TABLE OF CONTENTS

100 LOADS

Structures, to an agricultural engineer, means **buildings,** including the equipment and facilities in and around them.

An understanding of structures starts with knowing the loads imposed on them. All decisions made during design, selection, and construction, and all decisions made on appearance, durability, cleanability, etc., are made within the limits of supporting expected loads.

101 DEAD, SNOW, AND WIND LOADS

Introduction

Four general types of loads act on farm buildings—dead, live, snow, and wind loads. Dead loads are the weights of structural materials. Snow and wind loads depend on climate. Live loads include stored products such as silage, grain, and hay. Live loads vary both in intensity and in point of application. For live loads see chapters 102 and 103.

The intensity of loading in farm building design may differ from that for public and commercial buildings. Loading is usually lower in farm building design, because the consequences of failure are much less severe, e.g. in a machine shed than in a high school classroom. Even in farm buildings expected to last for many years, the owner is usually willing to accept greater risk of failure for lower initial cost.

Farm buildings are subjected to complex loads. Snow and wind loads vary with location, the height of the building, and its shape and roof slope. Recommended loads also vary with the expected life and use of the structure and an estimate of the risk to human life.

Snow and Wind Loads

This section discusses minimum snow and wind loads and methods of applying them for agricultural building design. It follows ASAE Engineering Practice EP288.4, Agricultural Buildings Snow and Wind Loads, which was approved late in 1986.

Terminology

Agricultural building:

A shelter for farm animals or crops; or when incidental to agricultural production, a shelter for processing or storing products of farm animals or crops, or for storing or repairing farm implements. An agricultural building is not for use by the public, for

Table 101-1. Weights of some common building materials.

Material	Weight* pcf	psf	plf
Reinforced concrete	150	...	...
Steel	490	...	...
Douglas fir	32	...	...
Oak	45	...	...
Yellow Pine	35	...	...
Insulation, per inch thickness	...	.1-.4	...
Steel roofing, 28 ga.	...	.8	...
Aluminum roofing, .024	...	.4	...
Wood shingles	...	2	...
Asphalt shingles	...	2.6	...
Softwood lumber			
1" lumber	...	2.4	...
3/8" plywood	...	1.1	...
1/2" plywood	...	1.5	...
2x4's 24" o.c.	...	0.7	...
2x6's 24" o.c.	...	1.1	...
Softwood lumber			
2x4	...	...	1.3
4x4	...	...	3.0
2x6	...	...	2.0
4x6	...	...	4.7
6x6	...	...	7.4
2x8	...	...	2.6
2x10	...	...	3.4
2x12	...	...	4.1

*pcf: pounds per cubic foot; psf: pounds per square foot; plf: pounds per linear foot.

Table 101-2. Approximate weights of trusses, psf.

Example: A 4-web truss for 4' spacing with 2x8 top chord and 2x6 bottom chord weighs about 1.3 + 0.7 = 2.0 psf. Dashed lines in table indicate example.

Chord size Top	Chord size Bottom	Truss spacing 2'	4'	8'
		Truss dead weight, psf		
2x4	2x4	1.6	0.8	0.4
2x6	2x4	2.0	1.0	0.5
2x6	2x6	2.4	1.2	0.6
2x8	2x6	2.7	1.3	0.7
2x10	2x4+2x4	3.3	1.6	0.8
2x12	2x4+2x6	4.0	2.0	1.0
2x12	2x6+2x6	4.4	2.2	1.1
Add the following for:				
2-&4-Web Truss		1.4	0.7	0.4
6 Web Truss		2.1	1.2	0.6

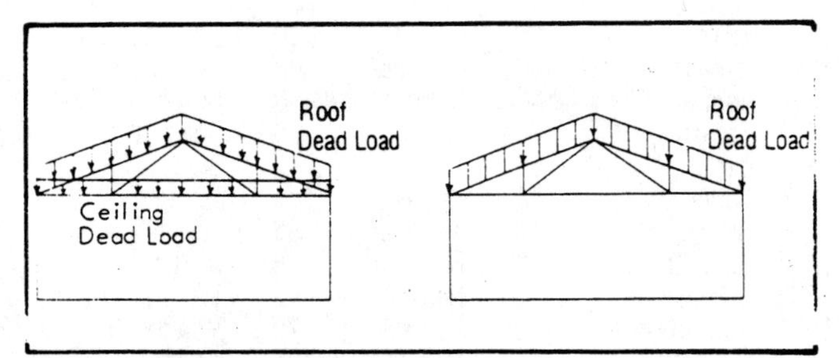

Fig 101-1. Basic dead and snow loads.
Psf-hp = lb/ft² – horizontal projection.

human habitation, or for commercial retail trade or warehousing.

Loads

EP288.4 does not include values for dead load and live load, nor the combinations used for building design.

Snow loads: Vertical loads from the weight of snow applied to the horizontal projection of a roof.

Wind loads: Loads from wind coming from any horizontal direction. Tornado winds are not included.

Other loads: Dead loads-Vertical loads due to the permanent weight of building construction materials such as floor, roof, framing, and covering. Live Loads-Both static and dynamic loads resulting from the use or occupancy of the building. These static loads result from the weight and/or pressure from equipment, livestock, and stored products and the materials used for construction and maintenance activities. The dynamic loads result from the dynamic effect of cranes, hoists and material handling equipment.

General Design Considerations

The values in the ground snow load and wind speed maps, Figs 101-2 to 101-4 and 101-7, have an annual probability of being exceeded equal to 0.02 (50-yr mean recurrence interval). Importance factors adjust the annual probability of the snow and wind loads to 0.04 (25-yr mean recurrence interval).

Minimum roof design load is 12 psf plus dead load. Exception: The minimum roof load need not apply to livestock sunshades, plastic-covered greenhouses, and similar structures.

Snow Loads

Ground snow loads, P_g

Design ground snow loads, P_g, for the contiguous United States are in Figs 101-2 to 101-4. Values for Alaska are in Table 101-3. In Alaska extreme local variations preclude state-wide mapping. Snow loads are zero for Hawaii.

Snow load values for shaded areas on the map are not for areas such as high country, where snow loads vary considerably. Consult local sources for data. Black areas on the maps have local variation in snow loads so extreme as to preclude meaningful mapping. Consult local sources for data.

Table 101-3. Ground snow loads, P_g, for Alaskan locations (psf).

psf		psf		psf	
20	Adak	65	Galena	130	Petersburg
45	Anchorage	60	Gulkana	45	St. Paul Isl.
75	Angoon	45	Homer	55	Seward
30	Barrow	70	Juneau	20	Shemya
60	Barter Island	55	Kenai	45	Sitka
35	Bethel	30	Kodiak	175	Talkeetna
60	Big Delta	70	Kotzebue	55	Unalakleet
20	Cold Bay	70	McGrath	170	Valdez
100	Cordova	55	Nenana	400	Whittier
55	Fairbanks	80	Nome	70	Wrangell
70	Ft. Yukon	50	Palmer	175	Yakutat

Balanced roof snow loads, P_s

Calculate the snow load, P_s, on the horizontal projection of an unobstructed roof with Eq 101-1.

Eq 101-1.

$$P_s = R \times C_e \times I \times C_s \times P_g$$

R = Roof snow factor (relates roof snow to ground snowpack).
= 1.0 for ground snowpack of 15 psf or less.
= 0.7 for ground snowpack of 20 psf or more in contiguous United States.
= 0.6 for Alaska.
= or the site specific coefficient of Table 101-8, Appendix.

C_e = exposure factor.
= 0.8 in windy areas with roof exposed on all sides with no shelter afforded by terrain, higher structures, or trees.
= 1.0 where wind cannot be relied on to reduce roof loads because of terrain, higher structures, or several trees nearby.
= 1.1 where terrain, higher structures, or trees will shelter the building increasing the potential for drifting.

I = Importance factor.
= 1.0 (.02 probability) for agricultural buildings that require a greater reliability of design to protect property or people.
= 0.8 (.04 probability) for agricultural buildings that present a low risk to property or people.

C_s = Slope factor (angles are from horizontal).
= 1.0 for 0° to 15° slope.
= 1.0 − (slope − 15°) ÷ 55 For 15° to 70° slope.
= 0 for slopes >70°.

P_g = Ground snow load from Figs 101-2 to 101-4.

Curved roofs: Consider slopes exceeding 70° free from snow load.

Multiple gable, sawtooth, and barrel vault roofs: Do not apply a reduction in snow load because of slope.

Example: Snow load on a low-risk, agricultural building 4/12 sloped roof in a sheltered area:

$$P_s = 0.7 \times 1.1 \times 0.8 \times 0.94 \times P_g$$
$$= 0.58\ P_g.$$

Unbalanced roof snow loads

Hip and gable roofs: For roofs with a slope less than 15° or more than 70°, unbalanced snow loads need not be considered. For slopes between 15° and 70°, design the structure to sustain an unbalanced uniform snow load on the lee side of 1.5 P_s while the windward side is free of snow.

Curved roofs: Consider the windward side and the portions of curved roofs having a slope exceeding 70° to be free of snow load. Determine unbalanced loads according to the loading diagrams in Fig 101-5.

If the ground or another roof abuts a Case II or Case III arched roof structure at or within 3′ of its eave, do not decrease the snow load below 1.5 P_s for the area between the 30° point and the eave, shown by dashed lines in Fig 101-5.

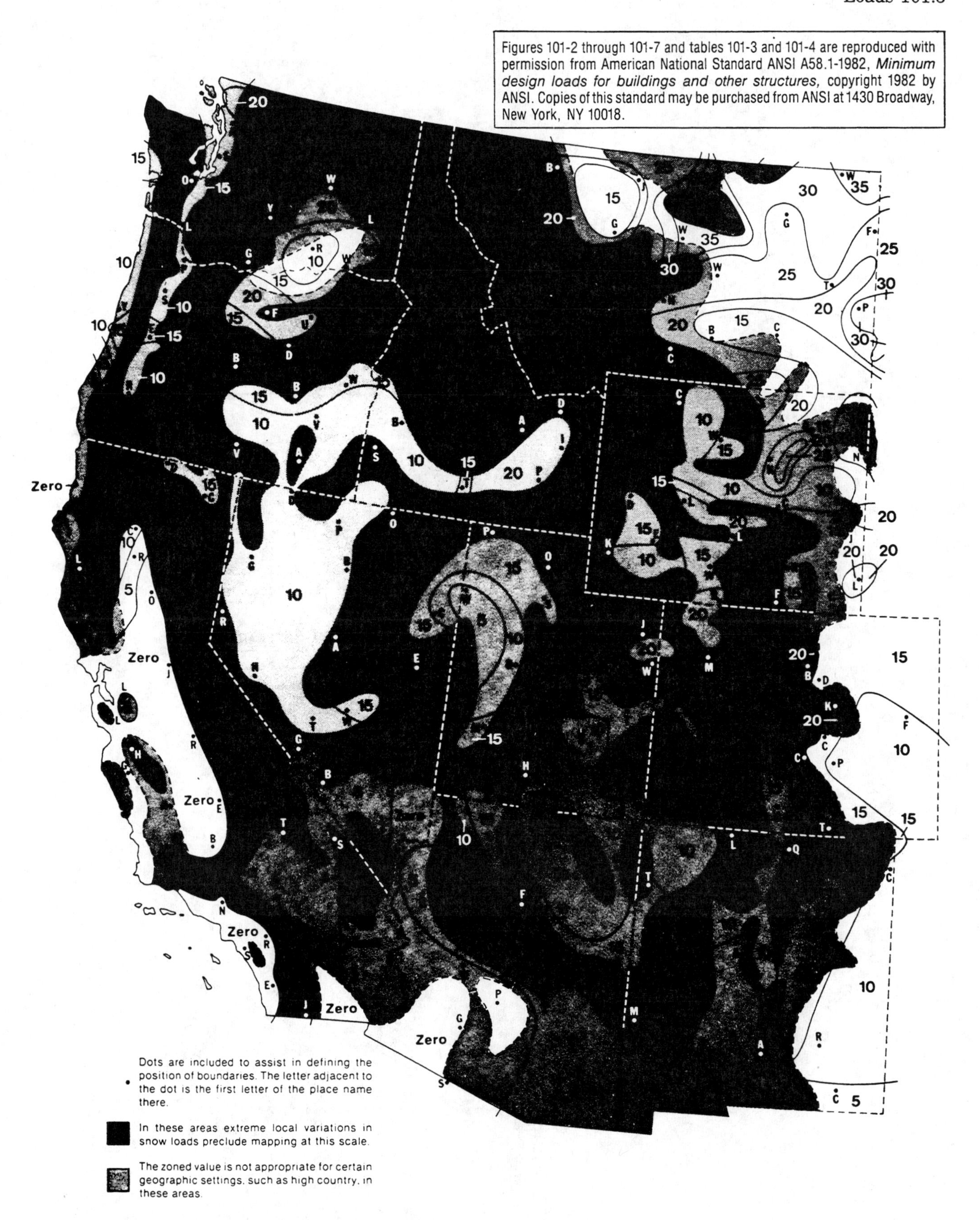

Fig 101-2. Western U.S. ground snow loads, P_g psf.
Values are loads associated with an annual probability of 0.02 (50-yr recurrence interval).
Source: ANSI A58.1-1982 modified for ASAE EP288.4 at MT and ND.

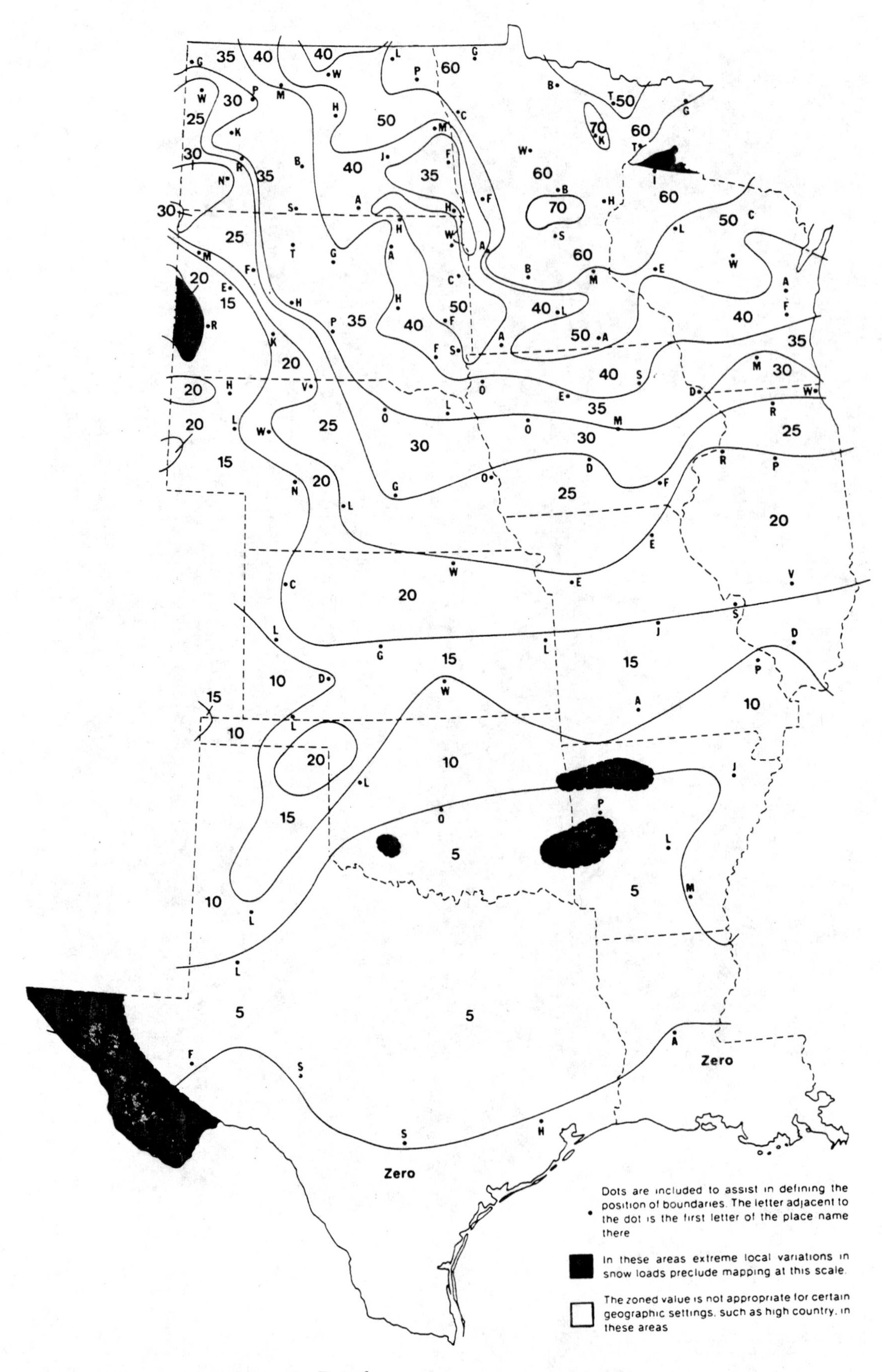

Fig 101-3. Central U.S. ground snow loads, P_g psf.
Values are loads associated with an annual probability of 0.02 (50-yr recurrence interval).
Source: ANSI A58.1-1982 modified for ASAE EP288.4 at ND, SD, MN, and WI.

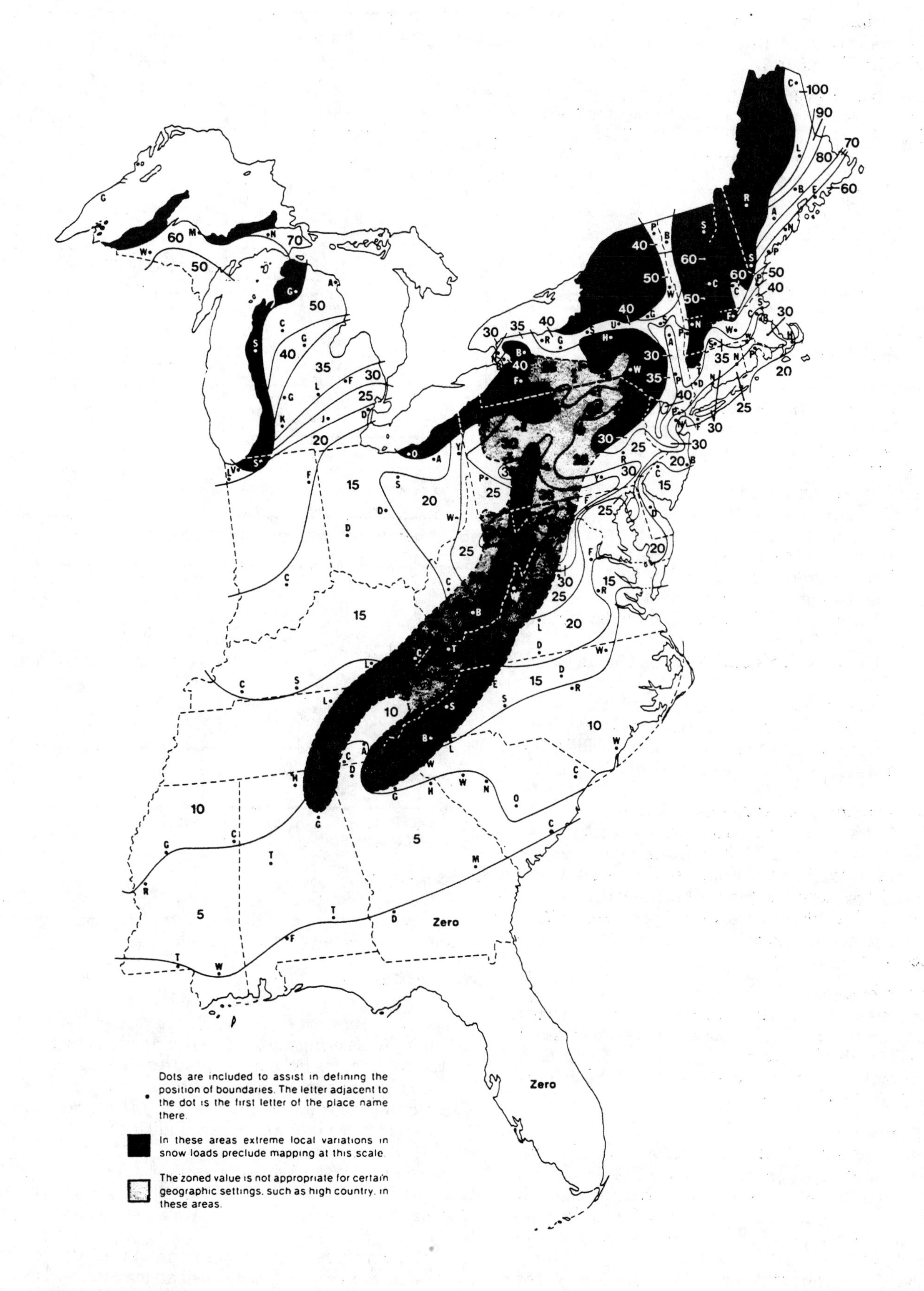

Fig 101-4. Eastern U.S. ground snow loads, P_g psf.
Values are loads associated with an annual probability of 0.02 (50-year recurrence interval).
Source: ANSI A58.1-1982.

Multiple gable, sawtooth, and barrel vault roofs: Consider the valleys as filled with snow of density d, as defined in Table 101-4.

Table 101-4. Densities for establishing drift loads.
Drifting is not considered where ground snow load, P_g, is 10 psf or less.

Ground snow load, P_g	Snow density, d
psf	pcf
11-30	15
31-60	20
>60	25

Drifts on lower roofs (aerodynamic shade)

Design roofs to support snow that drifts in the wind shelter of adjacent structures, terrain features, and higher portions of the same structure.

Lower roof of a structure

See dimensions in Fig 101-6. The geometry and magnitude of the drift load shall be determined as follows, except that drifting shall not be considered when $(H_r - H_b)/H_b < 0.2$.

Calculate drift height, H_d, ft $= [(W_b) \times P_s/(8xd)]^{0.5}$.

Where: $H_d + H_b$ cannot exceed H_r.

- W_b = horizontal dimension of upper roof normal to the line of the change in the roof level, ft.
- P_s = roof load on upper roof, lb/ft²
- d = density of snow, Table 101-4
- H_r = height differences between the upper and lower roofs, ft
- H_b = lower roof balanced snow load height, ft = $P_s \div d$

The width of the drift, W, is equal to 4 H_d or 4 $(H_r - H_b)$, whichever is less. Maximum intensity of the load, P_d, at the height change is as follows: $P_d = d\,(H_d + H_b)$ not to exceed $d \times H_r$

Adjacent structures and terrain features

Use Drifts on lower roofs to establish surcharge loads on a roof within 20′ of a higher structure or terrain feature that could cause drifting on it. However, the separation distance, s, between the two reduces drift loads on the lower roof. Apply the factor 1s/20 to the maximum drift load intensity, P_d, to account for spacing.

Roof projections

Consider as triangular the loads caused by a drift around a projection. Compute a surcharge load for all sides of any obstruction longer than 15′. Determine the drift surcharge load and the width of the drift as for Drifts on lower roofs.

Sliding snow

Snow may slide from a sloped roof to a lower roof, creating extra loads on the lower roof. Assume that snow accumulated on the upper roof under the balanced loading condition slides onto the lower roof. Determine the total extra load available from the upper roof by Eq 101-1.

Where a portion of the upper roof load is expected to slide clear of the lower roof, reduce the extra load on the lower roof accordingly.

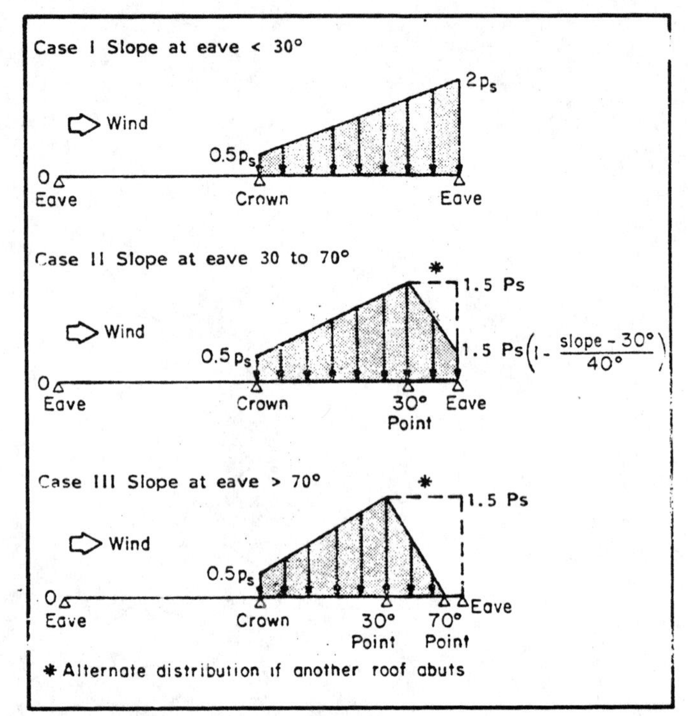

Fig 101-5. Unbalanced loading for curved roofs.

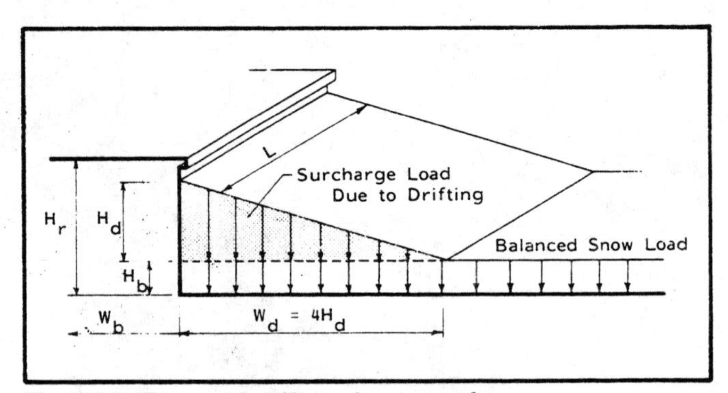

Fig 101-6. Shape of drift on lower roofs.
Source: MBMA Metal Building System Manual 1986.

Ponding loads

Consider roof deflections caused by snow loads when determining possible ponding of rain-on-snow or snow meltwater.

Wind Loads

Estimated extreme winds shown in Fig 101-7 are for open terrain with scattered obstructions less than 53′ high. Adjust map values for special winds such as the Santa Ana; for unusual exposures causing channeling or uplift such as ocean promontories, mountains, or gorges; and for exposures or elevations where wind records or experience indicate illustrated wind speeds are inadequate.

Recommendations in this section apply to agricultural buildings up to about 66′ tall and tall, slender structures (silos and grain legs) up to about 100′ in height.

Overturning moment due to wind is resisted by stabilizing moments of ⅔ of the dead load unless the structure is anchored to resist excess movement. When friction is insufficient to prevent sliding, anchor the building to resist sliding forces.

Wind velocity pressure:

Eq 101-2.

$$q = 0.00256 \times V^2 \times K_zG \times I^2$$

q = effective velocity pressure, psf.
V = wind velocity for extreme winds at 33′ elevation for a 0.02 annual probability (50-yr recurrence interval), m/s (mph), where V, m/s = 0.45V, mph, from Fig 101-7.
K_zG = combined building height, location exposure and gust factor from Table 101-3.
I^2 = importance factor.
= 1.0 (0.02 probability) for agricultural buildings that require greater reliability of design to protect property or people.
= 0.9 (0.04 probability) for agricultural buildings that present a low risk to property or people.

Example: Effective velocity pressure for a 48′ wide, 4:12 sloped roof, 16′ high side wall, low-risk building located near Des Moines, IA in an exposed location.

$$q = 0.00061 \times V^2 \times K_zG \times I^2.$$
$$q, psf = 0.00256 \times (80)^2 \times 1.14 \times 0.9 = 16.8\ psf.$$

Effective velocity pressure, q, in Eq 101-2, reflects dry air mass density at 59 F and 29.92″ mercury.

The building height and location exposure factor,

$$K_z = 2.6\ (z/z_g)^{2a}$$

applies both for sheltered (wooded areas or terrain with closely spaced obstructions) and exposed (flat, open country with scattered obstructions) locations.

where

z = design height of building in feet, equal to the mid-elevation of the roof. All heights are measured from the average grade line. Minimum z = 15′.
z_g = the gradient height at which the influence of surface friction is negligible. Gradient height, z_g, is assumed at 900′ for exposed and 1200′ for sheltered locations.
a = power law exponent. The recommended values of "a" for sheltered and exposed locations are 1/4.5 and 1/7, respectively.

The gust factor, G, is calculated by:

$$G = 0.65 + 3.65\ (T_z)$$

where:

T_z = a exposure factor which accounts for the degree of turbulence at height z and is expressed as:
$T_z = 2.35\ (D_o)^{0.5}/(z/h)^a$.
D_o = surface drag coefficient, use 0.01 for sheltered and 0.005 for exposed locations.
h = 33′.

Design pressure calculations

Calculate design pressure normal to the surface by multiplying the basic velocity pressure, q, by a dimensionless pressure coefficient, C_p, from Tables 101-6 or 101-7 or Fig 101-8. Positive pressure coefficients indicate inward pressure. Negative signs indicate outward pressure. Consider wind from all directions.

Table 101-5. Combined exposure and gust factors (K_zG).

Height ft	K_zG, Exposed	K_zG, Sheltered
15	1.07	0.62
20	1.14	0.68
30	1.24	0.77
40	1.32	0.84
50	1.39	0.91
60	1.45	0.96
70	1.50	1.01
80	1.54	1.06
90	1.58	1.10
100	1.62	1.14

Pressure coefficients in Fig 101-8 are for main structural systems such as rigid and braced frames, braced trusses, posts, poles, and girders.

Pressure coefficients for components and cladding, such as purlins, studs, girts, curtain walls, sheathing, roofing, and siding, are in Table 101-6.

An enclosed building is one that has a perimeter of solid walls. These "solid" walls can have openings, but openings must be protected by doors or windows.

An open building is one that has more than 20% of the windward surface open. A building or shed open on one side while the rest of the surfaces are fairly airtight, is an example.

Design pressure coefficients assume a uniform straight line pressure distribution except for ridge edges, perimeter eaves and canopy corners.

Consider higher wind loading at roof edges, ridge and corners due to turbulence. The area affected is 0.1 times the minimum width or 0.4 times the design height of the building, whichever is smaller, but not less than 3′ wide.

Coefficients for miscellaneous structures are in Table 101-7.

Table 101-6. Pressure coefficients, (C_p), for components and cladding.
Where two values are shown, check both cases.

Location	Enclosed building	Open building or overhang
Wall	+0.9; −1.0	+0.9; −1.5
Roof	−1.0	−1.5
Roof edges and ridge	−2.2	−2.2

Silo or tank Ht/dia.≤5	Angle B	Local C_p	Angle Local B	C_p
	0°	+1.0	105°	−1.5
	15	+0.8	120	−0.8
	30	+0.1	135	−0.6
	45	−0.8	150	−0.5
	60	−1.5	165	−0.5
	75	−1.9	180	−0.5
	90	−1.9		

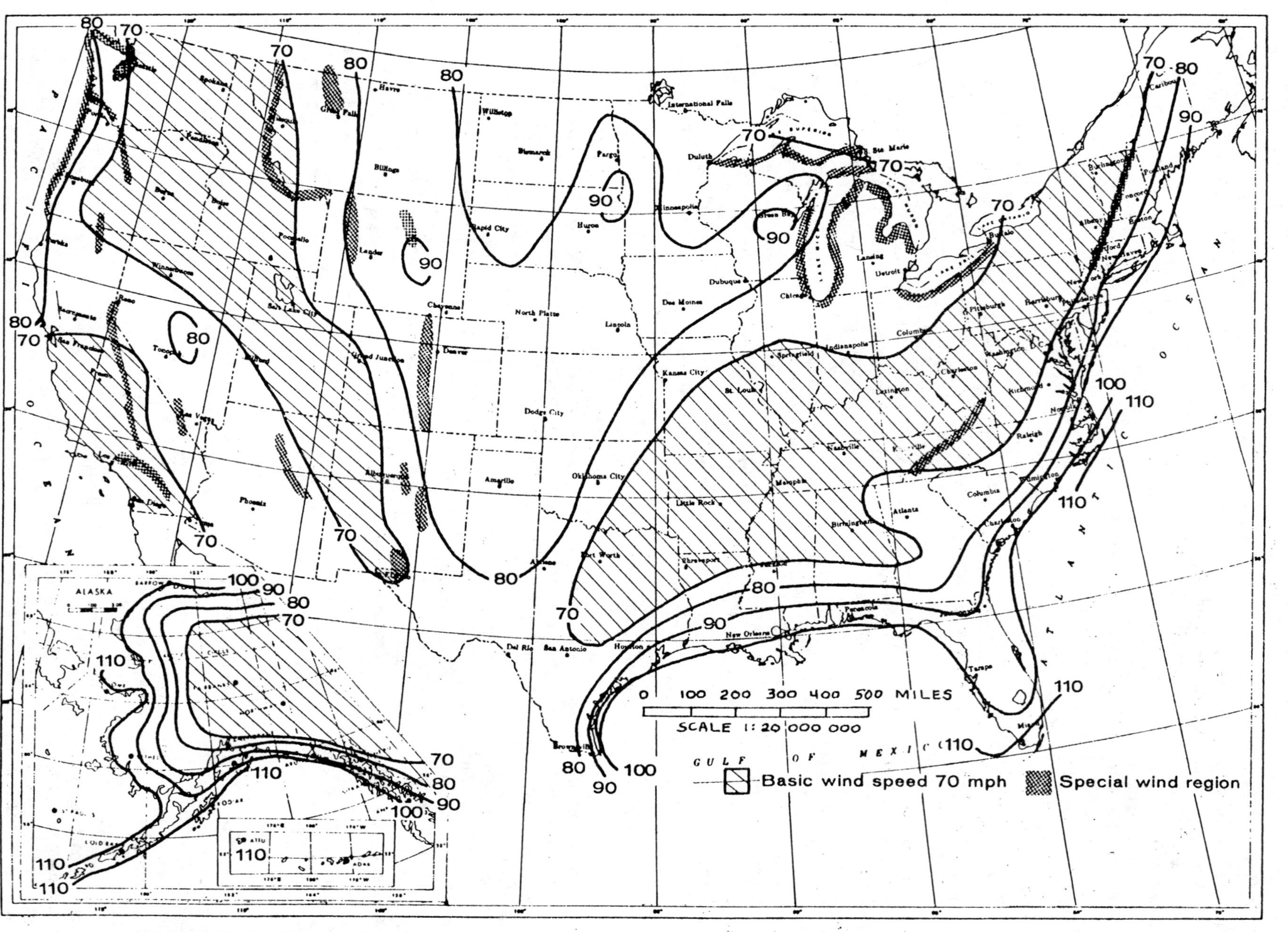

Fig 101-7. Basic wind speed, mph. Values are fastest-mile speeds at 33' above ground for rural exposure and an annual probability of 0.02 (50-yr recurrence interval).

Source ANSI A.58.1-1982.

Linear interpolation between wind speed contours is acceptable. Use caution with wind speed contours in mountainous regions of Alaska.

Wind direction →	Windward Wall	Windward Roof	Leeward Roof	Leeward Wall	End or side walls
Gable roof	Roof slopes from 10° to 30° (2:12 to 7:12) (Two cases illustrate wide variation in data for load on windward roof.)				
Case I	+0.7	+0.2	-0.7	-0.5	-0.7
Case II	+0.7	-0.7	-0.7	-0.5	-0.7
Gable Roof	+0.7	-0.7	-0.7	-0.5	-0.5
Open gable roof					
	—	-0.9	-1.2	-1.1	-1.4
	+1.3	+0.2	-0.2	—	-0.3
Monoslope roof					
	+0.7	-0.7		-0.7	-0.8
	+0.6	0		-0.6	-0.9
Monoslope Roof	+0.7	-0.7		-0.4	-0.6
Open monoslope roof					
	—	-1.3		-1.3	-1.3
	+1.1	+0.5		—	-0.4
Umbrella roof					
	—	+0.6	-0.6	—	—
Stored Product (h, 0.8 h)	—	-1.0	-1.1	—	—

Wind direction →	Windward Wall	Windward Lower roof	Windward Upper roof	Leeward Upper roof	Leeward Lower roof	Leeward Wall	End walls
Gambrel roof	+0.7	+0.5	-0.8	-0.6	-0.6	-0.4	-0.7

Multispan gable roof

h/w	Wind ward wall	First span a	First span b	Second span c	Second span d	Other span m	Other span n	wall
≈ 0.25	+0.6	+0.3	-1.0	-1.0	-0.6	-0.4	-0.5	-0.4
≈ 0.40	+0.6	-0.7	-1.0	-1.0	-0.6	-0.4	-0.5	-0.4

Arched roof

	Wind ward wall	Windward quarter	Roof Center half	Leeward quarter	Lee wall	End wall
On elevated structure	+0.8	2 h/w-0.4	-0.7	-0.6	-0.5	-0.7
Springing from ground	—	1.2 h/w	-0.7	-0.6	—	-0.7

Fig 101-8. Pressure coefficients, C_p, for agricultural structures.

The listed coefficients are the vector sum of both internal and external pressure. Positive coefficients are directed inward; negative coefficients are directed outward. 0.2 internal bursting coefficient is included.

Table 101-7. Pressure coefficients (C_p) for miscellaneous structures.

Structure or part	Description	C_p factor
Chimneys, tanks and silos	Square	1.4 any direction
	Round	0.6 any direction
Signs, flagpoles, lightpoles		1.4 any direction
Fences, latticed framework*	Flat sided members	1.7 any direction
	Round members	0.9 any direction

*(Ratio of solid area to gross area is 0.7 or less; for greater than 0.7, use 1.4)

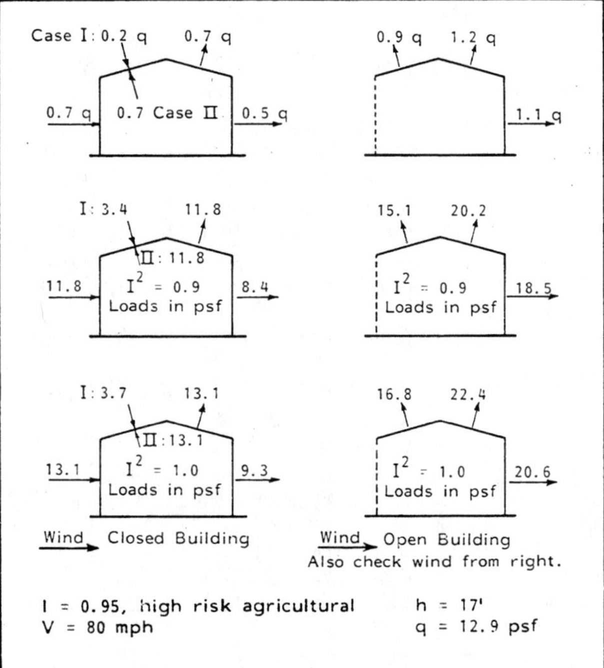

Fig 101-9. 80 mph wind loads on building frames.

Table 101-8. Roof snow factors.

Roof snow factors are probability sensitive and apply only with the importance factors listed.

Abstracted from M.J. O'Rourke and U. Stiefel, *Roof Snow Loads for Structural Design*, Journal of Structural Engineering, ASCE, Vol. 109, No. 7, July 1983, pp. 1527-1537.

Location	Importance factor 0.8	Importance factor 1.0	Location	Importance factor 0.8	Importance factor 1.0
Flagstaff AZ	0.576	0.597	Kalispell MT	0.568	0.587
Blue Canyon CA	0.592	0.616	Missoula MT	0.590	0.613
Mt Shasta CA	0.567	0.586	Grand Island NE	0.555	0.571
Alamosa CO	0.567	0.586	Norfolk NE	0.563	0.581
Colorado Spr CO	0.584	0.606	North Platte NE	0.579	0.600
Denver CO	0.606	0.633	Omaha NE	0.574	0.594
Grand Junction CO	0.556	0.573	Scottsbluff NE	0.587	0.610
Pueblo CO	0.617	0.647	Valentine NE	0.563	0.581
Bridgeport CT	0.572	0.591	Asheville NC	0.583	0.605
Hartford CT	0.606	0.634	Concord NH	0.574	0.594
New Haven CT	0.586	0.608	Atlantic City NJ	0.583	0.605
Wilmington DE	0.587	0.610	Neward NJ	0.583	0.605
Boise AP, ID	0.614	0.644	Albany NY	0.619	0.649
Pocatella AP, ID	0.632	0.655	Binghamton NY	0.593	0.618
Chicago/O-H, IL	0.599	0.625	Buffalo NY	0.574	0.594
Chicago IL	0.574	0.594	NYC-LA Guard NY	0.560	0.578
Moline IL	0.596	0.621	Rochester NY	0.594	0.618
Peoria IL	0.576	0.597	Syracuse NY	0.601	0.627
Rockford IL	0.568	0.587	Bismarck ND	0.569	0.588
Springfield IL	0.566	0.585	Fargo ND	0.563	0.581
Ft Wayne IN	0.571	0.591	Williston ND	0.580	0.601
Indianapolis IN	0.568	0.587	Clayton NM	0.586	0.609
South Bend IN	0.569	0.588	Elko NV	0.562	0.579
Burlington IA	0.577	0.598	Ely NV	0.638	0.673
Des Moines IA	0.569	0.588	Winnemucca NV	0.603	0.629
Dubuque IA	0.565	0.583	Akron-Canton OH	0.602	0.629
Sioux City IA	0.557	0.574	Cleveland OH	0.596	0.621
Waterloo IA	0.559	0.576	Columbus OH	0.592	0.616
Concordia KS	0.556	0.572	Dayton OH	0.587	0.610
Dodge City KS	0.575	0.595	Mansfield OH	0.572	0.592
Goodland KS	0.585	0.607	Toledo, Expr. OH	0.609	0.637
Topeka KS	0.563	0.580	Youngstown OH	0.615	0.645
Wichita KS	0.567	0.586	Burns OR	0.587	0.609
Covington KY	0.581	0.603	Allentown PA	0.584	0.606
Lexington KY	0.577	0.597	Erie PA	0.608	0.636
Caribou ME	0.586	0.609	Harrisburg PA	0.584	0.606
Portland ME	0.585	0.608	Philadelphia PA	0.566	0.584
Boston MA	0.595	0.620	Pittsburg PA	0.575	0.595
Nantucket MA	0.583	0.606	Scranton PA	0.597	0.623
Worcester MA	0.615	0.645	Williamsport PA	0.594	0.618
Alpena MI	0.593	0.617	Providence RI	0.587	0.610
Detroit MI	0.586	0.609	Aberdeen SD	0.565	0.583
Detroit AP, MI	0.587	0.610	Huron SD	0.552	0.568
Detroit WR, MI	0.557	0.573	Rapid City SD	0.586	0.609
Flint MI	0.568	0.587	Sioux Falls SD	0.560	0.577
Grand Rapids MI	0.569	0.588	Amarillo TX	0.570	0.589
Houghton Lk MI	0.575	0.596	Milford UT	0.589	0.613
Lansing MI	0.567	0.586	Salt Lake C UT	0.663	0.703
Marquette MI	0.680	0.725	Washington DC	0.608	0.636
Muskegon MI	0.581	0.603	Burlington VT	0.584	0.607
Sault S Marie MI	0.611	0.640	Spokane WA	0.562	0.580
Duluth MN	0.617	0.647	Stampede Pass WA	0.663	0.704
Int Falls MN	0.625	0.657	Yakima WA	0.555	0.571
Minn-St Paul MN	0.558	0.575	Green Bay WI	0.566	0.585
Rochester MN	0.554	0.570	La Crosse WI	0.552	0.567
St Cloud MN	0.562	0.580	Madison WI	0.555	0.571
Columbia Reg MO	0.556	0.572	Milwaukee WI	0.568	0.587
Kansas City MO	0.582	0.604	Charlestone WV	0.549	0.564
Springfield MO	0.574	0.594	Elkins WV	0.584	0.606
Billings MT	0.613	0.642	Huntington WV	0.571	0.591
Glasgow MT	0.579	0.600	Casper WY	0.615	0.645
Great Falls MT	0.616	0.646	Cheyenne WY	0.590	0.613
Havre MT	0.572	0.591	Lander WY	0.620	0.650
Helena MT	0.610	0.639	Sheridan WY	0.563	0.581

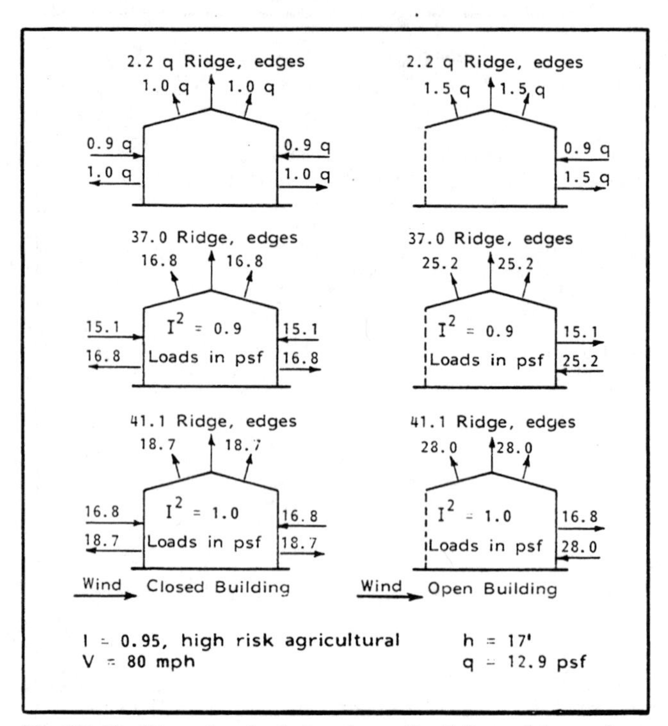

Fig 101-10. 80 mph winds loads on building girts and purlins.

102 FLOOR, SUSPENDED AND WASTE STORAGE LOADS

Live Floor and Suspended Loads

This section is based on ASAE EP378.2, 1981.

The minimum design load on floor area based on use of the area is listed in Table 102-1.

Slotted Floor

Where slats are interconnected between supports so that three or more slats must deflect together, design each span of slat between support and interconnection, and between two interconnections, for the recommended load per unit length of slat. Design the full span of each slat for one-half the recommended load per unit length.

Vehicle Storage

Vehicle storage (uniformly distributed): The minimum design load on a floor area used for farm machinery with traffic limited to access and egress is 150 psf. If loaded farm trucks or large farm tractors (those weighing more than 13,000 lb including mounted equipment) will occupy the area, the design load is 200 psf.

Table 102-1. Floor live loads.

An alternate requirement for slats which may dictate design is one concentrated load of 250 lb located first for maximum moment, then maximum shear, in slats and slat supports.

Design slotted floors in farrowing pens for a single concentrated load of 250 lb located first for maximum moment, then maximum shear, in slats and slat supports.

Design floors under bulk tanks in milkrooms according to the weight of the tank plus contents.

Where manure can accumulate, add 65 psf/ft of depth. Base design loads for floor-supported cages on weights and support intervals. Loads for suspended poultry cages are based on 4-row (double deck) or 6-row (triple deck) cages with 2 birds/8" cage, or 3 birds/12" cage, and 2" of manure accumulated on dropping boards under upper cages.

Calculate the design load for product storage on the basis of individual weights. The minimum load is 100 psf.

Livestock	Slats, per unit length plf	Solid floor and floor supports psf
Dairy and beef cattle	250	100
Calves to 300 lb.	150	50
stall area	250	60
Sheep—ewes, rams	120	50
—feeders	100	40
Swine		
to 50 lb	50	35
to 200 lb	100	50
to 400 lb	150	65
to 500 lb	170	70
Horses	250	100
Turkeys	25	30
Chickens, floor houses	15	20
Greenhouses		50
Manure		65

Vehicle storage (concentrated): In the absence of specific information, minimum design concentrated floor loads are:

- Tractors and implements: 5000 lb per wheel.
- Loaded trucks not exceeding 20,000 lb gross vehicle weight (GVW): 8,000 lb per wheel.
- Loaded trucks exceeding 20,000 lb GVW: 12,000 lb per wheel.

Vehicle loading and processing: In areas used for loading, unloading, or processing, increase minimum design loads by 50% to allow for impact or vibrations of machinery or equipment.

Waste Storage

Walls

The load on a tank wall from contained liquids is a simple hydrostatic load, increasing uniformly from top to bottom. The pressure at any depth is:

unit load, psf = unit weight x depth

The load imposed by soil on a submerged tank depends on the soil type and moisture content in addition to its weight and the depth. MWPS tank designs are based on 60 pcf Equivalent Fluid Density. Soil loads act inward, requiring reinforcing near the inside face of the wall.

Outward loads are assumed supported by the soil and coarse-grained backfill outside the wall for depths to 8'. Arbitrarily, tanks deeper than 8' are reinforced in both faces. Deep tanks are more heavily loaded, are more apt to extend below the water table, are a greater risk if failure occurs, and are more difficult to cast with consistently high quality and well placed concrete.

MWPS tanks are not designed for saturated fine sand (110 psf/ft). This material occurs in areas of high moisture and requires special consideration for design against flotation, floor failure, and overflow.

If vehicles must drive on soil within 5' of the tank walls, add a 100 psf uniform surcharge to the lateral load on exterior loads.

Table 102-2. Wall loads for waste storages.

	psf/ft of depth	Direction of load
Walls:	60	outward if above ground
	15	inward for good drainage
	30	inward for moderate drainage
	60	inward for high water table
	110	inward for saturated fine sand
Partitions:	60	each direction

Lid or Roof

Design loads for tank lids: 40 psf plus $0.8 \times P_g$ snow load on ground (Chapter 101) if outdoor tank lid is constructed at least 18" above ground level and is not accessible to livestock or equipment.

Select appropriate livestock or vehicle loads from this chapter.

103 GRAIN BIN LOADS AND PRESSURES

Chapter written by W.H. Bokhoven, PE, Curry Wille & Assoc., Ames, IA.

Definitions

Bin: a structure designed to contain free-flowing bulk grain and often including a hopper. Bin shape varies greatly, Fig 103-1. Bins are either free-standing or grouped together for efficiency. For design, bins are usually divided into two types: shallow and deep. There is no absolute or universally accepted definition of the difference between shallow or deep bins. One reasonable distinction is whether or not the rupture plane within the grain mass intersects the bin wall, Fig 103-1. In this definition, shallow bins have sidewall heights less than D/2 x tan (45 + ϕ/2) for circular bins or a/2 x tan (45+ ϕ/2) for rectangular or square bins. Certain long rectangular bins are both deep and shallow, depending on bin shape and surcharges. All other bins are deep bins.

Funnel flow: a condition in deep bins when grain flows from the top of the grain mass to the outlet through a channel that forms within the stagnant grain. The condition occurs when the hopper is not steep or smooth enough to permit the grain to slide along the walls. Under some conditions, mass flow occurs in the upper portion of a deep bin over a mass of grain at the bottom of the bin that is in funnel flow. Designs in this chapter do apply to this combined flow condition.

Mass flow: a condition when all of the grain in the bin is in motion whenever any grain is withdrawn. The condition occurs when the hopper is steep and smooth enough. Loads can be much greater than those predicted for funnel flow. Do not use this chapter for pressures in bins designed for mass flow.

Surcharge: the shape grain takes on the top of a pile when deposited from one or more points.

Hydraulic radius (R): R = (floor area/perimeter). Hydraulic radii for several bin shapes are in Fig 103-2. For odd shaped bins, such as pocket or fan bins, estimate hydraulic radius from the equivalent diameter.

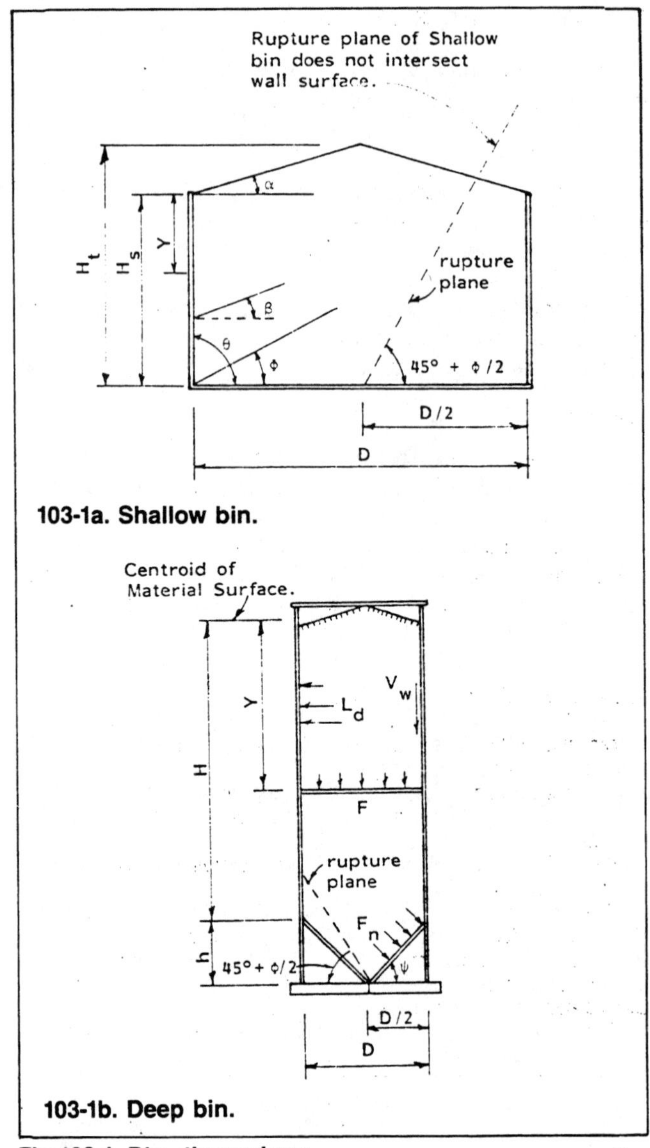

Fig 103-1. Bin dimensions.

Notation

A = area, or gross area of wall per unit width or height, ft^2

C_d = overpressure factor, used to predict the lateral static pressure increases during funnel flow, dimensionless. It converts static to expected dynamic pressures.

D = diameter of the horizontal cross section of a circular bin, ft

EFD = equivalent fluid density = w x k, lb/ft^3

F = vertical pressure created by stored grain at depth Y, psf

F_n = grain pressure normal to the sloping wall of a hopper, psf

H = height of grain to centroid of surcharge, ft, Fig 103-1b

H_s = height of grain at bin wall, ft, Fig 103-1a

H_t = height of grain to top of surcharge, ft, Fig 103-1a

L_e = lateral pressure increase due to eccentric discharge, psf

L_d = lateral design pressure created by dynamic conditions, psf, Fig 103-1b

L_s = lateral static pressure, psf

R = hydraulic radius of bin, ft

V_t = total vertical friction load on a unit width of wall above the point in question, lb/ft

Cross Section	Area A	Perimeter P	Hydraulic Radius R
Circle (D/2)	$\frac{\pi}{4} D^2$	πD	$\frac{D}{4}$
Rectangle (a, b)	ab	2(a+b)	$\frac{ab}{2(a+b)}$, Long side; $\frac{a}{4}$, Short side
Square (a)	a^2	4a	$\frac{a}{4}$
Regular Polygon N sides ($\frac{2\pi}{N}$, a, D/2)	$\frac{ND^2}{4} \sin\frac{\pi}{N} \cos\frac{\pi}{N}$	$ND \sin\frac{\pi}{N}$	$\frac{D}{4} \cos\frac{\pi}{N}$
	$\frac{Na^2}{4} \cot\frac{\pi}{N}$	Na	$\frac{a}{4} \cot\frac{\pi}{N}$
Interstice (D)	Approx. $(1-\frac{\pi}{4})D^2$	Approx. $\frac{\pi D}{2}$	Approx. 0.104D

Fig 103-2. Geometry of bin cross sections.

V_w = friction load on a unit area of wall or hopper surface, psf
Y = depth from level or sloping surface (in deep bins, from centroid of conical grain mass to point in question) ft, Fig 103-1b
a = length of short side of rectangle, ft
b = length of long side of rectangle, ft
e = eccentricity of outlet from center of bin, ft
h = depth of hopper or hopper-forming fill, ft, Fig 103-1b
k = ratio of lateral to vertical internal pressure, dimensionless
r = radius of bin, ft
w = bulk density of grain, pcf
α = filling angle of repose, usually less than the emptying angle of repose, degrees
β = angle of wall friction, equal to $\tan^{-1} \mu'$, degrees
ψ = angle of hopper wall from the horizontal, degrees, Fig 103-1b
ϕ = angle of internal friction, usually taken as the emptying angle of repose, degrees, Fig 103-1
μ' = coefficient of friction of grain on wall, dimensionless
θ = angle of wall from the horizontal, degrees, Fig 103-1a

Stored Material Pressures

Pressures created by free-flowing grain are predicted for normal conditions by the following equations for shallow and deep bins.

Shallow Bins

Predicted lateral pressures in shallow bins assume that the bin wall yields slightly through a distance sufficient to develop the full shearing resistance of the grain fill. If the wall is stiff or rigid, use deep bin theory to predict lateral pressures. Solutions for moments in shallow grain bin walls are in Table 103-3.

Lateral pressures in shallow bins have been predicted by two theories. Rankine's theory relates lateral to vertical pressure in noncohesive soils and comes from classical soils mechanics. It is valid for noncohesive soils, such as sand or gravel, which are similar in behavior to dry small grains. Coulomb's theory analyzes forces on a sliding wedge of material behind a wall. It yields identical values to Rankine's for a level pile where wall friction is neglected.

Lateral static wall pressure at depth Y below the surface:

Eq 103-1.

$$L_s = w \times Y \times k$$

where k is the lateral-to-vertical pressure ratio. The value of k varies with the angle of the wall from a horizontal plane, the angle of filling, the angle of internal friction, and the coefficient of friction between the wall and the grain mass. The generalized equation for k is:

Eq 103-2.

$$k = \sin^2(\theta-\phi) \div \sin^2(\theta) \sin(\theta+\beta) T^2$$
$$T = 1 + [\sin(\phi+\beta) \sin(\phi-\alpha) \div \sin(\theta-\alpha) \sin(\theta+\beta)]^{1/2}$$

This general equation reduces for conditions frequently encountered in bin design to one of the following:

• Vertical wall, level fill, and friction on wall neglected:

Eq 103-3.

$$k = (1-\sin(\phi)) \div (1+\sin(\phi))$$
$$k = \tan^2(45° - \phi/2) \text{ (Rankine's solution)}$$

• Vertical wall, sloping fill, filling angle less than emptying angle, and wall friction neglected:

Eq 103-4.

$$k = \sin^2(\theta-\phi) \div [1 + (\sin(\phi) \sin(\phi-\alpha) \div \sin(\theta-\alpha))^{1/2}]^2$$

• Vertical wall, filling angle equal to emptying angle, and wall friction neglected:

Eq 103-5.

$$k = \cos^2(\phi)$$

• Vertical wall, filling angle less than emptying angle, and filling angle equal to friction angle on wall:

Eq 103-6.

$$k = \cos(\alpha) [\cos(\alpha) - (\cos^2(\alpha) - \cos^2(\phi))^{1/2}] \div [\cos\alpha + (\cos^2(\alpha) - \cos^2(\phi))^{1/2}]$$

Determine k for other conditions with the generalized equation, Eq 103-2. To determine lateral pressures, wall friction is usually neglected, which results in a slightly conservative design.

Increases in lateral pressure from dynamic conditions do not usually occur within shallow bins. Therefore, do not take an increase in lateral pressure.

Lateral design pressure:

Eq 103-7.

$$L_d = L_s$$

Vertical pressure on a horizontal plane at depth Y below the surface:

Eq 103-8.

$$F = w \times Y$$

Note: F is maximum at height $Y = H_t$.
This condition occurs only at the center of the grain pile. This pressure is not necessarily over the entire floor—wall friction reduces floor pressure near the walls.

Vertical wall friction load:

Eq 103-9.

$$V_w = \mu' \times L_d$$

Total vertical wall friction load at depth Y below the surface:

Eq 103-10.

$$V_t = \mu' \times L_d \times Y \div 2$$

Deep Bins

Static vertical pressure at depth Y below the surface is predicted by Janssen's equation:

Eq 103-11.

$$F = (w \times R) \times (1\text{-}e^{-\mu' kY/R}) \div (\mu' \times k)$$

Vertical floor pressures need not total more than the weight of the grain. Vertical pressures predicted for static conditions are minimum: shock effects, such as the collapse of bridged or domed material above outlets, can create higher pressures. Examine each case for possible conditions and adjust values accordingly.

Lateral static unit pressure at depth Y:

Eq 103-12.

$$L_s = F \times k$$

During concentric unloading from deep bins, significant lateral pressure increases occur. Multiply static pressures by the appropriate overpressure factor, C_d, Fig 103-3, to obtain design pressures. Use lateral static pressures bin design only where the bin is unloaded from the top.

Lateral design unit pressure at depth Y:

Eq 103-13.

$$L_d = L_s \times C_d$$

Vertical friction load per unit wall area at depth Y:

Eq 103-14.

$$V_w = \mu' \times L_d$$

Total vertical friction load above depth Y per unit of wall perimeter:

Eq 103-15.

$$V_t = [(w \times Y) - (0.8 \times F)] \times R$$

While the total vertical friction load does not permit a static summation of forces, it is believed to more closely represent the actual wall load during withdrawal conditions.

Pressure on Sloping Floors—Shallow or Deep Bins

Grain pressure normal to a sloping surface:

Eq 103-16.

$$F_n = F \times \cos^2 \psi + L_d \times \sin^2 \psi$$

Friction force on sloping surface:

Eq 103-17.

$$V_w = \mu' \times F_n$$

Bulk Grain Properties

Common values for density, angle of repose, and coefficient of friction for grain and surface materials are in Tables 103-1 and 103-2. Densities and angles of repose vary depending on bulk grain properties and bin size, which affect the degree of compaction. Coefficients of friction also vary, and frequent use of a bin often causes lowered coefficients of friction. Select a range of values that predict the highest values for lateral and vertical loads.

Eccentric Discharge

Eccentric discharge from a bin occurs in several ways. Off-center discharge ports in the bin floor or sidewall cause grain to flow in a channel which is not in the bin center. Anti-dynamic tubes, installed along the bin wall, cause unsymmetrical top surfaces. Clean-out discharge ports in bin bottoms can leave large amounts of grain creating pressure around only a portion of the bin perimeter. Eccentric or multiple hoppers can also create unsymmetrical conditions.

Lateral pressures do change when grain is discharged through eccentrically located openings. Exact distribution of pressure increases or decreases has not been verified—test results are inconsistent. Estimate pressure increase and the design pressure at the opening by:

Lateral pressure increase due to eccentric discharge:

Eq 103-18.

$$L_e = 0.25\, L_s \times e \div r$$

Lateral design pressure with eccentric discharge:

Eq 103-19.

$$L_d = L_s \times C_d + L_e$$

The pressure increase is assumed constant for a height equal to D above the hopper or hopper forming fill and is then assumed to decrease linearly to the top of the bin.

Consider effects of uneven pressure distribution and unsymmetrical surfaces that cause bending moments and shear in bin walls, denting of thin-walled structures, and additional vertical wall loads.

Eccentric Filling

Loads from eccentric or off-center filling can be substantial, especially on large structures. Unsymmetrical top surfaces cause horizontal and vertical bending moments and additional vertical loads on bin walls. Avoid eccentric filling if possible, and design for the effects if necessary.

Thermal Effects

Differential dimensional changes between bin materials and grain result from external heating or cooling. For example, a metal bin filled on a hot day shrinks at night and creates a lateral pressure increase between the bin and the grain, resulting in strain in the bin wall.

The amount of lateral pressure increase depends on the radius, thickness, and material stiffness of the bin wall; the modulus of elasticity, Poisson's ratio, and bulk density of grain; and the ambient temperature decrease.

No theories exist on which design equations can be based. Conservatively, assume the grain mass is rigid and estimate the thermal strain in the bin wall. The result is a very conservative design, because some yielding of the grain mass undoubtedly occurs.

Moisture Effects

Lateral pressures increase significantly with increasing moisture content. Limited research data in-

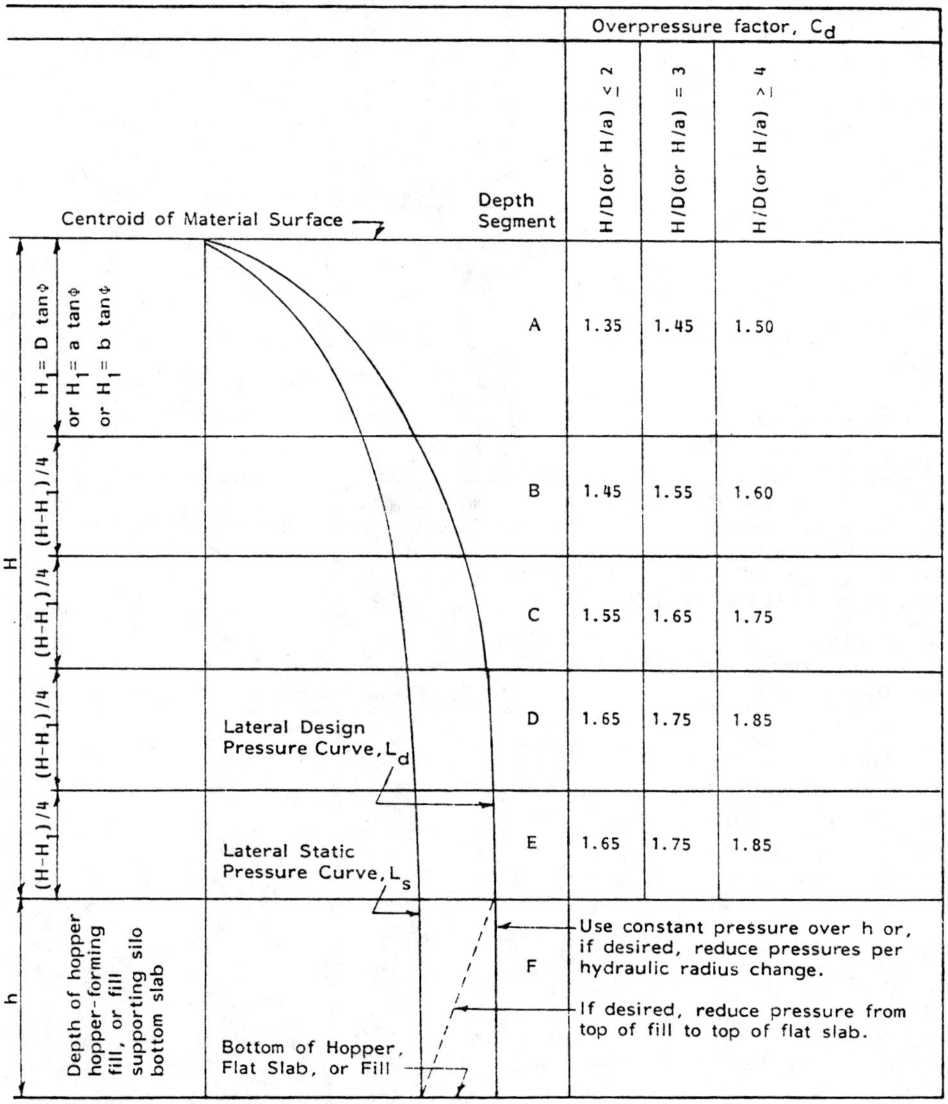
Centroid of Material Surface
H
h
$H_1 = D \tan\phi$
or $H_1 = a \tan\phi$
or $H_1 = b \tan\phi$
$(H-H_1)/4$
$(H-H_1)/4$
$(H-H_1)/4$
$(H-H_1)/4$
Lateral Design Pressure Curve, L_d
Lateral Static Pressure Curve, L_s
Depth of hopper hopper-forming fill, or fill supporting silo bottom slab
Bottom of Hopper, Flat Slab, or Fill
F
Use constant pressure over h or, if desired, reduce pressures per hydraulic radius change.
If desired, reduce pressure from top of fill to top of flat slab.

Depth Segment	Overpressure factor, C_d: H/D (or H/a) ≤ 2	H/D (or H/a) = 3	H/D (or H/a) ≥ 4
A	1.35	1.45	1.50
B	1.45	1.55	1.60
C	1.55	1.65	1.75
D	1.65	1.75	1.85
E	1.65	1.75	1.85

Fig 103-3. Minimum overpressure factors, C_d, for deep bins.
Interpolate C_d linearly between depth segments and H/D values shown.
H_1 = depth of segment A.
If $H_1 < H < 2H_1$ use C_d from segment B for total bin depth.
C_d values are minimum recommended values for free-flowing noncohesive grain and funnel flow conditions.
Do not use if mass flow conditions can occur.

dicates a sixfold increase in pressure as moisture content rises 4% and a tenfold increase with a 10% rise. Design and locate bins to prevent leaking, flooding, and soaking that can cause major load increases and possible failures.

Extraneous Loads

Objects in grain can cause excessive loads on portions of the bin. These objects include temperature cables, internal ladders, spouts, and stirring devices. Very little is known of the loads such objects exert.

Objects fixed in the grain mass, such as temperature cables and internal ladders, exert significant vertical loads on isolated portions of the bin. Discharging grain can create dynamic loads on temperature cables, depending on where the cable is located relative to the flow channel. Estimate vertical loads from the objects by assuming a load on the object

Table 103-1. Approximate bulk grain properties.

Grain	α Filling angle of repose, degrees	ϕ Emptying angle of repose, degrees	ω Bulk density, lb/ft³	EFD, lb/ft³ 1	2
Barley:					
Eastern	16	28	40.0	14.5	18.0
Western	16	28	43.2	15.6	19.4
Corn, shelled	16	27	48.0	18.0	22.5
Flaxseed	14	25	44.8	18.2	22.2
Oats:					
Central U.S.	18	32	33.6	10.3	12.9
Pacific NW and Canada	18	32	35.2	10.8	13.5
Rough Rice, American Pearl	20	36	41.6	10.8	13.6
Rye	17	26	46.4	18.1	23.3
Soybeans	16	29	48.0	16.7	20.5
Wheat:					
Soft red winter	16	27	48.8	18.3	22.9
Hard red spring	17	28	52.0	18.8	23.8
Hard red winter	16	27	51.2	19.2	24.0
Durum	17	26	52.0	20.3	26.1

1. For level fill and smooth vertical wall (wall friction neglected); use for shallow bins only.
2. For pile sloped at filling angle of repose and smooth vertical wall (wall friction neglected); use for shallow bins only.

Source: USDA Circular No. 835 & ASAE Data 241.2; EFD calculated.
Note: For Miscellaneous products, see ASAE Data 241.2

Table 103-2. Coefficients of friction (μ') for grains.
ASAE Paper 63-828 Brubaker & Pos, *Static Coefficients of Friction of Some Grains on Structural Materials.*
Valve in parentheses are from Table 1 of the paper and are from tests at 77 F and various relative humidities.

Material	Moisture Content (%)	CONCRETE Plastic Smooth Finish	CONCRETE Steel Trowel Finish	CONCRETE Wood Float Finish	WOOD Oak Grain Par.	WOOD Oak Grain Perp.	WOOD Douglas Fir Grain Par.	WOOD Douglas Fir Grain Perp.	PLASTIC Teflon	PLASTIC Poly-ethylene	METAL Mild Steel C. R.	METAL Galvanized Sheet Metal
Barley	10.7	.23	.56	.50	.23	.29	.27	.32	.17	.23	.20	.20
	14.3	.24	.57	.51	.21	.28	.30	.32	.13	.28	.23	.20
	16.4	.33	.62	.55	.30	.33	.37	.41	.11	.35	.21	.34
Oats	10.6	.28	.40	.43	.20	.23	.27	.29	.13	.20	.20	.22
	14.0	.33	.51	.42	.23	.25	.34	.36	.13	.28	.21	.18
	17.3	.50	.65	.64	.46	.48	.48	.50	.14	.50	.44	.32
Shelled Corn	7.5	.27	.41	.46	.24	.25	.27	.29	.17	.22	.23	.20
	9.9	.25	.59	.62	.28	.31	.31	.31	.18	.27	.20	.24
	13.9	.35	.64	.54	.29	.36	.37	.38	.12	.38	.24	.37
Soybeans	7.1 (7.0)	.25	.39	.39	(.29)	(.35)	.29	.31	(.23)	.25	(.32)	.21
	9.8 (11.6)	.31	.47	.37	(.34)	(.39)	.33	.31	(.19)	.29	(.41)	.18
	12.2 (15.4)	.36	.55	.52	(.42)	(.45)	.35	.44	(.17)	.43	(.45)	.20
Wheat	11.2 (9.7)	.36	.52	.51	(.30)	(.32)	.31	.35	(.19)	.27	(.33)	.10
	13.0 (11.9)	.46	.52	.55	(.28)	(.32)	.35	.38	(.17)	.35	(.33)	.14
	15.7 (15.1)	.56	.68	.69	(.35)	(.40)	.48	.50	(.12)	.45	(.38)	.33

similar to the vertical wall load. Compare coefficients of friction for the objects to those for the wall.

Objects that move within the grain, such as stirring devices, increase both lateral and vertical loads. Disturbing the static grain mass undoubtedly increases horizontal pressure, but exact amounts are unknown. Changes in vertical loads depend on the size and speed of the device and the type and quality of grain. Locate supports to avoid overstressing roof or sidewall members.

Table 103-3. Solutions for moments in shallow grain bin walls.

Units: any consistent units for weight and length.
E.g., EFD, equivalent fluid density = w x k = pcf
L, wall height = ft
M, bending moment = ft lb
H_s depth of fill = YA × L = ft
R, reaction = lb
E, elastic modulus = psf
I, moment of inertia = ft^4
Δ, deflection = ft

Moments and reactions are per unit length of wall. YA, YB, and YC are proportions of L.

SOLUTION I
Fixed Base
Bin Partially Full

SOLUTION II
Hinged Base
Bin Partially Full

SOLUTION III
Free Top, Bin Full
Optimum Tie Rod Location

See coefficients in table below.

MA = (coeff) x 0.06667 EFD × L^3
MB = (coeff) x 0.06667 EFD × L^3
YBL = (coeff) x L
R1 = (coeff) x EFD × L^2
R2 = (coeff) x EFD × L^2

MC = (coeff) x 0.06415 EFD × L^3
YCL = (coeff) x L
R3 = (coeff) x EFD × L^2
R4 = (coeff) x EFD × L^2

MD = MF = EFD × L^3
R5 = 0.318 EFD × L^2
R6 = 0.182 EFD × L^2
Δ = 0.0027 EFD × L^5/EI

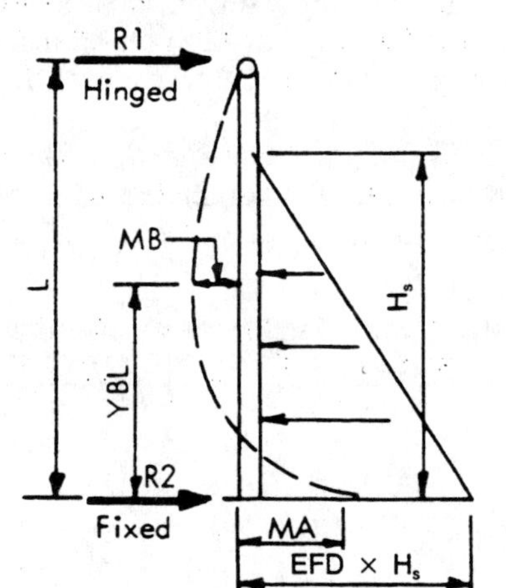

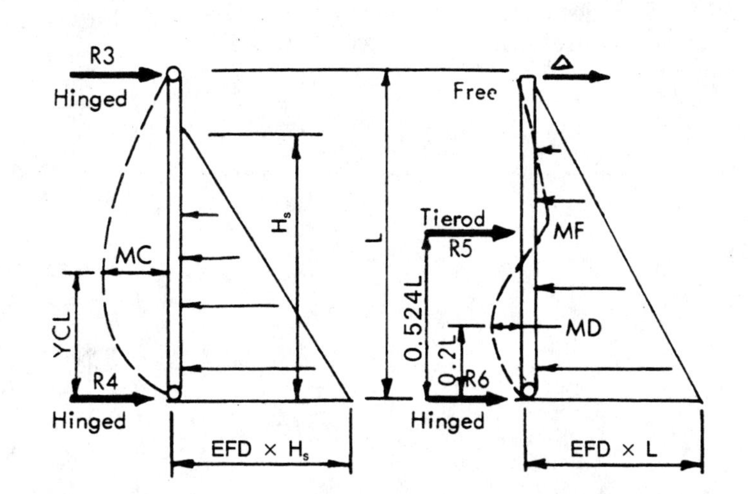

COEFFICIENTS FOR SOLUTION I

YA	MA	MB	YB	R1	R2
0.10	0.0023	0.0002	0.0951	0.00001	0.0050
0.20	0.0171	0.0023	0.1804	0.00019	0.0198
0.30	0.0532	0.0104	0.2564	0.00095	0.0440
0.40	0.1158	0.0288	0.3233	0.00294	0.0771
0.50	0.2070	0.0611	0.3814	0.00703	0.1180
0.60	0.3262	0.1096	0.4311	0.01426	0.1657
0.70	0.4703	0.1748	0.4728	0.02581	0.2192
0.80	0.6349	0.2552	0.5067	0.04301	0.2770
0.90	0.8137	0.3475	0.5333	0.06725	0.3377
1.00	1.0000	0.4472	0.5528	0.10000	0.4000

COEFFICIENTS FOR SOLUTION II

MC	YC	R3	R4
0.0024	0.0817	0.0002	0.0048
0.0173	0.1484	0.0013	0.0187
0.0535	0.2051	0.0045	0.0405
0.1160	0.2539	0.0107	0.0693
0.2066	0.2959	0.0208	0.1042
0.3249	0.3317	0.0360	0.1440
0.4682	0.3619	0.0572	0.1878
0.6324	0.3869	0.0853	0.2347
0.8118	0.4070	0.1215	0.2835
1.0000	0.4226	0.1667	0.3333

104 POTATOES

Potatoes exert pressures similar to other semi-fluids. These include horizontal wall pressures, vertical friction wall loads, and vertical floor loads.

Estimates of equivalent fluid density and center of force may be used to approximate pressure patterns, the maximum bending moment of the wall, reaction at the wall plate, and reaction at the wall sill.

A comparison between a typical measured potato pressure curve for vertical walls and calculated equivalent fluid density (EFD) line for bins that are wider than deep is shown in Fig 104-1. This shows that the EFD provides a reasonable prediction of the pressure curve. The pressure curve, however, exceeds the EFD line near the floor.

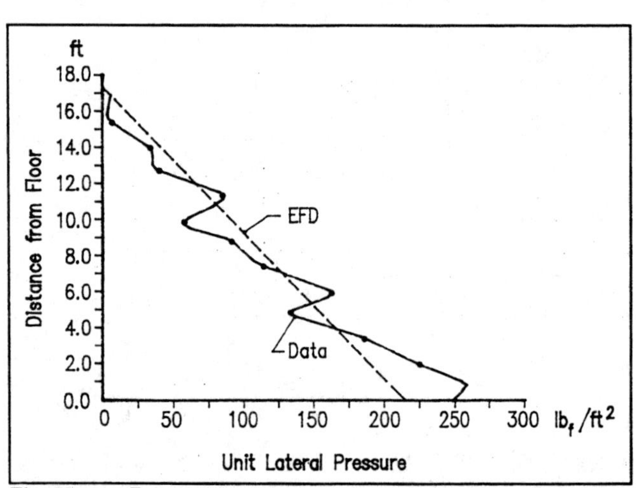

Fig 104-1. Potato pressure curve.
Typical measured potato pressure curve with calculated equivalent fluid density line for a shallow bin. Taken from ASAE Standard D446.

The center of force is the vertical distance measured from the bottom of the pile to the location at which the total force acts. It is often expressed as a fraction of the pile height (H). Table 104-1 shows the calculated equivalent fluid density and measured center of force (C) estimates for vertical potato bin walls.

Table 104-1. Potato bin walls.
Calculated equivalent fluid density (EFD) and measured center of force (C) estimates for vertical potato bin walls.

	Maximum calculated EFD				
	Shallow bin			Deep bin	
Pile depth, ft	Center of force, ft	Dry potatoes, EFD, lb/ft³	Wet potatoes lb/ft³	Center of ft	Dry potatoes lb/ft³
1	H/3.0	8	11	H/2.6	10
17.4	H/3.5	11	12	H/2.6	8

The equivalent fluid density of 13 lb/ft³ is commonly used for round potatoes. A lighter fluid density may be used for potatoes that are longer than round.

See the "Resource", page 104.2, for more information.

Tables 104-2 and 104-3 were developed using design procedures specified in the *Transactions of the ASAE*, Volume 26, No. 1, pp. 179-187, 1983.

Table 104-2. Potato storage wall forces on a vertical wall.
(EFD = 13 centroid adjusted. See *Transactions of the ASAE*, 1983, pp. 179-187)

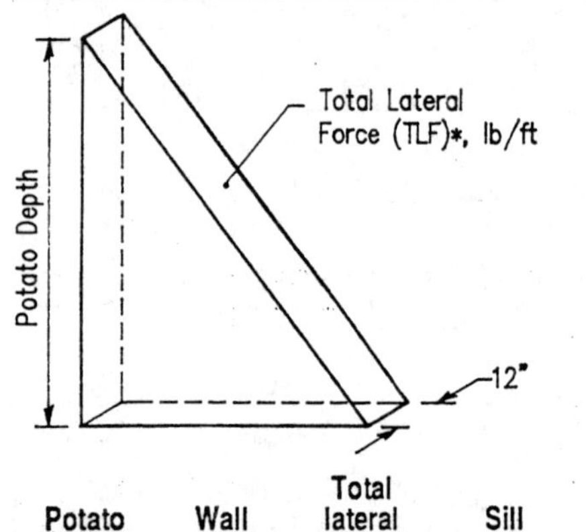

*TLF – The summation of the lateral pressures for a unit of wall length.

Potato depth ft	Wall height ft	Total lateral force lb/ft	Sill force lb/ft	Plate force lb/ft	Maximum bending moment in lb
10	12	650	469	181	10,000
12	14	936	669	267	20,100
14	16	1,274	902	372	31,500
16	18	1,664	1,171	493	46,200
18	20	2,106	1,474	632	65,250
20	22	2,600	1,812	788	88,300
22	24	3,146	2,185	961	116,900

Table 104-3. Potato storage wall forces with 4' leaner duct 2' from wall.

(EFD=13 centroid adjusted. See *Transactions of the ASAE*, 1983. pp. 179-187).

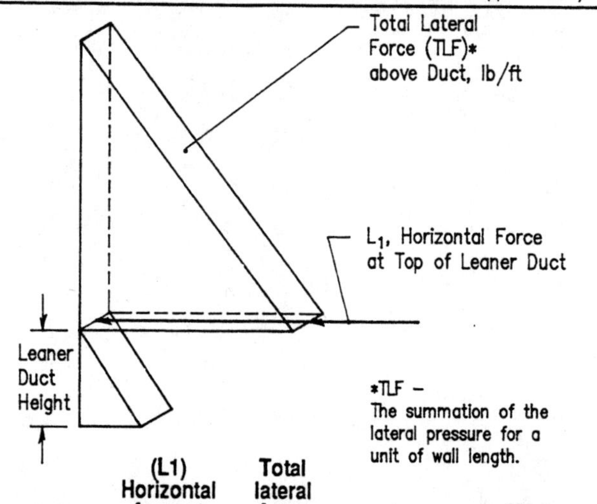

Potato depth ft	Wall height ft	(L1) Horizontal force at top of leaner lb/ft	Total lateral force above leaner duct lb/ft	Sill force lb/ft	Plate force lb/ft	Maximum bending moment in lb
10	12	272	275	338	209	14,550
12	14	333	470	507	296	22,975
14	16	394	717	711	400	34,570
16	18	454	1,016	949	521	49,520
18	20	515	1,367	1,222	659	69,100
20	22	575	1,770	1,530	815	92,600
22	24	636	2,225	1,873	988	120,980
24	26	696	2,732	2,248	1,179	154,675

The force on an inclined wall can be calculated as a reduced value of the force for a vertical wall with the relationship:

$F_1/F = ((a - b) \div (90 - b))3$

a = The angle that is included between the bin wall and the bin floor

b = Angle of internal friction. An average value is 32°.

F_1 = Total force on incline wall.

F = Total force on vertical wall.

Resource

For more information on lateral pressure of Irish potatoes stored in bulk refer to the American Society of Agricultural Engineers Standards ASAE D 446.

200 MATERIALS AND SELECTION

201 WOOD AS A MATERIAL

Wood's many characteristics differ greatly with such things as species, moisture content, and quality of the wood. Understanding these characteristics and knowing the relative ratings of the different species are helpful in choosing a species or quality of wood for a particular use.

Wood Terminology

Hardwoods and softwoods

All native species are divided into two classes—hardwoods and softwoods. In general, hardwoods have broad leaves and are **deciduous** (shed their leaves each season). Softwoods are **conifers** and have scalelike leaves as in cedars or needlelike leaves as in pines. "Hardwood" and "softwood" do not describe the wood's hardness or softness.

Heartwood and sapwood

Sapwood is the outer portion of a log, conducts sap, and has living cells. It varies in thickness with tree age and species, but it is commonly 1½"-2" thick.

Heartwood is inactive cells in the inner portion of a log. The heartwood may contain deposits that give it a darker color and usually make it more durable. Heartwood and sapwood of a given species are considered to be about equal in weight and strength.

Growth rings

In many species there is sufficient difference between spring and summer growth to produce well-marked annual rings.

Springwood is the inner part of the growth ring formed first in the growing season. Its cells have relatively large cavities and thin walls.

Summerwood is the outer part of the growth ring formed later in the season. Its cells have smaller cavities and thicker walls. When growth rings are prominent, as in southern yellow pine, springwood differs markedly from summerwood in physical properties. Springwood is lighter in color and weight, softer, and weaker than summerwood.

Direction

Many characteristics of wood vary with direction:

- Longitudinal—parallel to the long axis of the log or parallel to the grain of the wood.
- Radial—perpendicular to the growth rings.
- Tangential—tangent to the growth rings.

Plainsawed, flat-grained or slash-grained lumber is sawed tangent to the growth rings. The annual rings generally form angles of less than 45° with the wide face of the board.

Quartersawed, edge-grained or vertical-grained lumber is cut radial to the growth rings. The annual rings generally form angles of more than 45° with the wide face of the board.

Board measure

Lumber is usually priced by the board foot, a measure based on nominal dimensions. A board foot is 1'x1'x1". A 2x6 and 1x12 have one board foot per foot of length, and a 2x12 has 2 board feet per foot of length. To calculate the number of board feet in a piece of lumber, multiply width in inches times thickness in inches times length in feet and divide by 12 (board feet = w" x t" x L' ÷ 12).

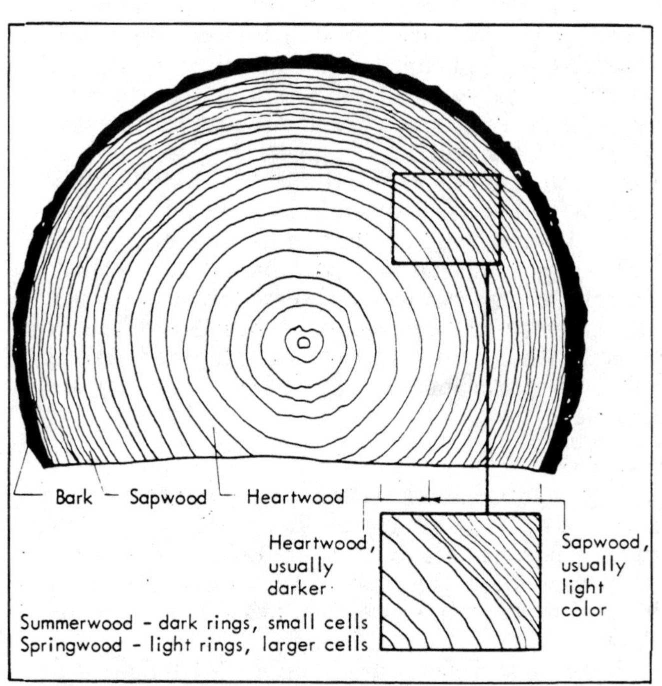

Fig 201-1. Tree growth rings.

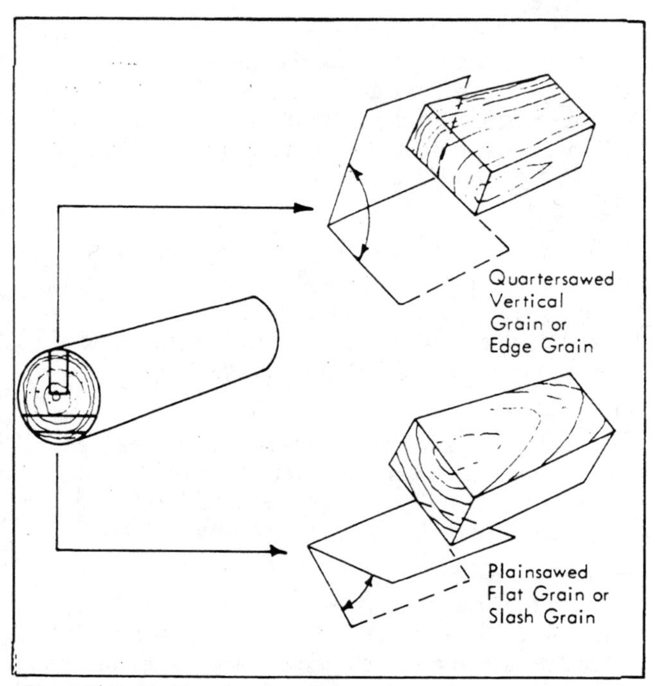

Fig 201-2. Cutting lumber from a log.

Table 201-1. Classification of characteristics and properties.
A, B, C and D are relative species ratings: A is high or desirable.

Kind of wood	Ease of Working	Paint Holding	Nail Holding	Heartwood Decay Resistance	Strength Bending	Strength Stiffness
Softwoods						
Cedar, inland red	B	A	C	A	C	C
Cedar, western red	A	A	C	A	C	C
Fir, Douglas	C	C	B	B	A	A
Fir, white	B	B	C	C	B	B
Hemlock, western	B	B	B	C	B	A
Larch, western	C	C	A	B	A	A
Pine, western white	A	A	A	B	B	B
Pine, lodge pole	A	A	B	B	B	B
Pine, Ponderosa	A	B	B	B	C	C
Pine, southern yellow	C	C	A	B	A	A
Pine, sugar	A	A	A	B	C	C
Redwood	B	A	B	A	B	B
Spruce, Englemanns	B	B	C	C	C	C
Spruce, Sitka	B	B	C	C	B	A
Tamarack	C	B	B	B	B	B
Hardwoods						
Ash, white	C	C	A	C	A	A
Birch, yellow	C	B	A	C	A	A
Cottonwood	B	B	C	C	C	B
Elm, rock	C	C	B	B	A	A
Hickory, true	C	C	A	C	A	A
Maple, hard	C	B	A	C	A	A
Oak, red or white	C	C	A	A	A	A
Walnut	B	C	B	A	A	A

Wood Characteristics

Ease of working

Wood that is easy to work looks better on the job and is easier to fabricate. Workability and machinability vary greatly. A general rating, Table 201-1, is based on hardness, texture, and character of the surface that can be obtained. Class A woods have soft, uniform texture and finish to smooth surfaces.

Paint holding ability

Paint holds better on edge-grain than on flat-grain. Knots in both white and yellow pines do not retain paint as well as the sound knots of the cedars, hemlocks, white fir, or western larch. The bark side of a flat-grained board is better than the pith side.

Nail holding power

Nail holding power is closely related to the density or specific gravity of the wood and to its splitting tendencies. Generally, the denser and harder the wood, the greater its holding power if it does not split.

Decay resistance

Wood that is continuously dry or continuously wet does not decay. But, wood at 21%-24% moisture content for appreciable periods does decay. The heartwood of some species has natural decay resistance.

Bending strength

The strength of a structural-size wooden member depends on the number, size, and type of defects and on the species, density, and moisture content. Relative strengths in bending are given for a number of species in Table 201-1. The bending strength measures the load-carrying capacity of such members as rafters, girders, and floor joists. Class A softwoods as given in the table dominate the structural field. Class B softwoods are used occasionally in heavy construction and extensively in light construction.

For a given species and member length, bending strength is proportional to the square of the depth, but is directly proportional to width. The geometric property that contributes to bending strength is called the section modulus. $S = bd^2/6$, for rectangular members.

Stiffness

Stiffness measures resistance to sag or deflection under load. It is an important characteristic for studs, joists, rafters, beams, shelving, ladder rails, etc. It depends on the cross section shape of its member and a property of wood called modulus of elasticity.

Relative stiffnesses are in Table 201-1. Softwoods in classes A and B are used where stiffness is the most

Table 201-2. Relative strength and stiffness.
For rectangular members, bending strength is proportional to member width and the square of its depth. Stiffness is proportional to member width and the cube of its depth.

Full size, on edge	2x4	4x4	2x8
Relative strength	1	2	4
Relative stiffness	1	2	8

Strength and stiffness decrease with span, depending on the type of load.

Span	10′	20′
Relative strength, uniform load	1	1/4
Concentrated load	1	1/2
Relative stiffness, uniform load	1	1/16
Concentrated load	1	1/8

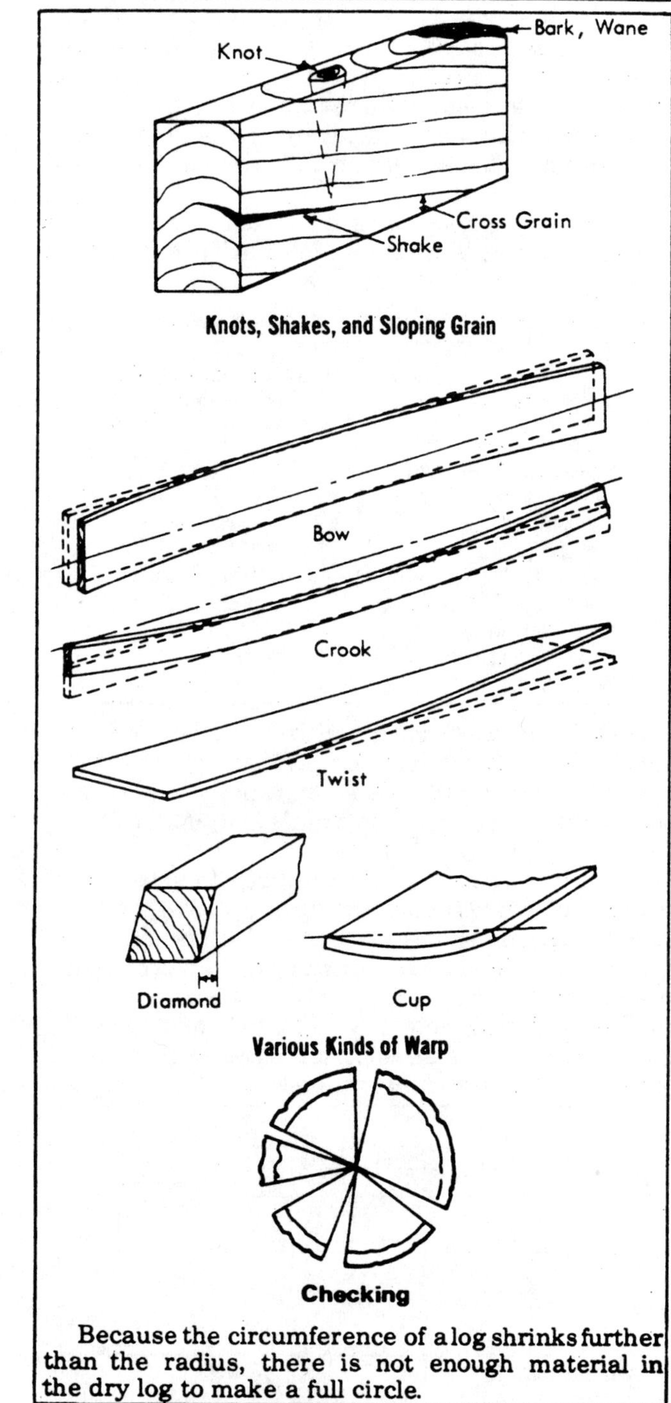

Fig 201-3. Defects in wood.

important requirement. Class A woods are usually heavier and harder than class B woods. Softwoods are often preferred over hardwoods because of their lighter weight. The geometric property that resists deflection is called the moment of inertia. $I = bd^3/12$, for rectangular members.

Wood defects

Some natural, drying, and milling characteristics result in defects in manufactured lumber. Some are illustrated in Fig 201-3. Others are:

- Decay: disintegration of wood by fungi.
- Holes: holes in wood from any cause.
- Pitch pocket: an opening between the annual rings caused by pitch accumulation.
- Stain: discoloration of wood from natural color.

Moisture content and shrinkage

Moisture content (% mc) in wood is the weight of water in the wood divided by the weight of the wood itself.

Eq 201-1.

$$mc\% = (Gwt - ODwt) \times 100 \div ODwt$$

mc% = % moisture content
Gwt = green weight of wood
ODwt = oven dry weight of wood

Green wood has "free water" in cell cavities and "absorbed water" in cell walls. As green wood dries, the cell walls remain saturated until the free water has been evaporated. The point at which cell walls begin to lose moisture is the "fiber saturation point" and is about 30% mc for most species, Fig 201-4.

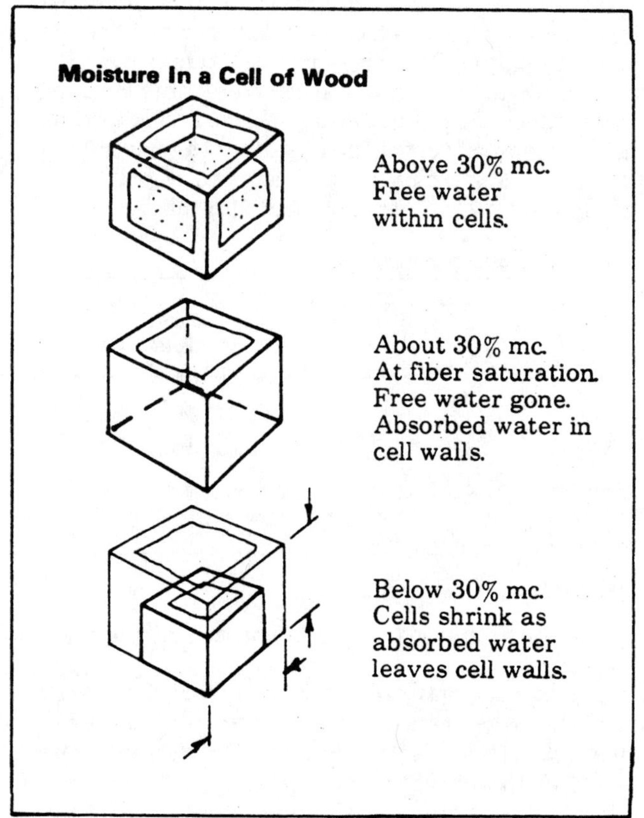

Fig 201-4. Moisture in wood cells.

Moisture content above the fiber saturation point does not affect the size or strength of wood. But as wood dries below fiber saturation and moisture begins to move from the cell walls, shrinkage begins and strength increases. The amount of shrinkage varies in the radial, tangential, and longitudinal directions, Table 201-3 and Fig 201-5.

If lumber is installed at about the % mc it will attain in service, only minor dimensional changes occur.

Table 201-3. Shrinkage of wood as % of green dimensions.
Total longitudinal shrinkage is 0.1%-0.3% of green size.

Species	Dried to 20% mc % Radial	% Tangential	Dried to 6% mc % Radial	% Tangential
Douglas Fir	1.7	2.6	4.0	6.2
Western Hemlock	1.4	2.6	3.4	6.3
Ponderosa Pine	1.3	2.1	3.1	5.0
Redwood	.9	1.5	2.1	3.5
Western Red Cedar	.8	1.7	1.9	4.0
Southern Yellow Pine	1.6	2.5	3.8	5.9

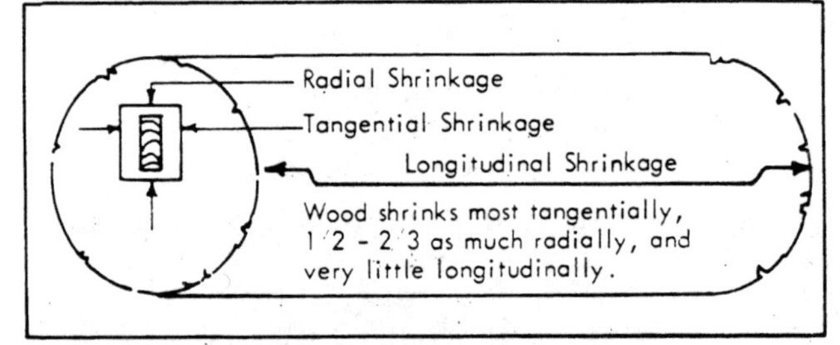

Fig 201-5. Relative shrinkage in three directions.

Typical lumber shrinkage problems

Shrinkages of house siding depend on initial moisture content, species, and climate. Excessive shrinkage breaks the paint seal at every board. If siding is at 20% mc, and due to climate will reach 6% equilibrium mc (emc) through part of each year, which species shrinks least? Compare 8" vertical grain siding of four species in Table 201-4. Western red cedar and redwood shrink and swell less than Douglas fir or southern pine.

Table 201-4. Siding shrinkage.

Wood	Radial Shrinkage Green to 6%	Green to 20%	Difference	Width of Board	Shrinkage
Western Red Cedar	1.9%	.8%	1.1%	8"	0.088"
Redwood	2.1	.9	1.2	8	0.096"
Douglas Fir	4.0	1.7	2.3	8	0.184"
Southern Yellow Pine	3.8	1.6	2.2	8	0.176"

Lumber

(Section prepared by Dr. J. O. Curtis, University of Illinois.)

Lumber is available in a number of species, strengths, appearance qualities, moisture contents, prices, etc. The objective is to obtain a satisfactory quality at the lowest possible cost.

Lumber is either ungraded (or native) lumber, or commerical lumber. Native lumber comes directly from local saw mills or is cut from the farmer's own logs. Commercial lumber is purchased via commercial trade channels and is available in standard strength and appearance grades.

Ungraded or Native Lumber

Native lumber includes some strong pieces, many pieces of adequate strength for light frame construction, and some weak pieces. It is a satisfactory material, especially if sorted to select the stronger pieces for members such as joists and rafters. The weaker pieces can be used for plates, sills, braces, studs in light buildings, etc. Native lumber is also recommended for subflooring, pens and partitions, and other nonstructural uses. When the stronger pieces are identified and used where strength is required, native lumber can often be substituted size-for-size for the commercial lumber normally shown on building plans.

The relative strengths of good clear samples of some common woods are shown to identify the strong and weak species.

Table 201-5. Relative strength of clear lumber.

Species	Relative strength in bending
Cottonwood	50%
Douglas-fir: coast type	100%
Elm: American	73%
Hemlock: western	86%
Hickory	127%
Maple: hard	100%
Oak: red and white	93%
Pine: southern yellow	100%
Spruce: Engleman	50%
Yellow-poplar	66%

Because native lumber can be used before it is thoroughly dry, it does not have its full strength at the time of fabrication. It dries fairly rapidly in place in the structure. Use ring-shank nails and tighten bolts after the lumber has dried.

Visually inspect and sort lumber to select the stronger pieces. Obviously those pieces that are nearly clear or that have only small knots are stronger. Select joists and rafters from pieces that have knots less than ⅓ the width of the piece in the middle ½ length. Knots may be up to ½ of the width near the ends of the piece. Discard pieces that show rot or that are noticeably undersize.

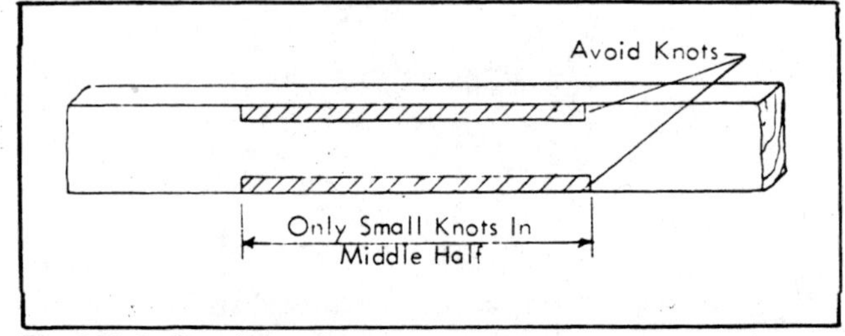

Fig 201-6. Small knots.

Commercial Lumber

Most commercial lumber in farm buildings is softwood lumber graded under the American Softwood Lumber Standard (ASLS). This standard is a common basis for uniform, industry wide inspection and grade marking practices for each piece of lumber produced and sold under its provisions. It also provides for coordinating the grades of species and preparing grading rules for each species. Lumber meeting the minimum size and grade provisions of the ASLS is graded under the rules of the various regional associations that have been approved by the Board of Review of the ASLS Committee.

Lumber is produced for various end uses, in a variety of sizes and moisture contents, and in several manufacturing classifications. Lumber is classified according to end use:

Yard lumber is for ordinary construction and general building purposes and is commonly available in local lumber yards.

Structural lumber is for framing requiring specific strength. Some structural lumber is also available from local lumber yards.

Factory and shop lumber is intended for remanufacturing.

Boards are less than 2″ in nominal thickness and 2″ or more in nominal width.

Dimension lumber is from 2″ to less than 5″ in nominal thickness and 2″ or more in nominal width.

Timbers are 5″ nominal or more in least dimension.

Lumber is classified according to nominal size, Table 201-6. It is also classified according to extent of manufacture:

Rough lumber has not been dressed (surfaced) but has been sawed, edged, and trimmed at least to saw marks showing in the wood on the four longitudinal surfaces of each piece for its overall length.

Dressed (surfaced) lumber has been planed to smooth surfaces and uniform size on one side (S1S), two sides (S2S), one edge (S1E), two edges (S2E), or a combination of sides and edges (S1S1E, S1S2E, S2S1E, S4S).

Worked lumber has been dressed, matched, shiplapped, or patterned.

Matched lumber has been worked to a tongue on one edge of each piece and a groove on the opposite edge for close tongue-and-groove joints by fitting two pieces together. When end-matched, the tongue and groove are also worked in the ends.

Shiplapped lumber has been worked or rabbeted on both edges of each piece for a close lapped joint between two pieces.

Patterned lumber is further worked to a pattern or molded form.

Lumber is surfaced either **dry** or **green. Dry** lumber has been seasoned or dried to a moisture content of 19% or less. **Green** lumber has a % mc over 19%. Lumber is surfaced green at a larger size than when surfaced dry so it is at the same size in service whether it was surfaced dry or green.

Table 201-6. Nominal and minimum dressed lumber sizes.
Thicknesses apply to all widths; widths apply to all thicknesses. Dressed sizes are for dry lumber.

	Thicknesses		Face widths	
Item	Nominal	Dressed	Nominal	Dressed
Boards	1	¾ in.	2	1½ in.
	1¼	1	3	2½
	1½	1¼	4	3½
			5	4½
			6	5½
			7	6½
			8	7¼
			9	8¼
			10	9¼
			11	10¼
			12	11¼
			14	13¼
			16	15¼
Dimension	2	1½	2	1½
	2½	2	3	2½
	3	2½	4	3½
	3½	3	5	4½
	4	3½	6	5½
	4½	4	8	7¼
			12	9¼
			12	11¼
Timbers (Dressed green)	5 and thicker	½ off	5 and wider	½ off

There are three regional associations that grade most softwood lumber: the Southern Pine Inspection Bureau (SPIB), the West Coast Lumber Inspection Bureau (WCLB), and the Western Wood Products Association (WWPA). Even though the grading rules of these three associations conform to the American Lumber Standard, there are variations in the names assigned to comparable grades.

Typical lumber grades and suggested uses for farm construction are in Table 201-7. For economy, select the lowest grade that is satisfactory for the intended use.

Most yard and structural lumber is grade stamped to help the consumer identify a given piece's grade. In addition to the grade, the stamp normally indicates the grading association, the mill that manufactured the lumber, the species, and the moisture content at which the lumber was surfaced. Typical grade stamps of two different grading associations are shown in Fig 201-7.

Designing With Lumber

(Reference: National Design Specification for Wood Construction, NFPA.)

Because strength is a function of moisture content, and member selection is based on both the size of the member and its strength, three elements are established before designing in wood: moisture content during use, size during use, and quality or allowable stresses.

Table 201-7. Grades of lumber.

Classification	Grades	Description or Use
Dimension		
2" to 5" thick 2" or more wide	**Structural Light Framing** (2"-4" thick, 2"-4" wide)	• For engineered use where higher strength is needed.
	Select Structural	• Use where high strength and stiffness and good appearance are needed.
	No. 1	• Use about the same as SEL STR, a little lower in quality.
	No. 2	• Recommended for most general construction uses.
	No. 3	• Use for general construction where appearance is not a factor.
	Light Framing (2"-4" thick, 2"-4" wide)	• Provides good appearance where high strength and high appearance are not needed.
	Construction	• Recommended and widely used for general framing purposes.
	Standard	• About same uses as CONSTRUCTION but a little lower in quality.
	Utility	• Used for studding, blocking, plates, etc. where economy and good strength are desired.
	Economy	• Suitable for crating, bracing and temporary construction.
	Studs (2"-4" thick, 2"-4" wide) Stud	• Only one grade; suitable for all stud uses.
	Structural Joists & Planks (2"-4" thick, 6" & wider)	• For engineering applications.
	Select Structural	• Use where high strength and stiffness and good appearance are needed.
	No. 1	• Use about same as SEL STR; a little lower quality.
	No. 2	• Recommended for most general co.. struction use.
	No. 3	• For use in general construction where appearance is not a factor.
	Appearance Framing (2"-4" thick, 2" & wider) A	• Use exposed in housing and light construction for high strength and finest appearance.
Timbers 5" or more in least dimension	Select Structural	• Use where superior strength and good appearance are needed.
	No. 1	• Similar uses to SEL STR; a little lower in quality.
	No. 2	• Recommended for general construction.
	No. 3	• Use for rough general construction.
Boards up to 1½" thick, 2" or more wide		• Graded for suitability for use in construction.
	SEL MER or 1	• Use in housing and light construction for exposed paneling, shelving, etc.
	CONST 2	• Use for subfloors, roof sheathing, etc.
	STD 3	• Used about the same as #2 but a little lower in quality.
	UTIL 4	• Combines usefulness and low cost for general construction purposes.
	ECON 5	• Use for low grade sheathing, crating and bracing.
Finish or Selects up to 1½" thick, 2" or more wide		• Graded on basis of appearance.
	B & BTR	• Nearly clear with only minor defects; suitable for natural finish.
	C	• More and larger defects than B & BTR but suitable for paint finish.
	D	• A little lower in quality than C but still suitable for paint finish.

WCLB = the grading association.
MILL 10 = the mill code number.
CONST = the lumber grade.
DOUG FIR = the lumber species.
S-DRY = lumber surfaced at 19% mc or less.

S-GRN = lumber surfaced green.
DOUG FIR-L = the species is mixed Douglas fir and larch.
WWP = the grading association.
12 = the mill code number.
SEL STR = the lumber grade.

Fig 201-7. Typical grade stamps.

Moisture Content

A green log contains free water in and between the cells and absorbed water in the capillaries of the walls of wood elements such as fibers and ray cells. The fiber saturation point (about 30% moisture for all species) is reached when all free water is removed and all absorbed water remains. For Douglas fir or southern pine, shrinkage occurs below 20% at about ⅕% of green size per 1% drop in moisture in edge grain, and ¼% per 1% in flat grain.

Moisture content of the wood during use affects not only the size and strength of the lumber, but also strength and perhaps durability of glue joints and performance of nails and other hardware fasteners.

Fig 201-8 relates relative humidity, dry bulb temperature, and equilibrium moisture content of wood. Even in livestock buildings, it is unlikely that relative humidities will remain high enough long enough to elevate the moisture content of lumber appreciably above 15%.

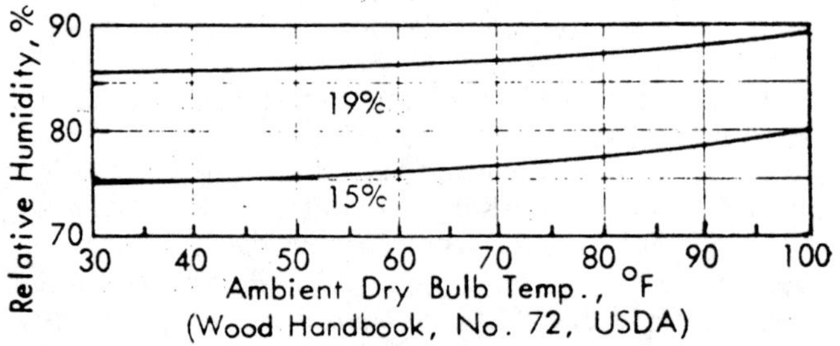

Fig 201-8. Equilibrium moisture content of wood.

Grading rules now increase allowable stresses for limited moisture content and set the thickness of 2" nominal lumber at 1½" for lumber dried to below 19% before milling.

The grading rules published by the WWPA (1970) establish three moisture levels with corresponding stresses and moduli of elasticity. The grading rules of the SPIB (1970) require kiln drying of all lumber to at least 19%.

Allowable stresses

Sample stresses for lumber grades and species commonly available are in the Design section. Engineers adjust these values for moisture content, duration of load, size and shape of cross section, fire-retardant treatment, and special conditions of shear and bearing.

Design

Stress

The term stress has several meanings. Stress, here, is the internal resistance of a member to external loads. In other words, stress is not the loads applied to the member, but rather the forces inside the member that hold up or resist the loads. Stress is measured in force per unit area, usually pounds per in², psi.

There are three basic types of stress—tension, compression, and shear. Tension stress occurs when two forces pull in opposite directions. Example: the rope in a game of tug-of-war is under tension stress. The bottom portion of a loaded floor joist is in tension. Compression stress occurs when two forces act toward each other. Example: 2x4 wall studs resisting a heavy roof load are under compression. Shear stress occurs when two forces, either toward or away from each other, do not act along the same line but are offset and tend to slide past each other. Example: a heavy load on a floor joist which in turn is resting on a basement wall. The basement wall pushes upward to resist the downward force of the load on the floor joist. At the edge of the basement wall, the floor joist tends to shear off and is under shear stress.

The design of wood members is rational—that is, principles of materials and how they behave are used to develop equations from which members and fasteners can be selected. The following derivation illustrates the procedure for a simple case.

More complete information is in Chapter 406.

Design example—simple beam

When a beam is loaded, it bends enough to support the load. Picture a 2x4 notched out, pinned at the centerline, and with two springs inserted near the outer edges. As the 2x4 bends, the top spring shortens—compression—and the bottom spring stretches—tension. The change in length of the springs is proportional to the load. The bigger the load, the bigger the change in length of each spring. Note that springs close to the hinge have little change in length.

In a wood beam, wood fibers act like springs. In a solid 2x4, the top fibers shorten a lot; near the centerline, they shorten a little. The same is true of the tension fibers.

The amount of compression is represented by the forces above the beam's centerline—a triangle of forces extending across the width. The amount of tension is below the centerline. C and T are at the centroids of their forces, ⅓ of the depth of the triangle, or d/6 from the outer edges of the 2x4.

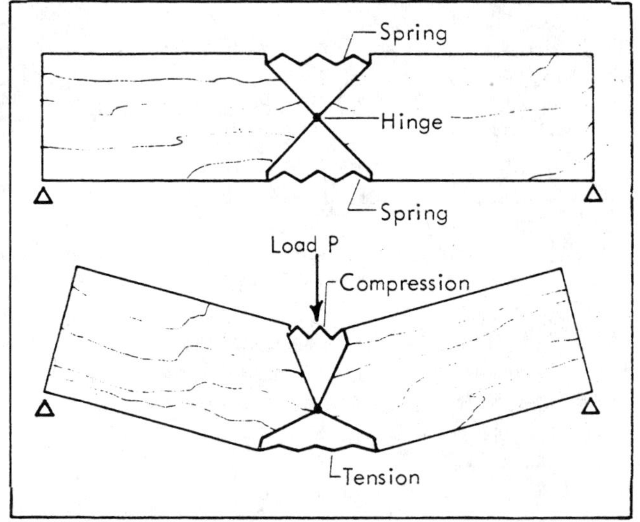

Fig 201-9. Schematic beam.

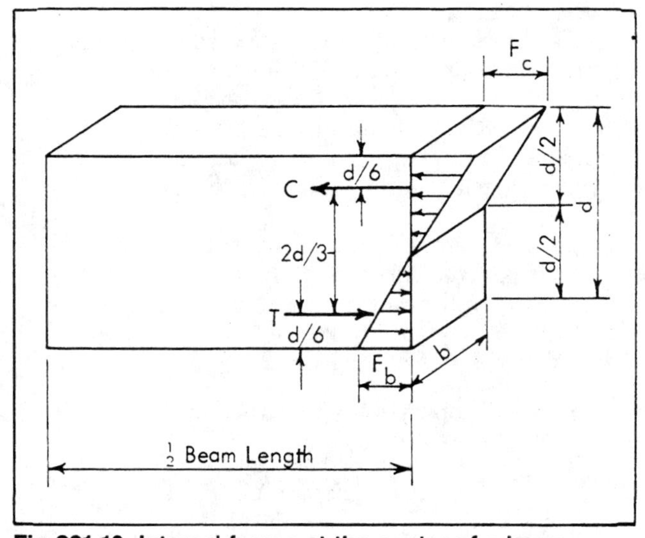

Fig 201-10. Internal forces at the center of a beam.

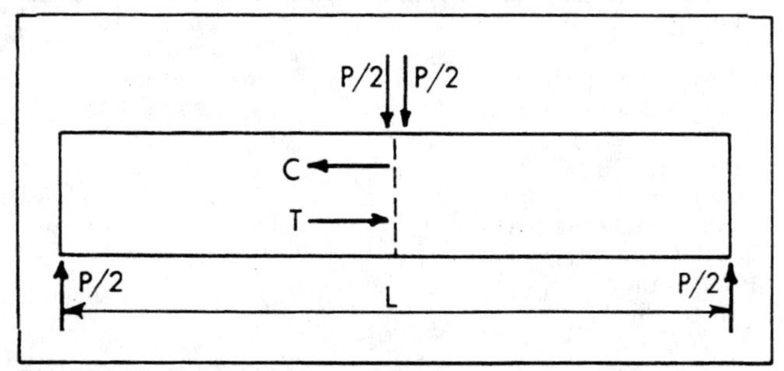

Fig 201-11. External beam forces.

$$T = C = (Fb \times b \times d/2) \div 2 \quad \text{Eq 201-2.}$$
$$= Fb \times bd \div 4$$

T = tension, lb
C = compression, lb
Fb = fiber stress, psi
b = member width, in.
d = member depth, in.

T and C are a couple, that is, two equal forces acting in opposite directions that tend to rotate the member. If you pushed-pulled on a 2x4 on a table, you would make the 2x4 rotate.

The couple inside the lumber is what resists the external forces on the beam. The load P is resisted by the two reactions, each P/2. There are two couples outside the beam: the forces P/2, L/2 apart.

The couple force times the couple distance is "bending moment," or moment, M. The external moment (Me) is:

$$Me = P/2 \times L/2$$
$$= PL/4 \text{ at center of beam}$$

The internal moment (Mi) is:

$$Mi = T \times 2d/3 \text{ (can use T or C)}$$

If the beam is in equilibrium (not moving or breaking), the internal couple is resisting the external couple, and the two couples are equal.

$$Me = Mi$$
$$PL/4 = T \times 2D/3 \text{ But,}$$
$$T = Fb \times bd/4$$
$$PL/4 = Fb \times bd/4 \times 2d/3$$
$$PL/4 = Fb \times bd^2/6$$

PL/4 is the moment due to load for a single load in the center of a simple beam. Moments for other cases are given at the end of the Design section.

Fb is the allowable bending stress, is a property of the material the beam is made of, and is given for some lumber species and grades in the Design section.

$bd^2/6$ is a property of the shape of the beam's cross section and is given for various lumber sizes in the Design section in the column headed "Section Modulus, S".

Therefore, M = FbS, or S = M ÷ Fb.

Usually, you know the lumber type, span, and load. Calculate the bending moment, M. Calculate the section modulus required: S = M/Fb. Find a member with an adequate S. Note: S is in^3, Fb is psi, M is in-lb. The load, w, is often lb/ft; divide it by 12. Use beam length in inches.

Design symbols

Section properties

For maximum strength of wood beams, the compression edge must have full length lateral support, the ends must be laterally supported, and bridging must support the tension edge.

Table 201-8. Bridging spacing for wood beams.
Bridging to reduce unbraced length to "short beam" for maximum allowable design stress.

Member depth	Maximum bridging spacing b=1½"	b=3"
3½"	33.5"	133.9"
5½"	21.3"	85.2"
7¼"	16.1"	64.6"
9¼"	12.6"	50.6"
11¼"	10.4"	41.6"

Poles

Preservative-treated poles are set in the ground as main structural members. Usually, the poles furnish both vertical support for the building and fixity at the bottom against rotation, to provide lateral bracing. Selecting and setting the poles is very important in pole building construction.

Wood species

Pole strength depends largely on the wood species. Table 201-9 lists the ultimate fiber stress in the outer fiber of the pole at failure in flexure as a cantilever beam. Note that weaker species are listed first. Other strength properties are also important, but the stress in bending is usually the limiting factor for poles used in pole buildings.

Table 201-9. Wood species for construction poles.
Code letters denote pole species.
Ultimate fiber stress, psi, is in the outer fiber of the pole at failure as a cantilever beam. The values are slightly reduced from average test values to provide for variability and for the effect of preservative treatment on strength. Reduce the stresses further for design to provide a safety factor depending on construction type.
ASAE-EP388

Species: Common Name	Botanical Name	Code Letters	Fiber Stress, psi†	Modulus of Elasticity, psi
Cedar, northern white	Thuja occidentalis	EC	4,000	880,000
Cedar, western red	Thuja plicata	WC	6,000	1,100,000
Pine, ponderosa	Pinus ponderosa	WP	6,000	1,100,000
Pine, jack	Pinus banksiana	JP	6,600	1,100,000
Pine, lodgepole	Pinus contorta	LP	6,600	1,320,000
Pine, red (Norway)	Pinus resinosa	NP	6,600	1,320,000
Douglas-fir, interior north	Pseudotsuga menziesii	DF	8,000	1,760,000
Douglas-fir, coast type	Pseudotsuga menziesii	DF	8,000	1,760,000
Larch, western	Larix occidentalis	WL	8,400	1,760,000
Pine, southern				
Loblolly	Pinus taeda	SP	8,000	1,760,000
Longleaf	Pinus palustris	SP	8,000	1,760,000
Shortleaf	Pinus echinata	SP	8,000	1,760,000
Slash	Pinus elliottii	SP	8,000	1,760,000

Wood quality

The average growth rate measured at the butt in the outer 2″ of poles should not be less than six rings/in., except that four rings are acceptable if 50% or more summerwood is present.

Prohibited defects are:

- Cross breaks, decay, dead streaks.
- Open or plugged holes, except plugged test holes up to ½″ diameter.
- Hollow butts, tops, and pith centers.
- Sap stain with softening or other wood disintegration.

Limited defects are:

- Firm red heart, with no softening or disintegration.
- Compression wood, if not visible in the outer 1″ of the pole.
- Insect damage, if no holes are over 1/16″ diameter or channeling is over 1/16″ deep.
- A knot, if not over 2″. The sum of knot diameters in 1′ should not exceed 4″ in poles to 16′ long, or 5″ in longer poles.
- Dead knots, if any decay is not associated with heart rot.
- Scars, if smoothly trimmed and not over 1″ deep.
- Shakes, if 2″ or more from the surface of the pole.
- Spiral grain, if not over one complete twist/30′.
- A split (or two checks terminating at the pith and separated at least 1/6 of the circumference), if not more than 6″ along the upper section of the pole.
- Shape defects are limited to sweep in only one plane and no "excessive" kinks or crooks. See Fig 201-12.

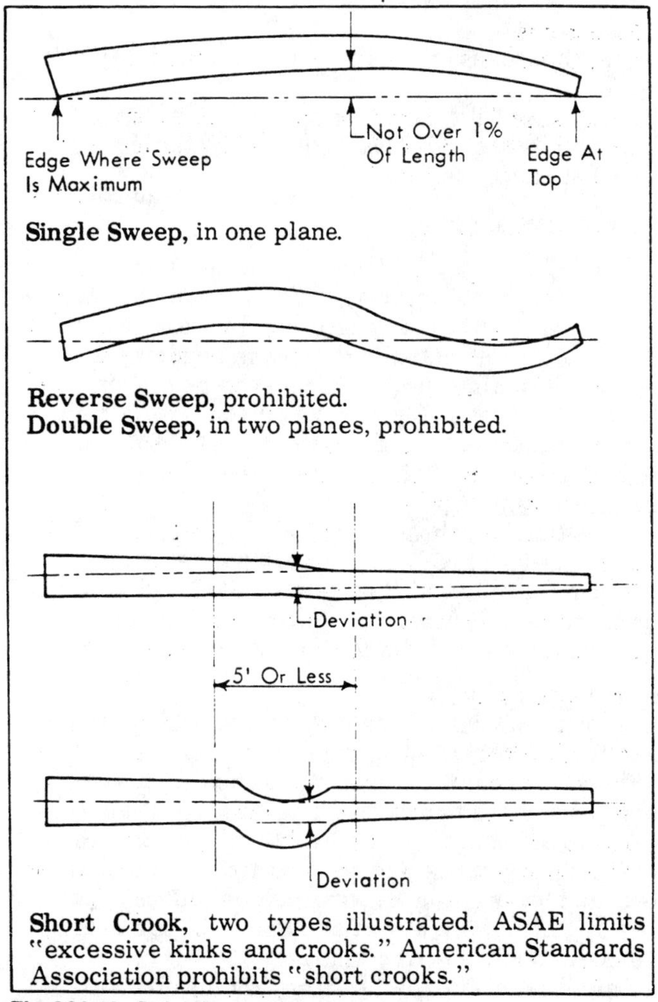

Edge Where Sweep Is Maximum
Not Over 1% Of Length
Edge At Top

Single Sweep, in one plane.

Reverse Sweep, prohibited.
Double Sweep, in two planes, prohibited.

Deviation
5′ Or Less
Deviation

Short Crook, two types illustrated. ASAE limits "excessive kinks and crooks." American Standards Association prohibits "short crooks."

Fig 201-12. Pole shape limitations.

Identification markings

For identification of graded poles, certain information is branded on the pole or included on a tag affixed to the pole. Obviously, it is better to brand poles on the side for better identification after construction. Brands or tags on pole ends are lost, because one end of the pole is put into the ground and the other may be cut off during construction.

Table 201-10. Dimensions of wood poles.
Breaking load is horizontal, 2′ down from the pole top. Groundline is for requirements on slabbing, scars, straightness, or other. For poles to 16′, 6000 psi, and 18′, 8000 psi: minimum circumference = top circumference + 0.075 in/ft taper; for 18′, 6000 psi, = 0.1 in/ft.

Class:		A=1	B=2	C=3	D=4	E=5	F=6	G=7	H=8	J=9
Breaking load, lb:		4,500	3,700	3,000	2,400	1,900	1,500	1,200	740	370
Min circumference at top, in.:		27	25	23	21	19	17	15	15	12
Length of pole, ft	**Groundline to butt ft**	**Min. circumference at 6′ from butt, in.**								
		Fiber stress = 6000 psi								
20	4	33.5	31.5	29.5	27.0	25.0	23.0	21.5	18.5	15.0
25	5	37.0	34.5	32.5	30.0	28.0	25.5	24.0	20.5	16.5
30	5.5	40.0	37.5	35.0	32.5	30.0	28.0	26.0	22.0	
35	6	42.5	40.0	37.5	34.5	32.0	30.0	27.5		
40	6	45.0	42.5	39.5	36.5	34.0	31.5	29.5		
		Fiber stress = 8000 psi								
20	4	31.0	29.0	27.0	25.0	23.0	21.0	19.5	17.5	14.0
25	5	33.5	31.5	29.5	27.5	25.5	23.0	21.5	19.5	15.0
30	5.5	36.5	34.0	32.0	29.5	27.5	25.0	23.5	20.5	
35	6	39.0	36.5	34.0	31.5	29.0	27.0	25.0		
40	6	41.0	38.5	36.0	33.5	31.0	28.5	26.5		

Dimensions

Length is measured between the ends and may be +6″ or -3″.

For species with minimum ultimate fiber stresses other than 6000 or 8000 psi, Table 201-10, the circumference at the groundline is:

Eq 201-3.

$$C = [wL/0.000264f]^{1/3}$$

C = groundline circumference, in.
w = breaking load 2′ from top of pole, lb
L = length between groundline and load, ft
f = ultimate fiber stress in bending, psi

The groundline circumferences can be converted to circumferences at 6′ from the butt by assuming a diameter taper of 0.1″/ft of length. When the poles are not perfectly round, the size is determined with a circumferential tape.

Classification: measure the circumference at 6′ from the butt. This dimension determines the true class of pole, provided that its top, measured at the minimum length point, is large enough. Otherwise the circumference at the top determines the class.

Round vs. sawn poles

For convenience in handling and attaching other structural members, sawn timbers are often used as columns instead of natural, round poles. However, round poles have two strength advantages. A round pole is 18% stronger in bending than a similar column with a rectangular cross section of the same area, and strength of a round pole is reduced very little by knots if limbs are not trimmed close enough to expose excessive cross grain.

Longer life is expected from round, pressure-treated poles because the sapwood is easier to treat than the exposed heartwood of sawn columns.

Poles are sometimes slabbed above the groundline on one or more sides before treating as a compromise between round and sawn. The butt section retains its thick envelope of thoroughly impregnated sapwood in the soil, where wood-destroying organisms are usually most active. Slabbing is a minimum cut to provide a single continuous flat face from groundline to top of intermediate wall poles, or two continuous flat faces at right angles from groundline to top of corner poles.

Depth of set (see Chapter 406)

Provide fixity at the base of the pole for structural stability against wind, snow, and ice loads. Both lateral and vertical design pressures vary considerably with soil conditions, reducing the practical value of equations for depth of set. Past experience is helpful in deciding on the depth of set and the material to be used around the pole. Usually, concrete provides necessary footing beneath the pole.

Wood Preservative Treatments

Organisms that destroy wood

The principal organisms that can degrade wood in farm buildings are fungi and insects.

The growth of fungi depends on mild temperatures, dampness, and oxygen. Termites are the major insect enemy of wood, but nationally are less serious than fungi.

Molds and fungi can stain sapwood different colors. Ordinarily they affect the strength only slightly. Stains can impair the use of wood where appearance is important.

Decay-producing fungi can attack either heartwood or sapwood. While the tree is alive, certain decay fungi attack the heartwood but rarely attack the sapwood. Most tree-attacking fungi die after the tree is cut. Other decay fungi destroy logs or manufactured products, such as sawed lumber.

Decay is relatively slow at temperatures below 50 F or above 90 F. Serious decay occurs only when the wood's moisture content is about 23%. The water vapor in humid air alone will not wet wood sufficiently to support significant decay. Wood can be too wet for decay as well as too dry. Water-soaked wood contains insufficient oxygen and will not decay.

Only the heartwood of decay resistant species contains natural preservative chemicals that retard fungi growth.

Insect damage. Bark and ambrosia beetles can damage freshly cut timber or rustic structures where the bark is left in place. Powder-post beetles may attack the sapwood of either freshly cut or seasoned timber. Prevent beetle damage by treating the wood with insecticides or fumigating existing buildings.

Subterranean termites cause most of the termite damage in the U.S. They develop colonies in the ground and build tunnels to a source of wood. Termites also require moisture. Termites do not infest buildings by being carried in lumber, but enter from ground nests. Signs of their presence are earthen tubes from the ground to the wood above.

The best protection from termites is to prevent them from gaining access to a building. Build the foundation of concrete or masonry that termites cannot penetrate. Use pressure treated wood where it is in contact with the ground or concrete. In crawl space foundations, keep floor joists at least 18″ and headers 12″ above the ground and provide good ventilation. In high termite hazard areas, chemically treat the soil around the foundation.

Carpenter ants sometimes damage poles, structural timbers, or buildings. They use wood for shelter rather than food. Carpenter ants may crawl directly into a building or be carried in firewood. Precautions that prevent termite attack and decay are usually effective against carpenter ants.

Wood preservatives

Wood can be protected from attack by applying selected chemicals. The degree of protection obtained depends on the kind of preservative used and the amount of penetration and retention. Some preservatives are more effective than others and some are more adaptable to certain use requirements.

Preservative oils. The wood does not swell from preservative oils, but it can shrink if it loses moisture during the treating process. Creosote and solutions with heavy petroleum oils help protect the wood from weathering outdoors but can adversely influence its cleanliness, odor, color, paintability, and fire resistance. Volatile oils or solvents with oil-borne preservatives that are removed after treatment leave the wood cleaner, but do not provide as much protection. Wood treated with preservative oils can be glued satisfactorily, but may require special cleaning to remove excess surface oils.

Coal-tar creosote is a black or brownish oil made by distilling coal tar. Its advantages include high toxicity to wood destroying organisms, relative low water solubility and volatility, high permanence, easy application, easily determined penetration, general availability and low cost, and a long record of satisfactory use.

Coal-tar creosote also has some disadvantages. It cannot be painted. The odor of creosoted wood is unpleasant to some persons. Creosote vapors are harmful to growing plants and some foodstuffs. Direct contact causes skin burns on some workers. Freshly creosoted wood can be easily ignited, readily burns, and produces dense smoke. After seasoning the more volatile oils disappear and creosoted wood burns like untreated wood.

Pentachlorophenol solutions for wood preservation generally contain 5% penta; the remainder is volatile solvents or heavy oils. The heavy oil solvents remain in the wood for a long time and usually do not produce a paintable surface. The volatile solvents are used with penta when the natural wood finish must be maintained or a paint finish is required.

Waterborne wood preservatives include acid copper chromate, ACC; chromated copper arsenate (types I, II, and III), CCA; ammonical copper arsenate, ACA; chromated zinc chloride CAC; and fluor chrome arsenate phenol, FCAP. These preservatives are often used when cleanliness and paintability are required. The CAC and FCAP formulations resist leaching less than preservative oils and are seldom used for wood in ground contact. The ACA and CCA treatments are now accepted for use on building foundations, building poles, utility poles, and piling. Waterborne preservatives leave the wood surface comparatively clean, paintable, and free from objectionable odor. These treatments decrease the danger of ignition and rapid flame spread, although formulations with copper and chromium stimulate and prolong glowing combustion of carbonized wood.

Recommended treatment levels

Preservative effectiveness is influenced by the type of chemical used, application method, penetration, and retention. Treating by brushing or dipping result in very low penetration and retention while pressure treating forces preservatives throughout the sapwood and can achieve high retention. Use only pressure treated material or decay resistant wood species in contact with the ground or manure.

Table 201-11. Recommended minimum preservative retentions, pcf.

Product and use	Creosote and creosote solutions	Penta-chlorophenol	Acid copper chromate, ACC	Ammoniacal copper arsenate, ACA	Chromated copper arsenate, CCA	Chromated zinc chloride, CAC	Fluor chrome arsenate phenol FCAP
Building poles (round)							
southern, Ponderosa pine,	7.5	0.38		0.60	0.60		
Red pine	10.5	0.53		0.60	0.60		
Coastal Douglas fir	9.0	0.45		0.60	0.60		
Jack pine, lodgepole pine	12.0	0.60					
Western larch, interior Douglas fir	16.0	0.80					
Fence posts							
Round, half or quarter-round less than 16′ long	8.0	0.40	0.50	0.40	0.40		
Lumber, timbers							
Above ground	8.0	0.40	0.25	0.25	0.25	0.45	0.25
Soil contact	10.0	0.50	0.50	0.40	0.40		
Major structural, membs.	12.0	0.60		0.60	0.60		
Foundations (home)				0.60	0.60		
Plywood							
Above ground	8.0	0.40	0.25	0.25	0.25	0.45	0.25
Soil contact	10.0	0.50	0.50	0.40	0.40		
Foundations (home)				0.60	0.60		
Plant structures							
Above ground				0.25	0.25		0.25
Soil contact				0.40	0.40		

Source: Fed Spec TT-W-00571J (AGR-AFS) and AWPS-C16.

202 WOOD GIRDER, POST, COLUMN, AND TRUSS SELECTION

This chapter presents design tables developed for selecting structural elements for farm building plans. Design loads are discussed in Section 100, and designing in Section 400.

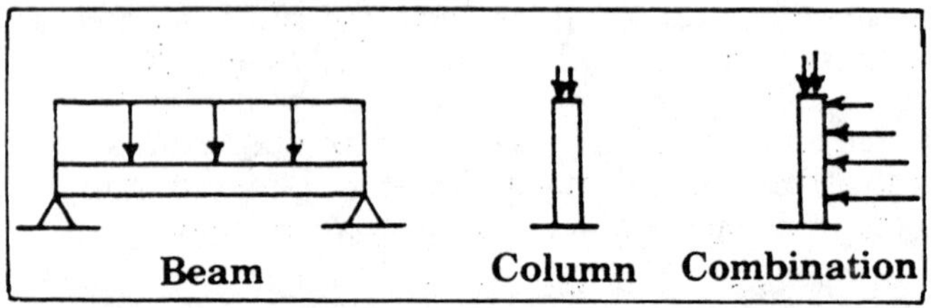

Fig 202-1. Basic structural loads.

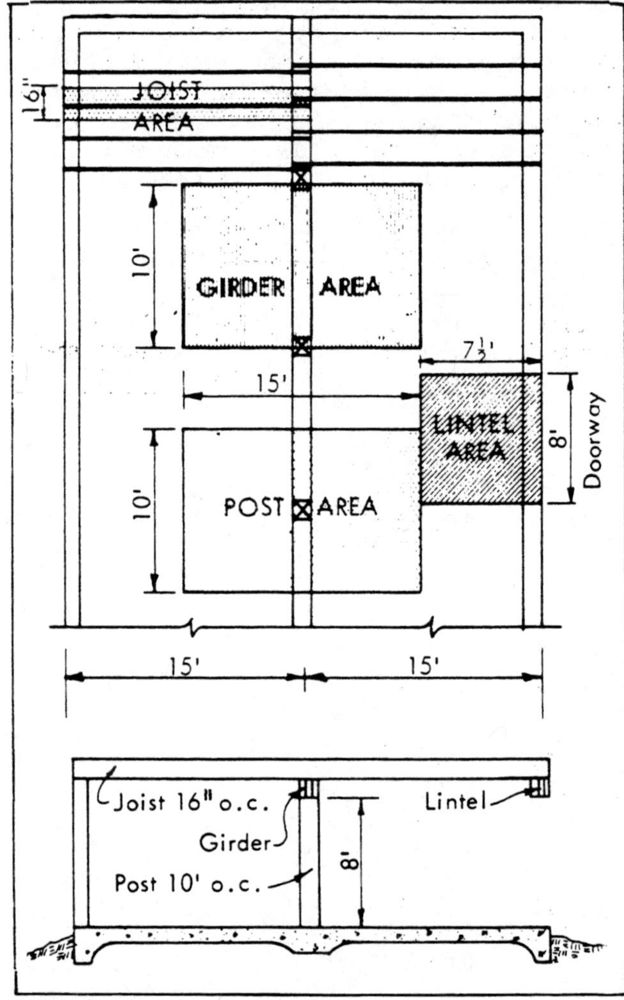

Fig 202-2. Loads on supports.

A beam supports loads that try to bend it—joists, lintels, a diving board, ladder rungs, etc.

A column supports loads that try to shorten it—studs or poles holding a roof, chair legs, ladder side rails, etc.

A support can be both—grain bin walls resist bending outward and also hold up the roof.

Rafters for low-slope roofs are commonly designed as beams, even though they carry some compression.

Beam Selection

Member Lengths

Beams: Use span between supporting columns or walls.

Columns: Use total length.

Example in Fig 202-2: joist = 15′; girder = 10′; lintel = 8′; post = 8′.

Loads on Members

Example total floor load is 40 psf.

Joist load, lb/ft
= (joist spacing, ft) x (floor load, lb/ft^2)
= (16″ ÷ 12 in/ft) x (40 psf) = 53.3 lb/ft

Girder load, lb/ft
= (girder spacing, ft) x (floor load, lb/ft^2. In this case, the girder spacing is half the building width.
= (15′) x (40 psf) = 600 lb/ft

Post load, lb
= (girder span, ft) x (joist span, ft) x (floor load, lb/ft^2)
= (10′) x (15′) x (40 psf) = 6000 lb

Lintel load, lb/ft
= (joist span ÷ 2, ft) x (floor load, lb/ft^2)
= (15′ ÷ 2) x (40 psf) = 300 lb/ft

Lumber Girt and Purlin Selection

Maximum spans and loads for purlins and girts are in Tables 202-1 to 202-4, Controlling loads are assumed to be snow for roof purlins and wind for wall girts.

Calculations assume purlins and girts are two-span continuous beams, and that bending is around one axis only, with rotational effects ignored. Rotational and minor axis bending effects become more critical as roof slope increases—use the table only for roof slopes of 4/12 or less.

Allowable bending and shear stresses are increased 15% (snow + dead) for purlins and 33% (wind), and also 15% for repetitive member bending with members 24″ o.c. or less. Maximum superimposable loads are listed for members both 24″ o.c. and more than 24″ o.c.

Table 202-1. Maximum spans for lumber purlins, ft.
#2 southern pine at 15% moisture content. Deflection limited to L/180′; "*" means deflection controls.

Purlin Size	Purlin Spacing (In)	Vertical Roof Load, Snow Plus Dead, PSF							
		15	20	25	30	35	40	45	50
		---- Maximum span, ft. ----							
Members Flat									
1x4	16	4.1*	3.7*	3.5*	3.3*	3.1	2.9	2.7	2.6
	20	3.8*	3.5*	3.2*	3.0	2.8	2.6	2.4	2.3
	24	3.6*	3.3*	3.0	2.7	2.5	2.4	2.2	2.1
	28	3.3	2.9	2.6	2.4	2.2	2.0	1.9	1.8
	32	2.1	2.7	2.4	2.2	2.0	1.9	1.8	1.7
	36	2.9	2.5	2.3	2.1	1.9	1.8	1.7	1.6
	40	2.8	2.4	2.2	2.0	1.8	1.7	1.6	1.5
	44	2.7	2.3	2.1	1.9	1.7	1.6	1.5	1.5
	48	2.5	2.2	2.0	1.8	1.7	1.6	1.6	1.5
									1.4
1x6	16	4.8*	4.4*	4.0*	3.8*	3.6*	3.5*	3.3*	3.2*
	20	4.4*	4.0*	3.8*	3.5*	3.4*	3.2*	3.1	2.9
	24	4.2*	3.8*	3.5*	3.3*	3.2*	3.0	2.8	2.7
	28	4.0*	3.6*	3.2	3.0	2.7	2.6	2.4	2.3
	32	3.8*	3.4	3.0	2.8	2.6	2.4	2.3	2.1
	36	3.7*	3.2	2.9	2.6	2.4	2.3	2.1	2.0
	40	3.5	3.0	2.7	2.5	2.3	2.1	2.0	1.9
	44	3.3	2.9	2.6	2.4	2.2	2.0	1.9	1.8
	48	3.2	2.8	2.5	2.3	2.1	2.0	1.8	1.8
2x4	16	8.2*	7.5*	7.0*	6.5*	6.2	5.8	5.5	5.2
	20	7.7*	7.0*	6.5*	6.0	5.5	5.2	4.9	4.6
	24	7.2*	6.5*	6.0	5.5	5.1	4.7	4.5	4.2
	28	6.7	5.8	5.2	4.7	4.4	4.1	3.9	3.7
	32	6.2	5.4	4.8	4.4	4.1	3.8	3.6	3.4
	36	5.9	5.1	4.6	4.2	3.9	3.6	3.4	3.2
	40	5.6	4.8	4.3	3.9	3.7	3.4	3.2	3.1
	44	5.3	4.6	4.1	3.8	3.5	3.3	3.1	2.9
	48	5.1	4.4	3.9	3.6	3.3	3.1	2.9	2.8
2x6	16	9.6*	8.7*	8.1*	7.6*	7.1	6.7	6.3	6.0
	20	8.9*	8.1*	7.5*	6.9	6.4	6.0	5.6	5.3
	24	8.4*	7.6*	6.9	6.3	5.8	5.4	5.1	4.9
	28	7.7	6.6	5.9	5.4	5.0	4.7	4.4	4.2
	32	7.2	6.2	5.6	5.1	4.7	4.4	4.1	3.9
	36	6.8	5.9	5.2	4.8	4.4	4.1	3.9	3.7
	40	6.4	5.6	5.0	4.5	4.2	3.9	3.7	3.5
	44	6.1	5.3	4.7	4.3	4.0	3.7	3.5	3.3
	48	5.9	5.1	4.5	4.1	3.8	3.6	3.4	3.2
Members on Edge									
2x4	16	14.5	12.5	11.2	10.2	9.5	8.9	8.3	7.9
	20	12.9	11.2	10.0	9.1	8.5	7.9	7.5	7.1
	24	11.8	10.2	9.1	8.3	7.7	7.2	6.8	6.5
	28	10.2	8.8	7.9	7.2	6.7	6.2	5.9	5.6
	32	9.5	8.3	7.4	6.7	6.2	5.8	5.5	5.2
	36	9.0	7.8	7.0	6.4	5.9	5.5	5.2	4.9
	40	8.5	7.4	6.6	6.0	5.6	5.2	4.9	4.7
	44	8.1	7.0	6.3	5.7	5.3	5.0	4.7	4.5
	48	7.8	6.7	6.0	5.5	5.1	4.8	4.5	4.3
2x6	16	20.8	18.0	16.1	14.7	13.6	12.7	12.0	11.4
	20	18.6	16.1	14.4	13.2	12.2	11.4	10.7	10.2
	24	17.0	14.7	13.2	12.0	11.1	10.4	9.8	9.3
	28	14.7	12.7	11.4	10.4	9.6	9.0	8.5	8.0
	32	13.7	11.9	10.6	9.7	9.0	8.4	7.9	7.5
	36	12.9	11.2	10.0	9.1	8.5	7.9	7.5	7.1
	40	12.3	10.6	9.5	8.7	8.0	7.5	7.1	6.7
	44	11.7	10.1	9.1	8.3	7.7	7.2	6.8	6.4
	48	11.2	9.7	8.7	7.9	7.3	6.9	6.5	6.1
2x8	16	27.4	23.8	21.2	19.4	18.0	16.8	15.8	15.0
	20	24.5	21.2	19.0	17.3	16.1	15.0	14.2	13.4
	24	22.4	19.4	17.3	15.8	14.7	13.7	12.9	12.3
	28	19.3	16.7	15.0	13.7	12.7	11.8	11.2	10.6
	32	18.1	15.7	14.0	12.8	11.8	11.1	10.4	9.9
	36	17.1	14.8	13.2	12.1	11.2	10.4	9.8	9.3
	40	16.2	14.0	12.5	11.4	10.6	9.9	9.3	8.9
	44	15.4	13.4	11.9	10.9	10.1	9.4	8.9	8.4
	48	14.8	12.8	11.4	10.4	9.7	9.0	8.5	8.1

Table 202-2. Maximum spans for lumber girts, ft.-15 psf load.
#2 southern pine at 15% moisture content. Deflection limited to L/120; "*" means deflection controls.

Spacing, in.	16"	20"	24"	28"	32"	36"	40"	44"	Dotted line is 48" example in text
Member size					Maximum span, ft.				
1x4 Flat	4.7*	4.4*	4.1*	3.6	3.4	3.2	3.0	2.9	2.7
1x6 Flat	5.5*	5.1*	4.8*	4.5	4.2	4.0	3.8	3.6	3.4
2x4 Flat	9.4*	8.8*	8.3*	7.2	6.7	6.3	6.0	5.7	5.5
2x6 Flat	11.0*	10.2*	9.6	8.3	7.7	7.3	6.9	6.6	6.3
2x8 Flat	12.0*	11.2*	10.5*	9.5	8.9	8.4	7.9	7.6	7.2

Table 202-3. Maximum superimposable loads for lumber purlins, plf.
#2 southern pine at 15% moisture content. Deflection limited to L/120; "*" means deflection controls. "v" means shear controls.

Span	1'	2'	3'	4'	5'	6'	8'		
Member size				Maximum superimposable load, plf					
1x4 Flat	352v	112	49	21*	11*	6*	2*	1*	Spaced 24" o.c. or less
	306v	97	43	21*	11*	6*	2*	1*	Spaced more than 24"
1x6 Flat	553v	176	78	34*	17*	10*	4*	2*	
	481v	153	68	34*	17*	10*	4*	2*	
2x4 Flat	703v	350v	199	112	71	49	21*	11*	
	612v	305v	173	97	62	43	21*	11*	
2x6 Flat	1105v	553v	262	147	94	65	34*	17*	
	961v	481v	228	128	82	57	32	17*	
2x4 Edge	703v	351v	234v	176v	141v	116	65	41	
	612v	305v	204v	153v	122v	101	56	36	

Span	4'	5'	6'	8'	10'	12'	14'	16'	18'	20'
Member size				Maximum superimposable load, plf						
2x4 Edge	176v	141	116	65	41	29	21	14*	10*	7*
	153v	122	101	56	36	25	18	14*	10*	7*
2x6 Edge	276v	221v	184v	135	86	60	44	33	26	21
	240v	192v	160v	117	75	52	38	29	23	18
2x8 Edge	364v	292v	243v	182v	146v	104	76	58	46	37
	317v	253v	211v	158v	127v	90	66	51	40	32

Table 202-4. Maximum superimposable loads for lumber girts, plf.
#2 southern pine at 15% moisture content. Deflection limited to L/180; "*" means deflection control. "v" means shear controls.

Span	1'	2'	3'	4'	5'	6'
Member size			Maximum superimposable load, plf			
1x4 Flat	406v	129	57	32	16*	9* Spaced 24" o.c. or less
	354v	112	50	28	16*	9* Spaced more than 24" o.c.
1x6 Flat	639v	203	90	50	26*	15*
	556v	177	78	44	26*	15*

Span	4'	5'	6'	8'	10'	12'
Member size			Maximum superimposable load, plf			
2x4 Flat	129	83	57	32	16*	9*
	112	72	50	28	16*	9*
2x6 Flat	170	109	75	42	26*	15*
	148	95	66	37	23	15*
2x8 Flat	225	144	100	56	34*	20*
	195	125	87	48	31	20*

Girder Selection

Girders are beams that support bending loads from other members, e.g., trusses between posts in a pole building, or rafters across a door opening in a stud wall building.

When selecting a girder, first determine how much load the girder must support and what its span is. Calculate a truss reaction with the following equation:

Truss reaction = (G) x (span/2) x (spacing)
Typical units: lb = (psf)(ft)(ft)
Where G = gravity loads: roof snow + roof dead + ceiling dead

Calculate the girder load from a uniformly distributed load with the equations in Load on Member, above.

Using the girder load and span, select a girder from Table 202-5. A girder may be simply supported or braced at each end as in Fig 202-3. Well fastened braces significantly increase the girder's strength.

Lumber girders

Table 202-5 is based on an allowable bending stress of 1500 psi, No. 1 Douglas fir or southern yellow pine. If lumber with a different allowable bending stress is used, multiply the tabulated loads by the ratio of its bending stress over 1500 psi.

Table 202-5. Allowable truss reactions on girders, lb.
1500 psi allowable bending stress. 15% increase for repetitive members, three or more in one beam, and load duration are included. Dressed lumber sizes. Deflection not limited. Braced girders have knee braces at each support.

Girder span	Member size	Section modulus	Simply supported girder: Uniform load lb/ft	Simply supported girder, Truss spacing, Truss reaction, lb: 2′ o.c.	4′ o.c.	8′ o.c.	Braced girder: Uniform load lb/ft	Braced girder, Truss spacing, Truss reaction, lb: 2′ o.c.	4′ o.c.	8′ o.c.
4′	moment		$wL^2/8$	$PL/4$			$wL^2/12$			
	deflection		$5wL^4/384EI$	$PL^3/48EI$			$wL^4/384EI$			
	2x6	7.56	543	1086			815			
	2x8	13.14	944	1888			1417			
	2-2x6	15.12	1087	2173			1630			
	2x10	21.39	1537	3074			2306			
	2-2x8	26.28	1889	3777			2833			
	2x12	31.64	2274	4548			3411			
	2-2x10	42.78	3075	6149			4612			
8′	moment		$wL^2/8$	$PL/2$	$PL/4$		$wL^2/12$			
	deflection		$5wL^4/384EI$	$19PL^3/384EI$	$PL^3/48EI$		$wL^4/384EI$			
	2x6	7.56	136	271	543		204			
	2x8	13.14	236	472	944		354			
	2-2x6	15.12	272	543	1086		408			
	2x10	21.39	384	768	1537		577			
	2-2x8	26.28	472	944	1888		708			
	2x12	31.64	569	1137	2274		853			
	2-2x10	42.78	769	1537	3074		1153			
	2-2x12	63.28	1137	2274	4548		1706			
	3-2x10	64.17	1326	2652	5304		1989			
12′	moment		$wL^2/8$	$3PL/4$	$PL/3$		$wL^2/12$	$0.4861PL$	$0.2222PL$	
	deflection		$5wL^4/384EI$	$11PL^3/144EI$	$23PL^3/648EI$		$wL^4/384EI$	$PL^3/64EI$	$10PL^3/1296EI$	
	2x10	21.39	171	341	768		256	527	1153	
	2-2x8	26.28	210	419	944		314	647	1416	
	2x12	31.64	253	505	1137		379	779	1705	
	2-2x10	42.78	342	683	1537		512	1054	2306	
	2-2x12	63.28	505	1010	2274		758	1559	3411	
	3-2x10	64.17	589	1178	2652		884	1819	3978	
	4-2x10	85.56	786	1571	3536		1178	2425	5304	
	3-2x12	94.92	872	1743	3922		1307	2690	5884	
16′	moment		$wL^2/8$	PL	$PL/2$	$PL/4$	$wL^2/12$	$0.6563PL$	$5PL/16$	$PL/8$
	deflection		$5wL^4/384EI$	$27PL^3/256EI$	$19PL^3/384EI$	$PL^3/48EI$	$wL^4/384EI$	$PL^3/48EI$	$PL^3/96EI$	$PL^3/192EI$
	2x10	21.39	96	192	384	768	144	292	614	1537
	2-2x8	26.28	118	236	472	944	177	359	755	1888
	2x12	31.64	142	284	568	1137	213	433	909	2274
	2-2x10	42.78	192	384	768	1537	288	585	1229	3074
	2-2x12	63.28	284	568	1137	2274	426	866	1819	4548
	3-2x10	64.17	331	663	1326	2652	497	1010	2121	5304
	4-2x10	85.56	442	884	1768	3536	663	1347	2828	7072
	3-2x12	94.92	490	980	1961	3922	735	1494	3138	7845
	4-2x12	126.56	653	1307	2615	5230	980	1992	4184	10460

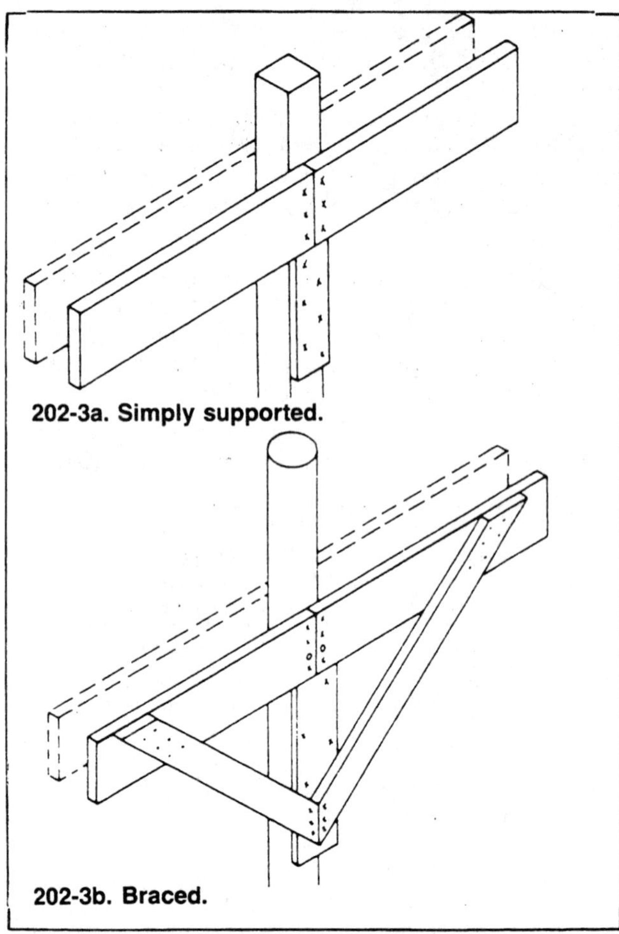

202-3a. Simply supported.

202-3b. Braced.

Fig 202-3. Girder support.

Table 202-6. Allowable loads on 5″ thick plywood glued laminated girders.
Repetitive members bending stresses of 1750 psi and 2050 psi for No. 1 and Select Structural Douglas fir or southern pine lumber respectively plus 15% increase for snow load, 2 months duration.

	Uniform load lb/ft	Truss spacing, o.c. 2′	4′	8′
No. 1 lumber and C-D plywood*				
24′ girder		Truss reaction, lb.		
18½″ deep	467	934	1924	4089
24″ deep	786	1573	3238	6882
32′ girder				
24″ deep	437	874	1776	3670
29¾″ deep	671	1342	2729	5640
Select structural lumber and structural 1 ext plywood				
24′ girder		Truss reaction, lb.		
18½″ deep	557	1114	2294	4874
24″ deep	937	1875	3860	8204
32′ girder				
24″ deep	520	1041	2117	4375
29¾″ deep	800	1600	3253	6723

*Use 80% of these values for No. 1 Hem-fir and 94% for Select Structural Hem-fir. Use C-D plywood.

Bending stress: 1000 1200 1400 1500 1600 1700
Multiply load by: 0.67 0.80 0.93 1.00 1.07 1.17

Example: Select a 16′ door header for a 40′ wide uninsulated pole machine shed with trusses 8′ o.c. located where snow load is 20 psf.

The girder span is 16′. The total roof load is 20 psf + 1 psf truss weight. The truss reaction is: (20 psf + 1 psf) x (40′ ÷ 2) x (8′) = 3360 lb. Assuming the girder is simply supported, Table 202-5 shows that 4-2x10′s or 3-3x12′s are adequate.

Laminated Girders

For long spans, more strength is required than is readily obtained with dimension lumber. Vertically laminating and gluing dimension lumber between layers of plywood produces strong straight girders. Hollow center box beams are similar, but need to be built in a jig to ensure straightness. When bar clamps are used to bring the narrow faces of two pieces of dimension lumber together, the resulting member is usually straight without the need for a jig.

Material Specifications

The girders listed in Table 202-7 are made of two high grades of lumber and plywood:

- No. 1 Douglas fir or southern pine lumber with ½″ C-D plywood, Identification Index = 32/16.
- Select Structural Douglas fir or southern pine lumber with ½″ C-C Exterior Structural I plywood.

Select a glue for the expected exposure. Casein, which is highly mold and water resistant but not waterproof, is adequate for most girders. Use it for girders that will stay dry throughout their life. Resorcinol resin is waterproof; use it and C-C EXT plywood for girders exposed to unusual moisture conditions. Read the label carefully before you buy—make sure you are getting the right glue.

Remove all dirt, oil, and sand from the lumber and plywood. Protect the girders from moisture for one week after fabrication. Temperatures below 70 F delay curing. Girders can be installed after 24 hr at 70 F, but require at least a week at 40 F.

Construction

Each girder has 6 layers. Locate members and joints as shown in Figs 202-4 and 202-5. The plywood may have horizontal joints if they are staggered at least 2″ from dimension lumber joints. Avoid horizontal joints along the neutral axis (mid-height) of the girder.

Assemble a girder in two pieces—layers 1, 2, and 3, and layers 4, 5, and 6. Clamp the narrow faces of the dimension lumber together.

For example, layer #2 = 2x8 + 2x12 = 2x20, 1½″x18½″. Spread glue on the plywood (layer #1).

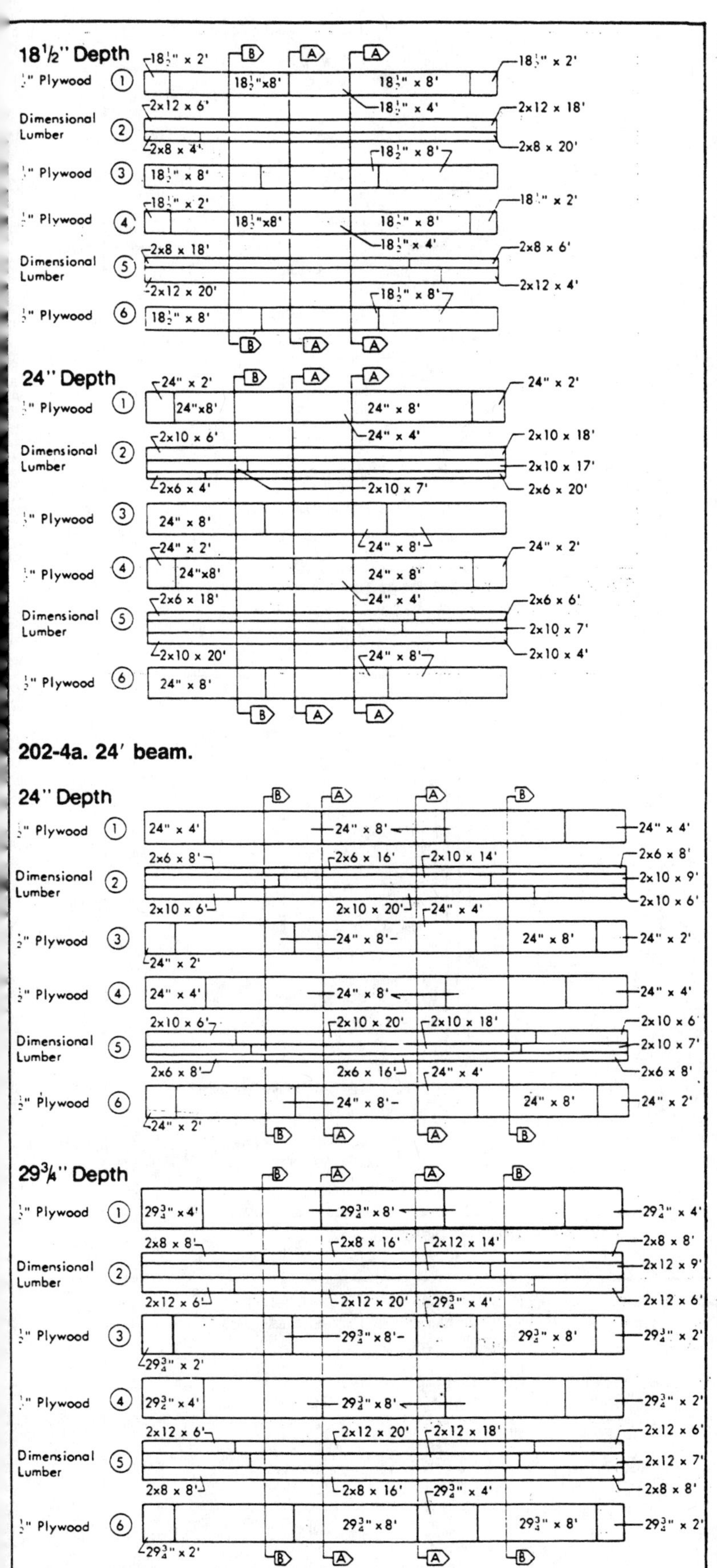

202-4a. 24′ beam.

202-4b. 32′ beam.

Fig 202-4. Cutting diagrams.

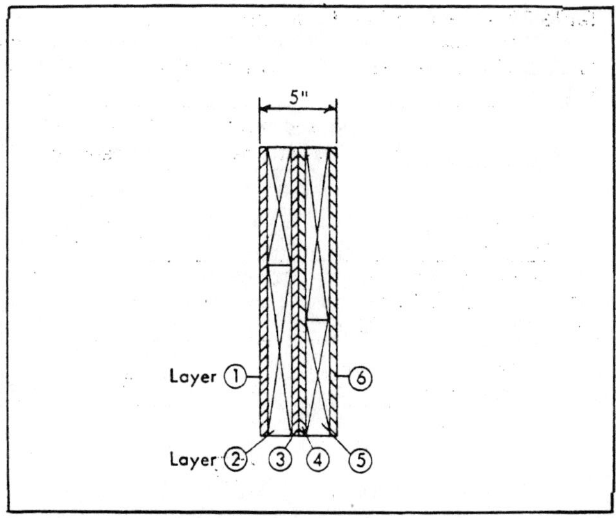

Fig 202-5. Laminated beam cross section.

Nail plywood to layer #2 with 6d nails, preferably galvanized or cement coated, 4″ o.c. both ways. Glue should squeeze out from the edges of the beam. Remove the clamps; similarly glue and nail layer #3 plywood to the other side of the dimension lumber. Then assemble layers #4, #5, and #6.

For final assembly, use method A or B.

A. Clamping method. When both halves of a beam have been assembled, apply glue to the two remaining inside surfaces. Place clamps about 2′ apart on the fully assembled beam and leave on for 24 hr. ½″ bolts can be used instead of clamps.

B. Weighting method. When both halves of the beam have been assembled, apply glue to the remaining inside surfaces. Lay the beam on a level surface. Place sufficient weight on the fully assembled beam to squeeze glue out from the edges of the beam. Leave on for 24 hr.

Pole Selection

Pole-to-Truss Fixity

To approximate a fixed joint at the top of a pole, fasten a knee brace at each pole, forming a triangle of at least 30″ on a side. The knee brace is 2″ dimension lumber as wide as the greatest width of the pole, e.g. a 2x6 for a 6x6 or a 2x8 for a 6x8. Securely anchor the knee brace to the pole and truss, Fig 202-6. If you leave the knee brace out, you have a typical pinned truss-to-pole connection.

Splicing Dimension Lumber Poles

If dimension lumber poles are made with pressure preservative treated wood in the ground and untreated wood above ground, the pole-to-truss joint must be fixed, preferably by a knee brace. Join the treated and untreated lumber at approximately ⅓ the wall height. Make the lap joint at least 12″ long and securely nailed or bolted. See Fig 202-7.

Table 202-7. Pinned-end columns.
Load = (snow load + dead load, psf) x ½ span x pole spacing. Circled numbers refer to examples. Tabulated values are based on the following assumptions:

	Allowable fiber stress	Modulus of elasticity
	psi	psi
Doug fir-larch	1250	1,800,000
Hem-fir	1000	1,500,000

Pole designs in Table 202-7 are for compression only, and do not allow for lateral wind stability. If either one or both ends of the column is fixed, select the column size from the Design section.

Pole dimension	Height ft	Load, lb Doug fir	Load, lb Hem fir
4x4 dressed	8'	9665	8084
	10'	6162	5133
	12'	4288	3577
	14'	3148	2622
4x4	8'	15,584	12,768
	10'	10,528	8,768
	12'	7,312	6,096
	14'	5,376	4,480
	16'	4,112	3,424
5x5 dressed	8'	21,850	17,719
	10'	16,808	13,973
	12'	11,705	9,761
	14'	8,627	7,189
	16'	6,581	5,488
	18'	5,204	4,334
5x5	8'	28,425	22,925
	10'	24,350	19,950
	12'	17,850	14,875
	14'	13,125	10,925
	16'	10,050	8,375
	18'	7,925	6,600
	20'	6,425	5,350
6x6 dressed	8'	35,453	28,496
	10'	32,125	26,076
	12'	24,955	21,568
	14'	19,269	16,063
	16'	14,732	12,251
	18'	11,586	9,680
	20'	9,438	7,865
6x6	8'	43,020	34,560
	10'	40,212	32,508
	12'	35,064	28,278
	14'	27,180	22,644
	16'	20,808	17,352
	18'	16,452	13,716
	20'	13,320	11,088

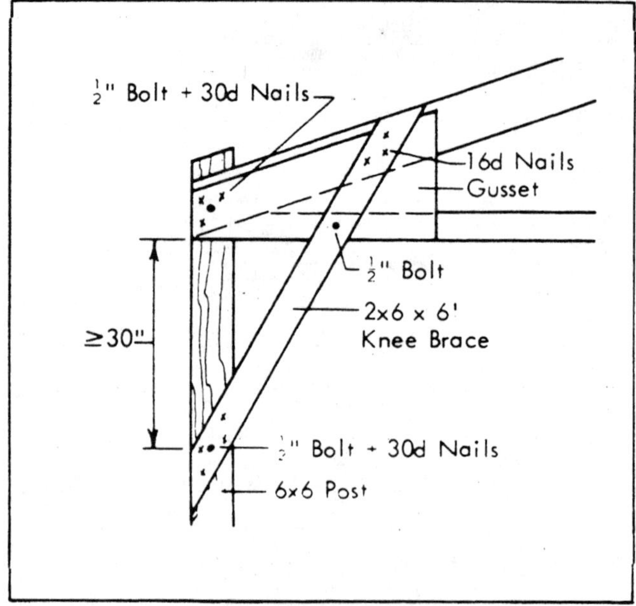

Fig 202-6. Typical fixed pole-to-truss connection.
Use ringshank hardened nails and bolts with washers.

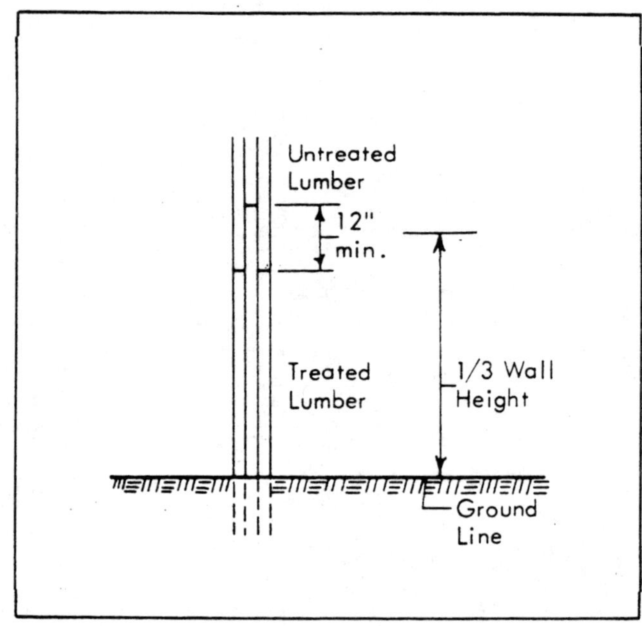

Fig 202-7. Laminated pole lap joint.

Table 202-8. Pole dimensions.

Sawn poles, in.

Nominal	Dressed
3x6	2½x5½
4x4	3½x3½
4x6	3½x5½
4x8	3½x7¼
4x10	3½x9¼
5x5	4½x4½
6x6	5½x5½
6x8	5½x7½
6x10	5½x9½
8x8	7½x7½
8x10	7½x9½
10x10	9½x9½

Round Poles, in.
Assumed dia. taper of 0.075 in/ft

ASA Class	Top dia.
1	8.6
2	8.0
3	7.3
4	6.7
5	6.0
6	5.4
7	4.8
10	3.8

2" Dimension laminated poles, in.
Repetitive member bending stresses used with 3-2x6's and up

Nominal	Laminated
2-2x4	3x3½
2-2x6	3x5½
2-2x8	3x7¼
3-2x6	4½x5½
3-2x8	4½x7¼
3-2x10	4½x9¼
4-2x8	6x7¼
4-2x10	6x9¼
4-2x12	6x11¼
5-2x10	7½x9¼
5-2x12	7½x11¼
6-2x12	9x11¼

Table 202-9. Douglas fir-larch or southern pine poles.

Lumber grades	Allowable fiber stresses, psi Bending F_b	Compression F_c	Tension F_t
Sawn poles (dressed and full size)			
Construction, 4x4	1050	1100	620
*No. 1, 3x6 & up	1200	825	800
Dimension lumber			
Const. grade @ 19%, 2x4	1050	1100	620
No. 1 @ 19%, 2x6 & up	1500	1250	1000
No. 1 @ 19%, 2x6 & up repetitive member	1750	1250	1000
Round poles (8000 psi ultimate strength; allowable stresses include 18% form factor)	2125	1400	1150

*For Southern yellow pine, 5x5 and up, bending stress can be 1300 psi; selection tables are in TR-6.

Truss Selection

Trusses are one roof framing systems for clear-span buildings. Clear-span construction provides for simple floor space arrangement. Halls, alleys, partitions, and equipment can be arranged and rearranged for best functional use of the space. There are no posts or bearing walls in the way. Machines and vehicles have maximum freedom of movement inside the building.

The bottom chords of trusses can support a ceiling. Head room clearance is restricted compared to rigid frame construction. Bird roosting can also be a problem.

Truss selection depends on:
- web pattern
- span
- roof slope
- truss spacing
- dead loads
- snow, wind, and other loads
- materials

Web Pattern

Pratt web pattern truss designs are good for buildings with or without ceilings. The number of webs increases with heavier loads and longer spans for maximum use of chord lumber.

Increasing the number of webs increases the number of pieces of lumber and plywood required, but can decrease chord size.

For buildings without ceilings, the compression webs may be perpendicular to the top chord rather than the bottom chord. The truss is not weakened if top chord panel points are not moved, and web gussets are made a little wider to cover the same amount of each web member.

Span

Span is measured as in Fig 202-8c.

Overhangs are limited to 2′ because they cause increased bending in the top chord at the heel joint. If an overhang must exceed 2′, use the top chord and heel gusset design for the next higher snow zone.

Trusses can extend beyond exterior walls, but not farther than the length of the bottom chord scarf cut.

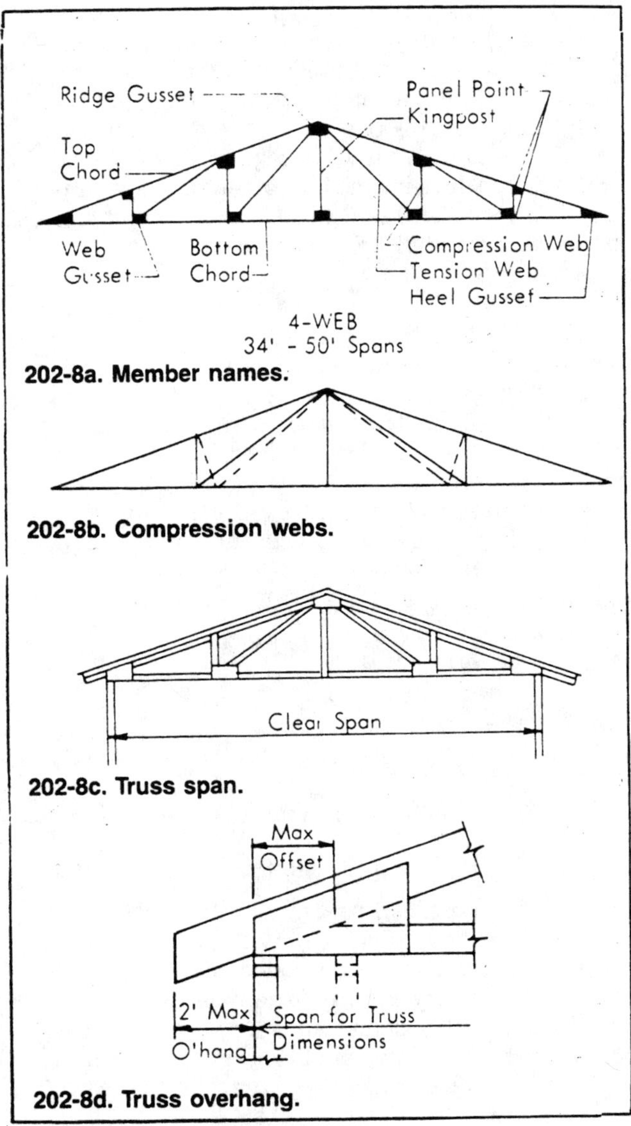

202-8a. Member names.

202-8b. Compression webs.

202-8c. Truss span.

202-8d. Truss overhang.

Fig 202-8. Truss parts.

Roof Slope

Roof slope significantly affects the forces in truss members. For a given span, spacing, and load, the member forces increase as throof slope is decreased.
- 3/12 slope or less is common in homes, commercial buildings, and some farm buildings in low snow load areas or for short spans and close spacings.
- 4/12 is the most common slope in farm buildings in all snow load areas.
- 5/12 slope significantly reduces member forces over lower slopes for buildings in areas of high snow loads or for long spans and wide spacings.

Truss Spacing

Roof and ceiling materials and wall framing influence truss spacing selection. In pole buildings, it is desirable to support trusses by a pole.
- 2′ spacing uses more truss material and labor than wider spacings. It is common in short spans and low roof slopes for buildings with ceilings, solid roof decks, and shingles.

- 4′ spacing supports many types of ceilings without special framing. It is common in insulated livestock buildings with metal roofs and in some machinery and equipment storage buildings.
- 8′ spacing requires the least truss material and labor. It is common in buildings that require no ceilings: machinery and equipment storage, uninsulated livestock buildings, etc. Total cost can be greater if a ceiling is needed.

Dead, Snow, Wind and Other Loads

See Chapter 101.

Dead loads include the weights of the roof, ceiling, trusses, braces, etc.

Add the weights of the truss, purlins or decking, roofing, and roof insulation to get the dead load on the top chord. Add the weights of the ceiling and its supports, insulation supported by the ceiling, and any fixtures hung from the ceiling to get the dead load on the bottom chord.

Example: Insulated livestock building.

Top chord dead load		
Truss weight—30′ span, 4′ spacing		1.5 psf
2x4 purlins—2′ o.c.		0.7
28 ga steel roofing		0.9
	Total	3.1 psf
Bottom chord dead load		
½″ plywood ceiling		1.4 psf
6″ insulation		2.4
	Total	3.8 psf

All MWPS trusses are designed to withstand winds of 80 mph against a building in suburban areas, towns, city outskirts, wooded areas, and rolling terrain. Winds tend to lift the roof and roll the building over. In flat, open country, open flat coastal belts, and in areas having winds higher than 80 mph, provide additional anchorage and bracing for the bottom chord and long diagonal webs.

MWPS trusses have not been designed for large concentrated loads. Avoid cranes and heavy roof-mounted equipment.

If crane loads are to be applied to trusses, select a truss designed for the load. Consider where the load is applied and its duration.

Materials

Lumber quality

Lumber for trusses must be clean, smooth, and of uniformly good quality. Where members of different thicknesses (variations in manufacture) come together at a joint, sand or plane the thicker member so the gusset can make full contact. Top chord lumber should be full-length if possible.

Plywood

Plywood quality is critical. Requirements include:

- C-C outer ply veneer grades. Or, CDX if building will remain dry, such as a home but not livestock housing—glue the C face.
- Exterior glue.
- ⅜″ thickness: Use 3-ply with 24⁄0 Identification Index (I.I.).
- ½″ thickness: Use 5-ply with 32⁄0 I.I. Or, 4-ply only if grade stamp states "Structural 1."
- Group 1 species in the outer plies. Some mills use weaker wood in the outer plies.
- Do not use sanded or other sheathing grades. Plywood stresses used in design are 250 psi shear through the heel joint, and 53 psi rolling shear for glue area.

Glue

The right glue to use:

Casein that is highly mold and water resistant, though not waterproof, is adequate for most trusses. Use it for trusses that will stay dry throughout their life. Resorcinol resin is waterproof; use it for trusses with any joint exposed to unusual moisture conditions. Read the label carefully before you buy—make sure you are getting the right glue.

The right way to use it:

Follow the manufacturer's specifications for mixing, pot life (the length of time the glue stays good after mixing), temperature during use, etc.

Joints

Gussets

The gussets in MWPS truss design tables fit over the members as shown. Apply a gusset to each side of each joint. All web gussets can be made of ⅜″ plywood. Heel gussets are the length and thickness specified in the selection tables from MWPS-9, *Designs for Glued Trusses.*

Heel joints

The heel gussets cover at least the scarf cut. **Face grain must be parallel to the bottom chord.**

Some pole builders fix the truss heel depth for sliding door header framing and offset the scarf cut. The heel gusset should cover the top chord a minimum of 4″, and 6″ when two layers of ½″ plywood are used, Fig 202-9.

Web gussets

Web gussets cover the top chord by 4″ and the entire width of the vertical web. Minimum size is 8″x8″ for 2x4 diagonal webs and 10″x10″ for 2x6s. They cover 4″ of single bottom chords and 6″ or 8″ of stacked bottom chords, Fig 202-10.

Ridge gussets

Ridge gussets cover the full depth of the top chord. Minimum ridge gussets are 8″x12″ for 2x4 diagonal webs and 8″x16″ for 2x6 diagonal webs, with the height increased as the top chord increases.

Laps

Lap joints connect the shortest web to the top chord and the king post to the bottom chord. Use 1″ lumber or ½″ plywood, as wide as the web or king post, glued and nailed to each side of each joint. Lap at least 4″ over each member of the joint.

Splices

Use full-length top chord and bottom chord lumber if possible.

When splices are necessary, locate them as shown in Fig 202-12. Do not locate chord splices at the panel points because the combinations of forces from the chords and webs overload the joints.

All splices: use 1″ lumber the same width as the pieces to be jointed. Glue-nail 16″ lengths for 2x4s, and 24″ lengths for 2x6s and up, on both sides of the joint. For 2x4 splices, ½″ plywood (I.I. 32/16) can be used instead of 1″ lumber; for 2x6s and up use ¾″ plywood (I.I. 48/24). Face grain must be parallel to the chord.

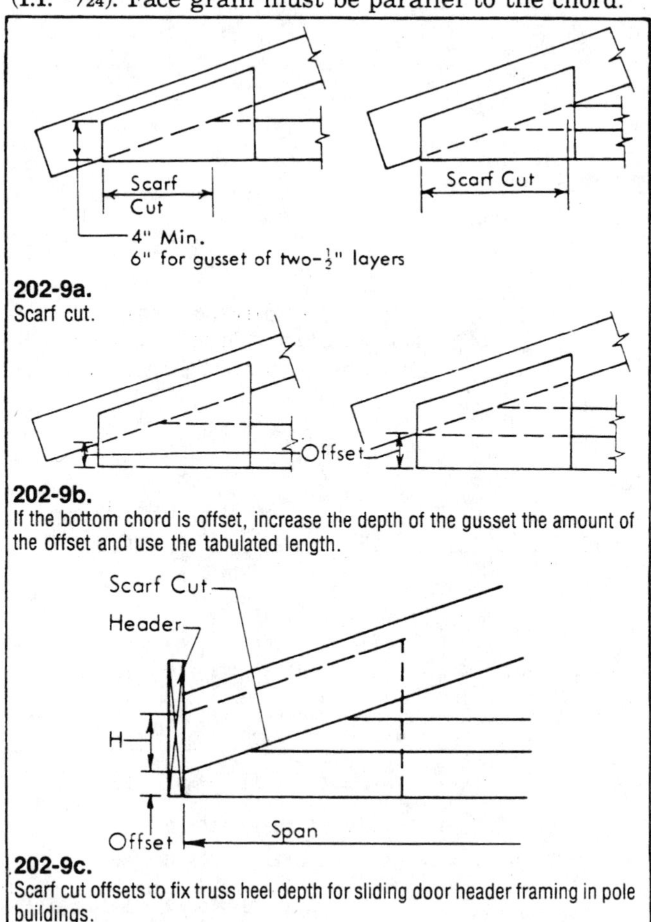

202-9a.
Scarf cut.

202-9b.
If the bottom chord is offset, increase the depth of the gusset the amount of the offset and use the tabulated length.

202-9c.
Scarf cut offsets to fix truss heel depth for sliding door header framing in pole buildings.

Fig 202-9. Heel gussets.

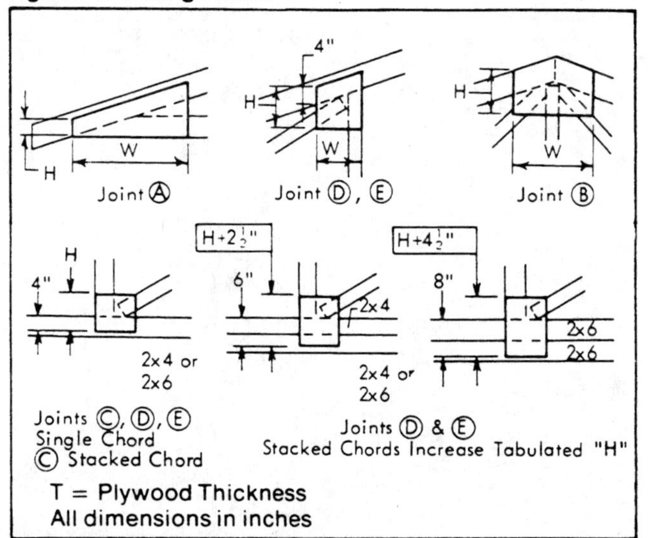

Fig 202-10. Tabulated gusset dimensions.
Larger gussets may be used.

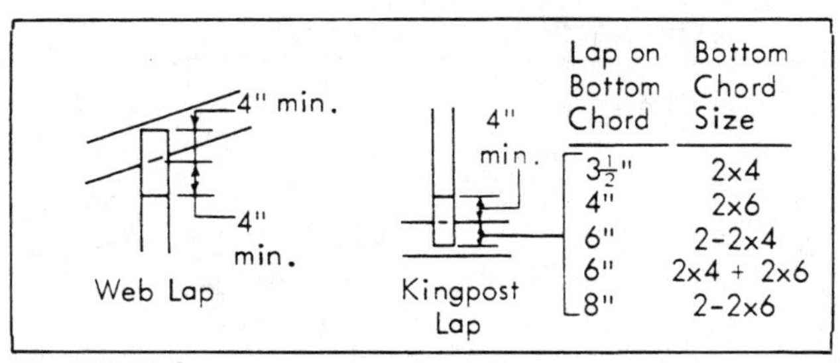

Fig 202-11. Laps.

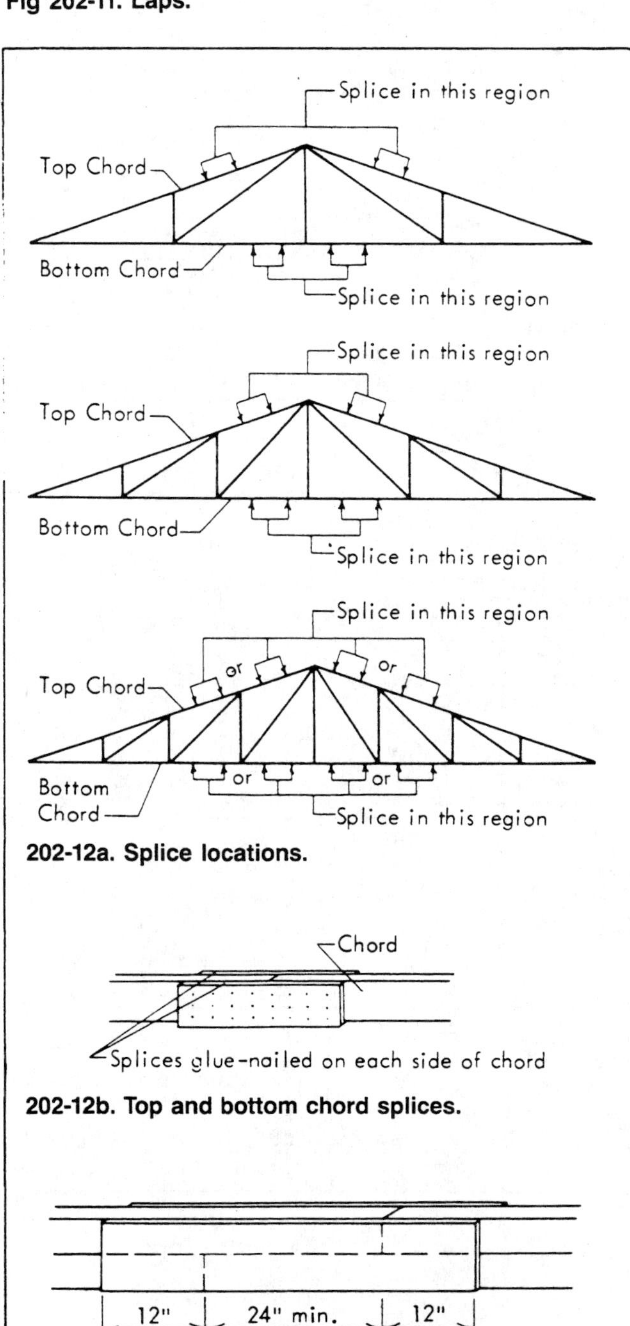

202-12a. Splice locations.

202-12b. Top and bottom chord splices.

202-12c. Splices for stacked bottom chords.
Plywood splices are recommended for stacked bottom chords—wide 1″ boards are apt to be cup-warped, which makes a good glue joint difficult with 4 pieces of lumber. Use ½″ plywood (32/16) for 2 - 2x4s and 3/4″ plywood (48/24) for 2x4 + 2x6 and for 2 - 2x6s.

Fig 202-12. Splices.

Table 202-10. Typical MWPS-9 selection table.

40' SPAN, 4-WEB

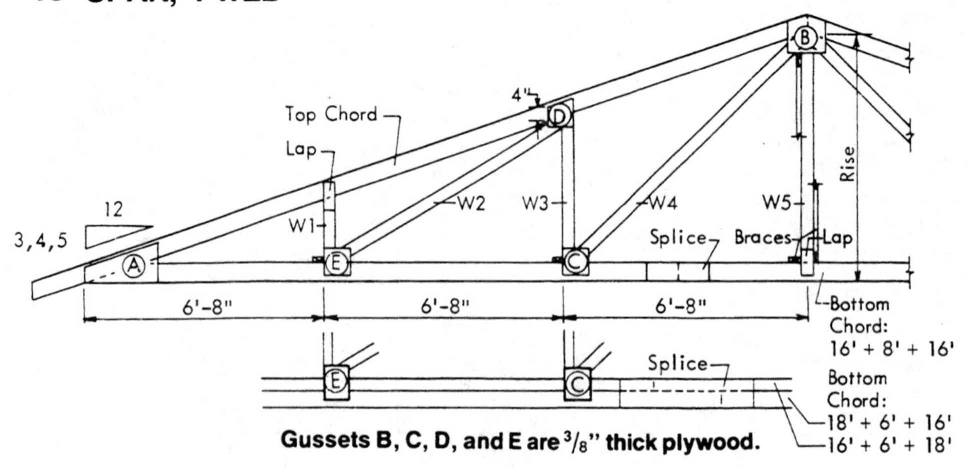

Gussets B, C, D, and E are 3/8" thick plywood.

Web Lengths

Roof Slope	Rise	Top Chord	W1	W2	W3	W4	W5
3/12	5'-0"	16'+5'	2'	7'	3'	8'	5'
4/12	6'-8"	18'+4'	2'	8'	5'	9'+8'	7'
5/12	8'-4"	18'+4'	3'	9'	6'	11'+10'	8'

4+4, 4+6, 6+6 indicates stacked lower chord. 4&4, 6&4, indicate double web; a 2x4 is attached to the web member to increase its stiffness.

1400f Lumber

			Truss spacing, ft. 2'			4'			8'			Web member sizes					Gusset Sizes, in.				
			Ceiling dead load, psf														A	B	C	D	E
	Top chord	Bottom chord	0	5	8	0	5	8	0	5	8	W1	W2	W3	W4	W5	T H W	H W	H W	H W	H W
			---Max. snow + roof dead load, psf---																		
3/12 Slope	2x4	2x4	27	25	23	0	0	0	0	0	0	2x4	2x4	2x4	2x4	2x4	3/8x3½x25	8x12	8x8	8x8	8x8
	2x6	2x4	37	34	32	0	0	0	0	0	0	"	"	"	"	"	½x4x16	10x12	"	"	"
	2x6	2x6	53	49	46	23	18	15	0	0	0	"	"	"	"	"	½x4x23	10x16	10x10	"	"
	2x8	2x6	55	51	49	24	19	16	0	0	0	2x4	2x4	2x4	2x4	2x4	½x4x24	12x16	10x10	8x8	8x8
	2x10	4+4	80	74	71	34	31	22	0	0	0	"	"	"	"	"	½x4x31	14x16	12x10	"	10x8
	2x12	4+6	100+	93	94	44	40	37	0	17	0	"	"	"	"	"	½x4x39	16x20	14x12	"	12x8
	2x12	6+6	-	98	99	46	42	39	23	19	16	"	"	"	4&4	"	½x4x44	18x20	18x12	8x10	14x8
4/12 Slope	2x4	2x4	31	29	28	0	0	0	0	0	0	2x4	2x4	2x4	2x4	2x4	3/8x3½x22	8x12	8x8	8x8	8x8
	2x6	2x4	49	47	43	0	0	0	0	0	0	"	"	"	"	"	½x4x17	10x12	8x10	"	"
	2x6	2x6	63	60	58	27	25	23	0	0	0	"	"	"	"	"	½x4x22	10x16	10x10	"	8x10
	2x8	2x6	75	70	67	33	29	25	0	0	0	2x4	2x4	2x4	2x4	2x4	½x4x23	12x16	10x10	8x8	8x8
	2x10	4+4	100+	98	98	45	41	31	0	0	0	"	"	"	"	"	½x4x33	14x20	12x12	"	10x10
	2x12	4+6	-	-	-	57	53	50	28	23	12	"	"	"	4&4	"	½x4x40	18x20	16x12	8x10	14x10
	2x12	6+6	-	-	-	60	55	52	30	25	23	"	"	"	"	"	½x4x38	18x20	18x12	"	16x10
5/12 Slope	2x4	2x4	34	33	32	15	13	0	0	0	0	2x4	2x4	2x4	2x4	2x4	3/8x3½x20	8x12	8x8	8x8	8x8
	2x6	2x4	62	57	52	27	13	0	0	0	0	"	"	"	"	"	½x4x16	10x16	10x8	"	"
	2x6	2x6	71	68	66	31	28	27	0	0	0	"	"	"	"	"	½x4x20	"	10x10	"	"
	2x8	2x6	94	88	84	41	37	32	0	12	0	2x4	2x4	2x6	2x4	2x4	½x4x24	12x20	10x12	8x10	8x8
	2x10	4+4	100+	100+	100+	56	51	39	28	14	0	"	"	"	4&4	"	2-½x6x18	16x20	14x14	8x12	10x10
	2x12	4+6	-	-	-	70	65	63	35	29	16	"	"	"	"	"	½x4x35	18x20	16x14	"	12x10
	2x12	6+6	-	-	-	71	66	63	35	31	29	"	"	"	"	"	½x4x38	20x24	20x14	"	14x10

203 PLYWOOD

Chapter prepared from publications of the American Plywood Association.

The number of species and manufacturing variations used in plywood has increased considerably in recent years. Design, selection, and use are therefore more complex—specified or needed grades may not be stocked locally.

Manufacture

Plywood is manufactured with a number of plies or veneers (thin sheets of wood). Some plywood is now made with an even number of plies. Softwood veneer for panels used in agriculture is usually "rotary peeled" rather than sliced or sawn.

Peeler logs are cut into "blocks" usually about 8½′ long. In a giant lathe, the blocks are rotated against a long knife that peels the wood off in a long, continuous, thin veneer. The veneer is cut to desired widths and dried to about 2%-5% moisture content. After careful grading, the veneer goes to the glue spreaders where the plywood panel is laid up.

The plywood then is cured, usually in a hydraulic press where both heat and pressure cure the glue in minutes. Panels are trimmed to size, and some grades are sanded. Plywood that conforms with U.S. Product Standard PS 1 carries a grade-trademark on every panel that permits easy identification.

Type

Plywood type, Interior or Exterior, depends on the resistance of the panels to moisture.

Some allowable stresses vary with panel type. Shear strength, however, varies with the kind of glue used.

Exterior type plywood has superior resistance to moisture and weather. It is made with waterproof glue and has no veneer grade below C. Plywood with some D veneers is marketed as "Interior plywood with exterior glue," or "Interior plywood with intermediate glue."

Interior type

Interior type plywood is used where its equilibrium moisture content will not continuously or repeatedly exceed 18% or where it is not exposed to the weather. Interior type plywood with exterior glue is used for sheathing where long construction periods are expected and for some protected exposures where a high moisture level might be reached.

Interior-type plywood has water-resistant glue and D veneers (not permitted in Exterior) for inner and back plies. Use Interior plywood where exposure to weather and moisture is limited.

An "intermediate" glueline is more durable than an interior, but is less durable than an exterior glueline.

Exterior type

Plywood to be exposed to weather or moisture contents continuously or repeatedly above 18% should be Exterior type.

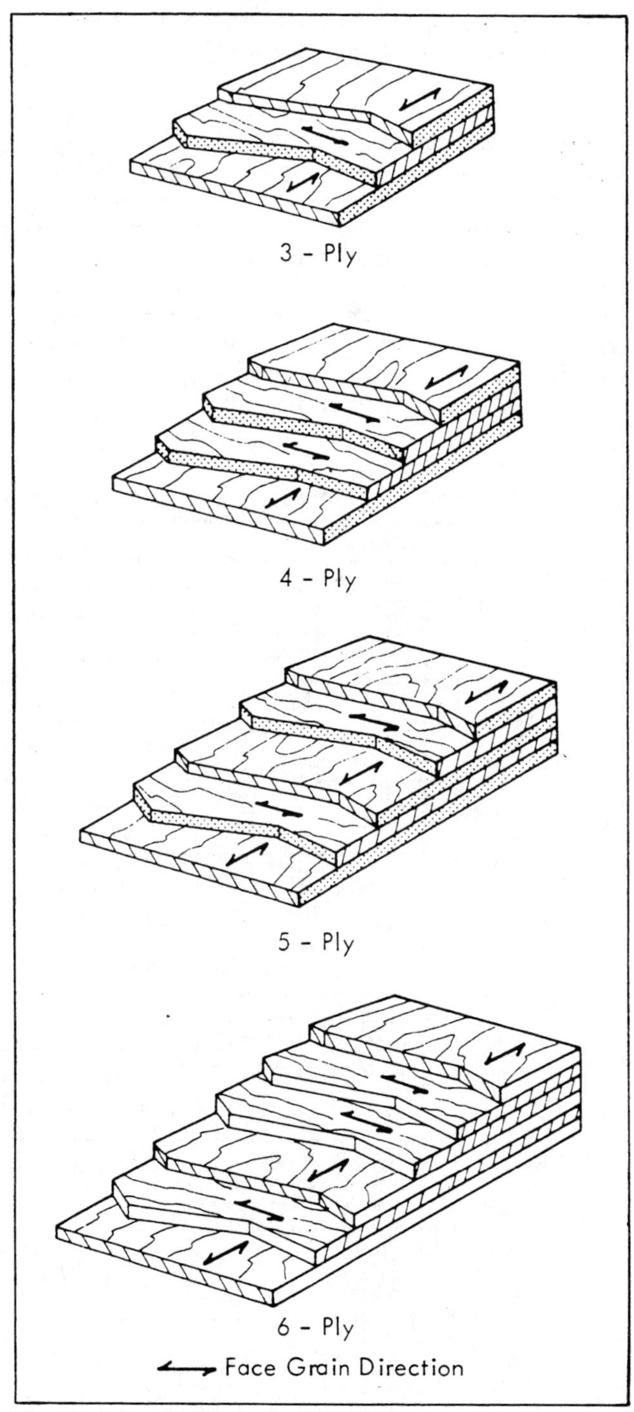

Fig 203-1. Typical 3- to 6-ply panels.

Interior type, exterior glue

Some grades of Interior plywood are made with exterior glue. They differ from EXT plywood because D-grade veneers are allowed. Three grades (C-D INT, and Structural I and Structural II C-D INT) when grade-stamped with exterior glue can be used instead of C-C EXT except under direct exposure to weather, wet manure, and frequent water. High humidities, such as condensation in livestock buildings, can cause delamination.

C-D INT with exterior glue (often called CDX) is good for sheathing, interior building linings, sheltered livestock equipment, and structural applications protected from direct weathering.

Grade

Plywood is graded into engineered or appearance grades. "Appearance Grades" are designated by the panel thickness, by the classification of face and back veneers, and by the species group of veneers. "Plyform" is an exception, where Class designates a species mix.

"Engineered Grades" are largely the unsanded sheathing panels designated C-D Interior or C-C Exterior. Either of these grades may be modified by the terms STRUCTURAL I or II. Use structural grades where strength is important. STRUCTURAL I is limited to Group 1 species. STRUCTURAL II allows Species Groups 1, 2 and 3. Both have exterior glue and some restrictions on knot size and repairs. Panels are designated by a thickness and an Identification Index, without reference to veneer species.

Identification Index

C-D INT, C-C EXT, and the Structural grades have an Identification Index (I.I.) as part of their grade stamps. It is an approximate measure of the panel's strength. Although developed for sheets to be used for sheathing or flooring, the Identification Index is useful in assuring that the right panels are used in a number of engineered applications.

The I.I. is two numbers separated by a slash—24/0, 32/16, etc. The first number is the maximum support spacing, in inches, when the panel is used for residential roof sheathing. The second number is support spacing for residential subflooring. For example, 32/16 plywood would be on rafters 32" o.c. or on joists 16" o.c. It is assumed that the sheets are laid with face grain perpendicular to the framing. For some farm buildings, somewhat less conservative designs are practical, so spans longer than the I.I. are sometimes recommended.

Table 203-1. Plywood grades for agricultural applications.

	Use these terms when you specify	Description and Most Common Uses	Veneer Grade			Most Common Thicknesses (inch) (3)					
			Face	Back	Inner Plies	1/4	5/16	3/8	1/2	5/8	3/4
APPEARANCE (1)	A-A EXT (2) (4)	Use where the appearance of both sides is important. Fences, built-ins, signs, cabinets, commercial refrigerators, tanks and ducts	A	A	C	✓		✓	✓	✓	✓
	A-C EXT (2) (4)	Exterior use where the appearance of only one side is important. Sidings, soffits, fences, structural uses, truck lining and farm buildings. Tanks, commercial refrigerators	A	C	C	✓		✓	✓	✓	✓
	B-C EXT (2) (4)	An outdoor utility panel for farm service and work buildings, truck linings, containers, tanks, agricultural equipment	B	C	C	✓		✓	✓	✓	✓
ENGINEERED	C-D INT w/ext glue (2)	A utility panel for use where exposure to weather and moisture will be limited	C	D	D		✓	✓	✓	✓	✓
	C-C EXT (2)	Unsanded grade with waterproof bond for subflooring and roof decking, siding on service and farm buildings. Backing, crating, pallets and pallet bins	C	C	C		✓	✓	✓	✓	✓
	C-C PLUGGED EXT (2)	For refrigerated or controlled atmosphere rooms. Also for pallets, fruit pallet bins, tanks, truck floors and linings. Touch-sanded	C Plgd	C	C	✓		✓	✓	✓	✓
	STRUCTURAL I & II C-D INT & C-C EXT	For engineered applications in farm construction. Unsanded. For species requirements see (4)	C	C or D	C or D		✓	✓	✓	✓	✓
SPECIALTY	HDO EXT (2) (4)	Exterior type High Density Overlay plywood with hard, semi-opaque resin-fiber overlay. Abrasion resistant. Painting not ordinarily required. For concrete forms, signs, acid tanks, cabinets, counter tops and farm equipment	A or B	A or B	C (5)		✓	✓	✓	✓	✓
	MDO EXT (2) (4)	Exterior type Medium Density Overlay with smooth, opaque resin-fiber overlay heat-fused to one- or both panel faces. Ideal base for paint. Highly recommended for siding and other outdoor applications. Also good for built-ins and signs	B	B or C	C		✓	✓	✓	✓	✓
	303 SIDING EXT inc. Texture 1-11 (2) (7)	Grade designation covers proprietary plywood products for exterior siding, fencing, etc., with special surface treatment such as V-groove, channel groove, striated, brushed rough-sawn	(6) C	C	C			✓	✓	✓	✓

Notes:

(1) Sanded both sides except where decorative or other surfaces specified
(2) Available in Group 1, 2, 3, 4, or 5 unless otherwise noted
(3) Standard 4 × 8 panel sizes, other sizes available
(4) Also available in STRUCTURAL I (all plies limited to Group 1 species) and II (limited to Groups 1, 2 and 3)
(5) Or C Plugged
(6) C or better for 5-plies. C Plugged or better for 3-ply panels
(7) Stud spacing is shown on grade stamp

Veneer grades

- N and A—highest grade level. No knots, restricted patches. Use N for a natural finish and A for painted surface.
- B—solid surface. Small round knots. Patches and round plugs are allowed. Most common use is in Plyform for concrete forms.
- C Plugged—special improved C grade used in underlayment.
- C—small knots, knotholes, patches. Lowest grade allowed in Exterior type plywood. For sheathing faces and inner plies in Exterior panels.
- D—larger knots, knotholes, some limited white pocket in sheathing grades. For inner plies and backs in Interior panels.

Wood Species

Woods for plywood are classified into five groups based on strength properties. Design stresses for a group are determined from clear wood specimens. Design stresses are published for Groups 1 to 4. All woods within a group are assigned the same working stress.

The group classification of a plywood panel is usually determined by the face and back veneer with the inner veneers allowed to be of a different group. Certain grades such as MARINE and the STRUCTURAL I have all plies of Group 1 species.

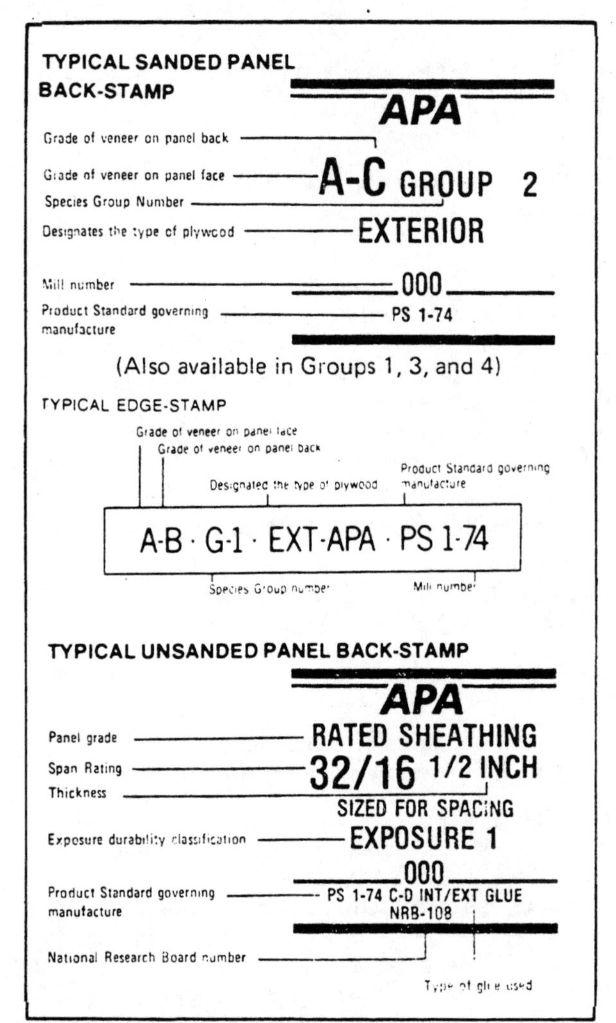

Fig 203-2. APA grade-trademarks.

Table 203-2. Classification of major species.

Group 1	Group 2		Group 3	Group 4	Group 5
Beech, American	Cedar, Port Orford	Maple, black	Alder, red	Aspen	Basswood
Birch: sweet, yellow	Cypress	Pine: Pond, red, Virginia, western white	Birch, paper	Cativo	
	Douglas fir 2		Cedar, Alaska	Cedar: incense, western red	Poplar, Balsam
Douglas fir 1	Balsam Fir: California red, grand, noble, Pacific silver, white		Fir, subalpine		
Larch, western		Spruce: red, Sitka, black	Hemlock, eastern	Cottonwood	
Maple, sugar		Sweetgum	Maple, bigleaf	Pine: eastern white, sugar	
Pine: Caribbean, Ocote southern loblolly longleaf, shortleaf, slash	Hemlock, western	Tamarack	Pine: jack, lodgepole, ponderosa, spruce		
	Lauan	Yellow-poplar			
			Redwood		
Tanoak			Spruce: Engelmann, white		

204 PLYWOOD SELECTION

Chapter prepared by Wendell Moore from publications of the American Plywood Association.

Farm Buildings and Equipment: Summary

What Grade to Buy

Most dealers stock only panels with exterior glue, except the "cabinet" grades A-B or A-D interior. The grades mentioned below are suggested minimums—A can be substituted for B, for example, if B face veneers are not stocked. Use C face veneer where appearance is not critical. Use B or C face veneer, preferably unsanded, for staining. Use B veneer or MDO if panels are to be painted.

Roof sheathing

Use "C-D INT with Exterior Glue" (CDX) touch sanded (or C-C EXT, unsanded).

Wall sheathing

Use CDX or C-C EXT, unsanded.

Roofing (small buildings or equipment)

Use B-C EXT and penetrating oil stain.

Exterior siding

Panels to be untreated or stained; if knot holes okay, use C-C EXT (unsanded).

Panels to be untreated or stained; knot holes undesirable, use C-C Plugged EXT (touch sanded), or B-C EXT (sanded).

Panels to be painted: MDO-EXT is best; B-C EXT may be somewhat cheaper.

Special patterns—303 Siding Ext.—See Residential Summary.

Ceilings and interior siding

Panels in high moisture buildings (livestock housing), use C-C EXT.

Panels in medium moisture buildings (open-front buildings), use CDX (or one of the Exterior Siding grades).

Panels in dry locations (shop, garage, feed and grain storages), use CDX.

Flooring

Subfloor, use C-C EXT on CDX unsanded.

Flooring without subfloor, use B-C EXT, or Sturd-I-Floor.

See Flooring, Residential Summary, for finished flooring.

Structural elements

For trusses, rigid frames, beams, braces, and other structural parts of equipment or buildings, use grade and thickness specified on the plan. If no grade is specified, CDX is a good minimum grade; B-C EXT has a face veneer without knot holes.

Partitions, gates, livestock equipment

Use C-C EXT.

What Thickness to Buy

Where face grain will be parallel to supports (studs, joists, etc.), **avoid** 3-ply 1/2" or 5/8" thicknesses—use 4-ply or 5-ply.

Structural applications, Identification Index

Sheathing grades are given an Identification Index (I.I.), which indicates strength. If you must substitute, select a panel with a larger I.I. e.g., if 30/12 is specified but unavailable, use 32/16. Some I.I.s apply to more than one thickness (thicker panel of weaker wood or thinner panel of stronger wood), so specify thickness if it matters (to match previously used sheets or your project's dimensions).

Table 204-1. Identification Index (I.I.).
The most commonly stocked panels are underlined.

Identification Index	Available thicknesses, in.
12/0	5/16"
16/0	5/16", 3/8"
20/0	5/16", 3/8"
24/0	3/8", 1/2"
30/12	5/8"
32/16	1/2", 5/8"
36/16	5/8"
42/20	5/8", 3/4", 7/8"
48/24	3/4", 7/8"

Table 204-2. Roof sheathing.
Face grain perpendicular to framing.

Total roof load, lb/ft²	Framing spacing 16"	24"	32"	48"
	- - Identification Index - -			
15 psf (South)	16/0	24/0	32/16	42/20
25 psf (Central)	"	"	"	"
35 psf (North)	"	"	"	48/24
45 psf (North, heavy)	"	"	"	"

Table 204-3. Nonstructural applications.

	Framing spacing 16"	24"	32"	48"
	- - - Panel thickness - - -			
Ceiling	1/4"	3/8"	1/2"	1/2"
Exterior siding on vertical studs				
Panels vertical	5/16"	3/8"	-	1/2"[a]
Panels horizontal	1/4"[b]	5/16"	3/8"	1/2"
Exterior siding on horizontal-wall girts				
Panels vertical	1/4"[b]	5/16"	3/8"	1/2"
Interior wall lining[c]				
Panels vertical	5/16"	3/8"	-	1/2"[a]
Panels horizontal	1/4"[b]	1/4"	3/8"	1/2"

[a]Block between studs 4' o.c.
[b]Use 1/4" only where not subject to physical damage—upper walls along livestock or feed alleys, lower walls in office or light-use areas.
[c]Assuming vertical studs.

Flooring or structural walls

Grades: C-D INT with EXT Glue (CDX), or C-C EXT.

Table 204-4. Face grain perpendicular to framing.
Allowable loads limited by strength. Deflections less than L/180.

Ident Index	12″	16″	24″	32″	48″
	Framing spacing				
	Uniform load, psf				
24/0	220	125	55	30	-
32/16	335	180	85	45	-
42/20	520	290	130	70	-
48/24	680	380	170	95	30
2-4-1	1365	765	340	190	65

Table 204-5. Face grain parallel to framing.
Allowable loads limited by strength. Deflections will exceed L/180: blocking end joints will prevent joints from opening up. If deflection may be a problem, redesign system so face grain crosses framing.

Ident Index	12″	16″	24″	48″
	Framing spacing			
	Uniform load			
24/0	45	25	-	-
32/16	75	40	15	-
42/20	130	70	26	-
48/24	325	180	65	15
2-4-1	1045	590	210	50

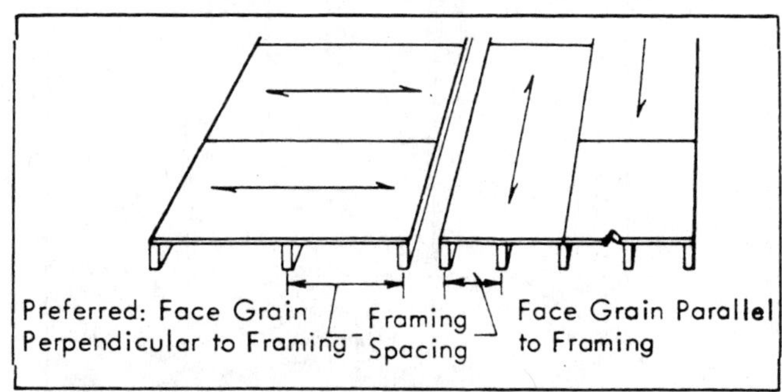

Fig 204-1. Plywood on framing.
For maximum strength and stiffness, apply panel face grain perpendicular to supports.

Installation

Install panels with a small space at each side and each end, generally 1/16″ at ends, 1/8″ at sides. Double these spacings for wet conditions. For a smooth floor, wet both sides of tongue-and-groove plywood and install with 1/16″ spaces all around. Similarly wet and space wall panels with shiplapped edges.

Use galvanized, aluminum, or nonstaining nails to avoid rust streaks. Ring-shank or screw-shank nails have more withdrawal holding power than common nails. Use 6d for 5/16″-1/2″ panels, 8d for 5/8″-7/8″ panels, and 10d for 1″ panel thickness. Space nails 6″ o.c. at panel edges and 12″ o.c. at intermediate supports; if spans are 48″ or more, nail 6″ o.c. at all supports.

Blocking under unsupported edges is desirable for floors and laterally loaded walls, unless tongue-and-groove plywood is used.

Residential Summary

What Grade to Buy

Buy interior (INT) grades for sheets that will stay dry; buy exterior (EXT) grades for sheets that will be exposed to high moisture conditions, including sheets stored outdoors during an extended construction period.

N veneers are highest grade; use them when a clear finish will be applied and best possible appearance is desired. A veneer is high quality, B is next, C has some small open knotholes, and D is lowest quality. Use the lowest grade that serves the purpose and is available—not all grades are stocked by all dealers.

The I.I. on engineered grades is a measure of strength. Larger numbers are stronger panels, so if you must substitute, use a panel with a larger I.I.—if 30/12 is specified but unavailable, use 32/16. Thicker panels with the same I.I. (e.g., 16/0 comes in 5/16″ and 3/8″ thicknesses) are not necessarily stronger; they can be of weaker wood species.

What Thickness and Grade to Buy

Where face grain will be parallel to supports (studs, joists, etc.) **avoid** 3-ply 1/2″ or 5/8″ thicknesses—use 4-ply or 5-ply.

Roof sheathing and subflooring

Use Identification Index (I.I.). After rafter and joist spacing are known: the left-hand number of the I.I. must be as large or larger than the rafter spacing, and the right-hand number of the I.I. must be as large or larger than the joist spacing. If rafters are to be 24″ o.c. and joists 16″ o.c., use 24/0 panels on the roof and 32/16 for subfloor.

Use unsanded grades with exterior glue for subflooring in bathrooms, laundry, and other areas exposed to high moisture conditions. Use Interior unsanded grades in other areas. Use Underlayment grades (or A-C EXT) if finish flooring is to be linoleum, tile, carpeting, or other non-structural material.

Apply roof sheathing and subflooring with the face grain of the panels perpendicular to the framing members.

Wall sheathing and siding

Use unsanded grades for sheathing and unsanded exterior if siding is to be stained. Use MDO for siding to be painted. Use 303 siding grades for special appearance effects.

Interior paneling

Consider special paneling materials.

In remodeling, 1/4″ thickness is adequate over plaster or other base. For new construction, use same thickness as for siding, Table 204-6.

Table 204-6. Wall sheathing and siding thicknesses.

Stud spacing	Plywood thickness Face grain on framing Parallel	Perpendicular
16″ o.c.	3/8″	3/8″
24″	1/2″	3/8″
32″	a	1/2″
48″	3/4″b	5/8″

[a]Not modular to 48″ wide sheet.
[b]Cross blocking 48″ o.c. between studs.
For Grade 303 Texture 1-11, install blocking 32″ o.c.

Discussion

Consider human occupancy factors, even in structures built outside building code jurisdiction. If the life hazard exposure is low, such as in crop storages, animal housing, or service buildings, greater support spacings than indicated by the I.I. can be used without reducing the structural adequacy of the building.

Construction

Use recommendations

Select plywood based on strength or appearance as given in the tables and recommendations in this chapter.

EXT plywood is recommended for farm buildings and equipment, both inside and outside. C-C EXT is most common, with CDX as an alternate where moisture content will be below 16%. CDX is also for sheathing not exposed to high moisture or the weather for an extended time. Even in livestock buildings, relative humidities are rarely high long enough for the moisture in lumber to get above 15%, Fig 201-8.

FRP plywood is coated with a fiberglass reinforced plastic for washability and durability.

Apply plywood panels with the face grain perpendicular to the supports for maximum strength and stiffness. When installing roof decking, wall sheathing, or subflooring, leave 1/16″ between panel ends and 1/8″ between panel edges to allow for slight expansion. Double these spacings if wet or humid conditions are expected. Tongue-and-groove or shiplapped plywood may be installed with only 1/16″ between panel edges.

Nailing

Use common nails in sheathing and subflooring. Use finish or casing nails in paneling and siding. Use ring-shank nails in underlayment and where extra holding power is needed. Galvanized nails are recommended. Nail no closer than 3/8″ to panel edges.

Single-wall construction

In single-wall construction, one layer of plywood siding is both sheathing and siding and is applied directly to the framing.

Minimum plywood thickness and installation recommendations are in Table 204-3. Place all panel edges over framing or blocking. Plywood graded 303 Texture 1-11 used vertically should have cross blocking at 48″ o.c. vertically when studs are spaced at 48″ o.c.

Fig 204-2 illustrates typical single-wall construction. Fig 204-5 shows siding joint details. These joints provide attractive and weathertight joints between plywood siding panels.

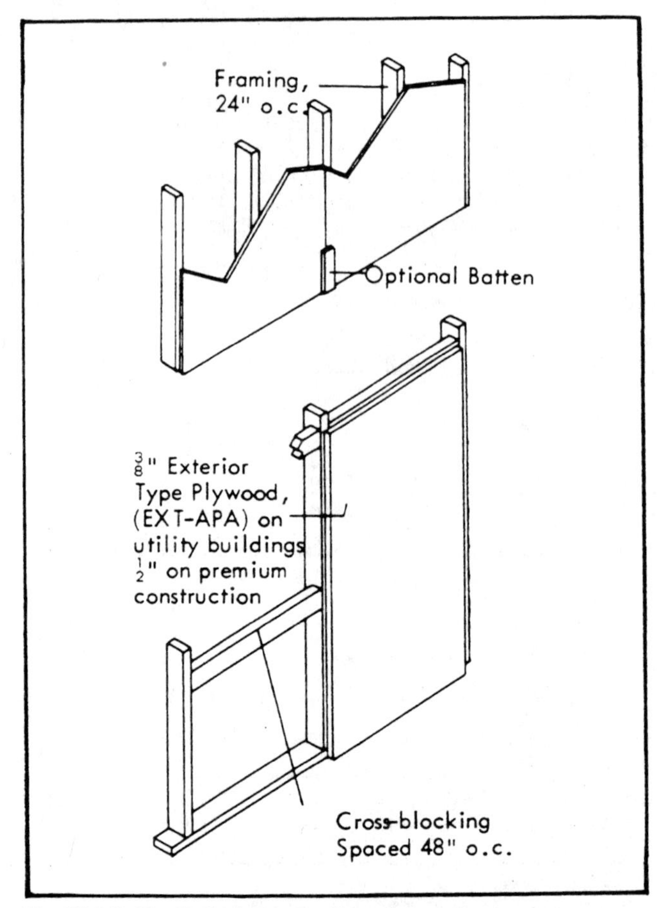

Fig 204-2. Typical single-wall construction.

Plywood and poles

C-C EXT makes a good siding for most farm service buildings. Properly applied, it can produce a neat, heavy-duty wall. Textured plywood surfaces can be considered for a premium appearance.

In most pole buildings, framing girts are 2″ dimension lumber nailed flat to the poles. These girts, spaced 32″ o.c., are adequate for 3/8″ EXT siding, Fig 204-4.

Roofs

Apply roof sheathing with the face grain across the supports. Materials and labor waste are reduced because the panels are accurately cut, sized, and squared. Fig 204-5 shows plywood over a truss or rafter roof. Fig 204-6 shows typical roof construction using plywood over purlins.

Minimum I.I. requirements for plywood roof sheathing are in Table 204-2.

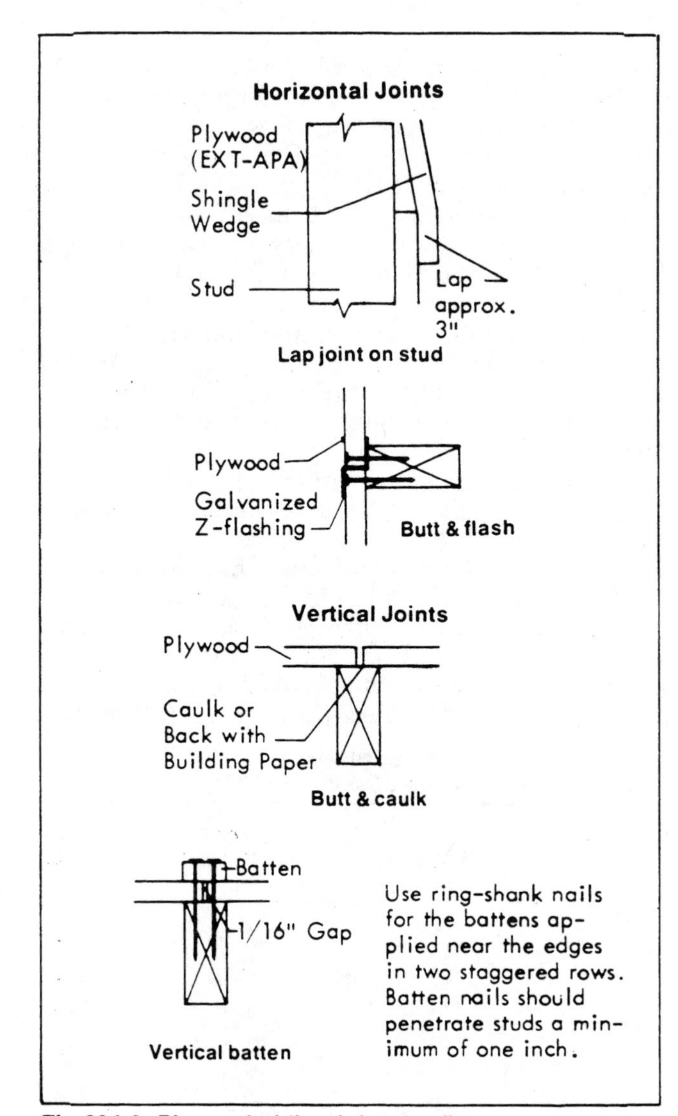

Fig 204-3. Plywood siding joint details.

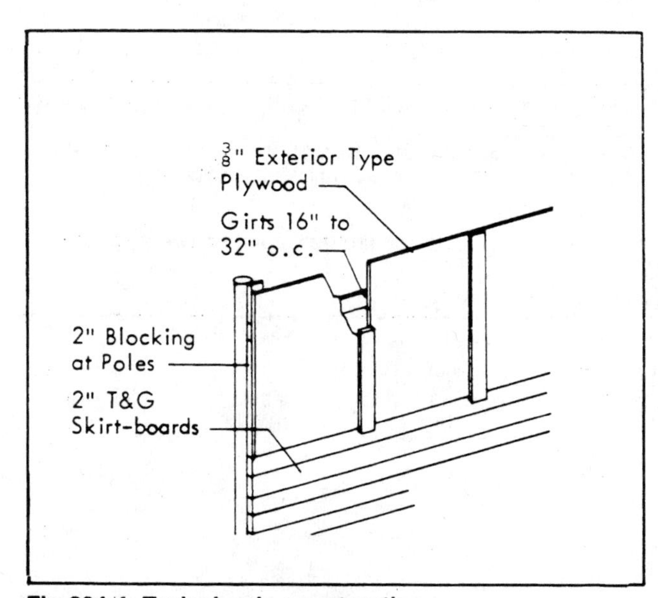

Fig 204-4. Typical pole construction.
Caulk the joint and use 1x3 or 1x4 battens.

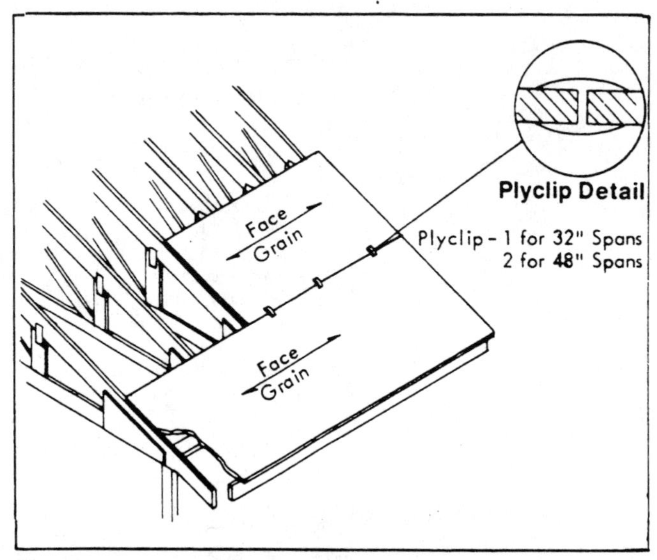

Fig 204-5. Plywood over trusses or rafters.

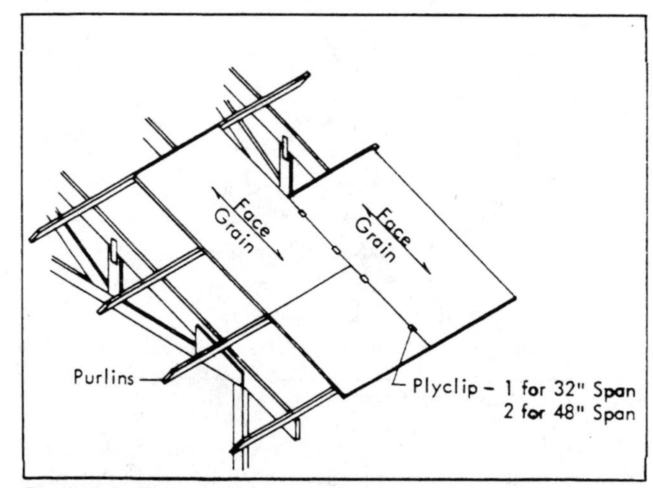

Fig 204-6. Plywood over purlins.

Ceilings

Typical ceilings are illustrated in Fig 204-7. A plywood ceiling can brace the bottom chords of trusses and support insulation above it.

An alternative to ½" T&G is square edge plywood with blocking or two plyclips along each panel edge. Also satisfactory is ⅜" plywood on trusses 32"-48" o.c. with 2x4 blocking at each edge.

Floors

Requirements for floors include strength; stiffness, especially at panel edges; resistance to concentrated loads, such as furniture and appliance legs; resistance to expected moisture conditions; and, for many installations, smoothness under nonstructural flooring such as linoleum or floor tile.

Use panels with exterior glue where high moisture conditions are expected: bathrooms, laundries, livestock buildings. Use adequate nailing. Plywood is usually covered with another flooring material, but some grades make good floors for portable livestock units and storage areas.

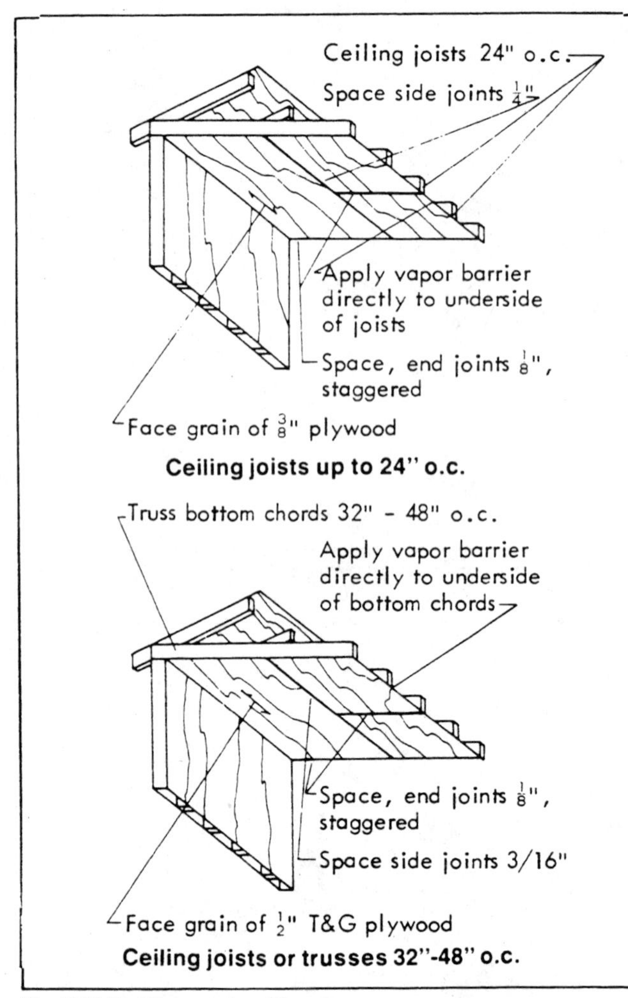

Fig 204-7. Plywood ceilings.

Subflooring is laid with the face grain across the supports, which have been most commonly 16″ o.c. in dwellings, but which are now often spaced 24″ or even 32″ or 48″ o.c. Place T&G wood or lightweight concrete flooring, or underlayment, over the subfloor.

Underlayment is a special material to provide a smooth surface for non-structural flooring materials such as linoleum or tile. Not only is the top veneer smooth, but the ply next to the top veneer has only limited defects to assure that small but heavy concentrated loads do not puncture the flooring and the underlayment. Underlayment depends on subflooring for much of its strength. Use underlayment to smooth an old floor before installing new flooring.

Sturd-I-Floor is combined subfloor and underlayment, usually with T&G edges.

Plywood flooring is adequate for many uses. B-D INT, for example, is sanded and has no holes in the top surface. C-D Plugged INT is only touch sanded but is adequate for attics or similar areas. C-D INT is adequate for floors of hay or grain storages. C-C EXT or CDX are adequate for portable livestock buildings and for storages where washing or other moisture conditions suggest using an exterior glue.

Select flooring panels from the information provided in the Agricultural or Residential Summary.

Granular storages

See Chapter 103 for grain bin loads.

Table 204-7 gives safe load data for plywood floors or other structural applications.

Tables 204-8 and 204-9 give recommendations for shelled corn or wheat bin floors and walls. Install the plywood with face grain across the supports.

Figs 204-8 and 204-9 show the required "pressure zone" for various combinations of material depth and equivalent diameter.

For round bins, equivalent diameter equals bin diameter. For rectangular bins, where the length is more than 1.5 times the width, it is the bin width. For rectangular bins with length less than 1.5 times width:

Equiv. diam. = 4 x floor area ÷ perimeter

The figures apply to other materials with similar properties. Note that in the figures, grain depths increase toward the **bottom,** and that the **lowest** numbered zones produce the highest pressures.

Each of the 14 zones corresponds to a particular support spacing and plywood description shown in Table 204-10. Sheathing is described by I.I.; sanded plywood is described by species group and thickness.

Table 204-7. Safe loads for plywood panels.
See also Tables 204-4, 204-5, and 204-10 to 204-13.

Support spacing in.	Identification Index	Panel thickness in.	Safe load psf[a]
12	24/0[b]	3/8, 1/2	220
	32/16	1/2, 5/8	335
	42/20	5/8-7/8	480
	48/24	3/4, 7/8	625
16	24/0[b]	3/8, 1/2	125
	32/16	1/2, 5/8	190
	42/20	5/8-7/8	290
	48/24	3/4, 7/8	380
24	32/16[b]	1/2, 5/8	85
	42/20[b]	5/8-7/8	130
	48/24	3/4, 7/8	170

[a]C-D Int w/exterior glue or C-C Ext plywood. Face grain across supports, deflection limited to span/180.
[b]Limited to light concentrated loads, such as foot traffic, and to uniform loads specified. For other conditions, specify greater Identification Index.

Table 204-8. Plywood for shelled corn or wheat bin floors.
Face grain across supports.

Support spacing in.	Maximum depth of level fill feet	Ident. Index	Panel thickness in.
12	8	32/16	1/2, 5/8
	12	42/20	5/8-7/8
	16	48/24	3/4, 7/8
16	6	42/20	5/8-7/8
	8	48/24	3/4, 7/8
	10	48/24*	3/4

*If stamped Structural I.

Table 204-9. Plywood for bin side walls.
For shelled corn or wheat. Face grain across supports.

Support spacing in.	Maximum depth of level fill feet	Identification Index
12	8	12/0
	10	20/0
	16	24/0
	20	32/16
16	8	24/0
	12	32/16
	20	42/20
20	8	32/16
	12	42/20
	20	48/24
24	8	42/20
	10	48/24
	12	48/24*

*If stamped Structural I.

Table 204-10. Plywood and support spacing.

Pressure zone	Identification Index	A-C Ext Group 1 in.*	Support spacing in.	Approx. pressure psf
14	24/0	3/8	24	50
13	42/20	5/8	32	65
12	32/16	1/2	24	75
11	48/24	3/4	32	80
10	24/0	3/8	16	110
9	42/20	5/8	24	115
8	48/24	3/4	24	145
7	32/16	1/2	16	165
6	24/0	3/8	12	195
5	42/20	5/8	16	255
4	32/16	1/2	12	295
3	48/24	3/4	16	325
2	42/20	5/8	12	450
1	48/24	3/4	12	575

*For each step the group number of sanded plywood departs from Group 1, take one step down to the next smaller zone number. For example, if Group 3 plywood is used, and the applicable figure shows Zone 11, use the plywood and support spacing shown for Zone 9.

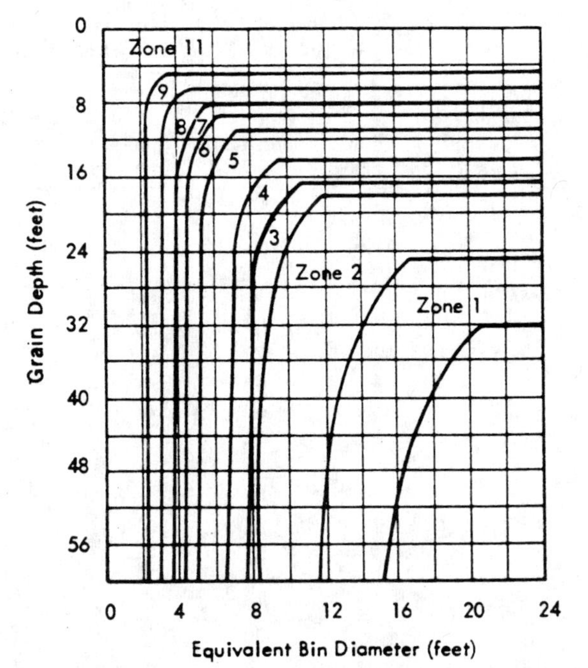

Fig 204-8. Shelled corn, wheat, flaxseed, rye, 48 pcf.

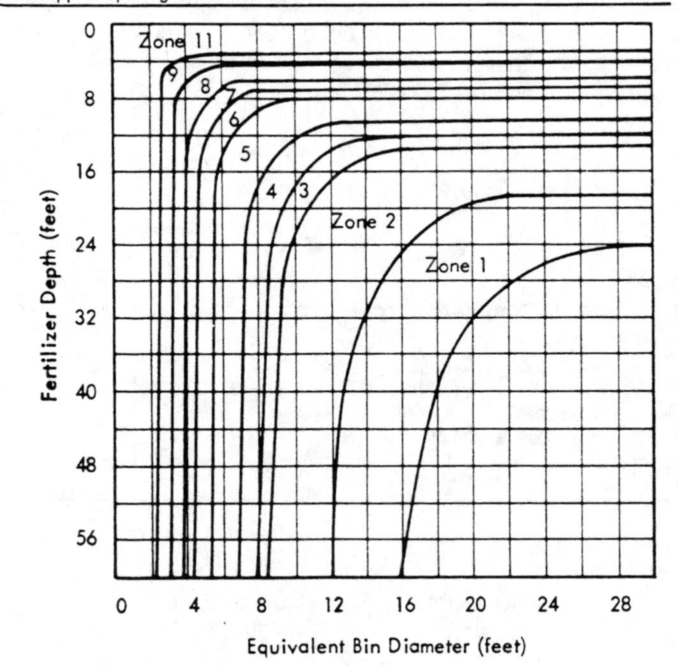

Fig 204-9. Fertilizer, 70 pcf.

To select the plywood thickness:
- Determine the equivalent diameter of the bin.
- Select the chart that applies to the material stored, Fig 204-8 or 204-9.
- Use the equivalent diameter and material depth to find the pressure zone from the chart.
- Select the plywood description and maximum support spacing from Table 204-10.

Example:

Select plywood walls for shelled corn storage 16′ wide, 28′ long, and 12′ high. A center partition, perpendicular to the 28′ wall, divides the building into bins of equal size. Plywood will be installed with its face grain across vertical studs spaced 16″ o.c.

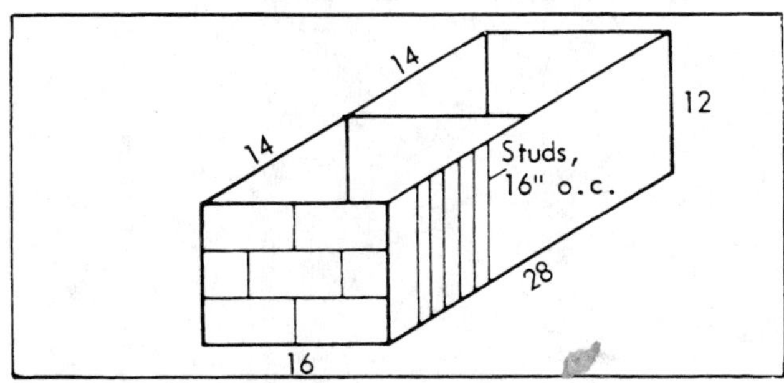

Fig 204-10. Example bin.

Each 16′x14′ bin is less than 1.5 times as long as it is wide:

Equiv. diam. = 4 x floor area ÷ perimeter
= 4x14x16 ÷ [2 x (14 + 16)] = 14.9′ = 15′

Because the bin will store shelled corn, find the pressure zone from Fig 204-8. For a 15′ equivalent diameter and 12′ depth of grain, Fig 204-8 shows Zone 5. Referring to Table 204-10, Zone 5 requires I.I. 42/20 or 5/8″ A-C EXT Group 1 with studs 16″ o.c.

Since the pressure decreases with increasing heights above the bottom of the bin, more economical plywood panels can be used toward the top. For an 8′ depth and 15′ equivalent diameter, Fig 204-8 shows Zone 8. For Zone 8, Table 204-10 gives plywood requirements only for supports 24″ o.c. Since the supports are spaced 16″ o.c., check the next-lower-numbered zone giving the plywood requirements for a 16″ support spacing. In this case, Zone 7 shows 32/16 or 1/2″ A-C EXT Group 1.

For a 4′ depth, Fig 204-8 shows Zone 11. Table 204-10 shows a plywood thickness for this zone based on a 32″ support spacing. Select 24/0 or 3/8″ A-C EXT Group 1 as indicated for Zone 10.

A similar procedure can be used to determine the allowable support spacing for a given zone and plywood thickness or I.I.

Install the plywood on the inside of the wall framing. If the plywood is installed on the outside of the framing, the fasteners will be loaded in withdrawal and require special design.

Table 204-11. Commercial roof decking—face grain across supports.

These values apply for C-D Int, C-C Ext, Structural I and Structural II C-D Int, and Structural I and Structural II C-C Ext grades. Plywood continuous over 2 or more spans.

5 psf dead load assumed; 15% stress increase used. Uniform load deflection limit: 1/180 span under live load plus dead load, 1/240 under live load only.

Panel Ident. Index	Max. span in.	Unsupported edge-max. length* in.	Allowable uniform live loads (psf) — Spacing of supports center-to-center in. 12	16	20	24	32	42	48
12/0	12	12	150						
16/0	16	16	160	75					
20/0	20	20	190	105	65				
24/0	24	24	250	140	95	50			
32/16	32	28	385	215	150	95	40		
42/20	42	32		330	230	145	75	35	
48/24	48	36			300	190	105	45	35
2-4-1	72	48				390	215	100	75
1 1/8″, Group 1&2	72	48				305	170	75	55
1 1/4″, Group 3&4	72	48				355	195	90	65

*Provide adequate blocking, tongue and grooved edges, or other suitable edge support such as Plyclips when spans exceed indicated value. Use two Plyclips for 48″ or greater spans and one for lesser spans.

Commercial and Residential Buildings

Roof decking

Allowable uniform loads for plywood roof decking are given in Table 204-11, which is based on plywood installed with face grain perpendicular to its supports.

Building roofs receive occasional foot traffic sufficient to cause differential deflection at unsupported panel edges. If edge support is desired, it can be achieved with Plyclips, tongue-and-groove plywood joints, or lumber blocking.

Table 204-12. Commercial roof decking[a]

Face grain parallel to supports.
5 psf dead load assumed.

Surface	Number & length of spans	Structural I Grades — Ident. Index	Structural I Grades — Allowable uniform live load psf	Grades other than Structural I — Ident. Index	Grades other than Structural I — Allowable uniform live load psf[b]
Unsanded	Four @12″	24/0	35	24/0[c]	45
				24/0[d]	
	Three @16″	32/16[d]	115	24/0[d]	65
		32/16[g]	75	32/16[d]	65
	Two @24″	32/16[g]	25	32/16[d,f]	25
		32/16[d]	40	32/16[h]	45
		42/20	80	42/20[h]	45
		48/24	110	42/20	50
				48/24	50
				48/24	120
Sanded[e]	One @48″	1 1/8	45	2-4-1	25
	Four @12″	3/8	80	3/8	50
				1/2	195
	Three @16″	3/8	30	1/2	80
		1/2	135	5/8	190
	Two @24″	1/2	45	5/8	65
		5/8	105	3/4	115
	One @48″	1 1/8	55	1 1/8	30

[a]All panel edged supported, as by Plyclips, tongue-and-groove edges or solid blocking.

[b]Deflection limitations: 1/180th for total load; 1/240th for live load only. Loads valid only for number of spans shown.

[c]3-ply and 4-ply construction, 1/2″ thick.

[d]5-ply construction only, 1/2″ thick.

[e]Sanded panel values other than Structural I are based on Group I panels. Reduce applicable loads by 5% for Group 2, 10% for Group 3 and 13% for Group 4 panels.

[f]Solid blocking recommended.

[g]4-ply construction, 1/2″ thick.

[h]5-ply construction only, 5/8″ thick.

Table 204-13. Roof live loads for APA.
Panel sheathing—face grain perpendicular to support.
Panel sheathing is a category of performance-rated panels that include waferboard, oriented strand board, and structural particleboard as well as veneer plywood.
10 psf dead load assumed.

Panel span rating	Panel thickness in.	Maximum span (in.) With edge support[a]	Maximum span (in.) No edge support	Allowable live loads (psf) Spacing of supports center-to-center (in.) 12	16	20	24	32	42	48
12/0	5/16	12	12	30						
16/0	5/16, 3/8	16	16	55	30					
20/20	5/16, 3/8	20	20	70	50	30				
24/0	3/8, 7/16, 1/2	24	20[b]	90	65	55	30			
24/16	7/16, 1/2	24	24	135	100	75	40			
32/16	15/32, 1/2, 5/8	32	28	135	100	75	55	30		
42/20	19/32, 5/8, 3/4, 7/8	42	32	165	120	100	80	55	35	
48/24	23/32, 3/4, 7/8	48	36	210	155	130	100	65	45	35

[a]Tongue-and-groove edges, panel edge clips (one between each support, except two between supports 48″ on center), lumber blocking, or other.

[b]24″ for 1/2″ panels.

205 FASTENERS

From the Forest Products Laboratory's *Wood Handbook*, USDA Handbook No. 72, and NDS-1982.

Nails and Wood

The following discussion outlines the effects of various factors on the strength of nailed joints.

Withdrawal Resistance

For bright, common wire nails driven into the side grain of seasoned wood that will remain dry, or unseasoned wood that will remain wet, the allowable withdrawal load is given by the formula:

Eq 205-1.

$$P = 1390\ G^{5/2}D$$

P = the allowable load (1/6 the ultimate load) per lineal inch of penetration in the member holding the nail point. The allowable loads in Table 205-7 are approximately = $1390\ G^{5/2}D$.

G = the specific gravity of the wood, oven-dry weight and volume.

D = the diameter of the nail in inches.

Dense, heavy woods generally have greater nail-holding power, but the less dense species do not split as readily. Increasing the diameter, length, and number of nails compensates for the wood's lower holding power.

Nails driven into green wood and pulled before any seasoning usually have the same withdrawal resistance as nails driven into seasoned wood and pulled soon after driving. However, common nails driven into green wood that season, or into seasoned wood subjected to cycles of wetting and drying, lose much of their withdrawal resistance. Even in seasoned wood not subjected to appreciable moisture content changes, the withdrawal resistance of nails may diminish with time.

Factors that affect withdrawal resistance

"Cement coating," properly applied, can double the resistance of nails to withdrawal immediately after they are driven into the softer woods. The increase is not permanent and drops to about 1/2 after a month or so for the softer woods.

Nails with special coatings such as zinc are used where corrosion and staining are important in permanence and appearance. Any increased resistance to immediate withdrawal is usually reduced after wetting and drying cycles.

Chemically etched or sand-blasted nails have somewhat higher withdrawal resistance than other coated nails and can retain it under varying moisture conditions, but not under impact loading.

Nail shanks are shaped to increase surface area without an increase in nail weight.

Their withdrawal resistance, except for some types of barbed nails, is generally greater than common nails in wood at a uniform moisture content. With changes in moisture content, some forms have greater withdrawal resistance than common wire nails, especially for nails driven into green wood that later seasons. In general, annular grooves increase withdrawal loads, and spiral grooves increase impact withdrawal.

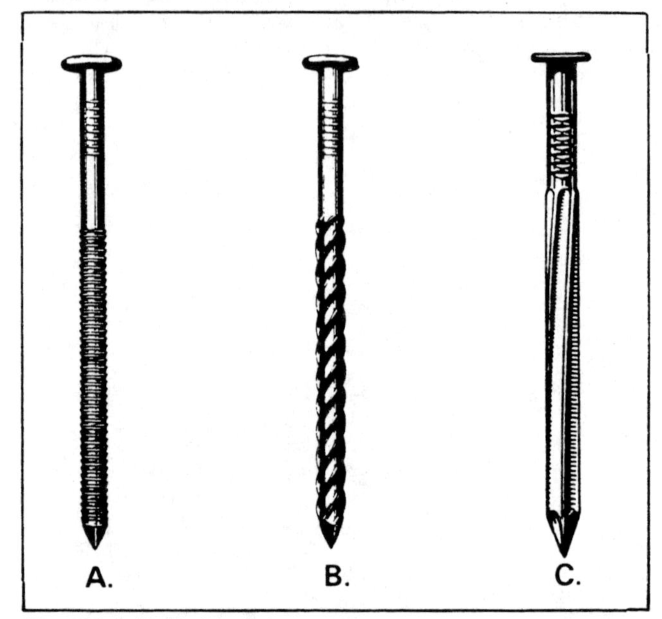

Fig 205-1. Nail shank types.
A, annular thread for general use, especially in softwoods. B, spiral thread for hardwoods such as flooring and pallets. C, shallow thread for masonry nails.

A long, sharp point usually has a higher withdrawal resistance, particularly in softer woods, than common nails, which usually have diamond points. Sharp points accentuate splitting. A blunt or flat point without taper reduces splitting and withdrawal resistance. Nails tapered to a blunt point split less than common nails and have about the same withdrawal resistance in heavier wood but less in lighter wood.

Withdrawal resistance is generally greatest for nails driven perpendicular to the grain of the wood. With end-nailing, withdrawal resistance is reduced in dense woods and can drop below 50% in softer woods.

Slant driving is usually superior to straight.

The maximum strength of toenailed joints is obtained by using the largest nail that will not cause excessive splitting, allowing an end distance of about 1/3 the nail length, driving the nail at a slope of 30° with the attached member, and burying the full shank of the nail but avoiding wood mutilation.

The allowable withdrawal load per nail in toe-nailed joints for all conditions of seasoning is ⅔ of calculated withdrawal resistance.

Lead holes prevent or reduce splitting and increase withdrawal resistance.

Clinching increases withdrawal strength up to 170%. Clinch nails across the grain.

Small nails in plywood cause less splitting when near an edge. Withdrawal resistance is 70%-85% that of solid wood. For plywood less than ½" thick, less splitting offsets the lower withdrawal resistance. The withdrawal load per inch decreases with increase in the number of plys. Face grain direction has little influence.

Lateral Resistance

The allowable load for a bright, common wire nail in lateral resistance when driven into the side grain (perpendicular to the wood fibers) of seasoned wood is:

Eq 205-2.

$$P = KD^{3/2}$$

P = the allowable lateral load, lb/nail
K = a factor allowing for wood species and for safety factor
D = the diameter of the nail, inches

Ultimate lateral nail loads in softwoods are about six times, and in hardwoods about 11 times, the load P. The side member and the member holding the nail point have about the same density, and nail penetration is not less than 10D for dense woods and 14D for light woods. When metal is held to wood, increase the allowable lateral nail load about 25%.

For end nailing, reduce P to about 60%.

Spacing

The NDS gives no guide on nail spacing. The following are from FPL report 0100, 1965.

Nail spacing, minimum:
- 1½" edge distance
- 1" across grain
- 2½" parallel to grain
- 2¼" end distance in tension members

Spikes

Common wire spikes are manufactured the same as common wire nails. They have either a chisel point or a diamond point and are made in lengths of 3"-12". For corresponding lengths (3"-6"), they have larger diameters than the common wire nails, and beyond 60d they are usually designated by inches of length.

The allowable withdrawal and lateral resistance formulas and limitations given for common wire nails are applicable to spikes, except that in calculating the withdawal load for spikes, the depth of penetration should be reduced by ⅔ the length of the point.

Fastener Selection

Much of this material is abstracted from the NDS and applies primarily to the more common structural woods. Refer to NDS for details for other species.

Table 205-1. Specific gravity, G, of woods.
Based on oven-dry weight and volume.

Group	Typical species	Specific gravity
I	Birch, Maple	0.66
	Ash	0.62
II	Douglas fir-larch	0.51
	southern pine	0.55
III	California Redwood; Idaho white, Ponderosa, sugar, and red pines	0.42
	Douglas fir, south; southern cypress	0.48
IV	Balsam fir, E. white pine	0.38

Nails and Spikes

Allowable loads are per nail and may be adjusted for duration of loading. For lumber pressure-treated with fire retardants, reduce nail and spike loads to:

90% in withdrawal if kiln dried after treatment.
22½% in withdrawal if not kiln dried after treatment.
66¾% in lateral resistance.

Table 205-2. Nail sizes.
"Steel nails" are threaded, hardened steel.

Penny-weight	Length in.	Wire diameter, in. Wire nails	Steel nails	Spikes
4d	1½"	0.098"		
6d	2	0.113	0.120"	
8d	2½	.131	.120	
10d	3	.148	.135	0.192"
12d	3¼	.148	.135	.192
16d	3½	.162	.148	.207
20d	4"	.192"	.177"	.225"
30d	4½	.207	.177	.244
40d	5	.225	.177	.263
50d	5½	.244	.177	.283
60d	6	.262	.177	.283
70d	7"		.207"	
80d	8		.207	
90d	9		.207	
5⁄16	7			.312
⅜	8½			.375

Table 205-3. Wire nails per pound.

Name	Nails/lb	Name	Nails/lb
8d common	106	8d floor brads	99
20d "	31	8d casing	145
60d "	11	3d brads	568
6d finish	309	3d fine	778
8d "	189	3d shingle	429
10d "	121	4d "	274

Table 205-4. Common nails in wood framing.

Connection	Nails
Joist to sill or girder, toe nail	3-8d
Bridging to joist, toe nail each end	2-8d
Ledger strip on girder	3-16d at each joist
1" x 6" subfloor or less to each joist, face nail	2-8d
Over 1" x 6" subfloor to each joist, face nail	3-8d
2" subfloor to joist or girder, blind and face nail	2-16d
Sole plate to joist or blocking, face nail	16d @16" oc
Top plate to stud, end nail	2-16d
Stud to sole plate, toe nail	4-8d
Doubled studs, face nail	16d @ 24" oc
Doubled top plates, face nail	16d @ 16" oc
Top plates, laps and intersections, face nail	2-16d
Continuous header, two pieces	16d @ 16" oc along each edge
Ceiling joists to plate, toe nail	3-8d
Continuous header to stud, toe nail	4-8d
Ceiling joists, laps over partitions, face nail	3-16d
Ceiling joists to parallel rafters, face nail	3-16d
Rafter to plate, toe nail	3-8d
1-inch brace to each stud and plate, face nail	2-8d
1" x 8" sheathing or less to each bearing, face nail	2-8d
Over 1" x 8" sheathing to each bearing, face nail	3-8d
Built-up corner studs	16d @ 24" oc
Built-up girders and beams	20d @ 32" oc along each edge

Table 205-5. Minimum nailing for plywood.

Plywood	Nail size	Nail spacing on supports: Panel edges	Intermediate
Roof and wall sheathing, paneling			
¼"	4d	6"	12"
5/16"-½"	6d	6"	12"
⅝"-⅞"	8d	6"	12"
When supports are 48" o.c. space all nails 6" o.c.			
Sub-flooring			
½"	6d	6"	10"
⅝"-⅞"	8d	6"	10"
1"-1¼"	10d	6"	6"
If resilient flooring is to be applied without underlayment, set nail heads 1/16"; use ring-shank nails.			
Underlayment			
¼"-½"	3d	6"	8" each way
⅝"	4d	6"	8" each way

Shingles to plywood

Asphalt	¾" or ⅞"-12 ga., ⅜" head nails
Wood	3d
Asbestos	1⅛"-12 ga. aluminum, screwthreaded shank, needlepoint

Table 205-6. Nail penetration for lateral resistance.
Data rounded up. Group I = 10 diameters, II = 11, III = 13, IV = 14. Minimum penetration = ⅓ tabulated. Reduce load by linear interpolation.

Common nails (inches penetration)

Group	8	10-12	16	20	30	40	60
I	1.4	1.5	1.7	2.0	2.1	2.3	2.7
II	1.5	1.7	1.8	2.2	2.3	2.5	2.9
III	1.7	2.0	2.2	2.5	2.7	3.0	3.5
IV	1.9	2.1	2.3	2.7	2.9	3.2	3.7

Threaded nails

Group	6-8	10-12	16	20-60	70-90
I	1.2	1.4	1.5	1.8	2.1
II	1.4	1.5	1.7	2.0	2.3
III	1.6	1.8	2.0	2.3	2.7
IV	1.7	1.9	2.1	2.5	3.0

Spikes

Group	10-12	16	20	30	40	50-60
I	2.0	2.1	2.3	2.5	2.7	2.9
II	2.2	2.3	2.5	2.7	3.0	3.2
III	2.5	2.7	3.0	3.2	3.5	3.7
IV	2.7	2.9	3.2	3.5	3.7	4.0

Withdrawal

Withdrawal from end grain is prohibited; avoid joints requiring strength in withdrawal from side grain. Reduce tabulated load for nails and spikes to 25% if driven into green wood that will season in use. Use ⅔ of the withdrawal load for withdrawal of toe-nails. See Table 205-7.

Lateral resistance

Strength in end grain for nails and spikes is ⅔ of the tabulated load.

In double shear, with nail fully penetrating all three members:

- Increase ⅓, each side member not less than ⅓ the thickness of the middle member.
- Increase ⅔, each side member equal to middle member.
- Double for wire nails to 12d with at least 3 diameters clinched over side members at least ⅜" thick. Similarly for hardened steel nails, but not clinched.
- Increase ¼ for wire nails or spikes with properly designed metal side plates.
- Decrease to ¾ for wire nails or spikes in unseasoned wood used wet or loaded before seasoning and for wood pressure-impregnated with fire-retardant chemicals.
- Increase 30% for diaphragm construction.

Bolts

Tabulated loads are for:

- normal load duration.
- one common bolt in double shear in 3-member joint.
- any grade of the wood species tabulated.
- wood seasoned to remain dry.
- bolt holes 1/32" to 1/16" larger than the bolt.
- loose nuts because of lumber shrinkage.
- washer, plate, or strap under head and nut.
- side plates ½ or more as thick as main member.
- bolt length in main member.
- grain of side and members in same direction.

Table 205-7. Allowable loads on nails and spikes.
Normal load duration loads for 6d-20d threaded, hardened steel nails are the same as for common wire nails:

Withdrawal loads,
lb/in. of penetration in side grain of seasoned wood or unseasoned that will remain wet. Normal duration.

Specific gravity G	Common nail, d							Threaded nail, d	
	8	10-12	16	20	30	40	60	30-60	70-90
	lb/in. penetration								
.66	64	72	79	94	101	110	128	94	110
.62	55	62	68	80	86	94	110	80	94
.54	39	44	48	57	61	67	78	57	67
.51	34	38	42	49	53	58	67	49	58
.48	29	33	36	42	46	50	58	42	50
.42	21	23	26	30	33	35	41	30	35
.38	16	18	20	24	25	28	32	24	28

	Common spike, d					
	10-12	16	20	30	40	50-60
	lb/in. penetration					
.66	94	101	110	119	128	138
.62	80	86	94	102	110	118
.54	57	61	67	72	78	84
.51	49	53	58	63	67	73
.48	42	46	50	54	58	62
.42	30	33	35	38	41	45
.38	24	25	28	30	32	35

Lateral loads (shear),
lb/nail. Penetration into member holding the point at least equal to requirements of Table 205-6.

Species group	Common nail, d							Threaded nail, d	
	8	10-12	16	20	30	40	60	30-60	70-90
	lb/nail								
I	97	116	133	172	192	218	275	172	218
II	78	94	108	139	155	176	223	139	176
III	64	77	88	114	127	144	182	114	144
IV	51	62	70	91	102	115	146	91	115

	Common spike, d					
	10-12	16	20	30	40	50-60
	lb/nail					
I	171	192	218	246	275	307
II	139	155	176	199	223	248
III	114	127	144	163	182	203
IV	91	102	115	130	146	163

Adjustments

The full load can be used for green lumber, even if lumber dries in service for wood side members with: a single bolt loaded parallel or perpendicular; single row of bolts loaded parallel; rows of bolts loaded parallel with separate splice plate for each row; or with metal gussets and a single row parallel to the grain in each member and loaded parallel or perpendicular. For other patterns, use 40% of tabulated load.

Reduce to 75% if exposed to weather, 67% if always wet.

With metal side plates, increase to 175% for parallel-to-grain loads.

For side members less than ½ the thickness of the main member, use tabulated load for main member twice as thick as the thinnest side member.

For single shear, two members of equal thickness, use ½ of the load for a piece twice the thickness of one member.

For single shear, two members of unequal thickness, use ½ of the load for the thicker member or for a piece twice the thickness of the thinner member.

For joints with more than three members of equal thickness, allow ½ tabulated load for each shear plane.

Placement

Member strength is computed for the net section after bolt holes are drilled.

Bolts spacings are computed from their centerlines.

Parallel to grain loading, minimum spacing equals 4 diameters.

Perpendicular to grain loading: if design load is less than bolt bearing capacity of side members, reduce spacing proportionally from 4 diameter minimum.

For rows of bolts, parallel loading, space rows at least 1½ diameters.

For rows of bolts, perpendicular loading, space rows at least 2½ diameters with L/d = 2; 5 diameters with L/d = 6 or more; straight line interpolation for L/d between 2 and 6.

Space no more than 5″ between rows of bolts paralleling the member, or use separate splice plates.

End distance, parallel loading: in tension, 7 diameters in softwoods, 5 diameters in hardwoods; in compression, 4 diameters.

End distance, perpendicular loading: 4 diameters.

Edge distance, parallel loading: 1½ diameters, except if L/d is greater than 6, use ½ row spacing.

Edge distance, perpendicular loading: 4 diameters to loaded edge, 1½ diameters to unloaded edge.

Staggered bolts, parallel loading: adjacent bolts computed as if at critical section unless spaced at least 8 diameter. Perpendicular loading: if design load of main member is less than bolt bearing capacity of side member, staggering can be employed.

Timber Fasteners

Special hardware develops structural strength in timber joints. Design loads given are examples only; use allowable loads and application and installation information furnished by the manufacturer. In some cases special nails are required. Split ring connectors require detailed design, as allowable loads vary considerably with edge and end spacing, slope of grain, ring spacing, and number of lumber faces affected.

Table 205-8. Bolt design values.
Design values, in pounds, on one bolt loaded at both ends (double shear) for following species.
L = length of bolt in main member, in.
D = diameter of bolt, in.
A = projected bolt area, LxD, in^2
P = lb/bolt parallel to grain
Q = lb/bolt perpendicular to grain

				Douglas fir larch (dense), southern pine (dense)		Ash, commercial white, Hickory		California Redwood (Close grain), Douglas fir-larch southern pine, southern cyress		Beech Birch sweet & yellow Maple black & sugar		Oak, red & white		Douglas fir south	
L	D	L/D	A	P	Q	P	Q	P	Q	P	Q	P	Q	P	Q
	½	3.00	.750	1100	500	1080	780	940	430	900	480	830	650	870	370
	⅝	2.40	.938	1380	570	1360	880	1180	490	1130	540	1050	730	1090	420
1½	¾	2.00	1.125	1660	630	1630	980	1420	540	1360	600	1260	820	1310	470
	⅞	1.71	1.313	1940	700	1910	1080	1660	600	1590	670	1470	900	1530	520
	1	1.50	1.500	2220	760	2180	1170	1890	650	1820	730	1690	980	1750	570
	½	5.00	1.250	1480	830	1450	1280	1260	720	1210	800	1120	1070	1290	620
	⅝	4.00	1.563	2140	950	2100	1460	1820	810	1750	900	1620	1220	1780	710
2½	¾	3.33	1.875	2700	1050	2650	1630	2310	900	2210	1010	2050	1360	2180	790
	⅞	2.86	2.188	3210	1160	3150	1790	2740	990	2630	1110	2440	1500	2550	860
	1	2.50	2.500	3680	1270	3620	1960	3150	1080	3020	1210	2800	1640	2920	940
	½	6.00	1.500	1490	970	1460	1460	1270	860	1220	960	1130	1130	1330	750
	⅝	4.80	1.875	2290	1130	2250	1750	1960	970	1880	1090	1740	1460	1980	850
3	¾	4.00	2.250	3080	1270	3020	1950	2630	1080	2520	1210	2340	1630	2560	940
	⅞	3.43	2.625	3760	1390	3700	2150	3220	1190	3080	1330	2860	1800	3040	1040
	1	3.00	3.000	4390	1520	4310	2350	3750	1300	3600	1450	3340	1960	3500	1130
	½	7.00	1.750	1490	1120	1460	1460	1270	980	1220	1090	1130	1130	1330	850
	⅝	5.60	2.188	2320	1310	2280	2020	1980	1130	1900	1270	1770	1690	2060	990
3½	¾	4.67	2.625	3280	1470	3220	2270	2800	1260	2690	1410	2490	1900	2820	1100
	⅞	4.00	3.063	4190	1630	4120	2510	3580	1390	3430	1550	3180	2100	3480	1210
	1	3.50	3.500	5000	1770	4920	2740	4270	1520	4100	1690	3800	2290	4050	1320
	⅝	8.80	3.438	2330	1890	2290	2150	1990	1410	1910	1570	1770	1770	3080	1220
	¾	7.33	4.125	3350	1930	3300	2980	2860	1880	2750	2100	2550	2490	2990	1630
	1	5.50	5.500	5930	2760	5830	4260	5070	2380	4860	2660	4510	3560	5270	2080
	⅝	12.00	4.688	2330	1210	2290	1870	1990	1260	1910	1410	1770	1560	2070	1100
	¾	10.00	5.625	3350	1930	3290	2710	2860	1820	2750	2030	2550	2260	2990	1580
7½	⅞	8.57	6.563	4560	2410	4480	3720	3900	2420	3740	2700	3470	3110	4060	2110
	1	7.50	7.500	5950	3090	5850	4760	5080	3030	4880	3390	4520	3980	5300	2640

Table 205-9. Rafter anchors.
Recommended safe working values.

Species Group	Resistance to Wind Uplift, pounds
I	965 lb
II	780
III	550
IV	410

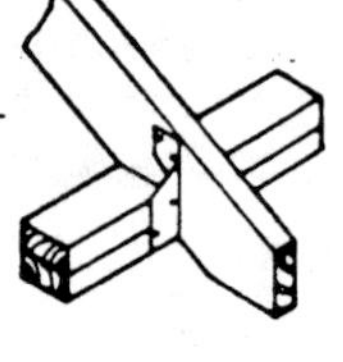

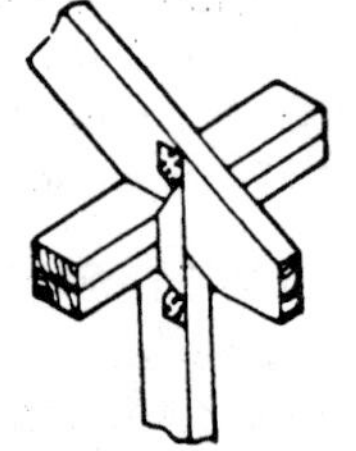

Table 205-10. Joist supports.

Joist or beam size	Steel ga.	Height	Seat depth	Safe load, lb 10d	16d
2x4	18	3¼″	1½″	425	755
2x6-8	18	4¾	2	635	805
2x10-12-14	18	7¾	2	1060	1270
4x6-8	16	5	2	1700	1070*
4x10-12	16	7	2	2200	1875*

*16d into header, 10d into joist.

Table 205-11. Framing anchors.
Recommended safe working values, lb/anchor.
Long term loading.
These values are for permanent load and can be increased from 33% to 50% for short duration loading such as wind and earthquake, depending upon local building code practice.

Direction of Load:	A	B	C	E	F
	—lb/anchor				
Short Term Loading (Wind or Earthquake)	450	825	420	450	675
Long Term Loading (Live & Dead Loads)	300	530	290	300	450

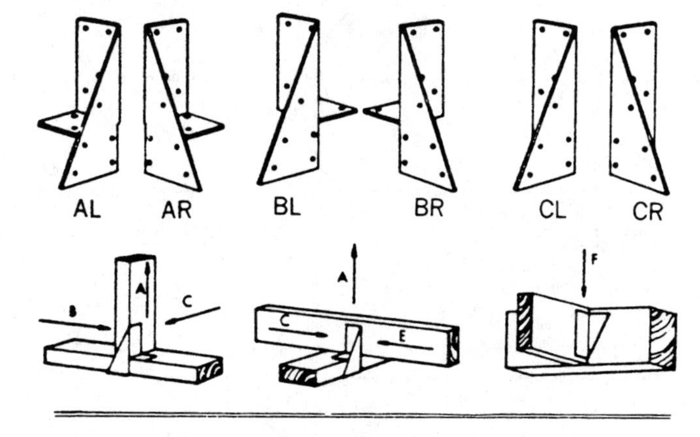

Direction of Load:	A	B	C	D
	—lb/anchor			
Short Term Loading (Wind or Earthquake)	600	450	180	600
Long Term Loading (Live & Dead Loads)	400	300	120	400

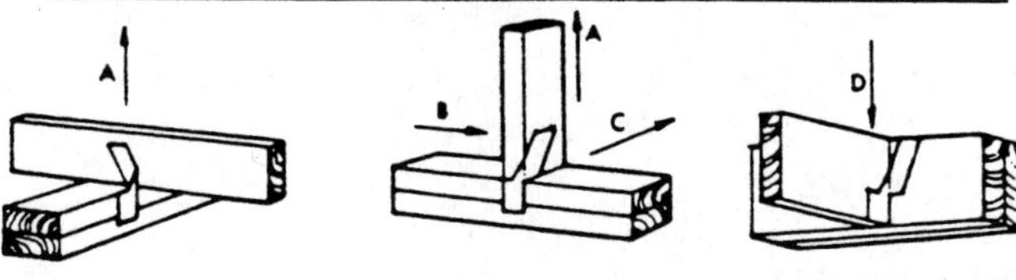

Table 205-12. Split rings.

Inside Dia.	Depth	Bolt Dia.	Lumber, Minimum Dimensions Ring in One Face	Rings in Both Faces
2 1/2″	3/4″	1/2″	1″ x 3 5/8″	1 5/8″ x 3 5/8″
4″	1″	3/4″	1″ x 5 1/2″	1 5/8″ x 5 1/2″

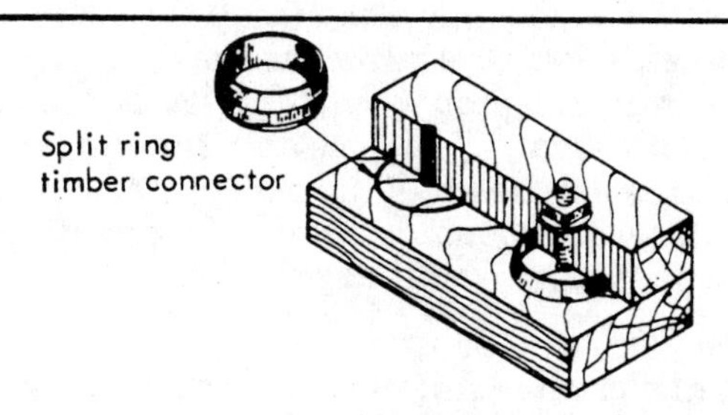

206 GLUING WOOD

Table 206-1. Glue properties.

PROPERTY	SYNTHETIC RESIN GLUES					PROTEIN GLUES	
	Resorcinol	Urea	Polyvinyl	Epoxy	Contact Adhesives	Casein	Animal
Needs Mixing	X	X		X		X	X
Crack-Filling			X	X	X	X	
Applied Hot							X
Applied Cold	X	X	X	X	X	X	
Colorless Glue Line		X	X		X	X	X
Dark Colored Glue Line	X						
Tends to Stain Certain Woods						X	
Pressed at 70°	X	X	X	X	X	X	X
Over 8-Hour Working Life		X	X				X
Low Moisture Resistance			X				X
Medium Moisture Resistance			X		X	X	
Good to High Moisture Resistance	X	X		X			
Low Temperature Resistance		X	X				
High Temperature Resistance	X			X	X	X	X
For Structural Gluing	X	X				X	
For Exterior Uses [1]	X			3			
For Interior Uses [2]		X	X	3	4	5	X

[1] Exterior uses include outdoor furniture, boats, and recreational equipment.

[2] Interior uses include furniture, cabinets, framing, and other shopwork that will be used in a moderately dry atmosphere.

[3] Excellent for bonding metal, plastics, and cloth to wood. No practical advantage on wood-to-wood gluing over resorcinol resin except it is a good joint filler.

[4] Covering counters and cabinets with leather, linoleum, and plastic laminates.

[5] Mold-resistant types are highly water resistant, though not waterproof.

Selection

Wood selection

Heavy woods are usually more difficult to glue than lightweight woods. Hardwoods are more difficult to glue than softwoods. Heartwood is more difficult to glue than sapwood.

Use dry, smooth wood free of dirt, oil, and other coatings. Most purchased lumber has been planed and is sufficiently smooth.

Clean dirt, paint, and other coatings from wood with a scraper, wire brush, steel wool, or other abrasive. Do not use wood that has oil or grease at joint locations.

For structural members and those expensive to replace, use wood pressure-treated with oil-base preservatives if it will be exposed to wet conditions, soil, or termite attack.

Generally, preservative-treated wood must be planed prior to gluing for maximum holding power. Wood treated with oil-base preservatives tends to bleed, so buy wood that has been steamed or otherwise cleaned until bleeding has stopped.

Moisture content

Wood gains and loses moisture with humidity and temperature changes, causing wood to change dimensions and stress joints.

Kiln-dried wood available at most lumber yards has about 15% moisture content. The moisture content of wood for interior home projects should be reduced to an average of 8% by storing it in an atmosphere similar to where it will be used. To assure uniform exposure to the air, place small blocks between the pieces. After several weeks the wood will usually be near an acceptable moisture level.

Exterior siding, trim, sheathing, and framing can usually be used as purchased. If the moisture content is above 15%, the wood should not be glued; store it in a drier atmosphere for several days.

Glue selection

No glue satisfies all requirements. Choose one for the requirements of one job. No glue is foolproof. Well-made joints require woodworking skill and control of gluing conditions. Improper use of glues may result in costly repairs or replacements.

Consider expected length of service. In permanent structures, the few added dollars needed to buy a more resistant glue and EXT plywood are well spent.

Recommended glue qualities for farm building construction:

- Water resistant for storage buildings and where the moisture content of the wood can be controlled (casein or resorcinol resin).
- Waterproof for livestock confinement buildings and where moisture problems are almost sure to occur, e.g., roof leaks, and outdoor furniture or equipment (resorcinol resin).
- Crack-filling (gap-filling).
- Easy mixing.
- Nail pressure for clamping.
- Cold setting—cures without heat.
- Moderate cost.

Two types are recommended—casein and resorcinol resin.

Recommended glue qualities for most interior uses:

- Easy mixing or pre-mixed.
- Moderate cost.
- Medium pressure requirements (clamps or nails).
- Cold setting.
- Does not stain the wood or leave a visible glueline.

Several glues usually have these qualities. Urea resin is a good structural glue for dry wood that will remain dry. Resorcinol resin and casein both leave visible gluelines.

Construction adhesives

An expanding array of adhesives useful in the construction—as opposed to prefabrication—industry are available. Products have been developed specifically for applying drywall panels, subfloors, decorative paneling, etc. Elastomeric construction adhesives do not set rapidly (as do contact cements, for example), retain some elasticity when cured, are water resistant (some may be waterproof), and adhere to wood, metals, fiberglass, concrete, brick, and other materials.

Construction savings result from subfloor panels being rigidly fastened to each other and to joists, from the ease of fastening materials to concrete floors and walls, from avoiding covering nail heads, and similar applications. Usually the adhesive is applied in a bead with a caulking gun—a single bead along a joist or stud, serpentine beads for panels, and intermittent beads for low-load members.

Gluing

Follow manufacturer's directions.

Have the room temperature above 70 F. Lower temperatures slow curing.

Follow mixing instructions, pot life (the length of time the glue stays usable after mixing), temperature during use, etc.

Foam, lumps, or too thick or too thin glue, lower joint quality. The glue film should have no air bubbles or foreign particles and should completely cover the joint area.

Before gluing, test for good fit; pieces should make contact at all points. Crack-filling glues can give a good bond despite small surface misfits. Glue a sample joint to test for staining.

The end grain of wood usually requires two coats. For farm construction, for heavy construction such as laminated timbers, or where the surfaces are rough and assembly periods may be long, spreading on both surfaces is desirable.

Apply enough glue. When pressure is applied, some glue should ooze out from around the joint. If it does not, you are not using enough glue. Do not skimp; the cost of the glue is a minor item in the total cost of construction.

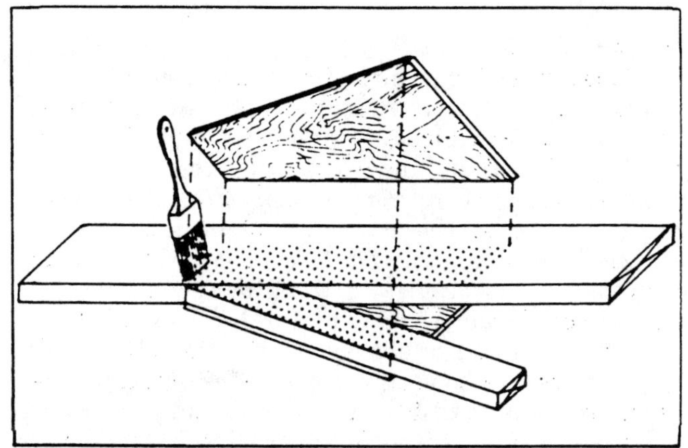

Fig 206-1. Applying glue to joint.

Apply glue with a brush, stick, or paint roller. Put pressure on the joint with clamps, nails, screws, or other fasteners before wiping off excess glue.

Assembly time

Assembly time is between spreading the glue and applying the pressure. "Open assembly" means the surfaces are open to air; "closed assembly" means the surfaces are in contact, but not clamped.

Permissible assembly time is reduced by: warm glue film temperatures, drier wood, the end of a batch of mixed glue, and only one surface coated with glue. Most glue containers have the assembly time and other precautions printed on the label.

Pressure

Pressure squeezes glue into a thin continuous film, forces air from the joint, brings glue and wood together, and holds them until the glue has set and cured.

Apply pressure uniformly over the entire joint. Keep joints under pressure at least until they have enough strength to resist separation—usually from 2 to 7 hr under favorable conditions. Leave pressure applied longer if possible.

Curing time

The time needed for curing depends on the glue type and the glueline temperature.

The higher the glueline temperature, the faster the glue will cure and the shorter the necessary clamping time. Room temperature means 70 F or above. The minimum temperatures for curing have not been well established, but below 70 F curing time is increased.

Durability

Durability depends on glue type, service conditions, gluing technique, and the design or construction of the joints.

Well-designed joints made with the proper glues will last indefinitely if the moisture content of the wood does not exceed approximately 15% and the temperature is within about -30 F to 110 F.

To increase the durability, use waterproof glue and preservative-treated wood, or use waterproof glue and apply preservatives to the wood after the joint is made.

Structural gluing

Wood must be clean and smooth and below 15% moisture. The surface of pressure-treated wood must be dry and free of bleeding preservative. Use unsanded (preferred) or sanded EXT plywood.

Use urea resin, casein, resorcinol resin, or a construction adhesive. Use resorcinol resin in treated wood. Use urea resin only for dry conditions. Casein can be used for moderately dry conditions; buy a glue that meets Federal Specification MMM-A-125 Type II. Apply glue at 40 F or above; 70 F is recommended. Apply pressure as soon as possible. Maintain pressure for two days at 40 F, 4 hr at 70 F, or 2 hr at 80 F.

Use resorcinol resin for either wet or dry conditions. Buy a glue that meets Military Specification Mil-A-46051. Apply glue at 70 F or above. Assemble but wait 5 to 10 minutes before applying pressure. Maintain pressure for 10 to 16 hr.

Pressure is often applied with staples or box, galvanized, or cement coated nails. Finish driving staples with at least one hammer blow. At the end of the pressing time, the glued elements can be moved, but full load should not be applied for about one week.

Structural joints

Joints connecting structural members are as critical as the members themselves. Use a nail for about each 8 in^2 of the joint. The location and size of nails are usually given in a designed plan or in the engineer's recommendations.

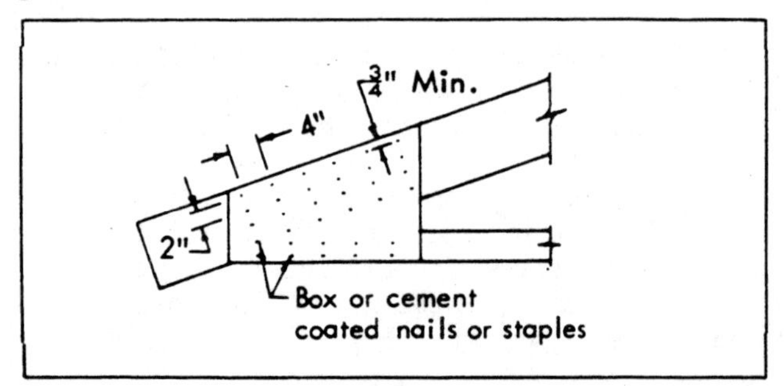

Fig 206-2. Typical truss joint.

Non-structural joints

Edge-to-edge joints can be strong as the wood itself. Tongue-and-groove lumber is easier to align, but the shapes may not fit snugly. A shallow tongue-and-groove is as effective as a deep one and is less wasteful of wood.

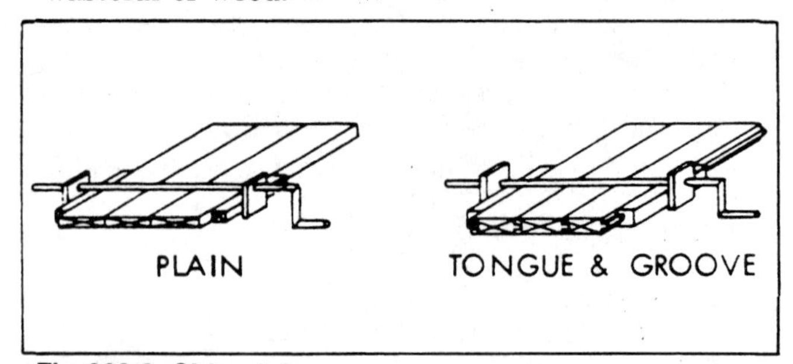

Fig 206-3. Side gluing.

End-to-end joints cannot be made to give the same strength as the wood itself. Use a shape that will give more glue surface area and, therefore, more strength.

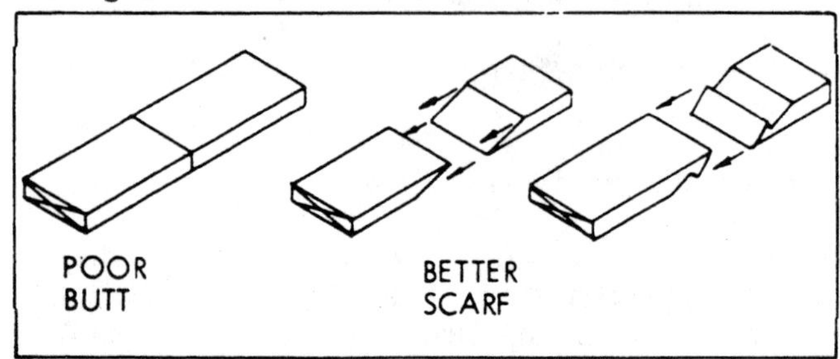

Fig 206-4. End gluing.

End grain-to-side grain joints are difficult to glue. They have high joint stresses because of unequal dimension changes in the two pieces. End-to-side joints should be tightly fitted and held together by dowels, tenons, gussets, or other reinforcing devices.

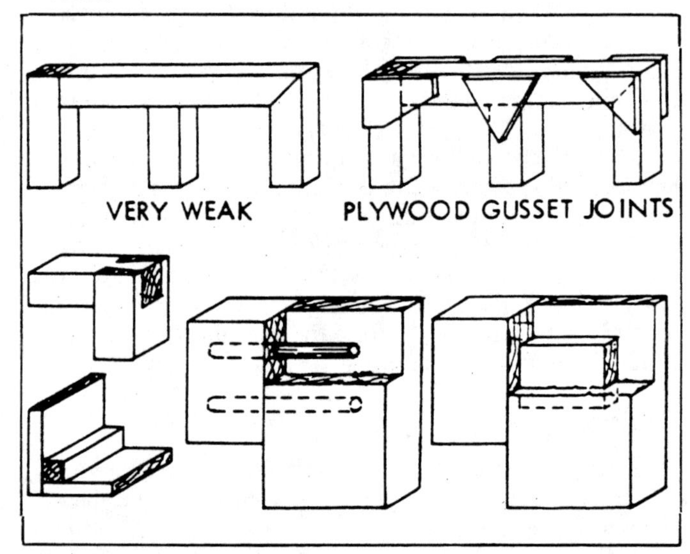

Fig 206-5. Cabinet joints.

207 CONCRETE AS A MATERIAL

Concrete is a mixture of Portland cement, water, air and aggregates. Portland cement is sold in bulk, or in bags of one cubic foot (94 lb). The aggregates provide volume at low cost, composing 66%-78% of the concrete.

Cement and water form a paste that hardens and glues the aggregates together. The quality of concrete is directly related to the binding qualities of this cement paste.

Selection

General properties

Concrete is very durable and resists attack by water, animal manures, chemicals such as fertilizers, and fire. Use high quality concrete around milk, silage, and animal manure.

Concrete is very weak in tension. Its strength in compression depends on the proportions of the mix. The compressive strength is 2 to 5 times that of wood. Most structural uses involve reinforced concrete, which depends on concrete's strength in compression and steel's strength in tension.

Concrete can be finished in either smooth or rough texture. It can be colored with pigments or painted.

Portland cement

Portland cement got its name from the Isle of Portland, near England, where it was first developed. This improvement over an ancient building material is made by burning limestone and clay, adding gypsum, and grinding the result to a fine consistent powder.

A number of variations have been developed for special purposes:

- Type I, normal Portland cement, is the general purpose type, the most common, and is usually furnished unless an alternative is specified.
- Type II is modified to release less heat during curing, and is therefore suitable in mass concrete—heavy retaining walls, or deadmen for suspension bridges. It is moderately high in resistance to sulphates. Type II is replacing Type I as the basic type in some areas.
- Type III is high-early-strength; it is very finely ground and sets very rapidly. It is useful for slipform construction and for cold weather jobs.
- Type IV is the lowest heat variety, and while suitable for large masses, develops strength relatively slowly. Its primary use is for mass concrete dams and other large volume structures.
- Type V is especially sulphate-resistant and cures more slowly than Type I.
- Types I, II, and III are available as Types IA, IIA, and IIIA. These are air-entrained Portland cements formulated with a compound that releases many tiny bubbles of air during curing. The resulting concrete is highly resistant to frost action and has some increased resistance to salts. Their use is recommended for all outdoor paving and for concrete exposed to animal wastes, even though slightly weaker than Type I, II, and III.

Air-entrained concrete

An air-entraining agent added to the cement produces millions of tiny bubbles in the concrete giving the concrete greater weathering resistance. Air entraining also reduces the strength and increases the concrete workability. Because the air-entraining agent increases the weathering resistance compared to its reduction of strength, air entraining is usually recommended for all concrete subjected to freezing and thawing. Table 207-1 relates the amounts of air entraining to aggregate size.

Other additives include pigments for coloring, gypsum to retard setting time, calcium to reduce setting time, and bentonite to improve workability.

Table 207-1. Air content for air-entrained concrete.

Max. aggregate size	Amount of air, %
1½", 2", or 2½"	4 to 6
¾" or 1"	5 to 7
⅜" or ½"	6½ to 8½

Water

Cement in concrete cures by the chemical combination of water and cement—not by drying like sheets in the sun. Use water for concrete that is essentially good enough to drink, without harmful chemicals, trash, or organic matter.

The strength of concrete is greatly dependent on the water-cement ratio. Enough water is needed for full curing, but excess water leaves voids when it evaporates. If a mix is too stiff to handle well, do not add water; reject the batch or add cement and water—reduce the amount of aggregates in subsequent batches.

Aggregates

Sand and gravel, the usual aggregates, are glued together by the cured cement paste. They must be free of, or very low in, silt and organic matter, and they must be hard and strong.

Particles up to ¼" are sand, and above ¼" are gravel. Well graded aggregates occupy most of the volume of the concrete, minimizing the amount of cement paste, which is the expensive ingredient. Using spheres for illustration, large balls have less surface area than small ones, requiring less cement. But small balls fit in between the large ones to help fill up the voids.

Table 207-2. Concrete mixes.
Make a trial batch to check for slump and workability. Wet sand occupies more volume than dry sand, so yield is not quite the same as in the example below.

	Max. size aggregate	[1]Gallons of water for each sack of cement, using: Damp[3] sand	Wet[4] (average) sand	Very[5] wet sand	[2]Suggested mixture for 1-sack trial batches: Cement, sacks ft^3	Aggregates Fine ft^3	Aggregates Coarse ft^3	Ready-Mix sacks cement per yard[6]
5-Gallon mix; use for concrete subjected to severe wear, weather, or weak acid and alkali solutions.	¾″	4½	4	3½	1	2	2¼	7¾
6-Gallon mix; use for floors (home, barn), driveways, walks, septic tanks, storage tanks, structural concrete.	1″	5½	5	4½	1	2¼	3	6¼
	1½″	5½	5	4½	1	2½	3½	6
7-Gallon mix; use for foundation walls, footings, mass concrete etc.	1½″	6¼	5½	4¾	1	3	4	5

[1]Increasing the proportion of water to cement reduces the strength and durability of concrete. Adjust the proportions of the batches without changing the water-cement ratio. Reduce gravel to improve smoothness; reduce both sand and gravel to reduce stiffness.
[2]Proportions vary slightly depending on gradation of aggregates.
[3]Damp sand falls apart after being squeezed in the palm of the hand.
[4]Wet sand balls in the hand when squeezed, but leaves no moisture on the palm.
[5]Very wet sand has been recently rained on or pumped.
[6]Medium consistency (3″ slump). Order air-entrained concrete for outdoor use.

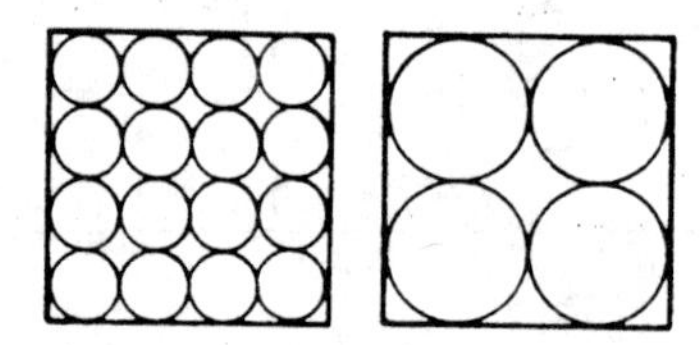

Fig 207-1. Aggregate size vs. surface area.

2″ cube holds 8-1″ diameter (d) balls;
volume $= 8(\pi d^3/6) = 8(0.5236)$
$= 4.189\ in^3$
surface $= 8(\pi d^2) = 8(3.14)$
$= 25.13\ in^2$
2″ cube holds 64½″ balls;
volume $= 64(0.06545) = 4.189\ in^3$
surface $= 64(0.785) = 50.26\ in^2$

Each size fills about one-half the volume, but the smaller ones have twice the surface area. Therefore, an economical mix is one that uses the largest practical size of aggregate, plus smaller aggregates to fill as many voids as possible.

The largest practical size of coarse aggregate is:

- About ⅕ the thickness of vertically formed concrete.
- About ⅓ the thickness of flat concrete work such as floors or walks.
- For reinforced concrete, not over ¾ the distance between bars. For example, if reinforcing steel is 1″ from the face of a slab, use ¾″ aggregate.

The moisture in the aggregate is part of the water needed to react with the cement.

Mixing and Placing

Trial mixes are recommended in Table 207-2.

Mix concrete at least one minute after all ingredients have been added for batches up to 1 yd. Add ¼ minute for each additional ½ yd or fraction.

Curing concrete

Concrete hardens by a chemical reaction called hydration. For this chemical reaction to occur, water must be present and the concrete temperature should be between 40 F and 90 F. This chemical reaction proceeds very rapidly initially and then slows down but continues for years. Concrete forms may be removed in 1 to 2 days in summer or 4 to 7 days in winter.

"Curing" usually means preventing mixing water from evaporating. To retain this water, cover fill with plastic, or soak it thoroughly before concrete is placed. After finishing, cover the concrete with plastic or spray with a curing compound. Other methods include ponding water or covering the concrete with straw, burlap, or sand, and sprinkling with water. Curing begins as soon as the concrete surface is hard enough not to be damaged by the water and should continue for 3 to 7 days.

Concrete finish

For a smooth finish, use a steel trowel. Concrete floors with a steel trowel finish are usually too slick for most farm animals. If the concrete is level, use wood float finish for animal floors. If the floor is inclined or it is an area where animals must have good footing, give the concrete a broom finish.

208 REINFORCED CONCRETE MANURE TANK SELECTION TABLES

Walls, lids, slats, beams, columns, footings

This chapter is intended primarily for engineers with an understanding of reinforced concrete design. While it includes design criteria and adequate reinforcing for selected sections, many essential details are not included, such as wall foundations, corner reinforcing, design of necessary openings in walls or lids, etc.

A plan based on this information is available from the MWPS at a nominal price; order MWPS plan 74303, *Liquid Manure Tanks.* See TR-3, *Concrete Manure Tank Design,* and TR-4, *Welded Wire in Concrete Manure Tanks,* for additional design tables.

One section of ACI 318-71 has not been rigidly followed. Paragraph 11.1.1(c) limits beams to 10″ deep without minimum shear reinforcement. Paragraph 11.1.1(d) states that if the allowable shear stress is cut in half, minimum shear reinforcement is not required. The Commentary explains that these requirements are for ductility. Ductility helps provide warning of failure and resistance to unexpected tension or catastrophic loads. Beams for slotted floors are supporting a relatively large dead load, almost one-half the total load in most cases, and are not subject to large, unexpected loads. Therefore, the beams in Tables 208-13, 208-14, and 208-17 use the full allowable shear stress for depths of up to 15″.

Hydrogen sulfide, H_2S, is given off from anaerobic waste storages. Where a tank has air over the liquids and condensation on the underside of the lid, hydrogen sulfide may lead to sulfuric acid. There is no general preventative to attack by the acid available to the farm tank designer or owner. Cement low in tri-calcium aluminate is naturally available in some areas and has some sulfate resistance, as does Type II cement, which is apt to be available in areas with water supplies naturally high in sulfates. Specification of a high quality air-entrained concrete is perhaps the best solution available to the engineer.

Procedure Summary

Materials

Obtain materials that meet or exceed the strength specifications of Table 208-1.

NOTE: Bar spacings are given to ¹⁄₁₀″; the designer can select designed rounding.

Tank Design

Tank walls (Fig 208-1)

Determine soil characteristics and choose the design load from Chapter 102.

Determine desired tank depth.

Select vertical steel from Table 208-5.

Select horizontal steel (minimum temperature and shrinkage steel) from Table 208-4.

Beams at the top of tank walls

Determine the distance between lateral restraints at the top of the tank wall.

Select extra reinforcing needed from Table 208-6 for the tank depth and design load, trying first a beam thickness equal to that of the wall itself before trying larger thicknesses.

Tank lids

Determine desired span of the slab.

Determine the design load from Table 208-7 for no vehicle traffic, or assume given tractor or manure wagon load.

Select principal reinforcing steel from Table 208-8 for the given tractor or manure wagon load or from Table 208-9 for the design snow and/or animal traffic load.

Select temperature and shrinkage steel for the chosen slab thickness from Table 208-3.

Slotted Floor and Support System Design

Slotted floor (Fig 208-2)

Determine load on slat from Table 208-11, second column.

Determine length of slat (usually not over 10′).

Select slat design from Table 208-12.

Floor-support girders (Fig 208-2)

Determine load per ft of girder.

Determine length of girder (usually not over 10′).

Deformed Bars

Select girder design from Table 208-13.

Columns (Fig 208-3)

Determine column load.

Determine column height (usually not over 8′).

Select column design from Table 208-14 or 208-16.

Column support (Fig 208-3)

Select adequate footing for column load from Table 208-14, 208-15 or 208-16.

Table 208-1. Allowable stresses, ultimate design.

Concrete		
Compressive strength	f'_c	3,500 psi
Punching shear	$3.4(f'_c)^{1/2}$	201 psi
Beam shear	$1.7(f'_c)^{1/2}$ or	100 psi
0.85	$1.9(f'_c)^{1/2} + 2500 \frac{(A_s)(V_u)}{(B)(M_u)}$	
Flextural tension	$3.25(f'_c)^{1/2}$	192 psi
Elasticity	$(1837)(33)(f'_c)^{1/2}$	3.6×10^6 psi
Steel		
Tension yield point	f_y	40,000 psi
Elasticity		2.9×10^7 psi
Development length	$\frac{(0.04)(A_b)(f_y)}{(f'_c)^{1/2}}$ but not less than $(0.0004)(d_b)(f_y)$ or 12″	

Table 208-2. Reinforcing steel.

No.	Area $in.^2$	Diam. in.	Weight lb/ft	Perimeter in.	Development Length in.
3	0.11	0.375	0.376	1.178	12.0
4	0.20	0.500	0.668	1.571	12.0
5	0.31	0.625	1.043	1.963	12.0
6	0.44	0.750	1.502	2.356	12.0
7	0.60	0.875	2.044	2.749	16.2
8	0.79	1.000	2.670	3.142	21.4
9	1.00	1.128	3.400	3.544	27.0
10	1.27	1.270	4.303	3.990	34.3
11	1.56	1.410	5.313	4.430	42.2

Bar spacing = (12)(bar area)/steel area required-in^2/ft; e.g. if A_s = 0.18 in 6" slab, #3 bars (0.11 in^2): spacing = 12x0.11/0.18 = 7.33"; To convert from one bar size to another: New bar spacing = (new bar area) (old bar spacing)/(old bar area), e.g. #4 bars (0.20 in^2) to replace #3's in example above: #4 spacing = (0.20)(7.33)/(0.11) = 13.33".

Table 208-3. Minimum steel in slabs and lids.
Temperature and shrinkage steel:
Area = 0.002bt both ways
Maximum spacing of bars is the lesser of 3 times the slab thickness or 18".

Slab Thickness	4"	5"	6"	7"
Min. A_s/ft	0.096	0.120	0.144	0.168
Bar Size	#3	#3	#3	#3
Bar Spacing	13.8"	11"	9.2"	7.8"
OR				
Bar Size	#4	#4	#4	#4
Bar Spacing	18.0"	18.0"	16.7"	14.3"

Slab Thickness	8"	9"	10"	11"	12"
Min. A_s/ft	0.192	0.216	0.240	0.264	0.288
Bar Size	#3	#4	#4	#4	#4
Bar Spacing	6.9"	11.0"	10.0"	9.1"	8.3"
OR					
Bar Size	#4	#5	#5	#5	#5
Bar Spacing	12.5"	17.2"	15.5"	14.1"	12.9"

Bar spacing = (bar area)/[(thickness)(minimum steel ratio)]

Table 208-4. Minimum steel in walls.
Temperature and shrinkage steel: Area = 0.0025bt horizontally and 0.0015bt vertically.
Maximum spacing of bars is the lesser of 5 times the slab thickness or 18".

Wall Thickness, inches	4"	5"	6"	7"
Vert. min. A_s/ft	0.0720	0.0900	0.1080	0.1260
bar size	#3	#3	#3	#3
bar spacing	18.0"	14.7"	12.2"	10.5"
Horiz. min. A_s/ft	0.1200	0.1500	0.1800	0.2100
bar size	#3	#3	#3	#4
bar spacing	11.0"	8.8"	7.3"	11.4"

Wall Thickness, inches	8"	9"	10"	11"	12"
Vert. min. A_s/ft	0.1440	0.1620	0.1800	0.1980	0.2160
bar size	#3	#3	#3	#4	#4
bar spacing	9.2"	8.2"	7.3"	12.1"	11.1"
Horiz. min. A_s/ft	0.2400	0.2700	0.3000	0.3300	0.3600
bar size	#4	#4	#4	#5	#5
bar spacing	10.0"	8.9"	8.0"	11.3"	10.3"

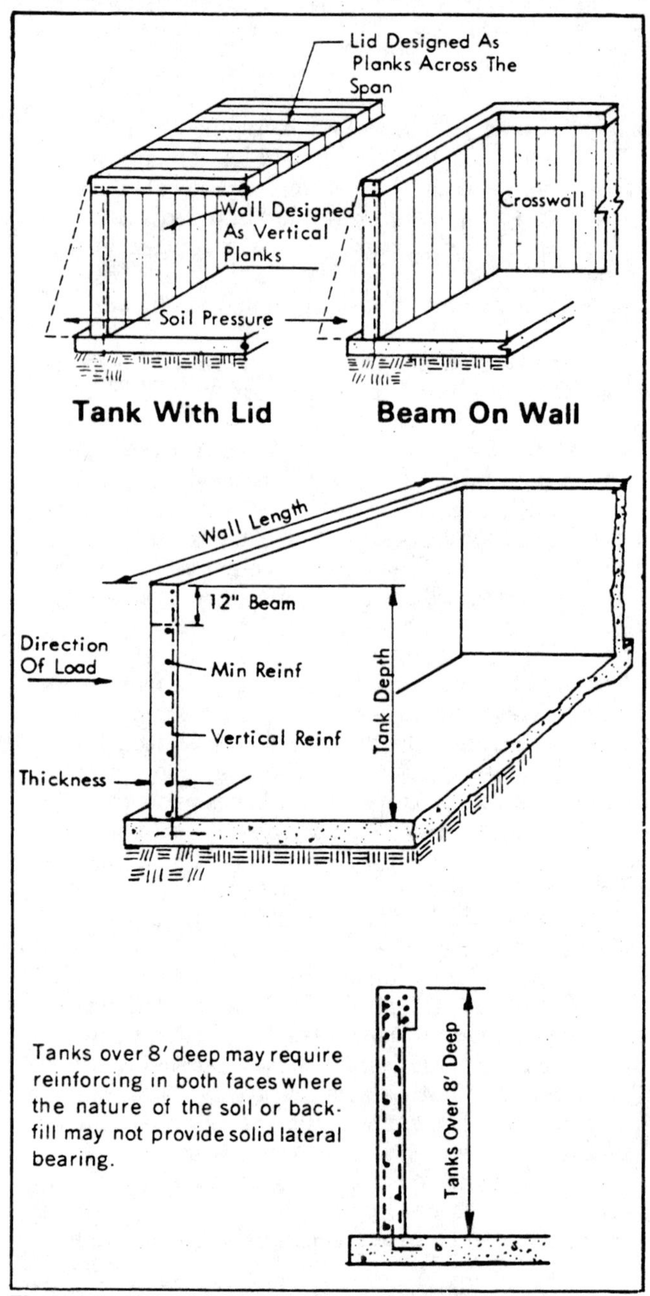

Fig 208-1. Manure tank design assumptions.

Tank Design

Assumptions

Tank walls—vertically reinforced

Ultimate Strength Design (ACI 318-71).

One-way slab design assumes pinned ends and continuous lateral support at top and bottom.

Maximum vertical steel spacing is five times the wall thickness or 18". (Arbitrarily set at 12" in Table 208-5.)

Triangular pressure distribution of the given design loads, except 100 psf vehicle surcharge.

Load factor = 1.7 on the design loads.
Allowable deflection = L/180.
Wall steel cover = 1½" for #5 or smaller.
= 2" for #6 or larger.

Table 208-5. Reinforced concrete tank walls.
Vertical reinforced design.
Tabulated reinforcing is vertical; minimum reinforcing (Table 208-4) is horizontal.

Hydrastatic Load

Bar Size, Spacing, Inches

Wall Thickness	Load, pcf	Tank Depth 6'	7'	8'	9'	10'
6"	15	3,12.2"	3,12.2"	3,12.2"	3,12.2"	3,10.7"
	30	"	"	3,10.5"	3, 7.3"	4, 9.6"
	60	"	3, 7.8"	4, 9.4"	5,10.0"	6, 8.9"
7"	15	3,10.5"	3,10.5"	3,10.5"	3,10.5"	3,10.5"
	30	"	"	"	3, 9.0"	4,11.9"
	60	"	3, 9.6"	4,11.6"	4, 8.0"	6,11.5"
8"	15	3, 9.2"	3, 9.2"	3, 9.2"	3, 9.2"	3, 9.2"
	30	"	"	"	"	3, 7.8"
	60	"	"	3, 7.6"	4, 9.6"	5,10.8"
9"	15	3, 8.2"	3, 8.2"	3, 8.2"	3, 8.2"	3, 8.2"
	30	"	"	"	"	"
	60	"	"	"	4,11.1"	4, 8.1"
10"	15	3, 7.3"	3, 7.3	3, 7.3"	3, 7.3"	3, 7.3"
	30	"	"	"	"	"
	60	"	"	"	3, 7.0"	4, 9.2"

Hydrastatic Load + 100 psf Surcharge For Vehicle Load

Wall Thickness	Load, pcf	Tank Depth 6'	7'	8'	9'	10'
6"	15	3,12.2"	3,11.0"	3, 8.0"	4,10.9"	4, 8.4"
	30	3,12.0"	3, 8.1"	4,10.4"	5,11.8"	6,11.0"
	60	3, 8.1"	4, 9.6"	5,10.2"	6, 9.0"	-
7"	15	3,10.5"	3,10.5"	3, 9.8"	3, 7.4"	4,10.3"
	30	"	3,10.0"	3, 7.1"	4, 9.5"	5,11.0"
	60	3, 9.9"	4,11.6"	4, 8.2"	5, 9.1"	7,11.6"
8"	15	3, 9.2"	3, 9.2"	3, 9.2"	3, 8.8"	3, 6.8"
	30	"	"	3, 8.4"	4,11.3"	4, 8.5"
	60	"	3, 7.8"	4, 9.8"	5,10.9"	6,10.5"
9"	15	3, 8.2"	3, 8.2"	3, 8.2"	3,10.2"	3, 7.8"
	30	"	"	"	3, 7.2"	4, 9.9"
	60	"	"	4,11.3"	4, 8.2"	5, 9.4"
10"	15	3, 7.3"	3, 7.3"	3, 7.3"	3, 7.3"	3, 7.3"
	30	"	"	"	"	4,11.2"
	60	"	"	3, 7.1"	4, 9.3"	5,10.7"

Beams at top of tank walls

Ultimate Strength Design (ACI 318-71).

Deflection was not restricted because the additional strength of the horizontal temperature and shrinkage steel was ignored in the design procedure.

Load factor = 1.7 on the triangular wall load to determine reactions carried by the beam.

Specified vertical and horizontal wall steel extends through the beam, so no minimum steel percentage is required for the beam.

Tank lids—reinforced across span

Tank lids for NO vehicle traffic

Ultimate Strength Design (ACI 318-71).

One-way slab design assumes pinned ends at each side wall.

Load factors = 1.7 on tabulated loads and 1.4 on deadweight of the slab.

Shear and an allowable deflection of L/180 checked for.

Lid steel cover = 1½" for #5 or smaller.
= 2" for #6 or larger.

Tank lids to support large tractor or manure wagon.

Ultimate Strength Design (ACI 318-71).

One-way slab design assumes pinned ends at each side wall.

Design load = 2-5000 lb loads 4′ o.c. plus deadweight of slab.

Load factor = 1.7 on axle load and 1.4 on deadweight of slab.

Design moment determined from the AASHO equation for bridge decks:

$$\text{Moment} = (1.7)(L + 2)P \div 32 + (1.4)WL^2 \div 8$$

L = span, ft; P = axle load of 10,000 lb; W = deadweight of slab.

Only bending moment checked.

Lid steel cover = 1½" for #5 or smaller.
= 2" for #6 or larger.

Tank walls

Chapter 102 includes a design load of 110 psf/ft for saturated fine sand. This soil type requires major modification of designs to allow for flotation, a structural floor, serious drainage and settling problems, and possible restrictions by building or other public authorities. Design elements are not included here.

Walls over 8′ deep, constructed where the natural soil or backfill may not provide solid laterial bearing, should also be reinforced in the outside face to resist full lateral hydraulic pressure (60 psf/ft) from stored liquid in the tank.

A 100 psf surcharge on the hydrostatic wall load is recommended where heavy equipment is within 5′ of the tank walls. See the bottom half of Table 208-5.

The wall designs assume continuous lateral support. The upper 12" of the wall may be reinforced as given in Table 208-6, to provide this lateral support and carry wall loads to a pilaster, partition, endwall, or cross beam.

Slat, and Floor or Lid Support Design

See Chapter 102 for design loads.

Reinforced concrete slats and girders

Ultimate Stress Design (ACI 318-71).

Clear distance between bars; not less than bar diameter, 1", nor 1⅓ times maximum size of coarse aggregate.

Cover = 1½" for #5 or smaller.
= 2" for #6 or larger.

Load factors = 1.4 on beam's weight and 1.7 on tabulated load.

Allowable Deflection = L/180.
Maximum steel percentage = 3.25%.
Minimum steel percentage = 0.50%.

The allowable load is reduced by 25% for steel percentages less than the minimum. Tabulated loads are live loads; dead loads have already been subtracted.

Masonry columns

Working stress design (specification for the design and construction of load-bearing concrete masonry, MCMA).

Compressive strength of masonry, f′m = 1100 psi.

Cores filled with grout.

Table 208-6. Beams at top of tank walls.
The wall designs assume continuous lateral support at the top and bottom of the walls. The upper 12" of the wall can be reinforced to provide this lateral support and carry wall loads to a pilaster, partition, endwall, or cross beam. Entries in Table 208-6 are the maximum wall length supported in feet. Reinforcing is horizontal.

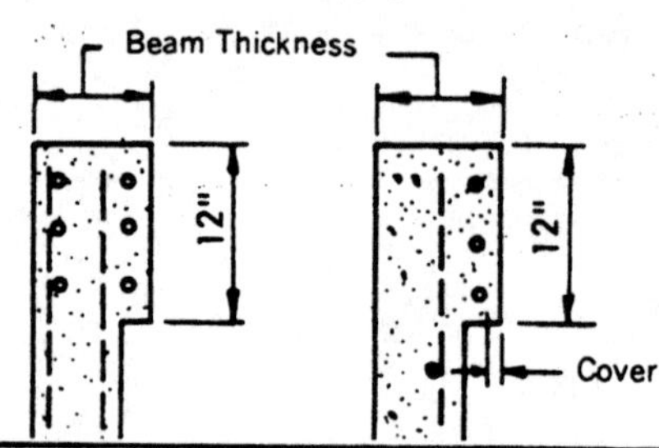

6" Beam Thickness

TANK DEPTH	LOAD pcf	1-#4	EXTRA REINFORCING IN TOP FOOT 1-#5	2-#4	2-#5	2-#6	2-#7
4'	15	17	21	24	28	-	-
	30	12	15	17	19	23	26
	60	8	10	12	14	16	18
5'	15	14	17	19	22	26	28
	30	9	12	13	15	18	21
	60	7	8	9	11	13	15
6'	15	11	14	16	18	21	25
	30	8	10	11	13	15	17
	60	5	7	8	9	10	12
7'	15	10	12	14	16	18	21
	30	7	8	9	11	13	15
	60	5	6	7	8	9	10
8'	15	8	10	12	14	16	18
	30	6	7	8	9	11	13
	60	4	5	6	7	8	9
10'	15	7	8	9	11	13	15
	30	4	6	6	7	9	10
	60	3	4	4	5	6	7

10" Beam Thickness

TANK DEPTH	LOAD pcf	1-#5	EXTRA REINFORCING IN TOP FOOT 2-#4	2-#5	2-#6	2-#7	2-#8
6'	15	20	22	27	28	-	-
	30	14	16	19	22	26	28
	60	10	11	13	16	18	20
7'	15	17	19	23	27	28	-
	30	12	13	16	19	22	25
	60	8	9	11	13	15	17
8'	15	15	17	20	24	27	28
	30	10	12	14	17	19	22
	60	7	8	10	12	13	15
9'	15	13	15	18	21	24	27
	30	9	10	12	15	17	19
	60	6	7	9	10	12	13
10'	15	12	13	16	19	22	25
	30	8	9	11	13	15	17
	60	6	6	8	9	11	12

8" Beam Thickness

TANK DEPTH	LOAD pcf	1-#4	EXTRA REINFORCING IN TOP FOOT 1-#5	2-#4	2-#5	2-#6	2-#7	2-#8
4'	15	21	26	28	-	-	-	-
	30	15	18	21	24	28	-	-
	60	10	13	14	17	20	23	26
5'	15	16	21	23	28	-	-	-
	30	12	14	16	19	23	26	28
	60	8	10	11	14	16	18	21
6'	15	14	17	19	23	27	28	-
	30	10	12	14	16	19	22	25
	60	7	8	9	11	13	15	17
7'	15	12	15	17	20	23	27	28
	30	8	10	12	14	16	19	21
	60	6	7	8	10	11	13	15
8'	15	10	13	14	17	20	23	26
	30	7	9	10	12	14	16	18
	60	5	6	7	8	10	11	13
9'	15	9	11	13	15	18	21	23
	30	6	8	9	11	12	14	16
	60	4	5	6	7	9	10	11
10'	15	8	10	11	14	16	18	21
	30	6	7	8	9	11	13	15
	60	4	5	5	7	8	9	10

12" Beam Thickness

TANK DEPTH	LOAD pcf	1-#5	EXTRA REINFORCING IN TOP FOOT 2-#4	2-#5	2-#6	2-#7	2-#8
8'	15	16	19	22	27	28	-
	30	11	13	16	19	22	25
	60	8	9	11	13	15	17
9'	15	14	16	20	24	27	28
	30	10	11	14	17	19	22
	60	7	8	10	12	13	15
10'	15	13	15	18	21	25	28
	30	9	10	12	15	17	20
	60	6	7	9	10	12	14

Table 208-7. Tank lid design loads.

Loading Condition	Design Load
human or poultry traffic	40 psf
livestock traffic or up to 15" of soil cover	150 psf
outdoor tanks with no livestock or vehicle traffic	40 psf + snow load on ground from Chapter 111.

Table 208-8. Tank lids for 2-5000 lb loads.
4' o.c. plus dead weight of slabs.

Span	Slab Thickness 5"	6"	7"	8"	9"	10"
5'	6,10.8"	5,11.9"	4, 9.4"	4,11.0"	4,11.0"	4,10.0"
6'	6, 9.1"	5,10.1"	4, 8.0"	4, 9.4"	4,10.7"	"
7'	7,10.6"	6,10.9"	5,10.7"	4, 8.1"	4, 9.2"	"
8'	7, 9.2"	6, 9.5"	5, 9.4"	5,10.9"	4, 8.0"	4, 9.0"
9'	8,10.5"	7,11.4"	6,10.6"	5, 9.7"	5,11.0"	4, 7.9"
10'	9,11.7"	7,10.1"	6, 9.4"	6,11.2"	5, 9.7"	5,10.8"
12'	-	8,10.7"	7,10.3"	6, 9.0"	6,10.4"	6,11.6"
14'	-	9,11.0"	8,11.2"	7,10.1"	7,11.6"	6, 9.5"
16'	-	10,11.5"	9,11.8"	8,11.2"	7, 9.7"	7,10.9"
18'	-	-	9,10.0"	8, 9.5"	8,10.9"	7, 9.2"
20'	-	-	10,10.8"	9,10.3"	9,11.8"	8,10.4"

Table 208-9. Reinforced concrete tank lids.
NO vehicle traffic.
One-way slab design adequate for load listed plus dead weight of slab.
All 4′ wide covers require only minimum reinforcing, except 4″ thickness, 150 psf = #3,11.7″.

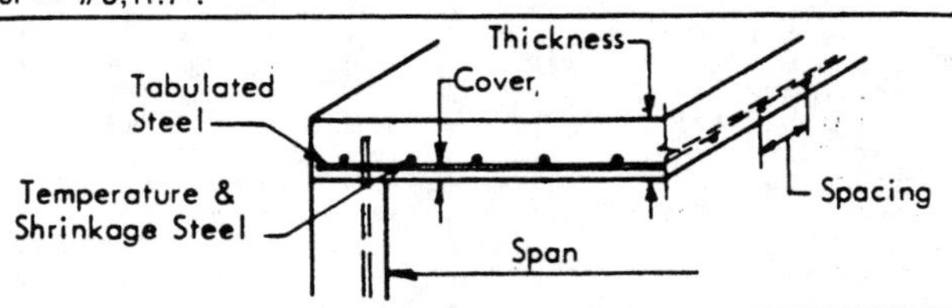

Lid Thickness	PSF	Lid Span 5'	6'	7'	8'	9'	10'	12'
4"	40	3,13.8"	3,13.8"	3,11.4"	3, 8.7"	3, 6.8"	4, 9.9"	-
	50	"	"	3,10.1"	3, 7.7"	4,10.9"	4, 8.7"	-
	60	"	3,12.5"	3, 9.1"	3, 6.9"	4, 9.8"	4, 7.8"	-
	70	"	3,11.3"	3, 8.2"	4,11.3"	4, 8.8"	-	-
	80	"	3,10.4"	3, 7.5"	4,10.4"	4, 8.1"	-	-
	90	"	3, 9.6"	3, 6.9"	4, 9.5"	5,11.5"	-	-
	100	3,12.9"	3, 8.9"	4,11.7"	4, 8.8"	-	-	-
	110	3,12.0"	3, 8.3"	4,10.9"	4, 8.2"	-	-	-
	150	3, 9.4"	4,11.7"	4, 8.5"	-	-	-	-
6"	40	3, 9.2"	3, 9.2"	3, 9.2"	3, 9.2"	3, 9.2"	3, 8.1"	4,10.1"
	50	"	"	"	"	3, 9.1"	3, 7.3"	4, 9.2"
	60	"	"	"	"	3, 8.3"	3, 6.7"	4, 8.4"
	70	"	"	"	"	3, 7.7"	4,11.3"	5,12.0"
	80	"	"	"	3, 9.1"	3, 7.1"	4,10.4"	5,11.1"
	90	"	"	"	3, 8.5"	3, 6.7"	4, 9.7"	5,10.3"
	100	"	"	"	3, 7.9"	4,11.3"	4, 9.1"	5, 9.7"
	110	"	"	"	3, 7.5"	4,10.7"	4, 8.6"	6,11.3"
	150	"	"	3, 7.9"	4,10.9"	4, 8.6"	5,10.7"	-

Lid Thickness	PSF	Lid Span 12'	14'	16'	18'	20'	22'	24'
8"	40	3, 6.8"	4, 9.0"	5,10.6"	6,10.8"	-	-	-
	50	4,11.3"	4, 8.3"	5, 9.7"	6, 9.9"	-	-	-
	60	4,10.5"	5,11.9"	5, 9.0"	6, 9.2"	-	-	-
	70	4, 9.8"	5,11.1"	6,10.9"	7,11.6"	-	-	-
	80	4, 9.2"	5,10.4"	6,10.2"	-	-	-	-
	90	4, 8.7"	5, 9.8"	6, 9.6"	-	-	-	-
	100	4, 8.2"	5, 9.2"	6, 9.1"	-	-	-	-
	110	5,12.0"	6,11.3"	7,11.7"	-	-	-	-
	150	5, 9.9"	6, 9.3"	-	-	-	-	-
10"	40	4,10.0"	4,10.0"	5,11.9"	5, 9.4"	6,10.0"	7,11.2"	-
	50	"	4, 9.4"	5,11.1"	5, 8.8"	6, 9.4"	7,10.5"	-
	60	"	4, 8.9"	5,10.4"	6,10.9"	6, 8.8"	7, 9.8"	-
	70	"	4, 8.3"	5, 9.8"	6,10.3"	7,11.2"	7, 9.2"	-
	80	"	4, 7.9"	5, 9.3"	6, 9.7"	7,10.6"	-	-
	90	"	5,11.6"	5, 8.8"	6, 9.2"	7,10.0"	-	-
	100	4, 9.7"	5,11.0"	6,11.1"	6, 8.7"	7, 9.5"	-	-
	110	4, 9.2"	5,10.4"	6,10.6"	7,11.3"	8,11.9"	-	-
	150	4, 7.7"	5, 8.8"	6, 8.8"	7, 9.4"	-	-	-

Reinforced columns, footings—plain and reinforced

Ultimate Stress Design (ACI 318-71).
Load factor on all loads = 1.7.
Minimum column steel area = 1% of column area.
Column steel cover = 1½″ for #5 or smaller.
= 2″ for #6 or larger.
4 bars per square tied column, minimum.
6 bars per round spiral column, minimum.
Modified R reduction factors for slenderness.
Space #3 ties no greater than 16 longitudinal bar diameters, least dimension of the column, or 16″.
Soil bearing = 3000 psf.
Cover between soil and footing steel = 3″.
Maximum bar spacing = 18″ or 3 times the footing thickness.

Design Example

Design a slotted floor system for 200 lb pigs.

Slats

Determine slat length from the building layout. For this example, use 8′. Determine the design load from Table 208-11: 100 plf. Select a slat design from Table 208-12 for a superimposable load of at least 100 plf and an 8′ span. One choice is a 4x5 with a #4 bar.

Girder

Determine the load in plf on the girders.
Load on girder, plf = [weight of floor + weight of animals] x girder spacing where:

Floor weight, psf = 12.5 x B″ x D″ ÷ (B″ + slot width)
Animal load, psf = value from column 2, Table 208-11.
Girder spacing, ft = slat length.
Load on girder = (50 psf + 50 psf) x 8′ = 800 plf.

Determine the girder length, which is equal to the column spacing from the building layout. For this example use 8′. From Table 208-13, select a girder design that supports 800 plf on an 8′ span. One choice is a 6x9 with 2-#5 bars.

Columns and Footings

Determine the column load. Each column supports one end of two girders, and therefore supports the total load on one girder plus the weight of the girder. The total load on one girder equals the girder load/ft x girder length.
Total girder load: 800 plf x 8′ = 6400 lb.
Weight of girder (lb)
= 1.04 x B″ x D″ x girder length (ft)
= 1.04 x 6 x 9 x 8
= 449.3 lb.

Total column load: 6400 lb + 449.3 lb = 6849.3 lb. Assume the column height in this example is 8′. Select a column design from Table 208-14 or 208-15. Select a column footing design from Table 208-14 or 208-15.

Table 208-10. Minimum member width, in.

Bar	Diameter	1 Bar	2 Bars	3 Bars
# 3	.375"	3.375"	4.750"	6.125"
4	.500	3.500	5.000	6.500
5	.625	3.625	5.250	6.875
6	.750	4.750	6.500	8.250
7	.875	4.875	6.750	8.625
8	1.000	5.000	7.000	9.000
9	1.128	5.128	7.384	9.640
10	1.270	5.270	7.810	10.350
11	1.410	5.410	8.230	11.050

Clear distance between bars not less than bar diameter, 1", nor 1 1/3 times max course aggregate

Table 208-11. Slotted floor design loads.

Livestock	Live Loads for Slat Design plf slat	Live Distributed Loads for Design of Girders and Columns psf floor area
Swine to 50#	50 plf	35 psf
to 200#	100	50
to 500#	170	70
Sheep	120	50
Calves	150	50
Beef, Dairy cattle	250	100

Table 208-12. Superimposable load on slats, lb/ft.
Width x total depth, B x D, in.; load, lb/ft.

Span, ft
Bar Size
Load, lb/ft

BAR SIZES		NO. 3							NO. 4							NO. 5						
SPAN		4	5	6	7	8	9	10	4	5	6	7	8	9	10	4	5	6	7	8	9	10
B	D																					
4																						
	4.0	192	118	78	53	37	27	19	324	186	128	96	70	52	40							
	4.5	239	147	97	67	48	34	25	410	257	173	112	76	55	42	529	338	196	123	82	57	41
	5.0	286	177	117	81	58	42	31	497	311	211	150	106	73	53	603	449	307	201	132	89	62
	5.5	333	206	137	96	69	50	37	583	366	248	177	131	100	72	680	524	366	263	177	121	85
	6.0	380	235	157	110	79	58	43	670	421	286	204	152	115	89	761	583	425	306	230	167	118
	6.5	426	265	177	124	89	66	49	756	476	323	232	172	131	102	847	645	434	349	262	202	157
	7.0	473	294	197	138	100	74	55	843	530	361	259	192	147	114	937	709	543	392	295	228	180
	7.5	*333	236	156	108	75	55	39	929	585	398	286	213	162	127	1031	776	602	435	327	253	200
	8.0	418	258	170	118	84	60	43	1016	640	436	313	233	178	139	1132	846	661	478	360	278	220
5																						
	4.0	192	117	76	51	35	24	16	332	206	138	96	70	51	38	467	260	156	104	76	62	54
	4.5	239	146	95	65	45	31	22	418	260	175	123	89	67	50	601	378	230	146	98	70	53
	5.0	285	175	115	78	55	39	27	504	315	212	150	110	82	61	702	463	315	209	139	96	68
	5.5	332	204	134	92	65	46	33	590	369	249	177	129	97	74	798	548	374	268	200	138	98
	6.0	378	232	153	106	75	54	38	676	423	286	203	149	113	86	899	634	472	311	232	177	131
	6.5	*311	189	123	82	57	39	26	762	478	323	230	169	128	98	1005	719	491	353	264	202	159
	7.0	345	210	137	92	63	44	30	848	532	360	257	189	143	110	1117	805	549	396	290	227	178
	7.5	380	231	151	102	70	49	33	935	586	397	283	209	158	122	1234	890	608	438	328	252	198
	8.0	414	252	165	112	77	54	37	1021	641	434	310	229	174	134	1359	975	667	481	360	277	218
6																						
	4.0	191	115	73	48	32	21	13	336	208	138	96	68	50	36	482	284	181	133	105	78	59
	4.5	237	143	92	62	42	28	16	422	262	174	122	88	64	48	616	386	261	168	114	83	64
	5.0	283	172	111	75	51	35	25	508	316	211	148	107	79	59	751	471	319	227	156	109	79
	5.5	*240	143	91	59	38	24	14	593	369	248	174	127	94	71	885	556	377	270	200	151	107
	6.0	274	164	104	68	45	29	17	679	423	284	201	146	109	82	1019	641	435	312	231	176	137
	6.5	307	185	118	78	51	33	21	765	477	321	227	166	124	94	1153	726	494	354	263	201	156
	7.0	341	205	131	87	58	38	24	850	531	358	253	185	139	105	1287	811	552	396	294	225	175
	7.5	375	226	145	96	64	43	27	936	585	394	279	205	154	117	1422	896	610	438	326	249	195
	8.0	409	247	159	105	71	47	30	1022	639	431	306	224	168	128	1556	981	668	480	358	274	214
7																						
	4.0	190	113	71	45	29	18	10	339	208	137	94	66	47	34	492	306	201	139	105	78	58
	4.5	235	141	89	58	38	24	15	424	261	173	120	85	62	45	626	391	263	181	128	99	77
	5.0	*203	119	73	46	28	16	7	509	315	209	146	104	76	56	760	475	321	228	167	121	89
	5.5	236	139	86	55	34	20	10	594	368	246	171	123	90	67	894	560	379	269	198	150	115
	6.0	270	159	100	63	40	24	12	680	422	282	197	142	105	78	1027	644	436	311	229	174	134
	6.5	303	180	113	72	46	28	15	765	475	318	223	162	119	89	1161	729	494	353	261	198	153
	7.0	336	200	126	81	52	32	18	850	529	354	249	181	134	100	1295	813	552	394	292	222	171
	7.5	370	220	139	90	58	37	21	690	425	281	195	133	100	72	1429	898	610	436	323	246	190
	8.0	403	241	152	99	64	41	24	753	465	308	213	152	110	80	1562	983	667	478	354	270	209
8																						
	4.0	188	110	68	42	26	15	7	339	207	135	92	64	45	31	499	309	206	144	104	76	56
	4.5	*167	95	57	33	18	8	0	424	260	171	117	83	59	42	632	393	264	185	135	100	75
	5.0	200	115	69	42	24	11	3	509	313	207	143	101	73	52	766	478	321	227	165	123	93
	5.5	233	135	82	50	29	15	5	594	366	243	168	120	87	63	899	562	378	268	196	147	112
	6.0	266	155	95	59	35	19	7	679	419	278	194	138	101	74	1032	646	436	309	227	171	130
	6.5	298	175	108	67	41	23	10	763	472	314	219	157	115	84	1166	730	493	350	258	194	149
	7.0	331	195	120	76	46	27	12	624	382	250	171	120	84	59	1299	814	550	392	288	218	167
	7.5	364	215	133	84	52	30	15	687	421	276	189	133	94	66	1432	898	608	433	319	241	186
	8.0	397	234	146	92	58	34	17	750	460	302	207	146	104	73	1566	982	665	474	350	255	204

BAR		NO. 6							NO. 7							NO. 8						
SPAN		4	5	6	7	8	9	10	4	5	6	7	8	9	10	4	5	6	7	8	9	10
B	H																					
5																						
	5.5	775	600	413	265	176	123	89														
	6.0	871	671	497	358	250	173	123														
	6.5	972	745	581	420	315	227	161	1068	822	666	534	390	268	190							
	7.0	1079	822	662	481	361	279	216	1175	899	726	607	467	354	252							
	7.5	1191	903	724	542	407	315	249	1288	980	788	658	531	412	327	1406	1073	865	723	620	510	371
	8.0	1309	986	789	603	454	351	278	1407	1063	852	709	595	463	368	1525	1157	930	775	663	576	460
6																						
	5.0	762	504	317	211	154	123	106														
	5.5	871	625	425	293	197	140	104														
	6.0	986	747	509	366	270	188	135	1079	833	649	462	306	211	150							
	6.5	1107	847	593	426	319	245	178	1200	922	746	552	415	287	204							
	7.0	1234	939	676	487	365	280	220	1328	1014	818	636	478	370	271	1442	1105	893	747	593	431	307
	7.5	1368	1035	760	548	410	316	249	1462	1110	892	720	542	420	333	1577	1201	967	808	678	527	404
	8.0	1509	1135	844	609	456	352	277	1603	1210	969	804	606	470	373	1719	1302	1044	870	744	594	473
7																						
	4.5	633	395	262	188	138	103	78														
	5.0	825	517	350	236	174	138	106														
	5.5	967	638	433	309	218	157	119	1058	809	551	358	240	168	123							
	6.0	1101	759	516	369	274	204	148	1191	919	666	480	334	231	165	1302	1006	813	561	371	255	181
	6.5	1241	880	599	430	320	244	190	1332	1022	781	563	422	307	219	1443	1110	899	698	504	347	247
	7.0	1389	1002	683	490	365	280	219	1480	1129	895	647	485	374	290	1592	1218	983	809	609	464	330
	7.5	1544	1123	766	551	411	315	247	1636	1241	996	730	548	424	334	1748	1330	1070	892	694	538	427
	8.0	1709	1244	849	611	456	350	275	1800	1357	1085	813	611	473	374	1913	1446	1159	965	778	604	480
8																						
	4.5	650	405	271	191	139	103	78														
	5.0	841	526	354	251	183	138	105	1006	663	418	280	207	169	140							
	5.5	1031	646	437	311	229	170	130	1151	829	564	387	261	186	139							
	6.0	1216	767	520	371	274	208	159	1304	995	678	487	355	247	178	1412	1090	837	586	389	269	192
	6.5	1376	888	603	431	319	243	188	1464	1122	792	570	426	327	236	1573	1208	978	714	531	367	261
	7.0	1544	1009	666	491	364	278	216	1632	1244	906	653	489	376	295	1741	1331	1073	825	620	480	349
	7.5	1721	1130	769	551	410	313	243	1810	1371	1021	736	551	425	334	1919	1458	1172	935	704	545	432
	8.0	1908	1250	851	611	455	348	271	1997	1503	1135	819	614	474	373	2106	1591	1274	1045	787	610	484

*Drop in strength is not an error; when % steel is below minimum allowable, load is decreased 25%.

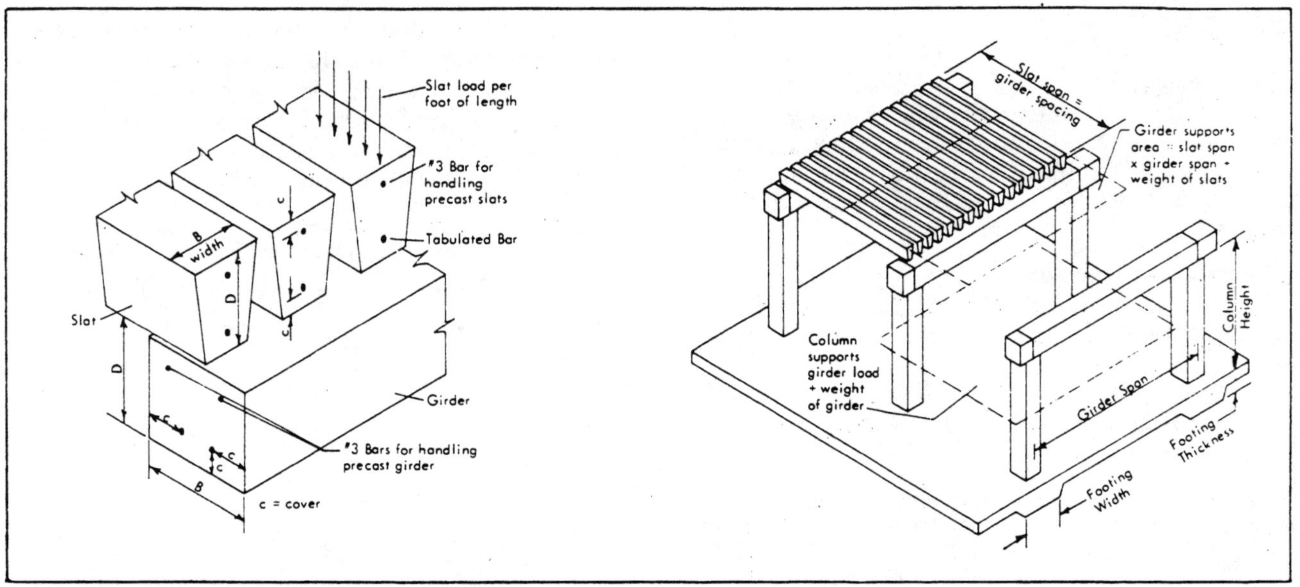

Fig 208-2. Slotted floor and support system.

Table 208-13. Live load on girders—efficient reinforcing. Width x total depth, B x D, in.; number and size of reinforcing bars; maximum efficient load on girder, lb/ft.

Member width	Size depth		Beam Span, Feet 6'		8'		10'		12'
6"	6"	2#5	776	2#5	446	2#5	219	2#5	116
	7		923		578		357		203
	8		1081		710		439		292
	9		1250		841		521		348
	10		1431		973		604		403
	11"		1626		1104		686		459
	12		1836		1236		769		515
	13		2063		1368		851		570
	14		2310		1499		933		626
	15		2578		1631		1016		682
7"	7"	2#5	1031	2#6	691	2#7	-	2#7	-
	8		1214		870		699		401
	9		1410		997		843		582
	10		1621		1130		940		694
	11		1847		1270		1042		805
	12		2092		1417		1148		917
	13		2356		1573		1259		1003
	14		2643		1737		1374		1090
	15		2955		1911		1495		1181
8"	8"	2#6	1341	2#7	1050	2#8	-	2#8	-
	9		1558		1195		1011		739
	10		1791		1347		1123		887
	11		2041		1507		1240	2#9	1083
	12		2309		1676		1361		1177
	13		2600		1854		1488		1274
	14		2914		2042		1620		1375
	15		3255		2241		1757		1479
9"	9"	2#6	1703	2#7	1294	2#8	1085	2#8	754
	10		1964		1465		1211		901
	11		2244		1644		1342	3#8	1252
	12		2545		1833		1478		1358
	13		2871		2033		1620		1468
	14		3224		2244		1768		1582
	15		3607		2467		1922		1700
10"	10"	3#5	2366	3#6	1636	3#7	1367	3#8	1219
	11		2660		1836		1512		1333
	12		3010		2047		1664		1450
	13		3388		2269		1822		1572
	14		3798		2504		1987		1698
	15		4244		2753		2159		1828
11"	11	3#6	2858	3#6	1974	3#7	1615	3#8	1413
	12		3228		2205		1782		1543
	13		3628		2450		1955		1676
	14		4061		2708		2136		1814
	15		4532		2981		2325		1957
12"	12	3#6	3464	3#7	2514	3#8	2042	3#8	1635
	13		3900		2781		2232		1780
	14		4371		3063		2430		1930
	15		4883		3361		2636		2086

Table 208-14. Unreinforced footings and masonry footings. Mortar f'm = 1100 psi; footing f'c = 3500 psi; soil pressure = 3000 psf; fill column cores with grout.

Nominal Column Size	Column Height	Allowable Load	Square Footing Width	Depth
8x8 in.	4'	11,405 lb	24"	6"
	6	11,152	24	6
	8	10,660	23	6
	10	9,849	22	6
	12	8,640	21	5
8x16 in.	4'	21,547 lb	33"	10"
	6	21,070	32	9
	8	20,140	32	9
	10	18,608	30	9
	12	16,323	28	8
16x16 in.	4'	44,609 lb	47"	12"
	6	44,495	47	12
	8	44,273	47	12
	10	43,907	46	12
	12	43,362	46	12

Masonry Column and Footing

From Table 208-14 select a masonry column that will support 6850 lb with an 8′ height. An 8x8 masonry column 8′ high will support 10,660 lb and is adequate.

Table 208-14 lists a 21 in^2 x 5″ thick footing for 8,640 lb; Table 208-16 lists a 19 in^2 x 4.1″ thick footing for 7500 lb—the latter footing is adequate and cheaper.

Reinforced Column and Footing

From Table 208-15 select a reinforced column that will support 6850 lb with an 8′ height. Either the minimum round or square columns are adequate. The minimum round column is 8″ diameter with 6-#3 bars and 1.46 in^3 of spiral reinforcing/ft of height. The minimum square column is 8″x8″ with 4-#4 bars and #3 ties spaced 8″ o.c.

The footings in Table 208-15 are for the maximum allowable load the given column will support and are excessively large for this low column load. The previously selected unreinforced footing for the masonry column is also adequate for the reinforced column.

Footing Designs

Masonry columns

The footings in Table 208-14 are adequate for the maximum loads each listed column will support.

If the masonry column is not fully loaded, a footing from Table 208-16, designed for the actual column load, is adequate.

Reinforced concrete columns

Each column in Table 208-15 is given with a matching reinforced footing that will support the column's maximum load.

If a reinforced column is not fully loaded, check for a smaller or unreinforced footing designed for the actual column load.

Table 208-15. Reinforced columns and footings.
Columns are assumed pinned end and concentrically loaded. Column dimension = diameter of a round column or face width of a square one; column load = kips, 1000 lb; spiral volume = volume of spiral reinforcing per ft of column height, in³/ft; width = side of a square footing, in.; depth = footing thickness, in., including 3" cover; As = footing steel area required in both directions, in²; bars = minimum number and corresponding size of equally spaced steel bars in both directions—a larger number of smaller bars may be used to meet the footing As required; #3 ties spacing = maximum vertical spacing, in., between #3 ties in square columns.

Square Columns

Column Dimension		4'	6'	8'	10'	#3 ties spacing
		Square columns with 4-#4 bars				
8"	Column Load:	60.5	57.0	51.2	45.4	8"
	Footing: Width	54	53	50	47	
	Depth	12	12	12	11	
	As	2.12	2.00	1.80	1.60	
	Bars	3#8	3#8	3#8	3#7	
		Square columns with 4-#5 bars				
8"	Column Load:	65.2	61.4	55.2	48.9	8"
	Footing: Width	56	55	52	49	
	Depth	13	13	12	11	
	As	2.28	2.15	1.94	1.73	
	Bars	4#7	4#7	3#8	3#7	

Round Columns

Column Dimension		4'	6'	8'	10'	12'	spiral volume
		Round columns with 6-#3 bars					
8"	Column Load:	44.6	44.1	40.6	37.0	33.4	1.46 in³
	Footing: Width	47	47	45	43	41	
	Depth	11	11	11	10	10	
	As	1.58	1.56	1.44	1.31	1.19	
	Bars	3#7	3#7	3#7	3#6	3#6	
		Round columns with 6-#4 bars					
8"	Column Load:	49.1	48.6	44.7	40.8	36.9	1.50 in³
	Footing: Width	49	49	47	45	43	
	Depth	11	11	11	11	10	
	As	1.73	1.72	1.58	1.45	1.31	
	Bars	3#7	3#7	3#7	3#7	3#6	

Footings in Table 208-14 for 8x8 or 8x16 columns (but not for 16x16 columns) can be used to support the same load from a reinforced column.

Footings in Table 208-16 can be used for Table 208-15 columns and do not require reinforcing if designed for the actual column load.

Table 208-16. Unreinforced square concrete footings.
8"x8" minimum column size; 3000 psf soil, 3500 psi concrete; load factor = 1.7.

Load, lbs	Size (sq.) inches	Depth inches
5000	15.5	3.0
7500	19.0	4.1
10000	21.9	5.2
12500	24.5	6.1
15000	26.8	7.0
17500	29.0	7.8
20000	31.0	8.5
22500	32.9	9.2
25000	34.6	9.9
27500	36.3	10.5
30000	38.0	11.1
32500	39.5	11.7
35000	41.0	12.3
40000	43.8	13.3
42500	45.2	13.8
45000	46.5	14.3

Welded Wire Fabric

Table 208-17. Minimum welded wire in walls.
Temperature and shrinkage steel: area = 0.002bt horizontally; area = 0.0012bt vertically.
Under special circumstances, the horizontal steel can be reduced allowing some vertical cracking. See Chapter 305.

Wall Thickness, in.	4"	5"	6"	7"
Vert. A_s/ft	0.0576	0.0720	0.0864	0.1008
Horiz. A_s/ft	0.0960	0.1200	0.1440	0.1680

Wall Thickness, in.	8"	9"	10"	11"	12"
Vert. A_s/ft	0.1152	0.1296	0.1440	0.1584	0.1728
Horiz. A_s/ft	0.1920	0.2160	0.2400	0.2640	0.2880

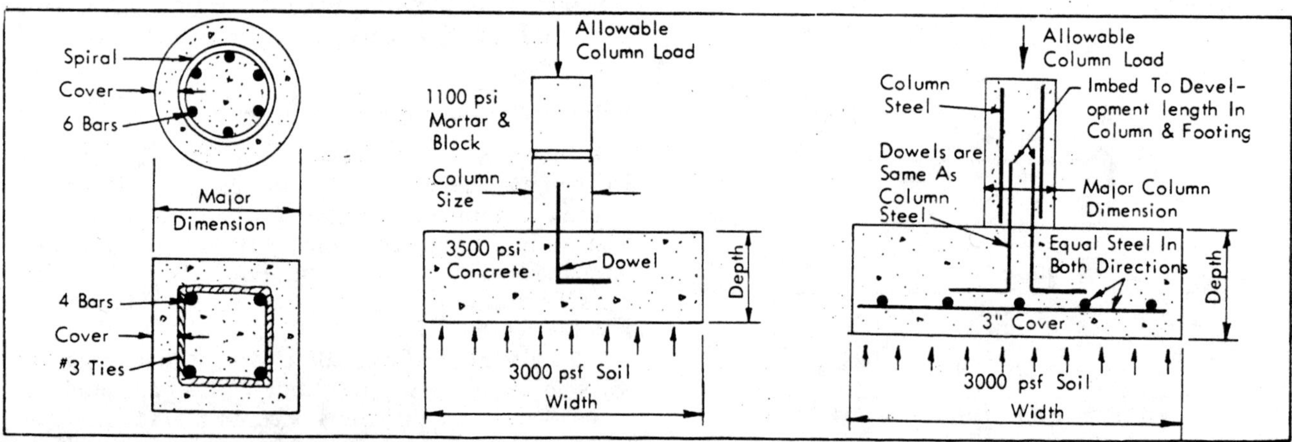

Fig 208-3. Column and footing design.

Table 208-18. Minimum welded wire in slabs.
Temperature and shrinkage steel: area = 0.0018bt in both directions.

Slab Thickness, in.	4"	5"	6"	7"
A_s/ft	0.0864	0.1080	0.1296	0.1512

Slab Thickness, in.	8"	9"	10"	11"	12"
A_s/ft	0.1728	0.1944	0.2160	0.2376	0.2592

Table 208-19. Wire size comparison.
"W" or "D" wire size = wire area in hundredths of a square inch; e.g. W4.5 has a cross-sectional area = 0.045 in².

W & D Size Number		Area	Nominal Diameter	AMERICAN STEEL & WIRE Gage
Smooth	Deformed	(sq in.)	(in.)	Number
W14	D14	0.122	0.394	4/0
W12	D12	.120	.390	
W11	D11	.110	.374	
W10.5		.105	.366	
		.103	.3625	3/0
W10	D10	0.100	0.356	
W9.5		.095	.348	
W9	D9	.090	.338	
		.086	.331	2/0
W8.5		.085	.329	
W8	D8	0.080	0.319	
W7.5		.075	.309	
		.074	.3065	1/0
W7	D7	.070	.298	
W6.5		.065	.288	
		0.063	0.283	1
S6	D6	.060	.276	
W5.5		.055	.264	
		.054	.2625	2
W5	D5	.050	.252	
		0.047	0.244	3
W4.5		.045	.240	
W4	D4	.040	.225	4
W3.5		.035	.211	
		.034	.207	5
W3		0.030	0.195	
W2.9		.029	.192	6
W2.5		.025	.177	7
W2.1		.021	.162	8
W2		.020	.159	
		0.017	0.148	9
W1.5		.015	.138	
W1.4		.014	.135	10

Table 208-20. Areas of welded wire fabric.
Area, in²/ft, of width for various spacings.

Wire Size Number		Nominal Diameter	Nominal Weight	Sectional Steel Area Wire Spacing, o.c.			
Smooth	Deformed	in.	lbs/lin ft	4"	6"	8"	10"
W12	D12	0.390	.408	.36	.24	.18	.114
W11	D11	0.374	.374	.33	.22	.165	.132
W10.5		0.366	.357	.315	.21	.157	.126
W10	D10	0.356	.340	.30	.20	.15	.12
W9.5		0.348	.323	.285	.19	.142	.114
W9	D9	0.338	.306	.27	.18	.135	.108
W8.5		0.329	.289	.255	.17	.127	.102
W8	D8	0.319	.272	.24	.16	.12	.096
W7.5		0.309	.255	.225	.15	.112	.09
W7	D7	0.298	.238	.21	.14	.105	.084
W6.5		0.288	.221	.195	.13	.097	.078
W6	D6	0.276	.204	.18	.12	.09	.072
W5.5		0.264	.187	.165	.11	.082	.066
W5	D5	0.252	.170	.15	.10	.075	.06
W4.5		0.240	.153	.135	.09	.067	.054
W4	D4	0.225	.136	.12	.08	.06	.048
W3.5		0.211	.119	.105	.07	.052	.042
W3		0.195	.102	.09	.06	.045	.036
W2.9		0.192	.098	.087	.058	.043	.035
W2.5		0.178	.085	.075	.05	.037	.03
W2.1		0.162	.070	.063	.042	.031	.025
W2		0.159	.068	.06	.04	.03	.024
W1.5		0.138	.051	.045	.03	.022	.018
W1.4		0.135	.049	.042	.028	.021	.017

Table 208-21. Areas of wire reinforcing.
Both bars and fabric are Grade 60 (60,000 psi) steel. Table 208-22 is based on equal allowable stresses. If comparing different grades of steel, the designer should compensate accordingly. For instance, fy of Grade 40 or 50 steel is 40,000 psi, whereas welded fabric and Grade 60 steel have a value of 60,000 psi for fy.
Six steel combinations equal 0.21 in²/ft: W7 or D7 wire 4" o.c., W10.5 or D10.5 wire 6" o.c., #4 bars 11½" o.c., or #5 bars 17½" o.c.

Wire Size Number (W or D Numbers) Spacing; in., o.c.			Steel Area in.²/ft.	Bar Spacing, in., o.c. Rebar Size Number		
4"	6"	12"		#3	#4	#5
	3.5	7	0.07 in.²/ft.	18"		
	4	8	0.08	16½		
3	4.5	9	0.09	14½		
	5	10	0.10	13		
	5.5	11	0.11	12		
4	6	12	0.12	11"		
	6.5		0.13	10	18"	
	7		0.14	9½	17	
5	7.5		0.15	9	16	
	8		0.16	8½	15	
	8.5		0.17	8"	14"	
6	9		0.18	7½	13	
	9.5		0.19	7	12½	
	10		0.20	6½	12	18"
*7	10.5		0.21		11½	17½
	11		0.22	6"	11"	16½"
			0.23		10½	16
8	12		0.24	5½	10	15½
			0.25		9½	15
			0.26	5		14
9			0.27		9"	13½"
			0.28	4½"	8½	13
			0.29			
10			0.30		8	12½
		31	0.31			12
			0.32	4"	7½"	11½"
11			0.33			
			0.34		7	11
			0.35			
12			0.36			10½

Selection tables in this section are provided only for manure tank walls and lids. Follow the Procedure Summary outlined previously but select steel areas for walls and lids from Tables 208-22 to 208-25.

Use flat sheets of welded wire fabric, not rolls. Welded wire fabric is made from cold drawn steel with a yield point of 60,000 psi. All other stresses are the same as those given in Table 208-1.

The development of at least two cross wires 2" or more beyond the point of critical section develops the full yield strength; embedment of one wire 2" or more develops ½ the yield strength.

Table 208-22. Typical wire sizes.
These items may be carried in sheets by various manufacturers in certain parts of the U.S. and Canada. The spacing, diameter, and sectional area of the welded wire fabric are the same in both the longitudinal and transverse directions.

Style Designation by Wire Gage	Style Designation by W-orD-Number	Spacing** in.	Diameter** in.	Sectional Area** sq in/ft	Weight lbs per 100 sq/ft
6x6-10x10	6x6-W1.4xW1.4	6	.135	.029	21
6x6-8x8	6x6-W2.1xW2.1	6	.162	.041	30
6x6-6x6	6x6-W2.9xW2.9	6	.192	.058	42
6x6-4x4	6x6-W4xW4	6	.225	.080	58
4x4-10x10	4x4-W1.4xW1.4	4	.135	.043	31
4x4-6x6	4x4-W2.9xW2.9	4	.192	.087	62
4x4-4x4	4x4-W4xW4	4	.225	.120	85

Table 208-23. Wire reinforced concrete tank walls.

Tabulated reinforcing is vertical; minimum reinforcing (Table 208-17) is horizontal. A 100 pcf surcharge on the hydrastatic wall load is recommended when installing or moving heavy equipment within 5′ of the tank wall. All 4′ deep tanks require only minimum temperature and shrinkage reinforcement.

Hydrastatic Load

Wall Thickness	Load, pcf	Tank Depth 6'	7'	8'	9'	10'
6"	15	0.0864	0.0864	0.0864	0.0864	0.0864
	30	"	"	"	0.1237	0.1714
	60	"	0.1163	0.1756	0.2540	-
7"	15	0.1008	0.1008	0.1008	0.1008	0.1008
	30	"	"	"	"	0.1379
	60	"	"	0.1413	0.2132	0.2824
8"	15-30	0.1152	0.1152	0.1152	0.1152	0.1156
	60	"	"	0.1184	0.1698	0.2350
9"	15-30	0.1296	0.1296	0.1296	0.1296	0.1296
	60	"	"	"	0.1459	0.2015
10"	15-30	0.1440	0.1440	0.1440	0.1440	0.1440
	60	"	"	"	"	0.1765
11"	15-60	0.1584	0.1584	0.1584	0.1584	0.1584
12"	15-60	0.1728	0.1728	0.1728	0.1728	0.1728

Hydrastatic Load + 100 psf surcharge for Vehicle Load

Wall Thickness	LOAD, pcf	Tank Depth 6'	7'	8'	9'	10'
6"	15	0.0864	0.0864	0.1131	0.1511	0.1967
	30	"	0.1113	0.1575	0.2157	0.2876
	60	0.1121	0.1712	0.2493	0.3509	-
7"	15	0.1008	0.1008	0.1008	0.1218	0.1581
	30	"	"	0.1269	9.1730	0.2295
	60	"	0.1378	0.1995	0.2786	0.3783
8"	15	0.1152	0.1152	0.1152	0.1152	0.1323
	30	"	"	"	0.1448	0.1915
	60	"	0.1155	0.1667	0.2318	0.3131
9"	15	0.1296	0.1296	0.1296	0.1296	0.1296
	30	"	"	"	"	0.1645
	60	"	"	0.1433	0.1988	0.2677
10"	15-30	0.1440	0.1440	0.1440	0.1440	0.1440
	60	"	"	"	0.1742	0.2341
11"	15-30	0.1584	0.1584	0.1584	0.1584	0.1584
	60	"	"	"	"	0.2982
12"	15-30	0.1728	0.1728	0.1728	0.1728	0.1728
	60	"	"	"	"	0.1875

Table 208-24. Tank lids for NO vehicles—wire reinforced.

One-way slab design adequate for load listed plus dead weight of slab; sectional area of steel required. All 5′ wide covers require only minimum reinforcing; except 4″ thickness, 150 psf = 0.0979. 10′ wide covers 8″ or thicker require only minimum reinforcing, except 8″ thickness, 150 psf = 0.1759.

Lid Thickness	PSF	Lid Span 6'	7'	8'	9'	10'	12'
4"	40	0.0864	0.0864	0.1068	0.1366	0.1709	-
	50	"	0.0913	0.1206	0.1545	0.1936	-
	60	"	0.1017	0.1345	0.1726	-	-
	70	"	0.1122	0.1485	0.1910	-	-
	80	0.0891	0.1228	0.1627	0.2097	-	-
	90	0.0967	0.1334	0.1771	-	-	-
	100	0.1044	0.1442	0.1917	-	-	-
	110	0.1121	0.1550	0.2064	-	-	-
	150	0.1434	0.1993	-	-	-	-
6"	40	0.1296	0.1296	0.1296	0.1296	0.1296	0.1628
	50	"	"	"	"	"	0.1794
	60	"	"	"	"	0.1345	0.1961
	70	"	"	"	"	0.1459	0.2129
	80	"	"	"	"	0.1573	0.2299
	90	"	"	"	0.1359	0.1688	0.2469
	100	"	"	"	0.1451	0.1804	0.2641
	110	"	"	"	0.1543	0.1920	0.2814
	150	"	"	0.1502	0.1917	0.2389	-

Lid Thickness	PSF	Lid Span 12'	14'	16'
8"	40	0.1728	0.1816	0.2391
	50	"	0.1969	0.2594
	60	"	0.2122	0.2797
	70	"	0.2276	0.3002
	80	0.1770	0.2431	0.3209
	90	0.1882	0.2586	0.3416
	100	0.1994	0.2742	0.3625
	110	0.2106	0.2898	-
	150	0.2560	0.3531	-
10"	40	0.2160	0.2160	0.2160
	50	"	"	0.2256
	60	"	"	0.2407
	70	"	"	0.2558
	80	"	"	0.2711
	90	"	0.2177	0.2863
	100	"	0.2292	0.3016
	110	"	0.2408	0.3170
	150	"	0.2874	0.3790

Table 208-25. Tank lids for wheel loads—wire reinforced.

Two 5000 lb loads 4′ o.c. plus dead weight of slabs. Sectional area of steel required.

Span	Slab Thickness 4"	5"	6"	7"	8"	9"	10"
4'	0.3539	0.2342	0.1782	0.1512	0.1728	0.1944	0.2160
5'	0.4325	0.2816	0.2138	0.1739	"	"	"
6'	0.5206	0.3321	0.2513	0.2046	0.1738	"	"
7'	-	0.3859	0.2910	0.2370	0.2015	"	"
8'	-	0.4433	0.3329	0.2711	0.2308	0.2024	"
9'	-	0.5049	0.3772	0.3069	0.2166	0.2298	"
10'	-	0.5711	0.4239	0.3447	0.2941	0.2586	0.2323
12'	-	0.7203	0.5255	0.4261	0.3639	0.3207	0.2889
14'	-	-	0.6393	0.5161	0.4407	0.3890	0.3512
16'	-	-	0.7677	0.6154	0.5249	0.4638	0.4194

209 TILT-UP CONCRETE CONSTRUCTION

Developed by Jay Runestad and other engineers of the Portland Cement Association, Skokie IL 60076; John Pedersen, MWPS; and E.A. Olson, University of Nebraska, retired.

You can build concrete walls for farm buildings by casting them in sections on a horizontal surface and then tilting or lifting them into position. Use solid walls for machinery or grain storage buildings and for horizontal silos. Solid walls also make good partitions in livestock buildings. Use insulated tilt-up walls for farrowing, nursery, or other "warm" buildings. For insulated walls, cast insulation in panels sandwiched between two concrete layers. The concrete protects the insulation from rodents, birds, livestock, fire, and physical damage. Custon-made precast insulated panels can be purchased in some areas.

Planning

See MWPS AED-22, *Tilt-Up Concrete Construction for Agriculture,* for details for casting and installing tilt-up panels.

Foundations

Establish the excavation lines and the foundation position by measuring from the building lines.

Some buildings are erected on pier foundations. Space the piers at the nominal panel length. Building loads are more concentrated than on a continuous foundation; make the piers large enough to prevent settling. Be sure the top surfaces of the piers are at the same level, so that the panels can be easily set plumb and the tops of adjacent wall panels will be at the same elevation.

Put footings at least 3′ deep, and in cold climates, to below the frost line.

For more information on pier and continuous footings and foundations, see *Foundations for Farm Structures,* from the Portland Cement Association.

Concrete Quality—see Chapter 207

Cement

Use not less than 6 bags of Portland cement per cubic yard of concrete with additives for 7% entrained air by volume.

Aggregates

Use well-graded, hard, dense, clean sand and coarse aggregate to make a medium or mushy mix; specify a 3″-4″ slump. If you have no vibrator, you may need to increase the slump, but do not exceed 6″. Increase slump (workability) by reducing the amounts of sand and rock, or by adding cement and water, not just by adding water. Once the water-cement ratio has been selected for the strength of concrete you need, do not upset that balance.

A stiffer mix is more economical because it has more aggregate. It also can be finished sooner, which can save labor. For economy, use aggregates well graded to the largest size permitted:

- ⅓ the thickness of an unreinforced slab on grade,
- ¾ the clear space between reinforcing bars and forms,
- or, ⅕ the minimum dimension of nonreinforced members. You can use 1½″ aggregate in an unreinforced 4″ slab, ¾″-1″ in solid wall panels, and ⅜″ in sandwich panels. If ⅜″ is unavailable, use the smallest size you can get. Crushed limestone is preferred because it is less apt to discolor.

Concrete Strength

Although durability is most important for most farm concrete, strength 3 to 7 days after casting is important for tilt-up wall panels. The strength needed to position the panels without cracking depends on the size of the panel and on the number and location of pick-up points. Select your panel size and concrete strength; then consult your concrete supplier for the mix design with the needed high early strength for lifting.

Table 209-1. Trial concrete mixes for tilt-up panels.
Water-cement ratio = 0.40.
Aggregates are assumed saturated but surface dried.
Reduce the amount of total water by the amount of water in the sand.

Max agg. size in.	Air content %	Water lb	Water gal	Cement lb	Sand lb	Coarse agg. lb
⅜″	7.5	340	40.9	850	1360	1150
¾″	6.0	263	31.7	658	2100	862

Use concrete mix with a water-cement ratio of 0.40 lb water/lb cement for adequate strength. The trial mixes in Table 209-1 should produce adequate strength in 3 days cured at 70 F, 5 days at 60 F, or 7 days at 50 F.

If the panels can be cured for a week or more at or above 70 F, a water-cement ratio of 0.53 should be adequate. Avoid a ratio greater than 0.53 lb water/lb cement.

To adjust for the water in the sand, weigh and dry a sand sample. Dry the sample until no surface water is present and then weigh the dried sample again. Then:

% moisture = 100 x loss of wt ÷ wet wt
lb water in sand = wt of sand x % moisture.
Subtract "lb water in sand" from the water required in the table; increase the amount of sand required by the same amount.

Example: Suppose a sample of sand weighed 1.0 lb when wet and 0.9 lb when dry. Loss of wt is 0.1 lb:
% moisture = 100 x 0.1 ÷ 1.0 = 10%.
For ⅜″ trial mix:
lb water in sand = 10% x 1360 = 136 lb.
Subtract from water: 340 - 136 = 204 lb water in trial mix.
Add to sand: 1360 + 136 = 1496 sand in trial mix.

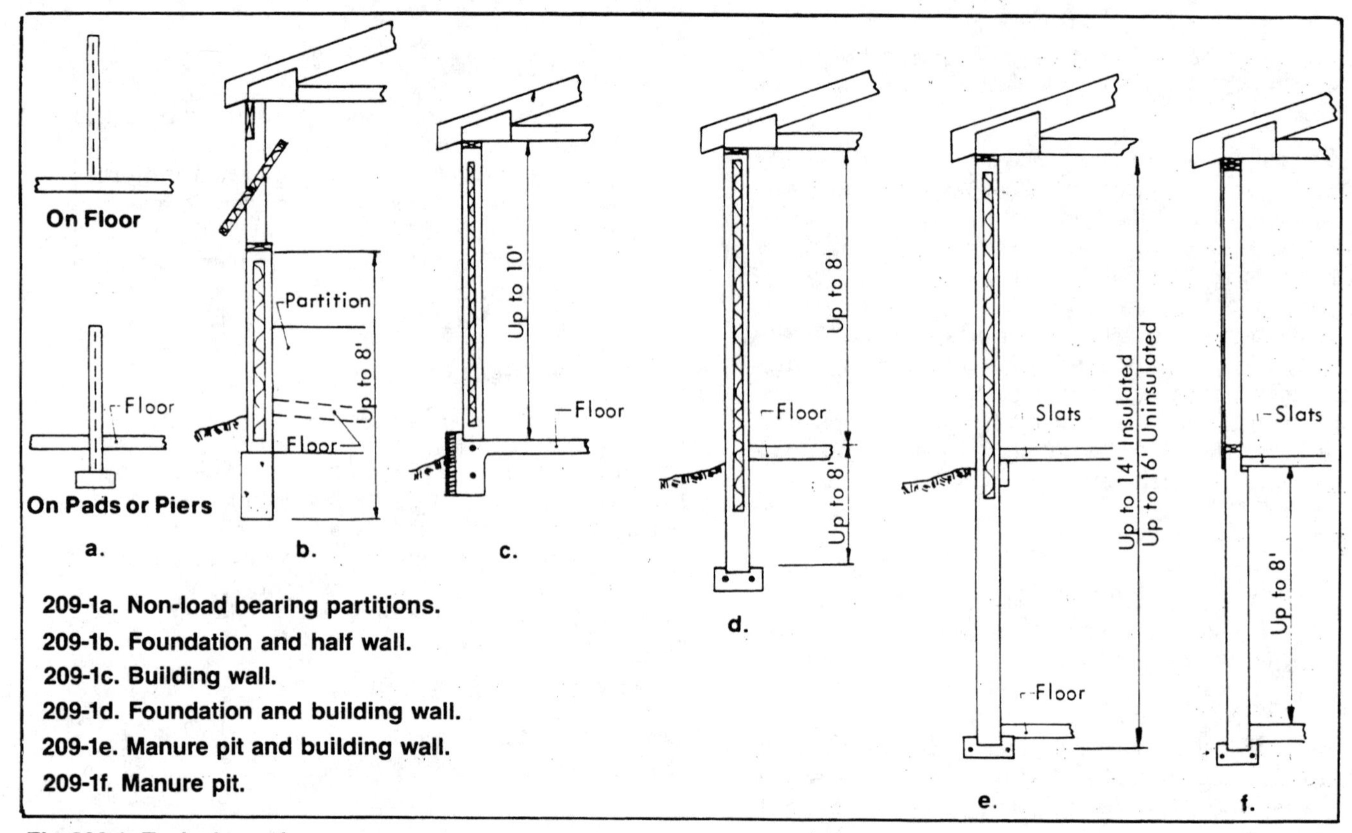

209-1a. Non-load bearing partitions.
209-1b. Foundation and half wall.
209-1c. Building wall.
209-1d. Foundation and building wall.
209-1e. Manure pit and building wall.
209-1f. Manure pit.

Fig 209-1. Typical panels.
Panels can be built with or without insulation.

Test cylinders taken of freshly mixed concrete can be properly cured and tested in a laboratory to assure adequate concrete strength.

Use a mixer large enough to make concrete for at least one panel per batch. Cast at least one whole panel at a time for uniform concrete in each panel and to avoid cold joints in the panels.

Begin curing immediately after the panels are cast and finished: provide a suitable covering, such as a spray or curing membrane or a clear polyethylene film covering so that the fresh concrete cannot dry out for 3 to 7 days.

Typical Panels

Examples of several applications for concrete wall panels are shown in Fig 209-1. Design and construction details are on the pages that follow.

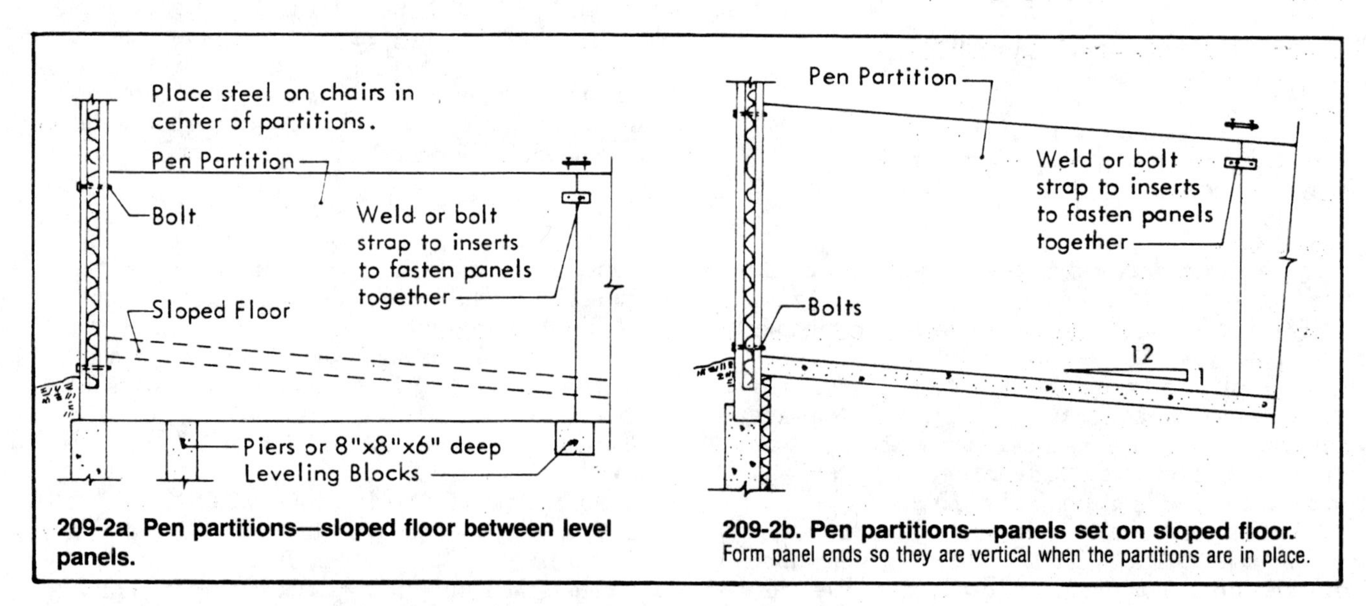

209-2a. Pen partitions—sloped floor between level panels.

209-2b. Pen partitions—panels set on sloped floor.
Form panel ends so they are vertical when the partitions are in place.

Fig 209-2. Pen partitions or non-loading bearing walls.

Table 209-2. Solid concrete pen partitions.
Concrete compressive strength is 2000 psi at time of lifting.

Solid panel thick.	Bar size	Reinforcing steel Horiz. & vert. spacing	Equivalent wire mesh	Max. panel height, ft Edge lift	Face lift
2½"	#3	24"	6x6 #6	7'	10'
3½"	#3	24"	6x6 #6	8'	12'

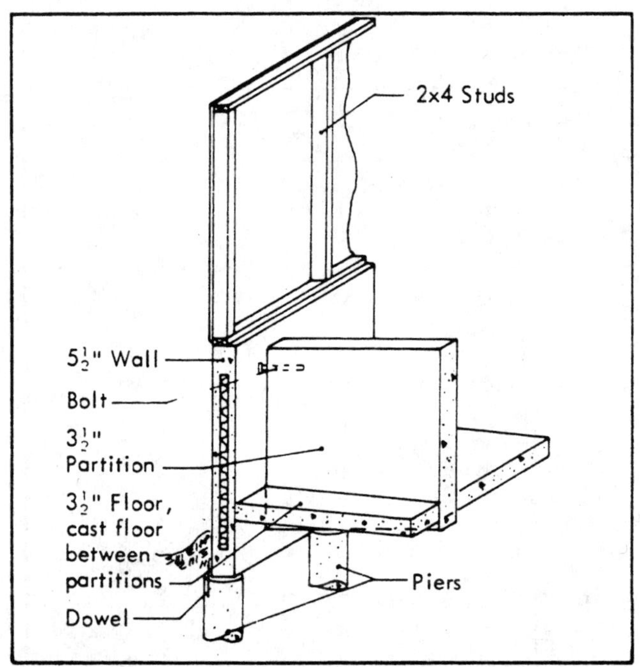

Fig 209-3. Half-wall construction with sandwich panels.

Partition walls (non-load bearing)

Solid pen partitions are a simple application of tilt-up construction. Cast panels about 4'x8' with right angle corners to rest on piers or leveling blocks. Then cast the sloped concrete floor against the panels, Fig 209-2a. Cast inserts in the panels for bolting them to wall panels or other partitions. Reinforcing is shown in Table 209-2.

Pen partitions can also rest on sloped floors. Form the ends so they are vertical when erected, Fig 209-2b. Bolt the panels to wall panels or other pen partitions. See Table 209-2 for reinforcing.

Insulated Half Walls

Insulated concrete sandwich panels, about 4' high and 8' long, can be the rugged yet insulated lower portion of exterior walls of livestock buildings. The upper walls and roof are conventional construction, Fig 209-2. See Table 209-3 and Fig 209-4 for reinforcing.

Wall Reinforcement

Masonry wall joint reinforcement, Fig 209-4, ties the wythes together, and its truss-like structure provides the required resistance to prevent one wythe from slipping past the other. That is, the joint reinforcement makes a sandwich panel resist bending somewhat like a solid panel.

Place the joint reinforcement between the strips of insulation to rest on the reinforcing steel. Then the upper wythe reinforcement can rest on the upper member of the joint reinforcement that protrudes up between the insulation. Position the joint reinforce-

Table 209-3. Reinforcing for building wall panels.
Concrete compressive strength at time of lifting is 2000 psi.
"Wythe" is the name for each of the two layers of concrete that enclose the insulation. Either wythe may be the face (outdoor) or back (indoor) layer.

Maximum panel width is 12'.
Openings for fans, windows, or doors up to ⅓ of the panel width can be cast into these panels.

Thickness (in.) t	t_1	t_2	t_3	Reinforcing in each wythe Horiz. & vert. size, spacing	Equivalent wire mesh	Maximum panel height, ft. Edge lift	Face lift
Insulated walls							
5½"	1¾"	2"	1¾"	#3, 24"	6x6 #6	7'	10'-6"
6	2	2	2	"	"	7'-4"	11'-0"
7	2	2	3	"	"	7'-6"	11'-4"
7¼	2	3	2¼	"	"	8'-4"	13'-4"
7¼	2	2	3¼	"	"	7'-6"	11'-4"
8	2	3	3	"	"	8'-6"	12'-9"
9¼	3	3	3¼	"	"	9'	13'-4"
Uninsulated solid panels							
3½"			#3, 24"	#3, 24"	6x6 #6	8'	12'-0"
4			#3, 20"	#3, 20"	6x6 #4	8'-6"	13'-0"
5½			#3, 16"	#3, 16"	4x4 #6	10'	15'-0"
6			"	"	"	10'-6"	15'-9"
7¼			#3, 12"	#3, 12"	4x4 #4	11'-6"	17'-6"
8			"	"	"	12'-4"	18'-6"

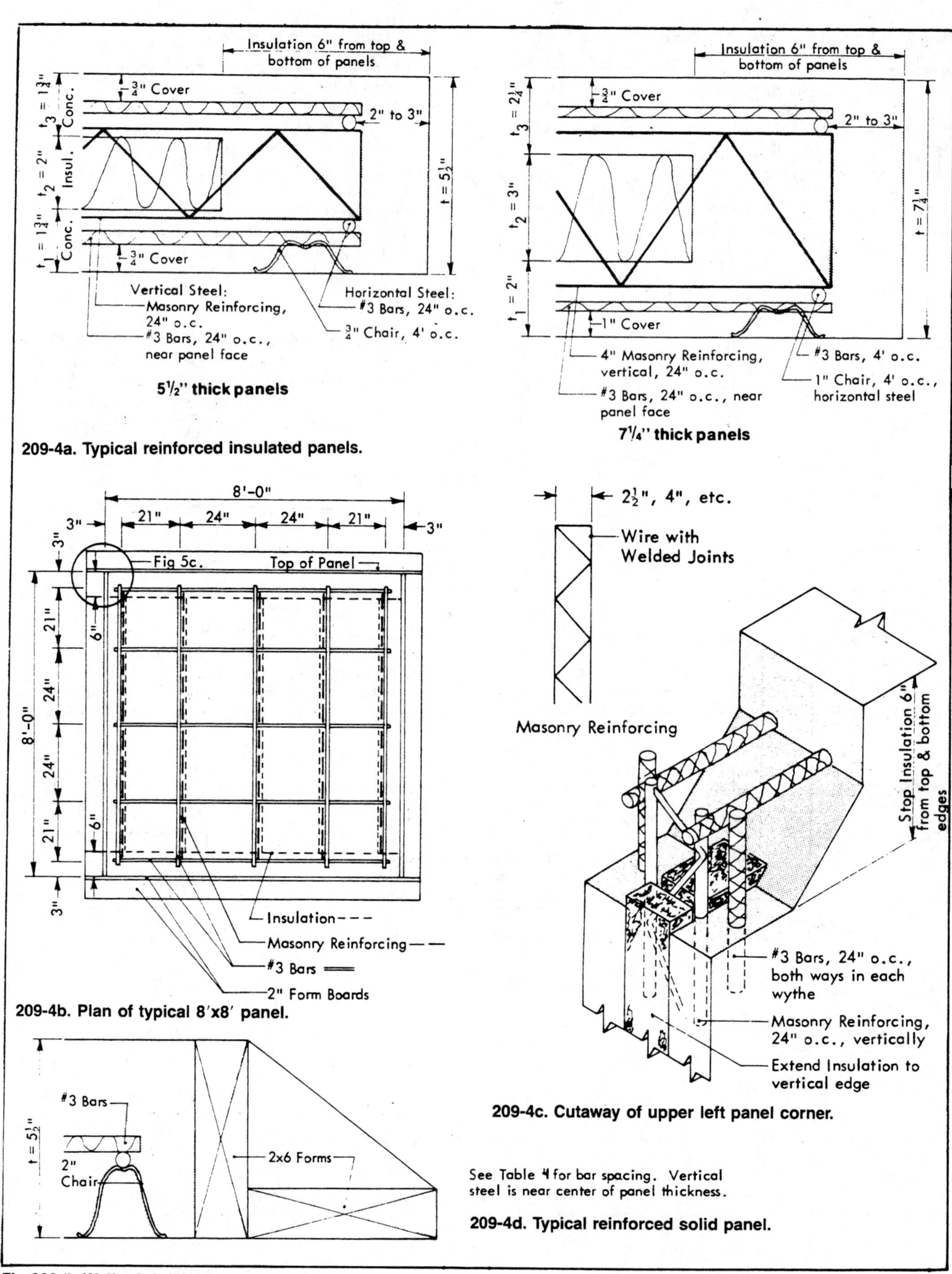

209-4a. Typical reinforced insulated panels.

209-4b. Plan of typical 8'x8' panel.

209-4c. Cutaway of upper left panel corner.

209-4d. Typical reinforced solid panel.

Fig 209-4. Wall reinforcing.
Vertical bars are close to panel faces. Insulation is 6″ away from top and bottom of panel, but extends to sides of panel.

ment as shown in Fig 209-4, 2″-3″ from each edge and then 2′ o.c. Use 2½″ wide joint reinforcement for 5½″ thick panels and 4″ joint reinforcement for 7¼″ thick panels.

Building Walls

Insulated panels have two rather thin layers of concrete (wythes) relative to their height. The wythes are connected so they act together to resist bending while being lifted to an upright position, Fig 209-4.

Table 209-4. Reinforcing for manure pit walls.
Concrete compressive strength is at least 3500 psi before backfilling.

Pit depth	Wall thick	Reinforcing steel Horizontal size, spacing	Reinforcing steel Vertical size, spacing	100 psf surcharge Vertical
6′	6″	#3, 7.3″	#3, 12.2″	#3, 8.1″
7′	″	″	#3, 7.8″	#4, 9.6″
8′	″	″	#4, 9.4″	#5, 10.2″
6′	7″	#4, 11.4″	#3, 10.5″	#3, 9.9″
7′	″	″	#3, 9.6″	#4, 11.6″
8′	″	″	#4, 11.6″	#4, 8.2″
6′	8″	#4, 10.0″	#3, 9.2″	#3, 9.2″
7′	″	″	″	#3, 7.8″
8′	″	″	#3, 7.6″	#4, 9.8″
6′	9″	#4, 8.9″	#3, 8.2″	#3, 8.2″
7′	″	″	″	″
8′	″	″	″	#4, 11.3″

Manure Pit Walls

Manure pit walls can be cast flat and then tilted up into position. Cast the wall panels with ledges to support slats and with cut-outs to support girders. Cut-outs can be in the center of the panel or centered over the joint between panels. See Table 209-4 for reinforcing details.

Combination Building and Manure Pit Walls

These building wall and foundation panels can be modified to serve as the building wall and manure pit wall. Extend the insulation to 2′ below grade as in the wall-foundation panels, with the rest of the manure pit wall solid.

The only changes needed are in the size and spacing of the reinforcing steel. Increase the amount of steel near the inside surface of the manure pit portion of the panels. See Table 209-4.

Adjust the vertical steel spacing in the upper portion of the panels (#3 @ up to 24″) so the bars lap every second or third bar in the manure pit portion.

Example 1: 6′ deep pit with 7″ wall, Table 5 calls for #3 rebars at 10.5″; therefore, place the vertical bars in the inside wythe 21″ apart so they lap alternate bars in the lower portion of the panel.

Example 2: 8′ deep pit with 7″ wall where machinery can drive close to the pit wall; vertical bars are #4 at 8.2″. Only slight adjustment is needed in the upper steel to lap every third bar in the lower steel. Lap rebars 12″.

210 CONCRETE MASONRY

Concrete masonry units (blocks) are adaptable, economical, durable, fire resistant, low maintenance, and readily available. They are used for buildings, manure tanks, or retaining walls.

Concrete blocks can be painted with portland cement or latex paint or be directly plastered. They are difficult to insulate adequately.

Types of Blocks

Concrete masonry units are designed as hollow load-bearing, solid load-bearing, or hollow non-load-bearing concrete blocks; concrete building tile; or concrete brick. They are made with either heavyweight or lightweight aggregates. Heavyweight 8x8x16 blocks made with portland cement and heavy aggregate—sand, gravel, crushed rock, or air-cooled slag—weigh 40 to 50 lb. The same size blocks made with portland cement and lightweight aggregate weigh 25 to 35 lb. Natural lightweight aggregates include volcanic cinders, pumice, and scoria; manufactured aggregates include coal cinders, expanded shale, clay, or slag. Lightweight blocks are easier to handle and have about 50% more insulating value, but are weaker than heavyweight blocks.

A "solid" concrete block has a core area of up to 25% of its gross cross-sectional area. "Hollow" blocks generally have core areas of 40%-50% of the gross area.

The 8x8x16 hollow block is the most common type. It is 7⅝" wide, 7⅝" high, and 15⅝" long. When laid with ⅜" mortar joints, total height is 8" and length is 16".

Blocks are 4", 6", 8", 10", and 12" wide. Some manufacturers make both 4" and 8" high blocks. Fig 210-1 shows some common types of blocks. Before starting a job, see your local supplier for sizes and shapes available.

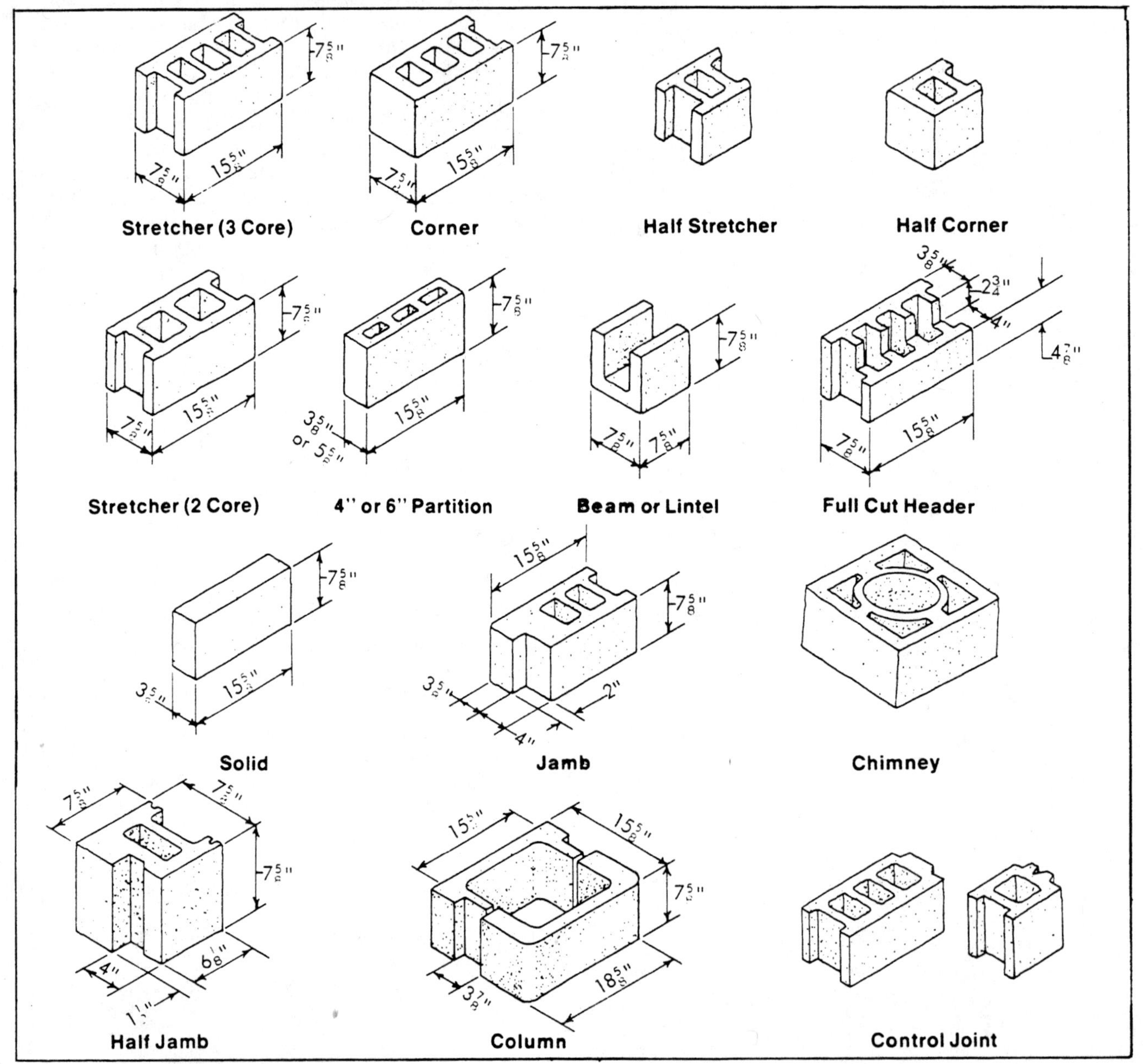

Fig 210-1. Typical concrete blocks.

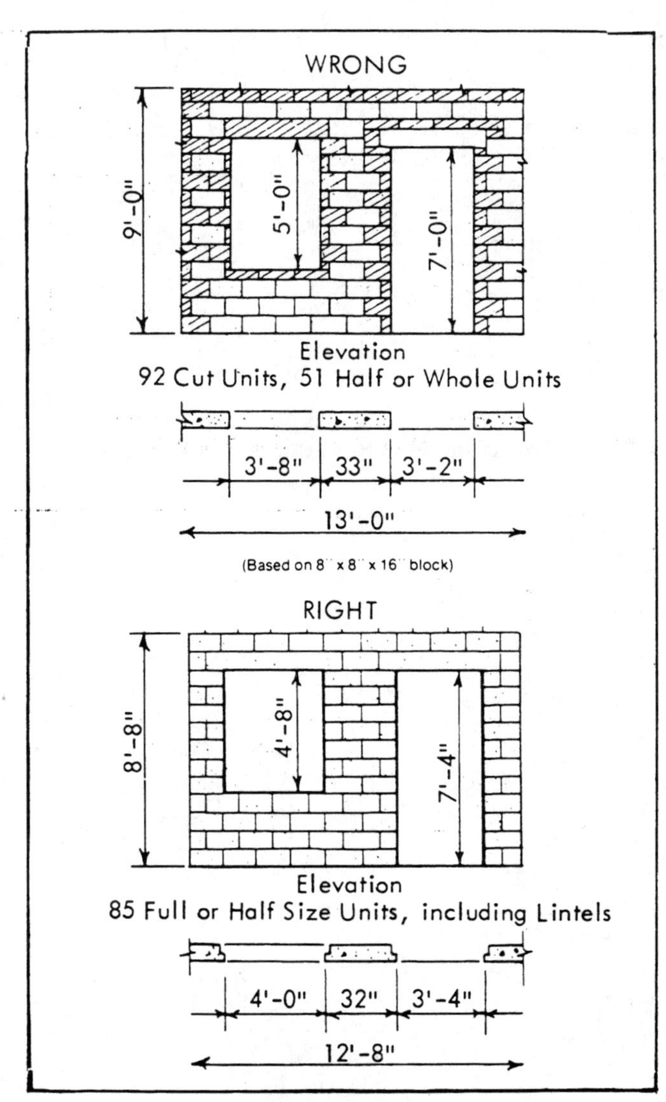

Fig 210-2. Planning block wall dimensions.

Laying Blocks with Mortar

Dimensions

Plan building dimensions to use full and half blocks to eliminate cutting, Fig 210-2. Make wall lengths, window and door widths, and window and door heights divisible by 8″: even ft, even ft + 8″, or odd ft + 4″ (8′-0″, 8′-8″, 9′-4″).

Mortar

Mortar must be workable to bond to concrete. Use properly graded sand, mortar with good water retention, and thorough mixing; avoid excess cement. Do not wet concrete block to control the absorption of water from the mortar.

Mortar is masonry cement (a mixture of portland cement and lime) and sand; or portland cement, hydrated lime, and sand. The proportions of each material identify the type of mortar. Use Type M mortar in construction requiring extra strength and frost resistance. For ordinary service, use Type N mortar, Table 210-1.

Avoid admixtures that lower the mortar's freezing point. The quantity needed to significantly lower the freezing point greatly lowers its strength and other desirable properties. Up to 1% calcium chloride added to masonry cement (2% to portland cement), or high-early-strength portland cement, increases curing rate and early strength.

Table 210-1. Recommended mortar mixes by volume.
Cement and lime sacks contains 1 ft³. Masonry cement is ASTM Specification C91 Type II.

Type & Service	Cement	Hydrated lime	Mortar sand in damp, loose condition
N: Ordinary Service	1-masonry cement	--	2¼ to 3
	or		
	1-portland cement	½ to 1¼	4½ to 6
M: Heavy loads or frost	1-masonry cement* plus 1-portland cement	--	4½ to 6
	or		
	1-portland cement	¼	2¼ to 3

Use mortar within 2½ hr after mixing when temperatures exceed 80 F and within 3½ hr when cooler. If the mortar stiffens from evaporation during this period, remix it and add a little water to restore workability.

Using standard 8″ blocks, about 13.5 ft³ of mortar is needed for 100 ft² of wall or 6 ft³ per 100 blocks. One ft³ of type N mortar requires about 0.16 sacks of portland cement and lime and 0.97 ft³ of sand.

1. Build corners first, 4 to 5 courses higher than the center of the wall. Lay first row of blocks in a full bed of mortar.
2. Make each corner course level and plumb.
3. Use a 1x2 with marks 8″ apart to locate the top of each course.
4. Use a mason's line stretched between corners to keep the wall straight and level.
5. Set blocks to grade and level by tapping with trowel handle.
6. Run an "O" or "V" shaped tool along joints after mortar has somewhat stiffened.

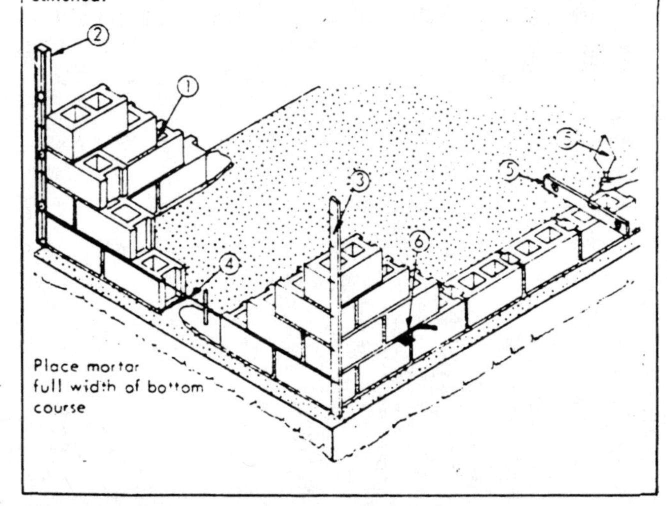

Fig 210-3. Steps in laying concrete block walls.

Laying Blocks

First lay out the building and construct level footings for the mason's line. Keep mortar beads uniform, set units firmly, and strike off excess mortar that squeezes to the outside of the wall. Joints can be rodded, raked, or flushed for the desired finish. Rodding to a concave shape or V-ing gives some resistance to water seepage. Follow the steps in Fig 210-3.

Control Joints

Control joints are continuous vertical joints in building walls that permit slight movements and prevent random cracking. Control joints relieve stress concentrations caused by expansion and contraction, unequal foundation settlement, or restraint by columns or intersecting walls.

The following are guides to control joint location, Figs 210-4, 210-5, and 210-6.

- In plain unbroken walls, space control joints no more than 2½ times the wall height.
- Continue foundation joints up through the wall.
- Extend roof and floor joints down through the wall. Place control joints at intersections of new and old construction.
- Place control joints at major openings. For openings less than 6′ wide put a control joint up one side; for wider openings build joints along both sides.
- Tie intersecting masonry together with a masonry bond only at corners. Terminate a butting wall with a control joint and connect it to the other wall with steel tiebars. Space tiebars not more than 3 courses apart. Fill cores to anchor tiebar ends.

Control joints are usually laid up with mortar like other joints. If the joint is to be caulked, rake stiffened mortar out of the joint to a depth of ¾″, prime with shellac or other sealant, and caulk.

Special tongue-and-groove control joint blocks provide excellent lateral stability for a wall. To use regular open-end block, place a strip of building paper over the core of one unit to break the mortar bond, then fill the core with mortar to form a key for lateral stability. Offset jamb blocks with a Z-type tie bar can also be used.

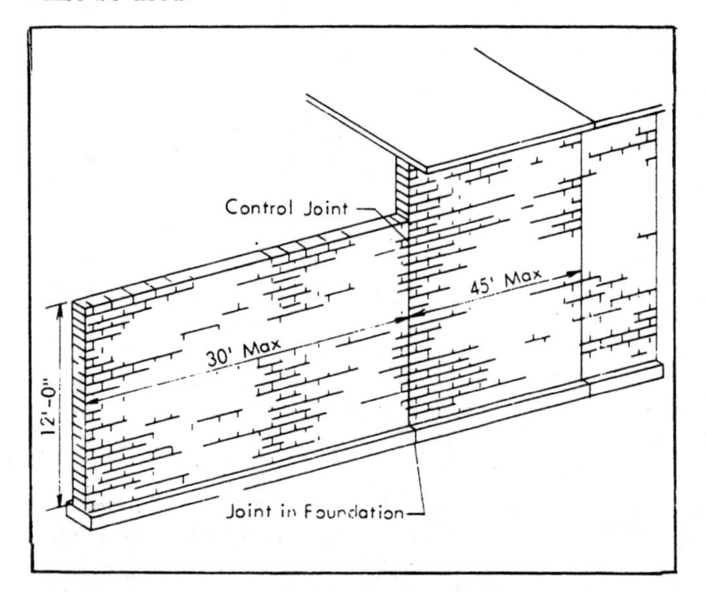

Fig 210-4. Control joints in masonry walls.

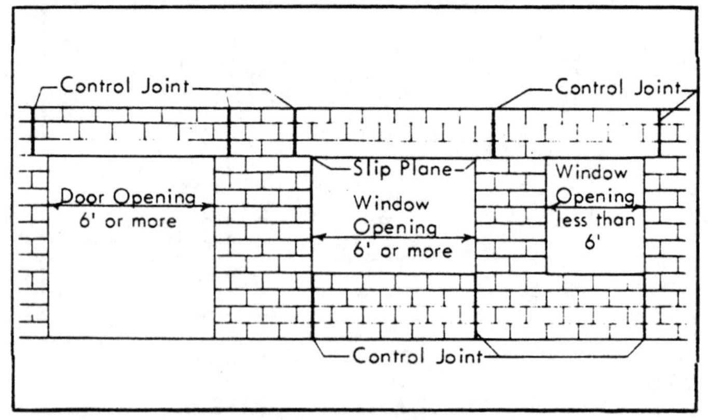

Fig 210-5. Control joints around doors and windows.

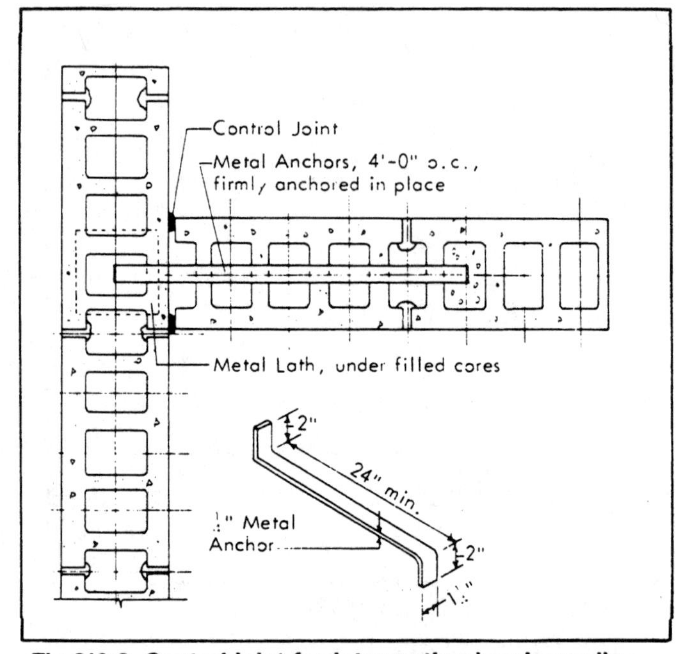

Fig 210-6. Control joint for intersecting bearing walls.

Reinforcing

Add steel reinforcing to some concrete block walls—bucking walls, retaining walls, and manure pit walls. Insert vertical reinforcing in block cores and fill with concrete. Reinforce horizontally with bars in the mortar joints—either one bar along each face or a prefabricated truss configuration, Fig 210-8.

Block walls can also be reinforced horizontally with bond beams. Bond beams are U shaped blocks with steel bars. They form one course of the wall and are filled with concrete after the reinforcing bars are added. Bond beams are put above and below windows, over doors, in retaining walls, or at the top of a wall with the plate anchored to it, Fig 210-9.

Block walls of hollow units are supported laterally by pilasters, buttresses, or cross walls spaced not more than 18 times the nominal wall thickness. Build pilasters with double corner blocks or pilaster blocks. For added strength, fill the block cores with concrete and add steel bars, Fig 210-10.

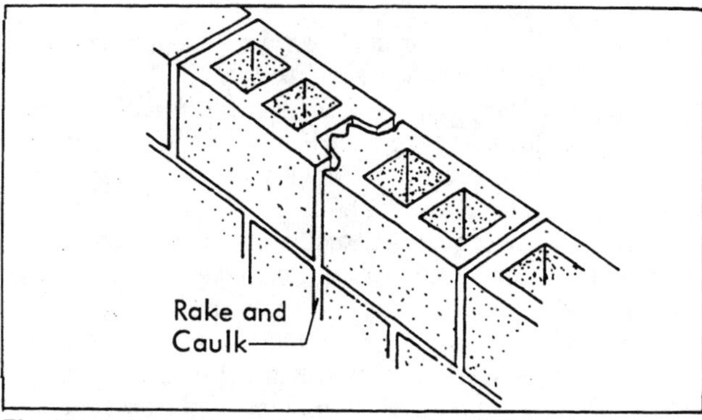

Fig 210-7. Tongue and groove control joint.

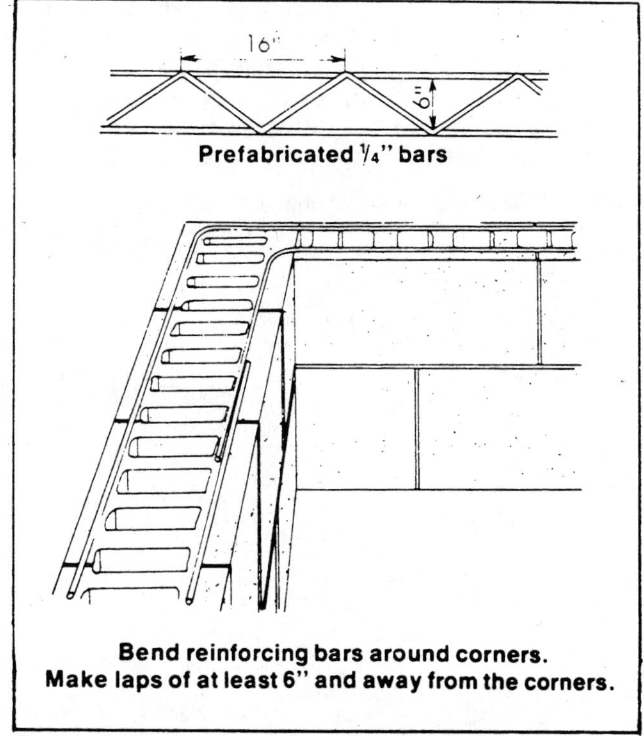

Fig 210-8. Horizontal masonry wall reinforcing.

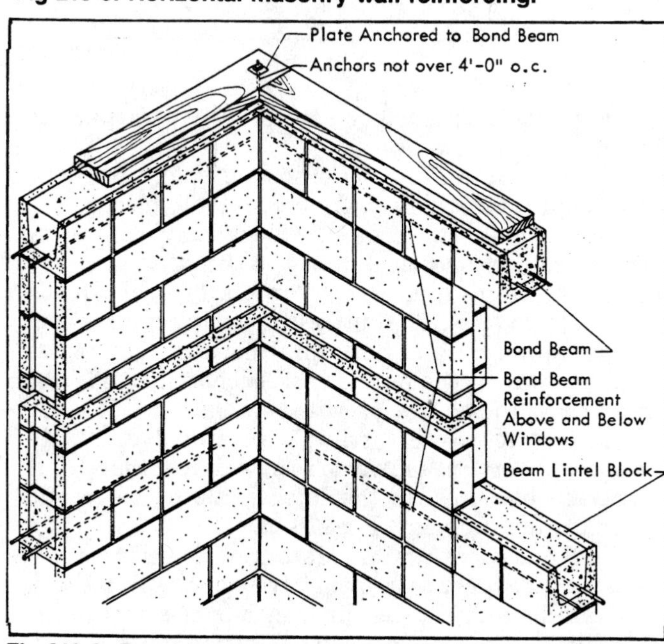

Fig 210-9. Bond beams.

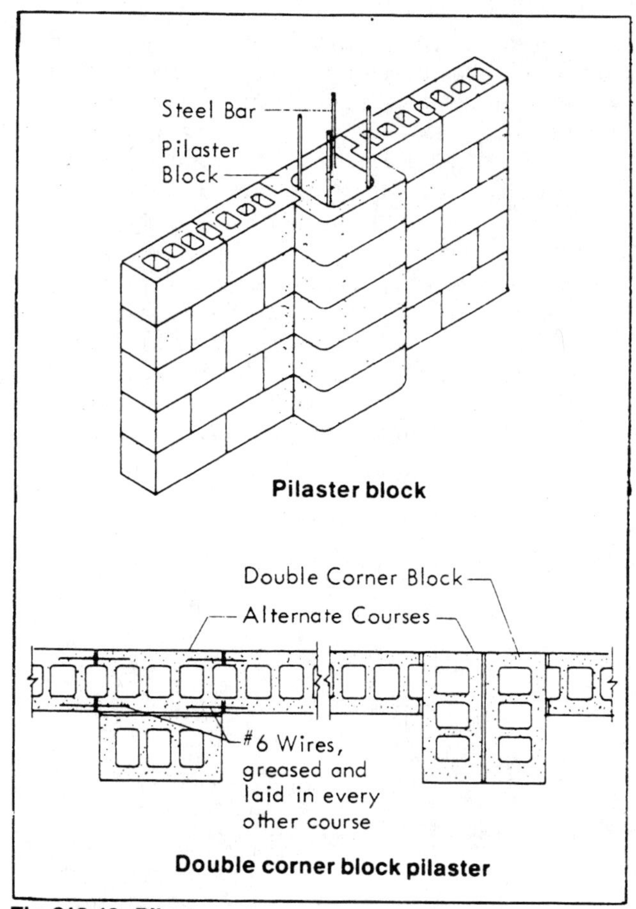

Fig 210-10. Pilasters

Insulation

See Chapter 631, Insulation and Vapor Barriers, for insulation values for concrete block walls. The problem of adequately insulating a concrete block wall is one of its main disadvantages. Lightweight aggregate blocks with the cores filled with insulation are moderately effective.

A cavity wall can be built of two walls separated by an air space and connected by metal ties. Fill the air space with foam or granular insulation. This type of construction is expensive and not common in farm buildings.

Concrete block basement walls can be insulated by nailing 2x2 furring strips to the inside of the wall, placing insulation between them, and covering with paneling. Plastic foam insulation boards can be glued directly to the wall and a finished surface glued to them. These methods are not practical in livestock housing because the durable inside wall surface, a main advantage of concrete, is lost.

Waterproofing

Water resistance of mortar joints is increased by proper tooling. Manure pit walls need not be waterproof. Seepage outward will be stopped by manure solids plugging the holes; seepage inward usually adds only minor amounts of water to the liquid manure.

A perimeter drain tile helps ensure dry basement walls and lower soil pressures against basement or manure pit walls (most unreinforced below-ground masonry walls are not strong enough to support high pressures from saturated soil). Place the drain tile beside the footing, slope it to daylight or a sump, and cover it with gravel or crushed rock. Sand below the floor is usually sufficient to relieve water pressure. Drain tile can also be laid in the sand where water problems are severe.

A good practice is to apply two ¼″ coats of portland cement plaster (called parging) on the exterior surface. Roughen the first coat before it drys to provide good bond for the second coat. Keep moist until it cures. A bituminous asphalt coating can be applied over the plaster for added protection, Fig 210-11.

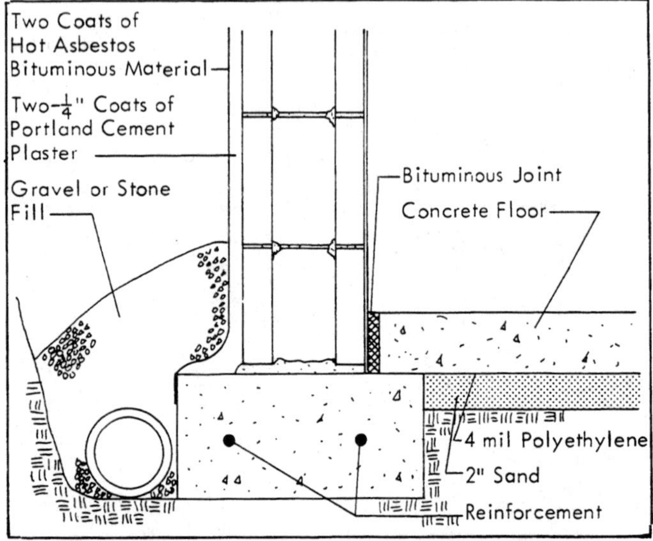

Fig 210-11. Waterproofing basement wall in wet soil.

Table 210-2. Maximum backfill depths on 8′ high unreinforced masonry walls.

Type of Block	Maximum depth below grade, ft Wall thickness, in		
	8"	10"	12"
Supporting frame bldg.			
Hollow block	5'	6'	7'
Solid block	5'	7'	7'
Supporting masonry bldg.			
Hollow block	5'	7'	7'
Solid block	7'	7'	7'

Basement walls

Backfill unreinforced basement walls only to the height given in Table 210-2. Avoid backfilling before the first floor is in place or brace the walls until the first floor is installed. Slope the finished grade away from the wall for good drainage. A typical basement wall is illustrated in Chapter 302 Foundations.

Building Walls

Design concrete masonry building walls according to the *American Standard Building Code Requirements for Masonry* (American National Standard A41.1953) or *Specification for the Design and Construction of Load-Bearing Concrete Masonry*(National Concrete Masonry Association).

The ratio of height or length to thickness for lateral support of load-bearing walls made of hollow core units is 18. An 8″ hollow core block wall needs lateral supports 12′ or less apart (8″ x 18 = 12′). Provide lateral support with cross walls, piers, buttresses, floors, or roofs. Non-load-bearing partitions can have lateral supports spaced up to 36 times the actual wall thickness.

The top story walls (12′ maximum) of a building less than 35′ high can be 8″ thick. Use 12″ block below the top story. Residence walls can be 8″ thick when not over 3 stories or 35′ high. Single story walls can be 6″ thick if not over 9′ high.

Non-bearing exterior walls can be 4″ thinner than bearing walls, but not less than 8″ thick except where 6″ walls are specifically permitted. Non-bearing interior walls can be 4″ thick.

Table 210-3. Reinforcing for cantilever retaining walls.
Soil pressure of 45 psf/ft, 3000 psi concrete, and 40,000 psi steel. Horizontal reinforcement: use two No. 4 bars in bond beams at 16″ o.c. or equivalent joint reinforcement at 8″ o.c.
H = height of wall; W = width of footing;
T = thickness of footing; A = distance to face of wall

Dimensions from Fig 210-13				Bar size & spacing	
H	W	T	A	Vertical wall rods	Horizontal footing rods
8″ walls					
3′-4″	2′- 4″	9″	8″	#3, 32″	#3, 27″
4′-0″	2′- 9′	9″	10″	#4, 32″	″
4′-8″	3′- 3″	10″	12″	#5, 32″	″
5′-4″	3′- 8″	10″	14″	#4, 16″	#4, 30″
6′-0″	4′- 2″	12″	15″	#6, 24″	#4, 25″
12″ walls					
6′-8″	4′- 6″	12″	16″	#6, 24″	#4, 22″
7′-4″	4′-10″	12″	18″	#7, 32″	#5, 26″
8′-0″	5′- 4″	12″	20″	#7, 24″	#5, 21″
8′-8″	5′-10″	14″	22″	#7, 16″	#5, 26″
9′-4″	6′- 4″	14″	24″	#8, 8″	#6, 21″

Retaining Walls

Fig 210-12 illustrates several types of concrete block retaining walls.

Gravity retaining walls depend solely on their own weight for stability. They commonly have a base thickness equal to ½ to ¾ of their height. Gravity walls are uneconomical over 4′-6′ high.

Buttressed walls span horizontally between the vertical supports. The buttresses are in compression. Major reinforcing is horizontal between buttresses and vertical in the buttresses. Anchor the wall and buttresses to the footing with dowels.

In cantilever walls the wall is reinforced to withstand the vertical tension on the inside surface. The concrete base resists the lateral sliding force and overturning moment, Fig 210-13.

Soil pressures against retaining walls can be considerably reduced by providing weep holes in the wall or running a drain tile the length of the wall in the backfill. Use 4″ tile for weep holes spaced 5′ o.c. hori-

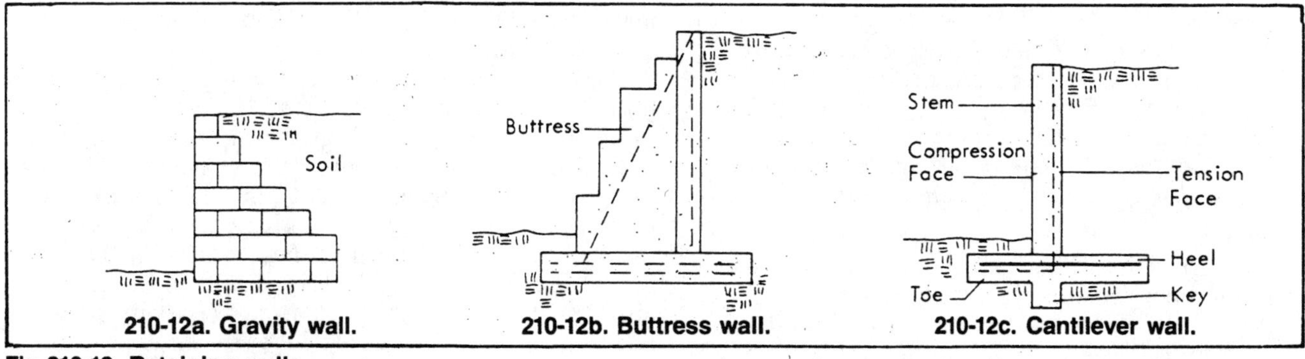

210-12a. Gravity wall. 210-12b. Buttress wall. 210-12c. Cantilever wall.

Fig 210-12. Retaining walls.

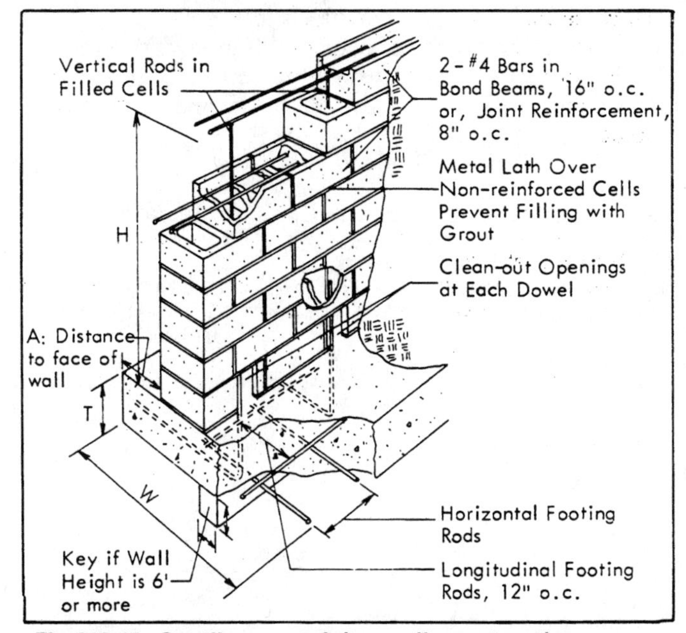

Fig 210-13. Cantilever retaining wall construction.
See Table 210-3 for steel required. Footing dowels are the same size and spacing as vertical wall rods. They extend to the toe of the footing and at least 30 bar diameters up into the wall. Clean out loose mortar at each dowel before filling cores.

zontally and 8" above finish grade on the low side. Place 1 ft³ of gravel in the backfill at each weep hole, Fig 210-14a. In poorly drained soils, use 4" tile for the longitudinal drain in the backfill located just above finish grade level and surrounded by gravel or crushed rock.

Fig 210-13 shows the construction of a reinforced concrete block retaining wall and Table 210-3 lists required sizes and reinforcing.

Surface Bonding

Surface bonding concrete block was introduced in the late 1960s. Concrete blocks are stacked without mortar between them to form walls, then both sides are coated with a thin layer of cement plaster containing fiber glass reinforcement. Surface bonding has several advantages over conventional concrete masonry construction:

- Less time and skill are required to build walls.
- It provides excellent resistance to rain penetration.

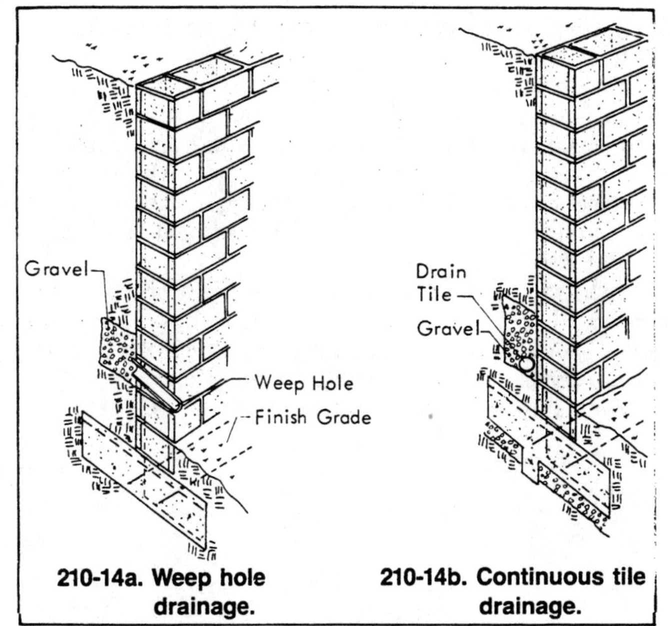

210-14a. Weep hole drainage. 210-14b. Continuous tile drainage.

Fig 210-14. Retaining wall drainage.

- Coloring pigment can be added to the mortar to eliminate painting.
- Surface bonded walls are as strong as conventional walls in bending.

Table 210-4. Wall and opening dimensions for surface-bonding.
Standard 16" blocks, 15⅝" long by 7⅝" high. Add ¼" for about each 10' of wall to allow for nonuniformity in block size. Make a trial stacking of blocks to determine the actual height of wall or opening before beginning construction.

Number of blocks	Wall length or opening	Number of courses	Wall height or opening
1	1'-3 5/8"	1	7 5/8"
2	2'-7 1/4"	2	1'-3 1/4"
3	3'-10 7/8"	3	1'-10 7/8"
4	5'-2 1/2"	4	2'-6 1/2"
5	6'-6 1/8"	5	3'-2 1/8"
6	7'-9 3/4"	6	3'-9 3/4"
7	9'-1 3/8"	7	4'-5 3/8"
8	10'-5"	8	5'-1"
9	11'-8 5/8"	9	5'-8 5/8"
10	13'-1/4"	10	6'-4 1/4"
11	14'-3 7/8"	11	6'-11 7/8"
12	15'-7 1/2"	12	7'-7 1/2"
13	16'-11 1/8"	13	8'-3 1/8"
14	18'-2 3/4"	14	8'-10 3/4"
15	19'-6 3/8"	15	9'-6 3/8"

Dimensioning

When blocks are laid using surface bonding and no mortar joints, wall dimensions are not in 8″ or 16″ modules. Wall lengths, heights, and openings for standard 8x16 blocks without mortar joints are given in Table 210-4. Add ¼″ to the wall length for every 10′ of wall to allow for nonuniform block size.

Construction

Start with a level footing, reinforced if recommended for your climate and soil conditions. Use ½″ steel dowels 48″ o.c. to tie the footing and wall together or run tie rods from the footing to the plate. Fill block cores containing reinforcing steel with mortar. Set the first course of block in a full mortar bed to obtain a good level base.

Stack the remaining courses dry. Butt the ends of the blocks tightly together. Use metal, mortar, plastic, or sand shims (not wood) to level uneven blocks. Block tops and bottoms can be ground to improve uniformity and wall compression strength. Use a mason's line and level on every third course to assure straight and plumb walls.

With no mortar bed joint, there is no space for horizontal joint reinforcing. Install a bond beam every 4′ vertically if horizontal reinforcing is needed. Provide control joints to prevent random cracking at locations similar to conventional block walls, but no more than 20′ o.c. in walls without openings or 60′ o.c. if bond beams are used.

Follow manufacturer's instructions when using premixed surface bonding mortar. Use clean masonry units. Wet the surfaces lightly with a hose prior to plastering. Apply a minimum thickness of 1/16″-⅛″ to **both** sides of the wall. Mix only small amounts that can be used before they set. Adding water and remixing stiff mortar can lower its bond strength. Trowel firmly to ensure a good bond and to achieve the desired finish, but do not over trowel. Wet again with a hose within 24 hr to prevent dehydration while mortar cures; protect it from heavy rain. Do not locate surface bonding joints over block joints.

Use the same types of door and window framing, pilasters, control joints, connecting wall ties, and roof anchorage methods as for conventional concrete block walls.

References

American National Standards Institute. 1954. American Standard Building Code Requirements for Masonry (A41.1), National Bureau of Standards Misc. Publication 211, ANSI, New York, N.Y.

National Concrete Masonry Association. 1968. Specification for the Design and Construction of Load-bearing Concrete Masonry TR75-B. NCMA, McLean, VA.

Randall, F. A., Jr. and W. C. Pancrease. 1976. Concrete Masonry Handbook for Architects, Engineers, Builders. Portland Cement Association, Skokie, IL.

211 CAULKS AND CAULKING

Caulking materials fill joints in buildings and fixtures and make them weatherproof or watertight. Caulks are usually required where materials change, such as wood siding abutting against brick chimneys or walls, joints between window frames and siding, or joints between walls and floors.

Caulks prevent water or water vapor from getting between, behind, or into construction materials—window frames, siding, floors—and therefore prevent deterioration. They can fill cracks and joints to prevent drafts and reduce heat requirements due to air infiltration. Caulking also helps keep out insects and vermin.

Caulks are available for almost all materials—masonry, wood, steel, glass, aluminum, and plastics.

Where to Use Caulks

When selecting a caulk for cracks, consider the width and depth of the crack and also if the crack is subject to movement.

Narrow cracks are up to about ¼″ wide and are usually up to about ¼″ deep.

Wide cracks are greater than ¼″ and often deeper than ¼″. Fill wide cracks with a filler (jute, okum, rope, etc.) before caulking. Most caulks do not work well in cracks larger than ¼″.

Select caulks that remain elastic for cracks subject to movement, such as joints between dissimilar materials, materials subject to frost heaving, or materials with different shrinkage characteristics. Examples are joints between a house and a sidewalk slab, between glass and a sash, between building siding boards, and the expansion-contraction joints in masonry walls and slabs. Be sure to select a caulk with suitable elastic properties; some caulks harden and will fail in a joint that moves.

If a crack between masonry and wood is more than ½″ deep, push filler into the crack before caulking. Squeeze the caulk from the tube and then force it into the crack with a putty knife. Acrylic latex caulk is recommended.

To seal between siding and door or window casings, clean out old caulk; clean and dust both surfaces. Hold the caulk gun at a 45° angle to the surfaces and apply the bead. Acrylic latex caulk is recommended.

Caulk around water pipes, wires, ducts, chimneys, etc., that go through walls and roofs to seal the openings against water, air, and insects that could cause extensive damage. To seal around chimneys or ducts that may be hot, select a caulk that can withstand high temperatures without smoking or starting a fire.

Seal around hose bibbs and electrical outlets to prevent water from seeping into a building. Force caulk into the joint around the fixture. If the opening is large, first stuff in urethane or other nonstaining filler. Acrylic latex caulk is recommended.

Caulk under or around the edges of flashing at fan outlets, vent pipes, chimneys, etc., to prevent mois-

Fig 211-1. Fill cracks or nailholes in wood or wood-base products.
Fill unsightly cracks and nailholes with caulk and then feather the edges smooth. Paint after removing any excess caulk. Acrylic latex caulk is recommended.

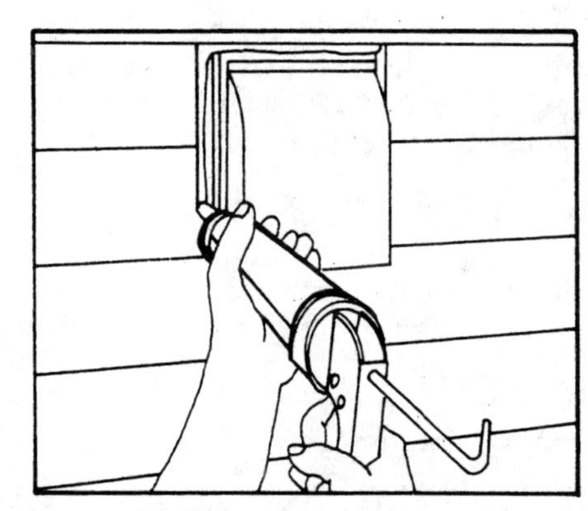

Fig 211-2. Seal cracks around ventilator caps and air conditioners.
Place a bead of acrylic latex caulk between the siding or roofing and the fixture. A tighter seal is possible by loosening the cap, placing a bead of caulk around the opening and resetting the cap in the caulk and tightening the cap down. Seal around window-mounted air conditioners.

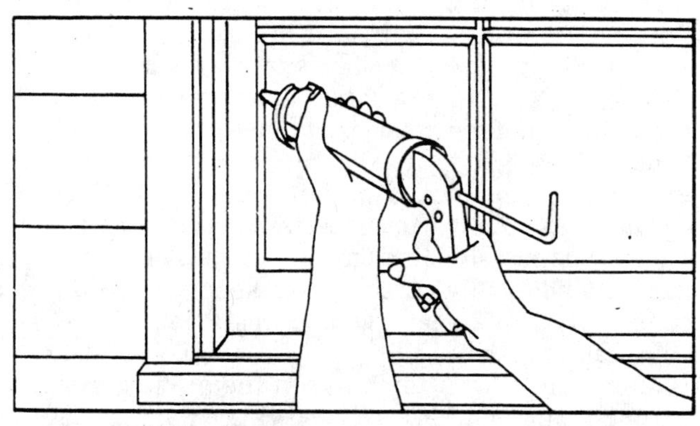

Fig 211-3. Glazing a window.
Caulking material makes quick, easy work of glazing projects. Remove all loose chips and dust; apply caulk to the glass-wood joint. Spread the caulk with a putty knife or wetted finger. Ocassionally, dip the knife in water to make spreading faster and easier. Acrylic latex caulk is recommended.

ture from getting between the flashing and siding or roofing material and causing deterioration. Caulk also stops leaks in rusted rain gutters.

Run a bead of caulk into the seam between loose chimney flashing and shingles to prevent water leakage into a house. If necessary, caulk between the flashing and the chimney. Butyl caulk is recommended.

To fix leaks in rusted rain gutters, be sure the surface is completely dry and free of rust and dirt. After running the bead with the caulking gun, tool the bead flat with a putty knife or wetted finger. Butyl caulk is recommended.

Relatively new but excellent silicone caulks are now available for sealing around bathtubs, wash basins, drain flanges, and other interior plumbing fixtures. Many of them remain soft and pliable after setting to withstand any movement without failing. Some bathtub and tile sealers contain fungicides to help prevent fungus and mildew growth.

Caulks

When you buy a caulking compound, consider more than initial cost. Good life expectancy and performance avoid the very time consuming job of removing poor quality materials and replacing them with new.

An ideal caulk:

- Squeezes easily out of a gun.
- Sticks to an unprimed surface.
- Becomes tack-free quickly to minimize dust pickup.
- Leaves a smooth surface.
- Cleans up with water.
- Shrinks little after curing.
- Remains plastic and rubbery for life.
- Is paintable.
- Does not require painting for maximum weather resistance.
- Lasts as long as the building it seals.

Because all caulking jobs are not alike, there is no one caulk that has all these properties.

Caulks come in cans, metal or plastic tubes, and 11 fl oz drop-in cartridges complete with a spout. The cartridges fit half-barrel caulking guns, and the spout can be cut to form different bead sizes. The cartridges and tubes are much easier to use than caulk from cans for most jobs.

Caulks cure either by evaporation of the solvents or by internal chemical reaction. Caulks that cure chemically usually perform better in the joint because they do not shrink from the loss of evaporated material. Chemically cured caulks keep well in the tube for only 1 to 2 years, so use fresh material.

Use caulks safely:

- Many caulks are toxic—keep them away from your mouth and eyes. These carry warning labels.
- Some caulks should not be used in closed quarters. These carry warning labels.
- Store caulks where children and animals cannot reach them.
- Follow label directions as to use.
- Some cleanup solvents are toxic and also flammable.

Types of Caulks

Table 211-1 lists the more common caulks and gives selection information. **Joint type** tells if the caulk can be used in a moving joint and gives limitations on joint width and depth. **Tack-free time** tells how much time is required before dust will not stick to the surface and it can be painted. **Cure time** shows how long it takes for the inner portion of the bead to completely set.

Oil base

Oil-base caulks give only a year or two of service before the oil dries out and they lose pliability. Then you must clean out the joint and re-caulk. Oil-base caulks are best sellers because of low cost, but price is all that recommends them.

Acrylic latex

The next step up in quality are the acrylic latex caulks. They are easy to use and nontoxic. They are easy to tool after placement and water can be used for cleanup. When they cure, the caulk remains flexible.

Do not use latex caulk in joints more than ¼" wide. Its expected life is much shorter than other high-performance products now readily available.

There are two types of latexes available: economy and quality. If the label does not say **acrylic,** then it is economy; some that say they are acrylic may be economy quality. The other latexes—ethylene vinyl acetate (EVA) and polyvinyl acetate (PVA)—are cheaper than good quality acrylic latex but are not as good in the joint, especially PVAs. Many of the low-cost latex caulks should be used only indoors.

The life of latex caulks in joints varies greatly. One manufacturer now guarantees its acrylic latex caulk for 20 years. Expect much shorter life from economy latexes. Most caulks are not guaranteed at all.

Butyl rubber

The lowest cost, most widely used flexible caulk does not become rigid and gives slightly if the joint moves. Home sealant is butyl rubber. Its price is about the same as a top-line latex, but it is more difficult to use. It cures by solvent evaporation. As the solvent evaporates, the caulk shrinks. Therefore, butyl's tolerance for joint-widening is limited. It is thus a low-performance flexible caulk and should not be used in moving joints.

Polysulfide rubber

Polysulfide was the first of the high-performance flexible caulks. It cures by chemical reaction, so there is no evaporation to make it shrink on the job.

It is toxic and difficult to apply. Because of its disadvantages, polysulfide is not recommended as a good caulk for homeowners.

Table 211-1. Caulk characteristics.

Caulk	Life (yrs)	Best uses	Adhesion	Shrinkage-free	Use primer on[1]	Joint type	Tack-free (hrs)	Cure (days)	Clean up with[2]	Paint	Available colors
Oil-base	1-7	Only price recommends it	Fair-good	Poor	Porous surfaces	Nonmoving to 1/4"w, 3/4"d	2-24	To 365	Paint thinner	Must	White, natural, gray
Acrylic-latex	2-10	Indoors & protected or painted	Excellent, except metal	Fair	Porous surfaces for best results	Nonmoving to 1/4" wide	1/4-1/2	3	Water	Best	White, black, gray, bronze
PVA tub/tile	3-10	Hole filling, indoors	Excellent	Poor	Usually none	Nonmoving to 1/4" wide	1/2	3	Water	Best	White
Acrylic latex tub/tile	5-10	Tub-and-tile	Excellent	Fair	None needed	Nonmoving to 1/4" wide	1/4	1-2	Water	Best	White
Butyl rubber	7-10	Narrow openings in metal, glass, wood, masonry, paint	Very good	Fair	None needed	Nonmoving to 1/4" x 1/4"	1/2-1 1/2	7	Paint thinner, naphtha	Best	White, clear, gray, black, brown, redwood, beige, bronze, sandstone
Neoprene rubber	15-20	Driveways, wall cracks	Excellent	Good	None needed	All 1/8"-1/4" wide	1	30-60	MEK, toluene, xyline	If desired	Aluminum, gray, black
Nitrile rubber	15-20	Small joints in metal, masonry	Excellent	Poor	None needed	All 1/8"-1/4" wide	1	30-60	MEK, toluene, xyline	If desired	Aluminum, gray, black
Polysulfide rubber	20+	Anywhere	Excellent	Excellent	Special primer on all but metal	All 1/2"x1/2"	24-72	7	TCE, toluene, MEK	If desired	White, black, gray, limestone, bronze, brown
Hypalon rubber	15-20	Anywhere	Very good	Good	None needed	All to 1/2"x 1/2"	24-48	60-90	Toluene, xylene, xylol, toluol	If desired	White, gray, bronze, black
Silicone rubber	20+	Outdoor metal, heat ducts, shallow joints	Good, excellent with primer	Excellent	Porous surfaces	All from 1/4" deep	1	2-5	Paint thinner, naphtha, toluol, xylol	Read label	White, black, clear, gray
Silicone rubber tub/tile	20+	Tub-and-tile	Good	Excellent	None needed	All from 1/4" deep	1	2-5	Paint thinner, naphtha, toluol, xylol	Read label	White, blue, pink, green, yellow, beige, gold
Urethane	20+	Anywhere. Pourable for horiz. joints	Excellent	Excellent	None needed	All to about 1/4" x 1/2"	24	4-14	MEK, acetone, lacouer thinner	If desired	White, gray, black, limestone, bronze; special colors when available
Weather strip/ caulking cord	To 20	Temporary draft-sealing and hole-plugging	None	Excellent	None needed	Nonmoving	--	None	Not sticky	No	Clear, gray

[1]"Porous" includes wood, wood products, concrete, and brick.
[2]MEK = methyl-ethyl-ketone, TCE = trichloroethylene

Silicone rubber

Silicone rubber is a high-performance flexible caulk, widely available in cartridges and tubes. It cures by chemical reaction to moisture in the air. It is the best-performing caulk.

It has three disadvantages:

- It is difficult to apply—tooling quickly before the bead begins to skin over.
- It is expensive.
- Some silicone caulks will not take paint.

Hypalon

Hypalon is similar to polysulfide rubber. Although it shrinks more, it is easy to run into cracks and joints and to tool smooth. It is also moderately priced.

Urethane

Urethane bonds with any surface without priming. It flows into the joint and smoothes easily. Urethane weathers well. It is nearly the ideal caulk.

The nonsagging type is for most sealing, including vertical joints; the pourable, self-leveling type is for horizontal cracks and joints. The one-part type is suitable for most applications, but use the two-part sealant where rapid and continual joint movement occurs.

A urethane caulk bead is easy to smooth out with a wetted finger. It is nontoxic. A dry cloth removes it from your hands if wiped immediately.

Weather strip/caulking cord

This is a one-of-a-kind, oil-free, nonshrinking plastic caulk made without solvents for easy application to nonmoving joints, both temporary and permanent. Adhesion of weather strip/caulking cord is low, so it should be used only where it will be held in place by the material on either side. Applied around storm windows or window air conditioners, it peels off clean at the end of the heating or cooling season. The cord collects dust so it should not be used where an attractive appearance is desired.

Special purpose

Many products on the market are made for one or two special jobs. Some of these are:

- Cement patching compound comes in cartridges like other caulks but is designed for filling horizontal or vertical cracks in concrete and masonry. It often has a butyl rubber base and is concrete gray or black in color.
- A low-cost asphalt-base product is available for sealing roof leaks. It is much like flowable roof cement but is much easier and neater to use. It should not be used on new work—once you use it no other material will stick.

Application

Follow the label instructions. Most caulks must be applied within a certain temperature range. Some caulks require a wet, a dry, or a primed surface.

For most applications, using a caulking gun and cartridge is easiest, cleanest, and most economical. The label gives proper instructions on how to cut the tip at the correct location and angle for your particular job.

In general, do not tool finish caulking if you can help it. Attempts at finishing usually result in a rough finish, smeared surfaces, and dirty tools. With a little practice, you can obtain a very good looking bead directly from the tube and further finishing with a tool will not be necessary.

Before applying any caulk or sealant, prepare the surfaces properly:

- Remove all loose, rotten, or rusty material.
- Remove dirt, dust, and rust.

Many caulks may have different or additional surface preparation requirements—**read the label.**

Table 211-2. Caulk joint coverage.

	Joint width, in.			
	1/8	1/4	3/8	1/2
Joint depth (in)	lineal feet per cartridge			
1/8	123	62	41	31
1/4	62	32	21	15
3/8	41	21	14	10
1/2	31	15	10	8
5/8	25	12	8	6
3/4	21	10	7	5

212 PAINTS AND STAINS

Paints, preservatives, and stains are all used on structures and equipment. Paints and enamels that contain a considerable amount of pigment (coloring) form a surface film subject to cracking and peeling. Paints and enamels are used on masonry, metal, wood, and wood products. Preservatives and semi-transparent stains that soak in and do not form a surface film are used only on wood and wood products. With no surface film, these products cannot peel as do paint and enamel. Clear finishes, such as varnish, that form a surface film are generally not as durable as paint containing pigment for color.

In selecting a durable finish system, consider the type and characteristics of the surface to be finished, the exposure and moisture conditions, and the characteristics of the finishing material.

Follow label directions. Follow the manufacturer's instructions for each material used—precautions, preparation, mixing, applying, drying, etc.

Much of this chapter defines material and ingredients. Three general terms are:

A **primer** is applied as the first coat of paint. Its main function is to provide a good bond between the topcoat and the surface to be painted and to prevent the topcoat from penetrating the surface producing an uneven gloss. It provides resistance to water and can help fill very small cracks and smooth the surface.

The **topcoat** produces the desired color, texture, and service characteristics you want. It is usually applied over a primer, although some topcoats are also used for the prime coat—read the label.

Gloss is luster or shine of a smooth painted surface. There are three degrees of gloss: shiny; semigloss; and flat, the least shiny. The lower the gloss the better the paint will hide surface irregularities. The following are some manufacturers' terms to indicate degree of gloss.

Gloss: high gloss, gloss, or ¾ gloss.
Semigloss: semigloss, eggshell, velvet, or satin.
Flat: dull, flat, or matte.

Solvent is the part of a paint that is volatile, i.e., that evaporates during drying.

Vehicle is the liquid portion—solvents, resins, and oils.

Latex paints are water soluble, and cleanup can be with soap and water. Most are water-base, but some have an alcohol solvent for lower viscosity and faster drying.

Many cleaners or washes are available for particular finishing jobs. Trisodium phosphate, if available, is a good general cleaner. Or mix 8 oz of chlorine bleach and 4 oz of a nonsudsing detergent per gal of water.

Types of Coatings

Interior primers

Alkali-resistant. A primer for previously unpainted plaster or interior masonry.

Alkyd. A blend of oils for all interior surfaces except paper-covered wallboard (drywall, sheet rock, gypsum board) or surfaces such as damp or partially cured masonry or plaster. It is slow drying, usually overnight before repainting.

Alkyd metal primer. A blend of solvents and corrosion-inhibiting pigments for metal surfaces. The pigments used depend on the type of metal to be covered.

Block filler. A relatively thick latex to fill the pores of masonry surfaces. It is best applied with a brush, but can be rolled or sprayed (airless). It can be applied to damp surfaces.

Cement grout. A thin mortar used to smooth rough concrete. Apply it with a brush or trowel. Grout is available as a powder for mixing or premixed.

Clear wood sealer. A sealer for the pores of wood without changing the wood's natural appearance. It is a quick drying first coat for floors if protected with one or more coats of varnish. It is usually used under a clear topcoat.

Enamel undercoaters. Primers for wood, plaster, or gypsum wallboard surfaces or previously painted surfaces. They are available in alkyd or latex types. They dry to hard, tight films that prevent enamel paints from penetrating into the surface.

Latex. A primer for brick, masonry, plaster, sheetrock wallboard, and some non- bleeding woods. It is quick drying and equipment cleanup is easy with soap and water.

Latex metal primer. A primer especially for metal. Metals may need to be penetrated. Protect the primed surface with a moisture resistant topcoat to prevent corrosion.

Primer-sealers or stain-blocking primers. Latex or alkyd primers to prevent water soluble color extractives in wood, particularly in western red cedar and redwood, from discoloring the topcoat.

Interior topcoats

Alkali-resistant enamel. Usually used to resist alkali on masonry surfaces. It can be a primer and topcoat. It can run on vertical surfaces and can bubble if applied to concrete floors with a roller.

Alkyd semigloss or gloss enamel. Usually used on primed plaster and wallboard, properly prepared wood trim and metal, and masonry. It is very washable and especially useful in kitchens and bathrooms. As an enamel, it has good gloss retention, grease and oil resistance, and better washability and resistance to abrasion than flat alkyd enamel.

Alkyd flat paint. A flat finish, almost free of sheen, as both a primer and topcoat on interior walls and ceilings of plaster, masonry, and similar surfaces. It is used in the same places as flat latex paint, but has better washability and abrasion resistance. It dries in about 4 hr and has little odor.

Alkyd floor enamel. A relatively fast drying enamel that is resistant to abrasion and impact. It is fairly resistant to water but not to the alkali on fresh concrete. Alkyd floor enamel blisters and peels if water comes through from behind the film on wood or masonry.

Fire-retardant paint. A specialized product for combustible surfaces such as wood where protection from fire is desired, such as furnace rooms or hallways leading to exits. It is usually a semi-gloss alkyd or latex paint. The paint expands at high temperatures to form a thick, insulating, non-flammable coat that retards flame spread. Some fire-retardant paints are susceptible to water. Prime and apply according to the manufacturer's instructions.

Latex semigloss or gloss enamel. An easy to apply and cleanup enamel for wallboard, wallpaper, wood, plaster, masonry, and metal. Advantages include rapid drying, low odor, the absence of harmful solvent fumes, and relatively good leveling. Do not use a latex enamel topcoat over a latex primer on metal because both coats are moisture permeable.

Latex flat wall paint. An interior paint for wallboard, wallpaper, plaster, and other porous absorptive materials. It is durable, has excellent coverage and good washability, dries quickly, and is easy to touch up. Water-base latex is safe to use and store and has practically no odor. It is often put over a sand or textured surface.

Latex floor enamel. Latex floor enamel has most of the advantages of latex wall paint plus fairly good abrasion resistance. It is useful in areas such as basements where ventilation may be poor and can be applied to damp but not wet surfaces. Standing water on a latex floor enamel may soften the film. Acid-etch concrete and metals for best results.

Urethane, epoxy ester, and polyurethane enamels and floor enamels. Two-part enamels mixed just before use, and with limited pot-life. They are difficult to use.

Moisture-cured urethane enamels require a specific relative humidity when applied. It is the most abrasion resistant of the three types. It is usually both the primer and topcoat.

Epoxy ester enamels are for high traffic areas requiring a wear-resistant coating and are sometimes both the primer and topcoat.

Polyurethane enamels are similar to urethanes.

Portland cement paint. A useful, low cost finish for rough masonry. It is used as a primer-topcoat. Dry portland cement powder is mixed with water just before application. The curing of the paint is caused by the reaction of the water with the portland cement powder. It must be kept damp until well cured. Portland cement paint does not make a good base for other types of coatings. It is not suitable for surfaces where abrasion is anticipated.

Portland cement paint for metal. An oil- or alkyd-based paint containing cement. Do not confuse it with dry portland cement paint for masonry. Portland cement paint is effective for use on galvanized steel.

Interior wood stains

Pigmented wiping stain. A stain to make woods, even different species, quite uniform in color. It can be used on floors and is available in many colors. It usually performs best on white pine and other softwoods.

Varnish stain. A varnish that stains and varnishes at the same time. It can be used on floors. A special protective topcoat may be needed. For large jobs, be sure all cans are from the same batch and mix them together if possible.

Interior clear finishes

Alkyd varnish. A good gloss finish for floors with light to moderate use if properly waxed and maintained. It is fairly resistant to stains and spots but shows scratches. Touch-ups or patched worn spots show.

Lacquer. A gloss finish less durable than varnish, but worn spots can be retouched with good results. Lacquer is difficult to apply unless sprayed because it dries rapidly. Use the right undercoat.

Moisture-cured urethane clear finish. A very abrasion-resistant finish for wood and ceramic tile floors. It must be applied when the humidity is at a certain level and is more expensive than a conventional clear finish.

Polyurethane clear finish. A highly abrasion-resistant finish for high traffic areas of wood or ceramic tile floors. It is more expensive and requires more care in application than a conventional clear finish.

Shellac. A transparent gloss finish with only moderate resistance to stains. It spots if liquids remain on it and is not alcohol resistant. When repainting with latex, seal stains (crayon, knot resin, grease spot) with shellac before retouching. It does not darken with age as quickly as varnish but does yellow and may remain tacky on some surfaces. It is quick drying and easy to apply.

Two-component epoxy and two-component urethane clear finish. Finishes for wood, concrete, and ceramic tile floors. Two parts are mixed before use and must usually be used within a few hours, before hardening can occur in the container. They are more abrasion resistant than polyurethane varnish and more expensive than conventional clear finishes.

Exterior surfaces

Paintable water-repellent preservative (WRP). A material that penetrates wood on the outside of houses and repels rain and dew. It helps wood resist decay, stain, fungus, swelling, shrinkage, and warping. It is relatively fluid. It leaves little or no film on the wood surface and does not protect against wear and abrasion.

Water-repellent preservative is an excellent pretreatment for wood to be painted. Reduce paint failure by brushing WPR over exterior windows, door, and trim, and into the joints prior to priming. Drying time depends on weather and the ingredients in the finish.

If you want to paint or stain over the preservative, see that the product label says the product you are purchasing is **paintable** WRP. **Silicone** water-repellent coatings are available, but are not paintable.

Paintable WRP is easy to apply by brush, roller, or dipping. It leaves the appearance of the wood relatively unchanged except for an initial slight darkening.

Where it is the only treatment applied to exterior wood, two coats are recommended. Best results are obtained when the first coat is applied to the back, face, edges, and ends of the wood before it is nailed into place. The second coat should be brushed over all exposed wood surfaces after installation.

The preservative's effectiveness lasts about 1 to 5 years. Darkening blotchy discoloration of the wood indicates that the treatment needs renewing.

Exterior primers

Alkali-resistant coatings. Topcoats on masonry walls or decks. These paints dry flat, the film chalks gradually, and they have higher moisture resistance than that provided by latex paints. They can be primers under less alkali-resistant topcoats. Apply two coats when used as a primer under alkali-sensitive topcoats. Their resistance to water, alkali, and acid is excellent, but their resistance to solvents is poor.

Alkyd-based metal primers. Primers that reduce corrosion on metal surfaces. Zinc dust primers are for galvanized steel. These primers are also available in an oil base. It is necessary to clean the surface more carefully when alkyd-based primers are used.

Alkyd primer. An opaque finish for wood. It is faster drying and harder than oil primers. It has greater resistance to bleeding, mold, and moisture.

Block filler. See Interior Primers.

Cement grout. See Interior Primers.

Knot sealer. An alternate to shellac or varnish to prevent resin from knots from bleeding through paint. It is good for knotty pine and plywood.

Latex metal primer. It is available in white, so it can be covered with one topcoat, usually within a couple of hours. Application and cleanup are easy. It gives good corrosion protection, but may show rust staining if not protected by a moisture-repellent topcoat.

Oil-base metal primers. Primers to prevent corrosion on most ferrous metal surfaces. Several pigments are available, including zinc chromate, zinc oxide, red lead, and iron oxide. Zinc dust primers are for galvanized steel.

Oil primer. This primer usually contains bodied oil to control penetration into wood.

Portland cement paint (dry). See Interior Topcoats.

Portland cement paint (oil). See Interior Topcoats.

Primer-sealers or stain-blocking primers. See Interior Primer.

Exterior topcoats

Acrylic clear finish. A rapid-drying lacquer for outdoor metal surfaces such as copper, brass, and bronze. It is easy to apply from a spray can because it dries rapidly.

Alkyd-based house paint. This paint dries quickly and is resistant to blistering. It has good durability and fair color retention.

Alkyd-based masonry paint. This masonry paint performs very well except when excessive moisture and alkali are present behind the film.

Alkyd trim enamel. An enamel for exterior wood, trim, windows, etc., and on properly primed metal. Brush application is easy. Drying time varies from a few hours to overnight. It has good retention of its gloss and bright colors. Apply it over an enamel undercoat on unpainted surfaces. It is not suitable where high resistance to acid, alkali, or other chemicals is required.

Aluminum paint, general purpose. A paint resistant to water and weather. It can be applied on new metal surfaces as a primer-topcoat system or as a topcoat over a rust-preventive metal primer.

Aluminum paint, special formula for wood. An excellent topcoat or a primer for latex or oil-base topcoats.

Barn paint. An oil-based paint for uniform appearance over poorly prepared and nonuniform surfaces. It has most of the characteristics of house paint. Red barn paint (non-oxide oil base) is one of the most durable paints for site application. Barn paint is also used on roofs, especially sheet metal, and on metal siding and trim. Most barn paint is red, although it is available in other colors.

Bituminous roof coatings. Asphaltic coatings for good weather resistance. Asbestos and other fillers prevent sagging on sloping roofs and permit applying thick coatings. Aluminum powder adds high reflectance for cooler temperatures inside the structure.

Colored aluminum roof coatings. Coating for house roofs in a limited range of colors with aluminum pigments added for heat reflectance.

Equipment or exterior enamel. Similar to alkyd trim enamel. Because it is used on metal, it may contain a rust-inhibitor.

Exterior oil-base metal paint. Some paints for exterior metal are still made with an oil base. They give good service and protection to structured steel but usually lose gloss earlier than similar alkyd products.

Gloss oil-base house paint. An easy to apply house paint with moderate flexibility for wood and metal.

Heat-resistant aluminum paint. The best are made with silicone resin—the more silicone, the greater the heat resistance. Substitutes for part of the silicone in some paints reduce heat resistance and durability. You cannot paint over this coating.

Latex house paint. A paint easy to apply, suitable for use on damp (not wet) surfaces, and easy to clean from equipment. It has good flexibility to avoid cracking on expanding or contracting wood. Latex house paints offer better color permanence than oil or alkyd paints and are more resistant to blistering or peeling. Most can also be used on masonry and primed metal. They are usually available in a low gloss to flat finish.

Latex masonry paint. This product is available in flat, semigloss, or gloss finish. Prime masonry with heavy chalk on the surface with a clear or lightly colored masonry surface conditioner.

Latex roof coating paints. Special latex paints are made for asbestos-cement, tile roofs, and roll roofing. Latex is not recommended for asphalt shingles, because it can cause the shingles to curl. Surface preparation and painting requirements can be difficult.

Latex trim enamel. A paint like latex house paint, available in a semigloss or gloss finish. The gloss may not be quite as high as the older alkyd trim gloss enamel, but the gloss is usually maintained much better. It can be used on any properly primed metal, but because it reacts to some solvent cleaners, it is not generally recommended for galvanized steel.

Exterior wood stains

Semi-transparent stain (penetrating stain). A latex or oil-base stain. A small amount of pigment changes the color of wood but does not hide its grain or texture. It is usually available in natural and wood-tone colors.

Two coats are recommended on new wood. For oil base, apply the second coat while the first is still damp. For latex, apply the second coat after the first is dry. Use a brush, roller, pad, or sprayer. The stain leaves a flat finish that will not peel.

The finish is worn away by the weather. Refinish when part of the wood shows through—one coat is usually enough.

Solid color stain (opaque stain). A latex or oil-base stain with more color and higher hiding power than a semi-transparent stain. It is more like paint and obscures the natural color of the wood.

It comes in many colors, from natural wood tones through pastel to deep tone colors. The latex type is easy to clean up, is slower wearing, and has better color retention. Light-colored latex stains can let the color extractives in the wood appear as a mottled red stain.

Usually only a single coat is applied; two coats may provide better and longer service or may lead to peeling. Brush application is best.

Exterior clear finishes

Clear finishes have a much shorter life than paint or stains. Ultraviolet rays from the sun penetrate clear coatings, changing the surface under the finish. The coating loses adhesion, causing discoloration, cracking, and peeling of the finish—perhaps in less than 2 yr. In time, the deteriorating finish must be completely removed. At least three coats of a clear wood finish are recommended.

Set up regular and strict maintenance for clear coatings on exterior wood. Apply a new clear coat before the old coat begins to deteriorate.

Alkyd varnish. A clear finish for outdoor walls, doors, and trim. It has moderate resistance to abrasion and wear.

Marine spar varnish. A finish for outdoor walls, doors, and trim. It has moderate resistance to abrasion and wear. If exposed to weather, it may fail in less than 2 yr if it is not repainted at least every 12 to 16 months.

Silicone water-repellent coating. A coating for masonry that repels water without changing the masonry surface appearance. Because it is transparent, use the recommended application rate on the label as your guide to apply the proper amount of coating. Water repellency will be less than expected with an inadequate film thickness. This coating cannot be topcoated until it has weathered for several years. No primer is required.

Urethane varnish. A clear, highly abrasion-resistant varnish for decks, porches, and other high traffic areas.

Matching the Finish and the Surface

Table 212-1 indicates the suitability and estimated life in years of some finishes on different types of wood and wood products.

The order of the finishes in the table, from left to right, is also the order in which they can be applied to wood. For example, a penetrating stain can be applied over a water repellent preservative. Paint can be applied over an oil-base penetrating stain finish. However, before painting, wipe off the previously stained surface with paint thinner to remove wax in the stain.

Table 212-2 and 212-3 give general finish recommendations for various surfaces.

Painting Wood

Steps for painting new wood

1. Water-repellent wood preservative (WRP): apply paintable water- repellent preservative with penta and wax to all bare wood and wood joints before priming.

Use on windows in new construction that have been treated by the manufacturer. Treat untreated lumber and the cut ends of treated lumber by brushing or dipping in a WRP solution; treat lap and butt joints of trim and siding liberally.

Let WRP dry for at least two warm, sunny days before painting with oil-base primer. The paint may develop a wrinkled pattern if painted too soon.

2. Apply the appropriate primer: apply the primer heavy enough to hide the grain of the wood. Use alkyd-oil base primers that are more resistant to moisture than latex primers on redwood and cedar that are susceptible to extractive staining.
3. Topcoats: apply **two topcoats** over the primer, especially on fully exposed areas. One topcoat may last 3 yr: two topcoats of good quality paint can last as long as 10 yr.

Follow the sun around the house if possible, painting the surfaces after the sun has warmed them.

Avoid painting late in the evening on cool spring and fall days when heavy dew is likely to form. The moisture will probably cause water marks on latex paint and may cause wrinkling and loss of gloss on oil-base paints.

Repainting

Repaint wood only when the old paint has weathered so much that it no longer covers or protects the wood. Often only the south and west sides of a building need to be painted. Use a putty knife or wood chisel to remove paint from around areas where spots of paint have peeled and exposed bare wood surfaces. Then use sandpaper to feather the broken paint edges around the bare spot.

Clean the old paint with trisodium phosphate or 8 oz bleach and 4 oz detergent per gal of water to remove chalk, dirt, or other contaminants that may keep the paint from sticking to the surface. Roughen glossy and unweathered surfaces with steel wool and wash them before repainting. **Rinse thoroughly.**

Clean the surface and then apply alkyd-oil base primer to severely weathered or chalked oil-base paint before applying latex paint—latex does not penetrate a chalky surface. Peeling between coats of paint is often due to inadequate preparation of old surfaces.

Treat bare wood with water-repellent wood preservative, as when painting new wood. WRP can be spread on the adjoining old paint. Spot prime the bare wood with housepaint primer before painting.

Table 212-1. Suitability and estimated life of finishes.

a. The first coat of water-repellent preservative (WRP) lasts 2 yr or less. When the surface starts to show spots of mildew resembling fly specks, treat the wood with another liberal brush or spray coat of WRP.
b. On smooth unweathered surfaces, apply only one coat. By the time the surface needs restaining, the wood will have weathered enough to absorb more stain, and the finish will last longer.
c. Apply two coats of penetrating stain less than 4 hr apart, or apply the second before the first coat has dried. This method allows both coats of stain to penetrate into the wood. If the first coat is dry before the second is applied, the surface is sealed, and the second coat cannot penetrate.
d. Applying a second topcoat will double the life of the paint job.
e. If the board is porous, the stain should penetrate well, and the finish will be durable.

	Water-Repellent Preservative Finish		Oil-Base Penetrating Stain Finish		Paint		
	Suitability	Estimated[a] life (years)	Suitability	Estimated life (years)	Suitability	Estimated life primer + 1 topcoat (years)	Typical cost
Wood							
A. Cedar and redwood lumber and bevel siding							
1. Smooth							
a. Bare	High	2	High	3 [b]	High	5 [d]	Moderate
b. Preprimed	-	-	-		High	5 [d]	Moderate
2. Rough	High	2	High	10 [c]	-	-	Moderate
B. No. 3 spruce or pine with knots (board and batten)							
1. Smooth	Moderate	1	High	3 [b]	Low	3 [d]	Low
2. Rough	High	2	High	10 [c]	-	-	Low
C. Cedar shingles and shakes	High	2	High	10 [c]	Low	3 [d]	High
Wood Products							
A. Exterior plywood							
1. D. fir and pine							
a. Smooth	High	2	Moderate	2 [b]	Low	3 [d]	Low
b. Textured, brushed, etc.	High	2	Moderate	2 [b]	Low	3 [d]	Moderate
c. Rough sawn	High	2	High	10 [c]	-	-	Moderate
d. Factory stained	-	-	-	2 to 5	-	-	Moderate
2. Cedar, redwood and lauan							
a. Smooth	High	2	High	3 [b]	Moderate	4 [d]	Moderate
b. Rough sawn	High	2	High	10 [c]	-	-	Moderate
c. Factory stained	-	-	-	2 to 5	-	-	Moderate
d. Textured, brushed, etc.	High	2	High	3 [b]	Moderate	4 [d]	Moderate
B. Medium density paper overlay/plywood	-	-	High	10 [c]	High	5 [d]	Moderate
C. Medium density hardboard siding materials							
1. Smooth or textured							
a. Base or untreated	-	-	Low to High	2 to 10 [e]	High	5 [d]	Moderate
b. Preprimed	-	-	-	-	High	5 [d]	Moderate
c. Prestained	-	-	-	2 to 10 [e]	-	-	Moderate

Table 212-2. Interior paints and finishes.

✓• Black dot indicates that a primer, sealer, or fill coat may be necessary before the finish coat, unless the surface has been previously painted.

	Semi-gloss latex	Flat paint - latex	Flat paint - alkyd type	Semi-gloss alkyd	Gloss enamel - alkyd	Rubber base paint (not latex)	Interior varnish	Shellac	Wax (liquid or paste)	Wax (emulsion)	Stain	Wood sealer	Floor varnish	Floor paint or enamel	Aluminum paint	Sealer or undercoater	Metal primer	Cement base paint	Clear urethane	Catalyzed enamel
Dry walls	✓•	✓•	✓•	✓•	✓•			✓				✓				✓				✓•
Plaster walls & ceiling	✓•	✓•	✓•	✓•	✓•	✓										✓				✓•
Wall board	✓•	✓•	✓•	✓•	✓	✓										✓				✓•
Wood paneling	✓•	✓•	✓•	✓•	✓•	✓	✓	✓	✓		✓	✓							✓	✓
Kitchen & bathroom walls	✓•	✓•		✓•	✓•	✓										✓				✓
Wood floors						✓		✓	✓	✓•	✓•	✓	✓•	✓•					✓	✓
Concrete floors						✓			✓•	✓•				✓					✓	✓
Vinyl & rubber tile floors									✓	✓										
Asphalt tile floors										✓										
Linoleum								✓	✓	✓			✓	✓						
Stair treads						✓		✓			✓	✓	✓	✓					✓	✓
Stair Risers		✓•	✓•	✓•	✓•	✓	✓	✓			✓	✓							✓	✓
Wood trim	✓•	✓•	✓•	✓•	✓•	✓	✓	✓	✓		✓					✓			✓	✓
Steel windows	✓•	✓•	✓•	✓•	✓•	✓									✓		✓			✓
Aluminum windows	✓•	✓•	✓•	✓•	✓•	✓									✓		✓			✓
Window sills	✓•	✓•			✓•	✓	✓													✓
Steel cabinets		✓•	✓•	✓•	✓•	✓											✓			✓
Heating ducts	✓•	✓•	✓•	✓•	✓•	✓									✓		✓			✓
Radiators & heating·pipes	✓•	✓•	✓•	✓•	✓•	✓									✓		✓			✓
Old masonry	✓	✓	✓	✓	✓	✓									✓	✓		✓		✓
New masonry	✓•	✓•	✓•	✓•	✓•	✓										✓		✓		✓•

Table 212-3. Exterior paints and finishes.

✓• Black dot indicates that a primer, sealer, or fill coat may be necessary before the finish coat, unless the surface has been previously painted.

	Ext. masonry paint - latex	House paint - latex (wood)	House paint - oil or alkyd	Transparent sealer	Cement base paint	Exterior clear finish	Aluminum paint - exterior	Wood stain, oil or latex	Roof coating	Roof cement	Asphalt emulsion	Trim paint	Awning paint	Spar varnish	Porch-and-deck enamel	Primer or undercoater	Metal primer	Latex types	Water repellent preservative
Clapboard siding		✓	✓•				✓	✓								✓		✓•	✓
Brick	✓		✓•	✓	✓		✓									✓		✓	
Cement & cinder block	✓•		✓•	✓	✓		✓									✓		✓	
Asbestos cement	✓		✓•													✓		✓	
Stucco	✓		✓•	✓	✓		✓									✓		✓	
Natural wood siding & trim						✓		✓						✓					✓
Metal siding		✓•	✓•				✓•					✓•					✓	✓•	
Wood frame windows		✓•	✓•				✓	✓				✓•				✓		✓•	✓
Steel windows		✓•	✓•				✓•					✓•					✓	✓•	
Aluminum windows		✓•	✓•				✓					✓•					✓	✓•	
Shutters & other trim		✓•	✓•					✓				✓•				✓		✓•	✓
Canvas awnings													✓						
Wood shingle roof								✓											✓
Metal roof		✓•	✓•														✓	✓•	
Coal tar felt roof									✓	✓	✓								
Wood porch floor															✓				
Cement porch floor															✓			✓	
Copper surfaces														✓					
Galvanized surfaces		✓•	✓•				✓•					✓•		✓			✓	✓•	
Iron surfaces		✓•	✓•				✓•					✓•					✓	✓•	

Painting Metals

Iron Base Metals

Painting metals successfully involves surface preparation, priming, and topcoating.

On new metal remove all dust, dirt, grease, and oil. When possible, use cleaners and preprimer treatments recommended by the primer manufacturer.

Remove loose and flaking paint from previously painted metals. Some paint and primer manufacturers recommend complete removal of all existing paint before applying their products. When possible, use a cleaner, primer, and topcoats made by the same manufacturer. Paint only when the surface is clean and dry.

Select a high quality primer intended for use on metal products and compatible with the topcoats. Apply enough primer to cover the metal, but avoid runs or sags. Allow recommended drying time.

Tractor and implement paint, enamel or lacquer, exterior enamel, and exterior house paint can all be used. Water-base or latex paints are less resistant to moisture than oil-base paints and depend on the primer to prevent rust. For interior use, interior paints and enamels can be used as topcoats over metal primer.

Galvanized Metal

Galvanizing covers the base steel with zinc. The zinc resists oxidation and helps protect the steel from rusting. However, the zinc eventually weathers away, the steel is exposed, and rusting starts. Painting galvanized metal helps maintain the zinc coating and extends the life of the metal.

If a new galvanized surface is to be painted immediately, pretreat it to the primer manufacturer's directions. Apply metallic zinc primer after the metal has been pretreated or has weathered 3 to 4 years.

Paint a galvanized surface before any rust appears or at the first sign of rust for best results. If galvanized metal is badly rusted, remove the loose rust and apply two coats of paint.

Metallic zinc dust paint weighs 20 lb or more per gal. Although relatively expensive, it covers 400 to 500 ft^2/gal and is economical when the extended life of the metal is considered.

It is usually battleship gray and may be used as a primer and a topcoat. To change color, apply a topcoat of exterior paint.

Wash off any grease or oil film according to label directions. Remove rust with a wire brush, sandpaper, or sandblasting.

Priming

Apply metallic zinc dust paint with a brush, roller, or sprayer when the temperature is above 40 F. Stir zinc paints often.

Aluminum

Wash bare aluminum with aluminum wool and a solvent wash. Avoid steel wool. Rinse thoroughly after washing.

Prime bare aluminum with zinc yellow primer. Apply enough to hide the aluminum, but avoid runs and sags.

Topcoat the primer with two coats of paint or machinery enamel.

For repainting, clean and rinse as described for bare aluminum. Lightly sand glossy areas with 100 to 150 grit open coat sandpaper for good paint adhesion. Glossy spots on buildings are likely to be in areas protected by eaves, overhangs, or other shade. Rinse or wipe off the sanding dust.

Occasionally chemical blisters form under the paint from engine exhaust fumes or other chemicals. Sand these areas with 80 or 100 grit sandpaper and then with 150 grit sandpaper to remove the paint from the blistered area. Remove both sanding dust and old chemicals.

If the primer is in good condition, the surface can be repainted with any paint similar to that originally used.

Paints for Masonry

Asbestos cement board

Wash all surfaces to be painted with trisodium phosphate or other nonsudsing, nonabrasive cleanser; rinse. Remove loose sand-like particles from weathered areas with a wire brush.

Select a system for **porous, paint-free** asbestos board recommended by one manufacturer and use his products for both paint and primer. Do not mix brands.

If asbestos cement siding has paint in good condition, remove excessive chalk with trisodium phosphate solution or bleach and detergent. Sand or steel wool glossy areas lightly for better adhesion. High quality latex house paint without primer should perform satisfactorily, if the surface is free of excessive chalk.

Apply two topcoats, especially on fully exposed areas, such as the south, east, and west sides of a building. One topcoat over the primer may last about 3 yr, but a two topcoat job with good quality paint may last 10 yr.

Concrete, brick, stucco and block

Select a filler, sealer, primer, and topcoat system produced by one manufacturer as needed for the condition of the surface. Latex paint is the most common topcoat.

Interior

Indoor finishing is similar to exterior, except useful topcoats include flat latex, latex enamel, oil or alkyd enamel, and epoxy. The two component epoxy system may be the most expensive but is the most durable, most resistant to stain, and the easiest to clean. Flat latex may be the least stain resistant and the first to wear away from vigorous cleaning.

Paint Problems

Paint failures are usually caused by moisture, improper surface preparation, or inappropriate or low quality materials.

Moisture

High moisture problems can be caused by:

- Inadequate vapor barrier.
- Inadequate ventilation in high moisture rooms and buildings.
- Seepage or leakage from roof, eaves, or plumbing.
- Ice dams in gutters causing backup under shingles and into the overhang and soffit.
- Water running behind siding through poorly caulked seams, especially gaps around windows and doors.

To reduce paint failure from moisture problems:

- Provide adequate ventilation, especially in high moisture areas such as bathrooms, kitchens, livestock buildings, etc.
- Add a paint vapor barrier on the interior of exterior walls. Two or three coats of vapor resistant coating reduce the moisture entering the wall.
- Check for roof leaks, defective flashing, and holes in gutters and downspouts. After a heavy rain, check for water standing in gutters, spouts, or flashing.
- Check caulking, especially the vertical joints.

Mildew

Mildew looks like dark spots of dirt. It may appear on surfaces partly or fully protected from sun, especially overhangs or where trees or shrubs are close to walls.

Do not paint over mildew. Wash immediately before repainting with trisodium phosphate mixed with household ammonia and water or with bleach and detergent. Rinse with clean water. Then paint with a latex paint or one with zinc oxide and a mildewcide.

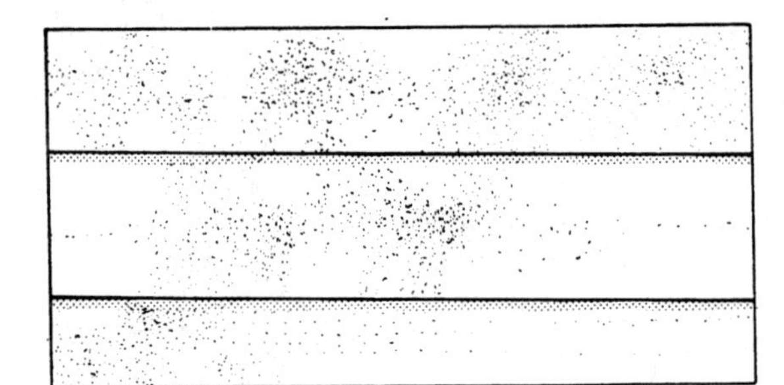

Fig 212-1. Mildew.

Wood pigment stains

Common building lumber has natural water soluble pigments. When moisture passes through the wood, it dissolves the pigment and carries it to the surface. The moisture bleeds through latex or breaks the paint film and dries out the surface, leaving a brown stain.

Before painting, locate and correct moisture source beneath the siding or in the building. Remove the stain with diluted chlorine laundry bleach. Repair breaks in the old paint carefully and sand smooth. Prime sanded areas and repaint.

Blistering

Blistering is usually caused by moisture coming through the wood and evaporating below the paint film. See Moisture. Before painting, correct the moisture problem.

Scrape off loose paint. Sand the bare wood and several inches around where the blister or peeling occurred. Dust and clean the area carefully. Apply water-repellent preservative. Prime the surface, then topcoat.

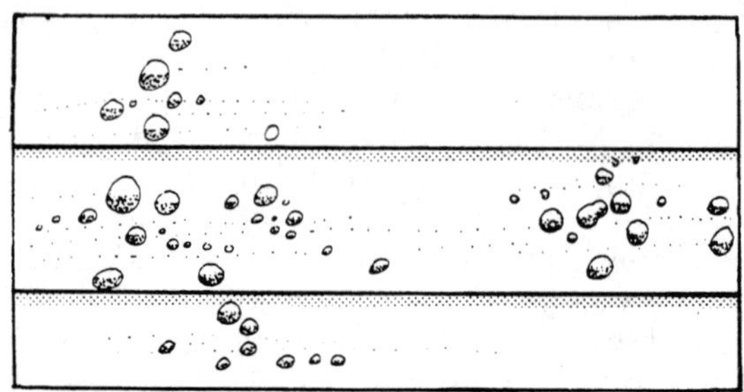

Fig 212-2. Blistering.

Peeling

Peeling is caused by:

- Applying the paint on greasy or oily surfaces.
- Inadequately prepared and primed surfaces.
- Applying oil-base paint over damp surfaces.
- Applying paint on woods with poor paint-holding characteristics.

Scrape and sand surfaces, apply WRP and primer, and apply a good quality paint.

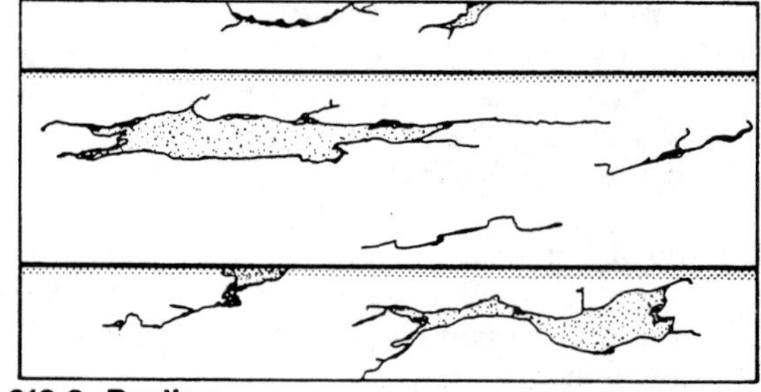

Fig 212-3. Peeling.

Checking

Checking is a network of small hairline cracks. They are parallel lines, in the shape of bird footprints or in small squares on the painted surface. They often extend down to bare wood and usually flake off as the paint ages and weathers. This condition is common on plywood veneers and is usually caused by expansion and contraction as the paint and wood get older.

Checking may be caused by:

- Applying oil paint over a damp surface or over a solvent or cleanser residue.
- Using an excessive amount of paint.
- Improperly mixing paint.

With a putty knife, determine if the paint is generally loosened enough to be removed entirely. If practical, scrape, sand, or use a paint remover. Sand the area smooth. If only partly removed to bare wood, blend the bare and sound areas together with fine sandpaper and fill deep cracks and joints with good-grade caulking compound. Remove dust and thoroughly clean the surface.

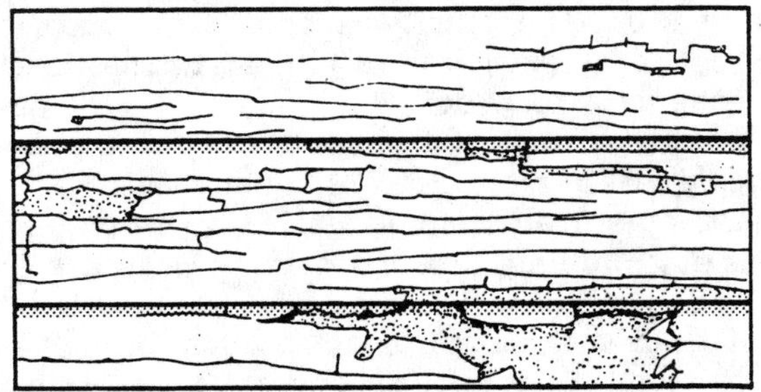

Fig 212-4. Checking.

Cracking and alligatoring

This problem is usually caused by improper primer and finish coat combination, or a thick coat over a soft or incompletely dried primer or WRP. It is also caused by multiple coats that are too thick, resulting in the underlying coats not drying completely. Cracks may extend down to the wood.

Before painting, remove the entire checked or alligatored surface and all paint that is not adhered. Sand surface smooth and blend with the surrounding sound areas. Dust, clean, and rinse. Spot prime bare wood or apply a general primer coat suited to the chosen finish. Topcoat only after allowing full drying time required by the primer.

Fig 212-5. Cracking and alligatoring.

Excessive chalking

Excessive chalking is most noticeable when it runs and stains other surfaces such as brick, masonry, or contrasting paint finishes. It can be caused by applying paint in light rain, fog, or mist or by applying the paint too thin.

Before painting, remove the chalk with a stiff bristle brush. Repaint with a nonchalking paint, or use one with a controlled chalking rate if it is desirable to maintain clean surfaces.

Stains

On rough-sawn or severely weathered surfaces, apply stain as soon as practical after installation or dip the wood before installing it.

On smooth lumber, apply the stain immediately, or wait through a summer before staining to allow checks and cracks to develop. Weathered wood absorbs more stain so the finish lasts longer.

Treat hard, dense woods (Douglas fir and southern pine) with a WRP. Allow the wood to weather for a year to develop cracks and checks, then stain.

On medium density hardboard, stain as soon as possible after installation.

In the Midwest, use an oil-base semi-transparent penetrating stain with the ingredients in about these percentages:

Midwest:
- Boiled linseed oil or oil-base varnishes—50%
- Mineral spirits paint thinner—15%-18%
- Pigments—6%-8%

Warmer and more humid climates:
- Boiled linseed oil or oil-based varnishes—about 30%
- Mineral spirits paint thinner—about 50%

Include a fungicide for mildew control and a small amount of wax (paraffin), which makes the stain water repellent.

One gal of stain covers 400 to 500 ft^2 with one coat on smooth surfaces. One gal covers 200 to 250 ft^2 with two coats on rough surfaces.

213 PLASTICS

Plastics are synthetic materials produced by the chemical industry. They contain basic organic elements—carbon, oxygen, hydrogen, and nitrogen—and will flow under heat and/or pressure.

American Society for Testing Materials (ASTM) defines plastics as "materials that contain as an essential ingredient an organic substance of large molecular weight, are solid in their finished state, and at some stage in their manufacture or processing into finished articles can be shaped by flow."

Characteristics

There are two types of plastics—thermoplastics and thermosets. Thermoplastics soften when heated and harden when cooled. Thermoplastics act like metal or wax. They can be reshaped many times. Polyethylene and PVC are thermoplastics. Thermosets are similar to concrete, because once they are set, they cannot be made to flow again. Melamine is a typical thermoset.

There are about 40 different commercially produced plastic families. A plastic may be soft (polyurethane foam); hard (melamine); clear (acrylic); heat-resistant (silicone); or heavier than iron (lead filled epoxy). There are hundreds of plastic types within the 40 families. It is difficult to make a choice for a particular application. ASTM standard test data or tests under actual use conditions are needed before final plastic selection.

Many plastics have desirable characteristics for the building trades. They can have virtually any finish or color; be transparent or opaque; can be molded, extruded, formed, or machined into intricate parts economically; and they combine with many other materials. Different plastics resist weathering, corrosion, abrasion, moisture, temperature, or fire. Plastics are available for insulating against electricity, heat, or sound. Some have high weight to strength ratio and most have low density. A range of hardness, flexibility, and creep characteristics is available.

Fillers for Plastics

Unless used in excessive amounts, fillers usually improve resin or composite properties. Fillers are nonreinforcing additives. They are inert extenders that displace part of the resin and serve the same purpose as air in a rigid foam.

Fibrous Reinforcements

Fibrous glass is the principle plastic reinforcement. Reinforcing fibers increase stiffness, strength, impact resistance, heat-deflection temperature, hardness, and dimensional stability. They decrease the coefficient of thermal expansion and change the electrical properties.

Asbestos fibers are used to reinforce PVC tile. "Glasboard" is a polyester-styrene copolymer reinforced with random chopped glass fiber.

Flame Retardants

Most plastics produce gases and smoke when heated by a fire. The toxicity depends on the type of fire and chemical composition. Cellulose nitrate plastics are the most hazardous.

In the past, fire resistance has been measured by small scale ASTM tests. Plastics passing these tests often burn vigorously in an actual fire. New testing methods that simulate a room corner are being developed.

Most flame retardants contain antimony, bromine, chlorine, nitrogen, phosphorous, or boron. Antimony trioxide is the most common, generally combined with bromine or chlorine containing compounds.

Rigid urethane foam insulation is usually flame-retarded by phosphorous-containing compounds.

General Properties and Applications

ABS Copolymers

ABS (Acrylonitrite-Butadiene-Styrene) is used for shoe heels, safety helmets, golf club heads, and waste, vent, and structural pipe and fittings.

ABS has high gloss and good rigidity, impact strength, hardness, chemical and moisture resistance, and electrical insulation properties. They are tougher, less subject to solvent attack, and resist creep rupture better than polystyrene. They ignite readily, soften and bubble while burning, and generate dense black smoke and clumps of airborne carbon. Cost is low.

Acetal Copolymers

Acetal copolymers are used in plumbing components, glass, auto parts, electronic equipment, and chromeplated plastic parts.

Acetals are tough, stiff, and have excellent impact resistance. They also resist hot water, organic solvents, inorganic salt solutions, lubricants, and hot air. They have hydrolytic stability and are one of the most creep resistant crystalline thermoplastics. Other properties are: fatigue endurance, dimensional stability, abrasion resistance, and excellent electrical properties.

Acrylics

Acrylics (methyl methacrylate polymers) are used for internally illuminated signs, transparent window glazing, adhesives, lacquers, and paints.

Unmodified acrylics are transparent, extremely stable against discoloration, and have desirable structural and thermal properties. Acrylics have excellent resistance to weathering, breakage, and chemicals. They are lightweight and combustible. They ignite readily and soften but do not usually drip. Acrylics have large coefficients of expansion but are not deformed by solar heat absorption. Acrylics are unaffected by alkalies, nonoxidizing acids, oils, grease, and household cleaners but are attacked by alcohols, strong solvents, and oxidizing agents.

Amino Resins

Amino resins are thermosetting urea formaldehydes and melamine formaldehydes.

Molded urea compounds are used for electrical wiring hardware. Melamine is used for dinnerware, wall panel and furniture laminates, adhesives, and binders for particle board and plywood.

Amino resin surfaces are resistant to moisture, oils, and solvents. They are tasteless, odorless, self extinguishing, good electrical insulators, and resist scratching and marring. Melamines are more resistant to chemicals, heat, and moisture than ureas. Urea formaldehyde is difficult to ignite. Melamine burns with a yellow flame. Both swell and crack as they burn.

Epoxy (thermosetting)

Brominated epoxies are fire-resistant. Cycloalphatic epoxies are used in electrical applications. Other uses include adhesives, electrical coatings, grouting compounds, and industrial maintenance coatings. Epoxy esters are used as baked on metal primers and chemically resistant floor paints.

PFA

PFA (perfluoroalkoxy, known as Teflon) is used for chemical liners in pumps, valves, pipes, and fittings. It is also used for electronic components, electrical wire insulation, and cookware liners.

PFA has good chemical resistance, anti-stick properties, a low coefficient of friction, good electrical characteristics, low smoke production, good flammability resistance, the ability to perform in temperature extremes, and good weatherability.

Phenolic Resins

Phenolic resins are in auto distributor housings, electrical plugs, and heat-resistant utensil handles. Some are adhesives for exterior plywood. Paper with phenolic resins and printed with wood grain is used for furniture and wall paneling. They are binders for particle board, sand paper, and grinding wheels. Phenolic resins can be foamed for thermal insulation.

Most phenolic resins are thermosetting. They are a combination of phenol and an aldehyde (e.g. formaldehyde). There are three basic groups: molding compositions; cast pehnolics; and adhesives and laminating solutions. Most phenolic base products have a high degree of fillers such as wood flour.

Polyethylene

High density polyethylene is used for corrugated agricultural piping, gas distribution piping, houseware, and pallets. Low density polyethylene film is used for packaging, vapor barriers, and insulation.

Polyethylene is manufactured in low, medium, and high densities. Desired toughness, resistance to acids, alkalies, and solvents, dielectric properties, and weatherability can be built in with additives. Polyethylene ignites readily, develops little smoke, and smells like paraffin.

Polyvinyls

Construction uses include flooring, pipe, conduit, pipe fittings, panels, siding, rainwater systems, windows, and trim.

Polyvinyls are thermoplastics derived from vinyl chloride, vinyl acetate, or vinylidene chloride. Rigid vinyls are low cost and can be designed to provide flame retardance, high modulus strength and toughness, good electrical and chemical resistance, water and gas impermeability, and weathering properties.

Foamed Plastics

Table 213-1 lists the engineering properties of foamed plastics used in agriculture.

ABS Foam

ABS foams are used as wood and metal replacements because of better strength-to-weight ratios and sound and thermal insulation properties.

ABS foams are similar to solid ABS. They have design flexibility, light weight, and rigidity. Standard and flame retardant grades are available.

Phenolic Foam

Phenolic foam insulation is used where fire resistance and low smoke production are important—roof decking, sandwich panels, and truck and trailer insulation.

Phenolic foam is a rigid, thermoset, cellular material. It has low flammability and flame spread characteristics. Phenolic foams are open cell products, but they have thermal insulating qualities comparable to other insulating materials.

Polyethylene Foam

Polyethylene foam has many uses in building construction. It is used for acoustical, insulating, and cushioning underlayment for vinyl and wood flooring.

Table 213-1. Foamed plastics properties.

PROPERTY	ASTM TEST METHOD	ABS- INJECTION MOLDING TYPE PELLETS	POLYETHYLENE- LOW DENSITY FOAM NONCROSS-LINKED SHEET	PHENOLICS FOAM IN PLACE				POLYURETHANE ISCYANURATE FOAMS POUR	POLYURETHANE ISCYANURATE FOAMS SPRAY	UREA FORMLDEHYDE BLOCK, SHRED, AND FOAM-IN-PLACE
1. Density, lb/cu ft.		40-56	2.1-3.3	1/3-1 1/2	2-5	7-10	10-22	1.5-3.0	2.0-3.0	0.8-1.2
2. Tensile strength, psi	D1623	2000-4000	35-100	3-17	20-54	80-130	-	25-75	20-60	Poor
3. Compression strength @ 10% deflection, psi	D1621	2300-3700 (1000 H)	3	2-15	22-85	158-300	300-1200	20-80	20-65	5
4. Maximum service temp, °F (dry)		176-180	160-180	-	Continued Service at 300°			300	300	120
5. Flammability Burning rate, in/min AEB in/ATB, sec.	D1692	-	2.2-3.0/31-57	-	-	-		-	-	-
6. Thermal resistivity (R_l value)	02326	1.72-0.48	3.57-2.94	4.76-3.57	5.0-4.55	4.17-3.57	-	9.09-5.88	9.09-5.88	5.56-4.76
7. Coefficient of linear expansion, 10^{-5} in/in-°F	D696	3.7-9.5	2.3	-	0.5	-	-	4	4	-
8. Water-vapor trans, Perm-in	C355	-	0.20- 0.40	-	2 lb/ft^3 5 lb/ft^3 2074g 1844g per day/m^2	-	-	3	3	28-35
9. Water absorption, % by	D2842							1.5	1.5	1.9

	ASTM TEST METHOD	POLYSTYRENE: PRODUCTS OR SHAPES MOLDED FROM EXPANDABLE BEADS OR FINISHED BOARDS		POLYSTYRENE: EXTRUDED BOARDS AND BILLETS				POLYURETHANE RIGID (CLOSED CELL): MOLDED PARTS: BOARDS, BLOCKS, SLABS: PIPE COVERING: ONE SHOT, TWO-AND THREE-PACKAGE SYSTEMS FOR FOAM IN PLACE, FOR SPRAY, POUR, OR FROTH-POUT TECHNIQUES				POLYURETHANE FLEXIBLE MOLDED (FOAM-IN- PLACE)
1. Density lb/cu ft.		2.0	1.0	5.0	1.5-2.0	2.0-2.6	2.0-5.0	1.3-3.0	4-8	9-12	13-18	1.2-20.0
2. Tensile strength psi	D1623	42-68	21-28	148-172	55-70	60-105	180-200	15-95	90-290	230-450	475-700	10-1350
3. Compression strength @ 10% deflection, psi	D1621	25-40	13-18	85-130	25-55 at 5%	25-60 at 5%	100-180 at 5%	15-60	70-275	290-550	650-1100	0.25-100 at 25%
4. Maximum service temp °F (dry)		165-175	165-175	165-175				200-250	200-250	250-275	250-300	150
5. Flammability Burning rate in/min AEB in/ATB, sec.	D1692	4.9 1.2-2.0 / 10-20(FR)	9.1 1.2-20/0-20(FR)	1.9 1.5 / 18(FR)	-	165-175 (aged)	165-175 (aged)	0.4-1.7 / 24-55	0.4-1.7 / 24-55	0.4-1.7 / 24-55	0.4-1.7 / 24-55	-
6. Thermal resistivity ft^2.h.°F/Btu.in (R value)	02326	4.35-4.17	4.0-3.85	4.07	40°F 4.76-3.45 70°F 4.35-3.85	40°F 5.88-5.26 70° 5.56-4.76	40° 5.88-5.26 70° 5.56-4.76	Blown with fluorocarbon: 9.09-5.88; Blown with CO_2: 4.35	Blown with fluorocarbon: 6.67-4.76; Blown with CO_2: 4.76-3.45	Blown with fluorocarbon: 5.26-40; Blown with CO_2: 3.22-2.86	Blown with fluorocarbon: 3.85-2.94; Blown with CO_2: 2.78-2.5	3.33 @ 2lb/cu ft density
7. Coefficient of linear expansion 10^{-5} in/in-°F	D696	3.0-4.0	3.0-4.0	3.0-4.0	-	3.0-4.0	3.0-4.0	4-8	4	4	4	-
8. Water-vapor trans, Perm-in	C355	0.6-1.2	1.2-3.0	0.4-0.6	-	0.3-1.1	0.3-1.1	0.6-4.0	0.9-2.0	-	-	-
9. Water absorption, % by	D2842	2-4	2-4	2-4	-	NIL to 1.0	NIL to 1.0	0.1-5.0	0.5	0.5	0.4	-

Low density extruded polyethylene foams are chemical, moisture, and solvent resistant; tough and flexible over a broad temperature range; electrical insulators; buoyant; and good thermal insulators.

Polystyrene Foam

Polystyrene foams are made with varying densities by different processing procedures. Expandable bead foams in the 1 to 2 psf range are used for packaging products, hot drink cups, and insulation board. Low density extruded polystyrene foams are stronger and have higher insulating properties than the expanded bead foams. Extruded foam from 5 to 20 psf is used for egg cartons, and meat and produce trays. High density structural foams, 25 to 50 psf, are used for window mouldings, frames, doors and door frames, and cabinets.

Urea-Formaldehyde Foam

Urea-formaldehyde foam is used primarily as insulating material. Urea-formaldehyde foam is a thermosetting plastic composed of urea-formaldehyde resin, air, and foaming agent. Expansion is mechanical rather than chemical and it sets in 5 to 120 sec. Before it sets it can be injected into forms or hollow walls. It drys from 2.5 psf to 0.75 psf in several days. Normal shrinkage is 7 to 15% which can be reduced to 2% with special formulations. Insulation and sound absorbent characteristics are good.

Urethane Foam

Urethane insulation boards have replaced polystyrene foam and fiber glass as insulation in refrigeration and cold storage. Foam insulation can also be sprayed on roofs, ceilings, walls, tanks, and pipes. It can be used for marine flotation.

Urethane varies from a tough hard solid used for insulation, molding, and structural applications to a very soft flexible material for cushions and bedding. Density varies from 0.5 to 60 psf. The heat insulation value of low density urethane is the highest of all available insulation materials. Polyurethane is combustible.

Isocyanurate Foam

Isocyanurate foam is used as low density, closed cell insulation in construction, refrigeration, transportation, industrial storage tanks, and marine applications.

They are second generation, isocyanurate-based foams that are chemically distinct from urethanes. They have superior thermal stability and improved flammability characteristics. Isocyanurate foam insulation has been produced with a flame spread rating of 25.

Vinyl Foam

Flexible foams are used for insulation, cushioned flooring, and weather stripping. Rigid foams are used as a wood replacement for trim, molding, and door frames.

Vinyl foam is produced by adding a gas to standard vinyl resin. PVC resins are used to produce both rigid and flexible foams.

300 CONSTRUCTION

301 HOW FARM BUILDINGS FAIL

In addition to the strengths of various building components (e.g. trussed rafters, pole embedment, and bracing), weaknesses or strengths also result from materials used, construction practices, or misuse of known design recommendations. Building failures caused by large snow and wind loads generally result from inadequate design or carelessness in construction. Failures are limited if an engineered plan is followed.

Truss Spacing

One obvious cause of failure is wide spacing of trusses, where trusses designed for 4′ o.c. spacing, for example, are installed 8′ o.c., cutting the roof strength in half.

Raised Lower Chords

Raising the lower chord in trusses to gain another 1′ or 2′ of head room is usually a mistake. The strength of the roof is limited to the strength of the upper chord from the plate up to the lower chord.

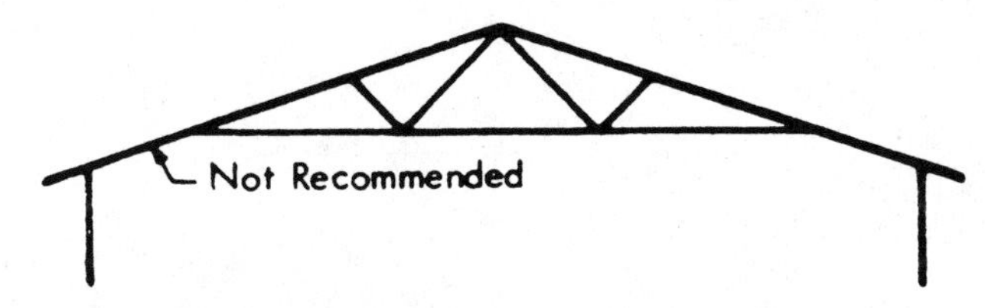

Fig 301-1. Beware of raised chords.

Raising the lower chord only 5″ above the plate line of a 40′ truss, can increase the upper chords from 2x8s to 2x12s. Raising the lower chord 2′ may increase a 2x8 upper chord to 3-2x12s!

Roof Load Transfer to Poles

In a 50′ wide building supporting a 25 psf roof load, each pole supports 5000 lb. Pole barn nails support about 160 lb each, and a ½″ bolt in single shear supports only about 600 lb. A couple of nails through a 2x6 scab are not enough to prevent failure.

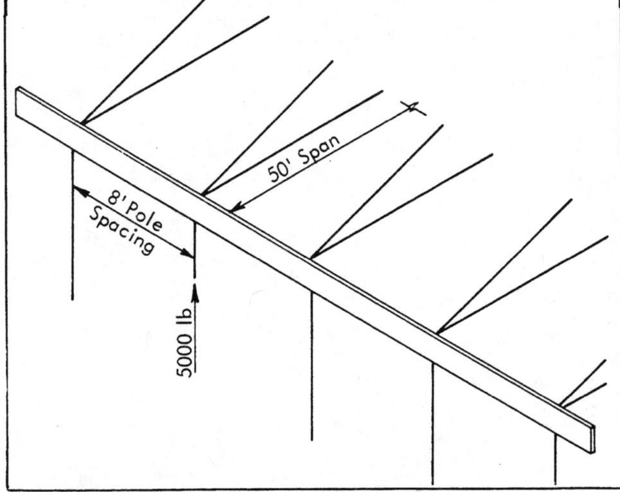

Fig 301-2. Transfer of roof load to pole.

With round poles, fastening the girders to the poles is the most critical joint in the building. Even with square poles, it is often difficult to design adequate joints.

Wind Loads

An 80 mph wind causes a net uplift load on each truss: about 1200 lb on 50′ trusses 4′ o.c. Trusses must be securely anchored.

A study of storm-damaged farm buildings found apparently substantial barns demolished because the builder had not considered fastenings such as adequate roof anchoring. Good fasteners properly applied add little to the cost of a building but can prevent complete building failure or at least reduce damage.

Metal roofing is often applied over roof purlins 2′ apart. Lifting wind forces attempt to pull the roof material off the purlins. Nail metal roofing carefully with screw- or ring-shank nails.

Purlin attachments to the trusses can be critically weak. With trusses 4′ apart and 2x4 purlins 2′ apart, each joint transfers wind loads from 8 ft^2 of roofing area. For the 80 mph wind, this load is at least 90 lb for each joint between purlin and truss—a 16d nail is needed at each joint. If the trusses are 8′ apart, rather than 4′, the uplift load per joint is 180 lb. Stagger the purlin joints and use a 60d nail in each purlin at each truss.

The spacing and bracing of trusses is critical, during construction and afterwards. If one truss buckles or fails, it can cause many or all of the others to fail.

Joints and Lumber Quality

Building joints are often particularly weak. Splices and joints often fail from inadequate fasteners—the strength of the members themselves is never fully used.

Initial failure in trusses from snow load are often in the bottom chord. Excess tension causes joints to fail or the member to fail at a knot or split. Use high quality lumber in the bottom chords. Do not splice chord members at the panel points. Make strong joints.

Simple rafters fail initially at similar weak points. Rafter supports and girders fail at joints or weak places in the members.

Foundation Failures

Anchor a building to its foundation with an adequate number and size of fasteners to keep it there.

Pole buildings that are otherwise well constructed may fail if the poles are not properly set. Shallow pole embedments are often the cause of building failure. Adequate imbedment and concrete backfill help resist both uplift and overturn of a building subjected to high winds.

In addition to buildings blowing off foundations, failures result from poor drainage, too-shallow foundations, or differential settlement of inadequate footings.

Building Material Failures

Lower grades of lumber than recommended are noted in many failed buildings. Knots and splits can cause main members to fail.

Failure of metal siding and roofing from wind is common. Failures occur where nail heads pull through the metal or with inadequate nailing.

Some failures result from the builder using interior-type plywood for exterior sheathing, gusset plates on livestock shelter roof trusses, and building braces.

Inadequate Construction Practices

Poor construction practices are found in many buildings. Knee-bracing can be poor or absent, building cross-tie members are poorly located, not enough nails were used, too small a surface area was used for joints, or framing members are too light for potential loads. Sometimes members are larger than necessary, but splices and joints are poorly made.

With contractors who know how to build a good building, supervision can be poor. The men actually driving the nails must understand potential problems in order to get a good job done.

302 FOUNDATIONS FOR FARM STRUCTURES

An adequate foundation is vital to the usefulness and life of a building. A foundation must resist the forces acting on it:

- The weight of the building, snow, and contents.
- Soil movement caused by a change in moisture content, settlement, or frost action.
- Wind forces that tend to lift and overturn the building.
- Horizontal soil pressure if it is used as a basement wall.
- Lateral forces from stored contents or from roof loads on rigid frames.

A good foundation distributes these forces so that movement of the building is **small** and **uniform.**

Frost Action

Heaving and settlement due to frost action can harm a building's foundation. For ice layers to form in the soil and cause heaving, there must be:

- Freezing soil temperatures.
- Ground water close to frost line.
- Soil that supports rapid movement of capillary water upward from the water table.

As the frost line penetrates the soil, water freezes in the soil pores. Freezing dries the soil, creates a capillary potential, and brings water upward to the frost line causing the ice particles to increase in thickness. A foundation can be lifted by a large ice lens, but the lens is more likely to saturate the soil and cause low soil bearing strength when it melts. See the chapter on climate data for maps of frost depth.

Prevent frost damage in slabs or shallow foundations by lowering the water table on sites with drainable subsoil, using coarse granular fill, or providing a layer of non-swelling clayey material with low capillary conductivity.

If suitable outlets are available, install tile drains around the building to lower the water table below the maximum frost penetration depth. Or raise the building grade to increase the distance to the water table.

Granular fill reduces frost damage by preventing capillary water movement. Use large gravel or crushed rock with the fines screened out. Drainage to below frost penetration is very desirable. Remove the subsoil to the maximum frost penetration depth if economically practical and replace with rock.

A capillary cut-off layer can also be used. Soil with a high clay content has high capillary potential, but very low permeability causing water to rise very slowly through it. Replace the subsoil with non-swelling clay having good particle size distribution.

On soil known to be susceptible to frost heaving, use one of the following construction techniques:

- Extend the foundation to below maximum frost penetration.
- Dig a perimeter trench to below frost penetration, backfill it with crushed rock, and then build a normal curtain wall foundation.
- Build a post and grade beam foundation with the posts extended to below frost penetration and the beam located above the ground surface.

In areas of deep frost penetration where frost heaving is not a problem, reinforce the top and bottom of shallow concrete footings for masonry walls or the top and bottom of shallow concrete foundations.

Soil Bearing Strength

If possible, select a building site that is well drained. The ideal foundation soil supports the building weight, neither swells nor shrinks excessively, and is not susceptible to frost action. Dry, well compacted, sandy clay is close to an ideal foundation soil. Compacted gravel or sand fill with good drainage makes a good foundation base.

Table 302-1. Allowable soil bearing strength.

Material	Bearing strength lb/ft^2
Gravel or coarse sand, well compacted	12,000
Dry, hard clay or coarse, firm sand	8,000
Moderately dry clay or moderately dry, coarse sand and clay	4,000 to 6,000
Ordinary clay and sand	3,000 to 4,000
Soft clay, sandy loam, or silt	1,000 to 2,000

Clay soils become soft when wet; improve their bearing strength with drainage or compaction. Before building on dry clay be sure it will not swell when wet. If a clay's capillary conductivity is high, it is subject to frost action.

Sand shrinks and settles when it dries and swells when wet, so provide drainage around the foundation. Spongy or peaty soils require special foundation design by an engineer.

A thin layer of firm soil or rock can lay on top of soft clay or loose sand. Foundation pressures transferred to this soft layer can cause undesirable settling; use large footings to reduce the pressure.

Avoid supporting foundations on fill dirt unless it is well compacted or has settled for a year. Uneven fill depth and soil types can result in differential settling. Extend foundations through fill dirt to undisturbed soil whenever possible.

Foundation Bearing Area

A footing increases a foundation's bearing area and lowers the load per ft^2 of soil.

Determine the amount of bearing area needed from building size, anticipated loads, and soil type. If possible, select a site with uniform soil quality. Most soil bearing capacities are determined from a specified allowable settlement after a given time period. The purpose is to prevent differential settlement—

one part of a foundation settling more than another—which is a major cause of foundation failures. It results from non-uniform foundation pressures and/or soil types around the building. If soil is uniform over the building site and foundations are sized for equal soil pressures, any settlement should be uniform, and typical allowable soil bearing strengths are very conservative.

Estimate the expected loads on the foundation from dead and live loads. See Loads. In storage structures such as granaries, use 100% of the expected weight of the building's contents. In homes where the design floor loads will not be present throughout the entire building at one time, use 50% of the design floor loads. For livestock buildings, use the distributed floor loads used to size the beams and columns supporting a slotted floor. See Fig 302-2 for an outline of calculating foundation loads.

Size perimeter wall and pier footings for the same soil pressures. Wall footings may be larger (for easy construction) than calculated from the allowable soil strength. Size pier footings for the wall bearing pressure, not the allowable soil bearing strength, to help minimize differential settlement.

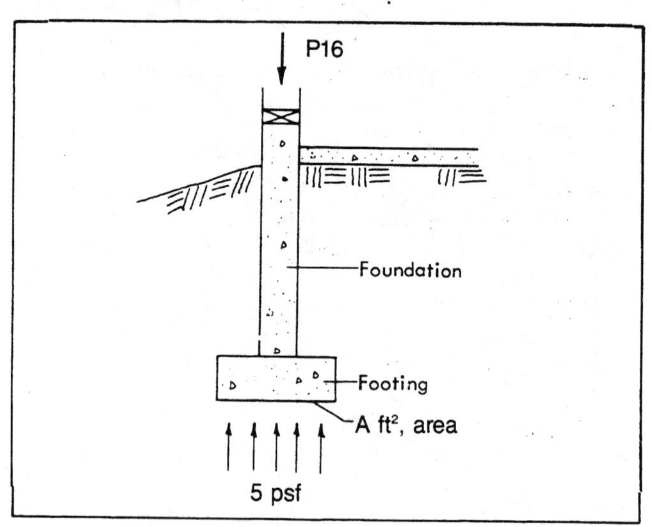

Fig 302-1. Footing load.

Required footing area is:

$A = P \div S$

A = footing area, ft^2

P = load, lb

S = soil bearing strength, psf

Lateral Soil Pressure

Basement walls hold back the soil and support the building above them. The pressure exerted on the wall depends on soil density, internal friction of the soil particles, and depth below the surface. Some soils with high cohesive forces support themselves along a vertical cut; saturated sands, with almost no cohesion, flow like water against a basement wall. An expansive clay can exert very high lateral pressures as it absorbs water.

Lower lateral soil loads with good drainage. Set basement walls on gravel or install perimeter drains covered with gravel that drain to a sump or free outlet. Build weep holes into retaining walls to relieve water pressure behind them.

Fig 302-3 illustrates typical triangular soil pressure against a basement wall. The soil pressure may be 15 to 110 psf/ft below the surface. Soil pressure can be calculated with the Rankine formula and equivalent fluid density. Most designs are based on 30 psf/ft with drainage, or 60 psf/ft without drainage.

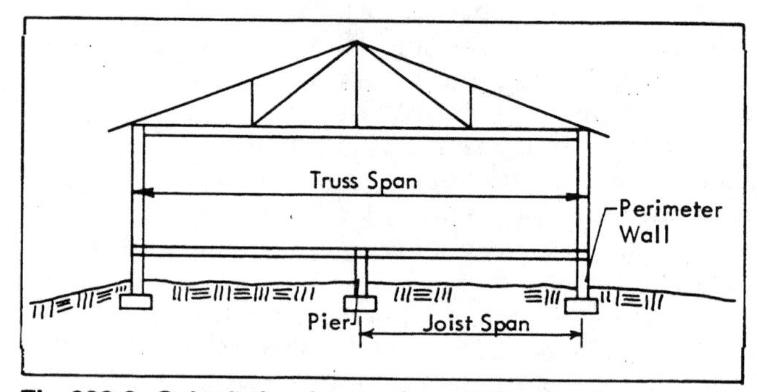

Fig 302-2. Calculating house foundation loads.

Wall load per foot of perimeter =
(Truss span) x 2 (snow + roof + ceiling loads) ÷ 2
+ wall dead load + (Joist span)
x (Floor live load + floor dead loads) ÷ 2
+ foundation and footing dead load
Pier load =
(Joist span) x (pier spacing)
x (floor live load + floor dead load) ÷ 2
+ pier and footing dead load

Floating Slab

A floating slab is both building floor and foundation. It is well suited for garages, shops, and homes without basements. The concrete floor and foundation are cast as one piece. The edge of a floating slab foundation is thickened and reinforced to form a grade beam bearing directly on the soil. The concrete is extended below grade only 12"-18" to prevent erosion from undermining the slab and rodents from burrowing under it. Tie the thickened edge to the floor with welded wire fabric. Anchor bolts through the sill hold the building on the foundation. See Fig 302-4.

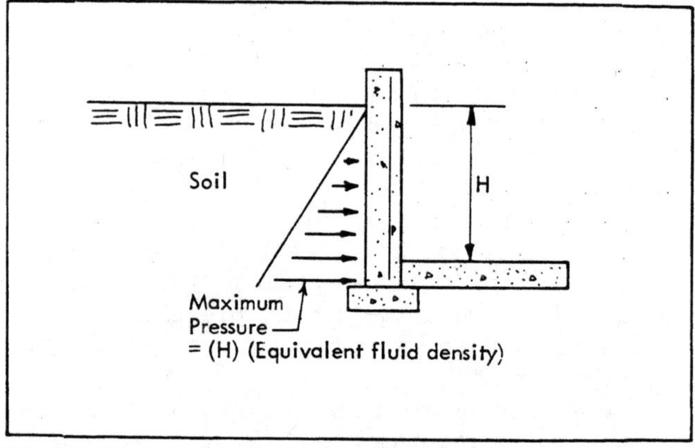

Fig 302-3. Lateral soil pressure.

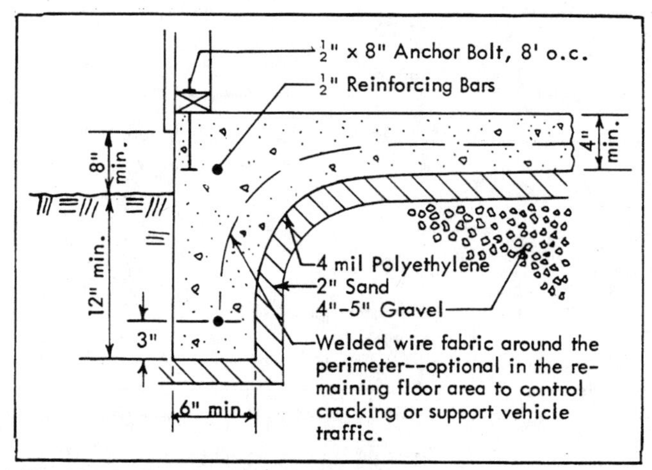

Fig 302-4. Floating slab foundation.

Level the building site and compact fill dirt to prevent settling and cracking. Cover fill with 4"-5" of gravel plus 2" of sand for good drainage. Lay 4 mil polyethylene plastic over the sand; it prevents the sand from drawing moisture from the fresh concrete, weakening it before it cures, and blocks moisture moving up through the floor.

See Frost Action for corrective measures if a floating slab is to be on soil susceptible to frost heaving.

In heated buildings, insulate the perimeter.

Raise the sills and wall lining above a frequently wetted floor to help prevent rotting, as in a milkhouse, Fig 302-5. Build a raised sill with longer anchor bolts and 4" concrete block around the slab.

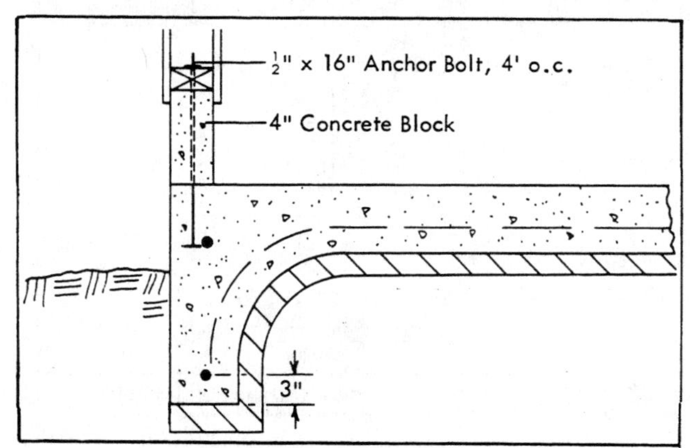

Fig 302-5. Floating slab with raised sill.

Curtain Wall Foundations

Curtain wall foundations are laterally supported by soil both inside and outside. In farm buildings, they support stud walls or rigid frames. An independent concrete floor can be poured inside the foundation. They may have a footing. In homes, they commonly support the walls and floor joists and form a crawl space under the house.

Curtain wall foundations support dead and snow loads and resist wind uplift; they usually have little or no lateral load. A wall supporting rigid frames must resist the horizontal frame reactions and may require bracing or a greater depth; see Fig 302-8.

Concrete

A concrete curtain wall can be formed and include a footing or poured in a trench with no footing. The traditional procedure is: dig a 4' wide trench around the building to the footing depth; form an 8"x16" footing; form an 8" wall to the desired height; place the concrete and anchor bolts, Fig 302-6.

A foundation wall reinforced to resist uplift and settlement need not extend to the frost line in areas where frost heaving is not a problem. Also, the foundation wall need not be wider than the sill if a footing provides adequate bearing area.

Another type of curtain wall is: dig a trench of the desired width and depth; place forms above the trench to obtain a straight, level wall; and pour the trench and forms full of concrete, Fig 302-7. Only cohesive soils, loess or high clay content, that can support themselves in a trench are suitable for this type of foundation. Determine trench width from the foundation load and soil bearing strength.

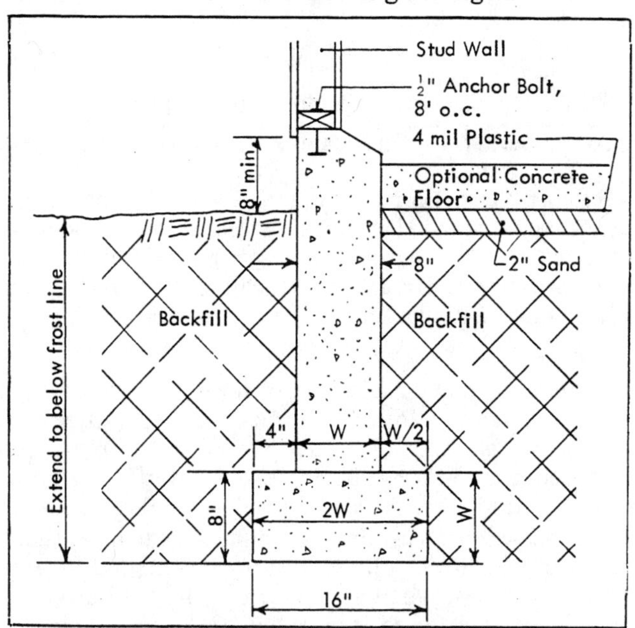

Fig 302-6. Typical curtain wall foundation.

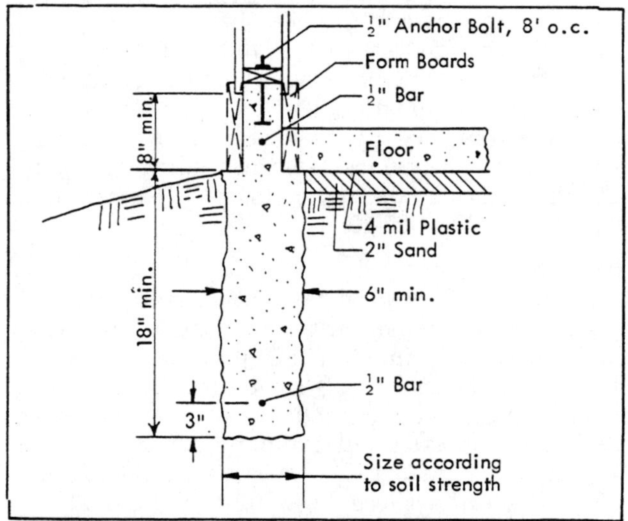

Fig 302-7. Trenched curtain wall foundation.

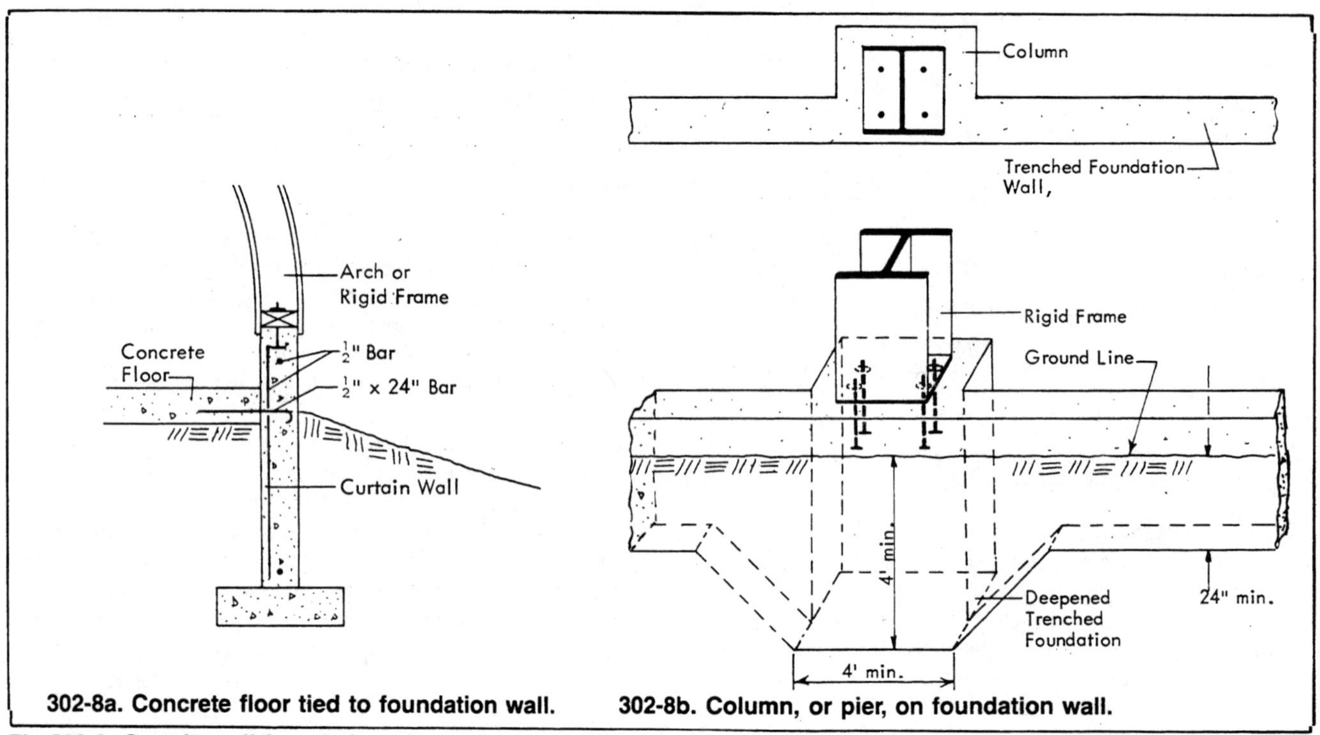

302-8a. Concrete floor tied to foundation wall. **302-8b. Column, or pier, on foundation wall.**

Fig 302-8. Curtain wall foundations for arches or rigid frames.

Special types of concrete curtain wall foundations are needed to resist the lateral forces from arches or rigid frames. The foundations can be braced with buttresses, tied to the floor with reinforcing bars, or strengthened with columns and deepened, Fig 302-8.

Concrete masonry

A concrete footing provides a level surface for a concrete masonry (block) foundation wall. Block walls are often adequate but are not as strong as reinforced concrete walls. Brace block walls under arches or rigid frames to support lateral forces.

In areas where frost does not penetrate to a depth of 18″, the ½″ bars in Fig 302-9 may be omitted in a footing on undisturbed soil. If the soil has uneven bearing strength or is compacted granular fill, always add the reinforcing bars. Also, include reinforcing bars if frost penetration is 18″-48″.

Where frost penetrates 4′ or more, possible damage from settling or heaving justifies strengthening a shallow foundation. Thickening the footing to 10″, adding reinforcing bars at the top, and adding dowels into the blocks increases the bending strength of the wall to bridge weak spots in the soil and resist frost heaving, Fig 302-10.

Treated wood

The American Plywood Association has developed a design procedure and construction details for wood foundations for homes. Their publication, *All-Weather Wood Foundation System,* contains member selection tables and construction and installation details. The system is suitable for many farm buildings. It can be easily insulated to prevent cold floors and high perimeter heat loss. Figs 302-11 and 302-12.

The wood foundation rests on a footing of gravel or crushed rock. Its main advantage is that no concrete

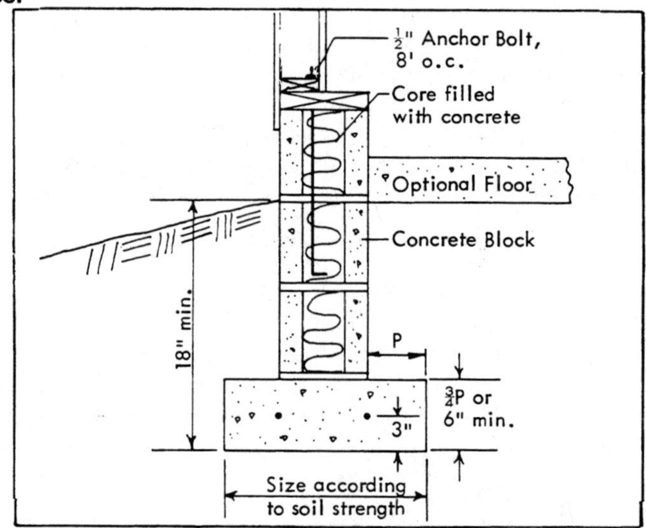

Fig 302-9. Typical concrete block curtain wall foundation.

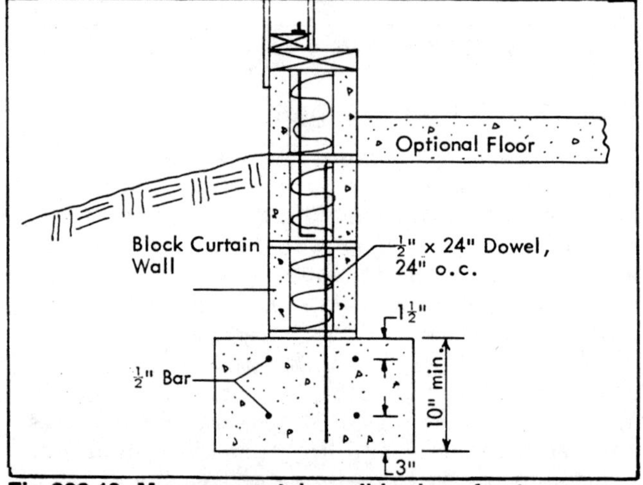

Fig 302-10. Masonry curtain wall in deep frost area.
Masonry reinforcing in the top mortar joint may replace the top two footing bars.

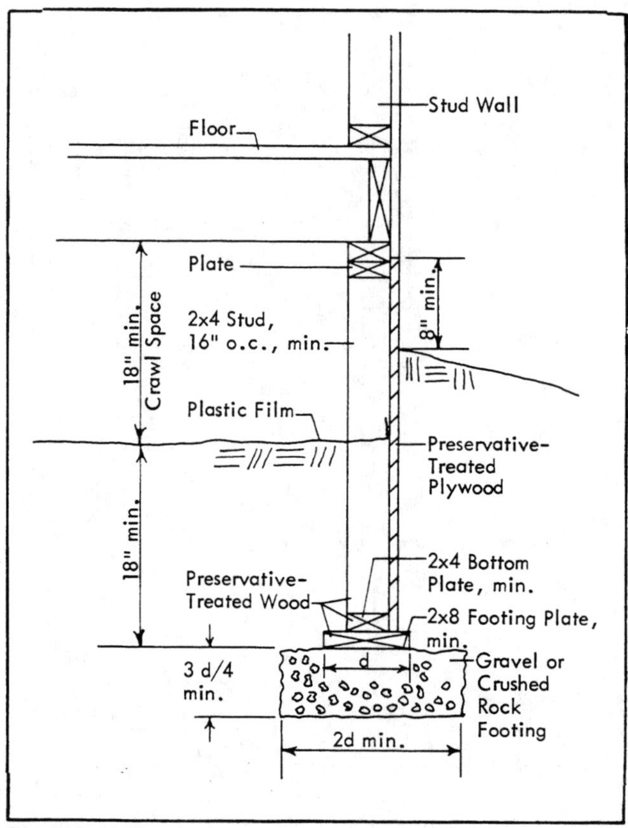

Fig 302-11. Treated wood foundation with a crawl space.

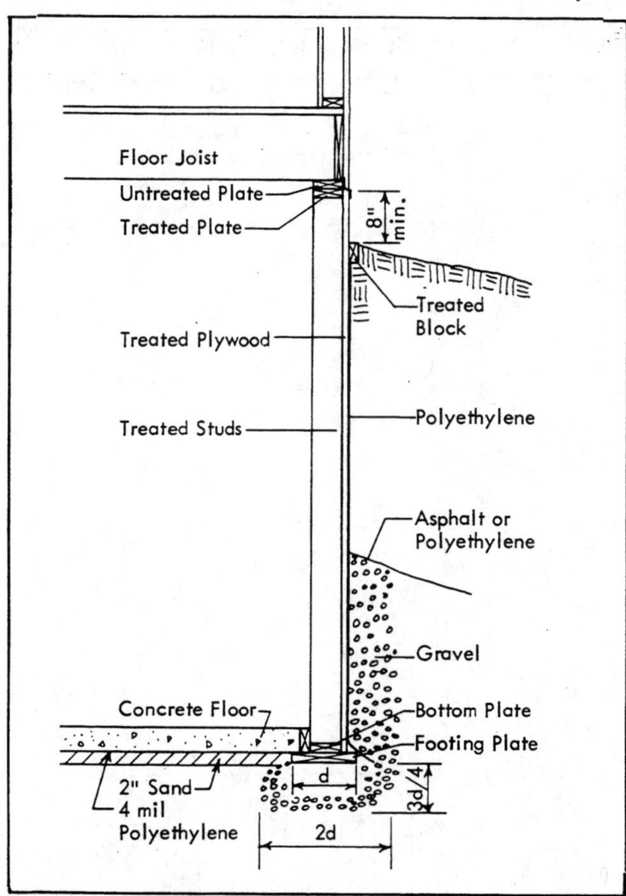

Fig 302-12. Preservative treated wood basement wall.
All wood below the first top plate is pressure preservative treated.

work is involved. It can be built in the winter when concrete would have to be protected from freezing. The allowable bearing strength of the gravel or crushed rock footing is assumed to be 3000 psf for sizing the footing plate. The downward load from the footing plate is assumed to be distributed outward through the footing at a 30° angle from vertical at the edge of the footing plate. If the width of the footing plate is d, the gravel footing is at least 2d wide and 3d/4 deep.

The wood foundation can support a stud wall without a crawl space. It would fit very well with a warm livestock building.

Manure Pit and Basement Walls

Manure pit and basement walls are similar to curtain wall foundations, except they must support lateral soil forces in addition to acting as a foundation. When the soil is at a higher elevation outside than inside the foundation, bending moments and outward horizontal forces at the top and bottom are created in the foundation wall. These forces greatly affect the design of a foundation wall. See the Selection section for manure pit wall designs. Lateral soil pressures are described earlier in this chapter. Install tile drains around walls over 4′ deep to lower soil pressure.

Reinforced concrete

Manure pit walls are usually reinforced concrete that is formed and cast in place. Precast and trenched pit walls are also used.

When the wall rests on a concrete floor, steel dowels transfer the horizontal soil reaction to the floor. The top of the wall must also be continuously braced, or reinforced, to provide lateral support between floor framing members.

Basement walls need a waterproof coating on the outside of the wall to ensure a dry basement. A perimeter tile drain lowers the possibility of water seepage.

Vertical steel carries bending moments created by lateral soil pressure. Horizontal steel controls cracking caused by temperature changes and shrinkage during curing.

A basement wall is like a pit wall, Fig 302-13, except lateral support at the top of the wall is provided by floor joists or partition walls.

To lower costs, some builders cast concrete manure pit walls in trenches and then excavate the soil where the pit is to be. Several problems can occur with this construction system:

- The trench walls may collapse before the concrete is placed.
- Accurate steel placement is very difficult.
- The pit wall may overturn when the pit is excavated if adequate lateral support is not provided.
- Drainage cannot be provided to lower the soil pressure.

The top of the wall must be formed to provide a straight, level foundation for the building or slotted floor. Concrete 3″ deep is needed between the steel

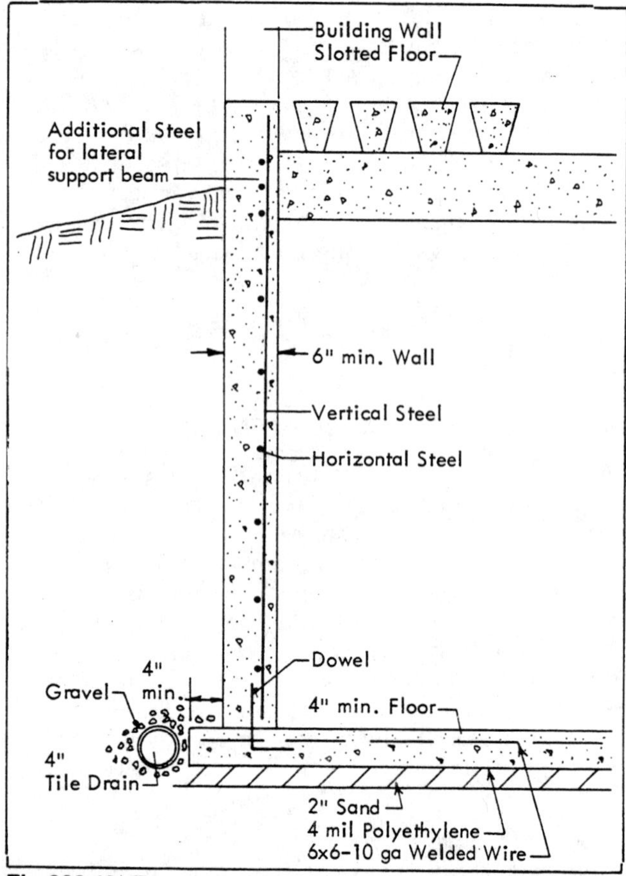

Fig 302-13. Formed concrete manure pit wall.

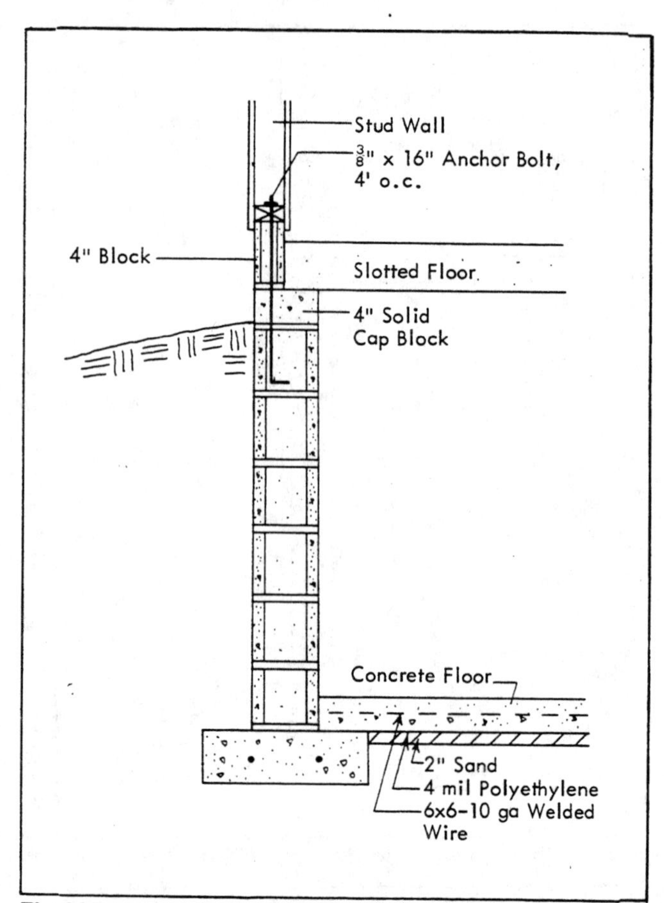

Fig 302-14. Shallow concrete block manure pit wall.

and soil, so the wall should be designed assuming the steel is at its center. The vertical bars can be driven into the soil at the bottom of the trench to hold the steel grid in place while the concrete is placed. This system is recommended only for shallow pits, 4′ or less, because it is more likely the trench walls will not collapse and reinforcing is not as critical.

Concrete masonry

Unreinforced concrete masonry has much less bending strength than reinforced concrete. Blocks are common for shallow manure pits or basements with good drainage. Block walls can be reinforced horizontally and/or vertically, or braced by pilasters or partition walls to increase their strength.

An unreinforced manure pit block wall should be continuously restrained at the top and bottom.

Use a curtain wall footing, Fig 302-9. If the footing and floor are cast together, add dowels to carry the lateral soil load from the wall to the floor.

Place a 4″ solid cap block at the top of the wall to seal the block cores and to provide bearing for a slotted floor.

For a raised sill, set a 4″ block on top of the cap block. Anchor the sill and stud wall with 3/8″ anchor bolts, 4′ o.c., extending through the cap block joints into the top 8″ block course, Fig 302-14.

Concrete block basement walls are commonly built to depths of 8′ without reinforcing, Fig 302-15. Provide good drainage for a dry basement and to lower soil pressure against the wall. An 8″ concrete

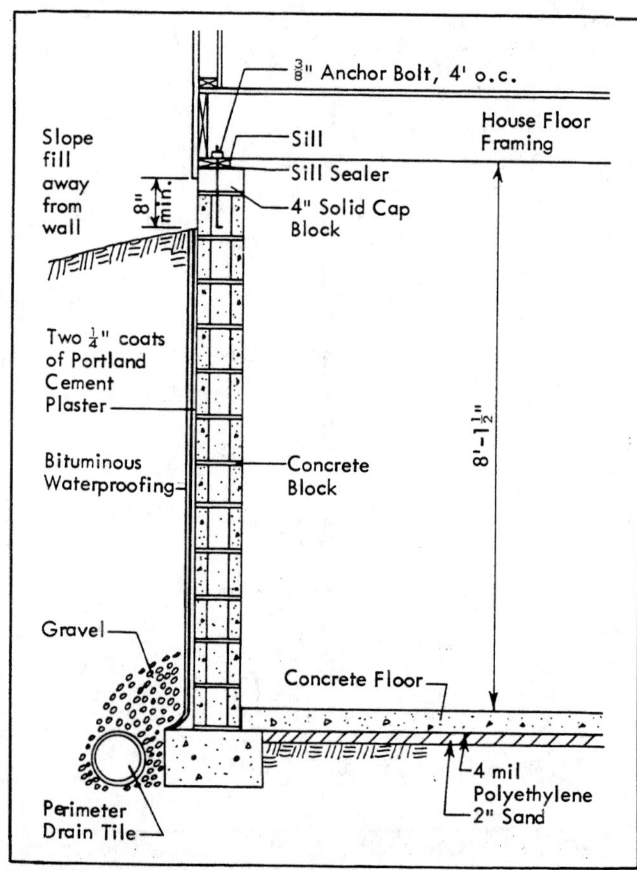

Fig 302-15. Concrete block basement wall.

block wall 8′ high under a frame house can support a soil depth of 5′. Reinforce the block wall if it must support more than 5′ of soil. Place horizontal reinforcing in the mortar joints of alternate courses or use bond beams. Vertically reinforce a block wall by adding pilasters or placing steel reinforcing bars in the block cores and filling with grout.

Treated wood

Treated wood is also a suitable material for manure pit or basement walls. See treated wood curtain walls. Use a sump under the basement floor to collect water moving through the gravel fill or install a drain tile to a sump or run to daylight. Check local building code requirements.

A wood basement wall can easily be insulated and finished to provide additional living space.

When treated plywood is used for a manure pit wall, omit the plastic. Building posts may replace the studs.

Post and Grade Beam

Post and grade beam foundations support buildings on stable soils below frost penetration when surface soils are susceptible to frost heaving or have low bearing strength. They are more economical than curtain wall foundations extended to the frost penetration depth in cold climates.

The post (column) supports the grade beam under the building. Size the post footings to support all gravity building loads. Select the posts and beams to resist both lateral and vertical building loads. Place 6″ of gravel under the grade beam for drainage and to prevent frost heaving from lifting the beam. Extend the beam at least 8″ above and 12″ below grade.

Concrete

Both the column and grade beam can be reinforced concrete. Dig the hole for the column with a post-hole auger and flare the end for a larger diameter footing if needed. Form the grade beam on top of the columns. Add steel reinforcing rods to both the column and grade beam. Use high quality air-entrained concrete, Fig 302-16.

This foundation is excellent for supporting a masonry wall. It can also support the floor and wall framing of a house with a crawl space.

Wood

Pressure preservative treated wood can support a stud wall building. New preservative treatments are very durable.

Set the posts on concrete footings below frost penetration. Construct the grade beam of treated dimension lumber. Securely nail the sill and wall to the grade beam, Fig 302-17.

The post can be extended above the ground and floor framing added to form a crawl space for a home.

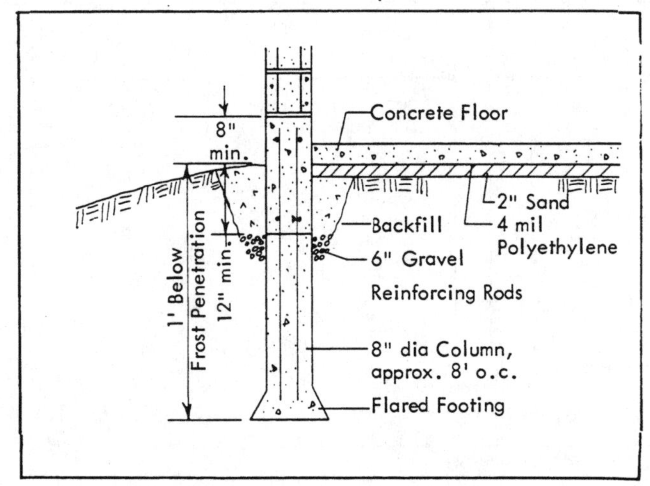

Fig 302-16. Concrete post and grade beam.

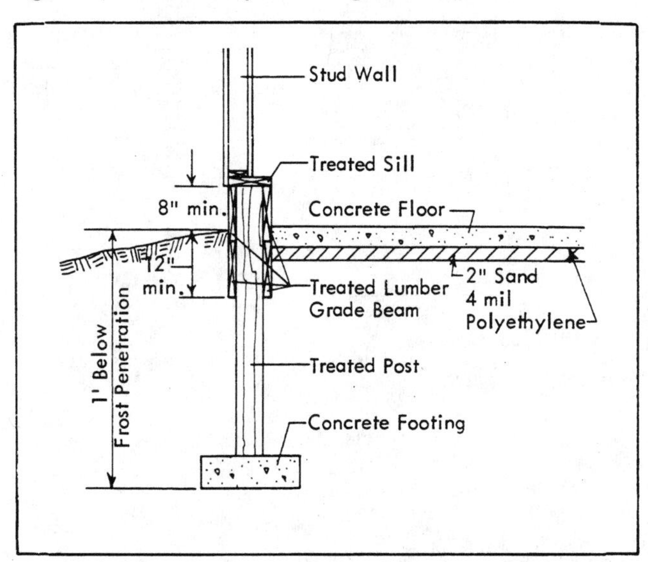

Fig 302-17. Treated wood post and grade beam.

Building Post Footings

Install footings under posts in post-frame buildings. Snow loads and wind uplift can both lead to building failures without adequate footings.

Size a concrete footing to support the snow load. The footing can be pre-cast, or a concrete mix placed in the hole dry which draws ground water from the soil and hardens. Increase footing size for posts next to door openings or in open walls where some of the posts are removed.

High uplift resistance is important for posts next to large door openings and in open front buildings. To increase the withdrawal resistance of posts above that provided by surface friction with the soil: nail treated lumber to both sides of the post; run a rod through the post and encase the butt in a concrete footing, or dowel the post to a concrete floor, Fig 302-18.

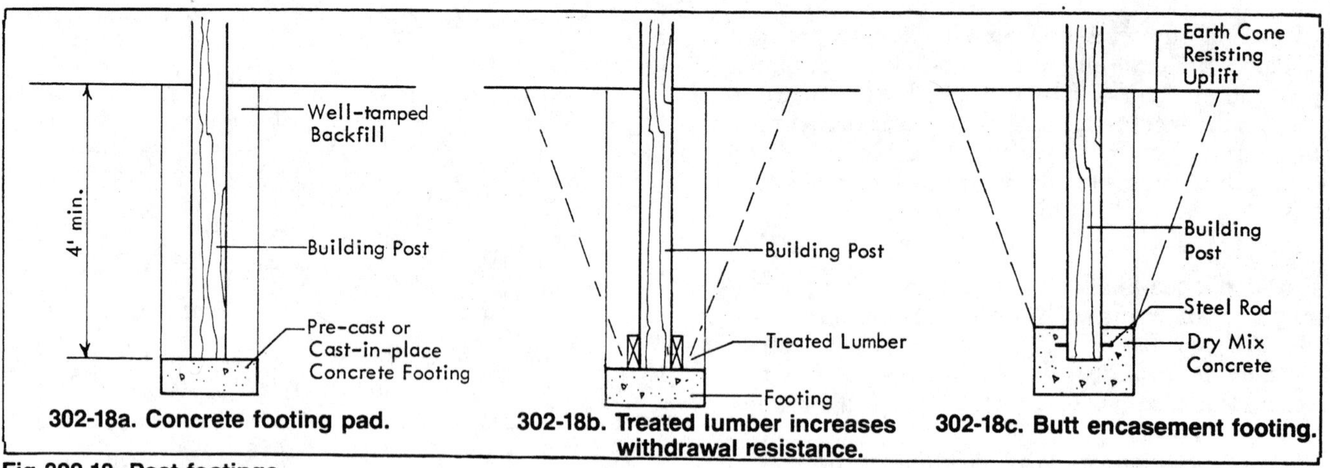

302-18a. Concrete footing pad.

302-18b. Treated lumber increases withdrawal resistance.

302-18c. Butt encasement footing.

Fig 302-18. Post footings.

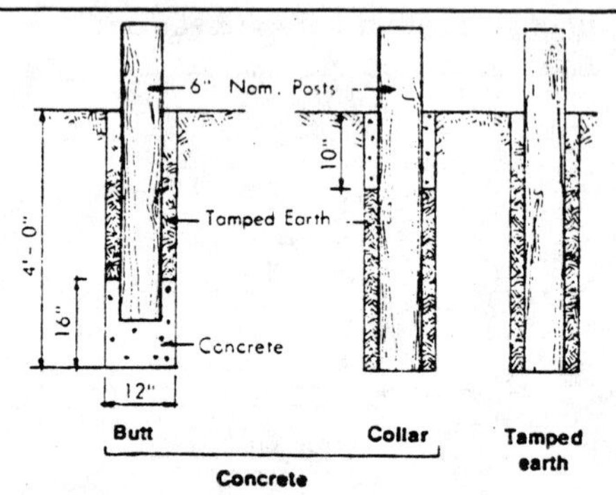

302-19.a. Tests A.

Each pole supported a 3200 lb load for 15 months; settling was recorded. Then the poles were pulled out.

	Concrete Butt	Concrete Collar	Tamped earth
Relative settlement:	1	3	3
Uplift load, lb:	5500	1200	1600
Relative deflection:	1	3	13

302-19.c. Tests C.

Overturn tests of 5″ top poles with various backfills. Three cycles of loads from 0 to 800 lb were applied laterally 14′ above grade. Earth backfill has 1½ times as much lateral deflection as the other backfills. Strengths were not reported for these tests.

	Butt	Concrete Collar	Encased
Depth of set:	2½′	3½′	5′
Relative deflection:	1	0.38	0.3

302-19.d. Tests D.

Settlement and uplift loads on the 5½″ poles are each the average of two tests.

Resistance to a horizontal load 4½′ above grade was, relative to tamped earth: 1.11 for 2x6 blocks; 1.78 for dry mix; 2.52 for pad and collar.

	Earth	Block	Concrete	Collar
Load at 1″ settlement:	1586 lb	2590 lb	5297 lb	3012 lb
Load at 1″ uplift:	242	563	4483	3500

4'-0"

Sand, earth, crushed rock   Pea gravel   Concrete

302-19.b. Tests B.

Withdrawal loads on poles with various backfills.

	Sand	Gravel	Concrete
Uplift:	1000 lb max.	2000 lb max.	6000 lb @ 0.2″
			9000 lb @ 1.5″

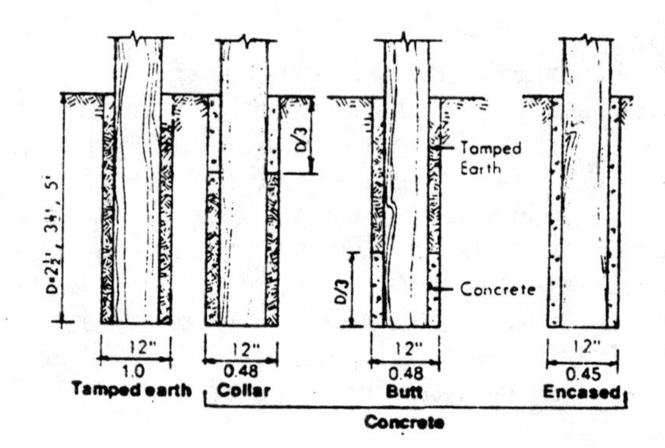

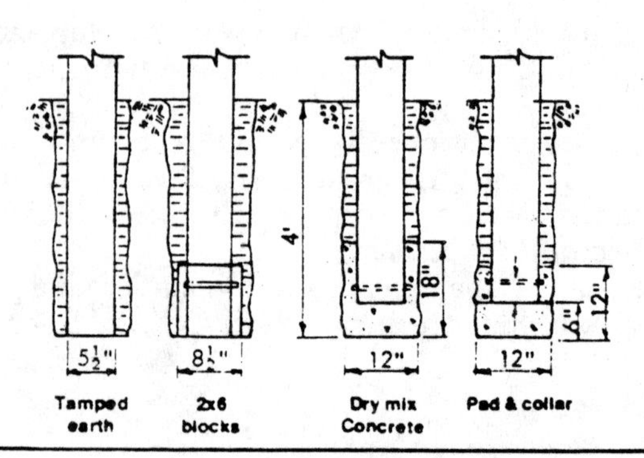

Fig 302-19. Pole anchorage.

Agricultural Engineers at Oklahoma State University (A,B,C) and the University of Illinois (D) tested the withdrawal and deflection of poles under load. The study results are valid only for their soils, but are reasonable for predominantly clay soils.

303 WALL AND FLOOR FRAMING

Stud Frame Construction

Platform Framing (Fig 303-1)

Platform framing is the most common framing for one-story houses. It is also used alone or in combination with balloon construction for two-story structures. The subfloor extends to the outside of the building and is a platform for the exterior walls and interior partitions.

Platform construction is easy to erect because it provides a flat work surface at each floor level. It also adapts easily to prefabrication. It is common to assemble wall framing on the floor and then tilt it into place.

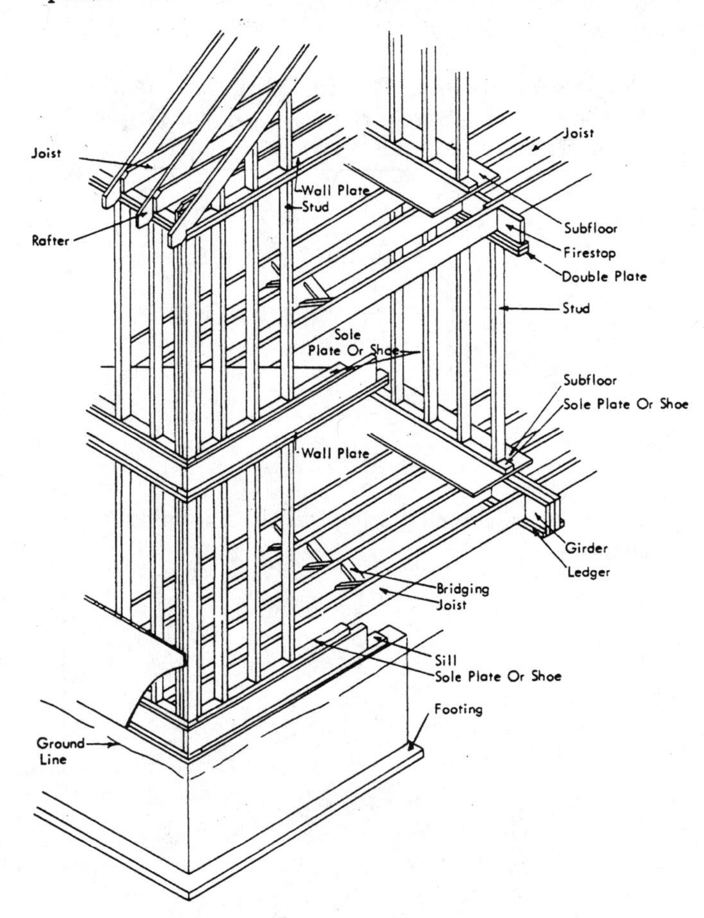

Fig 303-1. Platform frame construction.

Balloon framing (Fig 303-2)

Balloon framing is preferred for two-story buildings with an exterior veneer of brick, stone, or stucco, as there is less movement between the wood framing and the veneer. Both studs and first-floor joists rest on the anchored sill. Second-floor joists bear on a 1x4 ribbon (or ribband) let into the inside edges of the studs.

With solid masonry exterior walls, balloon framing for interior bearing partitions is desirable. It eliminates possible variations in settlement between exterior walls and interior supports.

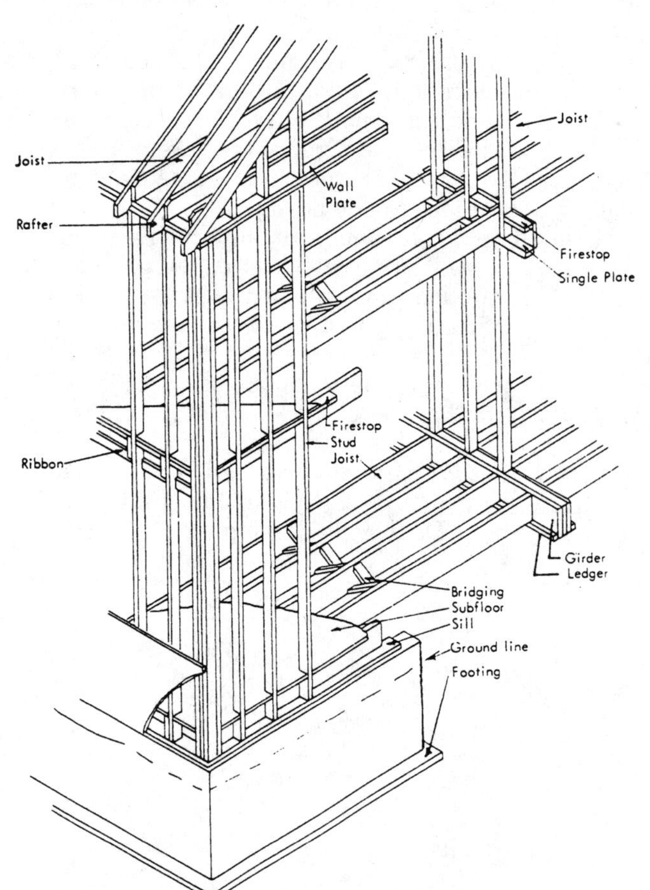

Fig 303-2. Balloon frame construction.

Floor framing

Floor framing includes sills, girders, joists and subflooring, with all members fastened together to support floor loads and exterior walls.

Sills on continuous foundation walls are usually 2″ lumber set on the walls to provide full and even bearing. They are anchored to foundation walls with ½″ bolts 8′ apart with at least two bolts in each sill piece, Fig 303-3. Embed bolts at least 6″ in poured concrete and at least 15″ in masonry units. Treated sill lumber is recommended.

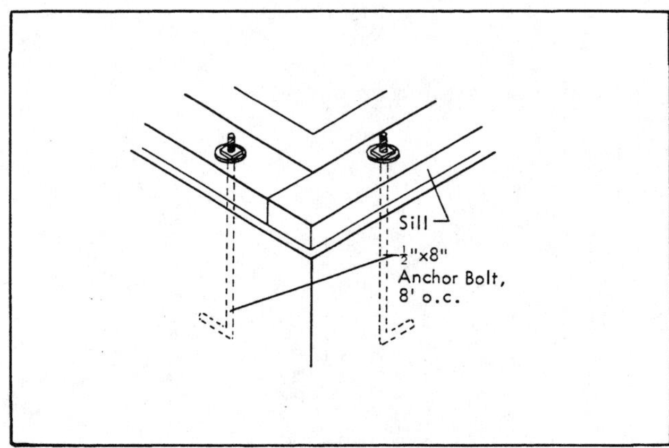

Fig 303-3. Anchoring sill to foundation wall.

Girders. Girders support joists. They are solid timbers or members built-up of 2″ pieces nailed together with the wide faces vertical, Fig 303-4. Girders are fastened together with one row of 20d nails 32″ o.c. along each edge. Glue-laminated members are also used, and flat-chord trusses are becoming more common. Beams and girders that are not continuous are tied together across supports. A bearing of 4″ on supports is recommended.

Joists. Joists are beams that support floors or ceilings. They are placed with top edges even for installing the subfloor and finished floor. Align the bottom edges of ceiling joists for an even ceiling. If joists carry both ceilings and subfloors, shims may be needed to align one or the other. Bearing for joists should not be less than 1½″ on wood or metal and 3″ on masonry.

Avoid notches for piping that exceed ⅙ the joist's depth in the top or bottom of joists and in the middle third of the span. Drill holes for piping or electric cables no closer than 2″ to the top or bottom of the joist; hole diameters should not exceed ⅓ of the joist depth. A notch at the end of a joist should not exceed one fourth the joist depth, Fig 303-4f.

Bridging. Brace joists to prevent them from buckling under load and so several joists tend to act together to support loads. Place diagonal struts or solid blocks between and at right angles to joists near each end and at up to 8′ o.c. Omit bridging at the ends of joists that are nailed to a header, joist band, or adjoining studs.

Framing floor openings (Fig 303-4e). Headers, trimmers, and tail joists frame openings, such as for stairs. Double trimmers and headers for spans over 4′. Support headers over 6′ long on joist hangers or framing anchors, unless they rest on a partition or wall. Support tail joists over 12′ long on framing anchors, or on ledger strips not less than nominal 2x2.

Partition support. Bearing partitions usually are placed directly above girders or walls that support the floor framing. In general, bearing partitions at

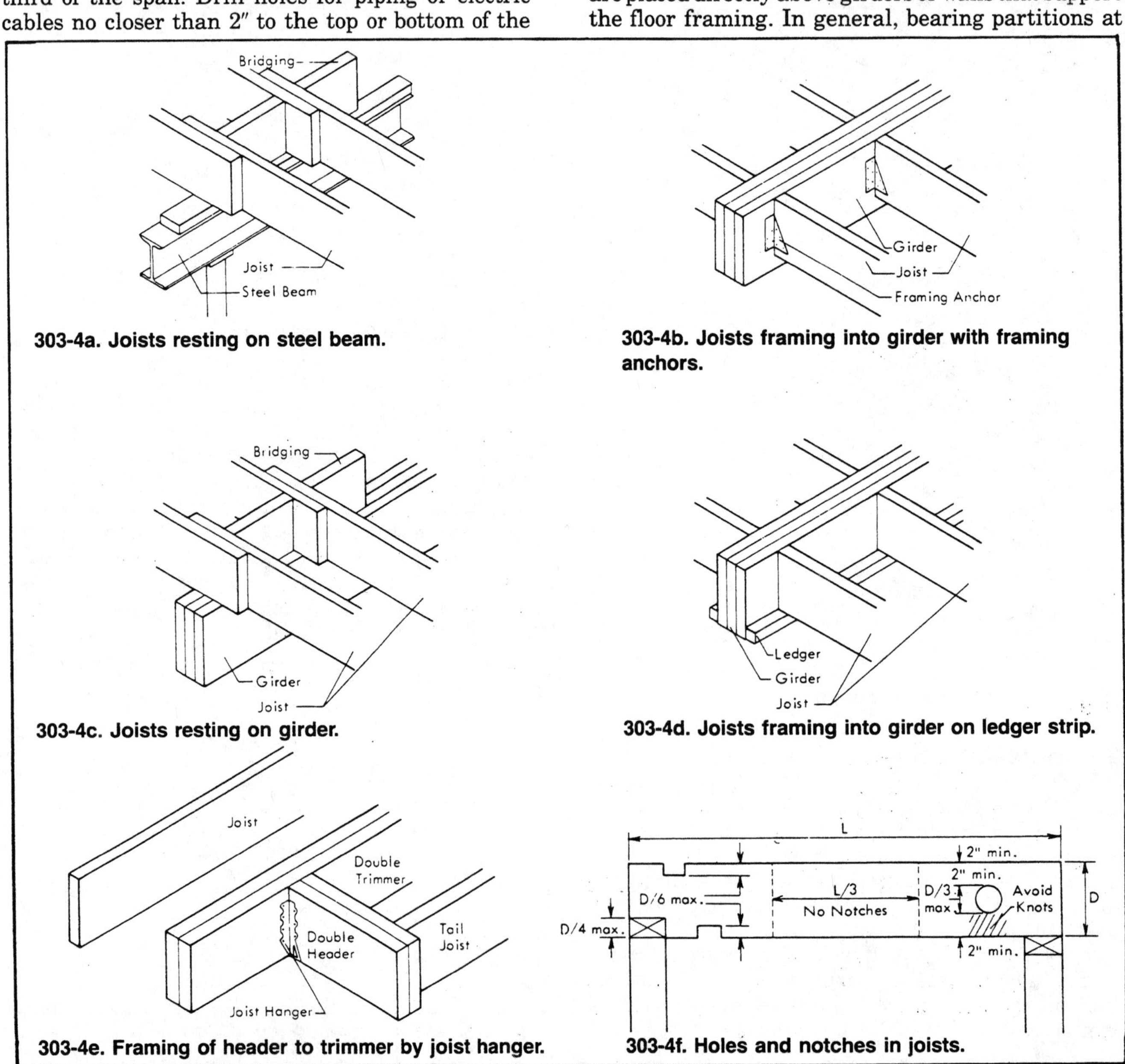

303-4a. Joists resting on steel beam.

303-4b. Joists framing into girder with framing anchors.

303-4c. Joists resting on girder.

303-4d. Joists framing into girder on ledger strip.

303-4e. Framing of header to trimmer by joist hanger.

303-4f. Holes and notches in joists.

Fig 303-4. Joist and girder framing.

right angles to joists should not be offset from main girders or walls more than the depth of the joist, unless the joists are designed to carry the extra load. Double the joists under non-bearing partitions that are parallel to joists.

Subflooring. Subflooring is usually plywood. Square-edge shiplap, or tongued-and-grooved (T&G) boards are also used. Plywood or diagonal subflooring permits the finish floor to be either parallel or at right angles to the joists, and provides lateral bracing and stiffness. Plywood subfloor may be faster to lay and gives even better stiffness.

Finish flooring

Subfloor preparation. Clean the subfloor of debris. Drive in raised nails and examine the floor for proper nailing. Lay building paper, such as 15 lb asphalt-saturated felt, over the subfloor.

Install finish T&G floor boards perpendicular to the joists. An expansion joint, covered later by the base board and base shoe, is left next to all vertical surfaces.

Start flooring strips square with the room against either sidewall. Face-nail the groove of the first piece next to the wall so moulding covers the nails.

Blind-nail other strips along the tongue with nails at an angle of 45°.-50°. with the floor, and deep enough to permit a tight match with the next piece. Square-edge strips are face-nailed, with nails set. Use 7d or 8d screw or cut steel nails spaced 10″-12″ apart.

If subflooring is other than wood boards, nails penetrate each joist and another nail is driven between joists. Nail each strip as close as practical to the end. For appearance, end joints should be well distributed.

Floor coverings (linoleum, tile, or carpet) may be placed directly on underlayment plywood.

Exterior wall framing

Exterior walls support floor and roof loads and resist lateral wind loads. Plates, studs, openings, bracing, and anchors are the basic components of wall framing. In balloon framing the bottom stud anchor is the sill. In platform framing the sole plate or shoe is the bottom anchor.

Sole plate or shoe. The sole plate is a 2″ piece nailed to the bottom of the studs in platform framing, often before the wall is tipped up into place; or the studs are toe-nailed to the sole plate.

Stud size and spacing. Studs in exterior walls are placed with the wide faces perpendicular to the wall. Studs are usually 2x4's in one- and two-story buildings. In three-story buildings, studs in the bottom story should be at least 2x6's. In one-story buildings, studs may be 24″ o.c. unless otherwise limited by the covering. In multi-story buildings, spacing should not exceed 16″ o.c. Stud spacing is measured center to center except at corners—measure from the outside of the corner stud to the center of the first interior stud. Multiple studs at the corners provide for attaching surface materials.

Gable-end walls. Studs in gable ends rest on wall plates and are notched and nailed to the end rafter.

Openings. A header carries loads over doors and windows. Support the header on trimmer studs, or by framing anchors for spans up to 3′. Where the opening width exceeds 6′, use triple studs with each end of the header resting on two trimmer studs.

Wall sheathing. Metal, plywood, and some wood siding can be applied in a single layer on building frames. Some plywood, building boards, and fiber board are rough sheathing requiring a finish siding. For weather-tight walls, cover rough sheathing on the outside with asphalt-saturated felt, or equivalent, weighing not less than 15 lb/100 ft². Sheathing paper must not be a vapor barrier. Lap felt 4″ at horizontal joints and 6″ at vertical joints. Install strips of sheathing paper, about 6″ wide, behind trim and around openings. The wall can then be completed with finish siding.

Bracing. Most frame walls are strong enough to support roof loads, and to resist wind pressures per-

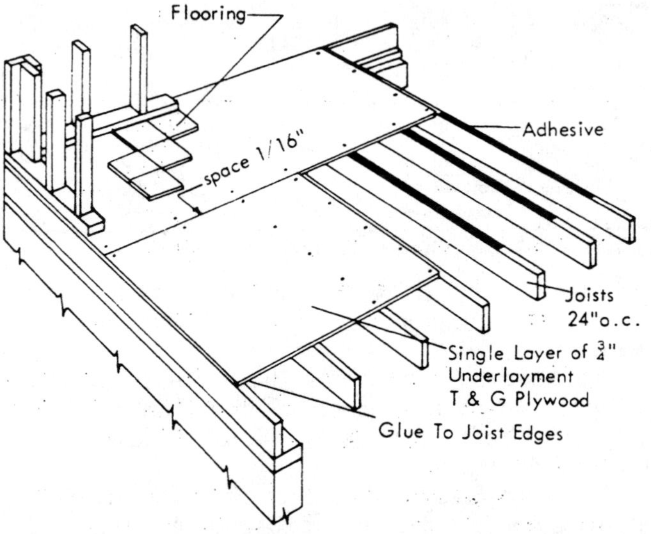

Flooring
Adhesive
space 1/16"
Joists 24"o.c.
Single Layer of 3/4" Underlayment T & G Plywood
Glue To Joist Edges

Joists— No. 2 or better
Douglas fir-larch lumber
2x10 repetitive members, 24″ o.c.

Grade	Moist	30 live + 10 dead sleeping & attic	40 live + 10 dead other areas
		Allowable span	
No. 1	15%	17′-0″	15′-5″
	19	16′-5″	14′-11″
No. 2	15%	16′-1″	14′-7″
	19	16′-1″	14′-7″

Underlayment— single layer of ¾″ plywood T&G (interior, or interior with exterior glue)
Stagger end joints
Leave 1/16″ between sheets at side and end joints
Site-applied elastomeric construction adhesive on joist faces and on T&G plywood joints
Nailing: 6d deformed shank, or 8d common, 12″ spacing along all bearings
If plywood is not T&G, block under all joints
Finish— Tile, carpet, linoleum or other non-structural flooring

Fig 303-5. Single underlayment floor.
A combination subfloor and base for finish floor is economical and practical.

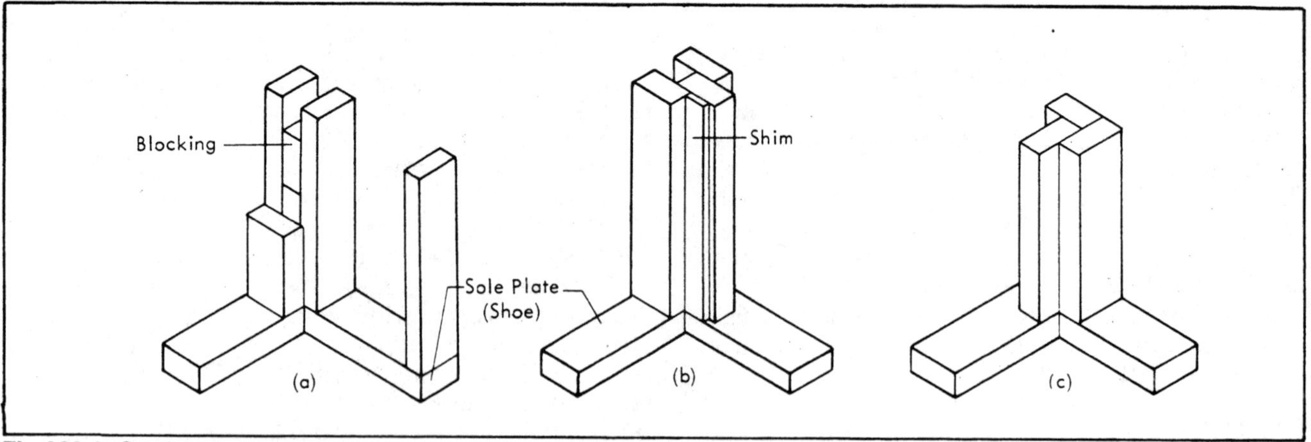

Fig 303-6. Stud corner framing.
In platform framing, each story starts with a sole plate as shown. In balloon framing, the studs start at the sill on the basement wall and extend to the plate at the top of the upper story.

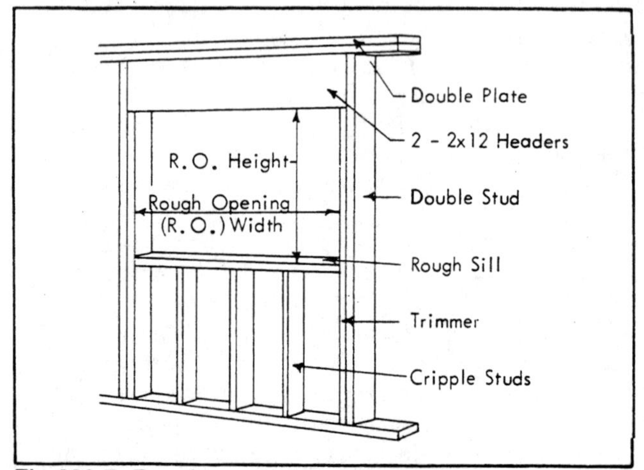

Fig 303-7. Framing exterior wall openings.
Omit rough sill and criple studs, and discontinue sole plate at doorways.

pendicular to them. Also important is resistance to end thrust or racking forces from the action of wind on adjacent walls. The factors most likely to affect resistance to racking are: the size and locations of wall openings; the type and size of framing, sheathing material, bracing, fasteners, and moisture content of the materials.

Fig 303-8 illustrates the relative stiffness and strength of sheathing and bracing systems.

Strength and stiffness may be provided by 1x4s let into the outside face of the studs at an angle of 45° and nailed to top and bottom plates and studs. Where plywood or other panel material is used, and where wood sheathing boards are applied diagonally, let-in braces are not necessary. Nail sheathing to sills, headers, studs, plates or continuous headers, and to gable end rafters. Plywood or fiberboard also provide good stiffness.

EFFECT OF SHEATHING (No Openings)		Relative Rigidity	Relative Strength
	2 8d Nails Per Stud Crossing	1.0	1.0
	2 8d Nails Per Stud Crossing	4.3	8+
¾" Fiberboard	8d Nails Spaced 3 Inches At Vertical Edges, 6 Inches On Intermediate Studs, 5 1/3 Inches On Plates	3.0	3.8
¼"	6d Nails Spaced 5 Inches On All Edges Of Plywood, 10 Inches On Intermediate Studs	4.2	5.2
	2 8d Nails Per Stud Crossing	0.7	0.8
Fiberboard	8d Nails Spaced 3 Inches At All Vertical Edges, 6 Inches On Intermediate Studs, 5 1/3 Inches On Plates.	1.6	2.1
¼"	6d Nails Spaced 5 Inches At All Plywood Edges And 10 Inches At Intermediate Studs	2.0	2.8

KEY: 1x8 Boards; Plywood ¼", ⅝"

Fig 303-8. Rigidity and strength of sheathing.

Partition framing (Fig 303-10)

Bearing partitions support floors, ceilings or roofs; non-bearing partitions carry only the weight of the partition.

Bearing partitions. 2x4 studs are set with the wide faces perpendicular to the partition and capped with a double top plate lapped or tied into exterior walls at the intersections. Studs supporting floors should be 16″ o.c.; those supporting ceilings and roofs may be 24″ o.c. Where openings occur, carry loads across the openings on headers like those for exterior walls.

Non-bearing partitions. Studs are 2x3 or 2x4 set with wide faces perpendicular or parallel to the partition. Stud spacing is 16″-24″, unless limited by the wall covering.

Firestopping. Concealed spaces in wood framing are firestopped with wood blocking, accurately fitted to fill the opening and arranged to prevent drafts from one space to another.

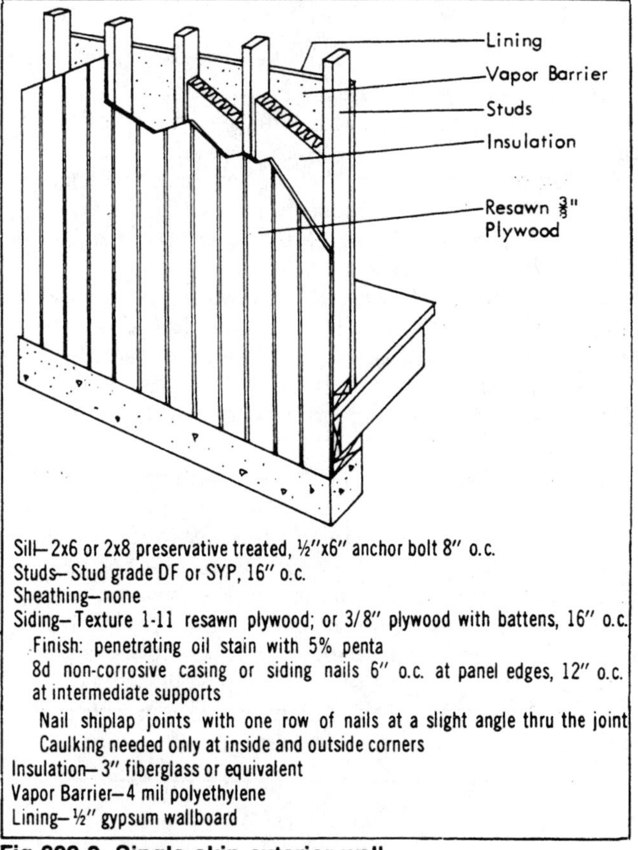

Fig 303-9. Single skin exterior wall.
A combined plywood sheathing and siding is economical and practical.

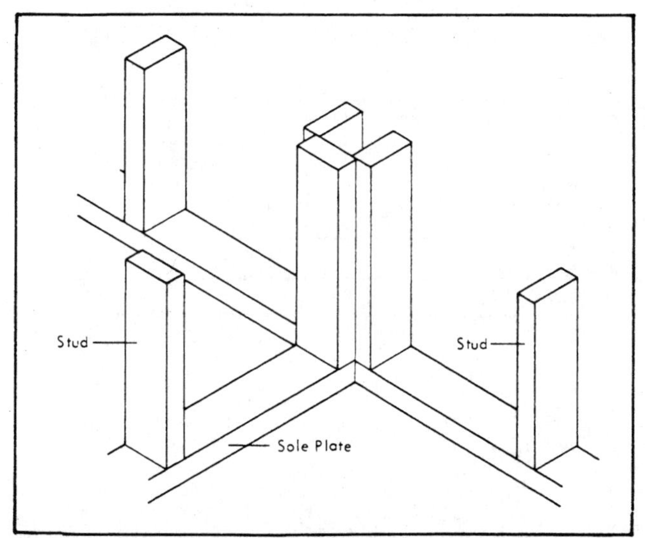

Fig 303-10. Framing where partition meets wall.

Pole-Type Construction

Pole-type construction uses treated round timber poles or sawn timber posts set in the ground as the main structural members. Girts, purlins, and rafters are attached to the poles to form a complete building frame. The poles serve as foundation, studs, and roof support.

Pole framing adapts to home construction as well as farm building construction. Girders attached to the poles above ground level can support joists, subflooring, and finished flooring over a crawl space.

Poles set to moderate depths in average soils support buildings safely even against high winds or heavy snow load.

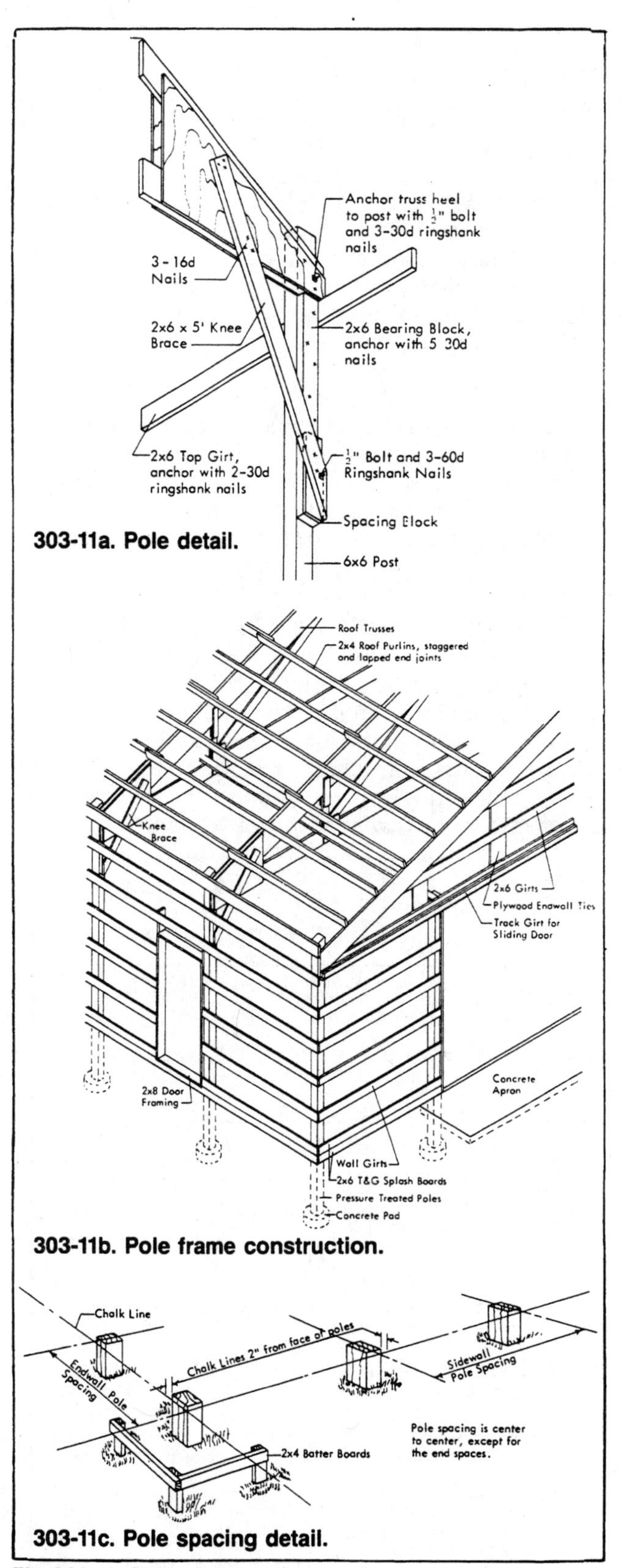

303-11a. Pole detail.

303-11b. Pole frame construction.

303-11c. Pole spacing detail.

Fig 303-11. Pole construction.

304 ROOF FRAMING

Roof Type

Fig 304-1 illustrates common roof shapes.

A shed roof is simple to build, especially on relatively narrow buildings. Gable and hip roofs adapt to many structures, particularly those with ceilings. The hip roof is used for its appearance and no gable-end maintenance. Gambrel, gothic, and gable rigid frames are used where high head room is needed. Monitors and half monitors have been used primarily for better lighting.

Common roofing terms:

Ridge: top line of roof.

Rake: exposed edge of roof, as at the gable end.

Eave: lower edge of roof.

Fascia board: trim board nailed to the fascia cut—the vertical cut at the end of a rafter.

Soffit or plancher board: trim board nailed under the eave on the plancher cut—the horizontal cut under the end of a rafter.

Roof slope: rise/run, usually ft/12′, e.g. $4/12$.

Roof pitch: rise/span, e.g. $4 \div 24 = 1/6$.

Gable roof. Fig 304-2 illustrates the parts of a gable roof. **Span** is the width between the outer edges of the plates. **Run** is one half of the span. **Rise** is the vertical distance from the top of the plate to the workline at the ridge. The run and the rise are the two legs of a right triangle; the rafter length along the workline is the hypotenuse. **Roof pitch** is the rise divided by the span. **Roof slope** is rise divided by the run.

Collar beams. Install 1x6 collar beams in the upper third of the attic space to every third pair of rafters on pitched roofs, to tie the ridge framing together during high winds.

Ridge pole or board. Ridge poles are 1″ or 2″ lumber placed between rafters at the ridge. They are 2″ deeper than the rafter to permit full contact. Rafters should butt directly opposite each other and be nailed to the ridge board. Notch rafters to fit exterior wall plates and fasten securely.

Hip roof. Components of the hip roof are shown in Fig 304-4.

Valley rafters at the intersection of two roof areas are twice as thick and 2″ deeper than the common rafters, to permit full bearing for the beveled end. Where ridges are at different elevations, provide vertical support for the interior end of the lower ridge board.

Hip rafters may be the thickness of common rafters, but should be 2″ deeper.

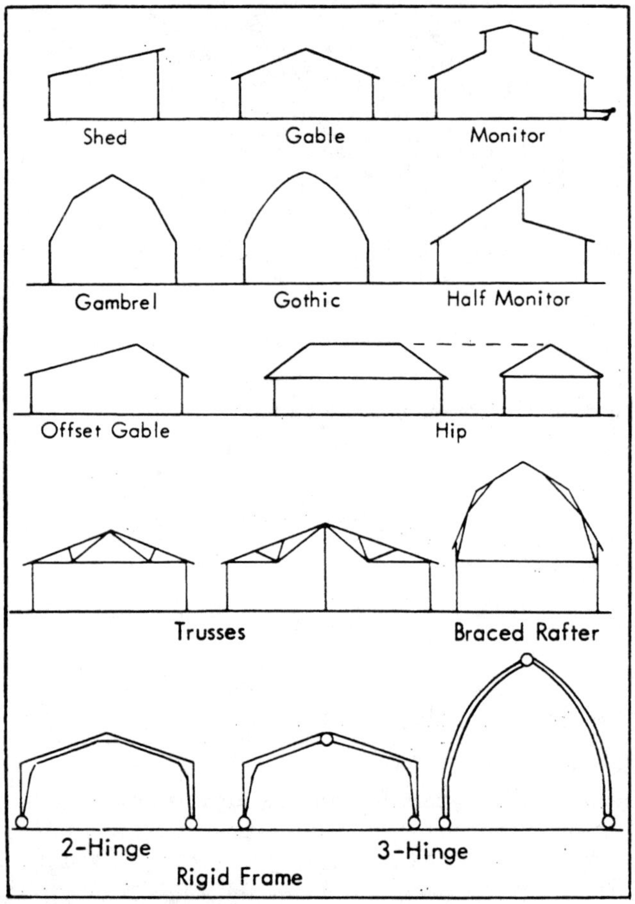

Shed
Gable
Monitor
Gambrel
Gothic
Half Monitor
Offset Gable
Hip
Trusses
Braced Rafter
2-Hinge
3-Hinge
Rigid Frame

Fig 304-1. Roof types.

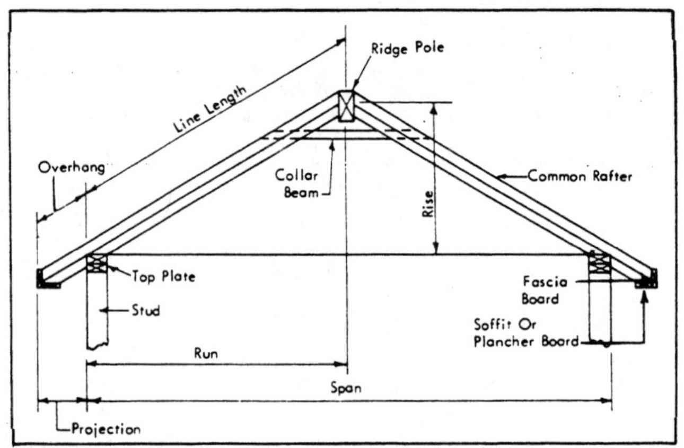

Fig 304-2. Gable roof.

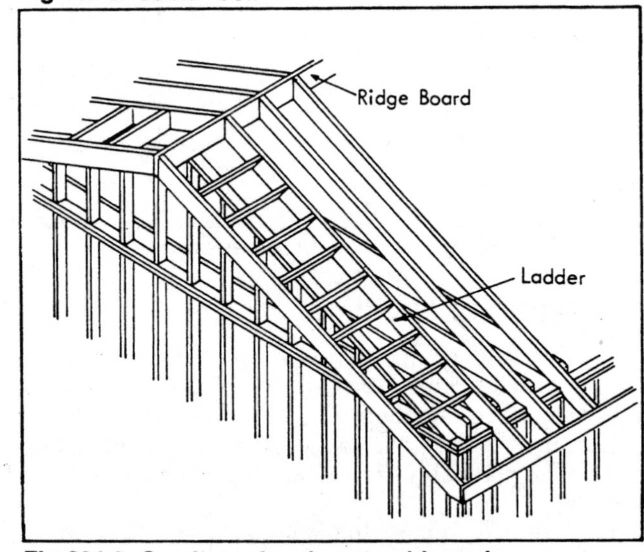

Fig 304-3. Overhang framing at gable end.

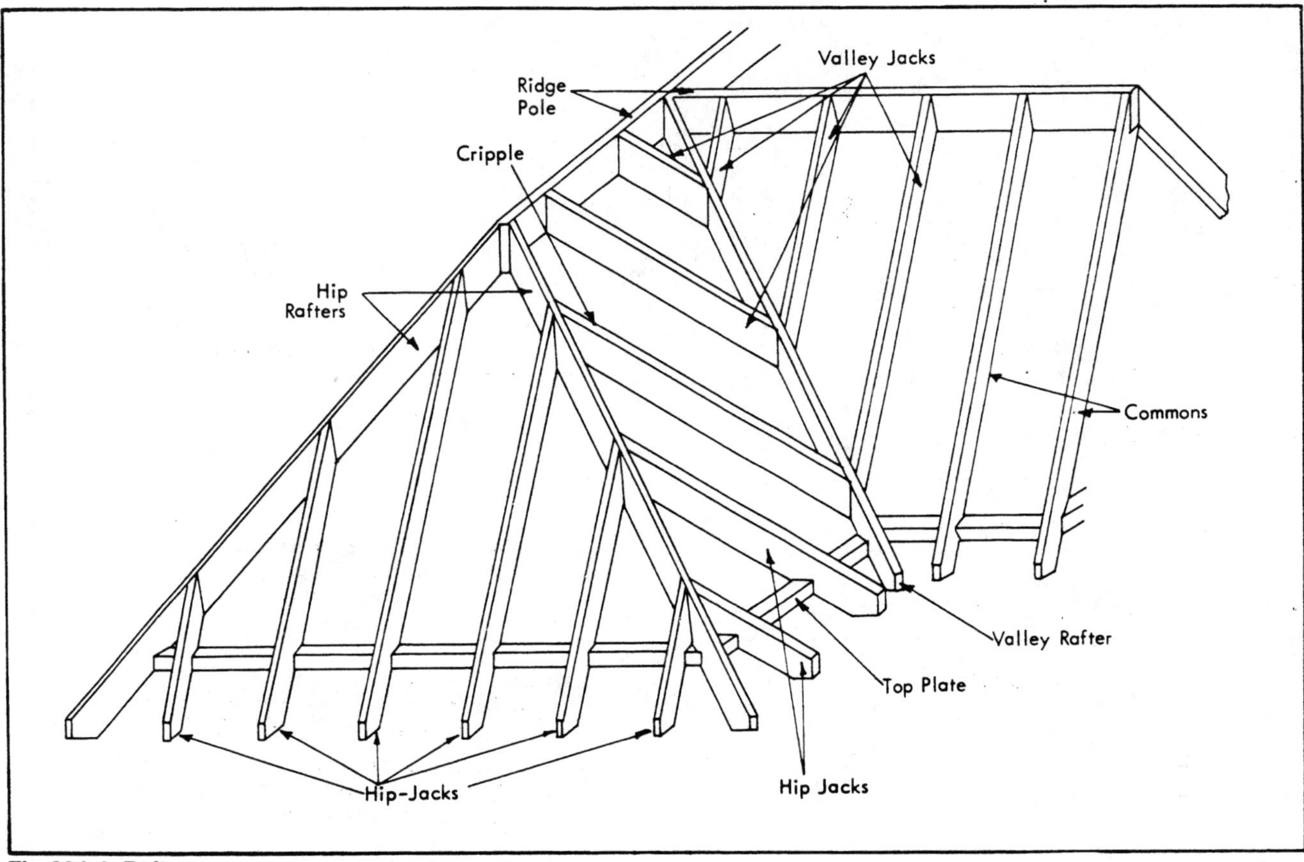

Fig 304-4. Rafters.

Flat roofs. In buildings with flat roofs the rafters or roof joists usually serve as ceiling joists for the space below. They are nailed to exterior walls. Where joists meet over interior partitions they should be nailed to plates and to each other. Avoid flat roofs as they are a major source of leaks; provide at least ¾"/ft slope if possible.

Common Rafters

Terms

- **Workline:** the line on the rafter extending from the point of the birdsmouth (outer edge of top of plate) to the ridge. It is parallel to the rafter length. The workline completes the triangle formed by the rise and run.
- **Plumb cut:** vertical cut at the top of the rafters.
- **Birdsmouth:** notch cut to fit the plate.
- **Fascia cut:** vertical cut at the lower end of the rafter overhang.
- **Plancher cut:** horizontal cut at the lower end of the rafter overhang.

Procedure

1. Lay out workline. Locate the workline on the rafter so the birdsmouth provides full bearing of the rafter on the plate. The top cut of the birdsmouth is as long as the width of the plate. Using appropriate settings on the square, and measuring from the inner edge of the rafter, locate the workline.
2. Determine square setting. A horizontal square setting of 12 is usually used for marking rafters. If roof pitch is given, the vertical square setting can be determined by using the roof pitch equation. The horizontal square setting, then, is the run. A sample calculation is shown in Fig 304-5 for a roof pitch of ¼ and a building span of 26′.
3. Mark plumb cut.
4. Measure rafter length on workline. The sample calculation applies.
5. Mark birdsmouth, fascia cut, and plancher cut. Length of overhang can be varied as desired.

Truss Construction

Have the trusses made by an experienced truss builder if possible. If you build them yourself, learn how to make good glued joints.

Assemble one truss

Select member sizes and lengths and gussets required.

A "camber" or slight raising of the bottom chord at the center of the truss provides for settlement from dead loads without the truss appearing to sag. Provide about ½" rise for 40′ trusses and 1" for 60′ trusses.

Draw the complete truss on the floor or a plywood base. Cut lumber to fit. Lightly nail **(don't glue)** the complete truss together. Check the dimensions of the truss; both sides of the truss should be the same (rise, slope, position of joints).

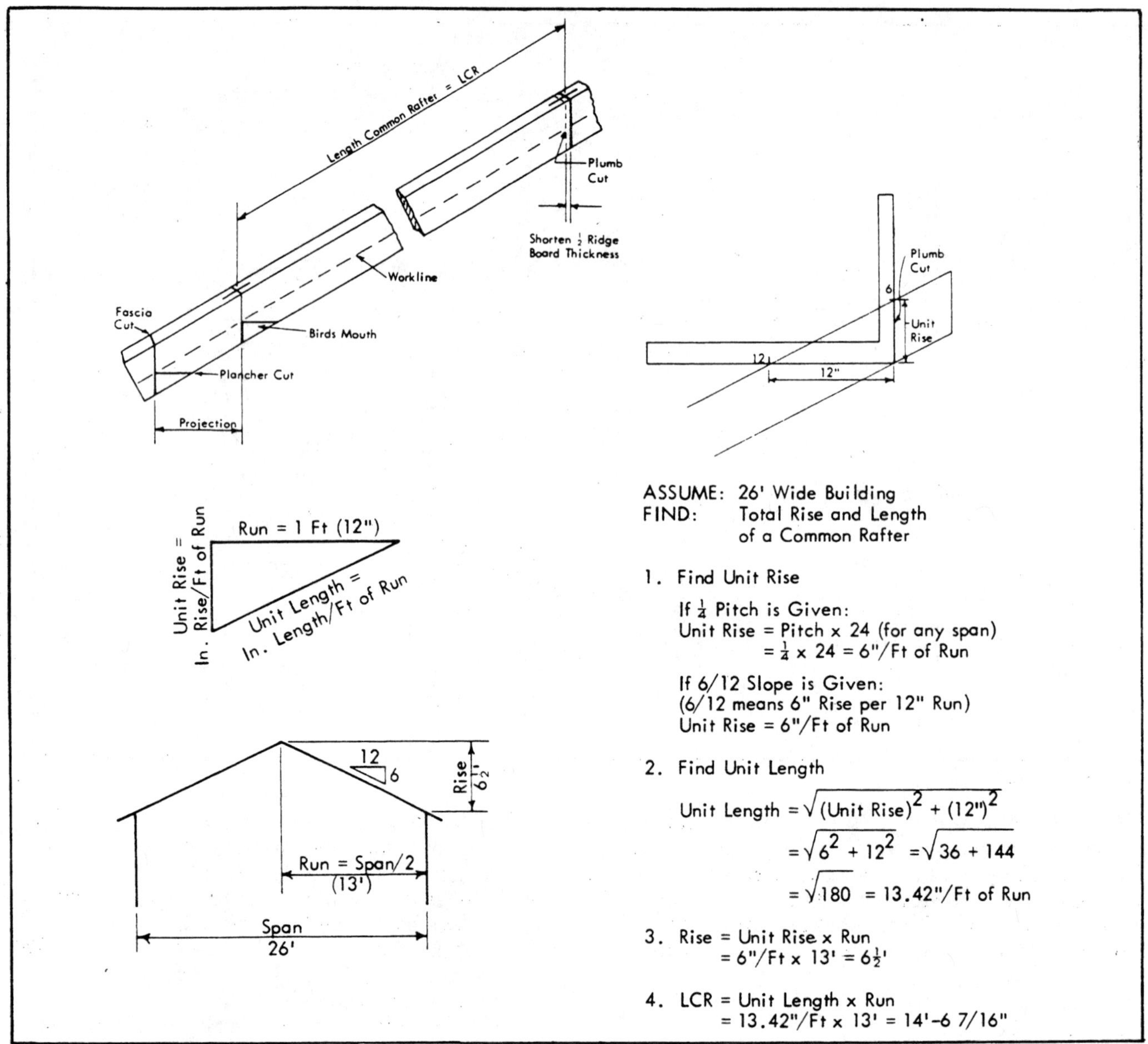

Fig 304-5. Rafter calculations.

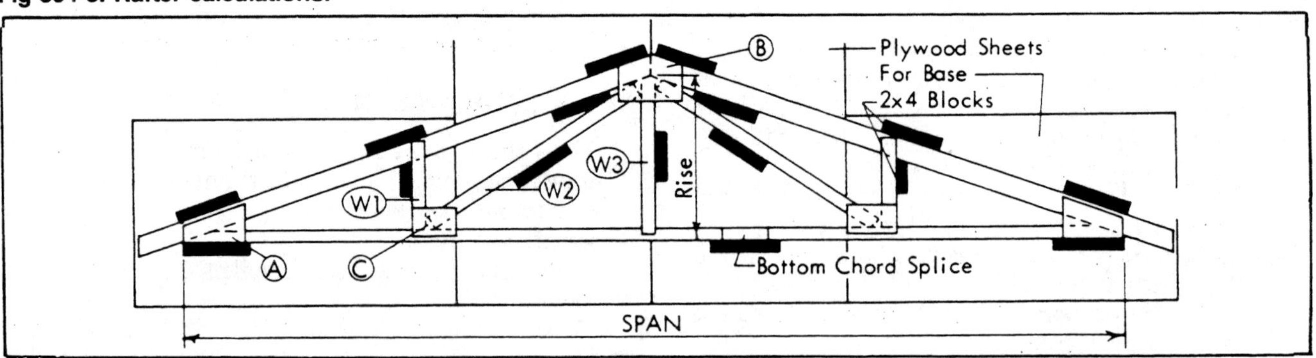

Fig 304-6. Truss jig.

Make a jig and cut other trusses

Nail 2x4 blocks to the base or floor around the lightly nailed truss. Mark off the 4" nail spacings along the jig at each joint.

Take the lightly nailed truss apart, and use its pieces as patterns for the rest of the trusses. Cut out the plywood gussets. The face grain of the heel gusset and any bottom chord splices **must** be parallel to the bottom chord. The direction of the grain in the other gussets is not critical.

The king post may be omitted in trusses that will not have crossbracing attached, although a king post is recommended for all trusses supporting ceilings.

Gussets are required on **both sides** of each joint.

Double webs

Buckling from forces created by snow loads or wind uplift is critical for web members in many large trusses—two 2x4s may be required to provide adequate stiffness.

Construct the truss with single web members: then if the plan calls for it, cut an additional 2x4 to fit between the gussets. Nail it securely with 10d nails about 6" apart.

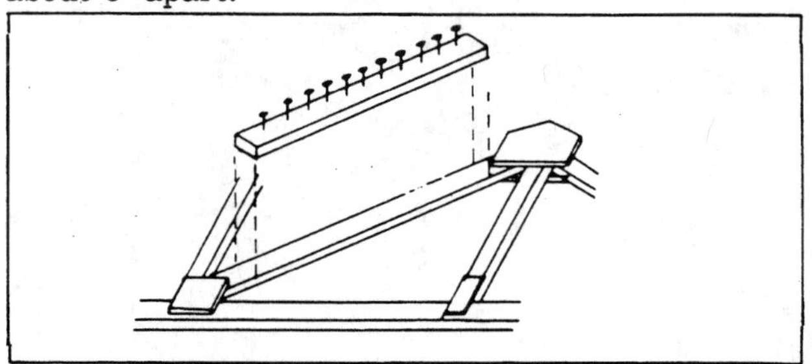

Fig 304-7. Nailing double webs.

In even larger trusses, both buckling due to wind uplift and tension due to snow loads are critical—a 2x6 and a 2x4 may be required. The 2x6 is required for tension and should be glued into the truss as the primary web member and the 2x4 added as a double web to provide adequate stiffness. The 2x4s used as double webs are stiffeners and do not need to be of high quality.

Assemble and glue

Use the jig to make uniform trusses.

Use plenty of glue—it's the cheapest part of the building. Spread the glue with a fiber glue brush or paint roller.

Apply to both surfaces to be joined. When you nail the gussets in place, glue should ooze out from around the joint—if it does not, you are not putting on enough glue.

Use 5d or 6d box nails, preferably galvanized or cement coated. Space rows of nails 2" apart across the lumber grain. Nail 4" apart with the lumber grain, starting with a middle row, and nailing outward to the edges of the lumber. You may machine nail, but as each join is nailed, hit each nail at least once with a hammer. Machines drive the fasteners so rapidly that the glue might not have time to spread—the final manual blow assures a tight joint. **Machine stapling is not recommended** because failures have occurred.

Raise the peak of the truss and turn the truss over, out of the jig. Glue and nail the gussets on this side, and nail on the double webs if required.

To eliminate errors in the pattern, which may result in an uneven ridge, mark one end—e.g. right end—of each truss and erect the trusses with that end on the same side of the building.

Move the truss out of the way and store flat for at least 24 hr. After the glue has cured, the truss is ready to be placed on the structure.

Precautions

Remove all dirt, oil, and sand from the lumber and plywood. This is necessary for a good glue bond.

Protect the glued joints from moisture for one week after fabrication. A plastic covering over stacked trusses is adequate.

Temperatures below 70 F delay curing. Trusses will be ready to erect in 24 hours at 70 F, but require at least a week at 40 F.

Above 85 F, pot life may be very short. Keep the glue in the shade, and close and nail joints as soon as possible after spreading. Consider keeping the pail of mixed glue in a bucket of ice in hot weather.

Endwall trusses

Endwall trusses are sometimes built without gussets on the outside face to simplify application of endwall siding. However, the endwall truss is not strong enough to support additional building length in any future expansion. If you might lengthen the building someday, put gusset plates on both sides of the joints and space plywood scraps along the truss members for siding application.

Truss Erection

A building must be strong enough to withstand the snow and wind loads that come after the building is complete.

The building walls support the trusses and anchor them against wind loads. Girders or lintels over openings should be strong enough to hold the trusses without excessive deflection. Consult an engineered plan for adequate design of the girder and its fastening.

Placing

Lighter trusses can be lifted by hand and placed on the walls upside down. Lift heavier trusses into place with a hoist. The trusses are positioned, anchored, and braced one at a time.

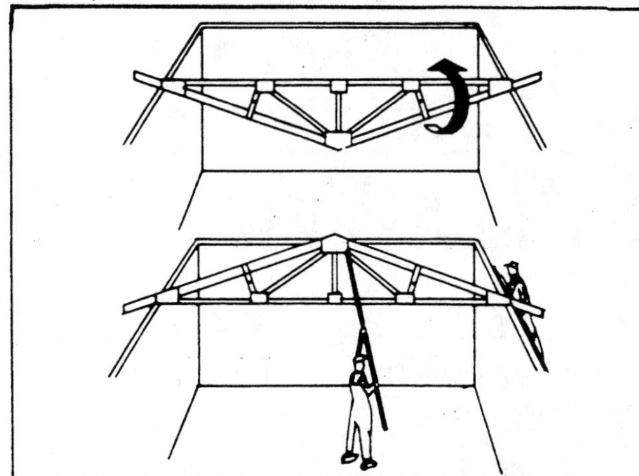

Fig 304-8. Placing trusses.

Windbracing

Brace and anchor the trusses as they are placed. The bracing helps align and hold the trusses in place while the roof and ceiling are applied.

Install stiffeners along the bottom chords at all panel points (where webs meet bottom chord) continuously from endwall to endwall unless a rigid ceiling is to be installed. The cross-bracing on the king posts is needed in all buildings.

Install cross-bracing for at least 16′ at each end of the building. Space additional 16′ braced sections not farther than 32′ apart.

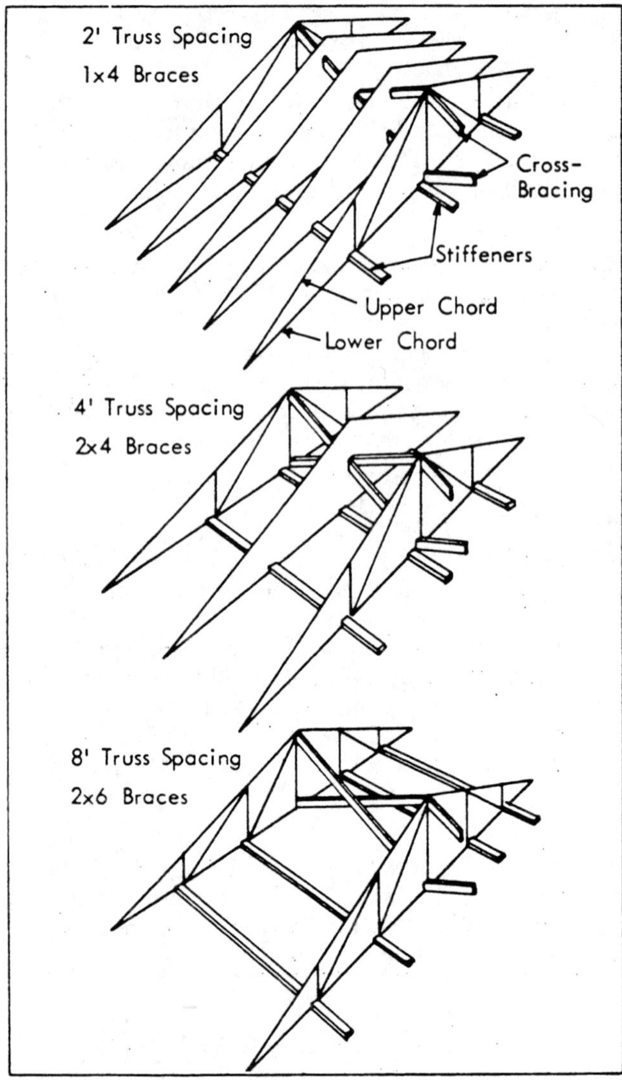

Fig 304-9. Windbracing.

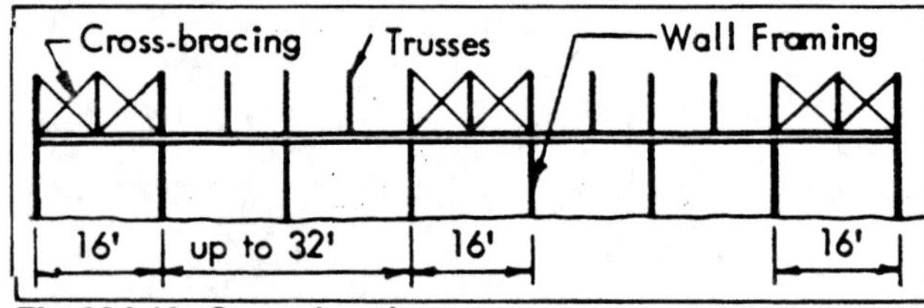

Fig 304-10. Cross-bracing.

Wind anchorage

The wall holds the truss up. Anchorage resists uplift on a roof due to the wind.

Table 304-1. Minimum fasteners for wind anchorage.
For both ends of each truss.
A = framing anchor
B = ½″ machine bolt
4-30d ringshank nails equal one ½″ bolt.

	---Truss Spacing---		
Truss Span	2′	4′	8′
20′-24′	1A or 1B	1A or 1B	2A or 1B
26′-30′	1A or 1B	1A or 1B	2A or 2B
32′-46′	1A or 1B	2A or 1B	3A or 2B
48′-50′	1A or 1B	2A or 1B	4A or 2B
52′-60′	1A or 1B	2A or 2B	4A or 3B

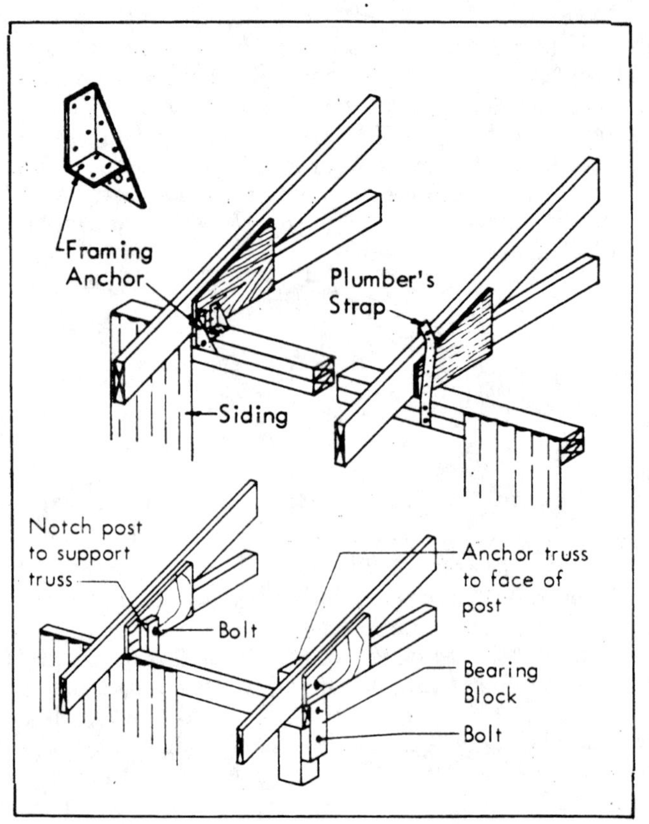

Fig 304-11. Truss anchorage.

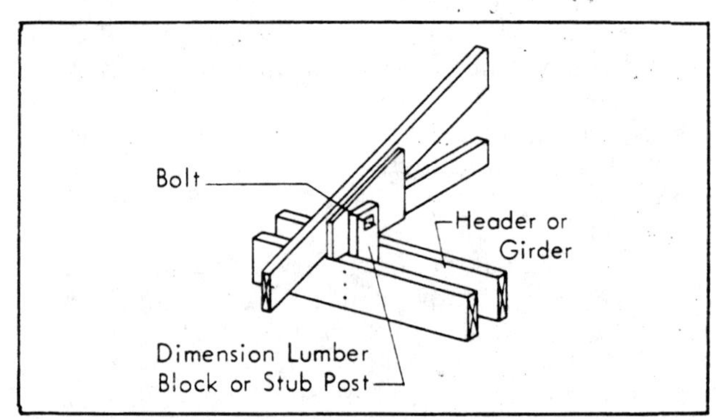

Fig 304-12. Truss to girder anchorage.

Knee bracing

In stud wall construction, knee braces stiffen the wall and roof joint to resist high winds. In buildings with rigid cross-partitions, such as homes, knee bracing is not required. In pole buildings with poles sized assuming rigid pole-to-truss joints, the knee-brace is part of the structural system. Knee bracing may be omitted in pole buildings with poles sized assuming pinned pole-to-truss joints.

In stud wall buildings where knee braces are not wanted, partition walls can be used for wind bracing. Brace walls approximately every 80′ by a partition wall of a structurally sound skin material: preferably plywood; possibly metal roofing or cement asbestos board; not insulation or fiber board. Anchor the wall to the floor and to the bottom chord of a truss as in endwalls. Some small doors will not seriously reduce the partition's stiffness.

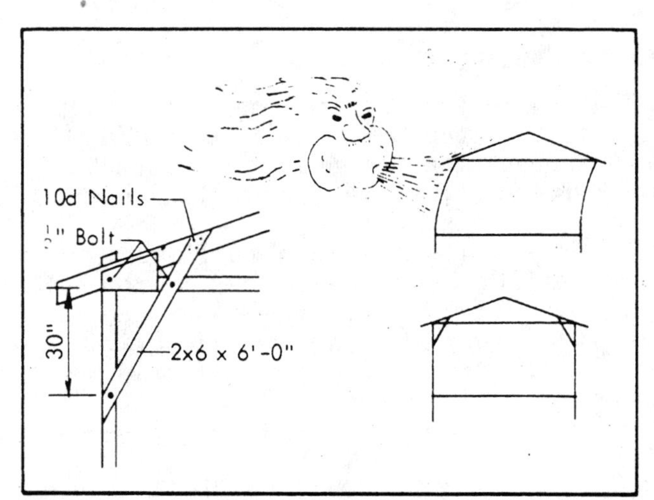

Fig 304-13. Typical knee brace—8′ o.c.

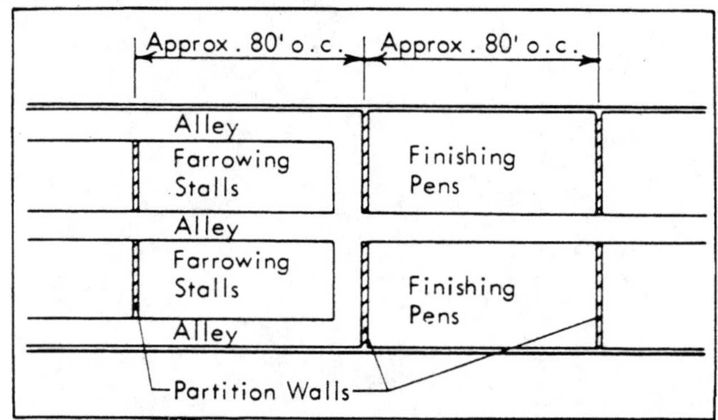

Fig 304-14. Windbracing with partition walls.

Roof Construction

After the trusses have been placed, anchored, and braced, the roof and ceiling can be applied. Select the type of roofing and ceiling material to be used. Framing and sheathing required depends on the roofing material and the truss spacing.

Sheathing

Many types of material can be used as sheathing. Plywood is common.

Anchor sheathing ½″ or less thick with 6d nails and thicker sheathing with 8d nails. Space nails 6″ apart along panel edges; 12″ apart along other supports. If panel supports are 48″ o.c., space all nails 6″ apart.

For support spacings of 32″ or more, use T&G plywood, blocking under joints, or plyclips on panel edges between supports.

Plywood may expand a little; leave 1/16″ space between roof or wall panel ends and ⅛″ between sides to allow for possible expansion.

Support 1″ lumber sheathing 24″ o.c., or 48″ o.c. if it is T&G, in both farm and residential construction. Anchor with 8d nails. Space supports for rolled steel and aluminum roofing according to manufacturer's recommendations.

Roof purlins

The maximum spacing between roof purlins depends on the sheathing or roofing materials used, and on the strength of the purlin itself.

Stagger purlin joints for continuity across the trusses. Purlins may be laid flat with 2′ and 4′ truss spacings. Butt joints over the trusses and use 8′, 12′, and 16′ lengths.

Alternate purlin lengths in pole buildings where the poles are spaced evenly and the trusses are not. For poles 8′ o.c., use 16′ and 18′ lengths with staggered lap joints if pairs of trusses are mounted on alternate sides of the poles.

Knee braces at sliding door and vent door locations may dictate altering this pattern.

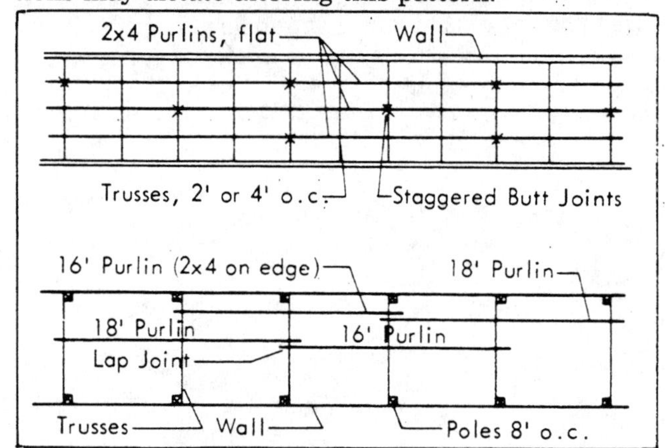

Fig 304-15. Staggered purlin joints.

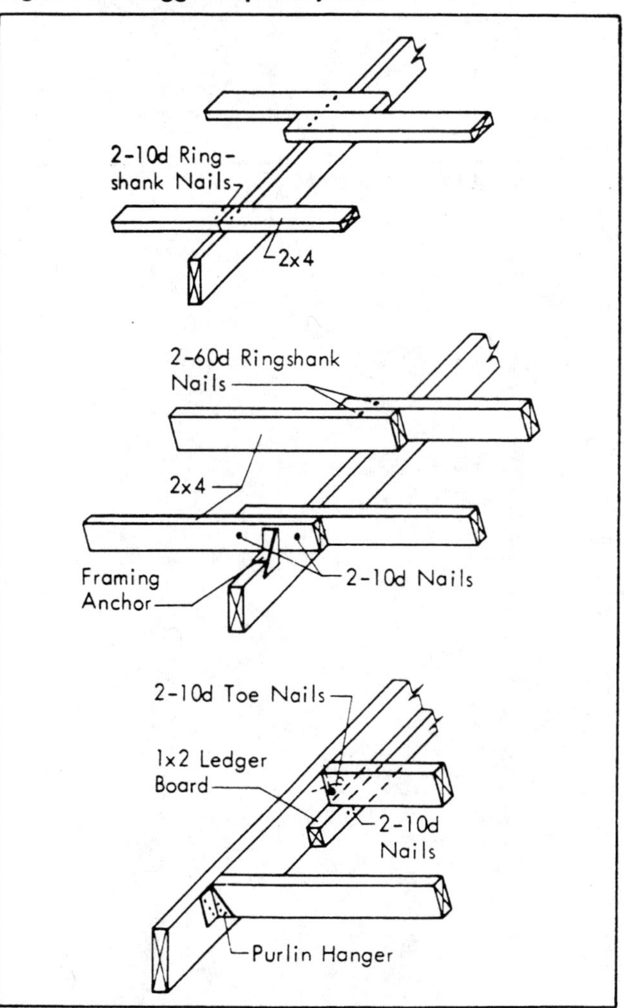

Fig 304-16. Roof purlin anchorage.

Rigid Frames

With this system two studs and two rafters are combined to form a structural unit with a gable shape similar to arched rafters. Plywood gussets are nailed and glued on each side of the eave joints and the ridge joint to form rigid frames.

Rigid frames are free of interior obstructions, and therefore are well suited for many types of farm buildings. Construction is simple because individual frames consist of a relatively small number of different sizes and shapes of framing members. The cost is no greater than for other systems of clear-span construction and may be less.

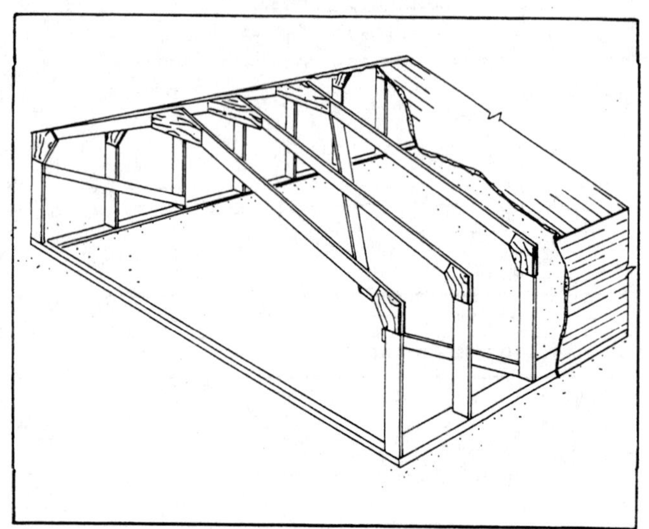

Fig 304-17. Rigid frame building.

Arched Rafters

Wooden arched rafters have been used for most kinds of farm production and storage buildings. Glued laminated bent rafters are commonly fabricated in factories while sawed (vertical laminated) rafters can be assembled on a job site.

Relatively complex framing is needed to put doors in the sides of these structures. Doors can be conveniently installed in the endwall. Ceilings may be installed, but many of the advantages of this framing system are lost. Arched buildings are especially useful where high ceilings are needed.

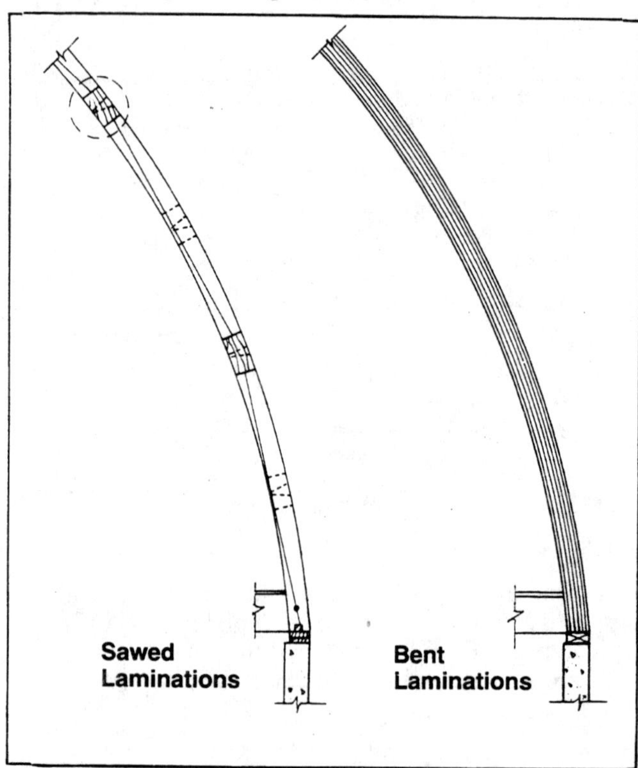

Fig 304-18. Laminated arches.

305 CONCRETE PAVEMENTS AND MANURE TANKS

Pavement Construction

Remove all sod and vegetable matter. The subgrade must provide uniform support and be easily drained. The top 6″ of subgrade should be sand, gravel, or crushed stone where subgrades may be water-soaked much of the time. Thoroughly dampen the subgrade before concrete is placed.

Use 6x6 #10 wire mesh in pavements poured over soft or spongy soils. In general, place the wire about 2″ from the top of 5″ and 6″ slabs and near the middle of thinner slabs.

Place the concrete near where it will be used to help prevent scaling and dusting from too much handling.

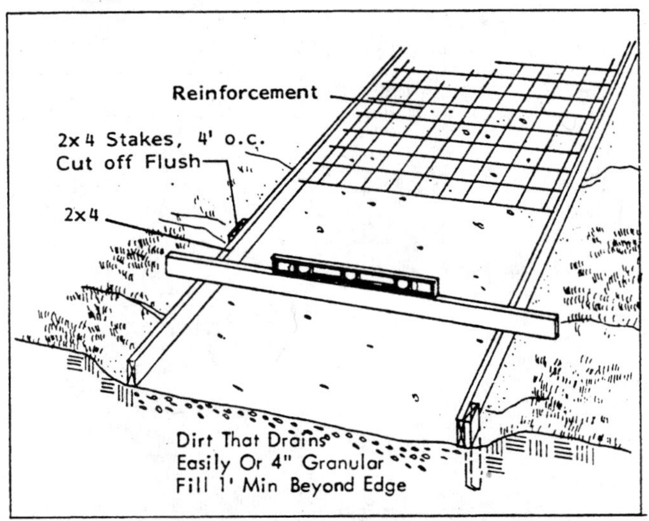

Fig 305-1. Preparing subgrade and laying reinforcing.

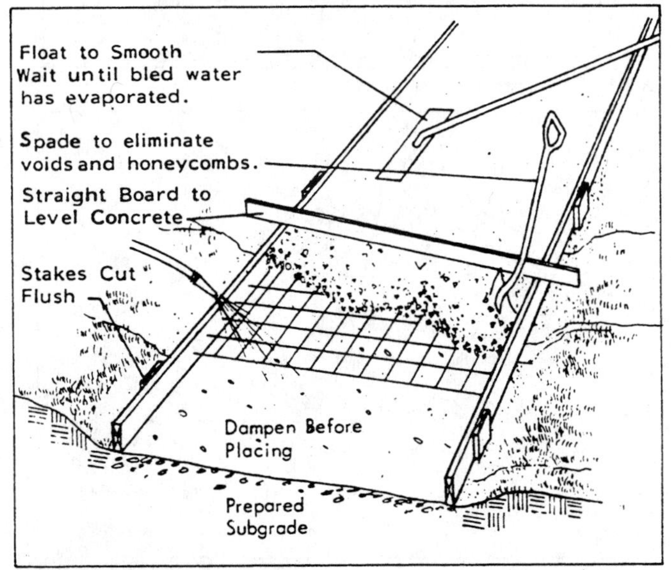

Fig 305-2. Placing concrete.

Remove concrete from wheelbarrows or mixers and spread as soon as possible. Spade or vibrate along the forms to eliminate voids or honeycombs. Strike it off to the proper grade, screed the surface, and then darby (using a darby or bull float to smooth and level the surface).

It is very important that the first operations of placing, screeding, and darbying be completed before any bleeding occurs. Bleeding is excess water in the concrete rising to the surface.

When all bleeding water has evaporated and the concrete has started to stiffen, start the other finishing operations, such as edging and jointing.

Cut control joints across a slab to control cracking. Cracks, if they occur, will usually be in the joints. Cut control joints soon after the concrete has been placed

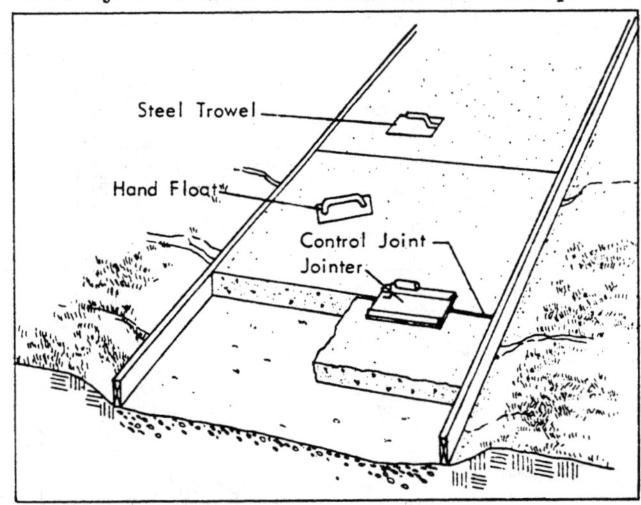

Fig 305-3. Control joints.
Space control joints at intervals equal to width of slab, but not more than 20′ apart.

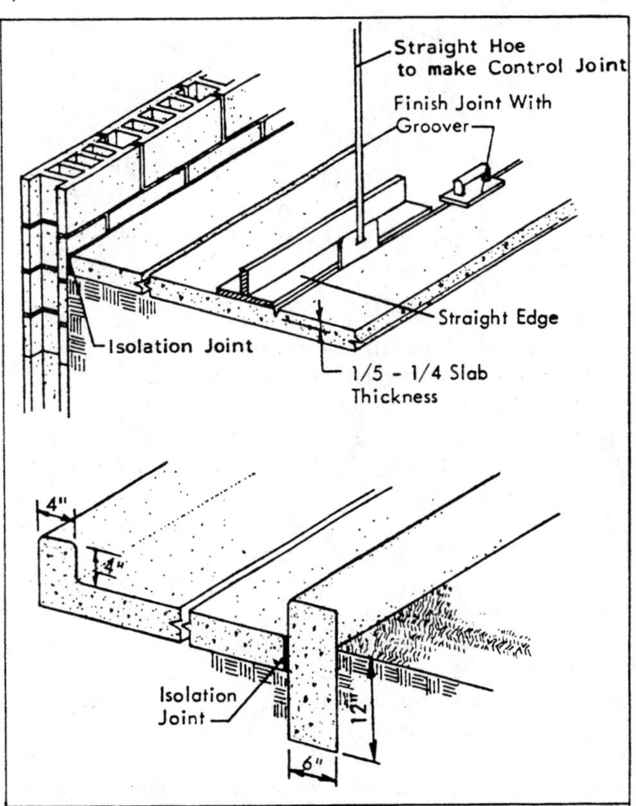

Fig 305-4. Isolation joints.
Place isolation joints along existing improvements such as buildings, water tanks, or paved drives to permit the slab to move with the earth.

to work coarse aggregates away from the joints. Space them about as far apart as the concrete is wide.

After edging and hand jointing, float the slab to embed large aggregates beneath the surface, remove slight imperfections and tool marks, and prepare the surface for other finishing.

Final smoothing is done with a steel trowel (use a magnesium trowel for air-entrained concrete) immediately following floating. If a rough surface is desired, final smoothing is omitted.

Isolation joints permit the slab to move with the earth. Place isolation joints along existing improvements such as buildings, concrete water tanks, or paved drives.

Expansion joints are the same as isolation joints and are installed in new walks and long drives to prevent buckling of the slab during hot weather.

Edges of slabs are thickened to reduce cracking and to allow for some erosion of adjacent soil. Construction joints key adjacent slab pours to each other.

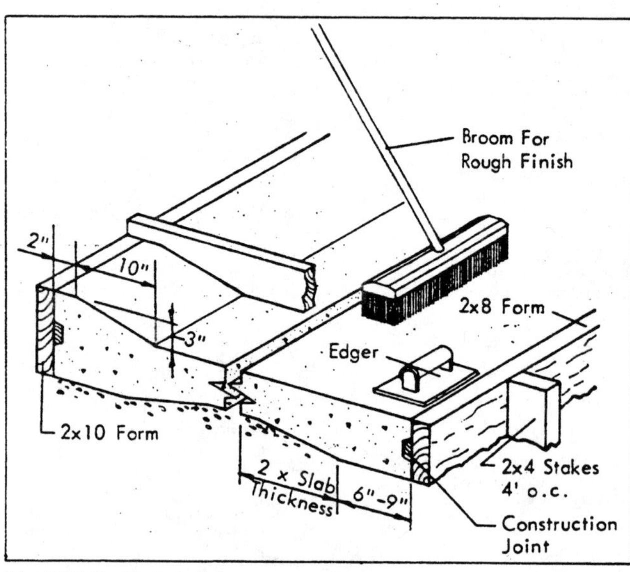

Fig 305-5. Slab joints.

Curing

Concrete does not dry—the paste sets by a chemical reaction between cement and water. Keep the surface of the concrete damp at least 5 days. Curing continues for months. Remove forms after about 5 days for slabs, 10 days for walls, and 28 days for structural elements.

Slip-Resistant Concrete Floors

Slippery concrete floors can result from concrete surfaces being finished with a steel trowel. Steel troweling brings fine aggregate and cement to the top, forming a glazed surface. Wood float and broom finished surfaces become smooth in time due to tractor scraping and constant animal traffic.

Select the degree of roughness based on the type of animal confined. Deep grooves make cleaning and disinfecting more difficult and can cause foot and leg problems with smaller animals.

New floors

To roughen new floors, score the surface with a homemade tool or add aluminum oxide grit after floating. Be certain the surface is hard enough not to let the concrete flow into the grooves or cover the grit.

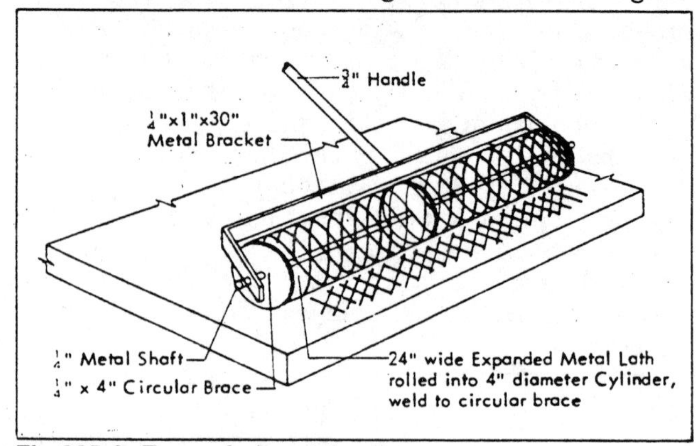

Fig 305-6. Expanded metal roller.

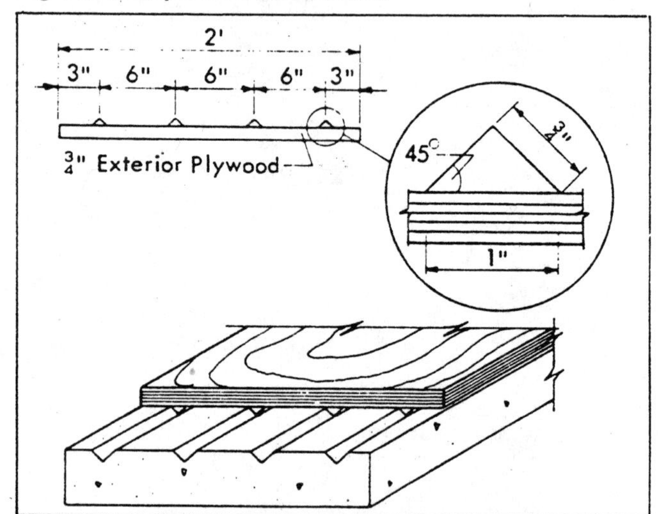

Fig 305-7. Wood groover.

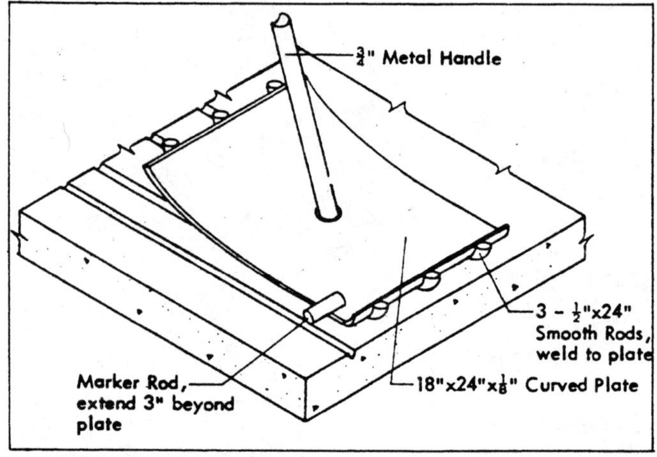

Fig 305-8. Steel groover.

Rough surfaces. Roughen the surface while the concrete is wet but firm. Make grooves diagonal to the direction of traffic to avoid catching scraper blades. Make crossing grooves for a diamond-shaped pattern.

Gritty surfaces. Develop gritty surfaces by applying aluminum oxide grit (as in sandpaper) at a rate of ¼ to ½ lb/ft² before the concrete sets. Two sizes of grit are generally available. The coarser grit (sized through a screen having 4 to 6 meshes/in.) is recommended because it is better anchored and less likely to pop out. The grit may cause extra wear on hoofs as less foot trimming has been experienced. Alley scraper blades wear faster when scraping a grit-surfaced alley.

Roughening hardened floors

There are several choices for roughening an existing concrete surface: chemical, mechanical, heat treatment, paint, or pouring a new 2″ layer over the existing slab. Treat the surface before it becomes so slippery the animals have difficulty walking.

Chemical. Chemically roughen the slippery surface with commercial muriatic (hydrochloric) acid. It is available at concrete ready mix plants and dairy supply, hardware, or drug stores. The acid attacks the lime in the concrete, removing the glaze and leaving a gritty surface.

Use extreme caution when handling acid. Wear protective gear such as goggles, rubber gloves, rubber boots, and a long-sleeved shirt. Wash any acid off your skin immediately. Do not breathe the fumes. Use a granite pail or crock to hold the water and acid mixture.

Acid deteriorates concrete, so carefully measure the amount used to limit chemical action. Floor surfaces vary in texture and hardness, so experiment where the results can be easily observed. One gallon of commerical muriatic acid roughens between 200 and 300 ft². Start with a solution of 1 part muriatic acid to 10 parts water. Work a liberal amount over the concrete with a long-handled scrub brush. The reaction between the cement and acid should cause the solution to fizz. If no fizzing occurs, gradually increase the acid. Do not exceed 3 parts acid to 10 parts water or excess deterioration of the concrete may result. Flush the surface with a generous amount of clean water after the fizzing stops—usually several hours. Carefully observe the amount of roughness you want, then treat the remaining area. Keep the acid out of sewer lines and liquid manure tanks—collect the drainage and haul it directly onto a field. In some areas, additional treatment may be necessary. Avoid excessive damage to the concrete.

Mechanical. Electric jack hammers can cut shallow grooves or chip the surface. It is usually more difficult and time consuming than the acid treatment. Rent the tools from a concrete contractor, highway maintenance department, or ready mix concrete supplier.

Pneumatic tools may be available with heads suitable for roughening, rather than breaking, the surface. A concrete saw or Carborundum wheel can make shallow cuts at close intervals to provide additional traction. Any of these measures requires special equipment which may not be readily available.

Concrete floors often become slippery after several years of scraping with a steel blade. Weld beads of special hard-surfaced welding rods to the under edge of the blade. The beads, at about 2″ intervals, score the concrete during scraping. Chains on the cleaning tractor tires also help break the surface glaze.

Heat can cause concrete surfaces to flake, leaving a rough, pitted surface. Although the method is slow, all you need is a welding torch, blow torch, or a weed burner.

Paint materials may not stick to existing concrete, and the cost is relatively high, but they are effective in areas of moderate human foot traffic.

A new 2″ layer can sometimes be poured over an existing floor. Thoroughly clean and dampen the surface before pouring. Use small aggregate. Roughen the floor as outlined above.

Outside Concrete Work in Cold Weather

Concrete cures very slowly at temperatures below 50 F. Water frozen in uncured concrete expands and damages the concrete. Cure concrete at least 48 hr before it freezes, but preferably 4 to 5 days.

Do not place concrete over frozen ground. Use Type IIIA Portland cement, or Type IA with calcium chloride dissolved in the mixing water at the rate of 2 lb per bag of cement.

When air temperatures are below 40 F:

- Heat the sand, gravel, and water to just below 150 F. Heat the sand and gravel in separate piles over culvert pipe, smokestack, or other firebox. Place a fire inside. Stir and rake the materials frequently to ensure even heating.
- Remove snow and ice from forms before placing concrete.
- Place concrete right after mixing, when the concrete temperature should be 60-80 F.
- Cover the concrete to retain as much heat as possible for 4 to 5 days. Canvas, straw or hay are often used as covers.
- Maintain concrete temperature at 70 F for 3 days or 50 F for 5 days. Do not allow it to freeze during the next 4 days. Use vented heaters indoors.
- Test for sufficient curing: pour hot water on the concrete. If frozen, the concrete will soften, but if properly cured, there will be no effect and forms may be removed.

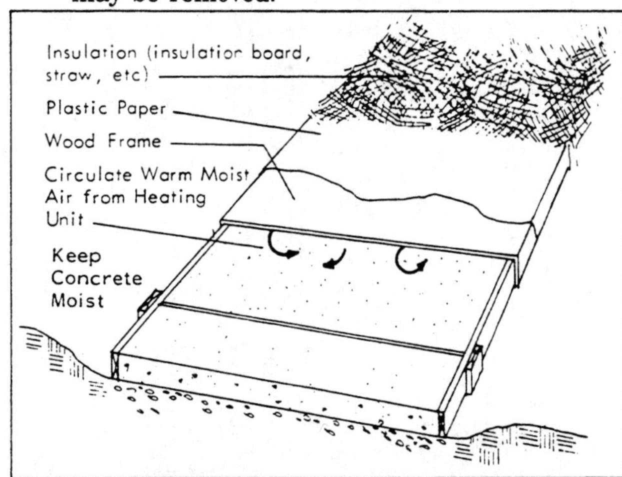

Fig 305-9. Protecting concrete in cold weather.

Outside Concrete Work in Hot Weather

As temperatures rise above 70 F, curing rates increase. Evaporation of water from the concrete also increases. A combination of wind, high temperature, and low humidity dries concrete too rapidly and weakens it.

Keep fresh concrete damp for at least 5 days. Retard evaporation with a sheet of 4 mil plastic over the dampened concrete.

In hot weather it may be necessary to reduce the temperature of the freshly mixed concrete. Stockpile aggregates in the shade, and cool mixing water. Start curing promptly to retard water evaporation.

Concrete Manure Tank Construction

See Chapter 208 for design criteria and selection tables. See MWPS TR-3, *Concrete Manure Tank Design,* and MWPS TR-4, *Welded Wire in Concrete Manure Tanks,* for additional design tables.

A liquid manure storage tank supports soil loads which tend to push the walls in. The walls are reinforced to prevent them from breaking and are keyed to the floor and top to prevent them from falling over. Partitions can have liquid on either or both sides, and so are doubly reinforced, Fig 305-11.

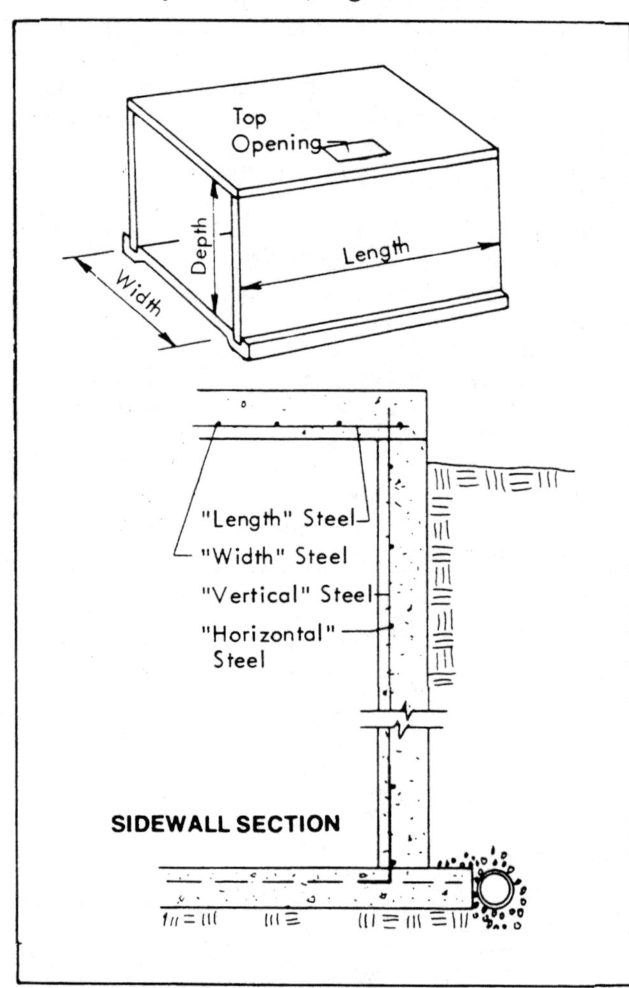

Fig 305-10. Manure tank construction.

If possible, locate the tank so that tractors, wagons, or other vehicle traffic cannot be on the tank top. If traffic cannot be avoided, select a tank top design adequate for vehicle traffic.

Design procedure

1. Select depth of tank required.
2. Select wall thickness and reinforcing. "Horizontal" steel depends on wall thickness. "Vertical" steel depends on tank depth. Use the 100 psf surcharge if heavy vehicles can drive within 5′ of the tank walls.
3. Select tank width.
4. Select the top thickness and reinforcing: "Width" steel depends on span and design load. "Length" steel depends on top thickness. For tanks under slotted floors, select the appropriate slat design.
5. If a column and girder system is to support the solid or slotted tank top, review Chapter 208, and consult an engineer.

The tanks are designed to have the walls supported by the top. Special provisions must be made for tanks under slotted floors, Figs 305-13 to 305-15.

Materials

Concrete: Air-entrained, f'_c = 3500 psi minimum 28 day strength.
Maximum aggregate size = ¾″.
Steel: Deformed reinforcing steel, f_y = 40,000 psi minimum

Locate steel accurately in forms; hold firmly in place with wire ties and with accessories such as slab bolsters and spacers.

Reinforcing steel must be covered with concrete for strength and protection against corrosion:

Cover = 1½″ for #5 bars or smaller
= 2″ for #6 bars or larger

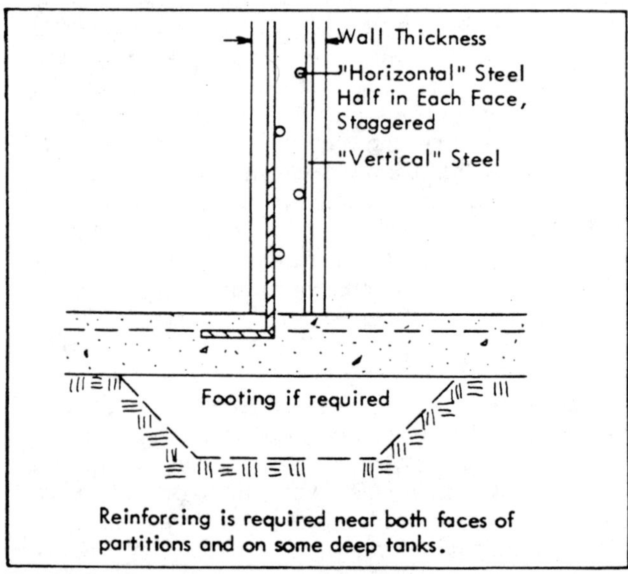

Fig 305-11. Partition wall construction.

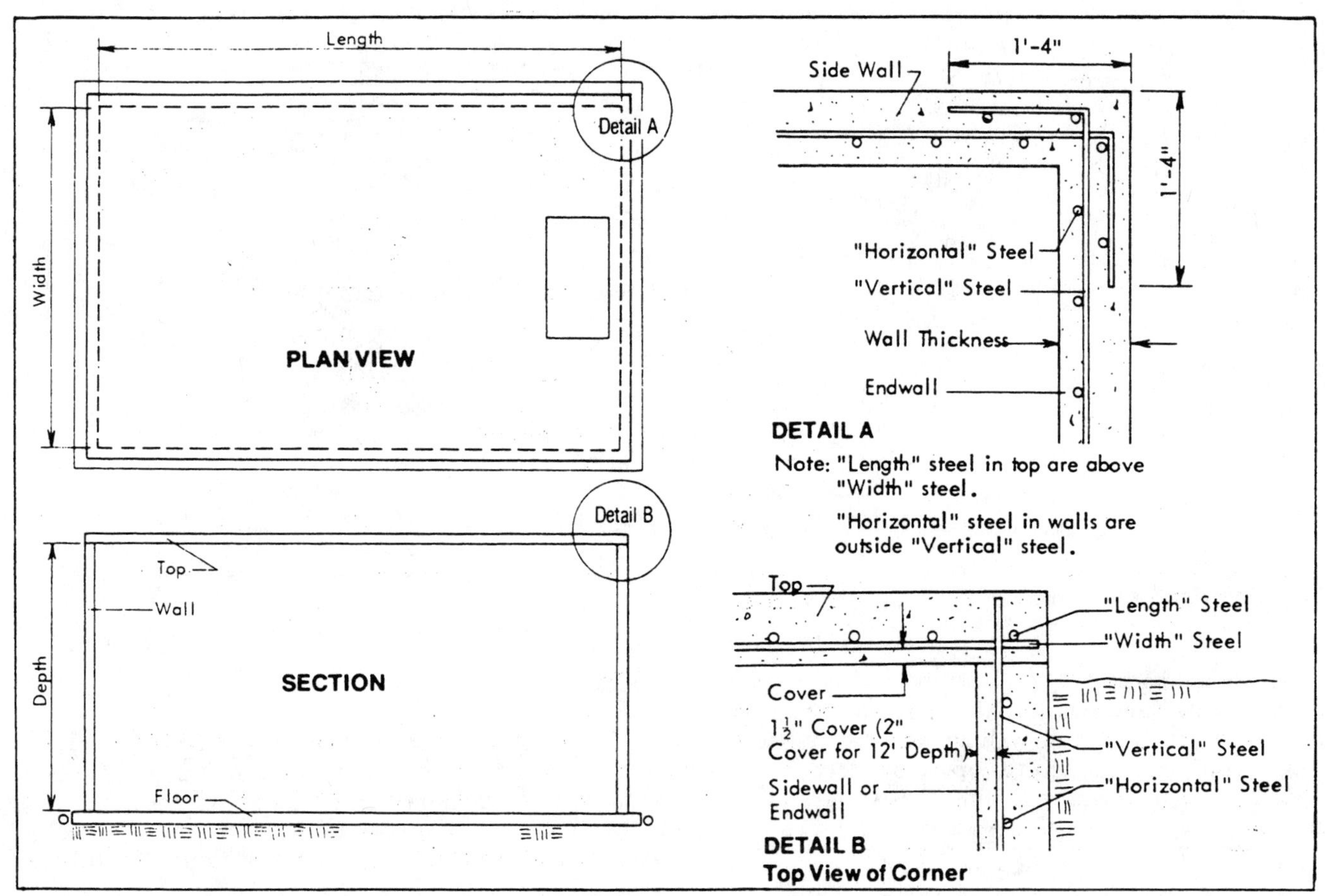

Fig 305-12. End wall construction.

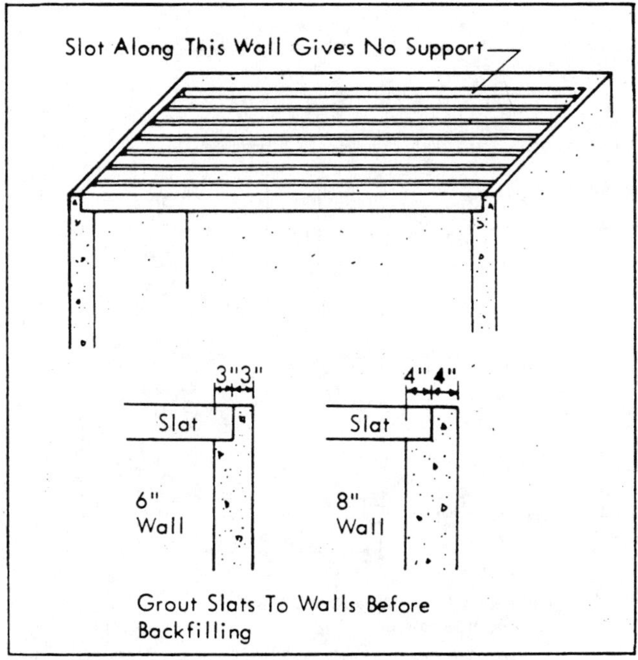

Fig 305-13. Supporting sidewalls with reinforced concrete slats.

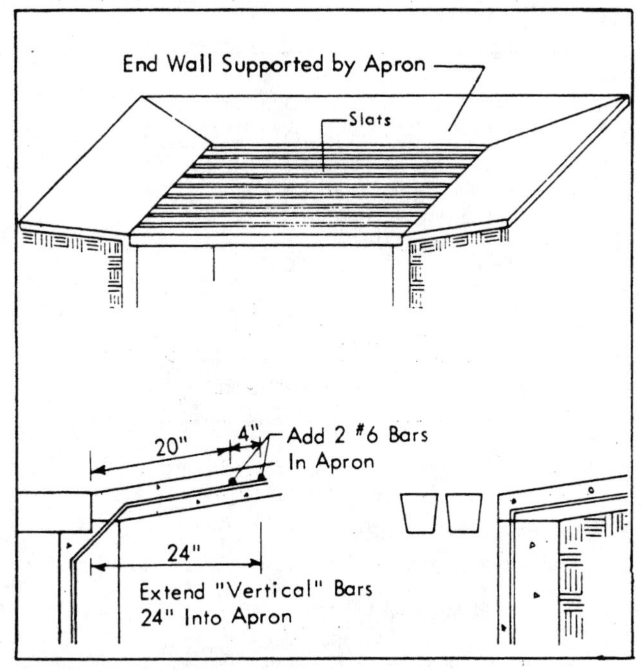

Fig 305-14. Paved apron to support the wall.

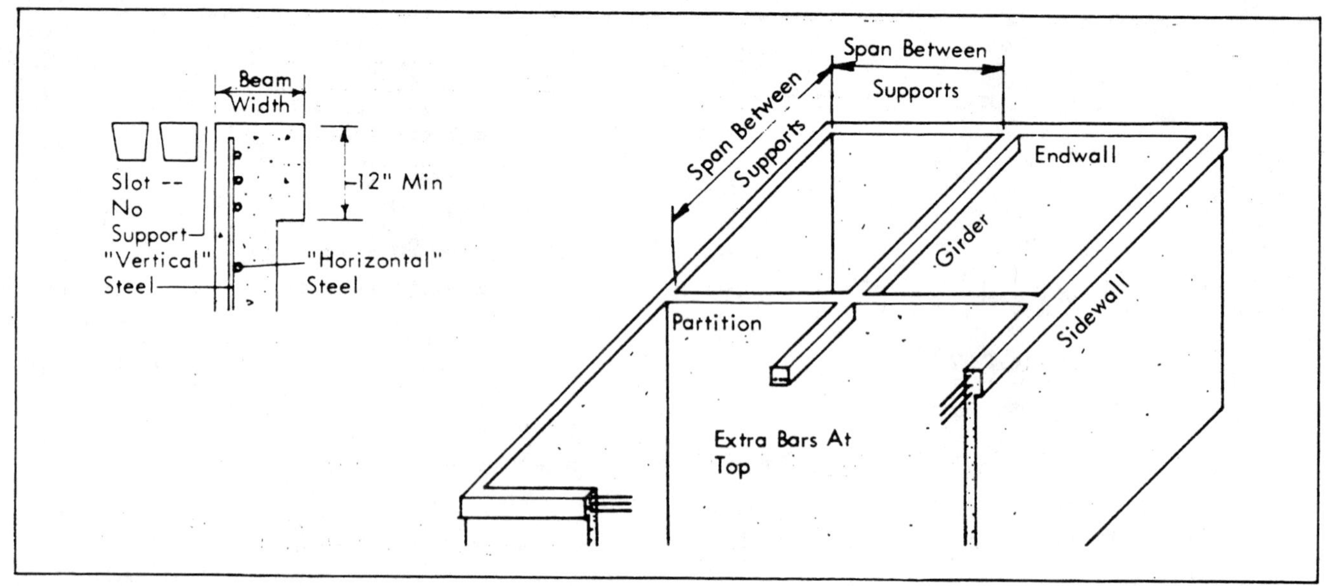

Fig 305-15. Top of wall reinforced between supports.
With a beam cast as part of the top of the wall.

Excavation and backfill

Slope excavations over 4′ deep no steeper than 1:1 above the 4′ level. Provide wall support before backfilling—reinforced concrete top or grouted slats. Backfill with free-draining, non-cohesive, granular materials.

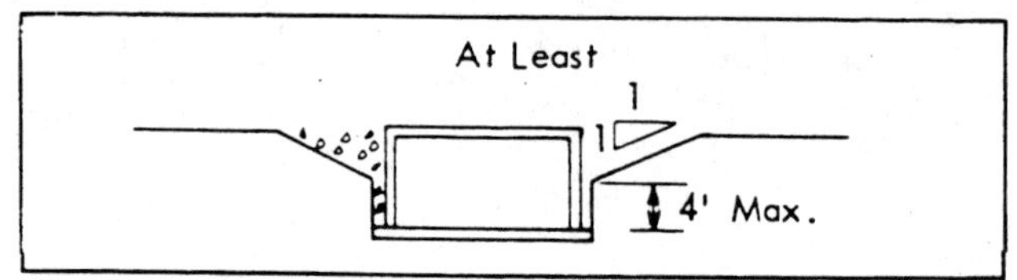

Fig 305-16. Excavation.

Floors

Footings—see Fig 305-17.

Construct floors at least 4″ thick over 4 mil polyethylene over 2″ sand, or 5″ thick over 2″ sand. Variations in floor thickness must be above these minimums. Use 6″x6″ 10 gauge in floors of tanks up to 8′ deep and 6″x6″ 6 gauge in floors of tanks up to 12′ deep.

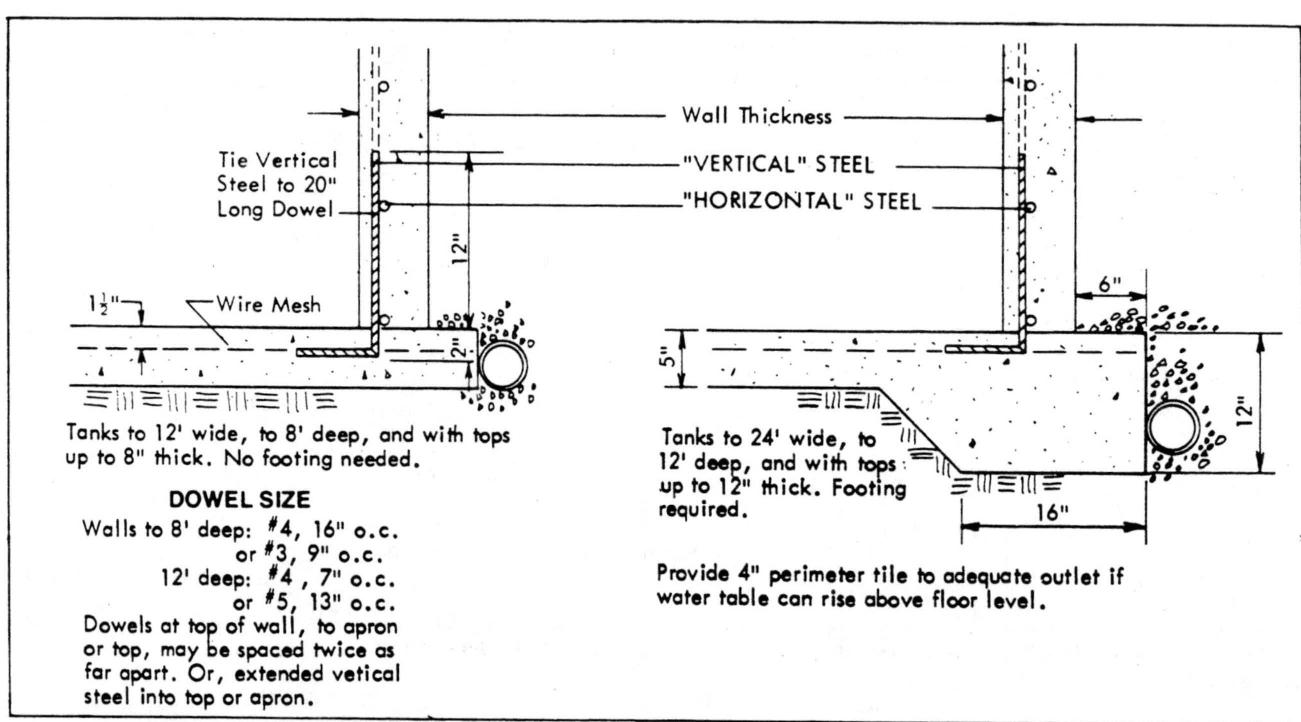

Fig 305-17. Footing construction.

Construction joints

In long tanks, install vertical construction joints in the walls about 100′ o.c. to avoid problems of differential settlement. Do not extend horizontal steel through the construction joint; insert a waterstop.

Install construction joints in floors with 10 gauge steel mesh not more than 50′ o.c. or not more than 100′ o.c. with 6 gauge mesh.

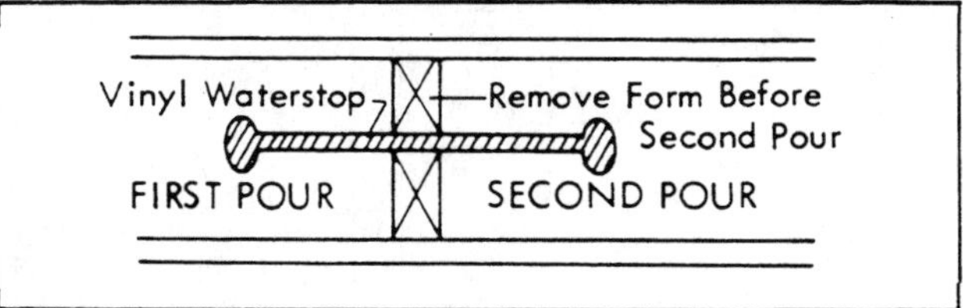

Fig 305-18. Waterstop.

Wall support

If the top of the tank is to be above grade:

Not more than 2½′ for 6″ thick walls, nor more than 3′ for thicker walls, reinforcing the inside face only is adequate.

If the top of the tank is to be farther above grade than 2½′ or 3′, the upper portion of the wall must be double reinforced. In the outside face, extend vertical steel (same size and spacing as in inside face) from the top to 1′ below grade; in the inside face use full-length bars. Half the horizontal steel in the upper portion of the wall should be moved to the outside face. See Fig 305-11.

Deep tanks—over 8′ deep

For tanks over 8′ deep built where the nature of the soil or the quality of backfill may not provide solid lateral bearing for the walls, reinforce the walls in the outside face to resist hydraulic pressure from stored liquids.

Build walls as partition walls, Fig 305-11, and where top beams are needed, reinforce against bending in both directions.

General Recommendations

Location

Obtain approval by appropriate regulatory agencies prior to starting construction. To avoid pollution of water supply and surface runoff:

- Locate the storage tank as far as feasible and downhill from the water supply, so that leakage or spillage will not adversely affect water supplies. Minimum distance is 100 ft.
- Avoid creviced limestone, shale, and bedrock sites which might allow direct ground water pollution.
- Avoid constructing tanks below the high-water table or in flood plains to prevent tank floatation and flooding.

Safety

External hydrostatic pressures cause uplifting forces on the tank. The tank is vulnerable to these uplifting forces when empty, particularly during construction. Add water to counteract bouyancy effects.

Design against accidents, asphyxiation, and possible over-exposure to toxic gases.

Protect necessary tank openings with grills and/or covers to prevent children, animals, equipment, and other objects from accidentally falling into storage tanks, Figs 305-19 to 305-21. Provide removable grills in only those openings used for stirring and pumping equipment. Design removable covers and grills to prevent their accidental loss into the tank and their unintentional removal, but for simple removal and replacement to encourage their use. Round openings are suggested. Use non-floating tank opening covers weighing at least 40 lb or covers protected from accidental removal.

Provide a permanent ladder or steps below all openings that have a least dimension of 15″, or larger, for emergency escape in case of accidental entry. Enclose open-top tanks with a fence at least 6′ high to prevent humans, livestock, or equipment from accidentally entering the tank.

Where gases may discharge into a building, provide necessary ventilation. Evacuate the building, if practical, during agitation prior to cleaning.

No person should enter a storage tank. However, if it is essential to enter, other persons must be present outside the tank with immediate means of removing the victim in case of any possible effect of dangerous gases. The person entering the tank should wear self-contained breathing equipment (a chemical reaction filter mask is not sufficient protection) and have one end of a rope secured around his body just below the arms, with the other end secured outside the tank.

Inspect the tank and its surroundings periodically for: leaks; vertical separation between ground and tank cover; deterioration of grills, covers, and ladders; and roof adequacy.

Tank Design Modifications

The following two modifications may be used where good construction practices are assured, and where all minimums will be met or exceeded. The modifications may not meet code or governmental requirements.

Reducing vertical steel for very good drainage

Where drainage is very good, tank walls may be designed for 30 pcf, or 30 pct + 100 psf vehicle load. Very good drainage means: surface drainage away from the tank site; coarse granular backfill that drains well near tank walls; tile around the tank floor perimeter that drains freely to an outlet. Where well-drained soils surround a tank, and where the tank walls are at least tank depth in from building walls, drain tile may not be needed.

Reducing horizontal steel in tank walls

Tank walls are designed with the main reinforcing steel vertical. Horizontal steel controls cracking due to shrinkage during concrete curing, and due to length changes during temperature changes. Horizontal steel also prevents cracking that might permit corrosion of vertical bars.

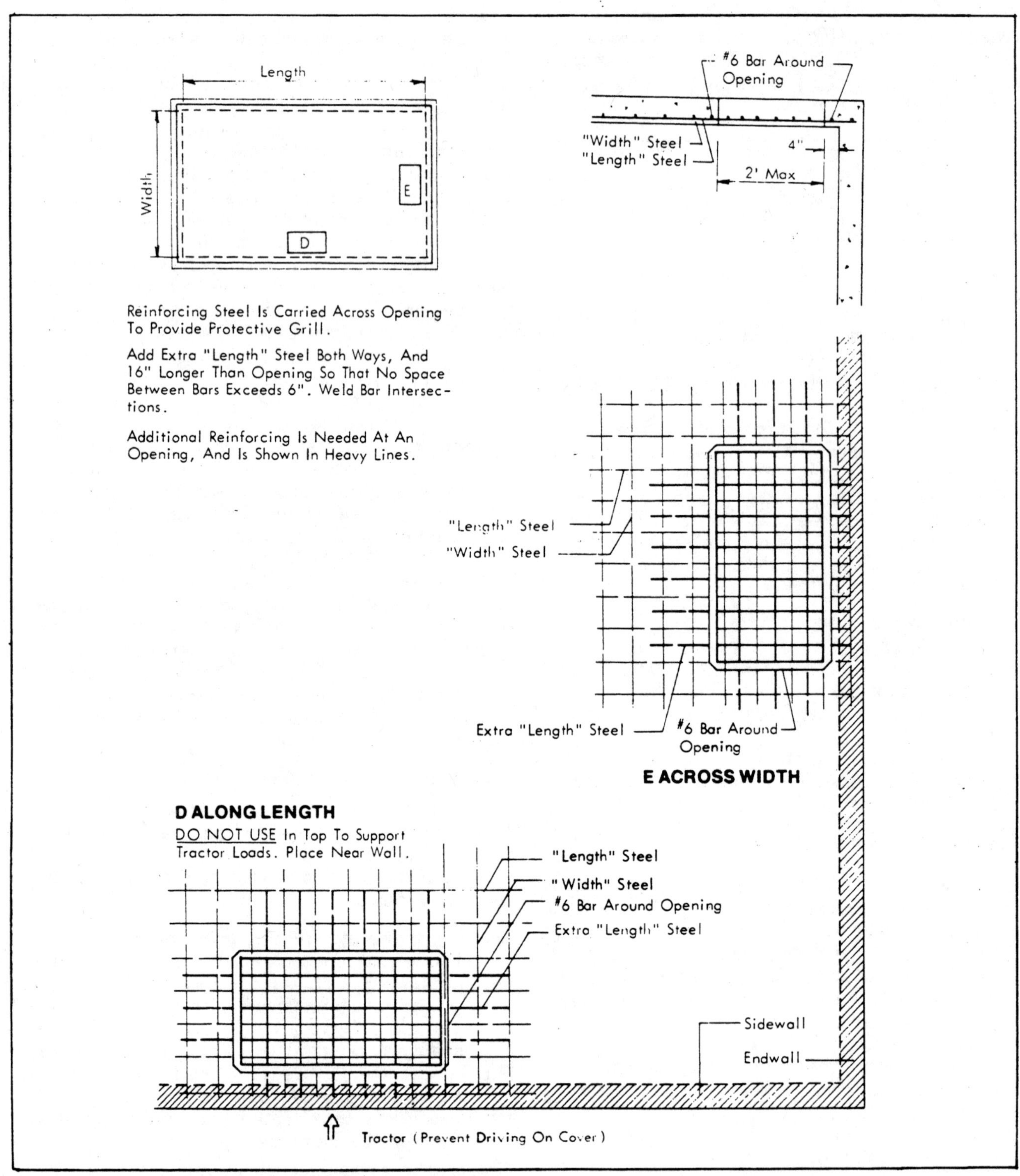

Fig 305-19. Slot openings for scrapings.

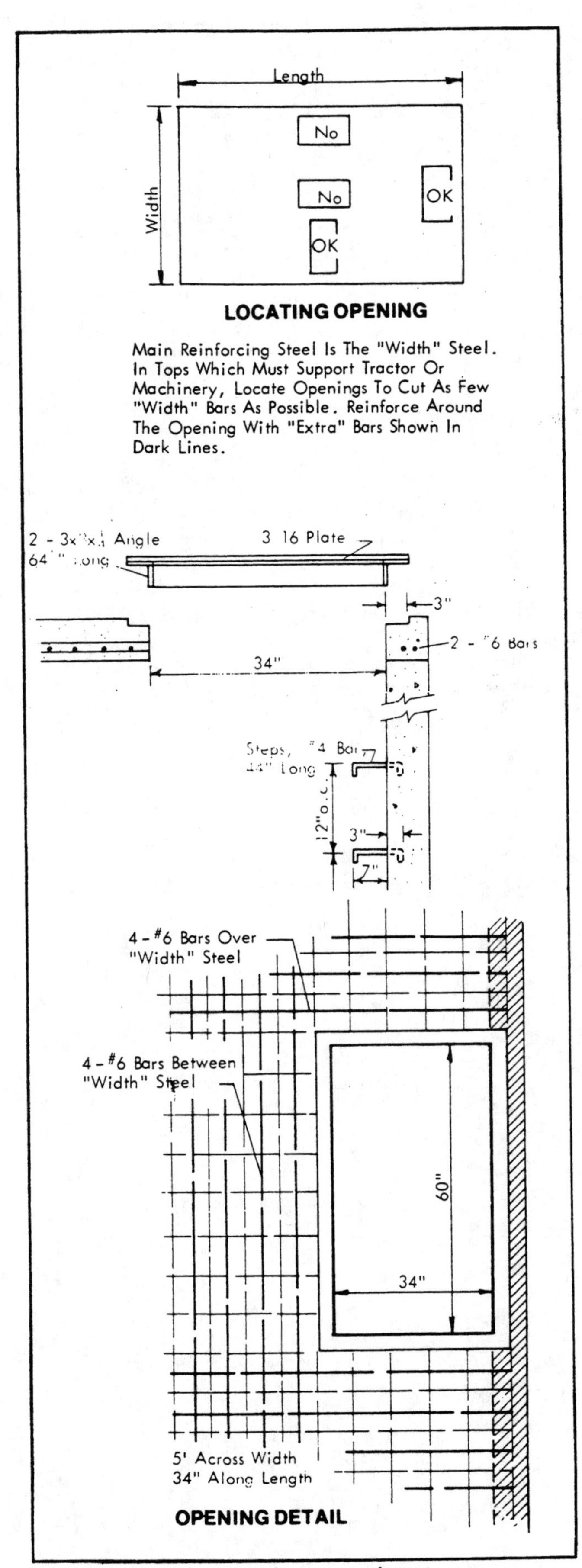

Fig 305-20. Pump and agitator openings.

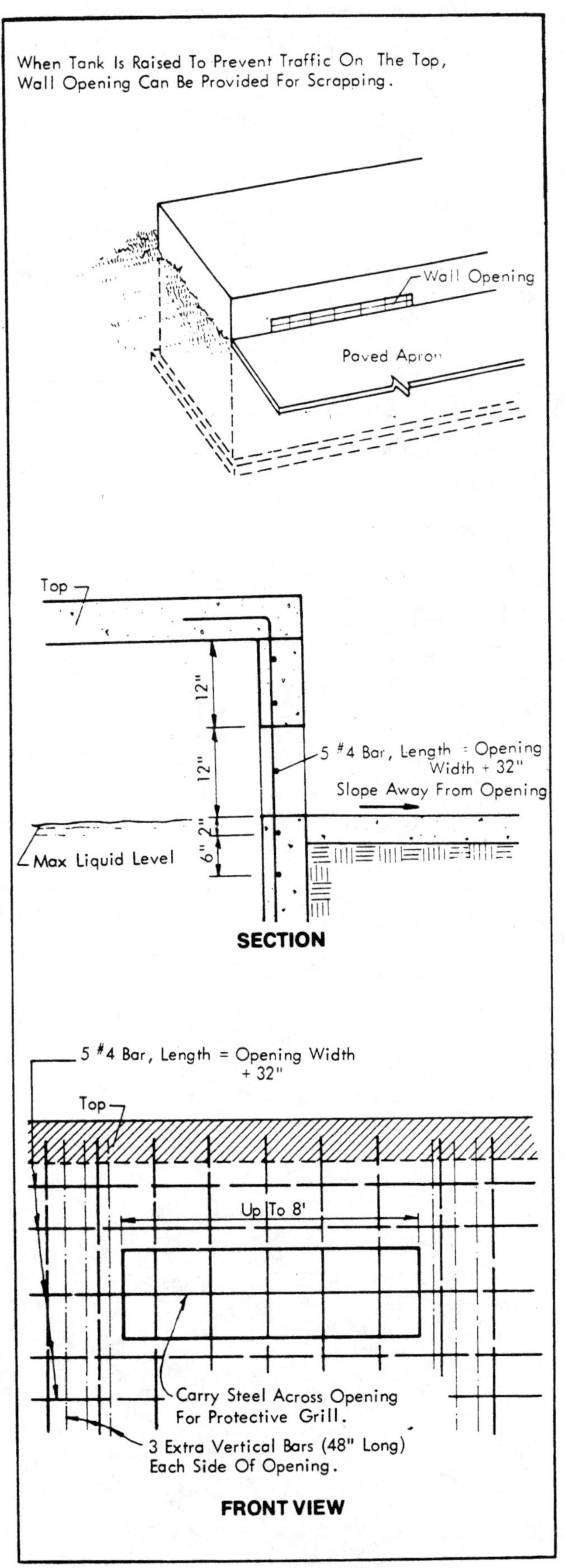

Fig 305-21. Scraping opening in wall.

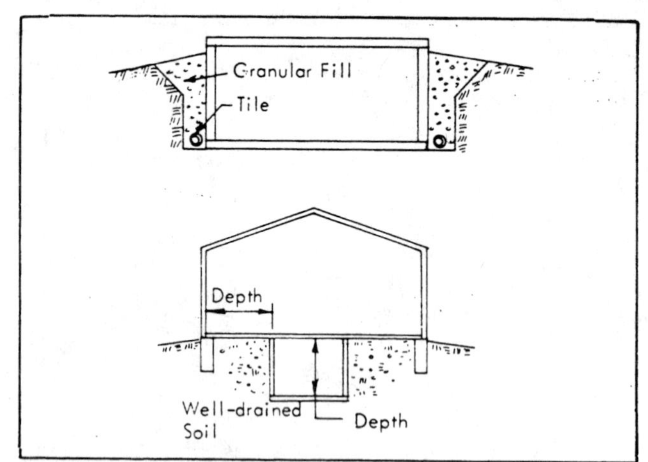

Fig 305-22. Good drainage.

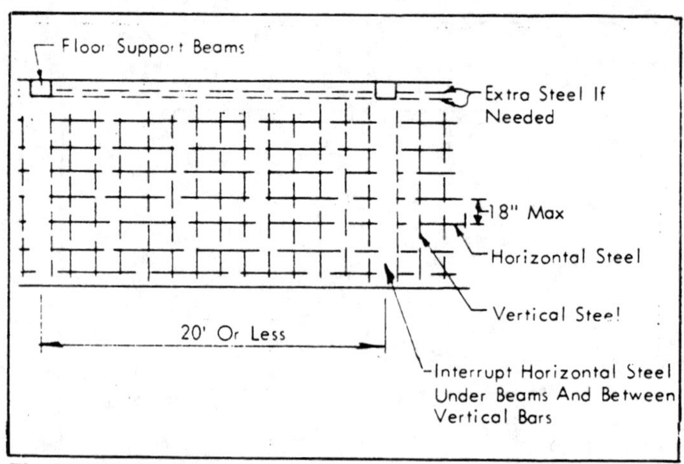

Fig 305-23. Reducing vertical steel.

Table 305-1. Horizontal steel permitting some vertical cracking.
Welded mesh assumed 60 grade; re-bars assumed 36 grade. Horizontal steel is about 0.001 bt for 36 grade.

Wall Thickness	Welded wire mesh size:	Maximum vertical spacing of re-bars #3	#4	#5
6″	4x4, 10g or 6x6, 8g	18″	18″	18″
7″	6x6, 6g	16″	18″	18″
8″	6x6, 6g	13″	18″	18″
9″	6x6, 4g	12″	18″	18″
10″	6x6, 4g	11″	18″	18″
12″	4x4, 6g	9″	17″	17″

Interrupt horizontal steel no farther apart than 20′. Position the interruptions between vertical steel bars. Provide horizontal steel spaced no farther apart than 18″ vertically.

Where vertical cracking is controlled, and where minor seepage can be tolerated, horizontal steel can be reduced below the minimums commonly specified. Use Table 305-1 to reduce horizontal steel.

306 FENCING

Lot Fencing

Panel fencing, Fig 306-1

Panel fencing is popular for fences for close confinement areas with frequent animal contact. They are usually 16′ long but can be cut to any length. The panels are easy to install with special 2½″ fence panel staples on posts 5′-4″ or 8′ o.c. Panels are usually galvanized and rigid and do not require stretching. Panel fencing is ideal for:

- Cattle, hog, and sheep lots.
- Sheep shearing pens.
- Calf holding pens.
- Finishing house dividers.
- Temporary pens.
- Haystack fencing.
- Horse corral.

Wooden fences

Wooden fences are often used when wood is plentiful or relatively inexpensive or where livestock crowding requires a strong fence. Figs 306-2 to 306-5 show typical wooden fences for livestock. Nail the boards to the posts with 2-16d spikes.

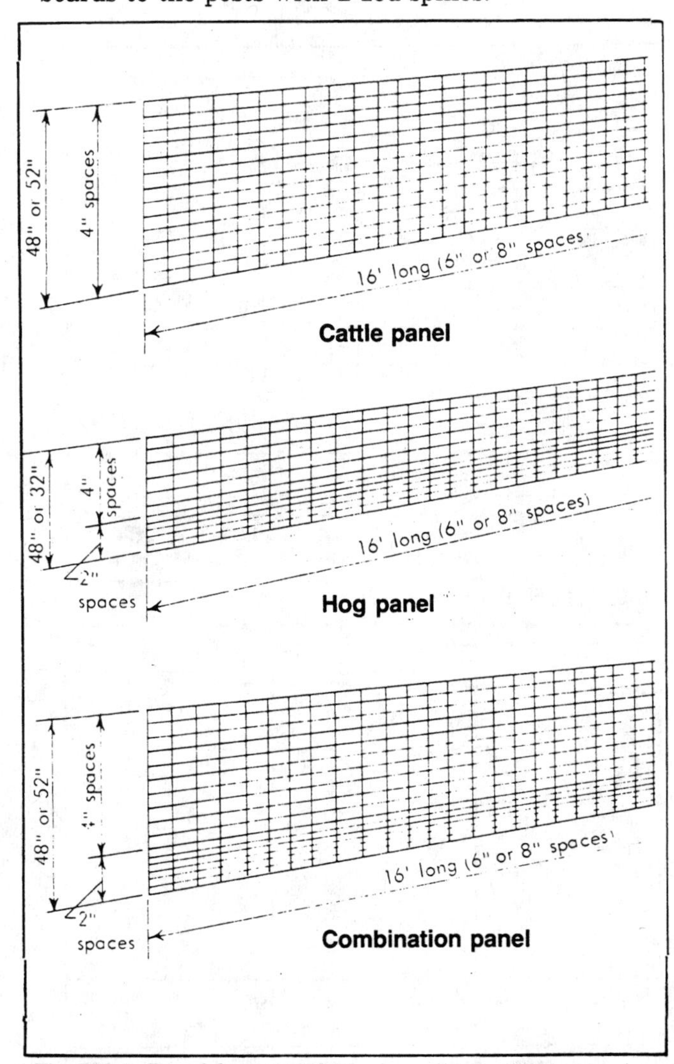

Fig 306-1. Panel fence styles.

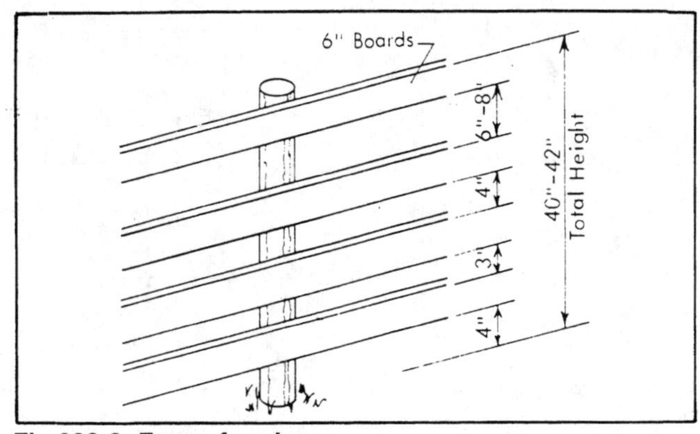

Fig 306-2. Fence for sheep.

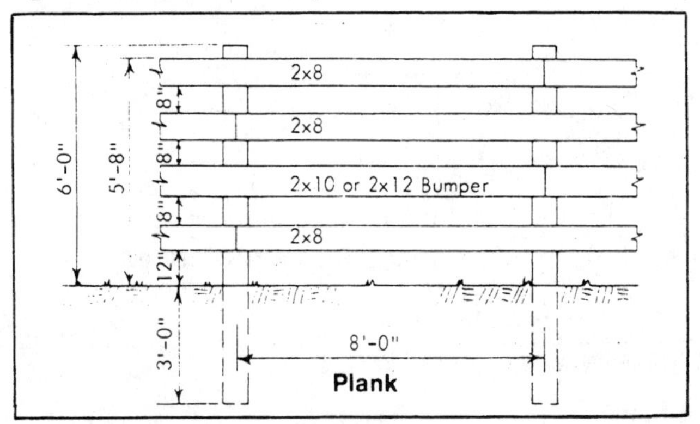

Fig 306-3. Line fence for horses.

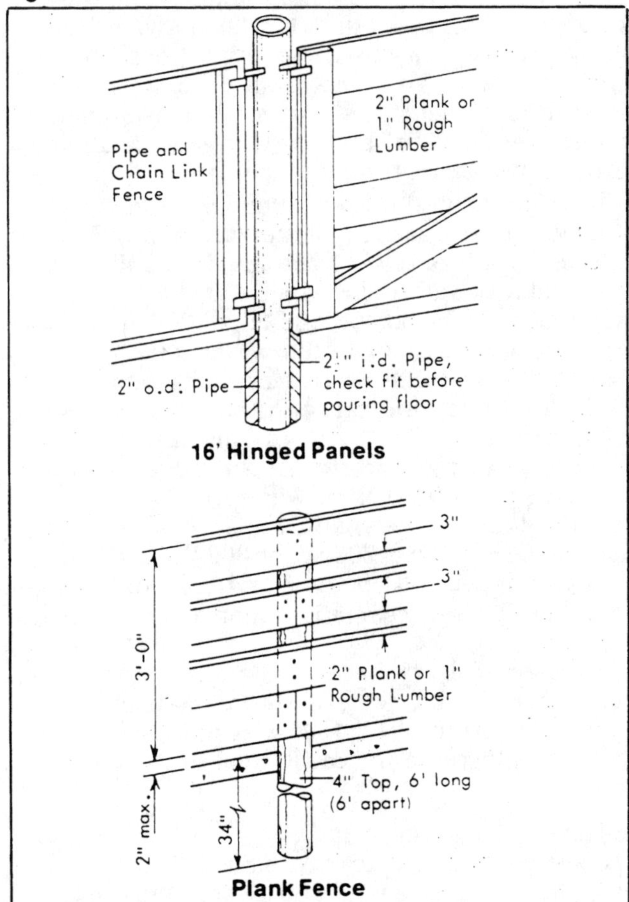

Fig 306-4. Wooden fences for hogs.

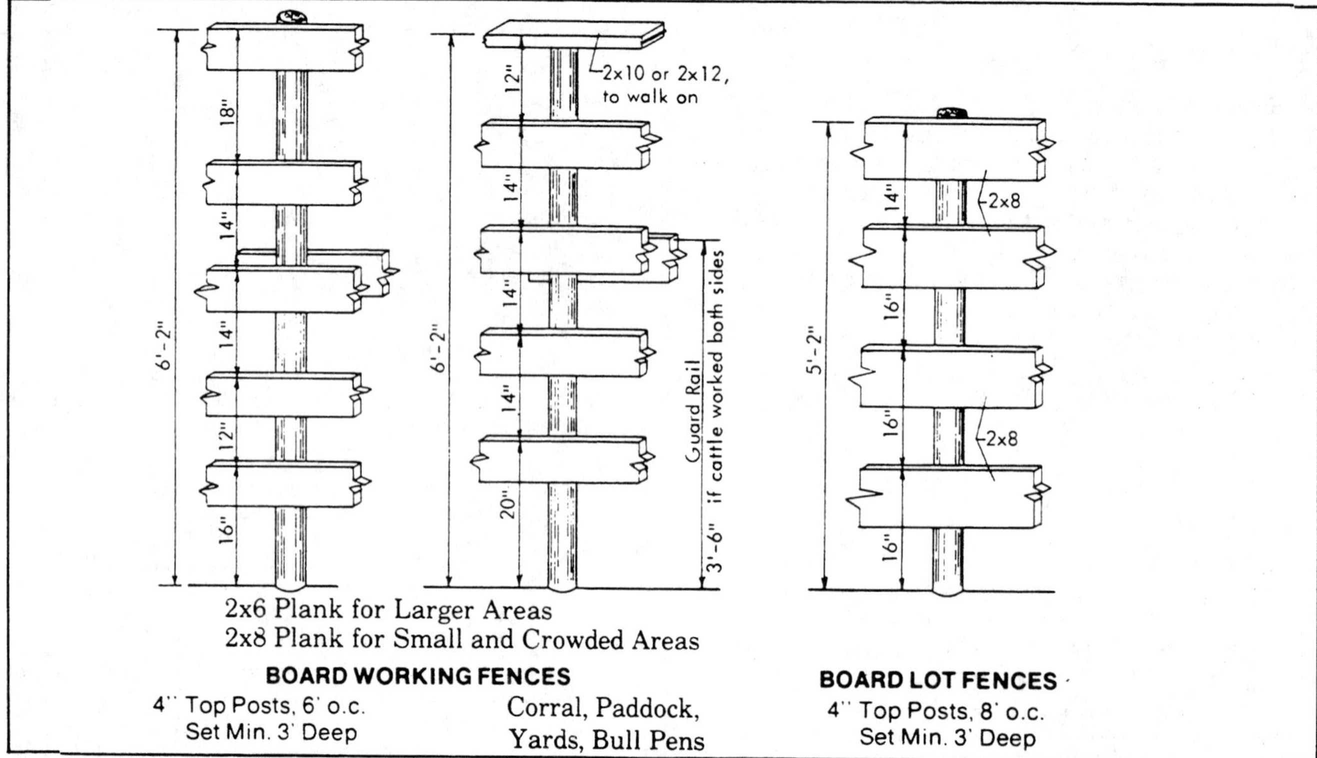

Fig 306-5. Wooden fences for cattle.

Wire fences

Barbed or woven wire are used alone or in combination for a number of different fence styles. Barbed wire above woven wire keeps large animals from leaning over and breaking down fencing. Below woven wire, it keeps small animals from digging under the fence. Wires are sized by gauge (ga), an arbitrary number representing diameter.

Fig 306-6 shows the most commonly used styles.

Cattle and horses: Use fence 306-6a or 306-6b. Use fence with 9 ga top and bottom wires and 11 ga fillers. Add a strand of barbed wire at the top.

Hogs: Use fence 306-6c, 306-6d, or 306-6e without the barbed wires above the woven wire. Styles 939 and 832 are available with a barbed bottom wire, or add a barbed strand below the woven wire. The barbed wire below the woven wire discourages animals from crawling or rooting under the fence. Use fence with 9 ga top and bottom wires and 11 ga fillers.

Cattle, horses, hogs, and sheep in the same field: Use fence 306-6c, 306-6d, or 306-7e. Use fence with 9 ga top and bottom wires and 11 ga fillers.

Fig 306-7 shows some common barbed wire fences.

For cow-calf operations, four strands of barbed wire are usually adequate if the animals are not crowded. Five strands of barbed wire work better for feeder cattle, because the cattle tend to rub on the fence.

High-tensile wire fencing

High-tensile wire fences have smooth, 11 to 14½ ga wire with a tensile strength of up to 200,000 psi and a breaking strength of up to 1,800 lb. A ten-strand

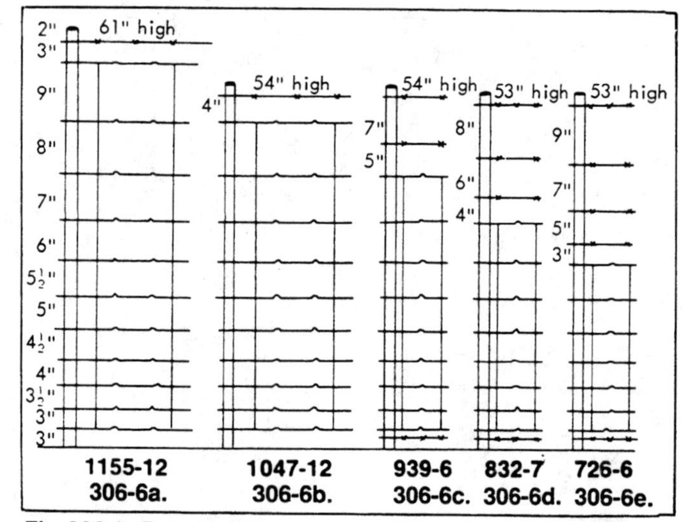

Fig 306-6. Fence styles.

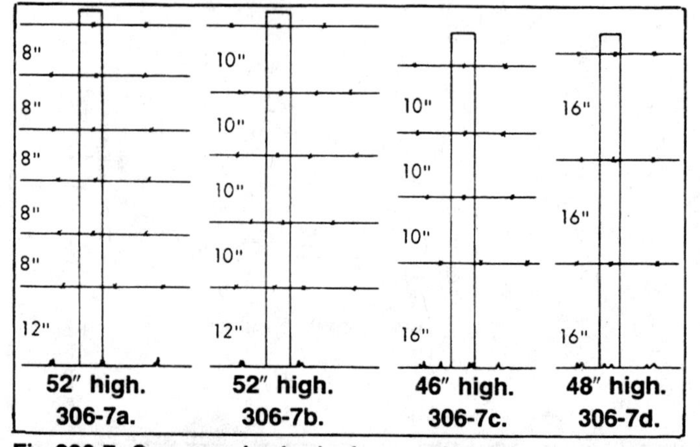

Fig 306-7. Common barbed wire spacings.
Fences 306-7a and 306-7b keep livestock from roads, railroads, or other problem areas. Fences 306-7c and 306-7d are line or field fences, so animals can occasionally get through.

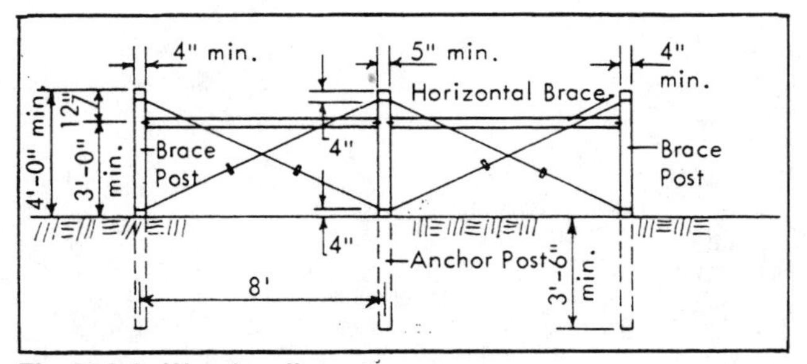

Fig 306-8. Wood pull post.
Post depths shown are minimum. For middle of long fence, place 40 rods apart.

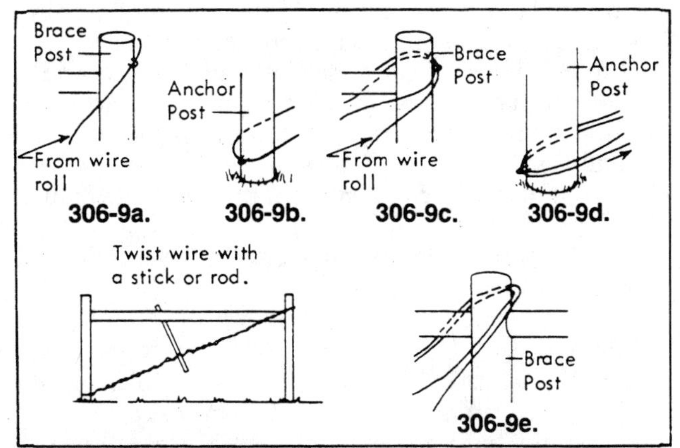

Fig 306-9. Threading 9-ga smooth wire diagonals.

high-tensile fence lasts longer and costs about two-thirds as much as an equivalent woven wire fence. The wire is also easier to handle, reduces damage to livestock and hides, has less sag, and requires little maintenance.

Wires are held in 200 to 250 lb tension along pressure-treated wood or fiberglass posts or a combination of posts and battens or droppers. The fences withstand over 1,200 pounds of livestock pressure or low temperature contraction without losing elasticity. A spring in each wire maintains tension.

Designs of six or more strand non-electric, and one or more strand electric fence, are available for all types of livestock and for deer and other predator protection.

High tensile wire **non-electric** fence is suitable for cattle, sheep, and horses. Specifications for non-electric fence are listed below. Ground the fencing for safety from lightning. Wire is galvanized 12½ ga with a 1300 to 1800 lb minimum breaking strength.

Typical spacing from ground to top wire:
- 10-strand livestock fence: 4", 4", 4", 4", 5", 5", 5", 5", 5", 5" (46")
- 10-strand cattle feedlot fence: 10", 4", 4", 4", 5", 5", 5", 5", 5", 5", (52")
- 8-strand cattle fence: 4", 5", 5", 5", 6", 6", 7", 8" (46")
- 6-strand cattle fence (not for calves): 14", 5", 6", 6", 7", 8" (46")

Permanent **electric fence** with up to five strands is suitable for holding sheep, hogs, goats, cattle, and horses. Wire is galvanized 12½ ga with a 900 lb minimum breaking strength.

Typical spacing:
- One strand 30" above ground (for milking cows).
- Two-strand wires at 17" and 38" above ground (for cows and calves, horses and foals).
- Three-strand wires at 17", 27", 38" above ground (for hard to hold cattle).
- Five-strand wires at 5", 10", 17", 27", 38" above ground (boundary fence for sheep, lambs, and cattle. Use where dogs are a problem).
- First and second wires up from the ground are charged or not, depending upon the conditions.

For complete planning and construction details, see NRAES-11, *High-Tensile Wire Fencing*, Northeast Regional Agricultural Engineering Service.

Corner and Pull Post Assemblies

After clearing the fence line of brush and other obstructions, locate and install corners and pull post assemblies. A fence is no better than its corners and braces, which often take up to half the total fence construction time.

Fig 306-8 shows good corner and pull post assembly construction. The double span assemblies are much stronger than the single span assemblies and are for fences over 200′ long. A post used as a corner post by more than one line fence requires a brace for each line fence. Post depths shown in Fig 306-8 are minimums. Use deeper settings for clay or wet soil conditions. The proper method of securing brace wire is shown in Fig 306-9.

Steel corner post and brace assemblies are available and can replace wooden assemblies. Anchor both posts and braces in concrete. The concrete anchor for corner posts should be 12" in diameter or square and extend 3½′ below ground. Braces can be anchored in 20" square blocks extending 2′ below the ground.

Fenceposts

Wood, steel, and fiberglass fenceposts are available in a variety of sizes and shapes. Base selection of posts for your fence on availability and cost of the post and on how long you want the fence to last.

Wood

Wooden posts may be more plentiful and less expensive than steel posts. For permanent fencing, select either a decay resistant variety such as Osage orange or posts that have been pressure preservative treated.

Wooden posts are available in 5½′-8′ lengths and in 2½" and larger top diameters. The larger the top diameter, the stronger the post. Line posts can be as small as 2½" but a minimum of 3½" provides a stronger, more durable fence. Use at least 5" top corner posts and gate posts. Brace corner posts or add strength to lines of steel posts with at least 3½" top posts.

Select post length from fence height and depth of set. Anchor posts should be set at least 3½′ into the ground. Line posts are set 2′-2½′ into the ground.

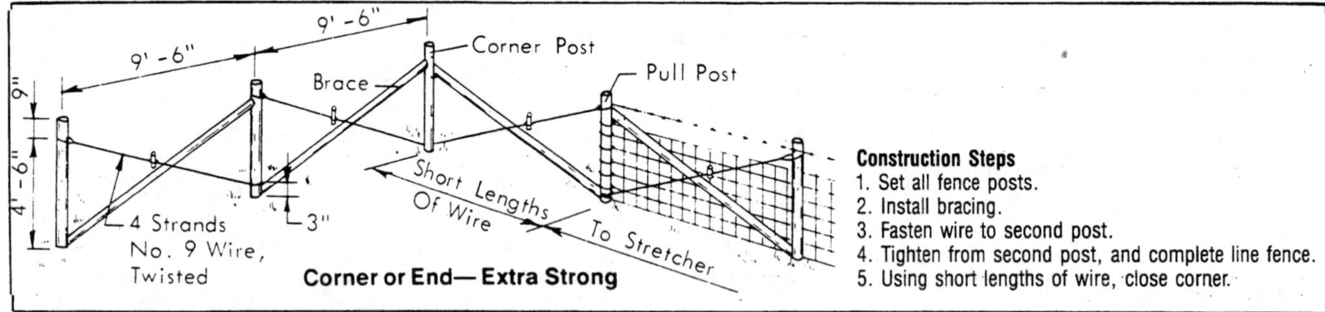

Fig 306-10. Recommended fence corner.

Steel, Fig 306-11

Steel posts have some advantages over wooden ones. They weigh less and are fireproof, extremely durable, and relatively easy to drive. They also ground fence against lightning when in contact with moist soil. Steel posts bend easily and usually cost more than wooden ones. Wooden anchor posts every 50′-75′ stiffen a fence with steel line posts.

Place steel posts so the anchor plate—usually a thin broad plate to help prevent bending from livestock pushing on the fence—is parallel to the wire. Drive steel posts so the top of the anchor plate is 4″-6″ below the soil surface.

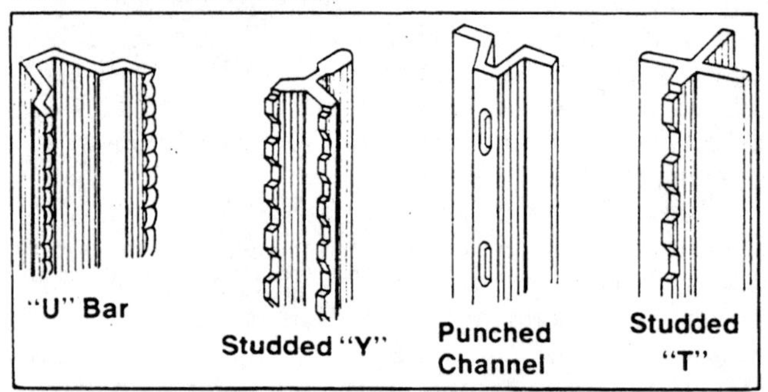

Fig 306-11. Common types of steel posts.

Fiberglass

Fiberglass posts are now available for either barbed or electric wire. They are 4′-7′ long and are as strong as steel but weigh about ¼ as much. They are very flexible and spring back from jolts. They do not rot or rust and can be installed with standard equipment. Insulators are not required with electric fence.

Line posts

Stretch a cord or wire between the completed corner and pull post assemblies to set line posts in a straight line.

Set line posts one rod (16½′) apart and at least 2½′ deep.

Orient steel line posts for correct wire placement when installed. Wires are usually on the animal side of the post except when fencing on contours.

Fence Installation

Fencing operations can result in painful and serious injury. The following **safety suggestions** are given to avoid difficulties:

- Wear heavy gauntlet leather gloves to protect hands and wrists, and boots or high shoes to protect legs and ankles. Tough, close-fitting clothing will prevent catching on wire.
- When stretching barbed wire and fence, stand on the opposite side of the post from the wire.
- Keep chains and wire stretching clamps in good condition.
- Carry staples in container or apron instead of in your pockets.
- If you handle treated posts, do not rub hands or gloves on face or skin.

General

1. Install and stretch in sections running from one corner or pull post assembly to the next.
2. Fasten the bottom wire first, then the next highest, etc.
3. Attach wire to the side of the post next to livestock except along curves.
4. Use galvanized staples or the clips that come with steel posts to attach wire to posts. Drive staples downward and at an angle. Leave the wire loose to move freely during temperature changes or under other stress.

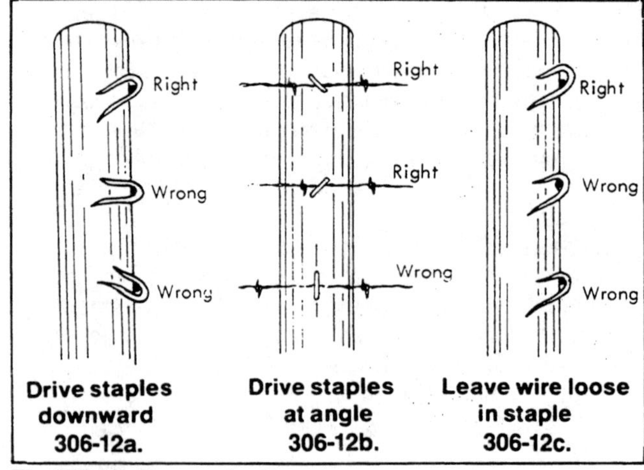

Fig 306-12. Fence stapling.

Woven wire

1. Attach the end of the wire to the pull post, removing a vertical wire and wrap line wires around the post, and splice them to themselves.

Fig 306-13. Wrap wire at pull post.

2. Unroll wire along the ground adjacent to the row of posts.
3. Prop the wire up against the line posts.
4. Attach the stretcher to the fencing, Fig 306-14. Use a single jack for fencing up to 35″ high and a double jack for higher fence. A block and tackle can also be used. Stretching fence with a tractor or truck is unwise, because it is too easy to overload the wire. The wire can be too tight or you can break the wire or pull the other end loose. A broken wire can be dangerous.
5. Stretch the fence slowly, making sure the wire doesn't snag on a post. Continue stretching until tension curves in the wire are straightened about ⅓, Fig 306-15.
6. Fasten wire to line posts starting at one end.
7. Fasten the end of the wire to the pull post.

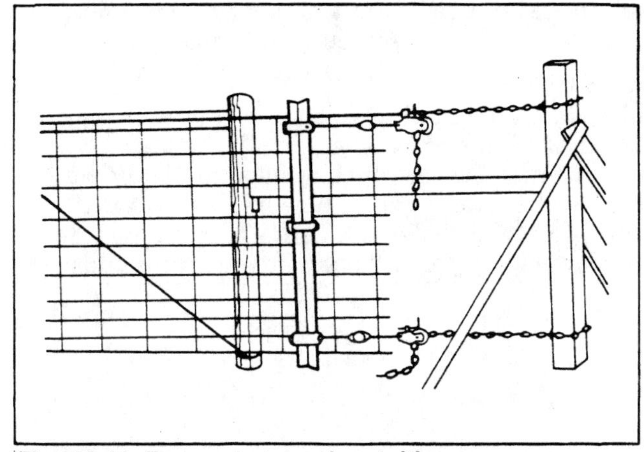

Fig 306-14. Fence stretcher in position.
Note braced "dummy" post at right.

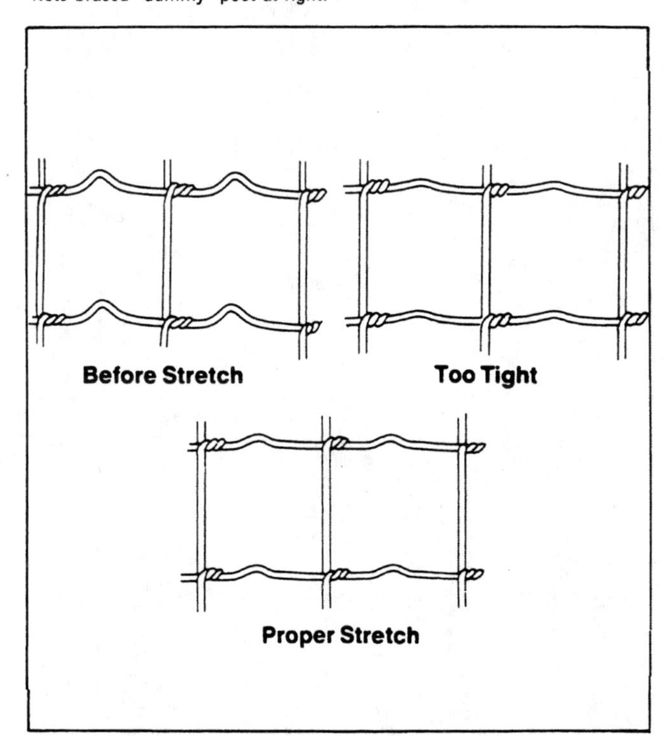

Fig 306-15. Typical tension curve in woven wire.
Stretch only until about ⅓ of the tension curve is removed.

Barbed Wire

Follow the same general procedure for barbed wire with the following precautions:

1. Unroll wire straight off the roll, not off the side.
2. Barbed wire does not have tension curves. Stretch wire until it is fairly tight; be careful not to break it.
3. Wear protective clothing and be careful with barbed wire.

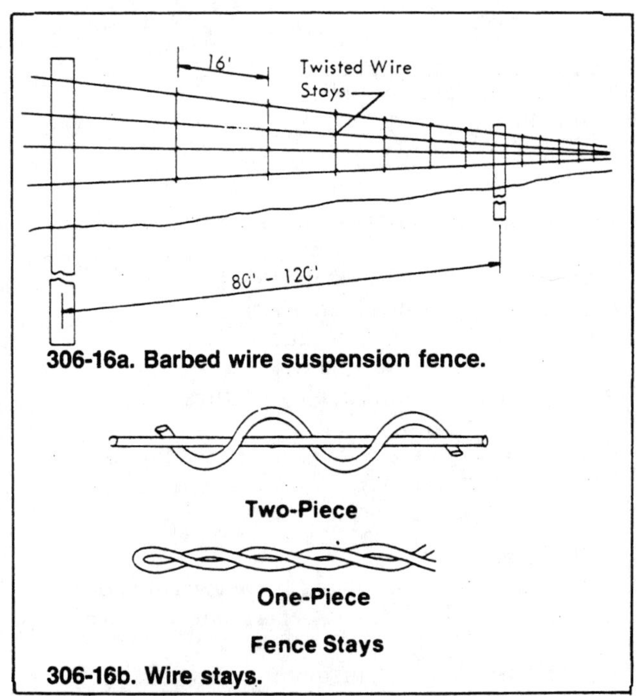

306-16a. Barbed wire suspension fence.

306-16b. Wire stays.

Fig 306-16. Suspension fences.

Suspension Fences

Barbed wire suspension fences are common on large cattle ranges, Fig 306-16. Support three to six strands of the wire with posts spaced 80′-120′ apart. Twisted wire stays, about 16″ apart, space the wires.

Fencing Across Waterways

Crossing low spots requires special precautions to prevent post withdrawal or washout. In locations not subject to frequent flooding, use extra length posts set 3½′-4′ deep or set posts 2½′ deep in concrete to prevent withdrawal, Fig 306-17. A hinged flood gate may be used in low spots which flood or when crossing streams with fences, Fig 306-18.

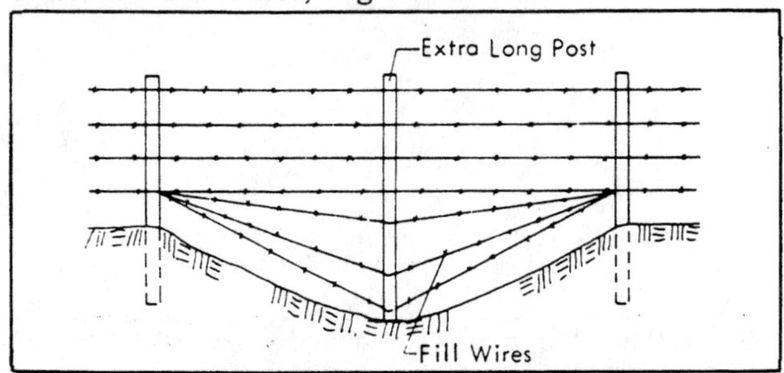

Fig 306-17. Line post in depression or low spot.

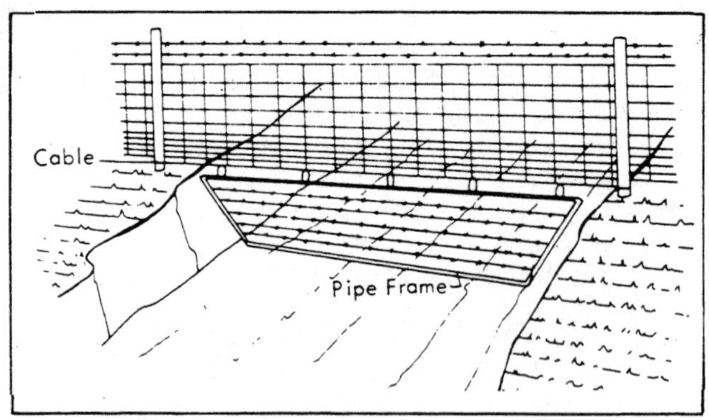

Fig 306-18. Hinged flood gate.
Use 1¼" pipe for gates up to 16′ wide. Use more than one gate for waterways over 16′ wide.

Electric Fences

Electric fencing is low cost and effective. It is ideal for:

- Managing pasture and rotation grazing.
- Salvaging of grain left in a field.
- Constructing temporary lanes.
- Extending life of old line fences.
- Adding safety to bull pen or pasture fences.
- Protecting terrace outlets.
- Protecting hay stacks or other feed supplies.
- Stopping animals from crowding fences.
- Lowering costs of feed lot fences.

Fence chargers

Fence chargers are powered by batteries or 120-volt current. Use the 120-volt type if power is available.

Fence chargers emit current intermittently, not continuously. The "on" time is usually ⅒ of a second 45 to 55 times a minute. The shock is sharp, but short and relatively harmless.

Select a charger approved by Underwriter's Laboratories (UL), or the U.S. Bureau of Standards. Notice of approval will be printed on the controller near the nameplate. Homemade chargers can and often do kill people or animals.

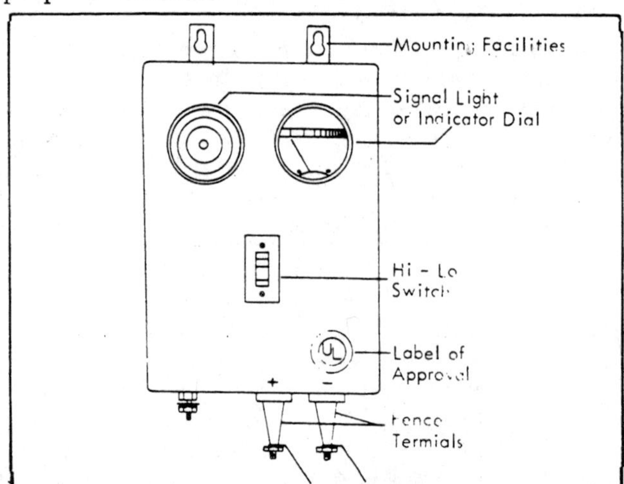

Fig 306-19. Desirable features of a fence controller.
(a) Signal light for showing when fence is shorted, or (b) a dial indicator that shows operating conditions. (c) A high-low switch for dry or normal conditions. (d) A method of mounting a controller so that it can be moved and mounted easily. (e) Label of approval.

Locate the 120-volt plug-in charger where it will stay dry, be accessible for inspection, and protected from livestock. Use 14 ga TW wire to lead the current from the charger to the fence and ground. Attach the black wire to the positive (+) terminal and the white wire to the ground (-) terminal of the charger.

Outside the building, attach the black wire to the electric fence. Attach the white wire to a ground rod with a steel clamp. The ground rod makes the earth a conductor to complete the circuit back to the controller. To be most effective, put the ground rod in moist soil or up to 8′ deep in dry soils. Effective electric fence operation requires good grounding.

Fence insulators

Fence insulators fasten electric fence wire to posts (not required with fiberglass posts) to prevent grounding the fence and making it ineffective. Insulators are available for any type of post and application. Fig 306-20 shows some typical insulators.

Construction

The success of an electric fence depends on good construction. Temporary fences can be smooth 12 or 14 ga galvanized wire, which is easy to handle but difficult for livestock to see. The wire is often broken by cattle being pushed or accidentally bumping into it. Use barbed wire for more permanent fences and to

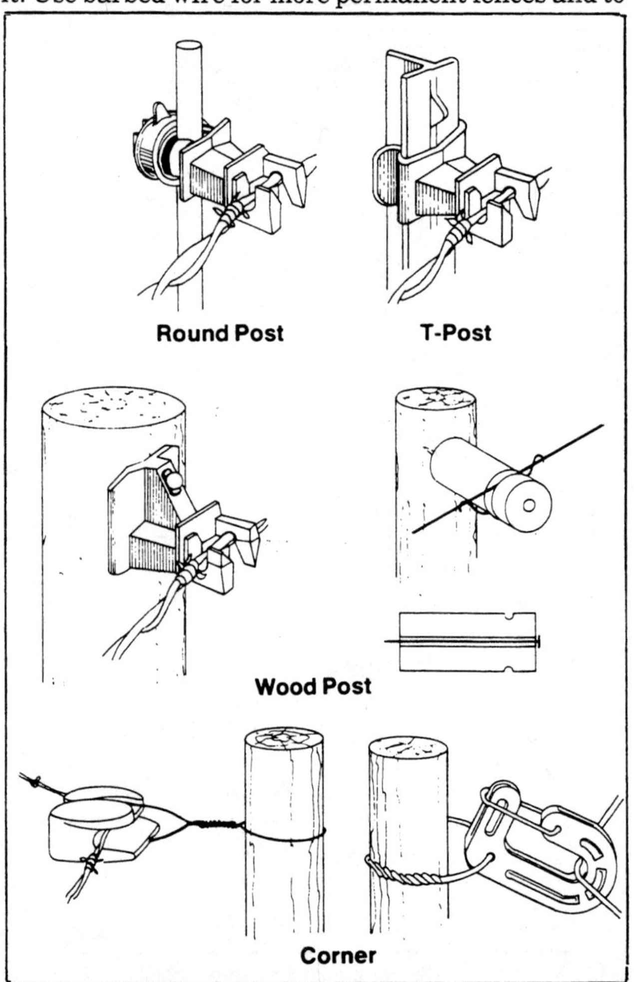

Fig 306-20. Electric fence insulators.

separate cattle from grain or forage crops. It is stronger, more visible to the livestock, and the barbs make it more effective.

Set 4" minimum diameter corner posts 36" deep and thoroughly tamp backfill. Fasten insulators to the corner posts. The wire between the insulator and the post is not charged. Set steel, wood, or fiberglass line posts spaced 50'-60' or more apart. With posts far apart, it is easy to mow under the wire with a sickle bar mower with the grass deflector removed.

Stretch the electric fence wire and attach it to insulators at each post. Install a small galvanized wire jumper at each corner to continue the charge around the field.

Make gates as shown in Fig 306-21c. Install an insulated handle that allows current to flow through a conductor in the center. Use a short length of plastic or rubber hose, old milker inflation, or a commercial handle. Make gates wide enough for the widest piece of equipment with plenty of clearance. Note that the current comes from the loop the handle hooks in to; when the gate is on the ground, the fence is not shorted out.

At swing gates, extend the posts on either side of the gate with 2x4s and string the electric fence wire overhead, Fig 306-22. Opening the gate does not short out any of the fence, but the gate is never charged.

Often a one-wire fence is sufficient, Fig 306-23. Height for a single wire is about ⅔ the animal height.

Gates

The more often you use a gate the better quality it must be to avoid frequent repairs.

Most farm gates must be wide enough for field machinery, vehicles, or equipment to maintain lots or pastures or haul crops, feed, or manure. Few gates are used only for livestock or people. A 10'-12' gate is wide enough for livestock and trucks or tractors. Gates up to 24' wide often are needed for field machinery. Allow extra width where large machinery will turn through a gate into a lane.

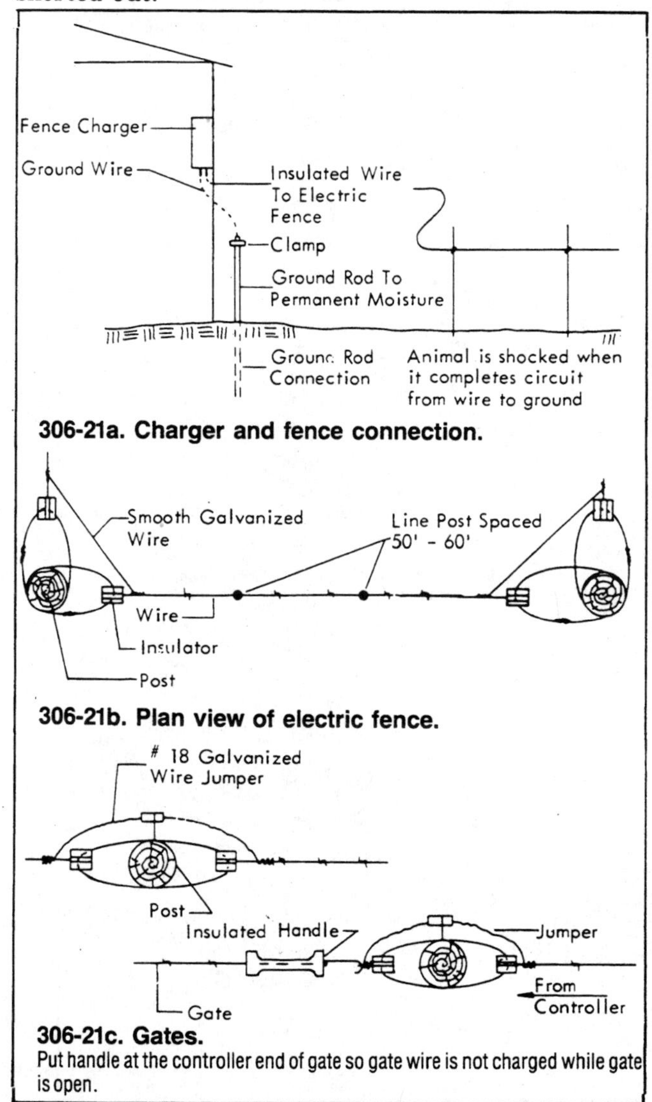

306-21a. Charger and fence connection.

306-21b. Plan view of electric fence.

306-21c. Gates.
Put handle at the controller end of gate so gate wire is not charged while gate is open.

Fig 306-21. Electric fence construction.

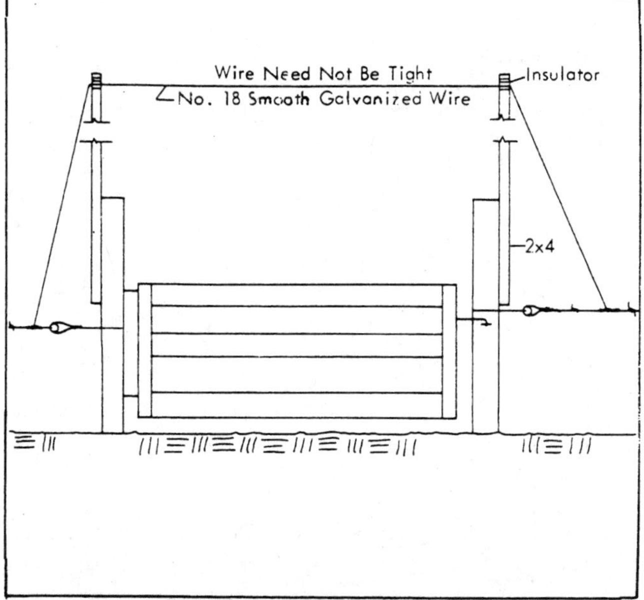

Fig 306-22. Electric fence above a swing gate.
Put the wire high enough to clear all machinery.

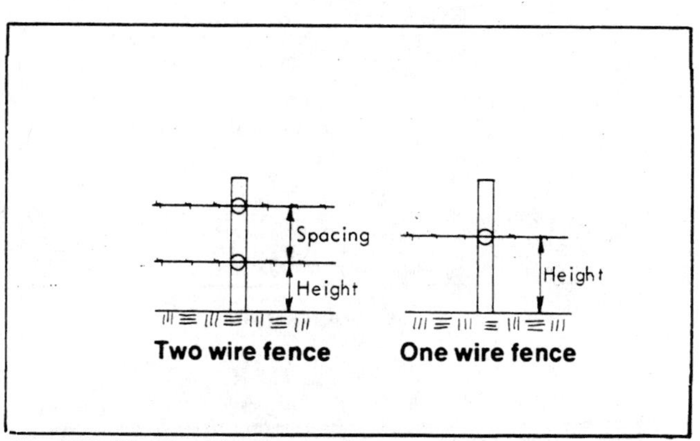

Fig 306-23. Electric fence wire spacing.

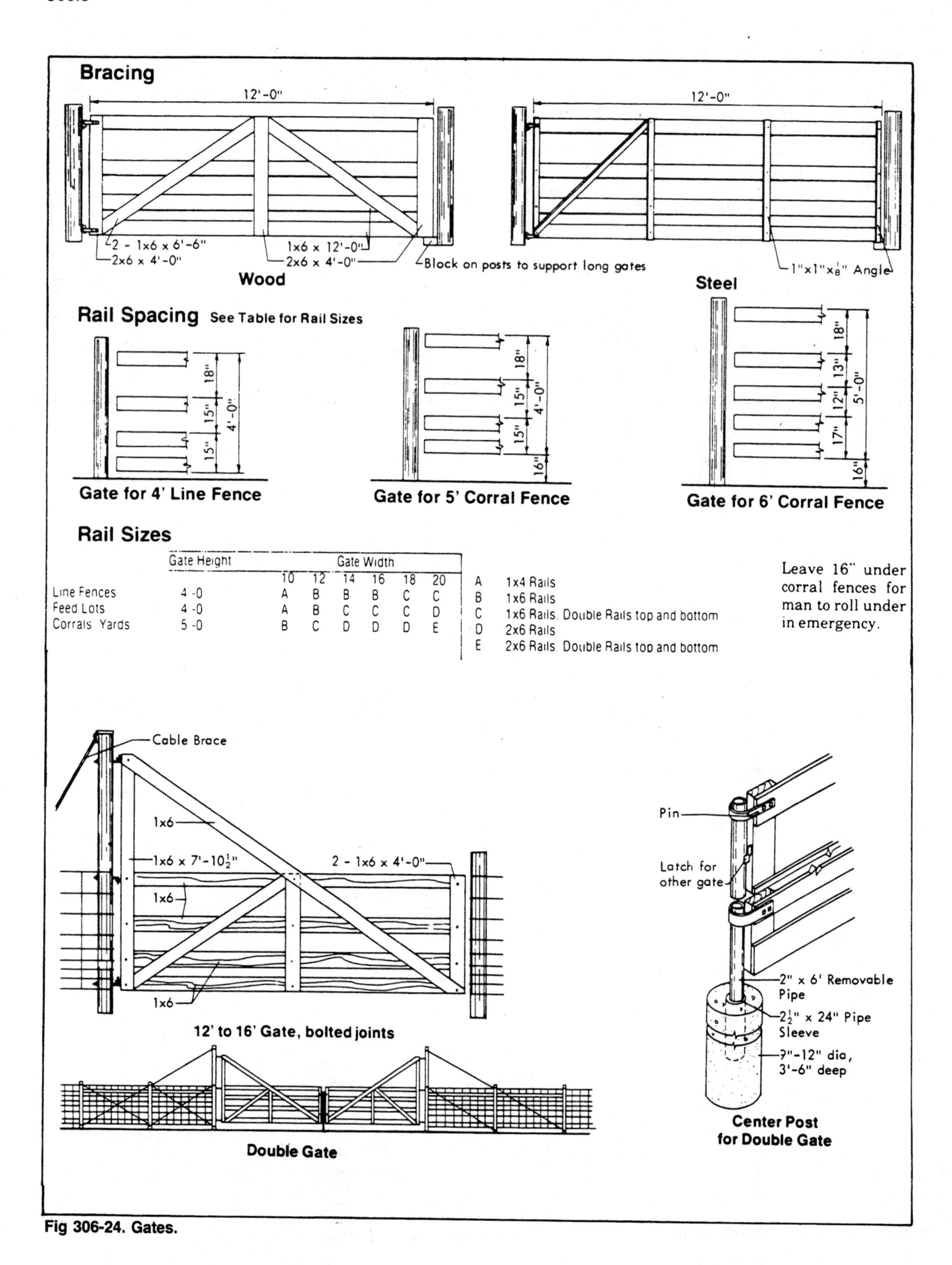

	Gate Height	Gate Width 10	12	14	16	18	20
Line Fences	4 -0	A	B	B	B	C	C
Feed Lots	4 -0	A	B	C	C	C	D
Corrals Yards	5 -0	B	C	D	D	D	E

A 1x4 Rails
B 1x6 Rails
C 1x6 Rails. Double Rails top and bottom
D 2x6 Rails
E 2x6 Rails. Double Rails top and bottom

Fig 306-24. Gates.

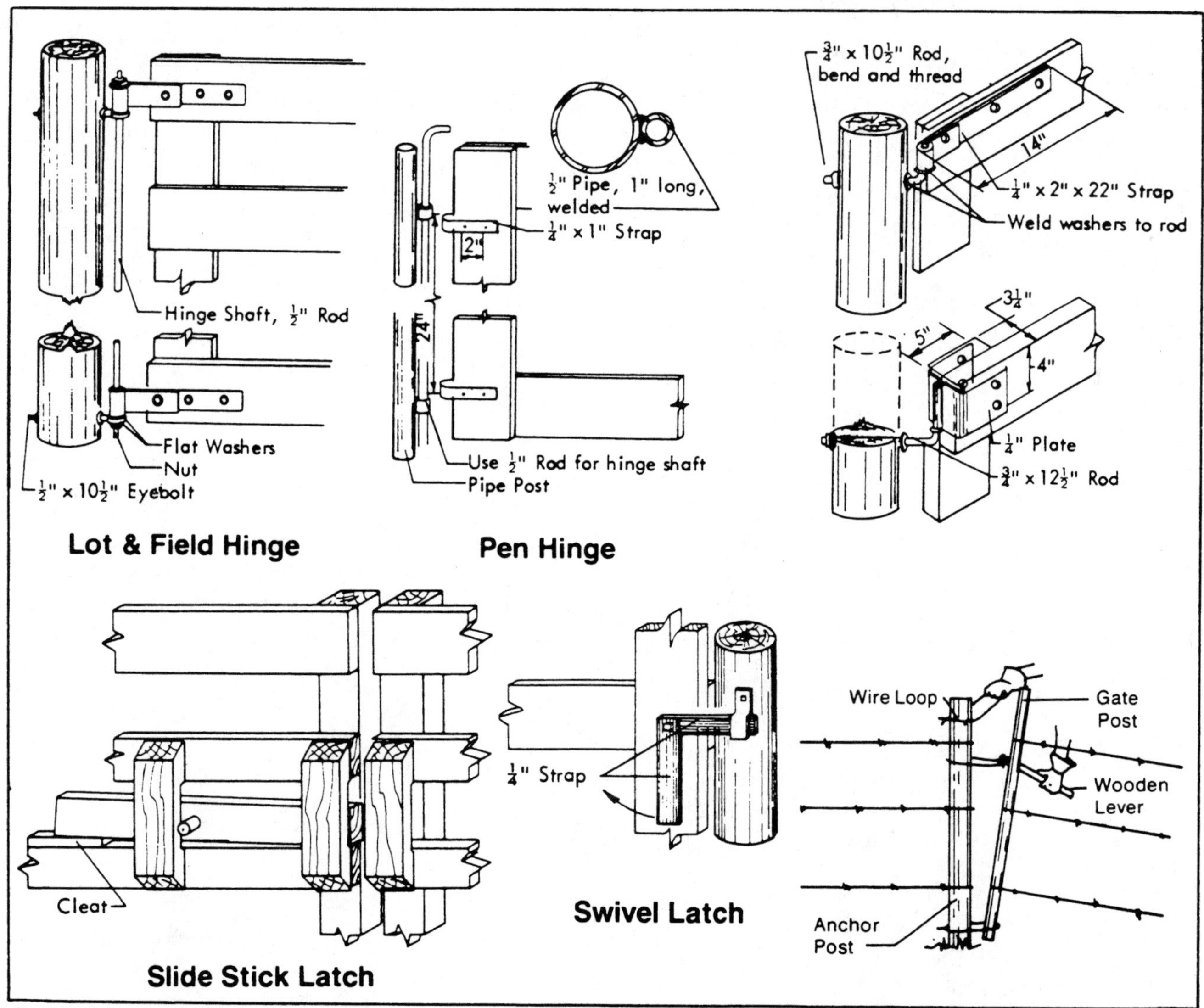

Fig 306-25. Gate hinges and latches.

400 DESIGN

401 STRENGTH OF MATERIALS

Fundamentals

Stress

External forces applied to a solid body produce cohesive or adhesive forces inside the body. The internal forces keep the body together; without them the body could not keep its shape or support the external loads.

Consider member AB, Fig 401-1, subjected to the tensile forces F and -F. The forces tend to pull member AB apart, so internal forces are necessary to keep the member together. Imagine the member cut in two parts at point C. To maintain the equilibrium of the free bodies AC and CB, a force -F must be applied to AC and a force F to CB. The two parts AC and CB were in equilibrium before the member was cut, so **internal forces** equivalent to these new forces must have existed in the member itself. The intensity of this internal force is **stress.** The magnitude of the stress is the force F divided by the cross-sectional area of member AB.

Stress = force/area

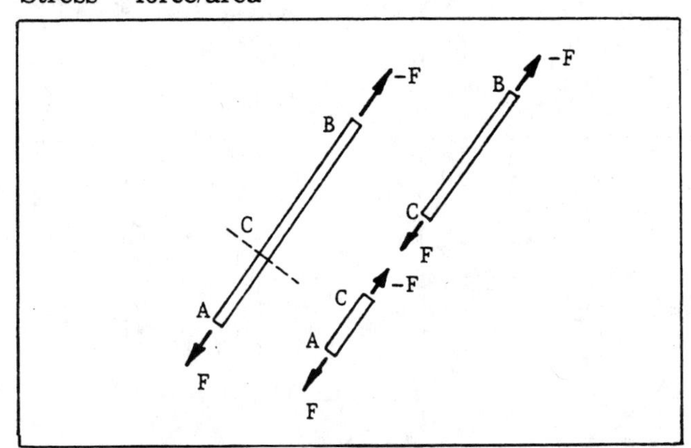

Fig 401-1. Member internal forces.

Normal stress is perpendicular to the cross section. **Shear stress** is parallel to the cross section. Normal stresses are either tensile or compressive. **Tensile stress** is from a pulling force and acts away from the cross section; **compressive stress** is from a pushing force and acts toward the cross section.

Normal stress, σ = dF normal/dA

Shearing stress, τ = dF parallel/dA

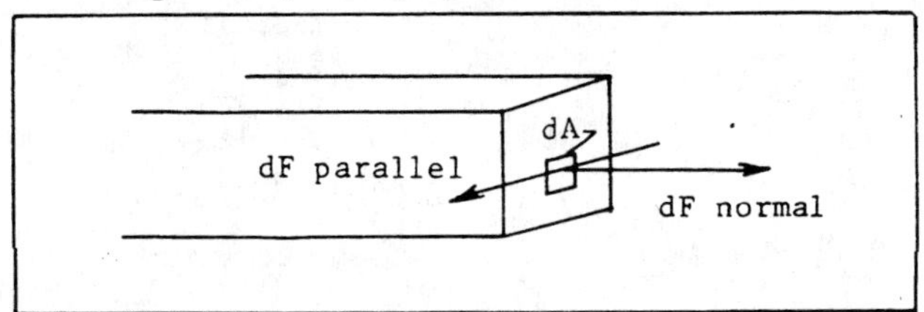

Fig 401-2. Normal and shearing stresses.

The dimensions of stress are force divided by length squared, usually lb/in^2, psi.

Note that a stress is associated with a plane area and is not a force vector. Stresses add algebraically while force vectors add by the parallelogram law.

The uniaxial state of normal stress in Fig 401-2 is its simplest form. Shear stress has the same magnitude on both the vertical and horizontal planes.

In more complex loading situations, elements of members can have normal stresses in two or three dimensions, and stress equations can be very complex.

Strain

External forces also cause dimension changes in a member that may or may not be visible. A longitudinal dimension change resulting from a normal stress is called a deformation, e. A deformation that increases length is an elongation or extension; one that decreases length is a contraction or compression. Angular distortion results in shear deformation.

Normal strain, ϵ, is the deformation per unit length:

$\epsilon = e_n/L$; $\epsilon = de_n/dL$

Shear strain, γ is the change in angle between two planes in the unstressed material. Shear strain can be a significant part of total deflection for deep beams with large loads.

$\gamma = e_s/L$; $\gamma = de_s/dL$

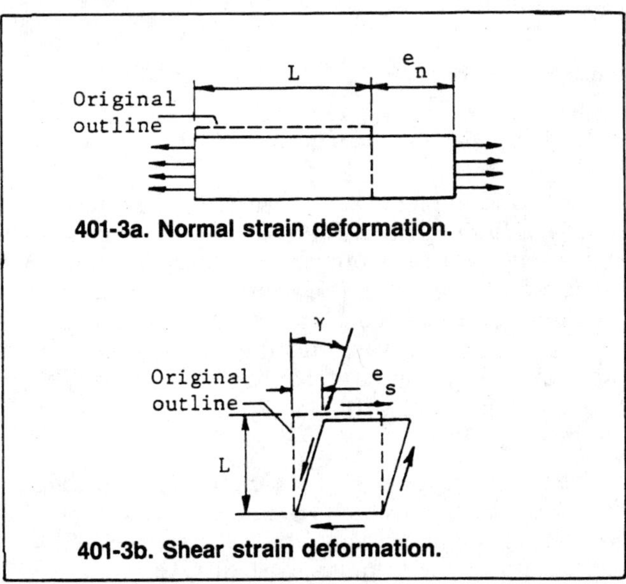

Fig 401-3. Normal and shearing strain.

Stress vs. Strain

A tensile test relates stress to strain. During the test, the load and the resulting elongation of a selected gauge length are recorded simultaneously. To plot stress vs. strain, convert load to stress by dividing it by the cross-sectional area of the member, and convert elongation to strain by dividing it by the original gauge length, Fig 401-4.

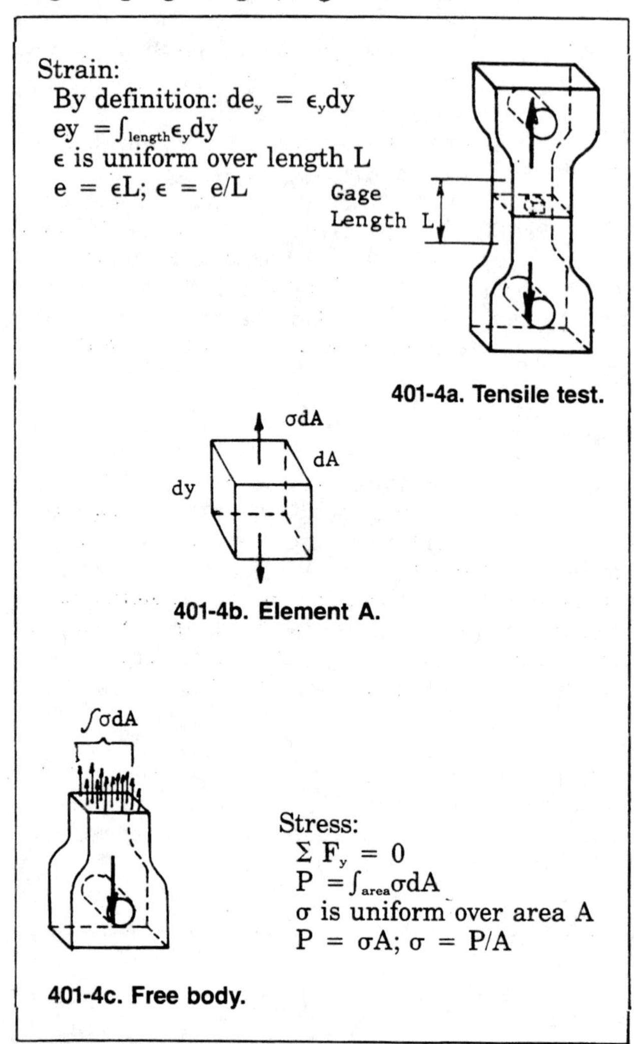

Fig 401-4. Stress-strain relationship.

Fig 401-5 is a plot of stress vs. strain for a tensile test on an aluminum rod. For most metallic engineering materials, stress is proportional to strain for the first portion of the curve. The stress at the upper limit of the initial straight portion of the curve is the proportional limit. Stress is related to strain below the proportional limit by the modulus of elasticity, E, a measure of material stiffness:

$E = \sigma/\epsilon$

The relationship between shear stress, τ, and shear strain, γ, is obtained by plotting torque on a cylinder vs. the resulting angle of twist. Shear stress is related to shear strain by the modulus of rigidity, G:

$G = \tau/\gamma$

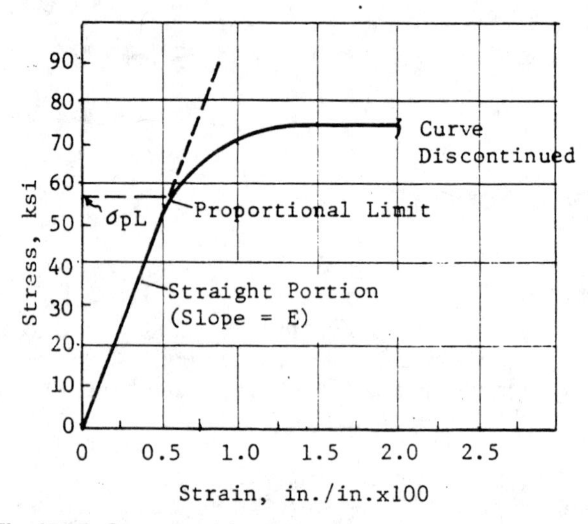

Fig 401-5. Stress-strain diagram.

Design Equations

Axially Loaded Members

Consider a member of constant cross section subjected to tensile force P parallel to the y axis, Fig 401-6. Cutting plane mm is passed normal to the line of action of external force P and the resulting free body diagram is drawn. Equilibrium requires that:

$\Sigma F_y = 0$

$\int_{area} \sigma dA - P = 0$

$P = \int_{area} \sigma dA$

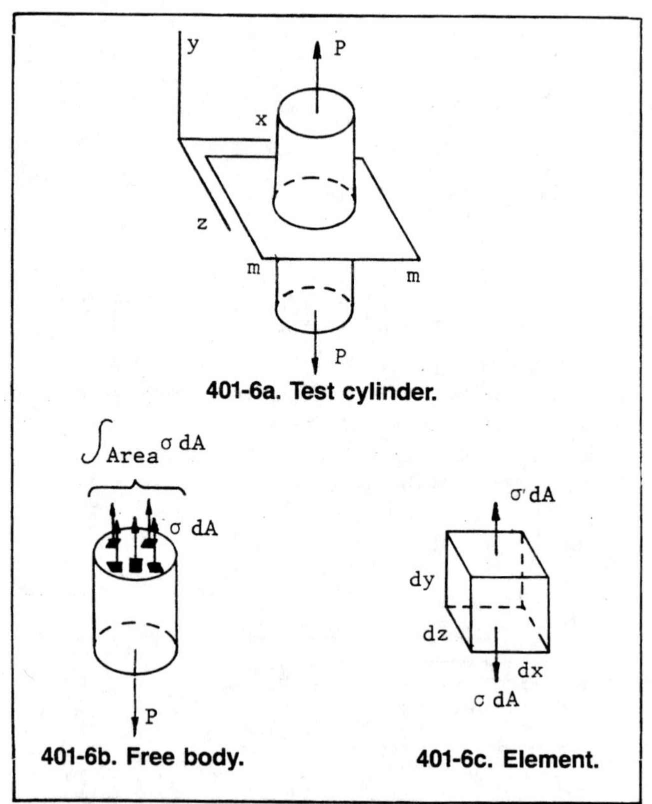

Fig 401-6. Axially loaded members.

The equation represents the sum of all of the cohesive forces on the cross section. It is the basic load-stress relationship for every axially loaded member.

If the axial stress is uniform, this equation simplifies to:

$P = \sigma \int dA$

$P = \sigma A$

$\sigma = P/A$

The equation is valid if:

- The portion of the member under consideration is straight.
- The portion of the member under consideration has a constant cross section.
- The member is of a homogeneous material.
- The member is loaded axially through its cross section's centriod.
- The load is not applied dynamically and is applied at a section well removed from the section where the stress is being calculated.

Buckling

Compressive loading of long members can result in buckling with large lateral deflections. Buckling is a stability failure and the resulting deflection is not linearly related to the applied load.

Subject a column to axial load P. If lateral force F is applied, the column "snaps" to a bent position. Depending on the value of P relative to the buckling load P_f, the column remains bent, or fractures and collapses. This behavior indicates that for axial loads greater than the buckling load, the straight position of a column is one of **unstable** equilibrium.

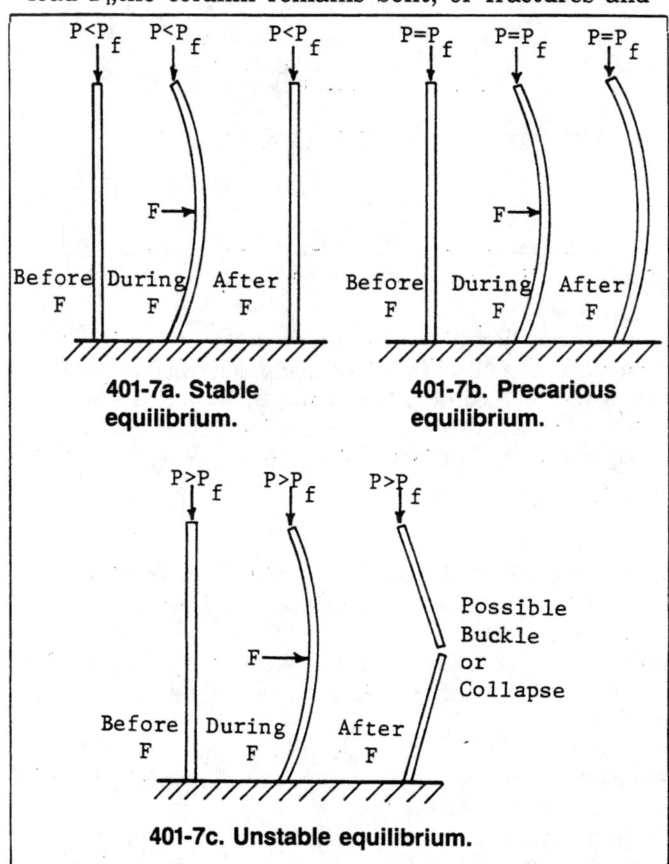

Fig 401-7. Buckling.

The following equation, the Euler equation, derived from the bending equation of a column, predicts the elastic buckling load of an ideal column.

$P_f = \pi^2 \times E \times A/(L/r)^2$

A = column cross-sectional area

r = radius of gyration about the bending or buckling axis

E = modulus of elasticity

This equation is valid if E is constant and the buckling load P_f causes an axial stress below the proportional limit at the start of buckling.

L/r is the column's slenderness ratio. A member with a slenderness ratio greater than 200 is too slender to carry much load. Members with slenderness ratios less than 30 have little tendency to buckle.

Equations and design procedures depend on the column material.

End Conditions

The Euler equation assumes zero bending moment at the column ends. Column length, L, in the equation is therefore the distance between two points of zero bending moment. Fig 401-8 shows four common end conditions. The effective length, L', is used in column design equations.

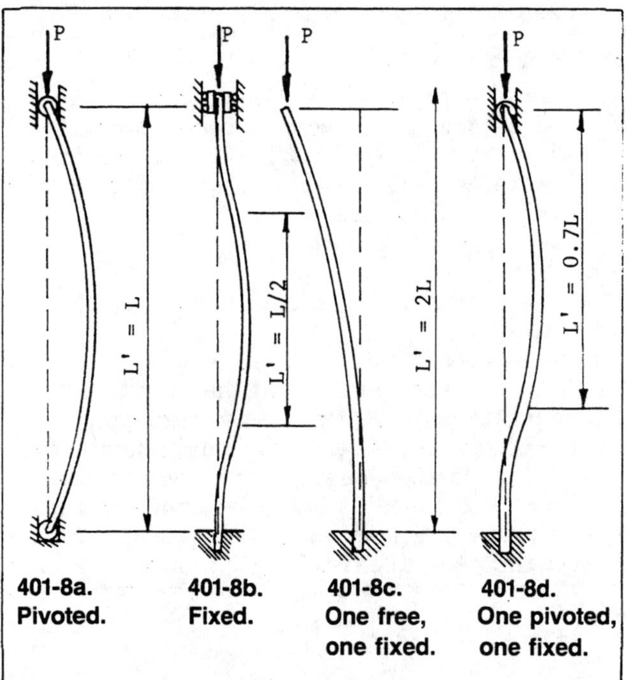

Fig 401-8. Effective column lengths.

Bending Members

Flexural Stresses

Longitudinal elements near the bottom of a beam are stretched and those near the top are compressed. There are both tensile and compressive stresses on transverse planes; they are fiber or **flexural** stresses.

Fig 401-9b is a free-body diagram of the beam in Fig 401-9a from the left end to plane a-a. Force V_r is the resultant of the shear forces at the section and moment M_r is the resultant of the normal stresses at the section. The magnitudes and directions of V_r and M_r are calculated from the equilibrium equations: $\Sigma F_y = 0$ and $\Sigma M_o = 0$.

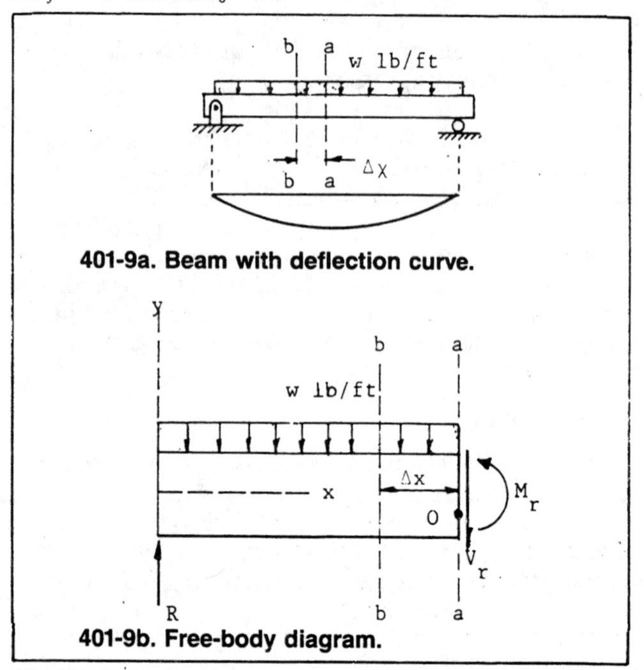

401-9a. Beam with deflection curve.

401-9b. Free-body diagram.

Fig 401-9. Typical beam.

A French engineer, C. A. Coulomb, discovered flexural stress distribution across a beam's cross section. His theory assumes that a plane section before bending remains a plane after bending. For this to be strictly true, the beam must:

- be bent only by couples.
- not buckle or twist.
- be of a homogeneous and isotropic material.
- be straight.
- have a constant cross section.

In Fig 401-10, fibers at the top of the beam shorten and those at the bottom stretch. At some distance, c, above the bottom of the beam, the longitudinal elements do not change length. The curved surface formed by these elements is the **neutral surface.** The intersection of the neutral surface with any cross section is the **neutral axis** of the section.

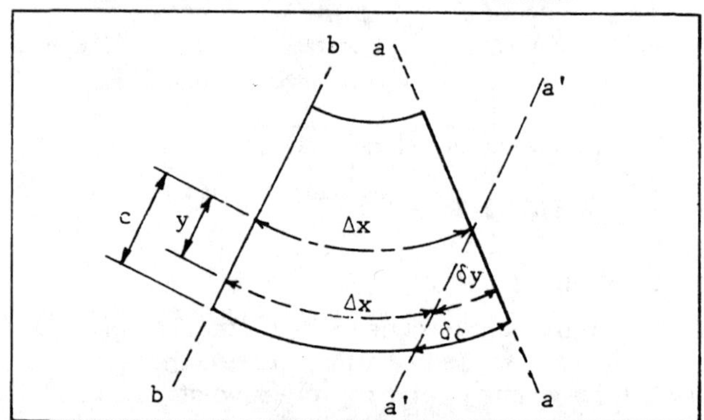

Fig 401-10. Flexural strain distribution.

The flexural deformation, δ, at distance y from the neutral axis is directly proportional to y:

$\delta_y / y = \delta_c / c$

Remembering that all elements had the same initial length, Δx, and that strain, $\epsilon = \delta/\text{length}$:

$\epsilon_y / y = \epsilon_c / c$

And, assuming that our homogeneous and isotropic material has a modulus of elasticity that is the same in tension and compression, and remembering that $\sigma = \epsilon E$:

$\sigma_y / y = \sigma_c / c$

$\sigma_y = y\sigma_c / c$

Stress increases linearly from zero at the neutral axis to a maximum at the outer fibers of the beam.

With the variation of flexural stress known, the free-body diagram of the beam section in Fig 401-9b can be redrawn as in Fig 401-11. The forces F_c and F_T are the resultants of the compressive and tensile flexural stresses respectively. The sum of the forces in the x direction must be zero; therefore $F_c = F_T$ and they form a couple of magnitude M_r.

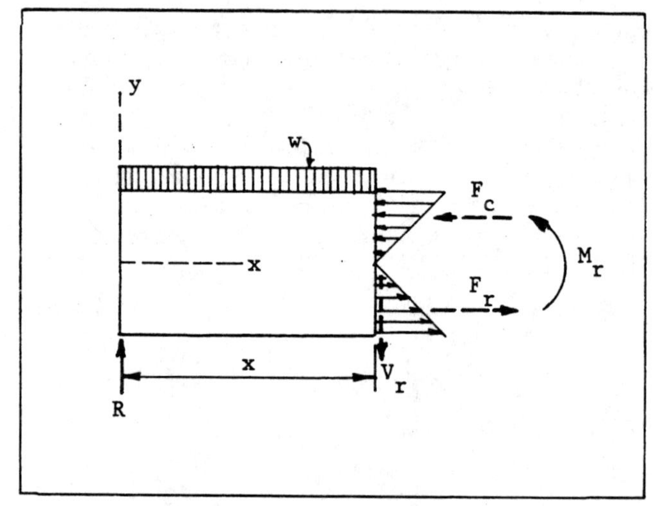

Fig 401-11. Flexural stress distribution.

To elevate flexural stresses, first locate the neutral surface. Fig 401-12 shows the elevation of part of a beam and the beam cross section. Differential force dF is the normal force action on differential area dA and equals σdA. The equilibrium equation $\Sigma F_x = 0$ gives

$\int_{area} dF = 0$

$\int_{area} \sigma y dA = 0$

where y is measured from the neutral surface to the area element, dA. Substitute $\sigma_y = (\sigma_c/c)y$:

$\sigma_c/c \int_{area} y da = 0$

But, the integral over the area of $ydA = y_m A$, where y_m is the distance to the centroidal axis. Therefore:

$\sigma_c/c \; y_m \; A = 0$

Because A, σ_c, and c cannot be zero, y_m must equal 0. Because $y_m = 0$, the neutral axis and the centroid of the area are the same.

The neutral axis moves away from the section's centroid in beams that are curved, subject to both axial and flexural stress, or subject to inelastic action on an unsymmetrical cross section.

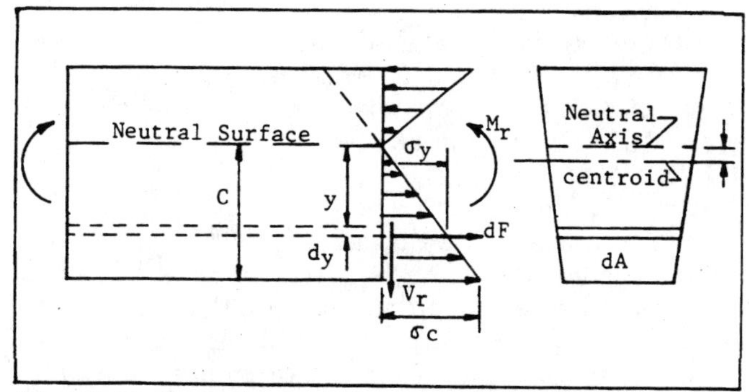

Fig 401-12. Neutral axis and stress distribution.

Flexure Formula

On the free-body diagram, Fig 401-12, the moment of the differential force, dF, around the neutral axis is ydF, and the moment of all such forces on the cross section is:

$M_r = \int_{area} ydF = \int_{area} y\sigma_y dA$

Substitute $\sigma_y = (\sigma_c/c)y$ and $\sigma_y/y = \sigma_c/c$:

$M_r = (\sigma_c/c) \int_{area} y^2dA = (\sigma_y/y) \int_{area} y^2dA$

The integral is the second moment of the cross-sectional area above the centroid, which is I, the moment of inertia:

$M_r = I\sigma_y/y$ and $\sigma_y = My/I$

This is the flexure formula, where σ is the normal stress at a distance y from the neutral surface and M is the resisting moment of the section.

At any section of the beam, the fiber stress is maximum at the surface farthest from the neutral axis:

$\sigma_c = Mc/I = M/S$

$S = I/c$ is the section modulus, a property of the shape of the cross section.

Shearing Stress Formula

An equation for horizontal shear stress is developed from the free-body diagram in Fig 401-13.

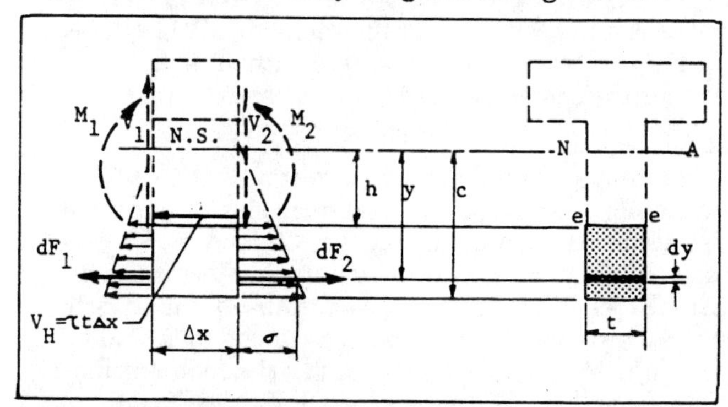

Fig 401-13. Horizontal shear stress, example.

Assume the bending moment increases with x, $M_2 > M_1$. Force dF_2 is the normal force acting on a differential area, dA, and is equal to σdA. The resultant of these differential forces is F_2 and $F_2 = \int\sigma dA$ integrated over the shaded area of the cross section. The fiber stress, σ, at distance y from the neutral axis is given by the bending equation $\sigma = My/I$. When the two equations are combined, force F_2 becomes:

$F_2 = (M_2/I) \int ydA = (M_2/I) \int_h^c ty\,dy$

Similarly, the resultant force on the left side of the element is:

$F_1 = (M_1/I) \int_h^c ty\,dy$

The equilibrium equation for the summation of forces in the horizontal direction in the free-body diagram yields:

$V_H = F_2 - F_1$

$= (M_2/I) \int_h^c ty\,dy - (M_1/I) \int_h^c ty\,dy$

$V_H = (\Delta M/I) \int_h^c ty\,dy$

The average shear stress is V_H divided by area tΔx along plane e-e. When length Δx approaches zero, ΔM also approaches zero.

$\tau = V_H/t\Delta x$

$It = (dM/Itdx) \int_h^c ty\,dy$

In this equation, dM/dx is the shear, V, at plane e-e. Also, the integral is the first moment about the neutral axis of the shaded portion of the cross section from plane e-e to the extreme fiber of the beam. The integral is called Q, and when V and Q are substituted into the equation for horizontal (or longitudinal) shear stress, it becomes:

$\tau = VQ/It$

This equation is accurate only if width t is not too great relative to the depth of the beam.

Vertical and horizontal shear stress are equal in magnitude at a given point.

Deflection

When a straight elastic beam is loaded, its longitudinal centroidal axis becomes an elastic curve. In regions of constant bending moment, the elastic curve is an arc of a circle of radius R, Fig 401-14, in which the portion AB of a beam is bent only with couples. The plane sections A and B remain planes after the moments are applied, and the deformation of the fibers is proportional to the distance from the neutral surface, which is unchanged in length.

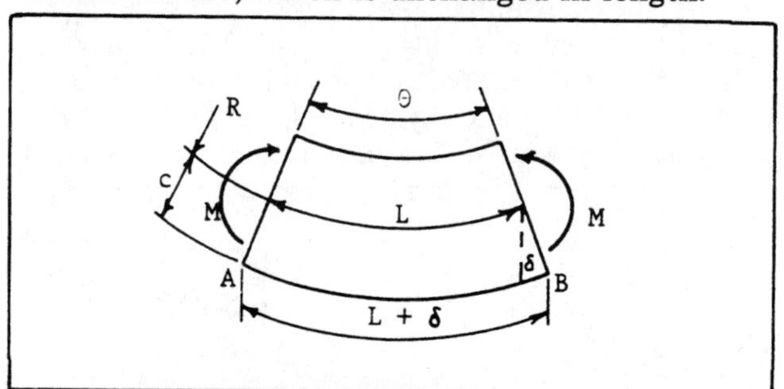

Fig 401-14. Elastic beam curvature.

From Fig 401-14,

$\theta = L/R = (L + \delta)/(R + c)$

from which $c/R = \delta/L = \epsilon = \sigma/E = Mc/EI$

and $1/R = M/EI$

This equation relates the radius of curvature of the neutral surface of the beam to the bending moment, M; the stiffness of the material, E; and the moment of inertia of the cross section, I. Quantity EI is sometimes referred as the flexural stiffness.

This equation is for constant bending moment for the portion of the beam involved. For most beams, the bending moment varies along the beam. The same equation can be derived from the curvature of an elastic curve from a calculus text:

$1/R = (d^2y/dx^2)/[1 + (dy/dx)^2]^{3/2}$

For actual beams, slope dy/dx is very small, and its square can be neglected in comparison to unity. With this approximation, the equation becomes:

$1/R = d^2y/dx^2 = M_x/EI$

$M_x = EI\ d^2y/dx^2$

This is the differential equation for the elastic curve of a beam with M dependent on its location along the beam.

The differential equation of the elastic curve can also be obtained from the geometry of the bent beam of Fig 401-15, where $dy/dx = \tan\theta = \theta$ (approximately, for small angles) and $d^2y/dx^2 = d\theta/dx$.

$d\theta = dL/R = dx/R$ (approximately for small angles)

therefore $d^2y/dx^2 = d\theta/dx = 1/R = M_x/EI$

or $M_x = EI\ d^2y/dx^2$

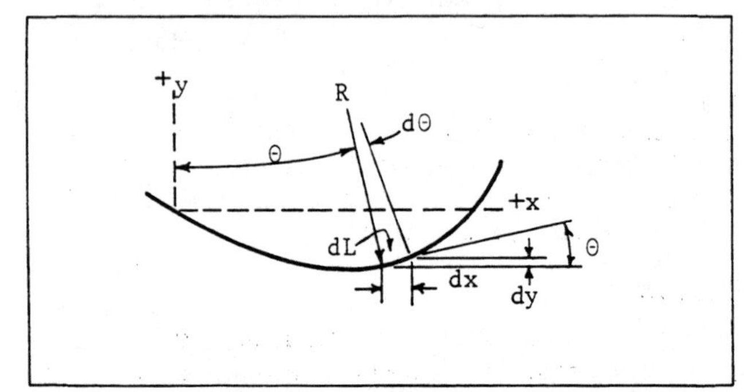

Fig 401-15. Elastic beam deflection curve.

The sign convention for bending moments established in the discussion of shear and moment diagrams is valid for this equation also. Both E and I are always positive; therefore the signs of the bending moment and the second derivative must be consistent. With the coordinate axes in Fig 401-16, the slope changes from positive to negative from A to B; therefore the second derivative is negative, which agrees with the sign of the bending moment. For the interval BC, both d^2y/dx^2 and M_x are positive.

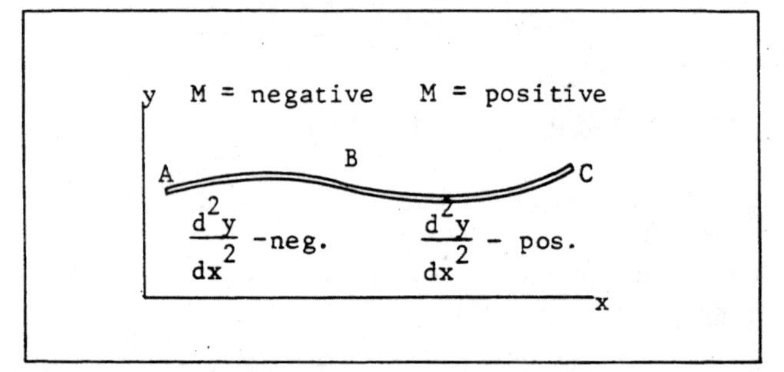

Fig 401-16. Bending moment sign convention.

The successive derivatives of deflection y of the elastic curve correlate with the physical quantities that they represent in beam action. Note that our sign convention for positive shear and moment leads to negative deflection and slope in Fig 401-17.

Deflection $= y$
Slope $= dy/dx$
Moment $= EI\ d^2y/dx^2$

For constant EI:
Shear $= dM/dx = EI\ d^3y/dx^3$
Load $= dV/dx = EI\ d^4y/dx^4$

To plot the slope of a member, substitute θ for slope dy/dx in the elastic curve equation:

$M = EI\ d\theta/dx$

$\int^B_A d\theta = \int^B_A Mdx/EI$

This equation shows that except for EI, the area under the moment diagram between two points along the beam is the change in slope between the same two points. Remembering that $dy/dx = d\theta$, the area under the slope diagram between two points along a beam is the change in deflection between these points.

Beam Diagrams

As an example, a complete series of diagrams is shown in Fig 401-17 for a simply supported beam with a concentrated load at the center of the span. To construct the diagrams, first calculate the reactions. The load is symmetrical and each reaction is P/2.

Shear diagram

The slope of the shear diagram equals the negative value of a distributed gravity load on the beam. In this example, the distributed load is zero. The change in shear equals the negative area under a distributed load. The upward reaction P/2 creates a positive shear in the beam from point A to B. At point B the load P creates a negative shear, added to P/2 from A to B yields a shear of -P/2 from B to C. The P/2 reaction at C causes the shear to close to zero.

Moment diagram

The slope of the moment diagram equals the shear value at the same point. The change in moment equals the area under the shear diagram. Because the beam is simply supported, the moments at the ends are zero. From A to B the shear is positive and the area is PL/4; the slope of the moment diagram is constant and positive, and the moment at B is PL/4. From B to C the shear is negative with an equal area causing the moment diagram to close to zero at C.

Slope diagram

The slope of the slope diagram equals the value of the moment diagram. The change in slope equals the area under the moment diagram divided by EI. Because of symmetry, the maximum deflection is at point B resulting in a slope of zero on the elastic curve. The slope diagram has zero slope at A and C. The slope at A plus the area under the moment diagram from A to B divided by EI, $PL^2/16EI$, equals zero; slope at A is $-PL^2/16EI$; slope at C is $PL^2/16EI$.

Deflection diagram

The slope of the deflection diagram equals the value of the slope diagram. The change in deflection equals the area under the slope diagram. The deflection diagram is zero at the supports and a maximum at B due to symmetry. The area under the slope curve from A to B is $-PL^3/48EI$, which is the deflection at B. The area from B to C is $PL^3/48EI$ causing the deflection to close to zero at C.

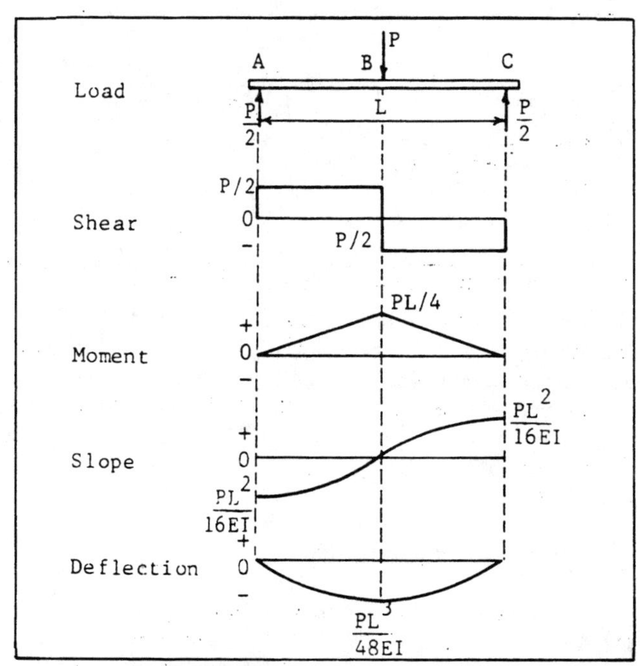

Fig 401-17. Beam diagrams.

The limitations to the flexure formula apply to calculating deflections, because the elastic curve equation was derived from the flexure formula. It was also assumed that:

- The square of the slope of the beam is negligible compared to unity.
- Beam deflection due to shear stress is negligible; a plane section remains a plane.
- E and I are constant; if either varies, it must vary as a function of x.

Deflections by Integration

When the assumptions in the previous section are essentially correct, and the bending moment can be expressed as an integrable function of x, the differential equation for moment can be solved for the deflection, y, of the elastic curve at any point, x, along the beam. Evaluate the constants of integration from the boundary or matching conditions.

A **boundary condition** is a known set of x and y, or x and dy/dx, at a point on the beam. One boundary condition can lead to only one constant of integration. A typical boundary condition is $y = 0$ at a reaction.

Many beams have abrupt load changes such as concentrated loads, reactions, or changes in distributed loads. The bending moment equations at any abrupt change in load are different functions of x. Write separate bending moment equations for each interval of the beam. Although the intervals are bounded by abrupt load changes, the beam is continuous and the slope and deflection at the junction of adjacent intervals must match. A **matching condition** is the equality of slope or deflection determined at the junction of two intervals from the elastic curve equation for both intervals. One matching condition leads to only one constant of integration. A typical matching condition is dy/dx is the same at each side of a point load.

Calculate the deflection of a beam by the double integration method in four steps.

1. Select the beam intervals; place coordinate axes on the beam with the origin at one end of an interval; indicate the range of values of x in each interval. For two adjacent intervals, x might be from 0 to L/2, and from L/2 to L.
2. List the boundary and matching conditions for each interval. Two conditions are required to evaluate the two constants of integration for each interval.
3. Express the bending moment as a function of x for each interval and equate it to EI d^2y/dx^2.
4. Solve the differential equations from step 3 and evaluate the constants of integration. Check the resulting equations for dimensional homogeneity. Calculate deflections as required.

Example: For the beam in Fig 401-18, locate and determine the maximum deflection between the supports.

Solution: From a free-body diagram of the beam and the equations of equilibrium, the left reaction is 7wL/12 upward. With $x = 0$ at the left support, the interval is x from 0 to L.

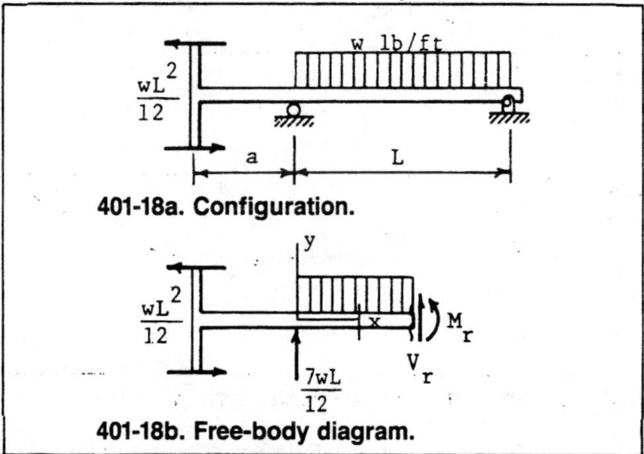

401-18a. Configuration.

401-18b. Free-body diagram.

Fig 401-18. Example beam.

The boundary conditions are $y = 0$ when $x = 0$ and $y = 0$ when $x = L$. The bending moment at any point in the interval is:

$EI\ d^2y/dx^2 = M_r = 7wLx/12 - wL^2/12 - wx^2/2$

Successive integration yields:

$EI\ dy/dx = 7wLx^2/24 - wL^2x/12 - wx^3/6 + C_1$ and

$EI\ y = 7wLx^3/72 - wL^2x^2/24 - wx^4/24 + C_1 x + C_2$

Substitute the boundary condition $y = 0$ when $x = 0$ into the second equation: $C_2 = 0$. Substitute the boundary condition $y = 0$ when $x = L$ into the same equation; $C_1 = -wL^3/72$. The elastic curve equation is:

$EIy = 7wLx^3/72 - wL^2x^2/24 - wx^4/24 - wL^3x/72$

The maximum deflection occurs where the slope, dy/dx, is zero:

$0 = 7wLx^2/24 - wL^2x/12 - wx^3/6 - wL^3/72$

The solution of this cubic equation in x gives the point of maximum deflection at $x = 0.541$ L between the supports. The deflection at that point is:

$y = -7.88\ wL^4/10^3EI$

y is negative, so deflection is downward.

Moment Area Method

The moment area method to determine deflection is semigraphical and relates successive derivatives of the deflection, y, to the moment diagram. For problems involving several changes in loading, the moment area method is usually much faster than double integration. The method is based on two theorems derived as follows.

From Fig 401-19 it is seen that for arc length dL along the elastic curve of a loaded beam:

$d\Theta = dL/R$

Remember $1/R = M/EI$ into $d\Theta = (M/EI)dL$. For actual beams, the curvature is so small that $dL = dx$ along the unloaded beam.

$d\Theta = M\ dx/EI$

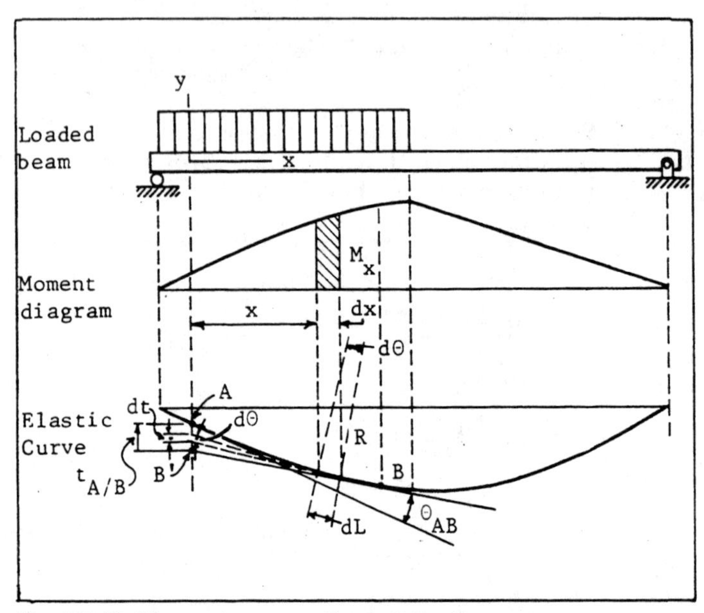

Fig 401-19. Moment area method derivation.

In Fig 401-19, $d\Theta$ is also the angle between tangents to the elastic curve at two points a distance dL apart. Select points A and B along the elastic curve; integrate the equation for $d\Theta$ between these limits:

$\int_A^B d\Theta = \Theta_B - \Theta_A = \int_A^B M\ dx/EI$

The integral on the left, designated $\Theta_A B$, is the angle between tangents to the elastic curve at A and B. The theorem can be written as:

$\Theta_{AB} = \int_A^B M\ dx/EI$

The integral is 1/EI times the area under the moment diagram when E and I are constants. If E or I varies along the beam, construct an M/EI diagram instead of a moment diagram. The **first moment area theorem** is: The angle between the tangents to the elastic curve of a loaded beam at two points along the curve equals the area under M/EI diagram of the beam between the same points.

Use the following sign convention to correctly interpret numerical results: areas under positive bending moments are positive. A positive area means that the angle is positive (counterclockwise) when measured from the tangent to the point on the left to the other tangent, Fig 401-20.

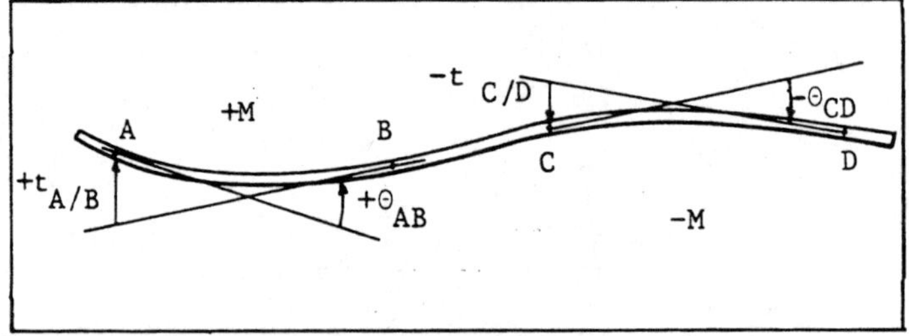

Fig 401-20. Moment area method sign convention.

To obtain the second moment area theorem, let B′ in Fig 401-19 be the intersection of the tangent to the elastic curve at B with a vertical from point A on the elastic curve. Because the deflections of a beam are small,

$dt = x\ d\theta = Mx\ dx/EI$

Integration yields

$\int_0^{t,A/B} dt = t_{A/B} = \int_A^B Mx\ dx/EI$

The left integral is the vertical distance $t_{A/B}$ of a point A on the elastic curve from a tangent drawn to another point B on the elastic curve. Call this distance the tangential deviation to distinguish it from the deflection, y. The integral on the right is the first moment, with respect to an axis through A parallel to the y axis, of the area through A parallel to the y axis, of the area under the M/EI diagram between points A and B.

The **second moment area theorem** is: The vertical distance (tangential deviation) **to** point A on the elastic curve of a beam **from** a tangent drawn at point B on the elastic curve equals the first moment, with respect to an axis at A, of the area under the M/EI diagram between points A and B.

The shape of the elastic curve is not always evident from the given loads; the following sign convention is recommended. The first moment of the area under a positive M/EI diagram is positive and gives a positive tangential deviation. A positive tangential deviation means that point A on the beam is above the tangent drawn from point B—the tangential deviation is in the positive y direction.

Moment diagram by parts

In deriving the moment area formulas, a moment diagram was drawn that included the effect of all loads and reactions. When calculating deflections, it is sometimes easier to draw a series of diagrams, each for the moment due to one load or reaction, Fig 401-21. Each moment diagram is drawn from the load, on the left, to an arbitrary reference section on the right. Select the reference section at the right end of the beam or at one end of a uniformly distributed load. Start with the leftmost load or reaction and draw its bending moment diagram ignoring other loads or reactions. Proceed toward the right, drawing a separate moment diagram for each load and reaction. Obtain the composite moment diagram by superposition (addition) of the component diagrams. Determine the deflection at any point on the beam by superposition of the effects of each moment diagram.

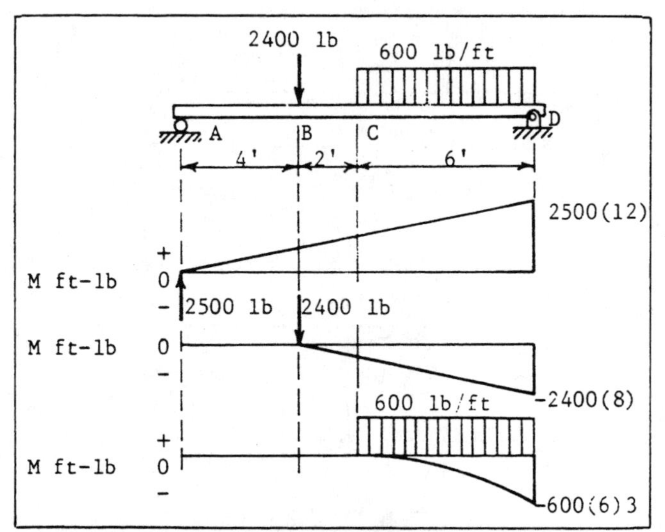

Fig 401-21. Moment diagram by parts.

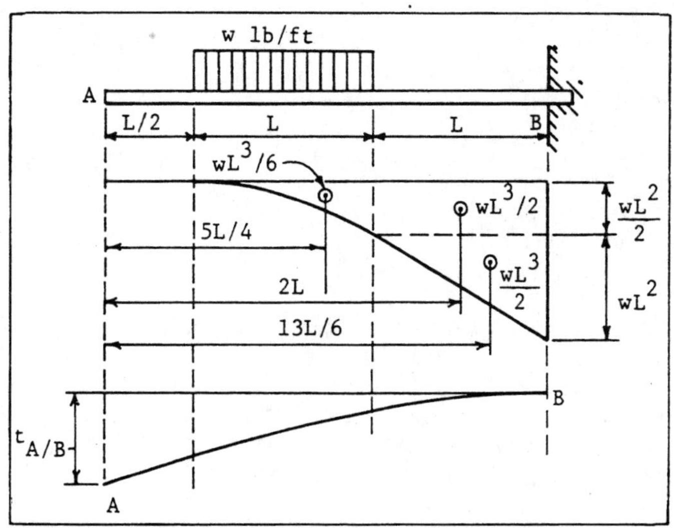

Fig 401-22. Deflections by moment area.

Deflections by moment area

The deflection of a specific point along the elastic curve of a beam can usually be determined with the second moment area theorem. Constructing the correct moment diagram and elastic curve sketch is essential:

1. Sketch the loaded beam, the moment or M/EI diagrams, and the elastic curve.
2. Visualize which tangent lines are most helpful, and draw them on the elastic curve. A tangent at a point of known zero slope may be useful.
3. By applying the second moment area theorem, determine the tangential deviation at the point where the beam deflection is desired and at other points as required.
4. From geometry, determine the perpendicular distance from the tangent line to the unloaded beam at the point where the beam deflection is desired. Using the results of step 3, solve for the desired deflection.

Example: Determine the deflection of the free end of the cantilever beam in Fig 401-22 in terms of w, L, E, and I.

Solution: E and I are constant, so use the bending moment diagram instead of an M/EI diagram. Divide the area under the moment diagram into rectangular, triangular, and parabolic parts to calculate areas and moments. The elastic curve is shown with the deflections greatly exaggerated.

Select points A and B at the ends of the beam because the beam has a horizontal tangent at B and the deflection at A is required. The vertical distance to A from the tangent at B, $t_{A/B}$, equals the deflection of the free end of the beam, y. Use the second moment area theorem to obtain the required deflection directly. The area of each of the three portions of the area under the moment diagram is shown in Fig 401-22 along with the distance from its centroid to the moment axis at the free end A. The second moment area theorem gives:

$$EIt_{A/B} = (-wL^3/6)(5L/4) - (wL^3/2)(2L) - (wL^3/2)(13L/6)$$

from which

$$y_A = t_{A/B} = 55wL^4/24EI \downarrow$$

Maximum deflection by moment area

When the location of the point of maximum deflection is unknown, use both the first and second moment area theorems to determine maximum deflection.

Fig 401-23 shows a typical situation where the maximum deflection, at point B on the elastic curve, is required. The deflection, y_B., or the tangential deviation, t_{AB} is wanted, but the distance x along the beam must be determined before the deflection can be obtained. Once $t_{C/A}$ is determined for the second moment area theorem, $t_{C/A}/L$ gives $\tan \Theta_A$ which is approximately equal to Θ_A. The angle Θ_A can also be determined, in terms of x, from the first moment area theorem because the tangent to the elastic curve at B is horizontal and $\Theta_A = \Theta_AB$. When these two equations for Θ_A are set equal to each other, the distance x is found. Finally $t_{A/B}$ is determined from the second moment area theorem.

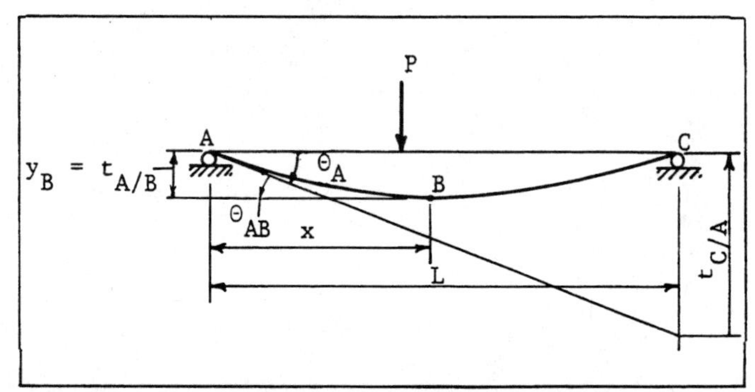

Fig 401-23. Maximum deflections by moment area.

Example: For the beam in Fig 401-24, determine the maximum deflection in terms of w, L, E, and I.

Solution: The equations of equilibrium give the left reaction as wL/2 upward and the right reaction as wL upward. The moment diagram by parts is in Fig 401-24. The elastic curve, very much exaggerated, of the loaded beam is shown in Fig 401-24 with tangents drawn at A and B. The second moment area theorem, with moment axis at C, gives the vertical distance to C from the tangent at A as:

$EIt_{C/A} = L(9wL^3)/4 - (2L/3)(wL^3) - (L/4)(wL^3/6)$
$= 37wL^4/24$
Because the deflections are small,
$\Theta_A = \tan \Theta_A = t_{C/A}/3L$
$= 37wL^3/72EI$
Because the tangent at B is horizontal, the first moment area theorem applied to the segment AB gives:
$\Theta_A = \Theta_{AB} = (1/EI)[(wL/2)(x)(x/2)$
$- (wL/2)(x-L)(x-L)/2]$
$= (wL/4EI)(2Lx - L^2)$
Equating the two expressions for Θ_A gives:
$37wL^3/72EI = (wL/4EI)(2Lx - L^2)$
from which $x = 55L/36$
Application of the second theorem to the segment AB, with the moment axis at A, gives the vertical distance from the horizontal tangent at B to A:
$-y_B = t_{A/B}$
$= (1/EI)[(2x/3)(wLx/2)(x/2)$
$- (L + 2(x-L)/3)(wL/2)(x-L)^2/2)]$
$= (wL^4/2EI)[(1/3)(55/36)^3 - (1/2)(19/36)^2$
$- (1/3)(19/36)^3]\ y_B$
$= -0.500\ wL^4/EI \downarrow$

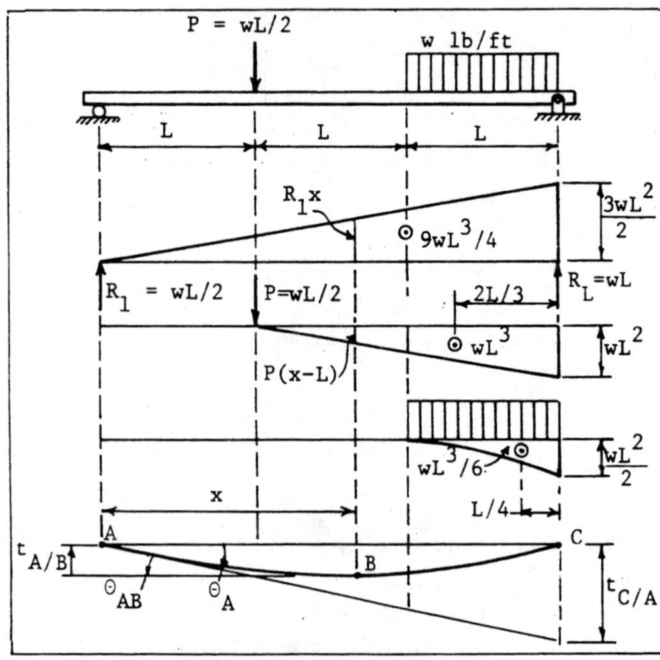

Fig 401-24. Maximum deflection example.

402 MOMENT AREA

The moment area method is outlined in the previous chapter. This chapter applies the method to the solution of indeterminate structures. The moment area method is easiest to use when analyzing a beam with one degree of indeterminacy. The methods of slope deflection and moment distribution are better suited for analyzing indeterminate frames and beams with more than one degree of indeterminacy.

Rigid-end beam

The beam in Fig 402-1 has four unknowns; there are only three equilibrium equations to solve for them, so it has one degree of indeterminacy. If one unknown force can be determined from the beam's deflections, then the other three unknown forces can be found with the three equilibrium equations. In this example, solve for the vertical force at C. First remove the vertical restraint at C so that there are only three unknowns. Then draw the moment diagram for the beam and calculate the deflection $t_{C/A}$.

The three equilibrium equations for the cantilever beam:

$$M_A = 10' \times 3000 \text{ lb} + 30' \times 1000 \text{ lb}$$
$$= 60{,}000 \text{ ft-lb} \circlearrowleft$$
$$R_A = 4000 \text{ lb} \uparrow$$
$$H_A = 0$$

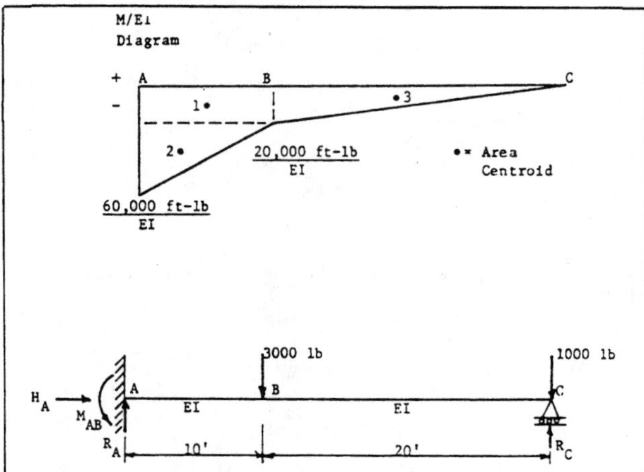

Fig 402-1. Example beam.

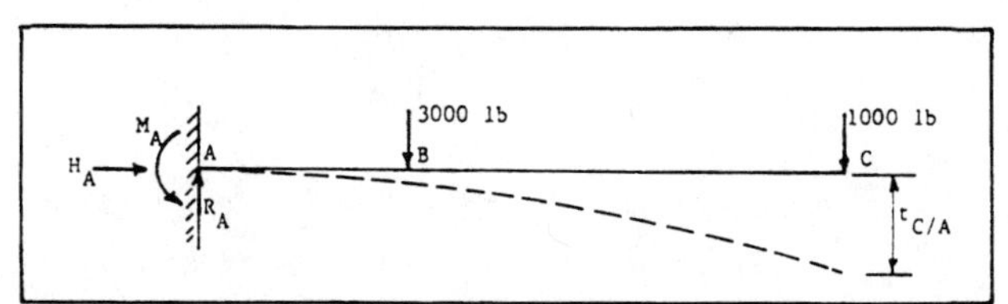

Fig 402-2. Rigid-end beam—restraint at C removed.

Using the second moment area theorem:

$t_{C/A}$ = moment of areas 1, 2 and 3 about point C

$$t_{C/A} = -(10)(20{,}000)(20 + 5)/EI$$
$$- (10)(40{,}000)(½)(20 + 20 \div 3)/EI$$
$$- (20)(20{,}000)(½)(40 \div 3)/EI$$
$$= 13{,}000{,}000 \downarrow$$

For total beam deflection to be zero at point C as required in the original problem, apply a force at point C such that $t_{C/A} + t'_{C/A} = 0$.

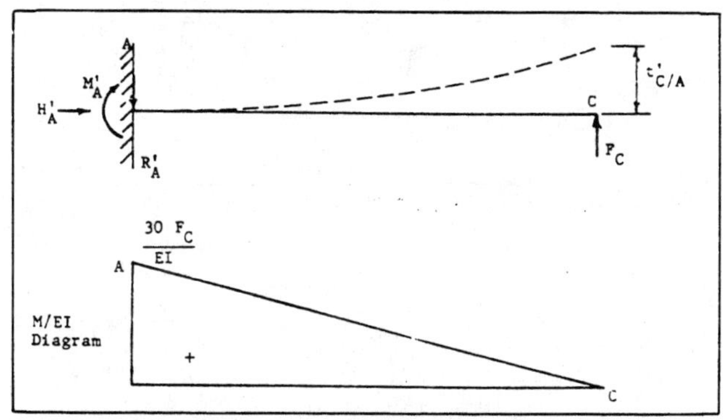

Fig 402-3. Rigid-end beam—force at C.

The original beam in Fig 402-1 is equal to the superposition of the two beams of Figs 402-2 and 402-3 if $t_{C/A} + t'_{C/A} = 0$.

From the three equilibrium equations:

$$M'_A = 30\ F_C \circlearrowright$$
$$R'_A = F_C \downarrow$$
$$H'_A = 0$$

Using the second moment area theorem:

$$t'_{C/A} = +(30F_C/EI)(30)(½)(20) = 9000F_C/EI \uparrow$$

Substituting into the compatibility equation for the deflection at C, $t_{C/A} + t'_{C/A} = 0$, yields:

$$(-13{,}000{,}000 + 9000F_C)/EI = 0$$
$$F_C = 13{,}000{,}000 \div 9000 = 1{,}444 \text{ lb} \uparrow$$

When the compatibility equation for deflection has been satisfied, then $R_C = F_C$ and $R_C = 1{,}444$ lb ↑.

Now the beam is determinate with three unknowns and three equilibrium equations.

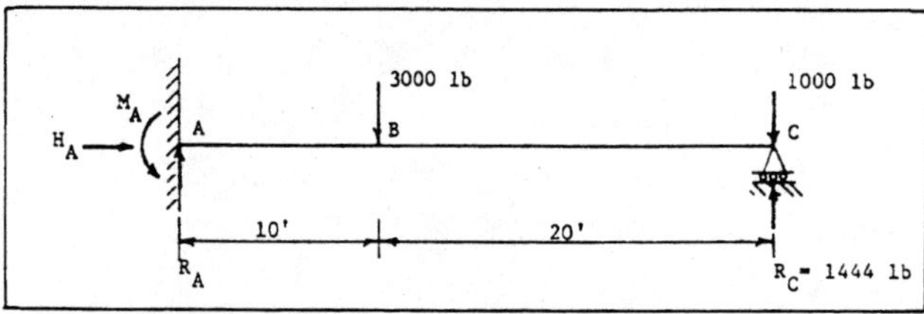

Fig 402-4. Rigid-end beam—reaction at C known.

Sum moments about point A:
$M_A = 16,680$ ft-lb ↺
Sum vertical forces:
$R_A = 6$ lb ↑
Sum horizontal forces:
$H_A = 0$

Continuous beam

Consider a beam continuous over three supports, Fig 402-5. Again there are four unknown forces and only three equilibrium equations, so the beam has one degree of indeterminacy. In this example remove the continuity of the beam over support B so that there are now two simply supported beams. The ends of the two segments rotate relative to one another at point B. Then apply sufficient moment to rotate the ends of the simple beams back to a position of no relative movement.

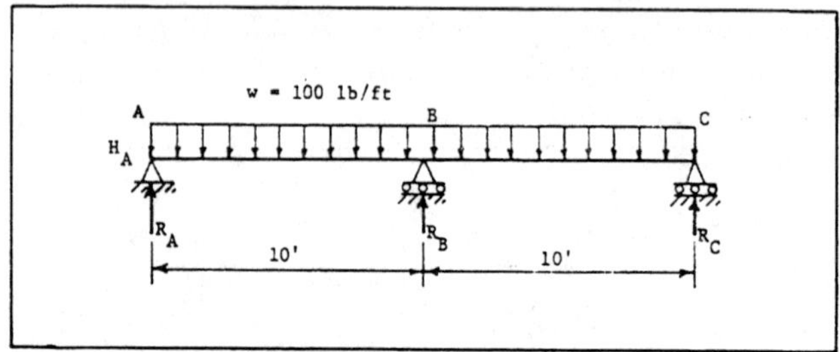

Fig 402-5. Continuous beam example.

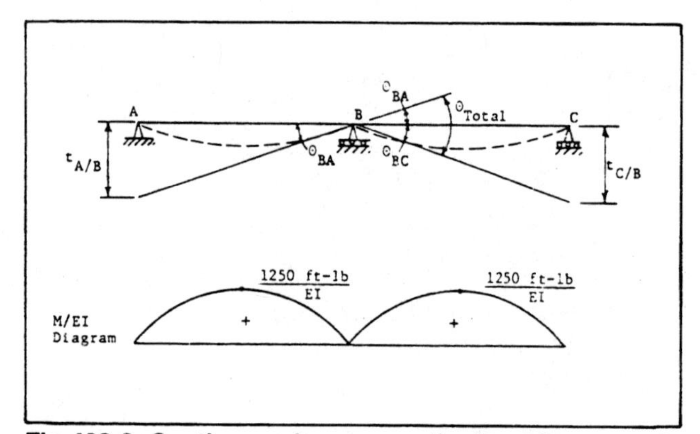

Fig 402-6. Continuous beam—continuity at B removed.

The moment diagram for the beam becomes:
$M_{max} = wL^2/8 = 1250$ ft-lb

The second moment area theorem:
$t_{A/B} = t_{C/B}$ from symmetry
$t_{A/B} = t_{C/B} = +(2/3)(10)(1250)(5)/EI$
$= 41,667/EI$ ↓
$t_{A/B} = (10)(\Theta_{BA})$
$t_{C/B} = (10)(\Theta_{BC})$
and $\Theta_{total} = \Theta_{BA} + \Theta_{BC}$
$\Theta_{total} = (2)(41,667)/(10)EI$
$= 8,333/EI$ ↻

If the beam is continuous, the angle between the two tangents to the elastic curve at support B is zero. Apply a moment, M_B, to the ends of member BA and BC at B to rotate them through an angle, $\Theta'_{BA} + \Theta'_{BC}$, equal to Θ_{total} in the opposite direction as shown in Fig 402-7.

The second moment area theorem:

$t'_{A/B} = t'_{C/B}$ due to symmetry
$t'_{A/B} = t'_{C/B} = -(M_B)(10)(1/2)(20 \div 3)/EI$
$= 33.3M_B/EI$ ↑
$t'_{A/B} = (10)(\Theta'_{BA})$
$t'_{C/B} = (10)(\Theta'_{BC})$
$\Theta_{total} = -\Theta'_{BA} - \Theta'_{BC}$
$\Theta_{total} = (2 \div 10)(33.3M_B/EI) = 8,333/EI$
$M_B = 1250$ ft-lb

Find the three reactions by drawing the free body diagrams of each half, BC and AB, and using the equilibrium equations.
$R_A = 375$ lb ↑
$R_B = 1250$ lb ↑
$R_C = 375$ lb ↑

These answers agree with the beam formulas given in Chapter 412.

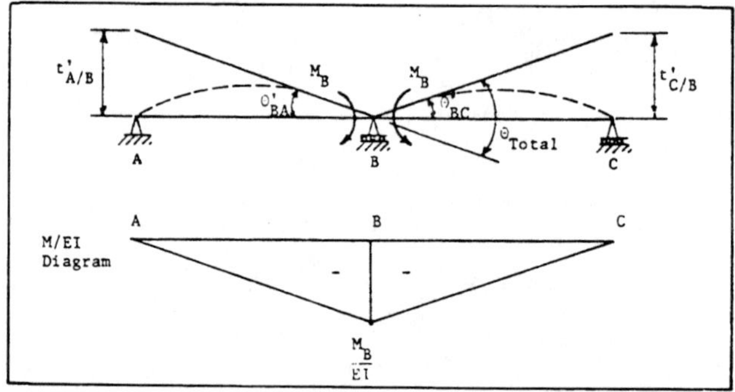

Fig 402-7. Continuous beam—moment applied at B.
Where $\Theta_{total} = -\Theta'_{BA} - \Theta'_{BC}$.

403 SLOPE DEFLECTION

Many indeterminate analysis sytems are adapted from slope deflection theory. It is easy to develop an understanding of the physical relationships with this system. Solutions involve a series of linear equations that lend themselves to computer solutions.

Continuous beam

Consider a continuous beam with three supports and two concentrated loads, P_1 and P_2., Fig 403-1.

The beam deforms under load, and at equilibrium appears as Fig 403-2.

The forces in the two sections of the beam, AB and BC, are shown in the free body diagrams in Fig 403-3.

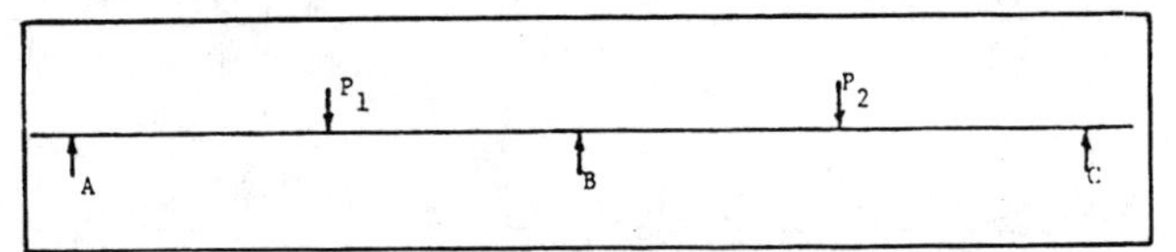

Fig 403-1. Continuous beam with two concentrated loads.

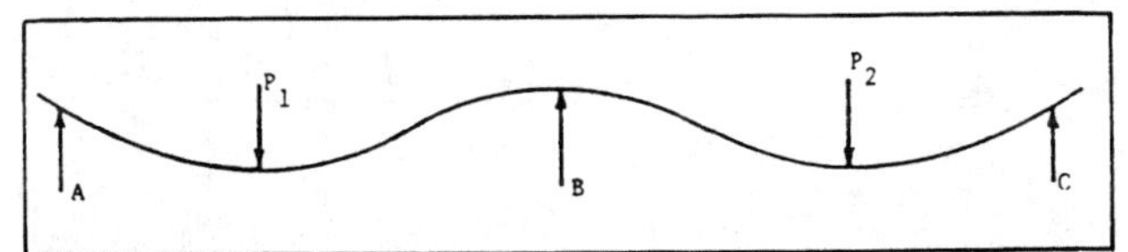

Fig 403-2. Deformed continuous beam.

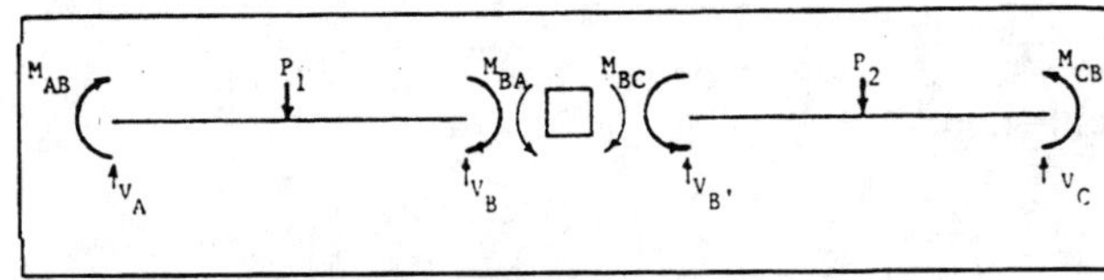

Fig 403-3. Continuous beam—free body diagram.

From the static equilibrium equations of the joint above support B, Fig 403-3, it follows that $M_{BA} + M_{BC} = 0$.

The moment in this joint can be induced by:

1. A load applied on the beam.
2. A rotation of one or both ends of the beam.
3. A differential deflection of the ends of the beam (if both ends deflect the same amount, no differential moment is introduced).

The moment at joint B contributed by each of these conditions can be added by superposition in the equilibrium equation for the joint; $M_{BA} + M_{BC} = 0$ where M_{BA} = Moment due to rotation at A, Θ_A, plus moment due to rotation at B, Θ_B, plus moment due to loads, plus moment due to differential deflection of A and B, δ.

The sign convention for slope deflection analysis usually considers clockwise rotation as positive.

The moment area theorems can be used to find the moment induced by a load, joint rotation, or differential joint settlement.

Fixed end moments: uniform load

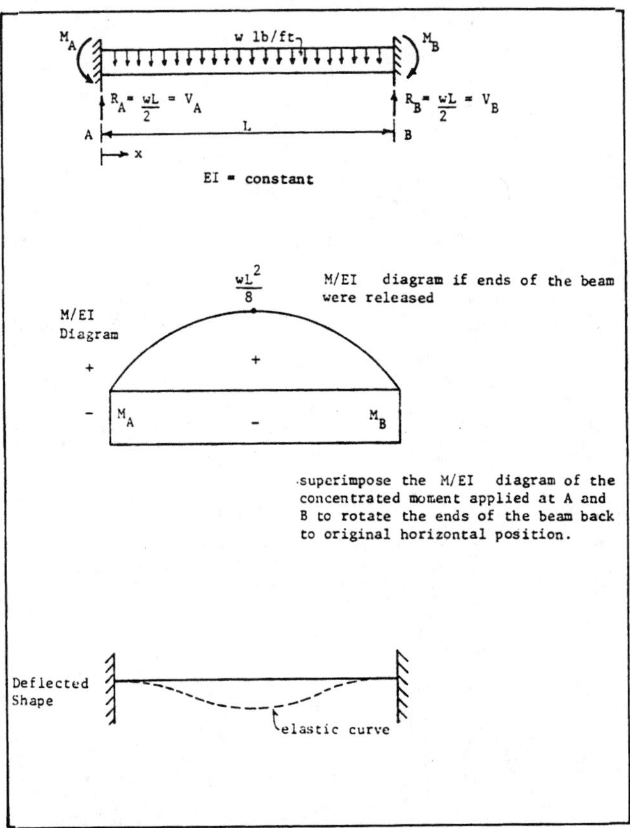

Fig 403-4. Beam with uniform load.

The first moment area theorem states that the angle between the tangent to the beam's elastic curve at A to the tangent to its elastic curve at B equals the area under the M/EI diagram.

$\Theta_{AB} = \int_A^B M/EI\, dx$

or,

$(\Theta_{AB})(EI) = \int_A^B M dx$

Because both ends of the beam are fixed and in line with each other, the tangents to the elastic curve at A and B must also be in line with each other and $\Theta_{AB} = 0$.

$$0 = \int_A^B M dx$$

The area under the parabolic curve for the moment due to the load when the ends of the beam are released is

$(2L/3)(wL^2/8) = wL^3/12$

so,

$$0 = \int_A^B M dx$$
$$= wL^3/12 - M_A(L) = 0$$
$$M_A = wL^3/12L = \pm\ wL^2/12 \circlearrowleft$$

Formulas for moments from loads on a beam with one or both ends fixed are in Chapter 412.

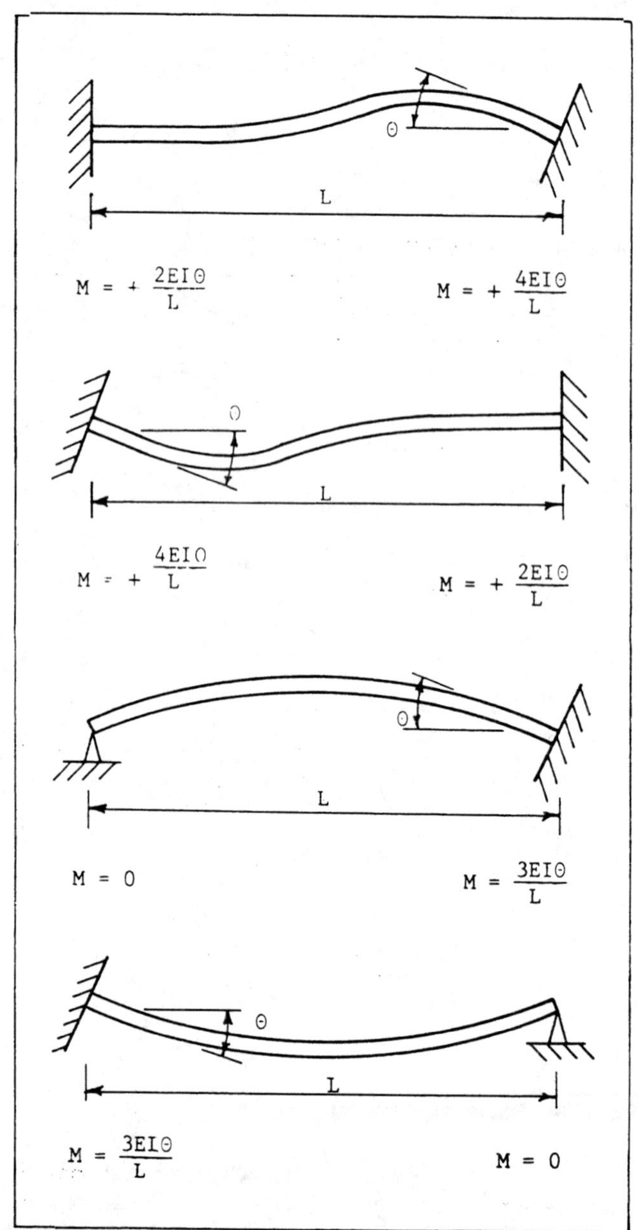

Fig 403-5. Fixed end beam—moments from rotating one end.

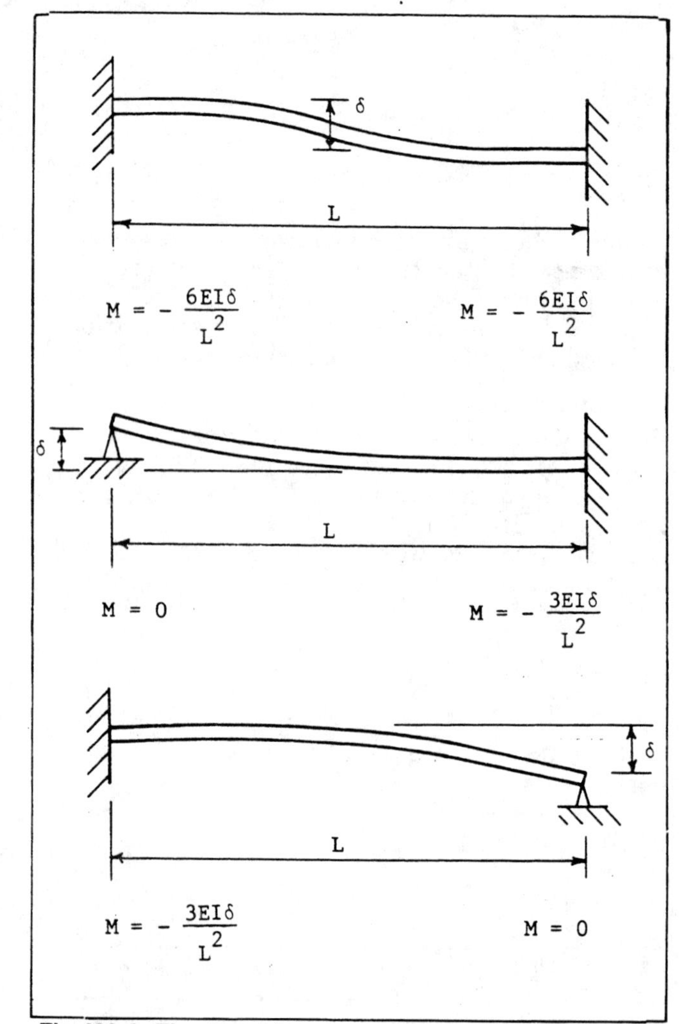

Fig 403-6. Fixed end beam—moments from differential settlement.

A derivation of some of these equations using moment area follows later in this chapter.

The following example illustrates determining forces in a continuous beam.

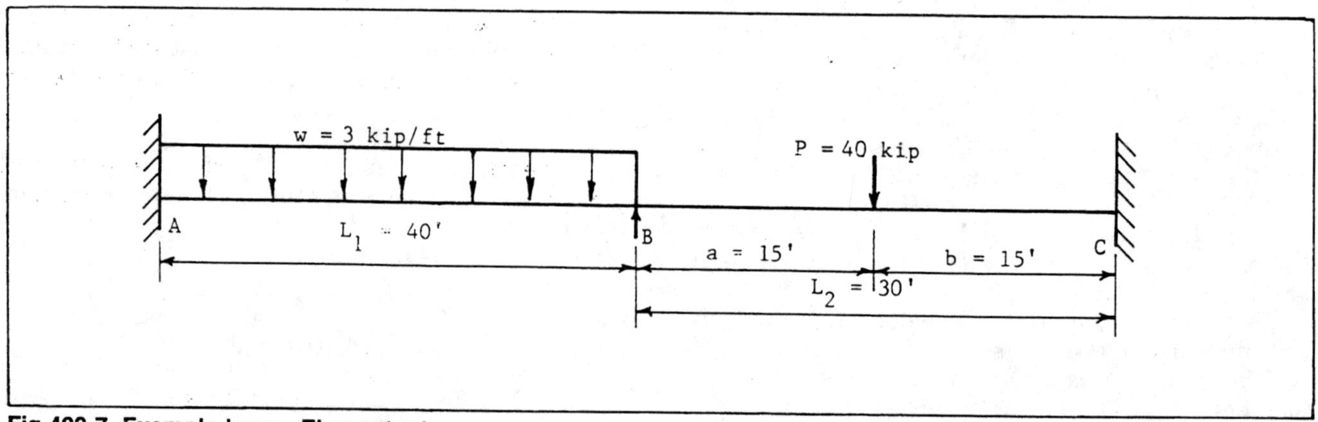

Fig 403-7. Example beam, EI constant.

Using the formulas given here and in Chapter 412, the moments are:

Table 403-1. End moments for example beam.

Moments due to		Load		Left end rotation		Right end rotation		Differential deflection
M_{AB}	=	$\frac{-wL_1^2}{12}$	+	$\frac{4EI\ \theta_A}{L_1}$	+	$\frac{2EI\ \theta_B}{L_1}$	-	$\frac{6EI\delta}{L_1^2}$
M_{BA}	=	$\frac{wL_1^2}{12}$	+	$\frac{2EI\ \theta_A}{L_1}$	+	$\frac{4EI\ \theta_B}{L_1}$	-	$\frac{6EI\delta}{L_1^2}$
M_{BC}	=	$\frac{-Pab^2}{L_2^2}$	+	$\frac{4EI\ \theta_B}{L_2}$	+	$\frac{2EI\ \theta_C}{L_2}$	-	$\frac{6EI\delta}{L_2^2}$
M_{CB}	=	$\frac{+Pa^2b}{L_2^2}$	+	$\frac{2EI\ \theta_B}{L_2}$	+	$\frac{4EI\ \theta_C}{L_2}$	-	$\frac{6EI\delta}{L_2^2}$

In this example, $w = 3$ kip/ft; $L_1 = 40'$; $\delta = 0$; $P = 40$ kip; $a = 15'$; $b = 15'$; $L_2 = 30'$; $\Theta_A = \Theta_C = 0$.

M_{AB} = -400 ft-kip + 2EI $\Theta_B \div 40$
M_{BA} = +400 ft-kip + 4EI $\Theta_B \div 40$
M_{BC} = -150 ft-kip + 4EI $\Theta_B \div 30$
M_{CB} = +150 ft-kip + 2EI $\Theta_B \div 30$

Sum moments around joint B:
$M_{BA} + M_{BC} = 0$
400 ft-kip + (4EI Θ_B/40) - 150 ft-kip + (4EI Θ_B/30) = 0
EI Θ_B = -1071.4
Substitute EI Θ_B into the previous equations:
M_{AB} = -453.6 ft-kip
M_{BA} = +293.0 ft-kip
M_{BC} = -293.0 ft-kip
M_{CB} = +79.0 ft-kip

Two-dimension frame

A second example illustrates finding moments due to differential deflection and moments for a member with one end fixed and one end pinned. Fig 403-8 is a two-dimensional frame with one degree of indeterminacy. Neglecting the change in length of member BC, it follows that $\delta_B = \delta_C$. As a result there are three unknowns of interest, Θ_B, U_C and δ_B.

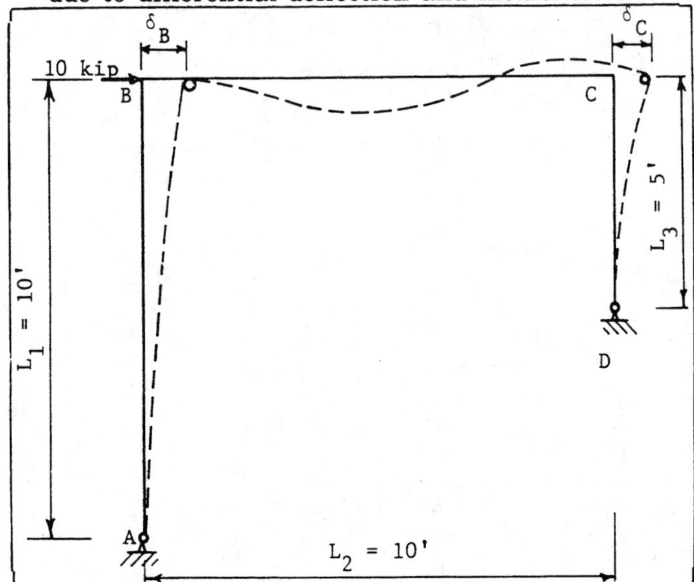

Fig 403-8. Two-dimension frame in sample problem.
$\delta_B = \delta_C$
EI = constant pinned support at A and D

The moment equations for the frame are in Table 403-2.

Table 403-2. Moment equations for two-dimension frame.
$L_1 = 10'$; $L_2 = 10'$; $L_3 = 5'$.

Moments due to		Load		Left end rotation		Right end rotation		Differential deflection
M_{AB}	=	0	+	0	+	0	+	0
M_{BA}	=	0	+	0	+	$\frac{3EI\ \theta_B}{L_1}$	-	$\frac{3EI\delta}{L_1^2}$
M_{BC}	=	0	+	$\frac{4EI\theta_B}{L_2}$	+	$\frac{2EI\ \theta_C}{L_2}$	+	0
M_{CB}	=	0	+	$\frac{2EI\theta_B}{L_2}$	+	$\frac{4EI\ \theta_C}{L_2}$	+	0
M_{CD}	=	0	+	0	+	$\frac{3EI\ \theta_C}{L_3}$	-	$\frac{3EI\delta}{L_3^2}$
M_{DC}	=	0	+	0	+	0	+	0

Sum moments around joint B and around joint C to determine two of the unknowns.
$M_{BA} + M_{BC} = 0$
3EI Θ_B/10 - 3EI δ/100 + 4EI Θ_B/10 + 2EI Θ_C/10 = 0
70 EI Θ_B + 20 EI Θ_C - 3EI δ = 0
$M_{CB} + M_{CD} = 0$
2EI Θ_B/10 + 4EI Θ_C/10 + 3EI Θ_C/5 - 3EI δ/25 = 0
5EI Θ_B + 25 EI Θ_C - 3EI δ = 0

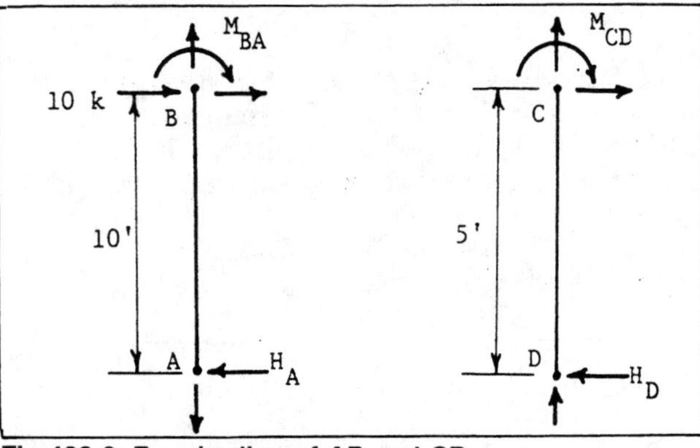

Fig 403-9. Free bodies of AB and CD.

Sum horizontal forces for the third equation. Find horizontal forces in terms of moments from free bodies of members AB and CD, Fig 403-9.

$\Sigma M_B = M_{BA} + 10H_A = 0$ ↻
$\Sigma M_C = M_{CD} + 5H_D = 0$ ↻
$H_A = -M_{BA}/10$
$H_D = -M_{CD}/5$

Sum horizontal forces for the whole frame as a free body.
$\Sigma F_H = 0 = 10$ kip $- H_A - H_D$
$\Sigma F_H = 0 = 10$ kip $+ M_{BA}/10 + M_{CD}/5$
$M_{BA} + 2M_{CD} = -100$
3EI Θ_B/10 - 3EI δ/100 + 6EI Θ_C/5 - 6EI δ/25 = -100
30 EI Θ_B + 120 EI Θ_C - 27 EI δ = -10,000

The equations for the three unknowns are:
$70\ EI\ \Theta_B + 20\ EI\ \Theta_C - 3EI\ \delta = 0$
$5EI\ \Theta_B + 25\ EI\ \Theta_C - 3EI\ \delta = 0$
$30\ EI\ \Theta_B + 120\ EI\ \Theta_C - 27\ EI\ \delta = -10{,}000$
Their solutions are:
$EI\ \Theta_B = 7.246$
$EI\ \Theta_C = 94.198$
$EI\ \delta = 797.06$

For the moments at joints B and C, substituting these values into the previous equations:

$M_{BA} = 3EI\ \Theta_B/10 - 3EI\ \delta/100$
$= -21.738$ ft-kip
$M_{BC} = 4EI\ \Theta_B/10 + 2EI\ \Theta_C/10$
$= +21.737$ ft-kip
$M_{CB} = 2EI\ \Theta_B/10 + 4EI\ \Theta_C/10$
$= +39.128$ ft-kip
$M_{CD} = 3EI\ \Theta_C/5 - 3EI\ \delta/25$
$= -39.129$ ft-kip

Gable frame

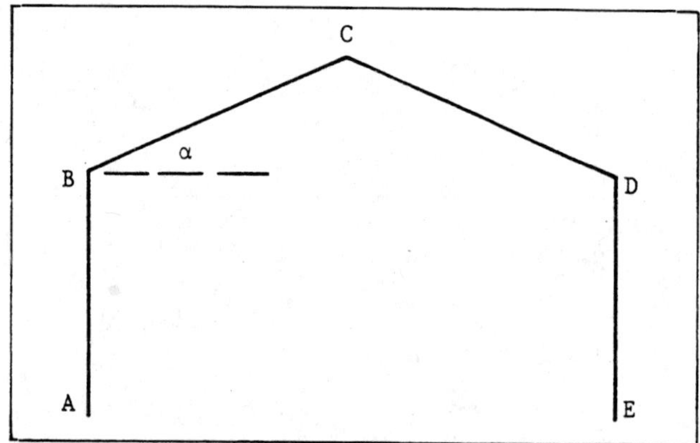

Fig 403-10. Gable type frame.

The forces in a gable frame, Fig 403-10, can be found with similar methods. The unknowns are: $EI\ \Theta_B$, $EI\ \Theta_C$, $EI\ \Theta_D$, $EI\ \delta_{BA}$,, $EI\ \delta_{BC}$, $EI\ \delta_{CD}$, $EI\ \delta_{DE}$.

It can be shown through geometrical relationships that:
$EI\ \delta_{BC} = (EI/2 \sin \alpha)(-\delta_{BA} + \delta_{DE})$
$EI\ \delta_{CD} = (EI/2 \sin \alpha)(\delta_{BA} - \delta_{DE})$

This reduces the number of unknowns from seven to five. Five equations are therefore needed. Sum moments for equilibrium at joints B, C, and D for three equations. Sum horizontal and vertical forces for the whole frame as a free body for the other two equations.

Stiffness coefficients using moment area theorems

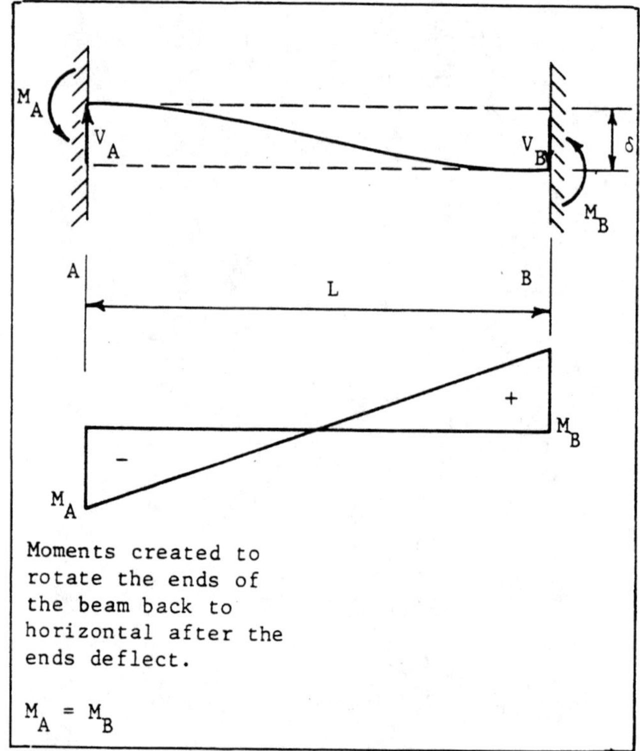

Fig 403-11. Fixed-end beam: settlement.

Using the second moment area theorem:
$t_{B/A} = \delta = \int^B_A Mx/EI\ dx$
Because EI is constant:
$\delta EI = \int^B_A Mxdx$
$\delta EI = -(M_A)(L/4)(L/2 + L/3) + (M_A)(L/4)(L/6)$
$= -(M_A)(L/4)(2L/3) = -M_A\ L^2/6$
$M_A = -6EI\ \delta/L^2$

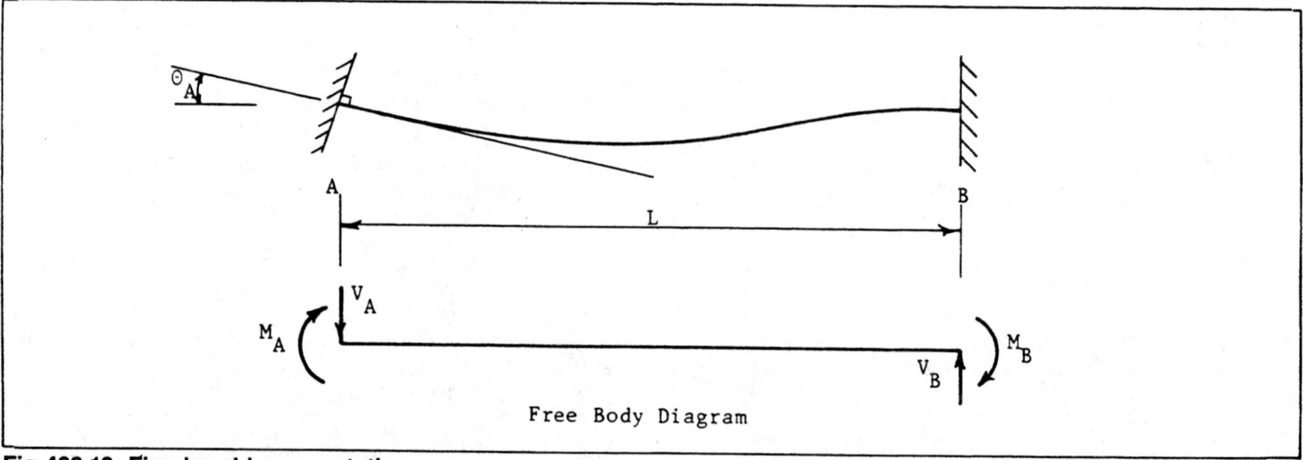

Fig 403-12. Fixed-end beam: rotation.

Using moment diagrams by parts, where the carry-over moment, M_B, is not known, and both moment area theorems:

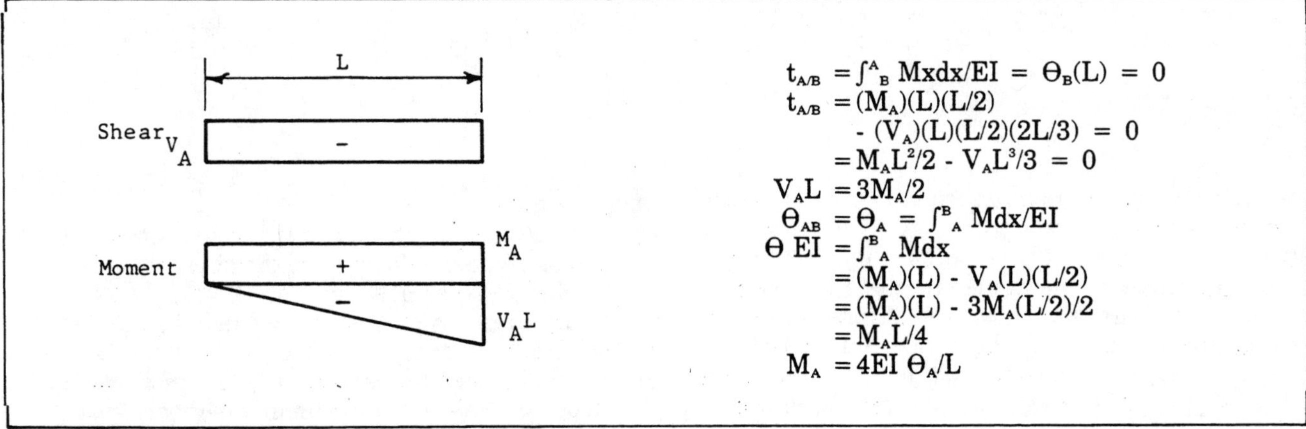

Fig 403-13. Moments and shears from A, Fig 403-12.

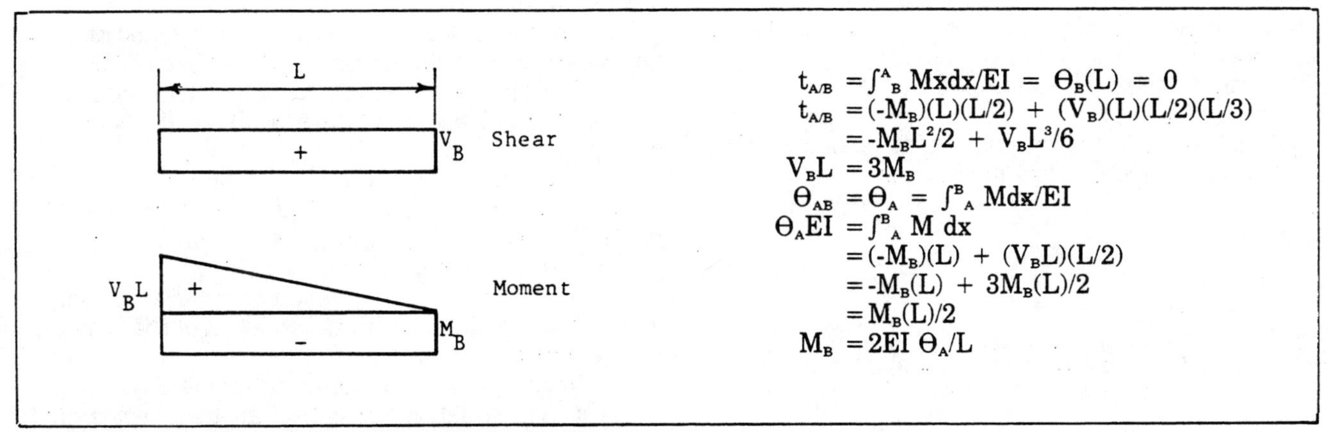

Fig 403-14. Moments and shears from B, Fig 403-12.

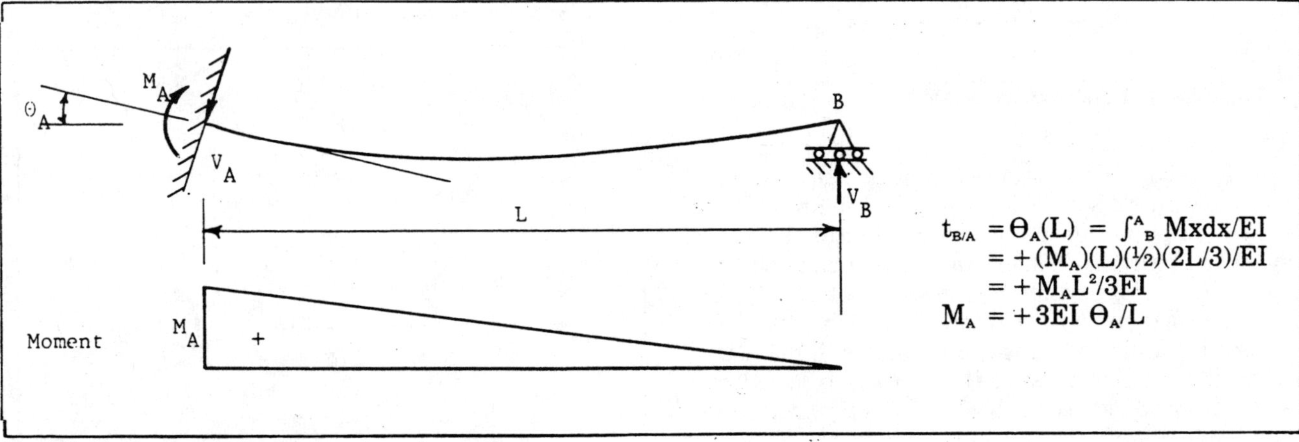

Fig 403-15. One end fixed, one pinned: rotation.

404 MOMENT DISTRIBUTION

Moment distribution is convenient for analyzing indeterminate rigid-joint structures. The slope deflection method may involve solving many simultaneous equations. Moment distribution involves no simultaneous equations, unless the structure moves laterally, and is often much shorter than slope deflection. A series of cycles converge to the final result and may be stopped at the degree of precision required.

Moment distribution involves the superposition of four effects:

- Moment due to the applied load.
- Moment due to the rotation of the near end of the member.
- Moment due to the rotation of the far end of the member.
- Moment due to the differential deflection of the joints at the end of the member.

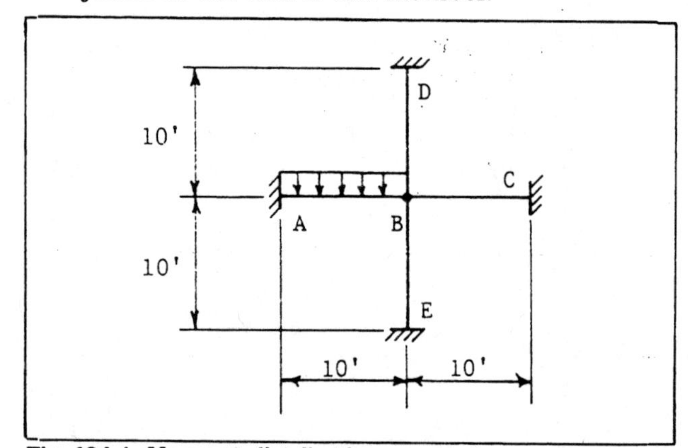

Fig 404-1. Moment distribution example.

Consider the structure in Fig 404-1. The supports are fixed and allow no rotation; joint B is rigid so all members framing into it undergo the same rotation; the load is applied to member AB. Obtain the moment at the ends of AB from the equations given in the section on slope deflection or in Chapter 412, Eq 412-47. Release joint B: it rotates until enough moment is built up in the ends of members AB, BC, BD, and BE at joint B to balance the moment brought to the joint from the load on AB when joint B was locked against rotation. Simultaneously moments develop at the fixed supports due to the rotation of joint B. When joint B is in equilibrium, the moment at the ends of the members is the algebraic sum of the moment due to the load when joint B was locked and the moment caused by the rotation of joint B. This procedure is essentially the moment distribution method.

Terminology

Fixed-end moments are those developed at the fixed ends of the members from the applied loads and joint deflections, as given by Eq 412-47.

Unbalanced moment is the algebraic sum of all of the fixed end moments at a joint. When a joint rotates under the unbalanced moment, moments are developed in the ends of the members meeting at the joint.

Distributed moments are those that restore the joint to equilibrium.

Carry-over moments are those developed at the far ends of each member as the joint rotates.

The **stiffness factor,** K, for a member is EI/L.

Clockwise moments are positive, as in slope deflection.

The **distribution factor** is the ratio of the stiffness of the member receiving moment to the sum of the stiffnesses of all members at the joint.

The **carry-over factor** is the ratio of the moment at the far fixed end of a member to the moment entering the member from the rotation of the joint at the near end.

Distribution of the unbalanced moment at a joint is based on the relative stiffness of the members framing into the joint. Draw a free body diagram of joint B and sum moments:

$$M_{BA} + M_{BC} + M_{BD} + M_{BE} + M = 0$$

M = unbalanced moment

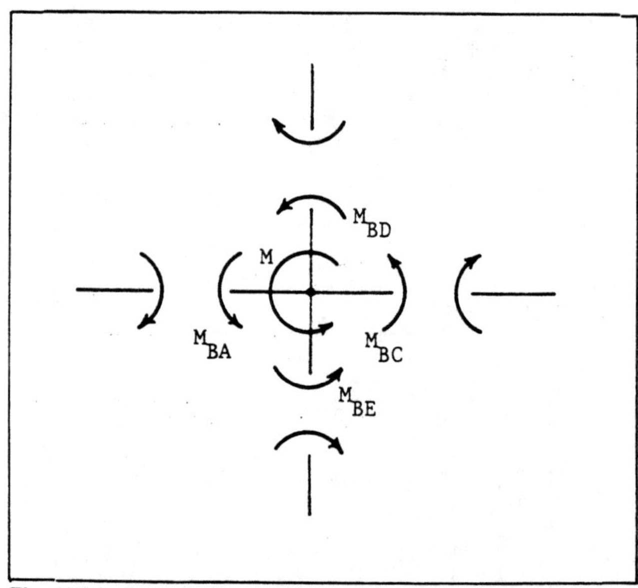

Fig 404-2. Free body diagram of joint B.

From the derivation of the effect of a joint rotation in the chapter on slope deflection:

$M_{BA} = 4(EI/L)_{BA}\ \Theta_B$
$M_{BD} = 4(EI/L)_{BD}\ \Theta_B$
$M_{BC} = 4(EI/L)_{BC}\ \Theta_B$
$M_{BE} = 4(EI/L)_{BE}\ \Theta_B$
$K_{BA} = 4(EI/L)_{BA}$
$K_{BD} = 4(EI/L)_{BD}$
$K_{BC} = 4(EI/L)_{BC}$
$K_{BE} = 4(EI/L)_{BE}$

Evaluate EI/L in each equation for the member in which the moment occurs. Substitute the moment equations into the equilibrium equation and simplify. For M_{BA}:

$$M_{BA} = -M\ K_{BA} \div (K_{BA} + K_{BD} + K_{BC} + K_{BE})$$

In general, for any member BM, the distributed moment is:

$$M_{BM} = -M\ K_{BM}/\Sigma K$$

The sum includes all of the members meeting at joint B where the unbalanced moment, M, occurs.

The **distribution factor** is defined as:

$$DF_{BM} = K_{BM} \div \Sigma K$$
$$M_{BM} = -DF_{BM}\ (M)$$

An equation for carry-over moments is also needed. If one end of a member, which is on unyielding supports at each end, is rotated while the other end remains fixed, the ratio of the moment at the fixed end to the moment producing the rotation is called the **carry-over factor.** Referring again to the derivation of the beam stiffness equations in the chapter on slope deflection:

$$M_{BM} = 4EI\ \Theta_B/L$$
and
$$M_{MB} = 2EI\ \Theta_B/L$$

$M_{MB}/M_{BM} = ½$ = carry-over factor for member BM for the distributed moment at Joint B.

In this case the carry-over moment is equal to ½ of its corresponding distributed moment and has the same sign.

Example I

A beam continuous over two supports between two fixed supports.

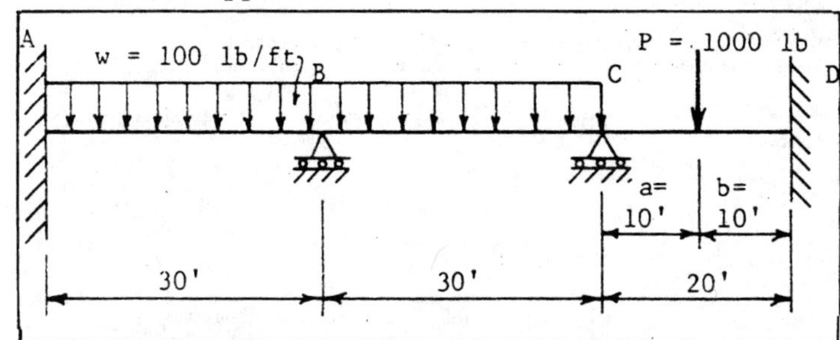

Fig 404-3. Fixed-end beam with two supports.

Step 1. Calculate fixed end moments for the three beam segments:

$FEM_{AB} = FEM_{BC} = -wL^2/12$
$= -(100\ lb/ft)(30')^2/12$
$= -7500$ ft-lb
$FEM_{BA} = FEM_{CB} = WL^2/12$
$= (100\ lb/ft)(30')^2/12$
$= 7500$ ft-lb
$FEM_{CD} = Pab^2/L^2$
$= -(1000\ lb)(10')(10')^2/(20')^2$
$= -2500$ ft-lb
$FEM_{DC} = Pa^2b/L^2$
$= (1000\ lb)(10')^2(10')/(20')^2$
$= 2500$ ft-lb

Step 2. Calculate distribution factors. The stiffness of the fixed supports is infinitely large, ∞.

$K_{AB} = 4EI/30$
$K_{BC} = 4EI/30$
$K_{CD} = 4EI/20$
$DF_{AB} = K_{AB}/(K_{AB} + \infty) = 0$
$DF_{BA} = K_{AB}/(K_{AB} + K_{BC}) = ½$
$DF_{BC} = K_{BC}/(K_{AB} + K_{BC}) = ½$
$DF_{CB} = K_{BC}/(K_{BC} + K_{CD}) = 0.4$
$DF_{CD} = K_{CD}/(K_{CD} + K_{BC}) = 0.6$
$DF_{DC} = K_{CD}/(K_{CD} + \infty) = 0$

Step 3. Calculate carry-over factors. Both ends of all of the members are continuous or fixed, so all carry-over factors are ½.

Step 4. Set up a table. It must start and end with a distribution of the unbalanced moments.

For the first distribution: The unbalanced moment at A is -7500; times (-DF) = 0. The unbalanced moment at B = 7500 - 7500 = 0, so there is no moment to distribute. The unbalanced moment at C is 7500 - 2500 = 5000; 5000(-.4) = -2000 in CB; 5000(-.6) = -3000 in CD. At C, unbalanced moment is 5000; distributed moment is 0.

The carry-overs are: Member AB, no distributed load to carry-over. End B of BC has no moment, but the ½ of the -2000 at end C carries over to end B. In CD, ½ of -3000 in end C carries to end D.

When adequate precision is reached, total the fixed end moments and the distributed and carry-over moments for each end of each member.

Table 404-1. Moment distribution for Example I.

	M_{AB}	M_{BA}	M_{BC}	M_{CB}	M_{CD}	M_{DC}
Dist. factor	0	1/2	1/2	0.4	0.6	0
Carry-over factor	1/2	1/2	1/2	1/2	1/2	1/2
Fixed end moment	-7500	7500	-7500	7500	-2500	2500
1st dist.	0	0	0	-2000	-3000	0
c.o.	0	0	-1000	0	0	-1500
2nd dist.	0	500	500	0	0	0
c.o.	250	0	0	250	0	0
3rd dist.	0	0	0	-100	-150	0
c.o.	0	0	-50	0	0	-75
4th dist.	0	25	25	0	0	0
c.o.	12.5	0	0	12.5	0	0
5th dist.	0	0	0	-5	-7.5	0
Total	-7237.5 ft-lb	8025 ft-lb	-8025 ft-lb	5675.5 ft-lb	-5657.5 ft-lb	925 ft-lb

For more precision the process could be lengthened to more than five distributions.

Example II

A beam continuous over one support and fixed at one end and free at the other.

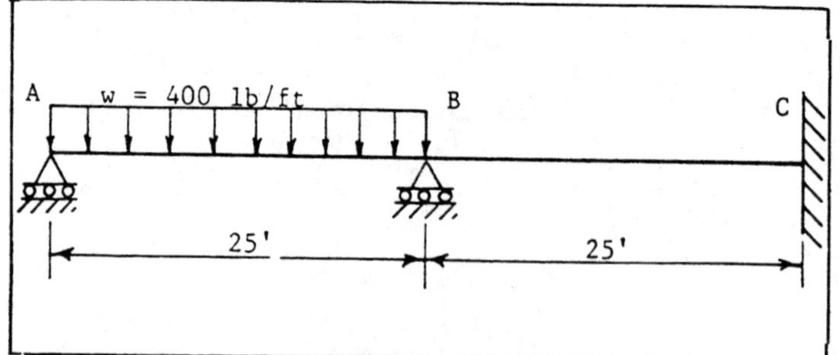

Fig 404-4. Two pinned, one fixed supports.

Step 1. Calculate fixed end moments for the two beam segments:

$FEM_{AB} = FEM_{BA} = wL^2/12$
$FEM_{BC} = FEM_{CB} = 0$
$FEM_{BA} = wL^2/12 = (400\ lb/ft)(25')^2/8$
$= 31{,}250$ ft-lb

Step 2. Calculate distribution factors. The stiffness of the fixed support is infinitely large, ∞, and that of the free end is zero. See last example, Chapter 403.

$K_{AB} = 3EI/25$
$K_{BC} = 4EI/25$
$DF_{AB} = K_{AB}/(K_{AB} + 0) = 1$
$DF_{BA} = K_{AB}/(K_{AB} + K_{BC}) = 0.43$
$DF_{BC} = K_{BC}/(K_{BC} + K_{AB}) = 0.57$
$DF_{CB} = K_{BC}/(K_{BC} + \infty) = 0$

Step 3. Calculate carry-over factors. For member BC with one end fixed and one end continuous the carry-over factors are ½ in both directions. For member AB the right end is continuous so the carry-over factor from A to B is ½, but the left end is free so the factor from B to A is zero.

Step 4. Set up a table. It must start and end with a distribution of the unbalanced moments.

Table 404-2. Moment distribution for Example II.
This example has reached a precise answer with two distributions.

	M_{AB}	M_{BA}	M_{BC}	M_{CB}
Dist. factor	1	0.43	0.57	0
Carry-over factor	1/2	0	1/2	1/2
Fixed end moments	0	31,250	0	0
1st dist.	0	-13,435	-17,812	0
c.o.	0	0	0	-8,906
2nd dist.	0	0	0	0
Total	0	17,812 ft-lb	-17,812 ft-lb	-8,906 ft-lb

Example III

A rigid frame with sidesway. EI = constant.

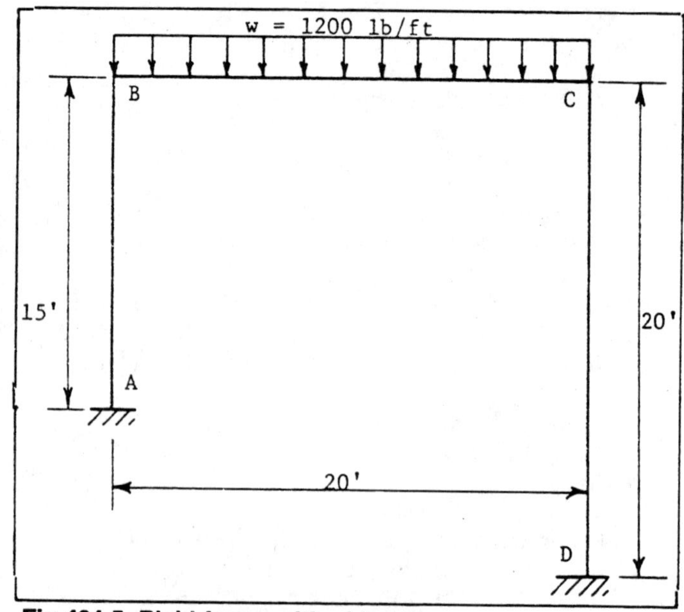

Fig 404-5. Rigid frame with sidesway.

Step 1. Calculate fixed end moments:

$FEM_{AB} = FEM_{BA} = FEM_{CD} = FEM_{DC} = 0$
$FEM_{BC} = -wL^2/12$
$= -(1200\ lb/ft)(20')^2/12 = -40{,}000$ ft-lb
$FEM_{CB} = wL^2/12$
$= (1200\ lb/ft)(20')^2/12 = 40{,}000$ ft-lb

Step 2. Calculate distribution factors. The stiffness of the fixed suppports is infinitely large, ∞.

$K_{AB} = 4EI/15$
$K_{BC} = 4EI/20$
$K_{CD} = 4EI/20$
$DF_{AB} = K_{AB}/(K_{AB} + \infty) = 0$
$DF_{BA} = K_{AB}/(K_{AB} + K_{BC}) = 0.57$
$DF_{BC} = K_{BC}/(K_{BC} + K_{AB}) = 0.43$
$DF_{CB} = K_{BC}/(K_{BC} + K_{CD}) = ½$
$DF_{CD} = K_{CD}/(K_{CD} + K_{BC}) = ½$
$DF_{DC} = K_{CD}/(K_{CD} + \infty) = 0$

Step 3. Calculate carry-over factors. Both ends of all of the members are continuous or fixed, so all carry-over factors are ½.

Step 4. Set up a table.

Table 404-3. Moment distribution for Example III.

	M_{AB}	M_{BA}	M_{BC}	M_{CB}	M_{CD}	M_{DC}
Dist. factor	0	0.57	0.43	1/2	1/2	0
Carry-over factor	1/2	1/2	1/2	1/2	1/2	1/2
Fixed end moment	0	0	-40,000	40,000	0	0
1st dist.	0	22,800	17,200	-20,000	-20,000	0
c.o.	11,400	0	-10,000	8,600	0	-10,000
2nd dist.	0	5,700	4,300	-4,300	-4,300	0
c.o.	2,850	0	-2,150	2,150	0	-2,150
3rd dist.	0	1,225	924	-1,075	-1,075	0
c.o.	612	0	-538	462	0	-538
4th dist.	0	307	231	-231	-231	0
Total	14,862 ft-lb	30.032 ft-lb	-30,033 ft-lb	25,606 ft-lb	-25,606 ft-lb	-12,688 ft-lb

Step 5. Calculate the unbalanced horizontal force in the frame. First draw free body diagrams of members AB and CD and calculate the shear at A and D.

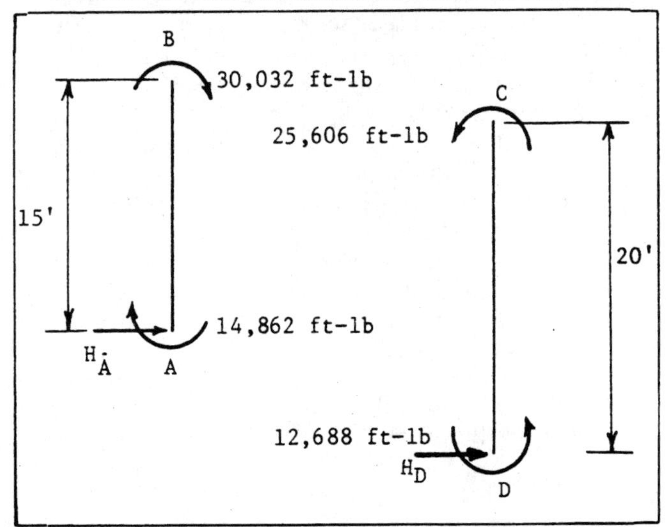

Fig 404-6. Example III: members AB and CD.

$$\Sigma M_B = 0 \circlearrowleft +$$
$$= 30{,}032 \text{ ft-lb} + 14{,}862 \text{ ft-lb} - (15')(H_A)$$
$$H_A = 2992.9 \text{ lb} \rightarrow$$
$$\Sigma M_C = 0 \circlearrowleft +$$
$$= -25{,}606 \text{ ft-lb} - 12{,}688 \text{ ft-lb} - (20')(H_D)$$
$$H_D = -1914.7 \text{ lb or } 1914.7 \text{ lb} \leftarrow$$

Calculate the unbalanced horizontal force that produces the sidesway.

$$\Sigma F_H = 2992.9 \text{ lb} - 1914.7 \text{ lb} = 1078.2 \text{ lb} \rightarrow \text{ unbalanced}$$

This means that the frame deflects to the right a distance δ. The unbalanced forces at A and D indicate that the frame would deflect to the left if a load greater than 1078.2 lb was applied at B or C in the negative direction.

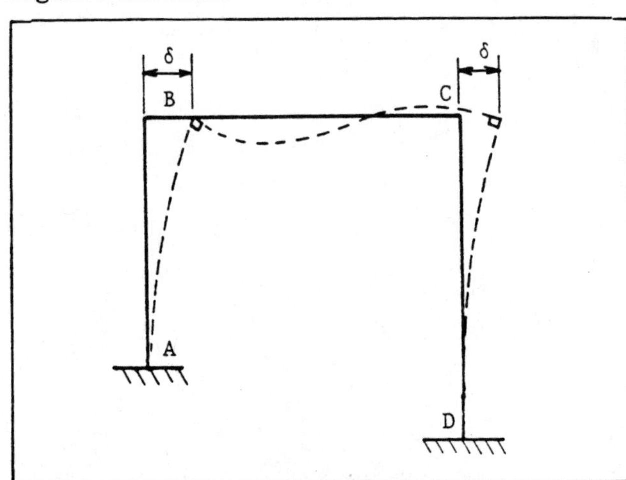

Fig 404-7. Example III: deformations.

Step 6. Assume some fixed end moments to counteract the unbalanced horizontal force. The moment created by a differential deflection of the joints was given in the slope deflection chapter.

$FEM'_{BA} = FEM'_{AB} = 6EI\ \delta/L^2 = 6EI\ \delta/15^2$
$FEM'_{DC} = FEM'_{CD} = 6EI\ \delta/L^2 = 6EI\ \delta/20^2$

For a given deflection, δ, there is a relationship between the moment generated in member AB and member CD.

$FEM'_{AB}/FEM'_{CD} = (20^2/6EI)/(15^2/6EI)$
Assume that FEM'_{BA} is -100 ft-lb.
$FEM'_{BA} = -100$ ft-lb
$FEM'_{CD} = -56.3$ ft-lb
$FEM'_{DC} = -56.3$ ft-lb

Step 7. Apply the assumed moments to the frame and go through the distribution process again. The same distributon and carry-over factors apply for this loading.

Table 404-4. Moment distribution for sidesway.

	M'_{AB}	M'_{BA}	M'_{BC}	M'_{CB}	M'_{CD}	M'_{DC}
Dist. factor	0	0.57	0.43	1/2	1/2	0
Carry-over factor	1/2	1/2	1/2	1/2	1/2	1/2
Fixed end moment	-100	-100	0	0	-56.3	-56.3
1st dist.	0	57	43	28.2	28.2	0
c.o.	28.5	0	14.1	21.5	0	14.1
2nd dist.	0	-8.0	-6.1	-10.8	-10.8	0
c.o.	-4.0	0	-5.4	-3.1	0	-5.4
3rd dist.	0	3.1	2.3	1.6	1.6	0
c.o.	1.6	0	0.8	1.2	0	0.8
4th dist.	0	-0.5	-0.3	-0.6	-0.6	0
Total	-73.9 ft-lb	-48.4 ft-lb	48.4 ft-lb	38.0 ft-lb	-37.9 ft-lb	-46.8 ft-lb

Step 8. Calculate the unbalanced horizontal force in the frame.

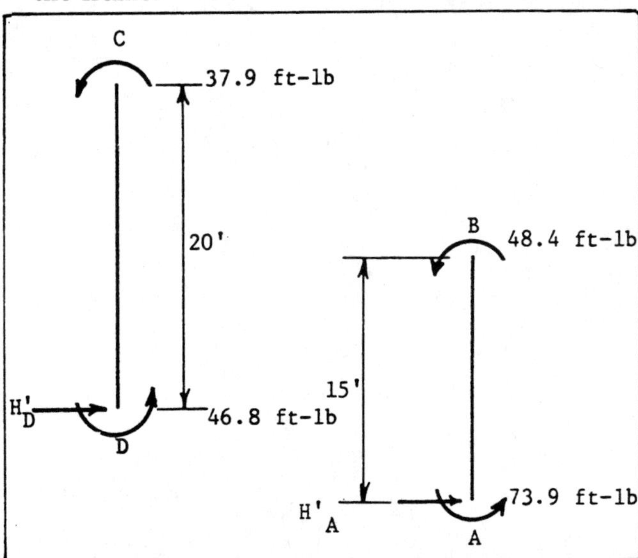

Fig 404-8. Sidesway in members AB and CD.

$$M'_B = 0 \circlearrowleft +$$
$$= -48.4 \text{ ft-lb} - 73.9 \text{ ft-lb} - (15')(H'_A)$$
$$H'_A = -8.15 \text{ lb or } 8.15 \text{ lb} \leftarrow$$
$$M'_C = 0 \circlearrowleft +$$
$$= -37.9 \text{ ft-lb} - 46.8 \text{ ft-lb} - (20')(H'_D)$$
$$H'_D = -4.24 \text{ lb or } 4.24 \text{ lb} \leftarrow$$
$$\Sigma F'_H = -8.15 - 4.24$$
$$= -12.39 \text{ lb or } 12.39 \text{ lb} \leftarrow$$

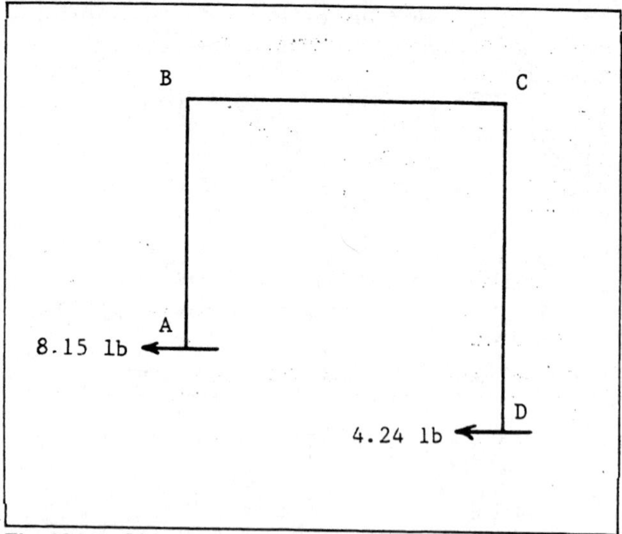

Fig 404-9. Sidesway reactions.

Step 9. Calculate the moments in the frame at equilibrium. The actual moments in the frame to counteract the initial unbalanced horizontal force are:
(1078.2 lb/12.39 lb) (Answers in Step 7) = (87.0) (Answers in Step 7)

Actual moments = (Answers in Step 4) + 87 (Answers in Step 7)

M_{AB} = (14,862 ft-lb) + 87 (-73.9 ft-lb) = 8,433 ft-lb

M_{BA} = (30,032 ft-lb) + 87 (-48.4 ft-lb) = 25,821 ft-lb

M_{BC} = (-30,032 ft-lb) + 87 (48.4 ft-lb) = -25,822 ft-lb

M_{CB} = (25,606 ft-lb) + 87 (38.0 ft-lb) = 28,912 ft-lb

M_{CD} = (-25,606 ft-lb) + 87 (-37.9 ft-lb) = -28,903 ft-lb

M_{DC} = (-12,688 ft-lb) + 87 (-46.8 ft-lb) = -16,760 ft-lb

405 LUMBER STRESSES & SIZES

Tables 405-1 and 405-2 are abstracts from Design Values for Wood Construction, March 1982 Supplement, National Forest Products Assn. There are other grades and several other factors that affect adjustments to design stresses.

Table 405-1. 1982 allowable unit stress for structural lumber, lb/in².
Allowable unit stresses are for normal loading conditions and lumber used at 19% m.c. max.

				- - - - - - - -Allowable unit stresses, lb/in²- - - - - - - -						
						Tension	Horiz	Compression		Modulus of
Species		Size		Bending, F_b		parallel	shear	Perp	Parallel	elasticity
Use	Grade	Thick	Wide	Engr	Repeat	F_t	F_v	F_{c1}	F_c	E
For lumber milled and used at 15% m.c., multiply stress by:				1.08	1.08	1.08	1.05	1.00	1.17	1.05
Douglas fir-larch (used at 19% m.c.; West Coast Lumber Inspection Bureau)										
Struct light framing	Select struct	2″-4″	2″-4″	2100	2400	1200	95	625	1600	1,800,000
	No. 1			1750	2050	1050	″	″	1250	″
	No. 2			1450	1650	850	″	″	1000	1,700,000
Light framing	Const.	2″-4″	4″	1050	1200	625	95	625	1150	1,500,000
	Stand			600	675	350	″	″	925	″
	Utility			275	325	175	″	″	600	″
Joist & plank	Select struct	2″-4″	5″ and wider	1800	2050	(See Table 405-2	95	625	1400	1,800,000
	No. 1			1500	1725		″	″	1250	″
	No. 2			1250	1450		″	″	1050	1,700,000
	Stud		5″&6″	725	850		″	″	675	1,500,000
Posts & timbers	Select struct	5″ or larger	Thick + 2″ and smaller	1500	-	1000	85	625	1150	1,600,000
	No. 1			1200	-	825	″	″	1000	″
Beams & stringers	Select struct	5″ or larger	Greater than thick + 2″	1600	-	950	85	625	1100	1,600,000
	No. 1			1350	-	675	″	″	925	″
Southern Pine (used at 19% m.c.; Southern Pine Inspection Bureau)										
Struct light framing	Select struct	2″-4″	2″-4″	2000	2300	1150	100	565	1550	1,700,000
	No. 1			1700	1950	1000	″	″	1250	″
	No. 2			1400	1650	825	90	″	975	1,600,000
	Stud			775	900	450	″	″	575	1,400,000
Light framing	Const.	2″-4″	4″	1000	1150	600	100	565	1100	1,400,000
	Stand			575	675	350	90	″	900	″
	Utility			275	300	150	90	″	575	″
Struct joist & plank	Select struct	2″-4″	5″ and wider	1750	2010	(See Table 405-2	90	565	1350	1,700,000
	No. 1			1450	1700		″	″	1250	″
	No. 2			1200	1400		″	″	1000	1,600,000
	Stud		5″&6″	725	850		″	″	625	1,400,000
Stress rated timber	No. 1 SR	5″x5″ and larger		1350	1350	875	110	325	775	1,500,000
	No. 2 SR			1100	1100	725	95	″	625	1,400,000

Table 405-2. 1982 allowable stresses in tension parallel to grain, lb/in².
Lumber 2" to 4" thick, wider than 4", and used at 19% m.c. max.

Use	Grade	Lumber width 5",6"	8"	10" or wider
Douglas Fir (WCLIB)				
Joist & plank	Select struct	1200	1080	960
	No. 1	1000	800	600
	No. 2	650	520	390
	Stud	375	-	-
Southern Pine (SPIB)				
Struct joist & plank	Select struct	1150	1025	920
	No. 1	975	780	585
	No. 2	625	500	375
	Stud	350	-	-

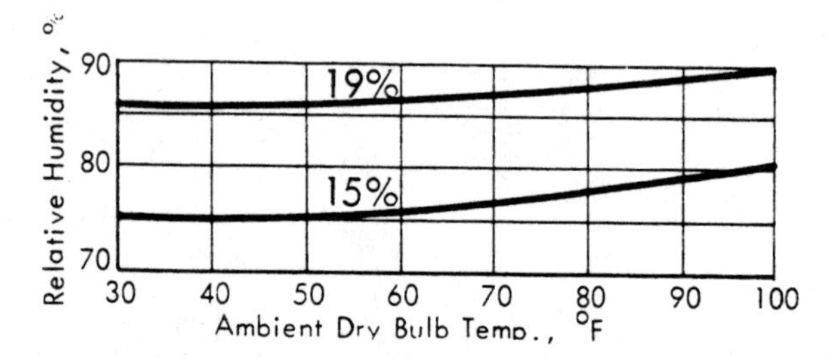

Fig 405-1. Equilibrium moisture content of wood.
Wood Handbook, No. 72, USDA.
Even in livestock buildings, it is unlikely that relative humidity is high enough long enough to raise the moisture content of lumber appreciably above 15%. Lumber shrinks during drying, so NDS stresses apply to moisture content and size at time of surfacing. NDS stresses still apply if the lumber is further dried, even though it shrinks.

Table 405-4. Lumber dimension properties.
Properties of the lumber used in truss designs. The bending properties, I and S, for stacked bottom chords assume the two members act independently, which is conservative. Equivalent B and D are the dimensions given the computer design program.
For 4 + 6, B and D were calculated to give the combined member the same A and S as the sum of the 2x4 and 2x6. The tabulated I was calculated from the equivalent B and D; I of 2x4 + 2x6 is 25.062.

2x6 4+6

Size	B In.	D In.	A In²	I In⁴	S In³	Equivalent B In.	Equivalent D In.
2x4	1.5	3.50	5.250	5.359	3.063		
2x6	1.5	5.50	8.250	20.797	7.563		
2x8	1.5	7.25	10.875	47.635	13.141		
2x10	1.5	9.25	13.875	98.932	21.391		
2x12	1.5	11.25	16.875	177.979	31.641		
4+4	1.5	7.00	10.500	10.718	6.126	3.0	3.5
4+6	1.5	9.00	13.500	25.062	10.626	2.86	4.72
6+6	1.5	11.00	16.500	41.594	15.126	3.0	5.5

Table 405-3. Lumber section properties.

Nominal BxD, in.	Dressed BxD, in.	Area A, in²	Inertia moment I, in⁴	Section modulus S, in³	Shear factor S_s, in²
1 x 2	¾x1½	1.13	0.21	0.28	0.75
x 3	x2½	1.88	0.98	0.78	1.25
x 4	x3½	2.63	2.68	1.53	1.75
x 5	x4½	3.38	5.70	2.53	2.25
x 6	x5½	4.13	10.40	3.78	2.75
x 7	x6½	4.88	17.16	5.28	3.25
1 x 8	¾x7¼	5.44	23.82	6.57	3.63
x 9	x8¼	6.19	35.09	8.51	4.13
x 10	x9¼	6.94	49.47	10.70	4.63
x 11	x10¼	7.69	67.31	13.13	5.13
x 12	x11¼	8.44	88.99	15.82	5.63
2 x 2	1½x1½	2.25	0.42	0.56	1.50
x 3	x2½	3.75	1.95	1.56	2.50
x 4	x3½	5.25	5.36	3.06	3.50
x 5	x4½	6.75	11.39	5.06	4.50
x 6	x5½	8.25	20.80	7.56	5.50
x 8	x7¼	10.88	47.63	13.14	7.25
2 x 10	1½x9¼	13.88	98.93	21.39	9.25
x 12	x11¼	16.88	177.98	31.64	11.26
3 x 2	2½x1½	3.75	0.70	0.94	2.50
x 3	x2½	6.25	3.26	2.60	4.17
x 4	x3½	8.75	8.93	5.10	5.84
x 5	x4½	11.25	18.98	8.44	7.50
x 6	x5½	13.75	34.66	12.60	9.17
x 8	x7¼	18.13	79.39	21.90	12.09
3 x 10	2½x9¼	23.13	164.89	35.65	15.42
x 12	x11¼	28.13	296.63	52.73	18.76
4 x 2	3½x1½	5.25	0.98	1.31	3.50
x 3	x2½	8.75	4.56	3.65	5.84
x 4	x3½	12.25	12.51	7.15	8.17
x 5	x4½	15.75	26.58	11.81	10.51
x 6	x5½	19.25	48.53	17.65	12.84
x 8	x7¼	x25.38	111.15	30.66	16.93
4 x 10	3½x9¼	32.38	230.84	49.91	21.59
x 12	x11¼	39.38	415.28	73.83	26.26
6 x 2	5½x1½	8.25	1.55	2.06	5.50
x 4	x3½	19.25	19.65	11.23	12.83
x 6	x5½	30.25	76.26	27.73	20.18
x 8	x7½	41.25	193.36	51.56	27.51
x 10	x9½	52.25	392.96	82.73	34.85
x 12	x11½	63.25	697.07	121.23	42.19
8 x 2	7¼x1½	10.88	2.04	2.72	7.25
x 4	x3½	25.38	25.90	14.80	16.92
8 x 6	7½x5½	41.25	103.98	37.81	27.51
x 8	x7½	56.25	263.67	70.31	37.52
x 10	x9½	71.25	535.86	112.81	47.52
x 12	x11½	86.25	950.55	165.31	57.53
10 x 2	9¼x1½	13.88	2.60	3.47	9.25
x 4	x3½	32.38	33.05	18.89	21.58
10 x 6	9½x5½	52.25	131.71	47.90	34.85
x 8	x7½	71.25	333.98	89.06	47.52
x 10	x9½	90.25	678.76	142.90	60.20
x 12	x11½	109.25	1204.03	209.40	72.87
12 x 1	12*x¾	9.00	0.42	1.13	6.00
x 2	x1½	18.00	3.38	4.50	12.01
x 3	x2½	30.00	15.63	12.50	20.01
x 4	x3½	42.00	42.88	24.50	28.01
x 5	x4½	54.00	91.13	40.50	36.02
x 6	x5½	66.00	166.38	60.50	44.02

*Full 12" wide.

406 WOOD POST FRAME AND TRUSS DESIGN

Chapter prepared by Dr. Dwaine Bundy, Agricultural Engineering Department, Iowa State University, Ames, Iowa.

Major references are:
National Design Specification for Wood Construction, 1985 Edition (NDS), National Forest Products Association.
Pole Building Design (AWPI), American Wood Preservers Institute.

Loads

Gravity

Lateral wind loads, rather than gravity loads, typically govern post size. The axial force in the posts from gravity loads is insignificant in selecting adequate post size. Therefore, even somewhat higher gravity loads than assumed for post selection do not affect post size.

Wind

From Chapter 101, a design speed of 90 mph for a 50-yr recurrence interval includes almost all the U.S. The velocity pressure includes the importance factor, wind velocity for extreme winds with a 50-yr recurrence, building height, and shape factor.

Truss and roofing materials are assumed to balance deflection (sidesway) in the windward and leeward posts.

The two loading cases of wind parallel and perpendicular to the ridge on a closed building determine the recommended post sizes.

Materials and Stress Selection

Materials Properties

Use the properties in Chapter 405 or the NDS Supplement for dimension lumber and sawn posts. Design values for round posts are from the round timber pile data in NDS.

Adjustments for Load Duration

Increase allowable stress, but not the modulus of elasticity, E, for short-duration loads. Load duration factors, LDF, are:

LDF = 0.9 for dead and other permanent loads
LDF = 1.15 for snow load
LDF = 1.33 for wind load

Post Analysis

A = area of member's cross section, in^2
b = width of the post, in.
C_f = form factor (for F_b only)
C_F = size factor (for F_b only)
C_s = slenderness factor for bending member
CUF = condition of use factor
d = depth of rectangular member, or least dimension of rectangular compression member, in.
D = embedment depth, in.
TabE = tabulated modulus of elasticity, psi
E = TabE × CUF, modulus of elasticity adjusted for use, psi
f_b = actual bending stress, psi
TabF_b = tabulated bending stress, psi
F_b = TabF_b × LDF × CUF × C_f, bending stress adjusted for load duration, use, and shape, psi
F'_b = design bending stress, adjusted for slenderness, psi
TabF_c = tabulated compression stress parallel to grain, psi
F_c = TabF_c × LDF × CUF, compression stress adjusted for load duration and use, psi
F'_c = design compression stress parallel to grain, adjusted for L_e/d ratio (short, intermediate, long column), psi
f_t = actual tensile stress, psi
TabF_t = allowable unadjusted tensile stress, psi
F_t = TabF_c × LDF × CUF
H = wall height, in
J = convenience factor for the interaction equation
K = smallest ratio of L_e/d for a long column
LDF = load duration factor
L_e = effective length, in.
L_u = unbraced length, in.
M_f = maximum bending moment, in-lb
M_f = maximum moment in post at D/3 below ground level, ft-lb
P = axial force, lb
R = horizontal truss reaction due to wind, lb
S = section modulus, in^3
S_1 = passive soil pressure, divide tabulated psf value by 144, psi
w = wind pressure, lb/ft^2

Forces on posts and symbols used are in Fig 406-1.

The maximum bending moment is at the upper soil reaction at about ⅓ of the embedment depth. The soil stress distribution and an allowable average soil passive stress of 2,500 psf are from AWPI for an average soil.

The top of the post can be designed as pinned or fixed. The maximum bending moments are in Eqs 406-1 and 406-2.

For top of post pinned (assume post is fixed at upper soil reaction at ⅓ of the embedment depth):

Eq 406-1.

$$M_f = R\,(H + D/3) + wH\,(H/2 + D/3)$$

For top of post fixed (minimum knee brace at each post is a 2x6 bolted to the truss and post forming a triangle at least 30″ on a side):

Eq 406-2.

$$M_f = R\,(H + D/3)/2 + wH\,(H/2 + D/3) - (H^2 + HD + D^2/3)\,(wH/6)/(H + D/3)$$

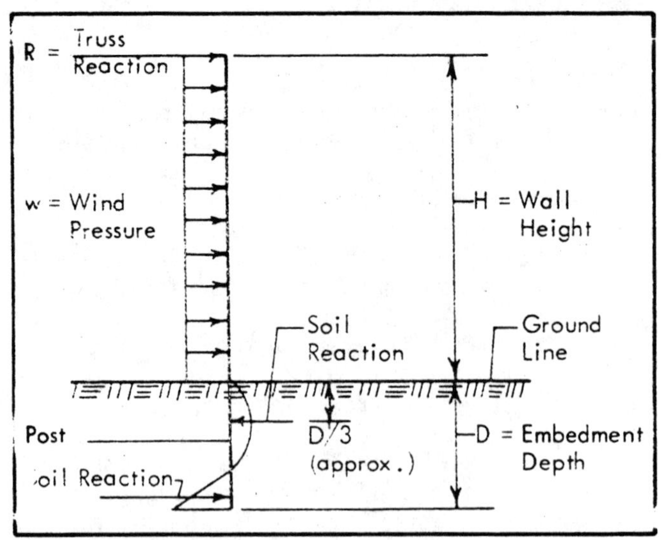

Fig 406-1. Pole load diagram.

Evaluate the interaction equations for moment and axial forces to select an adequate member size.

Tension + bending:

The interaction equations are:

Eq 406-3.

$(f_t \div F_t) + (M_f/S \div F_b) < 1.0$ and,

Eq 406-4.

$(f_b - f_t)/F'_b < 1.0$

The allowable buckling stress F'_b is no greater than F_b, which is adjusted for conditions other than the slenderness factor. The slenderness factor, C_s, is:

Eq 406-5.

$C_s = [(L_e \times d)/b^2]^{0.5} < 50$

For posts:

Eq 406-6.

Bending $L_e = 1.84\ L_u$

Where $L_u = H + D/3$ (conservative)

For short beams, $C_s < 10$:

Eq 406-7.

$F'_b = F_b \times C_F$

For intermediate beams, $10 < C_s < C_k$:

Eq 406-8.

$F'_b = F_b\ [1 - 0.333\ (C_s/C_k)^4]$

or $F'_b = F_b \times C_F$, whichever is smaller.

Eq 406-9.

in which $C_k = 0.811\ (E/F_b)^{0.5}$

For long beams, $C_k < C_s < 50$:

Eq 406-10.

$F'_b = 0.438E/C_s^2$

or $F'_b = F_b \times C_F$, whichever is smaller.

Compression + bending:

The interaction equation is:

Eq 406-11.

$[P/A \div F'_c] + [M_f/S \div (F'_b - JP/A)] < 1.0$

Eq 406-12.

$J = [(L_e/d) - 11] \div (K - 11)$; L_e for bending.

Eq 406-13.

$0 < J < 1.0$

if $L_e/d \leq 11$, $J = 0$

if $L_e/d \geq K$, $J = 1$

The allowable buckling stress, F'_c, is calculated by the equations below. Stress modifications are load duration and use (in F_c) and slenderness.

$L_e = 2.1\ L_u$ (pinned at top); L_e for compression (conservative).

$L_e = 1.2\ L_u$ (fixed at top); L_e for compression (conservative).

For short columns, $L_e/d < 11$:

Eq 406-14.

$F'_c = F_c$

For intermediate columns, $11 < L_e/d < K$:

Eq 406-15.

$K = 0.671\ (E/F_c)^{0.5}$

Eq 406-16.

$F'_c = F_c\ [1 - 0.333 \times (L_e/d \div K)^4]$

For long columns, $L_e/d > K$:

Eq 406-17.

$F'_c = 0.30\ E \div (L_e/d)^2$

Use a post's least dimension to check buckling stresses if the post has no girts providing lateral support in the weak direction (i.e., open wall).

The required embedment depth is:

Eq 406-18.

$D = \{1.47\ (R + wH) + [2.16\ (R + wH)^2 + 10.52\ S_1BM_f]^{0.5}\}/2S_1B$

Procedure and Example Problem

Design a post for 8′ spacing, 14′ height, and 48′ span. The building is an enclosed machine shed and is in Des Moines, Iowa. Assume a 3/12 slope (14°) gable roof and no ceiling.

1. Determine design loads.

   Determine design live and dead loads from Chapter 101. Investigate the load cases:

   Full snow plus dead loads for post axial load (both balanced and unbalanced loading).
   Dead and wind loads for post bending and uplift.
   Determine appropriate load duration factors.

   a. Design snow load. (See Chapter 101 for definition of terms.)

   R = 0.7
   C_e = 1.1

$I = 0.8$
$C_s = 1.0$
$P_g = 25$ psf
$P_s = R \times C_e \times I \times C_s \times P_g$
$= 0.7 \times 1.1 \times 0.8 \times 1.0 \times 25$
$= 15.4$ psf

b. Design wind load.
$I^2 = 0.90$
$V = 80$ mph
$K_zG = 1.14$
$q = 0.00256 \times V^2 \times K_zG \times I^2$
$= 0.00256 \times 80^2 \times 1.14 \times 0.9$
$= 16.8$ psf

c. Design dead load. See Fig 406-3.
Two load cases are in this example.

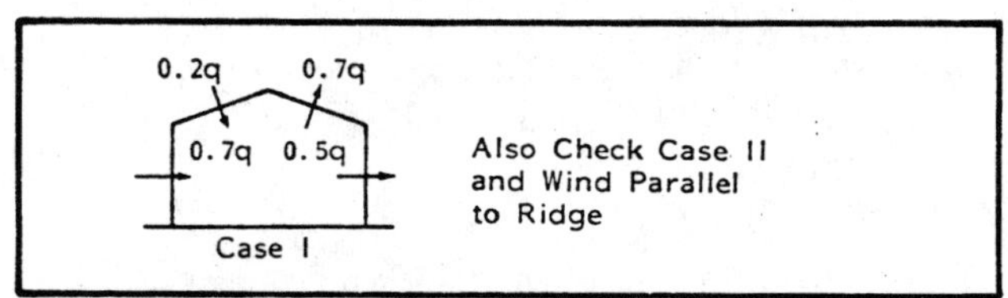

Fig 406-2. Wind load diagram.

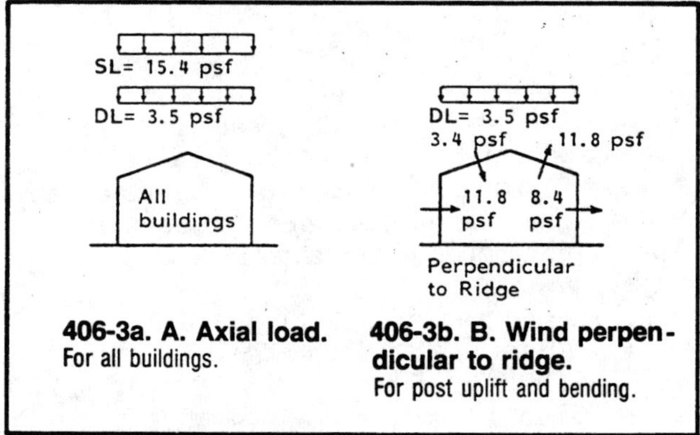

406-3a. A. Axial load. For all buildings.

406-3b. B. Wind perpendicular to ridge. For post uplift and bending.

Fig 406-3. Two load cases.

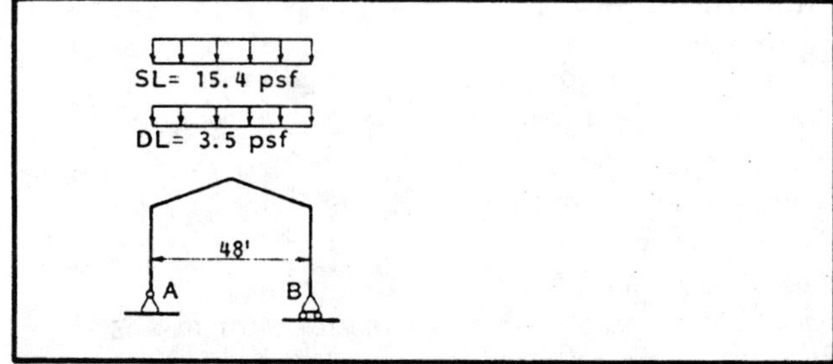

Fig 406-4. Axial post loads.

2. Evaluate soil at the building site.
Determine an allowable bearing strength and passive soil pressure from AWPI. For average soil and this example, use 2,500 psf passive soil pressure and 4,000 psf allowable bearing.
3. Determine allowable stresses.
Select lumber species and grade. Estimate the post's size classification. Look up the allowable unit stresses for a normal duration load. Fiber stresses for round, square, and laminated posts are in ASAE EP388. Determine allowable stress increases. Consider repetitive member bending, moisture content, and member orientation.
Assume No. 1 stress-rated southern pine, 5x5 or larger. The allowable stresses are:

Bending, F_b = 1,350 psi
Tension, F_t = 875 psi
Compression, F_c = 775 psi
Modulus of elasticity, E = 1,500,000 psi

4. Determine axial load for all cases.
   a. Determine axial load on a post for Case A and Case B for buckling.

   Case A. Sum moments about A and solve for reaction at B.
   Axial load = [3.5 psf × 8 × ½ × 48] + [15.4 psf × 8 × ½ × 48]
   = 672 lb + 2,957 lb = 3,629 lb
   b. Determine axial load and bending on post for Case B.
   Determine reactions at A and B; use the larger value in the interaction equation for determining post size.

   Case B. Wind perpendicular to ridge.
   Plus moments are ↻.

RB = axial load at B
48RB = [3.5 × 8′ × 48′ × 24′] + [(3.4 + 11.8) × 8 × 6 × (14 + 6/2)] + [(3.4 × 8 × 24 × 12)] − [11.8 × 8 × 24 × 36] + [(11.8 + 8.4) × 8 × 14 × 7]
48RB = 32,256 + 12,403 + 7,834 − 81,562 + 15,837
= −13,232 ft-lb
RB = −276 lb (tension in post)
RA = axial load at A
48RA = −[3.5 × 8′ × 48′ × 24′] + [(3.4 + 11.8) × 8 × 6 × (14 + 6/2)] − [(3.4 × 8 × 24 × 36)] + [(11.8 × 8 × 24 × 12)] + [(11.8 + 8.4) × 8 × 14 × 7]
48RA = −32,256 + 12,403 − 23,501 + 27,187 + 15,837
= −330 ft-lb
RA = −7 lb (compression in post)

Use axial load at B (larger of the axial loads). Post is in tension.

5. Evaluate fixity at the ends of the post.
Assume the post base is fixed at the centroid of the passive soil reaction ⅓ of the embedment depth below the surface. Assume the post top is pinned, or is fixed if a knee brace is used. For this example, assume the top of the post is pinned.
6. Estimate bending moment in the post.
Estimate embedment depth. Calculate bending moment at ⅓ of that depth for the post's end fixity and wind loads.
Assume embedment depth = 5′-0″, pinned top; calculate the bending moment:

Bending moment = M_f = R (H + D/3) + wH (H/2 + D/3)

For wind perpendicular to ridge:

R = horizontal truss reaction, lb
= (wind pressure on roof) × (post spacing) × (roof height) × ½
= (3.4 + 11.8) × 8 × 6/2
= 365 lb

w = uniform load on sidewall, lb/ft
= (wind pressure on walls × spacing × ½)
= (8.4 + 11.8) × 8 × ½
= 80.8 lb/ft

M_f = 365 lb × (14 + 5/3) + 80.8 × 14′ × (14/2 + 5/3)
= 15,522 ft-lb = 186,264 in-lb

7. Determine post size.
Use the estimated bending moment from Step 6 and the axial load from Step 4. Check both loading cases of wind parallel or perpendicular to the building ridge. Adjust the allowable stresses for load duration.
Estimate bending moment = 186,264 in-lb.

TabF_b = 1,350 psi
F_b = 1,350 × 1.33 × 1 × 1 = 1,795 psi
Trial: S = M_f/F_b
= 186,264/1795 = 103.8 in³

Try an 8x10.
S = 112.8 in³; A = 71.25 in²
TabF_t = 875 psi
F_t = 875 × 1.33 × 1 × 1 = 1,164 psi
Check interaction Eq 406-3: [(276/71.25) ÷ 1164] + [(186,264/112.8) ÷ 1795] < 1.0
0.003 + 0.92 < 1.0 OK
Also check Eq 406-4:
$C_s = (L_e d/b^2)^{0.5}$ = 7.64 < 10
Therefore, $F'_b = F_b$
$(f_b - f_t) \div F'_b < 1.0$
[(186,264/112.8) − (276/71.25)] ÷ 1795 < 1.0
0.92 < 1.0 OK

If the interaction equation value is larger than 1.0, try a larger post and check its interaction equation. If the value is much smaller than 1.0, try a smaller post (an 8x8 was too small).

8. Check the post for buckling under balanced snow load.
Determine the allowable compression stress for the approximate L_e/d using Euler equation.
Axial load (Step 4) is 3,629 lb.

S = 112.8 in³: Area = 71.25 in²
TabF_c = 775 psi
L_u = (14 + S/3) × 12 = 188″.
L_e/d = (2.1 × 188)/9.5 = 41.6
K = 0.671 [1,500,000/(775 × 1.33)]$^{0.5}$
= 25.6

Because K < L_e/d, this is a long column and F'_c is (Eq 406-17):

F'_c = 0.30 E/$(L_e/d)^2$
= 0.3 × 1,500,000/(41.6²)
= 260 psi

Allowable compressive stress, F'_c = 260 psi
Allowable column load = 260 psi × 71.25 in²
= 18,525 lb > 3,629 lb OK.
Therefore, column strength is adequate.

9. Determine post embedment depth, Eq 406-18:
Use the 8x10 post from Step 5, B = 7.50″.
Convert the passive soil pressure, S_1, to psi by dividing psf by 144: 2,500 psf = 17.36 psi

$$D = [1.47\ (R + wH) + [2.16\ (R + wH)^2 + 10.52\ S_1BM_f]^{0.5}] \div 2S_1B$$

To calculate, break the equation into terms:
D = [T1 + (T2 + T3)$^{0.5}$] ÷ T4

T1 = 1.47 [365 lb + (80.8 lb/ft × 14′)] = 2,199 lb
T2 = 2.16 (365 lb + 80.8 lb/ft × 14′)² = 4,835,407 lb²
T3 = 10.52 × 17.36 lb/in² × 7.5″ × 186,264 in-lb
= 255,126,546 lb²
T4 = 2 × 17.36 lb/in² × 7.5 = 260.4 lb/in
D = [2,199 + (4,835,407 + 255,126,546)$^{0.5}$] ÷ 260.4.
D = 70.4″ = 6′

The embedment depth is 6′. Recalculate the moment, post size, and embedment depth for all load cases.

10. Determine the post footing size.
Calculate the maximum vertical load on the post for full snow + dead loads.
Footing area = vertical load ÷ bearing strength
Footing diameter = (4 × footing area/π)$^{0.5}$
Footing thickness is about twice the distance from the face of the post to the edge of the footing.
Maximum vertical load = axial load on post + wall + post
Axial load on post = 3,629 lb (Step 4a)
Wall = 1.6 psf × 8′ × 14′ = 179 lb
Post = 39 lb/ft³ × 7.50″ × 9.5″ × 19.0′ × 1 ft²/144 in² = 367 lb
Maximum vertical load = 3,629 + 179 + 367 = 4,175 lb
Footing area = vertical load/soil-bearing strength
= 4,175 lb/4,000 psf = 1.04 ft² = 150 in²
Footing diameter (4 × 150/π)$^{0.5}$ = 13.8″ = 14″
Footing thickness: (14 − 7.5)/2 = 3.25″ overhang from side of post. Thickness is twice the overhang = 6½″.

11. Determine the post uplift anchorage.
Usually wind parallel to the ridge for a closed building or into the open front of a building is critical.
Calculate the maximum wind uplift minus dead load.
If this is more than a positive 1,000 lb, anchorage in addition to friction between post and soil is needed. A concrete collar dowelled to the base of the post would be more than adequate.

Post-to-Truss Fixity

To approximate a fixed joint at the top of a post, fasten a knee brace at each post, forming a triangle of at least 30″ on a side. The knee brace is 2″ dimension lumber as wide as the greatest width of the post, e.g., a 2x6 for a 6x6 or a 2x8 for a 6x8. Securely anchor the knee brace to the post and truss, Fig 406-5.

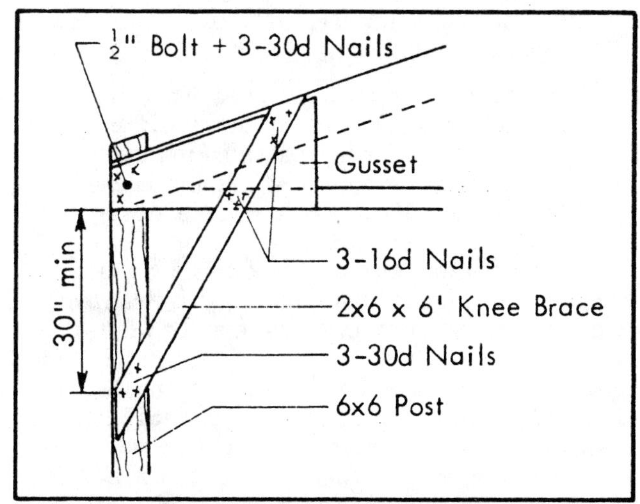

Fig 406-5. Typical fixed post-to-truss connection.
Use ring-shank hardened nails and bolts with washers.

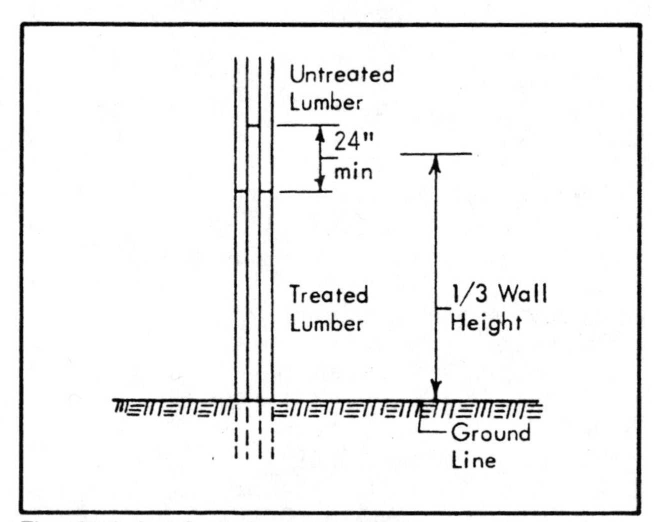

Fig 406-6. Laminated post lap joint.

Splicing Dimension Lumber Posts

If dimension lumber posts are made with pressure preservative-treated wood in the ground and untreated wood above ground, the post-to-truss joint must be fixed, preferably by a knee brace. Joint the treated and untreated lumber at about ½ the wall height, which is near the point of zero moment. Make the lap joint at least 12" long and securely nail or bolt it.

Wood Truss Design

This section discusses truss design assuming pinned joints with some modifications to allow for joint fixity with glued gussets. For designs based on rigid joint analysis, see MWPS-9, *Designs for Glued Trusses,* 4th edition, 1981, especially pages 72-76. Recommendations for the design of metal plate trusses may be found in *Design Specification for Metal Plate Connected Wood Trusses* by the Truss Plate Institute.

Two major references are:

National Design Specification for Wood Construction, (NDS), 1982 edition, National Forest Products Association.

Plywood Design Specification, (PDS), 1980, American Plywood Association.

Design Features

Members

Compute member forces assuming simple pinned joints. Compute bending moments for the top and bottom chords as $wL^2/12$, which assumes fixed ends, with L as the full panel point member length.

Effective member length has been studied by Suddarth (Purdue Plane Structures Analyzer). Effective member lengths in gravity load buckling are reduced from the panel point lengths to 0.8 for chords and 0.9 for webs (0.7 and 0.8, respectively, for wind). Top chords are laterally restrained by the roof construction. Bottom chords are laterally braced by a ceiling, when included in the dead loads, or by longitudinal braces at the panel points.

Evaluate the interaction equations, Eqs 406-3 and 406-4, and 406-11 through 406-13 for moment and axial forces to select adequate member sizes.

Allowable buckling stress, F'_b, is no greater than F_b, which is adjusted for conditions other than the slenderness factors. Determine slenderness factor, C_s, with Eq 406-5; calculate F'_b with Eqs 406-7 through 406-10.

Calculate buckling stress, F'_c, with Eqs 406-14 through 406-17.

Limit truss overhang and support offset as shown in Fig 202-8.

Heel joints

Heel joints resist bottom chord tension in shear perpendicular to the plies, with face grain of the heel gusset parallel to the bottom chord. The gusset glue area develops axial forces in the member framing into the joint. Bending moments and shear in the joints tend to change rolling and horizontal shear stresses to angles at which higher allowable stresses offset the effects of moment and shear.

Because ⅝" and ¾" thick plywood have little more effective thickness in shear than ½" sheets, two thicknesses of ½" plywood are recommended rather than single layers of the thicker panels.

Determine approximate minimum heel gusset thickness using beam shear in the gusset from the truss reaction, Fig 406-7. Assume that the critical section is halfway up the scarf cut, the two chords are the same size, and dimension lumber is effective in resisting shear.

The heel gusset must at least cover the scarf cut, which depends on member size and roof slope. See Fig 202-9.

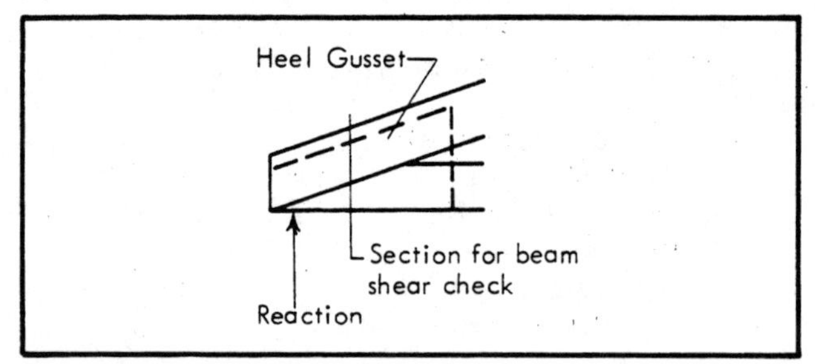

Fig 406-7. Beam shear check.
$F_v = 3V/2tD$
V is the vertical reaction and t is the truss thickness at the critical section.

The maximum length of heel gusset before going to the next larger thickness is arbitrary. Several factors apply:

- All gussets except the heel are ⅜", so use ⅜" where reasonable— trusses will stack better, only one plywood thickness need be purchased, and ⅜" with the correct number of plies may be more readily available than 5-ply ½".
- When the heel gusset is longer than the length of the scarf cut, the plywood between the upper face of the bottom chord and the lower face of the top chord is unsupported and tends to buckle as the top chord bends.
- The longer the heel gusset, the more "waste" because the additional height does not contribute to shear strength.
- Limit heel gussets to the length of the scarf joint plus about 14".

Minimum heel gusset height is measured at the outer end of the truss. The gusset extends from the bottom of the bottom chord to 4" up on the upper chord, but 6" up on the upper chord with gussets that are two layers of ½" plywood.

Other joints and splices

See Figs 202-10 through 202-12 and accompanying text.

Using allowable rolling shear for plywood and axial loads, find the minimum glue area needed at each end of each web member. For splices, also compare the tensile or compressive strength of the splice material with the member's axial load.

Procedure and Example of Truss Problems

Design a 3/12 slope gable truss for a building width of 30′ with trusses 4′ o.c. There is no ceiling in the building, which is in the Des Moines, Iowa, area.

Step 1. Determine the design loads.

See the post design example earlier in this chapter for load analysis. See Fig 406-3.

- Design snow load: P_s = 15.4 psf.
- Design wind load: q = 16.8 psf.
- Design dead load: 3.5 psf.

For wind uplift pressure, use wind parallel to the ridge.

Step 2. Determine the truss geometry.

Pick a slope, span, spacing and web configuration for the first trial design. Assign each member a reference number.

Assume a 2-web configuration with a king post and equal panel point spacing. See Fig 406-8.

Step 3. Select appropriate allowable stresses.

Pick a species and grade of lumber. Look up the allowable unit stresses for normal duration loads in the NDS Supplement for both 2x4s and larger members. (Note: Chapter 405 gives allowable stress for Douglas fir and southern pine.)

Determine allowable stress increases; load duration; repetitive member bending; moisture content adjustment; and orientation of the member, flat, or on-edge, etc.

Also, specify plywood for the gussets and determine its allowable rolling and horizontal shear stresses.

Assume the wood is southern pine at 19% MC. All chords are No. 1 grade and the webs are No. 2.

Allowable stresses are:

2x4s (#1)
- Bending = 1,700 psi, $TabF_b$
- Tension = 1,000 psi, F_t
- Compression = 1,250 psi, $TabF_c$
- Modulus of elasticity = 1.7×10^6 psi, TabE

2x4s (#2)
- Bending = 1,400 psi, $TabF_b$
- Tension = 825 psi, F_t
- Compression = 975 psi, $TabF_c$
- Modulus of elasticity = 1.6×10^6 psi, TabE

2x6s (#1)
- Bending = 1,450 psi, $TabF_b$
- Tension = 975 psi × 1, F_t
- Compression = 1,250 psi, $TabF_c$
- Modulus of elasticity = 1.7×10^6 psi, TabE

2x8s (#1)
- Bending = 1,450 psi, $TabF_b$
- Tension = 975 × 0.8 = 780 psi, F_t
- Compression = 1,250 psi, $TabF_c$
- Modulus of elasticity = 1.7×10^6 psi, TabE

Assume the gussets are ½" C-C Ext. plywood with span rating of 32/16. The plywood is unsanded, of mixed species, and has group 1 species in exterior plies. Allowable rolling shear stress is 53 psi; allowable horizontal shear stress is 190 psi; and effective shear thickness is 0.316"; tensile stress is 2,000 psi; effective tensile area is 2.5 in²/ft.

Step 4. Determine the length of each member.

Using trigonometry:
- Members 1, 2, 8, 9 = 92.77"
- Members 3, 4, 10, 11 = 90"
- Members 5, 13 = 22.5"
- Members 6, 12 = 100.62"
- Member 7 = 45"

Step 5. Determine members for gravity loads.

A. Calculate axial loads in the members assuming that the joints are pinned. After panel point

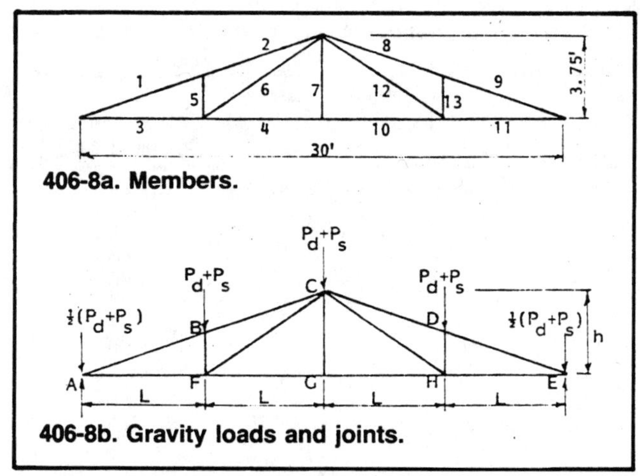

Fig 406-8. Truss geometry and gravity loads.

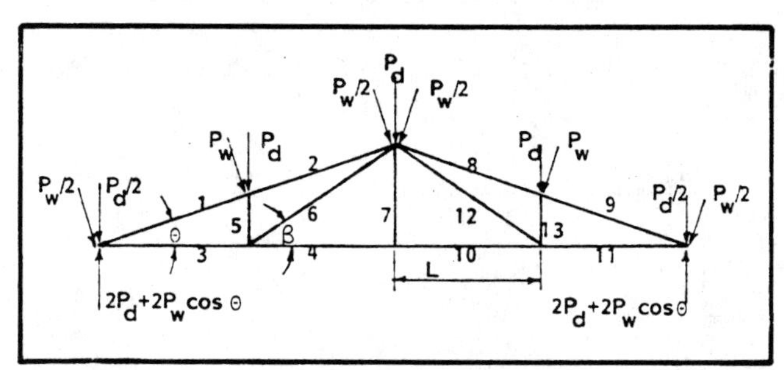

Fig 406-9. Truss gravity and wind loads.

loads are calculated, calculate member loads using the method of joints or sections or coefficients from a truss design handbook.

Balanced dead load at panel point is:

P_d = dead load × truss spacing × horizontal distance between panel points

$P_d = 3.5$ psf × 4′ × 7.5′
$= 105$ lb

Similarly,

$P_s = 15.4$ psf × 4′ × 7.5′
$= 462$ lb

B. Calculate estimated bending moments in the members. Tests at Purdue University showed that moments of about $wL^2/12$ are generated in the top chords by snow and roof loads and in the bottom chords by ceiling loads for nail-glue trusses. These moments are greater for split-ring, nailed, or metal plate trusses.

Using the methods of joints and assuming pinned ends, determine axial loads. The following equations are for the 2-web truss shown with M1 = member 1, etc.:

$M1 = -(1.5\ P_d + 1.5\ P_s)/\sin\phi$
$M2 = M1$
$M3 = M1 \cos\phi$
$M4 = 2L\ (P_d + P_s)/h$
$M5 = -(P_d + P_s)$
$M6 = -(P_d + P_s)/\sin\beta$
$M7 = 0$

C. Select minimum member size.
Effective member length, L_e, is 0.8 × the member length for chords and 0.9 × the member length for webs. For trusses with the top chord braced by purlins or sheathing, d may be taken as the depth of the top chord. Unless braced against buckling, the d for webs is the member width—usually 1½″. For bottom chords laterally braced at the panel points and no ceiling, d is the member width; with a structurally sound ceiling, d is the member depth.

The bending moment, M, in the top chords for the combined dead load and snow load is:
$M = wL^2/12$

Where

$wL = (P_d + P_s) \cos\phi$, and
L = length of top chord between panel points from step 4
$M = (P_d + P_s) \cos \phi \times L \div 12$
$M = (105 + 462) \cos 14.04 \times 92.77 \div 12$
$M = 4{,}253$ in-lb

With no ceiling, the bending moment in the lower chord is 0, Table 406-1.

D. Select minimum member size based on the interaction equation. For the top chord, where the compression edge of the bending member is supported, the slenderness ratio is 0. The top chord is sized in this example to illustrate the design procedure. For the top chord:

Try 2x6

$A = 8.25\ in^2$
$S = 7.56\ in^3$
$C_s = 0 < 10$ (short beam)
$F_b = TabF_b \times LDF \times CUF \times C_f$
$= 1{,}450$ psi × 1.15 × 1 × 1
$= 1{,}668$ psi
$L_e/d = 0.8 \times 92.77/5.5$
$L_e/d = 13.5 > 11$
$K = 0.671\ [1{,}700{,}000/(1.15 \times 1{,}250)]^{0.5}$
$K = 23.1$

Table 406-1. Member forces from dead and snow loads.
Negative axial loads are compression.

Member	Axial load lb	Bending moment in-lb
1	-3506	4253
2	-3506	4253
3	3401	0
4	2268	0
5	-567	0
6	1268	0
7	0	0
8	-3506	4253
9	-3506	4253
10	2268	0
11	3401	0
12	1268	0
13	-567	0

F'_c = TabF$_c$ [1 − 0.333 (L_e/d ÷ K)^4] × LDF × CUF
F'_c = 1,250 [1 − 0.333 (13.5/23.1)^4] × 1.15 × 1
F'_c = 1,381 psi

(Note: 0.8 converts chord length to effective length.) Effective length assumes pinned end times an adjustment factor. Check the interaction equation for member M1:

(P/A)/F'_c + (M/S)/[F_b − J (P/A)] < 1.
J = [(L_e/d) − 11] ÷ (K − 11)
J = (13.5 − 11)/(23.1 − 11)
J = 0.21

(3,506/8.25)/1,381 + (4,253/7.56)/(1,668 − 0.21 (3,506/8.25)) =
= .308 + .356
= 0.66 < 1 OK

The adjustment factors for buckling for other effective lengths for snow load and dead load are:
0.8 = M1, M2, M8, M9
0.8 = M3, M4, M10, M11
0.9 = M5, M6, M7, M12, M13

Unless laterally supported, d = member width.

Step 6. Wind load check.

A. Calculate the axial load as for the snow load. Determine panel point wind load, P_w.

P_w = wind pressure × truss spacing × length of top chord panel
P_w = −11.8 psf × 4′ × 92.77″ × 1′/12″
P_w = −365 lb

B. Calculate estimated bending moments for the members. Using the methods of joints and assuming pinned ends, determine axial loads for wind loads and dead loads. The following equations are for the 2-web truss shown:

M1 = −1.5(P_d + P_w cosφ)/sinφ
M2 = M1 − P_w tanφ
M3 = −(P_w sinφ + 2M1 cosφ)/2
M4 = 2P_d L/h + 2P_w cosφ(L/h) − P_w sinφ
M5 = P_d/2 + P_w cosφ/2 + M2 sinφ
M6 = −(P_d + P_w cosφ + 2 M2 sinφ)/2 sinβ
M7 = 0

C. Evaluate the interaction equation for wind load member size. If the interaction value is > 1, increase the member size until it is < 1. Check L_e/d ratios. In buckling due to wind loads, reduce effective length coefficients to 0.7 for chords and 0.8 for webs, to allow for less rigid control for short-term loads.

The bending moment (M) in the top chord for the combined dead load and wind load is:
M = wL^2/12

Where

wL = P_d cosφ + P_w
L = length of top chord between panel points
M = (P_d cosφ + P_w)L/12

Table 406-2. Truss member data for snow and dead loads.

Member	Member size	Section modulus in^3	Cross-sectional area, in^2	Interaction equation value
1	2x6	7.56	8.25	0.66
2	"	"	"	0.66
3	2x4	3.06	5.25	0.56
4	"	"	"	0.38
5	"	"	"	0.10
6	"	"	"	0.25
7	"	"	"	0.0
8	2x6	7.56	8.25	0.66
9	"	"	"	0.66
10	2x4	3.06	5.25	0.38
11	"	"	"	0.56
12	"	"	"	0.25
13	"	"	"	0.10

Table 406-3. Member forces from dead and wind loads.
Negative axial loads are compression.

Member	Axial load lb	Bending moment in-lb
1	1540	-2034
2	1632	-2034
3	-1450	0
4	-908	0
5	270	0
6	-606	0
7	0	0
8	1632	-2034
9	1540	-2034
10	-908	0
11	-1450	0
12	-606	0
13	270	0

M = [105 cos 14.04 + (−365)] × 92.77″/12
M = −2,034 in-lb

There is no ceiling, so the bending moment in the lower chord is 0.

D. Check the member size for wind load with the interaction equation. The procedure for determining member size is the same as Step 5D. The load duration factor is 1.33. The adjustment factor for buckling for other effective lengths for wind and dead loads are:
0.7 = M1, M2, M8, M9
0.7 = M3, M4, M10, M11
0.8 = M5, M6, M7, M12, M13

Unless laterally supported, d = member width. Check interaction equation value for each member.

Step 7. Determine splice and gusset size.

Bottom Chord Splice:

Member, F_t = 1,000 psi

Plywood F_t = 2,000 psi

Plywood F_s = 53 psi

For tension splice plates, reduce plywood F_t proportional to strip width, commencing with no reduction at 24″ and ranging to 50% at 8″ and less. Single strips less than 8″ wide that are used in stressed applications should be relatively free of surface defects. (Since human life is not endangered, the 50% reduction will not apply). Splice thickness required to develop full tension on both sides of joint:

Tension:

Member $F_t \times A = 1{,}000 \times 1.5 \times 3.5$

$= 5{,}250$ lb

Required Area of Plywood $= 5{,}250/2{,}000$

$= 2.63$ in^2

Available Area of Plywood $= 2 \times (2.5/12) \times 3.5 =$ 1.45 in^2 N.G.

Try 2 layers of plywood both sides.

Available area of plywood $= 4 \times (2.5/12) \times 3.5 =$ 2.92 in^2 OK.

Note: To avoid the 50% stress reduction and to improve joint predictability, the MWPS builds a splice from: the original members butted, a similar sized splice block parallel to the member, and two gussets. See Fig 406-11. Such a joint is designed for rolling shear between plywood and lumber.

Glue Area:
Refer to Fig 406-10.

$F = d \times L_1 \times 1.15 \times F_s \times 2$
$L_1 = 5{,}250/(3.5 \times 1.15 \times 53 \times 2)$
$L_1 = 12.3''$
$L_2 = 2 \times 12.3 = 24.6''$
L_1 = length of gusset on one member
L_2 = gusset length

Heel gusset:

Refer to Fig 406-12. The heel gusset should be long and thick enough to resist the load in the bottom chord in shear through the plies and have enough glue area on the top and bottom chords to develop the axial member loads. The heel gusset must at least cover the scarf cut.

Shear:
$\sin 14 = d/y$
$y = d/\sin 14°$
$= 3.5''/\sin 14°$
$= 14.5''$
$L_3 = 3.5/\tan 14°$
$= 14''$
$L_4 = 10''$
(Guess)

Maximum allowable tension available along scarf cut. Shear for plywood at an angle 15° from parallel is 1.3 F_v.

$T = (1.3 F_v \times t_e \times LDF \times 2y) + (F_v \times t_e \times LDF \times 2L_4)$

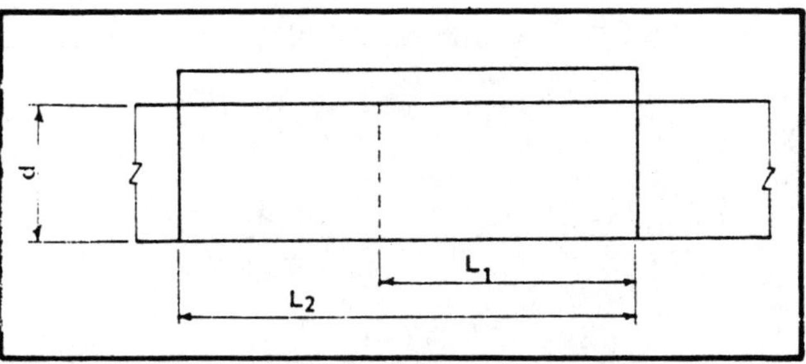

Fig 406-10. Plywood tension splice.

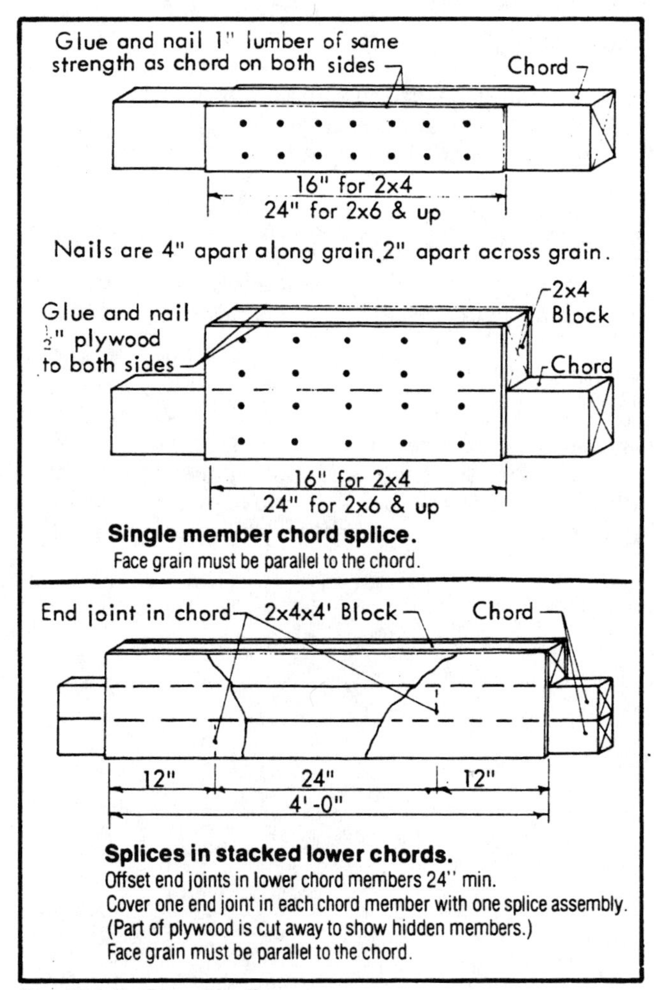

Single member chord splice.
Face grain must be parallel to the chord.

Splices in stacked lower chords.
Offset end joints in lower chord members 24" min.
Cover one end joint in each chord member with one splice assembly.
(Part of plywood is cut away to show hidden members.)
Face grain must be parallel to the chord.

Fig 406-11. Chord splices.

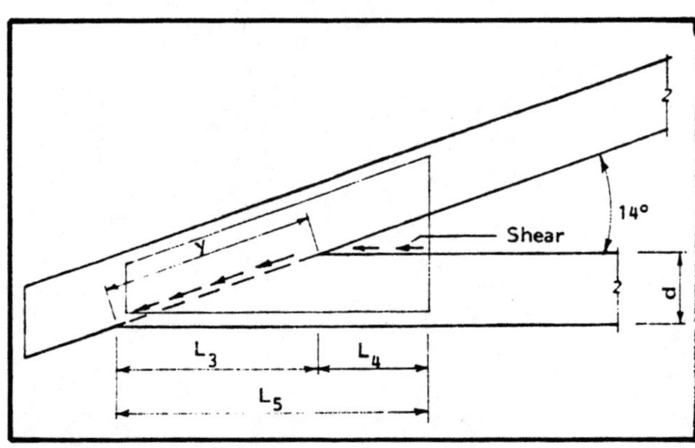

Fig 406-12. Heel joint.

$T = (1.3 \times 190 \times .316 \times 1.15 \times 2 \times 14.5) + (190 \times .316 \times 1.15 \times 2 \times 10)$
$T = 2{,}603 + 1{,}381 = 3{,}984 > 3{,}411$ OK

Glue area:

Available Area $= (d \times L_3/2) + (d \times L_4)$
$= (3.5 \times {}^{14}\!/_2) + (3.5 \times 10) = 59.5\ in^2$
$A_{req} = T_{req.}/(2 \times LDF \times F_s)$
$A_{req} = 3{,}411/(2 \times 1.15 \times 53)$
$A_{req} = 28.0\ in^2 < 59.5\ in^2$ OK.

The remainder of the gussets are left for the reader to design.

Step 8. Determine needed truss anchorage.

Reaction due to snow and dead load
= (design snow and dead load) × spacing × span/2
= (15.4 psf + 3.5 psf) × 4′ × 30′/2
= 1,134 lb

Uplift due to wind and dead load = (design wind load + dead load × spacing × span/2
= (−11.8 psf + 3.5 psf) × 4′ × 30′/2
= −498 lb

Size fasteners in the truss heel to withstand at least the uplift wind force. Size the bearing block and its fasteners for the snow load. Lateral load on a 30d ringshank nail = 139 lb. Lateral load on a ½″ bolt in single shear in a 1½″ truss heel = 438 lb. Lateral load on a ½″ bolt in single shear in a 1½″ bearing block = 635 lb.

Fasteners for uplift:

a. 1½″ bolt in truss = 1.33 × 438 = 583 lb > 498 OK.
b. 4 − 30d nails = 4 × 1.33 × 139 = 739 lb > 498 OK.

If the post is to be notched and the truss set on it to take the snow load reaction in direct bearing, no additional fasteners are required. If the post is not notched and a bearing block is used, determine the fasteners required for snow and dead load reaction:

½″ bolt, truss to post

= 1.15 × 438 = 504 lb

½″ bolt, block to post

= 1.15 × 635 = 730 lb

30d nails block or truss heel to post

= 1.15 × 139 = 159 lb

Number required for connection.

504 lb = 1 − ½″ bolt in truss

730 lb = 1 − ½″ bolt in block

318 lb = 2 − 30d nails in block or truss heel

Total = 1,552 lb > 1,134 lb OK.

407 PLYWOOD DESIGN

Chapter prepared with Wendell Moore of the American Plywood Association and from APA publications.

Plywood Section Properties

Section properties of plywood are needed to design the material as homogeneous orthotropic plates—plates with different properties in each direction. The multilayered makeup of plywood is allowed for in the "effective" properties. These properties are computed by transformed-section from the orthotropic nature, species groups, and manufacturing variables. Table 407-1 lists characteristics of engineered plywood grades.

Table 407-2 lists the species group for CD and CC grades marketed with the Identification Index and Sturd I Floor. From panel descriptions, select section properties from Tables 407-3 and 407-4 and design stresses from Table 407-5. Because different panel constructions are permissible within Grade, the properties of Tables 407-3 and 407-4 are generally the minimums that can be expected.

See Table 407-5 for guide to stress level selection. To qualify for stress level S-1, gluelines must be exterior and only veneer grades N, A, and C are allowed in either face or back. For stress level S-2, gluelines must be exterior and veneer grade B, C-plugged, and D are allowed on the face or back. Stress level S-3 includes all panels with interior or intermediate gluelines.

Stress levels are for 24" wide strips. Decrease 50% for strips 12" or less, 25% for strips 12"-24".

Face grain direction

The properties are per foot of panel width and depend on the face grain direction with respect to the applied stress. When face grain is across the supports, use **parallel** section properties. When face grain is parallel to the supports, use **perpendicular** properties.

Shear thickness

Panel thicknesses for shear-through-the-thickness allow for the reduced effectiveness of inner plies in mixed-species panels and glue shear strength. The values relate to veneer thickness, panel construction, and number of gluelines per inch of thickness.

Tension and compression areas

Effective areas for tension and compression are based on plies with grain parallel to the stress. Cross plies are allowed no tensile or compressive strength.

Moment of inertia

Do not use effective moments of inertia in bending stress calculations.

The effective moments of inertia are used with the modulus of elasticity of the face plies to calculate deflections. They were calculated with the reduced effectiveness of perpendicular plies shown in Fig 407-1. The modulus of elasticity of peeled wood veneer perpendicular to the grain is about 1/35 of its parallel modulus. The weakest permitted species group was assumed.

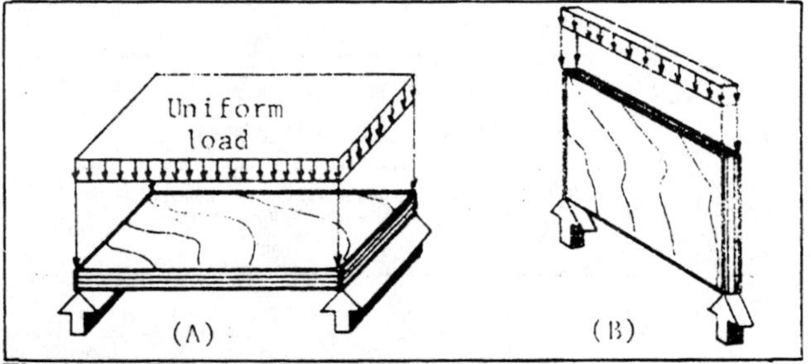

Fig 407-1. Plywood in bending.
Load applied perpendicular to the plane of the panel (A) and parallel to the plane of the panel (B).

Section modulus

Effective section moduli, KS, account for wood species, direction, and an empirical correction factor "K".

Use the tabulated KS values to calculate bending strength, rather than I/c. Section moduli perpendicular to the grain ignore the outermost tension ply.

Rolling-shear constant

Rolling-shear constants, Ib/Q, are convenient for calculating shear, $V = F_s(Ib/Q)$. For certain plywood constructions, the constant is omitted where rolling shear cannot control the design.

Plywood Stiffness and Design Stresses

Use the allowable unit stresses and moduli of elasticity in Table 407-5 to design with plywood. They may be modified for loading condition, treatment, or moisture content. They are used in both parallel and perpendicular directions. Plywood with adequate finger or scarf joints can carry full allowable stresses.

Use the allowable stresses assigned to the species group of the face plies regardless of inner ply species. Table 407-1 is a guide to using Tables 407-3 to 407-5.

Table 407-1. Guide to stress and section properties.

1. When exterior glue is specified, e.g. "interior with exterior glue", stress level 2 (S2) should be used.

2. Check local suppliers for availability of structural II and Plyform Class II Grades.

PLYWOOD TYPE	PLYWOOD GRADE	DESCRIPTION AND USE	TYPICAL GRADE-TRADEMARKS	VENEER GRADE			COMMON THICKNESSES	GRADE STRESS LEVEL (Table 407.5)	SPECIES GROUP	SECTION PROPERTY TABLE
				FACE	BACK	INNER				
INTERIOR TYPE PLYWOOD	C-D INT	sheathing grade for wall, roof, subflooring, and industrial applications. Also available with intermediate and exterior glue (1).	C-D 32/16 APA INTERIOR 000	C	D	D	5/16, 3/8, 1/2, 5/8, 3/4	S 3 (1)	See Table 2	Table 3 (unsanded)
	STRUCTURAL I C-D INT or STRUCTURAL II C-D INT (2)	Plywood grades to use where strength properties are of maximum importance. Made with exterior glue only. Structural I is made from all Group 1 woods. Structural II allows Group 3 woods	STRUCTURAL I C-D 42/20 APA INTERIOR 000 EXTERIOR GLUE	C	D	D	5/16, 3/8, 1/2, 5/8, 3/4	S 2	Structural I Use Group 1 Structural II Use Group 3	Table 4 (unsanded)
	UNDERLAYMENT INT	For underlayment or combination subfloor-underlayment under resilient floor coverings.	UNDERLAYMENT GROUP 3 APA INTERIOR 000 EXTERIOR GLUE	C plugged	D	C & D	1/2, 19/32 5/8, 23/32 3/4	S 3 (1)	As Specified	Table 3 (touch sanded)
	C-D PLUGGED INT	For built ins, wall and ceiling tile backing, NOT for underlayment.	C-D PLUGGED GROUP 2 APA INTERIOR 000	C plugged	D	D	1/2, 19/32 5/8, 23/32 3/4	S 3 (1)	As Specified	Table 3 (touch sanded)
	STRUCTURAL I or II (2) UNDERLAYMENT or C-D PLUGGED	For higher strength requirements for underlayment or built ins. Structural I constructed from all Group 1 woods. Made with exterior glue only.	STRUCTURAL I UNDERLAYMENT APA INTERIOR 000	C plugged	D	C & D	1/2, 19/32 5/8, 23/32 3/4	S 2	Structural I Use Group 1 Structural II Use Group 3	Table 4 (touch sanded)
	2.4.1 INT	Combination subfloor-underlayment. Available with tongue and groove	2·4·1 T&G GROUP 2 APA INTERIOR 000	C plugged	D	C & D	1-1/8"	S 3 (1)	Group 1	Table 3
	APPEARANCE GRADES	Generally applied where a high quality surface is required. Includes N-N, N-A, N-B, N-D, A-A, A-B, A-D, B-B, and B-D INT Grades.	A-D GROUP 1 APA INTERIOR 000	B or better	D or better	D	1/4, 3/8, 1/2, 5/8 3/4	S 3 (1)	As Specified	Table 3 (sanded)
EXTERIOR TYPE PLYWOOD	C-C EXT	sheathing grade with waterproof glue bond for wall, roof, subfloor and industrial applications.	C-C 36/16 APA EXTERIOR 000	C	C	C	5/16, 3/8, 1/2, 5/8, 3/4	S 1	See Table 2	Table 3 (unsanded)
	STRUCTURAL I C-C EXT or STRUCTURAL II C-C EXT (2)	For engineering applications in construction and industry where full exterior-type panels are required. Structural I is made from Group 1 woods only.	STRUCTURAL I C-C 32/16 APA EXTERIOR 000	C	C	C	5/16, 3/8, 1/2, 5/8, 3/4	S 1	Structural I use Group 1 Structural II Use Group 3	Table 4 (unsanded)
	UNDERLAYMENT EXT and C-C PLUGGED EXT	Underlayment for combination subfloor-underlayment or two layer floor under resilient floor coverings. Available with tongue and groove.	C-C PLUGGED GROUP 3 APA EXTERIOR 000	C plugged	C	C	1/2, 19/32, 23/32, 5/8, 3/4	S 2	As Specified	Table 3 (touch sanded)
	STRUCTURAL I or II (2) UNDERLAYMENT EXT or C-C PLUGGED EXT	For higher strength underlayment. All Group 1 construction in Structural I. Structural II allows Group 3 woods.	STRUCTURAL I UNDERLAYMENT APA EXTERIOR 000	C plugged	C	C	1/2, 19/32, 5/8, 23/32, 3/4	S 2	Structural I Use Group 1 Structural II Use Group 3	Table 4 (touch sanded)
	B-B PLYFORM CLASS I or II (2)	Concrete-form grade with high reuse factor. mill oiled unless otherwise specified. Available in HDO.	B-B PLYFORM CLASS I APA EXTERIOR 000	B	B	C	5/8, 3/4	S 2	Class I Use Group 1 Class II Use Group 3	Table 3 (sanded)
	MARINE EXT	Superior Exterior type plywood made only with Douglas-Fir or Western Larch. Special solid-core construction. Available with MDO or HDO face.	MARINE AA EXT APA	A or B	A or B	B	1/4, 3/8, 1/2, 5/8, 3/4	A face & back use S 1 B face or back use S 2	Group 1	Table 4 (sanded)
	APPEARANCE GRADES	Generally applied where a high quality surface is required. Includes AA, A-B, A-C, B-B, B-C, HDO and MDO EXT.	A-C GROUP 1 APA EXTERIOR 000	B or better	C or better	C	1/4, 3/8 1/2, 5/8 3/4	A or C face and back use S 1 B face or back use S 2	As Specified	Table 3 (sanded)

Grade Stress Level and Species Group

Allowable stresses are divided into three levels related to grade. Bending, tension, and compression stresses depend on veneer grade and species group. Shear stresses depend on glue type and species group. Modulus of elasticity and bearing strength depend only on species group.

Service Moisture Conditions

Table 407-5 lists allowable stresses for both wet and dry moisture conditions.

Use "dry" stresses for plywood continually at a moisture content of less than 16%. When equilibrium moisture content is 16% or greater, use "wet" stresses.

Table 407-2. Key to Identification Index and species group.
For panels with "Index" and thickness (or Sturd I Floor span) listed, use stress for species group given in table.
30/12 - 5/8", and 36/16 - 3/4" panels are also sometimes available. Use Group 4 stresses.

Thick. inch	Identification Index 12/0	16/0	20/0	24/0	32/16	42/20	48/24
5/16	4	3	1		Span rating Sturd I Floor		
3/8		4	3	1	16" o.c.	20" o.c.	24" o.c.
1/2				4	1		
5/8					3	1	
3/4						3	1
7/8						4	3

Use only plywood with exterior glue when equilibrium moisture content exceeds 18% or is exposed to the weather.

Stress Modifications

Direction of loading

Allowable unit stresses in Table 407-5 are for normal load duration—full loading either continuously or cumulatively for 10 years. Adjust allowable stresses for loading durations other than "normal". Do not adjust the modulus of elasticity. Duration increases for dimension lumber are applicable to plywood.

Pressure treatment

No adjustment is necessary for preservative treated plywood. Reduce the allowable stresses by 1/6 and modulus of elasticity by 1/10 for fire-retardant treated plywood.

Stresses

Modulus of elasticity

Use the modulus of elasticity for the species group of the face plies for both parallel and perpendicular to the face grain. Increase the modulus of elasticity 10% when calculating bending deflection if shear deflection is computed separately.

Table 407-3. Effective section properties.
Properties of 12" widths of panels. Face plies of different species group from inner plies. (Includes all Product Standard Grades except those noted in Table 407-4.)

			STRESS APPLIED PARALLEL TO FACE GRAIN				STRESS APPLIED PERPENDICULAR TO FACE GRAIN			
(1) NOMINAL THICKNESS in.	(2) APPROXIMATE WEIGHT psf	(3) EFFECTIVE THICKNESS FOR SHEAR in.	(4) A AREA in²/ft	(5) I MOMENT OF INERTIA in⁴/ft	(6) KS EFF. SECTION MODULUS in³/ft	(7) Ib/Q ROLLING SHEAR CONSTANT in²/ft	(8) A AREA in²/ft	(9) I MOMENT OF INERTIA in⁴/ft	(10) KS EFF. SECTION MODULUS in³/ft	(11) Ib/Q ROLLING SHEAR CONSTANT in²/ft
UNSANDED PANELS										
5/16-U	1.0	0.283	1.914	0.025	0.124	2.568	0.660	0.001	0.023	–
3/8 -U	1.1	0.293	1.866	0.041	0.162	3.108	0.799	0.002	0.033	–
1/2 -U	1.5	0.316	2.500	0.086	0.247	4.189	1.076	0.005	0.057	2.585
5/8 -U	1.8	0.336	2.951	0.154	0.379	5.270	1.354	0.011	0.095	3.252
3/4 -U	2.2	0.467	3.403	0.243	0.501	6.823	1.632	0.036	0.232	3.717
7/8 -U	2.6	0.757	4.109	0.344	0.681	7.174	2.925	0.162	0.542	5.097
1 -U	3.0	0.859	3.916	0.493	0.859	9.244	3.611	0.210	0.660	6.997
1-1/8 -U	3.3	0.877	4.621	0.676	1.047	10.008	3.464	0.307	0.821	8.483
SANDED PANELS										
1/4 -S	0.8	0.304	1.680	0.013	0.092	2.175	0.681	0.001	0.020	–
3/8 -S	1.1	0.313	1.680	0.038	0.176	3.389	1.181	0.004	0.056	–
1/2 -S	1.5	0.450	1.947	0.077	0.266	4.834	1.281	0.018	0.150	3.099
5/8 -S	1.8	0.472	2.280	0.129	0.356	6.293	1.627	0.045	0.234	3.922
3/4 -S	2.2	0.589	2.884	0.197	0.452	7.881	2.104	0.093	0.387	4.842
7/8 -S	2.6	0.608	2.942	0.278	0.547	8.225	3.199	0.157	0.542	5.698
1 -S	3.0	0.846	3.776	0.423	0.730	8.882	3.537	0.253	0.744	7.644
1 1/8 -S	3.3	0.865	3.854	0.548	0.840	9.883	3.673	0.360	0.918	9.032
TOUCH-SANDED PANELS										
1/2 -T	1.5	0.346	2.698	0.083	0.271	4.252	1.159	0.006	0.061	2.746
19/32 -T	1.7	0.491	2.618	0.123	0.337	5.403	1.610	0.019	0.150	3.220
5/8 -T	1.8	0.497	2.728	0.141	0.364	5.719	1.715	0.023	0.170	3.419
23/32 -T	2.1	0.503	3.181	0.196	0.447	6.600	2.014	0.035	0.226	3.659
3/4 -T	2.2	0.509	3.297	0.220	0.477	6.917	2.125	0.041	0.251	3.847
(2 4 1) 1 1/8 T	3.3	0.855	4.592	0.653	0.995	9.933	4.120	0.283	0.763	7.452

Table 407-4. Effective section properties.
Properties of 12″ widths of panels. All plies from same species group. (Includes Structural I and Marine.)

			STRESS APPLIED PARALLEL TO FACE GRAIN				STRESS APPLIED PERPENDICULAR TO FACE GRAIN			
(1) NOMINAL THICKNESS in.	(2) APPROXIMATE WEIGHT psf	(3) EFFECTIVE THICKNESS FOR SHEAR in.	(4) *A* AREA in ^{2}/ft	(5) *I* MOMENT OF INERTIA in ^{4}/ft	(6) *KS* EFF. SECTION MODULUS in ^{3}/ft	(7) *Ib/Q* ROLLING SHEAR CONSTANT in ^{2}/ft	(8) *A* AREA in ^{2}/ft	(9) *I* MOMENT OF INERTIA in ^{4}/ft	(10) *KS* EFF. SECTION MODULUS in ^{3}/ft	(11) *Ib/Q* ROLLING SHEAR CONSTANT in ^{2}/ft
UNSANDED PANELS										
5/16 - U	1.0	0.356	2.375	0.025	0.144	2.567	1.188	0.002	0.029	–
3/8 - U	1.1	0.371	2.226	0.041	0.195	3.107	1.438	0.003	0.043	–
1/2 - U	1.5	0.403	2.906	0.091	0.318	4.188	1.938	0.007	0.077	2.574
5/8 - U	1.8	0.434	3.464	0.155	0.433	5.268	2.438	0.015	0.122	3.238
3/4 - U	2.2	0.606	3.672	0.247	0.573	6.817	2.938	0.059	0.334	3.697
7/8 - U	2.6	0.776	4.388	0.346	0.690	6.948	3.510	0.192	0.584	5.086
1 - U	3.0	1.088	5.200	0.529	0.922	8.512	6.500	0.366	0.970	6.986
1 1/8 - U	3.3	1.119	6.654	0.751	1.164	9.061	5.542	0.503	1.131	8.675
SANDED PANELS										
1/4 - S	0.8	0.342	1.680	0.013	0.092	2.172	1.226	0.001	0.027	–
3/8 - S	1.1	0.373	1.680	0.038	0.177	3.382	2.126	0.007	0.078	–
1/2 - S	1.5	0.545	1.947	0.078	0.271	4.816	2.305	0.030	0.217	3.076
5/8 - S	1.8	0.576	2.280	0.131	0.361	6.261	2.929	0.077	0.343	3.887
3/4 - S	2.2	0.748	3.848	0.202	0.464	7.926	3.787	0.162	0.570	4.812
7/8 - S	2.6	0.778	3.952	0.288	0.569	7.539	5.759	0.275	0.798	5.671
1 - S	3.0	1.091	5.215	0.479	0.827	7.978	6.367	0.445	1.098	7.639
1 1/8 - S	3.3	1.121	5.593	0.623	0.955	8.840	6.611	0.634	1.356	9.031
TOUCH-SANDED PANELS										
1/2 - T	1.5	0.403	2.698	0.084	0.282	4.246	2.086	0.008	0.082	2.720
19/32 - T	1.7	0.567	3.127	0.124	0.349	5.390	2.899	0.030	0.212	3.183
5/8 - T	1.8	0.575	3.267	0.144	0.378	5.704	3.086	0.037	0.242	3.383
23/32 - T	2.1	0.598	3.337	0.201	0.469	6.582	3.625	0.057	0.322	3.596
3/4 - T	2.2	0.606	3.435	0.226	0.503	6.900	3.825	0.067	0.359	3.786

Bending stress

When loads are applied perpendicular to the plane of the panel, use the allowable bending stress with section modulus values, KS, from Tables 407-3 and 407-4, not with moment of inertia values. When loads are applied parallel to the plane of the panel, only the plies with face grain parallel to the stress direction are effective. Where end joints occur, modify the allowable stresses as allowed in the section on glued plywood end joints.

Tension or compression stress

The allowable stresses for tension, F_t, and compression, F_c, in the plane of the plies are used with the appropriate areas from Tables 407-3 and 407-4 to predict plywood strength.

Use $F_t/6$ for tension at 45° to the face grain. The allowable stress for compression at 45° to the face grain is $F_c/3$. If the panel is Structural I quality, use the full thickness. Use straight-line interpolation between 0° and 45°.

Bearing stress

Use the allowable bearing stress, $F_{c\perp}$, of the face plies for loads illustrated in Fig 407-3.

Shear stresses

The cross-laminated construction of plywood requires two different shear stresses.

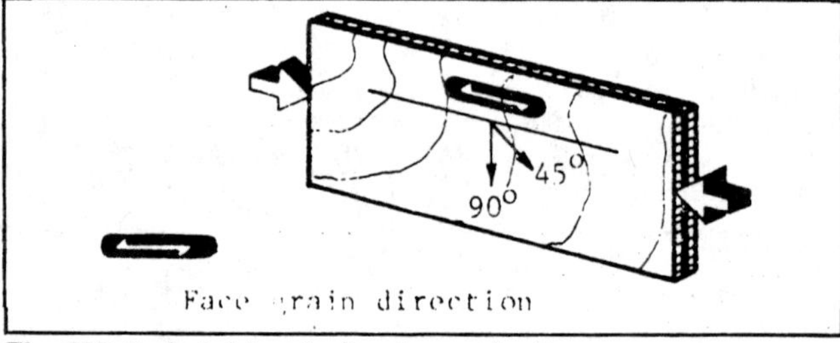

Fig 407-2. Axial loads in plane of plies.
Face grain assumed to be zero degrees.

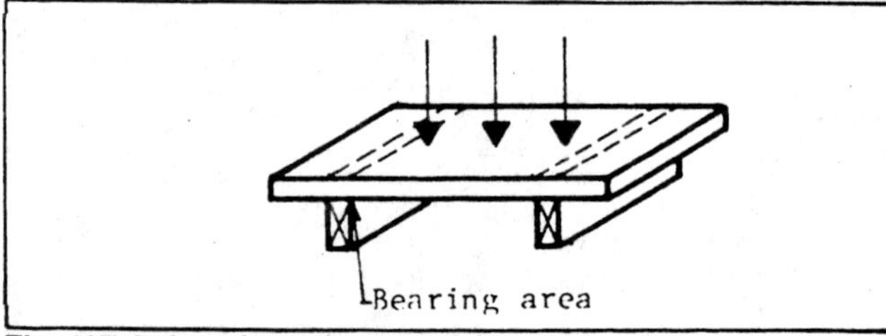

Fig 407-3. Load on the panel face.

Shear perpendicular to the plies

Use the allowable stresses for shear perpendicular to the plies, F_v, in Table 407-5 with the thicknesses in Column 3 of Tables 407-3 and 407-4. Double the allowable shear stress for shear at 45° to the face grain.

When plywood is attached to framing on only two sides, as in a Vierendeel truss, reduce shear stress to

Table 407-5. Allowable stresses for plywood.
Conforming to U.S. Product Standard PS-1-74 for Construction and Industrial plywood. Normal load basis in psi.

Type of Stress		Species Group of Face Ply	Grade Stress Level[1] S-1 Wet	S-1 Dry	S-2 Wet	S-2 Dry	S-3 Dry Only
EXTREME FIBER STRESS IN BENDING (F_b) TENSION IN PLANE OF PLIES (F_t) Face Grain Parallel or Perpendicular to Span (At 45° to Face Grain Use 1/6 F_t)	F_b & F_t	1	1430	2000	1190	1650	1650
		2, 3	980	1400	820	1200	1200
		4	940	1330	780	1110	1110
COMPRESSION IN PLANE OF PLIES (F_c) Parallel or Perpendicular to Face Grain (At 45° to Face Grain Use 1/3 F_c)	F_c	1	970	1640	900	1540	1540
		2	730	1200	680	1100	1100
		3	610	1060	580	990	990
		4	610	1000	580	950	950
SHEAR THROUGH THE THICKNESS Parallel or Perpendicular to Face Grain (At 45° to Face Grain Use 2 F_v)	F_v	1	155	190	155	190	160
		2, 3	120	140	120	140	120
		4	110	130	110	130	115
ROLLING SHEAR (IN THE PLANE OF PLIES) Parallel or Perpendicular to Face Grain (At 45° to Face Grain Use 1-1/3 F_s)	F_s	MARINE & STRUCTURAL I	63	75	63	75	—
		ALL OTHER[2]	44	53	44	53	48
MODULUS OF RIGIDITY Shear in Plane Perpendicular to Plies	G	1	70,000	90,000	70,000	90,000	82,000
		2	60,000	75,000	60,000	75,000	68,000
		3	50,000	60,000	50,000	60,000	55,000
		4	45,000	50,000	45,000	50,000	45,000
BEARING (ON FACE) Perpendicular to Plane of Plies	$F_{c\perp}$	1	210	340	210	340	340
		2, 3	135	210	135	210	210
		4	105	160	105	160	160
MODULUS OF ELASTICITY IN BENDING IN PLANE OF PLIES Face Grain Parallel or Perpendicular to Span	E	1	1,500,000	1,800,000	1,500,000	1,800,000	1,800,000
		2	1,300,000	1,500,000	1,300,000	1,500,000	1,500,000
		3	1,100,000	1,200,000	1,100,000	1,200,000	1,200,000
		4	900,000	1,000,000	900,000	1,000,000	1,000,000

(1) To qualify for stress level S-1, gluelines must be exterior and only veneer grades N, A, and C are allowed in either face or back. For stress level S-2, gluelines must be exterior and veneer grade B, C-Plugged and D are allowed on the face or back. Stress level S-3 includes all panels with interior or intermediate gluelines.
(2) Reduce stresses 25% for 3-layer (4-ply) panels over 5/8" thick. Such layups are possible under PS 1-74 for APA RATED SHEATHING, APA RATED STURD-I-FLOOR, UNDERLAYMENT, C-C Plugged and C-D Plugged grades over 5/8" through 3/4" thick.

0.89 x F_v when framing is parallel to the face grain, or 0.75 x F_v when it is perpendicular to the face grain.

See shear-through-thickness values for punching-shear.

Shear in the plane of the plies

Use the rolling shear stress, F_s, in Table 407-5 to calculate required contact areas. Reduce the allowable stress 50% for stress concentrations in structural assemblies such as the outside stringer of stressed-skin panels or the flange-to-web joint in box beams. Increase allowable rolling shear stress by ⅓ for shear at 45° to the face grain.

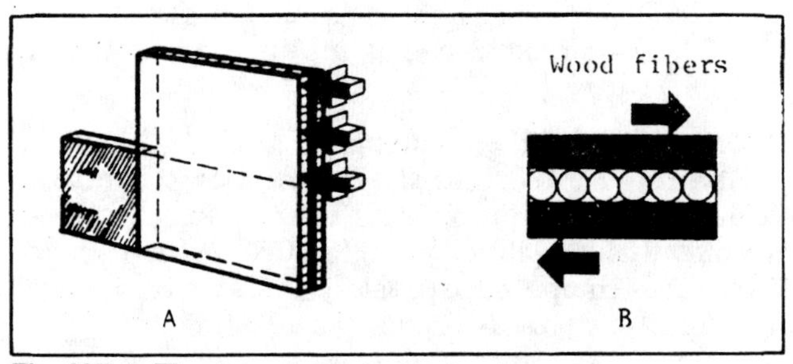

Fig 407-4. The two shear types in plywood.
Typical shear-through-thickness (A) and shear in the plane of the plies or rolling shear (B).

Design Loads and Formulas

Uniform Loads

With face grain perpendicular to supports, use the 3-span condition for spacings up to 32″. Use the 2-span condition for support spacings over 32″.

With face grain parallel to supports, use the 3-span condition for spacings up to 16″; the 2-span condition for spacings of 16″-24″; and 1-span condition for spacings over 24″.

Uniiform loads based on bending stress

These formulas assume one-way beam action and allow for the mixed length units noted.

Eq 407-1.

Single or double span:
$w_b = (96)F_b(KS) \div L^2$

Eq 407-2.

Three span:
$w_b = (120)F_b(KS) \div L^2$

w_b = allowable uniform load for bending stress, psf
F_b = allowable bending stress, psi
KS = effective section modulus, in^3/ft
L = support spacing, in. o.c.

Uniform loads based on shear stress

Eq 407-3.

Single span:
$w_s = (24)F_s(Ib/Q) \div L'$

Eq 407-4.

Double span:
$w_s = (19.2)F_s(Ib/Q) \div L'$

Eq 407-5.

Three span:
$w_s = (20)F_s(Ib/Q) \div L'$

w_s = allowable uniform load based on shear stress, psf
F_s = allowable rolling-shear stress, psi
Ib/Q = rolling shear constant, in^2/ft
L′ = support spacing minus support width, in

Uniform loads based on deflection

Bending deflection, D_b. For most cases a bending deflection calculation is sufficient using the effective modulus of elasticity, Table 407-5. When shear deflection is computed separately, increase the modulus of elasticity by 10% for bending deflection.

Eq 407-6.

Single span:
$D_b = w(L'')^4 \div (921.6)EI$

Eq 407-7.

Double span:
$D_b = w(L'')^4 \div (2220)EI$

Eq 407-8.

Three span:
$D_b = w(L'')^4 \div (1743)EI$

D_b = bending deflection, in.
w = uniform load, psf
E = modulus of elasticity, psi
I = effective moment of inertia, in^4/ft
L″ = clear span + sw, in.
sw = support-width factor, equal to 0.25″ for 2″ framing and 0.625″ for 4″ framing.

Shear deflection, D_s. Estimate shear deflection for all span conditions from:

Eq 407-9.

$D_s = w(C)t^2(L')^2 \div (1270)EI$

D_s = shear deflection, in.
w = uniform load, psf
C = constant: 120 for panels with face grain perpendicular to supports, 60 for parallel to supports.
t = nominal panel thickness, in.
E = modulus of elasticity, unadjusted, psi
I = effective moment of inertia, in^4/ft

Structural Assemblies

Adhesives

Adhesives for plywood-lumber structural assemblies provide both stiffness and strength.

For interior conditions where the moisture content will not exceed 18%, use water-resistant adhesives such as casein glue. When the assembly will be wet or used outdoors, use waterproof adhesives such as phenol and resorcinol resins.

Glued plywood end joints

Scarf joints 1 in 8 or flatter transmit full allowable tension or bending stress; joints 1 in 5 transmit 75% of allowable stress. Do not use scarf joints steeper than 1 in 5.

Butt joints backed with a glued plywood splice on one side transmit the stresses in Table 407-6. Shorter splices carry proportionally less stress. Use plywood for the splice equal to the grade of the plywood being spliced. Orient face grain perpendicular to the joint.

Butt joints across the face grain have 100% compressive strength when jointed by a 1 in 5 scarf joint or a splice of the length specified in Table 407-6.

Allowable Load Curves

Use the load span curves in Figs 407-5 to 407-9 for plywood under uniform loads. Evaluate the strength of a panel with the L/180, L/240, and strength curves. The plywood face grain is assumed perpendicular to the supports.

The curves are for plywood continuous over 3 spans for spacings up to 32″ and over 2 spans for larger spacings. The plywood is unsanded C-D Interior with exterior glue.

The following factors adjust the curves for other conditions:

Table 407-6. Butt joints—tension and flexure.
Glue plywood splice plate on one side of joint.

Plywood thickness	Length of splice plate in.	All Struct. 1 grades	1	Group: 2&3	4
		Maximum stress, psi			
1/4"	6	1500	1200	1000	900
5/16	8	"	"	"	"
3/8 sanded	10	"	"	"	"
3/8 unsanded	12	"	"	"	"
1/2	14	1500	1000	950	900
5/8 & 3/4	16	1200	800	750	700

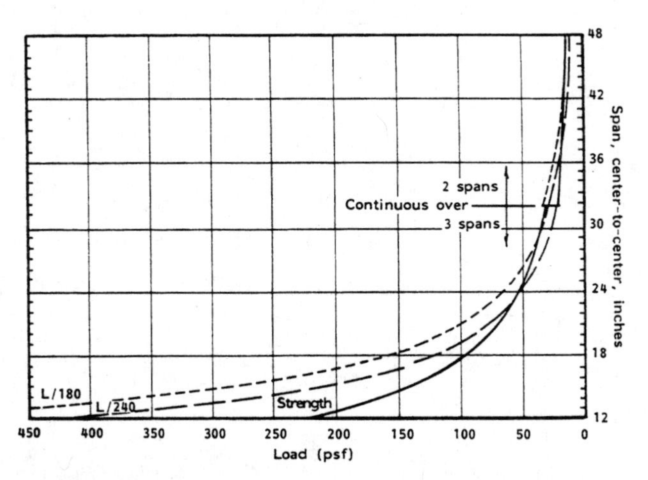

Fig 407-5. Unsanded C-D 24/0 Int with ext glue.
Face grain across supports.
Moisture content < 15%.
For face grain parallel to supports: use 5% of L/180 and L/240 loads; use 21% of strength load.

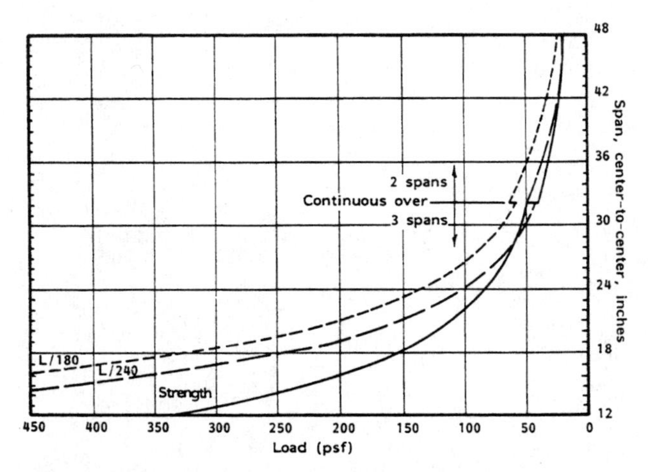

Fig 407-6. Unsanded C-D 32/16 Int with ext glue.
Face grain across supports.
Moisture content < 15%.
For face grain parallel to supports: use 6% of L/180 and L/240 loads; use 23% of strength load.

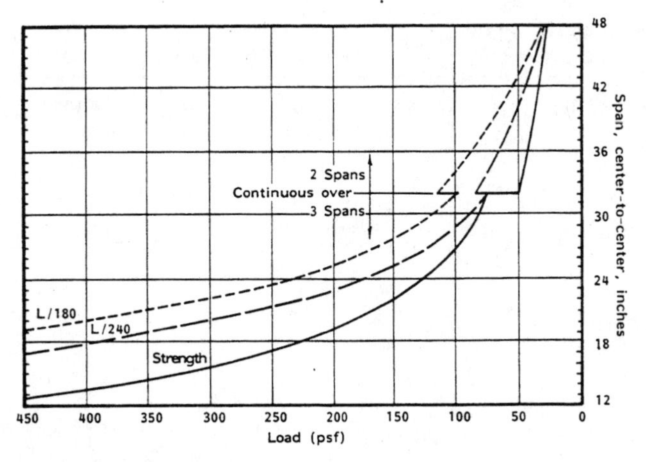

Fig 407-7. Unsanded C-D 42/20 Int with ext glue.
Face grain across supports.
Moisture content < 15%.
For face grain parallel to supports: use 7% of L/180 and L/240 loads; use 25% of strength load.

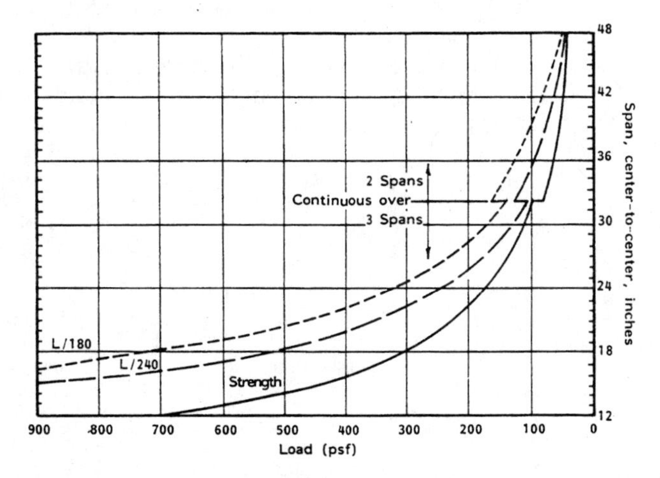

Fig 407-8. Unsanded C-D 48/24 Int with ext glue.
Face grain across supports.
Moisture content < 15%.
For face grain parallel to supports: use 16% of L/180 and L/240 loads; use 48% of strength load.

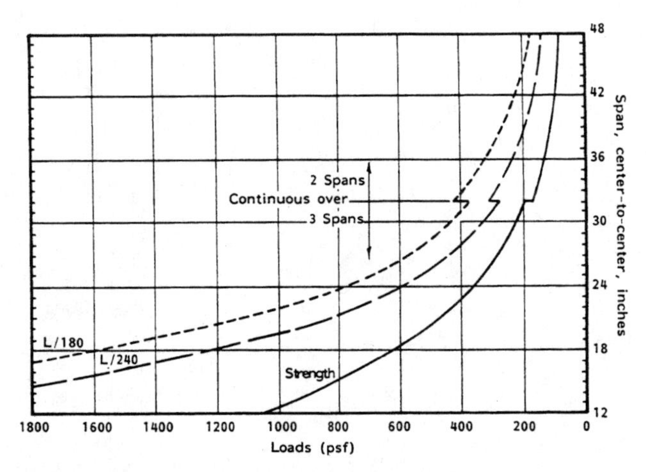

Fig 407-9. Unsanded 2-4-1 with ext glue.
Face grain across supports.
Moisture content < 15%.
For face grain parallel to supports: use 42% of L/180 and L/240 loads; use 93% of strength load.

Table 407-7. Plywood load curve factors.
Multiply allowable load from chart (Fig 407-5 to 407-9) by appropriate factors.

Condition in chart	Design condition	Adjustment factor
Strength		
3 spans	1 span	0.80
2 spans	1 span	1.00
CD-Int.	CC-EXT	1.17
	(Unsanded	
Deflection	L/180 or L/240)	
3 spans	1 span	0.53
2 spans	1 span	0.42
Moisture conditions—strength		
m.c. < 15%	m.c. > 15%	0.71
Moisture conditions—deflection		
m.c. < 15%	m.c. > 15%	0.83

Example:

What is allowable total load and deflection for unsanded C-C 48/24 Exterior plywood with face grain across supports spaced 24″?

- Select proper curve. See Fig 407-8 for C-D Interior with exterior glue plywood with a 48/24 Identification Index.
- Read load for strength = 170 psf. Read load for deflection of L/240 = 235 psf.
- Adjust values as required. From Table 407-7, the strength may be multiplied by 1.17 for C-C Ext.
  Load based on deflection of L/240:
  235 x 1.0 = 235
  Load based on strength: 170 x 1.17 = 199
  Allowable total load = 199 psf
- Deflection at allowable total load:

Eq 407-10.

$$\text{Defl} = \text{WTot} \times L \div (\text{WDef} \times 240 \text{ or } 180)$$

Defl = deflection, in.
WTot = allowable total load, lb
L = support spacing, in.
WDef = load at deflection limit
240 = deflection limit

With WTot = 199, L = 24, WDef = 235, and deflection limit of L/240:

$$\text{Defl} = 199 \times 24 \div (235 \times 240) = 0.0847''$$

408 REINFORCED CONCRETE

Chapter prepared by Dr. M.F. Brugger, Ohio State University.
Note: Tables 408-1, 408-2, and 408-3, and Fig 408-5 are from *Design and Control of Concrete Mixtures*, Portland Cement Association, 12th ed., 1979.

This chapter discusses fundamentals of structural concrete and strength method design for agricultural applications.

The American Concrete Institute's *Building Code for Reinforced Concrete,* ACI 318.77, is the basis for design. It is referred to here as ACI 318.

Definitions

Admixture: ingredient other than water, aggregate, or cement added to concrete before or during mixing to modify its properties.

Aggregate: inert material mixed with portland cement and water to produce concrete.

Aggregate, lightweight: aggregate having a dry, loose weight of 70 pcf or less.

Column: element that supports primarily axial compressive loads and has an unsupported height > 3 times its least lateral dimension.

Compressive concrete strength, f'_c: specified compressive strength of concrete, psi (ACI 318-4.3). Where f'_c is under a radical, the square root of the numerical value only is intended; the result is in psi.

Concrete: a mixture of portland cement, fine aggregate, coarse aggregate, and water.

Deformed reinforcement: reinforcing bars, deformed wire, welded wire fabric, and welded deformed wire fabric.

Development length: embedded reinforcement length required to develop the reinforcement design strength at a critical section.

Effective area of concrete: area of a section between the centroid of the tension reinforcement and the compression face of a flexural member.

Effective area of reinforcement: area of the right cross-sectional area of the reinforcement times the cosine of the angle between its direction and the direction for which the effectiveness is to be determined.

Embedment length: length of embedded reinforcement beyond a critical section.

Embedment length, equivalent (l_e): embedded reinforcement length to develop the same stress as a hook or mechanical anchorage.

End anchorage: length of reinforcement, a mechanical anchor, a hook, or a combination beyond the point of nominal zero reinforcement stress.

Load (dead, live, snow, or wind): see Chapter 101.

Load, design: load, multiplied by appropriate load factor, to proportion members.

Load, service: live and dead loads (without load factors).

Pedestal: upright compression member with unsupported height < 3 times its average least lateral dimension.

Plain concrete: concrete not meeting the definition of reinforced concrete.

Plain reinforcement: reinforcement not meeting the definition of deformed reinforcement.

Precast concrete: concrete cast in other than its final position in the structure.

Reinforced concrete: concrete containing reinforcement, including prestressing steel; design assumes that two materials act together in resisting forces.

Spiral: continuously wound reinforcement in the form of a cylindrical helix.

Stirrups or ties: lateral reinforcement, either open or closed units or continuously wound. Stirrups are usually used in horizontal members; ties are in vertical members.

Stress: intensity of force per unit area.

Yield strength or yield point, f_y: specified minimum yield strength or yield point of reinforcement, psi.

Principles

Plain concrete is a mixture of cement, aggregates, water, and frequently admixtures. Reinforced concrete has steel added in areas under tension to carry the tension stresses. The strength of reinforced concrete depends on many factors; the amount, type, and placement of steel; the proportions of concrete materials; and the conditions, including temperature and moisture, during placement and curing. See Section 200 for discussion of mixes, etc.

Strength & Deformation in Compression

The compressive strength of concrete is f'_c. It is the strength in psi of test cylinders 6" in diameter by 12" long measured on the 28th day after they are made. The water-cement ratio is the chief factor in determining strength, Fig 408-1.

The stress-strain relationship of the material indicates its performance under load. Typical stress-strain curves for specimens loaded at 28 days are in Fig 408-2. Low strength concrete has more ductility than high strength concrete. The stress-strain curve is influenced by:

- loading rate of the test specimens.
- water-cement ratio.
- specimen type.
- concrete age.
- curing.

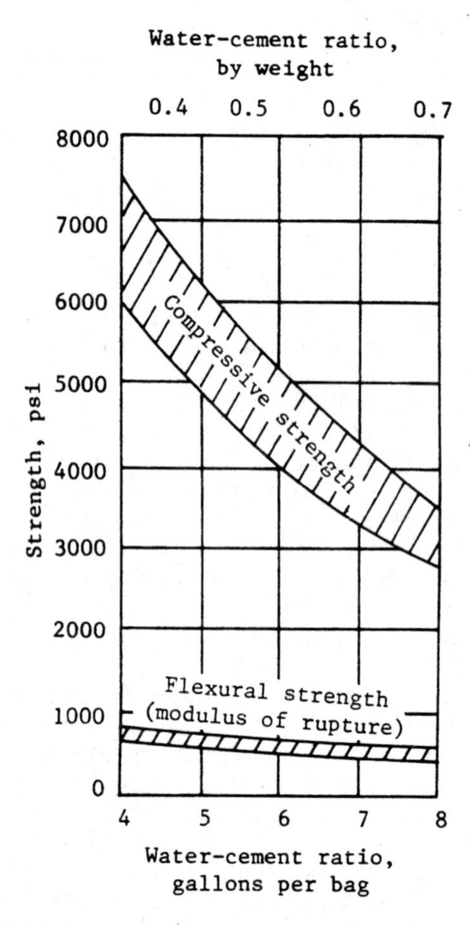

Fig 408-1. Water-cement ratio and strength.
Effect on 28-day strength in compression and flexure.

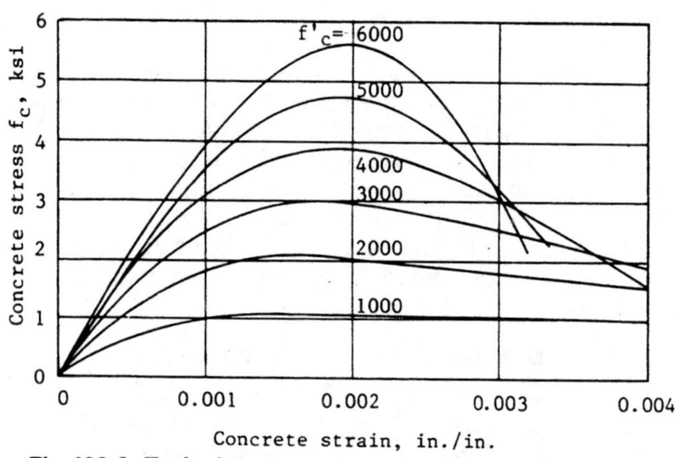

Fig 408-2. Typical concrete stress-strain curves.

The ultimate strain at crushing varies between 0.003 and 0.008 in/in. ACI 318 states: "The maximum usable strain at the extreme concrete compression fiber shall be assumed equal to 0.003."

The modulus of elasticity, E_c, is expressed by an empirical formula which may be considered the secant modulus for a compressive stress at service load level.

$$E_c = 33\ W^{1.5}\ f'^{\,0.5}_c$$

For normal-weight concrete, ACI 318-8.5.1 suggests:

$$E_c = 57{,}000\ f'^{\,0.5}_c$$

Creep

Creep is the property that causes continued deformation over time under constant load, Fig 408-3. This concrete sample was loaded at age 28 days with an instantaneous strain, ϵ_{inst}. The load was maintained for 230 days; creep caused the strain to almost triple. If the load is removed, the instantaneous strain, ϵ_{inst}, is removed, and some creep recovery occurs.

Creep deformations are proportional directly to applied stress and inversely to concrete strength. Sustained loads affect not only deformation but also concrete strength. Cylinder strength under sustained loading, f'_c, is only about 75% of the 28-day strength.

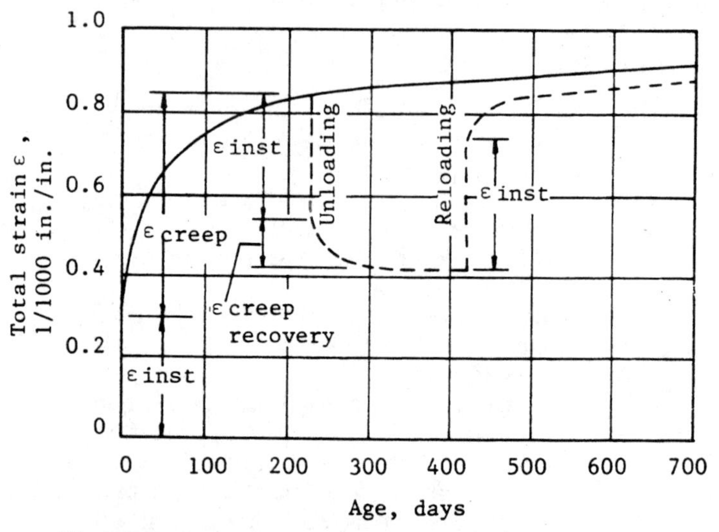

Fig 408-3. Creep in concrete.
Typical creep curve for concrete loaded to 600 psi at age 28 days.

Tension Strength

Concrete is best used where it is in compression, but its tension strength can also be significant. Shear and torsion strength of reinforced concrete beams depend on tension strength. Also, the formation of cracks on the tension side of flexural members depends on tension strength.

True tensile strengths are difficult to determine. Flexural tension properties have been measured in terms of the modulus of rupture by breaking unreinforced beams in flexure. The split-cylinder test is for axial tension.

The split-cylinder strength, f'_{ct}, is (6 to 7) $f'^{\,0.5}_c$ psi for sand-and-gravel concrete and (4 to 5) $f'^{\,0.5}_c$ psi for lightweight concretes. The true tensile strength, f'_t, for the former is (0.5 to 0.7) f'_{ct}, and the flexural tension, f_r, is (1.25 to 1.75) f'_{ct}. The smaller factors apply to higher strength concretes and the larger factors to lower strength concrete.

Volume Changes

All concrete mixes contain more water than is needed for hydration. If concrete is exposed to air, most of the free water evaporates. As concrete dries, it shrinks in volume. If dry concrete is submerged in water, it absorbs water and expands. When not adequately controlled, shrinkage can cause undesirable cracks or large harmful stresses. Final shrinkage of ordinary concrete is 0.0002 to 0.0007 in/in depending on water content (gal/bag), slump, type of aggregate, etc.

Concrete expands with increasing, and contracts with decreasing, temperature. The effects are similar to those caused by shrinkage. Concrete's coefficient of thermal expansion varies from 0.000004 to 0.000006 in/in-F.

Reinforcing Steel

Reinforcing steel is much stronger but more expensive than concrete. Steel and concrete are usually combined so that concrete resists compression and steel resists tension. Steel carries compression in some cases and minimum steel in compression members and beam faces guards against unexpected bending or tension.

Chemical adhesion, natural roughness of the surfaces, and surface deformations on the reinforcing bars bond the concrete and steel so they deform together. See illustration of standard hooks, Fig 408-9.

Steel's thermal expansion is 0.0000065 in/in-F, which is close enough to concrete to not cause problems. Steel has low fire and corrosion resistance, but it can be protected by concrete.

The most common reinforcement is round bars with diameters of ¼"-1⅜". Standard bar dimensions are in Table 208-2. Bars should meet appropriate ASTM standards.

The two important characteristics for steel performance are yield point and modulus of elasticity. Typical stress-strain curves are in Fig 408-4.

The most common reinforcing steel is Grade 40; yield strength, f_y, is 40,000 psi and ultimate strength is 70,000 psi. Grade 60, also widely used, has 60,000 psi yield strength and 90,000 psi ultimate strength.

Welded wire fabric is often used in slabs. It is cold-drawn steel wires welded together at right angles. Yield strength of cold-drawn welded wire is 64,000 psi. See Tables 208-19 and 208-20.

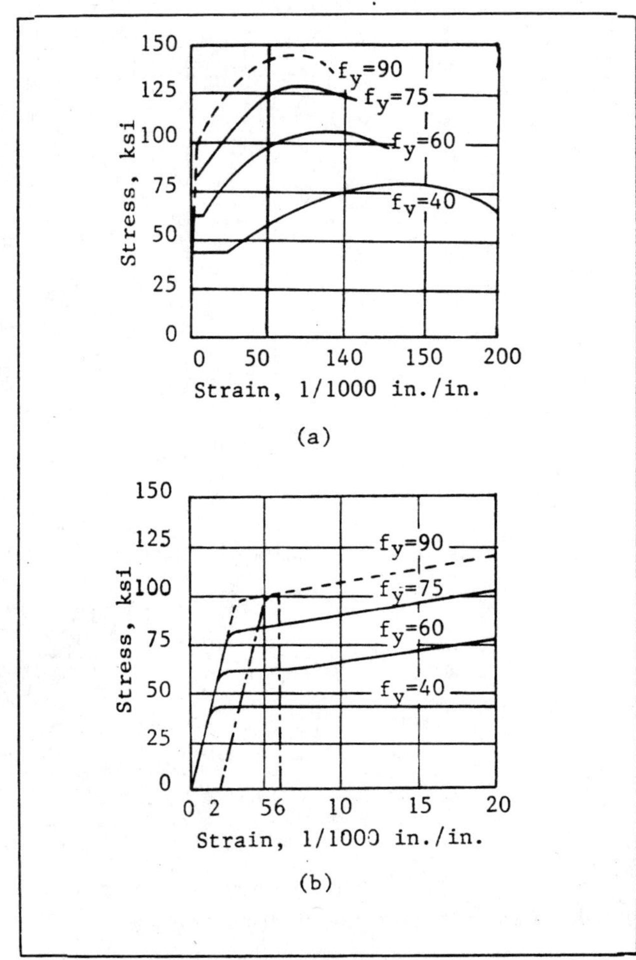

Fig 408-4. Typical steel stress-strain curves.

Concrete Quality

Proportioning Concrete for Strength

Base proportions, including water-cement ratio, on laboratory trial batches or field experience with the materials to be used. Select proportions for an average strength at the designated test age that exceeds f'_c by the amount indicated below, when both air content and slump are the maximums permitted by the specifications. The effect of the water-cement ratio and the type of cement on compressive strength is in Fig 408-5.

If the concrete producer has data on at least 30 consecutive strength tests on similar materials and conditions to those expected, select proportions for a strength that exceeds the required f'_c by at least:

400 psi if σ (standard deviation) is $<$ 300 psi
550 psi if σ is 300 to 400 psi
700 psi if σ is 400 to 500 psi
900 psi if σ is 500 to 600 psi
1,200 psi if σ is above 600 psi

Test laboratory trial batches for selecting concrete proportions according to *Method of Test for Compressive Strength of Molded Concrete Cylinders* (ASTM C 39). Prepare specimens according to *Method of Making and Curing Test Specimens in the Laboratory* (ASTM C 192). Establish a curve of the relationship between water-cement ratio (or cement content) and compressive strength. Select the maximum permissible water-cement ratio (or minimum cement content) from the curve.

Without suitable data from trial batches or field experience, permission may be granted to base concrete proportions on the water-cement ratio limits in ACI 318.

Proportioning Concrete for Exposure

Use Table 408-1 to select the water-cement ratio for various exposure conditions. Note that the quantities shown are the recommended **maximum** water-cement ratios without regard to strength.

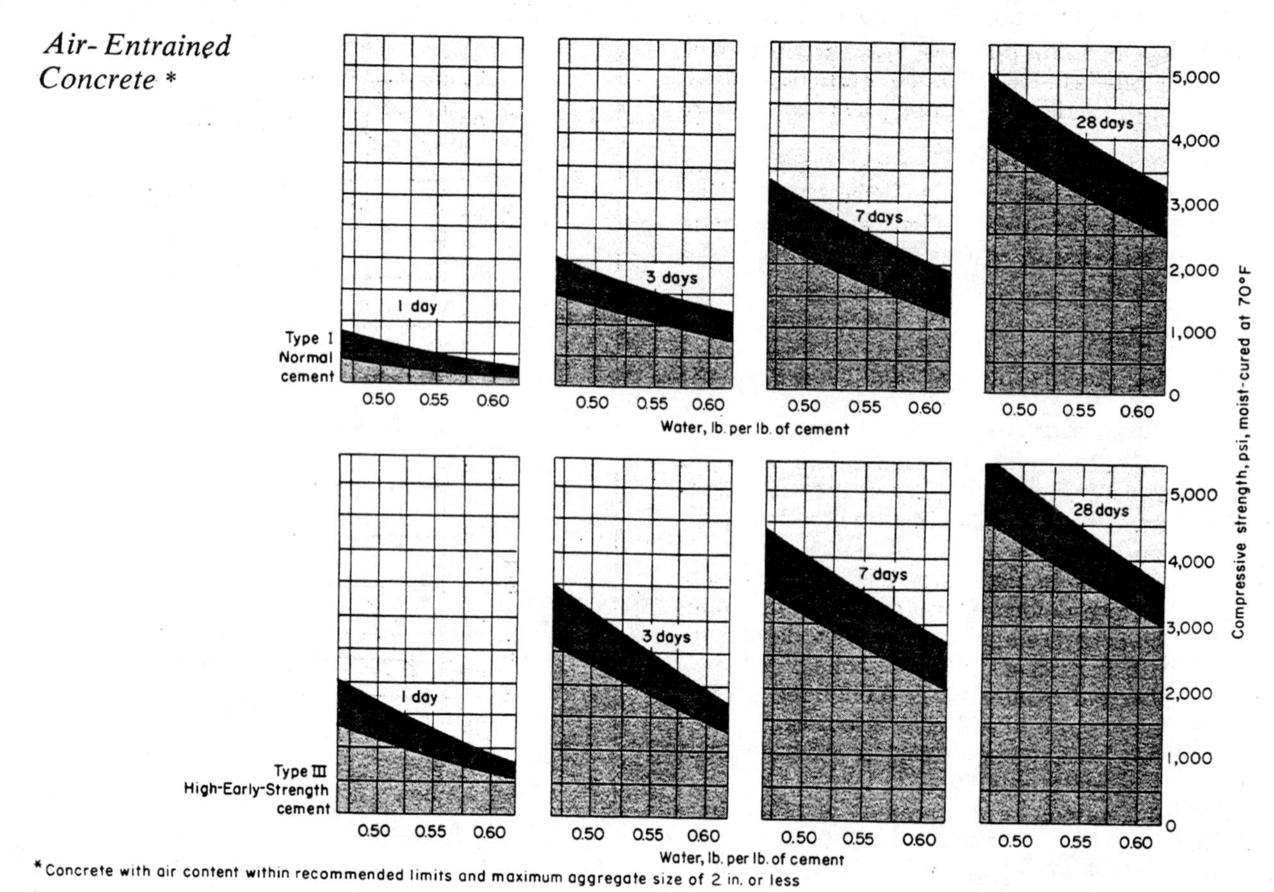

Fig 408-5. Water-cement ratio, strength, and age.

Table 408-1. Maximum water-cement ratio for exposure.

Exposure condition	Normal-weight concrete, absolute water-cement ratio by weight
Concrete protected from exposure to freezing and thawing or application of deicer chemicals	Select water-cement ratio on basis of strength, workability, and finishing needs.
Watertight concrete*	
In fresh water	0.50
In seawater	0.45
Frost-resistant concrete*	
Thin sections; any section with less than 2-in. cover over reinforcement and any concrete exposed to deicing salts	0.45
All other structures	0.50
Exposure to sulfates*	
Moderate	0.50
Severe	0.45
Placing concrete under water	Not less than 650 lb of cement per cubic yard (386 kg/m³)
Floors on grade	Select water-cement ratio for strength, plus minimum cement requirements.

*Contain entrained air within the limits of Table 2.

Table 408-2. Air content and aggregate sizes.

Nominal maximum size of coarse aggregate, in.	Total air content, percentage by volume
⅜	6 to 10
½	5 to 9
¾	4 to 8
1	3.5 to 6.5
1½	3 to 6
2	2.5 to 5.5
3	1.5 to 4.5

1 in. ≈ 25 mm

Proportioning Concrete for Consistency

Slump measures concrete consistency. Do not use it to compare mixes with very different proportions or kinds or sizes of aggregates. With different batches of similar mixtures, slump changes with materials, proportions, and water content. To avoid overly stiff or wet mixtures, see Table 408-3.

Aggregate and Steel Proportioning

Maximum aggregate size

Maximum coarse aggregate size depends on the sizes and shapes of concrete members and on the amounts and distribution of reinforcement. Generally, limit the maximum size to ⅕ the minimum dimension of the member or ¾ the clear space between

Table 408-3. Typical slumps.
For mechanically vibrated concrete. For hand methods, add 1" and increase the cement content in proportion to the water added. Minimum slump is 1".

Types of construction	Maximum slump
Reinforced foundation walls and footings	3"
Unreinforced footings and substructure walls	3"
Reinforced slabs, beams, and walls	5"
Building columns	4"
Pavements and slabs	3"

a bar and another bar or the forms. For unreinforced slabs on ground, limit size to ⅓ slab thickness. Smaller aggregates can be used if necessary or economical.

Reinforcement spacing

The clear distance between parallel bars in a layer should exceed nominal bar diameter, 1", or 1⅓ times the maximum aggregate size.

In walls and slabs, other than concrete joist construction, limit principal reinforcement spacing to three times the wall or slab thickness, with 18" maximum.

In spirally reinforced or tied compression members, the clear distance between longitudinal bars is at least 1½ times nominal bar diameter, 1½", and 1½ times maximum aggregate size.

Shrinkage and temperature steel

Provide reinforcement for shrinkage and temperature stresses normal to the principal reinforcement in structural floor and roof slabs and in walls where the principal reinforcement is in only one direction. For minimum steel, see Tables 208-3 and 208-4.

Design Methods and Requirements

Strength anaylsis of a reinforced concrete member is both empirical and rational, because the two materials act together.

Two design philosophies have been prevalent: working stress and ultimate strength. Only the working stress method was used until about 1960; the ultimate strength method has become popular since and is the method described here.

The working stress method (the "alternate design method" in ACI 318) requires that stresses in a structural element resulting from service loads and computed by the mechanics of elastic members not exceed some allowable value, e.g. $f_c = 0.45 f'_c$, $f_y = 20{,}000$ psi for grade 40 and 50 steel. The allowable values include a safety factor.

Some of the limitations of the working stress method:

- There is no simple way to account for different uncertainties among various loads, because the safety factor is in the limits on total stress.

Table 408-4. Protection for reinforcement.
(ACI 318-7.7) Cast-in-place concrete.

	Minimum Cover, in.
Cast against and permanently exposed to earth, footings and trenched walls	3
Exposed to earth, manure, or weather:	
#6 through #18 bars	2
#5 bars, W31 or D31 wire, and smaller	1½
Not exposed to weather or in contact with ground:	
Slabs, walls, joists:	
#11 and smaller	¾
Beams, girders, columns:	
Principles reinforcement, ties, stirrups, or spirals	1½
Precast concrete	
Exposed to earth, manure or weather:	
Wall panels:	
#11 or smaller	¾
Other members:	
#6 through #11	1½
#5 bars, W31 or D31 wire, and smaller	1¼
Not exposed to weather or in contact with the ground:	
Slabs, walls, joists:	
#11 or smaller	⅝
Beams, girders, columns:	
Principal reinforcement	dia. of the bar but not less than ⅝
Ties, stirrups, or spirals	⅜

- Creep and shrinkage are not easily accounted for.
- Stress is not proportional to strain up to crushing strength, so the inherent safety provided is unknown when allowable stress is a percentage of f'_c.

The ultimate strength method ("strength" in the ACI 318) requires increasing service loads by load factors to get design loads. The structural element is required to provide the desired ultimate strength. A code-defined ultimate strength value allows for concrete's nonlinear stress-strain behavior.

Safety factors allow for changes and uncertainty in service loads and also for possible undercapacity of the element because of construction error, e.g. concrete strength, size of element, or location of reinforcing steel.

Serviceability, such as deflection, is checked separately for service load conditions.

Safety Provisions

Two types of safety provisions are incorporated into the strength method. Overload factors are applied to service loads to give required strength, U. They account for the uncertainty and changes in service loads. Strength reduction factors, ϕ, are applied to the required strength, U/ϕ, to give the total theoretical strength to be designed for. They account for adverse variations in material strengths, dimensions, workmanship, quality control, and degree of supervision.

The basic overload equation (ACI 318-9.2.1) where wind and earthquake may be neglected is:

Eq 408-1.

$$U = 1.4D + 1.7L$$

U = required strength
D = dead load under service conditions
L = live load under service conditions

If wind load, W, is added, ACI 318-9.2.2 provides:

Eq 408-2.

$$U = 0.75\ (1.4D + 1.7L + 1.7W)$$

A more severe load situation may be only dead and wind loads. When live load is absent, Eq 408-2 becomes:

Eq 408-3.

$$U = 1.05D + 1.275W$$

If dead load has a gravity stabilizing effect in combination with wind, reduced dead load is checked using:

Eq 408-4.

$$U = 0.9D + 1.3W$$

Lateral earth pressure, H, is a live load, so:

Eq 408-5.

$$U = 1.4D + 1.7L + 1.7H$$

But, when dead load, live load, or both reduce the earth pressure effect, then:

Eq 408-6.

$$U = 0.9D + 1.7H$$

For fluid pressure, F, replace 1.7H with 1.4F in Eqs 408-5 and 408-6. The 1.4 instead of the 1.7 factor is used for fluids because liquid density is generally accurately known.

Any structural element is designed for the most severe load combination.

Table 408-5. Strength reduction factors, ϕ.
For combined compression and bending in d. and e., ϕ may be variable and increase to 0.90 as the axial compression decreases to zero. (ACI 318-9.3)

	ϕ Factors
a. Bending with or without axial tension	0.90
b. Axial tension	0.90
c. Shear and torsion	0.85
d. Compression members, spiral	0.75
e. Compression members, tied	0.70
f. Bearing on concrete	0.70
g. Bending in plain concrete	0.65

For example, applying the overload and undercapacity factors for moment in a beam with a moment due to dead load, M_D, and live load, M_L, gives the moment under factored loads, M_u, as:

$$M_u = 1.4\ M_D + 1.7\ M_L$$

The nominal strength requirement for flexure, M_n, is:

$$M_n > M_u/\phi$$

$$M_n > (1.4\ M_D + 1.7\ M_L)/0.90$$

Analysis and Design of Single Reinforced Beams

Understanding the basis of ultimate flexural strength is important. The conditons at theoretical strength in flexure are in Fig 408-6.

Actual concrete stress conditions may be replaced with any shape of stress block resulting in strength predictions agreeing with available test results. The rectangular stress block, Fig 408-6d, developed by Whitney is described as follows.

A compressive stress of $0.85\ f'_c$ is assumed uniformly distributed over an equivalent compression zone which is bounded by the edges of the cross section and by a line parallel to the neutral axis at a d istance $a = \beta_1 c$ from the fiber of maximum compression strain.

Eq 408-7.

$$\beta_1 = 0.85 \text{ for } f'_c < 4000 \text{ psi}$$

Eq 408-8.

$$\beta_1 = 0.85 - 0.05\ [(f'_c - 4{,}000)/1000]$$
$$\text{for } f'_c > 4{,}000 \text{ psi}$$

To compute nominal flexural strength, M_n, the following assumptions (ACI 318-10.2) are made in addition to the rectangular stress block:

- Applicable conditions of equilibrium and compatibility of strains are satisfied.
- Strain in the reinforcing steel and concrete are directly proportional to distance from the neutral axis.
- Maximum usable strain at the extreme concrete compression fiber is 0.003.
- Tensile strength of concrete is neglected (except for certain prestressed concrete conditions).
- Steel stress is E_s times strain up to f_y; steel stress equals f_y and is independent of strain at strains greater than that corresponding to f_y.

Nominal flexural strength is:

Eq 408-9.

$$M_n = T \text{ (or C) (arm)} = T \text{ (or C) } (d - a/2)$$

Assuming the steel yields before the concrete crushes, find the value of a:

Eq 408-10.

$$C = 0.85\ f'_c\ ba$$

Eq 408-11.

$$T = A_s f_y$$

Equating C = T gives:

Eq 408-12.

$$a = A_s f_y/(0.85\ f'_c b)$$

Therefore:

Eq 408-13.

$$M_n = A_s f_y\ (d - 0.59\ A_s f_y/f'_c b)$$

b = width of rectangular section
d = effective depth, or the distance from the extreme fiber in compression to the centroid of the tension steel area

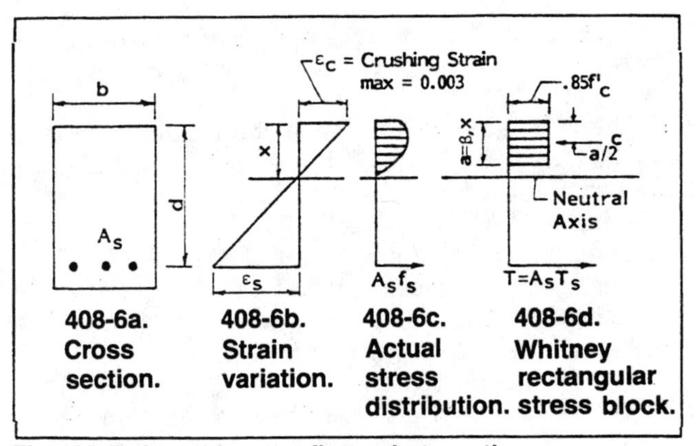

408-6a. Cross section. 408-6b. Strain variation. 408-6c. Actual stress distribution. 408-6d. Whitney rectangular stress block.

Fig 408-6. Conditions at flexural strength.

Example 1:

Determine the nominal flexural strength, M_n, for the rectangular section in Fig 408-7 if: f'_c = 5,000 psi, f_y = 40,000 psi, b = 12", d = 18.5", A_s = 4-#9 bars. Assuming the steel has yielded when strength is reached, the internal forces are:

$$C = 0.85\,f'_c ba = 0.85\,(5{,}000)\,(12)\,(a)$$
$$= 51{,}000\ a\ \text{(psi)}$$
$$T = f_y\,A_s = 40{,}000\,(4) = 160{,}000\ \text{(psi)}$$

For equilibrium, C = T; therefore:

$$51{,}000\ a = 160{,}000$$
$$a = 160{,}000/51{,}000 = 3.14''$$
$$\beta_1 = 0.80 \text{ for } f'_c = 5{,}000 \text{ psi, from Eq 408-8}$$

The neutral axis position is:

$$c = a/\beta_1 = 3.14/0.80 = 3.92''$$

Check to see if the steel does yield. When strain = 0.003 in the concrete, strain in the steel is, by proportion:

$$\epsilon_s = (0.003)(d - c)/c$$
$$= (0.003)\,(18.5 - 3.92)/3.92, \text{ Fig 408-6b}$$
$$= 0.011$$

Strain at the steel's yield point is:

$$\epsilon_y = f_y/E_s = 50{,}000/29{,}000{,}000 = 0.00172$$

The assumption that the steel yields is valid. Nominal flexural strength is:

$$M_n = C(d-a/2) \text{ or } T\,(d-a/2)$$
$$= 160{,}000\,(18.5 - 3.14/2)\,(1'/12'')$$
$$= 226{,}000 \text{ ft-lb}$$

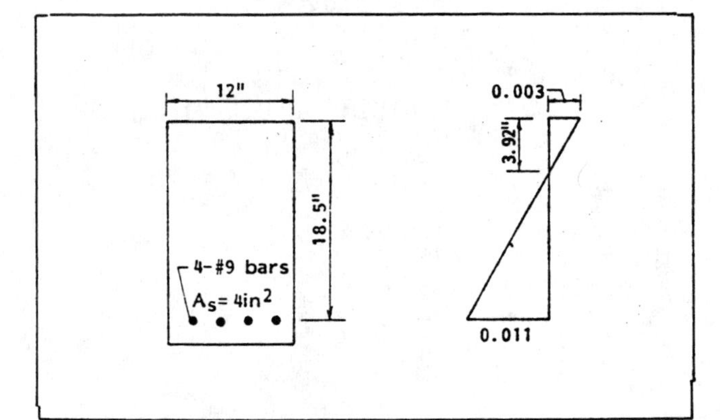

Fig 408-7. Beam for Example 1.

Example 2:

For the beam in Example 1, find the safe service moment, M_w, that may be applied according to ACI 318 if 40% of the moment is dead load and 60% is live load.

From Example 1:

$$M_n = \text{nominal strength} = 226{,}000 \text{ ft-lb}$$
$$M_u = \text{required strength} = 1.4\,M_D + 1.7\,M_u$$
$$\phi = 0.90 \text{ for flexure}$$

For safety:

$$M_n > M_u/\phi > (1.4\,M_D + 1.7\,M_L)/0.9$$

If $M_D = 0.4\,M_w$ and $M_L = 0.6\,M_w$:

$$M_n > [1.4\,(.40)(M_w) + 1.7\,(.60)(M_w)]/0.9 > 1.76\,M_w$$

The safe service moment is:

$$M_w = M_n/1.76 = 226{,}000/1.76 = 128{,}000 \text{ ft-lb}$$

A "balanced" condition exists when the maximum concrete strain reaches 0.003 as tension steel reaches $\epsilon_y = f_y/E_s$. The amount of tension steel is A_{sb} and the neutral axis distance is c_b.

When the actual $A_s > A_{sb}$, the beam is overreinforced. To maintain equilibrium as the tension force increases, the depth of the compression block, a, increases and the distance to the neutral axis, c, becomes greater than c_b. Under these conditions, concrete strain reaches 0.003 before the steel reaches yield. An overreinforced beam fails suddenly when concrete strain reaches 0.003. Because the concrete fails before the steel yields, there is little deformation to warn of impending failure.

When the actual $A_s < A_{sb}$, the beam is underreinforced. To maintain equilibrium as the tension force decreases, the depth of the compression block, a, decreases and the distance to the neutral axis, c, becomes less than c_b. Under these conditions, the steel yields before concrete strain reaches 0.003. Underreinforced beams fail gradually with noticeable deflection before the concrete strain reaches 0.003.

The amount of steel in a beam relative to balance significantly affects the failure mode. Therefore ACI 318 provides for ductile failure by limiting tension steel to not more than 75% of the amount in the balanced condition.

The relative amount of tension steel in a beam is ρ the reinforcement ratio or percentage:

Eq 408-14.

$$\rho = A_s/bd$$

The reinforcement ratio ρ_b for the balanced condition is determined by first looking at the strain condition:

Eq 408-15.

$$c_b/d = \epsilon_c/(\epsilon_c + \epsilon_y) = 0.003/(0.003 + f_y/29{,}000{,}000)$$
$$= 87{,}000/(87{,}000 + f_y)$$

The compressive force is:

Eq 408-16.

$$C_b = 0.85\,f'_c b\,\beta_1\,c_b$$

The tensile force is:

Eq 408-17.

$$T_b = A_{sb} f_y = \rho_b bd\,f_y$$

Equating $T_b = C_b$ gives:

Eq 408-18.

$$\rho_b = (0.85\, f'_c \div f_y)\, \beta_1\, c_b/d$$

Substitute c_b/d from Eq 408-15 (stresses f_y and f'_c in psi):

Eq 408-19.

$$\rho_b = (0.85\, f'_c/f_y)\,(\beta_1)\,[87{,}000/(87{,}000 + f_y)]$$

Example 3:

Does the beam in Example 1 meet the ACI 318 requirement of $\rho < 0.75\,\rho_b$?

From Example 1:

f'_c = 5,000 psi
f_y = 40,000 psi
β_1 = 0.80
A_s = 4 in^2
b = 12″
d = 18.5″

Substitute into Eq 408-19:

ρ_b = [(0.85)(5,000)/(40,000)](0.80) [(87,000)/(87,000 + 40,000)]
ρ_b = 0.058
$0.75\rho_b$ = 0.044

Calculate actual ρ with Eq 408-14:

ρ = 4.0/[(12)(18.5)] = 0.018

This is less than 0.044, so the ACI 318 requirement is met.

The design of a beam in bending is more involved than the analysis. There are only two equations but three unknowns: b, d, and A_s. Further, beam weight is often significant and must be approximated. The following additional equations aid in beam design. The two conditions of equilibrium are:

Eq 408-20.

$$C = T$$

and

Eq 408-21.

$$M_n = (C \text{ or } T)(d - a/2)$$

If the reinforcement ratio, ρ, is preset, then from Eqs 408-16 and 408-17:

$$0.85\, f'_c\, ba = \rho bd\, f_y$$

Eq 408-22.

$$a = \rho d\, [f_y/0.85\, f'_c]$$

Substituting Eq 408-22 into 408-21:

Eq 408-23.

$$M_n = \rho bd\, f_y\, [d - (\rho d/2)(f_y/0.85\, f'_c)]$$

or

Eq 408-24.

$$M_n = \rho bd^2\, f_y\, (1 - 0.59\, \rho\, f_y/f'_c)$$

For design, select ρ, for which case:

Eq 408-25.

$$bd^2 = M_n/[\rho f_y\, (1 - 0.59\, \rho\, f_y/f'_c)]$$

or, if b and d are preset, then:

Eq 408-26.

$$\rho = (0.85\, f'_c/f_y)\,[1 - (1 - 2\, M_n/0.85\, f'_c\, bd^2)^{0.5}]$$

The procedure for strength design of rectangular sections with only tension reinforcement involves the following steps:

1. Assume a value of ρ between the minimum ρ of $200/f_y$ (ACI 388-10.5) and the maximum ρ of $0.75\,\rho_b$.
2. Determine the required bd^2 from:
   $bd^2 = M_n/[\rho\, f_y\, (1 - 0.59\, \rho\, f_y/f'_c)]$
3. Choose b and d to approximate the required bd^2.
4. Determine the revised value of ρ with Eq 408-26.
5. Compute A_s from:
   A_s = (revised ρ) (actual bd)
6. Select the reinforcement.
7. Check the section to be certain that: $M_n > M_u/\phi$

Example 4:

Design a simply supported singly reinforced rectangular concrete beam to carry the floor load from a slatted floor holding area. Beam span is 20′, and the beams are 10′ apart. Floor design live load is 100 psf and the design dead load is 75 psf. Concrete, f'_c = 4,000 psi, steel, f_y = 40,000 psi. Calculate the ultimate moment:

M_u = 1.4 M_D + 1.7 M_L
M_L = $wL^2/8$ = 10′ x 100 psf x $(20')^2$ x 12 in/ft ÷ 8
M_L = 600,000 in-lb

Assume beam weight is 150 plf.
Total dead load is 150 plf + (10 ft^2/ft) (75 psf) = 900 plf.

M_D = 900 x $(20)^2$ x 12/8
M_D = 540,000 in-lb
M_u = 1.4 (540,000) + 1.7 (600,000) = 1,776,000 in-lb

Calculate the nominal moment for the beam,

M_n = M_u/ϕ
ϕ = 0.90
M_n = 1,776,000/.90 = 1,973,000 in-lb
ρ_b = (0.85 f'_c/f_y) β_1 [87,000/(87,000 + f_y)]
β = 0.85 for f'_c = 4,000 psi concrete
ρ_b = [(0.85) (4,000)/40,000] (0.85) [87,000/(87,000 + 40,000)]
ρ_b = 0.050

Maximum allowable ρ allowed by ACI 318:

ρ_{max} = 0.75 ρ_b = (0.75) (0.050) = 0.0375
ρ_{min} = $200/f_y$ = 200/40,000 = 0.005
ρ = 0.018 (choose middle of range)

Calculate bd^2 using Eq 408-25.

bd^2 = $M_n/[\rho\, f_y\, (1 - 0.59\, \rho\, f_y/f'_c)]$
bd^2 = 1,973,000/[(0.018) (40,000) (1 - (0.59) (0.018) (40,000/4,000))]
bd^2 = 2,770

Possible bd combinations:

b	d	bd^2
8″	18″	2592
10″	16″	2560
12″	15″	2700

Note that because of cover, the effective depth, d, is usually about 2½″ less than beam height, h.

Choose a 12″x18″ beam, which gives d = 15.5″ (approximately). Calculate required ρ from Eq 408-26.

$\rho = (0.85\ f'_c/f_y)\ [1 - (1 - 2\ M_n/0.85\ f'_c\ bd^2)^{0.5}]$

$\rho = (0.85 \times 4{,}000/40{,}000)\ [1 - (1 - 2 \times 1{,}973{,}000/(0.85 \times 4{,}000 \times 12 \times 15.5^2))^{0.5}]$

$\rho = 0.019$ (revised)

A_s = new $\rho \times bd$

$A_s = (0.019)\ (12'')\ (15.5'') = 3.53\ in^2$

Cover and bar spacing limit the number of bars in one layer for a given beam width, Table 408-6. Four #9 give $A_s = 4.00\ in^2$ and fit into a 12″ wide beam. Check the flexural strength of the beam.

$C = (0.85)\ (4{,}000)\ (12)\ (a) = 40{,}800a$ lb

$T = f_yA_s = (40{,}000)\ (4.00) = 160{,}000$ lb

$a = 160{,}000/40{,}800 = 3.92$ in.

$c = 3.92\ in/0.85 = 4.61$ in.

$\epsilon_s = (11.5 - 4.61)\ (0.003)/4.61 = 0.0045$

$\epsilon_y = 40{,}000/29{,}000{,}000 = 0.0014 < 0.0045$

Therefore steel has yielded:

$M_n = 160{,}000\ (15.5 - 3.92 \div 2)$

$M_n = 2{,}166{,}000$ in-lb

Calculate the required nominal moment using the new beam weight:

$Wt = 1' \times 1.5' \times 150\ lb/ft^3 = 225$ plf

$M_D = (975)\ (20)^2\ (12)/8 = 585{,}000$ in-lb

$M_n = [1.4\ (585{,}000) + 1.7\ (600{,}000)]/0.9$

$M_n = 2{,}043{,}000$ in-lb $< 2{,}166{,}000$ in-lb OK

Table 408-6. Bars per layer in beam.

Minimum beam width (in), ACI 318. Beam widths assume stirrups. For bars of different size, determine from table the beam width for smaller size bars and then add last column figure for each larger bar used. Maximum aggregate < ¾ of clear space between bars.

Bar size	Number of bars in one layer / Minimum beam width, in. 2	3	4	5	6	Width per added bar
#4	6.1	7.6	9.1	10.6	12.1	1.50
#5	6.3	7.9	9.6	11.2	12.8	1.63
#6	6.5	8.3	10.0	11.8	13.5	1.75
#7	6.7	8.6	10.5	12.4	14.2	1.88
#8	6.9	8.9	10.9	12.9	14.9	2.00
#10	7.7	10.2	12.8	15.3	17.8	2.54
#11	8.0	10.8	13.7	16.5	19.3	2.82

To select beam size, bar size, and bar placement:

- Assume beam weight, and design trial beam. Check the effect of the actual weight on design moment. Redesign the beam if the difference between assumed and actual moment is significant.
- Use ρ = ½ as the maximum value ($\rho = 0.0375\ \rho_b$). This seldom results in deflection problems.
- Maintain clear distances between bars within the beam, per ACI 318:
  1″ or nominal bar diameter, whichever is greater (ACI 318-7.6.1).
  For two layers, 1″ between layers (ACI 318-7.6.2).
- Allow for cover and stirrups (usually ⅜″) on each side of the beam. Measure cover from the face of the stirrup or tie to the form.
- Use whole inches for beam dimensions; ½″ is acceptable in floor slabs.
- For economy, beam depth is 1.5 to 2.0 times width.
- When the design yield strength for tension reinforcement exceeds 40,000 psi. Use ACI 318-10.6.4 to control flexural cracking in beams and one way slabs.

For reinforcing bars:

- Maintain bar symmetry.
- Use at least two bars whenever reinforcement is required.
- Use no more than two bar sizes and no more than two standard bar sizes apart for steel in one face at any location.
- Place bars in one layer, if practical, with two to six bars per layer.
- Place the larger of different sized bars in the layer nearest the beam face.

Reinforced Concrete Slab Design

A slab with one plan dimension significantly larger than the other is a one-way slab. Design a 1 ft wide typical strip across the least dimension as a simple beam with b = 12″. Provide minimum cover. Shear reinforcement is rarely required in slabs.

Example 5:

Design a one-way floor slab. Span is 20′. Design loads are 150 psf of live load, 40 psf dead load, and floor weight. Use $F'_c = 3{,}500$ psi and $F_y = 40{,}000$ psi.

Estimate floor thickness = 8″.

Floor weight = 1 x 1 x 150 x 8⁄12 = 100 psf

Calculate:

$M_n = M_u/\phi$

$M_u = 1.7\ M_D + 1.4\ M_L$ for 1′ strip.

$M_u = 1.7\ [(140)\ (20)^2/8]\ (12\ in/ft) + 1.4\ [150\ (20)^2/8]\ (12\ in/ft)$

$M_u = 268{,}800$ in-lb

$M_n = 268{,}800/0.90 = 298{,}700$ in-lb

Select $\rho = .3\ \rho_b$ to control deflection.

$\rho_b = [(0.85)\ (3{,}500)/40{,}000]\ (.85)\ 87{,}000/(87{,}000 + 40{,}000) = 0.043$

$\rho = 0.3\ (0.043) = 0.013$

Determine bd^2 (Eq 408-25):

$bd^2 = 298{,}700/[(0.013)\ (40{,}000)\ (1 - (0.59)\ (0.013)\ (3{,}500/40{,}000))]$

$bd^2 = 575\ in^3$

b = 12″, so:

$d = (575/12)^{0.5} = 6.9''$

Minimum cover = ¾″; try 8.5″ slab:

$d = 8.5 - 0.75 - .385 = 7.4''$

Calculate revised ρ (Eq 408-26).

$\rho = [(0.85)\ (3{,}500)/40{,}000]\ [1 - (1 - 2 \times 298{,}700/(0.85 \times 3{,}500 \times 12 \times 7.4^2)^{0.5})]$

$= 0.012$

$A_s = \rho\ bd = 0.012\ (12)\ (7.4) = 1.06$

Use Table 408-6 to select bar size and spacing. Select #6 bars at 5″ spacing; A_s = 1.06. Select steel perpendicular to main steel for temperature and shrinkage. Check the capacity of the final design (not shown here).

Table 408-7. Steel areas per foot of beam width.
Average area per foot of width by bar spacing.

	Bar spacing, in.								
Bar	3	4	5	6	7	8	9	10	12
	Area steel/ft beam width, in²								
3	0.44	0.33	0.26	0.22	0.19	0.17	0.15	0.13	0.11
4	0.78	0.59	0.47	0.39	0.34	0.29	0.26	0.24	0.20
5	1.23	0.92	0.74	0.61	0.53	0.46	0.41	0.37	0.31
6	1.77	1.32	1.06	0.88	0.76	0.66	0.59	0.53	0.44
7	2.40	1.80	1.44	1.20	1.03	0.90	0.80	0.72	0.60
8	3.14	2.36	1.88	1.57	1.35	1.18	1.05	0.94	0.78

Shear In Beams and One-Way Slabs

Checking for shear and the design of shear reinforcement is very important. Failures occur in pure shear but are more common in diagonal tension. No overall theory has been developed, so shear design is largely empirical.

A plain concrete beam behaves very much like a homogeneous elastic beam. Large tension stresses in the outer fibers lead to cracking and immediate beam failure. Shear has little effect on beam strength.

Tension reinforcement provides flexural tension strength after the concrete cracks and much higher loads are reached. Shear stress increases proportionately and creates significant diagonal tension stresses. Flexural tension steel does not reinforce concrete against diagonal tension stresses near a beam's neutral axis, and diagonal cracks open perpendicular to the local tension stresses.

Shear is transferred in reinforced concrete members by:

- shear resistance of uncracked concrete.
- aggregate interlock.
- dowel action of tension steel.
- arch action of deeper beams.
- shear resistance of stirrups.

In uncracked members, load capacity is limited by concrete strength. After inclined cracking, stirrups provide higher capacity. Stirrups are vertical U-shaped reinforcement anchored in the compression face of the beam and enclosing the flexural steel bars. They are usually #3, #4, or #5 bars.

Web reinforcing (stirrups) adds beam shear strength because:

- Part of the shear force is carried by the bars crossing a crack.
- The bars shorten the diagonal cracks, leaving more uncracked concrete to carry compression and shear.
- The stirrups tie the longitudinal steel to the rest of the beam and increase the force carried through dowel action in that steel.

The nominal shear force, V_n, is carried by the concrete and the stirrups:

Eq 408-27.

$$V_n = V_c + V_s$$

The nominal concrete shear stress, V_c, is specified (ACI 318-11.3) as:

Eq 408-28.

$$V_c = 2\ bd\ f'^{0.5}_c$$

or (ACI 318-11.6):

Eq 408-29.

$$V_c = (1.9\ f'^{0.5}_c + 2500\ \rho\ V_u d/M_u)\ bd$$

but not greater than 3.5 bd $f'^{0.5}_c$ with $V_u d/M_u < 1.0$.

- b = web width, in.
- d = effective depth, in.
- ρ = A_s/bd = reinforcement ratio
- V_n = nominal shear force, lb
- V_c = nominal concrete shear stress, psi
- V_s = nominal stirrup shear stress, psi
- V_u = factored shear force at section, lb
- M_u = factored moment at section, in-lb

Provide shear reinforcement when the factored shear force exceeds ½ of the concrete strength (ACI 318-11.5.5):

Eq 408-30.

$$V_u = \phi\ V_n > \phi V_c/2$$

except in:

- slabs and footings.
- concrete joist construction defined in ACI 318-8.11.
- beams with total depth not more than 10″, 2.5 times flange thickness, or ½ web width.

The factored shear force, V_u, within d of a support face is computed at d from the support, if the reaction in the direction of applied shear introduces compression into the end regions of the member (ACI 318-11.1).

The nominal shear strength added from vertical stirrups is (ACI 318-11.5.6):

Eq 408-31.

$$V_s = A_v f_y d/s$$

To find stirrup spacing, s, rearrange Eq 408-31 with $V_s = (V_u - \phi\ V_c)/\phi$ from Eq 408-27, and with stirrup area = A_v:

Eq 408-32.

$$s = \phi\ A_v f_y d/(V_u - \phi\ V_c)$$

Take V_u from the shear diagram; start at a distance d from the support face and then at convenient intervals until:

$V_u < \phi V_c/2$

Where torsional moment is small, minimum shear reinforcement is (ACI 318-11.5.5.3):

Eq 408-33.

$$A_v = 50\ bs/f_y$$

Or, the maximum spacing for a given A_v is:

Eq 408-34.

$$s = A_v f_y/50b$$

The maximum spacing of vertical stirrups is (ACI 318-11.5.4):

Eq 408-35.

s_{max} = d/2 or 24″ for $V_s < 4\ bd\ f'^{0.5}_c$

Eq 408-36.

s_{max} = d/4 or 12″ for $V_s > 4\ bd\ f'^{0.5}_c$

Place the first stirrup at s/2 from the face of the support; use the same s to at least s/2 past where the next V_u was calculated. As V_u decreases, s increases until s = s_{max}; then place the stirrups at s_{max} until $V_u = V_c/2$. The minimum number of stirrups for each value of s is 3.

Extend stirrups as close to the tension and compression surfaces as cover requirements and other reinforcement permit. Anchor ends of single leg or simple U stirrups as in ACI 318-12.14.2. For #3 to #5 Grade 40 bars, bending the bar ends in hooks at least 135° around longitudinal bars that are close to the compression face is adequate.

Example 6:

Design the stirrups for the beam in Fig 408-8 with f'_c = 4000 psi. Use #3 Grade 40 bars. The clear span is 19′. The factored load/ft of length is:

w_u = 1.4 (975) + 1.7 (1,000) = 3,065 plf

Draw the shear diagram as in Fig 408-8b. Depth d = total depth - (cover + stirrups + ½ bar diameter):

d = 18″ - (1.5″ + 0.375″ + 0.5″) = 15.6″

At d from the support or 21.6″ from the end:

V_u = 30,650 - 21.6 (3,065)/12 = 25,130 lb

Required:

$V_n = V_u/\phi$ = 25,130/0.85 = 29,560 lb

Available:

$V_c = 2\ bd\ f'^{0.5}_c$ = 2 (12) (15.6) $4{,}000^{0.5}$
= 23,680 lb

Stirrups are required if:

$V_u > 0.5\phi V_c$ = (0.85)(23,680) = 10,100 lb

Find point x where this shear is reached.

x = (30,650 - 11,840)/3,065
= 6.1′ or 73.6″ from free end
= 67.6″ (use 68″) from the support.

A_v for #3 = 2 x 0.11 = 0.22 in²

The required s from Eq 408-32 is:

s = [0.85 x 0.22 x 40,000 x 15.6] / [25,130 - (0.85 x 23,680)]
= 23.3″

The maximum spacing is d/2 = 15.6/2 = 7.8″

Use 7.5″

A_{vmin} = 50 bs/f_y = (50)(12)(7.5)/40,000
= 0.11 < 0.22″

Place each stirrup at about the center of the space, s, that it serves. Start the stirrups 3.5″ (about s/2) from the support and space them 7.5″. Use 10 stirrups at each end as in Fig 408-8c.

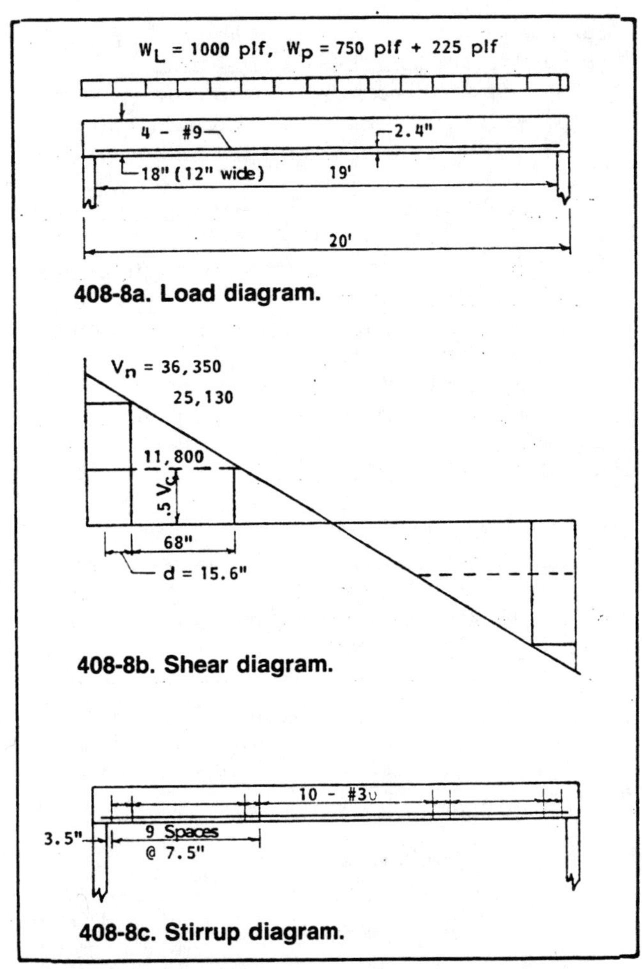

Fig 408-8. Beam for Example 6.

Reinforcement Development and Splices

To carry stress, a reinforcing bar must be properly anchored by development length, end anchorage, or a combination. Hooks can develop bars in tension. Development length and adjustments are in ACI 318-12.

Much is still unknown about failure mode and strength development in bars. The current code considers bar size d_b, f_y, and f'_c. Other factors being studied are bar spacing and cover and the influence of stirrups.

The development length of deformed bars and wire in tension is in ACI 318-12.2. The basic development length for #11 and smaller bars is:

Eq 408-37.

$l_d = 0.04\ A_b\ f_y/f'^{0.5}_c$

but not less than:

Eq 408-38.

$l_d = 0.0004\ d_b f_y$

and for deformed wires:

Eq 408-39.

$l_d = 0.03\ d_b f_y/f'^{0.5}_c$

with l_d = 12″ minimum except for lap splices and web reinforcement.

Multiply the basic development length by the following adjustment factors (ACI 318-12.2.3):

- 1.4 for top reinforcement (horizontal reinforcement with more than 12" of concrete below it).
- 2.0 - 60,000/f_y for reinforcement with $f_y > 60,000$ psi.

The following adjustments may be applied in addition to the above factors (ACI 318-12.2.4):

- 0.8 for reinforcement developed in length under consideration and spaced at least 6" o.c. with at least 3" clear from edge of bar to face of member in the direction of spacing.
- (A_s required) / (A_s provided) for reinforcement in flexural member in excess of that required by analysis.
- 0.75 for reinforcement enclosed in spiral reinforcement not less than ¼" in diameter and not more than ¼" pitch.

If space prevents sufficient development length for tension steel, a hook can add some capacity. A hook is useless in compression. See standard hooks (ACI 318-7.1) in Fig 408-9.

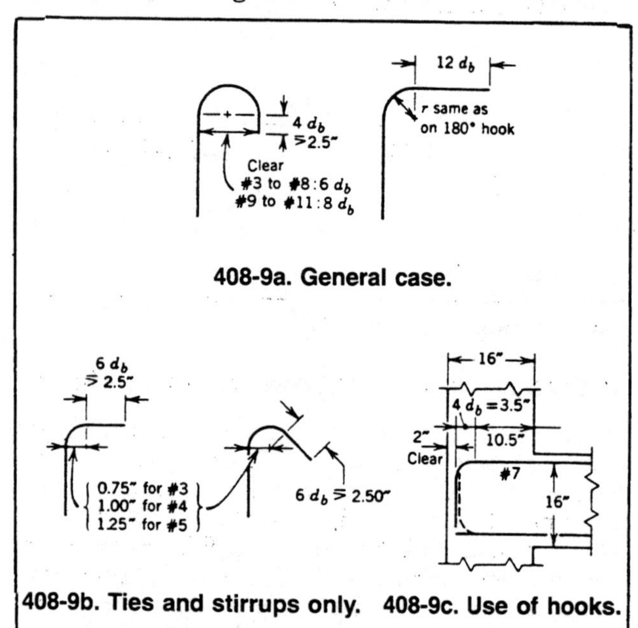

Fig 408-9. Standard hooks.
Radius and diameter are each to inside of hook.

With ξ limited to the values of Table 408-8, standard hooks are assumed to develop bar tensile stress to (ACI 318-12.5):

$$f_h = \xi f'^{0.5}_c \quad \textbf{Eq 408-40.}$$

Table 408-8. ξ values.

Bar size	f_y=60 ksi Top bars	f_y=60 ksi Other bars	f_y=40 ksi All bars
#3 to #5	540	540	360
#6	450	540	360
#7 to #9	360	540	360
#10	360	480	360
#11	360	420	360

Calculate an equivalent embedment length, l_e, for a standard hook by substituting f_h for f_y and l_e for l_d in Eqs 408-37, 408-38, and 408-39. A tie or stirrup hook (Fig 408-9) is valued as 0.5 l_d (ACI 318-12.14.2.1).

Example 7:

1. Calculate the development length for a #9 Grade 40 bar in a beam with f'_c = 4,000 psi.
   For a #9 bar, A_b = 1 in². From Eq 408-37:
   $l_d = 0.04\ A_b\ f_y/f'^{0.5}_c$
   $= 0.04\ (1)\ (40,000) \div 4,000^{0.5}$
   $= 25.3''$
   Check minimum l_d from Eq 408-38.
   $l_d = 0.0004\ (1.128)\ (40,000) = 18.0'' < 25.3''$
   Required l_d = 25.3"
2. If a standard hook is used at the end, what is the remaining length of embedment needed to develop full strength?
   From Table 408-8 for f_y = 40,000 psi:
   $\xi = 360$
   Using Eq 408-40:
   $f_h = 360\ f'^{0.5}_c = 360 \times 4,000^{0.5}$
   $f_h = 22,700$ psi
   Using f_h = 22,700 psi in Eq 408-37 gives the equivalent length for the hook as:
   $l_e = 0.04\ (1.0)\ (22,700)/f'^{0.5}_c$
   $l_e = 14.4''$
   The remaining length of embedment is 25.3 - 14.4 = 10.9"

Splices

Spliced tension bars with welds, Cadwell (fusion type) mechanical connectors, or laps are shown in Fig 408-10. The bars are commonly tied in contact with each other, but can be up to 6" apart, with a maximum spacing of ⅕ the lap length.

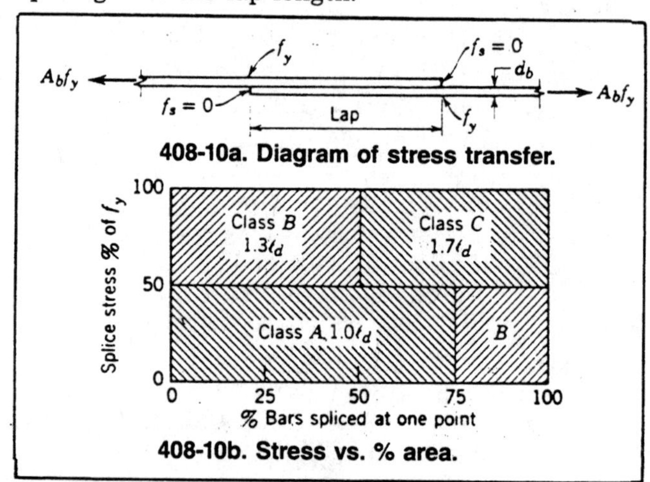

Fig 408-10. Tension lap splices.

Splice lap length is a multiple of l_d, with l_d based on the full f_y. Splices are in three classes by severity of conditions:

For stress always < f_y/2:
Class A if not over 75% of bars spliced within one lap length. Use 1.0 l_d.
Class B if more than 75% are spliced within one lap length. Use 1.3 l_b.

For stress > $f_y/2$:

Class B if not over 50% of bars spliced within one lap length. Use 1.3 l_d.

Class C if more than 50% are spliced within one lap length. Use 1.7 l_d.

- The minimum splice lap length is 12″. There is no minimum on l_d.

Columns

Columns are members that carry axial loads. In practice, columns are also subject to moment from eccentric loading or end rotation. All columns are designed to withstand a minimum moment.

Columns are reinforced with ties or spirals. In tied columns, the longitudinal bars are enclosed in a series of closed ties, Fig 408-11a. In spiral columns the longitudinal bars are closed in a continuous closely spaced spiral, Fig 408-11b.

Ties (ACI 318-7.10.5) are at least #3 for #10 or smaller bars, and #4 for #11 or larger and for all bundled bars. Limit tie spacing to 16 bar diameters, 48 tie diameters, or the least column dimension. Every corner and alternate bar is braced by the tie, and no bar is more than 6″ from such a laterally supported bar.

Limits are set on the steel ratio, $\rho_t = A_{st}/hb$, for longitudinal bars (ACI 318.10.9.1). Because of shrinkage and creep stresses on small areas, $0.01 < \rho_t < 0.08$. Crowding becomes a problem when ρ_t exceeds 0.08. Use at least 4 bars in rectangular and 6 bars in circular arrangements.

Columns are designed as "short" or "long." Lateral deflections are not significant for short columns but have an important effect on the strength of long columns.

In members braced against sidesway, the column is short if:

Eq 408-41.

$$K\ l_u/r < 34 - 12\ M_1/M_2$$

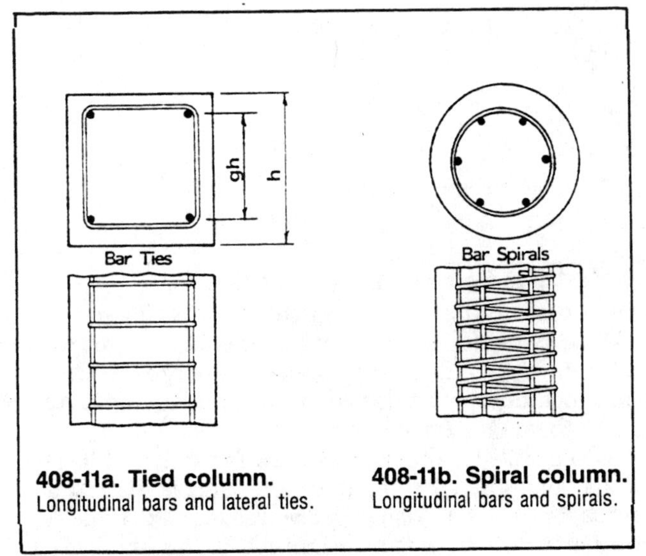

408-11a. Tied column. Longitudinal bars and lateral ties.

408-11b. Spiral column. Longitudinal bars and spirals.

Fig 408-11. Reinforced column types.

In unbraced members (sidesway permitted) the column is short if:

Eq 408-42.

$$K\ l_u/r < 22$$

K = effective length factor
l_u = unsupported column length
M = end moment acting on column, where $M_2 > |M_1|$.
M_1 is positive if member is bent in single curvature, negative if bent in double curvature.
M_2 is always positive.
r = 0.30 times overall dimension in the direction of stability for rectangular compression members
= 0.25 times the diameter of circular compression members

Short Column Design

Only the design for short columns is presented here. Consult a reinforced concrete text for the design of long columns.

The theoretical limit to the strength of an axially loaded column is:

Eq 408-43.

$$P_{no} = 0.85\ f'_c A_n + f_y A_{st}$$
$$P_n0 = 0.85\ f'_c\ (A_g - A_{st}) + f_y A_{st}$$

P_{no} = ultimate load capacity at yield point of tied column when eccentricity is zero
A_n = net area of concrete, $A_g - A_{st}$, in^2
A_g = gross area of concrete, in^2
A_{st} = area of vertical column steel, in^2
f'_c = standard cylinder strength of concrete, psi
f_y = yield point stress for steel, psi

Multiply P_{no} by ϕ for design strength. ϕ is 0.70 for a tied column and 0.75 for a spiral column.

The design strength is limited to 0.80 P_{no} for tied columns and 0.85 P_{no} for spiral columns. Maximum design strength for a tied column is (ACI 318-10.3.5.2):

Eq 408-44.

$$P_{n\text{-}max} = 0.80\ [0.85\ f'_c\ (A_g - A_{st}) + f_y A_{st}]$$

For a spiral column (ACI 318-10.3.5.1):

Eq 408-45.

$$P_{n\text{-}max} = 0.85\ [0.85\ f'_c\ (A_g - A_{st}) + f_y A_{st}]$$

These maximum load limits apply when the moment is small enough to keep the eccentricity under 0.10 h for a tied column and under 0.05 h for a spiral column, where h is the major column dimension. While these are approximate limits, if e is noticeably under them, the entire short column design can be based on maximum loads.

Example 8:

Design a short tied column to carry an axial dead load of 150,000 lb, an axial live load of 250,000 lb, and a live load moment of 200,000 in-lb. f'_c = 4,000 psi, f_y = 40,000 psi. Try for ν_t of about 0.03.

$P_u = 1.4 \times 150{,}000 + 1.7 \times 250{,}000 = 635{,}000$ lb
$M_u = 1.7 \times 200{,}000 = 340{,}000$ in-lb
$e = M_u/P_u = 340{,}000/635{,}000 = 0.54''$ eccentricity

From the loading, it appears that h much greater than 6" is needed and $e/h < 0.10$. Try the equation above for maximum load.

$P_{n\text{-}max} = P_u/\phi = 635{,}000/0.70 = 907{,}100$ lbs
$P_{n\text{-}max} = 0.80\,[0.85\,f'_c A_g + (f_y - 0.85\,f'_c)\,A_{st}]$
$= 0.80\,[0.85 \times 4{,}000\,A_g + (40{,}000 - 0.85 \times 4{,}000)\,0.03\,A_g]$
$= 3600\,A_g$
$A_g = 907{,}100 \div 3600 = 252$ in^2
$h = 252^{0.5} = 15.9''$

Use a 16"x16" column.
$A_g = 256$ in^2
Calculate A_{st}:
$907{,}100 = 0.80\,[0.85 \times 4{,}000 \times 256 + (40{,}000 - 0.85 \times 4{,}000)\,A_{st}]$
$A_{st} = 7.2$ in^2
Use 6-#10: $A_{st} = 7.62$ in^2
Use 3 bars on opposite faces.
For #3 ties, clear bar spacing =
$[16 - 2\,(1.5 + 0.38) - 3\,(1.27)]/2 = 4.2'' < 6''$ max
A single tie around all bars is adequate.
Tie spacing: $h = 16''$, 48 tie $d_b = 18''$,
16 bar $d_b = 20.3''$
Use #3 ties 16" o.c.

Short Eccentrically Loaded Columns

Replace an eccentric compressive load on a column by an equal load acting through the member centroid and a couple equal to $P_u e$, Fig 408-12a. Similarly, replace an axial load, P_u, and a bending moment, M_u, on a column by an equivalent system of forces, P_u, at an eccentricity, e, such that $M_u = P_u e$, Fig 408-12b. For eccentrically loaded columns, transform bending moments and/or axial loads into an equivalent axial load, P_u, at an eccentricity, e, from the plastic centroid of the column cross section.

The steel in an eccentrically loaded column is in compression on the side of the eccentricity and in tension on the opposite side. Parameters for tension steel are primed (A'_s, d', f'_s), and their compression steel counterparts are unprimed, Fig 408-13. At the limiting axial load for a given eccentricity, the concrete has just reached its ultimate strain of 0.003. The tension steel yields simultaneously as $\epsilon_c = 0.003$ (balance condition), the tension steel has already yielded (overreinforced—tension controls), or the steel has not yielded and $f'_s = E'_s \epsilon'_s$ (underreinforced—compression controls). In any of the three cases, the compression steel, A_s, may or may not have yielded.

The balance condition in the column is produced by a unique combination of load and eccentricity. The combination is the balanced load and the load and eccentricity are $P_n b$ and e_b, respectively.

The stress and strain distributions for the various possibilities of eccentricity are in Fig 408-14. The stress distribution at balance ($e = e_b$) is in Fig 408-14b. Note that A'_s has just yielded, but A_s may or may not have yielded. The magnitude of the compression block depth at balance, a_b, is calculated exactly as it was earlier for beams.

When $e > e_b$, the steel yields further, but T remains constant at $A'_s f'_y$ and the compression zone decreases ($a < a_b$). Also ϵ_s decreases, so C_s decreases. The net result is that as e increases, the axial load capacity, P_n, of the section decreases.

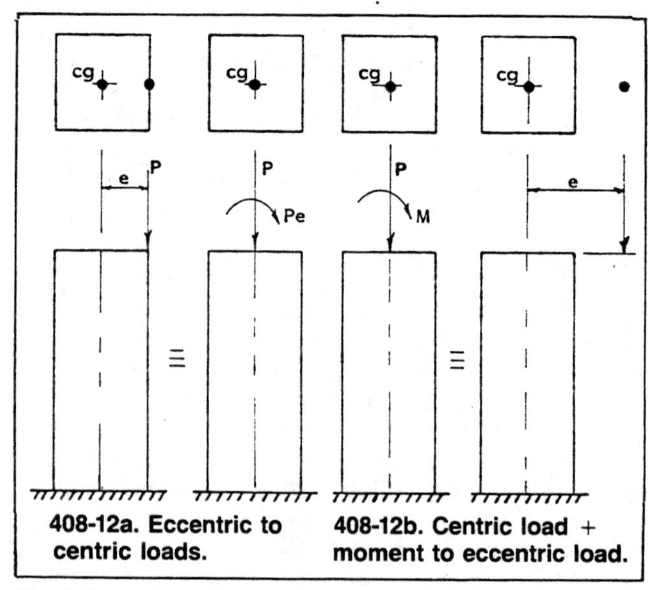

408-12a. Eccentric to centric loads.
408-12b. Centric load + moment to eccentric load.

Fig 408-12. Equivalent loadings for columns.

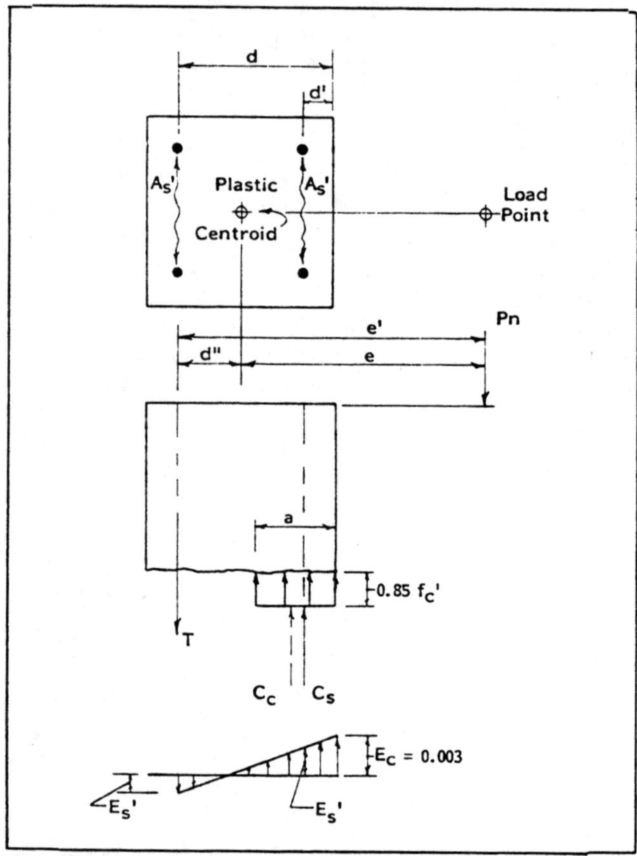

Fig 408-13. Variables for eccentric columns.

When $e < e_b$, A'_s no longer yields, compression block size increases ($a > a_b$) and ϵ'_s increases, but may still be less than ϵ'_y. The net result is that when $e < e_b$, $P_n > P_nb$ and increases to a limit value of $P_{n\text{-max}}$ as computed for a centrically loaded column ($e = 0$).

The eccentrically loaded column problem thus reduces to evaluating the combinations of P and e that it can carry and ensuring that $0.01 < \rho < 0.08$. A handy design aid for analyzing load capacities is the interaction diagram. It is a plot of the combinations of axial load, ϕ_n, and moment, $\phi_n e$, that induce failure. The column can safely carry any combination within the curve. Fig 408-15 is a general interaction diagram.

Equilibrium in reinforced columns (Figs 408-14 and 408-15) is more complex than in singly reinforced beams. Under axial load P_u, the compression and tensile resultants, C and T, are no longer equal and are no longer a simple couple. The equilibrium conditions for analysis are the sum of forces and the sum of moments about the tensile steel ($\Sigma M_T = 0$).

$$\Sigma F = 0$$

$$P_n = C_c + C_s - T$$

$$\Sigma M_T$$

$$P_n e' = C_c (d - a/2) + C_s (d - d')$$

$$1 = 0$$

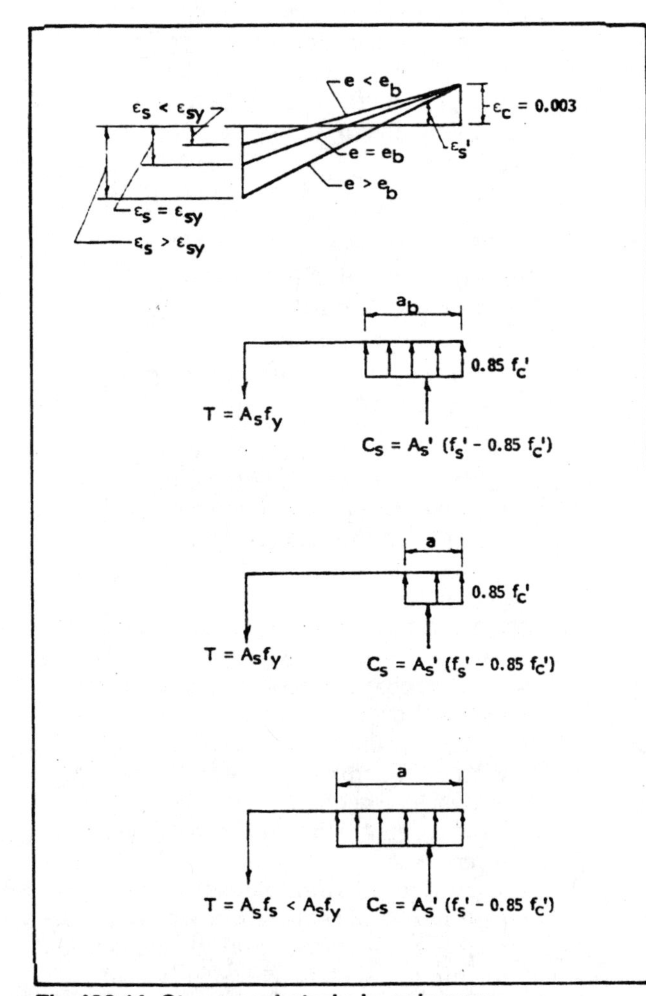

Fig 408-14. Stress and strain in columns.
For $e = e_b$, $e > e_b$, and $e < e_b$.

Summing moments about the plastic centroid is also useful when $e < e_b$ and A'_s does not yield.

$$M_{Pl} = 0 \text{ (about centroid)}$$

$$P_n e = Td'' + C_s (d - d'' - d') + C_c (d - d'' - a/2)$$

The interaction diagram for a rectangular column is in Fig 408-16. From this diagram, evaluate the combinations of axial load, ϕP_n, and moment, $M_n = \phi P_n \times e$, the column can safely carry. If the column load combinations fall within the envelope, the column can carry the load. Interaction diagrams for a variety of spiral and tied rectangular and circular columns are in the *Ultimate Strength Design Handbook: Part II, Columns.*

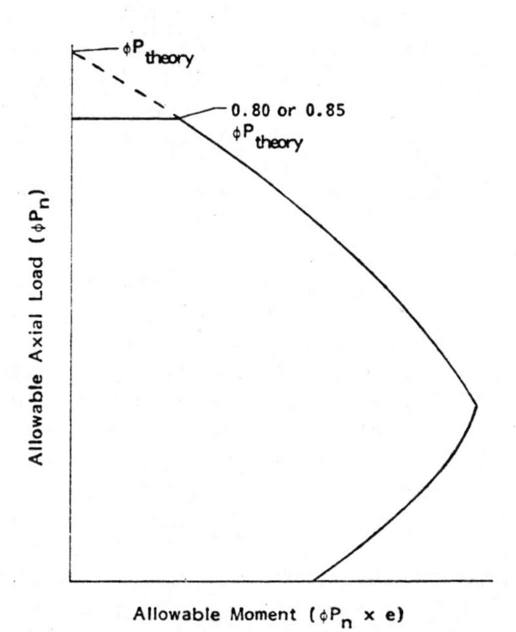

Fig 408-15. Interaction in eccentric columns.

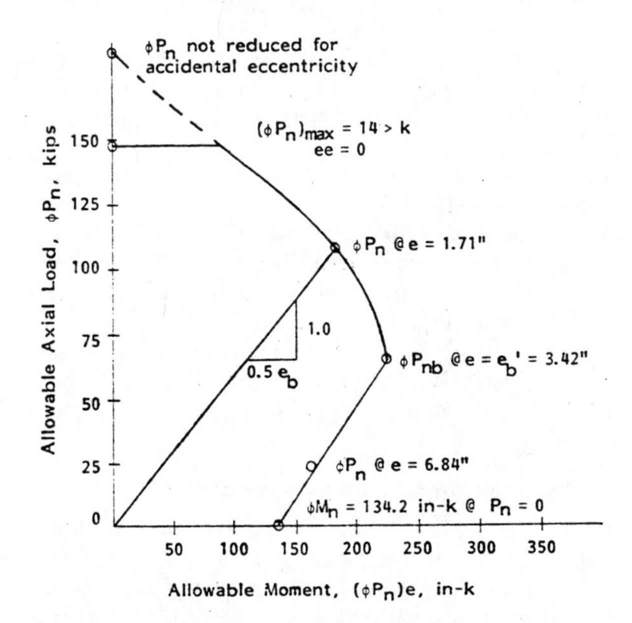

Fig 408-16. Interaction in typical columns.

409 STRUCTURAL STEEL DESIGN

This section was prepared by Dr. H.B. Manbeck, Pennsylvania State University; Dr. N.F. Meador, University of Missouri; and Dr. J.O. Curtis, University of Illinois.

Introduction

Structural steel is classified as hot-rolled or cold-formed, the main differences being the method of forming and the thickness of the elements of the structural sections.

Hot-rolled structural steel shapes include whole rolled shapes and those that are welded up from two or more hot-rolled plates. Cold-formed shapes are developed by: (1) cold rolling thin steel sheets—as with corrugated roofing and siding; (2) press brake operations; or (3) bend brake operations.

Hot-rolled elements are 0.1875″ to 4.9375″ thick. Cold-formed steel is typically 0.0149″ (28 gauge) to 0.25″ thick. Elements to 0.75″ thick can be cold-formed, but are not common.

The thickness of structural elements—webs, flanges, etc.—affects buckling. The thinner the element, the more susceptible it is to local buckling before the entire section reaches its elastic or plastic load capacity. Cold-formed shapes have thinner elements and are more susceptible to local buckling than thicker hot-rolled sections. Considerations of element thickness and local buckling cause major differences between hot- and cold-formed steel specifications.

The first two citations below are specifications for hot-rolled and cold-formed steel building design in the United States. The other three citations are excellent references for more information on steel design.

- *Manual of Steel Construction,* 8th Ed. American Institute of Steel Construction. Chicago, 1980. (Abbrev: AISC Manual)
- *Specification for the Design of Cold-Formed Steel Structural Members,* parts I to V. American Iron and Steel Institute (AISI). Washington, DC, 1968.
- *Steel Design Manual.* U.S. Steel Corp. Pittsburgh, PA, 1974.
- Salmon and Johnson. *Steel Structures.* Intext Educational Publishers. Scranton, PA, 1972.
- Yu, Wie-Wen. *Cold-Formed Steel Structures.* McGraw-Hill. New York, 1973.

The following material conforms to the AISC design specification but highlights only the most important aspects of steel design for agricultural structures.

Working Strength Design Philosophy

Structural steel is designed by the ultimate strength or the working strength method. Although the ultimate strength method is gaining acceptance, the traditional working strength method is still the more widely used and is usually a more familiar approach for the beginning student.

The working strength, or working stress, concept assumes that the steel is linearly elastic up to yield. Then, because theoretical and actual behavior do not quite agree, allowable design stresses are set at some level less than the steel's yield stress. The stresses in the structural elements are then predicted by the simple strength of materials equation, such as:
$f_b = Mc/I$, $f_t = P/A$, $f_v = VQ/It$.
The design specification sets the amount by which the yield stress is reduced for allowable design stresses for a variety of service conditions.

For example, in a flexural member, the actual bending stress is computed as $f_b = Mc/I$. But the allowable stress, F_b, equals kF_y, where k is a reduction factor from the design specification and F_y is the yield stress. The design condition to be satisfied is $f_b < F_b$, or $Mc/I < kF_y$. F_b depends on the steel alloy, heat treatment, rate of load application, temperature, cold working, repeated stresses, type of loading (tension, compression, flexure), and stability of the section elements (buckling).

List of symbols

A_e = net effective area of tension members
A_n = net area of tension members
A_w = area of girder web
C_b = bending coefficient dependent upon moment gradient

Eq 409-1.

$$C_b = 1.75 + 1.05(M_1/M_2) + 0.3(M_1/M_2)^2$$

C_c = column slenderness ratio between elastic and inelastic buckling

Eq 409-2.

$$C_c = (2\pi^2 E/F_y)^{1/2}$$

C'_c = column slenderness ratio where buckling stress equals yield stress
C_m = coefficient applied to bending term in interaction formula and dependent on column curvature caused by applied moments
C_t = tension member reduction coefficient
E = modulus of elasticity of steel (29,000 ksi)
F_a = axial stress permitted with no bending moment
F_{as} = axial compressive stress permitted with no bending moment for bracing and other secondary members
F_b = bending stress permitted with no axial force
F'_b = allowable bending stress in compression flange of plate girders as reduced for hybrid girders or large web depth-to-thickness ratios
F_c = column buckling stress
F'_e = Euler stress divided by safety factor

Eq 409-3.

$F'_e = 12\pi^2 E/23(kL_b \div r_b)^2$ if $KL/r > C_c$

F_p = allowable bearing stress
F_t = allowable tensile stress
F_u = specified minimum tensile stress
F_v = allowable shear stress
F_y = specified minimum steel yield stress, ksi. "Yield stress" is the specified minimum yield point for those steels that have one or specified minimum yield strength for those with no yield point.

Theoretical yield stress above which the section is not compact:

F'_y = due to flange buckling
F''_y = due to web crippling
F'''_y = under combined axial and flexural loading
K = effective length coefficient
L = span length (ft)
L_b = actual unbraced length in plane of bending (in.)
M = moment (k-ft)
M_1 = smaller moment at end of unbraced length of beam-column
M_2 = larger moment at end of unbraced length of beam column
M_D = moment produced by dead load
M_L = moment produced by live load
M_p = plastic moment
N = length of bearing of applied load (in.)
P = applied load (kips)
Q_a = ratio of effective profile area to the total profile of an axially loaded member
Q_s = axial stress reduction factor where width-thickness ratio of unstiffened elements exceeds limiting value. See Local Buckling.
R = reaction or concentrated transverse load on a beam (kips)
b_e = effective width of stiffened compression element
b_f = flange width of rolled beam or plate girder
c = distance from neutral axis to extreme fiber beam
d = depth of beam or bolt diameter
d_c = column web depth clear of fillets
f = axial compression load on member divided by effective area (ksi)
f_a = computed axial stress
f_b = computed bending stress
f_t = computed tensile stress
f_v = computed shear stress
g = transverse spacing between fastener gauge lines
k = distance from outer face of flange to web toe of fillet
r = governing radius of gyration = $(I/A)^{1/2}$
r_b = radius of gyration about axis of concurrent bending
r_y = lesser radius of gyration
s = spacing (pitch) between successive holes in line of stress
t = girder, beam, or column web thickness
t_f = flange thickness
x = subscript relating symbol to strong axis bending
y = subscript relating symbol to weak axis bending
ν = Poisson's ratio, may be taken as 0.3 for steel

Properties of Steel

Factors Affecting Mechanical Properties

Chemical composition

Chemical composition is the factor most affecting steel properties. In carbon steels, carbon and manganese influence strength, ductility, and weldability. Most structural carbon steels are 98%+ iron and about ½% carbon and 1% manganese by weight. Carbon increases hardness or tensile strength but decreases ductility and weldability, so small amounts of alloying elements are sometimes added to increase "hardenability" while keeping the benefits of a low percentage carbon.

Phosphorus and sulfur are limited because they reduce impact strength. Copper can increase corrosion resistance and silicon eliminates unwanted gases from the molten metal. Nickel and vanadium are also generally beneficial.

Some ductility is generally traded for increased strength, which is tolerable if enough ductility remains. What is important is the ductility of the fabricated structure, which is a function of material, design, and fabrication.

Heat Treatment and Strain History

Faster cooling rate, lower finishing temperature, and greater reduction of cross section during rolling increase the yield and tensile strength of hot-rolled steel. Some steels are quenched (heated above critical temperature and cooled rapidly) to increase strength and reduce ductility—and then tempered (heated below critical temperature and cooled slowly) to reduce cooling stress and restore part of the ductility at some loss of the strength gained by quenching.

Other steels are cold-rolled for higher strengths. Cold working strain hardens the material by, in effect, using up the lower part of the stress-strain curve. It is not quite that simple—for example, work-hardening in tension can reduce the yield strength in compression.

The grain size of steel, as affected by composition and heat treatment, is frequently important. Smaller grained steels are usually less brittle and notch sensitive and more impact resistant.

Heating during welding or burning also affects properties and residual stress patterns and is part of the total design problem.

Table 409-1. Some available structural steels.
Large sections (group 3 or higher) and thick sections (>1½") are omitted if they are not also available in smaller sizes.
Of the steels in Table 409-1, AISC approves A36, A242, A375, A440, A441, A514, A572, and A588 for structural application. The most common in agriculture buildings is A36.
From AISC.

Steel type	ASTM name	F_y	F_u	Available shape group	Bars & plates thickness	Use
Structural carbon	A36	36	58	All	All to 8"	Mild carbon steel for riveted, bolted, or welded bridges & buildings where strength/weight ratio not critical but stability is.
	A529	42	60	1	To ½"	Riveted, bolted, or welded when building. Common in joists & rigid frames.
Atmospheric corrosion resistance is about twice stru. carbon steel:						
High strength low alloy	A440	50	70	1,2	To ¾"	Manganese vanadium steel for welded, riveted, or bolted const. Where wt. is important.
	A572					
	Grade 42	42	60	All	½" to 6"	Manganese vanadium steel for riveted, bolted bridge const. & welded const. where dynamic loading & fatigue are important.
	Grade 50	50	65	All	To 2"	
	Grade 60	60	75	1,2	To 1¼"	
	Grade 65	65	80	1	To 1¼"	
Atmospheric corrosion resistance is 4-6 times stru. carbon steel for bare exposure to normal atmosphere:						
Corrosion resistant high strength, low alloy	A242	50	70	1,2	To ¾"	Welded, riveted or bolted const. where wt. & durability are important.
Atmospheric corrosion resistance is 4 times stru. carbon steel; for bare exposure to normal atmosphere:						
	A588	50	70	All	To 4"	Primarily welded const. where wt. & durability are important
Quenched & tempered alloy	A514	100	110		To 2½"	Primarily for welded bridges.
Pipe and structural tubing for welded, bolted or riveted const.						
Cold formed seamless & welded	A500 Grade B	46	58	Round, square, and rectangular		
Welded & seamless pipe	A53 Type B	35		Hot formed seamless and welded black hot dipped galvanized pipe.		
	Type S	35	60	Seamless pipe		
Hot formed welded & seamless	A501	36	58	Round, square, & rectangular		

Geometry, Temperature, Strain Rate

The size and shape of a member affect its stress distribution and structural behavior. Small elements tend to have higher unit strengths than larger ones, especially in fatigue and brittle fracture. See Table 409-1 for the allowable stresses and member thickness. Surface finish is also sometimes important. With fatigue, smoother surfaces give higher strengths.

Low temperature and high strain rate tend to increase yield and tensile strengths, but may reduce ductility. And both low temperature and high strain rate affect yield more than tensile strength, reducing the usual margin between the two strengths.

Table 409-2. Groups of structural shapes by tensile property.
Groups for structural tees are the group of the W, M, and S shapes from which they are made.
Group 4 = W14x233-550, W12x210-336
Group 5 = W14x605-730

Structural Shape	Group 1	Group 2	Group 3
W Shapes	W 14x22 −53	W 14x61 −132	W 14x145 −211
	W 12x14 −58	W 12x65 −106	W 12x120 −190
	W 10x12 −45	W 10x49 −112	
	W 8x10 −48	W 8x58, 67	
	W 6x9 − 25		
	W 5x16, 19		
	W 4x13		
M Shapes	to 20 lb/ft		
American Standard Channels (C)	to 20.7 lb/ft	> 20.7 lb/ft	
Angles (L), Structural & Bar-Size	to ½"	> ½ to ¾"	> ¾"

From AISC.

Structural Steel Categories

Stress-Strain Curves

Figs 409-1 and 409-2 illustrate the unique property of many steels that have a yield point. The elastic range is typically quite linear. Suddenly a change in the material gives a large increase in strain at no increase in stress (yield point). Thereafter, increased stress produces strain hardening from which the steel cannot return to the curve below the yield point.

Note that the strain range in Fig 409-2 goes only to 0.03 in/in.—it is a large-scale "map" of the first portion of Fig 409-1.

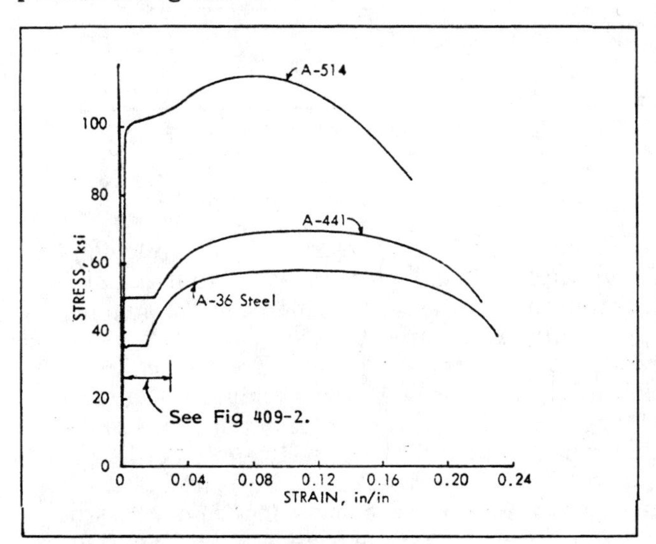

Fig 409-1. Stress-strain curves for structural steels.
With specified minimum tensile properties.

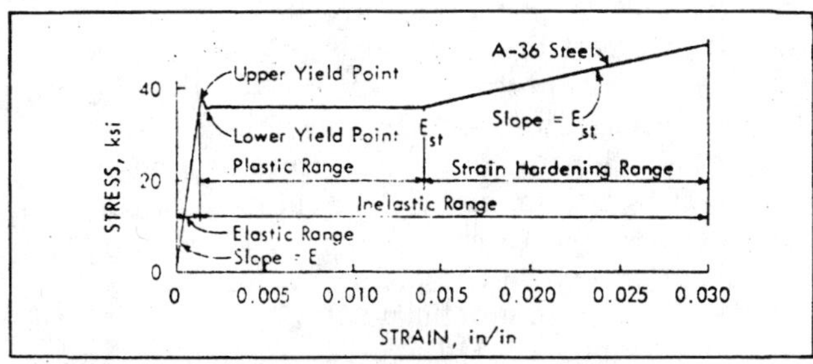

Fig 409-2. Initial stress-strain curve, A-36 steel.

Stability and Stiffened Compression Elements

Stability of Steel Sections

Instability (buckling) failures are important in steel design because the elements of sections are relatively slender. Designers consider both local buckling of elements and gross buckling of the entire structural member. The four most common buckling failures are local buckling of webs, local buckling of flanges, gross buckling of members, and lateral torsional buckling of members.

Local buckling of the webs (also called web crippling) is illustrated in Fig 409-3a. The load at which local web buckling occurs depends on the depth-to-thickness ratio of the web, d/t, and whether or not the element is stiffened. Web crippling is of more concern in beams than in columns.

Local flange buckling, Fig 409-3b, occurs in both flexural and compression members. The stress level for local buckling depends on the flange width-to-thickness ratio, b/t_f, and the element stiffening.

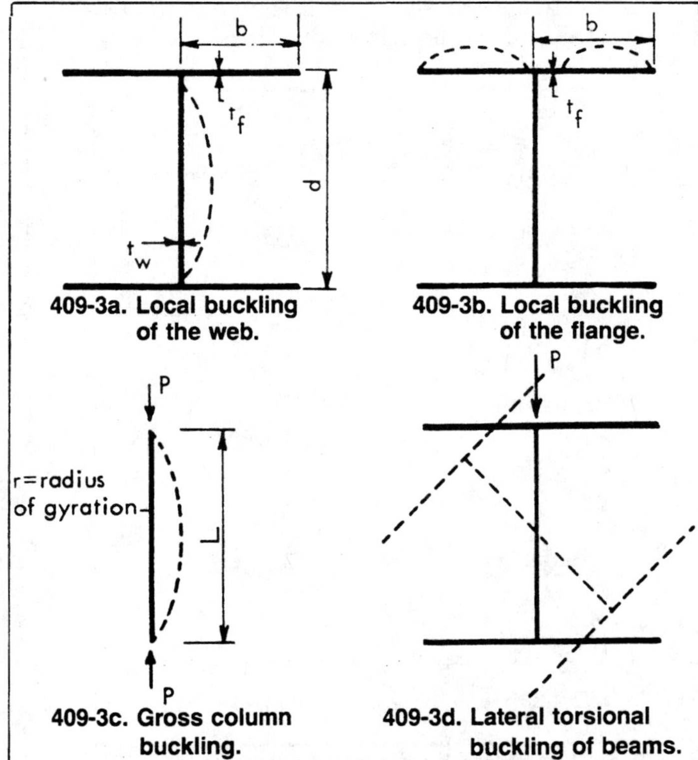

Fig 409-3. Instability failures in structural members.

Columns and beam-columns are both susceptible to gross buckling of the member, Fig 409-3c. It can occur before the compressive stress reaches the yield stress (elastic buckling) or after (inelastic buckling). The load at which gross buckling occurs depends on the slenderness ratio, L/r, (unsupported column length/radius of gyration), the degree of restraint (free, fixed, or pinned) at the ends of the member, and the steel's modulus of elasticity.

Lateral torsional buckling, Fig 409-3d, is instability in beams and beam-columns with long unbraced compression flanges. The flexural load at which the instability occurs depends on the beam's unbraced length, depth, and compression flange area. It is a gross buckling phenomenon in which a beam deflects laterally in the plane perpendicular to the applied load, i.e., the cross section twists or rotates out of its original plane. The load capacity of the member is usually seriously reduced.

The allowable stress level of a structural steel member must be low enough (with appropriate safety factors) to avoid instability failures or failure by yielding (with appropriate safety factor for material uncertainties). Designers carefully investigate member geometry, bracing, and loading to predict the first failure mode to occur before establishing an allowable stress for the member.

Stiffened and Unstiffened Compression Elements

The elements of a structural cross section are classified as either stiffened compression elements (SCE) or unstiffened compression elements (UCE). The classification indicates the degree of end restraint of each section element. If an element has one end restrained and one end free, it is an unstiffened element (UCE). If both ends of the element are restrained, it is stiffened (SCE).

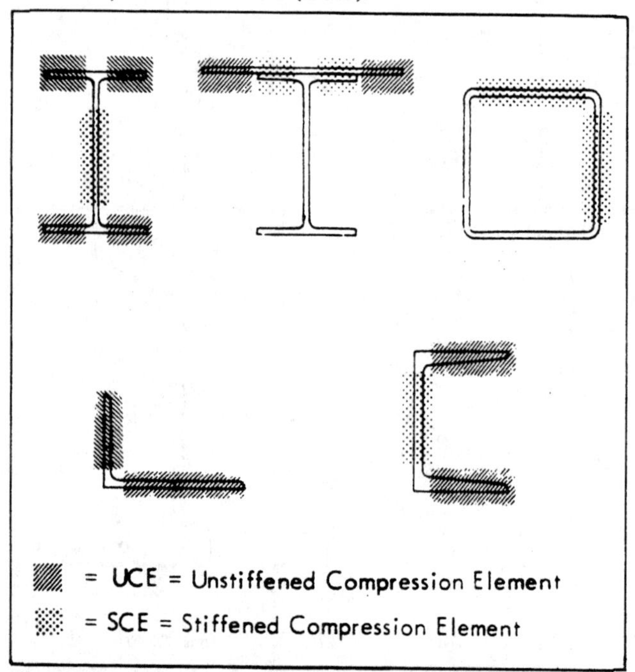

Fig 409-4. Stiffened and unstiffened compression elements.

Tension and Compression Members

The allowable stress, F_t, in tension members must not exceed 0.6 F_y on the gross area of the member nor 0.50 F_u on the effective net area. For pin-connected members, F_t must not exceed 0.45 F_y on the net area. For allowable stresses in threaded parts, $F_t = 0.33F_u$ on the nominal cross section area. The actual tensile stress, f_t, must not exceed F_t:

Eq 409-4.

$$f_t = P/A_e < F_t$$

P = tensile load

A_e = net effective cross section area

The net area of a section depends on the type, number, size, and placement of connectors. In welded connections, A_n is the gross section area. In bolted connections, A_n depends on: thickness, t; width, W_{gross}; bolt diameter, d; effective hole diameter (d + ⅛"); connector center-to-center spacing, s; and row center-to-center spacing, g. The bolt hole diameter is usually 1/16" larger than the bolt diameter, and 1/16" is added for damage to surrounding metal. Take the net area as no greater than 0.85 A_{gross}. For example, if a tensile member is fastened by two rows of bolts as shown in Fig 409-5, the net area is the smaller of:

Eq 409-5.

$$A_{nmax} = 0.85\ A_{gross}$$
$$A_{n1\text{-}1} = (t)(W_{gross} - 2(d + 1/8") + s^2/4g)$$
$$A_{n2\text{-}2} = (t)(W_{gross} - (d + 1/8")$$

The term $s^2/4g$ accounts for normal and shear forces on the diagonal area between bolts.

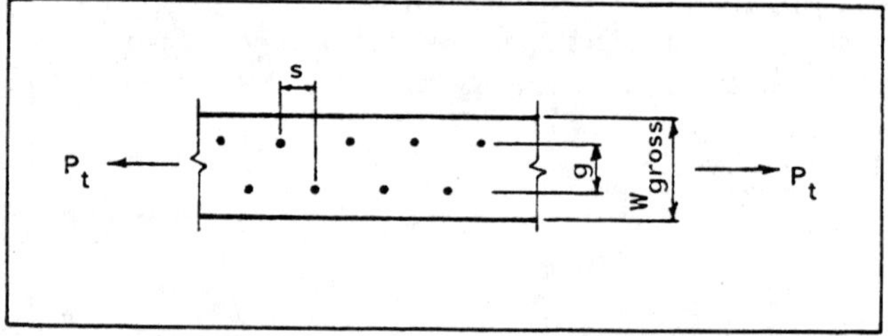

Fig 409-5. Net area of bolted tension members.

Eq 409-6.

$$A_e = C_t A_n$$

Compute the net effective area, A_e, for rolled and built-up tension members with bolts through some but not all the cross section elements of the member with a reduced net area to allow for non-uniform stress in the member at the joint. The reduction coefficient, C_t, is 0.9 for W, M, and S shapes with flange widths not less than ⅔ the depth, connections to the flange, and no fewer than three fasteners per line in the direction of the stress. For W, M, S, and built-up shapes not meeting the above conditions but still having no fewer than three fasteners per line, C_t = 0.85. All other members having only two fasteners per line have a reduction factor of 0.75.

Example 409-1:

What is the tension capacity of the member, not the connectors, for the A36 steel strap shown below?

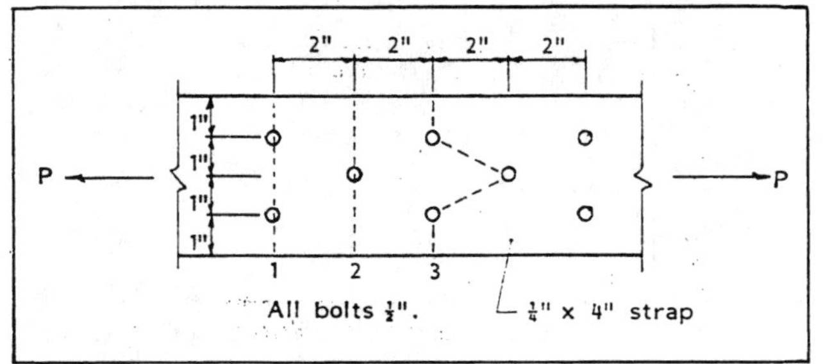

Fig 409-6. Joint of Example 1.

Solution:

Use Eq 409-5:

$A_{nmax} = 0.85 \times 0.25'' \times 4''$
$= 0.850 \text{ in}^2$
$A_{n1} = (0.25'')(4 - 2(½ + ⅛))$
$= 0.688 \text{ in}^2$
$A_{n2} = (0.25'')(4 - (½ + ⅛))$
$= 0.844 \text{ in}^2$
$A_{n3} = (0.25'')(4 - 3(½ + ⅛) + 2(2^2/(4x1))$
$= 1.031 \text{ in}^2$

The controlling net area is A_{n1}.

The allowable tensile force, P, is:

Eq 409-7.

$$P = A_{gross}\,(0.60\,F_y)$$

$= (0.25'' \times 4'')(0.60 \times 36 \text{ ksi})$
$= 21.6 \text{ k}$

or

$P = A_n\,(0.50 \times F_u)$
$= (0.688 \text{ in}^2)(0.50 \times 58 \text{ ksi})$
$= 19.95 \text{ k}$

The tension capacity of the steel strap, P, is 19.95 k.

Example 409-2:

What is the tension capacity of a threaded ½" A36 steel rod?

Solution:

Eq 409-8.

$$A_n = \pi D^2/4$$
$$A_n = \pi(0.5)^2/4 = 0.196 \text{ in}^2$$

D = nominal diameter
$P = A_n\,(0.33\,F_u)$
$= (0.196)(0.33)(58{,}000)$
$= 3{,}750 \text{ lb}$

For tension member (other than rods), slenderness ratios are limited: $L/r < 240$ in main members and < 300 in secondary members. This requirement prevents very slender members which are likely to vibrate excessively in service.

Compression Members

A column is a member subjected to axial compression stresses. Usually such members are also subject to some bending stresses but often can be safely designed as concentrically loaded columns. Although short columns may be stressed to their yield point, buckling occurs before the yield point is reached in most columns.

A **perfect column** is made of isotropic material, free of residual stresses, perfectly straight, loaded precisely at its centroid, and of uniform cross section. It is also assumed that the small deflection theory of ordinary bending is applicable, that shear may be neglected, and that twisting or distortion of the cross section does not occur during buckling.

If in addition, the column material has a linear stress-strain curve as in Fig 409-7a, the column starts to buckle at a stress predicted by Leonard Euler in 1757. He assumed that the member was slightly bent and then solved the differential equation for equilibrium. The **Euler equation** for a pinned-end column is:

Eq 409-9.

$$F_c = \pi^2 E/(KL/r)^2$$

F_c = stress at which buckling occurs
E = modulus of elasticity
K = effective length coefficient
L = column length between supports
r = radius of gyration $(I/A)^{1/2}$

In a perfect column, elastic buckling occurs when the Euler buckling stress is less than the yield stress. The relationship between the buckling stress and the effective slenderness ratio, KL/r, is shown in Fig 409-7b. The slenderness ratio at which buckling stress equals the yield stress is:

Eq 409-10.

$$KL/r = (\pi^2 E/F_y)^{1/2} = C'_c$$

At slightly lower slenderness ratios, the column usually fails by buckling after part of the cross section yields, which is inelastic buckling. At even lower slenderness ratios, yielding rather than buckling usually occurs.

Tangent Modulus

Euler's approach was generally ignored for design because columns of ordinary lengths were not as strong as his equation predicted. In 1889, Engresser realized that portions of those columns became inelastic before buckling and that Euler's E should be the slope of the stress-strain curve at the buckling stress. If the perfect column is of a material with a rounded portion in its stress-strain curve, Fig 409-7c, the buckling stress in the elastic range is given by the **Engresser equation:**

Eq 409-11.

$$F_c = \pi^2 E_t/(KL/r)^2$$

E_t, the tangent modulus, is the slope at stress F_c, and decreases with increasing stress. The buckling stress from Euler and Engresser equations is the same at the proportional limit, where $E_t = E$. The tangent modulus, while not useful in design because real columns are not perfect, is important in determining the strength of real columns.

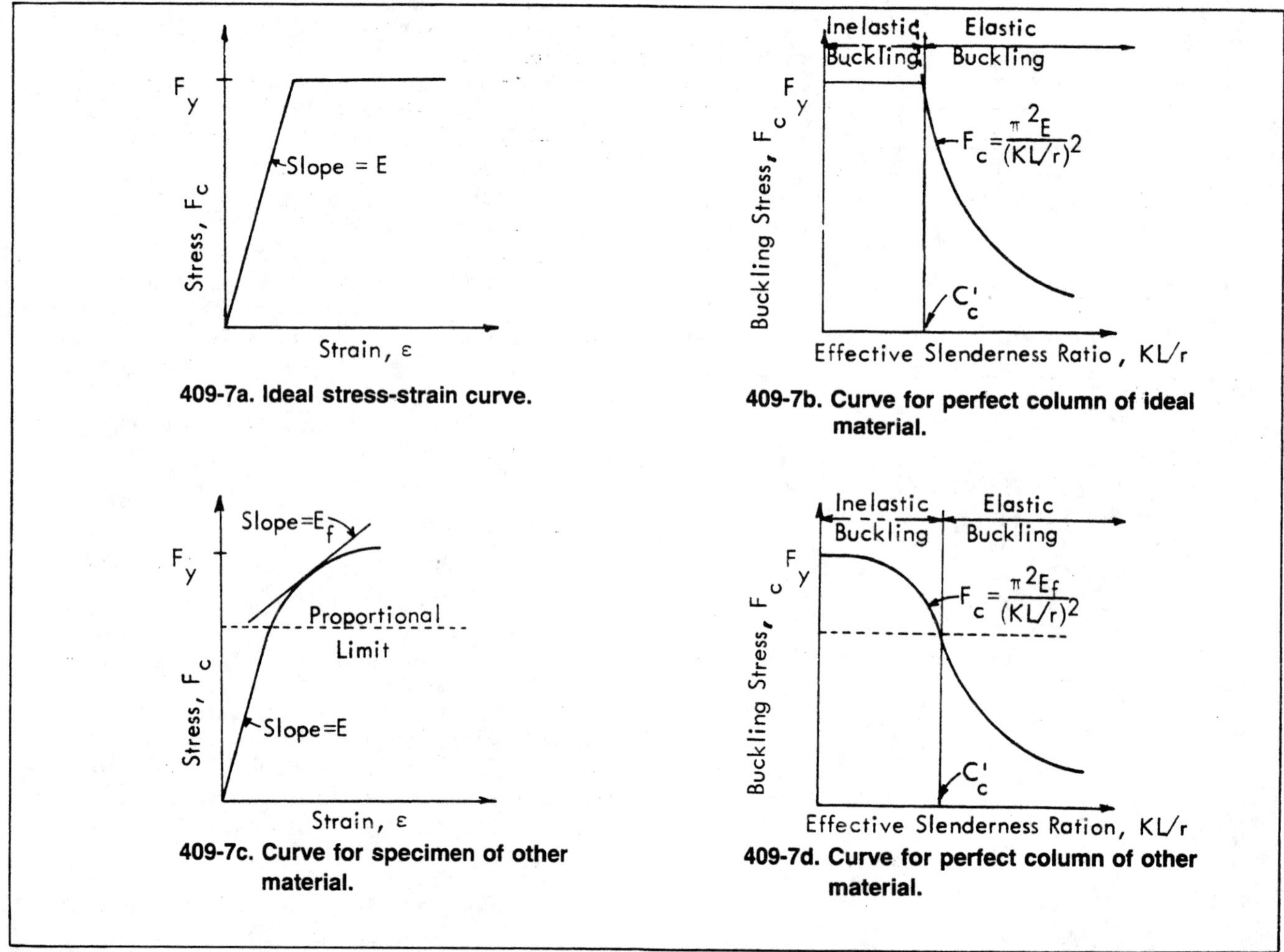

409-7a. Ideal stress-strain curve.

409-7b. Curve for perfect column of ideal material.

409-7c. Curve for specimen of other material.

409-7d. Curve for perfect column of other material.

Fig 409-7. Stress-strain and column curves.

Another valuable concept is the double modulus theory. With a perfect column, a stress-strain diagram as in Fig 409-7c, and a small deflection as Euler assumed, compressive stress on one side of the column is higher than on the other. As buckling proceeds, the stress difference increases rapidly. With a curved stress-strain diagram, E is lower on the higher stressed side and higher on the lower stressed side, so there are two tangent moduli. The tangent modulus, with appropriate safety factor, is used in design.

Residual Stress

Residual stresses are caused by differential rates of cooling in formed cross sections. Differential cooling induces large compressive stresses in parts of the section before any external load is applied. The maximum residual stresses, which average 12 ksi and are as high as 20 ksi, occur near the edges of the flanges and near the center of the webs, Fig 409-8a. The stress-strain curve, Fig 409-8b, and cross sections, Fig 409-8c, illustrate residual stress effect on the stress-strain behavior of the section. Because column buckling strength depends on E, and because residual stresses significantly alter the stress-strain curve, residual stresses also alter the strength of real columns. The AISC specification alters the Euler and tangent modulus equations for perfect columns to account for the discrepancy.

The Column Research Council (CRC) investigated the effect of residual stresses on column strength, Fig 409-9. The curves labeled "x-x" and "y-y" are observed column behavior. The "Basic Strength Curve" is the CRC strength curve for actual columns. It is slightly conservative for strong axis buckling but liberal for weak axis buckling in the inelastic range, which is more than compensated for by the specified safety factors. Note that the CRC strength curve is empirical and adequately defines column strength in the inelastic range.

Design Specifications

No Local Buckling

Most standard structural steel sections develop their yield stress before buckling locally. If the width-thickness ratios, b/t, of the compression elements of a section meet the following requirements, they yield before buckling locally. The width, b, is the full nominal dimension for plates, angles, channels, and Zs, and is ½ the width for I, H, and T flanges. The thickness, t, is the average flange thickness.

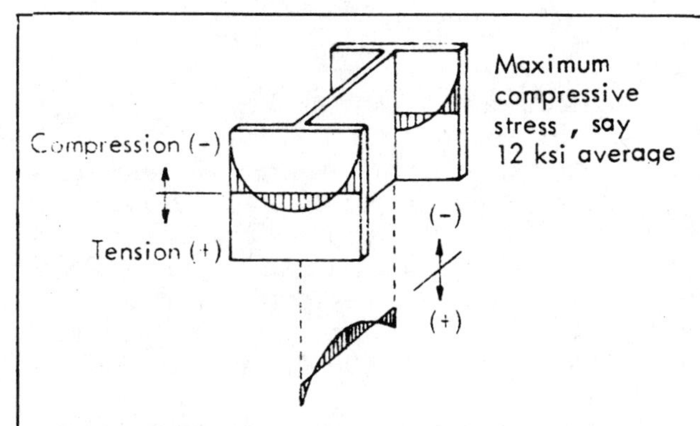

409-8a. Typical residual stress pattern on rolled shapes.

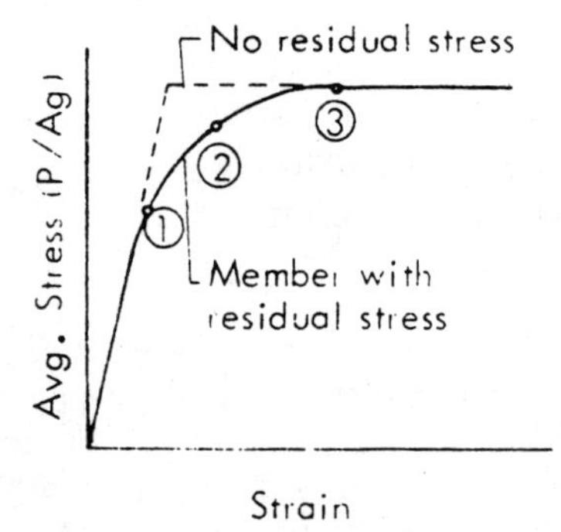

409-8b. Stress-strain curve for a real column.

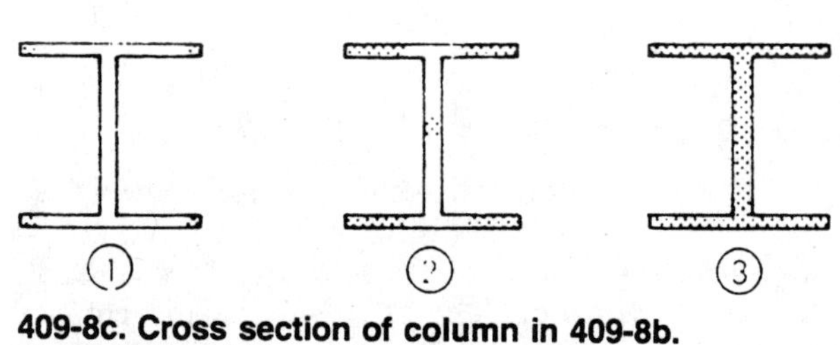

409-8c. Cross section of column in 409-8b.
Shaded portions yielded at indicated stress levels.

Fig 409-8. Residual stress in rolled shapes.

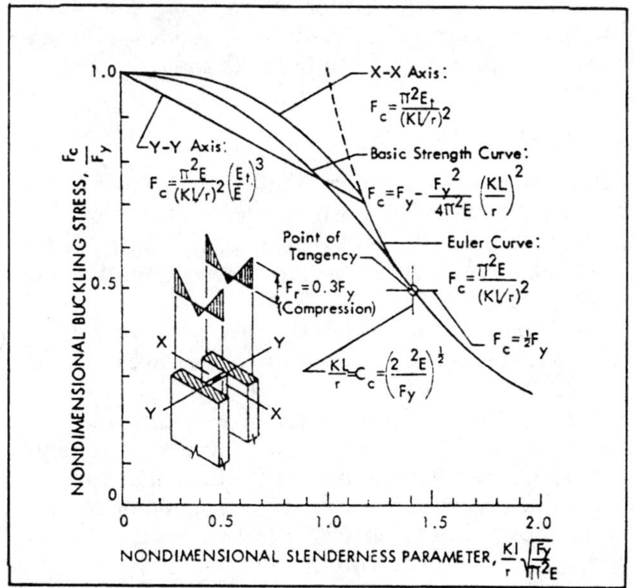

Fig 409-9. Effect of residual stress on column strength.

- Unstiffened Compression Elements (UCE)
  - Single and double angle struts with separators. $b/t < 76/F_y^{1/2}$
  - Double struts in contact; angles or plates projecting from girders, columns, or other compression members; and compression flanges of beam: $b/t < 95/F_y^{1/2}$
  - Stems of tees: $b/t < 127/F_y^{1/2}$
- Stiffened Compression Elements (SCE)
  - Square or rectangular box section: $b/t < 238/F_y^{1/2}$
  - Unsupported width of perforated cover plates: $b/t < 317/F_y^{1/2}$
  - Other uniformly compressed elements: $b/t < 253/F_y^{1/2}$
- Circular tubular
  - Outside diam./wall thickness $< 3300/F_y$

In the elastic buckling range, the column strength is still predicted by the Euler equation. Fig 409-9 shows a smooth transition between the CRC and Euler equations. The CRC equation was written to be tangent to the Euler curve when the buckling strength equals ½ the yield stress; i.e., when $F_c = P/A_{gross} = 0.5\ F_y$. Find the limiting slenderness ratio for elastic buckling by equating the CRC buckling strength equaton to $0.5\ F_y$:

Eq 409-12.

$$F_c = F_y - [F_y^2 \div (4\pi^2 E)][KL/r]^2 = 0.5\ F_y$$

Solve for KL/r:

$$(KL/r)_{lim} = (2\pi^2\ E/F_y)^{1/2} = C_c$$

The limiting slenderness ratio is C_c. If $KL/r > C_c$, the column buckles elastically and Euler's equation governs the design. If $KL/r < C_c$, the column buckles inelastically and the CRC equation predicts the column strength.

The AISC specification for columns has the CRC and Euler equations with appropriate safety factors. Specifically, for main structural members, and secondary members with $KL/r < 120$:

Eq 409-13.

If $KL/r < C_c$:

$$F_a = F_y[1-(KL/r)^2 \div 2C_c^2] \div [(5 \div 3 + 3(KL/r) \div 8C_c - (KL/r)^3 \div 8C_c^3]$$

If $KL/r > C_c$:

$$F_a = (12/23)[(\pi^2\ E)/(KL/r)^2]$$

For secondary members such as those used for bracing:

Eq 409-14.

If $KL/r > 120$:

$$F_{as} = Fa/(1.6 - L/200r)$$

where F_a is computed as for main members.

The safety factors for short columns increase from 1.67 (for stiff columns, the same as for tension members) to 1.92 as KL/r increases from 0 to C_c. Above $KL/r = C_c$, the safety factor is constant at 1.92.

Effective Slenderness Ratios

End restraint influences the buckling strength of columns. With other factors equal, a column with two fixed ends buckles at a theoretical load four times as large as a column with two pinned ends. The effective length factor, K, Fig 409-10, accounts for end restraint.

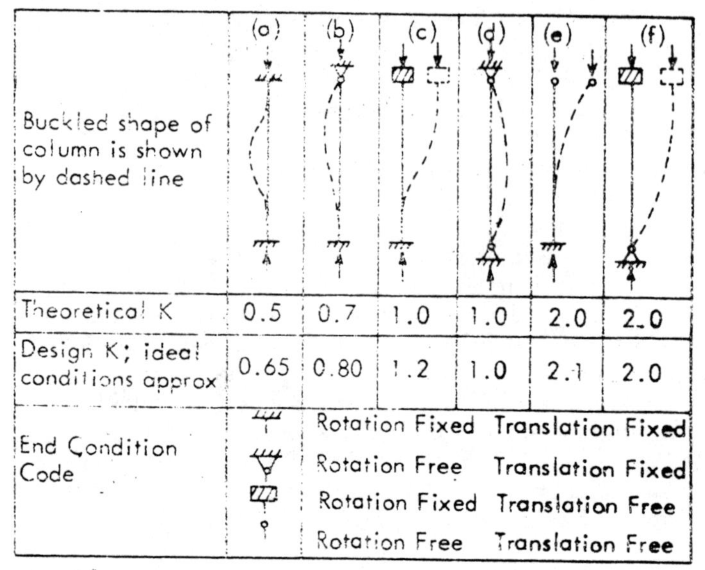

Fig 409-10. Effective length factors for columns.
From AISC.

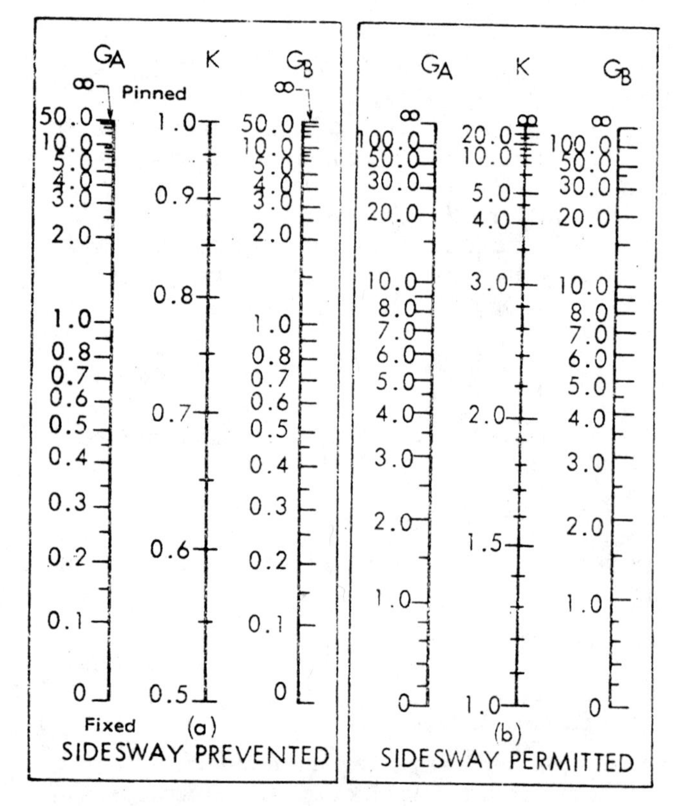

Fig 409-11. Effective lengths of columns.
From AISC. $G_{fixed} = 0$, $G_{pinned} =$ infinity.

K cannot always be estimated by the tabulated conditions. For example, in a rigid frame, the connection between the column top and the cross members is neither fixed, pinned, or free. Instead, it is partially restrained by the horizontal or inclined member. The degree of restraint depends on the stiffness of the column relative to the stiffness of the other members at the joint. For K of columns in frames, with and without sidesway, see Fig 409-11, the Jackson-Moreland charts.

In those charts, G is the ratio of the rigidity of the columns to the rigidity of the other members at the joint:

Eq 409-15.

$$G = \text{sum}(I_c/L_c) \div \text{sum}(I_g/L_g)$$

I = moment of inertia
L = member length
c = columns
g = girders

In Fig 409-11, the subscripts A and B refer to column ends. Two special cases are important. If one column end is supported but not rigidly connected to a footing, the ideal end condition is a pin and G is infinite. Practical design procedures use G = 10. If one column end is rigidly attached to a footing, the ideal end condition is fixed and G = 0. Use G =1.0.

Design Aids

Column Load Tables

Column load tables are included for selected W, M, pipe, and structural tubing of 36 ksi yield stress steel. They are typical of those found in the AISC *Steel Design Manual.* The tabulated loads comply with the AISC Specification.

Column Stress Tables

The allowable and Euler stresses for A36 steel columns are given in Table 409-7 for both main and secondary members and comply with the AISC Specification.

Columns

Tabular loads were computed for axially loaded members with the effective unsupported lengths, KL, indicated. KL is the actual unbraced length, L ft, multiplied by K, which depends on the column's end restraint, Fig 409-10.

Heavy horizontal lines in the tables indicate KL/r = 120; r is the radius of gyration. Loads are omitted when KL/r exceeds 200.

The W and M shape columns are assumed to fail around the y-axis. If a column is braced along the y-axis, the allowable load for x-axis failure is the tabulated value times r_x/r_y from the properties at the bottom of each table. Check for failure around the y-axis for the shortened column length.

Table 409-3. Axial column loads for W & M shapes.
F_y = 36 ksi.
Underlines indicate KL/r_y = 120. KL listed is for the least radius of gyration, r_y.

Name	W8				W6						W5		W4	M6	M5	M4
lb/ft	35	31	28	24	25	20	15	16	12	9	19	16	13	20	18.9	13
KL	kips															
0	222	197	178	153	159	127	96	102	77	58	120	101	83	127	120	82
6	201	178	155	133	136	109	81	96	72	54	115	97	78	122	114	78
7	197	174	150	129	131	105	78	92	68	51	111	94	75	119	111	74
8	191	170	144	124	126	100	75	87	64	48	107	91	71	116	106	70
9	186	165	138	118	120	95	71	82	60	45	103	87	67	112	102	65
10	180	160	132	113	114	90	67	76	55	41	99	83	62	107	96	60
11	174	154	125	107	107	85	62	69	50	37	93	79	57	103	91	54
12	168	149	118	101	100	79	58	62	44	33	88	74	52	98	85	48
13	162	143	111	95	93	73	53	54	38	28	82	69	46	92	78	42
14	155	137	103	88	85	67	48	46	31	23	76	64	39	87	71	35
15	148	131	95	81	77	60	43	38	26	19	70	58	33	81	64	29
16	141	124	87	74	69	54	38	32	22	16	63	52	28	74	56	24
17	133	117	78	66	61	47	33	27	18	13	55	46	24	68	48	21
18	125	110	69	59	54	42	30	23	16	12	48	40	20	61	42	18
19	117	103	62	53	49	38	27	20	14	10	42	35	18	53	36	15
20	109	95	56	48	44	34	24	18			37	31	16	47	32	
A in²	10.3	9.13	8.25	7.08	7.34	5.87	4.43	4.74	3.55	2.68	5.54	4.68	3.83	5.89	5.55	3.81
I_x in⁴	127	110	98.0	82.8	53.4	41.4	29.1	32.1	22.1	16.4	26.2	21.3	11.3	39.0	24.1	10.5
I_y in⁴	42.6	37.1	21.7	18.3	17.1	13.3	9.32	4.43	2.99	2.20	9.13	7.51	3.86	11.6	7.86	3.36
r_y in	2.03	2.02	1.62	1.61	1.52	1.50	1.45	0.966	0.918	0.905	1.28	1.27	1.00	1.40	1.19	0.939
r_x/r_y	1.73	1.72	2.13	2.12	1.78	1.77	1.77	2.69	2.71	2.73	1.70	1.68	1.72	1.84	1.75	1.77

From AISC.

Table 409-4. Axial column load in kips, extra strong pipe.
F_y = 36 ksi.

Diameter	6	5	4	3½	3
Thickness	.432	.375	.337	.318	.300
lb/ft	28.5	20.7	14.9	12.5	10.2
Effective length, KL, in feet with respect to radius of gyration					
6	166	118	81	66	52
7	162	114	78	63	48
8	159	111	75	59	45
9	155	107	71	55	41
10	151	103	67	51	37
11	146	99	63	47	33
12	142	95	59	43	28
13	137	91	54	38	24
14	132	86	49	33	21
15	127	81	44	29	18
16	122	76	39	25	16
17	116	71	34	23	14
18	111	65	31	20	12
19	105	59	28	18	11
20	99	54	25	16	
Properties					
Area A, in²	8.40	6.11	4.41	3.68	3.02
I, in⁴	40.5	20.7	9.61	6.28	3.89
r, in.	2.19	1.84	1.48	1.31	1.14

From AISC.

Table 409-5. Axial column load in kips, standard pipe.
F_y = 36 ksi.

Diameter	6	5	4	3½	3
Thickness	.280	.258	.237	.226	.216
lb/ft	18.9	14.6	10.7	9.1	7.5
Effective length, KL, in feet with respect to radius of gyration					
6	110	83	59	48	38
7	108	81	57	46	36
8	106	78	54	44	34
9	103	76	52	41	31
10	101	73	49	38	28
11	98	71	46	35	25
12	95	68	43	32	22
13	92	65	40	29	19
14	89	61	36	25	16
15	86	58	33	22	14
16	82	55	29	19	12
17	79	51	26	17	11
18	75	47	23	15	10
19	71	43	21	14	9
20	67	39	19	12	
Properties					
Area A, in²	5.58	4.30	3.17	2.68	2.23
I, in⁴	28.1	15.2	7.23	4.79	3.02
r, in.	2.25	1.88	1.51	1.34	1.16

From AISC.

Example 409-3:

Select the lightest weight A36 W or M section for an axial compressive load of 100,000 lb on a 10′ column.

- Weak axis: fixed both ends, braced at the third point.
- Strong axis: fixed at the bottom, pinned at the top with translation prevented.

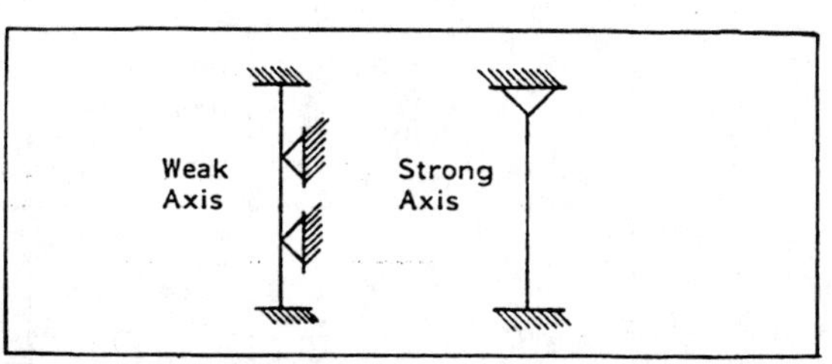

Fig 409-12. Weak and strong axis, Example 409-3.

Table 409-6. Axial load in kips, square structural tubing.
F_y = 46 ksi.

F_y = 46 ksi

COLUMNS
Square structural tubing
Allowable concentric loads in kips

Nominal Size		4 x 4					3 x 3		
Thickness		½	⅜	5/16	¼	3/16	5/16	¼	3/16
Wt./ft.		21.63	17.27	14.83	12.21	9.42	10.58	8.81	6.87
F_y		46 ksi							
Effective length in feet KL with respect to radius of gyration	0	176	140	120	99	76	86	71	56
	2	168	134	115	95	73	80	67	53
	3	162	130	112	92	71	77	64	50
	4	156	126	108	89	69	73	61	48
	5	150	121	104	86	67	68	57	45
	6	143	115	100	83	64	63	53	42
	7	135	110	95	79	61	57	49	39
	8	126	103	90	75	58	51	44	35
	9	117	97	84	70	55	44	38	31
	10	108	89	78	65	51	37	33	27
	11	98	82	72	60	47	31	27	22
	12	87	74	65	55	43	26	23	19
	13	75	65	58	49	39	22	19	16
	14	65	57	51	43	35	19	17	14
	15	57	49	44	38	30	16	15	12
Properties									
A (in.^{2})		6.36	5.08	4.36	3.59	2.77	3.11	2.59	2.02
I (in.^{4})		12.3	10.7	9.58	8.22	6.59	3.58	3.16	2.60
r (in.)		1.39	1.45	1.48	1.51	1.54	1.07	1.10	1.13

Nominal Size		6 x 6					5 x 5				
Thickness		½	⅜	5/16	¼	3/16	½	⅜	5/16	¼	3/16
Wt./ft.		35.24	27.48	23.34	19.02	14.53	28.43	22.37	19.08	15.62	11.97
F_y		46 ksi									
Effective length in feet KL with respect to radius of gyration	0	287	223	189	154	118	231	182	155	127	97
	6	257	201	171	140	107	200	159	136	111	86
	7	251	196	167	137	105	193	153	131	108	83
	8	244	191	163	133	102	186	148	127	104	80
	9	237	186	158	130	99	178	142	122	100	77
	10	229	180	154	126	96	169	135	116	96	74
	11	221	174	149	122	93	160	129	111	92	71
	12	212	168	143	117	90	151	122	105	87	67
	13	203	161	138	113	87	141	115	99	82	64
	14	194	154	132	108	83	131	107	93	77	60
	15	185	147	126	104	80	120	99	86	72	56
	16	175	140	120	99	76	109	90	79	66	52
	17	164	132	113	94	72	97	82	72	60	47
	18	153	124	107	88	68	87	73	64	54	43
	19	142	115	100	83	64	78	65	58	49	39
	20	131	107	93	77	60	70	59	52	44	35
Properties											
A (in.^{2})		10.40	8.08	6.86	5.59	4.27	8.36	6.58	5.61	4.59	3.52
I (in.^{4})		50.5	41.6	36.3	30.3	23.8	27.0	22.8	20.1	16.9	13.4
r (in.)		2.21	2.27	2.30	2.33	2.36	1.80	1.86	1.89	1.92	1.95

From AISC.

Table 409-7. Allowable stress for A36 compression members.

Main and Secondary Members Kl/r not over 120						Main Members Kl/r 121 to 200				Secondary Members* l/r 121 to 200			
$\frac{Kl}{r}$	F_a (ksi)	$\frac{Kl}{r}$	F_a (ksi)	$\frac{Kl}{r}$	F_a (ksi)	$\frac{Kl}{r}$	F_a (ksi)	$\frac{Kl}{r}$	F_a (ksi)	$\frac{l}{r}$	F_{as} (ksi)	$\frac{l}{r}$	F_{as} (ksi)
5	21.39	45	18.78	85	14.79	125	9.55	165	5.49	125	9.80	165	7.08
10	21.16	50	18.35	90	14.20	130	8.84	170	5.17	130	9.30	170	6.89
15	20.89	55	17.90	95	13.60	135	8.19	175	4.88	135	8.86	175	6.73
20	20.60	60	17.43	100	12.98	140	7.62	180	4.61	140	8.47	180	6.58
25	20.28	65	16.94	105	12.33	145	7.10	185	4.36	145	8.12	185	6.46
30	19.94	70	16.43	110	11.67	150	6.64	190	4.14	150	7.81	190	6.36
35	19.58	75	15.90	115	10.99	155	6.22	195	3.93	155	7.53	195	6.28
40	19.19	80	15.36	120	10.28	160	5.83	200	3.73	160	7.29	200	6.22

*K taken as 1.0 for secondary members.

From AISC.

Solution:

a. The effective lengths for the weak axis end sections:
$(KL)_y = 0.625(10 \div 3) = 2.08'$ (weak axis)
K from Fig 409-11 using G = 0 and 1.0, no sidesway
or for the middle section:
$(KL)_y = 0.78(10 \div 3) = 2.60'$ (weak axis)
(K from Fig 409-11 using G = 1.0 and 1.0, no sidesway.)
$(KL)_x = 0.7(10) = 7.0'$ (strong axis)
(K from Fig 409-11 using G = 0 and ∞, no sidesway.)

b. Whether weak or strong axis bending controls the design depends on the relative KL/r values.
$(KL)_x/(KL)_y = 7.0/2.60 = 2.69$
and r_x/r_y is less than 2.69 for many of the W and M sections shown. So, it appears that strong axis buckling controls.

c. Assuming a value for r_x/r_y of 2.0, enter the column load tables with:
$KL_y = (KL)_x/(r_x/r_y) = 7.0 \div 2 = 3.5'$
and P = 100 kips
A M5 x 18.9 is adequate, but $r_x/r_y = 1.75 < 2.0$

d. To check the section, enter the load tables with:
$KL = (7.0)/1.75 = 4.0'$
and P = 100 kips
The M5 x 18.9 section can carry 100 kips when KL = 4.0′.

e. To select a column using the stress tables, first assume a size column and then check the adequacy. Assuming an M5 x 18.9:
$(KL/r)_y = 2.60 \times 12/1.19 = 26.22$
$(KL/r)_x = 7 \times 12/(1.75 \times 1.19) = 40.34$
The X-axis controls.
From tables of allowable stresses:
$F_a = 19.16$ ksi, and
$P_{all} = F_a A = 19.16(5.55)$
$= 106.3$ kips > 100 kips

f. To select a column using the AISC equation, first assume a column, say M5 x 18.9, then:
$C_c = (2\pi^2 E/F_y)^{1/2}$
$= (2\pi^2(29 \times 10^6)/36)^{1/2} = 126.1$
$(KL/r)_{max} = 40.34 < C_c$

Thus, inelastic buckling occurs and:
$F_a = [1 - (40.34^2/2(126.1)^2]36/[(5/3) + 3(40.34)/8(126.1) - (40.34)^3/8(126.1)^3] = 19.16$ ksi
$P_{all} = F_a A = (19.16 \text{ ksi})(5.55 \text{ in}^2) = 106.3$ k

Flexural Members

The basic equation for flexure in steel beams is $M_{all} < F_b S$. Adjust the allowable bending stress, F, according to AISC to allow for discrepencies between theoretical and actual flexural behavior. F_b depends primarily on local buckling, lateral bracing, and the shape factor and type of section.

A **compact section** is proportioned so the cross section reaches its full plastic moment, M_p, before either local buckling of the elements or lateral torsional buckling occurs. A **non-compact** cross section has either local or lateral torsional buckling before reaching the plastic moment.

The following discussion of allowable stresses applies only to steels with yield stresses in the 36 to 50 ksi range and to beams built-up of only one steel grade.

Allowable Flexural Stresses

Compact Sections

The allowable flexural stress for a compact section that is also symmetrical about and loaded in the plane of its minor axis, is: $F_b = 0.66 F_y$. The requirements for compactness are moderately severe. Most standard sections of A36 steel, however, do meet the compact section requirements if the beam is adequately braced against torsional buckling.

A section is compact if:

1. The flanges are continuously connected to the webs.
2. The flange width-to-thickness ratio of UCEs is:

Eq 409-16.

$b_f/2t_f < 65/F_y^{1/2}$ (= 10.8 for A36)

3. The flange width-to-thickness ratio of SCEs is:

Eq 409-17.

$b_f/t_f < 190/F_y^{1/2}$ (= 31.7 for A36)

4. The depth-to-thickness ratio of the webs is:

Eq 409-18.

If $f_a/F_y < 0.16$:
$d/t_w < (640(1\text{-}3.74\ f_a/F_y)/F_y^{1/2}$ (= 106.7 for A36)
If $f_a/F_y > 0.16$:
$d/t_w < 257/F_y^{1/2}$ (= 42.8 for A36)
where f_a = computed axial stress in addition to bending stress

5. The compression flange of the beam is laterally supported at intervals, L_b, less than either L_c or L_u.

Eq 409-19.

$L_c = (76\ b_f/F_y^{1/2})$
If box or rectangular shaped members with depth < 6 x width and $t_f < 2t_w$:
$L_u = 20{,}000/[(d/A_f)(F_y)]$
If circular diameter/wall thickness < $3300/F_y$:
$L_u = (1950 + 1200\ M_1/M_2)b/F_y$

Condition (1.) assures continuity between the elements of the section. The other conditions assure reaching the plastic moment before flange buckling (2. and 3.), local web buckling (4.), or lateral buckling of braced beams (5.).

Nearly Compact Sections

If a cross section is compact **except** that:

Eq 409-20.

$65/F_y^{1/2} < b_f/2t_f < 95/F_y^{1/2}$,
then $F_b = [0.79\text{-}0.002(b_f/2t_f)\ F_y^{1/2}]F_y$

This equation decreases F_b from 0.66 F_y to 0.60 F_y as the UCE becomes more slender, avoiding an abrupt reduction in allowable stress when $b_f/2t_f$ exceeds $65/F_y^{1/2}$.

Minor Axis Bending and Solid Sections

Tension and compression on extreme fibers of doubly-symmetrical I and H members meeting compact requirements (1.) and (2.), and bent about their minor axis; solid round and square bars; and solid rectangular sections bent about their weaker axis is:

Eq 409-21.

$F_b = 0.75F_y$

Doubly-symmetrical I and H members bent about their minor axis and nearly compact as described above may be designed on the basis of an allowable bending stress:

Eq 409-22.

$F_b = F_y[1.075 - 0.005\ (b_f/2t_f)\ F_y^{1/2}]\ (1.5\text{-}5.b)$

Rectangular tubular sections meeting compact requirements (1.), (3.), and (4.) and bent about their minor axis may be designed on the basis of an allowable bending stress:

Eq 409-23.

$F_b = 0.66F_y$

Non-Compact Sections Because $L_b > L_c$ and/or $C_b.L_u$

If a section is compact except $L_b > L_c$, but less than C_bL_u, then $F_b = 0.60\ F_y$. The beams buckle laterally before reaching the plastic moment but not before reaching the yield moment.

If a non-compact section has adequate bracing, $L_b < L_c = 76/F_y^{1/2}$; if UCEs width-thickness ratio is less than $95.0/F_y^{1/2}$; and if SCEs width-thickness ratio is less than $253/F_y^{1/2}$; then $F_b = 0.60\ F_y$. The beams will not buckle locally or laterally before reaching the yield moment.

If a section is non-compact and inadequately braced because L_b is greater than the larger of L_c or C_bL_u, then F_b is less than 0.6 F_y because of lateral torsional buckling. To determine allowable stress:

1. Evaluate F_b for Euler-type buckling.
2. Evaluate F_b for St. Venant-type buckling.
3. Use the larger F_b, as it is unlikely that both Euler- and St. Venant-type buckling will occur simultaneously.

Euler buckling (warping) is a translation of the compression flange in the x-direction and twisting about the y-axis, Fig 409-13. Warping is resisted by the material stiffness and slenderness ratio of the flange plus ⅙ of the web area with respect to the y-axis, Fig 409-14. Torsional (St. Venant) buckling is

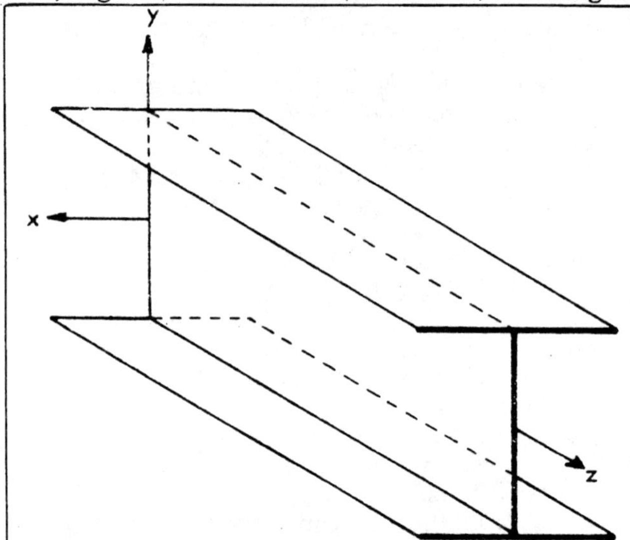

Fig 409-13. Axis orientation for a W-section.

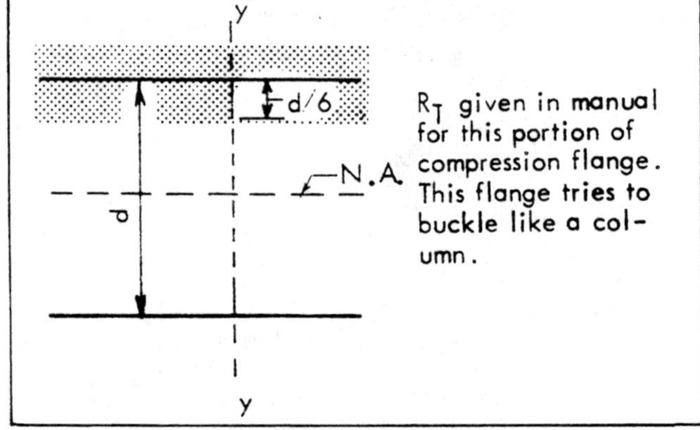

Fig 409-14. Definition sketch for R_T.

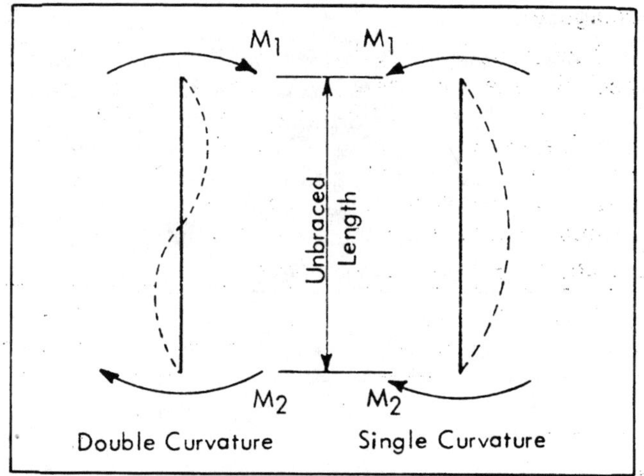

Fig 409-15. Definition sketch for C_b.

twisting about the z-axis. St. Venant buckling is resisted by torsional rigidity and the polar moment of inertia of the section.

Inelastic Euler-type buckling occurs at:

Eq 409-24.

$$F_b = [2/3 - (F_y(L_b/r_T)^2/1530 \times 10^3 C_b)]F_y$$

and occurs if:

$$(102{,}000\ C_b/F_y)^{1/2} < L_b/r_T < (510{,}000\ C_b/F_y)^{1/2}$$

where L_b/r_T is the slenderness ratio of the compression flange plus ⅓ the compression web area about the y-axis, Fig 409-14.

Elastic Euler-type buckling starts at:

Eq 409-25.

$$F_b = 170{,}000\ C_b/(L_b/r_T)^2$$

and occurs if: $L_b/r_T > (510{,}000\ C_b/F_y)^{1/2}$

If the compression area of a section is approximately rectangular, and if the tension and compression areas are nearly equal, torsional buckling will start at:

Eq 409-26.

$$F_b = 12 \times 10^3 C_b/[(d/A_f)L_b]$$

The allowable stress in bending for non-compact, inadequately braced sections is the larger of the values from Eqs 409-24, 409-25, or 409-26 but not larger than 0.60 F_y.

Comments

To clarify the various lengths that define the limits for lateral buckling:

1. If $L_b < L_c$ and the section is compact, the section can reach its plastic moment before lateral buckling; $F_b = 0.66\ F_y$.
2. If $L_c < L_b < C_b L_u$, the section can reach its yield moment before local buckling; $F_b = 0.60\ F_y$.
3. If $L_b > C_b L_u$, $F_b < 0.60\ F_y$.

C_b compensates for moment reversals (double curvature) between bracing points. A severe example is resisting lateral buckling under constant moment, because the full length of the compressive flange is subjected to the maximum stress.

But if the moments at the ends of the braced section are reversed, then only one point along the beam length is under maximum compression stress, and the tendency to buckle laterally is reduced. C_b is defined by:

Eq 409-27.

$$C_b = 1.75 + 1.05\ (M_1/M_2) + 0.3\ (M_1/M_2)^2 < 2.3$$

where (Fig 409-15):

$M_1 < M_2$, the moments at the ends of a braced section.

M_1/M_2 is negative when bent in single curvature.

M_1/M_2 is positive when bent in double curvature.

Note that C_b cannot exceed 2.3. $C_b = 1.0$ if the absolute value of an interior moment is greater than the absolute value of both M_1 and M_2.

The allowable stress for compact sections is 10% higher than for non-compact ones, because plastic moment develops in compact sections before local or torsion buckling. The 10% increase reflects that the plastic moment of typical structural shapes is about 1.12 times the yield moment.

Example 409-4:

Determine the allowable bending stress for an A36 W8 x 28 beam for 6; 10′, 15′, 20′ and 30′ spans laterally supported at the ends only.

The section properties from Table 409-15 for the section are:

r_t = 1.77″
d/A_f = 2.65
b_f = 6.535″
t_f = .465″
t_w = .285″
d = 8.06″

Checking compactness:

$b_f/2t_f = 7.0 < 10.83 = (b/2t)_{lim} = 65/(F_y)^{1/2}$
$d/t_w = 28.3 < 106.7 = (d/t)_{lim} = 640/(F_y)^{1/2}$
$L_c = 76 b_f/(F_y)^{1/2} = (76)(6.535)/36^{1/2}$
$= 82.78''$ or $6.90'$

If $L_b < 6.9'$, the section is compact.

$L_u = 20{,}000/(d/A_f)(F_y) = 20{,}000/(2.65)(36)$
$= 210'' = 17.5'$

a. If $L_b = 6' < L_c$,
then $F_b = 0.66\ F_y = 24$ ksi.

b. If $L_b = 10'$,
then $L_c < L_b < L_u$;
and $F_b = 0.6\ F_y = 22$ ksi.

c. If $L_b = 15'$,
then $L_c < L_b < L_u$;
and $F_b = 22$ ksi.

d. If $L_b = 20'$,
$L_b > L_u$; $F_b < 0.6\ F_y$
$L_b/r_t = 20 \times 12/1.77 = 135.6$
$(102{,}000\ C_b/F_y)^{1/2}$
$= (102{,}000 \times 1.0/36)^{1/2} = 53.2$
$(510{,}000\ C_b/F_y)^{1/2}$
$= (510{,}000 \times 1.0/36)^{1/2} = 119.0$
Because $L_b/r_b > 119$

$F_b = 170{,}000\ C_b/(L_b/r_t)^2$
$= 170{,}000 \times 1.0/(135.6)^2 = 9.25$ ksi.
Or $F_b = 12 \times 10^3 \times C_b/(d/A_f)(L_b)$
$= (12{,}0000 \times 1.0)/[2.65 \times (20 \times 12)]$
$= 18.87$ ksi > 9.25
Therefore, use $F_b = 18.87$ ksi.

e. If $L_b = 30' > L_u$, $F_b < 0.6\ F_y$
$(L_b/r_t) = 30 \times 12/1.77 = 203.4 > 119$
Therefore:
$F_b = 170{,}000 \times 1.0/(203.4)^2 = 4.11$ ksi
or $F_b = 12{,}000/[2.65 \times (30 \times 12)] = 12.58$ ksi
Therefore, use $F_b = 12.58$ ksi.

Shear Stresses

The allowable shear stress for a flexural member without web stiffeners between bearing points is $F_v = 0.4\ F_y$. The beam shear stress is:

Eq 409-28.

$$v = VQ/I_t$$

However, for standard structural shapes, approximate shear stress is calculated by the average shear stress over the web area, $d \times t_w$:

Eq 409-29.

$$v = V/(dt_w)$$

The AISC Specification requires that $v < F_v$.

If the depth-thickness ratio, d/t_w, of the web is lower than 62.6 for $F_y = 36$ ksi steel and 53.7 for $F_y = 50$ ksi steel, the web does not require web stiffeners. Most applications in agricultural structures meet this requirement. But web stiffeners are required in beams with very slender webs so the webs yield at $v = 0.4\ F_y$ before buckling.

Bearing Stresses

Web crippling occurs at end reactions or concentrated loads unless web stiffeners or properly sized bearing areas are provided. In agricultural applications, common loads and beam sizes do not require web stiffeners even at concentrated loads. But bearing plate dimensions must be:

• At interior concentrated loads.

Eq 409-30.

$$R/(t_w(N+2K)) < 0.75\ F_y$$

• At end-reactions.

Eq 409-31.

$$R/t_w(N+k) < 0.75\ F_y$$

R = concentrated loads
t_w = web thickness
N = length of bearing
k = distance from the outer face of flange to web toe of fillet

Deflection

Ordinary methods of elastic analysis are acceptable for evaluating deflections in steel beams. Allowable deflections are in building or structural codes.

Design Aids

Numerous design aids are available for structural steel. Steel fabricators are a source of design aids as in the AISC. The AISC *Manual of Steel Construction* contains section properties, allowable loads, tables, allowable moment tables, and other helpful data to assist the steel designer. Design aids for some sizes commonly used in agriculture are included here.

Table 409-8. Equivalent uniform loads.

Type of Loading: Equal Loads, Equal Spaces	Equivalent Uniform Load	Deflection Coefficient
P at center; L/2, L/2	2.00 P	0.80
P, P; L/3, L/3, L/3	2.67 P	1.02
P, P, P; L/4, L/4, L/4, L/4	4.00 P	0.95
P, P, P, P; L/5, L/5, L/5, L/5, L/5	4.80 P	1.01

Uniform Load Beam Tables

Table 409-11 contains uniform load constants to aid the designer in selecting steel beams. The assumptions made in developing the table are:

1. $L_b < L_c$.
2. Beam weight is included; subtract beam weight for net load.
3. The load is applied normal to the x-x axis.
4. $F_y = 36$ ksi.
5. Simple beam.

The terms are defined as follows:

W_c = uniform load constant, k-ft
where w = load/ft, L = beam span, ft
$= 2S_xF_b/3 > wL^2/2$

V = maximum web shear, kips
$= 0.4F_y dt_w > wL/2$

L_v = span length below which shear V in beam web governs, as compared to greater span lengths where flexure governs, ft

L_c = maximum unbraced length of the compression flange at which the allowable bending stress may be taken at $0.66\ F_y$ or as determined formulas for near compact members, ft

L_u = maximum unbraced length of the compression flange at which the allowable bending stress may be taken at $0.6F_y$, ft

L_b = unbraced length of compression flange, ft

W SHAPES

Designation	Area A	Depth d		Web Thickness t_w		$\frac{t_w}{2}$	Flange Width b_f		Flange Thickness t_f		Distance T	Distance k	Distance k_1
	In.²	In.		In.		In.	In.		In.		In.	In.	In.
W 12x 50	14.7	12.19	12¼	0.370	3/8	3/16	8.080	8⅛	0.640	5/8	9½	1⅜	13/16
x 45	13.2	12.06	12	0.335	5/16	3/16	8.045	8	0.575	9/16	9½	1¼	13/16
x 40	11.8	11.94	12	0.295	5/16	3/16	8.005	8	0.515	1/2	9½	1¼	3/4
W 12x 35	10.3	12.50	12½	0.300	5/16	3/16	6.560	6½	0.520	1/2	10½	1	9/16
x 30	8.79	12.34	12⅜	0.260	1/4	1/8	6.520	6½	0.440	7/16	10½	15/16	1/2
x 26	7.65	12.22	12¼	0.230	1/4	1/8	6.490	6½	0.380	3/8	10½	7/8	1/2
W 12x 22	6.48	12.31	12¼	0.260	1/4	1/8	4.030	4	0.425	7/16	10½	7/8	1/2
x 19	5.57	12.16	12⅛	0.235	1/4	1/8	4.005	4	0.350	3/8	10½	13/16	1/2
x 16	4.71	11.99	12	0.220	1/4	1/8	3.990	4	0.265	1/4	10½	3/4	1/2
x 14	4.16	11.91	11⅞	0.200	3/16	1/8	3.970	4	0.225	1/4	10½	11/16	1/2
W 10x 45	13.3	10.10	10⅛	0.350	3/8	3/16	8.020	8	0.620	5/8	7⅝	1¼	11/16
x 39	11.5	9.92	9⅞	0.315	5/16	3/16	7.985	8	0.530	1/2	7⅝	1⅛	11/16
x 33	9.71	9.73	9¾	0.290	5/16	3/16	7.960	8	0.435	7/16	7⅝	1 1/16	11/16
W 10x 30	8.84	10.47	10½	0.300	5/16	3/16	5.810	5¾	0.510	1/2	8⅝	15/16	1/2
x 26	7.61	10.33	10⅜	0.260	1/4	1/8	5.770	5¾	0.440	7/16	8⅝	7/8	1/2
x 22	6.49	10.17	10⅛	0.240	1/4	1/8	5.750	5¾	0.360	3/8	8⅝	3/4	1/2
W 10x 19	5.62	10.24	10¼	0.250	1/4	1/8	4.020	4	0.395	3/8	8⅝	13/16	1/2
x 17	4.99	10.11	10⅛	0.240	1/4	1/8	4.010	4	0.330	5/16	8⅝	3/4	1/2
x 15	4.41	9.99	10	0.230	1/4	1/8	4.000	4	0.270	1/4	8⅝	11/16	7/16
x 12	3.54	9.87	9⅞	0.190	3/16	1/8	3.960	4	0.210	3/16	8⅝	5/8	7/16
W 8x28	8.25	8.06	8	0.285	5/16	3/16	6.535	6½	0.465	7/16	6⅛	15/16	9/16
x24	7.08	7.93	7⅞	0.245	1/4	1/8	6.495	6½	0.400	3/8	6⅛	7/8	9/16
W 8x21	6.16	8.28	8¼	0.250	1/4	1/8	5.270	5¼	0.400	3/8	6⅝	13/16	1/2
x18	5.26	8.14	8⅛	0.230	1/4	1/8	5.250	5¼	0.330	5/16	6⅝	3/4	7/16

Nominal Wt. per Ft.	Compact Section Criteria $\frac{b_f}{2t_f}$	F_y'	$\frac{d}{t_w}$	F_y'''	r_T	$\frac{d}{A_f}$	Elastic Properties Axis X-X I	S	r	Axis Y-Y I	S	r	Torsional constant J	Plastic Modulus Z_x	Z_y
Lb.		Ksi		Ksi	In.		In.⁴	In.³	In.	In.⁴	In.³	In.	In.⁴	In.³	In.³
50	6.3	—	32.9	60.9	2.17	2.36	394	64.7	5.18	56.3	13.9	1.96	1.78	72.4	21.4
45	7.0	—	36.0	51.0	2.15	2.61	350	58.1	5.15	50.0	12.4	1.94	1.31	64.7	19.0
40	7.8	—	40.5	40.3	2.14	2.90	310	51.9	5.13	44.1	11.0	1.93	0.95	57.5	16.8
35	6.3	—	41.7	38.0	1.74	3.66	285	45.6	5.25	24.5	7.47	1.54	0.74	51.2	11.5
30	7.4	—	47.5	29.3	1.73	4.30	238	38.6	5.21	20.3	6.24	1.52	0.46	43.1	9.56
26	8.5	57.9	53.1	23.4	1.72	4.95	204	33.4	5.17	17.3	5.34	1.51	0.30	37.2	8.17
22	4.7	—	47.3	29.5	1.02	7.19	156	25.4	4.91	4.66	2.31	0.847	0.29	29.3	3.66
19	5.7	—	51.7	24.7	1.00	8.67	130	21.3	4.82	3.76	1.88	0.822	0.18	24.7	2.98
16	7.5	—	54.5	22.2	0.96	11.3	103	17.1	4.67	2.82	1.41	0.773	0.10	20.1	2.26
14	8.8	54.3	59.6	18.6	0.95	13.3	88.6	14.9	4.62	2.36	1.19	0.753	0.07	17.4	1.90
45	6.5	—	28.9	—	2.18	2.03	248	49.1	4.32	53.4	13.3	2.01	1.51	54.9	20.3
39	7.5	—	31.5	—	2.16	2.34	209	42.1	4.27	45.0	11.3	1.98	0.98	46.8	17.2
33	9.1	50.5	33.6	58.7	2.14	2.81	170	35.0	4.19	36.6	9.20	1.94	0.58	38.8	14.0
30	5.7	—	34.9	54.2	1.55	3.53	170	32.4	4.38	16.7	5.75	1.37	0.62	36.6	8.84
26	6.6	—	39.7	41.8	1.54	4.07	144	27.9	4.35	14.1	4.89	1.36	0.40	31.3	7.50
22	8.0	—	42.4	36.8	1.51	4.91	118	23.2	4.27	11.4	3.97	1.33	0.24	26.0	6.10
19	5.1	—	41.0	39.4	1.03	6.45	96.3	18.8	4.14	4.29	2.14	0.874	0.23	21.6	3.35
17	6.1	—	42.1	37.2	1.01	7.64	81.9	16.2	4.05	3.56	1.78	0.844	0.16	18.7	2.80
15	7.4	—	43.4	35.0	0.99	9.25	68.9	13.8	3.95	2.89	1.45	0.810	0.10	16.0	2.30
12	9.4	47.5	51.9	24.5	0.96	11.9	53.8	10.9	3.90	2.18	1.10	0.785	0.06	12.6	1.74
28	7.0	—	28.3	—	1.77	2.65	98.0	24.3	3.45	21.7	6.63	1.62	0.54	27.2	10.1
24	8.1	64.1	32.4	63.0	1.76	3.05	82.8	20.9	3.42	18.3	5.63	1.61	0.35	23.2	8.57
21	6.6	—	33.1	60.2	1.41	3.93	75.3	18.2	3.49	9.77	3.71	1.26	0.28	20.4	5.69
18	8.0	—	35.4	52.7	1.39	4.70	61.9	15.2	3.43	7.97	3.04	1.23	0.17	17.0	4.66

From AISC.

Table 409-10. Table of section properties.

Designation	Area A	Depth d		Web Thickness t_w		Web $\frac{t_w}{2}$	Flange Width b_f		Flange Thickness t_f		Distance T	Distance k	Distance k_1 / Grip	Max. Flge. Fas-ten-er
	In.^{2}	In.		In.		In.	In.		In.		In.	In.	In.	In.
W SHAPES														
W 8x15	4.44	8.11	8 1/8	0.245	1/4	1/8	4.015	4	0.315	5/16	6 5/8	3/4	1/2	
x13	3.84	7.99	8	0.230	1/4	1/8	4.000	4	0.255	1/4	6 5/8	11/16	7/16	
x10	2.96	7.89	7 7/8	0.170	3/16	1/8	3.940	4	0.205	3/16	6 5/8	5/8	7/16	
W 6x25	7.34	6.38	6 3/8	0.320	5/16	3/16	6.080	6 1/8	0.455	7/16	4 3/4	13/16	7/16	
x20	5.87	6.20	6 1/4	0.260	1/4	1/8	6.020	6	0.365	3/8	4 3/4	3/4	7/16	
x15	4.43	5.99	6	0.230	1/4	1/8	5.990	6	0.260	1/4	4 3/4	5/8	3/8	
W 6x16	4.74	6.28	6 1/4	0.260	1/4	1/8	4.030	4	0.405	3/8	4 3/4	3/4	7/16	
x12	3.55	6.03	6	0.230	1/4	1/8	4.000	4	0.280	1/4	4 3/4	5/8	3/8	
x 9	2.68	5.90	5 7/8	0.170	3/16	1/8	3.940	4	0.215	3/16	4 3/4	9/16	3/8	
W 5x19	5.54	5.15	5 1/8	0.270	1/4	1/8	5.030	5	0.430	7/16	3 1/2	13/16	7/16	
x16	4.68	5.01	5	0.240	1/4	1/8	5.000	5	0.360	3/8	3 1/2	3/4	7/16	
W 4x13	3.83	4.16	4 1/8	0.280	1/4	1/8	4.060	4	0.345	3/8	2 3/4	11/16	7/16	
M SHAPES													Grip	Max. Flge. Fas-ten-er
													In.	In.
M 14x18	5.10	14.00	14	0.215	3/16	1/8	4.000	4	0.270	1/4	12 3/4	5/8	1/4	3/4
M 12x11.8	3.47	12.00	12	0.177	3/16	1/8	3.065	3 1/8	0.225	1/4	10 7/8	9/16	1/4	—
M 10x9	2.65	10.00	10	0.157	3/16	1/8	2.690	2 3/4	0.206	3/16	8 7/8	9/16	3/16	—
M 8x6.5	1.92	8.00	8	0.135	1/8	1/16	2.281	2 1/4	0.189	3/16	7	1/2	3/16	—
M 6x20	5.89	6.00	6	0.250	1/4	1/8	5.938	6	0.379	3/8	4 1/4	7/8	3/8	7/8
M 6x4.4	1.29	6.00	6	0.114	1/8	1/16	1.844	1 7/8	0.171	3/16	5 1/8	7/16	3/16	—
M 5x18.9	5.55	5.00	5	0.316	5/16	3/16	5.003	5	0.416	7/16	3 1/4	7/8	7/16	7/8
M 4x13	3.81	4.00	4	0.254	1/4	1/8	3.940	4	0.371	3/8	2 3/8	13/16	3/8	3/4

Designation	Nom-inal Wt. per Ft	Compact Section Criteria $\frac{b_f}{2t_f}$	F_y'	$\frac{d}{t_w}$	F_y'''	r_T	$\frac{d}{A_f}$	Elastic Properties Axis X-X I	Axis X-X S	Axis X-X r	Axis Y-Y I	Axis Y-Y S	Axis Y-Y r	Tor-sional con-stant J	Plastic Modulus Z_x	Plastic Modulus Z_y
	Lb.		Ksi		Ksi	In.		In.^{4}	In.^{3}	In.	In.^{4}	In.^{3}	In.	In.^{4}	In.^{3}	In.^{3}
W SHAPES																
W 8x15	15	6.4	—	33.1	60.3	1.03	6.41	48.0	11.8	3.29	3.41	1.70	0.876	0.14	13.6	2.67
x13	13	7.8	—	34.7	54.7	1.01	7.83	39.6	9.91	3.21	2.73	1.37	0.843	0.09	11.4	2.15
x10	10	9.6	45.8	46.4	30.7	0.99	9.77	30.8	7.81	3.22	2.09	1.06	0.841	0.04	8.87	1.66
W 6x25	25	6.7	—	19.9	—	1.66	2.31	53.4	16.7	2.70	17.1	5.61	1.52	0.46	18.9	8.56
x20	20	8.2	62.1	23.8	—	1.64	2.82	41.4	13.4	2.66	13.3	4.41	1.50	0.24	14.9	6.72
x15	15	11.5	31.8	26.0	—	1.61	3.85	29.1	9.72	2.56	9.32	3.11	1.46	0.10	10.8	4.75
W 6x16	16	5.0	—	24.2	—	1.08	3.85	32.1	10.2	2.60	4.43	2.20	0.966	0.22	11.7	3.39
x12	12	7.1	—	26.2	—	1.05	5.38	22.1	7.31	2.49	2.99	1.50	0.918	0.09	8.30	2.32
x 9	9	9.2	50.3	34.7	54.8	1.03	6.96	16.4	5.56	2.47	2.19	1.11	0.905	0.04	6.23	1.72
W 5x19	19	5.8	—	19.1	—	1.38	2.38	26.2	10.2	2.17	9.13	3.63	1.28	0.31	11.6	5.53
x16	16	6.9	—	20.9	—	1.37	2.78	21.3	8.51	2.13	7.51	3.00	1.27	0.19	9.59	4.57
W 4x13	13	5.9	—	14.9	—	1.10	2.97	11.3	5.46	1.72	3.86	1.90	1.00	0.15	6.28	2.92
M SHAPES																
M 14x18	18	7.4	—	65.1	15.6	0.91	13.0	148	21.1	5.38	2.64	1.32	0.719	0.11	24.9	2.20
M 12x11.8	11.8	6.8	—	67.8	14.4	0.68	17.4	71.9	12.0	4.55	0.980	0.639	0.532	0.05	14.3	1.09
M 10x9	9	6.5	—	63.7	16.3	0.61	18.0	38.8	7.76	3.83	0.609	0.453	0.480	0.03	9.19	0.765
M 8x6.5	6.5	6.0	—	59.3	18.8	0.53	18.6	18.5	4.62	3.10	0.343	0.301	0.423	0.02	5.42	0.502
M 6x20	20	7.8	—	24.0	—	1.52	2.66	39.0	13.0	2.57	11.6	3.90	1.40	0.30	14.5	6.25
M 6x4.4	4.4	5.4	—	52.6	23.8	0.44	19.0	7.20	2.40	2.36	0.165	0.179	0.358	0.01	2.80	0.296
M 5x18.9	18.9	6.0	—	15.8	—	1.29	2.40	24.1	9.63	2.08	7.86	3.14	1.19	0.34	11.0	5.02
M 4x13	13	5.3	—	15.7	—	1.01	2.74	10.5	5.24	1.66	3.36	1.71	0.939	0.19	6.05	2.74

From AISC.

R = maximum end reaction for 3½″ bearing, kips. In cases where R would exceed V, the value of R is indicated by a dash.
= $0.75F_y t_w(3.5 + k)$

R_i = increase in R for additional in. of bearing, kips
= $0.75\ F_y t_w$

N_e = length of bearing to develop V, in.
= $(V/R_i - k)$

S_x = section modulus, X-X axis, in^3

D_c = uniform load deflection constants, in/ft²
= $(30/29)F_b/d$. Midspan defl = $D_cL^2/1000$, in.

For beams where $L_c < L_b < L_u$, reduce the tabular load by 22/24. The tables are valid only for unbraced lengths up to L_u. Midspan deflection assumes the extreme fiber stress is 0.66 F_y and therefore is the maximum deflection before exceeding F_b. Reduce the deflection by 22/44 if $L_c < L_b < L_u$.

Example 409-5:

Given: Using F_y = 36 ksi steel, select a 12″ deep simple beam to span 20′ and support three equal 7-kip concentrated loads at the quarter points. The compression flange is supported laterally at the ends and at the concentrated loads. Deflection must be L/240 or less.

Solution:

C_b = 1.0 for any unbraced section of this beam but if $L_b < L_c$ for the beam selected, then C_b is not a factor and the tables are valid.

Assume beam weight is 35 lb/ft.

Equivalent uniform load for moment = 4xP + dead load = 4x7 + 0.035 x 20 = 28.7 kips

wL x L = 28.7 kips x 20′ = 574 k-ft

wL/2 = 28.7 kips/2 = 14.35 kips

L_b = 5′

From Table 409-11, first examine a W12x30.

L_v = 6.6′ < 20′, therefore moment governs.

L_c = 6.9′ > L_b = 5′, therefore table value is valid.

W_c = 618 k-ft > wL^2, therefore moment resistance is adequate.

V = 46.5 kip > wL = 14.35 kips, shear strength is adequate.

R = 31.2 kips > 14.35 kips, therefore 3½″ bearing is adequate.

D_c = 2.0 therefore $D_{midspan}$
= (.95)(uniform load defl)
= $(.95)(2.0x20^2/1000)$
= 0.76″ < (20x12/240 = 1″), deflection is satisfactory.

Select W12x30.

Use of Beam Design Charts

The beam design charts in Fig 409-16 relate moment capacity to unbraced length, L_b, for W and M shape beams. These charts assumed C_b = 1, i.e., the absolute value of an interior moment is greater than the absolute value of both end moments. If C_b > 1.0, the assumption of C_b = 1.0 results in a conservative selection.

The curve for each beam applies the equation for that unbraced length. Up to the solid dot for unbraced length L_c, moment capacity is for F_b = 0.66 F_y for compact shapes. Unbraced length L_u is indicated by an open dot. Beyond this symbol, F_b < 0.60 F_y and the moment is for the largest allowable stress given by the three formulas that apply. The lines stop where F_b = 11 ksi.

The charts help select the lightest section. Dashed lines indicate that there is a beam that is lighter and stronger. To find the lighter beam, proceed upward and to the right until a solid curve is reached.

The curves ignore shear, bearing and deflection. Use these charts with the beam tables or formulas to check for other modes of failure.

Example 409-6:

Find the lightest A36 W section to carry a uniform 500 lb/ft load over a 20′ simply supported beam. Assume the beam is laterally supported.

Solution:

a. Maximum moment neglecting beam weight at this point $wL^2/8$
= $(0.5)(20^2)/8$ = 25.0 k-ft

b. C_b = 1.0, L_b = 0.

c. From the beam design charts at L_b = 0 and M = 25 k-ft, an adequate beam is a M6x20. But that curve is dashed, so there is a beam that is stronger and weighs less. Move upward and to the right and select a W12x14.

d. Now accounting for the beam weight, the maximum moment is:
M_{max} = $(0.5 + 0.014)(20^2)/8$ = 25.7 k-ft. OK.

e. Check shear, bearing, and deflection by either the basic equations or the beam design tables ($L_b < L_c$, so the tables can be used).

If the compression flange was supported laterally only at the end supports, then:
C_b = 1.0, L_b = 20′
Neglecting beam weight, M_{max} = 25.0 k-ft

From the beams chart at L_b = 20′ and M = 25.0 k-ft, W8x24 is the most economical choice. The curve is beyond the O symbol, so $L_b > L_u$ for the W8x24 beam. You can check shear, L_u, bearing, and deflection in the uniform load beam table, but you cannot select the beam for moment resistance since $L_b > L_u$.

Example 409-7:

Select a beam from the beam charts. Using 36 ksi steel, select the lightest W or M simple girder to support two 5.2 kip concentrated loads at the third points of a 30′ span. The compression flange is laterally supported at the ends and loads. Maximum moment is 52 k-ft in the 10′ between the loads.

Solution:

For the end spans of unsupported compression flange:

Eq 409-32.

$$C_b = 1.75 + 1.05\ (-0/52) + 0.3\ (-0/52)^2$$
$$= 1.75$$

Table 409-11. Beam uniform load constants.
Allowable beam moments, $C_b = 1$, $F_y = 36$ ksi.

BEAMS

$F_y = 36$ ksi

Uniform load constants
for beams laterally supported

For beams laterally unsupported, see page **2-51**

Shape	W_c	V	L_v	L_c	L_u	R	R_i	N_e	S	D_c
	Kip-ft.	Kip	Ft.	Ft.	Ft.	Kip	Kip	In.	In.^{3}	In./Ft.^{2}
W shapes										
W 12 x 50	1040	65.4	7.9	8.5	19.6	48.7	10.0	5.2	64.7	2.0
x 45	930	58.6	7.9	8.5	17.7	43.0	9.0	5.2	58.1	2.1
x 40	830	51.1	8.1	8.4	16.0	37.8	8.0	5.2	51.9	2.1
W 12 x 35	730	54.4	6.7	6.9	12.6	36.5	8.1	5.7	45.6	2.0
x 30	618	46.5	6.6	6.9	10.8	31.2	7.0	5.7	38.6	2.0
x 26	534	40.8	6.6	6.9	9.4	27.2	6.2	5.7	33.4	2.0
W 12 x 22	406	46.4	4.4	4.3	6.4	30.7	7.0	5.7	25.4	2.0
x 19	341	41.4	4.1	4.2	5.3	27.4	6.3	5.7	21.3	2.0
x 16	274	38.2	3.6	4.1	4.3	25.2	5.9	5.7	17.1	2.1
x 14	238	34.5	3.5	3.5	4.2	22.6	5.4	5.7	14.9	2.1
W 10 x 112	2020	124	8.1	11.0	53.2	110.0	20.4	4.2	126	2.2
x 100	1790	109	8.2	10.9	48.2	96.4	18.4	4.2	112	2.2
x 88	1580	95.1	8.3	10.8	43.3	83.7	16.3	4.2	98.5	2.3
x 77	1370	81.5	8.4	10.8	38.6	71.6	14.3	4.2	85.9	2.3
x 68	1210	70.9	8.5	10.7	34.8	61.9	12.7	4.2	75.7	2.4
x 60	1070	62.2	8.6	10.6	31.1	54.6	11.3	4.2	66.7	2.4
x 54	960	54.1	8.9	10.6	28.2	47.5	10.0	4.2	60.0	2.5
x 49	874	49.2	8.9	10.6	26.0	43.0	9.2	4.2	54.6	2.5
W 10 x 45	786	51.3	7.7	8.5	22.8	44.9	9.5	4.2	49.1	2.5
x 39	674	45.3	7.4	8.4	19.8	39.3	8.5	4.2	42.1	2.5
x 33	560	40.9	6.8	8.4	16.5	35.7	7.8	4.2	35.0	2.6
W 10 x 30	518	45.5	5.7	6.1	13.1	35.9	8.1	4.7	32.4	2.4
x 26	446	38.9	5.7	6.1	11.4	30.7	7.0	4.7	27.9	2.4
x 22	371	35.4	5.2	6.1	9.4	27.5	6.5	4.7	23.2	2.4
W 10 x 19	301	37.1	4.1	4.2	7.2	29.1	6.8	4.7	18.8	2.4
x 17	259	35.2	3.7	4.2	6.1	27.5	6.5	4.7	16.2	2.5
x 15	221	33.3	3.3	4.2	5.0	26.0	6.2	4.7	13.8	2.5
x 12	174	27.2	3.2	3.9	4.3	21.2	5.1	4.7	10.9	2.5
W 8 x 67	966	74.4	6.5	8.7	39.9	—	15.4	3.4	60.4	2.8
x 58	832	64.7	6.4	8.7	35.3	—	13.8	3.4	52.0	2.8
x 48	693	49.3	7.0	8.6	30.3	—	10.8	3.4	43.3	2.9
x 40	568	43.1	6.6	8.5	25.3	—	9.7	3.4	35.5	3.0
x 35	499	36.5	6.8	8.5	22.6	—	8.4	3.4	31.2	3.1
x 31	440	33.1	6.7	8.4	20.1	—	7.7	3.4	27.5	3.1
W 8 x 28	389	33.3	5.8	6.9	17.5	—	7.7	3.4	24.3	3.1
x 24	334	28.2	5.9	6.9	15.2	—	6.6	3.4	20.9	3.1
W 8 x 21	291	30.0	4.9	5.6	11.8	29.1	6.8	3.6	18.2	3.0
x 18	243	27.1	4.5	5.5	9.9	26.4	6.2	3.6	15.2	3.1

$F_y = 36$ ksi

BEAMS

Uniform load constants
for beams laterally supported

For beams laterally unsupported, see page **2-51**

Shape	W_c	V	L_v	L_c	L_u	R	R_i	N_e	S	D_c
	Kip-ft.	Kip	Ft.	Ft.	Ft.	Kip	Kip	In.	In.^{3}	In./Ft.^{2}
W shapes										
W 8 x 15	189	28.8	3.3	4.2	7.2	28.1	6.6	3.6	11.8	3.1
x 13	159	26.6	3.0	4.2	5.9	26.0	6.2	3.6	9.91	3.1
x 10	125	19.4	3.2	4.2	4.7	18.9	4.6	3.6	7.81	3.1
W 6 x 25	267	29.6	4.5	6.4	20.0	—	8.6	2.6	16.7	3.9
x 20	214	23.4	4.6	6.4	16.4	—	7.0	2.6	13.4	4.0
x 15	•152	20.0	3.8	6.3	12.0	—	6.2	2.6	9.72	•4.1
W 6 x 16	163	23.7	3.4	4.3	12.0	—	7.0	2.6	10.2	4.0
x 12	117	20.1	2.9	4.2	8.6	—	6.2	2.6	7.31	4.1
x 9	89	14.5	3.1	4.2	6.7	—	4.6	2.6	5.56	4.2
W 5 x 19	163	20.2	4.0	5.3	19.5	—	7.3	2.0	10.2	4.8
x 16	136	17.4	3.9	5.3	16.7	—	6.5	1.9	8.51	5.0
W 4 x 13	87	16.9	2.6	4.3	15.6	—	7.6	1.5	5.46	6.0
M shapes										
M 14 x 18	338	43.6	3.9	3.6	4.0	23.9	5.8	6.9	21.1	1.8
M 12 x 11.8	192	30.8	3.1	2.7	3.0	19.4	4.8	5.9	12.0	2.1
M 10 x 9	124	22.8	2.7	2.6	2.7	17.2	4.2	4.8	7.76	2.5
M 8 x 6.5	74	15.7	2.4	2.4	2.5	14.6	3.6	3.8	4.62	3.1
M 6 x 20	208	21.8	4.8	6.3	17.4	—	6.8	2.3	13.0	4.1
M 6 x 4.4	38	9.9	1.9	1.9	2.4	—	3.1	2.8	2.40	4.1
M 5 x 18.9	154	22.9	3.4	5.3	19.3	—	8.5	1.8	9.63	5.0
M 4 x 13	84	14.7	2.8	4.2	16.9	—	6.9	1.3	5.24	6.2

From AISC.

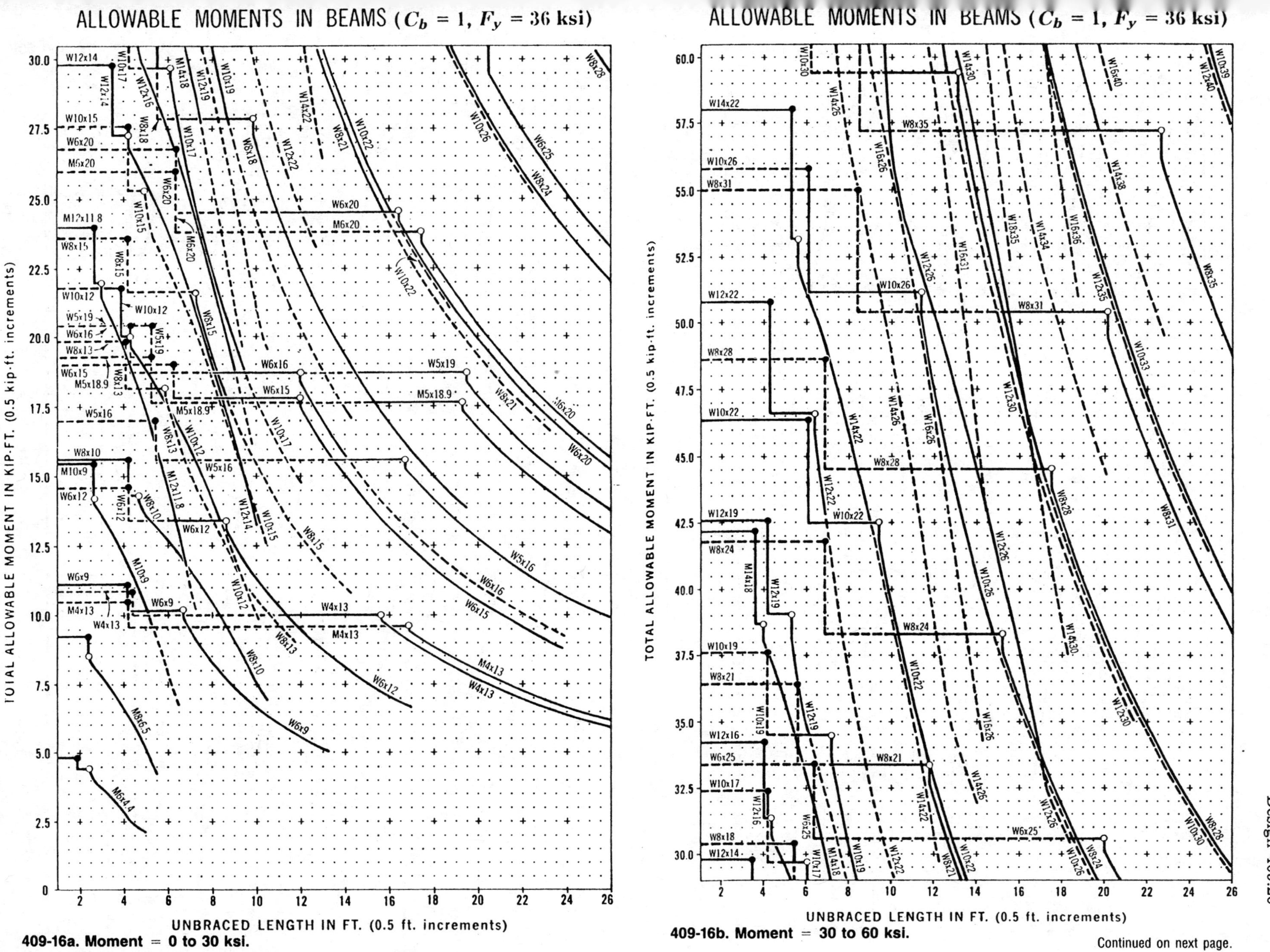

409-16a. Moment = 0 to 30 ksi.

409-16b. Moment = 30 to 60 ksi.

Fig 409-16. Allowable beam moments, C_b = 1, F_y = 36 ksi.

From AISC.

Continued on next page.

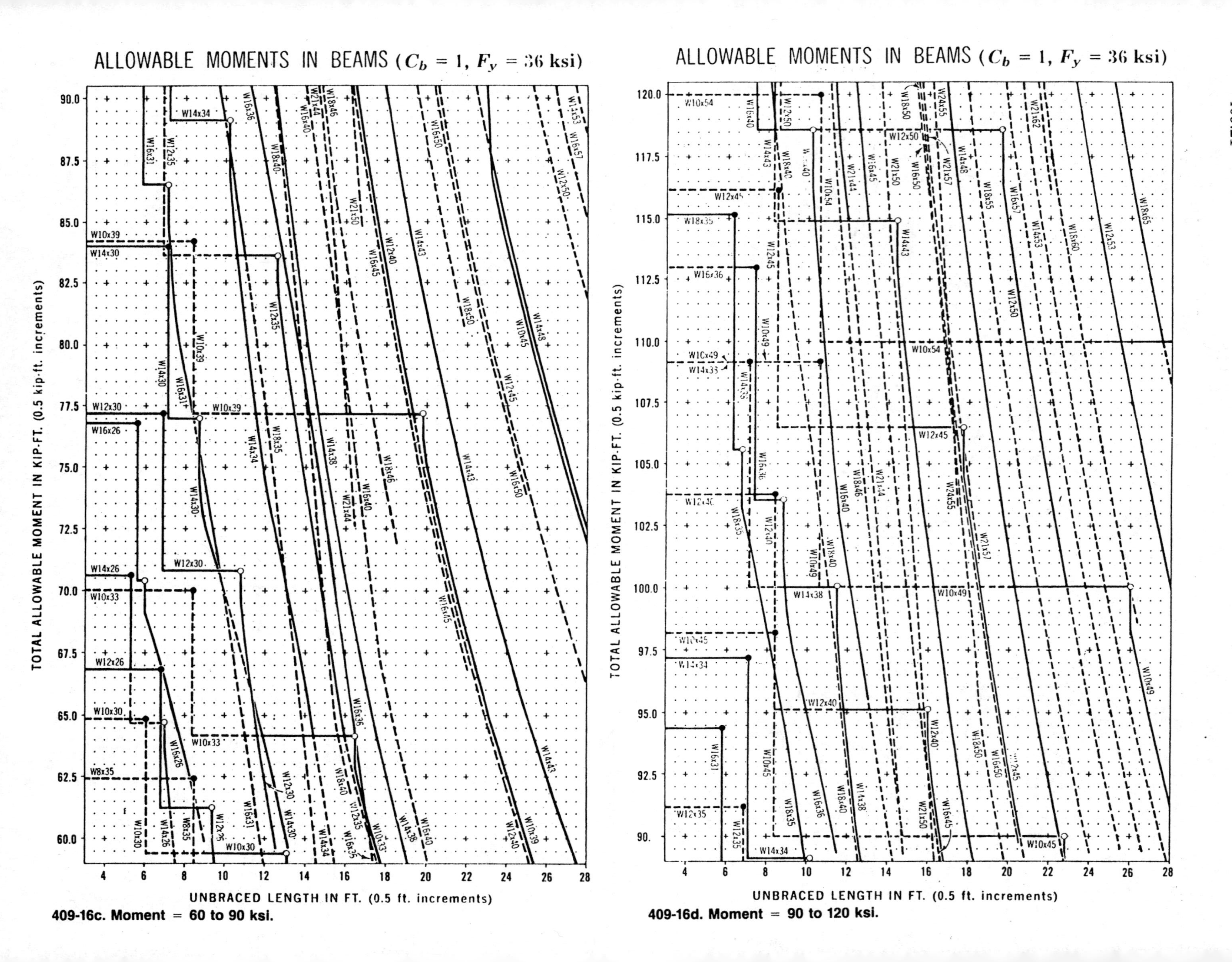

409-16c. Moment = 60 to 90 ksi.

409-16d. Moment = 90 to 120 ksi.

For the center span of unsupported compression flange:

Eq 409-33.

$$C_b = 1.75 + 1.05\,(-52/52) + 0.3(-52/52)^2 = 1.0$$

From the beam design charts, estimate beam size with C_b = 1.0 (because C = 1.0 where moment is maximum), L_b = 10′, and M = 52 k-ft.

Select a W12x26.

From the charts, $L_b < L_u$ so C_b does not affect beam selection.

Include the beam weight:

Eq 409-34.

$$M_{max} = 52 \text{ k-ft} + 0.026 \times 30^2/8 = 54.9 \text{ k-ft}$$

The W12x26 has adequate moment capacity. Check shear, bearing, and deflection in the uniform load beam tables.

V = 46.5 kips > (5.2 + 0.026 x 30/2) = 5.59 kips, shear OK.

R = 31.2 kips > 5.59 kips, so 3½″ bearing is OK.

D_c = 2.0 midspan deflection = (1.02)(uniform load deflection) = (1.02)(2.0 x 30^2/1000) = 1.84″ or L/196 OK.

Example 409-8:

A concentrated load of 10 kips is in the center of a 20′ simple beam. The concentrated load is applied over a 6″ length of the beam. Limit deflection to L/360. The end supports measure 2.5″ along the beam. What is the lightest A36 W10 or smaller section that can carry the load if L_b = 0′? L_b = 10′? L_b = 20′?

Solution:

For L_b = 0′ and neglecting beam weight:

Eq 409-35.

$$M_{max} = (5 \text{ kips})(10') = 50 \text{ k-ft}$$

Assume compact section:

Eq 409-36.

$$F_b = 0.66F_y = 24 \text{ ksi}$$

$S_{reqd} = M_{max}/F_b$ = 50(12 k-in)/24 ksi = 25in³

From the properties tables 409-9 and 409-10, W10x26 appears to be the lightest section. Confirm by checking Fig 409-16 and Table 409-11.

(W_c > (2x10 kips)20′ = 400).

Account for beam weight:

Eq 409-37.

$$M_{max} = 50 \text{ k-ft} + 0.026(20^2) \div 8 = 51.3 \text{ k-ft}$$

$S_{reqd} = M_{max}/F_b$ = 51.3 x 12 k-in/24 ksi = 25.65 in³ OK.

Therefore W10x26 has adequate moment resistance. Check shear stress:

$v = V/dt_w$ = 5.26 kips/10.33″ x 0.260 = 1.96 ksi

$F_v = 0.4F_y$ = 0.4 x 36 ksi = 14.4 ksi > 1.96 ksi. OK.

Verify in Table 409-16; shear strength of this beam is 38.9 kips.

Check bearing stresses to prevent web buckling:

Eq 409-38.

$R/(N+2K)t_w < 0.75\,F_y$ for interior loads

10K/[(6″+2x⅞″)0.260″] = 4.96 Ksi < (0.75)(36)

$R/(N+K)t_w < 0.75\,F_y$ for end supports

5.26K/[(2.5″+⅞″)x0.260″] = 5.99 ksi < (0.75)(36) OK.

Check deflection:

Eq 409-39.

$Defl_{max}$ = PL³/48EI = 10,000 lb(20x12)³/(48x29,000,000x144) = 0.69″

0.69″ = L/348 > L/360. Too high.

Deflection can also be calculated from Table 409-11.

Eq 409-40.

$Defl_{max}$ = (.8)(2.4x20²/1000) = 0.77″

which is greater than 0.69 because the example beam is not fully loaded.

Eq 409-41.

I_{reqd} = 360PL²/48E = 360x10,000 lb x 240²/48x29,000,000 = 149.0

From Table 409-10, it appears that W10x30 is the lightest beam to meet both the deflection and moment resistance criteria.

Check shear:

v = 5.30k/10.47″x0.30 = 1.69 < 14.4 ksi OK.

Buckling: adequate because k increased.

Select **W10x30.**

Solution if L_b = 10′:

Assume M_{max} = 50 k-ft + 0.030x20²/8 = 51.5.

From Fig 409-16b, with 51.5 k-ft and L_b = 10′.

Select a W10x30 and the charts show:

$L_c < L_b < L_u$

Therefore Fig 409-16b is valid even if C_b not equal 1.0.

The shear, buckling, and deflection are adequate as previously shown.

Select **W10x30.**

Solution if L_b = 20′:

Assume M_{max} = 51.5 k-ft.

From Fig 409-16b using 51.5 k-ft and L_b = 20′.

Select a W10x33 and the charts show:

$L_b > L_u$

Check C_b: find that C_b = 1.0, so the selection is valid. Check moment capacity using new beam weight:

Eq 409-42.

$$M_{max} = 50 \text{ k-ft} + 0.033 \times 20^2/8 = 51.6 \text{ k-ft}$$

W10x33 still adequate.

Check shear:
v = 5.33/9.73"x0.29" = 1.89 < 14.4 ksi. OK.
Buckling: OK because k is even larger than W10x30.
Deflection: OK because I is greater than with W10x30.
Select **W10x33.**

Combined Axial Compression and Flexure

The basic equation for combined stress is the interaction equation:

Eq 409-43.
Actual/allowable axial stress + Actual/allowable flexural stress < 1.0

The AISC Specification separates beam-columns into those for which f_a/F_a is greater than or less than 0.15. When $f_a/F_a > 0.15$, the moment magnification due to P-Δ effects become critical and are included in the interaction equation. When $f_a/F_a < 0.15$, these effects are neglected.

When $f_a/F_a < 0.15$, the beam-column must satisfy a stability requirement.

When $f_a/F_a > 0.15$, the beam-column must satisfy both a stability requirement and a strength requirement.

Connections

Bolted connections

A **friction-type** bolted connection is clamped together with sufficient bolt tension that the friction between the connected parts carries the design loads. Bearing failure of the connected parts or the bolt is not a concern. Only an "apparent" shear stress of the bolt is considered, and if it is exceeded, the connected parts slip and failure is assumed. High strength bolts (A325, A490, or A449 with hard-ended washers and A325 nuts) give adequate clamping.

Bearing-type joints are allowed to slip, so shear in the bolts and bearing stress in the connected parts carry the loads. The connections are not clamped to prevent slip between the members.

Allowable stresses for bolts are in Table 409-12. Allowable stresses are given for bearing connectors with threads or not in the shear plane. Threads do not affect friction connectors that do not fail by bolt shear.

Allowable bearing stress on the projected area of bolts in bearing connections must be less than 1.5 F_u, where F_u is the minimum tensile stress of the connected part. Bearing stress is not restricted in friction joints.

In bolted connections, the net area of the cross section with the connector holes must be large enough:

Table 409-12. Allowable stresses for bolts.

Bolt type (ASTM)	Tension (F_t) (ksi)	Shear (F_v) Friction type* (ksi)	Shear (F_v) Bearing type (ksi)
A325 and A449 threads in the shear plane	44.0	17.5	21.0
A325 and A499, threads not in the shear plane	44.0	17.5	30.0
A490, threads in the shear plane	54.0	22.0	28.0
A490, threads not in the shear plane	54.0	22.0	40.0
A307 threads in shear plane	20.0		10.0

*Standard size hole.

Table 409-13. Minimum edge distances for bolts (in).
Center of hole to edge.

Bolt diameter in.	At sheared edges in.	At rolled or gas-cut edges in.
½	⅞	¾
⅝	1⅛	⅞
¾	1¼	1
⅞	1½	1⅛
1	1¾	1¼

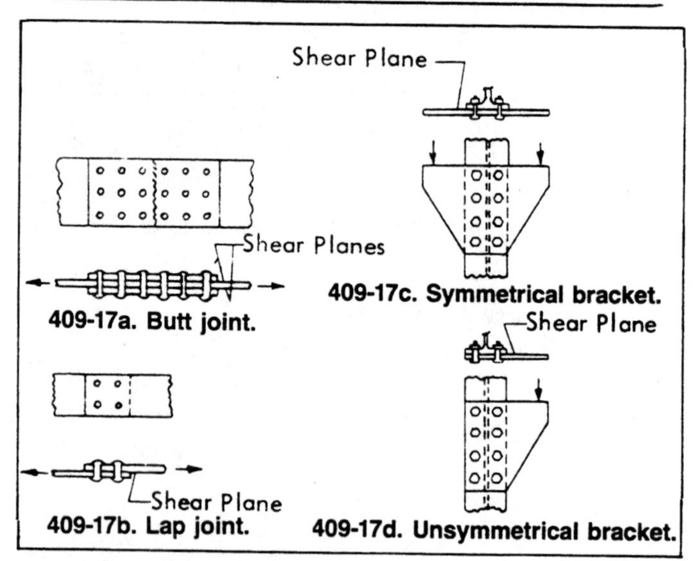

Fig 409-17. Bolts and rivets in shear.

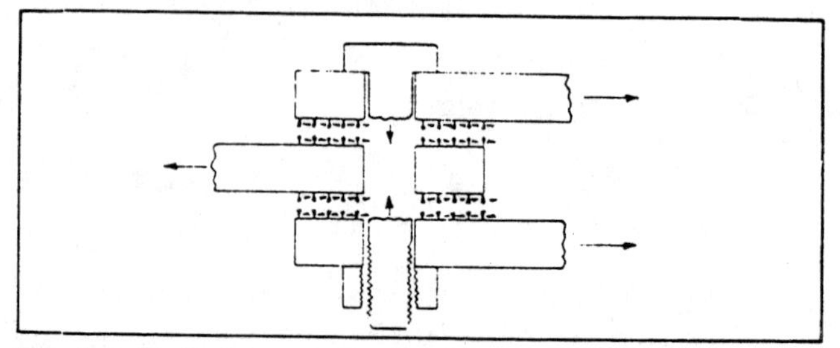

Fig 409-18. Bolt in tension in friction connection.

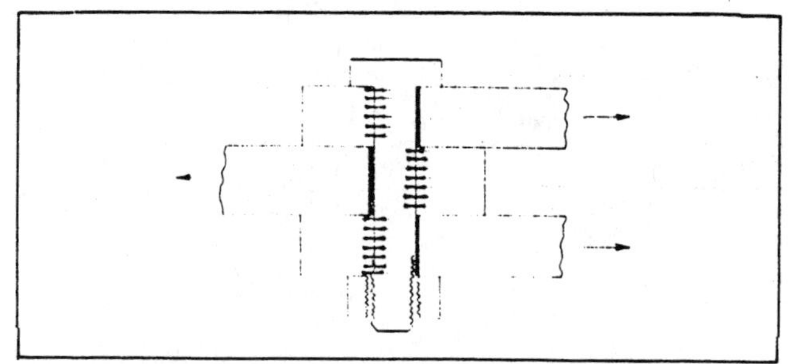
Fig 409-19. Bolt in shear in bearing connection.

Eq 409-44.

$F_t = P/A_{net} < 0.50\ F_u$

In connected parts carrying compressive stresses:

Eq 409-45.

$F_c = P/A_{gross} < F_a$

To develop the allowable stresses in Table 409-12, locate bolts properly.

- Space bolts 2⅔ times (preferably 3 times) the nominal bolt diameter.
- Provide the minimum edge distances in Table 409-13.
- Provide at least $2P/F_u$ + end distance for bearing connections in tension.

Spacing of bolts along the lines of transmitted force measured between hole centers must be:

$2P/F_ut + d/2$

Welded Connections

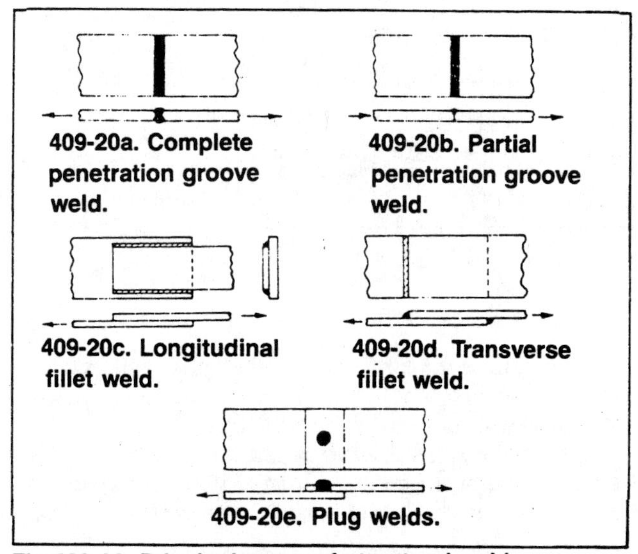

Fig 409-20. Principal types of structural welds.

Types of structural welds are in Fig 409-20. Weld strength depends on allowable stress, effective thickness, and weld length.

Allowable stresses for groove, plug, and fillet welds are in Table 409-14. Electrode designations E60XX and F6X-EXXX are for yield strengths of 60 ksi.

Effective Areas of Weld Metal

Effective area of groove and fillet welds is effective length times effective throat thickness.

Effective shear area of plug and slot welds is the nominal cross section of the hole or slot in the plane of the faying surface.

Effective length of fillet welds in holes and slots is the length of the weld centerline through the center of the plane through the throat. Effective area of overlapping fillets is limited to the nominal cross section of the hole or slot in the plane of the faying surface.

Effective length of a fillet weld is the overall length of the full-size fillet, including returns. Effective length of a groove weld is the width of the part joined.

Effective throat thickness of a fillet weld is the shortest root-to-face distance of the diagrammatic weld. For fillet welds made by the submerged arc process, effective throat thickness is the leg size for ⅜" and smaller fillet welds and the theoretical throat plus 0.11 for fillet welds over ⅜".

Fillet weld size is limited:

- If a joint has only fillet welds, minimum sizes are in Table 409-15.
- The maximum effective fillet size is material thickness to ¼" and weld along edges of 1⁄16" less than the material thickness.
- The minimum effective fillet length is four times the nominal size to develop its full strength. For shorter welds, use ¼ their length.
- If only longitudinal fillet welds connect ends of flat bar tension members, the length of each fillet is at least the perpendicular distance between them. The transverse distance between such welds is at least 8".
- Minimum lap in a lap joint is five times the thickness of the thinner part and not less than 1".
- Side or end fillet welds terminating at ends or sides of parts or members must, wherever practical, return continuously around the corners for at least twice the nominal weld size.

Plug and Slot Welds

Plug or slot welds transmit shear in lap joints, prevent buckling of lapped parts, and join components of built-up members.

Plug weld hole diameter is at least the part thickness plus 5⁄16", rounded up to the next odd 1⁄16", and not greater than 2¼ times the thickness of the weld metal.

On-center spacing of plug welds is at least four times the hole diameter.

Slot weld length is not more than 10 times weld thickness. Slot width is at least the part thickness, plus 5⁄16", rounded up to the next odd 1⁄16", and not greater than 2¼ times weld thickness. Slot ends are semicircular or have corners rounded to a radius not less than the part thickness, except those ends which extend to the edge of the part.

Table 409-14. Allowable stress on welds.

Type of Weld and Stress	Allowable Stress	Required Weld Strength Level
Complete-Penetration Groove Welds		
Tension normal to effective area	Same as base metal	"Matching" weld metal must be used.
Compression normal to effective area	Same as base metal	Weld metal with a strength level equal to or less than "matching" weld metal may be used.
Tension or compression parallel to axis of weld	Same as base metal	
Shear on effective area	0.30 × nominal tensile strength of weld metal (ksi), except shear stress on base metal shall not exceed 0.40 × yield stress of base metal	
Partial-Penetration Groove Welds		
Compression normal to effective area	Same as base metal	Weld metal with a strength level equal to or less than "matching" weld metal may be used.
Tension or compression parallel to axis of weld[e]	Same as base metal	
Shear parallel to axis of weld	0.30 × nominal tensile strength of weld metal (ksi), except shear stress on base metal shall not exceed 0.40 × yield stress of base metal	
Tension normal to effective area	0.30 × nominal tensile strength of weld metal (ksi), except tensile stress on base metal shall not exceed 0.60 × yield stress of base metal	
Fillet Welds		
Shear on effective area	0.30 × nominal tensile strength of weld metal (ksi), except shear stress on base metal shall not exceed 0.40 × yield stress of base metal	Weld metal with a strength level equal to or less than "matching" weld metal may be used.
Tension or compression parallel to axis of weld[e]	Same as base metal	
Plug and Slot Welds		
Shear parallel to faying surfaces (on effective area)	0.30 × nominal tensile strength of weld metal (ksi), except shear stress on base metal shall not exceed 0.40 × yield stress of base metal	Weld metal with a strength level equal to or less than "matching" weld metal may be used.

From AISC.

Table 409-15. Minimum size fillet welds.

Material thickness of thicker part joined in.	Minimum size of fillet weld in.
To 1/4 inclusive	1/8
Over 1/4 to 1/2	3/16
Over 1/2 to 3/4	1/4
Over 3/4	5/16

Minimum transverse spacing of lines of slot welds is four times slot width. Minimum on-center spacing along any line is two times slot length.

Plug or slot weld thickness in material up to 5/8″ is the thickness of the material. In thicker material, it is at least 1/2 the thickness of the material but not less than 5/8″.

410 EQUIVALENT FLUID DENSITY

Rankine's theory comes from classical soils mechanics. It relates lateral to vertical pressure in non-cohesive soils, such as sand and gravel, which are similar in behavior to small grains. It yields adequate equivalent fluid density values for designing shallow grain bin walls. See Chapter 103 for discussion of shallow vs. deep bins.

A companion theory, Coulomb's, analyzes forces on a sliding wedge of material behind a wall. It yields the same values as Rankine's for the angle of the failure plane and the ratio of vertical to lateral pressures when wall friction is neglected.

If a wall restraining the grain moves slightly away from the grain, the lateral pressure decreases, the shear stress increases, and the vertical pressure at the bottom of the grain mass remains unchanged. A plane of shear failure forms and the wedge of grain between it and the wall flows toward the wall as it moves. The failure plane that produces maximum horizontal forces when the pile is level and the wall is vertical is 45° + $\phi/2$ from the horizontal, Fig 410-1.

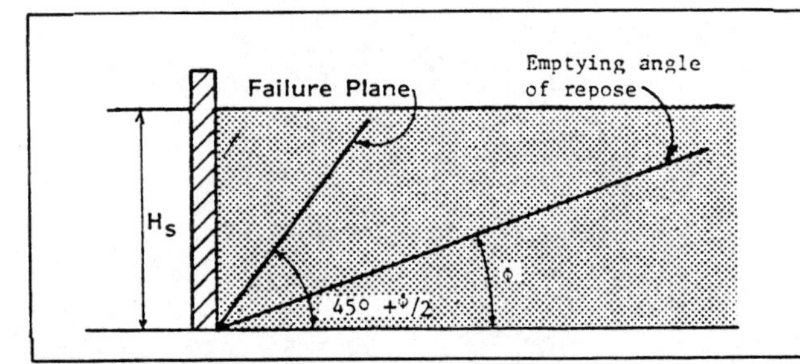

Fig 410-1. Failure plane and emptying angle of repose in grain.

Rankine's definition of the ratio of lateral to vertical pressure for the above condition is:

Eq 410-1.

$$k = \mathrm{Tan}^2(45 - \phi/2) = (1 - \mathrm{Sin}\ \phi)/(1 + \mathrm{Sin}\ \phi)$$

The equivalent fluid density of a grain is:

Eq 410-2.

$$\mathrm{EFD} = (\text{unit weight}) \times k = w \times k$$

Using corn as an example:

Its emptying angle of repose, ϕ = 27°, is the angle of internal friction.

Therefore:

$k = (1 - \mathrm{Sin}\ 27°)/(1 + \mathrm{Sin}\ 27°) = 0.38$

$\mathrm{EFD} = (48\ \mathrm{lb/ft^3}) \times 0.38 = 18\ \mathrm{lb/ft^3}$

The resulting load diagram on a wall is in Fig 410-2.

The case of an inclined grain surface is more complicated. Lateral pressure is assumed to act at an angle equal to the angle of the coefficient of friction between the wall and the grain mass. If this friction is neglected, which it usually is, then the direction of the thrust is horizontal.

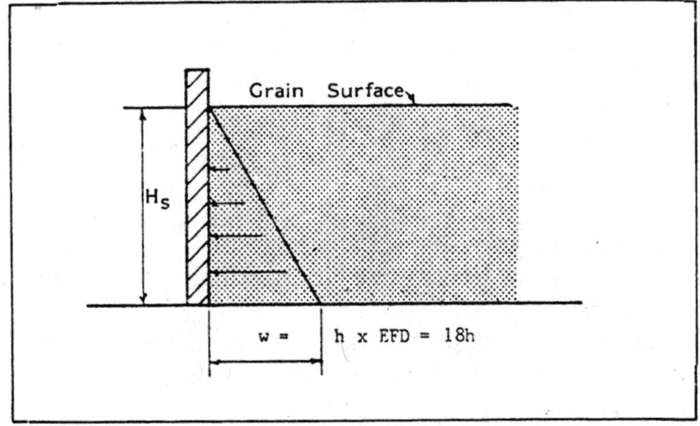

Fig 410-2. Load on a wall from a flat grain surface.

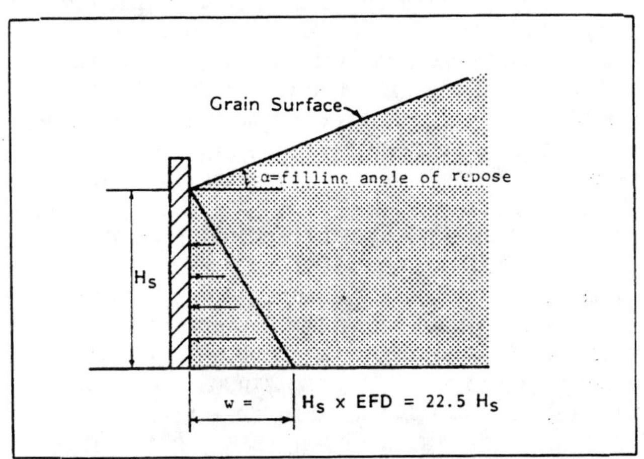

Fig 410-3. Grain surface at the filling angle of repose.

Use the more generalized equation for k if the grain surface or wall is inclined, Fig 410-3.

Eq 410-3.

$$k = \sin^2(\theta-\phi) \div \sin^2(\theta) \times \sin(\theta+\beta) \times T^2$$

$$T = 1 + [\sin(\phi+\beta) \times \sin(\phi-\alpha) \div \sin(\theta-\alpha) \times \sin(\theta+\beta)]^{1/2}$$

k = ratio of lateral to vertical internal pressure

ϕ = angle of internal friction, usually taken as the emptying angle of repose, degrees

θ = angle of wall from horizontal, degrees

β = angle of wall friction, degrees

α = filling angle of repose, degrees

Example, for corn:

$\phi = 27°$

$\theta = 90°$

$\beta = 0°$

$\alpha = 16°$

$T = [(\sin 27\text{-}0)(\sin 27\text{-}16) \div \sin(90\text{-}16)\sin(90+0)]^{1/2}$

$= 1.3$

$k = \sin^2(90\text{-}27)\sin^2(90)\sin(90) \times 1.3^2$

$= 0.4696$

$\mathrm{EFD} = (48\ \mathrm{lb/ft^3}) \times (0.4696)$

$= 22.5\ \mathrm{lb/ft^3}$

411 JANSSEN'S EQUATION

Many grain bins have significant wall friction with resulting lower vertical and lateral pressures than predicted by equivalent fluid density. The distinction between deep and shallow bins (those with and without significant wall friction) is discussed in Chapter 103.

The interaction of pressures on opposite bin walls has the effect of creating a "pressure dome," Fig 411-1b, which transmits pressure laterally to the sidewalls, creating reactions comparable to those in a structural arch. The lateral bin pressure is accompanied by vertical pressures in the bin walls. The static vertical component is equal to the static lateral pressure times the coefficient of friction between the grain and the bin wall.

Janssen's equation predicts static lateral and static vertical pressures within a grain mass. Because pressures during dynamic or unloading conditions are often greater than static pressures, do not use Janssen's equation directly for bin design. Overpressure factors and other modifications of these static pressures are discussed in Chapter 103.

A detailed derivation of Janssen's equation is illustrated in the following material.

Nomenclature:

ϕ = angle of internal friction of the grain (emptying angle of repose)
β = angle of friction of the grain on the bin walls
μ' = $\tan \beta$ = coefficient of friction of grain on the bin walls
w = weight of grain, lb/ft^3
F = vertical pressure, lb/ft^2
L_s = static lateral pressure, lb/ft^2
A = bin area, ft^2
U = bin circumference, ft
R = A/U = hydraulic radius of the bin, ft
H_s = height of the grain, ft
D = bin diameter, ft

The bin in Fig 411-1a has a uniform area, A, a constant circumference, U, and is filled with grain weighing w lb/ft^3 having an emptying angle of repose, ϕ. Let F be the vertical pressure and L the lateral pressure at any point, with both F and L assumed constant on any horizontal plane.

The weight of the grain between the sections of y and y + dy is (Aw)dy; the total frictional force acting upwards at the circumference is (LU)(tan β)dy; the total perpendicular pressure on the upper surface is VA; and total pressure on the lower surface is (F + dF)A.

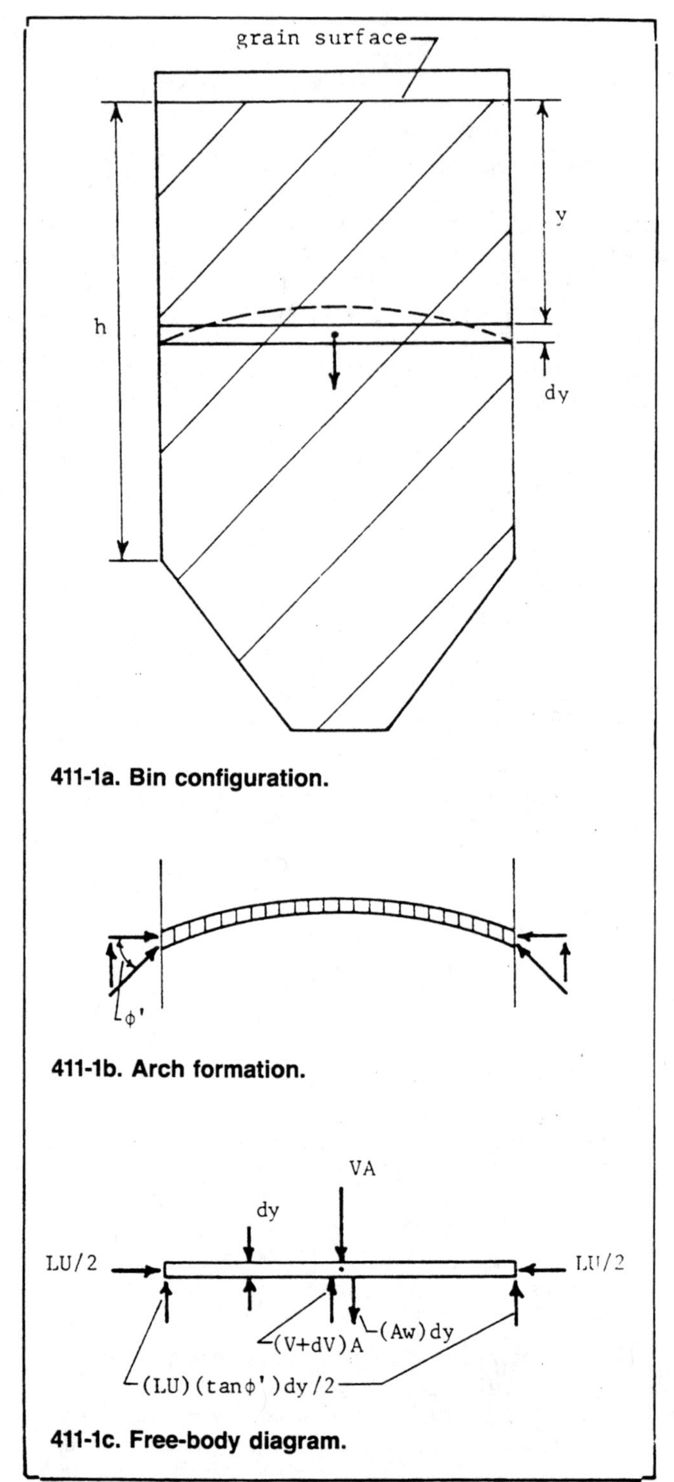

411-1a. Bin configuration.

411-1b. Arch formation.

411-1c. Free-body diagram.

Fig 411-1. Janssen's equation derivation.

The static vertical pressures must be in equilibrium with their sum equal to zero.

$$0 = FA - (F + dF)A + (Aw)dy - (LU)(\tan \beta)dy$$

Simplify:

$$dF = [w - L(\tan \beta)U/A]dy$$

In a granular mass, the static lateral pressure at any point is equal to the static vertical pressure at that point times a constant for the particular grain, k.

$$L = kV$$

The value of k may be determined by experiment, but the value generally used is the one derived by the Rankine method:

$$k = (1-\sin \beta)/(1+\sin \beta)$$

It is an approximation that ignores the friction between the grain and the bin walls.

Also let $R = A/U$ (the hydraulic radius) and $\mu' = \tan \beta$.

Substitute:

$$dF = (w - kF\mu'/R)dy$$

Let:

$$k\mu'/R = n$$

Substitute again:

$$dF = (w - nF)dy, \text{ or:}$$

$$dF/(w - nF) = dy$$

Multiply both sides of the equation by -n and integrate:

$$\ln(w-nF) = -ny + C$$

Evaluate the constant C at $y = 0$ where $F = 0$:

$$C = \ln(w)$$

Rearrange:

$$\ln (w - nF/w) = -ny$$

Take the exponential of each side of the equation:

$$(w - nF)/w = e^{-ny}$$

Solve for F:

$$F = w(1 - e^{-ny})/n$$

Substitute for n. The equation becomes:

$$F = (wR/k\mu')(1 - e^{-k\mu' Y/R})$$

412 BEAM FORMULAS

Equivalent Tabular Loads

Tabular loads are those listed in tables for specific materials and spans, and are usually expressed in total uniform load or in uniform load per unit length of beam. Tabular loads are calculated from the bending moment in a beam resulting from a uniform load, the size of a specified beam, and the allowable stress in the specified material.

Allowable load tables can be used for load patterns other than uniform load by converting the moment from the actual load pattern to its Equivalent Tabular Load (ETL), which is the uniform load that would result in the same maximum moment.

Equivalent Tabular Load (ETL)

1. Find the maximum bending moment resulting from the actual load pattern.
2. Convert units to those used in the load tables, usually feet and pounds.
3. If table lists total uniform load: ETL = 8 x maximum moment ÷ beam length.
   If table lists load per unit length: ETL = 8 x maximum moment ÷ (beam length)^2.
4. Select a beam that will support the ETL.
5. Check the selected beam for shear and deflection using equations for the actual load pattern.

Example: Simple beam with one concentrated load, P, at midspan. Maximum moment (Eq 412-4) = PL/4.

If load table lists total uniform load:
ETL = 8 x (PL/4) ÷ L = 2P lb.
If load table lists load per unit length:
ETL = 8 x (PL/4) ÷ L^2 = 2P/L lb/ft.
Select a beam to support 2P or 2P/L.

Superposition of loads

Given a beam with more than one load on it, the net effect of the loads at one point in the beam is equal to the sum of the effects of each load individually at that point. This is true for bending moments, shears, deflections, etc.

Symbol units

For moments in ft-lb, all lengths (L, a, b, c, x, etc.) are in ft and all distributed loads (w) are in lb/ft. Use lengths in inches and distributed loads in lb/in. for moments in in.-lb. For deflections in inches, the length and load units are inches and lb/in. Any compatible length and load units (in. and lb/in. or ft and lb/ft) yield correct shear and reaction values.

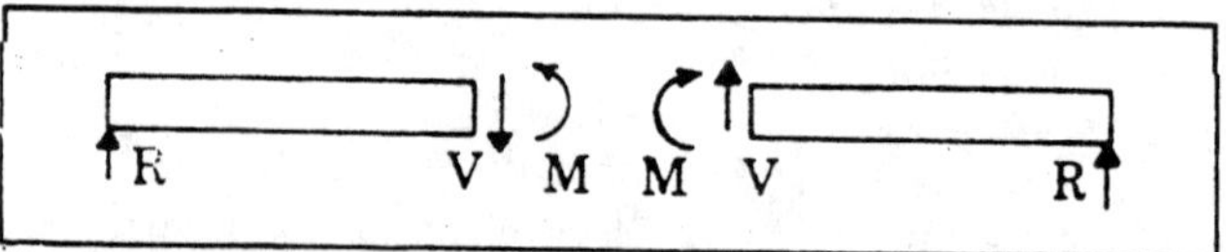

Fig 412-1. Sign convention.
Positive shears, moments, and reactions. Downward deflection is positive.

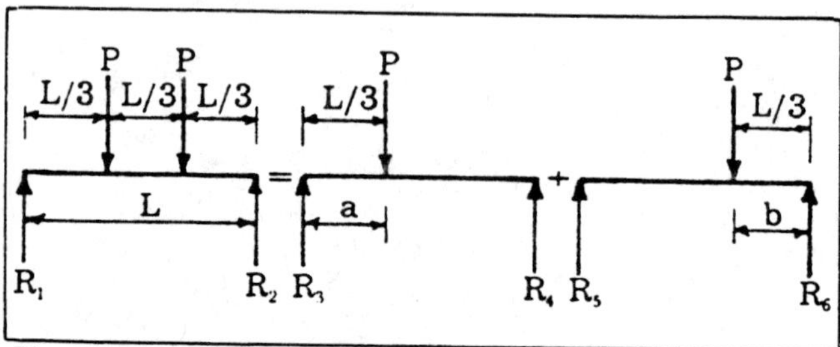

Fig 412-2. Sample problem.

The total effect of both loads is the sum of the effects of each load separately. From the equations for a single concentrated load at any point:

Moment at midspan:
Load left of center = Pa/2
= PL/6 at a = L/3
Load right of center = Pb/2
= PL/6 at b= L/3
Total Moment at midspan = PL/3

Check: see Eq 412-10; moment at midspan from 2 equal loads placed symmetrically at L/3 from each end = PL/3, which agrees with the calculation by superposition.

Similarly for end reactions:
$R_1 = R_3 + R_5$
$R_2 = R_4 + R_6$

And for deflection at midspan:
Load left of center
= Pa (L-x) (-a^2 - x^2 + 2xL) ÷ 6EIL
= 23PL^3 ÷ $(36)^2$ EI
at a = L/3 and x = L/2

Similarly for right load, and substituting b for a in the above equations,
Deflection at midspan
= 23PL^3 ÷ $(36)^2$ EI
Total deflections
= 46/PL^2 ÷ $(36)^2$ EI
= 23PL^2 ÷ 648EI

Check: see Eq 412-10; deflection at midspan
= Pa (3L^2 - 4a^2) ÷ 24EI where a = L/3
= 23PL^3 ÷ 648 EI which agrees with the computation above.

Simple Beams

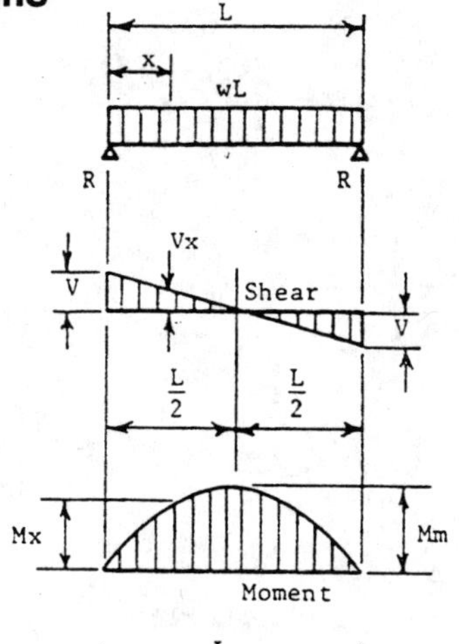

Eq 412-1

$R = V \ldots\ldots = \frac{wL}{2}$

$V_x \ldots\ldots = w(\frac{L}{2} - x)$

M_m [@ L/2] $\ldots\ldots = \frac{wL^2}{8}$

$M_x \ldots\ldots = \frac{wx}{2}(L - x)$

D_m [@ L/2] $\ldots\ldots = \frac{5wL^4}{384EI}$

$D_x \ldots\ldots = \frac{wx}{24EI}(L^3 - 2Lx^2 + x^3)$

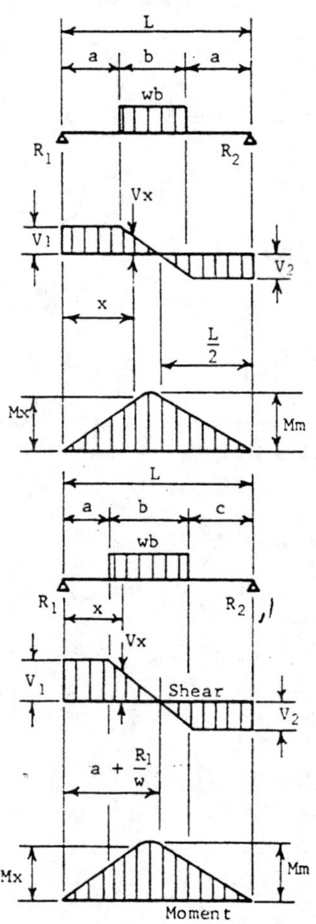

Eq 412-2

$R_1 = R_2 = V_1 = V_2 \ldots\ldots = \frac{wb}{2}$

V_x [@ a < x < (a + b)] $\ldots\ldots = \frac{wb}{2} - w(x - a)$

M_m [@ L/2] $\ldots\ldots = \frac{wb}{8}(2L - b)$

M_x [@ x < a] $\ldots\ldots = \frac{wbx}{2}$

M_x [@ a < x < (a + b)] $\ldots\ldots = \frac{wbx}{2} - \frac{w(x - a)^2}{2}$

D_m [@ x = L/2] $\ldots\ldots = \frac{5wL^4}{384EI} - \frac{wa^2}{48EI}(3L^2 - 2a^2)$

Eq 412-3

$R_1 = V_1$ [max. @ a < c] $\ldots\ldots = \frac{wb}{2L}(2c + b)$

$R_2 = V_2$ [max. @ a > c] $\ldots\ldots = \frac{wb}{2L}(2a + b)$

V_x [@ a < x < (a + b)] $\ldots\ldots = R_1 - w(x - a)$

M_m [@ $x = a + \frac{R_1}{w}$] $\ldots\ldots = R_1(a + \frac{R_1}{2w})$

M_x [@ x < a] $\ldots\ldots = R_1 x$

M_x [@ a < x < (a + b)] $\ldots\ldots = R_1 x - \frac{w}{2}(x - a)^2$

M_x [@ (a + b) < x < L] $\ldots\ldots = R_2(L - x)$

D_c [when a and c < L/2] $\ldots\ldots = \frac{5wL^4}{384EI} - \frac{w}{96EI}(3L^2a^2 - 2a^4 + 3L^2c^2 - 2c^4)$

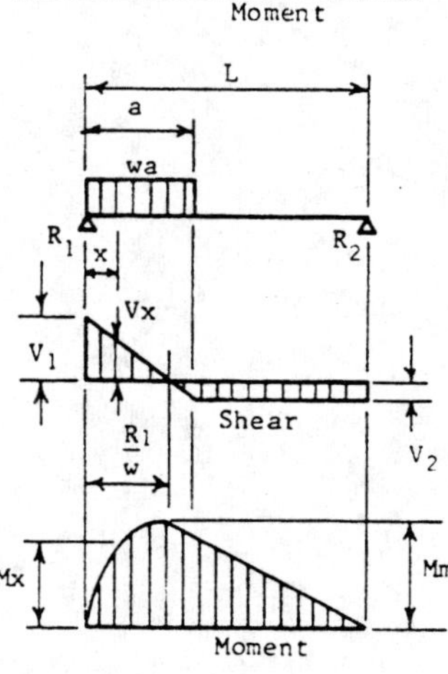

Eq 412-4

$R_1 = V_1 max. \ldots\ldots = \frac{wa}{2L}(2L - a)$

$R_2 = V_2 \ldots\ldots = \frac{wa^2}{2L}$

V_x [@ x < a] $\ldots\ldots = R_1 - wx$

M_m [@ $x = \frac{R_1}{w}$] $\ldots\ldots = \frac{R_1^2}{2w}$

M_x [@ x < a] $\ldots\ldots = R_1 x - \frac{wx^2}{2}$

M_x [@ a < x < L] $\ldots\ldots = R_2(L - x)$

D_x [@ x < a] $\ldots\ldots = \frac{wx}{24EIL}(a^2(2L - a)^2 - 2ax^2(2L - a) + Lx^3)$

D_x [@ x > a] $\ldots\ldots = \frac{wa^2(L - x)}{24EIL}(4xL - 2x^2 - a^2)$

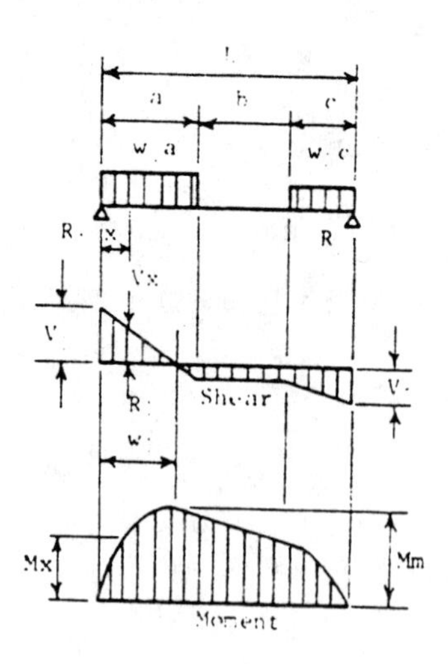

Eq 412-5

$R_1 = V_1$ . . . . $= \dfrac{w_1 a(2L - a) + w_2 c^2}{2L}$

$R_2 = V_2$ . . . . $= \dfrac{w_2 c(2L - c) + w_1 a^2}{2L}$

Vm $[@\ w_1 a > w_2 c]$ . . . . $= R_1$

Vm $[@\ w_1 a < w_2 c]$ . . . . $= R_2$

Vx $[@\ x < a]$ . . . . $= R_1 - w_1 x$

Mm $[@\ x = \dfrac{R_1}{w_1}\ @\ R_1 < w_1 a]$ . . . . $= \dfrac{R_1^2}{2w_1}$

Mm $[@\ x = L - \dfrac{R_2}{w_2}\ @\ R_2 < w_2 c]$ . . $= \dfrac{R_2^2}{2w_2}$

Mx $[@\ x < a]$ . . . . $= R_1 x - \dfrac{w_1 x^2}{2}$

Mx $[@\ a < x < (a + b)]$ . . . . $= R_1 x - \dfrac{w_1 a}{2}(2x - a)$

Mx $[@\ x > (a + b)]$ . . . . $= R_2(L - x) - \dfrac{w_2(L - x)^2}{2}$

D℄ [when a and c < L/2] . . . $= \dfrac{w_1 a^2}{96EI}(3L^2 - 2a^2) + \dfrac{w_2 c^2}{96EI}(3L^2 - 2c^2)$

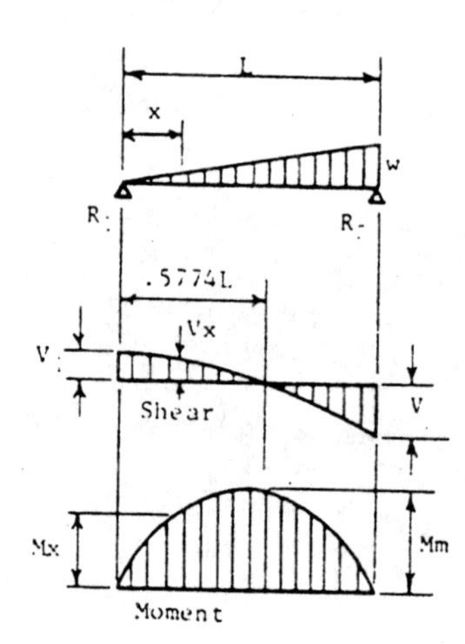

Eq 412-6

W . . . . $= wL/2$

$R_1 = V_1$ . . . . $= wL/6$

$R_2 = V_{2}$max. . . . . $= wL/3$

Vx . . . . $= \dfrac{wL}{6} - \dfrac{wx^2}{2L}$

Mm $[@\ x = .5774L]$ . . . . . . $= \dfrac{wL^2}{9\sqrt{3}} = 0.0642wL^2$

Mx . . . . $= \dfrac{wx}{6L}(L^2 - x^2)$

Dm $[@\ x = .5193L]$ . . . . . . $= 0.00652\ \dfrac{wL^4}{EI}$

Dx . . . . $= \dfrac{wx}{360EIL}(3x^4 - 10L^2x^2 + 7L^4)$

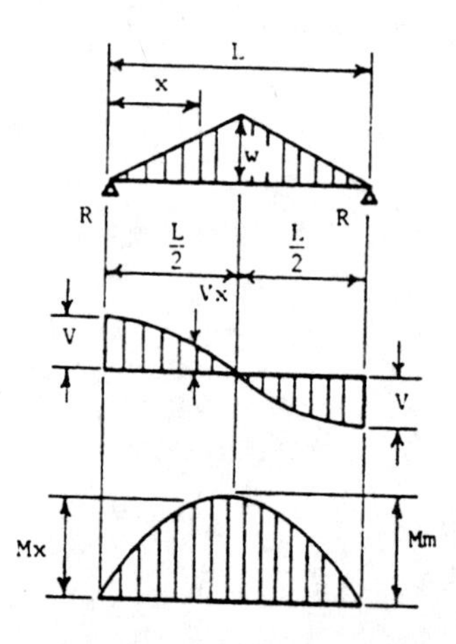

Eq 412-7

W . . . . $= wL/2$

R = V . . . . $= wL/4$

Vx $[@\ x < \dfrac{L}{2}]$ . . . . $= \dfrac{w}{4L}(L^2 - 4x^2)$

Mm [@ center] . . . . $= wL^2/12$

Mx $[@\ x < \dfrac{L}{2}]$ . . . . $= \dfrac{wx}{2}\left(\dfrac{L}{2} - \dfrac{2x^2}{3L}\right)$

Dm [@ center] . . . . $= \dfrac{wL^4}{120EI}$

Dx . . . . $= \dfrac{wx}{960EIL}(5L^2 - 4x^2)^2$

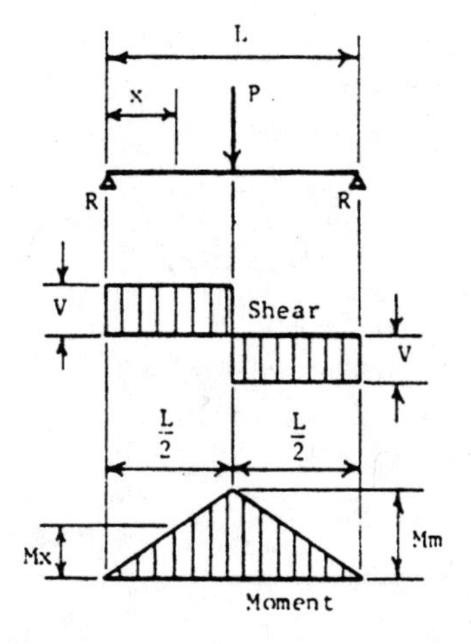

Eq 412-8

$R = V$ . . . . $= P/2$

M_m [@ load] . . . . $= PL/4$

M_x [@ $x < L/2$] . . . . $= Px/2$

D_m [@ load] . . . . $= \frac{PL^3}{48EI}$

D_x [@ $x < L/2$] . . . . $= \frac{Px}{48EI}(3L^2 - 4x^2)$

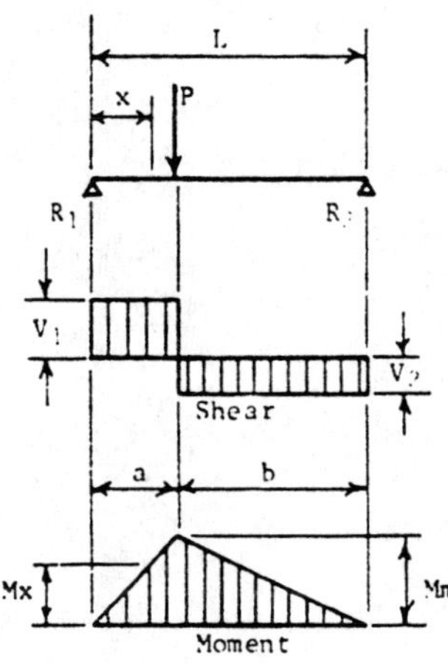

Eq 412-9

$R_1 = V_1$ [max. @ $a < b$] . . . . . $= Pb/L$

$R_2 = V_2$ [max. @ $a > b$] . . . . . $= Pa/L$

M_m [@ load] . . . . $= Pab/L$

M_x [@ $x < a$] . . . . $= Pbx/L$

M_x [@ $a < x < L$] . . . . $= \frac{Pa}{L}(L - x)$

D_m [@ $x = \sqrt{\frac{a(a + 2b)}{3}}$ @ $a > b$] . . $= \frac{Pab(a + 2b)\sqrt{3a(a + 2b)}}{27\ EIL}$

D_a [@ load] . . . . $= \frac{Pa^2b^2}{3EIL}$

D_x [@ $x < a$] . . . . $= \frac{Pbx}{6EIL}(L^2 - b^2 - x^2)$

D_x [@ $a < x < L$] . . . . $= \frac{Pa\ (L - x)}{6EIL}(-a^2 + 2xL - x^2)$

2 Equal loads

Eq 412-10 a = b = c = L/3

$R = V$ . . . . $= P$

M_m [between loads] . . . . . $= PL/3$

M_x [@ $x < L/3$] . . . . $= Px$

D_m [@ center] . . . . $= \frac{23PL^3}{648EI}$

D_x [@ $x < a$] . . . . $= \frac{Px}{6EI}\left(\frac{2L^2}{3} - x^2\right)$

D_x [@ $a < x < (L - a)$] . . . $= \frac{PL}{18EI}(3Lx - 3x^2 - L^2/9)$

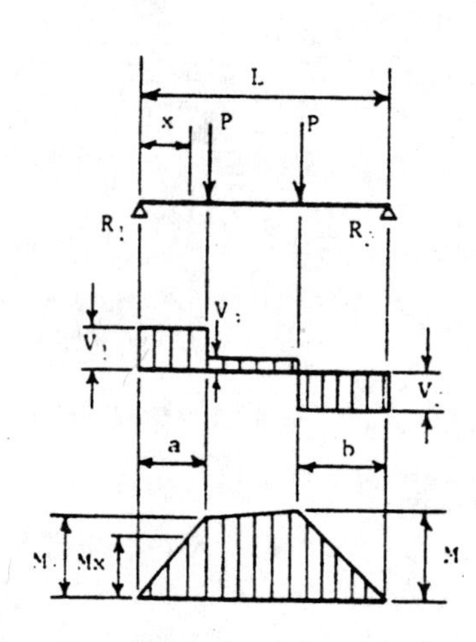

Eq 412-11 a = b ≠ c

$R = V$ . . . . $= P$

M_m [between loads] . . . . $= Pa$

M_x [@ $x < a$] . . . . $= Px$

D_m [@ center] . . . . $= \frac{Pa}{24EI}(3L^2 - 4a^2)$

D_x [@ $x < a$] . . . . $= \frac{Px}{6EI}(3La - 3a^2 - x^2)$

D_x [@ $a < x < (L - a)$] . . . . . $= \frac{Pa}{6EI}(3Lx - 3x^2 - a^2)$

Eq 412-12 a ≠ b ≠ c

$R_1 = V_1$ [max. @ $a < b$] . . . . . $= \frac{P}{L}(L - a + b)$

$R_2 = V_2$ [max. @ $a > b$] . . . . . $= \frac{P}{L}(L - b + a)$

V_3 . . . . $= \frac{P}{L}(b - a)$

M_1 [max. @ $a > b$] . . . . $= R_1 a$

M_2 [max. @ $a < b$] . . . . $= R_2 b$

M_x [@ $x < a$] . . . . $= R_1 x$

M_x [@ $a < x < (L - b)$] . . . . . $= R_1 x - P(x - a)$

M_x [@ $(L - b) < x < L$] . . . . . $= R_2(L - x)$

NOTE: For deflections use superposition of 2 single concentrated loads.

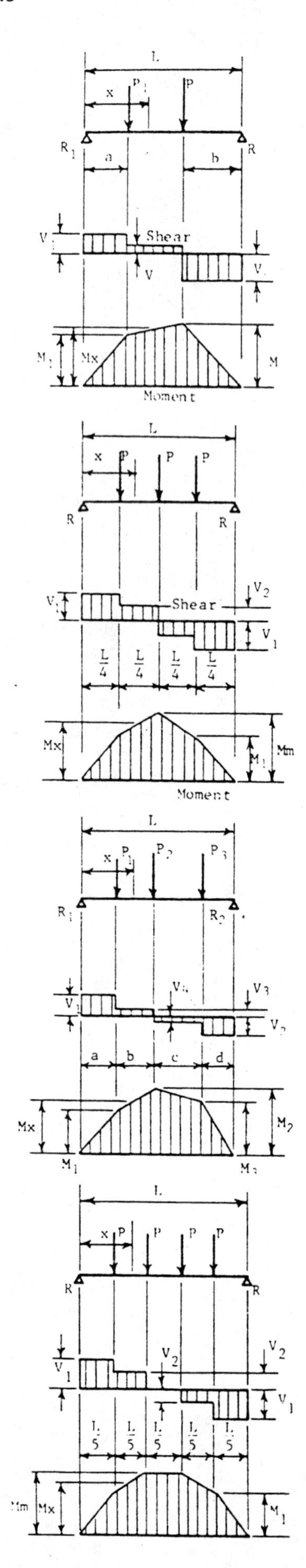

2 Unequal loads

Eq 412-13 $P_1 \neq P_2$

$R_1 = V_1$ . . . . $= \dfrac{P_1(L - a) + P_2 b}{L}$

$R_2 = V_2$ . . . . $= \dfrac{P_1 a + P_2(L - b)}{L}$

V_3 . . . . $= R_1 - P_1$

M_1 [max. @ $R_1 < P_1$] . . . $= R_1 a$

M_2 [max. @ $R_2 < P_2$] . . . $= R_2 b$

M_x [@ $x < a$] . . . . $= R_1 x$

M_x [@ $a < x < (L - b)$] . . $= R_1 x - P_1(x - a)$

NOTE: For deflections use superposition of 2 single concentrated loads.

Eq 412-14

$R = V_1$ . . . . $= 3P/2$

V_2 . . . . $= P/2$

M_x [@ $x < L/4$] . . . . . $= \dfrac{3Px}{2}$

M_x [@ $L/4 < x < L/2$] . . $= \dfrac{Px}{2} + \dfrac{PL}{4}$

M_1 [@ $x = L/4$] . . . . . $= \dfrac{3PL}{8}$

M_m [@ $x = L/2$] . . . . . $= PL/2$

D_m [@ $x = L/2$] . . . . . $= \dfrac{19PL^3}{384\ EI}$

Eq 412-15

$R_1 = V_1$ . . . . $= \dfrac{P_1(L - a) + P_2(c + d) + P_3 d}{L}$

$R_2 = V_2$ . . . . $= \dfrac{P_1 a + P_2(a + b) + P_3(L - d)}{L}$

V_3 . . . . $= R_1 - P_1$

V_4 . . . . $= R_1 - P_1 - P_2$

M_1 . . . . $= R_1 a$

M_2 . . . . $= R_1(a + b) - P_1 b$

M_3 . . . . $= R_1(L - d) - P_1(b + c) - P_2 c$

M_m [when $P_1 \geq R_1$] . . . . . . $= M_1$

M_m [when $P_1 + P_2 \geq R_1, P_2 + P_3 \geq R_2$] $= M_2$

M_m [when $P_3 \geq R_2$] . . . . . . $= M_3$

NOTE: For deflections use superposition of 3 single concentrated loads.

Eq 412-16

$R = V_1$ . . . . $= 2P$

V_2 . . . . $= P$

M_x [@ $x < L/5$] . . . . $= 2Px$

M_x [@ $L/5 < x < 2L/5$] . . . . $= Px + PL/5$

M_1 . . . . $= \dfrac{2PL}{5}$

M_m [@ $x = L/2$] . . . . $= \dfrac{3PL}{5}$

D_x [@ $x = L/2$] . . . . $= \dfrac{3.78\ PL^3}{60EI}$

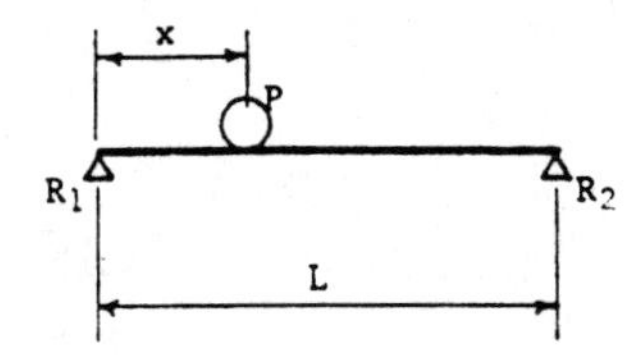

Eq 412-17

$R_1m = V_1m$ [@ x = 0] . . . . . $= P$

Mm [@ load, @ $x = \frac{L}{2}$] . . . . $= \frac{PL}{4}$

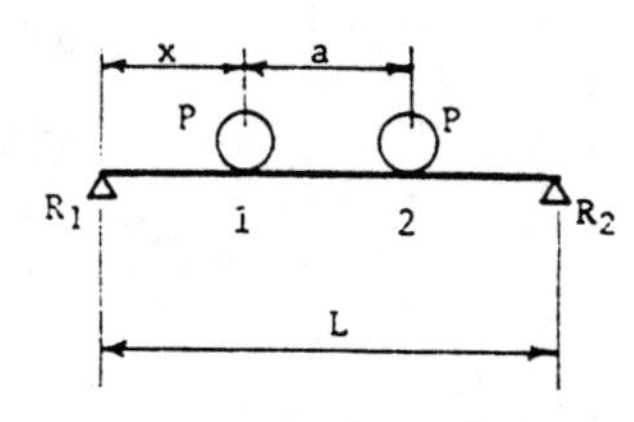

Eq 412-18

$R_1m = V_1m$ [@ x = 0] . . . . $= P(2 - \frac{a}{L})$

Mm
- [@ a < .586L, @ load 1 @ $x = \frac{1}{2}(L - \frac{a}{2})$] . . $= \frac{P}{2L}(L - \frac{a}{2})^2$
- [@ a > .586L with one load at center of span] . . $= \frac{PL}{4}$

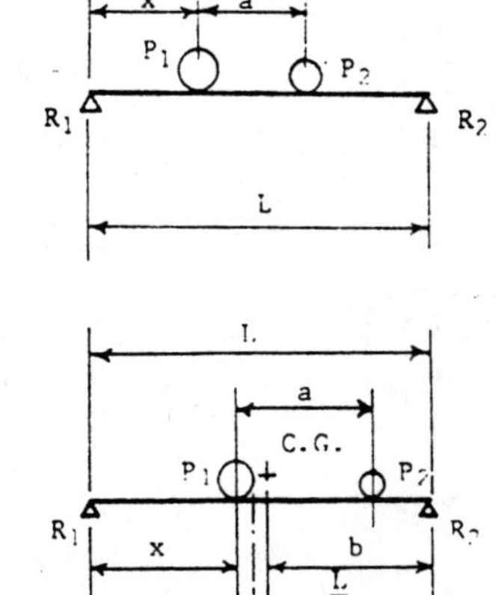

Eq 412-19

$R_1m = V_1m$ [@ x = 0] . . . . $= P_1 + P_2\frac{L - a}{L}$

Mm
- [under P_1, @ $x = \frac{1}{2}(L - \frac{P_2 a}{P_1 + P_2})$] . $= (P_1 + P_2)\frac{x^2}{L}$
- [Mm may be with P_1 at center of span and P_2 off span] . $= \frac{P_1 L}{4}$

Maximum shear is at one support when one of the loads is at that support. With several moving loads, locate them for maximum shear by trial.

Maximum moment is under one of the loads when that load is as far from one support as the center of gravity of all the moving loads on the beam is from the other support.

Mm is at P_1 when x = b, and when the span center line is midway between the center of gravity of loads and the nearest concentrated load.

Cantilever Beams

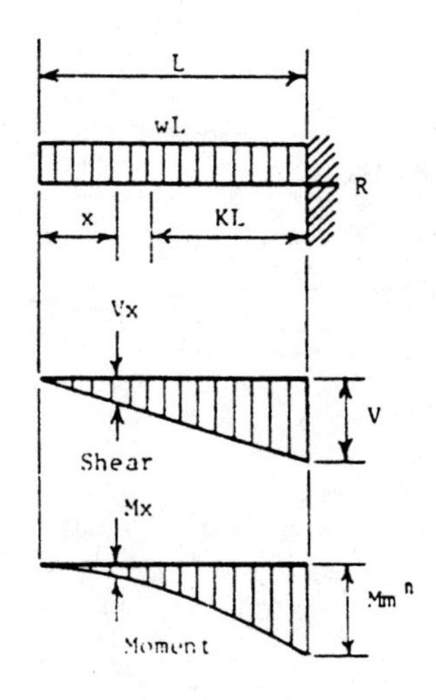

Eq 412-20

R = V . . . . $= wL$

Vx . . . . $= wx$

Mm [@ fixed end] . . . . $= \frac{-wL^2}{2}$

Mx . . . . $= \frac{-wx^2}{2}$

Dm [@ free end] . . . . $= \frac{wL^4}{8EI}$

Dx . . . . $= \frac{w}{24EI}(x^4 - 4L^3x + 3L^4)$

For partial load from support to distance KL, substitute KL for L. Measure x from left end of KL. Dm then becomes D_{KL}.

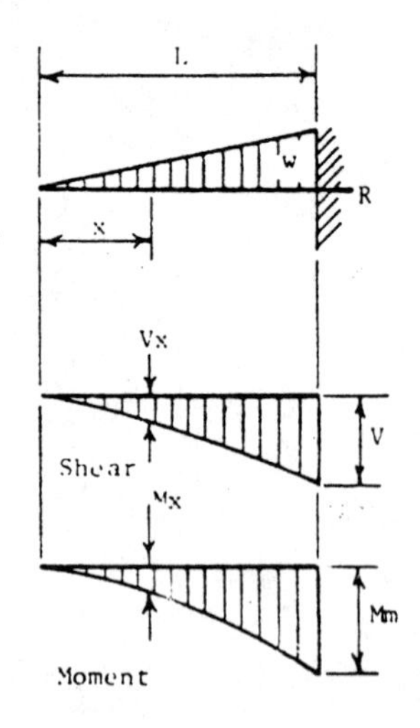

Eq 412-21

W . . . . $= wL/2$

R = V . . . . $= wL/2$

Vx . . . . $= wx^2/2L$

Mm [@ fixed end] . . . . $= -wL^2/6$

Mx . . . . $= -wx^3/6L$

Dm [@ free end] . . . . $= \dfrac{wL^4}{30EI}$

Dx . . . . $= \dfrac{w}{120EIL}(x^5 - 5L^4x + 4L^5)$

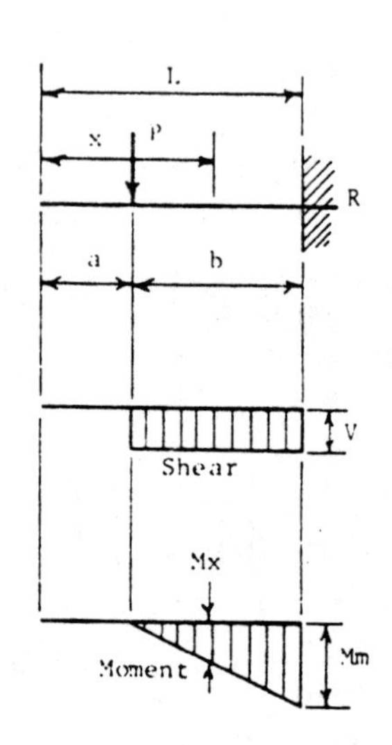

Eq 412-22 a = 0 (load at free end)

R = V . . . . $= P$

Mm [@ fixed end] . . . . $= -PL$

Mx . . . . $= -Px$

Dm [@ free end] . . . . $= \dfrac{PL^3}{3EI}$

Dx . . . . $= \dfrac{P}{6EI}(2L^3 - 3L^2x + x^3)$

Eq 412-23 a ≠ 0

R = V [@ a < x < L] . . . . . $= P$

Mm [@ fixed end] . . . . $= -Pb$

Mx [@ x > a] . . . . $= -P(x - a)$

Dm [@ free end] . . . . $= \dfrac{Pb^2}{6EI}(3L - b)$

Da [@ load] . . . . $= \dfrac{Pb^3}{3EI}$

Dx [@ x < a] . . . . $= \dfrac{Pb^2}{6EI}(3L - 3x - b)$

Dx [@ x > a] . . . . $= \dfrac{P(L - x)^2}{6EI}(3b - L + x)$

Overhang Beams

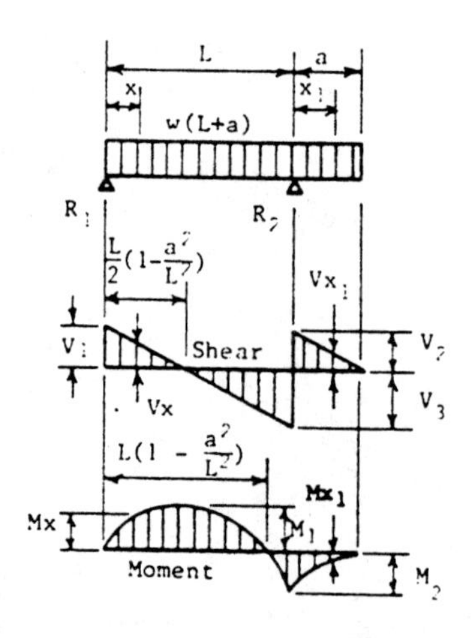

Eq 412-24

$R_1 = V_1$ . . . . $= \dfrac{w}{2L}(L^2 - a^2)$

$R_2 = V_2 + V_3$ . . . . $= \dfrac{w}{2L}(L + a)^2$

V_2 . . . . $= wa$

V_3 . . . . $= \dfrac{w}{2L}(L^2 + a^2)$

Vx . . . . $= R_1 - wx$

Vx_1 . . . . $= w(a - x_1)$

M_1 [@ $x = \frac{L}{2}(1 - \frac{a^2}{L^2})$] . . . . . $= \dfrac{w}{8L^2}(L + a)^2(L - a)^2$

M_2 [@ R_2] . . . . $= \dfrac{-wa^2}{2}$

Mx . . . . $= \dfrac{wx}{2L}(L^2 - a^2 - xL)$

Mx_1 . . . . $= -\dfrac{w}{2}(a - x_1)^2$

Dx . . . . . $= \dfrac{wx}{24EIL}(L^4 - 2L^2x^2 + Lx^3 - 2a^2L^2 + 2a^2x^2)$

Dx_1 . . . . . $= \dfrac{wx_1}{24EI}(4a^2L - L^3 + 6a^2x_1 - 4ax_1^2 + x_1^3)$

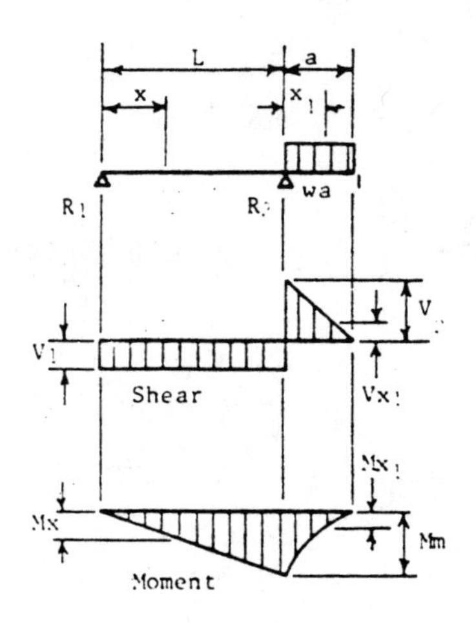

Eq 412-25

$R_1 = V_1$ . . . . $= -\frac{wa^2}{2L}$

$R_2 = V_1 + V_2$ . . . . $= \frac{wa}{2L}(2L + a)$

V_2 . . . . $= wa$

Vx_1 . . . . $= w(a - x_1)$

Mm [@ R_2] . . . . $= \frac{-wa^2}{2}$

Mx . . . . $= \frac{-wa^2x}{2L}$

Mx_1 . . . . $= -\frac{w}{2}(a - x_1)^2$

Dmx [@ $x = \frac{L}{3}$] . . . . $= -\frac{wa^2L^2}{18\sqrt{3}\,EI}$

$= -.03208\frac{wa^2L^2}{EI}$

Dmx_1 [@ $x_1 = a$] . . . . $= \frac{wa^3}{24EI}(4L + 3a)$

Dx . . . . $= -\frac{wa^2x}{12EIL}(L^2 - x^2)$

Dx_1 . . . . $= \frac{wx_1}{24EI}(4a^2L + 6a^2x_1 - 4ax_1^2 + x_1^3)$

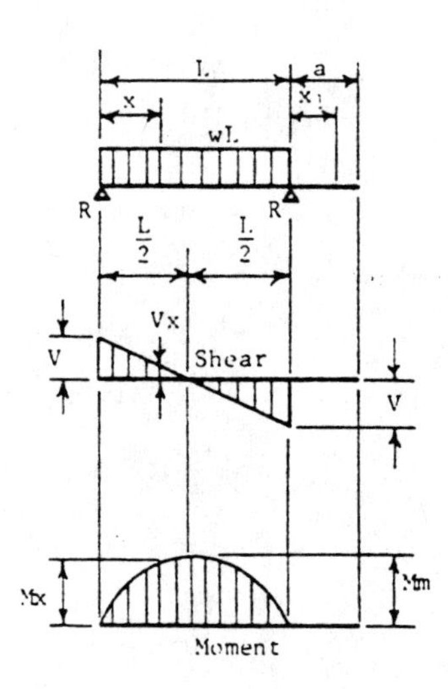

Eq 412-26

$R = V$ . . . . $= \frac{wL}{2}$

Vx . . . . $= w(\frac{L}{2} - x)$

Mm [@ center] . . . . $= \frac{wL^2}{8}$

Mx . . . . $= \frac{wx}{2}(L - x)$

Dm [@ center] . . . . $= \frac{5wL^4}{384EI}$

Dx . . . . $= \frac{wx}{24EI}(L^3 - 2Lx^2 + x^3)$

Dx_1 . . . . $= -\frac{wL^3x_1}{24EI}$

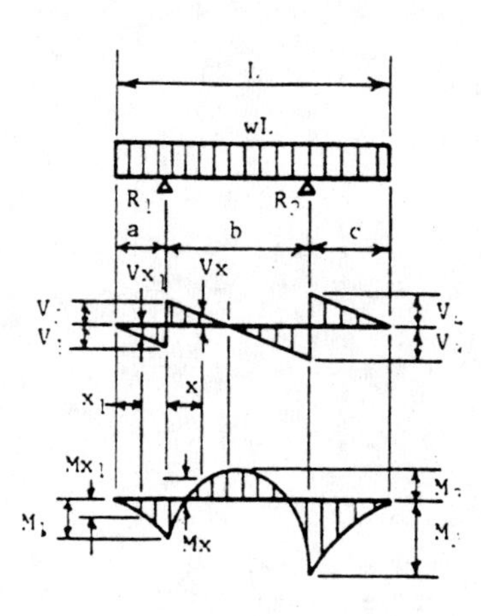

Eq 412-27

R_1 . . . . $= \frac{wL(L - 2c)}{2b}$

R_2 . . . . $= \frac{wL(L - 2a)}{2b}$

V_1 . . . . $= wa$

V_2 . . . . $= R_1 - V_1$

V_3 . . . . $= R_2 - V_4$

V_4 . . . . $= wc$

Vx_1 . . . . $= V_1 - wx_1$

Vx [@ $x < L$] . . . . $= R_1 - w(a + x_1)$

Vm [@ $a < c$] . . . . $= R_2 - wc$

M_1 . . . . $= -\frac{wa^2}{2}$

M_2 . . . . $= -\frac{wc^2}{2}$

M_3 . . . . $= R_1(\frac{R_1}{2w} - a)$

Mx [max. @ $x = \frac{R_1}{w} - a$] . . . $= R_1x - \frac{w(a + x)^2}{2}$

Mx_1 . . . . $= \frac{-wx_1^2}{2}$

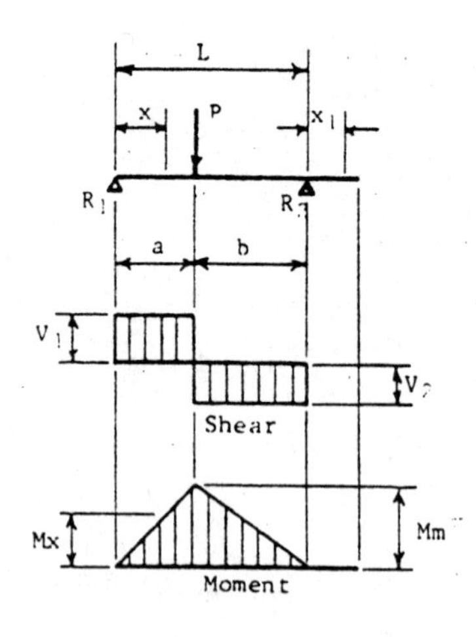

Eq 412-28

$R_1 = V_1$ [max. @ a < b] . . . $= \frac{Pb}{L}$

$R_2 = V_2$ [max. @ a > b] . . . $= \frac{Pa}{L}$

Mm [@ load] . . . . $= \frac{Pab}{L}$

Mx [@ x < a] . . . . $= \frac{Pbx}{L}$

Dm $\left[@\ x = \sqrt{\frac{a(a+2b)}{3}}\ @\ a > b\right] = \frac{Pab(a+2b)\sqrt{3a(a+2b)}}{27EIL}$

Da [@ load] . . . . $= \frac{Pa^2b^2}{3EIL}$

Dx [@ x < a] . . . . $= \frac{Pbx}{6EIL}(L^2 - b^2 - x^2)$

Dx [@ x > a] . . . . $= \frac{Pa(L - x)}{6EIL}(2Lx - x^2 - a^2)$

Dx_1 . . . . $= \frac{Pabx_1}{6EIL}(L + a)$

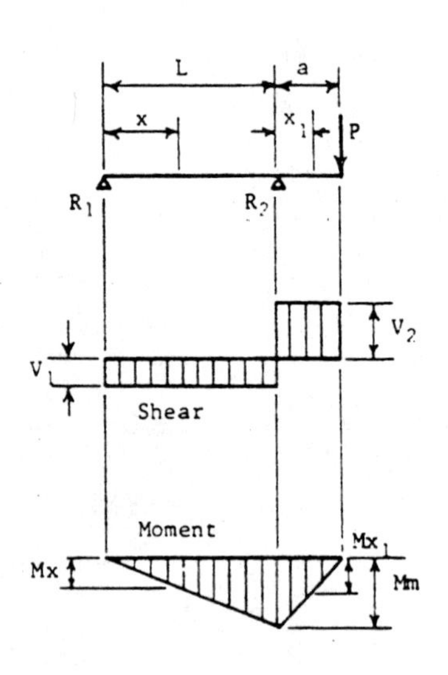

Eq 412-29

$R_1 = V_1$ . . . . $= \frac{-Pa}{L}$

$R_2 = V_1 + V_2$ . . . . $= \frac{P}{L}(L + a)$

V_2 . . . . $= P$

Mm [@ R_2] . . . . $= -Pa$

Mx [between supports] . . . . . $= \frac{-Pax}{L}$

Mx_1 [for overhang] . . . . . . $= -P(a - x_1)$

Dm $\left[@\ x = \frac{L}{\sqrt{3}}\right]$ . . . . $= -.06415\ \frac{PaL^2}{EI}$

Dm [@ $x_1 = a$] . . . . $= \frac{Pa^2}{3EI}(L + a)$

Dx [between supports] . . . . . $= \frac{-Pax}{6EIL}(L^2 - x^2)$

Dx_1 [for overhang] . . . . . . $= \frac{Px_1}{6EI}(2aL + 3ax_1 - x_1^2)$

Multi-Span Beams

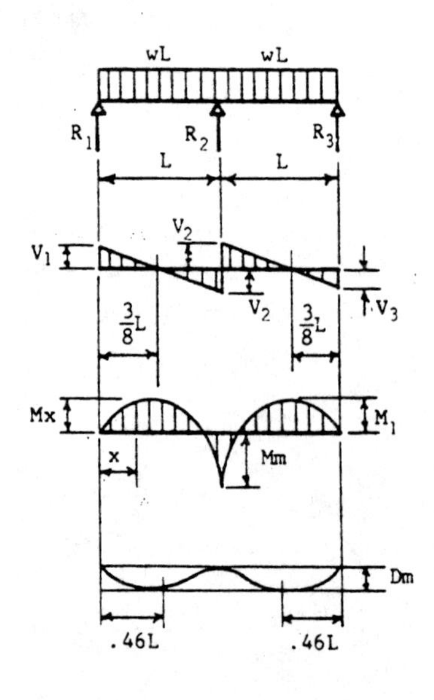

Eq 412-30

$R_1 = V_1 = R_3 = V_3$ . . . . $= \frac{3wL}{8}$

R_2 . . . . $= \frac{10wL}{8}$

$V_2 = Vm$ . . . . $= \frac{5wL}{8}$

Mm . . . . $= -\frac{wL^2}{8}$

M_1 $\left[@\ x = \frac{3L}{8}\right]$ . . . . $= \frac{9wL^2}{128}$

Mx [@ x < L] . . . . $= \frac{3wLx}{8} - \frac{wx^2}{2}$

Dm [@ x = approx. 0.46L] . . . . $= \frac{wL^4}{185EI}$

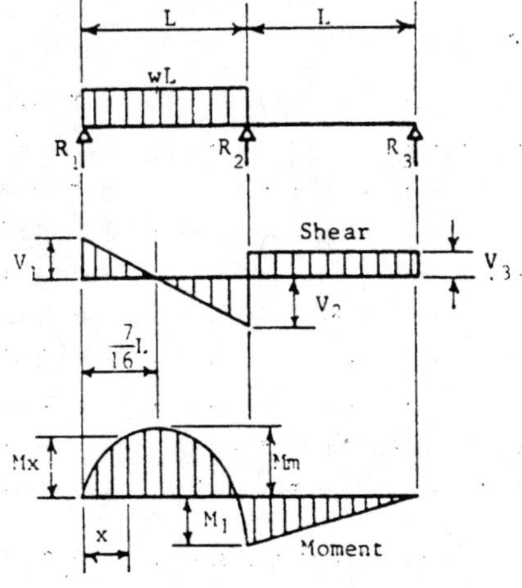

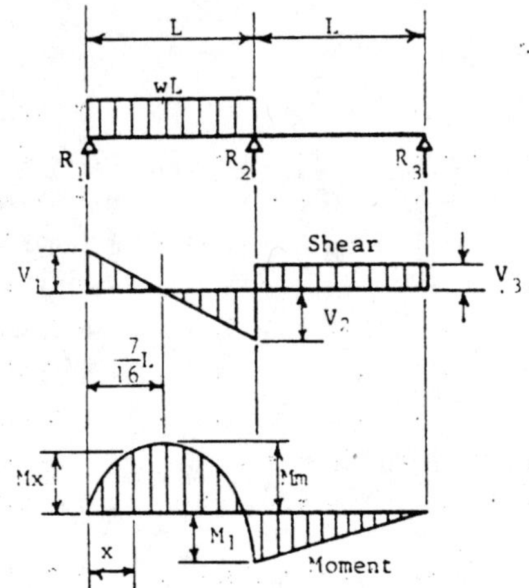

Eq 412-31

$R_1 = V_1$ . . . . $= \frac{7}{16}wL$

$R_2 = V_2 + V_3$ . . . . $= \frac{5}{8}wL$

$R_3 = V_3$ . . . . $= -\frac{1}{16}wL$

V_2 . . . . $= \frac{9}{16}wL$

Mm $[@\ x = \frac{7}{16}L]$ . . . . $= \frac{49}{512}wL^2$

M_1 $[@\ R_2]$ . . . . $= -\frac{1}{16}wL^2$

Mx $[@\ x < L]$ . . . . $= \frac{wx}{16}(7L - 8x)$

Dm $[@\ x = 0.472L]$ . . . . $= \frac{0.0092wL^4}{EI}$

Eq 412-32

$R_1 = V_1$ . . . . $= \frac{M_1}{L_1} + \frac{wL_1}{2}$

R_2 . . . . $= wL_1 + wL_2 - R_1 - R_3$

$R_3 = V_4$ . . . . $= \frac{M_1}{L_2} + \frac{wL_2}{2}$

V_2 . . . . $= wL_1 - R_1$

V_3 . . . . $= wL_2 - R_3$

M_1 . . . . $= -\frac{wL_2{}^3 + wL_1{}^3}{8(L_1 + L_2)}$

Mx $[@\ x < L_1$ max. $@\ x = \frac{R_1}{w}]$ . . . $= R_1x = \frac{wx^2}{2}$

Mx_1 $[@\ x_1 < L_2$ max. $@\ x_1 = \frac{R_3}{w}]$ $= R_3x_1 - \frac{wx_1{}^2}{2}$

Eq 412-33

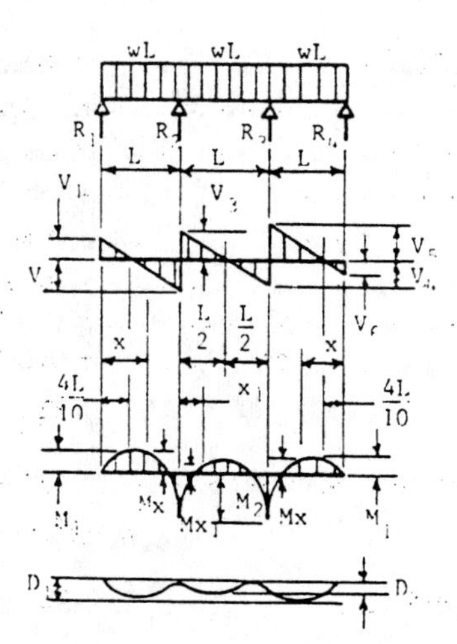

Max. Pos. Mom. M_1 $[@\ x = \frac{4L}{10}] = \frac{2wL^2}{25}$

Max. Neg. Mom. M_2 $[@\ x = L] = -\frac{wL^2}{10}$

$R_1 = R_4 = V_1 = V_6$ . . . . $= \frac{4wL}{10}$

$R_2 = R_3$ . . . . $= \frac{11wL}{10}$

$V_2 = V_5$ . . . . $= \frac{6wL}{10}$

$V_3 = V_4$ . . . . $= \frac{wL}{2}$

Mx $[@\ x < L$ max. $@\ x = \frac{4L}{10}]$ . . . . $= \frac{4wLx}{10} - \frac{wx^2}{2}$

Mx_1 $[@\ x_1 < L$ max. $@\ x_1 = L/2]$ . . . $= \frac{wLx_1}{2} - \frac{wL^2}{10} - \frac{wx_1{}^2}{2}$

$D_1 = Dm$ . . . . $= \frac{4wL^4}{581EI}$

D_2 . . . . $= \frac{wL^4}{1920EI}$

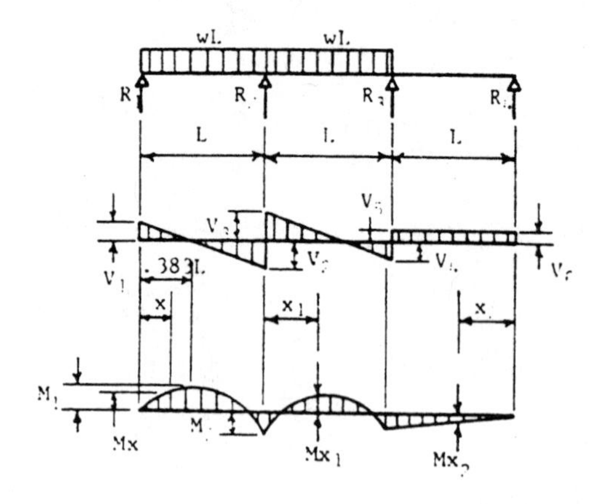

Eq 412-34

$R_1 = V_1$ . . . . = 0.383wL

R_2 . . . . = 1.20wL

R_3 . . . . = 0.450wL

R_4 . . . . = −0.033wL

V_2 . . . . = 0.617wL

V_3 . . . . = 0.583wL

V_4 . . . . = 0.417wL

$V_5 = V_6$ . . . . = 0.033wL

Max. Neg. Mom. M_2 [@ x = L] . . . . = $-0.1167wL^2$

Max. Pos. Mom. M_1 [@ x = 0.383L] . . = $0.0735wL^2$

Mx [@ x < L max. @ x = 0.383L] . . . = $0.383wLx - \frac{wx^2}{2}$

Mx_1 [@ x_1 < L max. @ x_1 = 0.583L] . . = $0.583wLx_1 - 0.1167wL^2 - \frac{wx_1^2}{2}$

Mx_2 [@ x_2 < L] . . . . = $-0.033wLx_2$

Dm [@ x = 0.430L] . . . . = $\frac{0.0059wL^4}{EI}$

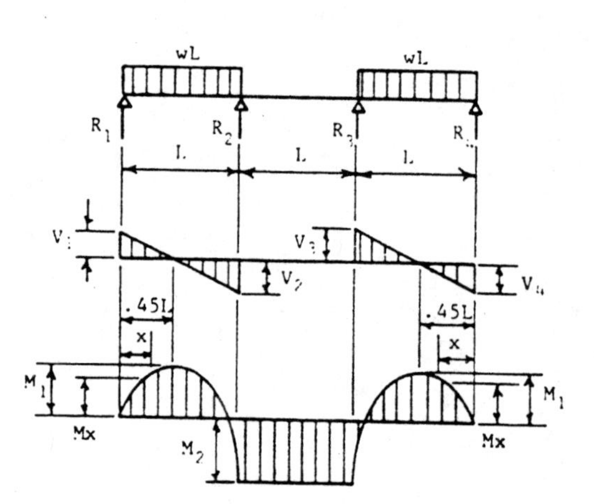

Eq 412-35

$R_1 = R_4 = V_1 = V_4$ . . . . = 0.450wL

$R_2 = R_3 = V_2 = V_3$ . . . . = 0.550wL

Max. Neg. Mom. M_2 [@ x = L] . . . . . = $-0.05wL^2$

Max. Pos. Mom. M_1 [@ x = 0.450L] . = $0.1013wL^2$

Mx [@ x < L] . . . . = $0.450wLx - \frac{wx^2}{2}$

Dm [@ x = 0.479L] . . . . = $\frac{0.0099wL^4}{EI}$

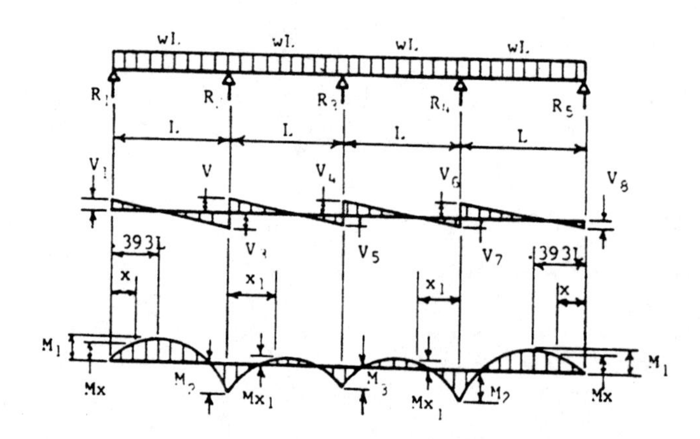

Eq 412-36

$R_1 = R_5 = V_1 = V_8$ . . . . = 0.393wL

$R_2 = R_4$ . . . . = 1.143wL

R_3 . . . . = 0.928wL

$V_2 = V_7$ . . . . = 0.536wL

$V_3 = V_6$ . . . . = 0.607wL

$V_4 = V_5$ . . . . = 0.464wL

M_3 . . . . = $-0.0714wL^2$

Mx [@ x < L max. @ x = 0.393L] . . . . . = $0.393wLx - \frac{wx^2}{2}$

Mx_1 [@ x_1 < L max. @ x_1 = 0.536L] . . . = $0.536wLx_1 - 0.1071wL^2 - \frac{wx_1^2}{2}$

Dm [@ x = 0.440L] . . . . = $\frac{0.0065wL^4}{EI}$

Max. Pos. Mom. M_1 [@ x = 0.393L] . . = $0.0772wL^2$

Max. Neg. Mom. M_2 [@ x = L] . . . . . . = $-0.1071wL^2$

$M_4 = 0.036wL^2$

Eq 412-37

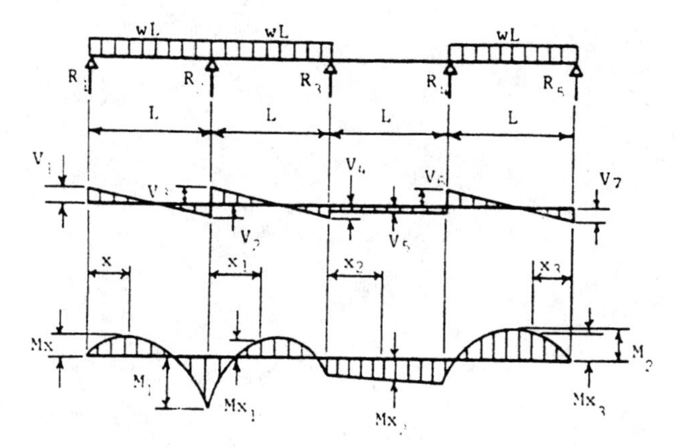

Max. Pos. Mom. M_2 [@ x_3 = 0.442L] = $0.0977wL^2$

Max. Neg. Mom. M_1 [@ x = L] . . . = $-0.1205wL^2$

$R_1 = V_1$ . . . . = 0.380wL

R_2 . . . . = 1.223wL

R_3 . . . . = 0.357wL

R_4 . . . . = 0.598wL

$R_5 = V_7$ . . . . = 0.442wL

V_2 . . . . = 0.620wL

V_3 . . . . = 0.603wL

V_4 . . . . = 0.397wL

V_5 . . . . = 0.040wL

V_6 . . . . = 0.558wL

Mx [@ x < L max. @ x = 0.380L] . . = $0.380wLx - \frac{wx^2}{2}$

Mx_1 [@ x_1 < L max. @ x_1 = 0.603L] . = $0.603wLx_1 - 0.1205wL^2 - \frac{wx_1^2}{2}$

Mx_2 [@ x_2 < L max. @ x_2 = L] . . . = $-0.04wLx_2 - 0.0179wL^2$

Mx_3 [@ x_3 < L max. @ x_3 = 0.442L] . = $0.442wLx_3 - \frac{wx_3^2}{2}$

Dm [@ x = 0.475L] . . . . = $\frac{0.0094wL^4}{EI}$

Eq 412-38

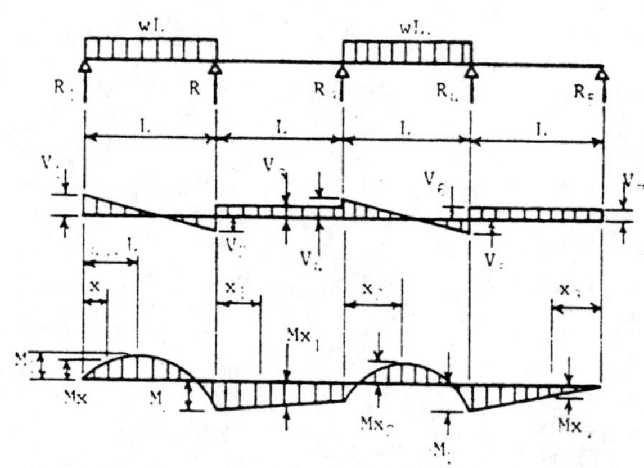

Max. Pos. Mom. M_1 [@ x = 0.446L] . = $0.0996wL^2$

Max. Neg. Mom. M_2 [@ x = L and @ x_2 = L] . . . . = $-0.0536wL^2$

$R_1 = V_1$ . . . . = 0.446wL

$R_2 = R_4$ . . . . = 0.572wL

R_3 . . . . = 0.464wL

R_5 . . . . = −0.054wL

V_2 . . . . = 0.554wL

V_3 . . . . = 0.018wL

V_4 . . . . = 0.482wL

V_5 . . . . = 0.518wL

$V_6 = V_7$ . . . . = 0.054wL

Mx [@ x < L max. @ x = 0.446L] . . = $0.446wLx - \frac{wx^2}{2}$

Mx_1 [@ x_1 < L max. @ x_1 = 0] . . . = $0.018wLx_1 - 0.0536wL^2$

Mx_2 [@ x_2 < L max. @ x_2 = 0.482L] . = $0.482wLx_2 - 0.0357wL^2 - \frac{wx_2^2}{2}$

Mx_3 [@ x_3 < L max. @ x_3 = L] . . . = $-0.054wLx_3$

Dm [@ x = 0.477L] . . . . = $\frac{0.0097wL^4}{EI}$

Eq 412-39

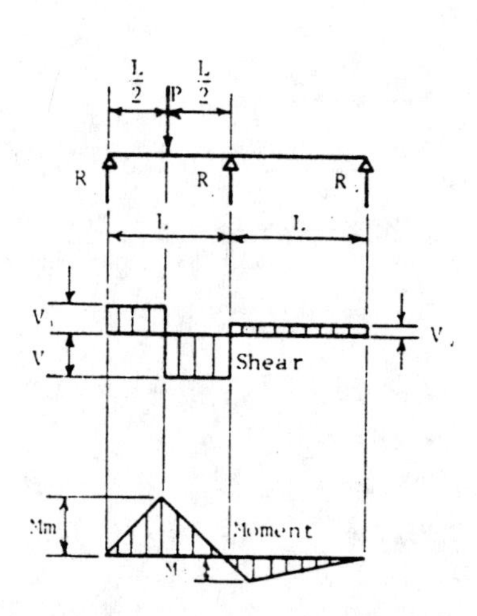

$R_1 = V_1$ . . . . = $\frac{13}{32}$ P

$R_2 = V_2 + V_3$ . . . . = $\frac{11}{16}$ P

$R_3 = V_3$ . . . . = $-\frac{3}{32}$ P

V_2 . . . . = $\frac{19}{32}$ P

Mm [@ load] . . . . = $\frac{13}{64}$ PL

M_1 [@ R_2] . . . . = $-\frac{3}{32}$ PL

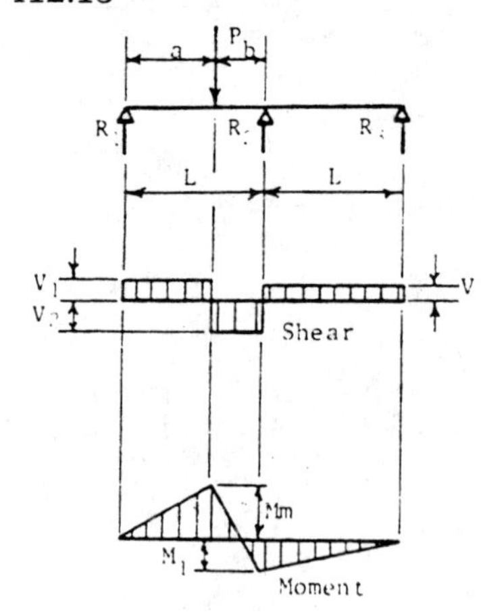

Eq 412-40

$R_1 = V_1$ . . . . $= \frac{Pb}{4L^3}(4L^2 - a(L + a))$

$R_2 = V_2 + V_3$ . . . . $= \frac{Pa}{2L^3}(2L^2 + b(L + a))$

$R_3 = V_3$ . . . . $= -\frac{Pab}{4L^3}(L + a)$

V_2 . . . . $= \frac{Pa}{4L^3}(4L^2 + b(L + a))$

Mm [@ load] . . . . $= \frac{Pab}{4L^3}(4L^2 - a(L + a))$

M_1 [@ R_2] . . . . $= -\frac{Pab}{4L^2}(L + a)$

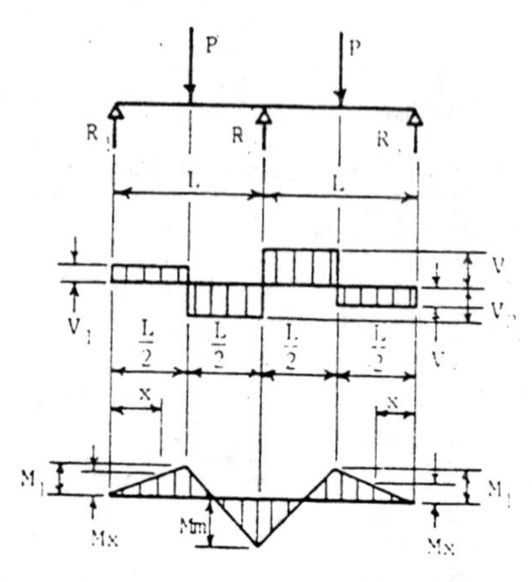

Eq 412-41

$R_1 = V_1 = R_3 = V_3$ . . . . $= \frac{5P}{16}$

$R_2 = 2V_2$ . . . . $= \frac{11P}{8}$

$V_2 = P - R_1$ . . . . $= \frac{11P}{16}$

Mm . . . . $= -\frac{3PL}{16}$

M_1 . . . . $= \frac{5PL}{32}$

Mx [@ $x < a$] . . . . $= R_1x$

Mx [@ $a < x < L$] . . . . $= R_1x - P(x - L/2)$

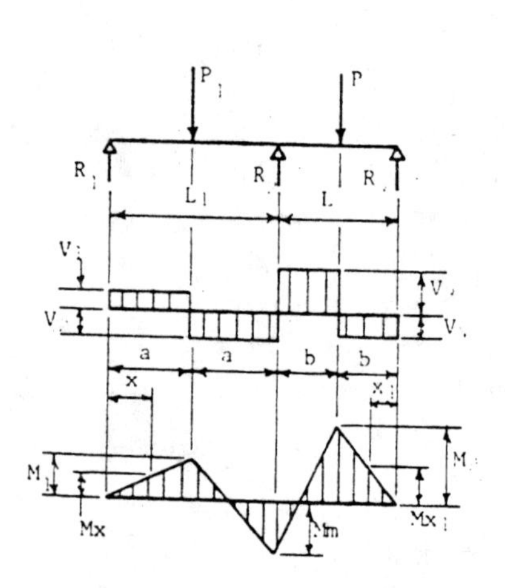

Eq 412-42

$R_1 = V_1$ . . . . $= \frac{M_3}{L_1} + \frac{P_1}{2}$

R_2 . . . . $= P_1 + P_2 - R_1 - R_3$

$R_3 = V_4$ . . . . $= \frac{M_3}{L_2} + \frac{P_2}{2}$

V_2 . . . . $= P_1 - R_1$

V_3 . . . . $= P_2 - R_3$

M_1 . . . . $= R_1a$

M_2 . . . . $= R_2b$

M_3 . . . . $= -\frac{3}{16}\left(\frac{P_1L_1^2 + P_2L_2^2}{L_1 + L_2}\right)$

Mx [@ $x < a$] . . . . $= R_1x$

Mx [@ $a < x < L$] . . $= R_1x - P(x - a)$

Mx_1 [@ $x_1 < b$] . . . $= R_3x_1$

Mx_1 [@ $b < x_1 < L$] . $= R_3x_1 - P(x_1 - b)$

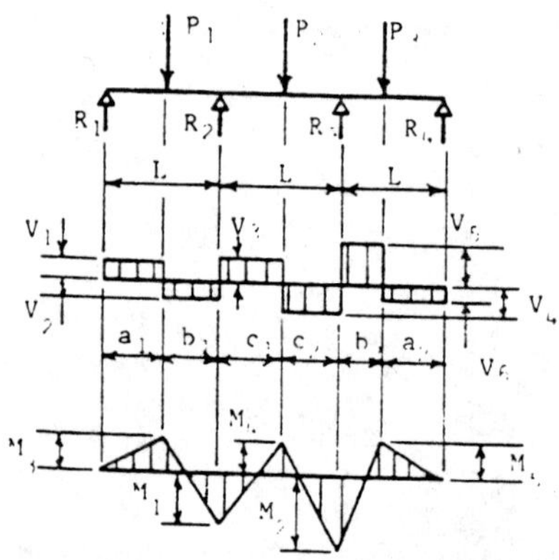

Eq 412-43

$R_1 = V_1$ . . . . $= \frac{M_1 + P_1b_1}{L}$

R_2 . . . . $= \frac{M_2 - 2R_1L + P_2c_2 + P_1(L + b_1)}{L}$

R_3 . . . . $= \frac{M_1 - 2R_4L + P_2c_1 + P_3(L + b_2)}{L}$

$R_4 = V_6$ . . . . $= \frac{M_2 + P_3b_2}{L}$

V_2 . . . . $= R_1 - P_1$

V_3 . . . . $= R_2 - V_2$

V_4 . . . . $= R_3 - V_5$

V_5 . . . . $= R_4 - P_3$

M_1 . . . . $= \frac{-4P_1a_1b_1(L + a_1) - P_2c_1c_2(7L - 5c_1) + P_3b_2a_2(L + a_2)}{15L^2}$

M_2 . . . . $= \frac{P_1a_1b_1(L + a_1) - P_2c_1c_2(2L + 5c_1) - 4P_3b_2a_2(L + a_2)}{15L^2}$

M_3 . . . . $= R_1a_1$

M_4 . . . . $= M_1 + V_3c_1$

M_5 . . . . $= R_4a_2$

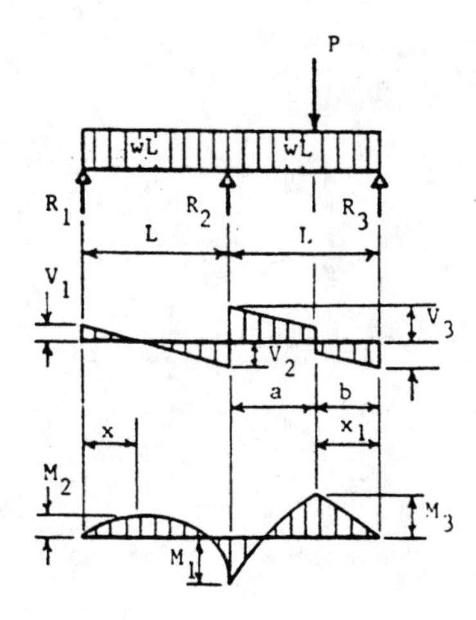

Eq 412-44

$R_1 = V_1$ . . . . $= \frac{M_1}{L} - \frac{wL}{2}$

R_2 . . . . $= 2wL + P - R_1 - R_3$

$R_3 = V_4$ . . . . $= \frac{M_1 + Pa}{L} + \frac{wL}{2}$

V_2 . . . . $= wL - R_1$

V_3 . . . . $= wL + P - R_3$

M_1 . . . . $= -\frac{wL^2}{8} - \frac{Pb(L^2 - b^2)}{4L^2}$

M_2 $[@\ x = \frac{R_1}{w}]$ . . . . $= R_1x - \frac{wx^2}{2}$

M_3 $[@\ x_1 = \frac{R_3}{w}$ when $R_3 < wb]$ $= R_3x_1 - \frac{wx_1^2}{2}$

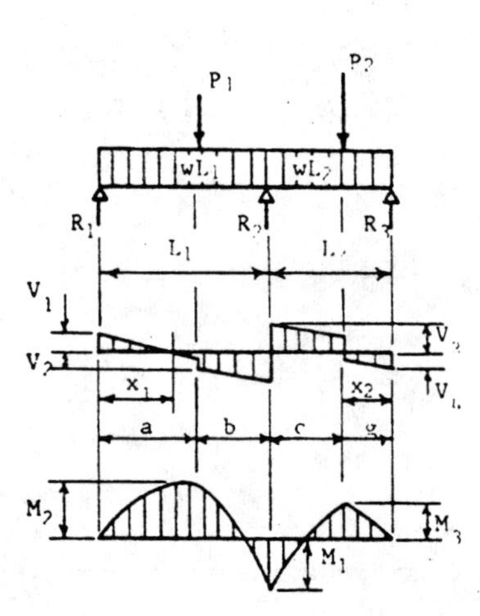

Eq 412-45

$R_1 = V_1$ . . . . $= \frac{M_1 + P_1b}{L_1} + \frac{wL_1}{2}$

R_2 . . . . $= wL_1 + wL_2 + P_1 + P_2 - R_1 - R_3$

$R_3 = V_4$ . . . . $= \frac{M_1 + P_2c}{L_2} + \frac{wL_2}{2}$

V_2 . . . . $= wL_1 + P_1 - R_1$

V_3 . . . . $= wL_2 + P_2 - R_3$

M_1 . . . . $= -\left[\frac{4P_1L_1^2(\frac{a}{L_1} - \frac{a^3}{L_1^3}) + 4P_2L_2^2(\frac{g}{L_2} - \frac{g^3}{L_2^3}) + wL_1^3 + wL_2^3}{8(L_1 + L_2)}\right]$

M_2 $[@\ \frac{R_1}{w}$ when $R_1 \leq wa]$ . . $= R_1x_1 - \frac{wx_1^2}{2}$

M_3 $[@\ \frac{R_3}{w}$ when $R_3 \leq wg]$ . . $= R_3x_2 - \frac{wx_2^2}{2}$

Eq 412-46

$R_1 = V_1$ . . . . $= \frac{wL}{2} + \frac{M_1 + P_1b_1}{L}$

R_2 . . . . $= 2wL + \frac{M_2 - 2R_1L + P_2c_2 + P_1(L + b_1)}{L}$

R_3 . . . . $= 2wL + \frac{M_1 - R_4L + P_2c_1 + P_3(L + b_2)}{L}$

$R_4 = V_6$ . . . . $= \frac{wL}{2} + \frac{M_2 + P_3b_2}{L}$

V_2 . . . . $= P_1 + wL - R_1$

V_3 . . . . $= R_2 - V_2$

V_4 . . . . $= R_3 - V_5$

V_5 . . . . $= P_3 + wL - R_4$

M_1 . . . . $= \frac{-4P_1a_1b_1(L + a_1) - P_2c_1c_2(7L - 5c_1) + P_3b_2a_2(L + a_2)}{15L^2} - \frac{wL^2}{10}$

M_2 . . . . $= \frac{P_1a_1b_1(L + a_1) - P_2c_1c_2(2L + 5c) - 4P_3b_2a_2(L + a_2)}{15L^2} - \frac{wL^2}{10}$

M_3 . . . . $= R_1a_1 - \frac{wa_1^2}{2}$

M_4 . . . . $= M_1 + V_3c_1 - \frac{wc_1^2}{2}$

M_5 . . . . $= R_4a_2 - \frac{wa_2^2}{2}$

Fixed-End Beams

Eq 412-47

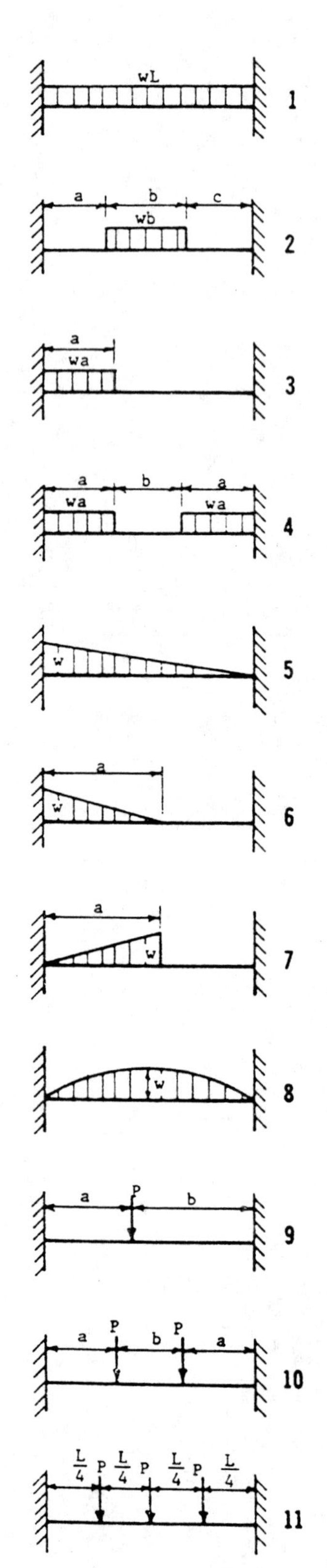

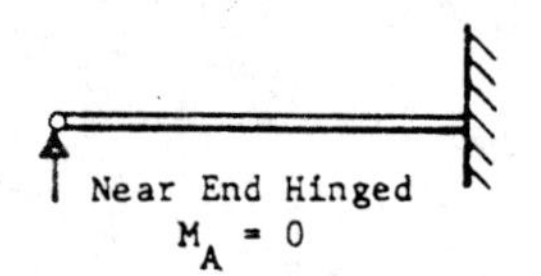

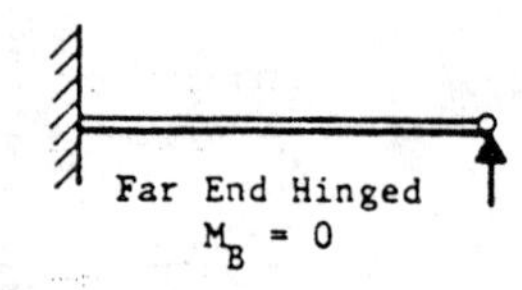

Both Ends Fixed

Fig #	M_A	M_B	R_A	R_B
1	$\frac{wL^2}{12}$	$\frac{wL^2}{12}$	$wL/2$	$wL/2$
2	$\frac{w}{12L^2}[(b+c)^3(4L-3b-3c) -c^3(4L-3c)]$	$\frac{w}{12L^2}[(a+b)^3(4L-3a-3b)-a^3(4L-3a)]$	$\frac{M_A-M_B+wb(c+b/2)}{L}$	$\frac{M_B-M_A+wb(a+b/2)}{L}$
3	$\frac{wa^2}{12L^2}(3a^2-8aL+6L^2)$	$\frac{wa^3}{12L^2}(4L-3a)$	$\frac{M_A-M_B+wa(L-a/2)}{L}$	$\frac{M_B-M_A+wa^2/2}{L}$
4	$\frac{wa^2}{6L}(3L-2a)$	$\frac{wa^2}{6L}(3L-2a)$	wa	wa
5	$\frac{wL^2}{20}$	$\frac{wL^2}{30}$	$\frac{7wL}{20}$	$\frac{3wL}{20}$
6	$\frac{wa^2}{60L^2}(3a^2-10aL+10L^2)$	$\frac{wa^3}{60L^2}(5L-3a)$	$\frac{M_A-M_B+\frac{wa}{2}(L-a/3)}{L}$	$\frac{M_B-M_A+wa^2/6}{L}$
7	$\frac{wa^2}{30L^2}(10L^2-15aL+6a^2)$	$\frac{wa^3}{20L^2}(5L-4a)$	$\frac{M_A-M_B+\frac{wa}{2}(L-2a/3)}{L}$	$\frac{M_B-M_A+wa^2/3}{L}$
8	$\frac{wL^2}{15}$	$\frac{wL^2}{15}$	$wL/3$	$wL/3$
9	$\frac{Pab^2}{L^2}$	$\frac{Pa^2b}{L^2}$	$\frac{Pb+M_A-M_B}{L}$	$\frac{M_B-M_A+Pa}{L}$
10	$\frac{Pa}{L}(L-a)$	$\frac{Pa}{L}(L-a)$	P	P
11	$\frac{5PL}{16}$	$\frac{5PL}{16}$	$3P/2$	$3P/2$

Near End Hinged, $M_A = 0$

Fig #	M_A	M_B	R_A	R_B
6	0	$\frac{wa^2}{120L^2}(10L^2 - 3a^2)$	$\frac{-M_B + \frac{wa}{2}(L - a/3)}{L}$	$\frac{M_B + wa^2/6}{L}$
7	0	$\frac{wa^2}{30L^2}(5L^2 - 3a^2)$	$\frac{-M_B + \frac{wa}{2}(L - 2a/3)}{L}$	$\frac{M_B + wa^2/3}{L}$
3	0	$\frac{wa^2}{8L^2}(2L^2 - a^2)$	$\frac{-M_B + (wa)(L - a/2)}{L}$	$\frac{M_B + wa^2/2}{L}$

Far End Hinged, $M_B = 0$

Fig #	M_A	M_B	R_A	R_B
1	$wL^2/8$	0	$5wL/8$	$3wL/8$
2	$\frac{w}{8L^2}(b^2 + 2cb)(2L^2 - 2c^2 - 2cb - b^2)$	0	$\frac{M_A + wb(c + b/2)}{L}$	$\frac{-M_A + wb(a + b/2)}{L}$
3	$\frac{wa^2}{8L^2}(2L - a)^2$	0	$\frac{M_A + wa(L - a/2)}{L}$	$\frac{-M_A + wa^2/2}{L}$
4	$\frac{wa^2}{4L}(3L - 2a)$	0	$\frac{M_A + waL}{L}$	$\frac{waL - M_A}{L}$
5	$wL^2/15$	0	$2wL/5$	$wL/10$
6	$\frac{wa^2}{120L^2}(20L^2 - 15aL + 3a^2)$	0	$\frac{M_A + \frac{wa}{2}(L - a/3)}{L}$	$\frac{-M_A + wa^2/6}{L}$
7	$\frac{wa^2}{120L^2}(40L^2 - 45aL + 12a^2)$	0	$\frac{M_A + \frac{wa}{2}(L - 2a/3)}{L}$	$\frac{-M_A + wa^2/3}{L}$
8	$wL^2/10$	0	$\frac{M_A + wL^2/3}{L}$	$\frac{-M_A + wL^2/3}{L}$
9	$\frac{Pab}{L^2}(b + a/2)$	0	$\frac{Pb + M_A}{L}$	$\frac{Pa - M_A}{L}$
10	$\frac{3Pa}{2L}(L - a)$	0	$\frac{M_A + PL}{L}$	$\frac{-M_A + PL}{L}$
11	$\frac{15PL}{32}$	0	$\frac{M_A + 3PL/2}{L}$	$\frac{-M_A + 3PL/2}{L}$

413 GABLE FRAME FORMULAS

Loads shown are positive and cause positive reactions. They cause bending moments that are plotted on the compression side of the member. Moments and reactions from superimposed loads are algebraically additive.

Each point on the frame has two coordinates. For symmetrical loads, a point on the rafter is x from the left, x′ from the ridge; a point on a leg is y from the base, y′ from the haunch. For unsymmetrical loads use x_1, x'_1, x_2, etc.

Two-hinge frames are statically indeterminate. Therefore the equations apply only to frames with uniform cross sections in both legs and in both rafters. Some increase in member size at the haunches and ridge, which are assumed rigid, has only a minor effect on moments and reactions.

Three-hinge frames are statically determinate. The members need not have uniform cross section—tapering does not affect moments or reactions.

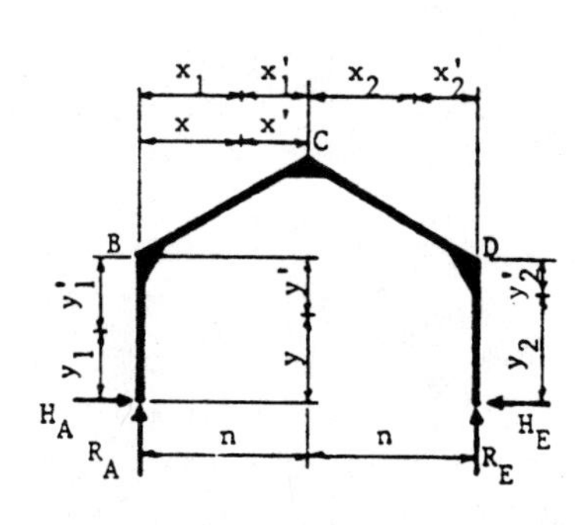

Frame coefficients for Eqs 413-1 to 413-7

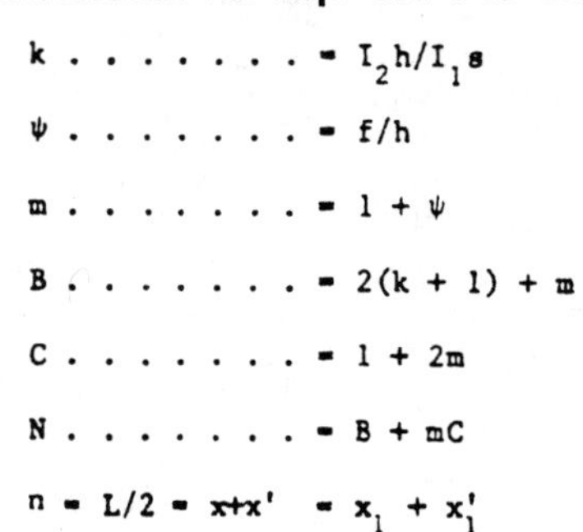

k . . . . $= I_2h/I_1s$

ψ . . . . $= f/h$

m . . . . $= 1 + \psi$

B . . . . $= 2(k + 1) + m$

C . . . . $= 1 + 2m$

N . . . . $= B + mC$

$n = L/2 = x + x' = x_1 + x'_1$

Symmetrical, 2-hinged, vertical legs

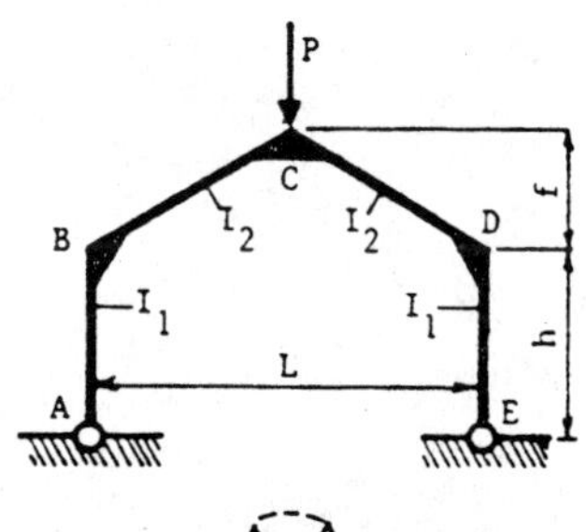

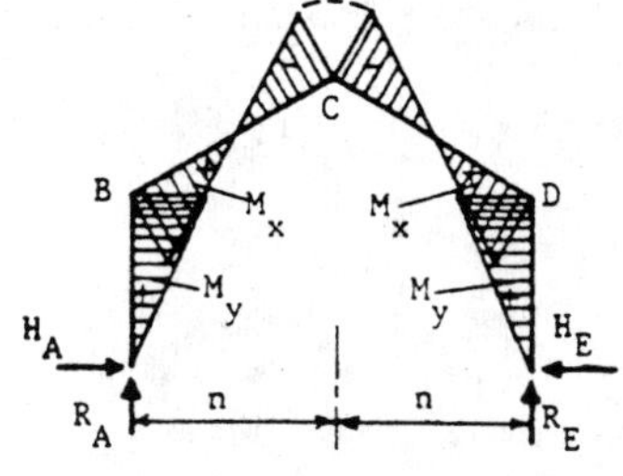

Eq 413-1

$M_B = M_D$ . . . . $= PLC/4N$

M_C . . . . $= -PLB/4N$

$R_A = R_E$ . . . . $= P/2$

$H_A = H_E$ . . . . $= M_B/h$

M_y . . . . $= yM_B/h$

M_x . . . . $= x'M_B/n + xM_C/n$

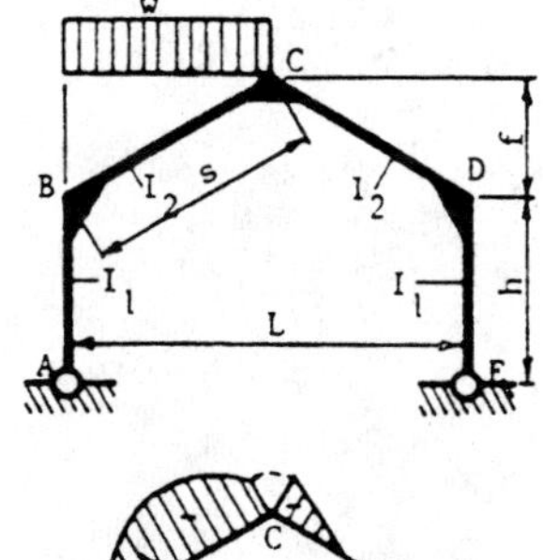

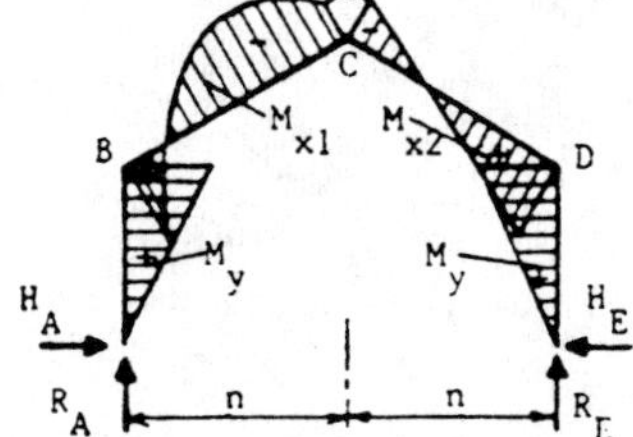

Eq 413-2

$M_B = M_D$ . . . . $= wL^2(3+5m)/32N$

M_C . . . . $= -wL^2/16 + mM_B$

R_A . . . . $= 3wL/8$

R_E . . . . $= wL/8$

$H_A = H_E$ . . . . $= M_B/h$

M_{x1} . . . . . . $= -wx_1x'_1/2 + x'_1M_B/n + x_1M_C/n$

M_{x2} . . . . . . $= x'_2M_C/n + x_2M_D/n$

M_y . . . . $= yM_B/h$

V_{x1} . . . . . . $= -wL^2(\frac{1}{2} - \frac{x_1}{n})/4s + (M_C - M_B)/s$

V_{x2} . . . . . . $= (M_D - M_C)/s$

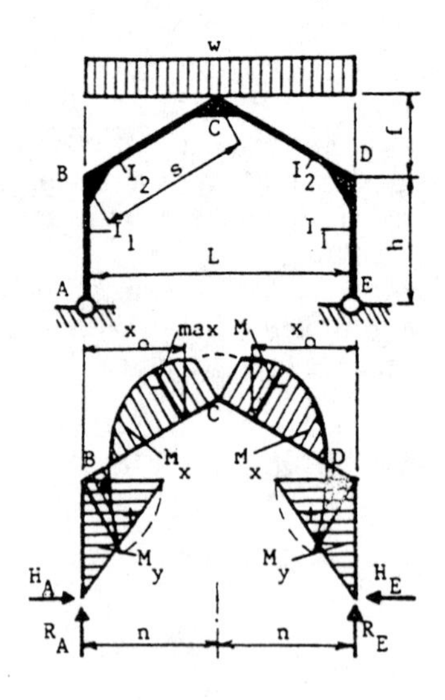

Eq 413-3

$M_B = M_D$ . . . . $= wL^2(3+5m)/16N$

M_C . . . . $= -wL^2/8 + mM_B$

$R_A = R_E$ . . . . $= wL/2$

$H_A = H_E$ . . . . $= M_B/h$

M_x . . . . $= -wxx'/2 + x'M_B/n + xM_C/n$

M_y . . . . $= yM_B/h$

V_x . . . . $= -wL^2(\frac{1}{2} - \frac{x}{n})/4s + (M_C - M_B)/s$

x_0 . . . . $= L/4 + (M_B - M_C)/wn$

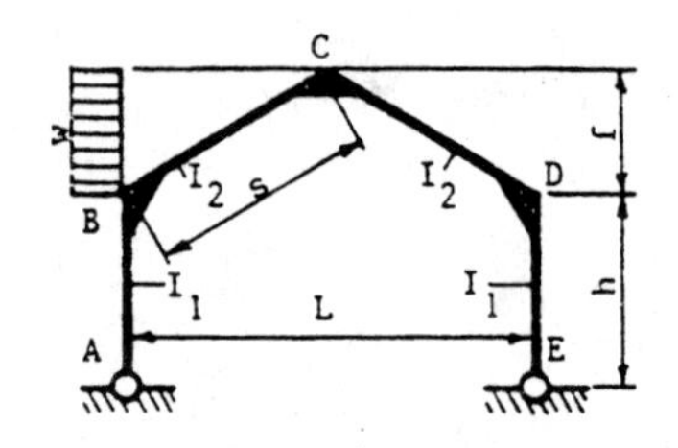

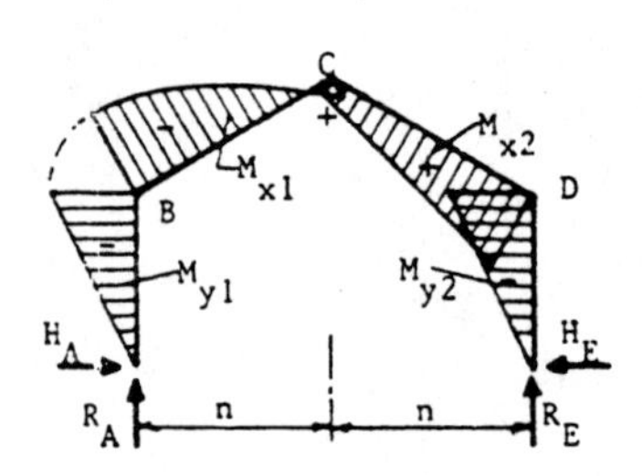

Eq 413-4

X . . . . $= wf^2(C+m)/8N$

M_B . . . . $= -X - wfh/2$

M_C . . . . $= wf^2/4 + mX$

M_D . . . . $= -X + wfh/2$

$R_A = -R_E$ . . . $= -wf(h+f/2)/L$

H_A . . . . $= -x/h - wf/2$

H_E . . . . $= -x/h + wf/2$

M_{x_1} . . . . . . $= -wf^2x_1x_1'/2n + x_1'M_B/n + x_1M_C/n$

M_{x_2} . . . . . . $= x_2'M_C/n + x_2M_D/n$

M_{y1} . . . . . . $= y_1M_B/h$

M_{y2} . . . . . . $= y_2M_D/h$

V_{x1} . . . . . . $= -wf^2(\frac{1}{2} - \frac{x_1}{n})/sn + (M_C - M_B)/s$

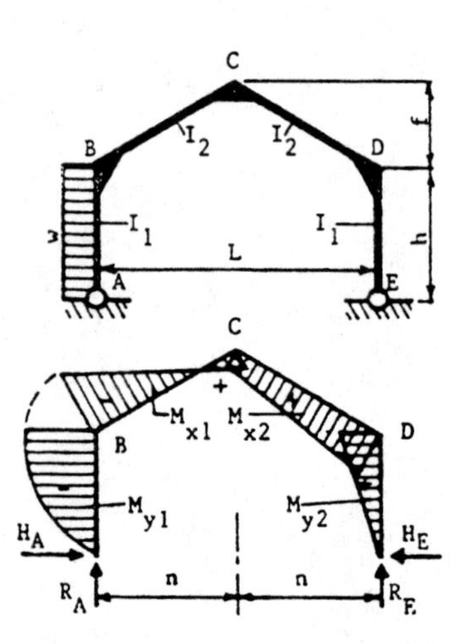

Eq 413-5

M_B . . . . $= -wh^2/2 + M_D$

M_C . . . . $= -wh^2/4 + mM_D$

M_D . . . . $= (wh^2)[2(B + C) + k]/8N$

$R_A = -R_E$ . . . $= -wh^2/2L$

H_A . . . . $= -(wh - H_E)$

H_E . . . . $= M_D/h$

M_{x_1} . . . . . . $= x_1'M_B/n + x_1M_C/n$

M_{x_2} . . . . . . $= x_2'M_C/n + x_2M_D/n$

M_{y_1} . . . . . . $= -wy_1y_1'/2 + y_1M_B/h$

M_{y_2} . . . . . . $= y_2M_D/h$

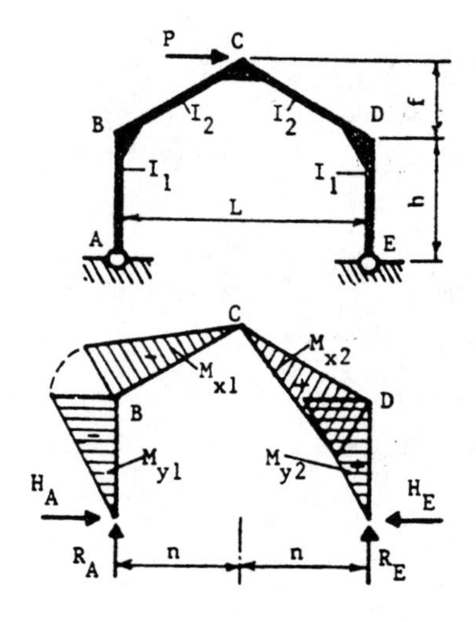

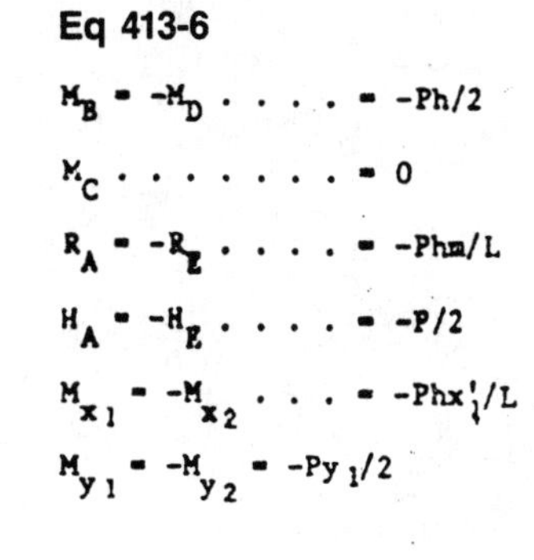

Eq 413-6

$M_B = -M_D \ldots\ldots = -Ph/2$

$M_C \ldots\ldots\ldots = 0$

$R_A = -R_E \ldots\ldots = -Phm/L$

$H_A = -H_E \ldots\ldots = -P/2$

$M_{x1} = -M_{x2} \ldots = -Phx'_1/L$

$M_{y1} = -M_{y2} = -Py_1/2$

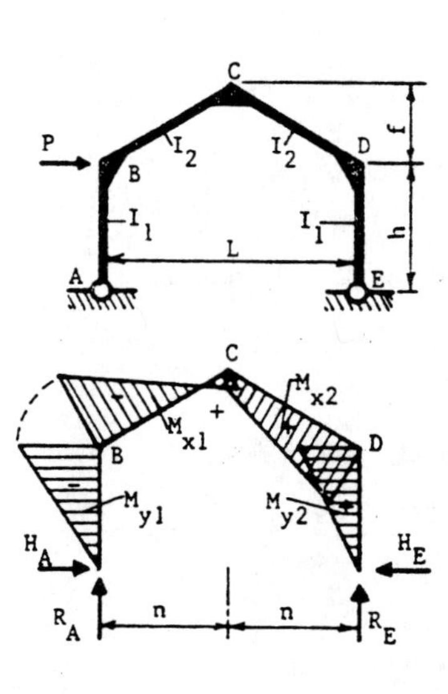

Eq 413-7

$M_B \ldots\ldots = -Ph + M_D$

$M_C \ldots\ldots = -Ph/2 + mM_D$

$M_D \ldots\ldots = Ph(B+C)/2N$

$R_A = -R_E \ldots = -Ph/L$

$H_A \ldots\ldots = -(P - H_E)$

$H_E \ldots\ldots = M_D/h$

$M_{x1} \ldots\ldots = x'_1M_B/n + x_1M_C/n$

$M_{x2} \ldots\ldots = x'_2M_C/n + x_2M_D/n$

$M_{y1} \ldots\ldots = y_1M_B/h$

$M_{y2} \ldots\ldots = y_2M_D/h$

Symmetrical, 3-hinged, vertical legs

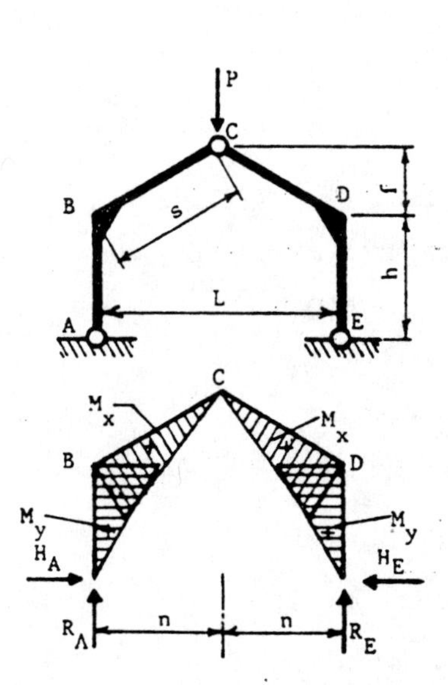

Eq 413-8

$M_B = M_D \ldots = PLh/4(f+h)$

$R_A = R_E \ldots = P/2$

$H_A = H_E \ldots = M_B/h$

$M_x \ldots\ldots = x'M_B/n$

$M_y \ldots\ldots = yM_B/h$

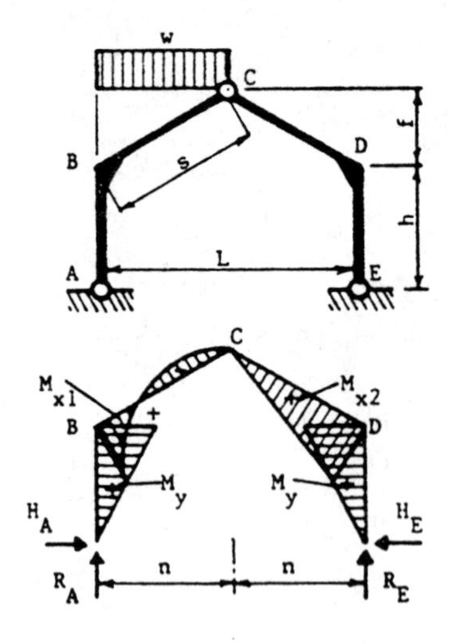

Eq 413-9

$M_B = M_D \ldots = wL^2h/16(f+h)$

$R_A \ldots\ldots = 3wL/8$

$R_E \ldots\ldots = wL/8$

$H_A = H_E \ldots = M_B/h$

$M_{x_1} \ldots\ldots = -wx_1x_1'/2 + x_1'M_B/n$

$M_{x_2} \ldots\ldots = x_2'M_D/n$

$M_y \ldots\ldots = yM_B/h$

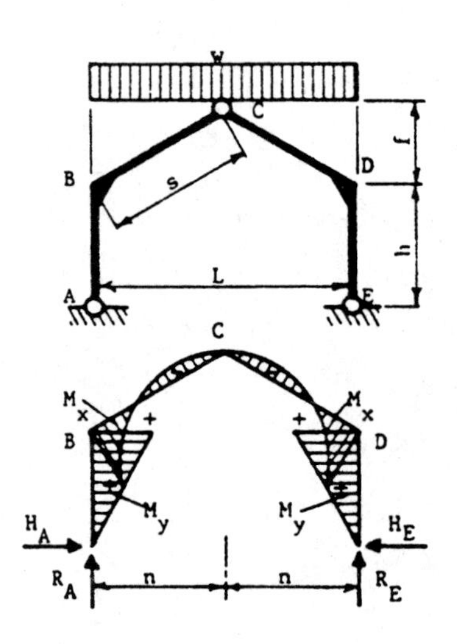

Eq 413-10

$M_B = M_D \ldots = wL^2h/8(f+h)$

$R_A = R_E \ldots = wL/2$

$H_A = H_E \ldots = M_B/h$

$M_x \ldots\ldots = -wxx'/2 + x'M_B/n$

$M_y \ldots\ldots = yM_B/h$

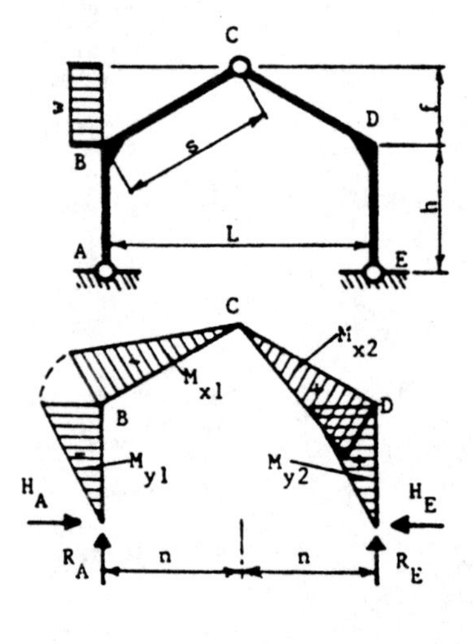

Eq 413-11

$M_B \ldots\ldots = -wfh(3f/4 + h/2)/(f+h)$

$M_D \ldots\ldots = wfh(3f/4 + h/2)/(f+h) + wfh$

$R_A = -R_E \ldots = -wf(h + f/2)/L$

$H_A \ldots\ldots = M_B/h$

$H_E \ldots\ldots = H_A + wf$

$M_{x_1} \ldots\ldots = -wf^2x_1x_1'/2n^2 + x_1'M_B/n$

$M_{x_2} \ldots\ldots = x_2M_D/n$

$M_{y_1} \ldots\ldots = y_1M_B/h$

$M_{y_2} \ldots\ldots = y_2M_C/h$

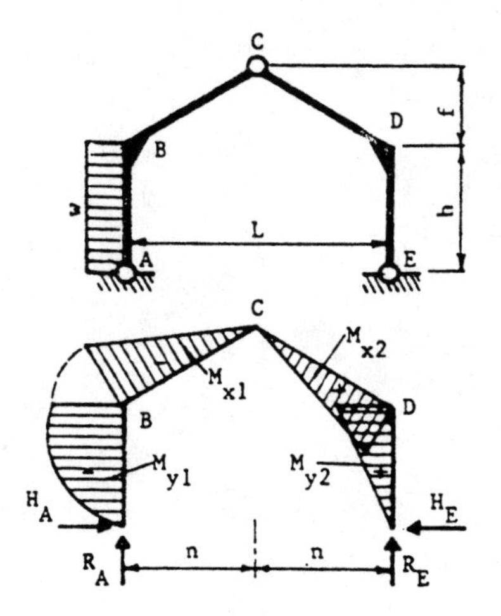

Eq 413-12

M_B . . . . . . $= -wh^2(3h/4 + f)/(f+h)$

M_D . . . . . . $= wh^3/4(f+h)$

$R_A = -R_E$ . . . $= -wh^2/2L$

H_A . . . . . . $= M_B/h - wh/2$

H_E . . . . . . $= wh + H_A$

M_{x1} . . . . . $= x_1' M_B/n$

M_{x2} . . . . . $= x_2 M_D/n$

M_{y1} . . . . . $= -wy_1 y_1'/2 + y_1 M_B/h$

M_{y2} . . . . . $= y_2 M_D/h$

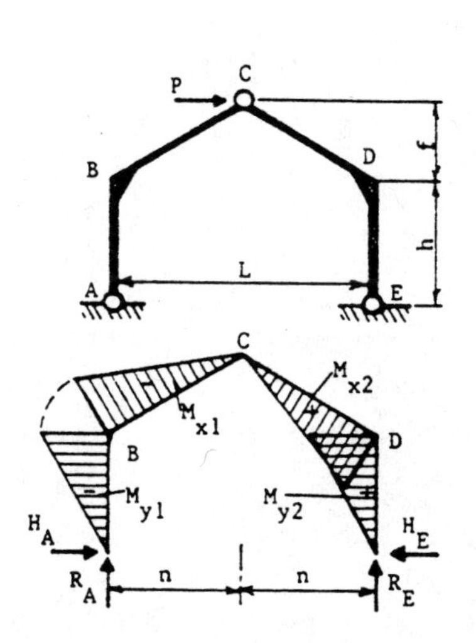

Eq 413-13

$M_B = -M_D$ . . . $= -Ph/2$

$R_A = -R_E$ . . . $= -P(f+h)/L$

$H_A = -H_E$ . . . $= -P/2$

$M_{x1} = -M_{x2}$ . . $= -Phx_1'/L$

$M_{y1} = -M_{y2}$ . . $= -Py_1/2$

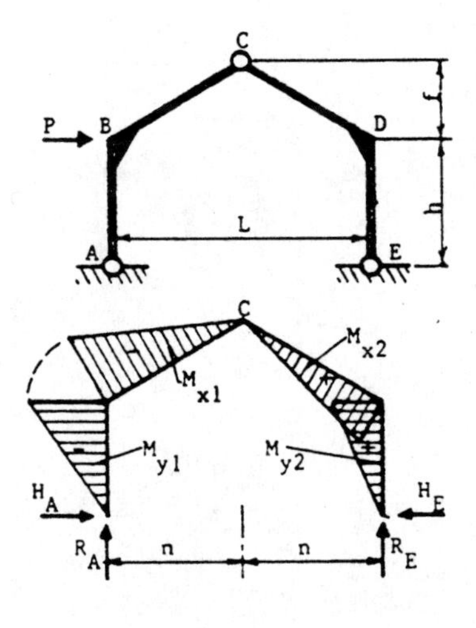

Eq 413-14

M_B . . . . . . $= -Ph(f + h/2)/(f+h)$

M_D . . . . . . $= Ph^2/2(f+h)$

$R_A = -R_E$ . . . $= -Ph/L$

H_A . . . . . . $= M_B/h$

H_E . . . . . . $= P + H_A$

M_{x1} . . . . . $= x_1' M_B/n$

M_{x2} . . . . . $= x_2 M_D/n$

M_{y1} . . . . . $= y_1 M_B/h$

M_{y2} . . . . . $= y_2 M_D/h$

500 PLANNING

501 FARMSTEAD PLANNING

Farmstead planning is setting goals, evaluating needs, mapping existing facilities, and making decisions. Planning objectives are expansion, improved performance, higher capacity, and better labor use. Careful planning includes reviewing the present, assessing the future, and providing flexibility for expansion.

Many factors determine the best plan, and while some are common sense, overlooking one can cause a poorly planned farmstead. Plan on paper, where mistakes can be easily corrected. It is less costly to correct a mistake during the planning stage than after construction begins. A building in the wrong place is a 20-year mistake that has long-term effects on farmstead planning.

Seek financial advice when making a cost/return analysis and evaluating a plan's feasibility.

Developing a Plan

Before starting a plan, prepare a detailed map of each site. Aerial photographs are helpful. Obtain photographs from the local ASCS and SCS offices or take your own. Take pictures toward the north with the sun only an hour or two above the horizon. Longer shadows make the terrain stand out. Contour maps are useful for planning building location, lot and drive slopes, and drainage patterns.

Plan from the farmstead map. Lacquer-type hair spray helps prevent pencil lines from smudging. Experiment with different ideas on tracing paper overlays to save the as-is plan. Take proposals to the field for a visual check. Lay out proposed roads, buildings, or distance zones to visualize the revised farmstead. Use stakes, marker flags, and twine to show boundaries, building walls, and drives.

Zone Planning

Zoning divides a farmstead into separate activity zones. For a farmstead with a family living area, the house is the center of the planning zones. For farmsteads without a house, center the activity zones around the farm court.

On tracing paper overlay, draw to scale circles 100′ apart using the house or farm court as the center. Activity zones help locate major activity centers, preserve the living environment, and encourage proper farmstead activity spacing. Although 100′ wide zones are used, wider zones are often desirable. See Figs 501-1 and 501-2.

Adjusting to changes in operator health, labor supply, or economics can be difficult unless space is available for expanded and new facilities. Plan for both grain and livestock facilities.

Zones

Locate the family living area in Zone 1. Include space for recreation, flower and vegetable gardens, and guest parking. Protect this area from noise, dust, and odors with trees and shrubbery.

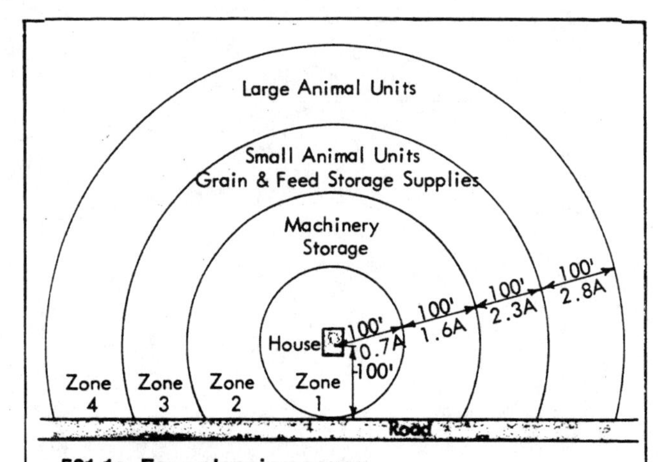

501-1a. Four planning zones.
Set the house back farther than 100′ from a busy road or if a tree windbreak is between the house and the road.

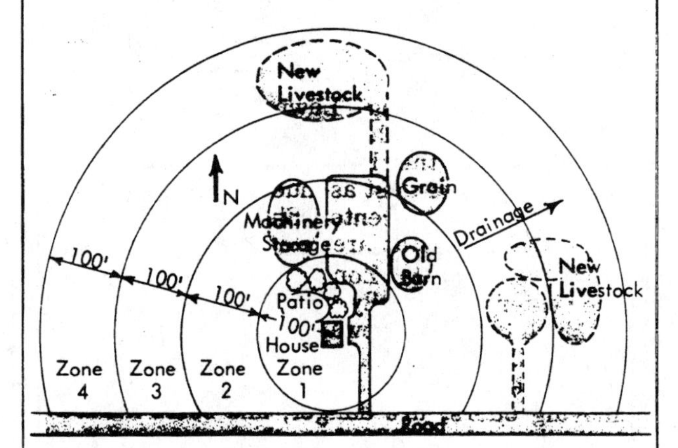

501-1b. Example: livestock enterprise north of the road.
Major centers: living, livestock. Secondary centers: machinery, grain. One driveway serves all centers; a separate drive could serve a new large livestock unit.

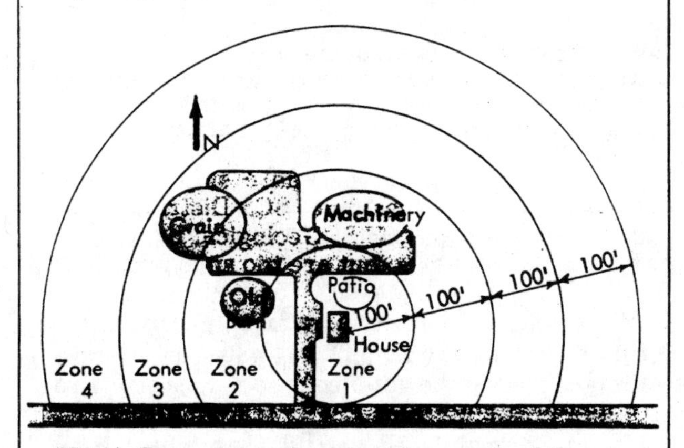

501-1c. Example: grain enterprise north of the road.
Major centers: living, grain. Locate machinery and supply areas for convenience and accessibility.

Fig 501-1. Farmstead planning zones.
The areas of the zones as shown are: Zone 1 = 0.7 acre; Zone 2 = 1.6 acres; Zone 3 = 2.3 acres; Zone 4 = 2.8 acres.

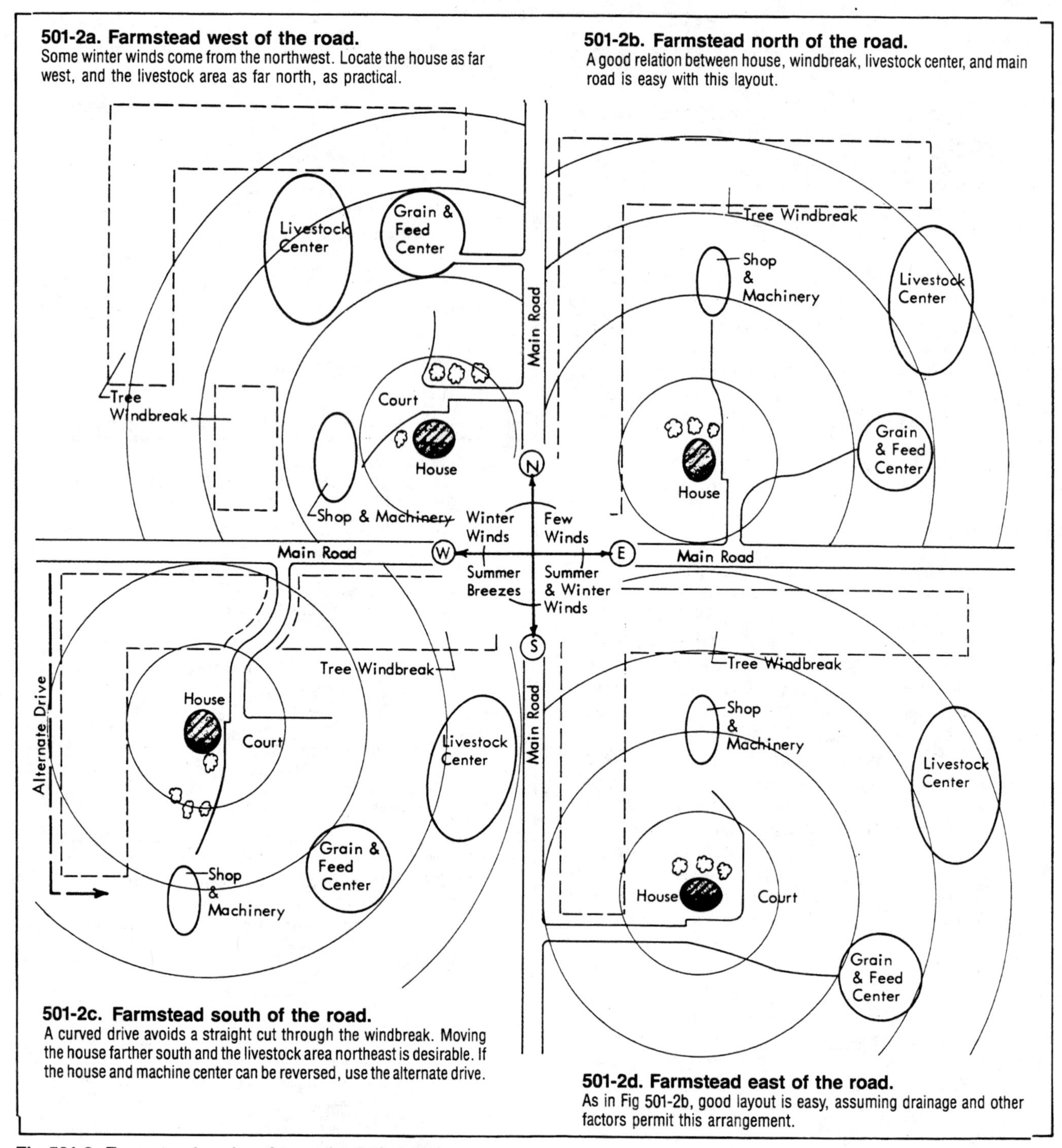

501-2a. Farmstead west of the road.
Some winter winds come from the northwest. Locate the house as far west, and the livestock area as far north, as practical.

501-2b. Farmstead north of the road.
A good relation between house, windbreak, livestock center, and main road is easy with this layout.

501-2c. Farmstead south of the road.
A curved drive avoids a straight cut through the windbreak. Moving the house farther south and the livestock area northeast is desirable. If the house and machine center can be reversed, use the alternate drive.

501-2d. Farmstead east of the road.
As in Fig 501-2b, good layout is easy, assuming drainage and other factors permit this arrangement.

Fig 501-2. Farmstead and main road relationships.
Only major activity centers are illustrated. See text for other factors.

Put the courtyard, farm shop, machinery storage, and related services in Zone 2. Keep fuel and chemical storages toward the outer edge to reduce odor, fire danger, and hazards to family members.

Zone 3 is for grain, feed, small livestock units, and pet housing. Provide good vehicle access and electric power to grain and feed handling and processing equipment. Keep heavy equipment, large dryers, and fire hazards away from the house.

Large livestock units need adequate space, drainage, waste management, vehicle access, loading facilities, feed distribution, and other services. Locate these facilities in Zone 4. Dust, odors, and noise are a problem, so locate livestock units as far from the family living area as possible.

Separation distance

Activity separation distances depend on:

- **Management needs.** Locate high-labor facilities at least 300′ and finishing and mature animal facilities at least 500′ from the family living area.

- **Operation size.** Larger operations create more noise, odors, dust, and traffic requiring greater separation distances. For example, a feed center for processing a few bushels a day has less impact on the living area than one handling large grain volumes and trucks.
- **Pollution hazards.** Livestock operations cause odor, dust, noise, and waste disposal problems. Odors from livestock operations can often be detected ½ mile downwind. Farm chemicals, fertilizers, pesticides, and fuels can also cause pollution problems if improperly handled.
- **Appearance.** A neat and attractive farmstead is important to many families. Consider locating easily landscaped facilities near the living area and less attractive facilities farther away.

When planning, assume that your operation will double in size. Provide space for new buildings, clearance between buildings, and expansion. Separate all buildings by at least 35′ for access, snow storage, and fire protection. Naturally ventilated buildings require at least 50′ clearance. Consider space needs for vehicle access and parking.

Topography, Climate, and Services

Topography and Drainage

Topography affects drainage, building location, access routes, and prevailing wind directions. Do not locate your farmstead on a flood plain, swale, low ground, peat soil, or very rocky soil.

Surface and subsurface drainage are an important feature of an effective farmstead. Natural drainage is preferred, but some improvements are usually needed. Intercept and divert surface water away from the farmstead with terraces. Construct terraces across the slope to carry runoff away. Terraces are usually grassed and maintained by mowing.

Good slopes for surface drainage are 2%-6%. If the site does not have 2%-6% slopes, locate buildings and grade as needed. A south slope exposure promotes drying and is good for livestock feeding areas.

Subsurface drainage is needed around building foundations, grain elevator pits, and below-ground storages to reduce frost heaving and soil pressures. Drainage from around grain pits and underground storages must not pollute streams or groundwater. Avoid sites with springs and high water tables.

Avoid locations near a river, section-line corner, rough topography, or other features limiting expansion.

Wind and Snow Control

Windbreaks help reduce winter winds and control snow. When wind speed is slowed, blowing snow and dust drop out and pile up. Take advantage of hills, trees, buildings, and haystacks for winter wind protection. Allow for summer air movement and drainage when locating windbreaks.

Tree Windbreak

Tree windbreaks reduce wind velocities 5 to 10 tree heights upwind and 10 to 20 tree heights downwind. Plant multi-row windbreaks of tall and short trees, evergreens, and shrubs for best protection. In the Midwest, tree windbreaks are usually planted on the north and west sides of the farmstead. Contact your local extension office for recommended tree species and spacing.

Locate buildings and activity centers at least 150′ downwind because of winter snow accumulation and reduced summer breezes.

Leave openings for lanes and roads. Plant windbreaks to avoid creating a wind tunnel. See Fig 501-3.

Windbreak Fences

Wind passing over a vertical barrier usually drops or swirls downward on the downwind side. A solid fence provides good wind protection for short distances but snow accumulates deeper on both sides of the fence. An 80% solid fence reduces wind speeds for a greater distance and spreads the snow out for faster melting, Fig 501-5.

Winter winds cause drifting and draft problems in open-front buildings. To protect them from wind and snow:

- Provide a continuous eave inlet on the closed

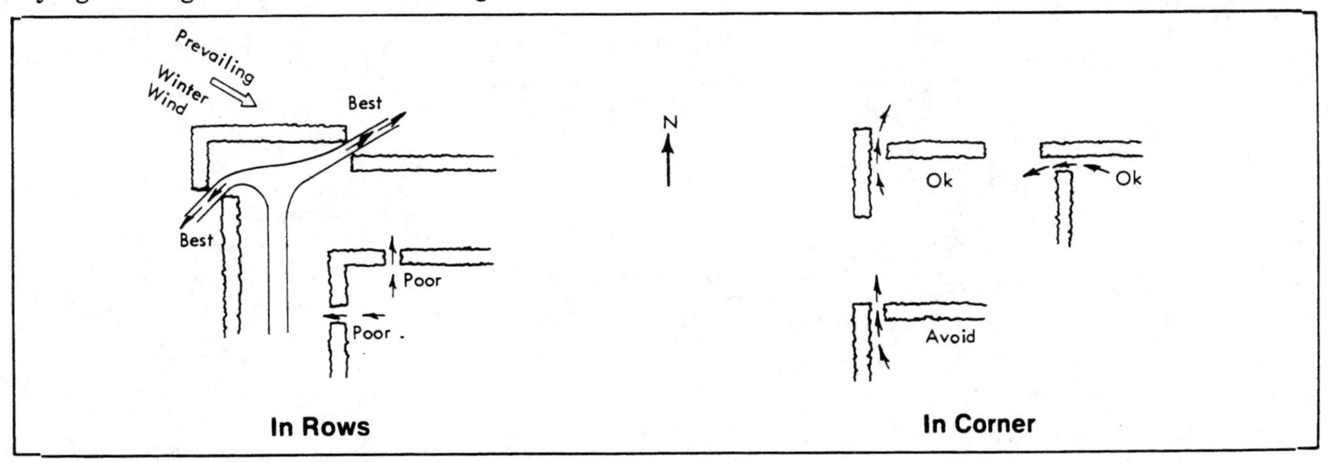

Fig 501-3. Traffic openings in tree windbreaks.
Arrange openings to reduce snow and wind problems.

side. Adjust the inlet to keep snow out during heavy snow storms.
- Do not build near a barn or silo. Provide at least a 30′ wind passage between buildings.
- Install windbreak fences at the ends to form a swirl chamber and reduce the wind in front of the building.
- Close part of the front wall at each end—up to ⅙ of the building length.
- Install solid cross partitions 50′ apart in long buildings to reduce drafts.

Snow fencing can provide temporary wind and snow control but is too porous to provide the best wind protection and is not heavy enough for use around livestock.

Windbreak fence construction

Horizontal or vertical slots in a slatted fence perform about the same. Close the opening below the fence with straw or snow in the winter to reduce drafts.

If fences are to be used with livestock, attach boards on the livestock side of the fence. Install a horizontal rub rail if livestock have access to both sides.

Sun

During cold weather take advantage of sun heat to warm buildings and dry surfaces. In the summer, protect buildings from the sun's heat with awnings, roof overhangs, trees, reflective roofing material, etc.

Orient buildings east-west.

Drives and Parking

When planning drives and parking consider:
- Surfacing for year-round access.
- Drainage to maintain the surface and control runoff.
- Area lighting for safety, convenience, and security.
- Space for maneuvering and parking.
- Space for snow storage.
- Safety and convenience when locating drives.

Entrance drive

Provide a single entrance drive for better traffic control and security. Exceptions are a second drive to a large livestock or grain center or a second house near the farmstead.

Locate drives either at the top of a hill or far enough from the top for safe visibility. Fast moving vehicles need time and distance to avoid a slower moving farm vehicle entering the farmstead. Provide turn-offs so drivers can wait safely for large equipment or opposing vehicles to clear the drive. Locate gates to the main drive at least 40′ (preferably 60′) from the road.

Make the drive surface about 16′ wide. Include at least 7′ of additional clearance on each side for overhanging equipment and snow storage. Drainage along the drive can be within this "right of way." Make drive straight or with gentle curves for easier movement.

Surface walks to parking areas and along drives.

Guest parking

Plan parking to encourage:
- Family guest to park near the house.
- Business visitors to park near the farm office.
- Delivery and shipping traffic to proceed to the court.

Provide 3 to 5 parking spaces and a direct walk to the guest entrance for family visitors. Screen the route to the service entrance and farm buildings with plantings.

Court

The farmstead court is usually an extension of the main drive. Plan for parking, maneuvering, and temporary storage of machinery and trucks.

A loop drive connecting the activity centers is common. Leave an open space in the center for overflow parking. Make circle drives with at least 110′ diameter.

A U-drive and court with two entrance drives is not recommended.

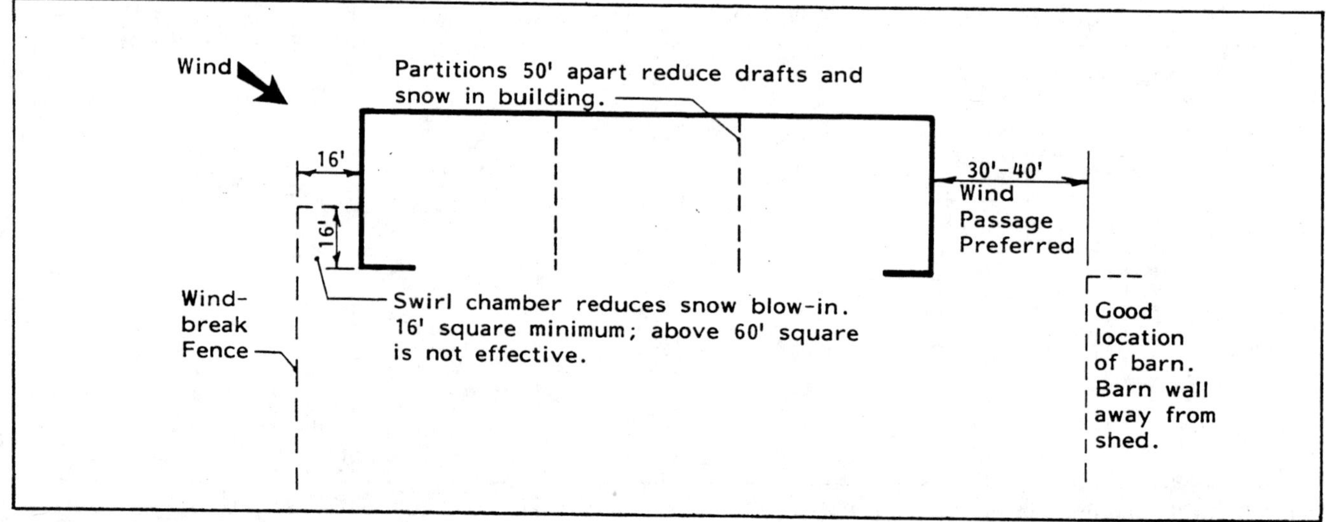

Fig 501-4. Protecting open sheds.

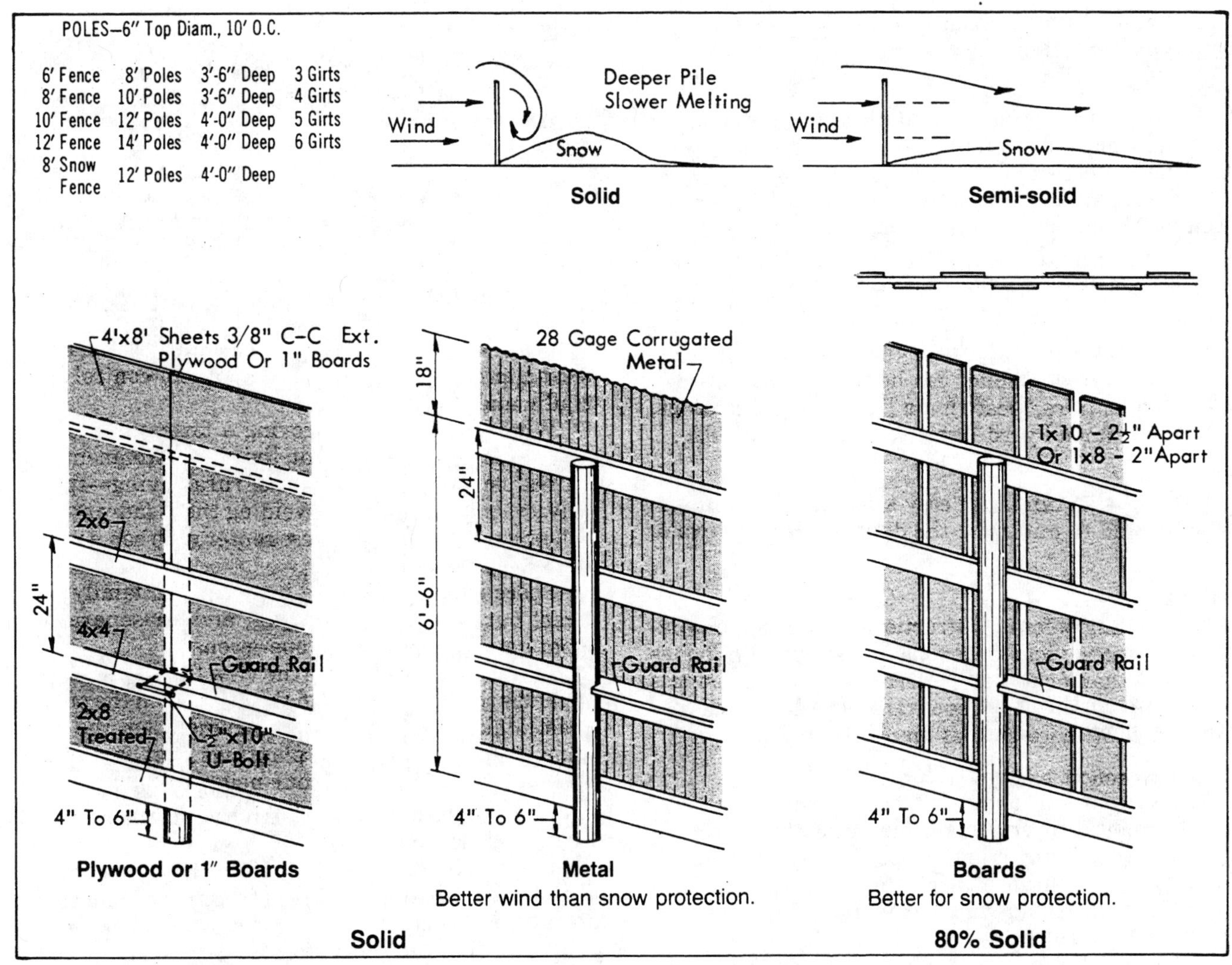

Fig 501-5. Windbreak fences.

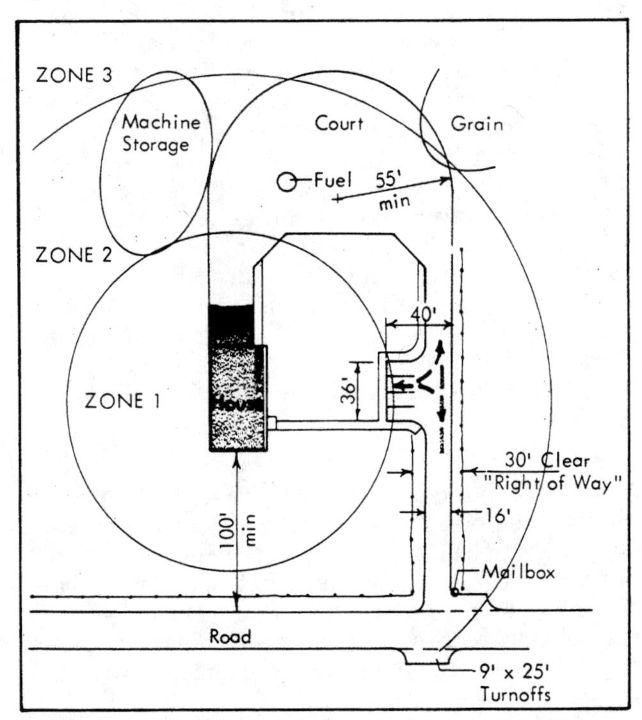

Fig 501-6. Overall dimensions for drives and parking.

Branch drives

Branch drives provide access to guest, office, and visitor parking; the family garage and parking; and small buildings.

Make straight drives at least 8′ wide (12′ is preferred). Plan curves with an 18′ inside and 30′ outside radius.

Utilities

Make a utilities overlay for your farmstead map showing the path of each overhead and buried utility. Include electric, gas, and telephone lines; water and sewer pipes; septic tanks; drainage field; and drains. Use colored pencils or pens and dashed, dotted, and solid lines to show each utility.

Utility lines are buried to improve appearance, avoid damage, and reduce hazards. Bury electric, fuel, gas, and telephone lines deep enough to avoid damage from heavy surface loads or tillage equipment. In firm soils with little or no surface grading, 18″ depth is enough. Bury water and sewer lines below frost penetration. See the Climate section for average frost penetration.

Record enough dimensions on the map to locate underground utilities. Indicate the type, size, and depth of each utility. Plan all utilities with your supplier. Be sure to meet insurance company and state regulations and limitations.

Water

Adequate water quality and quantity are essential to a farmstead. Water can be piped considerable distances if necessary. A remote storage tank filled with a relatively small pump can meet peak water demands. Each farm water system is unique, but if water is available, good design provides a dependable and high-quality system. See Chapter 801.

Electricity

Distribution policies vary with different power suppliers. Your supplier can help compare overhead and buried lines, locate main lines and power drops, and recommend yard lighting. See Chapter 802.

Gas

Locate LP tanks for easy servicing and maintenance, and to minimize the danger of fire or explosion.

Telephone

Planning before construction can reduce installation costs and yield more satisfactory service. Phones in more than one location permit handling business calls away from the house or calling a fire department with one or more phones inaccessible.

Fire prevention, safety, and security

Fire prevention includes adequate wiring, good housekeeping, properly maintained heating equipment, lightning protection, proper storage of fuels, and avoiding spontaneous ignition. Spacing buildings at least 50′ apart reduces fire spread and permits access for firefighting.

Equip each building and the fuel center with portable fire extinguishers at entrances and high-hazard areas. Provide water for fire department use from a pond or emergency cistern.

Protecting people and animals from fire, natural hazards, and accidents is part of all design decisions. A general awareness during planning is helpful; movements (machinery, vehicles, animals, falling or wind-blown objects), heights (ladders, pits), surfaces (ice, slippery floors), and materials (chemicals, fuels, bulk grain) suggest broad categories of risk.

Federal safety legislation requires owners to plan and operate their businesses as safely as possible. Consult state safety authorities for answers to specific questions and for applicability of the Occupational Safety and Health Act.

Security is a difficult problem—farms are vulnerable to theft, vandalism, arson, assault, etc. Be alert for protective measures and available equipment, and cooperate with local authorities. Consider fire and theft alarms and night lighting.

Planning Activity Centers

Try to overcome your familiarity with how things have been. We tend to work around that road, or this fence, or that mudhole without questioning.

Family Living Center

The family living area includes the house, lawn, patio, play areas, garden, garage, guest parking, and a portion of the drive.

Site

The house affects neighbors less than major production activities do. However, avoid nuisances from neighboring farms, and consider the home site relative to other land uses.

Locate the house upwind from livestock facilities to avoid odors carried by prevailing winds.

The house should be the first building seen, or at least approached, for appearance and for control over traffic and visitors.

When starting or improving a farmstead, locate the house first. A pleasant living environment requires using the advantages of rural living—space, privacy, and view—while avoiding the disadvantages of being near a busy business center with some undesirable byproducts.

Space is needed for the house, drive, family and guest parking, outdoor living, and for separation from neighbors and farm operations.

Locate the house where the view from the kitchen includes the farm court and the drive from the public road. Make the view from the living room the best your site offers; consider plantings that frame the desirable view and hide your neighbor's barn.

Provide for outdoor family activities as the living center is developed.

Shield a patio and play area from the farm court and the road with the house, the garage, fences, plantings, or any combination. Fencing or shrubbery can hide garbage or trash storage.

Easy access to a patio from the kitchen is desirable for outdoor meal preparation. A fenced play area for small children protects them from farm activities, stray animals, and possibly dangerous areas.

Consider an office in the house near the service entrance so you and business guests need not disrupt household activities.

Locate a second house as carefully as the first, with relations between two families an added concern. A completely separate site may be best if another desirable one is available on the farm. If two or more houses must be relatively close together, a separate drive gives some privacy and tends to separate visitors. Provide visual privacy for the homes and outdoor living areas using plantings and house orientation.

Distance

Quality of living is especially vulnerable to farm production centers. Zone planning of farmsteads including a living center is based on the position of the house. Generally, keep the house at least 100′ from a road to allow for possible road widening and to reduce noise and dust. More setback is needed if a tree windbreak is between the house and the road.

Drainage

Surface drainage from other areas may be polluted and odorous. Water around the house may prevent a dry basement and recreation areas.

Climate

Include the family living center inside the protected zone of a tree windbreak but outside the area of major snow catch. Most snow usually accumulates within 185′ of the windward row of trees. Provide additional wind protection with screening fences or plantings, but avoid blocking desirable summer breezes and causing snow drifts over walks and drives.

Water

Potable (or drinking quality) water is needed in the home; reasonable quality is needed for outdoor use. Special needs of farm families include family cleanup and washup near the service entrance and relatively large vegetable and landscaping gardens that may require watering.

Machinery Storage, Repair, and Service

The farm shop is usually near the machinery storage, either attached or in a separate building. A separate building reduces fire hazard. See Chapter 540.

Fuel, Chemical, and Fertilizer Storage

Consider a farm service station, with fuel storage and pumping, oil, water, and compressed air. For safety and because major equipment needs space to maneuver around the fuel pump, separate a fuel center about 50′ from the shop.

Provide gravel or hard surface access to the fuel pump from two or three sides. Slope the ground away from the fuel center so rainwater and fumes can drain away. Provide lights and a dry chemical fire extinguisher (fire type B) at the pump. Lock fuel pumps for safety and security.

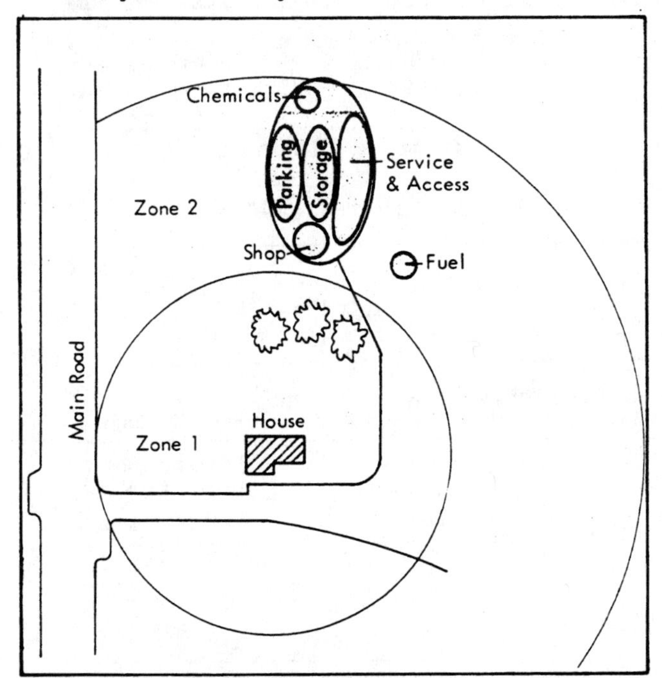

Fig 501-7. Machine shed layout.
Locate shop, machinery storage, chemical storage, and fuel center.

Fertilizers and pesticides create other storage problems. Ammonium nitrate is explosive if contaminated with fuel, oil, or other hydrocarbon or organic material. Some insecticides and herbicides can contaminate each other or fertilizer; some are flammable. State law may control storage, use, and disposal of some products.

A chemicals center is convenient if near the shop and machinery center, but move it as far as reasonable from the house because of odors and child safety.

A chemicals center requires water for mixing and an outside frostproof hydrant for filling sprayers, emergency flushing of eyes or skin, and rinsing utensils and containers. Electricity is needed for lights, ventilation fans, and perhaps heat for winter storage of liquids.

Do **not** provide batch storage containers for chemicals; store them in their original containers so they remain identified and with instructions for their care and use. Provide for disposal of washwater.

Grain Storage and Processing

A farm grain-feed center needs electric power, vehicle access, room for expansion, reasonable drainage, and proximity to livestock areas if the facility is for both storage and feed making.

Some water supply is needed for washing, dilution, and perhaps fire control.

Drainage must permit all-weather access. A deep dump pit requires good subsoil drainage.

Increased capacity or revised management may increase noise and dust levels and their impact on neighboring areas. The amount and kind of vehicle traffic may also change—consider road and bridge capacities required to support your needs.

Feed and grain facilities are in Zone 3, or even farther from the house, because of noise, dust, and traffic. Space around the center is needed for access and vehicles—often semi-trailer trucks.

Put a grain-feed center on a crowned or graded area with good surface drainage. Good subsurface drainage aids in designing an elevator pit, some con-

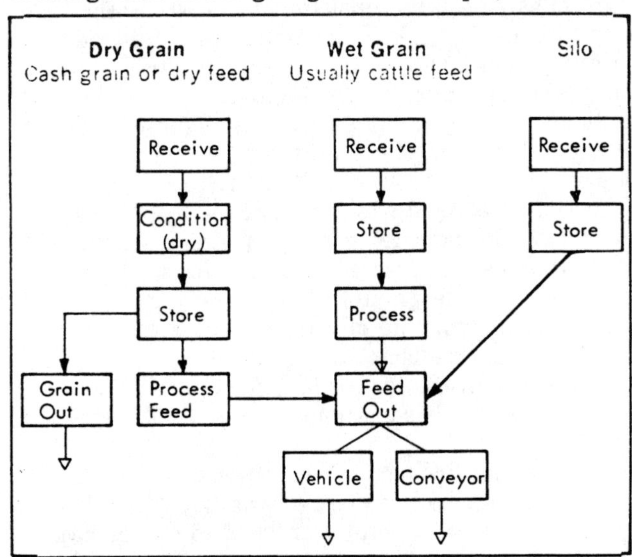

Fig 501-8. Feed center flow diagrams.

veyors, and foundations. Roads must be usable by heavy vehicles in all weather, so good surfaces and space to store plowed snow are needed.

On livestock farms, all grain may flow into a grain-feed center for storage and processing before being moved later by vehicle or mechanical conveyor. Large bulk delivery vehicles, even semi-trailers, may need access. Driving through, rather than backing up, is essential; ample turning space is needed.

The weight of grain transport trucks may affect locating the feed center to avoid light-load bridges or culverts. Putting a grain center in a remote location may mean developing and maintaining an extra road and more distance for providing utilities. But it also means less nuisance in the living area.

Especially on cash-grain farms handling over about 50,000 bu/yr, separate receiving and load-out drives may be desirable. Efficiency suffers when a loading truck blocks the receiving dump for extended times at harvest. If feed delivery wagons or trucks operate daily, a separate load-out lane avoids plugging grates and pits in the receiving alley and avoids cleaning equipment between ingredients.

Livestock Production

Essential factors

A new factor is added for livestock centers—waste disposal. Requirements, regulations, public demand, and simply good livestock management practices have made livestock waste management a subject for major farmstead planning decisions.

Water is essential for livestock. Optimum production requires adequate water; cool water is better than warm in hot weather. Management practices also often require water for cleaning (pens, milking equipment), sanitation spraying, waste removal, and summer cooling of animals or ventilation air.

Drainage is listed after water, because it is difficult to overcome a lack of water, but is usually possible to improve and even correct drainage. When engineers discuss livestock production or farmsteads, no subject is raised as consistently, as soon, and as fervently as drainage. Poor drainage has caused more problems, destroyed more otherwise successful layouts, and frustrated more developments than any other factor. Even many livestock waste management problems relate directly to drainage.

Generally, locate livestock facilities on high ground **—do not build in a hole.** Free fall below a livestock center permits gravity drainage of rain, snow melt, subsurface water around foundations and pits, and even some gravity waste handling.

Production volume of a major livestock enterprise affects, and is affected by, the whole farm and all operations. Utilities, feed, machinery, traffic, nuisances, space—all become more demanding as size increases.

Off-farm factors such as odors, dust, noise, and possible pollution by livestock wastes can affect your neighbors as well as your family. Neighbors can change from sympathetic farmers to less tolerant subdivision dwellers. While accurate long-range predictions may be impossible, observe current trends. Investigate zoning law developments with government offices, and avoid foreseeable major impacts on off-farm areas.

Most nuisances are air- or water-borne; locate livestock and waste facilities downwind from off-farm living centers as well as downwind from your own, and control runoff before it leaves your farm. It is usually easier in the long run to avoid complaints than to adequately respond to one already raised.

Wastes from a large modern livestock center can concentrate in such a small area that their handling and disposal become almost an "activity center." Waste management has become a critical factor because regulations can prohibit some developments, require or prohibit certain management practices, and constrain some planning decisions.

Planning factors

Separation distance for a livestock unit depends on the kind of animal; its age, size, or maturity; the size of the unit; the amount of care required; and factors such as housing in closed or open buildings, feeding in feedlots, or product processing in a milkhouse or egg room.

Table 501-1 compares livestock classes by their weight equivalents for distance planning. Total weight is a partial measure of amounts of feed and manure to be handled, amount of ventilation air, and approximate building size. Table 501-2 suggests minimum distances of livestock centers from the family living center. The numbers in the tables are more to help you understand separation than to suggest absolute values.

Animals in a closed, fan-ventilated building tend to create less nuisance than, for example, 200 feeder steers in a dirt lot. Baby animals tend to need more

Table 501-1. Enterprise sizes for livestock.
These numbers do not accurately measure the effect of enterprise size on either management or location but do suggest equivalent weights of different kinds of animals.

No. Animal Units	Maximum No. of Different Animals			
	1000# Beef	200# Hogs	1500# Dairy Cows	4# Hens
1 to 20	20	100	13	5,000
20 to 75	75	375	50	18,700
75 to 500	500	2,500	330	125,000

Table 501-2. Zones for livestock enterprises.
Consider the number of livestock and their size when planning a farmstead.

No. Animal Units	Farmstead Zone	Minimum Distance From House For Nearest Livestock Buildings
1 to 20	2 or 3	100′
20 to 75	3	200′
75 to 500*	4	300′

*Over 500 animal units may overwhelm other farmstead components. Consider a separate center isolated as far as practical from any residence.

care and create less nuisance than feeder or mature animals. They are also more apt to be in closed buildings. The milkhouse of a dairy center has little effect on family living except the visits of the bulk milk truck.

Topography determines overall drainage patterns. It can restrict the space available for a large center and can affect access routes and convenience to utilities and other farmstead centers.

Plan adequate space for livestock-related facilities: corral, hospital unit, access road, feed center, and processing and storage buildings (e.g. milkhouse, egg-handling room).

After reasonable space for the facility being planned is determined, double it—allow for twice the space needed soon. Extra space between buildings helps prevent fire spread, can increase access, and allows future changes without overcrowding.

Slopes of 2%-6% will drain but will not cause erosion in many soils. A south slope dries better and is especially preferred for unpaved lots. Earth moving can usually create desirable slopes and graded-up areas for buildings.

Existing Buildings—Remodel or Abandon?

When planning to expand, you may have to decide whether to remodel or abandon an existing building. Carefully consider your future, as well as current, needs. Conditions vary from farm to farm, but always evaluate these general factors.

- Compatibility of existing building to final setup.
- Structural integrity of existing building.
- Location of existing building.
- Cost of remodeling vs. new building.

Compatibility

The "best" remodeled building is one that requires few modifications. If an existing building must be changed drastically to meet your current needs, consider a new building.

Alternative Uses

First, decide if the existing building is more like another building you are planning to build in the future, rather than the building you are planning now. For example, it is easier to remodel an uninsulated machine shed for breeding-gestation than for a nursery or farrowing facility. Therefore, consider a new building for a nursery now. Remodel the machine shed into a gestation building later.

Factors affecting this decision include ceiling/roof height, location, adaptability to mechanical or natural ventilation, and insulating ease. Farrowing and nursery buildings require ceiling heights of 8′ or less, mechanical ventilation, and good insulation. Finishing hogs and gestating sows produce well in buildings with 8′-10′ high ceilings, natural ventilation, and less insulation.

Interior Space Arrangement

Determine if the existing layout adapts easily to the desired floor plan. Post locations frequently do not fit the new plan. It is possible to span existing posts with sub-beams or new posts and beams, but relocating a post usually requires a new footing. If you plan to use an existing concrete floor, new footings are costly and time-consuming.

Remodeled buildings usually make poor use of space—too much space in some areas and not enough in others. For example, a 20′x58′ building remodeled for farrowing in 20-5′x7′ stalls has only a 2′ center alley but 2′-3′ left over at one end. The narrow center alley is crowded and inefficient, and when considered over the life of the building, may make a new building desirable.

Environmental Control System

Determine if a good ventilating and heating system can be installed. For example, it is impractical to install floor heat in an existing concrete floor, so the building is not a good candidate for a farrowing building. Consider not only the initial cost of ventilating equipment, but also the cost of operating it in a less efficient remodeled building compared to a new building.

Check if air inlets and fans can be easily installed and properly located. See section on ventilation. Problems include:

- Ceiling obstructions. Exposed joists and beams block proper air distribution. A ceiling or fan and tube inlet system may be required.
- High ceilings. You may need a lower "false" ceiling to reduce drafts and heated space.
- Unplanned air inlets. Exhaust ventilation does not work well in a leaky building. Positive or neutral pressure ventilation may be required.

Manure Management

Will your manure handling system work in the remodeled building? It is difficult to add manure storages to existing buildings. If a flushing or scraping system is unsuitable, remodeling may be inadvisable. Special equipment to handle manure just for the remodeled facility is part of the remodeling cost.

Rodent Proofing

Can the existing building be made reasonably rodent proof? A flat slab or shallow footing (less than 2′ below grade) is inappropriate for livestock facilities, because rodents burrow and nest under the foundation.

Structural Integrity

Carefully evaluate the building's structure, whether it is for short- or long-term use. Check alignment and condition of foundations, strength and alignment of sidewalls, and condition of roof framing and covering. A building with a swayback ridgeline may have been overloaded or poorly built, and could need high future maintenance or repair costs.

Location

Evaluate location not only for suitability to the remodeled building but also for future buildings planned for the same site. Consider:

- Farm home and neighbors.
- Runoff and drainage. It may be impossible to maintain warm, dry floors in a building in a low area.
- Manure handling. Space must be available for manure handling facilities such as lagoons.
- Access roads. Does the present road provide all-weather access for heavy vehicles? Is building access convenient; can it be supervised?
- Room for expansion. Consider future buildings, feed centers, access roads, and manure handling facilities.

Do not compromise on location when deciding whether to remodel a building. A common argument is that the location may not be good, but the remodeled building is only temporary. After spending several thousand dollars to remodel a building, it often becomes the focal point for all future livestock building and feed handling construction—all in the wrong location.

Economic Considerations

Remodeling is not always the cheaper route, especially when future needs are considered. If remodeling cost is more than ½ to ⅔ of the new building cost, a new building is usually best. It may be possible to use some materials from the existing building for the new building. Remodeling costs include demolition of interior structural components and concrete floors.

Consider cost and availability of construction labor. Due to the many "unknowns" in remodeling, many farm builders are hesitant, if not unwilling, to accept remodeling jobs. If they accept a remodeling job, the contract is generally so heavily "padded" to compensate for possible "unknowns" that labor becomes very expensive.

Do not plan a new $100,000 swine facility around an existing $5000 building or $1000 concrete slab. Existing facilities may save money initially, but the long-term cost of restricted expansion and vehicle movement and reduced labor efficiency may offset the initial savings.

502 ENGINEERING ECONOMICS

This chapter was prepared by Dr. C.O. Cramer, University of Wisconsin.

In planning engineering projects, you may need to choose among several alternatives. Base your selection on economic factors, such as first cost or annual cost of ownership, and less tangible ones such as environmental or social effect, or personal preference. Engineering economics helps select engineering alternatives for maximum economy. The difference between alternatives is expressed in dollars.

Notation

- C = compound price change, inflation or deflation
- CAF = single payment compound amount factor
- CRF = capital recovery factor
- D = annual depreciation
- F = future sum of money, n interest periods from the present date
- f = declining balance fixed percentage
- i = interest rate per interest period
- i′ = effective interest rate
- I = total interest due
- k = number of years an asset has been utilized
- m = number of compounding periods per year
- n = estimated life in years, also number of interest periods
- P = a present sum of money, principal
- PWF = single payment present worth factor
- R = the end-of-period payment in a uniform series, annual net return
- SCAF = series compound amount factor
- SFF = sinking fund factor
- SPWF = series present worth factor
- V_a = value of asset when acquired
- V_p = asset book or present value
- V_r = replacement value of asset
- V_s = estimated salvage value

Time Value of Money

Money has a time value because of interest. Payments of different size can be equivalent if they are made on different dates.

If \$200 is put in a sock, at the end of 2 years the cash value will still be \$200. If the \$200 is deposited at 8% interest compounded annually, the cash value will be $\$200 \times (1 + .08)^2 = \$233.28$ after 2 years.

Therefore, \$200 today is equivalent to \$233.28 two years later at 8% earnings compounded annually. Similarly, to have \$200 in 2 years, deposit $\$200 \div (1 + 0.08)^2 = \$171.46$ at 8% interest now.

Money borrowed to finance an alternative has interest costs that are direct expenses against the alternative. An expense is also charged for using available funds, whether or not they are currently invested, for the interest income that could have been earned. For economic analysis, the cost of using funds is the same whether the money is borrowed or available.

Interest

Interest is either the cost of borrowed money or the earnings from loaned money. Because of interest, a dollar now is worth more than the prospect of a dollar next year.

Interest rate is a ratio between the interest payable at the end of a period of time, usually a year or less, and the money owed. Borrowers want to minimize the interest rate; lenders want to maximize it. The actual rate is a compromise based on the availability of capital, investment opportunities, and the degree of risk of losing capital.

Simple Interest

The amount of interest due is based on the initial value of the loan, the interest rate, and how long the loan lasts. It is usually payable at the end of each interest period.

Eq 502-1.

$$I = P \times i \times n$$

Eq 502-2.

$$F = P \times (1 + i \times n)$$

Compound Interest

Interest earned during the first period of the loan is added to the loan and interest is earned on this new total. The amount due at the end of the first period on which the second period's interest will be calculated is:

$$F_1 = P + P \times i$$

For the second period the amount due is:

$$F_2 = P + P \times i + (P + P \times i) \times i = P \times (1 + i)^2$$

At the end of the nth year the amount due is:

Eq 502-3.

$$F_n = P \times (1 + i)^n$$

Nominal Interest Rate

A year is often divided into several interest periods. Interest is earned and compounded during each period. The total interest figure can be quoted as an annual rate and the number of interest periods per year. For example, interest at 4% per half year can be given as 8% compounded semiannually. The nominal interest rate is 8%. For \$1,000:

$$F_{6mo} = P + P \times i = \$1{,}000 + (\$1{,}000 \times .04) = \$1{,}000 + \$40 = \$1{,}040$$

$$F_{1yr} = \$1{,}040 + (\$1{,}040 \times .04) = \$1{,}040 + \$41.60 = \$1{,}081.60$$

Effective Interest Rate

The way the interest rate is quoted affects the actual or effective rate. Comparing the future value of \$1,000 compounded at 8% annually (\$1,080) and semiannually (\$1,081.60):

- For $1,000 at 8% compounded annually:
  Effective interest rate =
  (F - P) ÷ P = ($1,080 - $1,000) ÷ $1,000 = .08 = 8%
- For $1,000 at 8% compounded semiannually:
  Effective interest rate =
  (F - P) ÷ P = ($1,081.60 - $1,000) ÷ $1,000 = .0816 = 8.16%
- The effective annual interest rate, i′, can be determined from this expression:

  $i' = (1 + i)^m - 1$

Compound Interest Factors

The money consequences of an alternative usually occur over an extended period. Translate cash flow for all the alternatives to a particular time by finding its future or present worth.

A future worth calculation converts a single sum or a series of values to an equivalent amount at a later date. Present worth calculations convert a single future sum or a series of future values to an equivalent amount at an earlier date, usually the present.

For equivalent values, calculate the compound amount of each sum for each period. Avoid this tedious task with the following compound interest factors, which can be selected from a table or easily computed with a hand calculator.

One type of factor converts a single amount to a future or present value. The other converts a series of regularly deposited amounts.

Single payment compound amount factor (CAF)

If a single sum is deposited at compound interest, it grows at the interest rate to a future sum. Calculate the future sum (F) based on the principal (P), interest rate (i), and number of interest periods (n).

Eq 502-4.

$$CAF = (1 + i)^n$$

$$F = P \times CAF \text{ (Same as Eq 502-3)}$$

If the future sum, F, is known, the size of the needed deposit or principal is:

Eq 502-5.

$$P = F \div CAF$$

Eq 502-5 is often expressed as P = F x PWF, where PWF is the **present worth factor** in many interest tables. PWF is simply 1 ÷ CAF.

Series compound amount factor (SCAF)

If a uniform sum, R, is deposited regularly over n periods, the total amount will be the sum of the compound amounts of each deposit. The money deposited at the end of the first period earns interest for (n-1) periods. The second period payment earns interest for (n-2) periods; the third period, (n-3) periods; and so on until the last payment which earns no interest. Use the SCAF to calculate the future sum, F.

Eq 502-6.

$$SCAF = [(1 + i)^n - 1] \div i$$

Eq 502-7.

$$F = R \times SCAF$$

A future sum, F, derived from a series of uniform payments that earn interest is called a sinking fund.

To find the size of regular payment needed to develop a desired fund, use:

Eq 502-8.

$$R = F \div SCAF$$

Eq 502-8 is often expressed as R = F x SFF, where SFF is the **sinking fund factor** in many interest tables and is 1 ÷ SCAF.

Series present worth factor (SPWF)

The SWPF is used to find the present value, P, of a future series of uniform payments, R, made over n periods at interest rate i.

Eq 502-9.

$$SPWF = [(1 + i)^n - 1] \div [i \times (1 + i)^n]$$

Eq 502-10.

$$P = R \times SPWF$$

The **capital recovery factor,** CRF, which is 1 ÷ SPWF, is the factor most often used in engineering economy problem solutions. The CRF when multiplied by a present debt gives the uniform end-of-year payment necessary to repay the debt in n periods with an interest rate of i. An example is the mortgage on a house. The appropriate CRF multiplied by the mortgage sum, P, yields the amount of the uniform end-of-period payment, R, to repay the loan with interest in n periods.

Eq 502-11.

$$R = P \times CRF = P \div SPWF$$

Table 502-1. Compound interest factors.

To find	Given	Use this compound interest factor
F	P	$CAF = (1 + i)^n$
P	F	$PWF = 1/CAF$
F	R	$SCAF = [(1 + i)^n - 1]/i$
R	F	$SFF = 1/SCAF$
P	R	$SPWF = [(1 + i)^n-1]/[i(1 + i)^n]$
R	P	$CRF = 1/SPWF$

Engineering Alternatives Comparison

Four steps are needed to solve an engineering economics problem:

1. Define the alternatives.
   - List all feasible alternatives.
   - Select a plan, process, or method and prepare plans with sufficient detail to estimate costs.
   - Select materials—low cost, shorter life materials vs. more expensive, longer life materials.
2. Reduce alternatives to consistent time and money differences considering:
   - Initial investment—the estimated cost of each alternative. This capital investment must be recovered with interest over the life of the asset. This is called capital recovery and is used instead of depreciation.
   - Operating costs—annual expense of repairs, maintenance, taxes, and insurance. Also con-

sider fuel and labor costs. These are called annual disbursements.
- Operating returns such as the fertilizer value of manure.

3. Assume an interest rate to make alternative money-time series comparison.
4. Reduce alternatives to a comparable basis with one of the following methods—annual cost, present worth, or prospective rate of return on extra investment. Each method yields the same solution when selecting the best alternative. However, some problems may be more easily solved by one method than another.

To demonstrate these three comparison methods, compare two building alternatives, Table 502-2.

Table 502-2. Building alternative comparison.

	Building X	Building Y
Initial cost	$14,000	$20,000
Estimated life	15 yr	25 yr
Estimated salvage value	$0	$3,000
Annual expenses	$800	$500

Annual Cost Method

This is the most common method for engineers to compare a nonuniform series of payments. Nonuniform payments are converted to an equivalent uniform series. The uniform payments recover capital investment with interest and charge off annual disbursements.

Calculate the SPWF from Eq 502-9 or select it from interest tables. The amount of capital recovery (CR) is:

Eq 502-12.

$$CR = (V_a - V_s) \div SPWF + V_s \times i$$

Table 502-3 shows that Building Y has an advantage over Building X using this method. Annual cost of Building Y is $103.20 less than Building X.

Table 502-3. Annual cost example.
Use 8% interest.

Building X			
CR @ 8%	$14,000 / 8.559	=	$1,635.71
Annual expenses		=	800.00
Annual cost		=	$2,435.71
Building Y			
CR @ 8%	$17,000 / 10.675 + 3,000 x .08	=	
	1,592.51 + 240	=	1,832.51
Annual expenses		=	500.00
Annual cost		=	$2,332.51

Present Worth Method

Two or more engineering alternatives with different money-time series can be compared by converting each money-time series to an equivalent single sum at some specified date such as the present. To an investor, the present worth of future payments is the present investment needed to secure those future payments with interest. Two features of present worth studies are:

- Salvage values are negative payments.
- The same number of years of service is used for each alternative.

Calculate the present worth, PW, for each renewal using Eq 502-5. The future sum, F, equals the initial value, V_a, minus the salvage value, V_s. Use the compound amount factor, CAF, for the number of periods at each renewal. Compound interest factors can be selected from a table or for values not in the table easily calculated with a hand calculator.

$$PW = (V_a - V_s) \div CAF_n$$

Each renewal is a single event so CAF or PWF is used. The annual expenses are a series of payments so SCAF or SPWF should be used.

When comparing alternatives use the least common multiple of the products' estimated life and calculate renewals with constant initial costs, salvage values, product lives, and annual disbursements. For the example use a 75-yr comparison—5-15 yr periods for Building X and 3-25 yr periods for Building Y.

The example in Table 502-4 indicates that Building Y has a $1,284.42 advantage in present worth.

Table 502-4. Present worth example.
Use a 75-yr comparison (5-15 yr periods; 3-25 yr periods).

	Building X		
Initial cost			$14,000.00
PW, 1st renewal	$(V_a-V_s)/CAF_{15}$ $14,000 / 3.172	=	4,413.62
PW, 2nd renewal	$(V_a-V_s)/CAF_{30}$ $14,000 / 10.063	=	1,391.24
PW, 3rd renewal	$(V_a-V_s)/CAF_{45}$ $14,000 / 31.920	=	438.60
PW, 4th renewal	$(V_a-V_s)/CAF_{60}$ $14,000 / 101.257	=	138.26
PW, annual expense	expense x $SPWF_{75}$ $800 x 12.461	=	9,968.80
Total PW for 75 yr			$30,350.52
	Building Y		
Initial cost			$20,000.00
PW, 1st renewal	$(V_a-V_s)/CAF_{25}$ ($20,000-$3,000)/6.848	=	2,482.48
PW, 2nd renewal	$(V_a-V_s)/CAF_{50}$ ($20,000-$3,000)/46.902	=	362.46
PW, annual expense	expense x $SPWF_{75}$ $500 x 12.461	=	6,230.50
Total PW for 75 yr			$29,075.44
Less PW, salvage value	V_s/CAF_{75} -$3,000 / 321.205	=	-9.34
Total PW			$29,066.10

Comparing the present worth method with the annual cost method, Building Y's annual cost advantage was $103.20. Multiply this by the series present worth factor ($SPWF_{75}$ = 12.461) and it equals

$1,285.98, the PW advantage of Building Y ($1.56 rounding error).

Prospective Rate of Return On Extra Investment

This method compares two money-time series by finding the interest rate that makes them equivalent to one another. The objective is to find the interest rate that yields equal annual costs or present worths for both alternatives.

Two or more interest rates are selected, annual cost or present worth values are calculated, and the rate of return is determined by interpolation.

When comparing two alternatives, the unknown interest rate may be thought of as the prospective rate of return on the extra investment from the alternative with the higher initial cost.

Determine the rate of return on the extra investment of Building Y using annual costs. Present worth can also be used. See Table 502-5 for annual costs of Building X and Y.

By interpolation, the rate of return on this extra investment is 9.5%.

$$i = 8\% + [103.20 \div (103.20 + 32.24) \times (10\% - 8\%)] = 9.5\%$$

The investor must decide if it is worthwhile to spend an extra $6,000 (difference in initial costs) now to save $300/yr (difference in annual expenses). If the rate of return on the extra investment is attractive, the more expensive alternative would be selected.

Table 502-5. Rate of return example.

	Annual cost		Annual cost Advantage	
i	Building X	Building Y	Building X	Building Y
8%	$2,435.71	$2,332.51	—	$103.20
10%	$2,640.65	$2,672.89	$32.24	—

Depreciation

The previous section presented methods for selecting the most economical engineering alternative but did not consider income taxes. Depreciation of capital equipment used in a business is an important component in computing income taxes.

Depreciation is the decrease in value of an asset due to physical wear and tear, obsolescence, or depletion of resources. Depreciation is occasionally confused with maintenance, because both deal with physical deterioration resulting from age and service.

Maintenance and repairs are routine activities to keep property in reasonably good condition, for example, painting a building or replacing a worn-out roof. Both maintenance and repair costs are operating expenses. Occasionally extensive repairs restore value lost through depreciation. Consider extensive repair expenditures as new investments.

Depreciation is the loss in value that occurs despite prudent maintenance and repairs. Consider depreciation in an economic study to recover capital invested in physical assets and to charge depreciation costs as production costs.

Estimate the life of an asset to determine the amount of annual depreciation. For tax purposes, the shortest allowable life is generally most advantageous. Unless reliable information suggests a shorter useful life, use the values suggested by the U.S. Internal Revenue Service (IRS). Depreciation regulations as they affect the computation of federal taxes change with the economic and political conditions of the country. Changes in depreciation schedules are sometimes used to encourage investment. Examples of schedule changes are a 20% additional first year depreciation on qualifying property and writing off the value of an asset in periods shorter than its useful life.

Depreciation Accounting Methods

In depreciation studies, depreciation charges are a series of payments to replace the asset. The charge is used for tax purposes and the actual money is handled as working capital. However, in an engineering economic study, depreciation charges are assigned to each alternative to recover the initial cost with interest. This is similar to capital recovery which was discussed in the interest section.

The book or present value of an asset is the original cost minus the accumulated depreciation charges. Book value can differ from market or sale value depending upon the depreciation method used.

Include salvage values for assets with significant value at retirement. Assets must not be depreciated below their salvage value, or in other words, the book value at retirement is equal to the salvage value.

There are a number of methods for computing annual depreciation and book value. Some of the more common methods will be discussed.

Straight line method

This is the simplest and most widely used depreciation method. The annual depreciation is constant. The cost-less-salvage value uniformly decreases over the prospective life. Disadvantages are:

- Does not discount future services, which are less valuable as the service becomes more remote.
- Is not the most favorable for tax purposes.
- Ignores the actual increase in rate of value loss toward the end of the useful life.

To compute annual straight line depreciation:

Eq 502-13.

$$D = (V_a - V_s) \div n$$

Present value of an asset using straight line depreciation is:

Eq 502-14.

$$V_p = V_a - [k \div n \times (V_a - V_s)]$$

Sinking fund method

An imaginary sinking fund is built by uniform end-of-year annual deposits throughout the life of the asset. Assume the deposits draw interest and build a fund equal to the asset's cost minus its estimated salvage value.

The gradual early loss and more rapid later loss are more consistent with the real loss in value. However, this method usually writes off costs too slowly for favorable income tax deductions.

Table 502-6. Interest tables, compound interest factors.

	Single Payment	Uniform Annual Series		Single Payment	Uniform Annual Series		
n	CAF	SCAF	SPWF	CAF	SCAF	SPWF	n
		4.%			6.%		
1	1.040	1.000	0.962	1.060	1.000	0.943	1
2	1.082	2.040	1.886	1.124	2.060	1.833	2
3	1.125	3.122	2.775	1.191	3.184	2.673	3
4	1.170	4.246	3.630	1.262	4.375	3.465	4
5	1.217	5.416	4.452	1.338	5.637	4.212	5
6	1.265	6.633	5.242	1.419	6.975	4.917	6
7	1.316	7.898	6.002	1.504	8.394	5.582	7
8	1.369	9.214	6.733	1.594	9.897	6.210	8
9	1.423	10.583	7.435	1.689	11.491	6.802	9
10	1.480	12.006	8.111	1.791	13.181	7.360	10
11	1.539	13.486	8.760	1.898	14.971	7.887	11
12	1.601	15.026	9.385	2.012	16.870	8.384	12
13	1.665	16.627	9.986	2.133	18.882	8.853	13
14	1.732	18.292	10.563	2.261	21.015	9.295	14
15	1.801	20.024	11.118	2.397	23.276	9.712	15
16	1.873	21.824	11.652	2.540	25.672	10.106	16
17	1.948	23.697	12.166	2.693	28.212	10.477	17
18	2.026	25.645	12.659	2.854	30.905	10.828	18
19	2.107	27.671	13.134	3.026	33.759	11.158	19
20	2.191	29.778	13.590	3.207	36.785	11.470	20
25	2.666	41.646	15.622	4.292	54.864	12.783	25
30	3.243	56.085	17.292	5.743	79.057	13.765	30
35	3.946	73.652	18.665	7.686	111.433	14.498	35
40	4.801	95.025	19.793	10.286	154.758	15.046	40
45	5.841	121.029	20.720	13.764	212.738	15.456	45
50	7.107	152.667	21.482	18.420	290.328	15.762	50

	Single Payment	Uniform Annual Series		Single Payment	Uniform Annual Series		
n	CAF	SCAF	SPWF	CAF	SCAF	SPWF	n
		8.%			10.%		
1	1.080	1.000	0.926	1.100	1.000	0.909	1
2	1.166	2.080	1.783	1.210	2.100	1.736	2
3	1.260	3.246	2.577	1.331	3.310	2.487	3
4	1.360	4.506	3.312	1.464	4.641	3.170	4
5	1.469	5.867	3.993	1.611	6.105	3.791	5
6	1.587	7.336	4.623	1.772	7.716	4.355	6
7	1.714	8.923	5.206	1.949	9.487	4.868	7
8	1.851	10.637	5.747	2.144	11.436	5.335	8
9	1.999	12.488	6.247	2.358	13.579	5.759	9
10	2.159	14.487	6.710	2.594	15.937	6.145	10
11	2.332	16.645	7.139	2.853	18.531	6.495	11
12	2.518	18.977	7.536	3.138	21.384	6.814	12
13	2.720	21.495	7.904	3.452	24.522	7.103	13
14	2.937	24.215	8.244	3.797	27.975	7.367	14
15	3.172	27.152	8.559	4.177	31.772	7.606	15
16	3.426	30.324	8.851	4.595	35.949	7.824	16
17	3.700	33.750	9.122	5.054	40.544	8.022	17
18	3.996	37.450	9.372	5.560	45.599	8.201	18
19	4.316	41.446	9.604	6.116	51.158	8.365	19
20	4.661	45.762	9.818	6.727	57.274	8.514	20
25	6.848	73.106	10.675	10.835	98.346	9.077	25
30	10.063	113.283	11.258	17.449	164.491	9.427	30
35	14.785	172.316	11.655	28.102	271.019	9.644	35
40	21.724	259.056	11.925	45.258	442.583	9.779	40
45	31.920	386.504	12.108	72.889	718.887	9.863	45
50	46.901	573.768	12.233	117.388	1163.877	9.915	50

Continued on next page.

Table 502-6. Interest tables, compound interest factors, continued.

	Single Payment	Uniform Annual Series		Single Payment	Uniform Annual Series		
n	CAF	SCAF	SPWF	CAF	SCAF	SPWF	n
		12.%			14.%		
1	1.120	1.000	0.893	1.140	1.000	0.877	1
2	1.254	2.120	1.690	1.300	2.140	1.647	2
3	1.405	3.374	2.402	1.482	3.440	2.322	3
4	1.574	4.779	3.037	1.689	4.921	2.914	4
5	1.762	6.353	3.605	1.925	6.610	3.433	5
6	1.974	8.115	4.111	2.195	8.535	3.889	6
7	2.211	10.089	4.564	2.502	10.730	4.288	7
8	2.476	12.300	4.968	2.853	13.233	4.639	8
9	2.773	14.776	5.328	3.252	16.085	4.946	9
10	3.106	17.549	5.650	3.707	19.337	5.216	10
11	3.479	20.655	5.938	4.226	23.044	5.453	11
12	3.896	24.133	6.194	4.818	27.271	5.660	12
13	4.363	28.029	6.424	5.492	32.088	5.842	13
14	4.887	32.393	6.628	6.261	37.581	6.002	14
15	5.474	37.280	6.811	7.138	43.842	6.142	15
16	6.130	42.753	6.974	8.137	50.980	6.265	16
17	6.866	48.884	7.120	9.276	59.117	6.373	17
18	7.690	55.750	7.250	10.575	68.393	6.467	18
19	8.613	63.440	7.366	12.056	78.968	6.550	19
20	9.646	72.052	7.469	13.743	91.024	6.623	20
25	17.000	133.333	7.843	26.462	181.868	6.873	25
30	29.960	241.332	8.055	50.949	356.781	7.003	30
35	52.799	431.662	8.176	98.098	693.559	7.070	35
40	93.051	767.088	8.244	188.879	1341.996	7.105	40
45	163.987	1358.224	8.283	363.670	2590.501	7.123	45
50	289.000	2400.004	8.304	700.214	4994.387	7.133	50

	Single Payment	Uniform Annual Series		Single Payment	Uniform Annual Series		
n	CAF	SCAF	SPWF	CAF	SCAF	SPWF	n
		16.%			18.%		
1	1.160	1.000	0.862	1.180	1.000	0.847	1
2	1.346	2.160	1.605	1.392	2.180	1.566	2
3	1.561	3.506	2.246	1.643	3.572	2.174	3
4	1.811	5.066	2.798	1.939	5.215	2.690	4
5	2.100	6.877	3.274	2.288	7.154	3.127	5
6	2.436	8.977	3.685	2.700	9.442	3.498	6
7	2.826	11.414	4.039	3.185	12.141	3.812	7
8	3.278	14.240	4.344	3.759	15.327	4.078	8
9	3.803	17.518	4.607	4.435	19.086	4.303	9
10	4.411	21.321	4.833	5.234	23.521	4.494	10
11	5.117	25.733	5.029	6.176	28.755	4.656	11
12	5.936	30.850	5.197	7.288	34.931	4.793	12
13	6.886	36.786	5.342	8.599	42.218	4.910	13
14	7.988	43.672	5.468	10.147	50.818	5.008	14
15	9.266	51.659	5.575	11.974	60.965	5.092	15
16	10.748	60.925	5.668	14.129	72.938	5.162	16
17	12.468	71.673	5.749	16.672	87.067	5.222	17
18	14.462	84.140	5.818	19.673	103.739	5.273	18
19	16.776	98.603	5.877	23.214	123.412	5.316	19
20	19.461	115.379	5.929	27.393	146.626	5.353	20
25	40.874	249.213	6.097	62.668	342.598	5.467	25
30	85.850	530.309	6.177	143.368	790.935	5.517	30
35	180.313	1120.707	6.215	327.991	1816.616	5.539	35
40	378.719	2360.743	6.233	750.362	4163.117	5.548	40
45	795.439	4965.242	6.242	1716.641	9531.340	5.552	45
50	1670.693	10435.570	6.246	3927.249	21812.490	5.554	50

The annual depreciation expense is the sinking fund deposit plus interest earned on the imaginary fund.

$$D = (V_a - V_s) \times i \div [(1 + i)^n - 1] + i \times (V_a - V_s) \times [(1 + i)^{k-1} - 1] \div [(1 + i)^n - 1]$$

Eq 502-15.

$$D = (V_a - V_s) \times i \times (1 + i)^{k-1} \div [(1 + i)^n - 1]$$

or by using the interest tables:

Eq 502-16.

$$D = (V_a - V_s) \times SFF_n \times (1 + i \times SCAF_{k-1})$$

The present value at any time is the first cost of the asset minus the current amount accumulated in the imaginary (sinking) fund.

The present value at the end of k years of an asset with a life of n years using the sinking fund method is:

Eq 502-17.

$$V_p = V_a - (V_a - V_s) \times [(1 + i)^k - 1] \div [(1 + i)^n - 1]$$

or

Eq 502-18.

$$V_p = V_a - (V_a - V_s) \times SFF_n \times SCAF_k$$

Declining balance method

The annual depreciation charge is a fixed percentage of the declining book value. If the fixed percentage is 2 times the straight line rate based on average service life, over 50% of the cost of the asset is written off during the first one third of its life. Twice the straight line method is called double declining balance depreciation.

This method gives rapid depreciation write-off for income tax purposes. It is more difficult to calculate than straight line, because the depreciation charges are different each year. Salvage values complicate this method and are neglected here.

Calculate depreciation for year k by:

Eq 502-19.

$$D = f \times V_a \times (1 - f)^{k-1}$$

where f = fixed percentage depreciation charge (use 2 x straight line rate or rate allowed by the IRS) so that f = 2/yrs expected life.

The present asset value at the end of k years is:

Eq 502-20.

$$V_p = V_a \times (1 - f)^k$$

Sum of digits method

This method gives larger depreciation during the early years of ownership. Advantages and disadvantages of this method are similar to the declining balance method. The annual charge is a ratio of the sum of the remaining years of life, (n - k + 1), to the sum of the entire life, (1 + 2 + 3 + ... + n), multiplied by the difference between the acquired and salvage value, $(V_a - V_s)$.

$$D = (n - k + 1) \times (V_a - V_s) \div (1 + 2 + 3 + ... + n)$$

Eq 502-21.

$$D = 2 \times (n - k + 1) \times (V_a - V_s) \div [n \times (n + 1)]$$

The present asset value at the end of k years is:

Eq 502-22.

$$V_p = 2 \times (V_a - V_s) \times [1 + 2 + ... + (n - k)] \div [n \times (n + 1)] + V_s$$

Example 1:

A dairy barn constructed in 1971 cost $80,000. Its probable useful life is 25 years with zero salvage value. Use 10% interest rate.

Determine:

- Annual depreciation based on new cost in 1971.
- Present value of the barn at the end of 1983 (12 years old).

Solution:

a. Straight line depreciation method:

$$D = V_a \div n = \$80{,}000 \div 25 \text{ yr} = \$3{,}200/\text{yr}$$

Present value, V_p in 1983 is:

$$V_p = V_a \times (1 - k \div n) = \$80{,}000 \times (1 - 12 \div 25) = \$41{,}600$$

b. Sinking fund method:

$$D_1 = (V_a - V_s) \times i \times (1 + i)^{k-1} \div [(1 + i)^n - 1] = \$80{,}000 \times .10 \times (1 + .10)^0 \div [(1 + .10)^{25} - 1] = \$8{,}000 \div 9.835 = \$813.42$$

$$D_{12} = \$80{,}000 \times .10 \times (1 + .10)^{11} \div [(1 + .10)^{25} - 1] = \$22{,}824.96 \div 9.835 = \$2{,}320.79$$

Or, using the interest tables:

$$D_1 = (V_a - V_s) \div SCAF_{25} = \$80{,}000 \div 98.347 = \$813.45$$

$$D_{12} = (V_a - V_s) \times (1 + i \times SCAF_{11}) \div SCAF_{25} = \$80{,}000 \times (1 + .10 \times 18.531) \div 98.347 = \$2{,}320.84$$

Present value in 1983 is:

$$V_p = V_a - (V_a - V_s) \times SCAF_{12} \div SCAF_{25} = \$80{,}000 - (\$80{,}000 \times 21.384) \div 98.347 = \$62{,}605.27$$

c. Declining balance method:
depreciation rate, f = 2 x 1 ÷ 25 = .08

$$D_1 = f \times V_a \times (1 - f)^{k-1} = .08 \times \$80{,}000 \times (1 - .08)^0 = \$6{,}400.00$$

$$D_{12} = .08 \times \$80{,}000 \times (1 - .08)^{11} = \$2{,}557.68$$

Present value in 1983 is:

$$V_p = V_a \times (1 - f)^k = \$80{,}000 \times (1 - .08)^{12} = \$29{,}413.31$$

d. Sum of digits methods

$$D_1 = (V_a - V_s) \times (n - k + 1) \div (1 + 2 + ... + n) = \$80{,}000 \times (25 - 1 + 1) \div (1 + 2 + ... + 25) = \$80{,}000 \times 25 \div 325 = \$6{,}153.85$$

$$D_{12} = \$80{,}000 \times (25 - 12 + 1) \div 325 = \$3{,}446.15$$

Present worth in 1983 is:

$$V_p = 2 \times [1 + 2 + ... + (n - k)] \times (V_a - V_s) \div [n \times (n + 1)] + V_s$$

$$V_p = 2 \times (1 + 2 + ... + 13) \times \$80,000 \div (25 \times 26) + 0$$
$$= 2 \times 91 \times \$80,000 \div 650 = \$22,400.00$$

Inflation

Inflation is a substantial rise of prices. It is a decrease in the value or purchasing power of money.

Price changes may or may not need to be considered in economic analysis. However, costs and benefits must be computed in comparable units.

Inflation has a significant effect on taxable income. Taxable income is based on inflated current dollars, but depreciation deductions on fixed assets are based on the original investment made with more valuable dollars. The net result is an increase in taxable income.

Impact of Inflation Before Taxes

Inflation increases the cost of fixed assets so that at the end of an asset's life, the replacement cost will be higher. Compute the increase in cost for each year. Use the average inflation rate of each period.

Price increases are similar to interest compounded annually. Each increase is added to the original cost, and the following increases are based on the compounded increases.

The replacement cost, V_r, is a future sum, F, and the factor used to determine V_r is similar to the single payment compound amount factor, CAF. Assuming that the initial cost, V_a, equals the principal, P, and that the average inflation rate, C, equals the interest rate, i,

$$F = P \times CAF = P \times (1 + i)^n$$

becomes

$$V_r = V_a \times (1 + C)^n.$$

The present worth of future inflated dollars can also be calculated. See Example 3.

Example 2:

A dairy barn costs $80,000 to build. Its estimated useful life is 25 years.

Determine the replacement cost at the end of its 25-yr life. Assume an average inflation rate, C, of 6%.

$$V_r = V_a \times (1 + C)^n$$
$$= \$80,000 \times (1 + .06)^{25} = \$343,350$$

Example 3:

Assume that you wish to purchase an annuity to pay you $6,000 a year for the next 8 years. You want an 8% return above the rate of inflation. It appears that inflation will average 6% per year.

How much should you pay at this time for the annuity?

- Calculate the equivalent interest rate, i', to yield the desired rate of return plus inflation.

$$F = P \times (1 + C)^n \times (1 + i)^n$$
$$= P \times (1 + C + i + C \times i)^n$$

$$i' = C + i + C \times i$$
$$= .08 + .06 + (.08 \times .06)$$
$$= .1448$$

- The amount you should pay for the annuity is:

$$P = R \times SPWF$$
$$= R \times [(1 + i')^n - 1] \div [i' \times (1 + i')^n]$$
$$= \$6,000 \times [(1 + .1448)^8 - 1] \div [.1448 \times (1 + .1448)^8]$$
$$= \$27,715$$

- If the inflation rate was zero, you could pay more for the annuity. The equivalent interest rate would be 8%.

$$P = \$6,000 \times [(1 + .08)^8 - 1] \div [.08 \times (1 + .08)^8]$$
$$= \$34,480$$

Impact of Inflation After Taxes

Taxes adversely affect real income during an inflationary period. Even though the returns from an investment may follow the inflation rate, the real returns after taxes are reduced because the inflationary portion of the return is also taxed.

In addition, inflation diminishes the real value of the depreciation deduction on a fixed asset. The returns from this investment may increase, but the depreciation schedule does not change, which increases taxable income.

Example 4:

A piece of equipment was purchased in 1982 for $10,000. It has a useful life of 5 years with zero salvage value and will produce an estimated return of 12% per year.

a. With a zero inflation rate and i = 12%:

- Annual net return, R, is:

$$R = P \times CRF_5 = P \div SPWF_5$$
$$= \$10,000 \div 3.605 = \$2,774$$

- With straight line depreciation, the annual depreciation deduction, D, is:

$$D = (1 \div n) \times V_a$$
$$= (1 \div 5) \times \$10,000 = \$2,000/yr$$

- Taxable income = R - D = $2,774 - $2,000 = $774

b. With 6% annual inflation and returns increasing with inflation, determine the taxable income for the third year, 1985.

- 1985 net return = $R \times (1 + C)^n = \$2,774 \times (1 + .06)^3 = \$3,304$
- The depreciation deduction remains constant at $2,000.
- Taxable income = $3,304 - $2,000 = $1,304 in 1985 dollars.

Or in 1982 dollars, taxable income = $\$1,304 \div (1 + .06)^3 = \$1,095$.

The taxable income has increased 68%, [($1,304 - $774) ÷ $774], in terms of 1985 dollars or 41%, [($1,095 - $774) ÷ $774], in terms of 1982 dollars.

Example 4 illustrates the unfavorable effect of inflation on the after-tax rate of return. Equipment with a substantial salvage value that increases with inflation could offset some of the inflationary loss.

510 SWINE

Summary of Design Data

	Weight lb	Ventilation, cfm/hd: Cold weather rate	Mild weather rate	Hot weather rate	Winter room temperature F	Supplemental heat, Btu/hr/hd: Slotted floors	Bedded or scraped floors	Manure, ft³/hd Liquids + solids + 15% extra
Sow and litter	400	20	80	500	80	4000	—	0.66
					70	3000	—	
					60	—	3500	
Prenursery pig	12-30	2	10	25	85	350	—	0.03
Nursery pig	30-75	3	15	35	75	350	—	0.07
					65	—	450	
Growing pig	75-150	7	24	75	60	600	—	0.14
Finishing pig	150-220	10	35	120	60	600	—	0.24
Gestating sow	325	12	40	150*	60	1000	—	0.20
Boar	400	14	50	300	60	1000	—	0.25

*300 cfm for gestating sows in a breeding facility.

Slot Widths

For slatted floors. Wire mesh, metal, or plastic slats preferred in farrowing and prenursery.

	Slot widths in.	Concrete slat widths, in.
Sow and litter	⅜	4
Prenursery pig	⅜	Not recommended
Nursery pig	1	4
Growing-finishing pig	1	6-8
Gestating sows or boars:		
Pens	1	6-8
Stalls	1	4

Water Requirements

Animal type	Gal/hd/day
Sow and litter	8
Nursery pig	1
Growing pig	3
Finishing pig	4
Gestating sow	6
Boar	8

Space Requirements

Enclosed housing:

Pigs	Weight lb	Area ft²
Prenursery[a]	12-30	2-2½
Pig-nursery[b]	30-75	3-4
Growing[b]	75-150	6
Finishing[b]	150-220	8

[a]Avoid concrete slats, slats over 2" wide, and partly slotted floors for prenursery pigs.
[b]For slotted, flushed, or scraped floor.

Shed With Lot:

More lot area is often provided to facilitate manure drying.

	Weight lb	Inside ft²/hd	Outside ft²/hd
Nursery pig	30-75	3-4	6-8
Growing/finishing pig	75-220	5-6	12-15
Gestating sow	325	8	14
Boar	400	40	40
Sow in breeding	325	16	28

Feeder Space:

Sows: 1'/self-feed sow, 2'/group-fed sow.

Pig (12-30 lb):	2 pigs/feeder space
Pig (30-50 lb):	3 pigs/feeder space
Pig (50-75 lb):	4 pigs/feeder space
Pig (75-220 lb):	4-5 pigs/feeder space

Space Requirements

Pasture Space:
Depends on rainfall and soil fertility.

10 gestating sows/acre
7 sows with litters/acre
50 to 100 growing-finishing pigs/acre

Shade Space:

15-20 ft²/sow
20-30 ft²/sow and litter
4 ft²/pig to 100 lb
6 ft²/pig over 100 lb

Waterer Space:

Minimum of 2 waterers per pen.
Pig (12-75 lb): 10 pigs/waterer
Pig (75-220 lb): 15 pigs/waterer

Floor and Lot Slopes

Slotted floors: usually flat
Solid floors:
Farrowing:
¼"-½"/ft without bedding
¼"/ft with bedding

Pigs:
½"/ft without bedding
¼"/ft with bedding

Paved lots: ½"/ft
Paved feeding floors:
Indoors: ¼"/ft
Outdoors: ½"/ft

Building alleys:
½"/ft crown or side slope
⅛"/ft to drains

Scheduling

Facility scheduling is one of the most important and often the most difficult task in managing an intensive production system. Production system capacity is usually based on desired farrowing frequency, sow herd size, farrowing building size, or number of pigs finished per year. Once you have a value for two or more of these factors, determine **general** facility sizes from Table 510-1. Make a separate **detailed** scheduling analysis of your particular system before starting construction.

Table 510-1 shows the average number of sows, boars, and pigs in various production schedules, based on a 10-stall farrowing building (40 stalls for a weekly schedule). For a larger or smaller farrowing

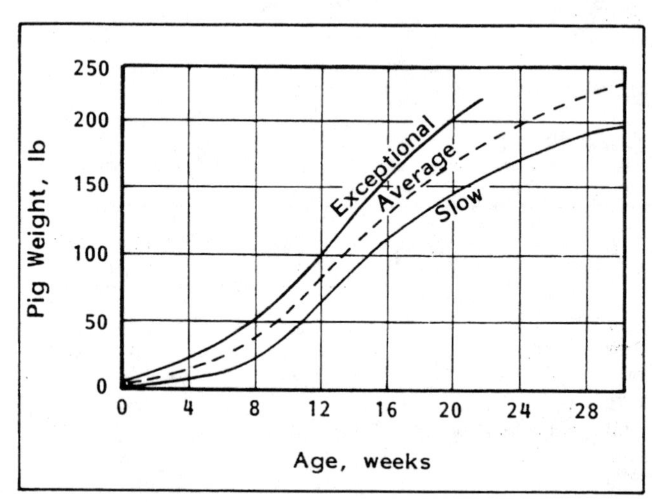

Fig 510-1. Typical growth of swine.

Table 510-1. Swine scheduling table.
Animal capacity per 10 sows farrowed per period.

					Use "Multiplier" on these values:									
						Sow capacity in:				Pig capacity in:				
Row	Farrowing frequency A	Farrowing interval, weeks B	Number of sow groups C	Farrowing stalls D	Total sows E	gestation F	gilt pool F	breeding F	Boars	prenursery G	nursery G	growing G	finishing G	Pigs per year
1	6x	8	3	10	39	24	12	15	4	—	96	—	160	520
2	7x(1st)	7	3	10	39	24	12	15	5	—	96	80	80	595
3	8x	6	4	10	51	36	12	15	4	—	96	80	160	695
4	10x(1st)	5	4	10	51	36	12	15	5	—	96	160	160	830
5	12x(1st)	4	5	10	63	48	12	15	6	96	80	160	160	1040
6	Weekly(1st)	4	20	40	243	168	12	45	7	340	320	640	680	4160

A. 6x = farrowing about 6 times per year.
(1st) = sows bred during first heat after weaning.
B. Time per sow group for farrowing and cleaning.
C. Weekly farrowing requires 5 sow groups for each of 4 rooms = 20 groups.
D. Group size = number of farrowing stalls. Weekly farrowing requires at least 4 rooms of 10 stalls each = 40 stalls. Five rooms (50 stalls) are preferred to allow one week for cleaning between farrowing.
E. Sows = sows + gilts. Gilt pool not included.
Total sows = (group size x 1.2) x (number of sow groups + ¼). The 1.2 allows for 80% conception. The ¼ allows for open sows returned from gestation.
F. Building capacity needed—not necessarily number of sows at all times. In most cases, space for one group is needed only when cleaning farrowing room and is not used after the group is moved to farrowing. It may be possible to provide short term housing for this group elsewhere on the farm. Gilt pool = 6 x typical number of sows to be replaced in each sow group (20% replacement assumed). Breeding area allows extra capacity for low conception rates during warm weather.
G. Assumes 8 pigs marketed per litter. Pigs marketed at the average age of 26 weeks. Capacity for first group moved to prenursery or nursery is increased by 20% to account for overfarrowing and slow growers.

building, calculate a multiplier by dividing the desired number of farrowing stalls by the table value. Apply the multiplier to the table entries in the same row as the desired farrowing frequency. To base the capacity on number of pigs produced per year, divide desired number of pigs per year by the table value to get the multiplier.

Example A

A producer wants to farrow 7 times a year in a 24-stall farrowing building. Determine the size of the other facilities and the number of finishing hogs produced per year.

Row 2 in Table 510-1 is for 7 farrowings per year. Calculate the multiplier by dividing the desired number of farrowing stalls by the table value: 24 ÷ 10 = 2.4. Multiply the other values in row 2 by 2.4:

- sows in entire herd = 2.4 x 39 = 93.6 (round up to 94).
- sows in gestation unit = 2.4 x 24 = 57.6 (58).
- boars in breeding unit = 2.4 x 5 = 12.
- pigs in nursery unit = 2.4 x 96 = 230.4 (231).
- pigs in growing unit = 2.4 x 80 = 192.
- pigs in finishing unit = 2.4 x 80 = 192.
- pigs marketed per year = 2.4 x 595 = 1428.

Example B

A producer wants a new weekly farrowing complex. He needs to produce about 5000 pigs per year to utilize his labor and grain.

Row 6 is for weekly farrowing. Determine the multiplier by dividing the desired number of pigs per year by the table value: 5000 ÷ 4160 = 1.2. Multiply the other values in row 6 by 1.2:

- farrowing stalls = 1.2 x 40 = 48 (12 stalls per room).
- sows in entire herd = 1.2 x 243 = 291.6 (round up to 292).
- sows in gestation unit = 1.2 x 168 = 201.6 (202).
- boars in breeding unit = 1.2 x 7 = 8.4 (9)
- pigs in prenursery unit = 1.2 x 340 = 408.
- pigs in nursery unit = 1.2 x 320 = 384.
- pigs in growing unit = 1.2 x 640 = 768.
- pigs in finishing unit = 1.2 x 680 = 816.
- pigs marketed per year = 1.2 x 4160 = 4992.

Example C

A producer has two 20-sow farrowing rooms. He wants to farrow one sow group every 3 weeks (i.e. each room will be occupied for 6 weeks per farrowing). This is equivalent to two separate 20-sow farrowing systems, so figure the capacity of one room and double it.

Row 3 is for 6 weeks per farrowing. The multiplier is the actual number of sows per room divided by 10: 20 ÷ 10 = 2.0. Multiply the other values in row 3 by 2.0:

- number of sow groups = 4 per room (8 for two rooms).
- total sows = 2.0 x 51 = 102 (204 for two rooms).
- sows in gestation = 2.0 x 36 = 72 (144 for two rooms).
- sows in breeding = 2.0 x 15 = 30 (60 for two rooms).*
- boars needed = 2.0 x 4 = 8 (8 for two rooms).*
- pigs in nursery = 2.0 x 96 = 192 (384 for two rooms).
- pigs in growing = 2.0 x 80 = 160 (320 for two rooms).
- pigs in finishing = 2.0 x 160 = 320 (640 for two rooms).
- pigs produced per year = 2.0 x 695 = 1390 (2780 for two rooms).

* This example assumes second heat breeding (3½ weeks after weaning), so space for two groups of sows is required in breeding building. However, the two sow groups are bred 3 weeks apart, so you need only enough boars for one sow group.

Farrowing

Typical Farrowing Schedules

- **One Litter per Year.** Gilts are farrowed once a year in mild weather, usually on pasture. Building and equipment investment must be small, because the cost is charged to only one group of sows and litters per year.
- **Two Litters per Year.** One group of sows is farrowed twice a year, often on pasture. Farrowings are scheduled to avoid weather extremes and periods of high labor demand. Overhead for buildings and equipment is prorated to two litters per year.
- **Multiple Litters per Year.** Two or more groups of sows farrow twice a year, allowing 4 to 12 farrowings a year. Many producers farrowing 8 or more times per year end up with continuous farrowing because of the difficulty in getting groups of sows to farrow over a short time. Environmentally controlled facilities are common for this schedule, which increases overhead costs but reduces labor per pig. Facility costs are charged to many pigs, lowering the cost per pig.
- **Farrowing Each Week.** Some large producers farrow a set number of sows (usually 8 to 12) each week in separate, small farrowing rooms, Fig 510-2. Weekly farrowing provides better overall management and full use of specialized labor and facilities, resulting in groups of sows bred weekly, litters born within the same week, and rooms that can be emptied and disinfected between groups. Provide at least 4 farrowing rooms for 3-week weaning—5 rooms for 4-week weaning. An additional room allows time for more convenient cleanup and space for farrowing excess or late sows.

Building Design Factors

Temperature

Sows and young nursing pigs require different temperatures. A newborn pig needs a dry, draft-free environment at a temperature of 90-95 F the first three days of life. In contrast, a sow is most comfortable at 60-65 F.

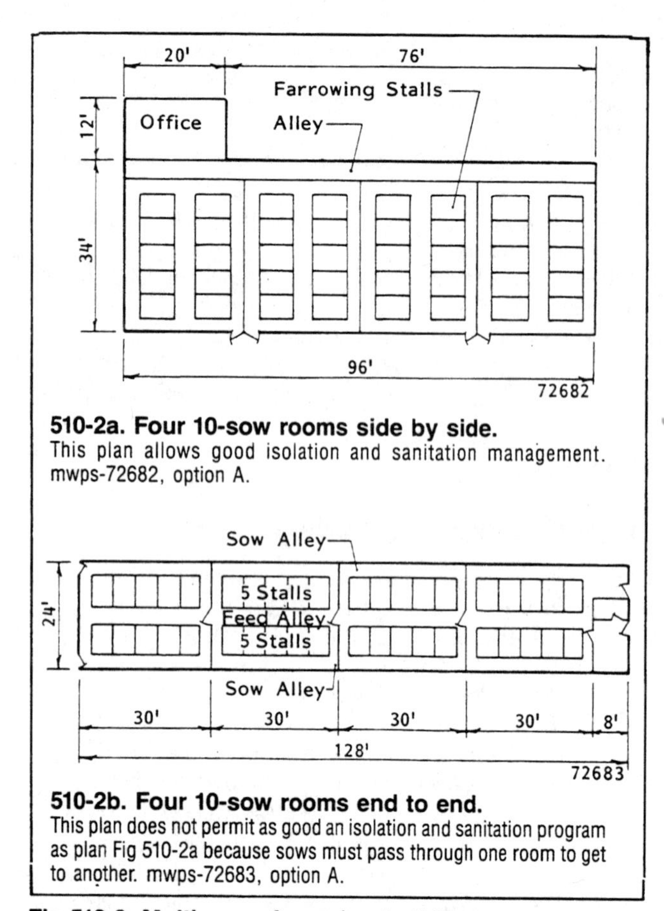

510-2a. Four 10-sow rooms side by side.
This plan allows good isolation and sanitation management. mwps-72682, option A.

510-2b. Four 10-sow rooms end to end.
This plan does not permit as good an isolation and sanitation program as plan Fig 510-2a because sows must pass through one room to get to another. mwps-72683, option A.

Fig 510-2. Multi-room farrowing buildings.

Provide these two temperatures by keeping the farrowing building at 65-75 F and providing supplemental heat in the creeps. Creep heaters may be lamps, catalytic gas heaters, electric floor pads, electric heating cables, hot water pipes in the floor, or some combination. See section on ventilation. Provide creep heat for at least the first week after birth. Room temperature can be 60 F if bedding is used. If the building is not environmentally controlled, use draft barriers, bedding, and hovers.

Sanitation

Properly designed farrowing units are easy to keep clean. Avoid porous or rough surfaces that can harbor bacteria. Rough floors are abrasive to the feet and knees of nursing pigs and retain moisture and manure. Smooth surfaces drain and dry more rapidly, are easier to clean and disinfect, but can be slippery. Slick floors also result in injured feet and legs. Avoid excessive troweling of concrete floors—power troweling frequently results in slippery floors.

Slope solid floors for proper drainage. Slats made of slick materials (aluminum, stainless steel, plastic) should have ribs down the center or punched slots for better footing. Avoid slats with continuous, parallel ribs which can trap moisture and manure, resulting in unsanitary conditions.

Establish a schedule for breeding and follow it so you can completely empty the building and clean and disinfect between sow groups.

Stalls or Pens

The main differences between farrowing stalls and pens are the amount of exercise for the sow and the degree of protection of the pigs from crushing. With stalls, pigs are protected from the sow. Heated creeps further protect pigs by attracting them away from the sow when not nursing. Stalls require less bedding and labor than do pens. It is more difficult to catch pigs in a stall but the operator is protected from the sow when handling pigs.

Farrowing pens allow more sow movement but require more cleaning labor and floor space. A sow in an open pen must be restrained for any physical treatment. Pens may be preferred for late weaning to give the larger pigs more space. Convert farrowing pens with feeders and waterers to small nursery pens by removing the guard rails.

Stalls are usually 5′ wide by 7′ long (some are 6′ or 6½′ long with the feeder mounted on the outside of the stall). The width includes an 18″ pig area on both sides of a 24″ sow stall, Fig 510-3. Most commercial stalls are adjustable for large or small sows. Some special farrowing stalls are only 4½′ or even 4′ wide. The narrower stalls work better for gilts, smaller sows, and for pigs weaned by 3 weeks.

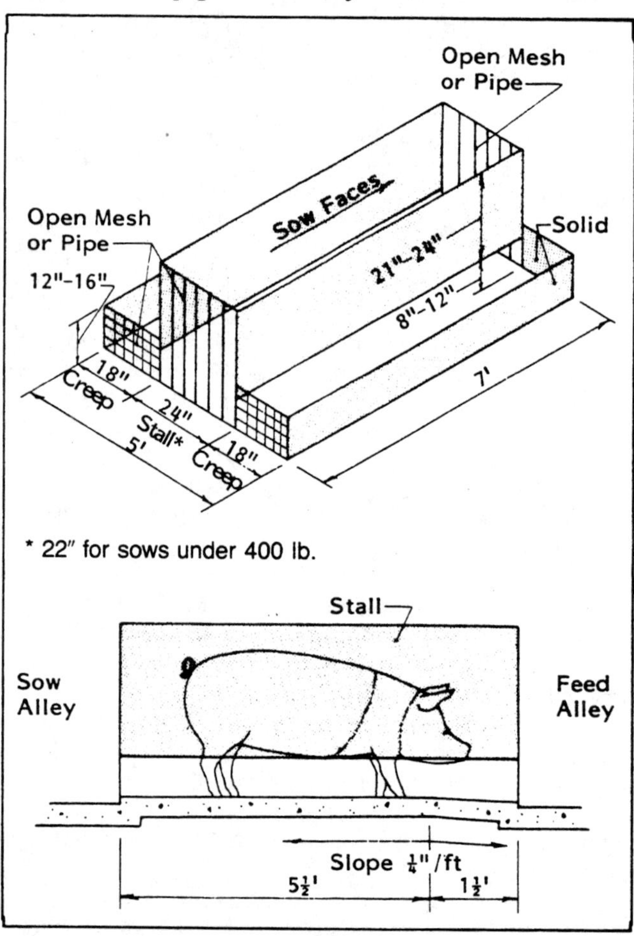

Fig 510-3. Farrowing stall (or crate).
Farrowing stalls are often 1″ lumber, ¾″ exterior plywood, or 1″ galvanized or aluminum pipe. Solid partitions reduce drafts. Make the front and side creep panels solid. Make the rear creep panels and the front and rear sow stall panels open mesh or pipe.

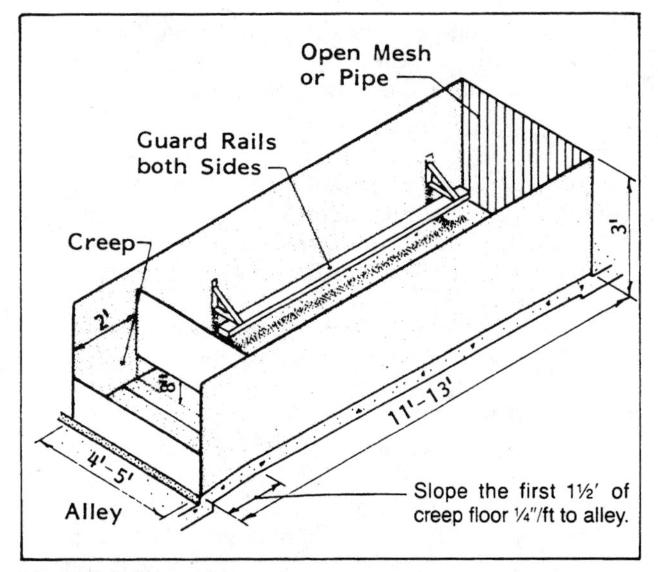

Fig 510-4. Farrowing pen.
The two-slope floor allows drainage from the creep area directly to the alley instead of across the pen area.

Typical pens are 4′ to 5′ wide by 11′ to 13′ long, Fig 510-4. The front 18″ to 24″ of the pen is the creep area. Guard rails, about 6″ out from the wall and 8″ off the floor, reduce pig crushing. The rear 4′ to 5′ is often slotted. Cover slats during, and for the first few days after, birth. Locate waterer over slats.

Some producers use farrowing pens as "free stalls." Sows are free to come and go from the pens to a feeding floor for feed and water. Keep pigs in the pens with a low gate behind the sow, Fig 510-5.

Floors

Slotted floors

Slotted floors greatly reduce labor to remove manure. They separate the animal from its manure and reduce internal parasite problems. They also result in drier floors.

Buildings with totally slotted floors require more environmental control than buildings with partly slotted or solid floors. Use a solid covering in the creep area of totally slotted floors for about one week after farrowing to increase pig comfort and prevent hoof injuries.

Slat spacing is critical in a farrowing unit. Space slats either a uniform ⅜″ apart **or** 1″ apart. Pig's legs may get caught in spaces **between** ⅜″ and 1″. Cover any spacing over ⅜″ during, and for 3 days after, farrowing. Space concrete and wood slats 1″ apart behind the sow to improve cleaning. A 2″x4″ opening behind the sow is useful for manual manure removal, but keep it covered except during use. Maximum slat width is 4″.

Place slats parallel to the sow to provide better footing for the sow to get up and for the pigs while nursing.

Solid floors

Slope solid floors to alleys and drains. Use a two-slope floor if waterers are installed at the front of stall or pen, Fig 510-3.

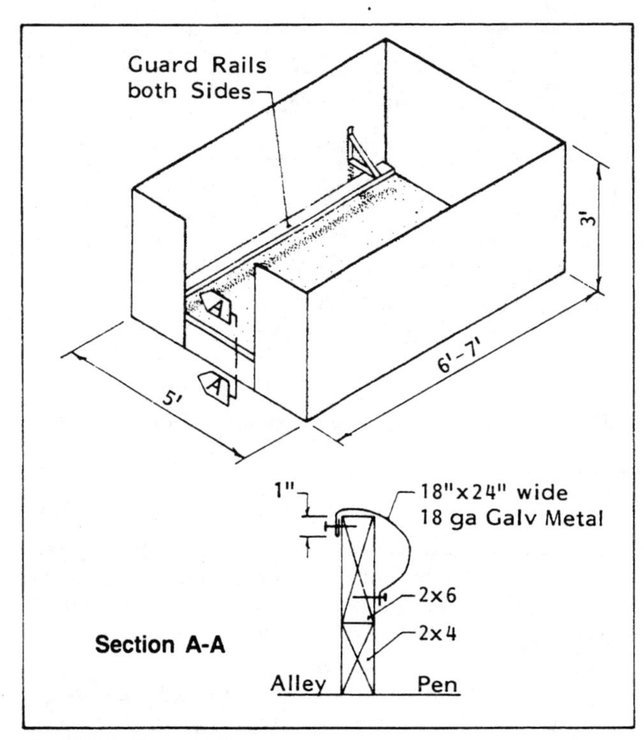

Fig 510-5. Free stall.
One type of homemade pig stop is shown; commercial units are available.

Floor Plans

Farrowing buildings commonly have 2 rows of farrowing stalls facing the center alley for feeding ease and sow interaction, Fig 510-6. Hand feed sows from a feed cart or feeding stations.

Fig 510-7 shows sow gestation stalls in the same room as farrowing stalls. Some producers hold sows in the farrowing building for 6 to 7 weeks before farrowing to encourage pig immunity to scours. This building will rarely be empty for total room cleaning.

Farrowing stall floor arrangements are shown in Fig 510-8. Fig 510-8a shows a two-slope solid floor, which often contains floor heat. Sows face the outside wall in this arrangement. Turn sows out twice a day for feed and water, to increase sow interaction, and to reduce cleaning of solid floors. A mechanical gutter cleaner just outside the rear of the stall can facilitate cleaning.

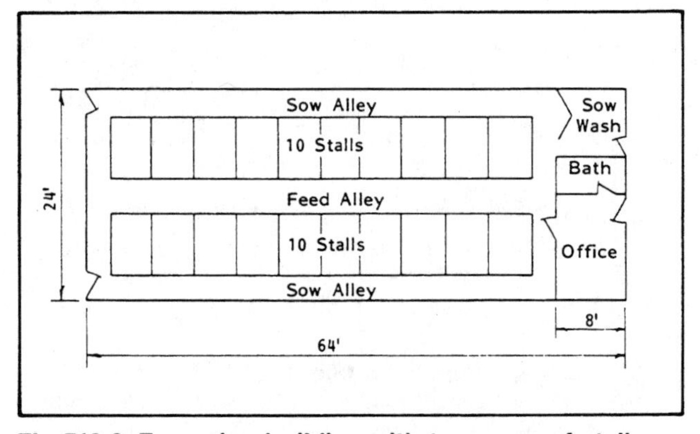

Fig 510-6. Farrowing building with two rows of stalls.
mwps-72680.

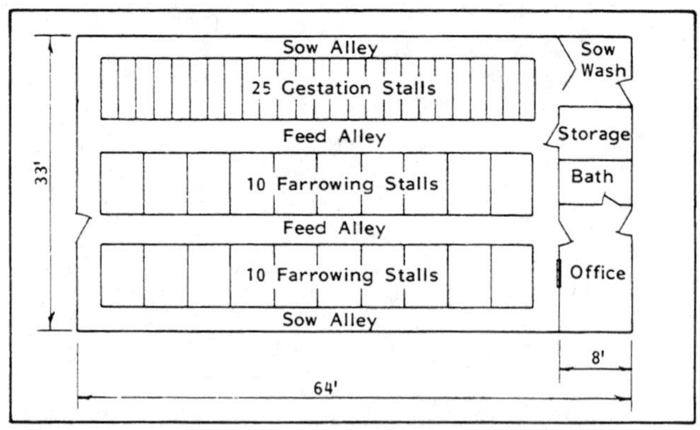

Fig 510-7. 20-sow farrowing building with 25 gestation stalls.
mwps-72699.

Figs 510-8b, 510-8c, and 510-8d show floors with slats at the rear of the stall. The concrete portion of the creep area often contains heat. Slope the front portion of the floor forward to drain spilled water. Dunging at the front of the stalls is a problem with pigs older than 3 weeks.

Fig 510-8e shows a floor with about 12" slotted in the front and 30" slotted over a gravity drain gutter at the rear.

Figs 510-8f and 510-8g show totally slotted floors under the stalls or the entire building. Scrape or flush manure toward center alley.

Nursery

A nursery is for weaned pigs or for sows with litters. There are three types:

- **A late-wean nursery (pig-nursery)** for raising pigs weaned at 5 to 8 weeks from about 30 lb up to 75 lb.
- **An early-wean nursery (prenursery)** for raising pigs weaned at 3 to 4 weeks from about 12 lb up to 25-30 lb. After 3 to 4 weeks, move pigs to a pig-nursery.
- **A sow-pig nursery** for relocating sows and their 3-day to 3-week-old litters from the farrowing unit. After the sows are taken out, the pigs stay until they are moved at about 30 lb to a pig-nursery or at about 75 lb to a growing unit. Labor for moving sows with pigs is a disadvantage.

Temperature

For 3-week-old pigs, provide 85 F temperature at pig level for the first few days after weaning. Lower

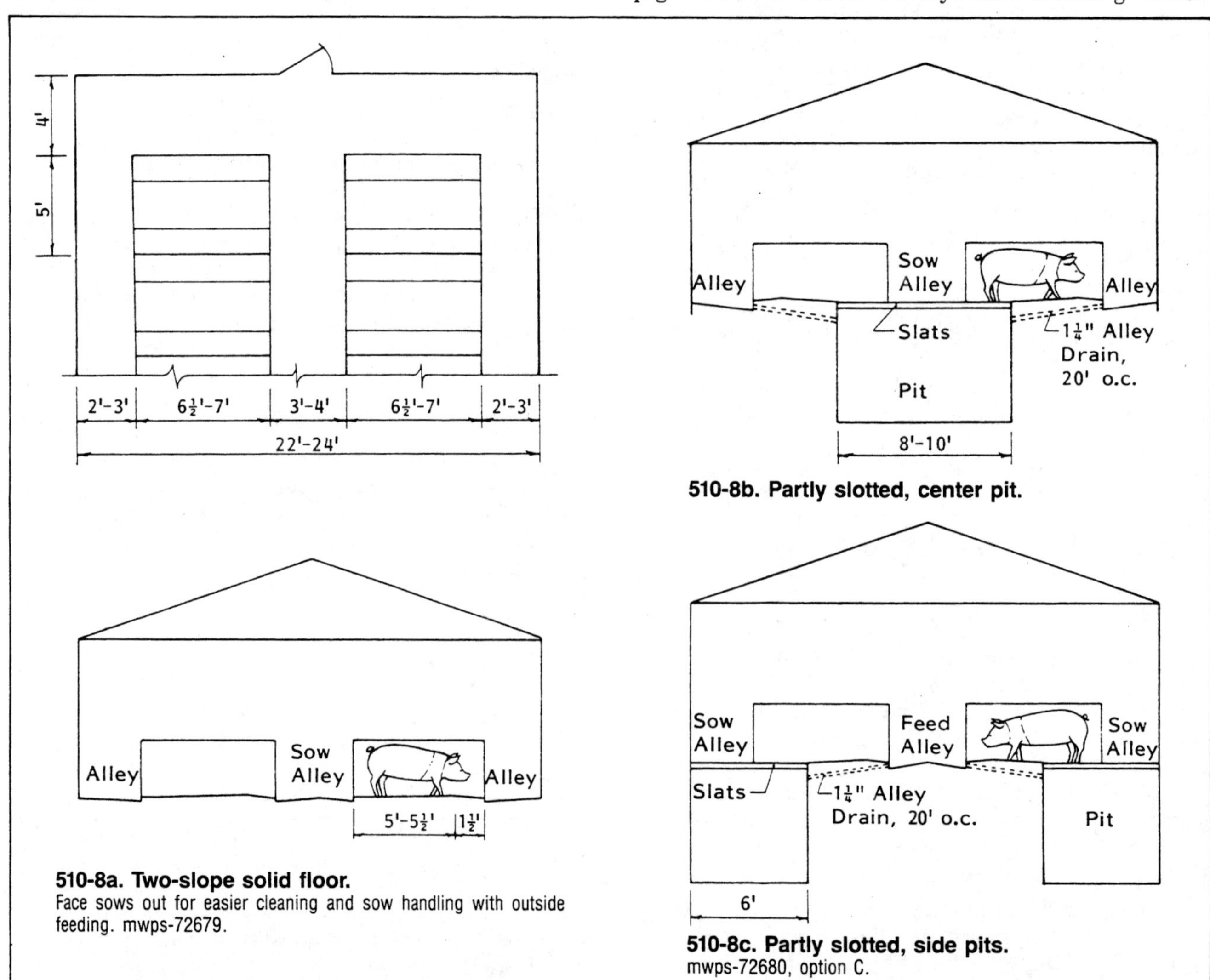

510-8a. Two-slope solid floor.
Face sows out for easier cleaning and sow handling with outside feeding. mwps-72679.

510-8b. Partly slotted, center pit.

510-8c. Partly slotted, side pits.
mwps-72680, option C.

Fig 510-8. Farrowing building floor arrangements.

Continued on next page.

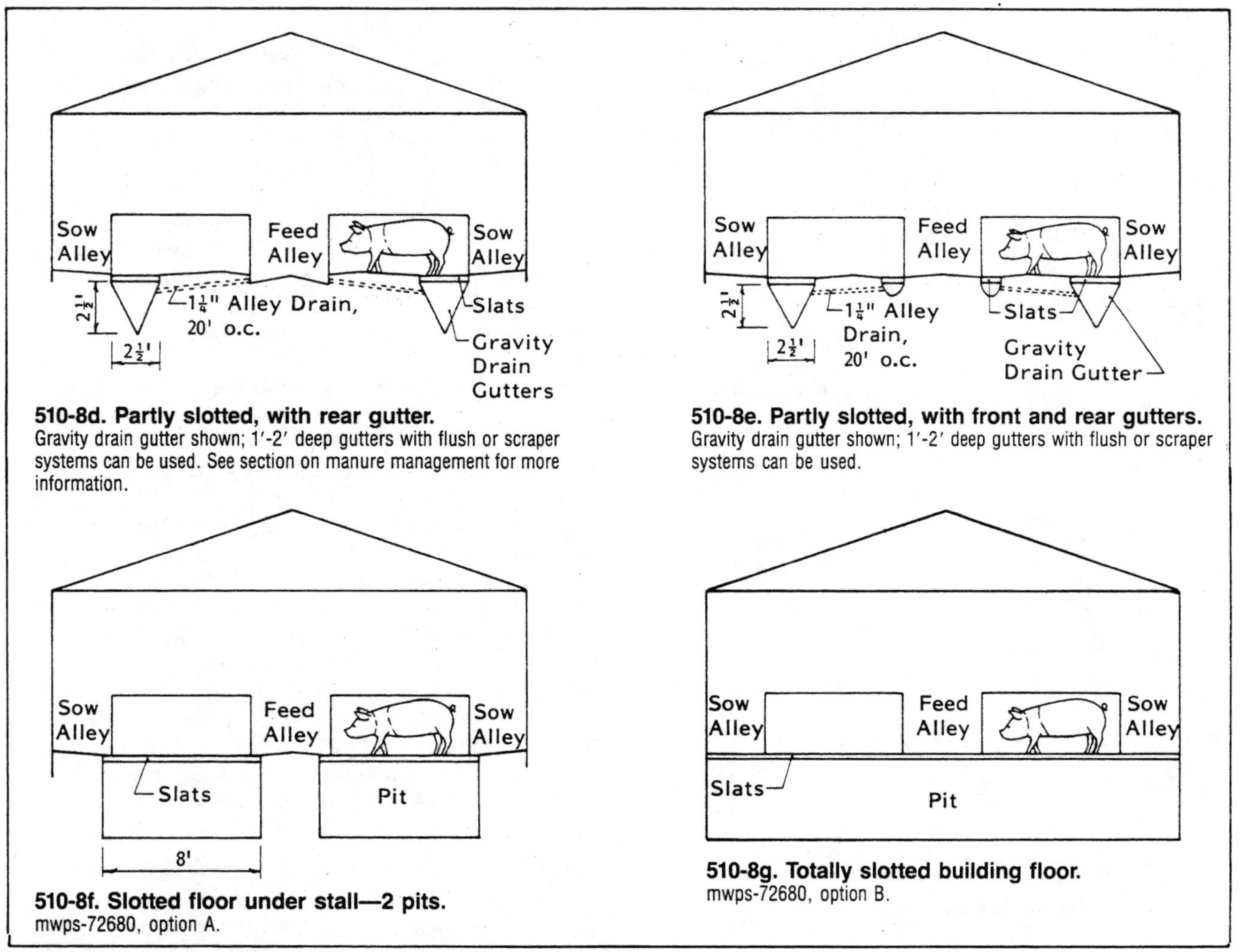

510-8d. Partly slotted, with rear gutter.
Gravity drain gutter shown; 1'-2' deep gutters with flush or scraper systems can be used. See section on manure management for more information.

510-8e. Partly slotted, with front and rear gutters.
Gravity drain gutter shown; 1'-2' deep gutters with flush or scraper systems can be used.

510-8f. Slotted floor under stall—2 pits.
mwps-72680, option A.

510-8g. Totally slotted building floor.
mwps-72680, option B.

Fig 510-8. Farrowing building floor arrangements, continued.

the temperature 3 F per week to a minimum of about 70 F for 8-week-old pigs. Provide warm floors with infrared heaters, heating pads, or floor heat. With only space heaters, you may have to set the thermostat higher than the desired room temperature to maintain warm floors. For prenursery, consider preheating ventilation air to reduce drafts.

In a non-bedded sow-pig nursery, two different environmental conditions are required. Keep the room temperature at 65-75 F with supplemental heat in the creeps. Room temperature can be 60 F if bedding is used.

Space

In a pig-nursery, provide 3 to 4 ft²/pig for totally or partly slotted pens. Size each pen for 16 to 20 pigs. Keep litters together as much as possible, but sort to maintain size uniformity.

In a prenursery, provide 2 to 2½ ft²/pig in totally slotted pens with woven wire floors or slats less than 2" wide. Avoid concrete slats, slats over 2" wide, and partly slotted pens. These space allowances assume that the pigs enter at about 12 lb and leave at about 30 lb. Size each prenursery pen for no more than 16 pigs; preferably no more than 10 pigs.

In totally or partly slotted sow-pig nurseries, minimum pen size is 5'x10' for one sow and litter; 8'x10' for two. Minimum pen width is 5' and minimum pen length is 10'. Provide 1 ft² of heated creep area per pig in each pen. Allow an additional 40 ft² of open lot per sow and litter in shed and lot systems. Do not combine more than two sows and litters per pen.

Facilities

Use fenceline feeders in prenursery pens and decks, and in partly slotted pig-nurseries. Round or fenceline feeders can be used in totally slotted pig-nurseries.

Section the openings of feeders for early weaned pigs so the pigs cannot crawl into them. Leave at least 6" between feeder and pen partitions to prevent dunging in end feeder holes.

Provide one waterer for each 10 prenursery or pig-nursery pigs, with a minimum of 2 waterers per pen. Place nipple height at about 10" for 10 lb pigs; about 12" for 12 lb. Raise accordingly as pig grows.

Make pen partitions 24" high in prenurseries and 32" high in pig-nurseries.

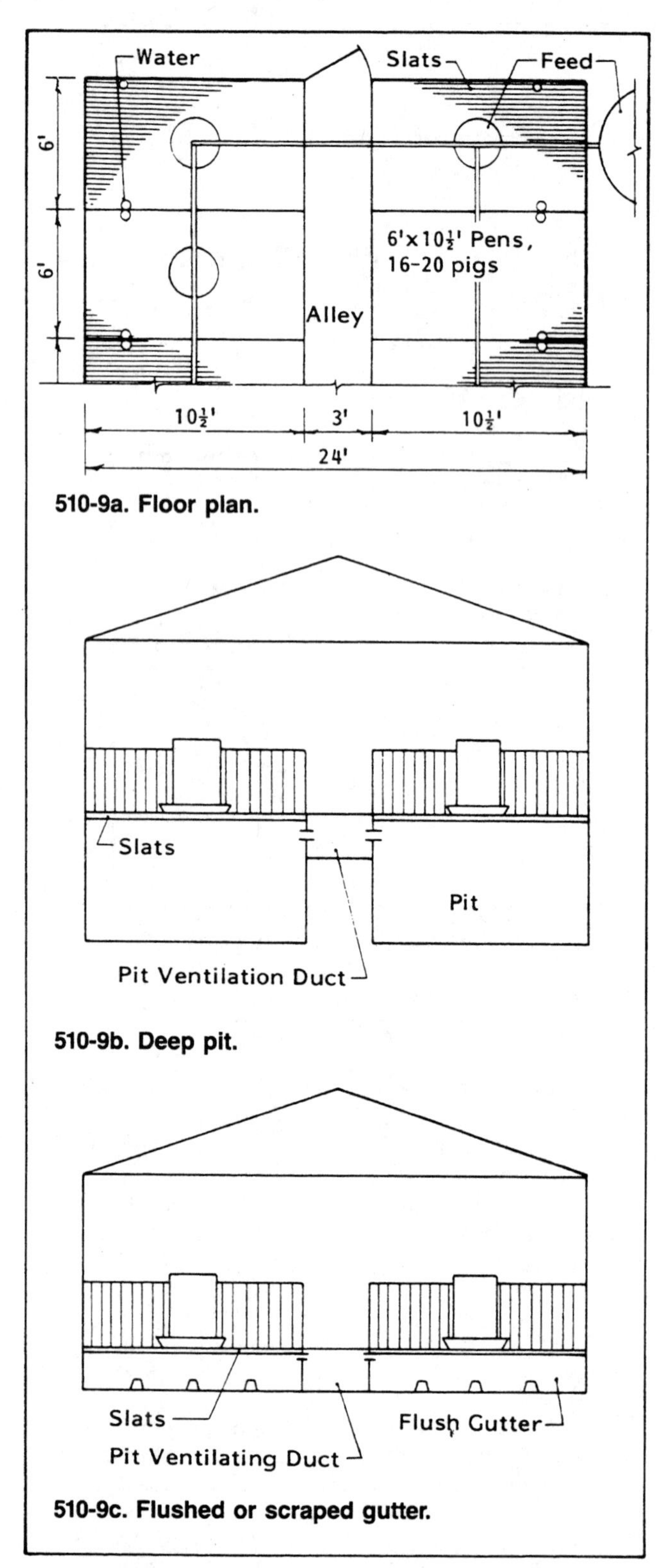

510-9a. Floor plan.

510-9b. Deep pit.

510-9c. Flushed or scraped gutter.

Fig 510-9. Pig-nursery with totally slotted floor.
mwps-72686.

Decks

Advantages include:
- Increased stocking density per building.
- Lower operating costs per pig.
- Fewer pigs per group in smaller pens.

Disadvantages include:
- Pig handling in upper decks is difficult.
- Design and management of environmental control system is more critical.
- Maintenance of upper deck equipment may be more difficult if not arranged for easy access.
- Pigs in lower pens can be dirtier than normal.

Pigs are usually put in and removed from upper decks by hand. Remove pigs before they reach 50 lb, limit deck depth to 4′, and use removable front gates to ease pig handling. Arrange feeders, solid partitions, and overlays in both decks to facilitate observation.

Consider the higher pig density and the influence of the top decks on airflow patterns when selecting a ventilating system. Never locate pigs in the direct path of the ventilating air. Reduce drafts with solid pen partitions and solid floor overlays in the sleeping area. Monitor room temperature and airflow patterns closely because fluctuations can lead to scours or poor performance in small pigs.

Locate the waterer at rear of pens to train pigs to dung there. Solid floor overlays near the feeder also improve dunging patterns and reduce feed wastage.

All-in, all-out management reduces disease risk, because it allows thorough cleaning and disinfecting between pig groups. Keep facilities as clean as possible, but avoid washing pens when occupied by pigs. Dry scraping is best when pigs are present.

Floor Surfaces

Slotted or perforated floors greatly reduce cleaning labor and help separate manure quickly from the pigs. Totally slotted floors are highly recommended for nursery pigs.

For partly slotted floors, slot at least two-thirds of the floor in prenurseries and at least 40% in pig-nurseries. It is difficult to train nursery pigs to sleep on solid floors and dung on slats, so consider the following:

- Use a long, narrow pen shape (length 2 to 4 times the width). Minimum pen width is 4′ for prenursery (5′ for pignursery) with partly slotted floor.
- Use solid partitions over solid floors and open partitions over slotted areas. In buildings with ceiling slot air inlets, use open partitions along alleys for cross ventilation in summer.
- Place feeders on solid floors and waterers over slats.
- Provide a 1″-2″ step up from slotted floors to solid floors to separate dunging and sleeping areas and to keep manure off the solid floor.
- During cold weather, provide supplemental zone heat in solid floor areas for small pigs.
- Use hinged or removable hovers over the sleeping areas to reduce drafts and retain heat.
- Wet down the slotted area just before putting pigs in the pen.
- Feed on the solid floor for the first few days.

Avoid concrete slats for pigs under 30 lb—select nonabrasive slats. Slope solid floors where bedding is used about ¼″/ft for drainage. If no bedding, slope floors ½″/ft.

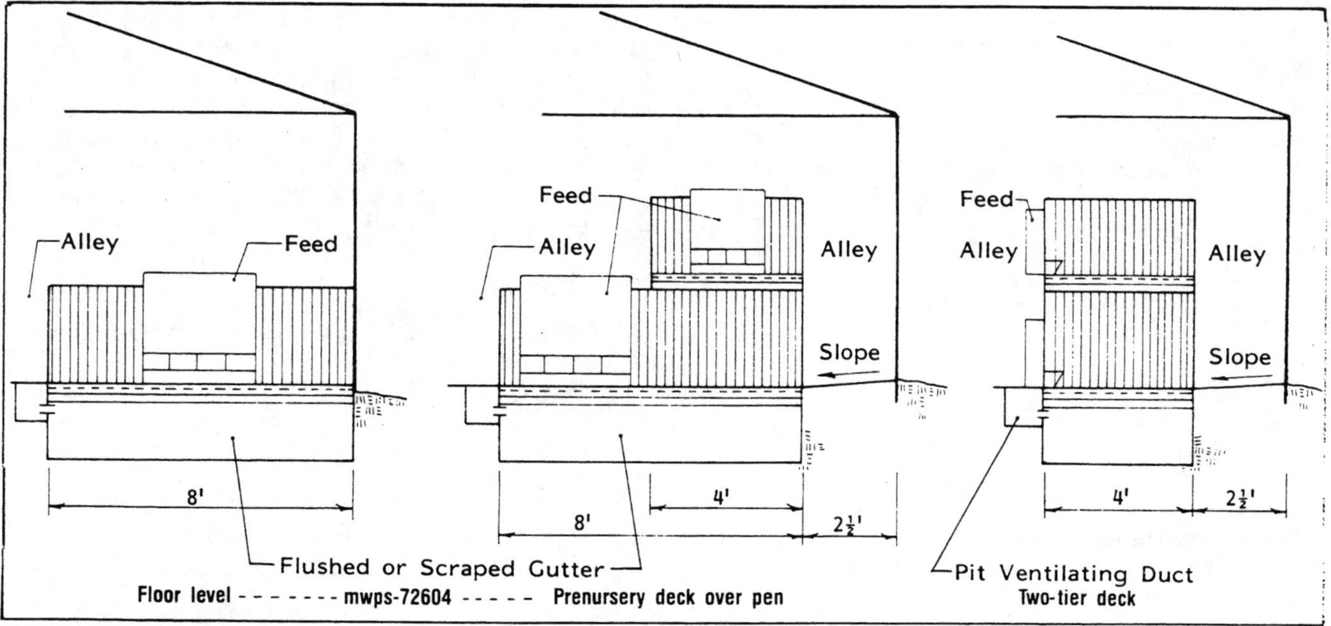

Fig 510-10. Prenursery pens.
Floors and decks are ⅜"x1¼" woven wire mesh. Pen width is 4'.

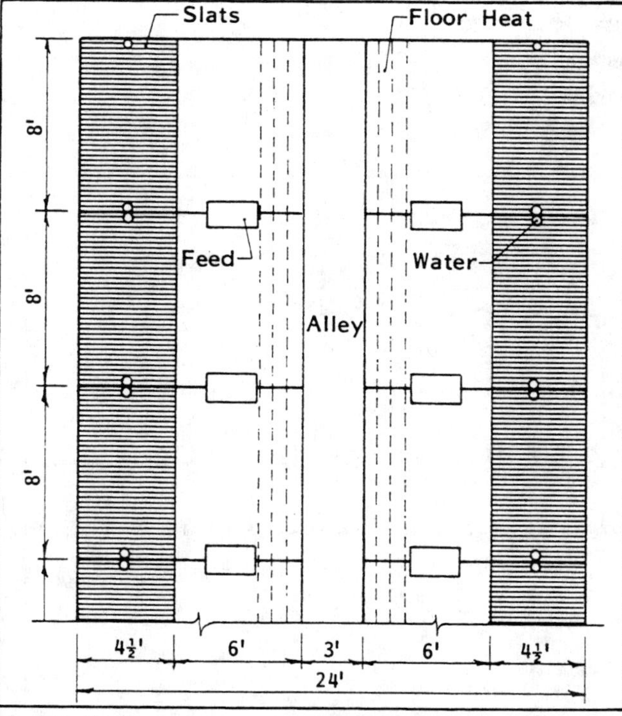

Fig 510-11. Sow-pig nursery with partly slotted floor.
Two sows and litters per pen. mwps-72685.

Growing and Finishing

In this handbook the growth stages of pigs are:
- Growing stage - 75 to 150 lb.
- Finishing stage - 150 to 220 lb or market weight.

Temperature

For growing-finishing pigs, a temperature of 60-70 F is recommended. Although finishing swine can thrive in lower temperatures, they grow faster and with less feed in the recommended range.

Space

Space requirements depend on animal size and type of housing. Overcrowding causes animal discomfort, poor performance, and increases stress and disease susceptibility. For environmentally controlled or modified open-front buildings, provide 6 ft^2/growing pig (75 to 150 lb), 8 ft^2/finishing pig (150 to 200 lb), and 9 ft^2/finishing pig during hot weather or if over 200 lb. As pigs grow, move each group to a larger pen to maintain recommended space per pig. For open-front sheds with lot, provide 5-6 ft^2/growing-finishing pig indoors and 12-15 ft^2/pig outdoors.

Pen size and pigs/pen vary with management goals. As the number of pigs/pen increases, competition affects performance. Good management, including adequate feed, water, and floor space per pig, reduces the adverse effect of large groups. A reasonable maximum is 20 to 25 pigs/pen (3 litters) in buildings with partly or totally slotted floors.

Facilities

Provide one drinking space for each 15 pigs with a minimum of 2 waterers per pen. Place permanently mounted nipple waterers at 26" and install a 4" step for smaller pigs. Adjust movable nipple waterers at about the height of the pig's back.

Use fenceline feeders on partly slotted floors and round or fenceline feeders on totally slotted floors. Provide one feeder space per 4 to 5 pigs.

In buildings with partly slotted floors, use solid partitions over solid floors and gated partitions over slotted areas to promote good dunging habits. Use hog-proof materials on walls and partitions exposed to swine. Concrete partitions are recommended on solid floors. Over slats, consider vertical pipe spaced 6" o.c. to prevent pigs from climbing. Make pen partitions at least 32" high.

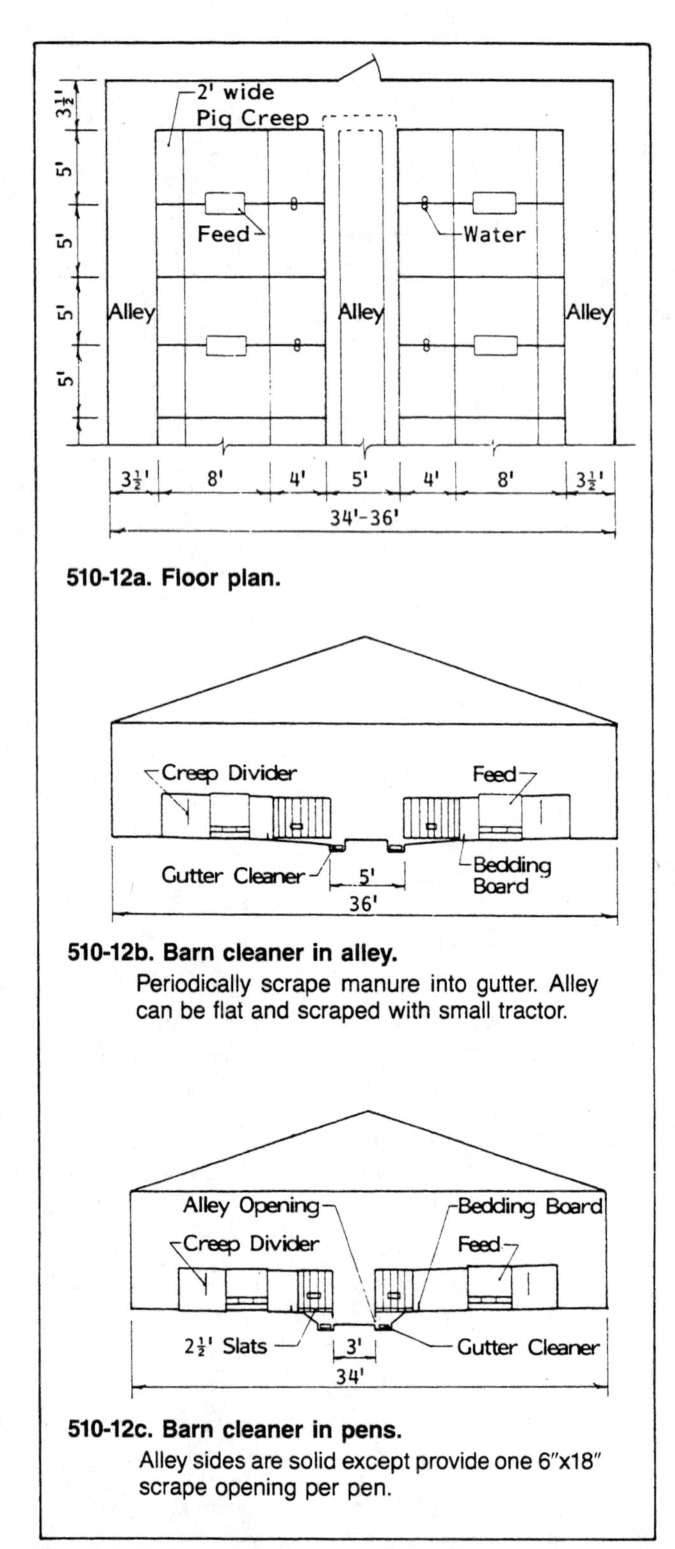

510-12a. Floor plan.

510-12b. Barn cleaner in alley.
Periodically scrape manure into gutter. Alley can be flat and scraped with small tractor.

510-12c. Barn cleaner in pens.
Alley sides are solid except provide one 6"x18" scrape opening per pen.

Fig 510-12. Sow-pig nursery with bedded pens and barn cleaner.

Floor Surfaces

Make at least 30%-40% of the pen floor slotted. For concrete slats, use 5"-8" wide slats with 1" slots. See section on manure management.

Partly slotted floors work best when the pen is about 20′ long (8′-10′ slotted) and 6′-10′ wide. Partly slotted buildings with a single row of pens (alley along front or back) generally result in good dunging patterns. Partly slotted buildings, with pens on both sides (center or off-center alley), require good management to train pigs to dung on the slotted area. See nursery section for other design factors which help toilet train pigs on partly slotted floors. Pen shape is not important on totally slotted floors.

Building Types

Growing-finishing pigs are housed in three types of buildings, Table 510-2:

- Environmentally controlled—fan ventilated in winter; often naturally ventilated in summer.
- Modified open-front—naturally ventilated.
- Open-front with outside lot.

Table 510-2. Finishing building types.
Summer performance of the three types is about the same. Approximate rank.

Building type	Initial cost	Winter performance	Operating cost	Labor requirements
Environmentally controlled	Higher	Higher	Higher	Lower
Modified open-front	Lower to medium	Higher	Lower	Lower
Open-front/ outside lot	Lower	Lower	Lower	Higher

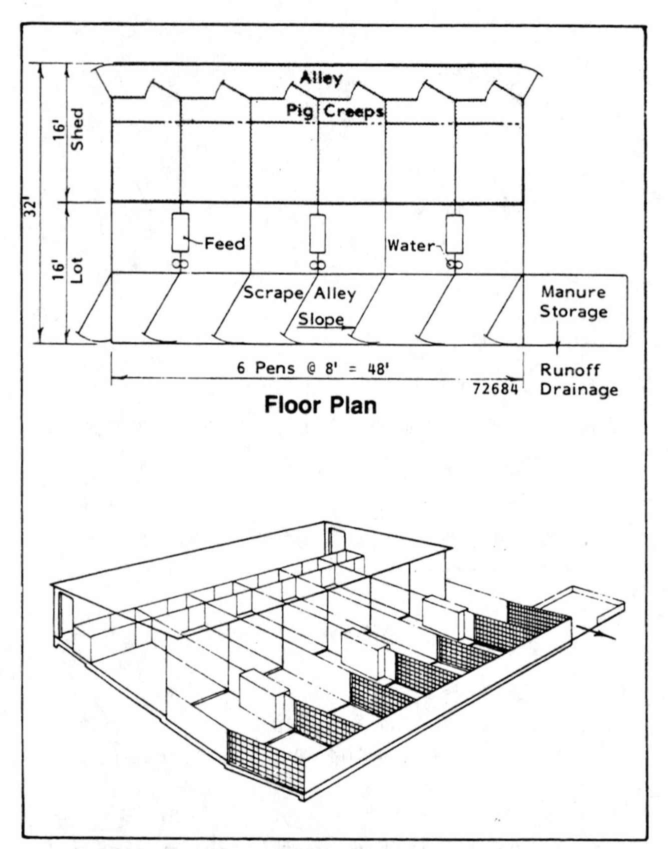

Fig 510-13. Sow-pig nursery with tractor-scraped open lot.
Face open front south or east. mwps-72684.

Environmentally Controlled Units

Environmentally controlled (EC) buildings are usually the most expensive, Fig 510-14. An EC building is usually 32′-44′ wide with two rows of pens. They are often totally slotted, because bad dunging habits and inadequate manure storage can be a problem with partly slotted floors.

An EC building has an insulated ceiling, fan ventilation in winter, and often natural ventilation in summer with sidewall vent doors. Vent doors must seal tightly for effective winter ventilation. Odor levels in EC buildings with pit manure storage are high but can be reduced with pit ventilation. See section on ventilation.

Modified Open-Front Building

A modified open-front (MOF) building is insulated and naturally ventilated, Fig 510-15. The building is usually 28′-33′ wide with a single row of pens. The floor can be partly slotted or have an open gutter with a flush or scrape system. Floors can be totally slotted in mild climates where building temperatures stay above 60 F. Make partly slotted floors about 30%-40% slotted.

Open ridge vents and adjustable sidewall vent doors provide ventilation. Adjust vent doors with manual or automatic controls. For best natural ventilation, orient the building perpendicular to prevailing summer winds. In most of the Midwest, orient the long axis east-west. Adequate insulation reduces

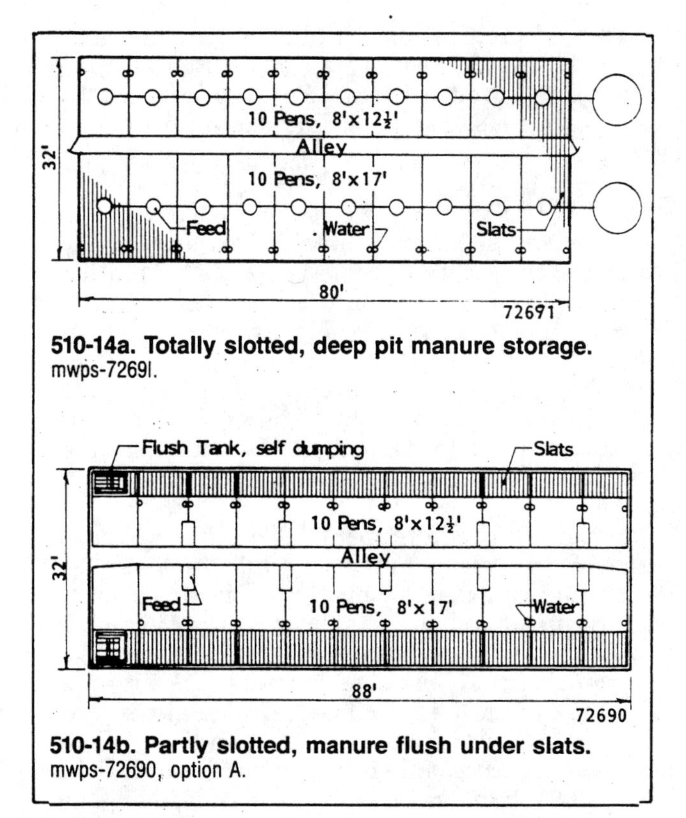

510-14a. Totally slotted, deep pit manure storage.
mwps-72691.

510-14b. Partly slotted, manure flush under slats.
mwps-72690, option A.

Fig 510-14. Environmentally controlled growing-finishing buildings.
Off-center alleys and two pen sizes allow you to move growing pigs across alley to finishing pens.

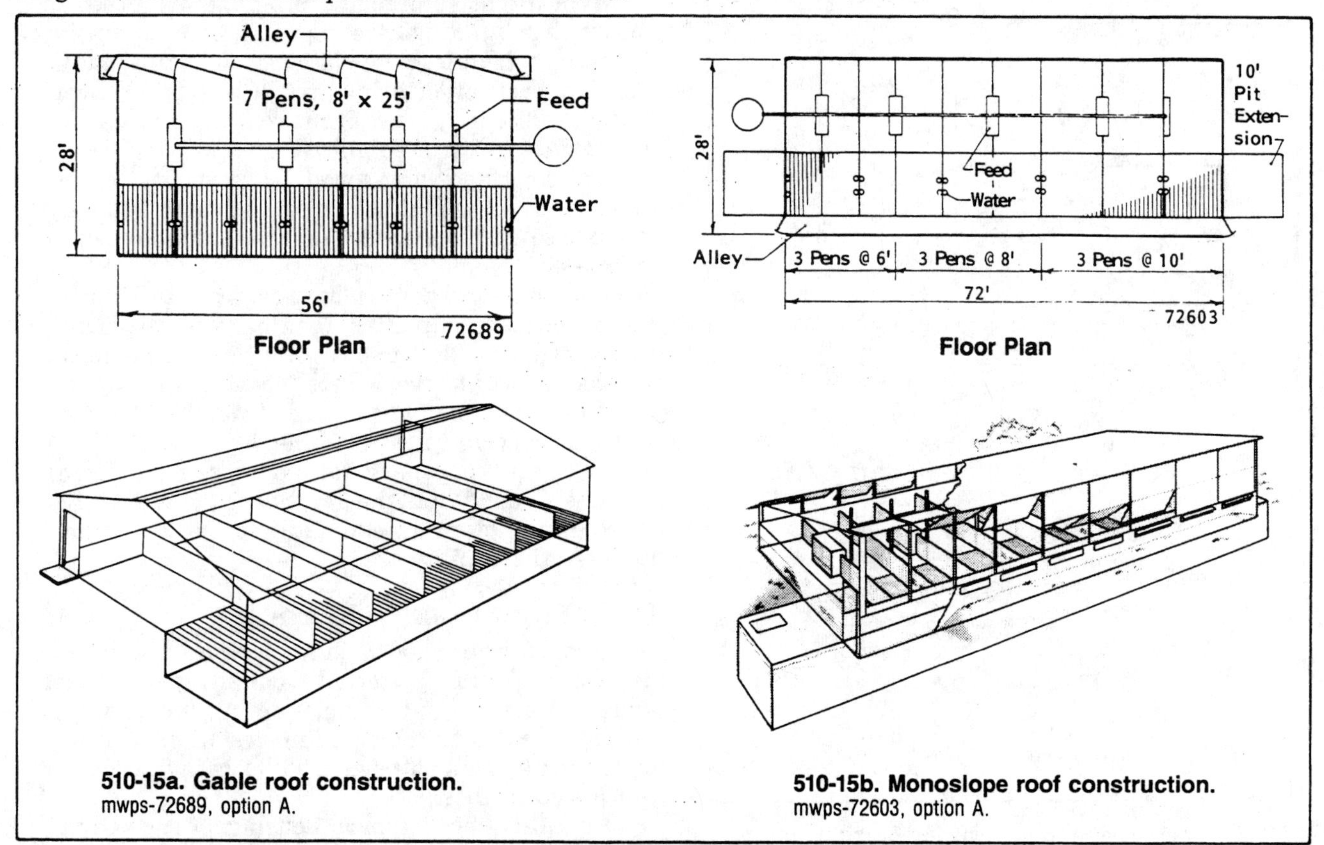

510-15a. Gable roof construction.
mwps-72689, option A.

510-15b. Monoslope roof construction.
mwps-72603, option A.

Fig 510-15. Modified open-front, growing-finishing buildings.
These buildings are partly slotted with long narrow pens. Deep pits and flush or scrape gutters can be used under the slats.

winter condensation and summer radiation heat gain, and conserves heat to maintain suitable winter temperatures. Provide overhead or floor zone heat in the growing pens to start pigs moved in from a nursery.

Open-Front Shed and Lot

Open-front sheds with outside lots are usually lowest cost of the three types, Fig 510-16. Ventilation is through the open front and doors in the back wall. The ceiling is insulated to control winter condensation and provide summer shade. Construction requires less skill because most of the concrete is in flat slabs rather than manure pits. The smaller growing pigs do not perform as well in this building type during cold weather, but may perform better in warm weather.

Outside lots require more labor than buildings with slats. Handle manure as a solid with a scraper, loader, and spreader. Channel surface runoff away from the unit and slope lots at ½"/ft. Control lot runoff to prevent pollution of surface or ground water. A lot is a "point" discharge and may require runoff control structures to meet environmental regulations. See section on manure management. Open lots produce more odor (because of their larger surface area) than EC or MOF buildings, so their use may be limited by nearby neighbors. Bait and spray to control flies.

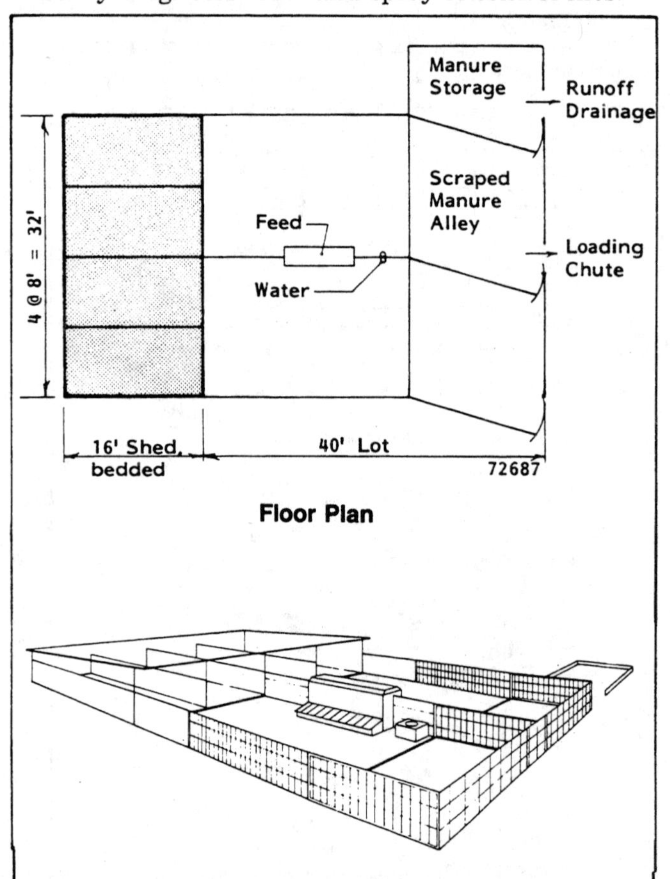

Fig 510-16. Open-front and lot growing-finishing unit.
The 16′ open-front shed provides a covered sleeping area. Pigs feed, water, and dung outside on the paved lot. mwps-72687.

Gestation and Breeding

Alternatives for gestation and breeding facilities range from pasture systems with limited shelter to environmentally controlled facilities with gestation stalls or tethers.

Advantages of environmentally controlled (EC) housing for breeding-gestation include:

- Better control of mud, dust, and manure.
- Reduced labor for feeding and sow handling.
- Better supervision of breeding program.
- Improved control of internal and external parasites.
- More control over sow feed intake.
- Smaller land requirements.
- More efficient use of boars.
- Improved operator and animal comfort and convenience.

Disadvantages include:

- Higher initial investment.
- Possible delayed sexual maturity, lower conception rates in gilts, and lower rebreeding efficiency in sows with average management.
- Requires more intensive management and daily attention to details.

Facility Sizing

Provide space for gestating sows and gilts, replacement gilts, recently weaned sows, boars, and breeding pens.

In an intensive breeding program, overbreed to ensure that buildings are kept full. Some intensive farrowing schedules require that sows be bred on the first heat cycle after farrowing. With an 80% conception rate, 20% of the sows do not conceive and are sold and replaced with bred gilts.

For some farrowing schedules, sows that do not conceive after the first heat are added to the following group of sows. Sows open after the second heat cycle are culled.

Also, some low performance sows are culled from the farrowing group after weaning. With all factors considered, provide 3 to 6 times as many gilts in the gilt pool as will be needed for replacement at breeding. The larger number is for hot weather breeding when conception rates are lower.

The required space for boars and gestating females depends on the farrowing schedule. See scheduling section for estimating the size of breeding-gestation facilities.

Open-Front Shed With Lot

An open-front shed with lot provides a well-bedded sleeping area protected from adverse weather and an outside lot for feeding, watering, and dunging. This system moves the breeding-gestation operation off pasture and out of the mud with low-investment or remodeled facilities.

Fig 510-17 shows typical housing for sows from a 20-sow farrowing building with 6 or 7 farrowings per year (3 sow groups). Allow about 8 ft^2 of housing area and about 14 ft² of lot area per sow in gestation pens. Provide 40 ft² of housing area and 40 ft² of lot area per boar in individual or group pens. Double these areas in the breeding pens. Group up to 12 sows per gestation pen and no more than 6 sows (plus boar) per breeding pen.

Environmental control for a shed is limited. Use solid partitions between pens in the shed and use doors covering the top half of the open front to reduce drafts. Leave a 2" crack above the door for escape of warm, moist air. During cold weather, provide plenty of dry bedding in the shed.

Handle manure scraped from the lot as a solid. Provide a manure storage area and a conventional manure spreader. Channel lot runoff to an approved management system, such as a settling basin or infiltration channel. See section on manure management.

Enclosed Sow Housing

Options for enclosed sow housing include:

- Mechanical ventilation year-round; often with evaporative cooling to relieve heat stress.
- Mechanical ventilation in winter and natural ventilation in summer.
- Natural ventilation year-round in modified open-front buildings.

Space requirements are listed in Table 510-3. Group **no more** than 12 sows per pen and **no more** than 6 (plus boar) in breeding pens. Sow groups of 5 or 6 reduce competition and increase flexibility in grouping sows of equal size and compatible temperament.

Table 510-3. Space requirements of housed breeding swine.

Animal type	Weight lb	Group pens: Solid floor ft²	Group pens: All or partly slotted floor* ft²	Animals per pen	Stall size
Breeding					
Gilts	250-300	40	24	up to 6	
Sows	300-500	48	30	up to 6	
Boars	300-500	60	40	1	2'-4"x7'
Gestating					
Gilts	250-300	20	14	6-12	1'-10"x6'
Sow	300-500	24	16	6-12	2'-0"x7'

* Use this column for flushed open gutter. Open gutter not recommended in breeding because of slick floors.

Fig 510-18 shows naturally ventilated modified open-front gestation buildings. Either floor feed daily on the solid portion of the floor or feed sows individually in freestalls along the side of the pen. Fig 510-19 shows a naturally ventilated gestation building with pens and freestalls for individual sow feeding.

Freestalls are either full size, 2'x6'-7', or abbreviated 1½'x2' deep. Arrange freestalls on partly slotted floors so sow manure is deposited on the slats.

Fig 510-20 shows mechanically ventilated gestation buildings with 2, 3, and 4 rows of stalls. Each sow is fed individually in her stall. These buildings can be naturally ventilated in summer.

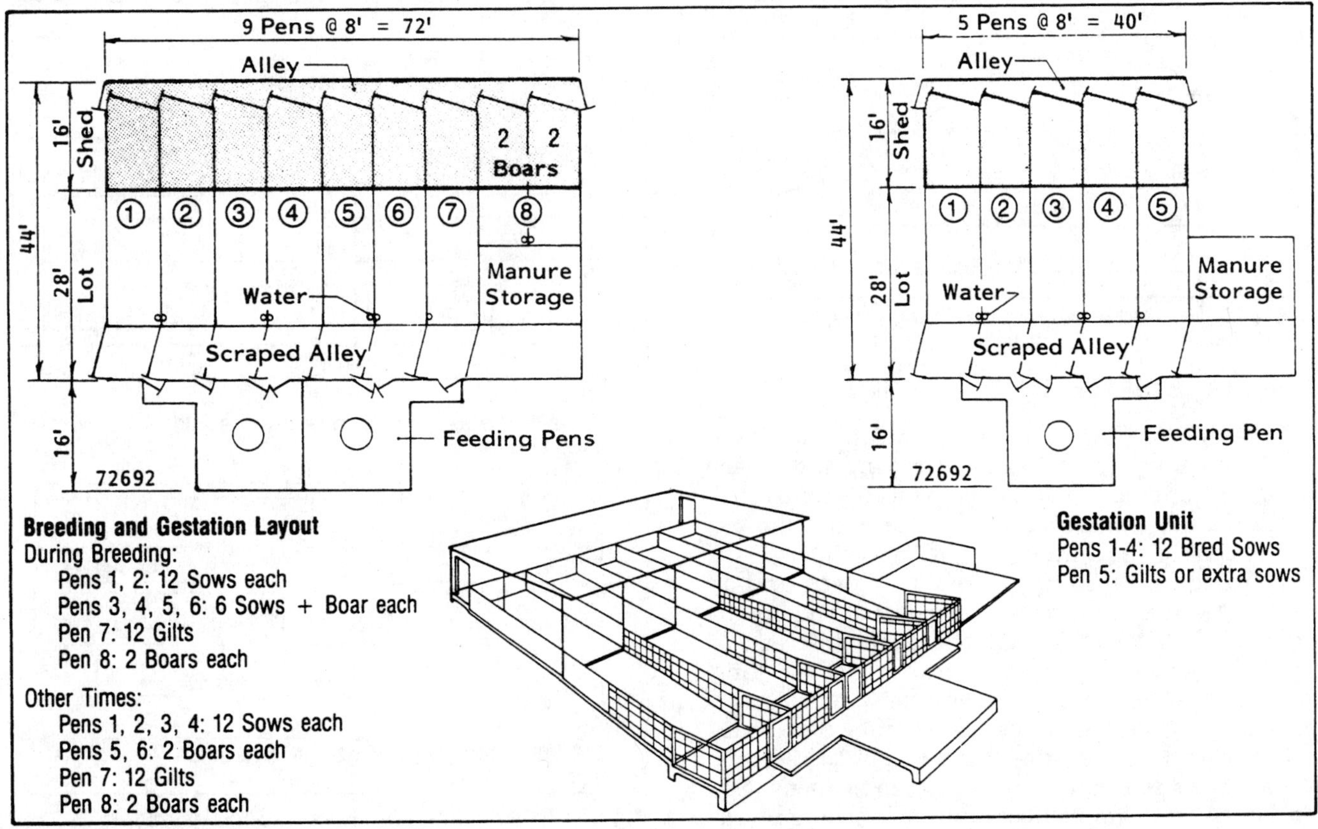

Fig 510-17. Open-front and lot gestation unit.
mwps-72692.

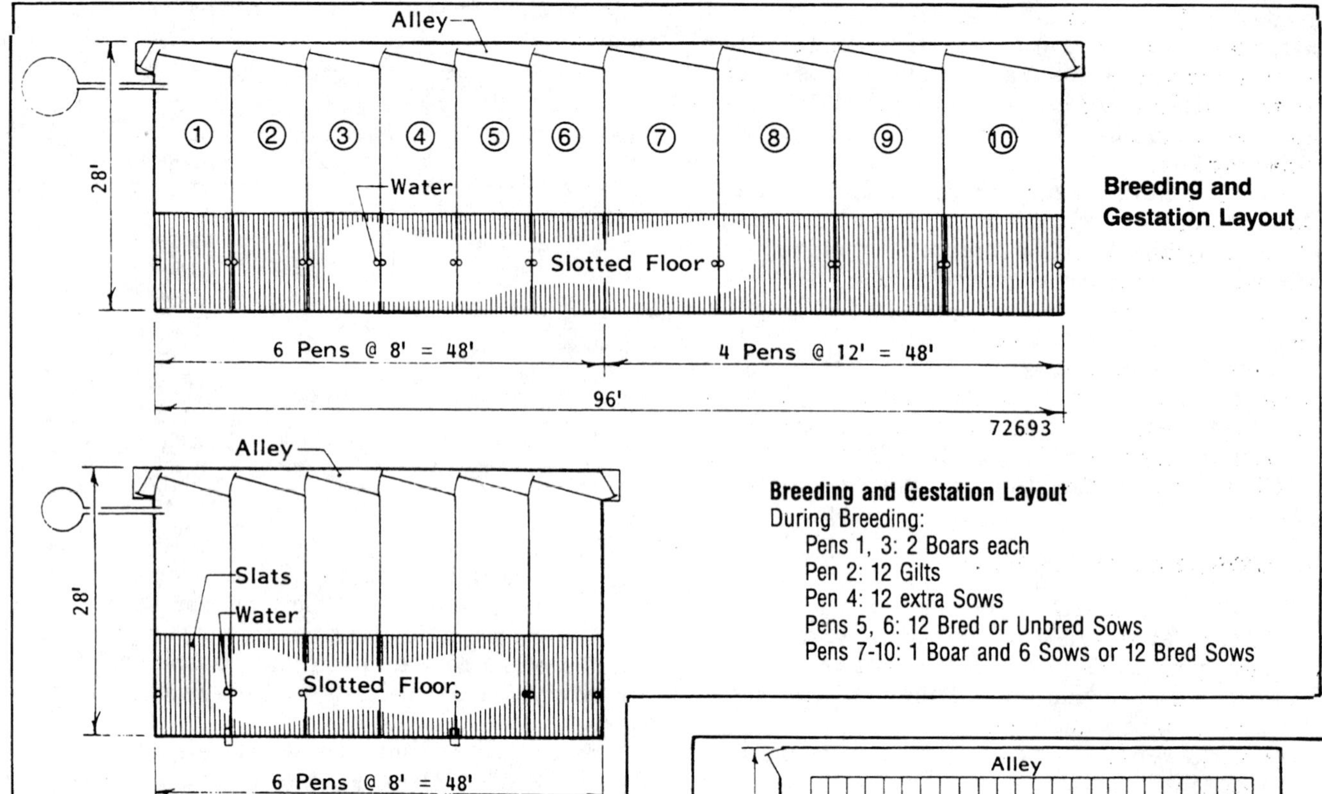

Fig 510-18. Modified open-front gestation unit.
mwps-72693, option A.

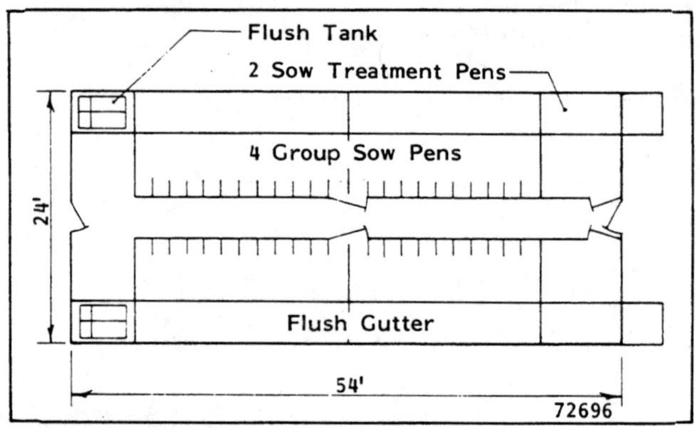

Fig 510-19. Free stall gestation building.
mwps-72696.

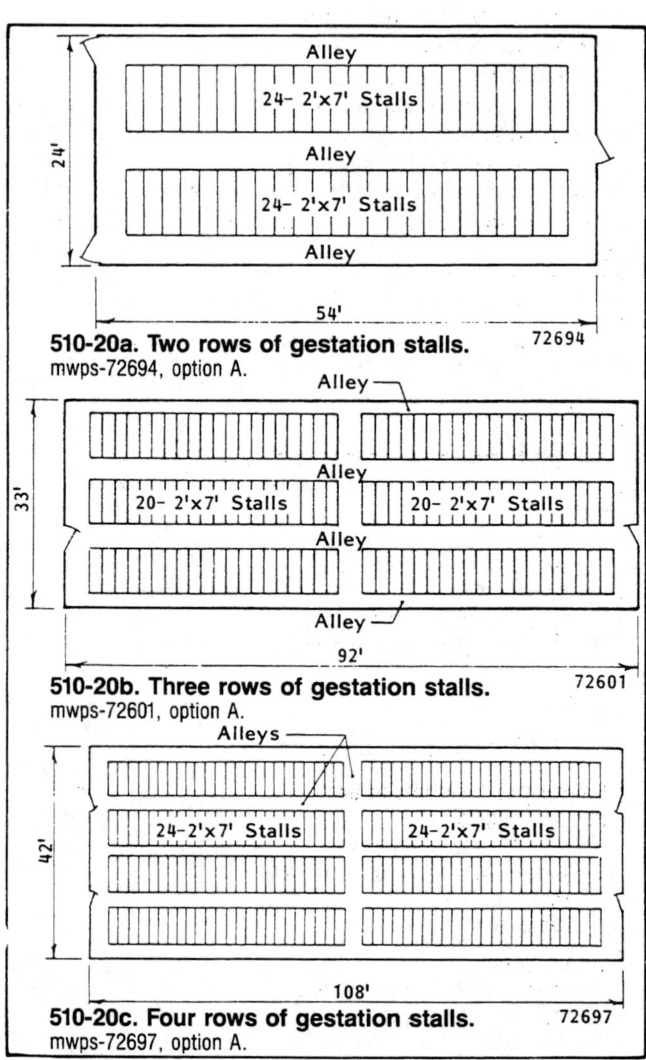

510-20a. Two rows of gestation stalls.
mwps-72694, option A.

510-20b. Three rows of gestation stalls.
mwps-72601, option A.

510-20c. Four rows of gestation stalls.
mwps-72697, option A.

Fig 510-20. Environmentally controlled gestation buildings.

Breeding Buildings

Breeding buildings are economically justified only for weekly or continuous farrowing schedules. Breeding buildings are environmentally controlled to eliminate hot and cold weather breeding slumps. See section on ventilation.

A breeding building has holding pens and/or stalls or tethers, boar pens and/or stalls, and breeding pens. Pen layout affects the social interaction between boars and females, a vital part of breeding performance. Alternate boar pens with female pens and use open pen partitions to encourage social contact. Locate boars as close as possible to recently weaned sows to induce strong heat indications. See Fig 510-21 for example floor plans of breeding buildings.

Developing and breeding gilts in environmentally controlled buildings presents special problems. Gilts are most fertile when housed in small groups with adequate floor space, Table 510-3. Exposure to boars hastens the onset of puberty, so put boars in adjacent pens.

Sows are usually held in stalls (called stress, dry-off, or stimulus stalls) to prevent fighting for the first 2 to 3 days after weaning. Weaning within 6 weeks of farrowing triggers the estrus (heat) cycle 4 to 8 days after pigs are removed. After holding in stimulus stalls, move sows to a holding pen where it is easier to detect heat. Locate this pen close to the boars, because fenceline contact between sows and boars induces a stronger indication of heat.

An alternative to stimulus stalls are pens holding 4 to 6 sows after weaning. Keep a mature boar in the pen for the first 48 hours to reduce fighting among newly weaned sows and to stimulate the onset of estrus. After 48 hours, move the boar to an adjacent pen for fenceline contact.

After heat is detected, place the sow in a breeding pen with the boar. The actual heat period (when the sow stands and allows the boar to mount) lasts about 2½ days. Breed 22 to 26 hr after the onset of standing heat (16 to 20 hr for gilts). After breeding, return the female to a holding pen; breed again 8 to 12 hr later. Breed in the boar pen or a separate neutral pen. A neutral pen provides better footing because the flooring has a chance to dry between uses. Slick floors result in poor breeding performance. Improve footing by roughening concrete slats.

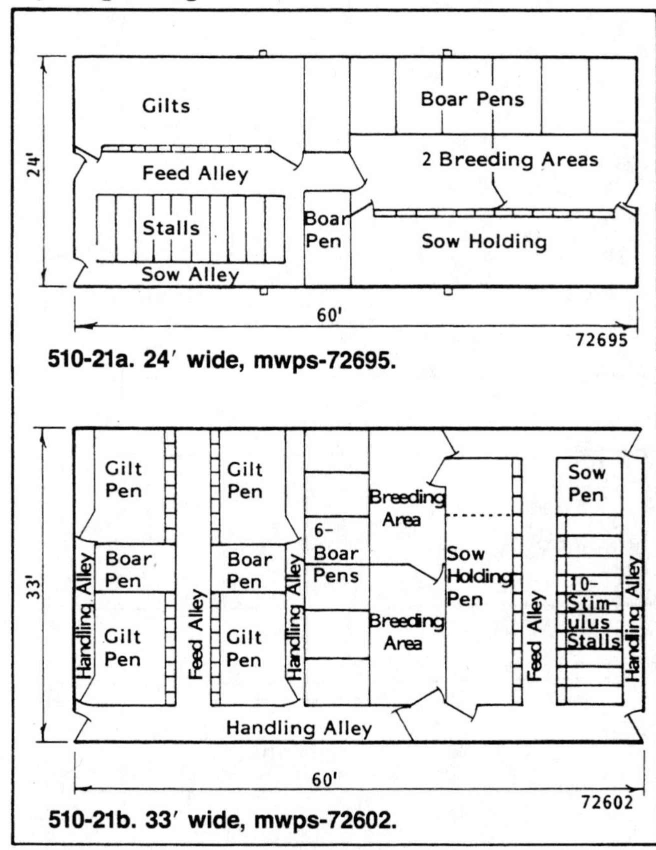

510-21a. 24′ wide, mwps-72695.

510-21b. 33′ wide, mwps-72602.

Fig 510-21. Breeding buildings.

See Table 510-3 for space requirements in breeding buildings. Make breeding pens at least 8′x8′ to allow animals room to mount. Use pen partitions at least 42″ high. Use vertical planks or bars to keep animals from "climbing" partitions. Arrange pens and alleys so you can see all animals from adjacent walkways.

Environmental requirements of breeding swine are critical. Temperatures above 85 F with high humidities reduce boar fertility, which may take 3 months or longer to correct. Maintain temperatures below 85 F for sows during the first 2 to 3 weeks of gestation to ensure maximum litter size and during the last 2 to 3 weeks to reduce stillborns and abortions. Temperatures below 55 F affect operator comfort. See section on ventilation.

Boar Housing

House boars in individual stalls or small individual pens. Grouping boars leads to fighting, homosexual activity, and injury. Rotate boars daily between breeding pens to reduce the influence of inactive or sterile boars.

Provide separate housing for new boars and isolate them for at least 60 days before exposure to the sow herd. This helps you detect disease and prevent herd infection.

Provide at least one boar for each four sows weaned/bred per week, which assumes that each sow is bred twice. Allow one or two additional boars for injuries or sickness. Another way to compute the number needed is to use a mature boar no more than twice a day and no more than ten times a week (use young boars no more than once a day and six times a week).

Provide additional boars if sows are weaned and bred as a group. Refer to Table 510-1.

Combining Buildings

When combining buildings into a swine production system, consider the movement of three major products—feed, pigs, and manure. Animal movement includes:

- Sows from gestation-breeding to farrowing and back again.
- Pigs from farrowing to nursery, growing, and finishing.
- Gilts from growing-finishing to gestation-breeding.
- Market or feeder pigs and culled sows to market.

Place gestation-breeding on one side of farrowing and pig growing on the other side. Sows move easily between gestation-breeding and farrowing; pigs move easily from farrowing to the nursery, growing, and finishing buildings.

Pasture Production

Producing swine on pasture requires relatively low investment in buildings and equipment. Year-round production is difficult in cold climates.

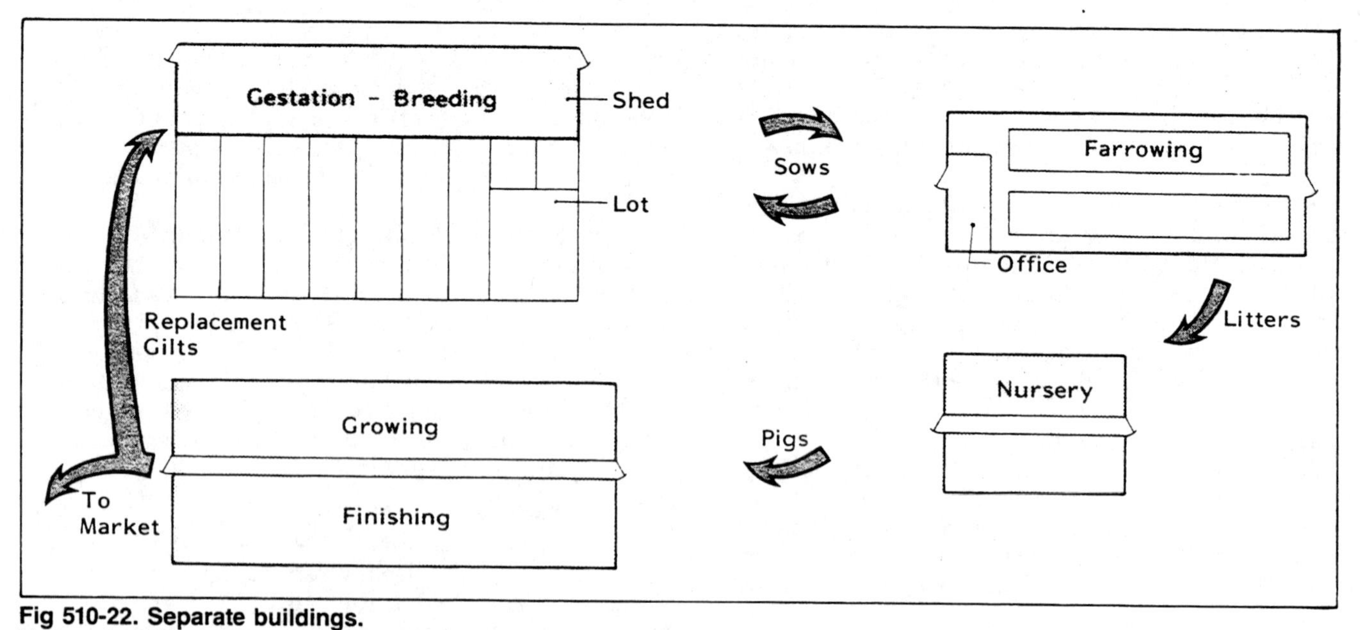

Fig 510-22. Separate buildings.
Moving animals from building to building is difficult. Expand by adding onto building lengths or duplicating entire building system.

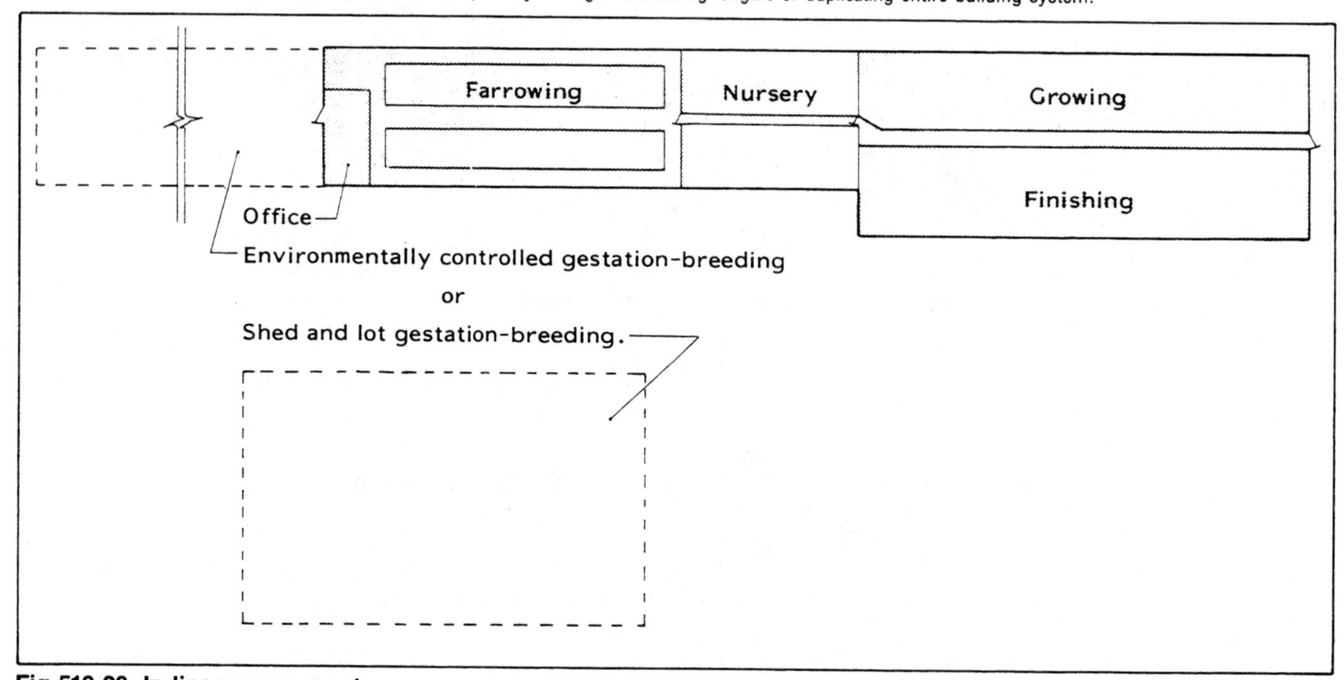

Fig 510-23. In-line arrangement.
Buildings are end to end so animals can be easily moved within the building. Gestation-breeding may be a separate facility or attached to the farrowing building.

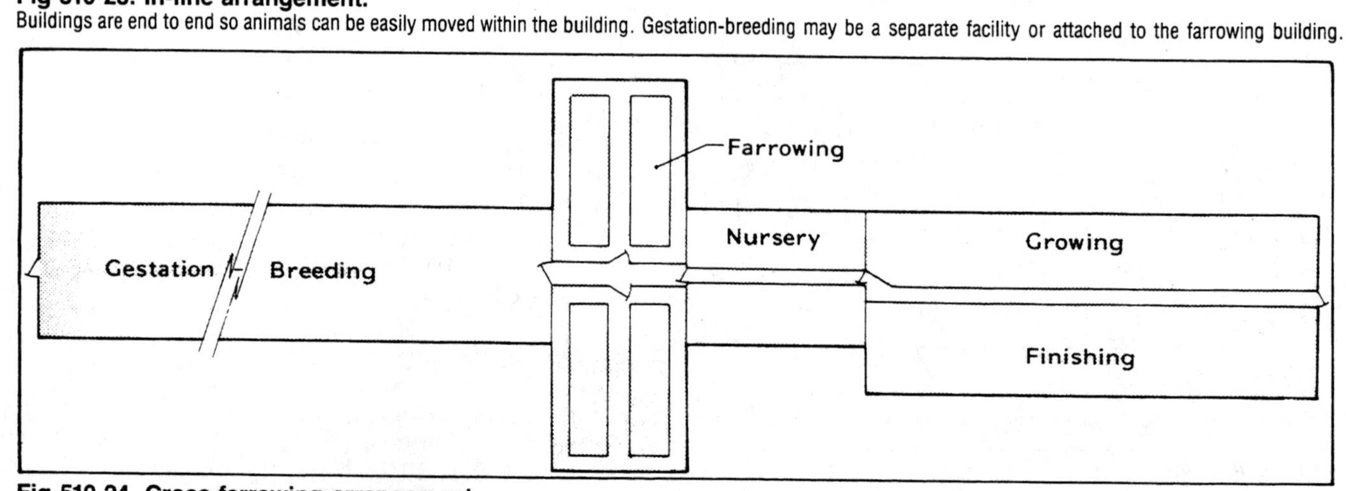

Fig 510-24. Cross-farrowing arrangement.
Similar to the in-line arrangement except farrowing is turned crossways to reduce overall building length.

Recommendations for pasture production:

- Provide shade and shelter. Move portable houses periodically.
- Locate feeders at a well-drained site, preferably on pavement or wooden platforms.
- Arrange pasture lots along a permanent road. Prevent wallows near waterers by forming a curb around the slab. Slope the slab and surrounding area for quick drainage.

Swine Handling

The faster and quieter hogs are sorted, treated, and loaded, the less weight gains are affected. If possible, sort a few hours before loading to allow hogs to calm down. A good handling facility includes a holding pen, a sorting alley and gates, a squeeze chute or other treatment space, and a loading chute. Other features such as scales and crowd gates are also useful.

Provide sick pens in each building for about 1% of animal capacity. Allow about 8 to 10 ft² / 100 lb of swine in sick pens. Locate sick pens near access doors or holding areas for convenient servicing.

Loading chutes are often the bottleneck in handling facilities. Pigs are reluctant to walk on dirt or wood floors if they were raised on concrete. If the pigs were raised on concrete floors just before shipping, consider a concrete floor for the chute. Keep chute floor clean and dry for good footing. Install steps (4" rise per 12" run preferred) or cleats 8" o.c. Construction must be sturdy because pigs are reluctant to walk up a shaky chute.

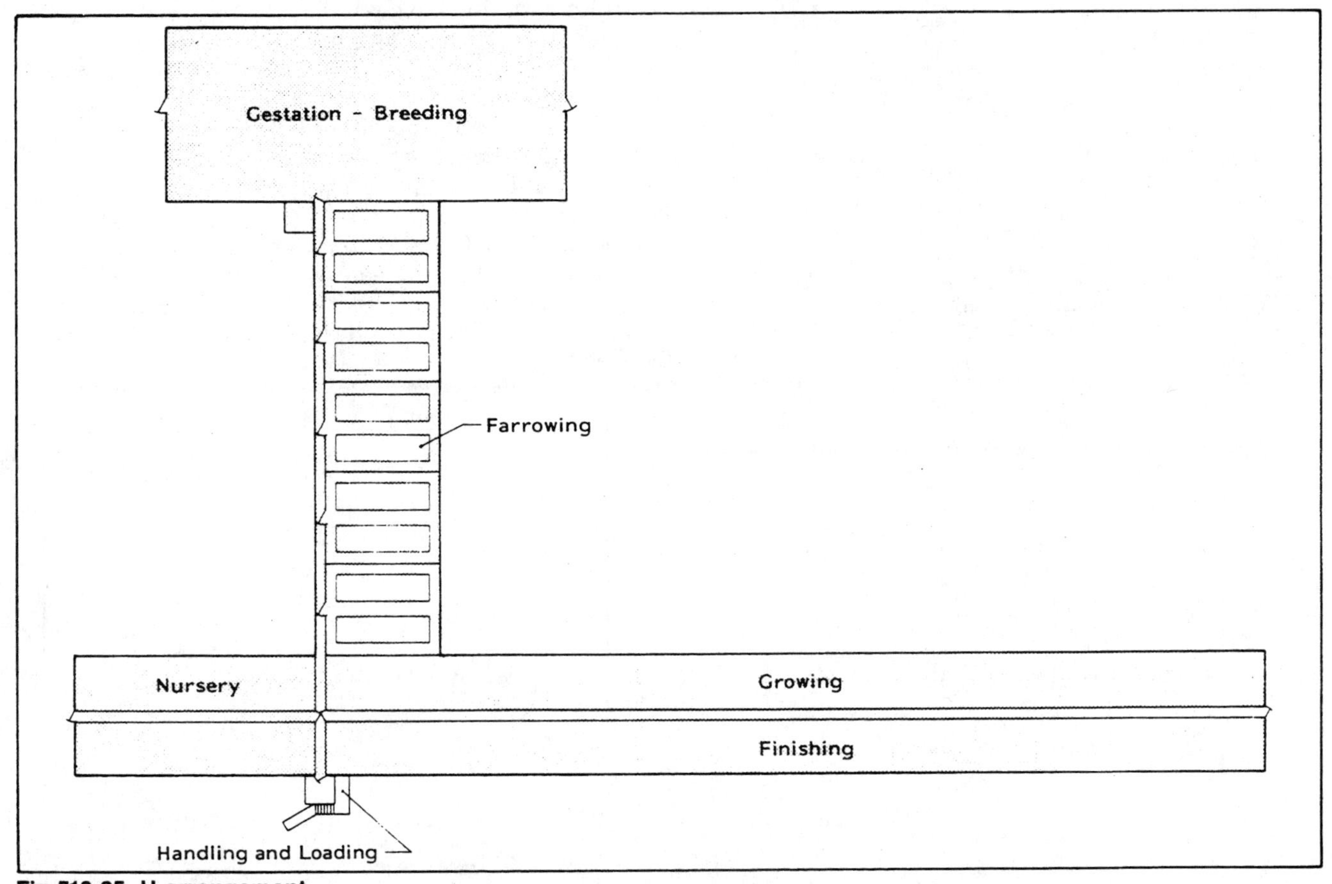

Fig 510-25. H arrangement.
Overall building complex length is reduced by placing farrowing lengthwise between gestation-breeding and pig growing.

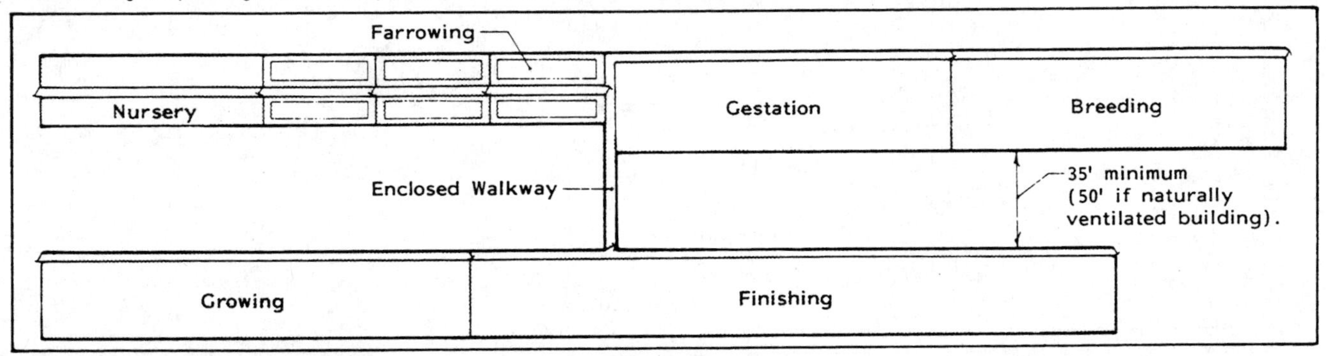

Fig 510-26. Parallel arrangement.
Parallel buildings are connected with enclosed walkways for easier animal movement.

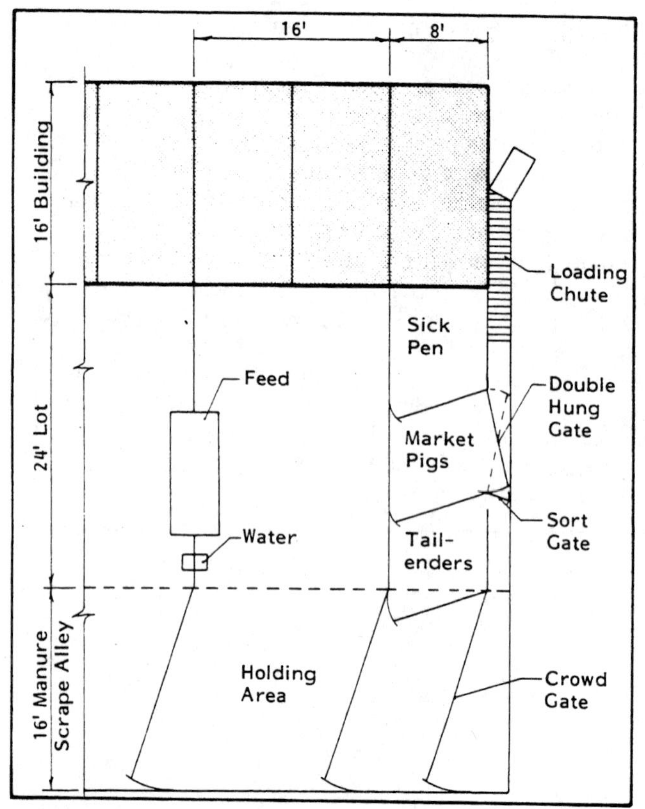

Fig 510-27. Handling layout for shed and lot.
An extra pen is required for sorting and for sick pigs.

Restrict chute width to 22″ or less to reduce turnarounds. Unloading chutes can be 4′-5′ wide. Slope chutes no more than 20° (about 4½′ rise per 12′ run). Pigs load better if the chute sides are solid and they can enter the truck from a flat area rather than off a sloping chute. Turn the chute about 45° at the top of the ramp and construct a flat floor for 3′-4′ before hogs enter truck, Fig 510-30.

A one-way gate at the top of the chute keeps animals in the truck. A self-aligning dock bumper reduces injuries as animals enter the truck. Provide a catwalk along the chute to help you keep hogs moving.

Hogs don't like to go into bright sunlight from a dark building or vice versa. Provide uniform and diffuse light (no bare light bulbs) with no shadows across the handling area.

Build handling fences and gates (in the chute and working alley) at least 42″ high (32″ if used only for finishing hogs) with 3″ clearance between the base of the fence and the floor to prevent manure accumulation. Make the working alleys 16″ wide and use solid panels so hogs can only see and move straight ahead. Construct the circular crowd gates with open-wire hog panels, because hogs are reluctant to enter a pen with solid walls. Likewise, avoid making sow washes and other handling areas with three solid sides. Hinge sorting gates (2′ wide) at both ends to allow sorting of both incoming feeder pigs and outgoing market pigs.

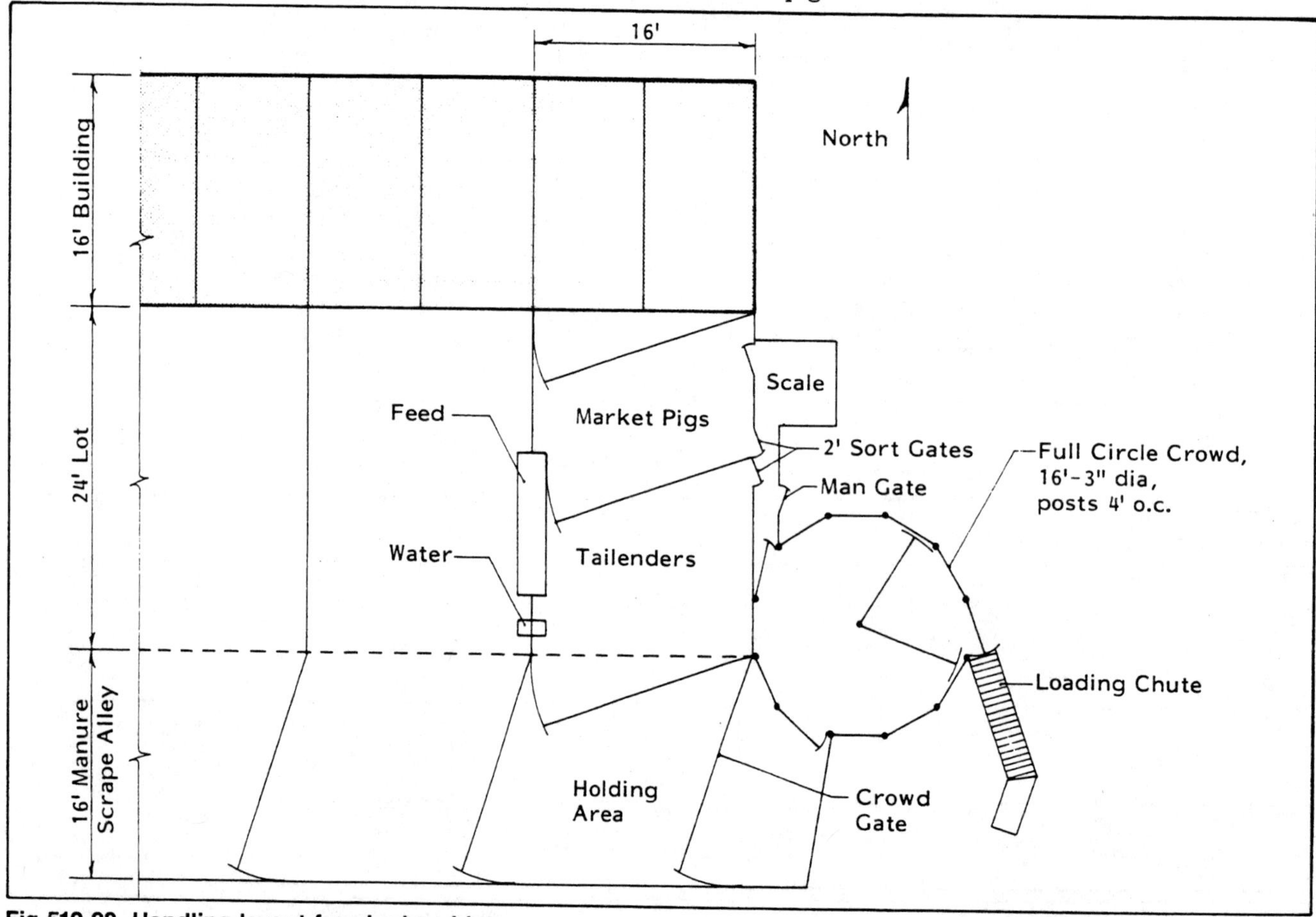

Fig 510-28. Handling layout for shed and lot.
End feeding pen is used for sorting. Shut up feeding pigs in building before bringing in the sorting pigs.

Pigs can be killed or crippled in transport, usually due to overcrowding. Allow 1¼ ft^2 per 40 lb pig, 1½ ft^2 per 60 lb pig, 3¾ ft^2 per 200 lb pig, and 4¼ ft^2 per 250 lb pig on trucks.

Livestock Conservation, Inc., suggests the following guidelines for transporting swine.

- Make trailers and loading facilities free of protruding hardware and boards.
- Load as quickly and as quietly as possible and leave immediately.
- Check animals about one hour after loading, then keep moving to reduce travel time.
- Provide good footing to reduce injuries.
- Provide bedding—use straw on dry sand if below 60 F; wet sand if above 60 F.
- Use partitions to avoid piling due to sudden stops or cold weather.
- Keep vehicles as open as possible in summer to maximize air movement over animals. Keep vehicles as closed as possible in winter.
- In summer, load and haul during the cool of the day.
- If the temperature exceeds 80 F, wet down animals and increase floor space per pig by one-third.
- Put 25 lb ice blocks on truck floor in very hot weather.

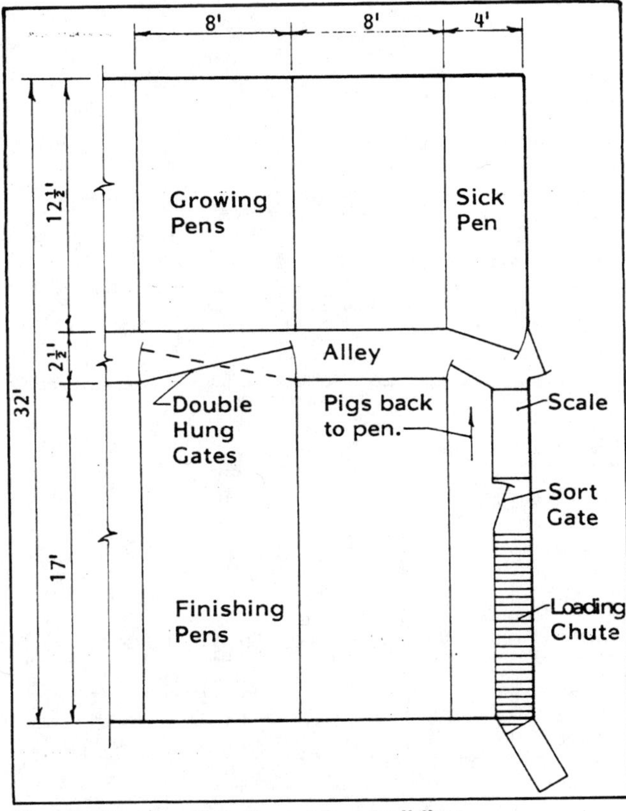

Fig 510-29. Handling layout in a building.

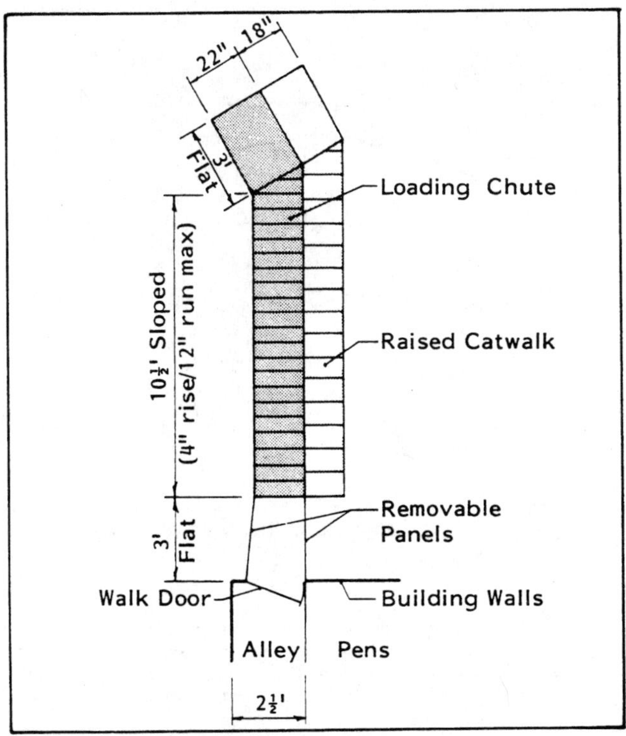

Fig 510-30. Permanent loading chute.

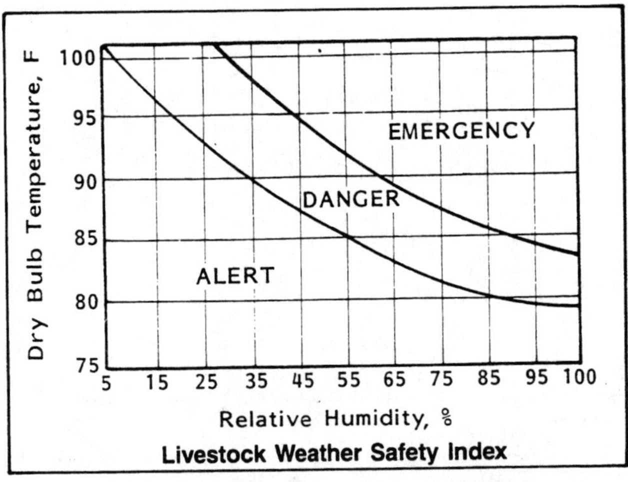

Fig 510-31. Weather guide for safe swine shipping.
From Livestock Conservation, Inc.

Dimensions for Swine

Reference Points (See Fig 510-32)

L = length from behind to base of tail
H = height from floor to top of back
D = depth from lowest point on belly to top of back
W = width at widest point near middle of hog. The width across shoulders and hams was found to be nearly equal to this dimension.

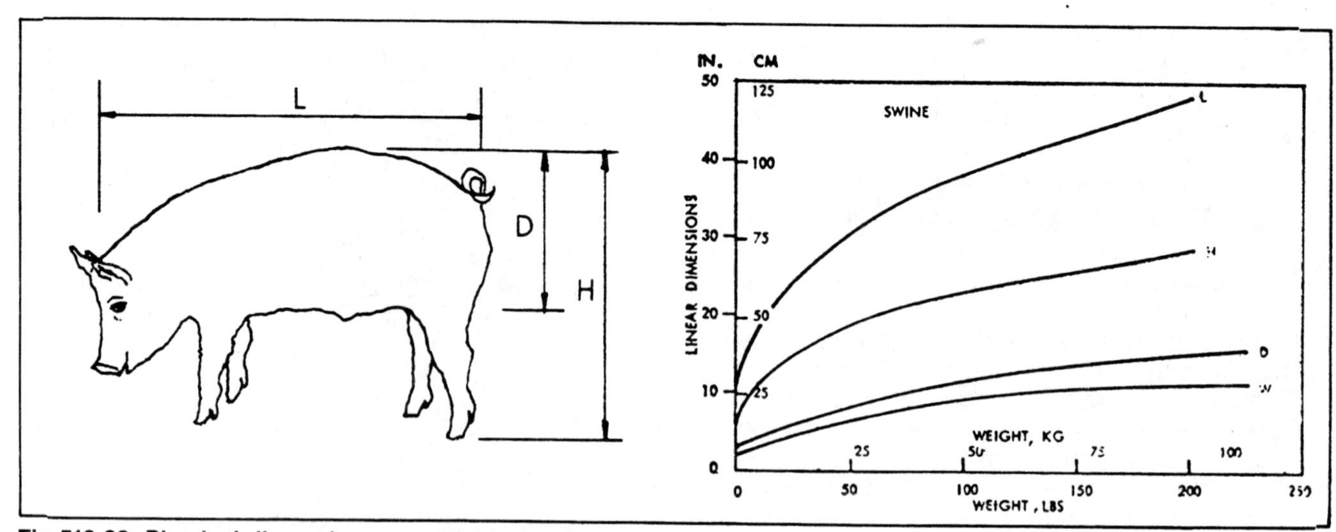

Fig 510-32. Physical dimensions vs. weight of swine.

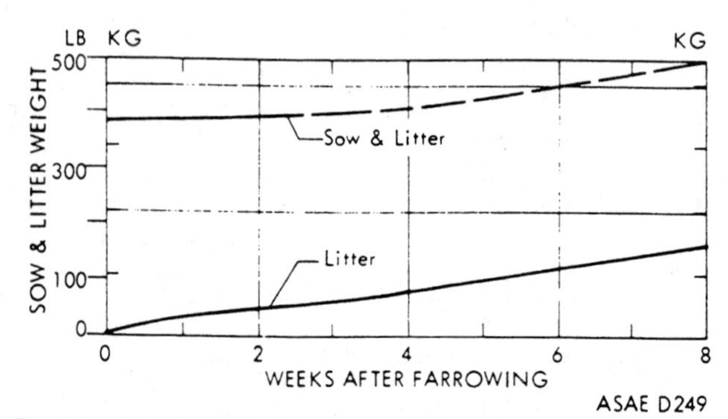

Fig 510-33. Weight of a sow and litter.

Feed and Water

Table 510-4. Daily water consumption.

Weight of Hog or Condition	Daily Water Consumption lb	kg
30 lb (13.6 kg)	5-7	2.3-3.2
60 to 80 lb (27.2-36.3 kg)	7-10	3.2-4.5
75-125 lb (34.0-56.7 kg)	10-16	4.5-7.3
200-380 lb (90.7-172.4 kg)	12-30	5.4-13.6
Pregnant sows	30-38	13.6-17.2
Lactating sows	40-50	18.1-22.7

Table 510-6. Daily feed for swine.

Live Weight lb	kg	Expected Daily Gain lb	kg	Total Air-Dry Feed lb	kg
		Growing finishing pigs			
10-25	4.5-11.3	0.6	0.3	1.2	0.5
25-50	11.3-22.7	1.0	0.4	2.5	1.1
75-125	34.0-56.7	1.6	0.8	5.2	2.4
125-175	56.7-79.4	1.7	0.8	6.7	3.0
175-225	79.4-102.0	1.9	0.9	7.8	3.5
		Bred gilts and sows			
300	136.1	1.0	0.4	5.5	2.5
500	226.8	0.7	0.3	6.5	2.9
		Lactating gilts and sows			
350	158.8	-	-	11.0	5.0
450	204.1	-	-	12.5	5.7
		Young and adult boars			
300	136.1	1.0	0.4	6.0	2.7
500	226.8	-	-	7.5	3.4

Table 510-5. Complete rations for swine.
Extracted with permission from *The 1969 Agricultural Engineers Yearbook.* The American Society of Agricultural Engineers, Box 229, St. Joseph, MI 49085.

Ingredient	Pig Starter lb	kg	Pig Grower lb	kg	Hand-Fed Gestation lb	kg	Gilt Growing & Self-Fed Gest. lb	kg	Lactation lb	kg	Growing-Finishing 50-75 lb (22.7-34.0 kg) lb	kg	Growing-Finishing 75-100 lb (34.0-45.4 kg) lb	kg	Growing-Finishing 150-200 lb (68.0-90.7 kg) lb	kg
Gr. yellow corn	1200	544.3	1500	680.4	1100	498.9	600	272.2	1200	544.3	1600	725.8	1700	771.1	1800	816.5
Gr. oat groats	200	90.7	-	-	-	-	-	-	-	-	-	-	-	-	-	-
Gr. oats	-	-	-	-	500	226.8	600	272.2	200	90.7	-	-	-	-	-	-
Alfalfa hay, wheat bran or midds	-	-	-	-	-	-	600	272.2	100	45.3	-	-	-	-	-	-
Pig supplement	600	273.1	500	226.8	-	-	-	-	-	-	-	-	-	-	-	-
All-purpose supplement	-	-	-	-	400	181.4	200	90.7	500	226.8	400	181.4	300	136.1	200	90.7
Total	2000	907.2	2000	907.2	2000	907.2	2000	907.2	2000	907.2	2000	907.2	2000	907.2	2000	907.2

511 BEEF

Data Summary

Table 511-1. Space requirements.

	Cow-calf	Backgrounding to 600 lb	Finishing 600 lb-market
Feeding lot, ft²/hd			
paved	60-75	40-50	50-60
unpaved (w/mounds)		150-300	250-500
unpaved (w/o mounds)	350-800	300-600	400-800
Mound, ft²/hd	40-45	20-25	30-35
Open front shelter, ft²/hd	25-30	15-20	20-25
Cold confinement barn, ft²/hd			
solid floor, bedded	40-50	20-25	30-35
flume or slotted floor			18-20
Shade, ft²/hd			20-25
Calving pen (1 pen/12 cows), ft²	120 min		
Isolation & sick pen, ft²		40-50	50-60
no. of pens/100 hd		2-5	2-5
Feeders, in/hd			
self feeder (feed always available)		3-4	4-6
fed daily	calf 14-18 cow 24-30	18-22	22-26
calf creep	3″-5″/calf		
Handling facilities, ft²/hd			
total area	60-80	40-50	50-60
holding pen	20-25	12-15	15-20
crowding pen	12-15	6-8	8-12

Table 511-2. Waterer space.
Water space = 1 cup waterer or 18″ of tank perimeter.

Dry lot	Max 25 hd/waterer
Pasture	Max 15 hd/waterer

Table 511-3. Floor and lot slopes.

Handling facilities	¼″-½″/ft
Feeding lots	
earth	½″-¾″/ft
paved	⅛″/ft min.
Mound	
side slope	4:1 to 5:1
longitudinal slope	½″/ft max.
Slotted floors	Usually flat
Solid feeding floors	¾″-1″/ft
Bunk apron	¾″-1″/ft

Table 511-4. Handling facilities dimensions.
Use dimensions for over 1,200 lb for cow-calf operation.

	To 600 lb	600-1200 lb	Over 1200 lb
Holding area sq ft/head	14	17	20
Crowding pen sq ft/head	6	10	12
Working chute with vertical sides			
width	18″	22″	26″
Desirable length (min.)	20′	20′	20′
Working chute with sloping sides			
Width at bottom inside clear	15″	15″	16″
Width at top inside clear	20″	24″	26″
Desirable length (min.)	18′	18′	18′
Working chute fence			
Height—solid wall	45″	50″	50″
Depth of posts in ground	36″	36″	36″
Overall height (7′ clear min. below cross ties to walk under)			
Top rail, farm cattle	55″	60″	60″
Top rail, range cattle	68″	72″	72″
Corral fence			
Recommended height	60″	60″	60″
Depth of posts in ground	30″	30″	30″
Loading chute			
Width	26″	26″	26″-30″
Length (min.)	12′	12′	12′
Rise in/ft	3½	3½	3½
Ramp height for:			
Gooseneck trailer 15″			
Pickup truck 28″			
Van type truck 40″			
Tractor-trailer 48″			
Double deck 100″			

Table 511-5. Bunk design.

Throat height (max)	
calves	18″
feeders	22″
mature cows	22″
Bunk width (max 60″)	
both sides feeding	
cow-calf	48″-60″
backgrounding	36″
finishing	48″-60″
one side feeding	18″
auger feeder (Fig 511-14)	Add 6″-12″ up to max width
Bunk floor height above apron	8″-12″
Step along bunk	
height	6″-8″
width	12″-16″

Cow-Calf Facilities

Beef cows usually need few facilities. Cows bred to calve in the spring can be wintered outdoors. Trees or natural terrain provide desirable and satisfactory shelter. Confine cows near the farmstead prior to calving for observation. Provide a calving barn for protection during the first few days after birth and for sick cow care. An open-front building with electric hot air blowers is adequate for winter calving. Use portable calf shelters on pasture.

Provide separate feedyard areas for mature cows, first calf heifers, bulls, steers, and heifer calves.

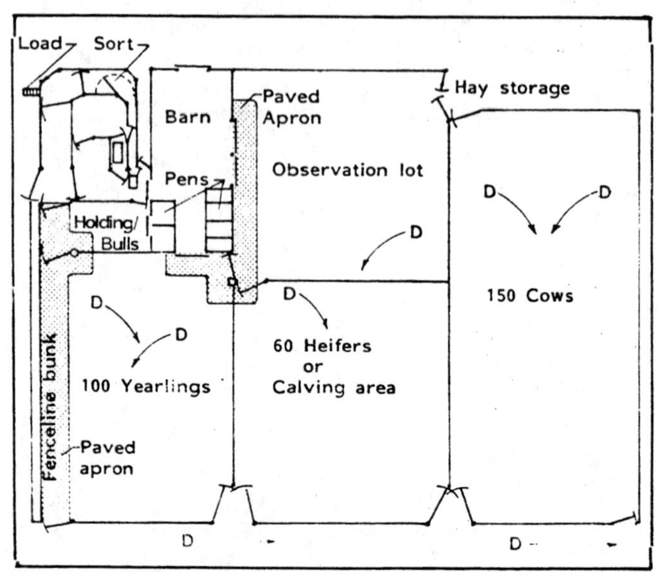

Fig 511-1. 150 cow-calf feedyard.

Use movable pen dividers for easier cow and calf handling and barn cleaning. Provide a frost-proof hydrant for water supply during winter calving.

During the winter, warm chilled calves and dry newborn calves in 4'x4'x2½' warming boxes. The inside temperature is 70-110 F. Use a heated hover to gradually adjust calves to winter temperatures. With movable calf shelters, return cows and calves to an outdoor lot in 2 or 3 days. Provide ventilation to remove moisture.

Cattle Feeding Facilities

Barn and Feedlot

Buildings to protect cattle and feeding equipment are common in cold, wet, and windy climates to improve feed efficiency. Open-front barns and lots with mechanical or fenceline bunks are common in the Midwest.

Face the long, open side away from prevailing winter winds, usually to the east or south. Provide panels in the closed walls to be opened for summer ventilation. Eave troughs keep rain water out of the manure handling system. Connect a paved apron along the open side of the barn to the paved feedbunk apron. Extend aprons 4'-6' into buildings or under roof overhangs to prevent mud from blown-in snow and rain. Make unpaved floors inside barns at least 12", and paved floors at least 6" above outside grade. Slope floors toward the open side.

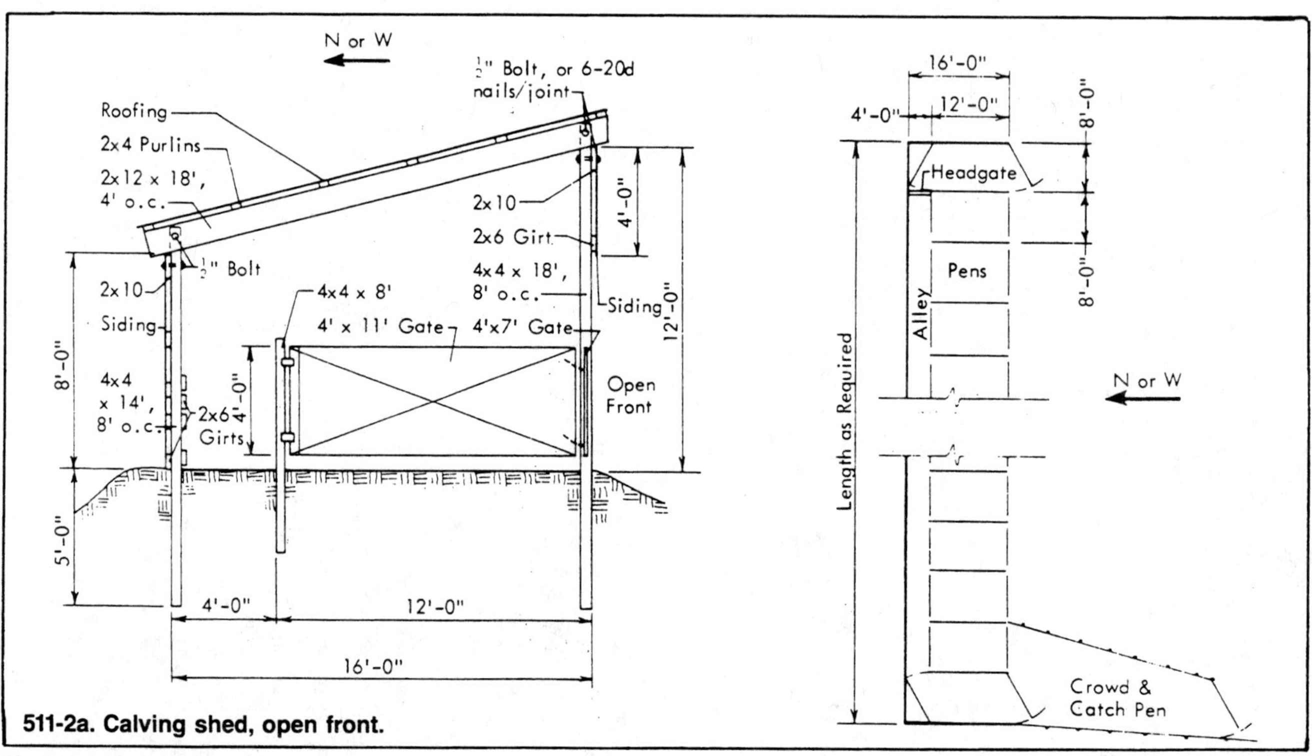

511-2a. Calving shed, open front.

Fig 511-2. Calving sheds.

Continued on next page.

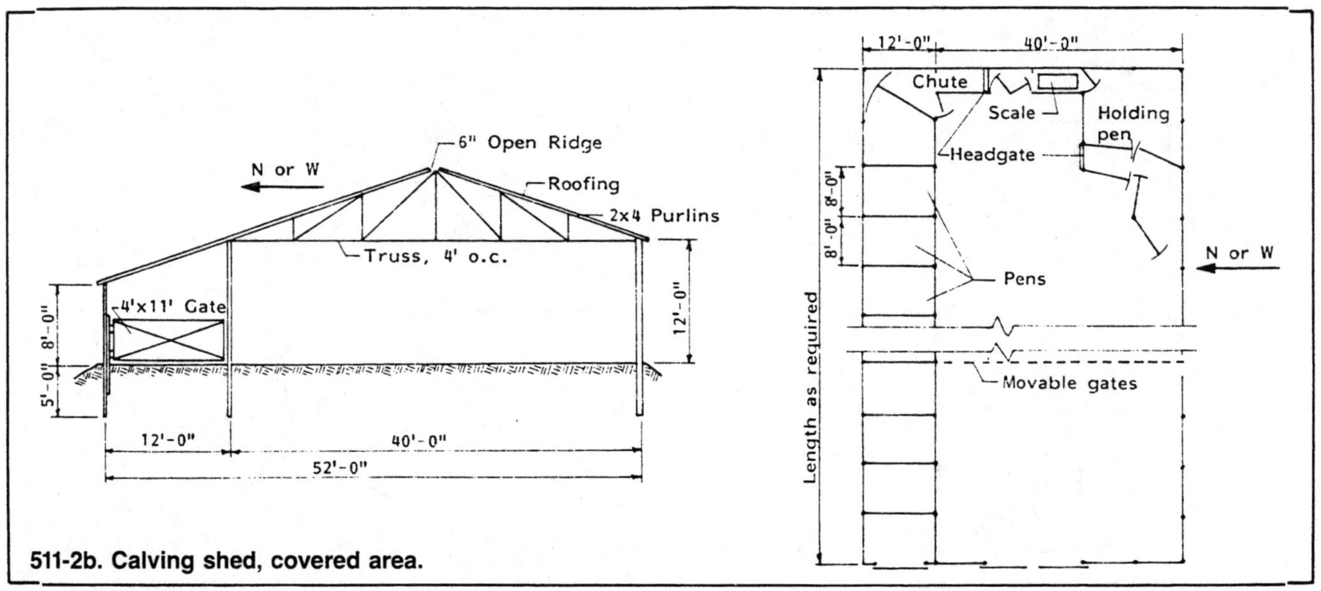

511-2b. Calving shed, covered area.

Fig 511-2. Calving sheds, continued.

Plan for controlling feedlot runoff and cattle handling facilities. Fig 511-3 shows barn and lot layouts.

Feeding Barn and Lot

The building is for feeding, summer shade, and winter shelter. The open lot is for cattle resting and exercise.

Advantages of a feeding barn are:

- Feed and equipment are protected from wind, rain and snow.
- Cattle feed intake is more consistent, especially during severe weather.

Open Feedlot

Open feedlots are usually unpaved lots that require more land area. Weather protection is often limited to a windbreak fence in winter and/or a sun-

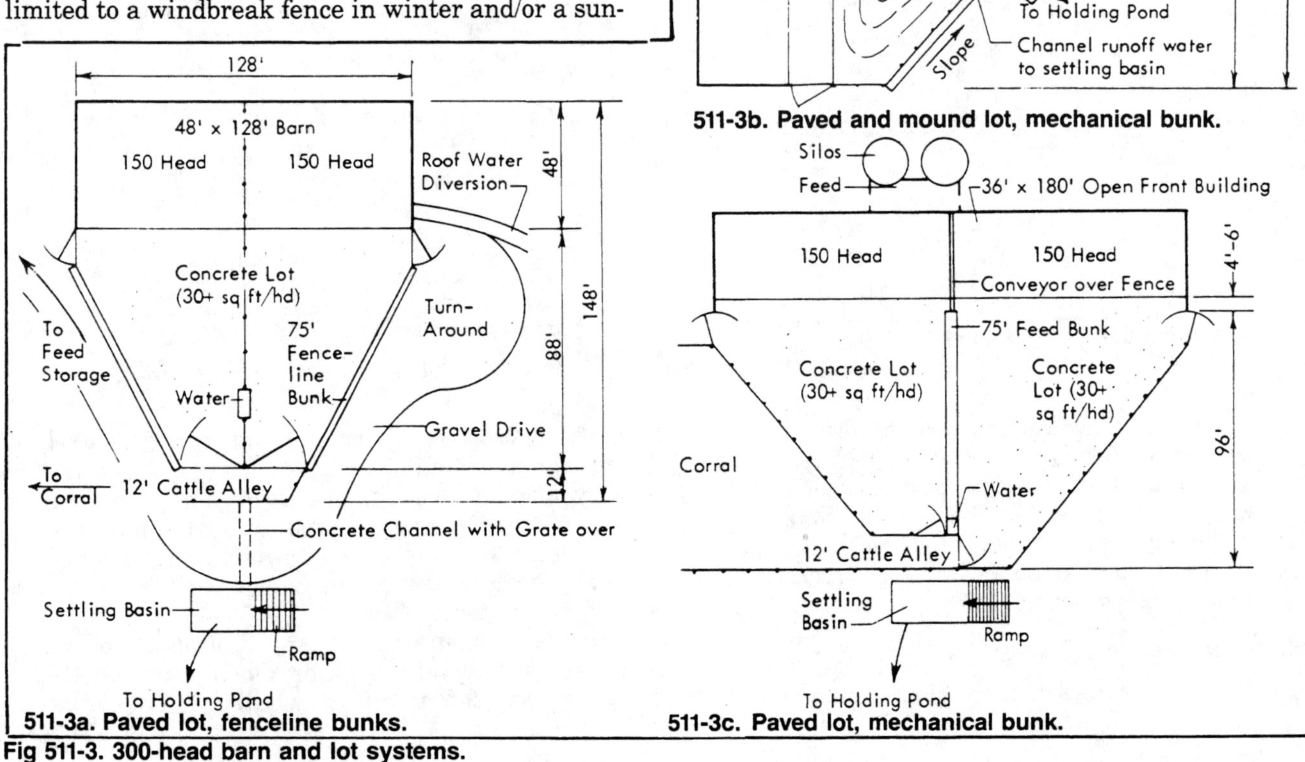

511-3b. Paved and mound lot, mechanical bunk.

511-3a. Paved lot, fenceline bunks.

511-3c. Paved lot, mechanical bunk.

Fig 511-3. 300-head barn and lot systems.

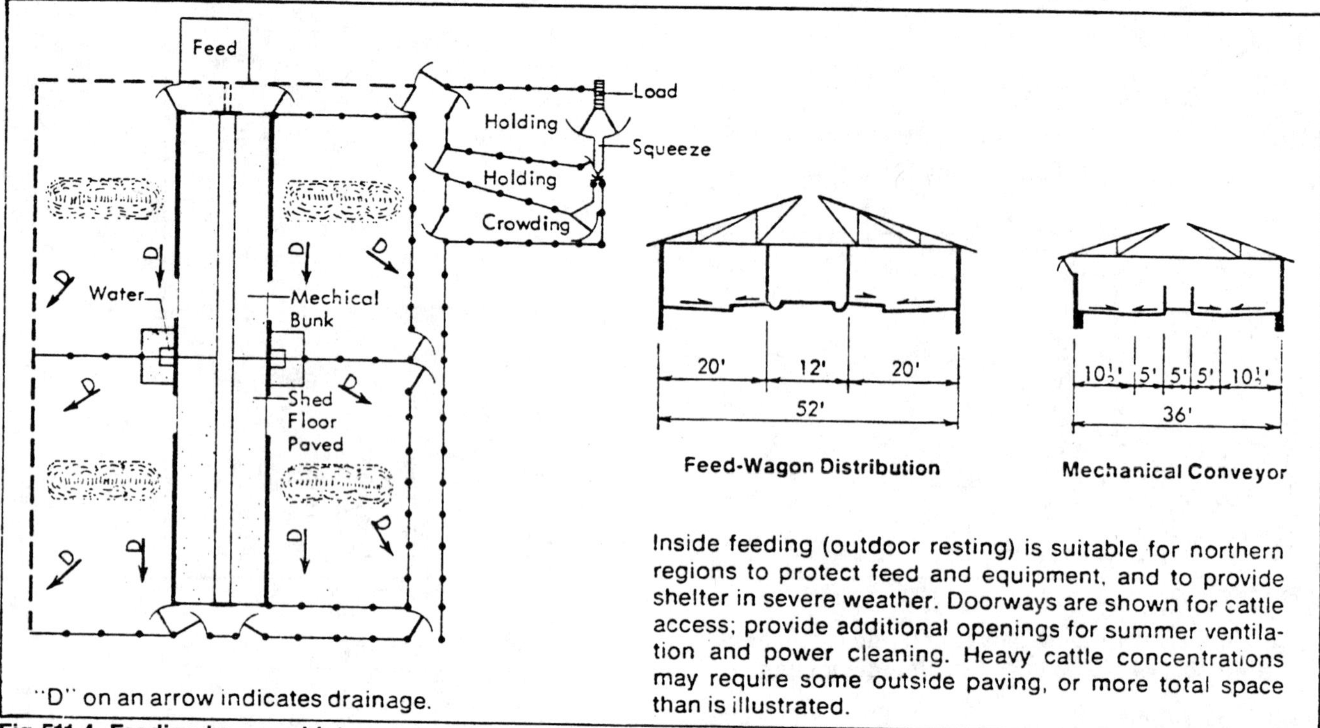

Fig 511-4. Feeding barn and lot system.

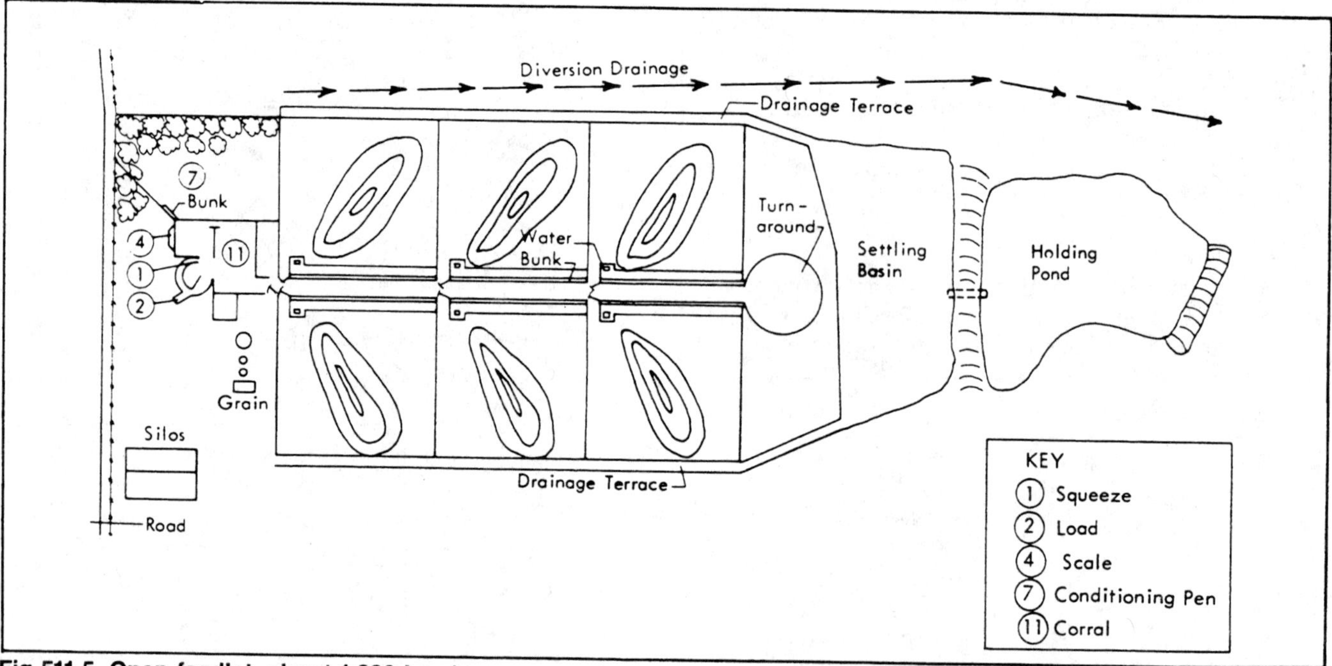

Fig 511-5. Open feedlot, about 1,200 head.
This lot is on a very gentle slope; mounds are only about 3′ high. Because drainage is poor at the back of the lots, cattle are handled in the feed alleys. Each lot is 250′ square.

shade in summer. Pave an apron along feed bunks and around water tanks.

Fig 511-5 shows an example of a real situation where some compromises have been made. Note that lots are not always rectangular. If possible, plan separate cattle and vehicular traffic lanes and receiving, shipping, and treatment facilities.

Confinement Feeding Barns

Confinement barns solve feedlot problems caused by drifting snow, severe wind, mud, runoff control, and mound maintenance. A cold confinement barn is usually open on the south or east, depending on prevailing winds. Feeding is in mechanical or fenceline bunks. Solid, semi-solid, and liquid manure systems are used with confinement barns. See the Waste Management section.

Provide an open ridge and open eaves to control moisture during cold weather. Open doors on the north or west side ¼ to ½ of the length for better cross-ventilation during the summer. Inside temperatures during the winter are usually 5-10 F above outdoors. Circulating or heated waterers are required to prevent freezing. Birds, flies, and frozen manure can be problems.

Pen size depends on building dimensions, management program, and number of cattle purchased or sold at one time. Maximum pen capacity is usually 120 head. Plan pens so cattle are of about uniform weight.

Wall heights in beef confinement buildings are usually 10′-14′. Higher walls improve summer comfort but are not a substitute for a well designed and properly managed ventilation system.

Confinement buildings can also be completely enclosed, insulated, and mechanically ventilated. These are more expensive to construct and operate but do eliminate some of the problems with open, naturally ventilated buildings.

Barn location

Locate an open-sided, confinement barn with the open side facing away from prevailing winds—usually south or east. Locate buildings for maximum natural air movement during warm months. Provide space for feed storage and handling, manure storage, and cattle handling.

Locate buildings with storage and scraper systems on level ground. Keep the bottom of the storage above the ground water table and at least 100′ from the water supply.

Locate a barn with a flushing system on a slope about the same as the gutter or flume slope. Build the outlet higher than the lagoon or storage surface so that a line from the barn to the storage slopes at least 1%. If the outlet is not higher than the storage surface, a lift pump is needed.

Floors

Slotted floors reduce labor for manure removal and keep cattle cleaner. Concrete slats are typical for beef buildings. Slat spacing is usually 1½″-2″ and slats are up to 12″ wide.

Ventilation is needed in confinement barns to control moisture, modify temperatures, and remove gases and odors.

Feedlot Planning

When planning a feedlot, consider:

- Truck access to receiving, loading, and feed processing areas.
- Arranging lots in a U shape around the central activity area.

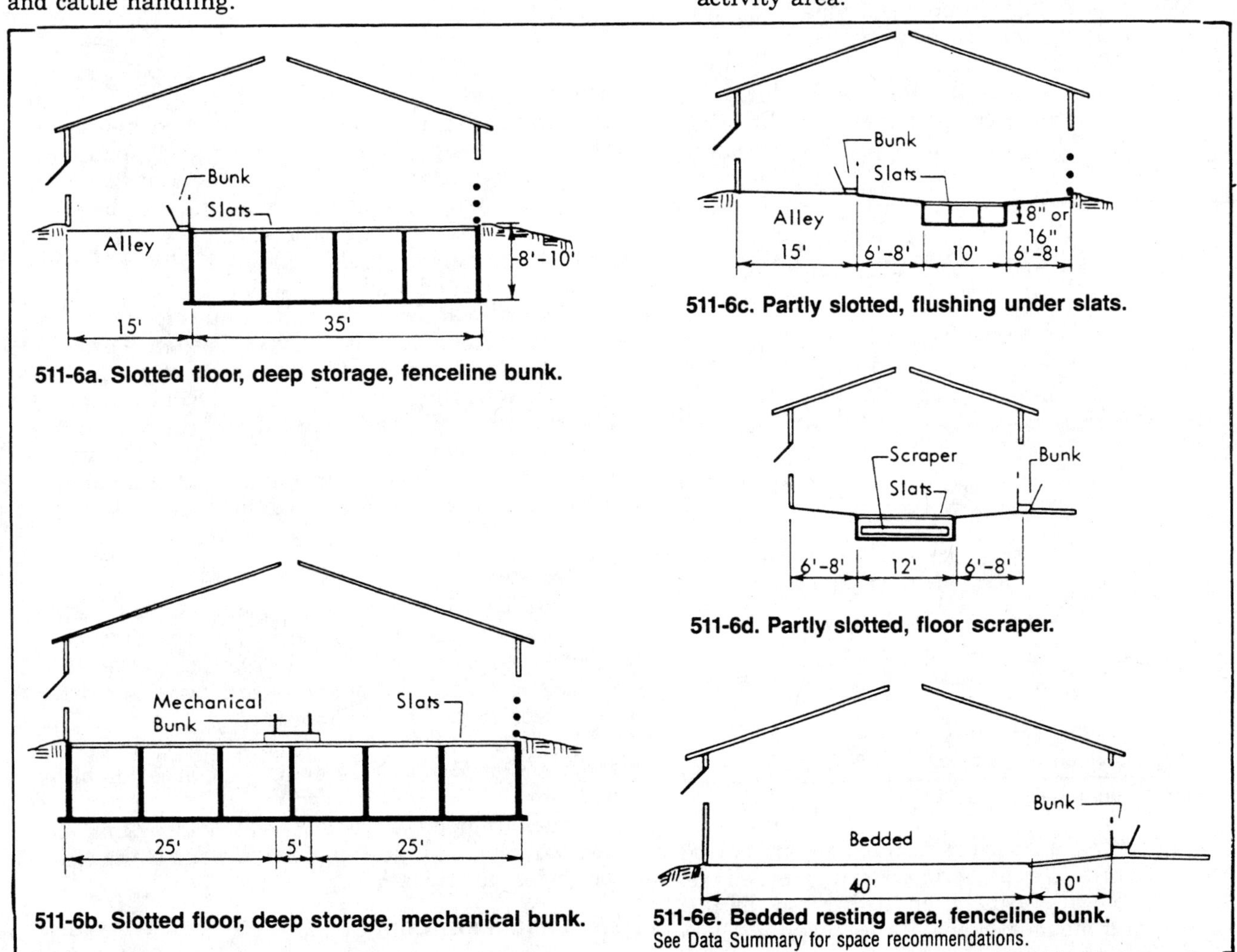

511-6a. Slotted floor, deep storage, fenceline bunk.

511-6c. Partly slotted, flushing under slats.

511-6d. Partly slotted, floor scraper.

511-6b. Slotted floor, deep storage, mechanical bunk.

511-6e. Bedded resting area, fenceline bunk.
See Data Summary for space recommendations.

Fig 511-6. Confinement barn cross sections.

- Locating feeding pens near the feed mill to reduce travel.
- Orienting traffic lanes and feed bunks north-south to maximize snow melting and drying.
- Rounding lot corners for easier turning at intersections.
- Space for a service area, cattle working facilities, separate conditioning lots, and hospital facilities.
- Potential for expansion. Assume your operation will double in size and plan accordingly.
- Consider winter and summer airflow patterns, runoff control, snow accumulation and removal, and manure management.

Plan for 120 or 240 head per lot. These are multiples of an average truckload of 60 head. Plan lots so cattle are of about uniform weight. Smaller, uniform size groups grow more efficiently and are easier to manage.

Lot area depends on cattle population and size and extra space needed for manure and snow storage, dust and mud reduction, and equipment use. Extra space needed varies with the amount of paved space, soil type, drainage, annual rainfall, and freezing and thawing cycles. See the Data Summary for recommended lot areas.

Use mounds in lots to improve drainage and provide areas that can quickly dry.

Provide vehicle traffic lanes at least 14′ wide and cattle alleys up to about 12′ wide. Keep vehicle and cattle traffic separate when possible to simplify work routines. Use a 30′-60′ wide alley between two parallel feedbunks.

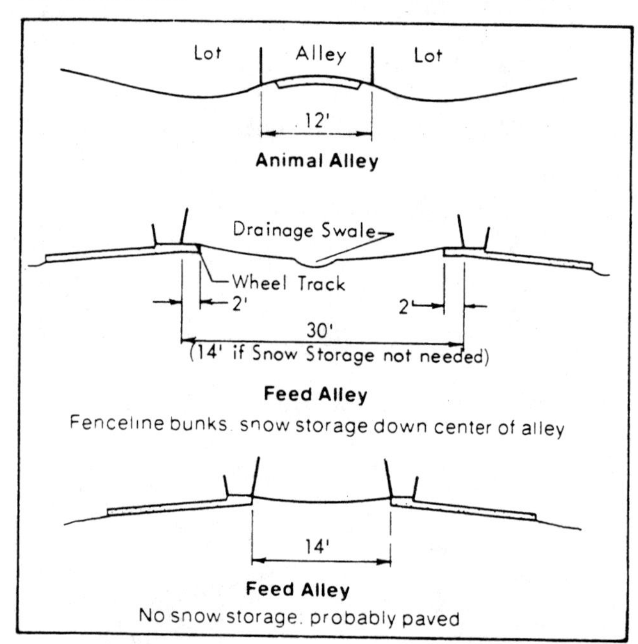

Fig 511-7. Feedlot alleys.

Pave heavy traffic areas, around waterers, and along feedbunks. Extend aprons 4′-6′ into buildings. Pave an 8′-10′ walkway between the barn and bunks, waterers, and mounds. Slope all aprons away from bunks, sheds, and waterers—1″/ft is nearly self-cleaning and ½″/ft is minimum.

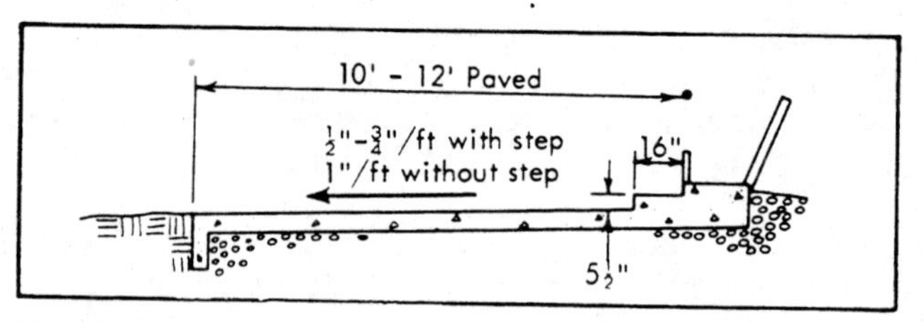

Fig 511-8. Bunk aprons.
Use 12′ width if the soil below the apron will be muddy or drifted with snow part of the year.

Topography and Drainage

Slope open lots 4%-6% away from winter prevailing winds, usually south or east. Unpaved lots sloped over 8% have excessive erosion. Complete land grading before starting construction.

Slope lots away from buildings and feedbunks. Avoid drainage from one lot to another. Direct water away from the lot with eave troughs and downspouts on barns. Use diversion ditches or terraces to intercept surface water from above the lot.

Locate feedbunks along the high side of the lot or up and down the slope for best lot drainage. A north-south or northwest-southeast orientation is preferred.

Mounds

Orient mounds parallel to the lot slope so runoff does not pond. Begin mounds at feedbunks and slope them away. Round mound tops to avoid holes and promote drainage. Yearly maintenance is required.

Stabilize the upper half of the mound surface with chopped bedding, a soil manure mixture, or by disking in lime. Barn lime reacts quicker than agricultural limestone. Start by adding 1 lb/ft^2; add more lime until mound surface is stable.

A windbreak fence on the mound ridge is recommended in the northern Midwest. The fence permits cattle to move from one side to the other depending on weather conditions. Build the fence 8′-12′ high and 80% solid (e.g. 6″ boards spaced 1½″ apart). Provide minimum mound space on each side of the windbreak. Slope lots away from fences to control manure accumulation and insect breeding.

Service Area

Plan for a business office, lounge, and parking and access area. An office is for maintaining herd records and conducting daily business activities. Computer-based feeding and record systems require protection from dirt and moisture.

In the lounge area, provide a toilet, sink, and shower. Consider lockers and a rest area for hired workers.

Provide employee, visitor, and truck parking. Locate shipping and receiving facilities near the office for easier supervision.

Wind and Snow Control

See Farmstead Planning for wind and snow control.

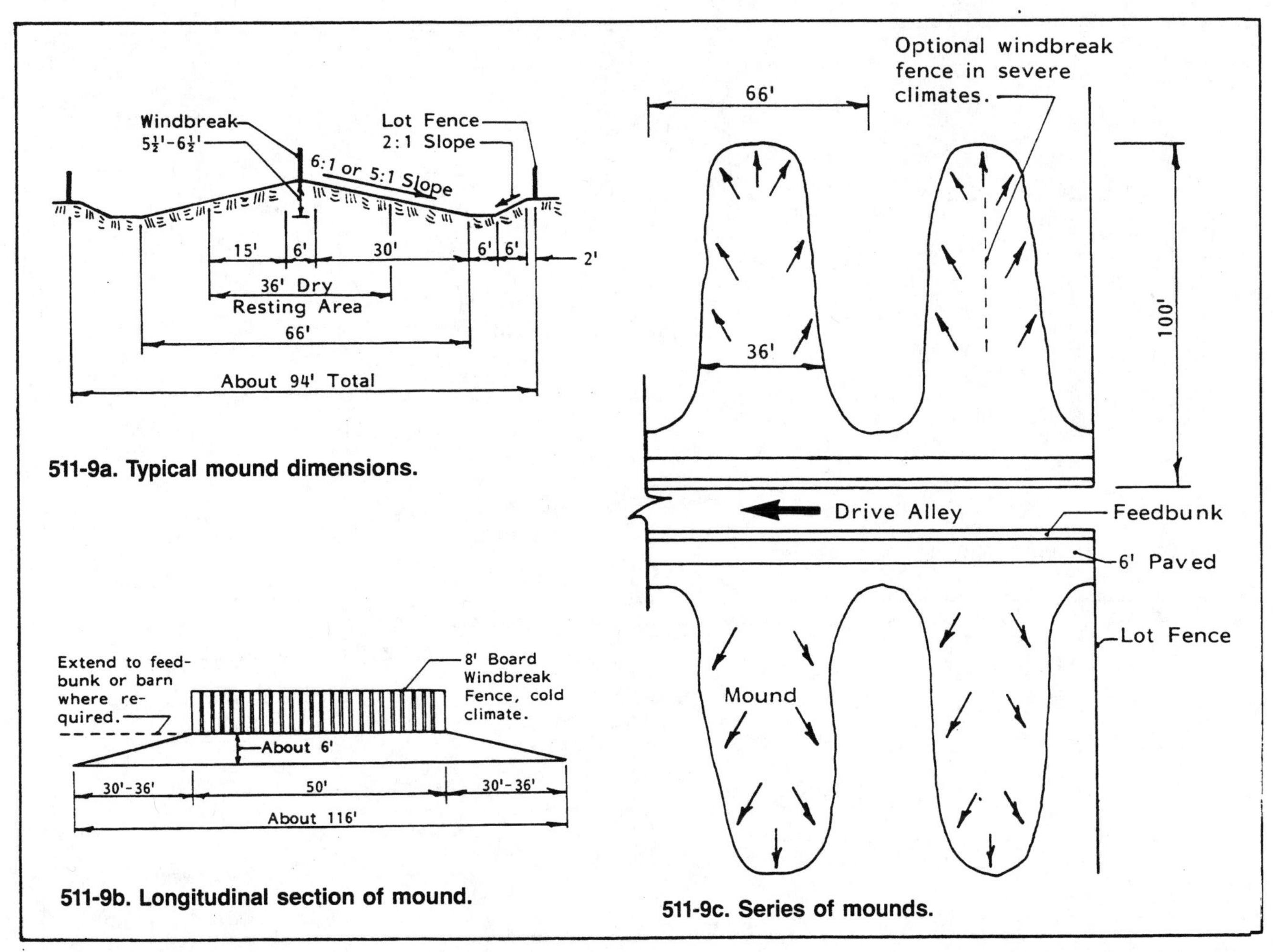

511-9a. Typical mound dimensions.

511-9b. Longitudinal section of mound.

511-9c. Series of mounds.

Fig 511-9. Mound construction.

Handling Facilities

Cattle handling facilities are needed for treating, handling, and sorting cattle. Locate handling facilities on a well drained site near barns, lots, and an access road.

Provide protection from prevailing winter winds. Trees near a handling area provide summer shade and winter wind protection. Consider the effect on snow accumulation. Roofed handling facilities improve cattle handling.

Complete cattle handling facilities include:

- Conditioning, holding, sorting, and crowding pens.
- Working, cutting, and loading chutes.
- Squeeze and headgate.
- Dipping facilities.
- Scale.
- Hospital facilities.

Conditioning Lots

Conditioning (or receiving) lots are for incoming cattle before they go to the feedlot. New cattle are usually stressed from weaning, removal from range, crowding in trucks and rail cars, motion sickness, thirst, hunger, fright, and pain. Many have never eaten from a bunk or drunk from a tank or waterer.

To help calves adapt, provide:

- Lots near the receiving corral for up to 60 head, with 100 ft²/hd in unpaved lots. Avoid overcrowding.
- About 2′ bunk space/head. Offer hay for 8 to 10 hr before starting other roughages. Start grain or silage gradually during the conditioning period, which can be 3 to 4 weeks.
- Waterers that can be turned off—calves tend to over-water at first. A running hose attracts calves that know the sound of water.

Constructing Facilities

Locate handling facilities farther from the service yard or house to reduce noise and dust. When cattle are handled in all types of weather, consider indoor cattle handling. Provide enough space for holding pens, feed and equipment storage, chutes, and treatment pens.

Provide one holding pen/250 head for up to 1,000 head, 1 pen/500 head for up to 5,000 head, and one for each additional 2,000 head. Make holding pens about 600 ft² (20′x30′) with two or more larger pens (1,000 ft²) for feedlots of 5,000 head or more. Provide 12′ working alleys between pens with several gates to improve cattle movement and worker safety.

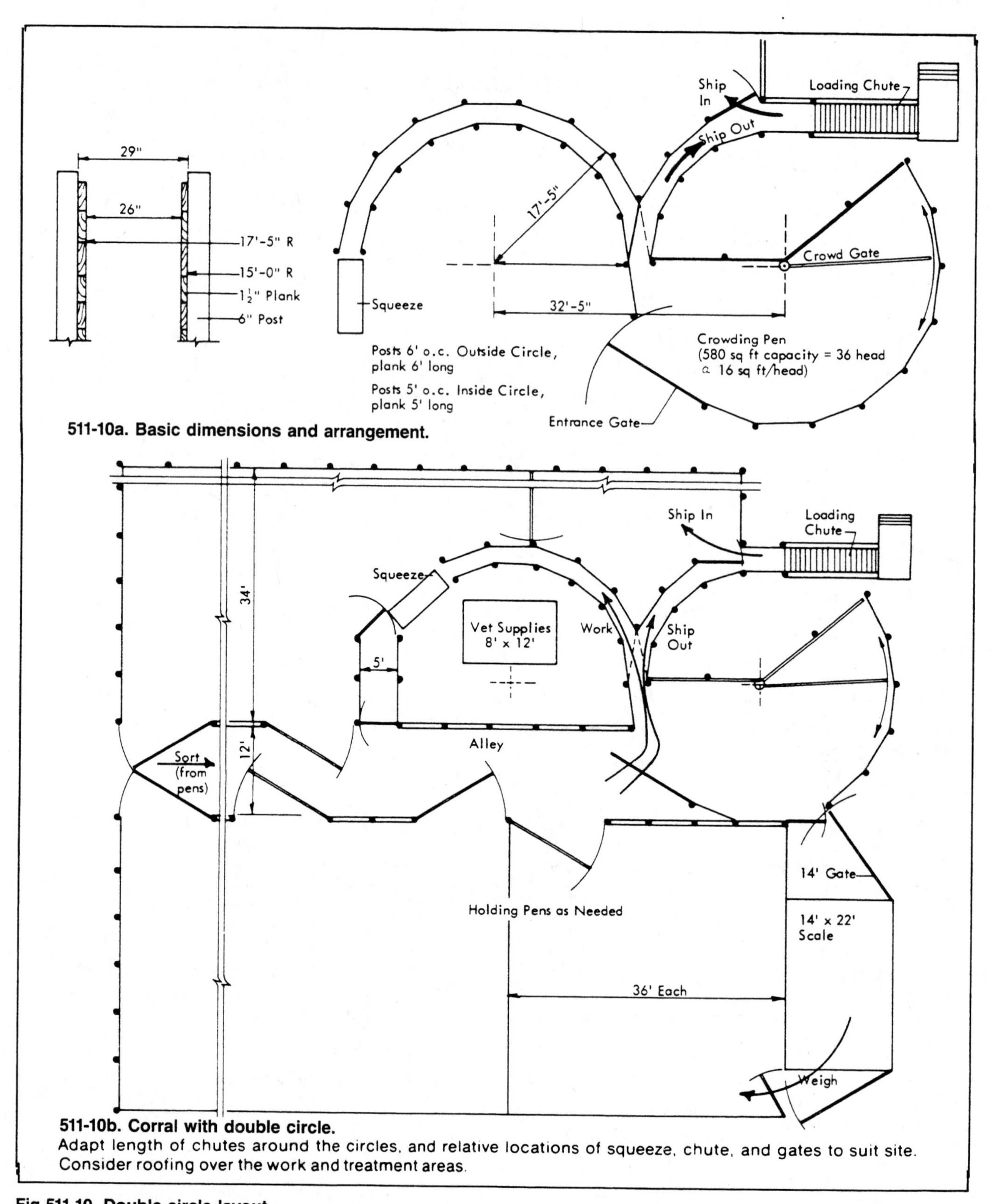

511-10a. Basic dimensions and arrangement.

511-10b. Corral with double circle.

Adapt length of chutes around the circles, and relative locations of squeeze, chute, and gates to suit site. Consider roofing over the work and treatment areas.

Fig 511-10. Double circle layout.

Crowding pen, loading chute, and squeeze are combined into an efficient unit.

A crowding pen is used to crowd small groups of cattle into the chute. It can be used for spraying and some injections. Taper the pen from 12′ to about 2′ at the chute entrance. Use solid-sided fences to improve cattle movement. Circular crowding pens are becoming more popular. Provide at least a 6′ working space on each side of the working chute. Use a solid, 2′ wide cutting gate in the working chute for sorting.

The headgate and squeeze restrains cattle for treatment, branding, etc. A curved neck headgate prevents up-down neck movement. Choking is a problem. A tilting table lowers cattle to a horizontal position for hoof trimming and treatment.

Use an approved scale to weigh cattle and keep production records. Offset the scale from the working chute so cattle can bypass the scale.

Locate an inclined and a trailer type loading chute for easy access during wet weather. A north-south orientation south of the buildings promotes drying.

Hospital Facilities

Provide one hospital area for every 6,000 head, close to working corrals and conditioning lots. Hospital pens must be well drained. Use roughened concrete sloped ¼"/ft or more to a drain in outdoor hospital pens. A well ventilated barn reduces stress and speeds recovery. Provide medical supply storage, hot water, a refrigerator, records center, and a sling for animal support.

Provide 40 to 50 ft^2/hd of hospital space for 2%-5% of the finishing cattle. Avoid locating treatment facilities where incoming cattle can be contaminated. Do not overcrowd treatment facilities.

Thoroughly clean treatment and handling areas as needed and in early summer. Use white wash containing cresol (a disinfectant) for washing walls, posts, and other barn surfaces. Cresol controls ring worm and lice. Heavily spread lime on indoor floor areas about 30 days before use.

Heat and mechanically ventilate tight, well-insulated rooms and intensively used barn areas. See the Ventilation section.

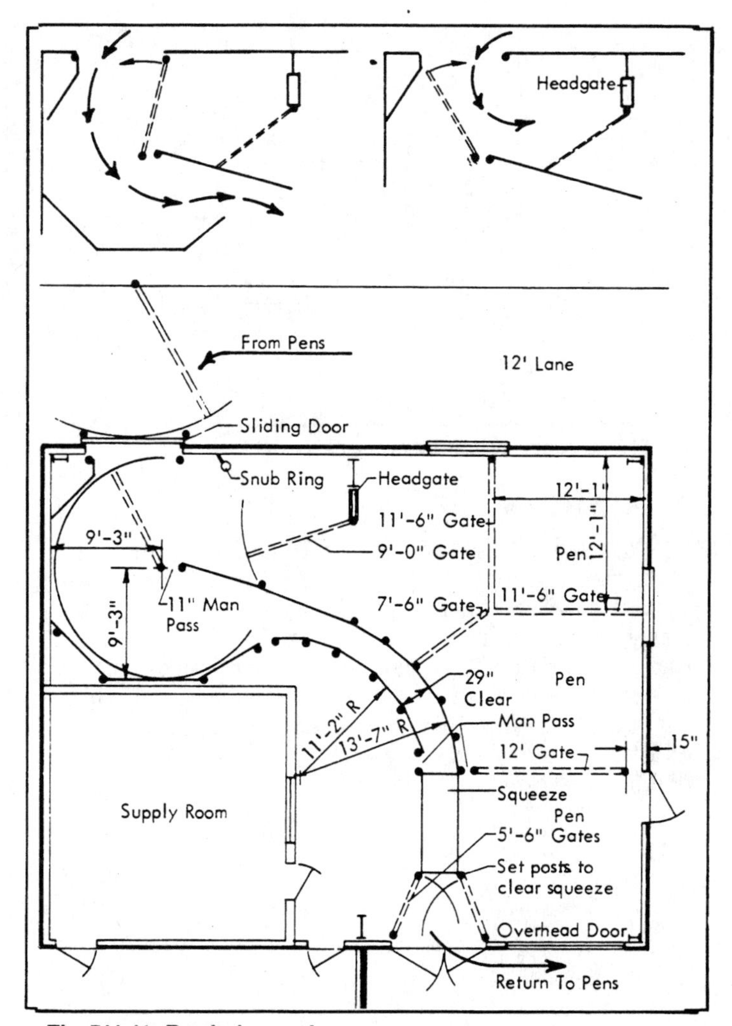

Fig 511-11. Roofed corral.
This layout can be a hospital center. Gates swing out of the way for tractor cleaning. Mount floodlights to light head, sides, and tail of cattle in squeeze or headgate, and in the corner pen for care during calving.

Bunk Planning

Feeding space and bunk planning depend on cattle size and the number that eat at one time. Tables 511-1 and 511-5 in the data summary give the space and bunk size needed.

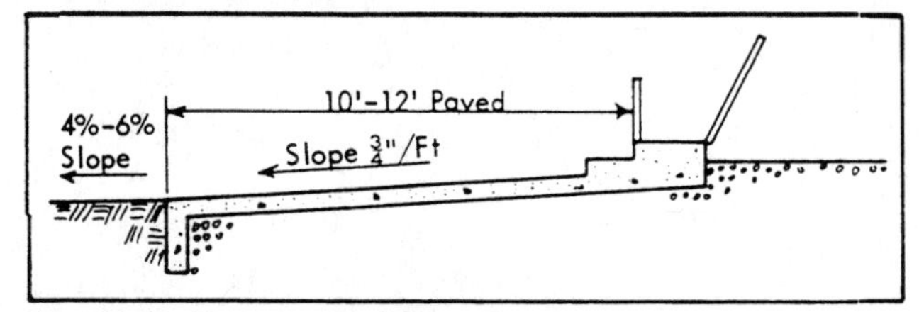

Fig 511-12. Pavement along a bunk.

Fenceline Bunks

Cattle feed from one side. Bunks are usually filled with a self unloading wagon or truck. Provide all-weather year-round access. Fenceline bunks require twice as much bunk length (but not twice the cost) as bunks that feed from both sides.

Mechanical Bunks

Mechanical bunks are usually wider because cattle eat from both sides. If the feed conveyor divides the eating space, make the bunk wider. See Fig 511-14 and Table 511-5.

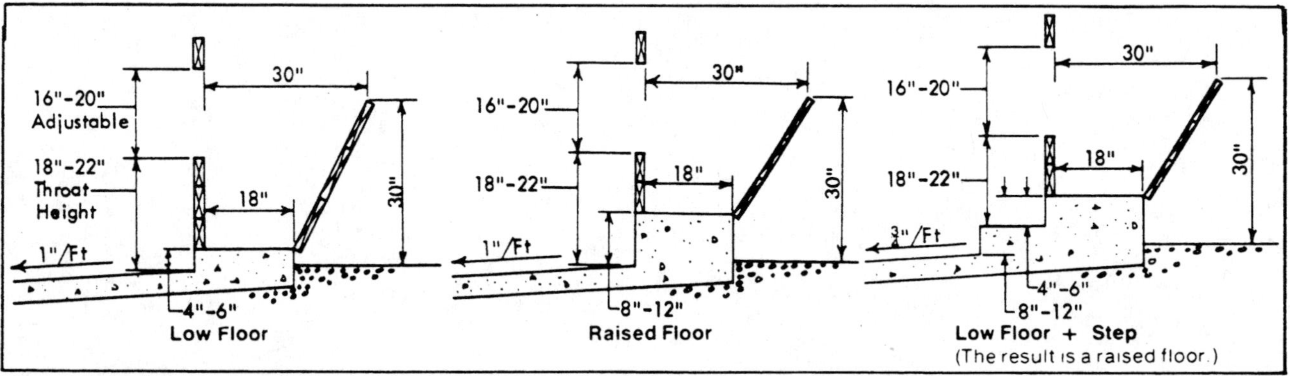

Fig 511-13. Fenceline bunks.

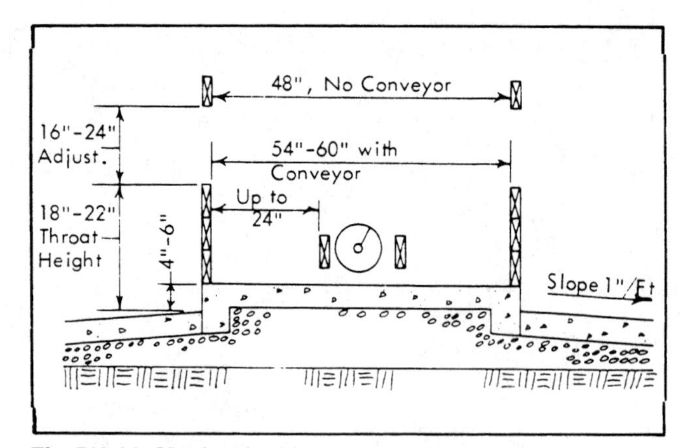

Fig 511-14. Mechanical bunk.

Covered Bunks

Covered bunks protect feed and mechanical equipment. They can also provide some shade.

Use a narrow roof above the apron for bunk protection and minimum shade, Fig 511-15a. A wider roof high enough to clear cleaning equipment can provide bunk and cattle shelter and summer shade, Fig 511-15b. Wide-roofed covered bunks can cause additional snow drifting and prevent thawing in northern regions.

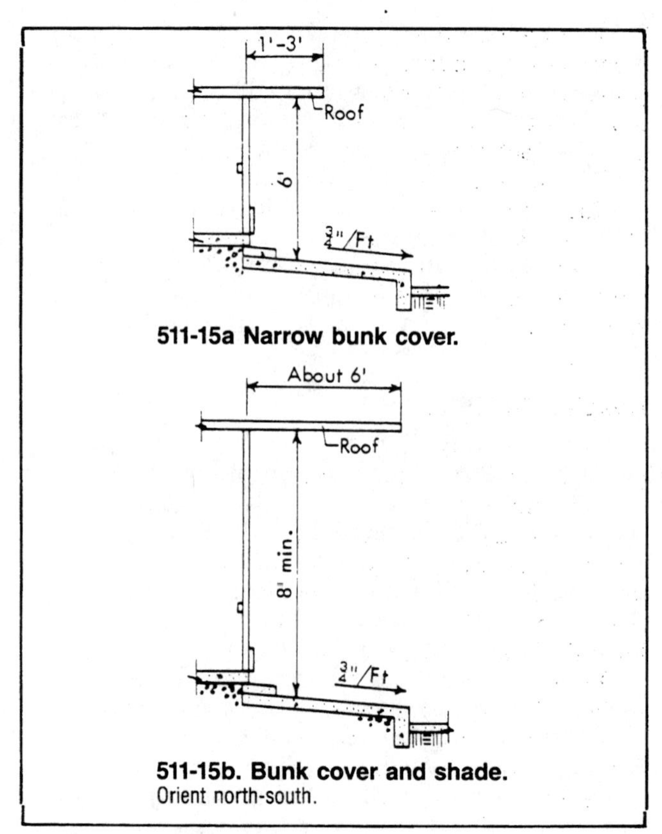

511-15a Narrow bunk cover.

511-15b. Bunk cover and shade.
Orient north-south.

Fig 511-15. Covered bunks.

512 DAIRY

Data Summary

Table 512-1. Typical herd makeup.
Replacement numbers assume uniform calving throughout the year, 12-month calving interval, no death loss or culling, 50% male and 50% female calves, and all males sold at birth.

	Number				Weight lb.
Cows milking	33	62	83	208	1400
Dry cows	7	13	17	42	1550
Total mature cows	40	75	100	250	1450
Heifers					
16-24 mo.	15	28	38	95	1050
13-15 mo.	5	9	12	30	800
9-12 mo.	7	13	17	43	600
5-8 mo.	7	13	17	42	400
3-4 mo.	3	6	8	20	250
Calves 0-2 mo.	3	6	8	20	150
Total replacements	40	75	100	250	

Table 512-2. Annual feed requirements.
Approximate needs per dry and milking cows and replacements. If using high moisture grain, adjust figures accordingly.

	% dry matter	Lbs milk/cow-year 12,000	14,000	16,000	18,000
		---- quantity per cow ----			
Medium level of silage:					
Hay silage, T	40-50	11.0	11.6	12.3	12.9
or hay, T	80-85	4.3	4.5	4.7	4.9
Corn silage, T	30-35	12.0	12.0	12.0	12.0
Shelled corn, bu		72	84	102	118
High level of silage:					
Hay silage, T	40-50	6.7	6.9	7.1	7.5
or hay, T	80-85	2.6	2.7	2.8	2.9
Corn silage, T	30-35	16.5	16.5	16.5	16.5
Shelled corn, bu		52	64	86	106

Source: Chore Reduction for Free Stall Dairy Systems, *Hoard's Dairyman,* Fort Atkinson, WI.

Table 512-3. Estimated cropland.
Necessary to produce the annual feed required per cow and replacements. If using high moisture grain, adjust figures accordingly.
Values based on Table 512-2 and yields shown.

	% dry matter	Lbs milk/cow-year 12,000	14,000	16,000	18,000
		--- acres/cow and replacements ---			
Medium level of silage:					
Hay silage 6 T/A	40-50	1.8	1.9	2.1	2.2
or hay 3 T/A	80-85	1.4	1.5	1.6	1.6
Corn silage 15 T/A	30-35	0.8	0.8	0.8	0.8
Shelled corn 80 bu/A		0.9	1.1	1.3	1.5
High level of silage:					
Hay silage	40-50	1.1	1.2	1.2	1.3
or hay	80-85	0.9	0.9	0.9	1.0
Corn silage	30-35	1.1	1.1	1.1	1.1
Shelled corn		0.7	0.8	1.1	1.3

Table 512-4. Housing space requirements.

Housing type	Age, months 3-4	5-8	9-12	13-15	16-24
	---------- ft²/animal ----------				
Resting area & paved outside lot	20 30	25 35	28 40	32 45	40 50
Total confinement					
Solid floor	20	25	30	40	60
Slotted floor	11	12	13	17	25
	--------- dimensions ---------				
Free stall &	2'x4'-6"	2'-6"x5'	3'x5'-6"	3'-6"x6'-6"	3'-6"x7'
manure alley width	6'	8'-10'	8'-10'	8'-10'	8'-10'

Table 512-5. Feeder space requirements.

	Age, months 3-4	5-8	9-12	13-15	16-24
	------- in/animal -------				
Self feeder (hay or silage)	4	4	5	6	6
Once-a-day feeding	12	18	22	26	30

Table 512-6. Cow stall platform sizes.
Use electric cow trainers. Dimensions from edge of curb to edge of gutter.

	Stanchion stalls		Tie stalls	
Cow weight	Width	Length	Width	Length
Under 1200 lb	4'-0"	5'-6"	4'-0"	5'-9"
1200 to 1600 lb	4'-6"	5'-9"	4'-6"	6'-0"
Over 1600 lb	Not recommended		5'-0"	6'-6"

Table 512-7. Recommended stall barn dimensions.

Alley width	
Flat manger-feed alley	5'8"-6'6"
Feed alley with step manger	4'0"-4'6"
Service alley with barn cleaner	6'0"
Cross alley[a]	4'6"
Manger width	
Cows under 1200 lb	20"
Cows 1200 lb or more	24" – 27"
Gutters	
Width[b]	16" or 18"
Depth, stall side	11" - 16"
Depth, alley side	11" - 14"

[a]Taper the end stalls inward 6" at the front for added turning room for a feed cart.
[b]Or as required for barn cleaner.

Table 512-8. Free stall dimensions.
Stall width measured center-to-center of 2" pipe dividers. For wider divider dimensions, increase stall width accordingly. Stall lengths are measured from front of stall to alley side of curb.

	Width x Length
Calves	
6 weeks to 4 months	2'-0" x 4'-6"
5 to 7 months	2'-6" x 5'-0"
Heifers	
8 months to freshening	3'-0" x 5'-6"
Cows (average herd weight)	
1000 lb	3'-6" x 6'-10"
1200 lb	3'-9" x 7'-0"
1400 lb	4'-0" x 7'-0"
1600 lb	4'-0" x 7'-6"

Table 512-9. Milking parlor cow throughputs.[a]

	Side-opening		Herringbone				Trigon			Polygon			
Mechanization[c]	D-2	D-3	D-4	D-6	D-8	D-10	12-stall	16-stall	18-stall	16-stall	20-stall	24-stall	32-stall
	Cows per hour[b]												
None	25-35^{1}	50-63^{2}	29-42^{1}	50-66^{2}	64-80^{2}	80-89^{2}	53-74^{2}	68-89^{2}	76-95^{2}	71-97^{2}	86-112^{2}	101-127^{2}	121-157^{3}
Crowd gate	28-38^{1}	52-65^{2}	34-47^{1}	55-71^{2}	69-87^{2}	88-97^{2}	59-79^{2}	74-96^{2}	83-103^{2}	78-104^{2}	94-120^{2}	110-136^{2}	131-167^{3}
Prep stalls	34-44^{1}	52-65^{2}											
Crowd gate & feedgates			34-47^{1}	58-74^{2}	72-90^{2}	92-101^{2}	62-83^{2}	78-100^{2}	87-107^{2}	83-109^{2}	98-126^{2}	117-143^{2}	139-175^{3}
Crowd gate & prep stalls	38-48^{1}	56-69^{2}											
Detachers			33-46^{1}	49-65^{1}	60-78^{1}	72-81^{1}	50-70^{1}	63-85^{1}	67-87^{1}	68-94^{1}	75-101^{1}	79-107^{1}	117-153^{2}
Detachers & crowd gate	36-46^{1}	44-57^{1}	37-50^{1}	54-70^{1}	68-84^{1}	79-88^{1}	56-76^{1}	71-92^{1}	75-94^{1}	76-102^{1}	83-109^{1}	90-116^{1}	129-165^{2}
Detachers, crowd gate, & feedgates			39-52^{1}	57-73^{1}	70-88^{1}	83-92^{1}	59-78^{1}	73-95^{1}	78-98^{1}	81-107^{1}	89-115^{1}	96-122^{1}	137-173^{2}
Detachers, crowd gates & prep stalls	40-50^{1}	50-63^{1}											

[a]Low range value is for cows producing about 60 lb /day; high value for cows producing about 38 lb/day.
[b]Steady-state throughputs; parlor setup and cleanup and group changing not included.
[c]Assume parlors have power-operated entrance and exit gates.
[d]Superscripts (1,2,3) denote number of operators.

Table 512-10. Utility room space.

	Area, ft²
Milk vacuum pump	6-9
Compressor	8-10
Water heater	4-6
Furnace	3-5
Storage	3-5
Work alleys	20-30
Refrigerator	6-9
Total	50-75

Table 512-11. Milking system components.

	Milking system	
Components	**Bucket**	**Pipeline**
Vacuum pump(s)	X	X
Balance tank/header	X	X
Vacuum controller(s)	X	X
Vacuum line and stallcocks	X	
Pulsation line and stallcocks		X
Vacuum supply line to header/trap		X
Bucket unit with claw, shells and liners	X	
Pulsator—electric, master or pneumatic	X	X
Milk pipeline		X
Weigh jars (optional)		X
Receiver, pump, trap		X
Strainer or milkveyor plus filter	X	
Filter system		X

Table 512-12. Milking center water heater sizes.

Parlor size	Washing cows 115°F	CIP water 165°F
Double 4	50 gal	80 gal
Double 8 or milking barn	80 gal	120 gal

Total Dairy Facility

When planning new construction or major modification of a dairy system, consider:

- Calf, heifer, dry cow, and milking cow housing.
- Feed types and feeding system.
- Manure handling method.
- Milking system and equipment.
- Labor requirements.
- Building environment.
- Animal handling, maternity, and treatment facilities.
- Building selection and construction.
- Sanitary and pollution control regulations.
- Future expansion.

In this chapter, herd size is the total number of both dry and milking cows. Include calves and heifers in the herd makeup if you raise replacements. Table 512-1 gives typical herd makeup assuming uniform calving year-round.

Maximum herd size may depend on farm size. Estimate cropland and storage needs based on your rations or Table 512-2.

Cropland estimates are affected by milk production, ration, forage choice, crop yields, etc. See Table 512-3. If all feeds except supplements are raised on the farm, needed cropland is 2.62 to 4.43 acres/cow; 3 to 4 acres per cow is a good estimate.

Feed storage space is discussed later.

Provide housing for the different animal groups based on Table 512-1. More than one group can be housed in the same building, but allow for managing each group separately. Management includes feeding, animal treatment, etc.

Housing Types

Housing is often based on building environment.

A **cold barn** has indoor winter temperatures about the same as outdoors. It is usually uninsulated with unregulated natural ventilation.

Cold barns are the least expensive to build, but freezing and condensation on inside walls during severe winter weather can be a problem. Heated waterers are needed. Adjust rations to maintain sufficient body heat production.

A **modified environment barn** has indoor winter temperatures higher than outdoors. It is usually above 32 F. A modified barn is insulated and designed for controlled natural or fan ventilation.

A modified barn has fewer problems with frozen manure. Insulation increases the cost of modified housing but helps keep the barn cooler in summer and reduces condensation in winter. Natural ventilation must be properly managed.

Warm barns have uniform inside winter temperatures of 40 F or above. They have insulation, mechanical ventilation, and supplemental heat for environmental control. Warm barns have more comfortable working conditions but are more expensive to build.

Replacement Animal Housing

House all replacement animals separate from the milking herd. Separation in a free stall housing system allows feeding these animals as a special group. Suit space, equipment, environment, rations, and care to the age group being handled. When planning replacement animal housing, provide:

- Adequate space.
- Dry, draft-free resting area.
- Fresh air.
- Adequate feed and water space.
- Animal groups based on size or age.
- Proper sanitation.

The number of replacement animals is proportional to the number of milking cows in the herd, Table 512-1. As the herd size expands, increase replacement numbers. Increase space for these animals to avoid crowding.

Calf Housing

Calf housing and management depend on age. The sucking instinct persists in unweaned calves (0 to 2 months) for about 30 minutes after feeding.

Individually pen calves, because mouth-to-mouth contact can transmit disease, and calf sucking can cause udder damage. House unweaned calves separate from the milking herd and maternity area.

Calves in warm housing can be in 2'x4' floor level tie stalls with a 2' wide feed alley in front and a 3'-4' wide service alley in the rear. Disadvantages of elevated tie stalls are cleaning difficulties, drafts, and bruises to small calves' legs. With elevated stalls, provide solid sides, a solid floor in the front half, and shields to prevent facial contact.

With cold housing systems, 4'x8' bedded pens with 4' partitions work well. Solid partitions minimize drafts and calf-to-calf contact. Provide a combined feed and manure alley.

Unweaned calves grow well in outdoor calf hutches. Hutches are 4'x8' enclosed shelters with a 4'x6' outside run. Separate hutches to prevent facial contact through the fence of the outside run. Move hutches after each use to improve sanitation. Provide winter wind protection.

Experience shows that young calves can be housed in a stanchion barn with a milking herd of less than 40 cows. With larger herds, calf losses tend to increase because of:

- Crowding.
- Inadequate sanitation.
- Reduced animal observation and individual care.
- Calf-to-calf contact.
- Drafts.

Sanitation

Build pens so they are easy to clean. Proper sanitation to reduce disease in calves includes:

- Regular manure removal—daily from elevated stalls.
- Washing stalls, pens, and hutches after animal removal.
- Resting or drying out pens and hutches for at least a week between use, which requires more facilities.
- Moving hutches to a new site to break disease cycles.

Put newborn calves on fresh bedding and add bedding as needed to keep them dry.

Heifer Housing

Weaned calves can be in group pens. Young heifers (2 to 6 months) are susceptible to drafts, so use a building with four walls instead of one with an open side. For calves up to 6 months of age, limit group size to 5 to 8 animals. See Table 512-4 for space needs.

Plan for ½ T of bedding/calf-year. If bedding is limited, consider totally slotted floors, free stalls, or sloped floors.

Six- to 24-month-old heifers can withstand the stress of larger groups. Use Table 512-4 to determine space needs. When possible, move animals as a group to keep uniform group sizes.

Bred Heifers and Dry Cows

First calf heifers and dry cows are often housed together, especially when heifers are within three months of calving.

Tie stalls are expensive for bred heifers and dry cows; group housing is adequate. Observe them frequently for calving signs and move to maternity pens before calving. Locate the maternity area near the dry cow and bred heifer group for easy access.

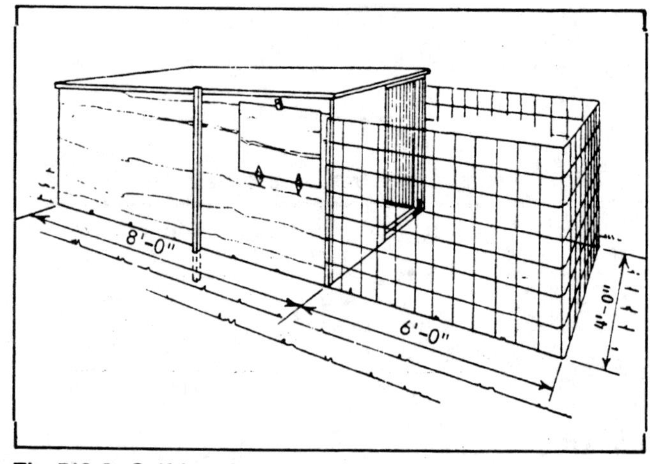

Fig 512-2. Calf hutch.
Face calf hutches south, on a well drained base. Provide a windbreak to the north and west.

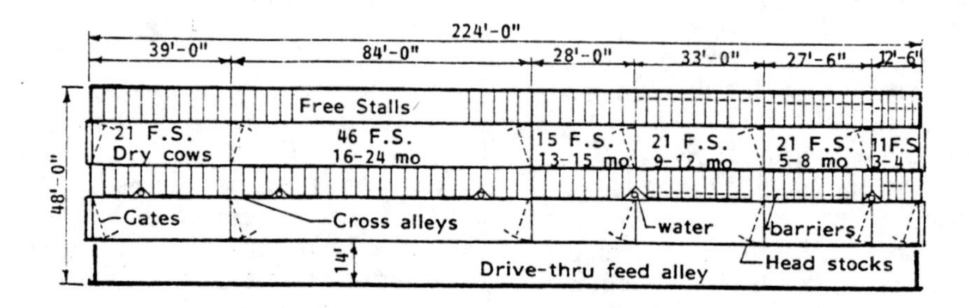

Fig 512-1. Heifer and dry cow barn.
Open-front loafing barns are flexible and work well with yard feeding. Manure accumulates on the pack and in the yard and requires more bedding than slotted floors.

Milking Herd Facilities

Stall Barns

Tie stall (stanchion) barns are the most common housing system for herds up to about 80 cows. A stall barn can be as mechanized as a free stall barn, except for stooping during milking. Tie stalls permit greater individual attention than free stalls.

Before you build or remodel a barn, determine:

- Number of cows. Base milking herd size on farm size, available labor, and production goals.
- Stall arrangement, size, and type.
- Stall floor covering or bedding type.
- Number of pens needed.
- Insulation and ventilation type.
- Type of milking system.
- Manure management system.
- Expansion possibilities.

Stanchion stalls

Stanchion stalls have a metal yoke fastened at the top and bottom that is free to swing from side to side, Fig 512-3. Usually cows are released individually, but lever stalls open and close a group of stanchions which reduces labor. On some stanchions, the yoke adjusts forward and backward to help position cows so that manure falls in the gutter. These adjustments have little value. Use proper platform size, well adjusted cow trainers, and good management to keep platforms clean.

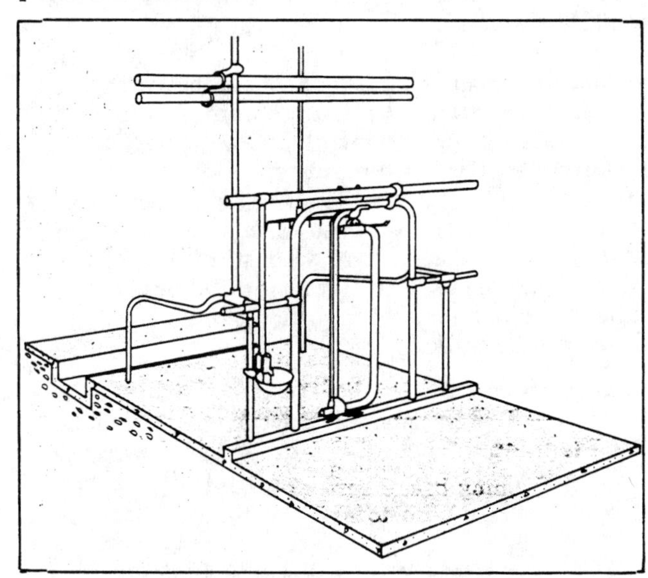

Fig 512-3. Stanchion stalls.
Water bowls are usually between every other cow over the manger. The water line can be in the floor, part of the stanchion frame, or overhead.

Tie stalls

Each cow has a neck chain or a strap to prevent her from backing out, and a pipe to prevent her from walking into the manger. Cows are fastened and released individually, requiring more labor than with lever stanchions. Generally, tie stalls are preferred over stanchion stalls because of greater cow comfort. Proper stall platform size is also important to cow comfort and cleanliness.

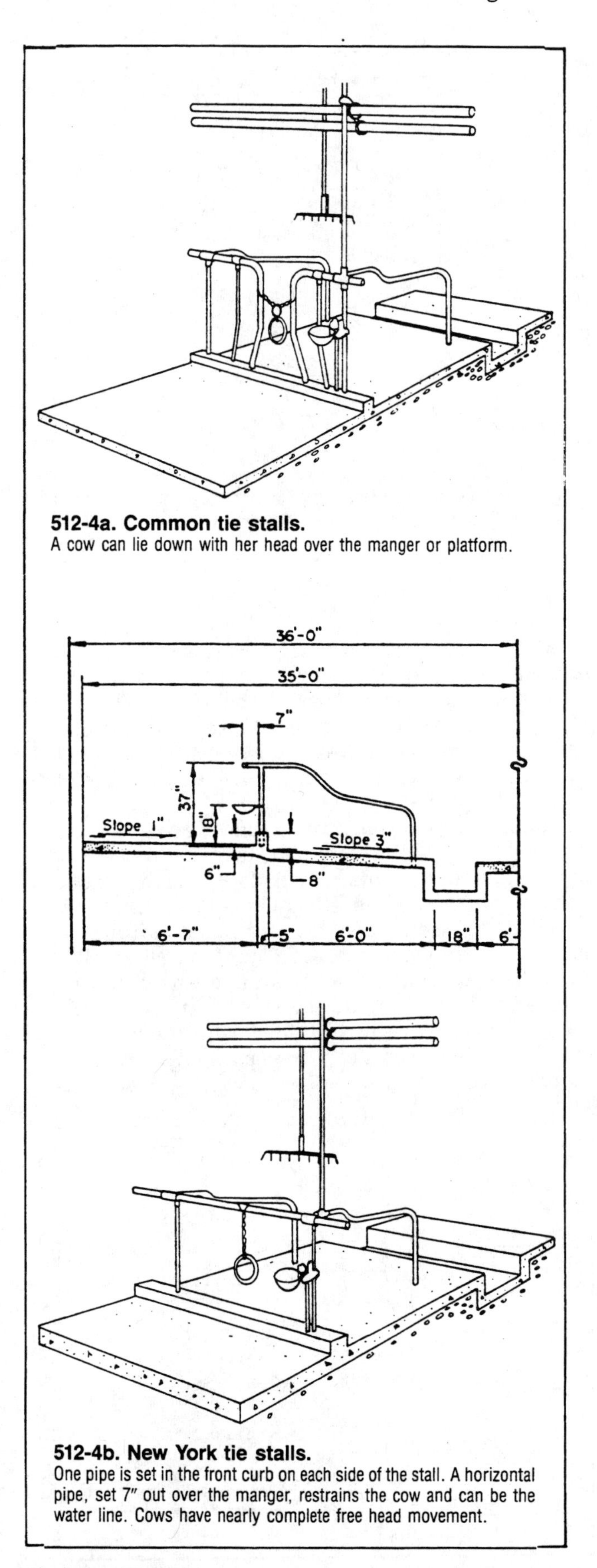

512-4a. Common tie stalls.
A cow can lie down with her head over the manger or platform.

512-4b. New York tie stalls.
One pipe is set in the front curb on each side of the stall. A horizontal pipe, set 7" out over the manger, restrains the cow and can be the water line. Cows have nearly complete free head movement.

Fig 512-4. Tie stalls.
More labor is required to tie and untie cows than with stanchion stalls.

Continued on next page.

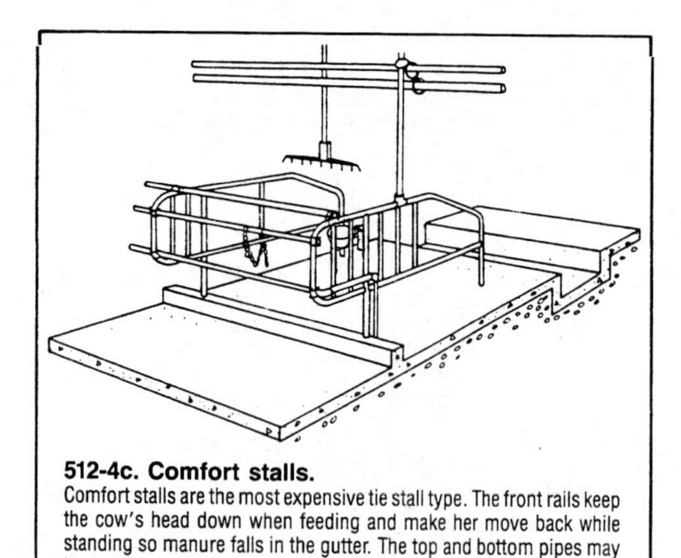

512-4c. Comfort stalls.
Comfort stalls are the most expensive tie stall type. The front rails keep the cow's head down when feeding and make her move back while standing so manure falls in the gutter. The top and bottom pipes may be the vacuum and water lines. A flat manger is easier to clean than a step manger.

Fig 512-4. Tie stalls, continued.

Stall mats

Many producers want an alternative to bedding on concrete. Bedding is often scarce, expensive, and can interfere with liquid manure handling. A highly absorbent, expendable, and good insulating alternative is hard to find. Indoor-outdoor carpeting is not recommended because it wears badly and is difficult to hold in place.

Rubber mats are softer than concrete but cows are cleaner with bedding. Manure, urine, and bedding accumulate under mats held in place with lag screws and regular cleaning is difficult and time consuming. To reduce these problems, lay rubber mats within one hour of pouring the concrete. Cut each mat smaller than the stall platform by about 6″ at the front and back and about 3″ along the sides. Fill around the mats with concrete finished flush with the mat surface, Fig 512-5. Eventually these mats loosen and are more difficult to clean under than mats placed on hardened concrete.

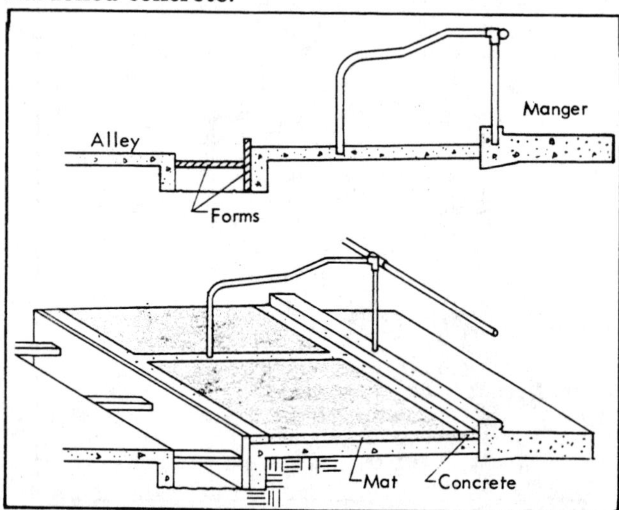

Fig 512-5. Stall mats.
Cast the alley, manger, and curb with stalls in place. Cast the cow platform concrete to a level even with **bottom** of the rubber mat. Wait one hour, place the mats, add concrete, and finish the border flush with the mat surface.

Mats do not keep cows clean. Use a small amount of finely-chopped bedding and install cow trainers to make cows deposit manure and urine in a gutter. Because mats insulate and destroy the electrical ground for trainers, use tie chains in tie stalls rather than leather straps. In stanchion stalls there is usually enough ground through the yoke.

Rubber mats reduce the wear on cow hooves so periodic trimming is required.

Stall size

Make stalls long enough so cows can lie down with their udders on the platform. Adequate stall width is also important for cow and milker comfort. See Table 512-6. Small stalls cause teat and udder injuries which can ruin a cow. Injuries have led to unjust criticism of stall housing.

Even with properly-sized stalls, some manure may drop on the platform. Adjust cow trainers to force cows to move backward and arch their backs for proper evacuation. Trainers that are located too far back on the front shoulders can cause problems in the reproductive tract because of incomplete evacuation.

Arrangement

Studies show that more of an operator's time is spent behind the cows than in front. Therefore, arrange stalls in two rows facing out for better labor efficiency. Only one litter alley is needed and a gutter cleaner can be economical. Walking during milking is reduced, and a pipeline milker can be more economically installed.

Feeding is easier from a common feed alley when cows face in, but milking is less convenient with two litter alleys. With this arrangement, a pipeline milker is more difficult to install and walls next to the litter alleys spatter with manure.

With a pipeline milker, slope the barn floor about 1″ in 10′ toward the milkhouse so that the milk line height above the floor is nearly constant. Locate the milkhouse at the end of the barn. A milkhouse at the middle of the barn makes it difficult to properly install a pipeline milking system.

Barn dimensions

Modern stall barns usually have an outside width of 36′. A 34′ wide barn is sometimes used with medium-size cows. Provide a 6′ wide litter alley. Use a flat manger feed alley for easier cleaning and equipment movement.

Manure gutters are usually 16″ or 18″ wide, Fig 512-7. Less urine splashes on the service alley with 18″ wide gutters. A depth of 16″ on the platform side and 14″ on the alley side is common. Some producers prefer the stall and service alley at the same elevation with a 12″ gutter. Steel grates over gutters help keep bedding out of liquid manure systems.

Slope stall platforms at least 1″ to the gutter. With minimum bedding, slope platforms 3″ for rapid draining of liquids.

Free Stall Barns

Free stall barns are practical and common for herds of about 60 or more and are usually used for

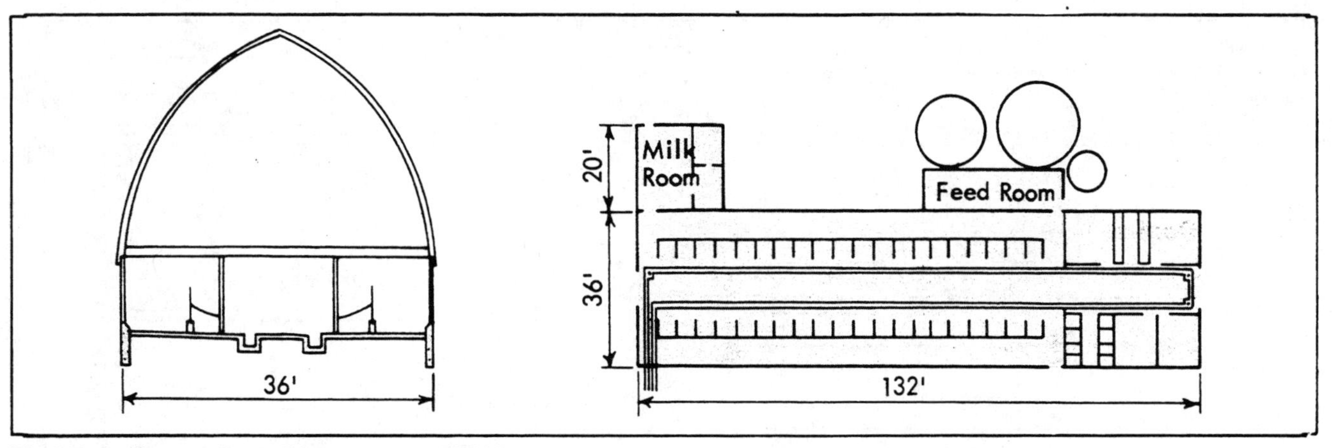

Fig 512-6. 40-cow tie stall barn, mwps-72326.
Two maternity pens and two calf pens are separated from the main barn by a partition and have their own heating and ventilation systems.

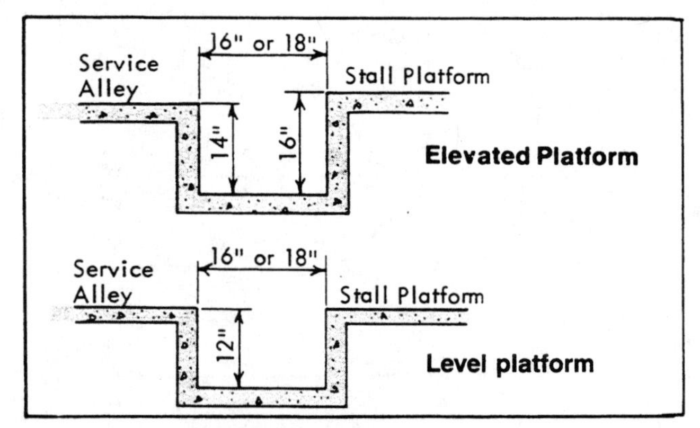

Fig 512-7. Typical barn cleaner gutters.
Gutter depths of 14" next to the service alley and 16" next to the stall platform are common.

herds of 100 or more. Free stalls were first used to replace deep manure packs in loose housing systems. The stalls reduce bedding needs and keep cows cleaner.

With properly designed free stalls, most manure is deposited in alleys. With solid floor alleys, clean alleys daily with a tractor-mounted blade, mechanical alley scraper, or flushing. Move manure to a cross conveyor, pump, storage area, or spreader. Mechanical scrapers and flushing can remove manure while cows are in the stall area.

Slotted alleys eliminate scraping. Manure passes through slots into a storage below. See Waste Management.

Free stall barns are either "warm" (well insulated and mechanically ventilated) or "cold" (little or no insulation and naturally ventilated). Feeding is usually in the barn or an adjacent feeding area.

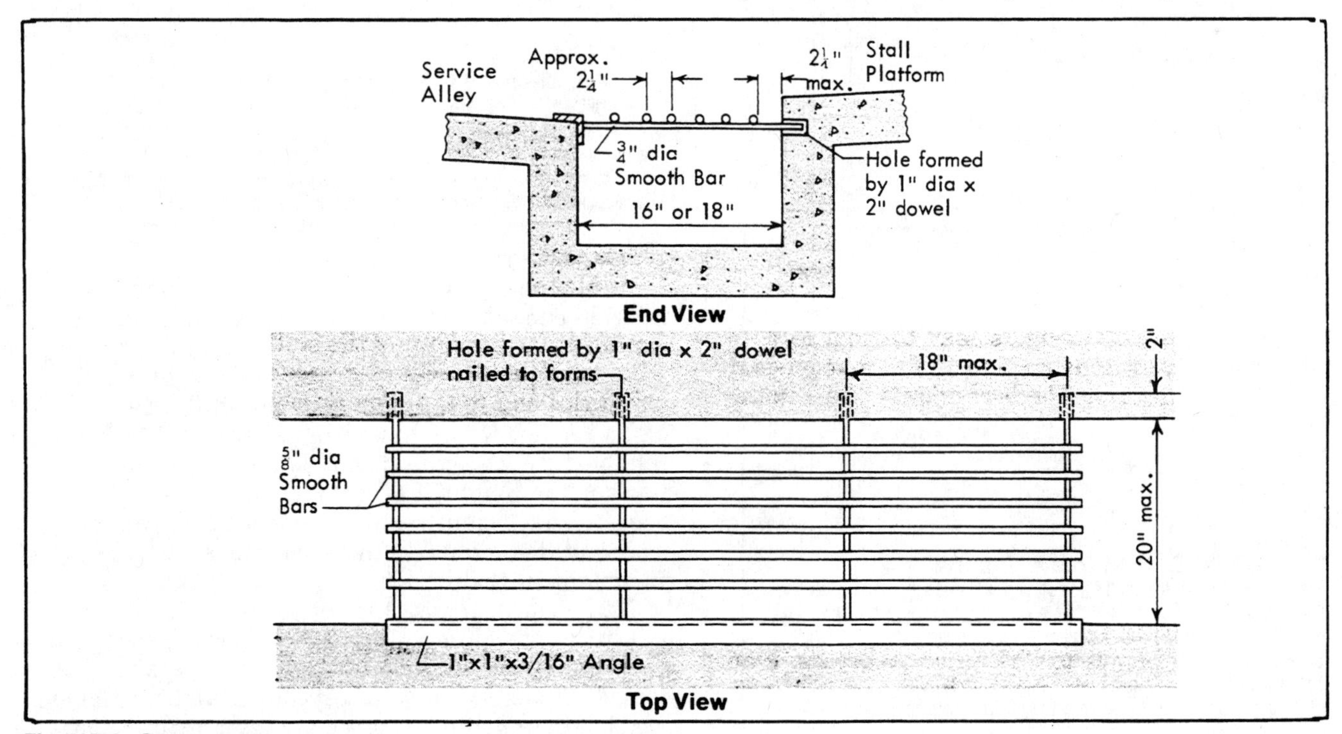

Fig 512-8. Gutter grates.
Use steel angles on both sides if cow platform is not higher than service alley.

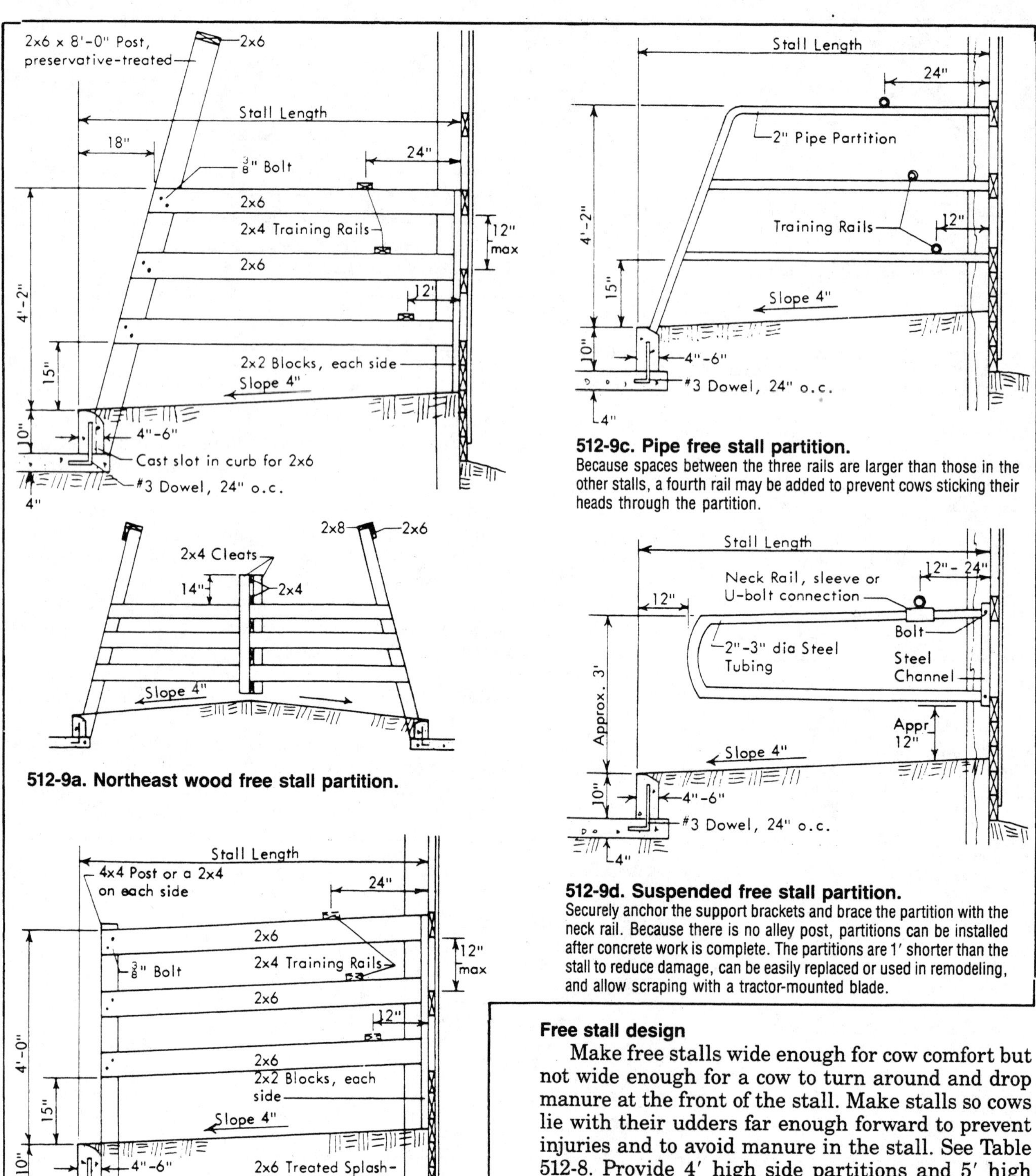

512-9c. Pipe free stall partition.
Because spaces between the three rails are larger than those in the other stalls, a fourth rail may be added to prevent cows sticking their heads through the partition.

512-9a. Northeast wood free stall partition.

512-9d. Suspended free stall partition.
Securely anchor the support brackets and brace the partition with the neck rail. Because there is no alley post, partitions can be installed after concrete work is complete. The partitions are 1′ shorter than the stall to reduce damage, can be easily replaced or used in remodeling, and allow scraping with a tractor-mounted blade.

512-9b. MWPS wood free stall partition.
Set rails parallel to the stall floor slope for constant partition height. Facing free stalls can have either a center supporting partition and sloped rails or horizontal rails and a suspended stall front like Fig 512-9a. Space wall girts 12″ or less to protect the siding from the cow's head. Anchor splashboards securely outside the building posts or install them inside to prevent the fill from pushing the wall out.

Fig 512-9. Typical free stall construction.
Install a 2x8 bottom partition board, except in Fig 512-9d, if cows work fill and bedding from stall to stall creating deep holes.

Free stall design

Make free stalls wide enough for cow comfort but not wide enough for a cow to turn around and drop manure at the front of the stall. Make stalls so cows lie with their udders far enough forward to prevent injuries and to avoid manure in the stall. See Table 512-8. Provide 4′ high side partitions and 5′ high front partitions so cows do not hang their heads over. Increase the height if necessary.

Make partitions open for good air movement. Space rails less than 6″ or more than 12″ so cows do not get caught in them.

Neck boards across the top stall rail force cows back when standing up so manure is dropped in the alley. Adjust the neck board forward or backward to keep stalls clean. Cows can get caught under a single neck board and injure themselves. Two or three horizontal rails at the front allow a cow to lie down comfortably, Fig 512-9, but she will not get caught under the top rail.

For a comfortable resting area, place bedding even with the curb at the rear and 2″-4″ above the curb at the front. Sawdust, chopped straw or corn stover, and sand are common bedding materials. Bedding type is usually determined by available materials and limitations of the manure handling system.

One stall is usually provided for each cow. For certain housing systems and management practices, more cows can be kept in the barn than there are free stalls. Cows do not always use the same stall or all lie down at the same time.

Detecting cows in heat is no more problem than in other housing types. Although heat periods are obvious, they often occur at night and outward signs are not detected. Contact between cows and producer is reduced with free stall barns.

Finish floors to reduce slipping and avoid injuries. See section on Concrete Pavements and Manure Tanks.

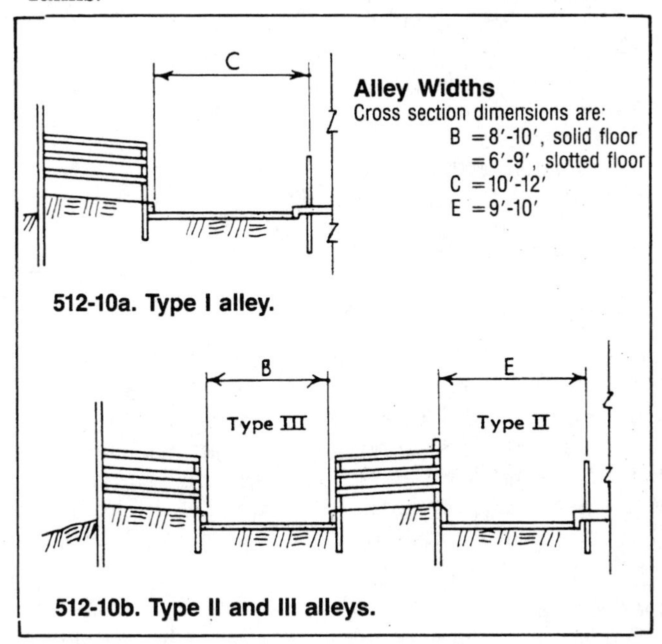

512-10a. Type I alley.

512-10b. Type II and III alleys.

Fig 512-10. Alley types.

Barn layout

Free stall barns commonly have 2, 3, or 4 alleys. Most designs use alleys for both feed and stall access. A type I alley is for both feed and stall access. A type II alley is between a bunk and the front of a stall row. A type III alley is between the backs of two stall rows. Avoid dead-end alleys. Barn width is directly related to alley widths, Fig 512-10. Make crossover alleys 4′ wide, or 8′ wide if cows are moved as a group. With mechanical scrapers, use narrower resting alleys. Slope solid floor alleys to storage areas or cross conveyors.

Free stall arrangements for at least three production groups are common. Separate high and low producers and dry cows and feed them accordingly.

In solid floor barns, use curbs to control manure collection. Make the curbs straight, and locate them to protect walls, free stalls, bunks, and waterers from damage during scraping. Elevate fences, waterers, and cow cross alleys from collection areas.

Barn plans

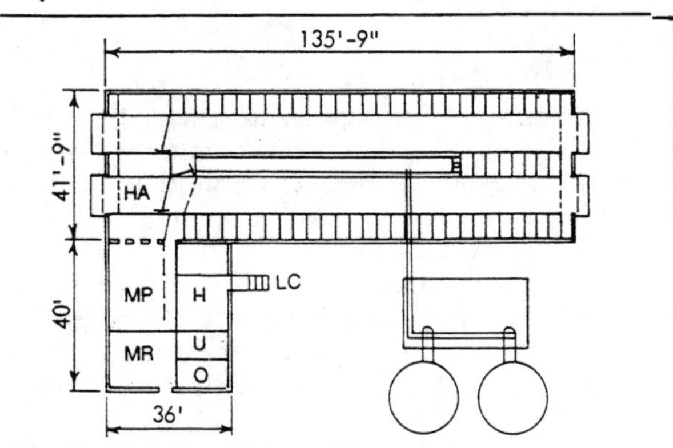

512-11a. 62 free stall barn (2 groups), mwps-72352.

Use a tractor scraper or automatic scraper to move manure to a storage area. Space is included for a double-4 herringbone parlor. If the feed center is located across the barn from the milking center, expand the barn to 124 stalls by building a similar free stall area to the left and a second return alley from the parlor.

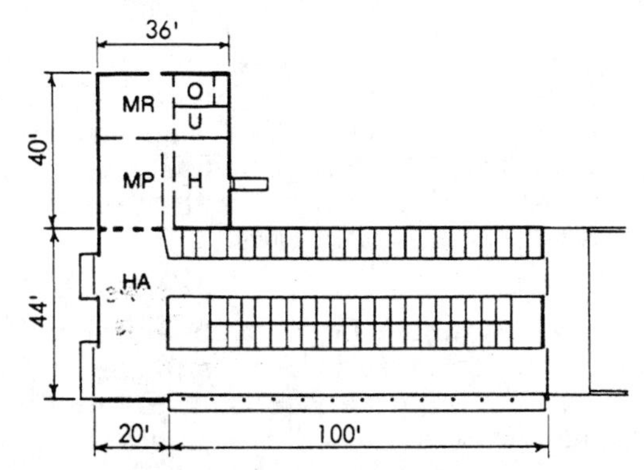

512-11b. 64 free stall barn.

The fenceline bunk is under the roof overhang for outside filling. Manure is scraped to one end of the building for removal. Space is included for a double-4 herringbone parlor. Expand the barn to 128 stalls by building a similar free stall area to the left and a second alley from the parlor.

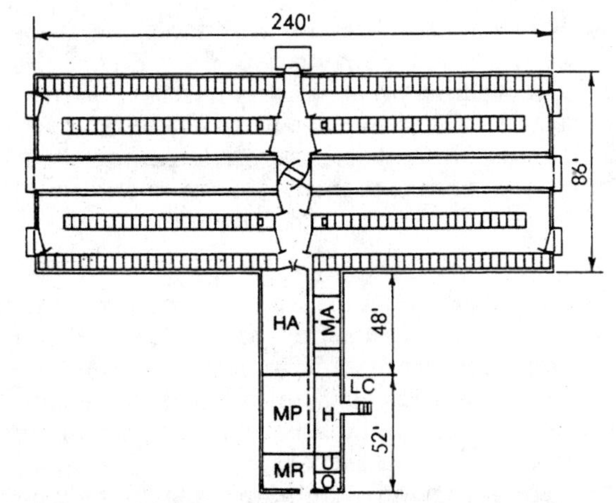

512-11c. 86′ barn with 206 free stalls (4 groups), mwps-72355.

Manure is scraped to the center of the building and then moved to a storage area opposite the milking center. Space is included for a double-8 herringbone parlor.

Fig 512-11. MWPS cold dairy barn plans.

These pole buildings are naturally ventilated. Design building layouts for adequate ventilation. The herringbone milking centers include a milking parlor, milkhouse, utility, shower, and office.

Milking Centers

A milking center for free stall housing includes milking parlor, holding area, utility room, milk room, lounge area, office, and treatment area. A stall barn milking center has a utility room, milk room, lounge area, and office. Cows are usually milked in the stalls with bucket units or an around-the-barn pipeline.

The U.S. Public Health Grade A Pasteurized Milk Ordinance requires approval of a plan for each farm before construction or remodeling begins. Contact your milk buyer about approval procedures before building.

Milking center dimensions depend on equipment selection, milking stall type, cow size, milk tank size and type, compressor location, and type of washing and milk handling equipment. Coordinate construction with your building contractor and equipment supplier.

Building

Interior walls

Interior walls should be a light color, washable, and kept clean. The bottom 4″ should be impervious to water. A bump railing 36″ above the floor helps protect painted surfaces from cows.

Lighting

Provide two continuous rows of 40 W cool white flourescent tubes the length of the milking parlor. Install the rows over the operator's pit 10″ from the platform edge. Avoid equipment that interferes with lighting.

Provide one fluorescent fixture with two 4′-40 W cool white fluorescent tubes for each 100 ft^2 of floor area in the milkhouse and an additional fixture over each wash vat. Do not locate fixtures so broken glass can fall into bulk tank openings.

With white ceiling and wall surfaces, no light reflectors are needed. Provide reflectors if surfaces are dark. Use weatherproof lighting fixtures fastened directly to the ceiling to prevent rust and dust accumulation. Install weatherproof electrical outlets 5′ above the floor to protect them from cows and wash water.

Drains

Use deep seal floor drains. Install sewers, drains, fittings, and fixtures according to the state plumbing code. Use only materials approved in that code.

For easy access and good sanitation, do not locate floor drains under the bulk tank or directly under its outlet valve. Use at least 4″ drain lines. Size waste disposal systems to handle the water used in cleaning and sanitizing operations. See Milking Center Waste Disposal. Cast iron sewer piping with approved seals is recommended. Pipe wash and rinse vat water directly into the disposal system.

In the milking parlor, provide a 96 in^2 (8″x12″) water seal drain behind each cow or a grated gutter and one drain per line of cows.

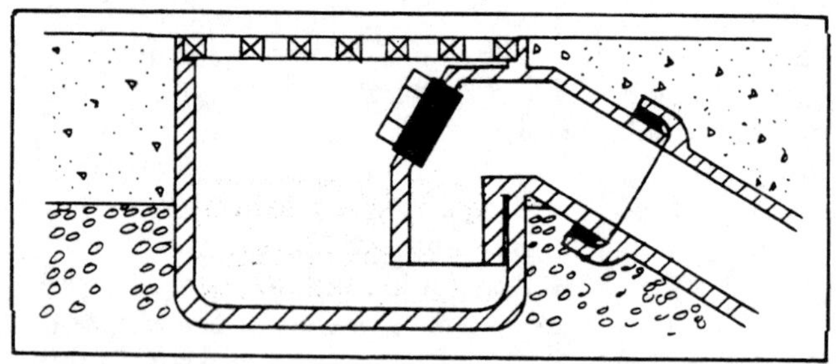

Fig 512-12. Drain.

Milking Parlors

Parlor type influences:
- Milking center size, layout, and location.
- Cow traffic patterns.
- Milking routine.
- Amount of mechanization.

Parlor size depends on:
- Initial mechanization.
- Plans for future improvements.
- Number of cows milked.
- Available labor.
- Milking time available.
- Milk production level.

Parlor selection also depends on:
- Herd expansion.
- Initial investments.
- Annual costs.
- Personal preference.

Flat barn

Flat barns were used primarily in the West for milking only. In the Midwest and Northeast, 6 to 10 stall flat barns have been used as an alternative to more expensive elevated parlors. A flat barn can also be an interim facility until an elevated parlor can be afforded. Put stalls in a room the size of the future elevated parlor.

Side-opening

Side-opening parlors, Fig 512-13, have been used since the 1920's. Layouts are usually double-2, -3, or -4 and for herds of less than 250 cows. The advantage of side-opening parlors is individual cow attention. Their main disadvantage is a greater distance between udders compared to the herringbone. This distance becomes important when the parlor is mechanized. With mechanization, more cows are needed in the parlor to keep the milker busy and effectively use the equipment. Pit length increases by 8′-10′ for each two milking stalls.

Herringbone

Herringbones are the most common elevated parlor. They range from double-4 to double-10, Fig 512-14. Larger herringbones do not increase cow throughput or milking quality.

The walking distance between udders varies from 36″-48″ depending upon the manufacturer. With a herringbone, the parlor is shorter and the milker can recognize and attend to problems faster, such as accidental machine drop-off.

Cows are handled in batches. A disadvantage of a larger herringbone (double-12 or larger) is that a slow-milking cow can hold up all the stalls on that side. Cow movement improves when cows are handled in groups compared with being handled individually in side-opening and rotary parlors.

Polygon

The polygon parlor, Fig 512-15, combines some of the advantages of the herringbone and side-opening parlors. Distance between udders is minimized, and with fewer cows per side, fewer cows are held up by a slow-milking cow.

Usually, polygons have 4 sides with 6 cows per side. The number of cows per side depends on the operator's routine, degree of mechanization, and management. Polygons with 4, 5, 6, 8, or 10 cows per side are in use.

Trigon

A trigon is a three-sided polygon, Fig 512-16. It was designed for the medium-sized dairy (250 to 500 cows). Trigons have 12, 16, 18, 22, and 24 stall capacities.

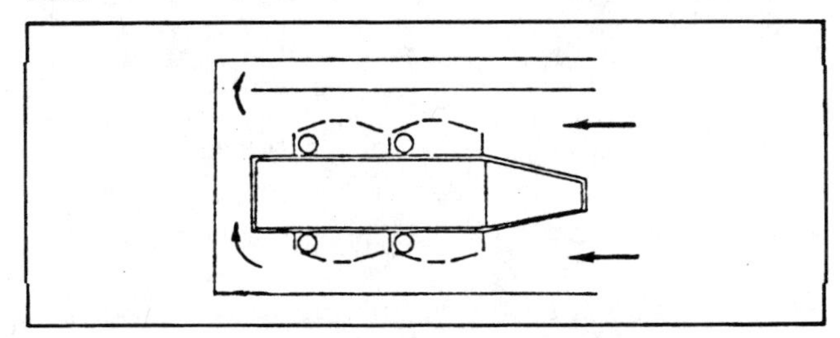

Fig 512-13. Double-2 side-opening parlor.

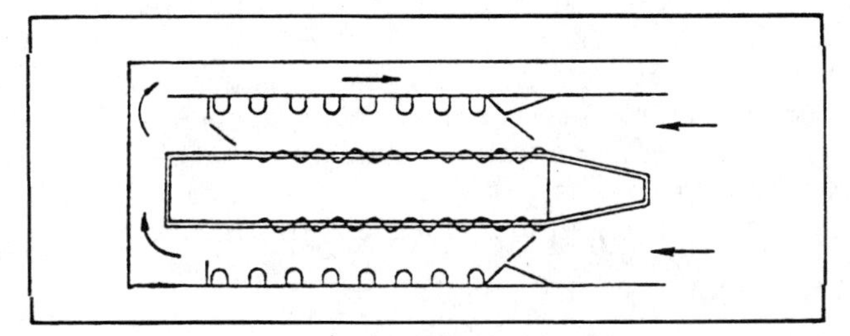

Fig 512-14. Double-8 herringbone parlor.

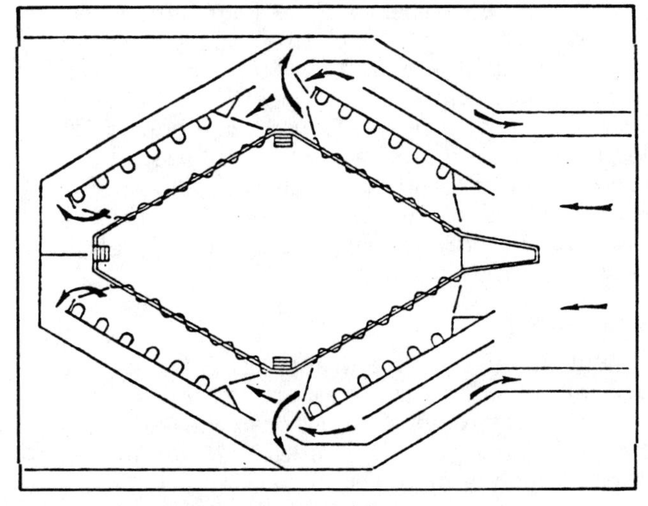

Fig 512-15. 24-stall polygon parlor.

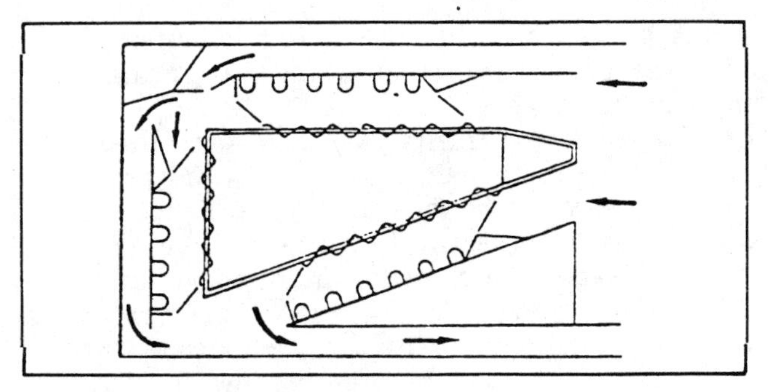

Fig 512-16. 16-stall trigon parlor.

Holding Area

Holding pens confine cows ready to enter the milking parlor. Cover the holding area to protect animals from sun, rain, snow, and wind. Provide a crowding gate to encourage cows to enter the parlor.

With confinement housing, the holding area is usually connected to the barn and all traffic areas are covered. Part of a resting or feeding alley can be used as a holding area in some arrangements, but use of a crowd gate, future expansion, and division of the herd into production groups is limited. Alley scraping can also be difficult.

If milking more than one group of cows, a holding area separate from the housing unit is preferred. Cow movement patterns, feeding coordination, manure removal, and milking are simplified.

With an enclosed holding area, replace the wall between the holding area and milking parlor with an overhead garage door. Movement into the parlor is improved when cows can see parlor activity while waiting in the holding area. If the parlor opens directly into a cold barn, do not remove the wall because the parlor can be very cold in the winter.

The holding area does not need to be insulated. Cows brought in for milking quickly warm the holding area so that opening the overhead door does not lower the parlor temperature except during very cold weather. Provide an open ridge and open eaves for winter moisture removal. Add sidewall openings for summer ventilation.

If the wall is removed and no door is used, insulate the holding area to conserve heat in the winter. Use heat rejected from the bulk tank refrigeration system to heat the parlor and holding area.

Provide 15 ft^2/cow in the holding area. Consider parlor throughput when planning the holding area. Hold cows for no more than 2 hrs at each milking (1 hr in hot climates or where cows are milked three times daily). For parlors that continue milking while changing groups, increase the holding area by 25% to allow for overlap.

Parlor Layout

A straight entrance from the holding pen into the milking parlor is preferred. Turns at the entrance can cause slower cow movement and more frequent operator interruption.

Avoid steps or ramps at the parlor entrance. If regulations require a step, keep step height to a minimum.

A straight exit from the parlor is also preferred. But if cows must be turned, make it at the exit rather than at the entrance.

Provide return lanes from the parlor to the housing unit. A single return lane is common—one group of cows crosses over the front of the parlor to exit. Return lanes outside the building can be wide enough to scrape with a tractor. Make inside lanes 33"-36" wide to prevent exiting cows from turning around. Lanes can be hosed down or hand-scraped. Allow for sorting and possibly treating cows from the return lane.

Milk Room

The milk room often contains the bulk milk tank, a milk receiver group, a filtration device, in-line cooling equipment, and a place to wash and store milking equipment.

Some milk cooling devices are:

- Conventional "compressor/condensor" bulk tank cooler. Size compressors according to milk flow per hour entering the tank.
- Combination of heat exchanger and bulk tank cooler. Plate coolers and tube coolers are installed between the receiver jar and the bulk tank. Well water runs countercurrent to milk flow through the heat exchanger. Milk temperature decrease depends on heat exchanger surface area, water temperature, and water and milk flowrates.
- Heat exchanger plus ice bank. Ice water is circulated through the heat exchanger in place of well water. Milk enters the bulk tank at approximately 36 F. The bulk tank maintains milk temperature and does not cool the milk.

Milk room size depends on bulk tank size. Plan for a larger tank and possible expansion.

Check local milk codes for required separation distances between the bulk tank and equipment or a wall. Recommended distances are 24" from the rear and end of the tank and 36" from the outlet valve and working ends. Gravity-type milk filtration systems or large bulk tanks may require ceiling heights of 10' or more.

Many larger bulk tanks are designed so that a major portion of the tank extends through one wall (bulkheaded) to the outside or to an adjacent utility room, Fig 512-17. This reduces milk room size and the cost of building materials.

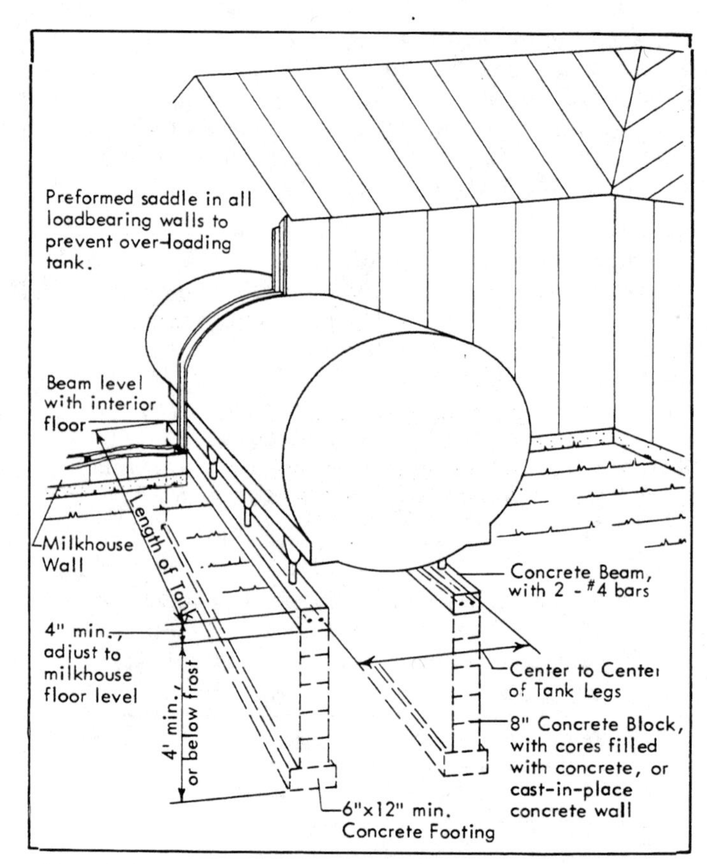

Fig 512-17. Masonry footing for bulkheaded milk tank.

Water Supply

Provide 105 F water in the milking parlor for washing and sanitizing udders. Use 165 F water in the milk room and for washing milk handling equipment. Two separate water heaters are recommended, but 105 F water can be obtained from 165 F water through a mixing faucet. Insulate all hot water pipes with preformed insulation.

Pipe hot and cold water to all wash and rinse vats in the milk room. Provide a mixing faucet with a hose and nozzle for cleaning the bulk tank, floor, and walls. Dispense acidified water and/or sanitizer through this hose and nozzle to improve sanitation and milk quality. CIP (clean-in-place) cycle rinse water can be retained in a storage tank and used to wash the floor.

For some installations, a hot water booster heater is needed. A calrod heater can be set in an open wash tank or installed in a stainless steel jacket as part of the hot water line. These heaters are controlled by the automatic washing equipment and a thermostat.

Utility Room

Provide space for milking and processing equipment and supply storage. Store cleaning compounds and other supplies on shelves off the floor. Store medical supplies according to label directions, preferably not in the milkhouse. Install good lighting, ventilation, and a floor drain.

Provide electrical service for vacuum pumps, milk cooler, water heater, ventilating fans, furnace, grain meters, and electric fence controllers. Evaluate your power needs to determine the entrance capacity. Locate the electric entrance panel on an interior wall to prevent condensation and rusting.

Lounge Area

Provide a toilet, sink, and showers in the lounge area. Consider lockers and a rest area for hired workers. For sanitary reasons, do not open the bathroom door into the milk room. See Milking Center Wastes for waste handling methods.

Office

An office is for maintaining herd health and production records and is sometimes the main farm office. Computer-based feeding and record systems require protection from dirt and moisture.

Handling and Treatment

Equipment to restrain cattle for examinations, treatment, hoof trimming, artificial insemination, etc., is important to a dairy operation. The milking parlor is often used for examination and treatment in free stall barns. This is inconvenient and cows may hesitate to enter the milking parlor because of previous treatment.

In stall barns many of these activities are conducted while the cow is in the stall. Usually there is no space to properly restrain the animal or perform surgical or treatment procedures.

There are four types of handling areas—separation, treatment, maternity, and loading.

Separation Area

Cows with problems such as sore feet or bad legs and joints need to be separated from the herd. Pen them near the milking parlor, the feed bunk, waterer, and manure alley. A dirt floor can help. Provide one 10'x14' or 12'x12' pen per 100 cows. Include one or more stanchions in the pen for restraints. Remove cows with long-term illnesses from the herd.

Treatment Area

Confine cows in heat and for artificial insemination, postpartum examination, pregnancy diagnosis, examination, and surgery. Provide 3'-5' in front of and behind a cow to perform a complete examination.

Locate the treatment area in or near the milking center. Provide hot and cold water, heat, and dry and refrigerated storage. Provide the same number of stalls as are on one side of the milking parlor—an 8-stall treatment area for a double-8 herringbone parlor. Use lever-operated stanchions so cows have little freedom and can be restrained simultaneously.

For large herds, consider a separate veterinary cattle handling facility.

Maternity Area

Freshening pens must be wide enough so the cow can turn around easily and long enough to use a calf puller when needed. Maternity pens are 12'x12' or 10'x14'. Place a stanchion diagonally across one corner of the pen. Provide a concrete curb between each stall to improve sanitation. For cows insecure on concrete, provide a pen with a dirt floor. In some cases, use deep bedding on concrete floors to prevent cows from slipping.

Locate pens for cleaning with a tractor or loader. Use movable gates or partitions that swing against the wall for cleaning.

Loading Chutes

Provide a solid side chute for receiving and shipping cows. Locate the chute so a pen can be used to hold cows prior to loading. Provide steps or cleats on the chute floor so the cows do not slip.

Dairy Dimensions

Information extracted with permission from *The 1982 Agricultural Engineers Yearbook,* The American Society of Agricultural Engineers, Box 229, St. Joseph, Michigan 49085.

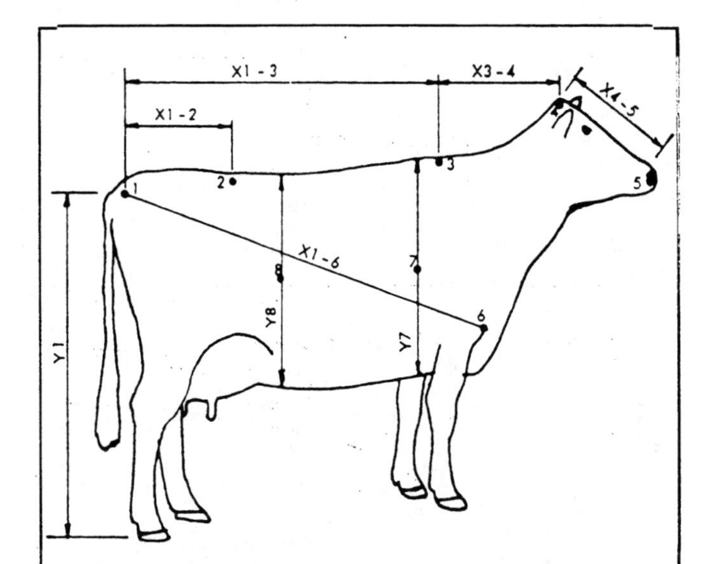

Reference Points (See Fig 423-22).

Height above floor (Y)

Y1 Height to pinbones (Y1), hips (Y2), and withers (Y3) are nearly the same for all breeds and ages. The pinbones are generally lower by up to 5 percent, especially in mature animals.

Lengths (X)

X1-6 Pinbone to front of shoulder (body)
X1-2 Pinbone to hip (rump)
X1-3 Pinbone to wither (back)
X3-4 Wither to pole (neck)
X4-5 Pole to muzzle (face)

Depths (Y)

Y7 Chest, at heart girth
Y8 Barrel

Depth at chest and barrel are nearly the same at birth. By six months of age the following approximations apply:

Y7 = Y8-3 cm for Holstein and Guernsey
Y7 = Y8-2.5 cm for Jerseys
Y7 = Y8-4.5 cm for Ayrshire

Circumferences (girths) (C)

C5 Muzzle
C7 Chest, at heart girth
C8 Barrel

Widths (Z)

Z1 Pinbones
Z2 Hips
Z4 Forehead
Z7 Chest, at heart girth

Fig 512-18. Dimensions of dairy cattle.

References:

Comparative Measurements of Holstein, Ayrshire, Guernsey, and Jersey Females from Birth to Seven Years, Nebraska Agricultural Experiment Station Research Bulletin 179, 1956.

Brody, S., Bioenergetics and Growth, Reinhold Publishing Corp., 1945.

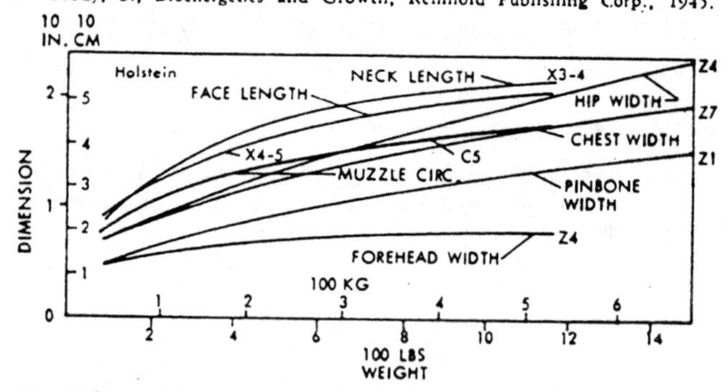

Fig 512-19. Weight vs. dimension.

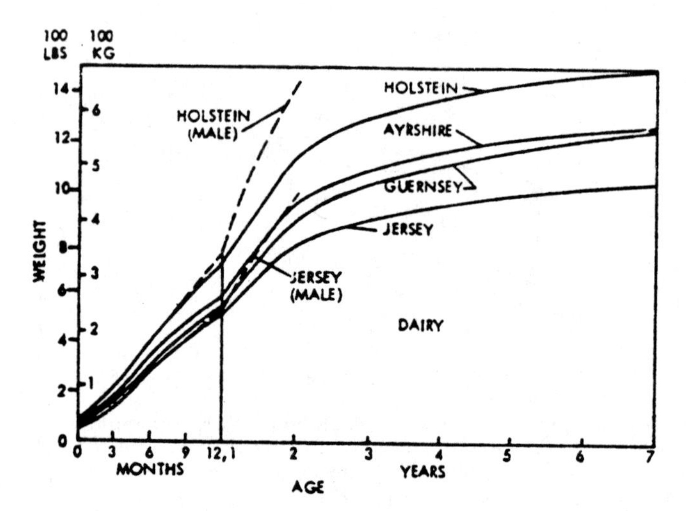

Fig 512-20. Weight vs. age of dairy cattle.

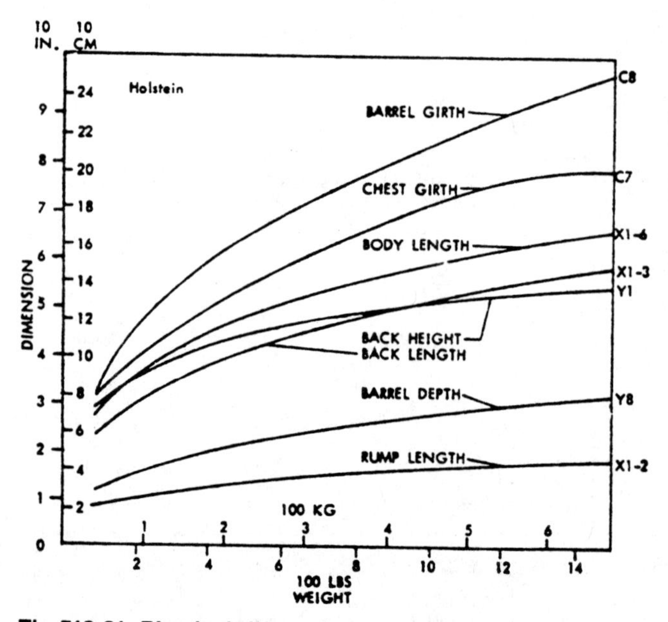

Fig 512-21. Physical dimensions vs. weight.

513 SHEEP

Rearing a higher percentage of lambs born helps justify investment in facilities. Lamb losses are mainly from chilling, starvation, and pneumonia. Facilities to reduce drafts and keep lambs dry, and pens for ewes with newborn lambs play a key role in reducing lamb death loss. Facilities are needed for pregnant ewes, ewes with newborn lambs, orphan lambs, mixing pens, and lactating ewes and lambs.

Saving labor is an important function of facilities. Facilities are needed for feeding, sorting, treating, lambing, shearing, and waste disposal. Carrying feed long distances, hand-feeding, or struggling through the flock to feeders wastes labor. Plan for efficient feed distribution. Relatively simple, inexpensive corrals, alleys, and chutes improve labor efficiency. For lambing, include lambing pens (jugs), mixing pens, creeps, and a drop area. Modify the environment to improve lamb performance.

Design facilities to be flexible to allow for expansion and improved production efficiency. Traditionally, sheep producers have lambed once each year. Some use the same building and equipment for up to three or four lambings annually, or lambing as well as lamb feeding.

Data Summary

		Rams 180-300 lb	Dry ewes 150-200 lb	Ewes with lambs	5-30 lb	Feeder lambs 30-110 lb
Building floor space (ft^2)/hd	Solid	20-30	12-16	15-20[1]	1.5-2 ft^2 of creep space per lamb	8-10
	Slotted	14-20	8-10	10-12[1]		4-5
Lot space (ft^2/hd)	Dirt	25-40	25-40	30-50		20-30
	Paved	16	16	20		10
Feeder[2] space (in./hd)	Limit-fed	12"	16"-20"	16"-20"	2"/lamb creep	9"-12"
	Self-fed	6"	4"-6"	6"-8"		1"-2"
Water (head/bowl or nipple)[3]		10	40-50	40-50	water available	50-75
(head/ft)	Tank	2	15-25	15-25		25-40
(gal/hd-day)[4]		2-3	2	3	0.1-0.3	1.5
Manure/day	(lb)	10	6	7		4
	(ft^3)	0.15	0.1	0.12		0.065
		plus bedding and spilled water				
Supplemental heat		—	—	100-200 Btu/100 lb plus 50-250 watt heat lamp		
Ventilation		Cold barns: provide ridge openings and adjustable wall openings. Warm barns: provide adjustable ceiling fresh air intakes and exhaust fans for 25 to 200 cfm/1000 lb.				
Wool produced (lb/yr)		6-18	5-14	—		4-7
Approximate feed needed[5] (lb/day per animal)	Hay	4-7	2.5-4	4-7 + grain		1-2 + grain
	Haylage	8-10	5-7	8-10 + grain		2-4 + grain
	Corn silage	12-20 + supp	7-9 + supp	12-18 + supp		4-6 + supp
	Grain	0.5-2.5	0.0-0.75	0.75-2.5		1.0-3.0
	Supplement	0.0-0.25	0.12-0.25	0.25-0.5		0.25-0.5

[1]For lambing rates above 170%, increase floor space 5 sq ft/hd.
[2]Feeder space/animal depends on: animal size, shorn vs unshorn, breed, pregnancy stage, number of times fed/day, and feed quality.
[3]Use heated or circulating type in cold buildings.
[4]Water requirements vary considerably with time of year and ration. Use clean water and keep waterer clean. Maintain water above 35 F in winter and below 75 F in summer.
[5]Approximate rations for 3 optional forages. Data are only for computing feed storage and handling needs.

Provide lambing pens (jugs) for about 10% of a 100 ewe flock; 7-9% of a 600 ewe flock; 4-6% of a 1000 ewe flock. 4'x4'x32" (min) or 5'x5'x36" (large ewes).

Group Sizes

Limit the number of sheep per pen:

- Young lambs in large groups can easily get separated from their mothers.
- Downed or sick sheep are not easily seen in large groups.

The information in Table 513-1 is given for planning building layout and pen arrangement. For large flocks, it may be more practical to use larger groups and suffer the losses associated with larger groups.

Table 513-1. Recommended group sizes for sheep.

Ewes with lambs before weaning	Birth to 1 day	2 to 4 days	5 to 7 days	8 to 14 days	14 days to weaning
	maximum group size				
singles	lambing jugs	10 ewes + lambs	20 ewes + lambs	40 ewes + lambs	50-100 ewes + lambs
twins	lambing jugs	5 ewes + lambs	10 ewes + lambs	20 ewes + lambs	50-100 ewes + lambs
Pregnant ewes	**200**				
Ewes about to lamb	**50**				
Early weaned lambs	**50**				

Sheep Production Systems

Many production systems can be described depending upon flock size, time of lambing, facilities, etc. The following four systems illustrate the types available and the facilities needed. Affect of flock size is indicated.

Winter Confinement Lambing

Ewes lamb in January and February. Lambs are creep-fed before weaning and self-fed until market. Ewes are housed until weaning and are kept on drylot, range, or pasture after weaning until the next winter lambing season. This system requires lambing and lamb feeding facilities, a corral, productive ewes, rapid growing lambs, top sanitation, good management, available labor, grain supply, and limited range or pasture. Slaughter lambs are marketed in late spring. The facility selected depends upon flock size and climate.

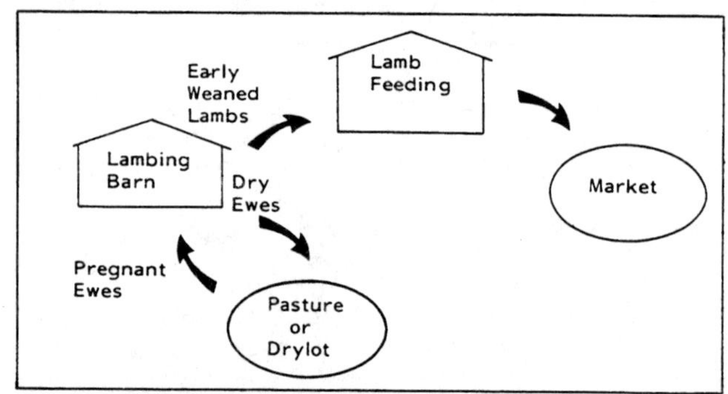

Fig 513-1. Winter lambing.

Early Spring Confinement Lambing

Lambing facilities are needed but probably not lamb feeding facilities. Ewes lamb in early spring and are grazed with their lambs until weaning. Lambs may be sold as slaughter lambs at weaning, or feeders may be grazed or fed in a drylot until marketed. Lambs are not creep fed. This system requires lambing facilities (less expensive than winter lambing), good management and labor, good sanitation, productive ewes, good pasture management, limited grain, a corral, and adequate range or pasture. The facility selected depends upon flock size, pasture available, and climate.

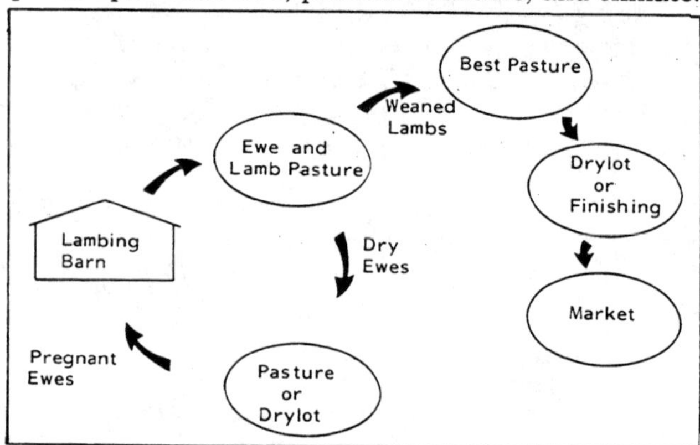

Fig 513-2. Early spring lambing.

Late Spring Pasture Lambing

This system may not require lambing facilities but may require lamb feeding facilities. Ewes lamb on range or pasture after cold wet spring weather has passed. Ewes and lambs graze together until weaning. Ewes continue to graze and lambs are grazed separately, fed in drylot, or sold as feeder lambs or slaughter lambs. Well-protected lambing pastures are necessary. Other requirements include ewes with good maternal traits, adequate range or pasture, maybe a feedlot, a corral, some grain, top sheep and pasture management, and good markets for both choice grade slaughter lambs and feeder lambs. Facility requirements are minimal.

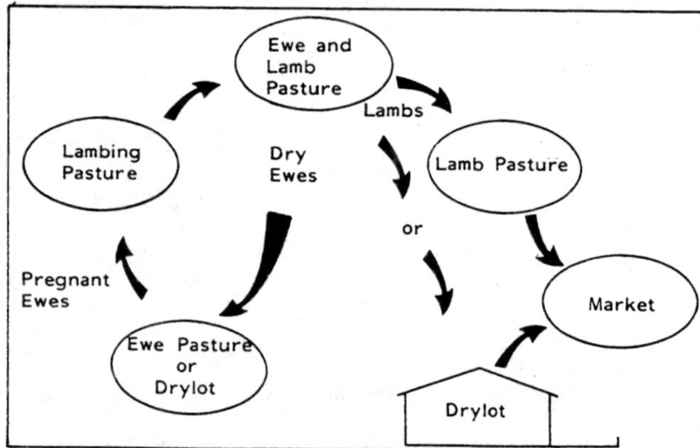

Fig 513-3. Late spring pasture lambing.

Phase and Accelerated Lambing

In **phase lambing** each ewe lambs only once annually, but the flock is separated into different breeding groups, lambing at different times. This system essentially combines two or all three of the first systems described. It allows flock expansion to maximize facility use, especially lambing and feeding facilities.

Lambing three times in two years is called **accelerated lambing,** Fig 513-4. Breeds with longer than normal breeding seasons are necessary. Lambing is followed by early weaning and immediate rebreeding. Lambs are usually fed in drylot while ewes are grazed, fed in drylot, or both. This intensive system requires lambing and lamb feeding facilities, a corral, drylot facilities, reproductively efficient ewes, skilled management and labor, sanitation, fast growing lambs, grain supply, forage supply, some pasture, and stable markets for slaughter lambs. Kind and size of facilities depends on flock size, frequency of lambing, available pasture, climate, feed supply, and feeding system selected. All ewes will not lamb on an accelerated schedule.

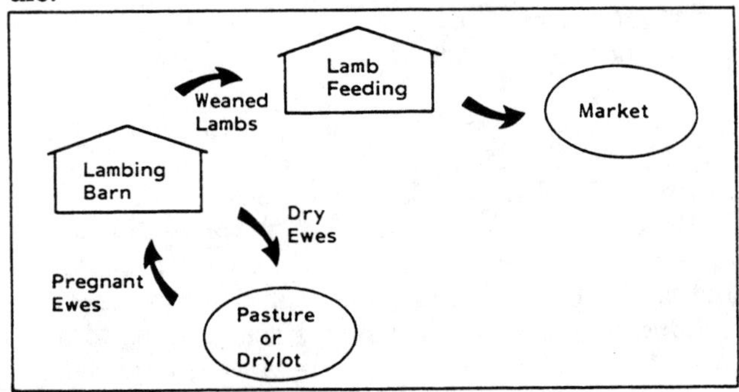

Fig 513-4. Accelerated lambing.

Barn and Lot Layouts

Three types of housing are common for sheep:

- Open-front barn with lot.
- Solid floor confinement barn.
- Slotted floor confinement barn.

Open-front Barn With Lot

Select open-front barns with clear-span construction to permit rearranging space without major building changes. Natural ventilation, or "cold housing," usually keeps inside temperatures a few degrees cooler than outside temperatures in hot weather and a few degrees warmer in cold weather. See Ventilation.

Slope the floor ¼"/ft toward the open front and the concrete apron ½"-1"/ft away from the building. A concrete floor is desirable in heavy traffic areas where mud causes problems, but is **not** desirable for inside pens; 2"-4" deep crushed limestone works well and is easily replenished. Sand, gravel or compacted clay are also used.

Place a barn with the open side away from prevailing winter winds.

Although dirt lots work best in areas with less than 30" annual rainfall, they are successful in all parts of the North Central Region with proper lot slopes, mounding and windbreaks. Sheep are relatively intolerant of mud and need year-round dry-surface access to shelter, feed, and water.

Figs 513-5 and 513-6 show open-front barns. Roof insulation reduces winter condensation and summer heat load.

Fig 513-5 shows grain self-feeders for feeder lambs in the building to protect both feed and lambs. To mechanize filling self-feeders, run an overhead conveyor the length of the barn with a feed drop at each feeder.

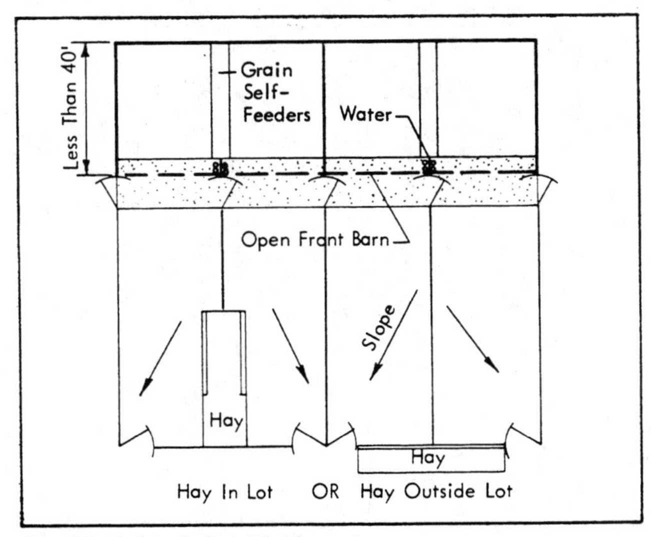

Fig 513-5. Lamb feeding layout.
Use 6 to 8 ft²/lamb barn space; 15 to 20 ft²/lamb lot space; and 1"-2"/lamb self-feeding space.

Fig 513-6 shows ewe feeding layouts. Fenceline bunks are shown in Fig 513-6a and 513-6b, and a mechanical bunk is shown in Fig 513-6c. Provide bunk space for all animals to eat at one time (16"-20"/ewe). Set the fenceline bunk out slightly from the barn for feed-wagon access. Leave space at the low end of the lot for surface drainage. A central drive through the building allows a shorter lot, but prevents blocking the ewes away from the feeding area inside the barn, Fig 513-6b.

Adequate bunk space to feed a large flock mechanically and all at one time is expensive. Fig 513-6c shows an arrangement for mechanically feeding one pen at a time. Feed one pen of ewes at a time in the feeding area. This method uses the same equipment to feed several groups, called **group-feeding.** Moving ewes requires more labor than fenceline feeding.

Pavement

Pave heavy traffic areas, around waterers, and along feedbunks. Additional pavement pays for itself if sheep are otherwise in mud for extended periods. Concrete is usually the most satisfactory paving material.

Paved 4' wide sidewalks provide good access routes for sheep between bunks, waterers, shelter, and mounds.

Slope aprons away from bunks and waterers—1"/ft slope is nearly self-cleaning.

To reduce mud in unpaved areas, use stone, gravel, or disk in barn lime at 1 lb/ft², or build mounds.

Topography and Drainage

Slope open lots a uniform 4-6% away from prevailing winter winds, usually south or east. Unpaved lots with slopes over 10% erode. Direction of lot slope can be changed, but may be expensive. Consider moving from present lot site to obtain desirable slope. Complete land grading before starting construction.

Use diversion ditches or terraces as drainageways to intercept surface water from above the lot. Slope lots away from barns and feedbunks. Avoid drainage from one lot to another. Install eave troughs and downspouts on barns and direct the water away from the lot.

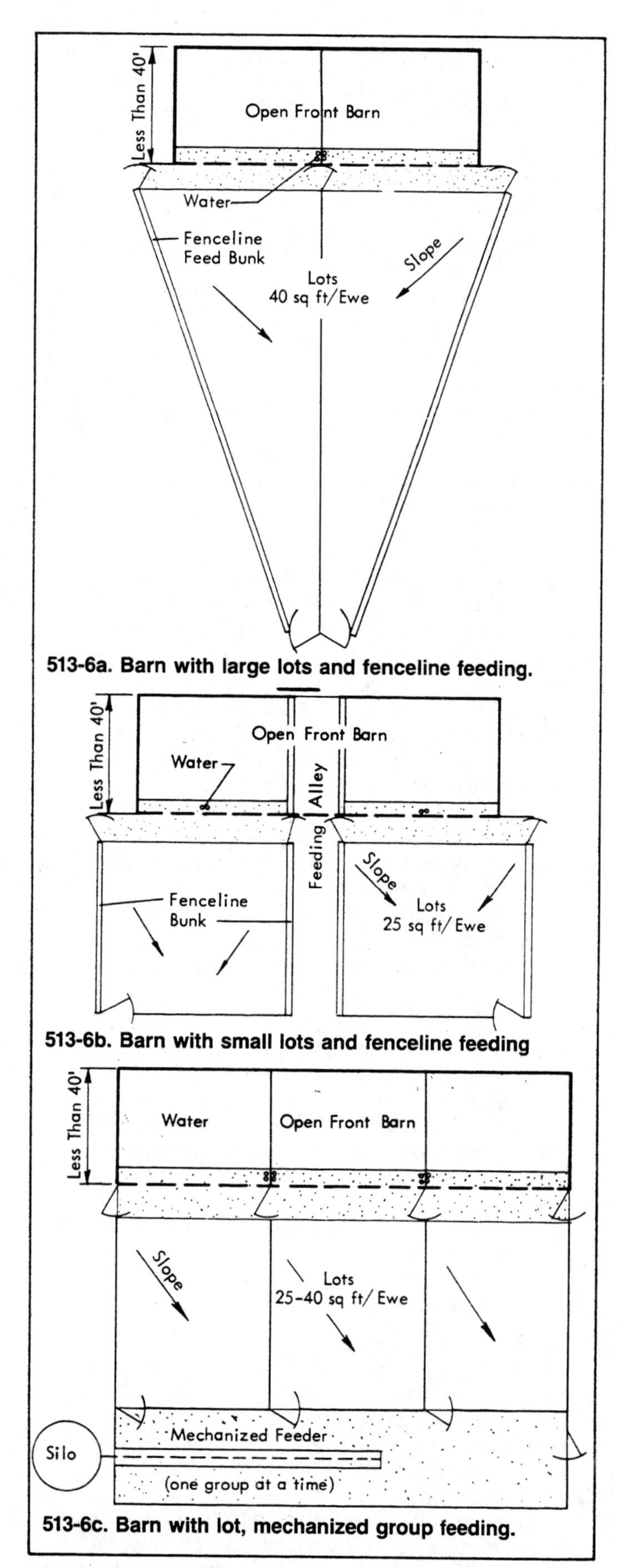

513-6a. Barn with large lots and fenceline feeding.

513-6b. Barn with small lots and fenceline feeding

513-6c. Barn with lot, mechanized group feeding.

Fig 513-6. Ewe feeding layouts.
Use 10 to 12 ft²/ewe barn space; 16"-20"/ewe feeding space; and 40 ft²/ewe lot space with 25" or more annual rain or 25 ft²/ewe with good lot drainage and less than 25" annual rain.

Orient feedbunks to provide best lot drainage, either at the high side of the lot or up and down the slope. When compatible with drainage, a N-S or NW-SE orientation is usually preferred, because the sun melts ice and dries along both sides of the bunk.

Mounds

Consider mounds in dirt lots with limited drainage and where mud is a problem. General design recommendations include:

- Mound area/animal: about 25 ft²/head; if there is a windbreak along the top of the mound, provide 25 ft² on each side of the mound so sheep can move with the sun or wind.
- Mound height: 4'-6'.
- Top width: 6' rounded.
- Top length: at least 50'.
- Side slopes: 5:1 or 6:1. (One foot of rise for each five or six foot horizontal.)
- Longitudinal slope: 6% max (16:1).

Build mounds or ridges parallel to the prevailing lot slope so runoff water is not ponded in the lot. Begin mounds at feedbunks and slope them away. A rounded-top mound avoids holes in the surface leading to a complete breakdown. Yearly maintenance is required.

Stabilize the upper half of the mound surface with chopped bedding or soil manure mixture, or by disking in lime. Barn lime is preferred over agricultural limestone because it reacts much more quickly. Start by adding one lb/ft²; if mound surface isn't stable enough, add more lime until satisfied.

A windbreak fence on the ridge of the mound is recommended in the northern part of the North Central Region. The fence permits sheep to move from one side to the other depending on weather conditions. Build the fence 8' high and 80% solid (e.g. 6" boards spaced 1½" apart). Spaces can be either vertical or horizontal.

Windbreaks

Windbreak fences and tree shelterbelts change wind direction but do not **stop** the wind. Consider local experience to determine the distance facilities should be placed from windbreaks. The windward side of a tree shelterbelt is usually 100'-300' away from protected areas. The shorter distance is suitable where snow drifting is less severe. Remember that shelterbelts also reduce air movement around barns and feedyards in summer.

Solid Floor Confinement

A confinement building offers several advantages over outside lot systems: feed in open lots gets wet or blows away; snow and ice collect in feedbunks; grain, hay and silage freeze; outside space for lot arrangements may be inconvenient and expensive; soil type, slopes, and climate or pollution from lot runoff may prohibit satisfactory lot operation.

Fig 513-7a shows open-front confinement with four self-feeders for each pen of 50 ewes. A mechanical feedbunk through the barn can replace the conveyor and self-feeders if silage is fed.

Fig 513-7b shows open-front confinement with an inside feed alley the length of the barn. Silage is distributed with an unloading wagon. With mechanical feeding, this floor arrangement allows temporary lambing pens in the center alley.

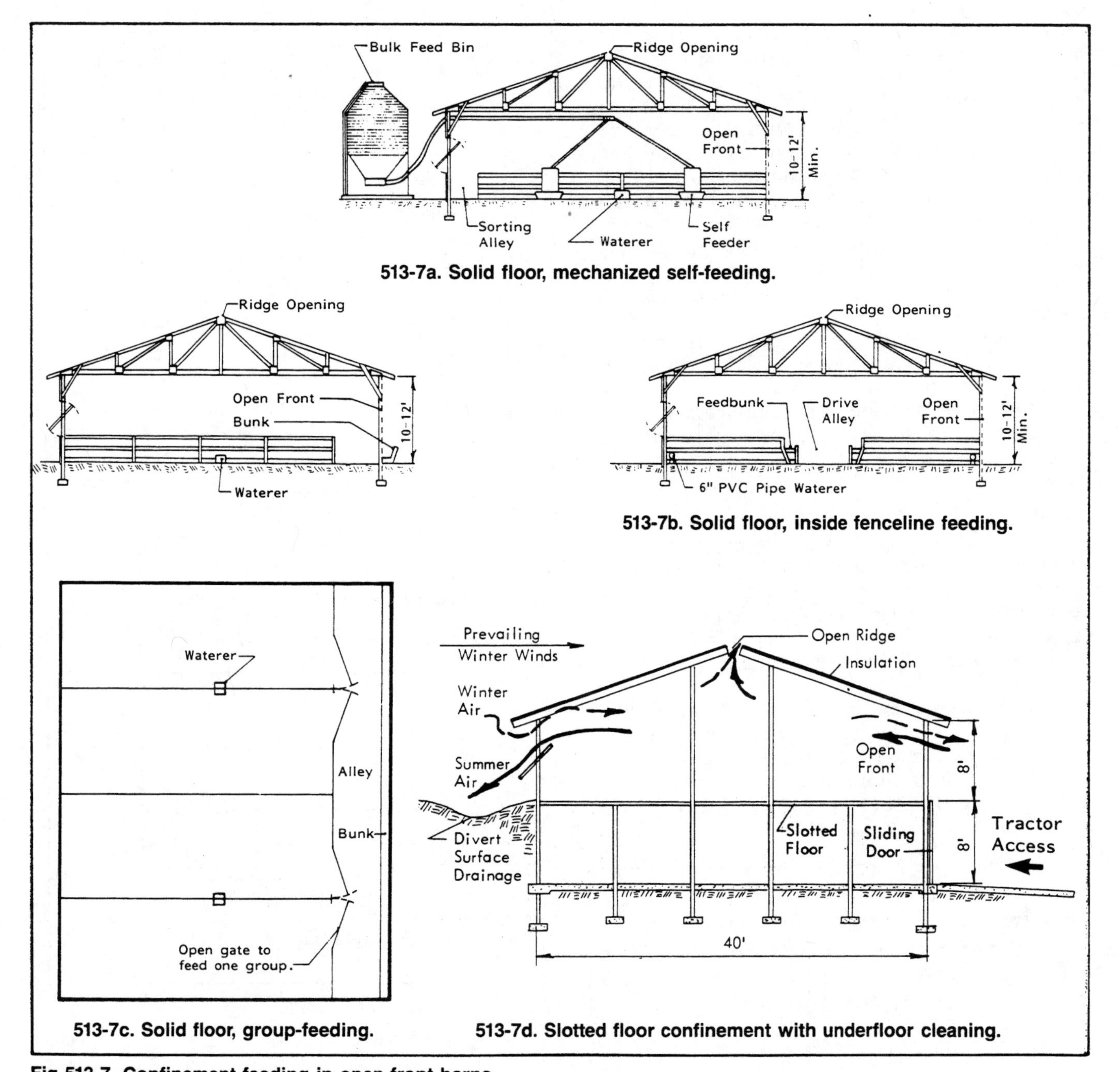

513-7a. Solid floor, mechanized self-feeding.

513-7b. Solid floor, inside fenceline feeding.

513-7c. Solid floor, group-feeding.

513-7d. Slotted floor confinement with underfloor cleaning.

Fig 513-7. Confinement feeding in open-front barns.
Slotted floor space: 8 to 10 ft^2/ewe, 10 to 12 ft^2/ewe and lamb, 4 to 5 ft^2/feeder lamb. Solid floor space: 12 to 16 ft^2/ewe, 15 to 20 ft^2/ewe and lamb, 8 to 10 ft^2/ feeder lamb. Group-feeding space: 16"-20"/ewe, 9"-12"/feeder lamb.

It is difficult to provide adequate bunk space for group-feeding in confinement. Consider self-feeding or separate feed bunks from pens with sheep fed in groups. Additional labor is required to open gates and move animals at each feeding. The maximum number of pens that can be managed depends on the time an operator is available to move animals. With each group at the bunk for about 4 hours, three to four groups can be fed in an 8-12 hour period with the last group remaining at the bunk over night. Fig 513-7c.

Slotted Floor Confinement

Slotted floor confinement provides clean, dry floors with minimum cleaning labor. About twice the animals can be housed on slotted floors as an equivalent size solid floor, so the added investment is spread over more animals. Drafts through the floor can be a problem, especially for small animals. Unflattened or rough expanded metal works well for sheep. See Manure Management.

Fig 513-7d shows an open-front slotted floor sheep barn with a drive-in manure pit. Locate the barn on a south or east slope; the pit can be entered from the ground level for cleaning, but the high wall is protected from winter winds. During hot weather, open manure doors for additional ventilation.

Slotted Floors

Slotted floors significantly reduce daily labor requirements for manure handling. Many types of slotted floor materials have been used: green, undressed, native hardwood (oak); concrete; steel grids (used in some slotted floor hog units); flattened, expanded metal X-plate; unflattened, expanded metal; metal and plastic slats; and kiln-dried pine. Floors and fleeces stay cleaner on more open floors. Sheep are more sure-footed and hooves wear more evenly on expanded metal than on other materials. Hardwood floors have irregular openings between slats that can trap lambs' feet. Pine floors soon show heavy wear.

Space 3"-4" wide concrete slats about ½" apart. A smooth slightly crowned slat is more self-cleaning and causes less wool staining, but makes animals less sure-footed.

Run slats in the short pen direction to reduce foot damage (makes no difference with ¾" expanded metal) and parallel to feeders.

X-plate with 4½" long openings can damage young lambs' feet, so use it only for larger animals.

Floor Plans

Most sheep barns serve different functions at different times of the year. As a result, pen partitions, feeders, creeps, waterers, etc. should be movable. Figs 513-8 to 513-14 are layouts for producers with ewe-breeding flocks. Provisions are included for ewes during gestation, lambing, and nursing. Figs 513-15 to 513-18 show layouts for lambing, or lamb or ewe feeding.

Note: Sizes of rooms and pens are shown within the recommended ranges and are approximate. Where room or pen sizes are labeled, dimensions are in feet with the horizontal dimension shown first. For example, in Fig 513-8 the work room is "8x16"—it is 8' left-to-right and 16' top-to-bottom.

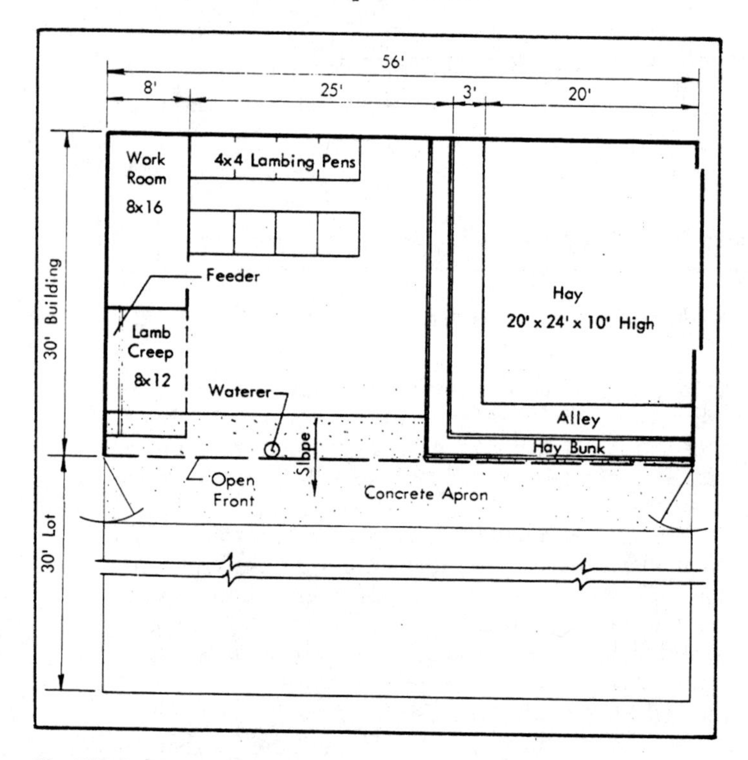

Fig 513-8. 50 ewes, inside feeding and hay storage.

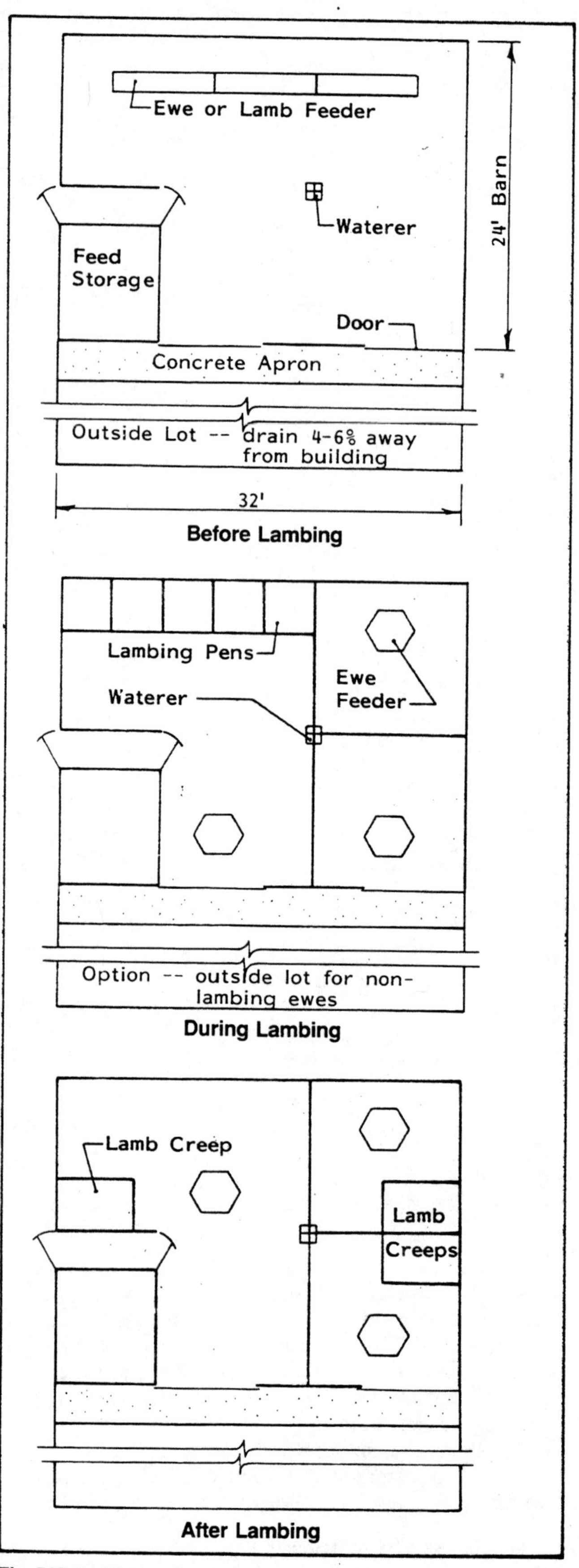

Fig 513-9. 40-ewe lambing shed, mwps-72509.
Movable pens, creeps, and feeders permit flexible space use. Separate hay storage required.

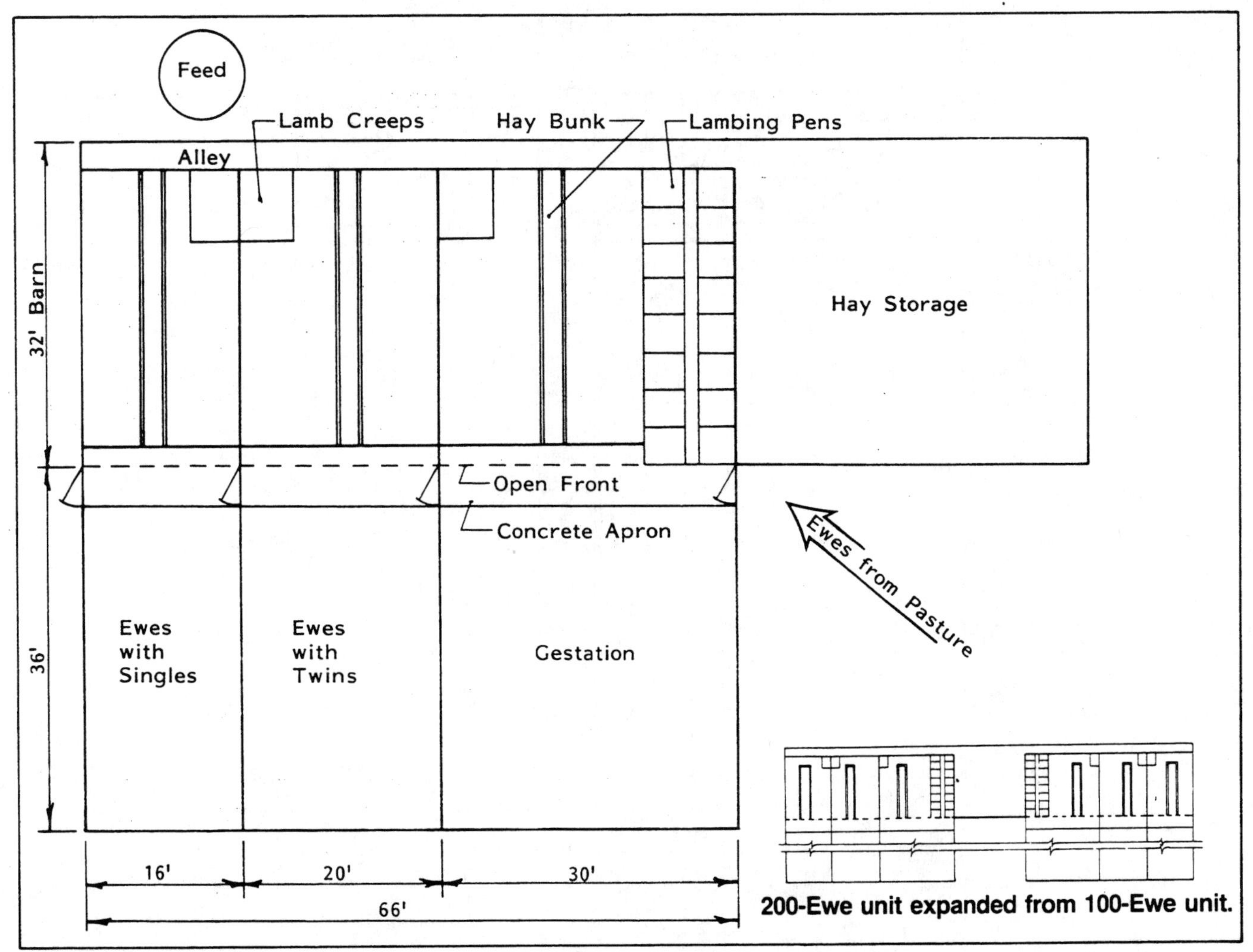

Fig 513-10. 100 ewes, inside feeding, separation of twins and singles.
200-ewe unit expanded from 100-ewe unit.

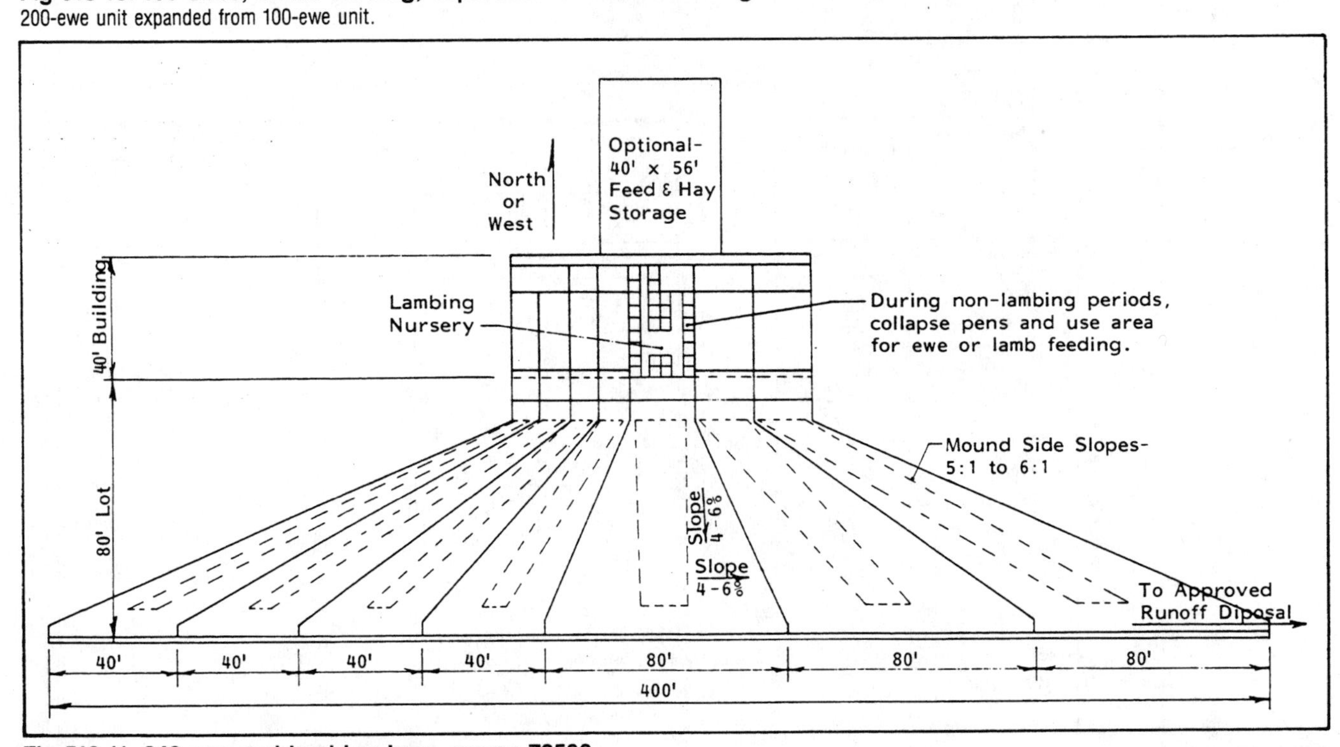

Fig 513-11. 240-ewe and lambing barn, mwps-72506.
Outside feeding, open-front, feed and hay storage in adjacent barn.

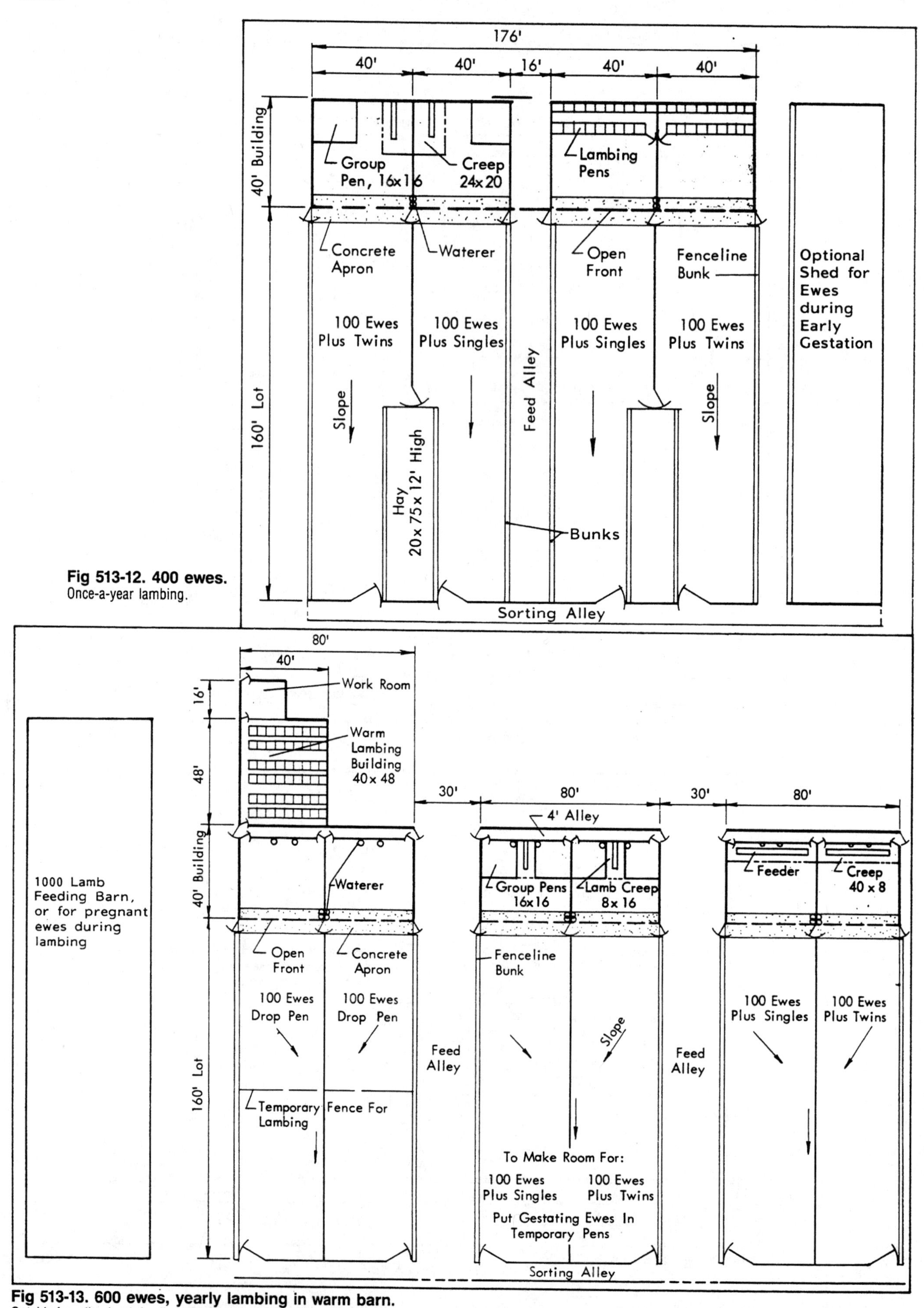

Fig 513-12. 400 ewes.
Once-a-year lambing.

Fig 513-13. 600 ewes, yearly lambing in warm barn.
Outside fenceline bunk feeding. Breed ewes in groups of 100 and mark when bred. Less lambing area is required with longer lambing period. Zero pasture assumed. Additional lamb feeding or ewe maintenance facility required. Group pens and creep feeders installed as needed.

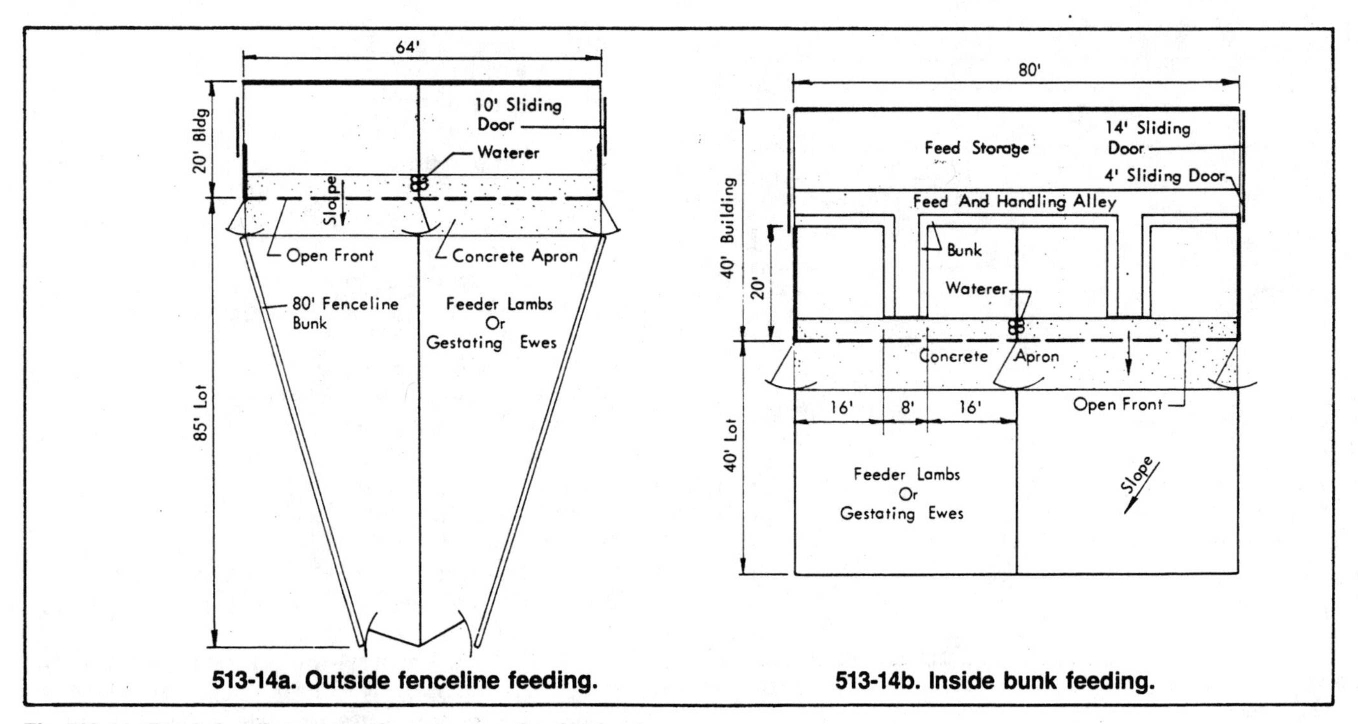

513-14a. Outside fenceline feeding. 513-14b. Inside bunk feeding.

Fig 513-14. Bunk feeding, gestating ewes or feeder lambs.
Each pen holds 55 to 65 ewes or 80 to 100 feeder lambs. Separate facilities are needed for lambing.

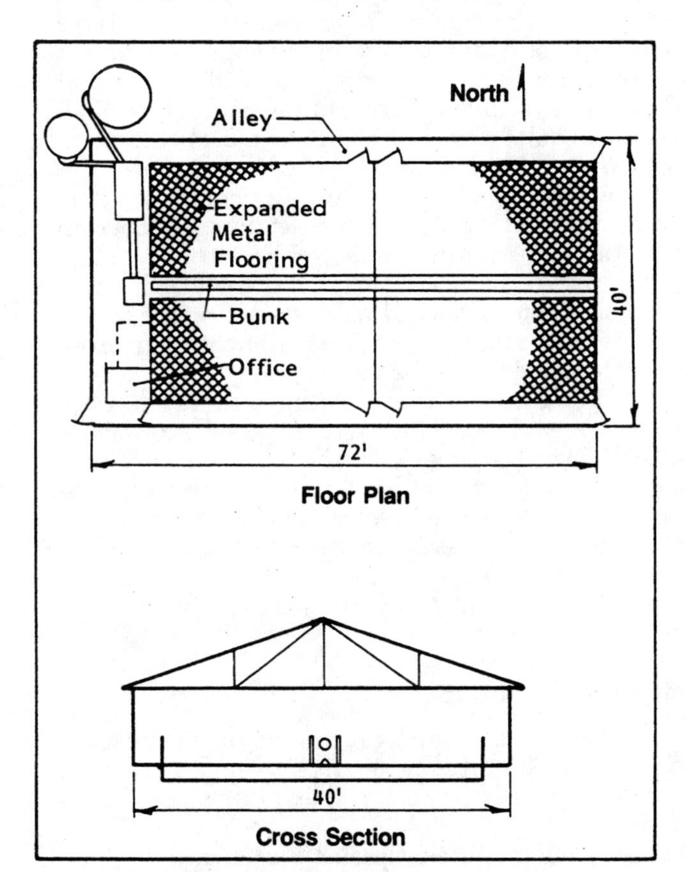

Fig 513-15. Slotted floor confinement barn, mwps-72505.
Removable panels provide slotted floor over shallow pits cleaned with mechanical scrapers. Alternate is tractor-scraped with access from one end. Capacity: up to 200 ewes at 8 to 10 ft²/ewe ; up to 400 lambs at 4-5 ft²/lamb.

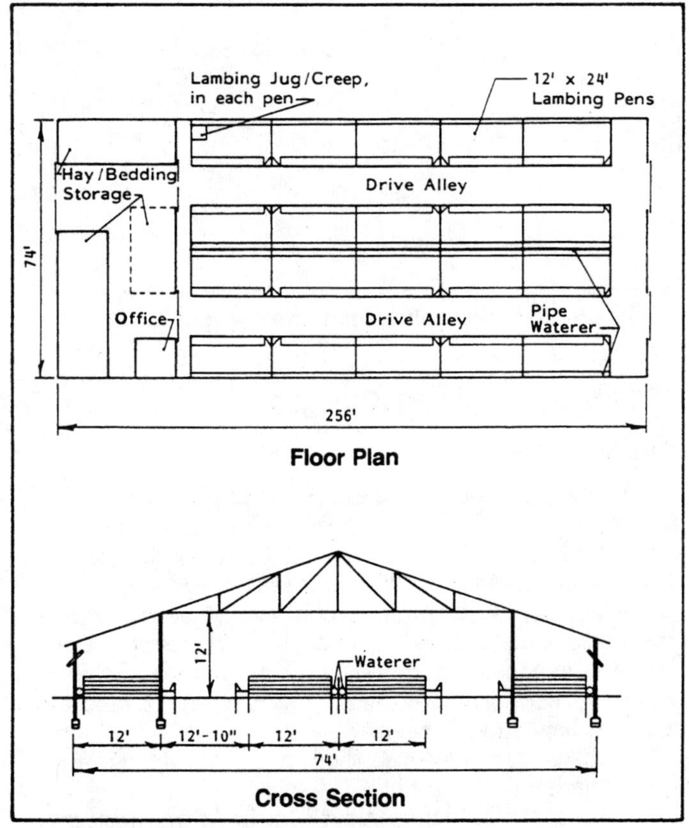

Fig 513-16. 500 ewes or 1000 ewes on accelerated lambing, mwps-72507.
Inside lambing and bunk feeding. Separate ewe or lamb feeding required.

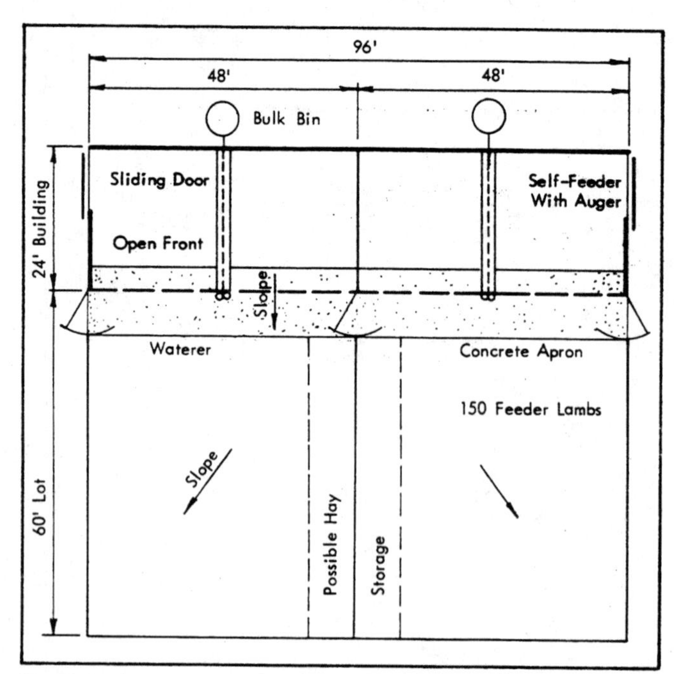

Fig 513-17. 300 feeder lambs.
Inside self-feeding.

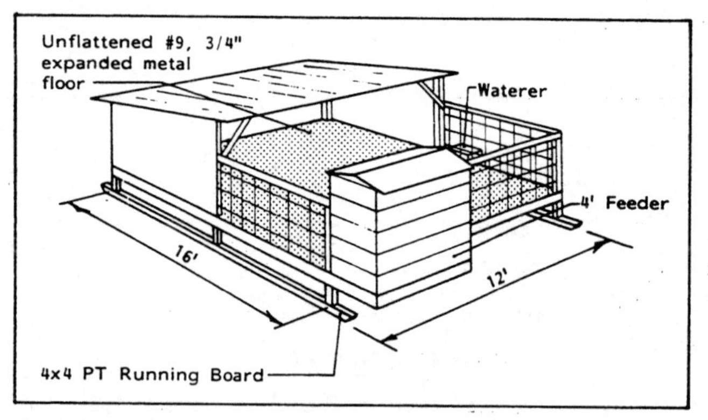

Fig 513-18. Portable lamb feeding unit, mwps-72508.

Treating and Handling

Planning

First determine the number of animals to be handled at one time. For sheep confined to a relatively small area or barn, consider **one** site for handling facilities. For flocks spread over a large area, consider two or more facilities, or portable equipment. Consider future increases in flock size—build larger facilities or leave space for expansion. Some facilities may also be used for cattle with proper planning. Take advantage of the natural behavior of sheep:

- Sheep move toward other sheep and follow one another.
- Sheep prefer to move up a slope and toward open spaces.
- Sheep tend to move through corners where they cannot see what lies ahead.

Corrals

Corrals are penned areas for treating, handling, and sorting animals. Locate a corral:

- near the sheep and barns.
- near a serviceable road.
- on a well-drained site.
- where electricity and water are available.

Plan for proper disposal of spray chemicals. Trees near a corral can provide shade during summer and wind protection during winter.

A corral includes:

Gathering area. Two pens are common—a reception pen and a crowding pen. The reception pen is for sheep from the field or barn. Size is based on the number of animals to be handled at one time. Typical pen sizes are 4-5 ft^2/sheep and 7 ft^2/ewe and lamb.

Provide a large entrance gate that can be latched open for clear access and visibility to sheep entering the pen. Use rectangular (not square) pens with wide gates to the working area, creating a funnel effect.

Force sheep from the reception pen to the working area through small pens or a circular crowding pen. Circular crowding pens have two center-hung gates that swing 360 degrees, Fig 513-19.

Working area. The working area is used for sorting, tagging, foot bath, worming, vaccinating, etc.

In many cases one chute (or race) serves all functions. Sheep enter the crowding pen in groups of 6-8. Gates are closed at each end and the operator treats sheep from the side of the chute. Place a portable wood or fiberglass foot bath in the chute. Important features are shown in Fig 513-20. Expanded metal flooring can improve footing in the chute. Install an enclosed shelf for holding equipment, medicines, etc., near the working chute for easy access.

Some operators like wider working chutes with separate footbaths, Fig 513-21, which let the operator get in with the sheep. Normal chute width is about 3½'. The pen is filled to capacity and the shephard works through and keeps treated and untreated animals separate. With two chutes, a continuous flow of animals can be maintained to the sorting chute.

Holding area. Sheep from the working chute are transferred or sorted into holding pens. Allow about 4-5 ft^2/sheep. Larger areas make handling more difficult. Arrange for cycling sheep through pens and back to the gathering area or working chute for further treatment. A perimeter passage allows groups of sheep to be passed, Fig 513-24. Also, individual animals can be drawn from various groups into the passage.

Related Facilities

Weighing. A useful location for a weigh scale is in the sorting chute before the sorting gate. Consider a removable panel in the side of the chute for access to the scale.

Loading facilities. A portable loading chute is useful for multi-pasture set ups. Consider a permanent loading ramp for large flocks and to reduce animal injury. Provide all-weather access for vehicles and lighting.

Shearing. Good facilities speed animal movement. Handling pens next to a shearing barn make sorting easier and quicker and hold sheared sheep. Small catch

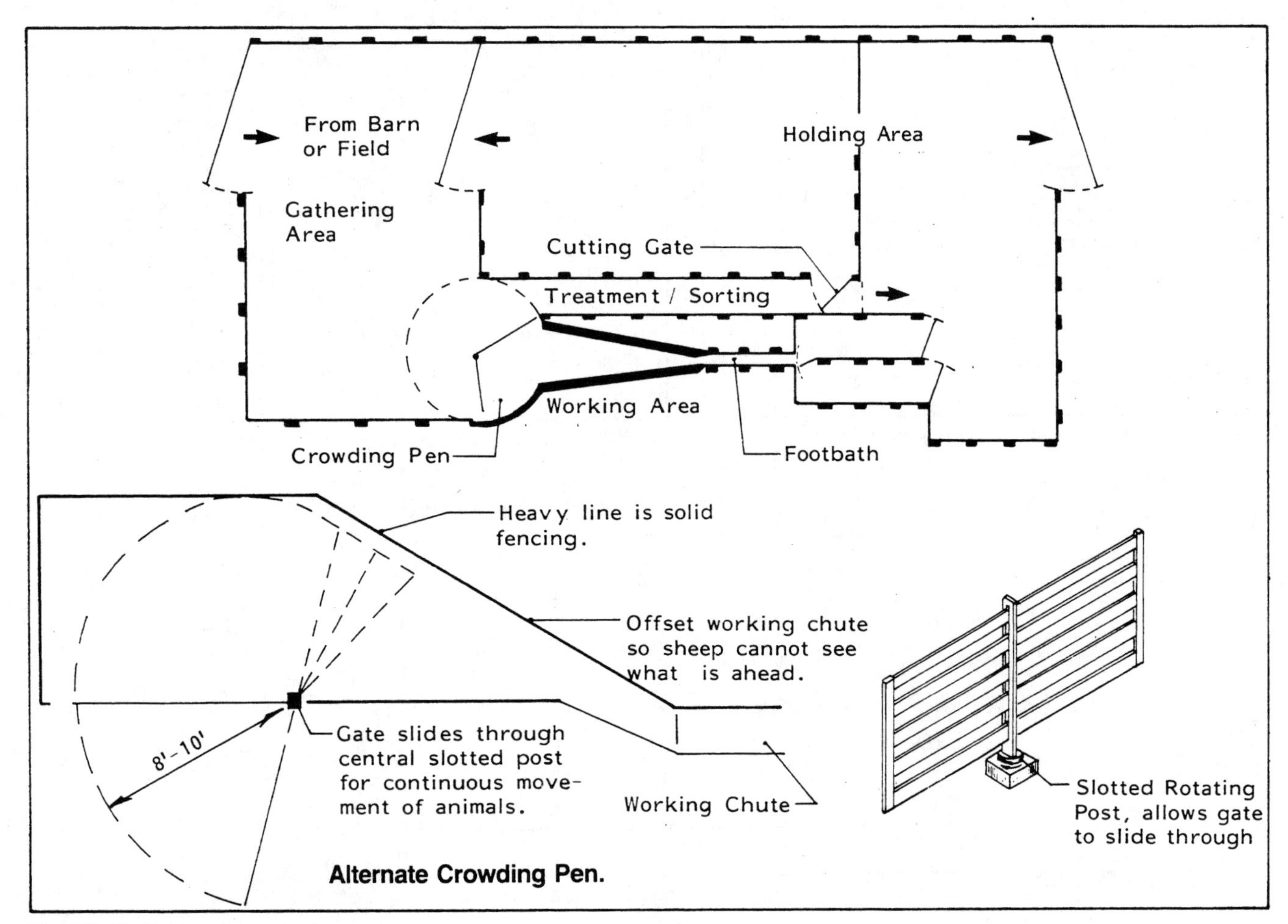

Fig 513-19. Parts of a corral.

pens are necessary for shearing. Construct floors that can be kept clean—slats are often used. Wool bagging equipment and storage space may be necessary.

Handling Equipment Construction

All softwood should be preservative-treated, especially for use outdoors. See Building Materials. Prefabricated metal handling equipment is available in some areas. With proper finish and upkeep, it should outlast wood equipment. Avoid sharp edges. Noise can be a problem.

Improving Working Conditions

A roof over the working area keeps sun and rain off the operator, sheep, and handling equipment. Consider combining an open-front barn with the handling unit. Install good lighting to encourage sheep to enter.

Office

Locate an office near the handling and lambing areas for convenience. Consider the following equipment:

water heater	desk and record system
refrigerator-freezer	stove or hot plate
storage cabinets	sink
space heater	toilet and shower
exhaust fan	wood storage (separate room)
milk replacer mixer	couch

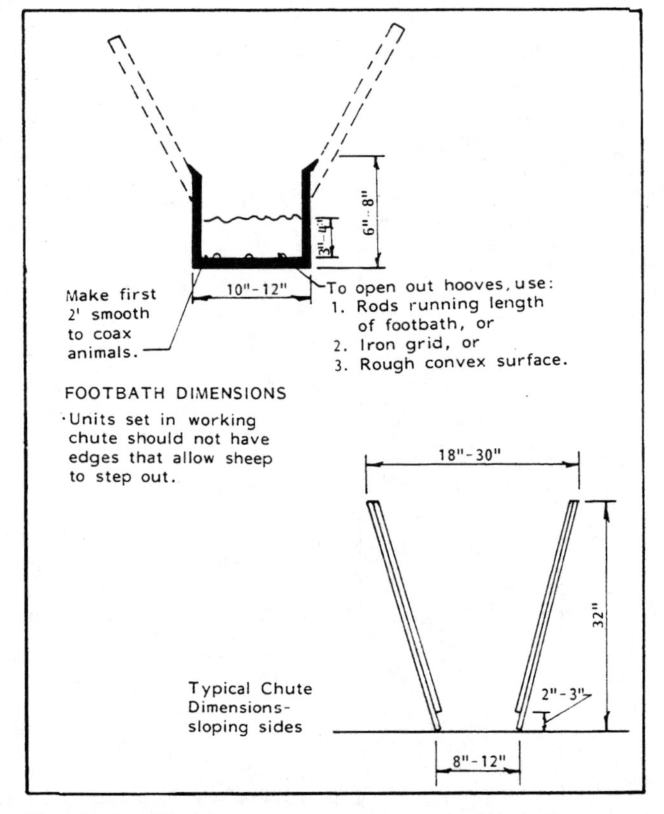

Fig 513-20. Working area for sheep—single chute.
Length 15'-20'.

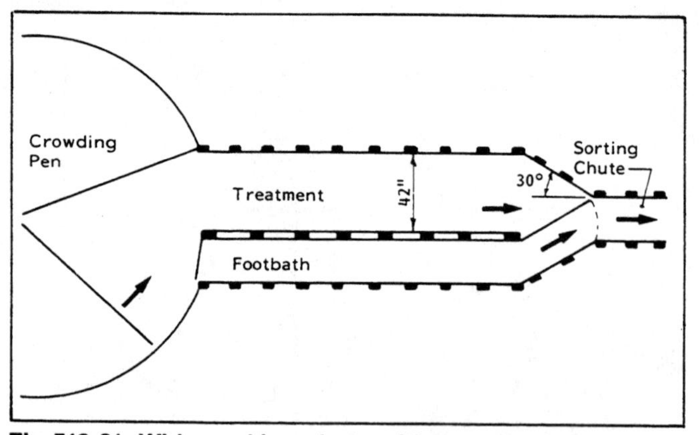

Fig 513-21. Wide working chute with separate footbath.

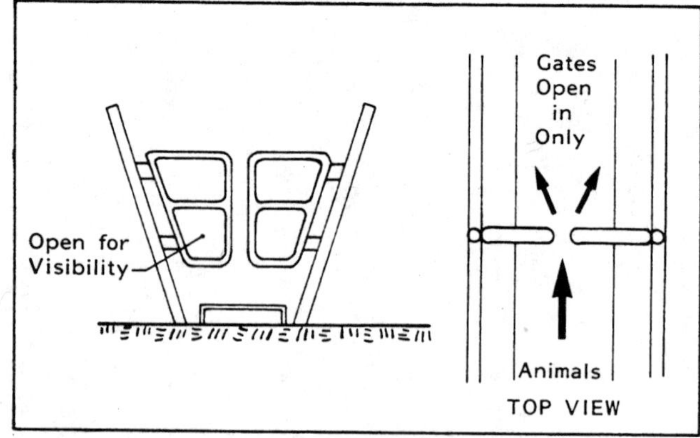

Fig 513-22. One-way gate in working chute.

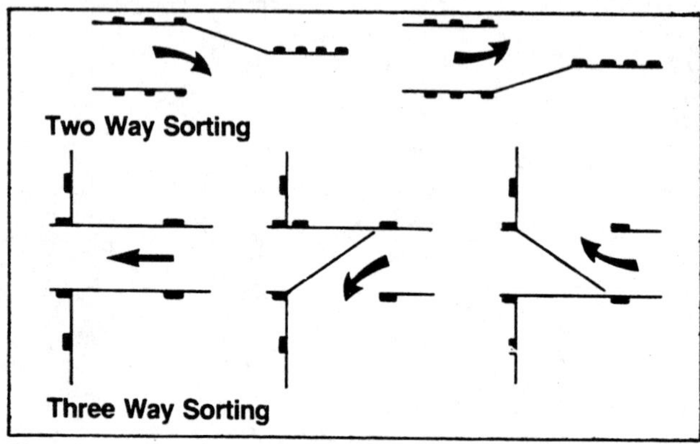

Fig 513-23. Sorting gate layouts.

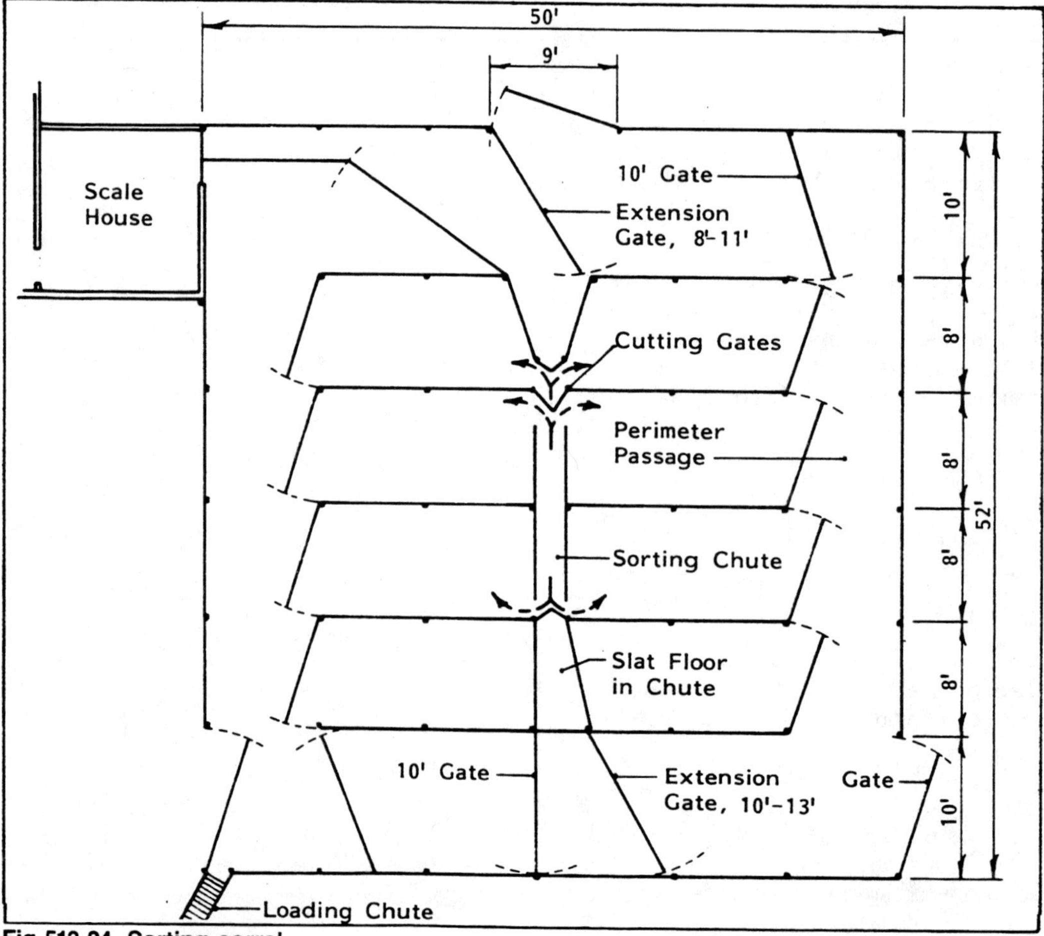

Fig 513-24. Sorting corral.
Source: University of Wisconsin.

Raising Orphan Lambs

Extra or orphan lambs occur in almost all sheep flocks at lambing time. These are lambs that ewes cannot rear and result from quadruplets, triplets, twins from ewes with borderline milk production, or from ewes that die. As the sheep industry adopts the use of more prolific sheep breeds and increased lambing frequency, the number of extra or orphan lambs will increase.

Adoption Equipment

Often, a ewe with only one lamb but adequate milk to rear two lambs is not identified soon enough to use either slime or wet grafting. Or, a ewe with twins and an adequate milk supply loses one lamb. Unable to use the extra milk, producers have reared the extra lambs artificially, with labor-intensive milk replacer. Adoption equipment is a less expensive alternative.

Build in-pen adoption stanchions, Fig 513-25 and 513-26 out of lumber or metal tubing and pipe. They fit 4'x4' lambing jugs so the ewe has access to water and can be easily fed. The stanchions are in the pens only when they are needed.

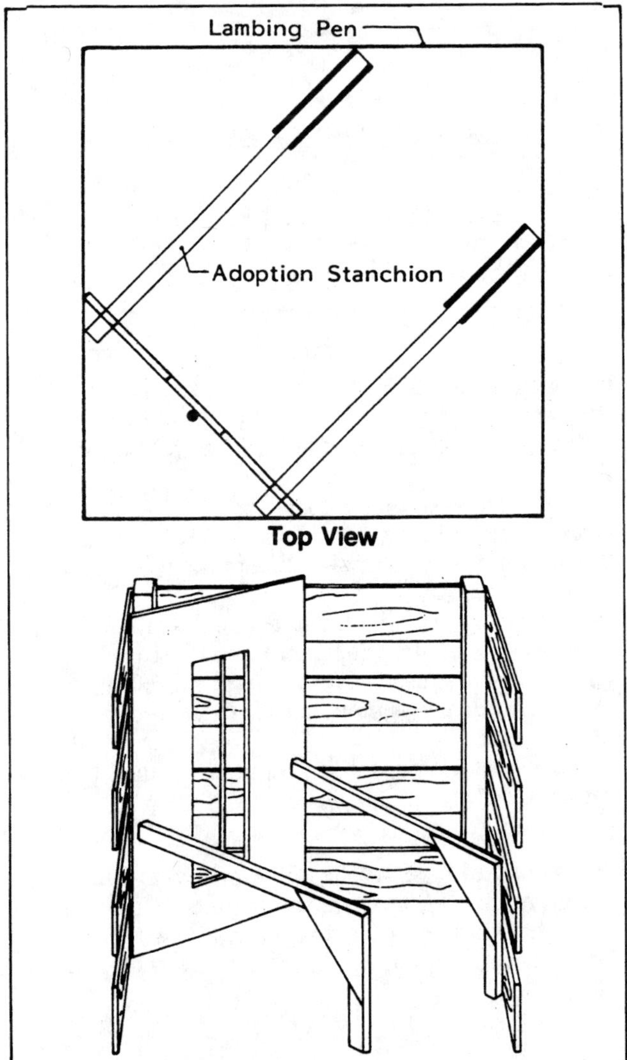

Fig 513-25. Lumber adoption stanchion.

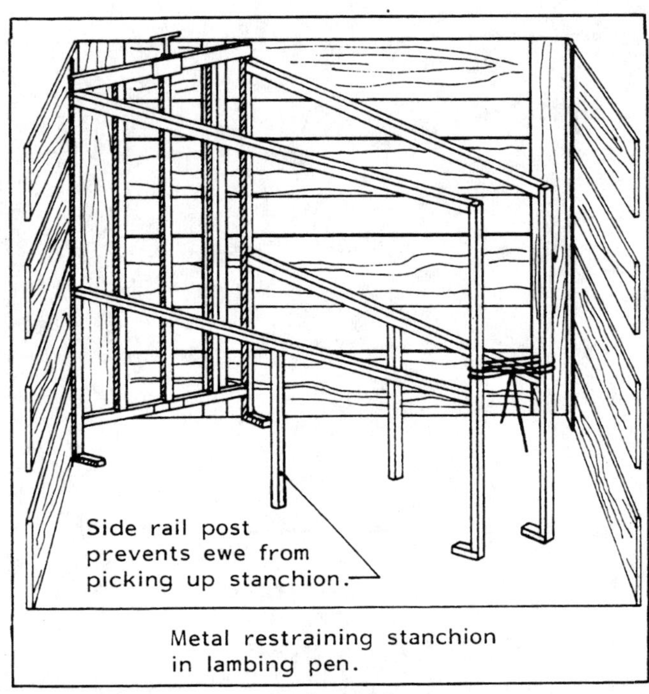

Fig 513-26. Metal adoption stanchion.

Match extra lambs needing milk as best you can for size and vigor with the ewe's own lamb. You do not have to modify the smell of the extra lamb, nor is suckling training usually necessary.

The stanchions permit the ewes to lie down or stand as they desire, but hold their heads so they cannot see or smell the lambs. Other types of adoption crates and fostering pens are shown in Figs 513-27 to 513-29.

Artificial Rearing

Newborn lambs can be successfully reared on milk replacer diets, low-labor self-feeders, and early weaning from liquid diets. Young lambs require a liquid milk diet until stomach development permits solid feeds. Additional information is available from the U.S. Sheep Experiment Station, Dubois, Idaho 83423.

Feeding

For a few lambs, mix milk powder and water with an egg beater or electric hand mixer. For many lambs, use an agitator-type washing machine or electric paint stirrer. Commercial mixers are also available. One day's milk solution can be prepared and refrigerated until needed. Equipment requires regular cleaning.

For minimum labor and maximum growth, feed lambs free-choice. Nipples attached to tubes leading from an insulated chest is a satisfactory feeder, Fig 513-30. Place nipples 12"-15" above the floor. Have cold milk available in the tubs at all times—water frozen in plastic jugs can cool the milk in warm weather. Nipples are attached to a metal strap mounted on the side of the pen partition. Fig 513-31. Use a smaller insulated jug with a few nipples in the training pen. A shallow container holds the milk near the nipple level making it easier for the lambs to obtain milk.

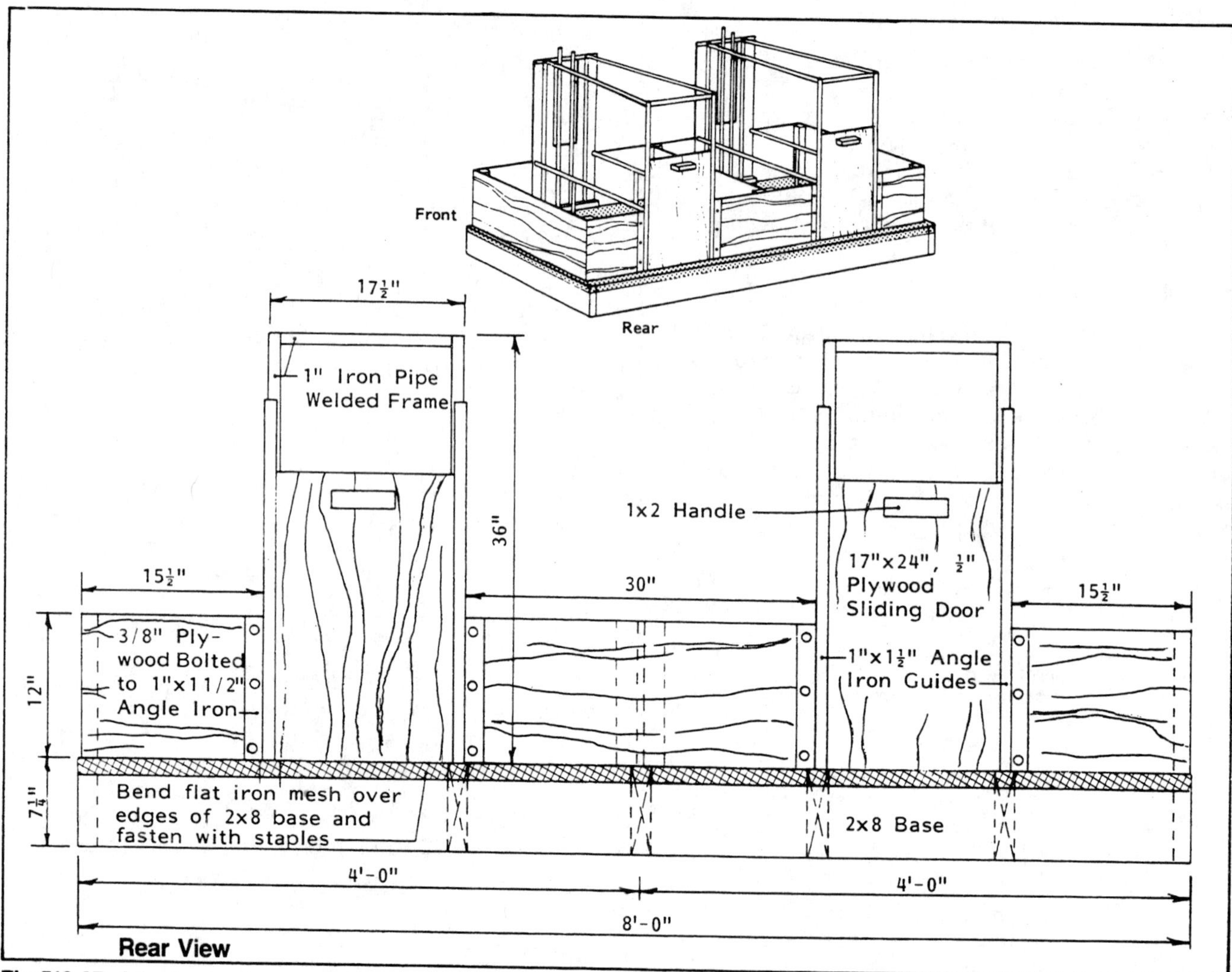

Fig 513-27. Adopt-a-lamb ewe crate.

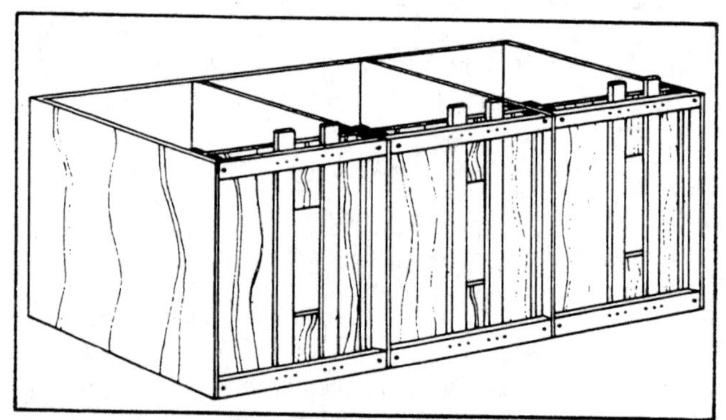

Fig 513-28. Three-pen adoption stanchion.

Another milk feeder for lamb pens has nipples and tubes leading from a 3"-4" polyvinylchloride (PVC) pipe, which is the milk reservoir, Fig 513-32. The PVC pipe is fed from a refrigerated bulk milk tank. Regulate the flow with a liquid-level control valve. Select simple, easy to clean valves.

Allow three to five lambs/nipple. Lambs can be fed in groups of 25, but performance is more satisfactory in groups of 15 or less.

Nipples

Two types of nipples are available. The self-primed type has a valve and an attached plastic tube. Lambs suck with minimum effort, but the valve tends to clog, restricting milk flow. This type is also difficult to clean. The other type is simply a nipple attached to a plastic tube.

Train lambs to suck on the same type of nipple to be used throughout training. Lambs trained to suck on a self-primed nipple and later switched to the other type may refuse to suck.

Mount either type of nipple just above the milk level in the reservoir to require the lambs to suck the milk up through the plastic tubes. When the lambs finish sucking, the milk in the tubes flows back into the reservoir, preventing waste.

To reduce coccidiosis, select and place feeders and waterers to keep out lambs' feet.

Heat

Provide infrared (110 volt, 250 watt) heat lamps with reflectors about 30" above the pen floor. Use one lamp for each 15 lambs. Or, install quartz-type infrared heat lamps according to manufacturer recommendations.

Space

Bed solid floor pens with straw or wood shavings. Expanded metal or slotted floors reduce pen-cleaning requirements when 2-4 ft^2 of floor space is allowed per lamb. Use 6-8 ft^2/lamb on solid floors.

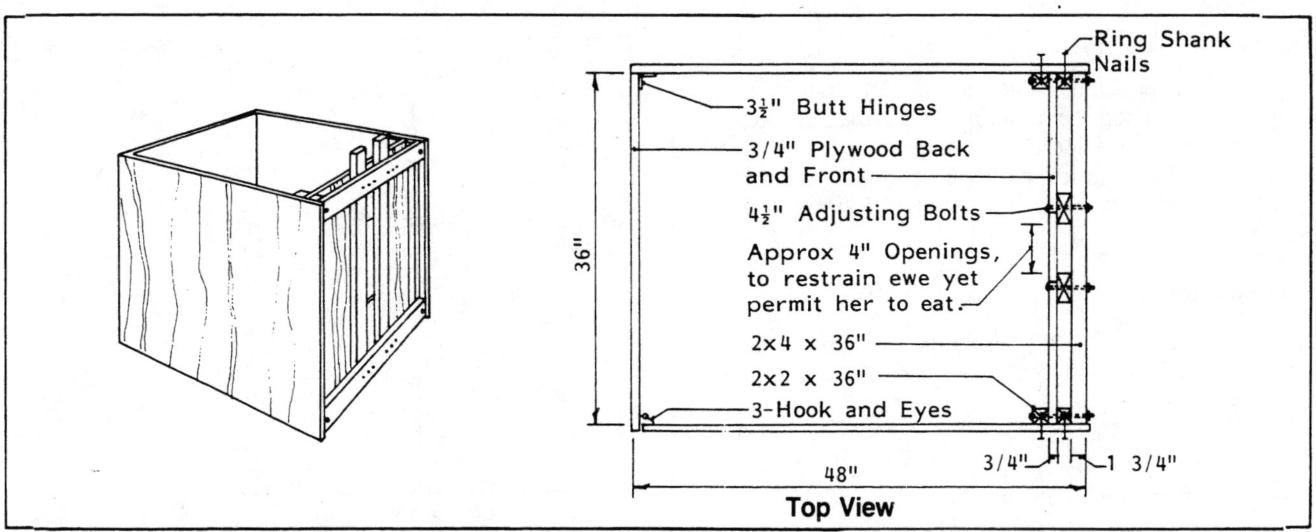

Fig 513-29. Single-pen adoption stanchion.

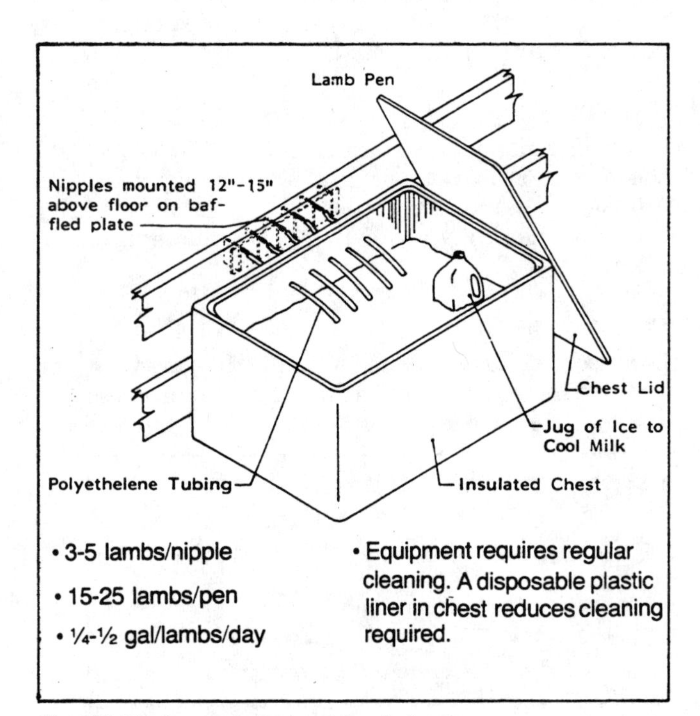

Fig 513-30. Insulated chest lamb feeder.

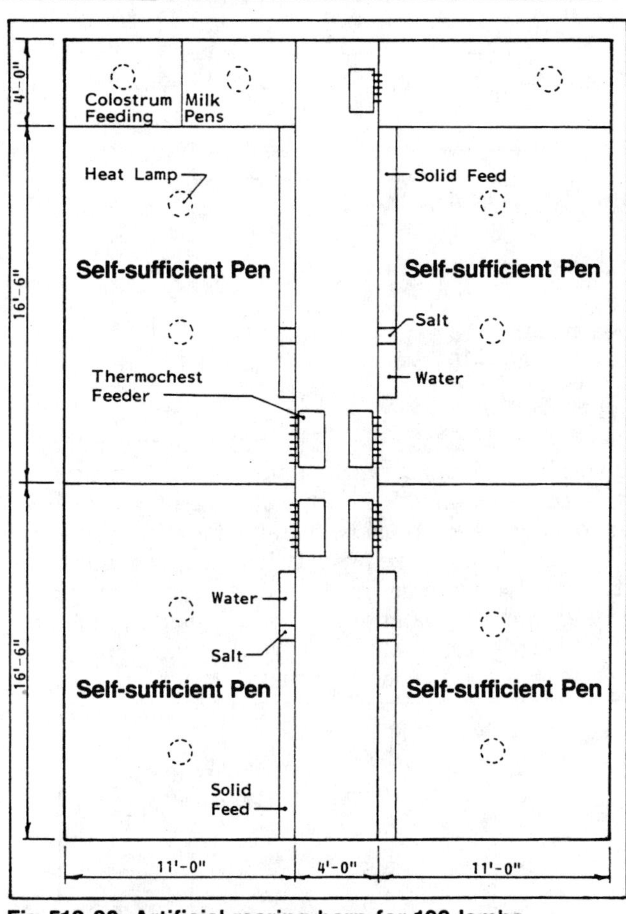

Fig 513-32. Artificial rearing barn for 100 lambs.

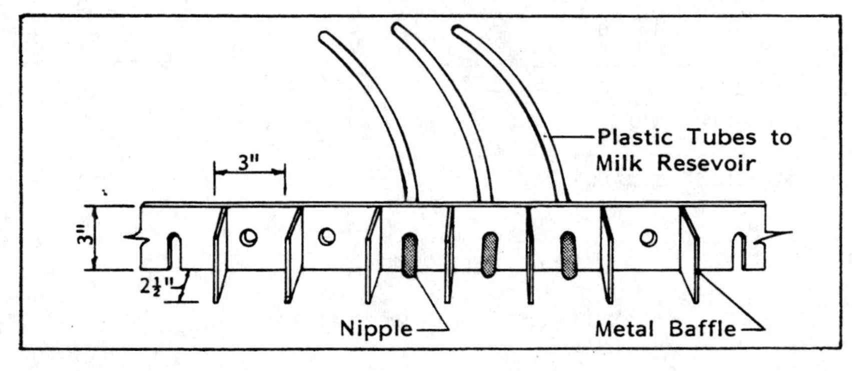

Fig 513-31. Strap-mounted nipple feeder. Mount to pen partition. Tubes extend to milk resevoir. Baffles prevent lambs from chewing off nipples.

Bunk Planning

Feeding space per sheep is determined by their size, whether they are shorn or unshorn, and the number that eat or drink at one time. If all sheep eat at the same time, allow 16″-20″ of bunk space/ewe, and 9″-12″/feeder lamb. If feed is always available, allow 4″-6″/ewe; and 1″-2″/lamb.

Mechanical Bunks

Sheep usually feed from both sides. The bunks can be lot dividers. The feed center should be near the feeding line.

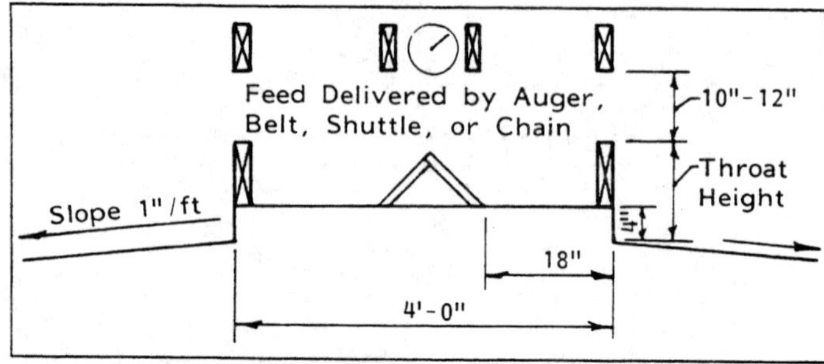

Fig 513-33. Mechanical bunk.

Throat Height, Maximum

ewe—15″
feeder lamb—13″
creep feeder—10″

Fenceline Bunks

Sheep feed from one side. The bunks are usually filled with an unloading wagon or truck. They can be lengthened to fit any size operation. Feeding drives take a lot of space and must be surfaced for all-weather use. Fenceline bunks require twice as much bunk length (but not twice as much cost) as bunks where sheep feed from both sides. A high-tensile electric fence the length of a fenceline bunk keeps lambs in but allows feeding with a side-unloading wagon.

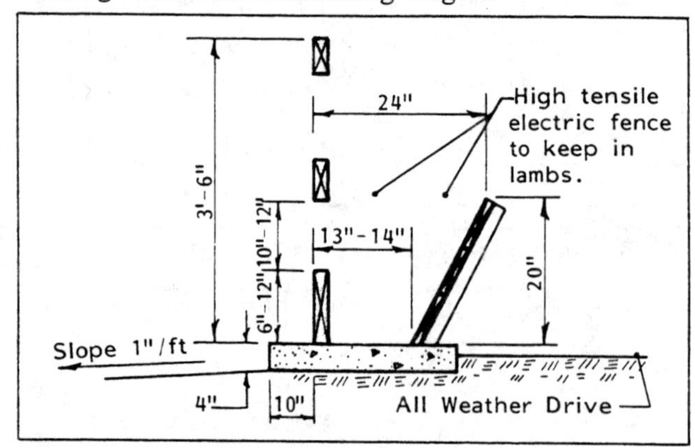

Fig 513-34. Fenceline bunk.

Hand-Fed Bunks

Small flocks are often hand-fed. Locate feeders in an area that can be kept dry.

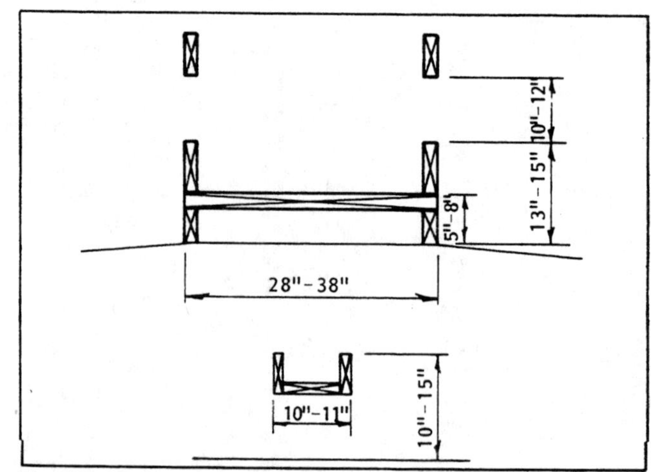

Fig 513-35. Hand-fed bunk.

Apron

Provide a well-drained apron sloped 1″/ft away from feeders and waterers. If paved, a 6′ width is adequate. If the apron is gravel or clay, use 20′.

Floor Height

Indoor hay manger floors are commonly 1″-2″ above the sheep's feet. Raise outdoor bunks to help keep sheep and dirt out of bunks and prevent sheep standing on soil, snow or manure while eating below the level of their feet.

Low Floor

Low floors are usually used indoors. They are used outdoors where the area next to the bunk is more or less clean. Low floors are least cost and easiest to build.

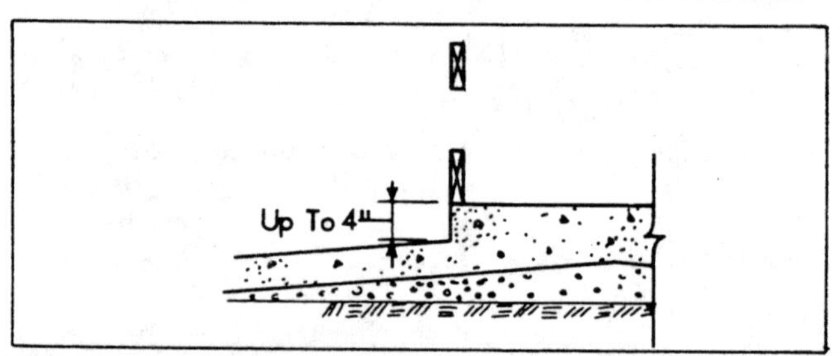

Fig 513-36. Low floor bunk.

Raised Floor

Raised floors are used where frozen manure and ice may collect or where wood framing makes a higher bunk floor easier to build.

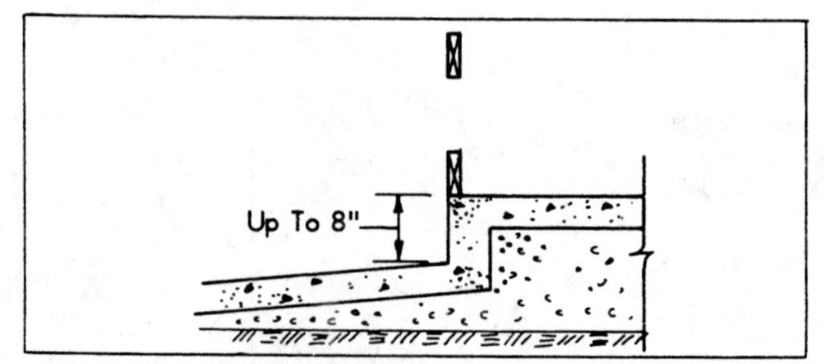

Fig 513-37. Raised floor bunk.

514 HORSES
Data Summary

Table 514-1. Hay manger and grain box dimensions.
Dimensions for inside the stall.

	Hay Manger[1]	Dimensions	Grain Box
All Mature Animals (Mares, Geldings, Brood Mares, Stallions)	30″ - 36″	Length	20″ - 24″
	38″ - 42″	Throat Height	38″ - 42″
	20″ - 24″	Width	12″ - 16″
	24″ - 30″	Depth	8″ - 12″
Foals and 2-year olds	24″ - 30″	Length	16″ - 20″
	32″ - 36″	Throat Height	32″ - 36″
	16″ - 20″	Width	10″ - 16″
	20″ - 24″	Depth	6″ - 8″
Ponies	24″	Length	18″
	32″	Throat Height	32″
	18″	Width	10″
	20″	Depth	6″ - 8″

[1] Wall corner hay racks are often used instead of mangers. Five feet is the usual distance between the floor and bottom of the rack. Many horsemen feed hay on the stall floor in both box and tie stalls and use a wall-mounted grain box in the corner of the stall.

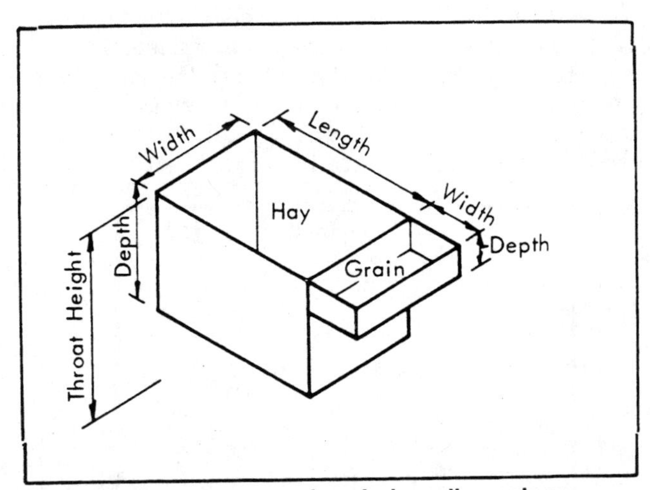

Fig 514-1. Hay manger and grain box dimensions.

Table 514-2. Light horse rations.

Age, Sex and Use	Live Weight lb	Live Weight kg	Daily Allowance (lb per 100 lb live weight) Grain	Hay	Ration Type[2]
Stallions in breeding season	900-1400	408-635	3/4-1 1/2	3/4-1 1/2	1, 2, 3
Pregnant mares	900-1400	408-635	3/4-1 1/2	3/4-1 1/2	4, 5, 6
Foals before weaning	100-350	45-159	1/2-3/4	1/2-3/4	4, 7, 8
Weanlings	350-450	159-204	1-1 1/2	1 1/2-2	8, 9, 10
Yearlings	450-700	204-318	–Summer Pasture–		
in training	450-700	204-318	3/4-1 1/4	3/4-1 1/4	3, 4, 8, 10
Yearlings, or rising 2 year olds	700-1000	318-454	1/2-1	1-1 1/2	3, 4, 11
Light horses at work					
hard use	900-1400	408-635	1/4-1 1/2	1-1 1/4	3, 12, 13
medium use	900-1400	408-635	3/4-1	1-1 1/4	3, 12, 13
light use	900-1400	408-635	1/2	1-1 1/4	3, 12, 13
Mature idle horses	900-1400	408-635	–	1 1/2-1 3/4	–

[1] From Light Horses, USDA Farmer's Bulletin 2127.
[2] Refer to Table 425-6.

Table 514-3. Door and ceiling dimensions.

Door	Width	Height
Stalls	4′	8′
Small wheeled equipment	10′	10′
Horse and rider	12′	12′
Large equipment, horse and rider	16′	14′

Ceiling	Height minimum
Horse	8′
Horse and rider	12′

Table 514-4. Space requirements for horses in buildings.
In open-front shelters, plan for 60 to 80 ft²/1000 lb liveweight.

Dimensions of Stalls including Manger	Box Stall Size		Tie Stall Size
Mature Animal (Mare or Gelding)	10′ x 10′	small	
	10′ x 12′	medium	5′ x 9′
	12′ x 12′	large	5′ x 12′
Brood Mare	12′ x 12′	or larger	
Foal to 2-year old	10′ x 10′	average	4½′ x 9′
	12′ x 12′	large	5′ x 9′
Stallion[1]	14′ x 14′	or larger	
Pony	9′ x 9′	average	3′ x 6′

[1] Work stallions daily or provide a 2-4 acre paddock for exercise.

Table 514-5. Grain rations for light horses.

Feed Material	1	2	3	4	5	6	7	8	9	10	11	12	13
	lb of Feed Material per 100 lb of Ration												
Barley	–	–	–	–	45	–	–	30	–	–	35	–	30
Corn	–	35	–	–	–	–	–	–	–	–	–	70	–
Oats	55	35	100	80	45	95	50	30	70	80	35	30	70
Wheat	20	15	–	–	–	–	–	30	–	–	–	–	–
Wheat bran	20	15	–	20	10	–	40	–	15	–	15	–	–
Linseed meal	5	–	–	–	–	5	10	10	15	20	15	–	–

[1] From Light Horses, USDA Farmer's Bulletin 2127.

Table 514-6. Water requirements.

Type and age of animal	Gallons per head per day
Mature (mare or gelding)	8-12
Brood mares	8-12
Foals to 2-yr olds	6- 8
Stallions	8-12
Ponies	6- 8

Planning

Before you begin detailed planning and construction of your horse facility, take time to:

- Study the chapter on Farmstead Planning.
- Collect ideas on horse facilities from publications, tours, visits to existing layouts, and from experienced horsemen and professional managers.
- Choose a style for the barn or other buildings that fits the site and is in harmony with the surroundings.
- Determine the total amount of space needed under roof to stable and care for the animals. Basic space needs are for stalls, traffic lanes, and feed, tack, and equipment storage. Include special-purpose features such as wash rack, trailer storage, breeding area, show stable, riding arena, and resting space.
- Figure the amount of space needed under roof for separate but associated buildings such as open shelters, riding arena, sales barn, training barn, equipment storage, exercise area, hay and feed storage, office and lounge, and other areas (including living quarters if needed).
- Determine the amount of space needed for open and fenced areas such as lanes and roads, outside lots and corrals, exercise areas, parking space for trailers and other vehicles and equipment, and special events. Also include distance between buildings for pleasing appearance, fire protection, and future expansion in your space allotment.

The size of the total layout is determined by the land area covered by the necessary buildings and directly related open and fenced areas.

Site Selection

The building site should be well drained, accessible, and slope about 5′/100′ away from the building in all directions to assure good surface drainage. Grade and fill for a well-drained site, using only clean soil, sand, gravel, or crushed rock. Refer to the Farmstead Planning chapter for factors influencing building site selection.

Housing

A horse barn, large or small, should be well planned, durable, and attractive. Its basic purpose is to provide an environment that protects horses from temperature extremes, keeps them dry and out of the wind, eliminates drafts through the stables, provides fresh air in both winter and summer, and protects them from injury.

Stables and other facilities are usually in two basic building shells—open-front buildings for minimal shelter and completely enclosed buildings for optimum shelter. The usable space in open-front or enclosed basic shells may be confined to the ground floor level. Depending on wall height and roof design, additional space overhead for hay, bedding, and feed storage can be provided.

Open-Front Buildings

When horses are kept outside, provide free-choice shelter. An open-front shed usually provides adequate protection. If a large building is needed, provide a building at least 32′ wide for moderate climates and 40′ wide for cold climates, with 60 to 80 ft^2 of floor space per 1000 lb of animal weight using the shelter. Additional room may be needed in the shelter for hay and bedding storage, foal creep, pens, etc.

Face the open side of the building away from prevailing winter winds. Provide a 10′ high clearance on the open side. Feed hay in racks to conserve bedding. Provide enough fresh clean water, perhaps in a heated stock waterer for winter use. A single heated water bowl serves 8 to 10 horses.

Allow for space taken up by walls and other structural members that reduce the usable space inside the building. Less space and smaller facilities than actually needed can result from planning with minimum space requirements. Two rows of 10′x10′ box stalls separated by a 10′ wide alley do not fit in a 30′ wide building, but do fit with very little room to spare in one that has a 32′ outside dimension.

Typical building width (outside measurement) for an enclosed structure and box stall arrangement is illustrated in Fig 514-2. Building width for tie stalls is determined by stall length and alley width. If the building has both box stalls and tie stalls, the size of the box stalls usually determines barn width.

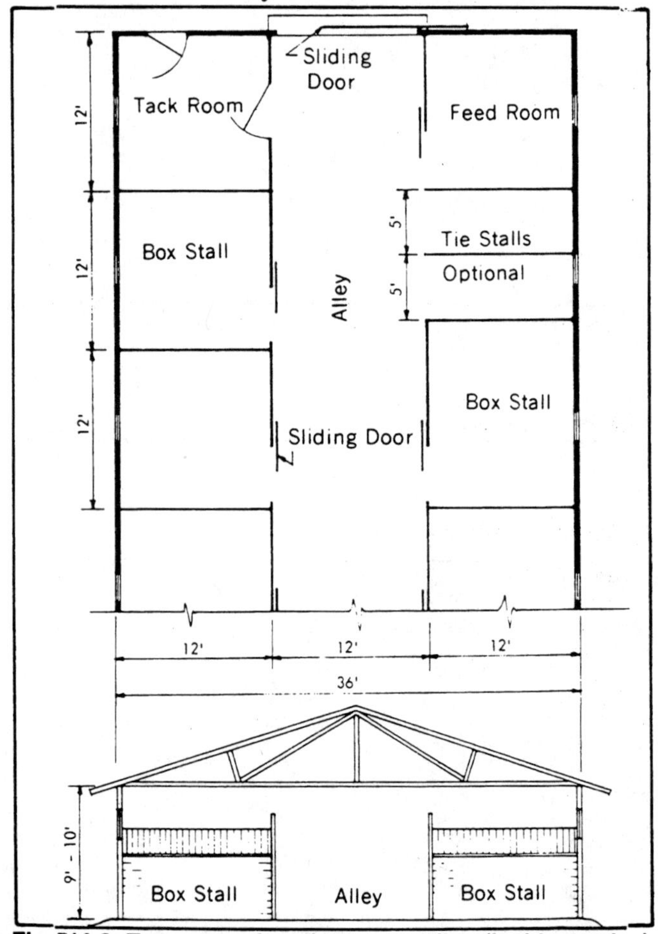

Fig 514-2. Two rows of stalls, center alley (inside service).

Other Horse Facilities

Provide wide clear-span structures or outside arenas for activities involving strenuous exercise. Covered facilities and their widths based on use include:

- Alleys separating two rows of stalls in box stall barns are for handling and resting animals, and for limited exercise and training. The alley can be as narrow as 10′ in barns 32′ wide, and as wide as 28′ in barns 50′ wide.
- Clear-span structures 36′-50′ wide are for exercising and training horses and for sales barns. Riding is limited.
- Clear-span structures 50′-100′ wide are for exercise, training, and riding arenas. Widths less than 60′, though usable for riding, are less desirable than widths of 60′ or more; 80′ or more is spacious. A clear-span width of 110′ is recommended for an indoor show ring.
- Roofed, open-front housing structure widths are variable. A width of 32′ is minimal for animal shelters, storage, and utility buildings. Clear-span structures are usually more desirable than post and beam construction.

Barn Construction

Ceiling Height

The minimum distance from the floor to the underside of roof or ceiling framing is 8′ for a horse, and 12′ for a horse and rider. Low ceilings interfere with ventilation, make the barn dark, and are a safety hazard for people and horses.

Common ceiling heights are 9′-10′ for stall barns and 14′-16′ for riding areas.

Alleys

Provide 10′ or wider litter and work alleys between stalls for convenience, safety, animal traffic, and vehicles.

Provide at least a 6′ litter alley when it is at the back of a single row of stalls. Narrow alleys require more hand labor to clean and are difficult to move horses through.

Use a 4′ minimum width for feed alleys for convenient feeding, especially with baled hay. Cross alleys are also 4′. Paving alleys is optional.

Doors at the ends of litter alleys are either sliding or overhead type. Size them for the traffic using them.

Stalls

The upper 2′-2½′ of a stall partition is often a rugged open panel or guard to aid ventilation and observation. The alley wall may be 5½′ high with no guard.

Many barns have only box stalls, but few have only tie stalls. In most horse barns some box stall space is necessary for sick animals, mares at foaling time, and colts.

Table 514-7. Stall characteristics.

Item	Tie Stalls	Box Stalls
Water	out-of-stall	in-stall
Feed	in-stall	in-stall
Manure	less carrying	more carrying
Bedding	less required	more required
Exercise	out-of-stall	limited in-stall
Space	45 to 60 sq ft	100 to 320 sq ft
Floor	clay or plank	clay or plank
Partitions	strong & tight	strong & tight
Top Guard	manger end	on partitions

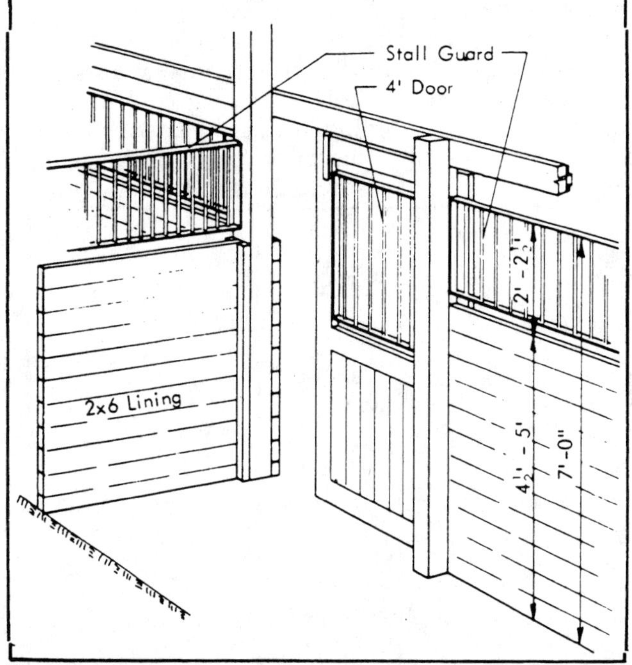

Fig 514-3. Typical box stall construction and arrangements.

Stalls are usually 2″ hardwood planks which can withstand rough treatment, chewing, and kicking. Concrete masonry walls are widely used in areas where climatic conditions can be controlled by waterproofing the outside stall walls and filling the block with insulation material to reduce heat loss. Blocks are smooth and withstand rough treatment and chewing by horses. Concrete masonry partitions in horse barns are becoming more common.

Box stalls

Box stalls are preferred to tie stalls for pleasure horses. Use 7′ high box stall partitions to prevent fighting. The lower 4½′-5′ of the 7′ stall partitions and inside walls are usually made tight in cold climates to prevent drafts and protect the animals.

For riding horses, the minimum box stall is 10′x10′, 12′x12′ is more common, and stalls 16′x16′ or larger are not uncommon. If the barn layout permits, a stall 16′x20′ or larger is useful for foaling mares. Box stalls for ponies may be smaller, depending on breed. A larger stall can be obtained by removing the common partition between adjoining box stalls.

Equip box stalls with rugged Dutch doors, full length sliding doors, or swinging gates hung with heavy hardware. Horses can open doors, so to be safe, install door fasteners that the attendant can operate from inside and outside the stall and that the horse cannot open. A stay roller or guide outside the bottom of sliding doors holds them in place.

Provide a convenient and safe stall arrangement. Consider both the attendant and the horse. Include provisions for watering; feeding hay, grain, salt and minerals; and for cross-tying the animals.

Either built-in hay-grain box combination mangers, similar to those used in tie stalls, or individual wall-mounted hay racks and grain boxes are used in box stalls. Hay can be fed on the stall floor.

Tie stalls

Tie stalls are less common for pleasure horses than box stalls. However, they do provide for animals that are restless in box stalls. Tie stalls require about half the area, less bedding, and are easier to clean than box stalls. They can often be constructed in existing buildings unsuitable for box stalls.

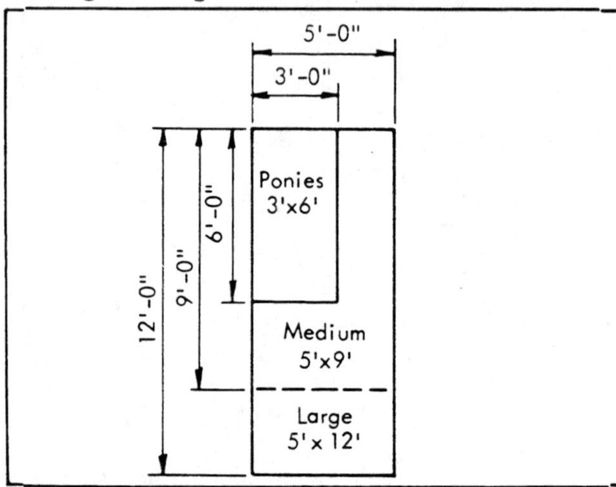

Fig 514-4. Tie stall dimensions.

The typical tie stall is 5'x9' (3'x6' for ponies), although lengths up to 12' are common. Measure stall length from the front of the manger or grain box to the rear of the stall partition. The top of a built-in wooden manger is about 2' wide. The standing platform is 7' or longer. The built-in manger slopes forward at the bottom on the horse's side to protect its knees. Individual grain boxes and hay racks are sometimes used in tie stalls instead of built-in wood mangers. Hay is occasionally fed on the floor. Horses are usually removed from the stalls for water and exercise, although watering in the stall is not uncommon.

A litter alley at least 6' wide with a gutter is required behind a single row of tie stalls. A 10'-12' litter alley with gutters separates two rows of stalls facing out. A 3½'-4' feed alley in front of each row of stalls is desirable. But, feeding can be handled satisfactorily from the rear of a stall not serviced by a front feed alley.

Provide partitions at least 6' high at the front or manger end to keep animals from fighting. Use a height of 4½' at the rear.

The amount of bedding needed depends upon the type of floor, type of stable, and the weather. Provide 8 to 15 lb of bedding per day per 1000 lb animal weight in occupied stalls.

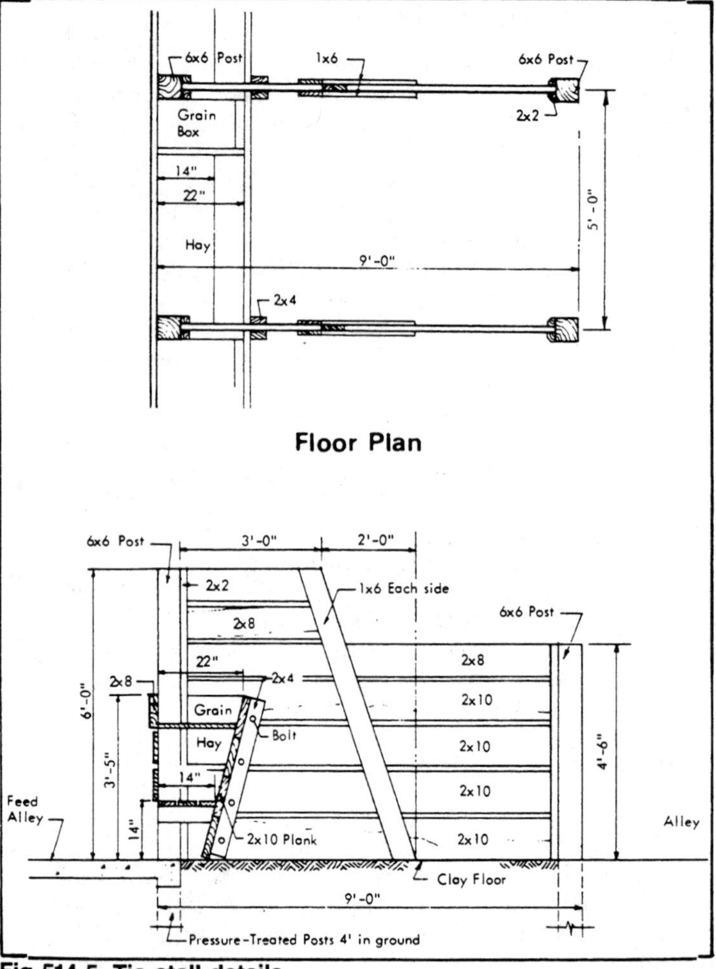

Fig 514-5. Tie stall details.

Feed Room

Feed rooms are seldom larger than 12'x12'. Organize the feed room for convenience and easy housekeeping. Plan storage for feed materials, equipment, and tools.

Store hay in an overhead loft or on the ground floor. Small rooms and narrow doors are inconvenient and add to hay handling labor.

Store grain feed in vermin-proof bins or containers such as large garbage cans or wood bins with tight lids. Hopper-bottom bins with mechanical unloading are also used either inside or outside the barn.

Provide storage space for about 18 lb of hay and 10 lb of grain per day per 1000 lb of liveweight.

Tack Room

A well organized and maintained tack room that is enclosed, dry, and free of dust is important to good stable management. In smaller stables it is simply an equipment storage area. In larger stables the tack room has also traditionally been management headquarters: office, service shop for cleaning and maintaining tack, and meeting and lounging place.

The tack room can contain all or some of the following: saddle racks, tack box, bridle and halter racks, shoeing box, first aid kit, clothes closet, storage cabinets, shelves, filing cabinets, desk, chairs, miscellaneous furniture, refrigerators, sink, working

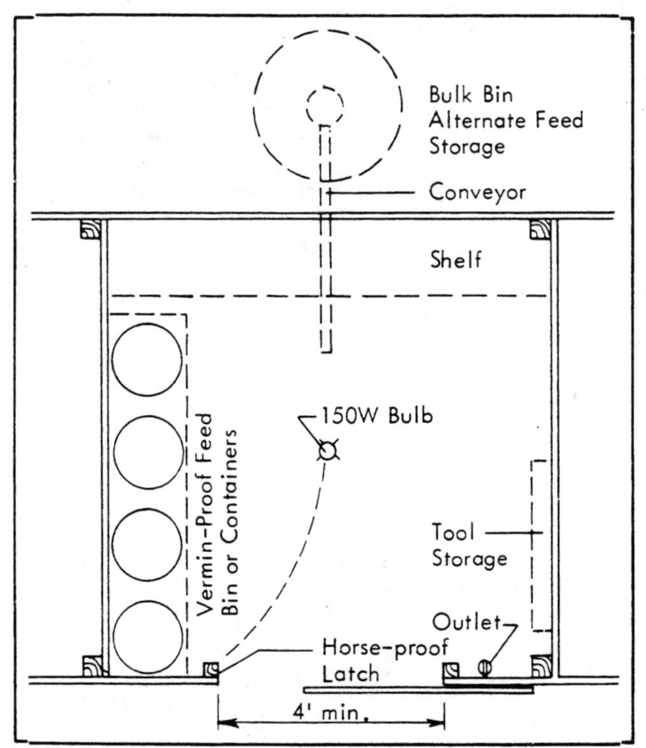

Fig 514-6. Typical feed room.

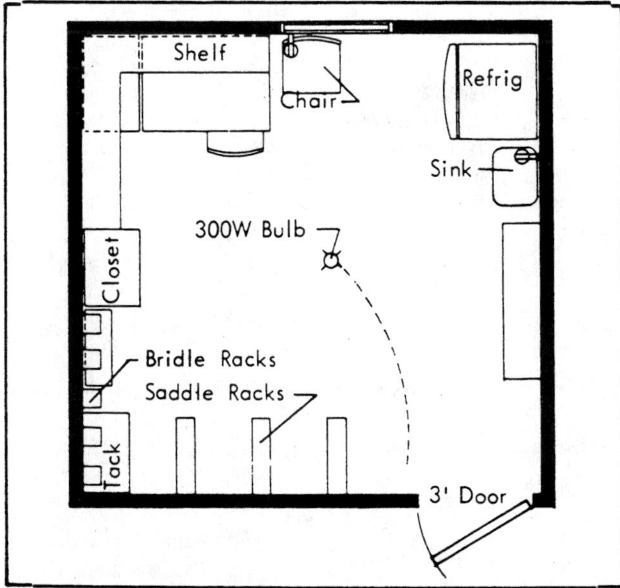

Fig 514-7. Typical tack room.

and lounging areas, heating equipment, hot and cold running water, and cooking facilities. Toilet and shower facilities can be adjacent.

Floors

Packed or puddled rock-free clay on a well-drained base makes one of the best floors for stables. It is usually easy to obtain. However, it is difficult to keep clean and has to be renewed from time to time. Wood plank stall floors or wood block floors on concrete are very difficult to keep dry and odor-free. Concrete floors are the least desirable for stalls but are preferred for the wash area, feed room, feed alleys, and tack room. Use plenty of bedding on concrete stall floors to prevent stiffness. Artificial turf (tartan) can also be used for stall floors and alleys.

Mangers and Racks For Hay and Grain

Construct mangers and grain boxes from 2" hardwood or equivalent material using the dimensions in Table 514-2. Corner or wall hay racks may replace built-in wood mangers. Corner hay racks extend about 2' each way from the stall corner and usually have a circular front. Wall hay racks are about 4' wide with straight or circular fronts. Build mangers so they are convenient for horses to eat from and for attendants to service.

Separate grain boxes are used in both box and tie stalls, especially when hay is fed in wall racks or on the floor. They are usually in a stall corner for convenient feeding.

Windows

An adjustable 2'x2' window in each box stall provides light and ventilation but is not essential if other lighting and ventilation are provided. Allow at least 6' between the window sill and stall floor. Protect windows that can be reached by horses with heavy 1"x2" welded wire or a steel grating. Make the protective guard removable for window cleaning.

Translucent roof and wall panels can admit daylight. However, they can become a heat source in the summer and are subject to severe moisture conditions and frosting in cold weather.

Water

A healthy horse needs 8 to 12 gal/day of fresh water of a quality suitable for humans.

Most horses are hand-watered. Locate in-stall waterers where spillage can drain directly from the stall—they may need filling more than once a day in a cold barn because of freezing. A frostproof hydrant or electrically heated water bowl supplies fresh water in freezing weather. One heated waterer serves two stalls or 8 to 10 horses in a lot.

Manure Handling

Check local regulations for storage and disposal recommendations. If regulations exist, follow them. In the absence of regulations:

- Dispose of manure daily when possible.
- Provide temporary storage for manure that cannot be disposed of daily—about 12 ft^2 of fenced or enclosed storage per horse.
- Locate the storage in an approved or safe area for convenient removal, away from any water source and out of natural drainage channels.
- Empty the storage at least weekly during fly breeding season (spring temperatures above 65 F until the first killing frost in the fall).
- Drain storage areas away from surrounding facilities and lots.

515 POULTRY AND RABBITS

Chickens and Turkeys

Space Requirements

Chicks require ½ ft^2/bird up to 4 weeks and 1 ft^2/bird for 4- to 10-week-old birds. Chickens 10 weeks to market need about 2 ft^2/bird. For mature laying hens, provide 1¼ to 3 ft^2/hen and 3 ft^2/bird for broiler hens.

The confined floor space required for turkeys is 1¼ to 1½ ft^2/turkey up to 8 weeks and 4 to 6 ft^2/turkey from 8 weeks to market. The confined space necessary for breeding turkeys is 6 to 8 ft^2/bird.

For breeding turkeys build roosts 1½'-2' off the ground. Provide 12"-16"/bird on roosts. Growing chickens (0 to 20 weeks) need 4"-6"/bird on roosts spaced 8"-12" apart. Provide 8"-10" of roost length for mature birds on roosts 12"-15" apart.

Range space depends upon soil, crop, and climate conditions. Limit chickens to 500 birds/acre. With turkeys, provide 1 acre/1000 birds and move to a new location each week. Shelter size is based on roost space. Plan for 1½ to 2 ft^2/turkey or ½ ft^2/chicken of range shelter space.

Environment

Keep young turkeys and chicks warm with hovers 3" above the floor. Provide one hover for every 300 birds or 4 heat lamps/150 birds. The starting temperature under a hover is 95 F, reduced 5 F each week.

A 40 to 60 watt (W) incandescent bulb is required for each 200 ft^2 of floor area. Chicks require 15 W per 200 ft^2 for 24 hr/day until they are 3 weeks old to prevent pileups.

In work areas or for observation, space fluorescent lighting 10' apart.

Consider a standby generator for large flocks—it is a must for windowless houses.

Debeak chicks at 1 day of age and before they are housed. Debeak turkeys at 3 to 5 weeks to control cannibalism.

Equipment

The edge of feeders and waterers should be level with the birds' backs. Birds should not have to walk more than 10' to feed, water, or nests. See Table 515-1 for space requirements. To keep birds out of feed and water troughs, place one live and one ground wire 1" apart over the trough.

Provide 1 nest for every 4 to 5 turkey or chicken hens. Nest size is 1½'x2' for small breeds of turkey and 2'x2' for larger breeds. Nests 9" wide are adequate for laying hens, but use 11"-12" wide nests for broiler hens.

For chickens, provide nests on the range or house the pullets before they start to lay. Use the same type of nests on the range as in the laying house.

An adequate electric confinement fence is ⅜"x40" rod posts 16' o.c. with two #12 smooth wires 9" and 16" above the ground. Mowing or chemical weed control near the wires is required.

Table 515-1. Feeder and waterer space per 100 birds.

	Age (weeks)	Feeding troughs (linear ft)*	Watering troughs (linear ft)*
Chicks	0-3	8	6
	3-6	10	6
	6-12	12	6
	12-16	16	8
	16-20	25	8
Laying hens		30	12
Broilers	0-3	8.5	6
	3-6	17	6
	6-10	20	6
	10-16	25	6-8
Turkeys	0-4	25	6-8
	4-16	40	6-8
	16-24	50	18

*On double-sided troughs, count the length of both sides.

Table 515-2. Daily water consumption per 100 birds.

	Age, week	lb	gal
Chickens	1-3	5-15	.6-1.8
	3-6	10-25	1.2-3.0
	6-10	25-35	3.0-4.2
	9-13	30-40	3.6-4.8
	Pullets	25-40	3.0-4.8
	Non-laying hens	40	4.8
	Laying hens (mod. temp.)	40-60	4.8-7.2
	Laying hens (90 F)	70	8.4
Turkeys	1-3	65-150	7.8-18.0
	4-7	215-490	25.8-58.8
	9-13	500-800	60.0-96.0
	15-19	1000	120.0
	21-26	850	102.0

Table 515-3. Feed and time required to obtain average live weights.

Average live weight (lb)	Quantity of feed (lb) required per bird		Age (weeks) at which live weights are reached	
	Female	Male	Female	Male
0.5	.7	.7	2.0	1.8
1.1	1.6	1.6	3.5	3.3
2.2	3.8	3.5	5.6	5.2
3.3	6.6	5.7	7.0	6.9
4.4	9.4	8.6	8.3	7.6
5.5	12.3	11.2	9.7	8.7
6.6	15.9	14.0	11.0	9.6
8.8	23.4	20.1	13.5	11.2
11.0	33.1	27.2	16.3	13.1
13.2	45.1	34.8	19.2	14.9
15.4	58.7	42.4	23.1	16.8
17.6	—	51.0	—	18.6
19.8	—	60.0	—	20.5
22.0	—	70.0	—	22.3

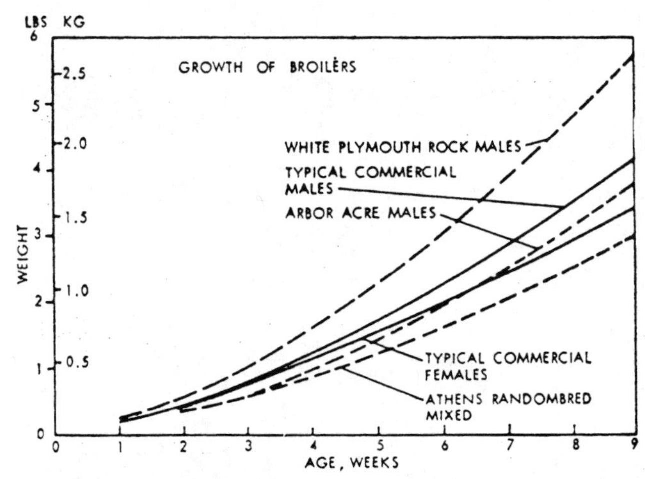

Fig 515-1. Weight vs. age of broilers.

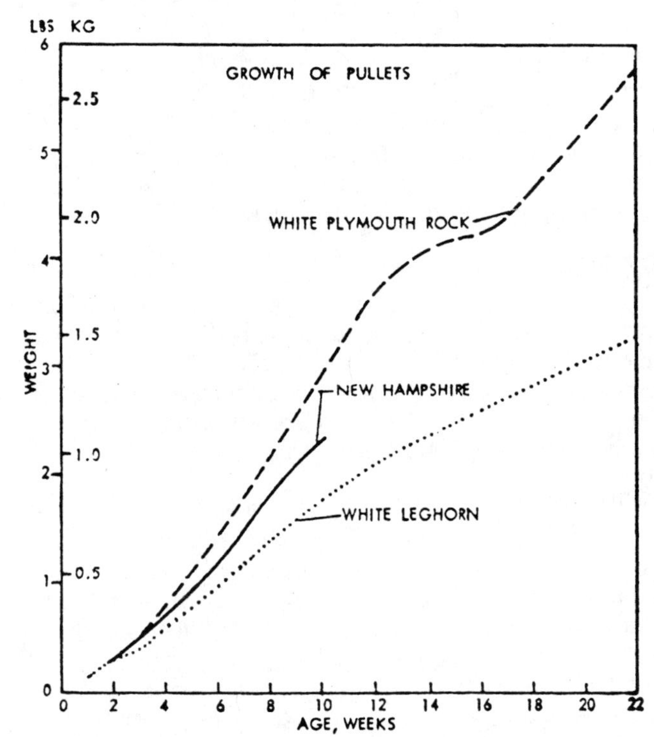

Fig 515-2. Weight vs. age of pullets.

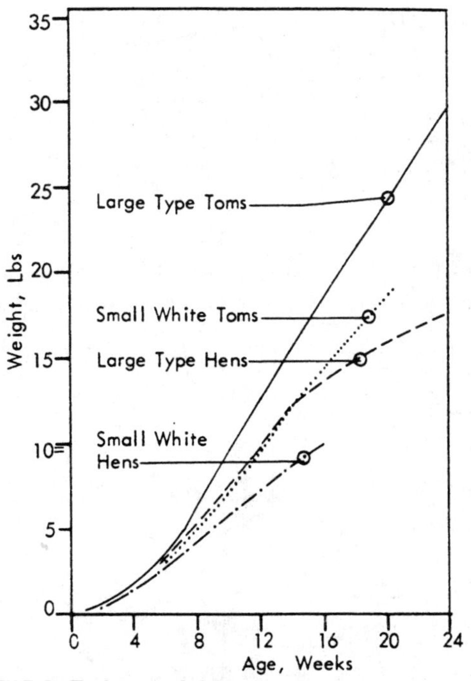

Fig 515-3. Turkey weight vs. age.

Extracted with permission from *The 1982 Agricultural Engineers Yearbook,* The American Society of Agricultural Engineers, Box 229, St. Joseph, Michigan 49085.

Table 515-4. Sizes for egg cooling rooms.
1 case/100 hens/week

Flock	3000-3600	3600-4200	4200-6000	6000-9600
Inside Size	6′ x 6′	6′ x 7′	7′ x 8′	8′ x 9′
Cases	24	28	40	64
Baskets	15	19	21	21
Refrigeration	4500 BTU	5100 BTU	6900 BTU	9900 BTU

(1 H.P. refrigeration approx. = 9000 to 12,000 BTU/hr.)

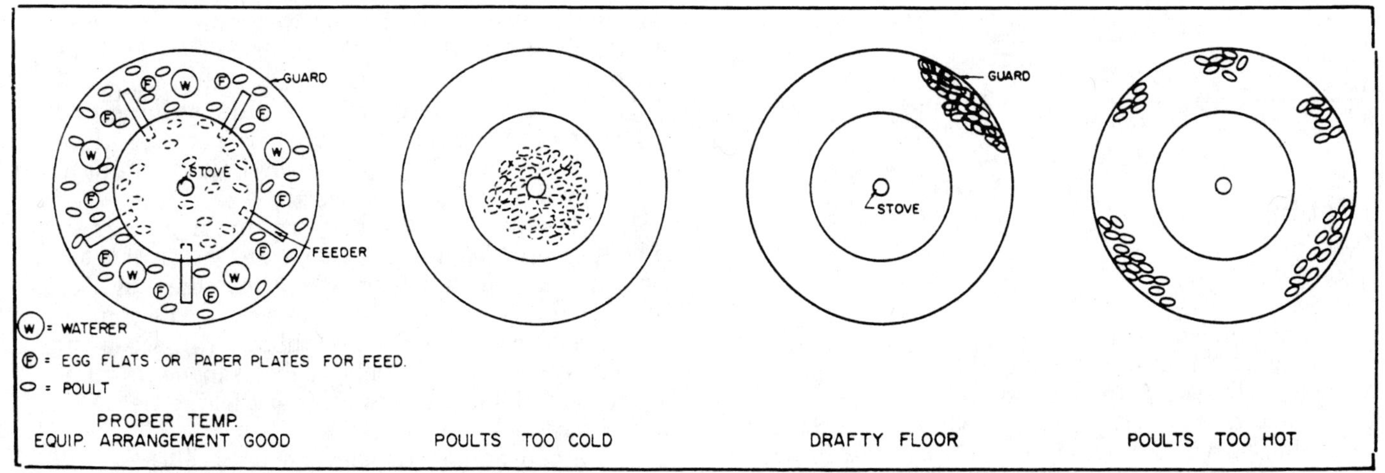

Fig 515-4. Poult or chick guide to hover temperature.

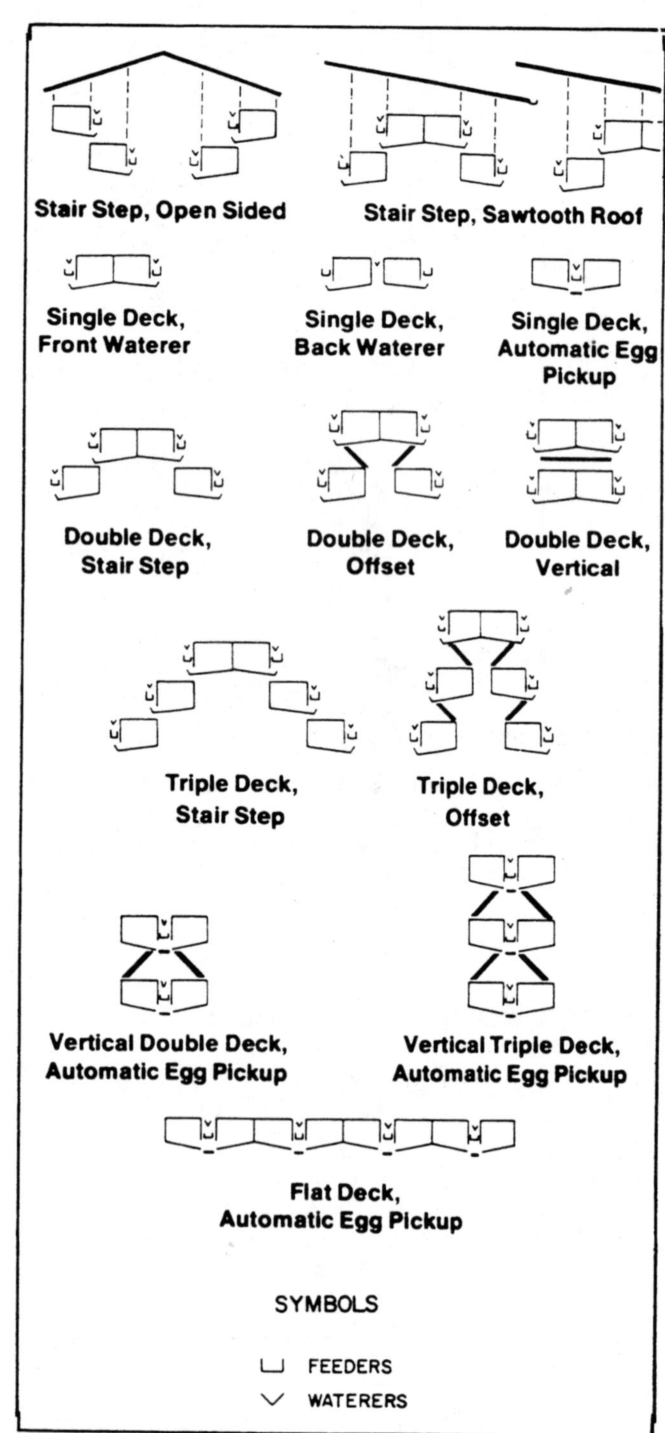

Fig 515-5. Cage arrangements.

Rabbits

Mating

Light breeds: 4 to 5 months old.
Medium breeds: 6 to 7 months old.
Giant breeds: 9 to 12 months old.

The estrus cycle is 15 or 16 days, with conception possible with non-pregnant does at any time. Provide one buck per 10 does. Gestation is 31 to 32 days.

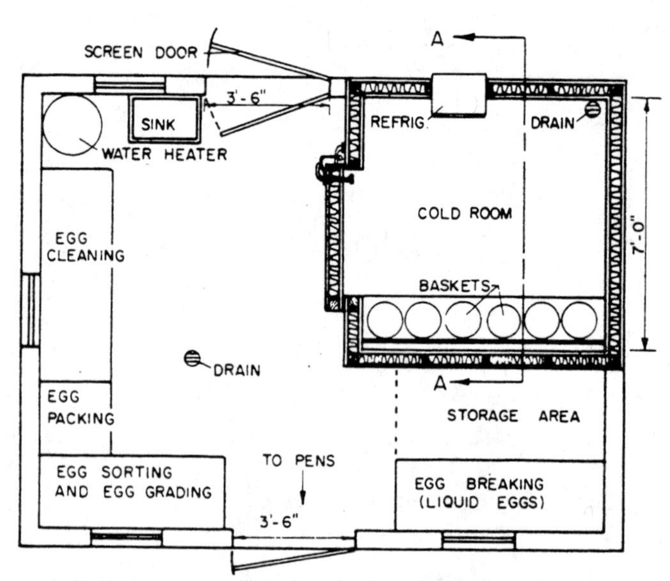

Fig 515-6. Typical floor plan for egg handling room.

Kindle 2 or 3 litters/year for show animals and 4 litters for commercial breeding. To kindle means to give birth. Some strains can be mated 6 weeks after kindling and produce 5 litters/year.

Kindling

Provide a nest box in the hutch 27 days after mating. Does will pull fur to cover the litter. If they do not, pull fur from the doe and arrange the bed. Litter may be kindled on the hutch floor. The doe needs to be undisturbed a day or two before kindling and until restlessness stops after kindling.

Young Litters

On the day after kindling, carefully remove any dead or deformed young from the nest. Does can care for 6 to 8 young. Baby rabbits can be transferred to a foster mother. Mate several does to kindle at the same time.

Wean young from the does at 6 to 8 weeks. At this age the young will have started eating other feed.

Equipment

Individual hutches for mature rabbits are up to 30″ deep and 24″ high. Waist-high, single-tier hutches are preferred. Wire floors require no bedding. Wire should be galvanized after welding and any sharp points removed. In colder climates, close the back and ends. Slope solid or slat floors to the back (rabbits will chew exposed wood).

Build crocks, troughs, hoppers, and hay mangers to hold several feedings. A lip on the inside of these containers prevents feed waste. Anchor smaller units. Dividers keep young rabbits out of the trough. Separate each kind of feed, or rabbits will scoop out less desirable feeds to eat other components first. Complete pellet feed is now common, which means no hay or hay rack is needed.

Use a half-gallon crock or anchored coffee can for a waterer.

532 LOW TEMPERATURE DRYING

Low temperature drying, as used in this handbook, is a process of slow drying with natural air or air heated only a few degrees (2 F to 10 F). Grain is dried and stored in the same bin so the process is sometimes referred to as in-storage drying. It is also called deep-bed drying and is one method of in-bin drying.

The bin usually is equipped with a fully perforated floor. Provide a fan capable of supplying at least 1 cubic foot of air per minute per bushel (cfm/bu) and exhaust vents of 1 sq ft/1000 cfm of fan capacity. Equipping each bin with a grain distributor to spread fines is recommended, Fig 1. A stirring device is sometimes installed, but is not essential.

Drying depends on the air's ability to evaporate water so its relative humidity is a key factor. The lower the relative humidity of the drying air, the more water the air evaporates from the grain, resulting in a lower final grain moisture content. The air's ability to evaporate water is referred to as **drying potential.** It is influenced by temperature as well as relative humidity.

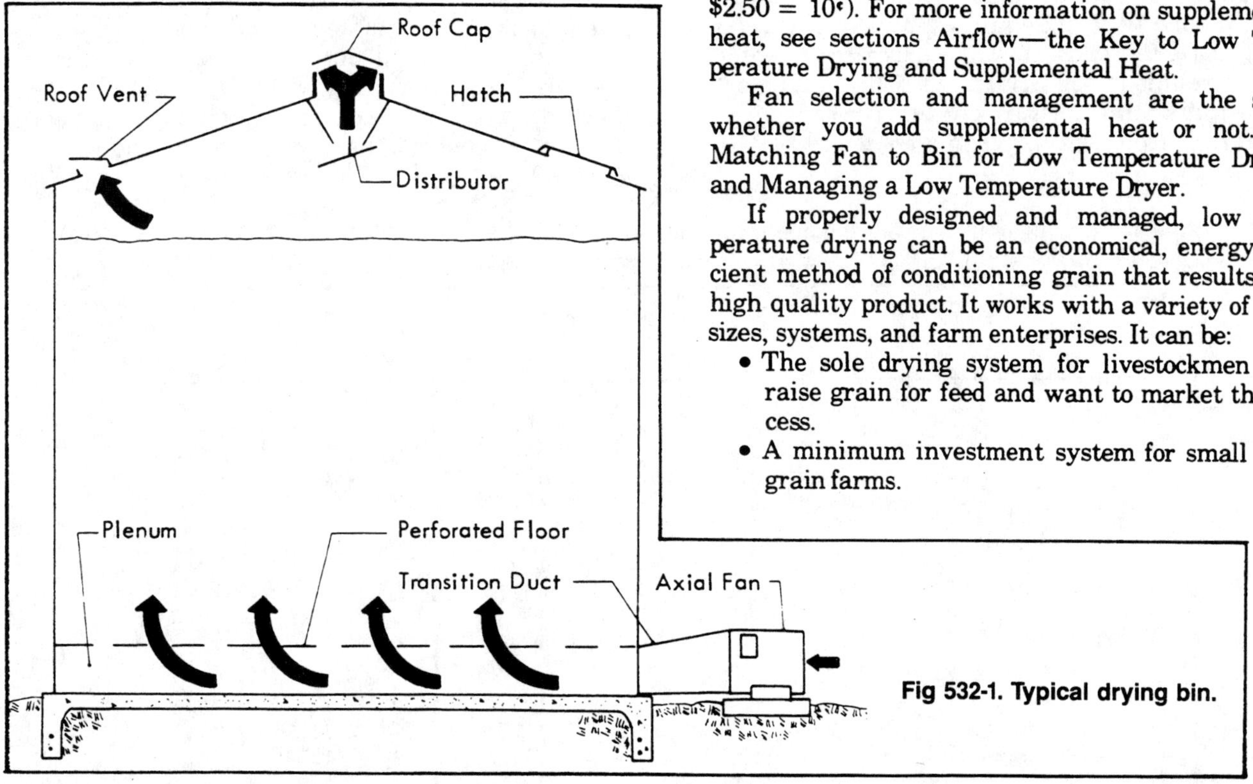

A drying fan heats the air and reduces its relative humidity. As fan power increases, heat added to the air increases. Supplemental heat, sometimes from a solar collector, has often been added.

Whether you use supplemental heat or not depends on the drying potential of the air in your area and the final moisture content you want. Often, the natural air along with heat from the fan is sufficient for drying corn** in the North Central region. (For more information, see Equilibrium Moisture Content.) Corn dries to a moisture content for safe winter storage (18% or less) 9 years out of 10 during the fall drying season. Spring drying is needed some years to reach a safe long-term storage moisture content (13% to 14%).

Supplemental heat causes costly overdrying in low temperature dryers without stirring machines: **1**) You do not sell the maximum amount of water allowed for #2 yellow corn, and **2**) You pay for the energy to overdry.

Estimate the cost of overdrying at 2% of corn market price/point of drying below 15.5%/bushel. That is, corn dried to 13.5% and sold for $2.50/bu costs you 10¢ per bushel (2% x 2 points x $2.50 = 10¢). For more information on supplemental heat, see sections Airflow—the Key to Low Temperature Drying and Supplemental Heat.

Fan selection and management are the same whether you add supplemental heat or not. See Matching Fan to Bin for Low Temperature Drying and Managing a Low Temperature Dryer.

If properly designed and managed, low temperature drying can be an economical, energy efficient method of conditioning grain that results in a high quality product. It works with a variety of farm sizes, systems, and farm enterprises. It can be:

- The sole drying system for livestockmen who raise grain for feed and want to market the excess.
- A minimum investment system for small cash grain farms.

Fig 532-1. Typical drying bin.

*Abstracted from MWPS-22, "Low Temperature and Solar Grain Drying Handbook."

**Most research has been directed to low temperature drying of corn so the text is specific for corn. Although principles for the low temperature drying of other crops are similar, recommendations are scarce. For available information see Table 532-9.

- One method of increasing the capacity of a high temperature dryer by using a combination high temperature/low temperature system.
- The major drying system for a large cash grain farm with a well designed multi-bin set-up.

Advantages of low temperature drying include:

- Minimum grain handling.
- Good quality grain (few stress cracks, high test weight).
- Less dependence on petroleum based fuels.
- High drying efficiency (few purchased Btu/lb of water removed) because it uses atmospheric heat.

Disadvantages of low temperature drying include:

- Initial moisture content limitations.
- High electrical power demand (connected kilowatts).
- Weather dependency.
- Possible limitation on fill rate.

Principles of Low Temperature Drying

Equilibrium Moisture Content

Low temperature grain drying works because grain readily gives up its moisture to the air. Corn dries in the field while still on the stalks—the principle is the same in a bin except air must be forced through the grain with a fan. When air is forced through grain, the air evaporates moisture from the grain.

For each temperature and relative humidity combination of the drying air there is a corresponding moisture content for the grain. If the air is kept at constant conditions, the grain eventually reaches a moisture content close to that indicated in Table 1—the grain is in equilibrium with the air. Table 1 gives **equilibrium moisture contents** for corn.

In grain drying, the air temperature and relative humidity are not constant. The average temperature /relative humidity combination throughout the drying period determines the final moisture content. Table 1, however, gives you an idea of the final grain moisture content that can be reached using only natural air.

Example 1. If the outside air averages 50 F and 70% relative humidity, to what moisture content will the air dry the corn? Answer: 15.4%.

Notice the effect of temperature. As air gets warmer, it dries grain to a lower moisture content for the same relative humidity. High humidity drying air dries wet grain (moisture content about 24%) as long as temperature is above freezing.

Supplemental heat increases the drying air temperature, and reduces relative humidity. Use Table 1 to estimate the final moisture content when heat is added. Reduce the relative humidity by 10 points for each 5 F of heat added.

Example 2. In example 1, the outside air was 50 F, 70% relative humidity. If 5 F is added, what will be the final corn moisture content?

Answer: The new temperature is 55 F and relative humidity is reduced to 60%: the corn will dry to 13.5% moisture content.

Table 1 is also useful in determining air conditions within the grain mass. Knowing these conditions allows you to evaluate the storability of the grain. The section, Mold Growth and Spoilage , gives more information.

Example 3. If corn is 20% moisture and 50 F, what is the relative humidity within the grain mass? Answer: about 90%.

Grain can rewet if the air passing through it is at a high relative humidity for an extended time. However, the average relative humidity and temperature throughout the drying season usually pre-

Table 532-1. Equilibrium moisture content for shelled corn.
Dashed lines indicate examples in text. (Developed from Chung-Pfost equation, ASAE paper 76-3520.)

Temp.	Relative humidity, percent								
deg. F	10	20	30	40	50	60	70	80	90
	Equilibrium moisture content, percent								
20	9.4	11.1	12.4	13.6	14.8	16.1	17.6	19.4	22.2
25	8.8	10.5	11.9	13.1	14.3	15.6	17.1	19.0	21.8
30	8.3	10.1	11.4	12.7	13.9	15.2	16.7	18.6	21.1
35	7.9	9.6	11.0	12.3	13.5	14.8	16.3	18.2	20.8
40	7.4	9.2	10.6	11.9	13.1	14.5	16.0	17.9	20.5
45	7.1	8.8	10.2	11.5	12.8	14.1	15.7	17.6	20.5
50	6.7	8.5	9.9	11.2	12.5	13.8	15.4	17.3	20.2
55	6.3	8.2	9.6	10.9	12.2	13.5	15.1	17.0	20.0
60	6.0	7.9	9.3	10.6	11.9	13.3	14.8	16.8	19.7
65	5.7	7.6	9.0	10.3	11.6	13.0	14.6	16.5	19.5
70	5.4	7.3	8.7	10.0	11.4	12.7	14.3	16.3	19.3
75	5.1	7.0	8.5	9.8	11.1	12.5	14.1	16.1	19.1
80	4.9	6.7	8.2	9.6	10.9	12.3	13.9	15.9	18.9
85	4.6	6.5	8.0	9.3	10.7	12.1	13.7	15.7	18.7
90	4.4	6.3	7.7	9.1	10.4	11.9	13.5	15.5	18.5
95	4.1	6.0	7.5	8.9	10.2	11.7	13.3	15.3	18.4
100	3.9	5.8	7.3	8.7	10.0	11.5	13.1	15.1	18.2

vents grain from rewetting above about 17%. You can expect corn to rewet to within about 1 percentage point below the equilibrium moisture contents given in Table 1. If air would dry wet grain to 18%, the same air would rewet dry grain to 16% or 17%.

Rewetting can be helpful in low temperature grain drying. For instance, air is drier in early fall and temperatures are warmer. This combination causes grain in the lower part of the bin to overdry.

As winter approaches, relative humidity increases and air temperature decreases. You would expect little drying under these conditions. However, as the air passes through the lower layers of the overdried grain, moisture is taken out of the air. Drying continues in the upper layers. Overdried grain also dehumidifies the air on rainy and foggy days and evens out the day-night differences in relative humidity to allow drying to continue during the night. Therefore, operate fans continuously during drying if there is wet corn (greater than 18% moisture content) in the top of the bin.

Mold Growth and Spoilage

Mold is the major cause of spoilage in grain. There are two groups of mold that affect grain quality: field molds and storage molds.

Field molds invade kernels while grain is still in the field. Generally, they grow in high moisture grain (greater than about 20%) at temperatures of 30 F to 90 F. The high moisture content and cool temperatures of grain in the top of a low temperature drying bin resemble field conditions so some spoilage in a low temperature dryer results from field molds. Fall temperatures of 40 F to 50 F reduce mold activity, and prevent major spoilage. Once the grain is dried, these molds die, or become inactive.

Storage molds dominate in storage facilities when grain moisture content is too low for field molds (less than about 20%). The moisture and temperature requirements of these molds determine the recommended safe storage moisture contents given in Table 533-1. Because the effects of temperature and moisture content on mold growth are interrelated, grain moisture content can be higher when grain temperature is kept below about 40°F with proper aeration, management, and close observation. For summer long-term storage, safe storage moisture contents are lower.

By controlling moisture content and temperature, mold growth is restricted and grain can be dried without significant spoilage. Grain temperature and moisture content determine the **allowable storage time (AST)**—how long grain can be kept before it spoils. Table 2 gives the AST of shelled corn. (AST has not been developed for other grains.) The AST gives an estimate of how long you have to dry grain before it spoils and how long you can maintain grain quality in storage. The AST is the time it takes for the grain to drop one market grade due to moldy kernels developed under the particular temperature and moisture conditions of the grain.

Notice that as grain moisture content increases for a given temperature, the AST for drying and storing decreases. Also, as temperature increases, AST decreases. The AST is approximately cut in half for each 10 F increase in temperature, and cut in half for each 2 percentage points increase in moisture content. Mechanical damage also affects AST. Clean grain and whole seeds are more resistant to mold.

The AST is used to design low temperature drying systems. The recommended airflow rate is chosen so that drying finishes within the AST. Some examples show how to use AST to estimate how long you have to dry grain.

Example 4. Grain is harvested at 24% moisture and grain temperature is 50 F. How much time do you have before the grain drops one market grade?

Answer: about 25 days, Table 2. If conditions in the bin remain constant, the grain will drop one market grade in about 25 days, the AST at 24% moisture content and 50 F.

Example 5. If at the end of 7 days, the corn in Example 4 has dried to 20% moisture and is at 40 F, how does AST change?

Answer: During the first 7 days, 7/25 of the AST from Ex 4 was used up; 18/25 remain. The AST for the new conditions, 20% and 40 F, is 142 days, Table 2. But, only 18/25 of that time remains: 142 x 18/25 = 102 days. The AST is extended by drying.

Table 532-2. Allowable storage time for shelled corn.
Dashed lines indicate examples in text. (Developed from Thompson, Transactions of ASAE 333-337, 1972.)

Grain temp. deg. F	Corn moisture, percent						
	18	20	22	24	26	28	30
	- - - days - - -						
30	648	321	190	127	94	74	61
35	432	214	126	85	62	49	40
40	288	142	84	56	41	32	27
45	192	95	56	37	27	21	18
50	128	63	37	25	18	14	12
55	85	42	25	16	12	9	8
60	56	28	17	11	8	7	5
65	42	21	13	8	6	5	4
70	31	16	9	6	5	4	3
75	23	12	7	5	4	3	2
80	17	9	5	4	3	2	2

How Drying Occurs in a Bin

In a low temperature drying system without stirring, grain is dried in zones as shown in Fig 2. Air flows up through the grain, evaporating and carrying away water. Drying takes place in a drying zone 1' to 2' thick. The drying zone advances through the grain at a rate largely determined by airflow and the drying potential of the air.

The characteristics of three zones can be identified. The first zone is the **dry grain.** It is about the temperature of the drying air. The moisture content is close to that given in Table 1.

In the **drying zone,** grain is drying and the moisture content is less than the moisture content when the bin was filled, but the grain has not yet reached moisture equilibrium with the air. The grain temperature is less than the drying air temperature because evaporation of water causes cooling.

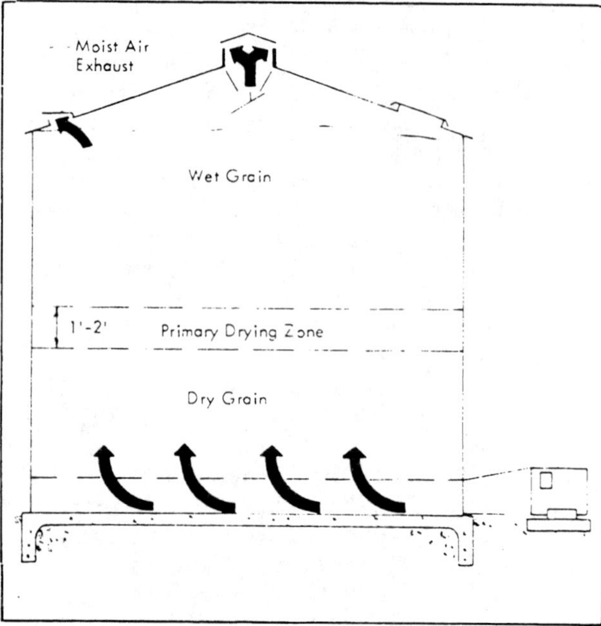

Fig 532-2. Three zones in drying grain.

The **wet grain** above the drying zone is near the initial moisture content and is subject to rapid spoilage if the conditions within the bin are suitable for mold growth. Close observation of the surface grain is critical to the success of low temperature grain drying. This grain must be dried within the allowable storage time (AST). If the surface grain begins to spoil, it must be removed and dried by some other method.

The AST is somewhat longer than expected because the air leaving the drying zone is cooler than the entering air which extends the AST of the wet grain.

In example 3, conditions within the bin were 90% relative humidity for 20% moisture content corn at 50 F. These conditions are favorable for mold growth. The temperature in the top layer, however, can be as low as 40 F—this 10 degree difference can extend AST by about 50%.

Airflow

Airflow—the Key to Low Temperature Drying

Airflow rate largely determines drying time. It is measured in cubic feet per minute (cfm); recommendations are in airflow per bushel (cfm/bu). The air's ability to evaporate water from the grain (drying potential) mostly affects final moisture content, but also affects drying time.

There are two ways to reduce drying time. Note that the two are not equal in effect.

- **Increase the airflow rate**—more air moving through the grain carries out more water. Drying speed is proportional to airflow. If airflow increases 25%, drying rate also increases by 25%.
- **Raise drying air temperature** to increase water-carrying capacity, Ex 2. During an extended period of damp, cold weather, added heat permits drying to continue. During more normal weather, added heat increases the speed of the drying front, but very little if corn moisture content is below about 24%. Supplemental heat also increases the amount and rate of water removal, but primarily from the lower part of the bin, resulting in overdrying. See section on Supplemental Heat.

Compare the progress of the drying front in the example bins in Fig 3. Each bin is the same: 30' diameter with 17.5' of shelled corn (about 9,900 bu). The bins differ only in the way they are equipped and their energy requirements. Initial corn moisture content is 22%. Drying begins Oct. 15.

Notice that bin #1, with 1.1 cfm/bu, finishes drying in 8 weeks. When 2.5 F of supplemental heat is added, Bin #2 finishes in 7 weeks and the final average moisture content is lower because of overdrying in the bottom.

When airflow is increased to 1.4 cfm/bu, Bin #3 finishes in the fifth week. The final average moisture content is also lower than that of bin #1. The lower final moisture content is mainly because drying is completed earlier in the fall when the drying potential of the air is usually better.

The example bins show that airflow is the key to low temperature drying. The more air that is delivered, the faster the drying front moves through the grain, the greater the amount of water removed, and the more reliable the system is. Supplemental heat does not permit reducing airflow, nor does it make up for inadequate airflow. Because grain in the top does not dry much faster, added heat does not reduce the chance of spoilage. The section, "Supplemental Heat", gives more information.

Airflow Resistance and Static Pressure

There are practical limits to how much airflow can be delivered to a full bin because the grain acts as a barrier. This airflow resistance, which must be overcome by the fan, depends on a number of factors.

- Resistance depends on the type of grain. Larger grains like corn, soybeans, and sunflower seeds provide bigger open spaces between seeds and offer less resistance to airflow. Small grains like oats, rough rice, or wheat pack tighter in a bin and offer greater resistance.
- Broken grain and foreign material increase resistance. They concentrate under the spout and reduce airflow to the "dirty" area. Clean the grain and use a spreader to minimize this problem.

Fig 532-3. Progress of drying front.

Based on 28 years of average weather data for Des Moines, IA. Temperature shown in top of bin is wet-grain temperature.

Bin #1	Bin #2	Bin #3
10 Hp	10 Hp	20 Hp
1.1 cfm/bu	1.1 cfm/bu	1.4 cfm/bu
3.13" SP	3.13" SP	4.63" SP
2.5° from fan	2.5° from fan	3.8° from fan
	2.5° supplemental heat (electric)	

Oct 22 Avg Temp 52 F Avg RH 62 %
Bin #1: 47.7 F; 19.4-21.7%; 16.0 %; 13.4 %
Bin #2: 48.8 F; 20.9-21.8%; 18.5%; 12.4-14.7%
Bin #3: 48.4 F; 19.7-21.7%; 16.6%; 12.6-13.8%

Oct 29 Avg Temp 49 F Avg RH 63%
Bin #1: 45.6 F; 20.0-21.7%; 17.2%; 13.0-14.6%
Bin #2: 46.5 F; 20.5-21.7%; 17.9%; 12.1-14.7%
Bin #3: 46.0 F; 20.2-21.7%; 17.7%; 12.5-14.9%

Nov 5 Avg Temp 43 F Avg RH 66 %
Bin #1: 40.8 F; 19.6-21.6%; 17.1%; 13.0-14.9%
Bin #2: 42.0 F; 20.3-21.5%; 18.0%; 12.1-15.1%
Bin #3: 41.8 F; 19.0-20.7%; 16.4%; 12.5-14.5%

Nov 12 Avg Temp 41 F Avg RH 69 %
Bin #1: 39.9 F; 20.1-21.1%; 15.9-18.1%; 13.0-14.7%
Bin #2: 42.1 F; 20.5%; 15.8-18.4%; 12.1-14.0%
Bin #3: 42.7 F; 16%

Nov 19 Avg Temp 36 F Avg RH 69 %
Bin #1: 35.8 F; 19.1%; 17.0%; 13.1-15.3%
Bin #2: 38.5 F; 16.6%; 12.1-14.6%
Bin #3: Nov 13; Dry

Nov 26 Avg Temp 32 F Avg RH 69 %
Bin #1: 33.6 F; 16.3%; 13.3-15.1%
Bin #2: Nov 22; Dry

	Bin #1	Bin #2	Bin #3
Date dry	Nov. 30	Nov. 22	Nov. 13
MC range	13.3% to 15.5%	12.1% to 15.2%	12.5% to 15.4%
Avg. final MC	14.2%	13.2%	13.4%
Temp. range	34.5 F to 34 F	41 F to 39.2 F	44.8 F to 42.7 F
Fan hours, drying	1104	912	696
Kilowatt hrs. used	9,685	16,027	12,212

- For the same amount of grain and cfm/bu, a larger diameter bin offers less resistance than a smaller one.
- Resistance increases as grain is packed into a bin. A grain spreader may pack grain, but its advantage in spreading fines and leveling outweigh the disadvantage of packing.

Static pressure is the driving force a fan provides to push air through a mass of grain. It is measured in inches of water.

Select a fan that can overcome the resistance in your particular bin system and deliver the required airflow. The section, Matching Fan to Bin for Low Temperature Drying , gives more information.

Fans

Types of Fans

There are two types of fans for grain drying: axial-flow (propeller) and centrifugal (squirrel cage).

Air passes through an axial-flow fan in a direction generally parallel to the fan axis. Most axial-flow fans have blades mounted directly on the motor shaft. The fan is in a cylindrical housing.

Most centrifugal fans used in grain drying are the backward-curved type—the rotor blades curve back away from the direction of rotation. Centrifugal fans are quieter, but may be more expensive to buy.

If an axial-flow and a centrifugal fan of the same horsepower rating are compared, the axial-flow fan usually delivers more air at less than 3.5" of static pressure. Generally, a centrifugal fan performs better when higher static pressure (greater than 4.5") is required. For static pressures between 3.5" and 4.5", consider both types. Axial-flow fans have an advantage in layer filling because the fan operates a significant portion of the time when the bin is less than full (low static pressure).

Fans heat the drying air and increase its drying potential. A centrifugal fan increases drying air temperature 1.5 F to 3.5 F. An axial-flow fan usually adds more heat (2 F to 4 F) because its motor is mounted in the airstream.

Fan Performance

Fan performance is based on the airflow (cfm) a fan delivers at a given static pressure.

Laboratory tests predict better performance than installed fans can provide. Pick a fan that delivers about 25% more than the recommended airflow.

Fan size is designated by the horsepower (Hp) of the fan motor. But, don't buy a fan on the basis of Hp alone because output airflow for fans of the same Hp varies. Also, Hp gives little indication of the airflow delivered at a given static pressure. Know the static pressure required for your system and compare performance data of several fans before selecting one. The section Matching Fan to Bin for Low Temperature Drying, gives more information on fan selection.

Rated Hp is not a good indicator of the power required to operate the fan motor, either. Most air-over fan motors (motors mounted in the airstream) operate above their rated Hp which means they draw more electricity than you would expect.

If possible, compare fans for **cfm/watt** as well as airflow delivery. Cfm/watt is a measure of a fan's electrical efficiency much like miles per gallon measures a car's fuel efficiency. Drying costs are directly related to cfm/watt. The cost of drying with a fan that delivers 1 cfm/watt is half the cost of drying with a fan that delivers only 0.5 cfm/watt. Select an axial-flow fan that delivers at least 1 cfm/watt at 3" of static pressure. A centrifugal fan should deliver at least 1 cfm/watt at 4" of static pressure.

Although cfm/watt data for fans is seldom published, the manufacturer should have it. Insist the sales representative try to get this rating for you.

Management Decisions Affecting Fan Selection

Before selecting a fan you need to have an idea of the following:

- Kind of grain and how much will be dried with low temperature
- Harvest capacity
- Bin size (diameter and depth) and static pressure
- Average harvest moisture from year to year
- Power availability

Bin size

The shallower the depth of grain, the less static pressure and therefore the less horsepower required to deliver a given airflow through that grain. Decrease maximum depth to maintain reasonable fan horsepower levels at high airflows. Select large diameter bins when possible.

Harvest moisture content

Lower harvest moisture makes low temperature drying work better, and reduces energy use and drying costs. Manage planting/harvesting/drying practices to "control" moisture content going into a low temperature bin as much as possible.

Take advantage of short season varieties to get a lower harvest moisture content. Some early-maturing varieties produce yields comparable to long-season varieties.

Power availability

Fan horsepower may be limited by the local electric power supplier. Find out the approximate fan size that the supplier will accept and the rates and demand charges. Shallower grain depths permit higher airflows per horsepower. At 1.25 Hp/1000 bu of bin capacity, a fan delivers about:

1 cfm/bu through 20' of corn
1.25 cfm/bu through 16' of corn
1.5 cfm/bu through 13' to 14' of corn

Matching Fan to Bin for Low Temperature Drying

Tables in this section help select a fan for a new system, and estimate airflow in an existing bin.

Select a fan that delivers the airflow given for your location in Fig 4. Provide more airflow if you can. The eastern and southern portions of the region need higher airflows because of higher humidities and somewhat warmer temperatures at harvest. If you follow the recommendations for bin filling and dryer management, you should be able to dry grain to a safe storage moisture content before the AST is exceeded. Do not try to dry grain with less than 1 cfm/bu because the airflow is not sufficient to cool grain and prevent major spoilage.

More airflow than the minimums in Fig 4 increases drying speed, gives an added margin of safety, and reduces the need to finish drying in the spring. Table 3 gives static pressure for various airflows and grain depths. Use this table as a guide to practical maximum airflows. To reduce drying costs, design your system for less than 4.5" of static pressure. See Table 4 for bin capacities.

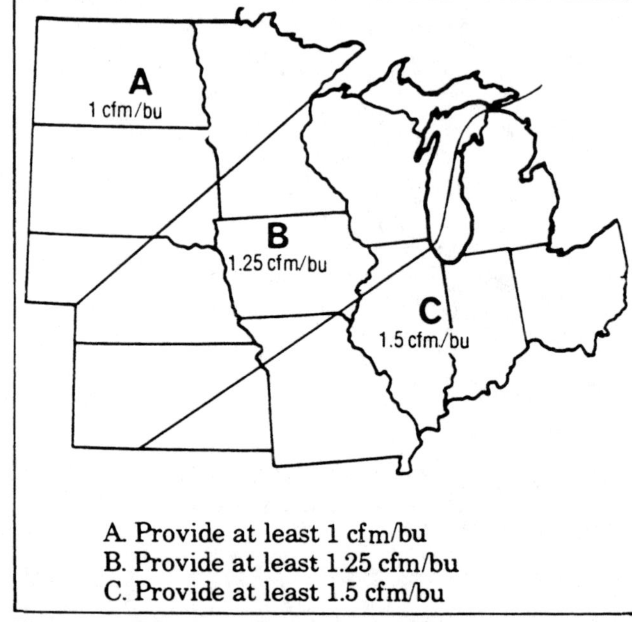

Fig 532-4. Full-bin airflow for fan selection.

Table 532-3. Static pressures for shelled corn.

This table is based on center filling or use of a drop-type distributor with wet shelled corn. With a sling-type distributor, multiply the static pressure by 1.3. With a sling-type distributor and a stirring machine, use static pressure values in the table.
Avoid static pressures above 5.0''.

Grain depth ft	Airflow rate, cfm/bu 0.5	0.75	1.0	1.25	1.5	1.75	2.0	2.25	2.5	2.75	3.0
	static pressure, inches of water										
2							0.1	0.1	0.1	0.1	0.1
3			0.1	0.1	0.1	0.1	0.1	0.1	0.2	0.2	0.2
4		0.1	0.1		0.2	0.2	0.2	0.3	0.3	0.4	0.4
5	0.1	0.1	0.2	0.2	0.3	0.3	0.4	0.5	0.5	0.6	0.7
6	0.1	0.2	0.2	0.3	0.4	0.5	0.6	0.7	0.8	0.9	1.0
7	0.2	0.2	0.3	0.5	0.6	0.7	0.9	1.0	1.2	1.3	1.5
8	0.2	0.3	0.5	0.6	0.8	1.0	1.2	1.4	1.6	1.8	2.1
9	0.3	0.4	0.6	0.8	1.0	1.3	1.6	1.8	2.1	2.4	2.8
10	0.3	0.5	0.8	1.0	1.3	1.7	2.0	2.4	2.8	3.2	3.6
11	0.4	0.7	1.0	1.3	1.7	2.1	2.5	3.0	3.5	4.0	4.6
12	0.5	0.8	1.2	1.6	2.1	2.6	3.1	3.7	4.3	5.0	5.7
13	0.6	1.0	1.4	1.9	2.5	3.1	3.8	4.5	5.3	6.1	7.0
14	0.7	1.2	1.7	2.3	3.0	3.7	4.6	5.4	6.4	7.4	8.4
15	0.8	1.4	2.0	2.7	3.5	4.4	5.4	6.5	7.6	8.8	
16	0.9	1.6	2.3	3.2	4.2	5.2	6.4	7.6	8.9		
17	1.1	1.8	2.7	3.7	4.8	6.1	7.4	8.9			
18	1.2	2.1	3.1	4.3	5.6	7.0	8.5				
19	1.4	2.4	3.5	4.9	6.4	8.0	9.8				
20	1.5	2.7	4.0	5.5	7.2	9.1					

Axial fans: above solid line. Centrifugal fans: below dashed line.

Example 7. Design a new system.

1. How much grain in each bin? 10,000 bu of corn
2. Recommended full-bin airflow rate for your area, Fig 4? 1.25 cfm/bu
3. Recommended airflow (1 x 2)? 12,500 cfm
4. Total required airflow? Add 25% to #3: 15,625 cfm
5. Select possible bin sizes, Table 4.
   Bin A 30'd x 18'h
   Bin B 33'd x 15'h
6. Find static pressure of selected bins, Table 3.
   Bin A 4.3"
   Bin B 2.7"
7. Select a fan for bin A that delivers 15,625 cfm at 4.3" of static pressure. At this static pressure consider both axial-flow and centrifugal fans. Remember, axial-flow fans have an advantage in layer or controlled filling. Compare cfm/watt ratings if available.
8. Repeat #7 for bin B. Select a fan that delivers 15,625 cfm at 2.7" of static pressure.
9. Compare possible fan/bin combinations for fan and bin costs, projected annual drying costs, and cfm/watt. Remember that shallow grain depths are better for low temperature drying. Also, consider dealer service and maintenance.

Example 8. Estimate full-bin airflow.

The bin is 30'd x 18'h, holds 10,000 bushels, and is equipped with a 10Hp fan.

1. Get fan performance data for your fan. Fig 5 is the fan curve for this example.
2. Calculate cfm/bu for several static pressures on the graph (or table): cfm/bu = airflow ÷ bin capacity; at 1" of static pressure, 17,800 cfm ÷ 10,000 bu = 1.78 cfm/bu.

Static pressure Fig 5	Total airflow Fig 5	Cfm/bu Fig 5
1"	17,800	1.78
2"	15,300	1.53
3"	12,800	1.28
4"	10,800	1.08

3. For about the same static pressures, find the approximate cfm/bu for the 18' deep bin. In Table 3, the 18' depth shows 1.2" of static pressure at 0.5 cfm/bu.

Static pressure Table 3	Cfm/bu Table 3
1.2	0.5
2.1	0.75
3.1	1.0
4.3	1.25

4. Plot the cfm/bu data on a graph as in Fig 6. The fan operates at the static pressure where the cfm/bu values are equal (the lines cross). The graph is actually a fan curve with airflow in cfm/bu. The example fan is operating between 1 and 1.25 cfm/bu (between 3" and 4" of static pressure).
5. Because this procedure only estimates airflow, round down. (Use 1 rather than 1.25 cfm/bu.) Assume the system operates at the lower airflow rate when planning a filling strategy.

Table 532-4. Capacities of round bins.
This chart is based on 1 cu ft = 0.8 bu and does not involve test weight, moisture content, or shrinkage.

Grain depth ft	Bin diameter, ft 15	18	21	24	27	30	33	36	39	42	48	60
						- - - - bushels - - - -						
1	142	204	278	363	460	568	687	818	960	1113	1453	2271
2	284	409	556	727	920	1136	1374	1635	1919	2226	2907	4542
3	426	613	835	1090	1380	1703	2061	2453	2879	3338	4360	6813
4	568	818	1113	1453	1840	2271	2748	3270	3838	4451	5814	9084
5	710	1022	1391	1817	2299	2839	3435	4088	4798	5364	7267	11355
6	852	1266	1669	2180	2759	3407	4122	4905	5757	6677	8721	13626
7	994	1431	1947	2544	3219	3974	4809	5723	6717	7790	10174	15897
8	1136	1635	2226	2907	3679	4542	5496	6541	7676	8902	11628	18168
9	1277	1840	2504	3270	4139	5110	6183	7358	8636	10015	13081	20439
10	1419	2044	2782	3634	4599	5678	6870	8176	9595	11128	14535	22710
11	1561	2248	3060	3997	5059	6245	7557	8993	10555	12241	15988	24981
12	1703	2453	3338	4360	5519	6813	8244	9811	11514	13354	17441	27252
13	1845	2657	3617	4724	5978	7381	8931	10628	12474	14466	18895	29523
14	1987	2861	3895	5087	6438	7949	9618	11446	13433	15579	20348	31794
15	2129	3066	4173	5450	6898	8516	10305	12264	14393	16692	21802	34065
16	2271	3270	4451	5814	7358	9084	10992	13081	15352	17805	23255	36336
17	2413	3475	4729	6177	7818	9652	11679	13899	16312	18919	24709	38607
18	2556	3679	5008	6541	8278	10220	12366	14716	17271	20030	26162	40878
19	2697	3883	5286	6904	8738	10787	13053	15534	18231	21143	27616	43149
20	2839	4088	5564	7267	9198	11355	13740	16351	19190	22256	29069	45420

Summary of Design Recommendations

- Equip bin with full perforated floor.
- Install a grain distributor.
- Provide a minimum of 1 cfm/bu.
- Provide roof openings of 1 sq ft/1000 cfm.
- Design for a static pressure of 4.5" or less.
- Limit bin depth to 20'; 14' or 15' is preferred.

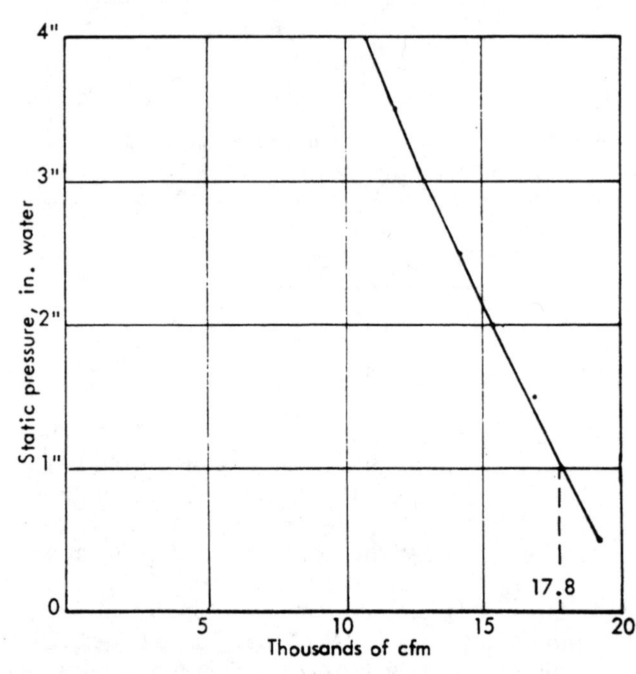

Fig 532-5. Performance data for fan in Example 8.
Dashed lines indicate example in text.

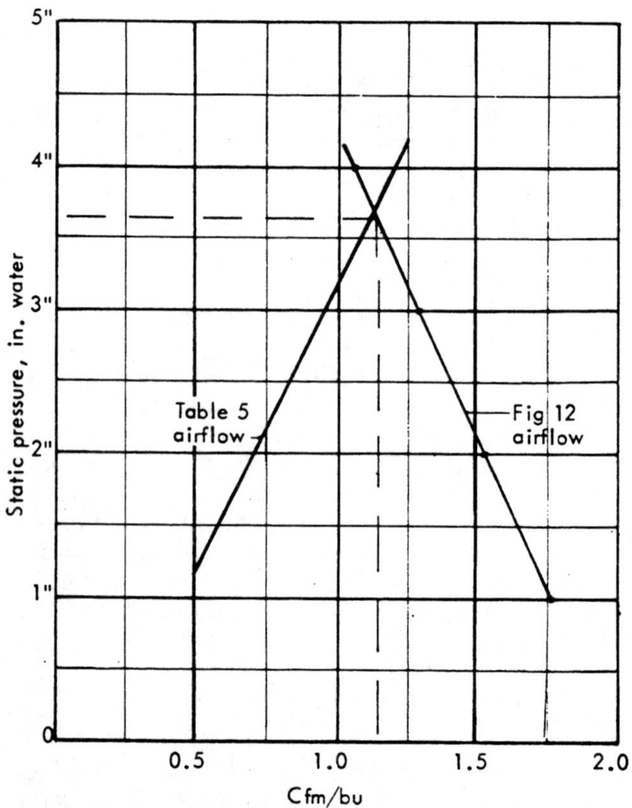

Fig 532-6. Estimating airflow for Example 8.
Lines cross at estimated airflow.

Managing a Low Temperature Dryer

Filling Strategies

Do not "overload" the system with wetter grain than it is designed to dry. If you do, the risk of spoilage increases sharply. Choose from 3 filling strategies to handle harvest moisture conditions for an individual year.

- **Single filling:** Sometimes called fast filling. The bin is filled in 1 to 3 days.
- **Layer filling:** Grain is added in layers over a period of time. There are two ways to decide when to add the next layer: **1)** Let the drying front come through before adding more, or **2)** Fill on a set time period (eg. weekly). Drying front may or may not be through the top. Layer filling is a safe way to dry corn, but can slow harvest when drying conditions are good, relative to controlled filling.
- **Controlled filling:** A method of managing layer filling. Harvest proceeds as fast as drying conditons permit because you don't wait for the drying front to come through the surface grain.

Single filling

Harvest moisture content is the limiting management factor in a single filling situation. Choose varieties known for early field dry down, use combination drying, or delay harvest and let grain dry in the field.

Fig 7 gives the maximum initial moisture content for fast filling systems depending on harvest date and airflow rate. Following these recommendations, corn should dry to a safe storage moisture content before the allowable storage time (AST) is exceeded 9 years out of 10. The initial moisture content is lower early in the season because the warmer temperatures of early fall shorten AST. The recommended initial moisture content drops by mid-November because the drying potential of the air has dropped.

Layer filling

Usually the first grain harvested is also the wettest. Limiting grain depth to get a higher airflow rate is one way to handle higher moisture corn than the system can handle on a full-bin basis.

Airflow is higher in a partially filled bin because of lower grain depth and static pressure. The actual airflow rate depends on individual fan performance, but in a bin designed for 1 cfm/bu on a full-bin basis, the airflow rate is at least 4 cfm/bu if the bin is ¼ full (10,000 cfm ÷ 2,500 bu = 4 cfm/bu), Fig 8.

When layer filling, you load a fraction of the bin's capacity each week. While the first layer starts to dry in the bin, the standing corn continues to dry down in the field. The second layer needs less airflow when it is loaded into the bin.

The maximum moisture contents given in Fig 7 can be increased if you spread out filling. If you harvest grain wetter than the moisture limits of Fig 7, layer fill to compensate:

- Add grain every 5 to 7 days

Before Nov. 1:

- Add ¼ of bin depth if 3 points too wet
- Add ⅓ of bin depth if 2 points too wet
- Add ½ of bin depth if 1 point too wet

After Nov. 1:

- Add ¼ of bin depth if 2 points too wet
- Add ⅓ of bin depth if 1 point too wet

Zone	Full-bin airflow cfm/bu	Harvest date 9-1	9-15	10-1	10-15	11-1	11-15	12-1
		- - Initial moisture content, percent -						
A	1.0	18	19.5	21	22	24	20	18
	1.25	20	20.5	21.5	23	24.5	20.5	18
	1.5	20	20.5	22.5	23	25	21	18
	2.0	20.5	21	23	24	25.5	21.5	18
	3.0	22	22.5	24	25.5	27	22	18
B	1.0	19	20	20	21	23	20	18
	1.25	19	20	20.5	21.5	24	20.5	18
	1.5	19.5	20.5	21	22.5	24	21	18
	2.0	20	21	22.5	23.5	25	21.5	18
	3.0	21	22.5	23.5	24.5	26	22	18
C	1.0	19	19.5	20	21	22	20	18
	1.25	19	20	20.5	21.5	22.5	20.5	18
	1.5	19.5	20	21	22	23.5	21	18
	2.0	20	21	22	23	24.5	21.5	18
	3.0	21	22	23.5	24.5	25.5	22	18
D	1.0	19	19.5	20	21	22	20	18
	1.25	19	19.5	20.5	21	22.5	20.5	18
	1.5	19	19.5	21	22	23	21	18
	2.0	19.5	21	21.5	23	24	21.5	18
	3.0	20.5	21.5	23	24	25	22	18

Fig 532-7. Maximum corn moisture for single-fill drying.
Developed by Thomas L. Thompson

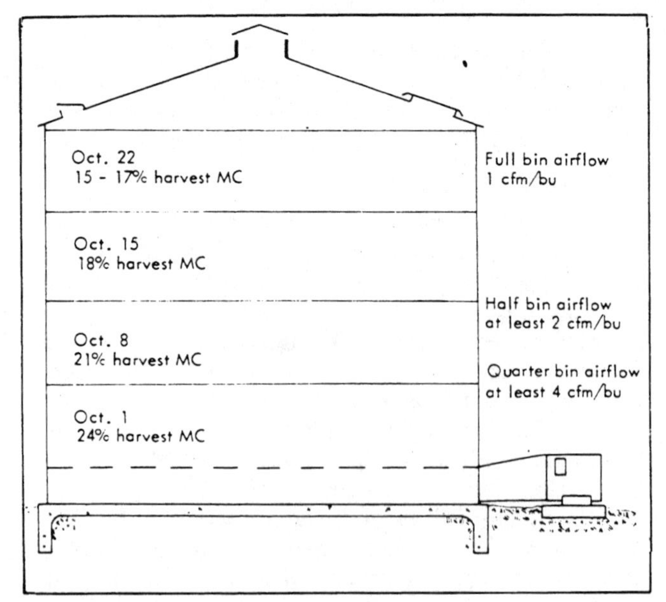

Fig 532-8. Example of layer filling.
The higher airflow rates early in the filling permit a higher initial moisture content to be loaded.

Controlled filling*

Controlled filling is similar to layer filling except the drying front does not come through the top before another layer is added. Filling proceeds as fast as drying conditions permit. If properly managed this strategy greatly reduces risk of spoilage. Systems designed for 1 to 1.5 cfm/bu with an axial-flow fan are recommended.

Controlled filling requires monitoring the drying front progress and information about the moisture content of the grain in the field. With these two pieces of information you decide how much grain can be harvested without overloading the system.

To manage the controlled filling strategy you need to know:

- The quantity of corn per foot of bin depth, Table 5.
- The amount of airflow in each bin in cfm/sq ft, Table 5.
- The moisture content of the grain in the field. A portable moisture meter is accurate enough.
- The maximum safe depth of wet corn, Table 6.
- Rules of controlled filling, Table 7.
- Progress of the drying front. See "Locating the Drying Front", below.

Tables 5 to 8 explain how controlled filling works. Once you understand the filling procedure, you need only Tables 6 and 7.

Locating the Drying Front

You can walk on the surface of wet grain (20% mc or higher); at most your feet sink 2" to 3". In a bed of dry grain your feet sink 5" to 10". It is this difference in firmness that you feel when using a drying front probe.

*Developed from ASAE paper 79-3032, Van Ee and Kline.

To locate the drying front slowly push the probe into the grain until you feel the end slip into the drying front. Using the 2' spacings of the extension couplers, estimate the distance to the drying front from the grain surface.

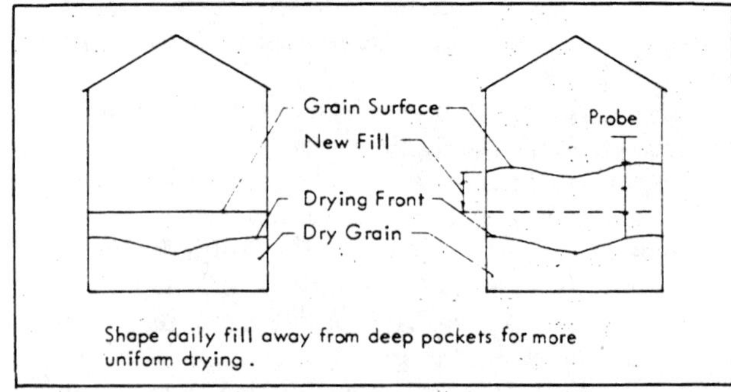

Fig 532-9. Locating the drying front.

Progress of the drying front may not be uniform throughout the bin because of fines accumulation and reduced airflow. Often it moves slower through the center of the bin. Probe several areas of the bin at least once a week to be sure you find the deepest area of the front. During filling, you can vary the thickness of a layer slightly to encourage uniform progress of the front.

Planning Before Harvest

1. Determine quantity of grain per foot of bin depth. Table 5.
2. Determine total airflow rate at 3" of static pressure for each fan. See manufacturer's data. Then, from Table 5, figure cfm/sq ft for each fan.
3. Mark the performance of your system on Table 6. This estimates system performance for the intermediate stage of drying when grain depth is 6' to 12' (static pressure is about 3").
4. The next higher level of airflow rate on Table 6 estimates system performance in the beginning stage of filling 0' to 6'. And, the next lower level of airflow estimates performance in the final stages, 12' to 18'.
5. Paint a grain depth scale in 1' increments in 2 or 3 places around the inside of the bin.
6. You are ready to fill. Table 8 outlines a typical harvest/fill schedule and shows how the 6 rules of controlled filling (Table 7) affect harvest.

Example 9. Planning for controlled filling.

1. A 30' diameter bin holds 570 bu per foot of grain depth.
2. 10 Hp fan, 12,000 cfm at 3" of static pressure: 16.9 cfm/sq ft
3. Underline the wet grain depth for the average airflow of 16 cfm/sq ft and label **Intermediate 6' to 12'.**
4. Underline the next higher airflow, 18 cfm/sq ft, and label it as **Beginning 0' to 6'.** Underline the next lower airflow, 14 cfm/sq ft, and label as **Final 12' to 18'.**

Table 532-5. Bin capacity and average airflow.
Dashed lines indicate examples in text.

Fan discharge cfm	Bin diameter, ft. 18	21	24	27	30	33	36	42	48
	Quantity of corn, bu/ft								
	200	280	360	460	570	690	820	1110	1450
	Average airflow, cfm/sq ft								
1,000	3.9								
2,000	7.8	5.7	4.4						
3,000	11.8	8.6	6.6	5.2	4.2				
4,000	15.7	11.5	8.8	7.0	5.6	4.6	3.9		
5,000	19.6	14.4	11.0	8.7	7.0	5.8	4.9	3.6	
6,000	23.6	17.3	13.2	10.4	8.5	7.0	5.9	4.3	
7,000	27.6	20.2	15.4	12.2	9.9	8.1	6.9	5.0	3.8
8,000	31.4	23.1	17.6	13.9	11.3	9.3	7.8	5.7	4.4
9,000		26.0	19.8	15.7	12.7	10.5	8.8	6.5	4.9
10,000		28.8	22.1	17.4	14.1	11.7	9.8	7.2	5.5
12,000		34.6	26.5	20.9	16.9	14.0	11.7	8.6	6.6
14,000			30.0	24.4	19.8	16.3	13.7	10.1	7.7
16,000			35.3	27.9	22.6	18.7	15.7	11.5	8.8
18,000				31.4	25.4	21.0	17.6	13.0	9.9
20,000				34.9	28.3	23.3	19.6	14.4	11.0
22,000					31.1	25.7	21.6	15.8	12.1
24,000					34.9	28.0	23.6	17.3	13.2
26,000						30.4	25.5	18.7	14.3
28,000						32.7	27.5	20.1	15.4
30,000							29.4	21.6	16.5
34,000							33.4	24.5	18.7
36,000								27.4	20.1
42,000								30.3	23.2
46,000									25.4
50,000									27.6
54,000									30.9

Table 532-6. Maximum grain depth above the drying front.
Dashed lines indicate examples in text.

Average airflow cfm/ft^2	Incoming grain moisture, percent 18	19	20	21	22	23	24	25	26
	Max. wet grain depth, ft								
6	15	7.5	5.0	3.7	3.0	2.5	- Don't fill		
8		10	6.6	5.0	4.0	3.3	2.5	2.0	
10		12	8.3	6.2	5.0	4.1	3.1	2.5	2.1
12		15	10	7.5	6.0	5.0	3.7	3.0	2.5
14 Final (12'-18')		18	11	8.7	7.0	5.8	4.4	3.5	2.9
16 Intermediate (6'-12')			13	10	8.0	6.6	5.0	4.0	3.3
18 Beginning (0'- 6')			15	11	9.0	7.5	5.6	4.5	3.7
20			17	13	10	8.3	6.2	5.0	4.1
24				15	12	10	7.5	6.0	5.0
28 20 feet depth				17	14	11	8.7	7.0	5.8
32 max. recommended					16	13	10	8.0	6.6
36					18	15	11	9.0	7.5
40						17	12	10	8.3

Table 532-7. Rules for controlled filling.

Rules 4 and 5 must both be met before more grain filling is allowed.

1. Begin harvest as soon as the corn moisture in the field is 26% or less.
2. For corn 24% to 26% moisture content, the maximum daily filling depth is 2'.
3. For corn less than 24% moisture content, the maximum filling depth is 4'.
4. For corn 22% moisture content and higher, the initial drying front must be at least halfway up the grain profile before you can add more wet grain.
5. For all incoming corn moistures, the filling depth may not exceed the maximum recommended wet grain depth, Table 8.
6. Controlled filling may be skipped on any day the quantity allowed is too small to be practical or other harvest activities take precedence.

Table 532-8. Example of controlled filling.

Oct. 10: corn moisture 26%; harvest begins (Rule #1). Table 532-6 shows that in the beginning stage of filling, the example bin can handle 3.7′ of wet grain, but Rule #2 controls and reduces fill to 2′.
Oct. 11: corn moisture 25.5%; drying front must be halfway through the grain that is in the bin (Rule #4). It is not; do not fill.
Oct. 12: corn moisture 25%; drying zone is halfway through grain profile. Add 2′.
Oct. 16: corn moisture 23%; maximum daily fill is 4′ if drying zone is halfway through existing grain. Table 532-6 shows 6.6′ is the maximum depth of wet grain allowed. 3.6′ is added (6.6′ - 3′ = 3.6′). 4′ maximum is not violated.
Oct. 19: corn moisture 21.5%; maximum wet grain depth (Table 532-6) is 9′; probing shows 5.1′ of wet grain already in the bin; add 3.9′ (Rule #5).
Oct. 20: corn moisture 21%; 8.5′ of wet grain in the bin, can have maximum of 8.7′. Choose not to fill (Rule #6).
Oct. 23: filling completed after 14 days.
Nov. 11: drying completed after 33 days. During the early stages of filling, Rules 2, 3, and 4 usually control. During intermediate stages, Rules 4 and 5 usually control. For the final stage of filling, Rule 5 usually controls.

Date	Grain moisture in field percent	Bin condition, beginning of day: Total grain bu	Grain depth ft	Dry front position ft	Depth of wet grain ft	Max. wet grain depth[a] ft	Quantity to add ft[b]	bu[c]
Beginning								
Oct. 10	26.0					–	–	–
Oct. 11	25.5	1140	2.0	0.5	1.5	4.1	–	–
Oct. 12	25.0	1140	2.0	1.0	1.0	4.5	2.0	1140
Oct. 13	24.5	2280	4.0	1.5	2.5	5.1	–	–
Oct. 14	24.0	2280	4.0	2.0	2.0	5.6	2.0	1140
Oct. 15	23.5	3420	6.0	2.5	3.5	6.5	–	–
Intermediate								
Oct. 16	23.0	3420	6.0	3.0	3.0	6.6	3.6	2052
Oct. 17	22.5	5472	9.6	3.5	6.1	7.3	–	–
Oct. 18	22.0	5472	9.6	4.0	5.6	8.0	–	–
Oct. 19	21.5	5472	9.6	4.5	5.1	9.0	3.9	2223
Final								
Oct. 20	21.0	7695	13.5	5.0	8.5	8.7	0.2[d]	–
Oct. 21	20.5	7695	13.5	5.5	8.0	9.8	1.8	1026
Oct. 22	20.25	8721	15.3	6.0	9.3	10.4	1.1[d]	–
Oct. 23	20.0	8721	15.3	6.5	8.8	11.0	2.2	1254
Oct. 24	*	9975	17.5, full	7.0	10.5	*	*	*
Oct. 25	*	9975	17.5	7.5	10.0	*	*	*
Oct. 26	*	9975	17.5	8.0	9.5	*	*	*
.	.	.	.	.	.	.	.	.
Nov. 11		9975	17.5	17.5	– – Drying completed – – –			

a. From Table 532-6.
b. From Table 532-7.
c. Calculated from bu/ft, Table 532-5.
d. Choose to disregard increments less than 1.5′.
* Filling completed.

Combination Drying

Combination drying fits with all three filling strategies, but is particularly helpful when using single or layer filling because of their initial moisture content limitations. Combination drying uses a high temperature dryer to dry to the moisture content the system can handle. Any type of column dryer, or a bin equipped with a stirring machine and heat can be used.

Fan Management

Regardless of filling strategy, follow the fan management practices given below. Do not shut off the fan during periods of high humidity, or foggy or rainy days. Air movement cools and controls heating of the grain even if drying slows. It also equalizes moisture differences (top to bottom).

Fall Drying

Turn on the fan(s) as soon as grain covers the floor of the bin. **Do not** turn the fan(s) off until one of the following occurs:

- Complete drying: all grain in the bin is dried to 15.5% moisture content or less. Aerate to cool for winter storage and to maintain quality.
- Late in the season if the drying zone is through the top and the moisture content is 18% or less. Begin aeration and winter management. Plan to spring dry.

Aeration—see Managing Dry Grain in Storage

Spring Drying

If spring drying is needed, start the fan when outdoor temperatures average 40 F to 45 F and run continuously until top grain is at 15.5%. Because overdrying is difficult to avoid in the spring, average grain moisture content will likely be 13% to 14%.

Handling Wet Corn in the Wet Season

An unusually wet fall, when grain does not dry down in the field, presents some problems for low temperature drying. Drying potential of the air is low because it is damp. Increasing drying potential with supplemental heat helps little in this type of drying season. Often, drying cannot be finished in the fall (corn moisture to 18% or less). Some options for handling wet grain in the wet season are:

- Use a high temperature dryer to predry to a moisture content your system can handle.
- Feed or sell the grain at harvest.
- Cool and hold the wet grain with aeration. Feed or sell it within the allowable storage time.
- If the grain is above 20%, mix it with dry grain so that grain moisture averages less than 18% and aerate to assure uniform grain temperature throughout the bin.

Summary of Management Tips

- Clean grain to reduce resistance to airflow and improve storability.
- Use grain distributor to spread remaining fines and foreign material, and to improve uniformity of airflow.
- Keep grain level to promote uniform progress of drying front.
- Start fan as soon as bin floor is covered.
- Open all roof hatches and vents when drying to provide adequate escape for exhaust air.
- Run fan continuously until drying is completed, or until average daily temperature is consistently below freezing.
- Cover fans when not in use to prevent air currents through the bin.
- Check grain periodically for signs of heating and/or spoilage.

Supplemental Heat

The drying ability in a low temperature dryer comes mainly from the air, not the added heat as in a high temperature dryer. If adequate airflow is provided through fan size and/or layer filling, corn dries to a safe moisture content for winter storage almost every year by the end of fall. Spring drying is necessary many years to reach a moisture content for safe long term storage.

By adding supplemental heat you cannot decrease the recommended airflow or load the system with wetter corn than recommended, nor will the grain dry much faster. Supplemental heat increases energy use and/or fixed costs at a given airflow. Supplemental heat reduces grain moisture content in the lower layers of the bin, but does not reduce moisture content in the top of the bin.

The controlled addition of supplemental heat (eg. electric heater on a humidistat, continuous fan operation) helps achieve the desired final moisture content and prevent rewetting during an extended period of damp, cold weather. Add heat only when the natural air cannot dry the grain (when relative humidity is above about 75%) to avoid overdrying.

If you want a lower final average moisture content by the end of fall:

- Increase airflow, or
- Add supplemental heat.

Between adding heat or increasing airflow, more air (shallower bins, larger fan) is usually the better choice. Additional airflow has more advantages (reliability, drying capacity) than increasing the drying air temperature.

Increased fan power:

- Speeds drying so that it is finished earlier in the fall when relative humidity is lower and results in a lower average final moisture content.
- Results in increased fan heat which increases temperature and reduces relative humidity of the drying air.
- May require excessive static pressure. Remember that doubling the airflow through a given depth of grain requires about 5 times as much power. Select large diameter, shallow bins to reduce static pressure at high airflows.
- Requires less energy than when heat is provided with an electric resistance heater. See Fig 3.

If you choose to add supplemental heat to an unstirred low temperature dryer, consider a solar collector. A solar collector can provide adequate supplemental heat for low temperature drying because:

- The longer drying period provides an extended period for collecting solar energy.
- Short periods of cloudy weather, and nights seldom cause problems because grain stores energy.
- High airflow requirements along with low temperature rises mean that simple, relatively inexpensive collectors can be used.

Pick a collector that heats drying air temperature an average of 2 F to 3 F over a 24-hour period. Design the collector, ducting, and transition to minimize additional static pressure. For more information, see the section on collector sizing.

Note that the additional static pressure from the collector, ducting, and transition can reduce airflow 10% to 15%. Select a fan that delivers the minimum recommended airflow despite the increased static pressure. Also, increased electrical demand and costs of the fan reduce potential benefits from the collector. And, keep in mind that solar heat is not a controlled heat source. It is not available on damp, cloudy days when it is needed. It is useful for reaching a lower average final moisture content if that is what you want. Providing heat with a solar collector reduces the amount of purchased energy compared to supplying heat with electricty.

Stirring and Low Temperature Drying

Although a stirring machine is not necessary in a properly designed and managed low temperature drying bin, it can provide some management advantages when drying conditions are unusually difficult. In a low temperature dryer a stirring machine:

- **Reduces overdrying.** Stirring mixes wet and dry grain and reduces moisture differences top to bottom. You can stop drying when the average moisture content is at the desired level instead of waiting for a top layer to dry. If natural air in your climate consistently dries below 15.5%, consider a stirring machine. Balance the cost of the stirrer against the cost of overdrying. Estimate the cost of overdrying at 2% of the market price of the corn/point of moisture below 15.5%/bushel. That is, if corn is dried to 13.5% and the market price is $2.50 then, 2% x 2 points x $2.50 = 10¢/bu. If supplemental heat causes overdrying, turn it off before deciding to buy a stirrer.

- **Increases airflow.** Stirring loosens grain which increases airflow. Airflow may be increased as much as 30% if a slinger-type distributor is used. Increased airflow decreases drying time and improves reliability of a low temperature dryer.
- Breaks up pockets of slow-drying grain.
- Eliminates the drying front. Grain does not dry in layers so there is a gradual moisture reduction in all of the grain and less chance of spoilage in the top.
- **Increases costs.** The total yearly cost of a stirrer (fixed and variable) is about 10¢/bu. This is between 25% and 35% of the yearly cost of a bin without the stirring machine. (Cost based on a 10,000 bu bin at $13,000 construction cost without a stirrer, 10% interest rate, and a $2,500 stirrer depreciated over 7 years.)
- Occupies about 2' of bin space. Install an extra ring to avoid loosing storage capacity.
- May be less reliable than other components of a low temperature dryer because of its complexity.

If you install a stirring machine in a low temperature drying bin, **continue to follow recommended procedures for low temperature drying, such as harvest moisture content and airflow rate.**

Table 532-9. Static pressure.
Data for wheat, barley, and soybeans from *Grain Drying on the Farm*, an extension publication from North Dakota State University (NDSU). Sunflower data from L. F. Backer, NDSU. Sorghum values estimated at 2.5 times corn values given in Table 532-3.

Grain Depth ft.	Airflow, cfm/bu 1	2	3
Wheat	- inches water -		
4	0.5	0.8	1.1
6	0.9	1.6	2.4
8	1.4	2.5	4.3
10	2.0	4.3	7.8
Barley			
4	0.4	0.6	0.8
6	0.6	1.0	1.5
8	0.9	1.5	2.7
10	1.3	2.2	4.3
Soybeans			
4	0.3	0.4	0.4
6	0.4	0.6	0.8
8	0.5	0.8	1.2
10	0.6	1.3	2.0
12	0.9	1.8	2.9
Sunflowers			
4	0.1	0.2	0.3
6	0.2	0.4	0.7
8	0.3	0.7	1.2
10	0.4	1.1	1.9
12	0.6	1.6	2.8
14	0.8	2.2	3.8
16	1.1	2.9	5.0
18	1.4	3.6	6.4
20	1.7	4.5	7.9
Sorghum			
4	0.3	0.5	1.0
6	0.5	1.3	2.5
8	1.3	2.5	5.3
10	2.0	4.3	9.0
12	3.0	6.5	14.3
14	4.3	9.3	21.0

533 MANAGING DRY GRAIN IN STORAGE

Although storage problems are common during the bad harvest year, many also result from poor dry-grain management practices. Proper aeration and insect control along with adequate observation minimize these dry grain problems.

Storage Problem Causes

If grain has been dried correctly for the storage period intended, problems with grain condition usually result from:

- Improper grain cooling.
- Inadequate observation of the stored grain.
- Poor initial grain quality.
- Improper insect control.

Each of these problems can be minimized with good management.

See Table 533-1 for safe grain moisture contents for storage. Reduce the moisture 1% below table values for poor quality grain resulting from drought, frost, blight, harvest damage, etc.

Table 533-1. Safe storage moisture.
Good quality aerated grain.

Grain	Maximum safe moisture content
Shelled corn and sorghum	
To be sold as #2 grain or equivalent by spring	15½%
To be stored up to 1 year	14%
To be stored more than 1 year	13%
Soybeans	
To be sold by spring	14%
To be stored up to 1 year	12%
Wheat	13%
Small grain (oats, barley, etc.)	13%
Sunflowers	
To be stored up to 6 months	10%
To be stored up to 1 year	8%

Grain Temperature and Moisture Migration.

More dried grain goes out of condition because grain temperatures are not controlled than for any other reason. Improper control of temperature causes moisture to move or migrate from one part of the grain mass to another, where the moisture can accumulate and cause grain spoilage problems.

Although moisture migration problems can occur any time grain temperatures vary considerably in different parts of the bin, the most critical time occurs when warm grain is stored into cold winter temperatures. Grain is typically put into storage when the grain temperature is 50°-80°F, and perhaps higher.

A bin of grain has 30%-60% air space within the grain mass, and this air surrounding the grain is at the same temperature as the grain.

By late fall or early winter, average outside temperatures fall to 20°F and colder in the Corn Belt. This drop in temperature causes the grain and air near the bin walls to cool. Because grain has fairly good insulating qualities, most of the large center mass of grain and air in the bin remains at about the same temperature as it was when placed in storage.

These temperature differences create a slow movement of both moisture and air. This natural air circulation is called convection currents.

Convection currents develop because the grain and air near the bin walls cool. Cooling causes the air to become heavier and settle toward the bin floor. As the air moves toward the floor and then into the center of the bin, it becomes warmer and less dense or lighter. This causes the air to rise through the warm grain where it continues to increase in temperature. As the air increases in temperature, its moisture-holding capacity increases and it begins to absorb small amounts of moisture.

The slowly moving air rises into the cooler grain mass in the upper portion of the bin where the air is cooled. Some of the moisture in the air is deposited in the grain by both moisture condensation onto the cold grain surfaces and by moisture diffusion into the cooler grain.

Fig 533-1 shows this natural movement of air and moisture, called moisture migration, which is the most common cause of problems in stored grain.

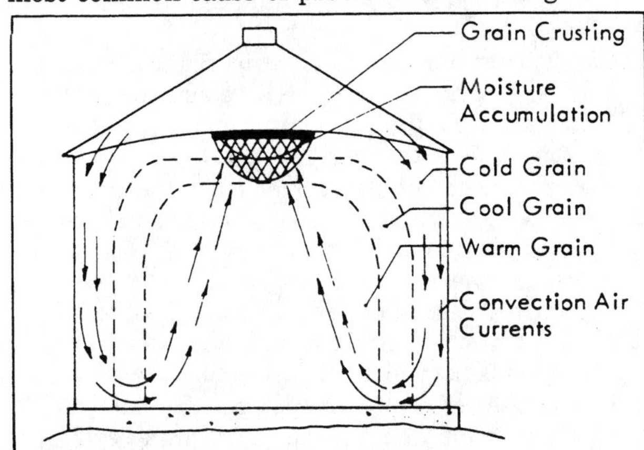

Fig 533-1. Moisture migration in grain.
Relative winter temperatures and moisture movement.

Moisture migration problems are usually first evident when a crust begins to form on the grain surface. The crust is seen as a thin surface layer of kernels that are slightly wet, slimy or tacky. The kernels tend to stick or even freeze together.

Crusting forecasts a potentially severe spoilage process. It is almost a sure sign of undesirable temperature differences in the grain.

If crusting occurs, the surface should be stirred or, in extreme cases, removed, and aeration started immediately. If the top surface is permitted to seal over, severe spoilage is imminent.

Often, minor moisture migration problems develop into severe spoilage only during spring warm-up. Moisture can also migrate when cold grain is stored into the warm or hot months.

*By B. A. McKenzie, Purdue University, and L. D. VanFossen and H. J. Stockdale, Iowa State University.

Check for condensation or frost on the underside of the roof surface on cold, crisp mornings before the sun warms the roof. Such condensation is almost always a sure sign of moisture migration and often indicates poor grain condition.

Regardless of what time of the year grain is stored, the best rule is to maintain the grain temperature within 15°-20°F of the average daily temperature.

Aerate for Temperature Control

In the past, elevator operators turned grain from bin to bin to equalize temperatures, cool the grain, and blend the grain mass. Although turning effectively interferes with natural moisture migration and breaks up hot spots, turning requires an empty bin, time and labor. It also tends to damage more grain.

Modern grain management uses aeration to control grain temperature. Aeration forces air through the grain either continuously or intermittently. Aeration is **not** drying although small moisture changes do occur with a change in temperature. Aeration may be used to maintain quality in wet grain and to cool hot grain from a dryer, but these practices require procedures not discussed here.

Aerate to cool stored grain in the fall so there is no warm grain mass in the center of the storage. Aerate to warm the grain in the spring if storage is to be continued into the hot summer months.

Fig 533-2 shows how grain is tempered (cooled or warmed) by either negative or positive aeration systems. With either system, a tempering (cooling or warming) zone moves through the grain.

The tempering zone starts at the top of the grain in a negative pressure or suction type aeration system and moves down. The last grain to cool or warm is normally next to the floor.

In a positive pressure system, the tempering zone starts at the bottom of the bin and moves up. The grain at the top is normally the last to cool or warm.

The movement of the tempering zone completely through the grain is one cooling or warming cycle. Once a cycle has been started, operate the fan continuously until the zone moves completely through the grain. The time required to complete each cycle depends almost entirely on the aeration airflow rate, as shown in Table 533-2.

Grain drying or rewetting is usually insignificant during grain aeration. Because the cooling (or warming) front moves through the grain about 50 times faster than a drying or wetting front, only a small fraction of the grain is rewetted during an aeration cycle, even with high humidities. As a precaution, operate the aeration fan only long enough to accomplish the grain cooling or warming cycle. This is particularly important with higher capacity aeration fans.

Although round bins are shown here, the basic storage and aeration principles are the same for flat grain storage.

Aeration Airflow Rates

Successful aeration rates for farm-type storage usually range from 1/20 cfm/bushel (cfm = cubic feet of air per minute) to over 1 cfm/bushel in bins equipped for rapid grain cooling or drying. About 1/10 cfm/bushel is most common in farm bins, with higher rates more often selected for new bins. The aeration rate for a particular bin depends on the type of bin, air distribution system, desired storage moisture content, and the management procedures to be practiced.

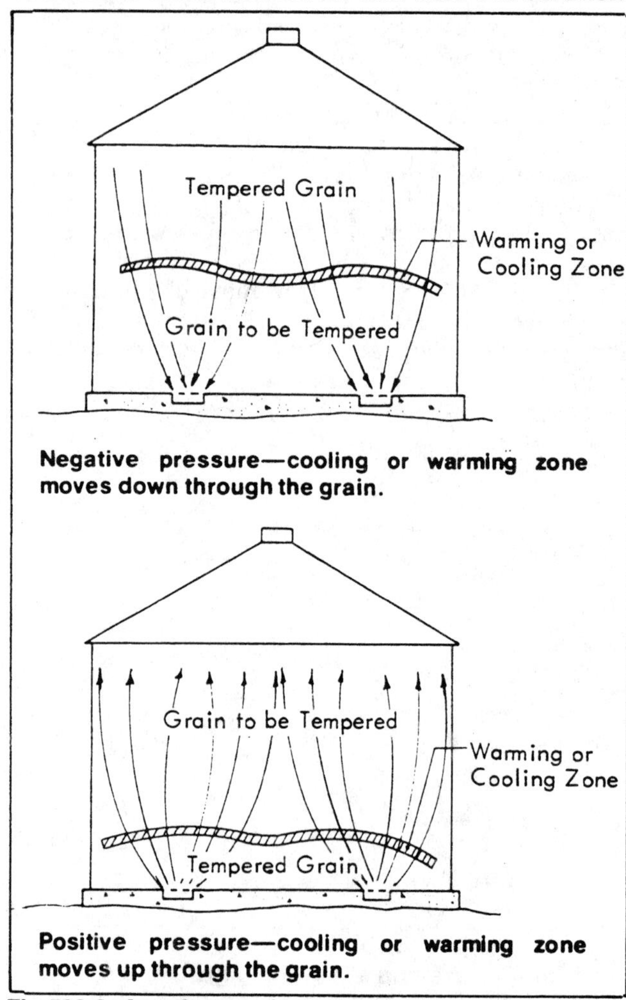

Fig 533-2. Aerating to change grain temperature.
Grain behind the zone of cooling or warming has been tempered. Grain in the zone is changing and approaching outside air conditions. Grain in front of the zone has changed very little. One temperature change or cycle is the time needed to move the tempering zone completely through the stored grain mass.

The time for one cooling or warming cycle to completely pass through grain depends on the aeration rate (cfm/bushel) and the time of the year. The number of hours is estimated by:

$$\text{Hours} = \frac{15}{\text{cfm/bu}} \quad \text{(fall)}$$

$$\text{Hours} = \frac{20}{\text{cfm/bu}} \quad \text{(winter)}$$

$$\text{Hours} = \frac{12}{\text{cfm/bu}} \quad \text{(spring)}$$

Example: During the fall, 1/10 cfm/bushel moves one cooling cycle through the grain in approximately 150 hours, or about 6¼ days (150÷24), or

$$\frac{15}{1/10 \text{ cfm/bu}} = 150 \text{ hours}$$

The hours for several aeration rates have been calculated for you in Table 533-2.

Usually two or three cooling cycles are needed to cool or warm the grain to the desired storage temperatures.

Table 533-2. Grain cooling or warming times.
Times are based on 60-lb bushels in the Midwest and 10°-15° temperature changes.

Airflow rate cfm/bu	Fall cooling hours	Winter cooling hours	Spring warming hours
1/20	300	400	240
1/10	150	200	120
1/5	75	100	60
1/4	60	80	48
1/3	45	61	36
1/2	30	40	24
3/4	20	27	16
1	15	20	12
1¼	12	16	10
1½	10	13	8

Cooling Grain for Winter Storage

Aerate to cool the grain to 35°-40°F for winter storage in most of the Midwest. In the northern parts of the Midwest (South Dakota, North Dakota, Minnesota, and Wisconsin), stored grain should be cooled to 30°F to 35°F because the average winter temperatures are colder. Start an aeration cycle whenever the average day-night air temperature is 10°-15°F cooler than the grain. Because grain from a high-temperature dryer is usually 10°F or more above air temperature at the same time grain is placed in a bin, start aerating immediately or at least as soon as the bin is full.

Not operating an aeration fan long enough is a bad decision that causes grain spoilage. As an example of the time to cool grain during the fall, assume the fan capacity on a bin is 1/10 cfm/bu. Table 533-2 estimates 150 hr (nearly a week) are needed to cool grain 10-15 F in the fall. But roughly 1 to 2 weeks after this cycle is completed, outside air temperatures will have dropped another 10-15 F on an average. Consequently, the cycle must be repeated to bring the grain temperature to the colder outside air temperature. Usually a maximum of three cooling cycles totaling possibly 500 hr of fan operation (150 hr + 150 hr + 200 hr) are needed to properly cool the grain for winter storage. Fig 533-3 shows how outside temperatures and grain temperatures typically lower during the fall with three 1/10 cfm/bu aeration cycles.

The number and length of cooling cycles depend on fan capacity, when and at what temperature grain was binned, and how fast average air temperatures cool during the fall. For example, at 1 cfm/bu rather than 1/10, cooling proceeds 10 times faster and only 15 hours is required for one cycle. Some operators prefer only two cooling cycles to conserve energy. The grain is cooled to 50°-60°F on the first cycle. Additional aeration is then delayed until outside air temperatures will cool grain to 35°-40°F for the final cycle.

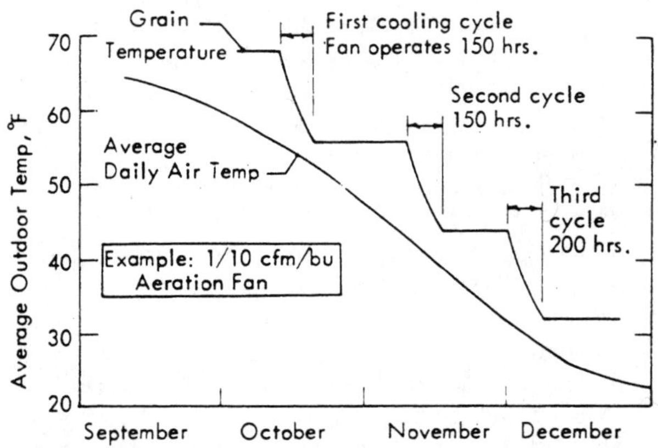

Fig 533-3. Example of aeration cooling cycles.

With low airflow rates (less than 1/5 cfm/bu), you can largely ignore outdoor humidity. Cooling the grain is important. Keep the fan running even if outdoor air has high humidity for a day or two. Any rewetting during these wet periods helps offset the unwanted drying during previous good weather periods. Many operators with airflow rates of 1/10th cfm/bu operate the fan continuously from the time the initial fill is placed in the bin at harvest, until outside air temperatures of 35°-40°F have prevailed for at least 1-2 weeks. With higher airflow rates (more than 1/4 cfm/bu), each cooling cycle is short enough that aeration can be delayed a day or two to avoid warm, high humidity air conditions. **But if there are any signs of heating or hot spots, no matter what the season or the weather, run the fan continuously until no heating can be detected.**

Be sure to continue each aeration cycle until the cooling front has moved completely through the grain. This minimizes the chances for a moisture front within the grain mass that can cause spoilage.

You need to know how fast your aeration fan will move a cooling or warming cycle through a bin. If you don't know the aeration rate (cfm/bu), contact the dealer who furnished the fan.

You can also tell when a cycle has passed through the grain by taking temperatures. With a suction (negative pressure) aeration fan, take the temperature of the exhaust air with a thermometer suspended on the protective screen over the fan. With a pressure aeration system, place a thermometer 6" to 12" into the grain at the top of the bin. When the cooling or warming zone is through the grain, the exhaust air temperature drops suddenly. It will be about the average outdoor temperature when the aeration cycle was started.

Be aware that the thermometer in an airstream does not tell the complete story. For example, air passes around a pocket of fines, so that the temperature which may be quite high, is not reflected at all. Also, unless there is a center duct in the bin floor, the tempering zone may not have reached the grain near the center of the bin floor.

Take the temperature of the first gallon or so of grain removed from the unloading auger to check the bottom grain. Take the temperature of the next gallon or two to check the critical center mass. If any of the temperatures are higher than the airstream temperature, run the aeration fan until these grain areas are cooled.

Taking the temperature in several locations in the grain surface in a pressure aeration system shows when the entire cooling or warming zone has passed through the grain.

Air does not pass through pockets of fines. These pockets can start heating and spoil. Although not completely reliable, taking temperatures, particularly numerous temperatures in a pressure aeration system, can help detect these hot spots.

Record grain temperature readings along with the date and the air temperature taken in the shade near the bin site. Cross check the time for a cooling or warming cycle with Table 533-2. If cooling or warming is taking much longer than expected, either your airflow is less than you thought (which is useful information for next year) or many fines may be blocking airflow. Once you establish this time for one aeration cycle, it will always take about the same time to move a temperature change through the depth of grain for the same bin and aeration equipment.

Freezing grain slightly decreases the potential for spoilage but is not needed for grain that is properly dried, aerated and managed. Because of possible problems, freezing grain is not encouraged.

Condensation during aeration can be a problem in grain cooled well below freezing. It may be difficult to warm grain in the spring without condensation immediately freezing into ice. Frozen chunks of grain block aeration warming cycles and grain unloading. In the winter, operate aeration fans in frozen grain only with relatively dry air that is as cold or colder than the grain.

In the spring, start warming the grain as soon as air temperatures are about 10°F warmer than the grain to avoid massive condensation and freezing. High airflow aeration for warming grain is advantageous because the faster warming reduces the need for aerating in undesirable (particularly high humidity) weather.

Observation and Management of Stored Grain

Observe dry grain in storage weekly during the critical fall and spring months when outside air temperatures are changing rapidly, and during the summer. Check at least every 2 weeks during the winter. Establish a **regular** day of the week and time of the day to check grain to avoid forgetting.

Observe grain by climbing into the bin. If you have filled the bin to the roof, sell or feed enough grain so you can safely get into the grain surface. You are searching for small changes, indicators of problems to come. Check the surface for indications of crusting—wet, sticky, or frozen grain. Observe the roof for evidence of present or past condensation. Thrust your arm into the grain and see if you can detect any heating. Grasp a handful of grain from arm depth and the surface, and smell for musty, moldy odors. Run the fan to smell exhaust air. Keep a small diameter, stiff rod, such as a long endgate rod with a loop for a handle, in each bin. Poke the rod into the mass, down the center, outward toward the sides. Feel for any hard, compacted, or moist masses. Leave the rod thrust into the grain when you leave. Tie it to the roof! Withdraw it quickly next time you check the grain to feel if you can detect warmth. See Table 533-3 for a summary of observations and corrective actions in managing grain.

Once the grain has been cooled, run the aeration fan to smell the exhaust air for bad odors coming from grain that might be going out of condition and to dry condensation in the fan motor. When you run the fan, pick a clear, crisp day when the air and the grain are at about the same temperature. Because this day may not always occur on your chosen, regular inspection day, it may not be feasible to always run the fan when you check grain condition.

It is best to cover the fan intake when not aerating to block the air draft that is drawn in through the fan and up through the grain by the natural chimney effect of tall bins. Covering the fan also prevents rodent and pet access.

If additional aeration is needed to cool small hot spots, run the fan as long as outdoor air is not more than about 10°F above the grain temperature. If necessary, turn off the fan during the warm part of the day. If serious heating is occurring, run the fan day and night regardless of weather and temperature until the heating has stopped.

If the heating can't be stopped, the best action will likely be to remove the grain for drying, cleaning, breaking up hot spots, or for selling. Remember, it's better to sell grain with a minor storage problem, even at low grain prices, than to permit a bin to go completely out of condition.

Warming Grain in the Spring

Consider warming the grain with the aeration fan if:

- The grain will be stored into July and August.
- The grain was accidentally or intentionally frozen.

Spring warm-up prevents moisture migration with resulting high-moisture pockets that may spoil. Proper warming thaws any frozen grain that may interfere with aeration or handling.

Start the fan when the average outdoor temperature is 10°-15°F above the grain temperature. Always run the fan continuously for a complete warming cycle; i.e., until the warming front has moved completely through the grain and the average grain temperature has been raised 10°-15°F. Stopping midway almost guarantees a deposit of condensed moisture that will encourage spoilage. Repeat the warming cycles as needed to bring the average grain temperature up to 50°-60°F.

Conduct weekly grain inspections. Warmer weather and warmer grain temperatures encourage mold and insect activity.

Summer Management

Do not warm the grain to high summer temperatures (80°-90°F). Aerate as needed if heating occurs, but stop spring warm-up at about 50°-60°F.

Although some good grain managers are storing corn at 15½% into summer in the colder areas of the Corn Belt, the grain should be down to the moisture contents in Table 533-1 to minimize risk. All of the grain in the storage must be at or below the maximum. Pockets of higher moisture grain will likely spoil.

Other Aeration Tips

Suction aeration systems, especially with high airflow, should not be operated until several feet of grain are in the bin to avoid pulling fines toward the duct. Also, a perforated floor and metal perforated ducts get cold while the fan is off. Consequently, when suction aeration starts, warm, moist air from the grain can cause condensation on the floor and in grain near the floor. The wet grain can spoil and plug the aeration system. Also, warm, moist air drawn through uncovered aeration fans when they are not operating can cause moisture condensation on cold metal ducts and perforated floors. An advantage for suction systems is that grain temperatures and odors are more convenient to check than with pressure systems.

A pressure system is more likely to have under-roof condensation than a suction one. Because the grain at the top of the bin is the last to change during an aeration cycle, the grain most likely to spoil is the easiest to observe.

Commercial automatic aeration control systems are available that use a humidistat and a thermostat to cycle the fan according to preset conditions. Follow the same observation and management procedures to assure proper performance. Do not use automatic controls during spring warm-up because continuous fan operation is essential. Automatic controls are useful and can relieve you of some management, but monitor them regularly to make sure the system is working properly. The grain value is worth it!

Aeration Costs

Operating a grain aeration system is very inexpensive. Operating cost = hours x kilowatts(kW) x ¢/kilowatt-hour(kWh). For estimating, assume that 1 horsepower = 1 kilowatt. Therefore, operating cost is approximately equal to fan horsepower x hr x ¢kWh. To properly cool or warm a bin of grain seldom exceeds 1/10¢ per bushel for electricity. A very small cost compared with the benefits!

Peaking Grain in Storage

Most dry grain peaks at an angle of 16°-20° for center filling without a distributor. The volume of grain that can be stored in the peak can be estimated by cubing the bin diameter and multiplying by 0.035—for a 24-foot bin, 24 x 24 x 24 x 0.035 = 484 bushels. Although it is tempting to store those extra bushels, keep in mind they interfere with uniform aeration and add to the migration problems.

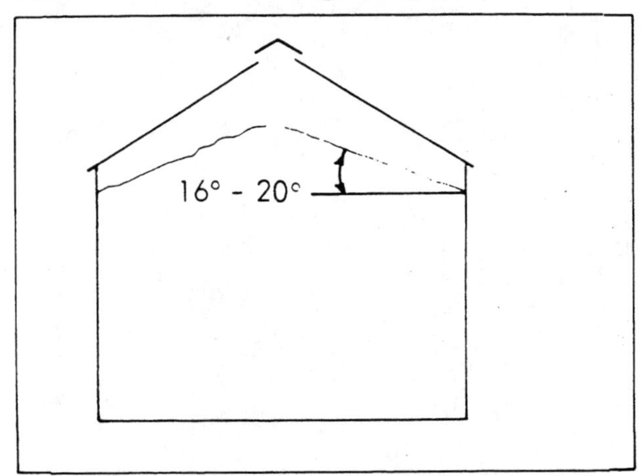

Fig 533-4. Natural filling angle for most dry grains.

Peaking also makes it difficult and dangerous to enter the bin for observation. Because of dust and high temperatures during the summer, entering a small space between roof and grain should be prohibited. Shifting grain may block the exit.

If the grain is peaked during harvest, withdraw the peak immediately after harvest. Lowering the center core of the bin improves airflow through the center of the bin, and probing and sampling are made easier and safer. Some fines are also removed.

Managing Fines in Storage

Broken grain and foreign material, or fines, can create two problems in stored grain, particularly when they accumulate in pockets. First, broken kernels are more susceptible to spoilage than unbroken ones. Second, airflow from aeration fans tends to go around pockets of fines so they cool more slowly. These pockets often develop into hot spots that result in spoiled grain.

Strive to minimize the fines produced from harvesting, drying, and handling, rather than later trying to solve resulting storage problems. Consider three grain storage management techniques that avoid problems from fines that do go into the storage:

- The most common technique is to use a grain spreader to minimize concentration of fines.
- Grain can also be cleaned before binning to improve storability. Unfortunately, cleaned grain may have no greater market value. Fines do have value, both as livestock feed and as added weight in market grain. It makes sense for livestock farmers to clean all their feed grain because they can feed the fines. Unless fines are causing serious storage problems or grade reduction, cash grain farmers may lose money by cleaning grain

Table 533-3. Observations and actions in managing dry stored grain.

Observation	Probable Cause	Solution/Recommended Action
Musty or spoiled grain odor.	Heating, moisture accumulation in one spot.	Run the fan. Smell the exhaust while in the bin or in front of the exhaust fan. Run the fan to cool any hot spots. If damage is severe, remove the grain.
Hard layer or core below grain surface.	High moisture or spoiled, caked grain mass.	Run the aeration or drying fan. Check to see if caked or compacted mass blocks airflow. Cool and dry if airflow is adequate, otherwise unload to remove all spoiled grain.
Warm grain below the top surface.	Moisture content too high.	Run the fan regardless of weather conditions until the exhaust air temperature equals the desired grain temperature.
Surface grain wet or slimy. Perhaps grain sticking or frozen together.	Early signs of moisture migration, often noticeable only 1-2 weeks after binning.	Run aeration fan. Cool grain until exhaust temperatures equal desired grain temperature or outside air temperatures.
Hard surface crust, caked, and blocking airflow. Possibly strong enough to support a man.	Severe moisture migration and condensation in the top surface.	Remove the spoiled layer. **Wear a dust mask to filter mold spores.** Run the fan to cool grain after spoilage is removed. Sample grain with probe to determine condition throughout center mass below the crust. Consider marketing grain to arrest further spoilage.
Under-roof condensation dripping onto surface.	Warm grain in cold weather, severe convection circulation and moisture migration.	Aerate until exhaust air temperature equals outdoor air temperature at beginning of aeration cycle.
Wet or spoiled spots on grain surface outside center point.	Condensate drip from bolt end or under roof fixture that funnels condensate flow; possible roof leak.	Check grain for heating. Check roof under surface at night. Check for caulking around roof inlets and joints.
Wet, spoiled spot directly under fill cap.	Leaking roof cap or condensed water from gravity spout.	Check bin cap seal and hold down. Block or disconnect gravity spout so air from bin and grain cannot flow up tube. Marginal solution: hang bucket under spout inlet and check bucket for water accumulation.
No Air flow through grain with aeration fan running.	Moldy, caked grain mass blocking flow; possible moldy grain layer immediately above aeration duct or perforated floor on suction system.	Try to determine location and scope of spoilage. Unload storage and market or re-bin good grain.
White dust visible whenever grain is stirred.	Mold on grain but not sufficient spoilage to seal top surface.	**Wear dust mask in working grain.** Evaluate grain condition throughout bin where possible. Observe caution in continued storage because grain condition has deteriorated to some degree.
Cooling time required much longer than usual.	Increased fines in grain resisting and reducing airflow; increased fines can cause airflow resistance to increase as much as 2-4 times over that of clean grain.	Run the fan longer time. Operate fan until grain and exhaust air temperature readings indicate grain is at desired temperature, **regardless of the fan time required.**
Exhaust air temperatures in center of bin surface warmer than those away from center.	Fine material accumulation in storage center resisting airflow; airflow through center mass grossly reduced compared to relatively clean grain around outside of storage.	Run the fan sufficient time to cool the center irrespective of the outside grain temperatures. Draw down the bin center to remove fines and decrease the grain depth for easier air passage in the center core.
Unknown grain conditions in the bin center.	Too deep to probe; bin too full to access; no temperature sensing cables installed.	Withdraw some grain from all bins to feed or market. Observe (look, feel, smell) **first** grain to flow with each withdrawal, since it was in the center core. Withdraw any storage filled above level full, as soon as possible following harvest, to reduce moisture migration tendencies and permit access for observation and sampling.

unless they can sell the gleanings to a livestock farmer or grain elevator.

When planning a new grain storage and drying system, plan for a way to include grain cleaning now or in the future. The grain grading rules may change to make on-farm grain cleaning profitable.

Another possibility is no grain spreading coupled with frequent withdrawal of the center core of grain to remove accumulated fines. Direct the flow of grain toward the exact center of the storage. The grain drops onto the grain pile and rolls down toward the outside wall. Remove grain at regular intervals (daily or more often) during filling to remove the peak. The goal is to draw the surface daily to an inverted cone with a diameter of 4'-10'. Repeated withdrawals remove many of the fines that accumulate in the center core. See Fig 533-5.

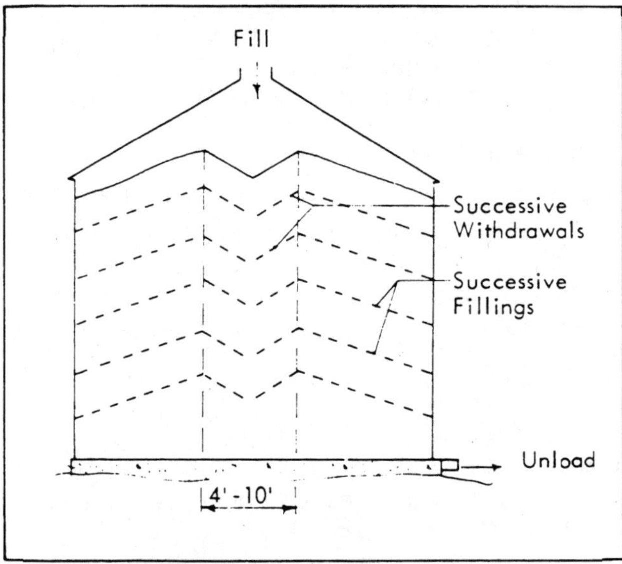

Fig 533-5. Withdrawals during filling remove fines.

A secondary benefit of nondistributed filling is that the grain rolling down the cone of grain results in a looser fill than when using a grain distributor. This loose fill reduces the resistance to airflow through the grain. However, with nondistributed fines not removed by withdrawal, the aeration airflow patterns in the bin may not be uniform.

Temperature Sensing

Consider installing temperature sensing units in large grain storages. Fig 533-6 illustrates locating four cables with temperature sensors 4'-5' apart along each cable. Suspend the cables from the roof. Mount the center cable to one side of the center of the bin to reduce the drag on the cable when unloading grain. Check with the bin manufacturer to be sure the cables, supports, and roof can withstand the drag from grain filling and unloading. Anchor the bottom of the cables to assure alignment, but allow for a sweep unloader. Bring the circuits from the top of the cables to a central point for attaching the read-out device.

Temperature sensors accurately trace the progress of aeration cooling or heating cycles. They help identify hot spots within the grain mass. They also indicate overall heating and approximate average grain temperature.

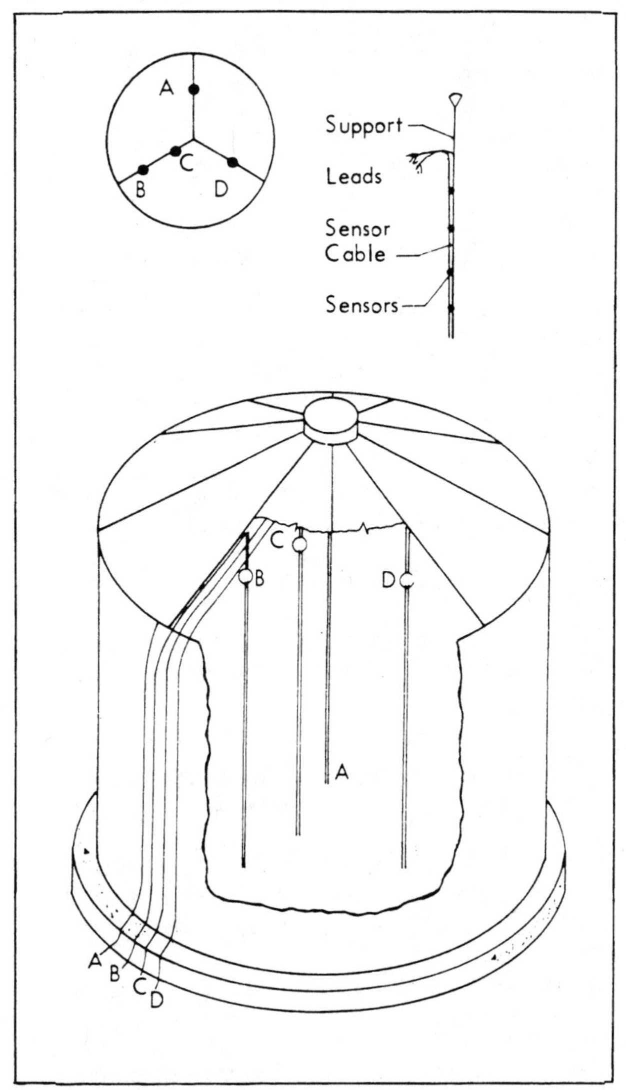

Fig 533-6. Bin temperature sensing system.
Four sensing cables are suspended from the roof: 3 are half way from wall to bin center, one is close to the center.

Temperatures may change only 1°-2°F per week so read and record them accurately. A continual increase in temperature is a warning that must be heeded, especially if one spot in the bin is heating faster than the bin as a whole. Experience indicates that once heating starts, it continues to increase at an increasisng rate until cooling is applied.

Insect Control in Stored Grain

Insect infestations in storage can come from grain residues in combines, handling equipment, and old grain left in storage. Correctly drying, aerating and managing stored grain minimizes the risks of insect infestation and damage. Insect activity goes with moisture accumulation and grain heating. Where yearly insect problems are common, practice regular preventive treatment. Where infestations are more often found with poor quality grain, or where there is an occasional slip in overall management, emphasize observation and control. Look for insect activity during every storage visit. Follow this preharvest checklist:

1. Clean all debris from harvesting, handling and drying equipment (trucks, augers, elevators).
2. Sweep old grain, grain particles and dust from inside the bin. If possible, remove debris under perforated floors and dispose of the sweepings by burning, burying, etc. or saturate this debris with 2½% malathion or 2½% methoxychlor at least 3 weeks before using the fan.
3. Repair bin if any signs of water leakage are found (spoiled grain on the floor, holes in the roof, etc.).
4. Apply a 2½% methoxychlor or 2½% malathion-water emulsion spray to all surfaces of clean, empty bins.
5. Remove piles of boards, spilled grain, and vegetation from around the bin that attract rodents. If there is no chance of runoff to adjacent fields, gardens, etc., spray a 2' band of soil sterilant around the foundation so vegetation will not grow.
6. Don't put new grain on top of old grain. Just a few insects in the old grain can contaminate the entire bin.

Preventive Insect Control

For frequent insect problems, apply a grain protectant in addition to using a residual insecticide spray.

- Spray grain being augered or elevated into the bin with a protectant—1 pint 57% malathion in 3 to 5 gallons water per 1,000 bushels of grain.
- After the bin is full, apply a cap-out malathion spray. Mix ½ pint 57% malathion in 1 to 2 gal of water/1,000 sq ft and apply to grain surface. The cap-out spray is a barrier against insects entering or feeding on the grain surface. Each time the grain surface is disturbed by taking a sample, etc., repair the barrier with malathion.
- Indian meal moths are resistant to malathion, so hang dichlorvos resin strips over the grain to destroy the moths before they lay eggs (the strips do not control larvae). Install the strips in early spring, before temperatures reach 50°F. Use one strip/1,000 cu ft air space above the grain.

See Table 533-4 for the number of strips to use in bins full to the eave.

Table 533-4. Use of dichlorvos resin strips.

Bin diameter ft	Number of dichlorvos resin strips
To 24'	1
27' - 30'	2
32' - 33'	3
36' to 37'	4
40'	5
42'	6

If the bin is not full, add one strip/1,000 cu ft of cylindrical volume between the grain and eave. This volume is about:

V = (bin dia.)^2 x (feet of exposed sidewall) x (0.8)

Example: 27'-diameter bin with 2½' of sidewall exposed:

V = (27)^2 x 2½ x 0.8 = 1458 cubic feet

Use two strips for 1458 cubic feet plus two strips for the attic space. Replace these strips monthly until temperatures are below 50°F in the fall.

Controlling an Infestation

You have two alternatives for controlling a stored grain infestation:

- Fumigate with a liquid, solid or gas grain fumigant in storage or as the grain is being turned. **Fumigants are toxic and must be applied with proper safety precautions and equipment.** Each fumigation job is different; hire a commercial applicator.
- If Indian meal moth is not present and the grain can be turned into a nearby empty bin, apply a malathion grain protectant. Also, passing grain over a screen removes many insects and debris.

Safety Practices

Take time to review safety measures with workers and all family members. "Better safe than sorry" is very real around grain and hazardous machinery.

Absolutely forbid entry into a bin or gravity unload vehicle when grain is flowing. It is a major cause of accidental death when handling and unloading grain.

With modest flow rates of a 6" auger, you are helpless only two to four seconds after stepping into the cone of flowing grain. You are totally submerged within 20 seconds at a grain flow rate of only 1,000 bu/hr.

A child submerges even more quickly. Even in gravity-unload vehicles not as deep as the child's height, the massive outflow rate drags him down into the discharge cone, folding his legs and shortening the height necessary for complete submergence and suffocation.

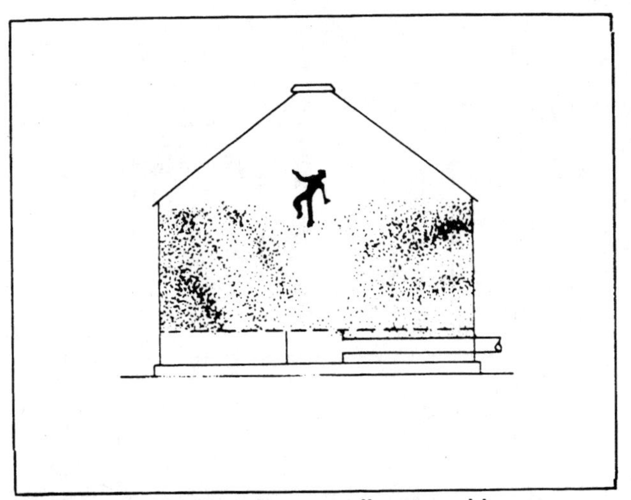

Fig 533-7. Bridged grain can collapse and bury you.
Beware of bridged or crusted grain.

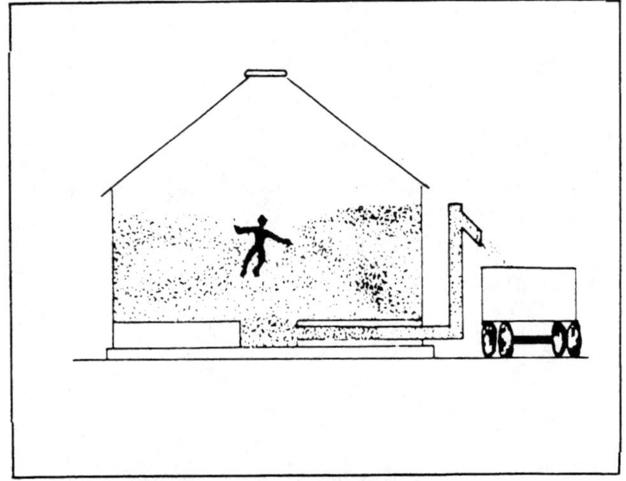

Fig 533-8. Flowing grain can suffocate you.
Flowing grain can trap and bury you in seconds.

Safety Precautions

- Don't enter a bin of flowing grain.
- Don't enter a bin to break a crust or remove a blockage when unloading equipment is running, whether or not grain is flowing. Restarted flow can trap you.
- Before entering a bin or cleaning or repairing conveyors:
  - Lock out the control circuit on automatic unloading equipment, as on a wet holding bin.
  - Flag the switch on manual equipment so someone else doesn't start it.
- Don't enter a bin unless you know the nature of previous grain removal, especially if any crusting is evident.
- Beware of walking on any surface crust.
- Don't depend on a second person—on the bin roof, on the ground, or at some remote point—to start or stop equipment on your shouted instructions.
  - Equipment noise can block out shouts for action or assistance.
  - That person may fall or over-exert in the panic and haste of getting off the bin or running to the control point.
- Be wary and alert while working with grain that has gone out of condition—there may be molds, blocked flow, cavities, cave-ins or crusting.
- When entering a questionable bin or storage, have two outside and one inside workers. Attach a safety rope to the man in the bin with the two men outside capable of lifting him out without entering the bin. One man outside cannot do this and cannot go for help while giving first aid.
- Always wear a respirator capable of filtering fine dust to work in obviously dusty-moldy grain. Never work in such conditions, even with protection, without a second person on safety standby.
- Parents, watch your children.
  - Keep them away from bins and vehicles with flowing grain.
  - Small hands and feet can penetrate even properly shielded augers, belts and PTOs.
- If a grain bin is peaked close to the roof, be extremely cautious. Crawling between roof and peak can cave grain and block the exit.
- Maintain proper and effective shields and guards on hazardous equipment.

534 GRAIN, FEED, SILAGE, AND BEDDING DATA

Grain Drying

Wet vs. Dry Basis

To convert moisture on a wet basis to moisture on a dry basis, and vice versa, use:

$M_d = (100 \times M_w) \div (100 - M_w)$
$M_w = (100 \times M_d) \div (100 + M_d)$

M_d = moisture content, dry basis
M_w = moisture content, wet basis

Example:

If corn is 15½% moisture content on a wet basis, what is it on a dry basis?
$M_d = (100 \times 15½) \div (100 - 15½) = 18.3\%$

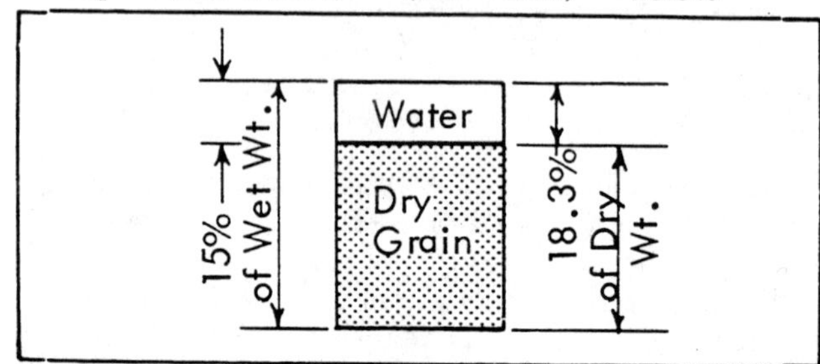

Fig 534-1. Wet vs. dry basis.

Moisture Loss During Drying

To determine the weight of grain after drying in lb, bu, ft³, etc., use:

DRY = Wet x $(100 - P_w) \div (100 - P_d)$

DRY = dry grain weight
WET = wet grain weight
P_w = percent moisture content, wet
P_d = percent moisture content, dry

Example:

If you dry 50 lb of grain at 28% moisture content to 15½% moisture content, how many lbs of grain will you have?

Solution:

DRY = 50 x (100 - 28) ÷ (100 - 15½) = 42.6 lb
WATER REMOVED = 50 - 42.6 = 7.4 lb

Table 534-1. Moisture content for safe storage.
Lower limit moisture content at which fungi damage occurs in stored grain. ASAE DATA D 244-1.

Fungus	Lower moisture content, percent (wet basis)		
	Wheat, barley, oats, corn	Soybeans	Flaxseed
Aspergillus restrictus	13.0-13.5	12.0-12.5	10.0-10.5
A. repens, amstelodami, ruber	13.5-14.0	12.5-13.0	10.5-11.0
A. candidus, orchraceus	15.0-15.5	13.5-14.0	12.0-12.5
A. flavus	17.5-18.0	15.5-16.0	14.5-15.0
Penicillium	16.0-16.5	14.5-15.0	13.0-13.5
Pullularia, yeasts	20.0-21.0	17.0-18.0	15.0-16.0

*These limits apply at tmeperature range of 5-10C to 30-35C. Growth of these fungi is much slower at 5-10C than at 25-30C. 5-10C = 41-50°F. 25-30C = 77-86°F. 30-35C = 86-95°F.

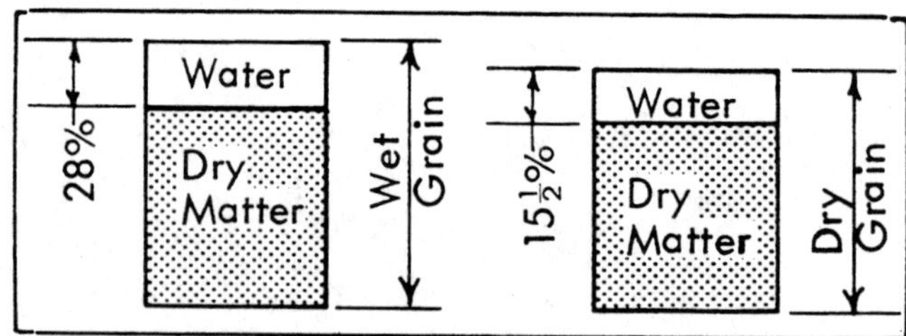

Fig 534-2. Drying moisture loss.

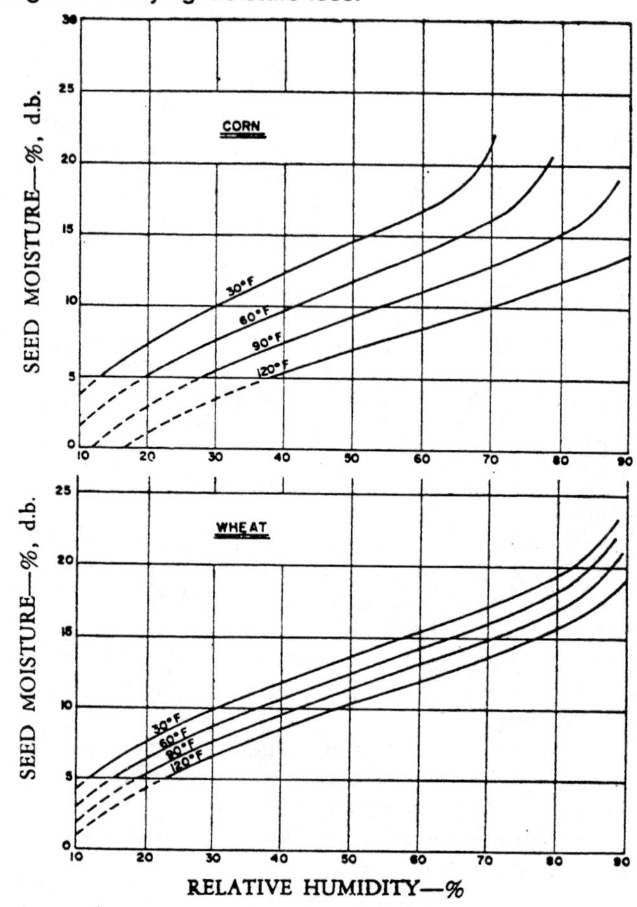

Fig 534-3. Corn and wheat equilibrium moisture content.

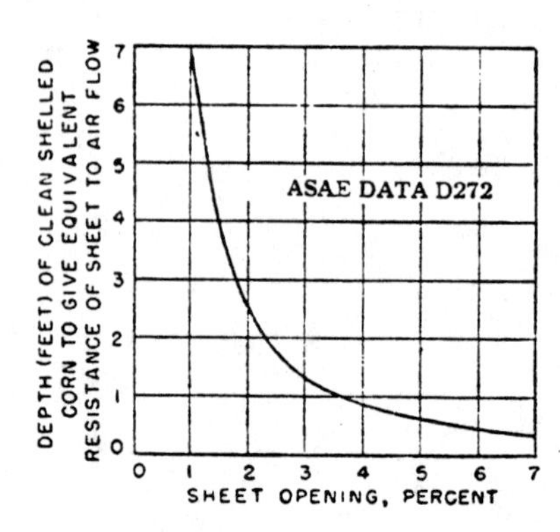

Fig 534-4. Air flow resistance through perforated metal sheets.
Several small perforations are preferred to only a few large perforations. The curve shown is based on tests with 0.04" - 0.13" perforation widths.

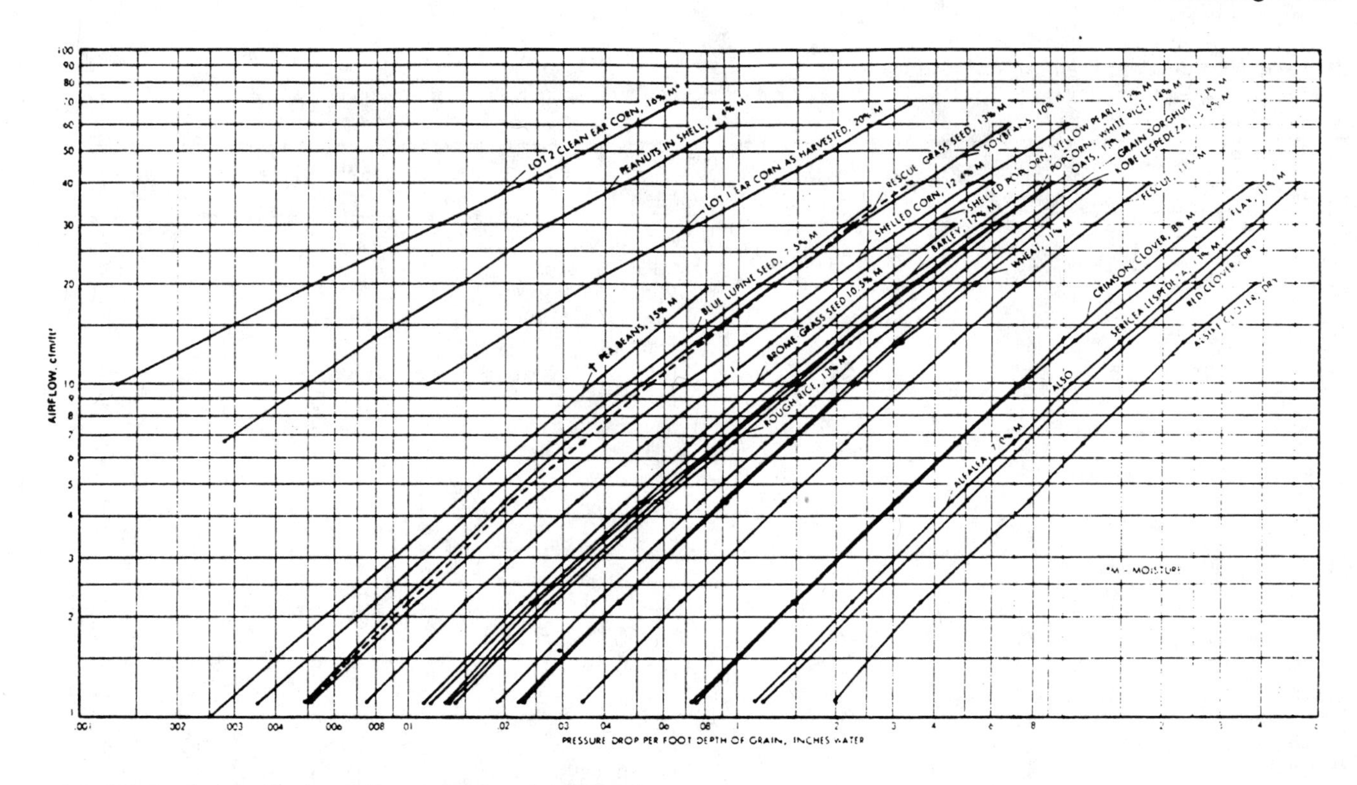

NOTES: This chart gives values for a loose fill (not packed) of clean, relatively dry grain.
For a loose fill of clean grain having high moisture content (in equilibrium with relative humidities exceeding 85 percent), use only 80 percent of the indicated pressure drop for a given rate of air flow.
Packing of the grain in a bin may cause 50 percent higher resistance to air flow than the values shown.
When foreign material is mixed with grain no specific correction can be recommended. However, it should be noted that resistance to air flow is increased if the foreign material is finer than the grain, and resistance to air flow is decreased if the foreign material is coarser than the grain.
White rice is a variety of popcorn.

ASAE Data: ASAE D272.

Fig 534-5. Air flow resistance through grains.

Feeds and Grains

Table 534-2. Storage requirements of feeds.

Material	lb/cu ft	cu ft/ton	Material	lb/cu ft	cu ft/ton
Alfalfa meal	15	134	Malt sprouts	15	134
Alfalfa, chopped	12	170	Mixed mill feed		
Barley meal	28	72	(bran & middlings)	15	134
Beet pulp, dried	15	134	Molasses	78	26
Bran	16	125	Molasses beet pulp	20	100
Brewers grains, dry	15	134	Meat scraps	34	59
Buckwheat bran	15	134	Oats, ground	18-20	111-100
Buckwheat middlings	23	88	Pellets, mixed feed	35-40	57-60
Coconut meal	38	53	Pellets, ground hay	38-45	53-44
Concentrates, typical	45	45	Rice bran	20	100
Corn meal	38	53	Rice polish	31	65
Corn & cob meal, dry	36	56	Rye meal	38	53
Cotton seed meal	38	53	Salt, fine	50	40
Distillers' grains, dry	15	134	Soybean meal	42	48
Gluten feed	33	61	Tankage	32	63
Gluten meal	46	45	Wheat bran	14	154
Hominy meal	28	72	Wheat feed, mixed	15	134
Kaffir meal	27	74	Wheat, ground	45	46
Lime	60	33	Wheat middlings (Std)	20	100
Linseed meal	23	88	Wheat screenings	27	77

SOURCE: Based primarily on California Agricultural Circular 517.

Table 534-3. Storage requirements of grains.
About 2 bushels of husked ear corn require 70 lb to yield 1 bu or 56 lb of shelled corn. A bag or sack (beans, rice) is typically 100 lb.
Source: ASAE data: ASAE D241.1, 1967.

Material	Approximate lb/bu	Weight lb/ft³
Alfalfa	60	48.0
Barley, 15%	48	38.4
Beans:		
Lima, dry	56	44.8
Lima, unshelled	32	25.6
Snap	30	24.0
Other, dry	60	48.0
Bluegrass	14-30	11.2-24.0
Broomcorn	44-50	35.2-40.0
Buckwheat	48-52	38.4-41.6
Castor beans	46	36.8
Clover	60	48.0
Corn, 15½%		
Ear, husked	70	28.0
Shelled	56	44.8
Green sweet	35	28.0
Cottonseed	32	25.6
Cowpeas	60	48.0
]Flaxseed, 11%	56	44.8
Grain sorghums, 15%	56 & 50	44.8 & 40.0
Hempseed	44	35.2
Hickory nuts	50	40.0
Hungarian millet	48 & 50	38.4 & 40.0
Katir	56 & 50	44.8 & 40.0
Kapok	35-40	28.0-32.0
Lentils	60	48.0
Millet	48-50	38.4-40.0
Mustard	58-60	46.4-48.0
Oats, 16%	32	25.6
Orchard grass	14	11.2
Peanuts, unshelled:		
Virginia type	17	13.6
Runners, Southeastern	21	16.8
Spanish		
Southeastern	25	19.7
Southwestern	25	19.8
Perilla	37-40	29.6-32.0
Popcorn:		
On ear	70	28.0
Shelled	56	44.8
Poppy	46	36.8
Rapeseed	50 & 60	40.0 & 48.0
Redtop	50 & 60	40.0 & 48.0
Rice, rough	45	36.0
Rye, 16%	56	44.8
Sesame	46	36.8
Sorgo	50	40.0
Soybeans, 14%	60	48.0
Spelt (p. wheat)	40	32.0
Sudan grass	40	32.0
Sunflower	24 & 32	19.2 & 25.6
Timothy	45	36.0
Velvet beans (hulled)	60	48.0
Vetch	60	48.0
Walnuts, black	50	40.0
Wheat, 14%	60	48.0
Walnuts, English	24-26	20-22

Table 534-4. Storage requirements of silage and high moisture corn.

Material Description	lb/cu ft	cu ft/ton
Corn shelled:		
25% moisture	43.1	46
30% moisture	39.7	51
Corn and cob meal:		
30% moisture	38.5	52
Silage:		
upright silo	40*	50
horizontal silo	35*	60
spread in bunk	25*	80

Table 534-5. Beef feed storage.
Steers are fed for 10 months from 450 to 1,050 lb. Assume 2 lb gain/day and 1% grain ration.
Yearlings fed from 650 to 1,100 lb in 6 to 7 months. Plan for 1 or 2 batches/ year. Assume 2.2 lb gain/day and 1% grain ration.
Corn silage is 65% moisture content.

Feed component	Unit	Number of animals					
		1	25	50	100	300	500
		Storage requirements for feeding period					
Steer with silage							
Corn silage	Ton	3.5	87.5	175	350	1,050	1,750
Corn[b] 30%	Bu	43	1,075	2,150	4,300	12,900	21,500
	Ton	1.5	37.5	75	150	350	750
or 15%	Bu	43	1,075	2,150	4,300	12,900	21,500
	Ton	1.2	30	60	120	360	600
40% supplement[c]	Ton	0.23	5.63	11.25	22.50	67.50	112.5
Steer with hay							
Hay[d,e]	Ton	0.88	21.96	43.75	87.5	262.5	473.5
Corn[b] 30%	Bu	63	1,575	3,150	6,300	18,900	31,500
	Ton	2.14	53.55	107.1	214.2	642.6	1,071
or 15%	Bu	63	1,575	3,150	6,300	18,900	31,500
	Ton	1.76	44	88	176	528	880
40% supplement[c]	Ton	0.23	5.63	11.25	22.5	67.50	112.5
12% hay[f]	Ton	0.063	1.56	3.13	6.25	18.75	31.25
Yearling with corn silage							
Corn silage	Ton	2.5	62.5	125	250	750	1250
Corn[b] 30%	Bu	35	875	1,750	3,500	10,000	17,500
	Ton	1.2	30	60	120	360	600
or 15%	Bu	35	875	1,750	3,500	10,500	17,500
	Ton	1.0	25	50	100	300	500
40% supplement[c]	Ton	0.158	3.9	7.9	15.75	47.25	78.75
Yearling with hay							
Hay[d]	Ton	0.63	16	31	62.5	187.5	312.5
Corn[b] 30%	Bu	50	1,250	2,500	5,000	15,000	25,000
	Ton	1.7	42.5	85	170	510	850
or 15%	Bu	50	1,205	2,500	5,000	15,000	25,000
	Ton	1.4	35	70	140	420	700
40% supplement[c]	Ton	0.13	3.3	6.6	13.25	39.75	66.25
12% hay[f]	Ton	0.05	1.25	2.5	5	15	25

Cow herd
Allow 25 lb hay / cow per day during dry period.
Allow 30 lb hay / cow per day after calving.
If limit fed, 2 tons / head (0.35 tons / head per month) for 6-month winter hay feeding period.
If not limit fed, 2½ tons (0.45 tons / head per month).
Adjust hay storage amounts for the length and severity of the winters in your area.

[a]Use 90% of these amounts for heifers fed for 10 months (400-900 lb at 1.8 lb gain / head per day). *Source: Purdue University*

[b]If ground ear corn replaces shelled corn, feed 25% more corn and 20% less corn silage or ⅓ less hay.

[c]If 64% supplement is used, multiply tons by 0.6.

[d]If haylage is used, multiply tons by 2.0.

[e]Increase hay requirements by ⅓ to cover losses if cattle have free access to big package hay. Restricted feeding cuts hay losses.

[f]Reduced supplement required with good legume hay of at least 12% protein.

Table 534-6. Dairy annual feed requirements.
Use these values to estimate storage needs.

	% dry matter	Lbs milk/cow-year 12,000	14,000	16,000	18,000
		- - - - quantity per cow - - - -			
Medium level of silage:					
Hay silage, T	40-50	11.0	11.6	12.3	12.9
or hay, T	80-85	4.3	4.5	4.7	4.9
Corn silage, T	30-35	12.0	12.0	12.0	12.0
Shelled corn, bu		72	84	102	118
High level of silage:					
Hay silage, T	40-50	6.7	6.9	7.1	7.5
or hay, T	80-85	2.6	2.7	2.8	2.9
Corn silage, T	30-35	16.5	16.5	16.5	16.5
Shelled corn, bu		52	64	86	106

Source: Chore Reduction for Free Stall Dairy Systems, *Hoard's Dairyman*, Fort Atkinson, WI.

Table 534-7. Pounds of water per bushel of grain.
One bushel = 56 lb shelled corn or grain sorghum at 15½% moisture (47.32 lb dry matter), 60 lb of wheat or soybeans at 14% (51.6 lb dry matter), or 32 lb of oats at 14½% (27.4 lb dry matter). The pounds of grain needed to get a bushel of dried grain = lb dry matter + lb of water.

Grain Moisture content %	Shelled corn, grain sorghum	Wheat, soybeans	Oats
	Pounds	water per bushel	
35	25.4	27.8	14.6
30	20.2	22.1	11.7
28	18.4	20.1	10.6
26	16.6	18.2	9.6
24	14.9	16.4	8.6
22	13.3	14.6	7.7
20	11.8	12.9	6.8
18	10.4	11.4	6.0
16	9.0	9.8	5.2
14	7.7	8.4	4.4
12	6.5	7.0	3.7
10	5.3	5.8	3.0
8	4.1	4.5	2.3

SOURCE: USDA FB2214

Table 534-8. Pounds of water in ear corn at harvest.

Kernel Moisture Content (percent)	Amount of Water in a Bushel of Ear Corn[1] In Kernels	In Cobs	Total
	Pounds	Pounds	Pounds
35	25.4	12.4	37.8
30	20.2	9.9	30.1
28	18.4	8.8	27.2
26	16.6	7.8	24.4
24	14.9	6.7	21.6
22	13.3	5.5	18.8
20	11.8	4.4	16.2
18	10.4	3.2	13.6
16	9.0	2.1	11.1
14	7.7	1.4	9.1
12	6.5	0.9	7.4
10	5.3	0.5	5.8

[1] A bushel of ear corn is defined here as the quantity that will yield 56 pounds of shelled corn at 15.5 percent moisture.

Table 534-9. Rate of harvesting corn.*

Machine	Acres/Hr	Bushels Harvesting at Yield of 75 bu/A	100 bu/A	125 bu/A
One-row picker	.6 to .7	50	65	75
Two-row picker	1.1 to 1.3	95	120	140
Two-row picker sheller	1.1 to 1.4	105	125	145
Two-row combine	1.2 to 1.4	110	135	140
Four-row combine	2.3 to 2.6	200	240	275

* Allows for down time for changing wagons, unloading bins, major adjustments and turning, and reasonable driving speed to minimize field losses.

The rate of harvest will vary with field conditions, moisture content of corn harvested and yield per acre.

SOURCE: Robert G. White, Michigan State University

Hay, Straw, and Bedding

Table 534-10. Storage requirements of hay and straw.

Materal	ft³/ton	lb/ft³
Loose		
Alfalfa	450-500	4-4.4
Non legume	450-600	3.3-4.4
Straw	670-1000	2-3
Baled		
Alfalfa	200-330	6-10
Non legume	250-330	6-8
Straw	400-500	4-5
Chopped		
Alfalfa, 1½" cut	285-360	5.5-7
Non legume, 3" cut	300-400	5-6.7
Straw	250-350	5.7-8

Table 534-11. Water absorbing capacity of bedding.

lbs of water/lb of bedding	
	WOOD
4.0	Tanning bark
2.5	Dry fine bark
3.0	Pine chips
2.5	sawdust
2.0	shavings
1.0	needles
1.5	Hardwood chips, shavings or sawdust
	CORN
2.5	Shredded stover
2.1	Ground cobs
	STRAW
2.6	Flax
2.8	Oats, threshed
2.5	combined
2.4	chopped
2.2	Wheat, combined
2.1	chopped
3.0	HAY, chopped mature
	SHELLS, HULLS
2.7	Cocoa
2.5	Peanut, cottonseed
2.0	Oats

Storage

Grain

Figuring grain storage capacity

1 bu ear corn = 70 lb = 2.5 ft^3 (15½% moisture)
1 bu shelled corn = 56 lb = 1.25 ft^3 (15½% moisture)
1 ft^3 ear corn = 0.4 bu (15½% moisture)
1 ft^3 shelled corn = 0.8 bu (15½% moisture)

Rectangular or square crib or bin

$V = W \times L \times H$
V = volume, ft^3
W = width, ft
L = length, ft
H = height, ft

Table 534-12. Ear corn crib storage capacity.
Based on 2½ ft^3/bu. Includes ½ cone space with no deduction for center tunnel. Roof slope 1:1.

Rectangular			Round[2]		
Width feet	Height feet	Bu per 10 ft Length	Diameter feet	Height feet	Capacity bu
4	12	188	12	12	540
	16	256		16	720
	20	320		20	900
6	12	288	14	12	740
	16	384		16	980
	20	480		20	230
8	12	384	16	12	960
	16	512		16	1280
	20	638		20	1610
10	12	480	18	12	1220
	16	640		16	1620
	20	800		20	2030

Example:
Capacity of a 6′ wide, 12′ high, and 40′ long ear corn crib is:
6 x 12 x 40 = 2,880 ft^3 x 0.4 bu/ft^3 = 1,152 bu

Round cribs, bins, or silos

$V = 3.1416 \times R^2 \times H = 0.785 \times D^2 \times H$
V = volume, ft^3
R = radius, ft
D = diameter, ft
H = height, ft

Example:
A 14′ high, 14′ diameter, round ear corn crib's capacity = .785 x $(14')^2$ x 14′ = 2,152 ft^3 x 0.4 bu/ft^3 = 861 bu.
A 14′ high, 14′ diameter, round shelled corn bin's capacity = .785 x $(14')^2$ x 14′ = 2,152 ft^3 x 0.8 bu/ft^3 = 1,722 bu.

Table 534-13. Round steel feed tank storage capacity.
Tank has a 60° hopper and 24″ slide valve clearance.

Description	Approx. Overall Height-ft	Material Weight 30#/ cu ft	40#/ cu ft	50#/ cu ft	Total Capacity cu ft	Bu
		Tons capacity				
6′ diameter-center draw off	10	2.0	2.7	3.4	135	108
	16	4.2	5.7	7.1	285	228
	20	6.2	8.6	10.4	415	322
6′ diameter-side draw-off	15	2.8	3.7	4.7	187	150
	20	5.0	6.7	8.4	338	270
	25	7.3	9.7	12.1	487	390
9′ diameter-center draw off	17	8.4	11.2	14.0	561	413
	22	13.5	18.0	22.5	900	720
	28	18.6	24.7	30.9	1236	990
12′ diameter-center draw off	20	16.3	21.7	27.1	1085	870
	28	29.7	39.6	49.5	1980	1585
	36	43.1	57.5	71.9	2875	2300
	42	52.0	69.4	86.8	2472	2780

Table 534-14. Round bin storage capacity.
Values based on 1 ft^3 = 0.8 bu. Does not involve test weight, moisture content, or shrinkage.

Grain depth ft	Bin diameter, ft: 15	18	21	24	27	30	33	36	39	42	48	60
	- - - - bushels - - - -											
1	142	204	278	363	460	568	687	818	960	1113	1453	2271
2	284	409	556	727	920	1136	1374	1635	1919	2226	2907	4542
3	426	613	835	1090	1380	1703	2061	2453	2879	3338	4360	6813
4	568	818	1113	1453	1840	2271	2748	3270	3838	4451	5814	9084
5	710	1022	1391	1817	2299	2839	3435	4088	4798	5564	7267	11355
6	852	1266	1669	2180	2759	3407	4122	4905	5757	6677	8721	13626
7	994	1431	1947	2544	3219	3974	4809	5723	6717	7790	10174	15897
8	1136	1635	2226	2907	3679	4542	5496	6541	7676	8902	11628	18168
9	1277	1840	2504	3270	4139	5110	6183	7358	8636	10015	13081	20439
10	1419	2044	2782	3634	4599	5678	6870	8176	9595	11128	14535	22710
11	1561	2248	3060	3997	5059	6245	7557	8993	10555	12241	15988	24981
12	1703	2453	3338	4360	5519	6813	8244	9811	11514	13354	17441	27252
13	1845	2657	3617	4724	5978	7381	8931	10628	12474	14466	18895	29523
14	1987	2861	3895	5087	6438	7949	9618	11446	13433	15579	20348	31794
15	2129	3066	4173	5450	6898	8516	10305	12264	14393	16692	21802	34065
16	2271	3270	4451	5814	7358	9084	10992	13081	15352	17805	23255	36336
17	2413	3475	4729	6177	7818	9652	11679	13899	16312	18919	24709	38607
18	2556	3679	5008	6541	8278	10220	12366	14716	17271	20030	26162	40878
19	2697	3883	5286	6904	8738	10787	13053	15534	18231	21143	27616	43149
20	2839	4088	5564	7267	9198	11355	13740	16351	19190	22256	29069	45420

Hay

Table 534-15. Hay shed capacity with 20′ sidewall.

Shed Width	Capacity per foot of Length tons		
	Baled	Chopped	Loose
24'	2.0	1.9	0.8
30'	2.6	2.3	1.0
36'	3.1	2.8	1.2
40'	3.4	3.1	1.4

Silage

Determining silo size

The capacity of a silo depends on height and diameter. Crop moisture is also important, but silo capacity on a dry matter basis is consistent between 50%-70% moisture content.

Height

Minimum height = amount of silage removed, ft/day x feeding period, days + 10′. Silage removal is in ft/day for easier calculating. See Table 534-16. Provide 10′ of extra height for settling and silo unloader storage.

Diameter

Determine the tons of dry matter.

$$DM = 5 \times FP \div 2{,}000 \text{ lb/ton} \times DMF$$

DM = tons of dry matter
S = silage fed, lb/day
FP = feeding period, days
DMF = dry matter factor, Table 534-17

The dry matter factor converts moist silage to dry matter content. Estimate the percent moisture of silage and select the dry matter factor from Table 534-17. Grass or corn silage is usually stored at about 65% and haylage at about 50%.

Select the maximum diameter from Table 534-21 using the calculated minimum height and tons of dry matter. Any silo with a greater height and smaller diameter meeting the needed storage can also be used.

Table 534-16. Silage removal.

	Silage removed, ft/day	
	Weather	
	Cold	Warm
Whole corn	2″	4″-6″
Alfalfa-brome	2″	3″-4″
Chopped ear corn	2″	2″
Cracked shelled corn	4″	4″

Table 534-17. Dry matter factor.

	Moisture content, %					
	45%	50%	55%	60%	65%	75%
Factor	1.8	2.0	2.2	2.5	2.8	3.3

Table 534-18. Horizontal silo maximum exposed surface area.
Remove at least a 4″ vertical slice per day.

Feeding Rate lb per cow	Square Feet per cow
20	2
30	3
40	4
50	5
60	6
70	7
80	8

Table 534-19. Tower silo, pounds of dry matter in 2″ layers.

Silo Diameter feet	Pounds Dry Matter* in 2" layers
12'	282
14	384
16	502
18	636
20	785
22'	950
24	1131
26	1329
28	1539
30	1767

*Assuming 15 lb dry matter/cu ft.

Table 534-20. Horizontal silo capacity per 10 ft of length.
Level full at 50 ft³ = 1 ton. Silage at about 65% moisture content.

Depth ft		Silo floor width, ft 20'	30'	40'	50'	60'	70'	80'	90'	100'	Closed end ratio
10'	Bu	1800	2600	3400	4200	5000	5800	6600	7400	8200	1/8
	Tons	45	65	85	105	125	145	165	185	205	
12'	Bu	2208	3168	4128	5088	6048	7008	7968	8928	9888	1/7
	Tons	55	79	103	127	151	175	199	223	247	
14'	Bu	2632	3752	4872	5992	7112	8232	9352	10472	11592	1/6
	Tons	66	94	122	150	178	206	234	262	290	
16'	Bu	3064	4344	5624	6904	8184	9464	10744	12024	13304	1/5
	Tons	77	109	141	173	205	237	269	301	333	
18'	Bu	3528	4968	6408	7848	9288	10728	12168	13608	14048	1/5
	Tons	88	124	160	196	232	268	304	340	376	
20'	Bu	4000	5600	7200	8800	10400	12000	13600	15200	16800	1/4
	Tons	100	140	180	220	260	300	340	380	420	

Example: 50' wide, 12' deep, 120' long, 1 end closed.
Capacity per 10', from table = 127 tons.
Capacity per 120' = 12 x 127 = 1524 tons.
Closed end holds additional silage.
Multiply closed end ratio for silage depth times the capacity/10': 1/7 x 127 = 18 (about).
Open end is less than full. Deduct for capacity lost:

Slope of silage	Deduction
1/4	closed end capacity
1/2	2 x " " "
1/1	4 x " " "

Assume slope of 1/2.
Closed end capacity = 18 tons.
Deduct 2 x 18 = 36 tons.
Total capacity = 1524 + 18 - 36 = 1506 tons.

Table 534-21. Approximate dry matter capacity of silos, tons*.

Silo height, ft	Silo diameter, ft										
	10	12	14	16	18	20	22	24	26	28	30
20	8	12	16	21	27	33	40	47	56	65	74
24	11	15	21	27	34	43	52	61	72	83	96
28	13	19	26	35	44	53	64	76	90	104	119
32	16	23	32	41	52	65	78	93	109	127	145
36	19	28	37	48	62	76	92	109	129	150	172
40	22	32	44	57	72	89	107	127	150	173	199
44		37	50	65	82	102	123	147	172	200	229
48		42	56	74	93	115	140	166	195	226	260
52			64	83	105	129	157	186	219	254	291
56			71	93	117	144	174	207	243	282	324
60			78	102	129	159	192	228	273	309	357
64					142	174	210	250	298	340	391
68					155	190	228	272	324	370	425
72								293	350	400	458
76								314	376	427	489
80								334	392	455	520

*Capacities allow one foot unused depth for settling in silos up to 30 ft high, and one additional foot for each 10 ft beyond 30 ft height.
Tons of moist silage = tons of dry matter × dry matter factor
Dry matter factor = 100/(100-percent moisture)

Source: ASAE D252.1, 1983.

Table 534-22. Reinforcing rods for concrete stave and cast-in-place silos.
For storing legume and grass silage and high moisture or dry shelled corn.

Diameter	10'0"		12'0"		14'0"		16'0"		18'0"		20'0"	
Silo Type[1]	S	C-P	S	C-P	S	C-P	S	C-P	S	C-P	S	C-P
Rod Diameter	9/16"	1/2"	9/16"	1/2"	9/16"	1/2"	9/16"	5/8"	9/16"	5/8"	9/16"	5/8"
Distance From Top in Feet	Spacing of Horizontal Rods, in Inches											
0-5	30	24	30	24	30	24	30	24	30	24	30	24
5-10	30	24	30	24	30	24	30	24	15	24	15	24
10-15	30	24	15	24	15	20	15	24	15	24	15	22
15-20	15	22	15	18	15	15	15	20	10	18	10	16
20-25	15	17	15	14	15	12	10	17	10	15	10	13
25-30	15	14	15	12	10	10	10	14	10	12	7½	11
30-35	15	12	10	10	10	9	10	12	7½	11	7½	9
35-40	10	10	10	9	7½	8	7½	10	7½	9	6	8
40-45			10	8	7½	7	7½	9	6	8	6[2]	7
45-50			7½	7	7½	6	6	8	6	7	6[2]	6
50-55					6	5	6	7	5	6	6[2]	6
55-60					6	5	7	7	5	6	5[2]	5

[1] S = Stave Silo, Round Rods -- Rolled Threads -- Galvanized

C-P = Cast-In-Place Silo, 2 foot rod overlap at splices, 6 inch wall thickness, reinforcing rods centered in wall. Additional vertical steel is needed for all silo sizes: 3/8" round bars 30" apart.

[2] 5/8" diameter rod.

If the reinforcing in your silo is not exposed and you cannot find out what it is, it would be wise to add reinforcing bands on the outside in accordance with the recommendations in the table.

SOURCE: "Building and Remodeling Silos for Legumes and Grass Silage," Concrete Information, No FB-7, Farm Bureau, Portland Cement Association, Chicago 1954.

Conveyors

Vertical Elevator Height

Bucket elevator height includes boots and hoppers, distributor or valve, and head clearance for the drive wheel at the top.

The effective elevator height is the distance between the hopper and the bottom of the distributor. Spouting height is:

- For a 45° slope, 1 ft/ft between leg and bin fill-point.
- For a 37° slope, ¾ ft/ft between leg and bin fill-point.
- For a 60° slope, 1¾ ft/ft between leg and bin fill-point.

Example:

If a bin fill-point is 30′ above grade, a 45° spouting is used, and the elevator is 24′ from the center of the bin, the effective elevator height must be: 30′ (bin) + 1 x 24′ (spout) = 54′

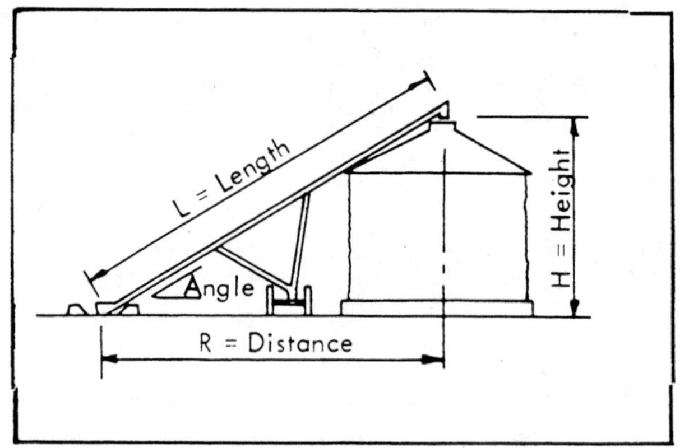

Fig 534-6. Conveyor length, distance, and discharge height.

Table 534-23. Conveyor length and discharge distance and height.

Height Of Discharge H, ft	Length & Distance, feet 6:12 26½° L	R	10:12 40° L	R
20	45	40	28	24
22	50	44	31	26
24	54	48	34	29
26	59	52	37	31
28	63	56	39	34
30	68	60	42	36
35	79	70	49	42
40	90	80	54	48

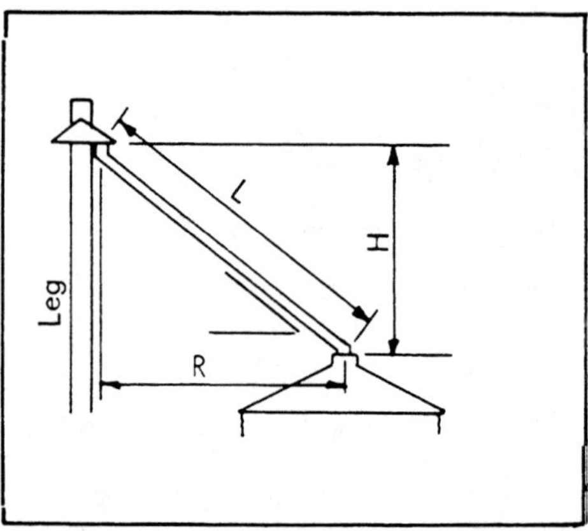

Fig 534-7. Conveyor length and leg height for gravity flow.

Table 534-24. Elevator discharge height and conveyor length.

Distance To Bin R, ft	Length & Height, feet Dry Grain 37° L	H	Wet Grain 45° L	H	Feed 60° L	H
16	20	12	23	16	32	28
18	22½	13½	25½	18	36	31
20	25	18	28½	20	40	35
22	27½	16½	31	22	44	38
24	30	18	34	24	48	42
26	32½	19½	37	26	52	45
28	35	21	40	28	56	48½
30	37½	22½	42½	30	60	52
35	44	26½	49½	35	70	61
40	50	30	57	40	80	70
45	57	34	64	45	90	78
50	63	38	71	50	100	87

Table 534-25. Gravity flow spout angles and floor slopes.
Angles less than 30° are not recommended. Bulky or damp grain can block rusty pipe.

Material	Spout Angle or Floor Slope	Slope Equivalents Rise:Run
Grains, dry	37°	3/4:1
Grains, wet or tough	45° min	1:1
Pellets	45°	1:1
Meal	60°	1 3/4:1

Table 534-26. Auger conveyor capacities.
Horsepower is that required at auger drive shaft. Drive train horsepower loss must be added to determine total horsepower needs. Assume a 10% power loss and add to table values. To estimate between values, note that capacity decreases and horsepower increases as angle of incline increases.

534-26a. 4" conveyer, dry shelled corn.
Bushel weight, 54 to 56 lb; moisture content, 13.2%-14.2% wet basis.

Auger Speed	Length of Exposed Helix At Intake	Angle of Elevation 0°		45°		90°	
RPM	inches	bu/hr	hp/10'	bu/hr	hp/10'	bu/hr	hp/10'
	6	140	.11	110	.13	40	.10
	12	150	.12	120	.15	60	.11
200	18	150	.13	120	.17	70	.12
	24	150	.14	120	.18	80	.13
	6	270	.23	180	.25	90	.19
	12	290	.29	220	.29	130	.24
400	18	290	.33	240	.32	150	.26
	24	300	.38	240	.36	160	.27
	6	410	.33	280	.40	160	.29
	12	470	.43	350	.52	220	.41
700	18	480	.51	380	.64	250	.47
	24	480	.60	380	.76	270	.49
	6	490	.41	320	.61	200	.46
	12	650	.63	460	.81	310	.67
1180	18	740	.85	530	1.01	360	.79
	24	770	1.08	560	1.21	380	.88

534-26b. 4″ auger conveyor, dry soybeans.
Bushel weight, 54.5 to 56 lb; moisture content, 11%-11.2% wet basis.

Auger Speed	Intake Exposure	Angle of Elevation of Screw									
		0°		22.5°		45°		67.5°		90°	
RPM	inches	bu/hr	hp/10'	bu/hr	hp/10'	bu/hr	hp/10'	bu/hr	hp/10'	bu/hr	hp/10'
300	6	210	.15	180	.21	150	.22	100	.17	80	.17
	12	215	.16	190	.22	160	.24	140	.23	110	.19
	18	220	.21	190	.25	160	.26	150	.25	120	.21
	24	220	.21	190	.26	170	.27	150	.26	130	.24
500	6	330	.23	280	.31	230	.34	170	.32	130	.27
	12	340	.27	300	.39	260	.43	200	.40	160	.33
	18	350	.35	310	.45	270	.47	230	.43	180	.35
	24	360	.38	310	.48	280	.49	250	.46	220	.41
700	6	420	.28	360	.41	290	.45	210	.41	170	.37
	12	450	.37	400	.54	350	.60	270	.57	210	.47
	18	470	.49	410	.63	380	.66	290	.60	240	.49
	24	500	.53	435	.66	380	.71	330	.65	290	.58
900	6	465	.33	400	.49	330	.55	240	.51	200	.45
	12	520	.47	470	.67	410	.74	310	.71	250	.60
	18	570	.62	520	.81	460	.85	350	.77	300	.63
	24	640	.69	550	.87	470	.93	400	.84	350	.73
1100	6	490	.38	420	.56	340	.64	265	.60	220	.55
	12	600	.55	530	.78	460	.86	320	.82	280	.71
	18	690	.77	610	1.00	530	1.03	390	.92	340	.81
	24	780	.84	650	1.06	540	1.14	450	1.01	400	.92

534-26c. 6″ auger conveyor, dry shelled corn.
Bushel weight, 54 to 56 lb; moisture content, 14.5% wet basis.

Auger Speed	Intake Exposure	Angle of Elevation of Screw									
		0°		22.5°		45°		67.5°		90°	
RPM	inches	bu/hr	hp/10'	bu/hr	hp/10'	bu/hr	hp/10'	bu/hr	hp/10'	bu/hr	hp/10'
200	6	590	.20	520	.30	370	.33	280	.31	220	.25
	12	590	.38	550	.41	500	.44	400	.44	280	.32
	18	620	.32	570	.43	510	.47	430	.45	310	.36
	24	630	.44	590	.50	550	.55	470	.54	350	.40
400	6	970	.35	850	.52	650	.60	480	.57	380	.46
	12	1090	.56	1010	.82	850	.88	690	.83	520	.70
	18	1170	.74	1070	.92	940	1.02	720	.92	560	.80
	24	1190	.97	1110	1.13	1010	1.18	830	1.07	660	.92
600	6	1210	.49	1050	.72	820	.82	590	.77	490	.64
	12	1510	.84	1400	1.22	1160	1.28	910	1.16	740	1.05
	18	1650	1.17	1500	1.42	1270	1.52	1010	1.42	800	1.23
	24	1700	1.47	1570	1.74	1440	1.80	1140	1.60	920	1.40
800	6	1320	.58	1100	.86	890	.95	640	.92	540	.77
	12	1760	1.07	1660	1.54	1370	1.62	1080	1.46	890	1.32
	18	1990	1.57	1790	1.96	1510	2.08	1220	1.94	1000	1.64
	24	2140	1.95	1910	2.32	1740	2.39	1360	2.12	1100	1.89

534-26d. 6″ auger conveyor, dry soybeans.
Bushel weight, 54 to 56 lbs; moisture content 11%-12% wet basis.

Auger Speed	Intake Exposure	Angle of Elevation of Screw									
		0°		22.5°		45°		67.5°		90°	
RPM	inches	bu/hr	hp/10'	bu/hr	hp/10'	bu/hr	hp/10'	bu/hr	hp/10'	bu/hr	hp/10'
	6	490	.30	410	.41	320	.41	240	.38	180	.34
	12	500	.40	430	.53	360	.57	290	.50	220	.40
200	18	520	.50	500	.60	440	.66	360	.60	240	.45
	24	540	.60	520	.67	470	.68	390	.64	290	.52
	6	880	.52	710	.71	570	.77	400	.70	310	.60
	12	990	.84	830	1.14	690	1.20	540	1.04	390	.79
400	18	1110	.98	1030	1.18	880	1.29	740	1.23	460	.95
	24	1180	1.36	1040	1.62	900	1.63	800	1.54	560	1.14
	6	1080	.68	890	.96	700	1.07	510	1.00	390	.87
	12	1350	1.20	1130	1.61	930	1.71	710	1.48	500	1.10
600	18	1620	1.45	1510	1.74	1280	1.94	10[illegible]0	1.88	660	1.47
	24	1690	2.13	1520	2.52	1320	2.51	1100	2.32	790	1.76
	6	1180	.78	960	1.12	740	1.28	550	1.22	420	1.10
	12	1610	1.51	1310	1.98	1080	2.10	820	1.84	640	1.50
800	18	1980	1.93	1840	2.29	1530	2.54	1230	2.44	810	1.98
	24	2020	2.93	1850	3.43	1640	3.48	1320	3.24	1000	2.56

534-26e. 6″ auger conveyor, dry vs. wet shelled corn.
Moisture content, 14% and 25% wet basis; 12″ exposed helix at screw inlet.

Auger Speed	Corn Moisture	Angle of Elevation of Screw Conveyor									
		0°		22.5°		45°		67.5°		90°	
RPM	percent	bu/hr	hp/10'	bu/hr	hp/10'	bu/hr	hp/10'	bu/hr	hp/10'	bu/hr	hp/10'
200	14	594	.28	552	.41	498	.44	402	.44	276	.32
	25	372	1.37	318	1.40	282	1.31	204	.97	156	.32
400	14	1086	.56	1008	.82	852	.88	690	.83	516	.70
	25	696	1.84	618	1.89	510	1.78	402	1.45	300	.70
600	14	1512	.84	1404	1.22	1164	1.28	906	1.16	744	1.05
	25	948	2.32	822	2.34	678	2.27	516	1.92	378	1.09
800	14	1764	1.07	1656	1.54	1368	1.62	1080	1.46	888	1.32
	25	1098	2.80	948	2.85	774	2.75	582	2.44	474	1.55

Table 534-27. Shelled grain elevator capacity and horsepower.
Belt-type elevator.

Bucket	Bucket Spacing	Belt Speed	Capacity	Horsepower
3 x 2	8		50	.075
	4		100	
4 x 3	8		200	0.10
	6	270	300	0.125
6 x 4	4½	270	500-550	0.20
	4½	335	700	0.25
7 x 5	8		900	0.30
	6	335	1200	0.33
9 x 5	7	265	1600	0.5
	6	300	1800	0.5
9 x 6	12	385	1500	0.625
	6	385	3000	1.25

Table 534-28. Flight conveyor specifications.

Flight Width (in)	HP Required For Various Lengths (ft)							Flight Speed fpm	Approximate Capacities bu/hr
	5-15	16-20	21-25	26-30	31-35	36-40	41-50		
5	1/4-1/3	1/2	3/4	3/4				290-450	300-500
6	1/4-1/2	1/2- 3/4	3/4-1	1				290-450	500-600
8	1/2-3/4	1/2-1	1/2-1	3/4-1	3/4-1	3/4-1		150-365	400-600
14	3/4	3/4	1	1	1 1/2	1 1/2-3	2	150-225	
18		1/2-1	3/4-1	1-1 1/2	1-2	1 1/2-3	2-5	150	
20			3/4-1	1-2	1-2	1 1/2-3	2-5	150	1200-1400

535 SILAGE AND HAY

Types of Silos

Upright silo capacities are in Table 534-21. Bottom-unloading upright silos should take material from the silo center, unless the structure has been specifically designed for side withdrawal. Most cylindrical storages depend on uniform outward pressures that are disturbed by side unloading.

Upright gas-tight silos have lower storage losses (see the table below) but higher initial cost. They are also more flexible, because refilling with other materials in almost any combination and season does not interfere with the unloader or feeding.

Conventional silos need tight walls and sealed doors to be airtight. Sealing the top with plastic can keep storage losses below 5%. Once the silo is opened, feed out at least 3"/day in warm weather. Silo walls must be stronger for grass silage or wet grain than for corn silage. A roof is strongly recommended.

Horizontal silos cost less per stored ton than uprights, but have higher storage losses. Bunkers are above ground; trenches are below ground; stacks are temporary storage on the ground. Most trenches are built into a bank so the floor can drain out through the open end. Concrete walls are preferred below ground, plank or concrete above ground; adequate strength and water tightness are important.

Table 535-1. Silo spoilage.
Estimated averages with good management.
Moisture content = 15%-60%.

Type of silo	Percent of loss: Average	Range
Gas tight	5%	1-11%
Concrete stave	6	2-12
Horizontal (bunker)	15	10-25
Stack	25	15-30

Horizontal silos require more demanding management than uprights. Minimize surface exposure during filling; fill one end to full depth quickly, Fig 535-1. Compact silage thoroughly during filling and beware of the risk of the tractor tipping over bunker walls and on the sloped face of the pile. Shape the silage surface to shed water; install and anchor a surface cover daily, usually plastic. Unload 2"-4" daily and from only a portion of the surface if necessary, Fig 535-2. Disturb the exposed face as little as possible. Remove the surface cover week by week; maintain a hard surface floor sloped to drain, to control wastage.

Horizontal silos are widely accepted and successful. They adapt readily to feed-wagon distribution and to high speed unloading with tractor loaders or horizontal silo unloaders. Many are wide enough for the feed vehicle to circle in the silo, load, and return to the lots.

Although good for small herds where a tractor loader moves silage directly from storage to a bunk, horizontal silos have relatively high losses because of more surface area per volume and because shallow depths reduce compaction. If silage depth will be less than 16', consider the increased cost but decreased spoilage of an upright silo.

Stack silage only for emergency or temporary storage or for quantities of several thousand tons. Corn silage stacks are fairly successful with material at 70% moisture, wet basis. Make stacks as small and as deep as is safe and practical to improve quality.

Silage Moisture Content

For silage, wet basis moisture content should be 60%-70%; use the bottom of the range for bottom-unloading uprights, the middle of the range for top-unloading uprights, and the top of the range for horizontals. Too much moisture increases seepage, odors, and unloading problems. If too dry, the

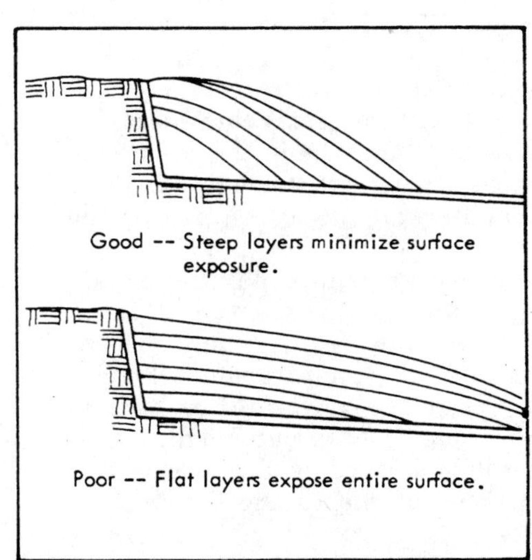

Fig 535-1. Horizontal silo surface exposure.

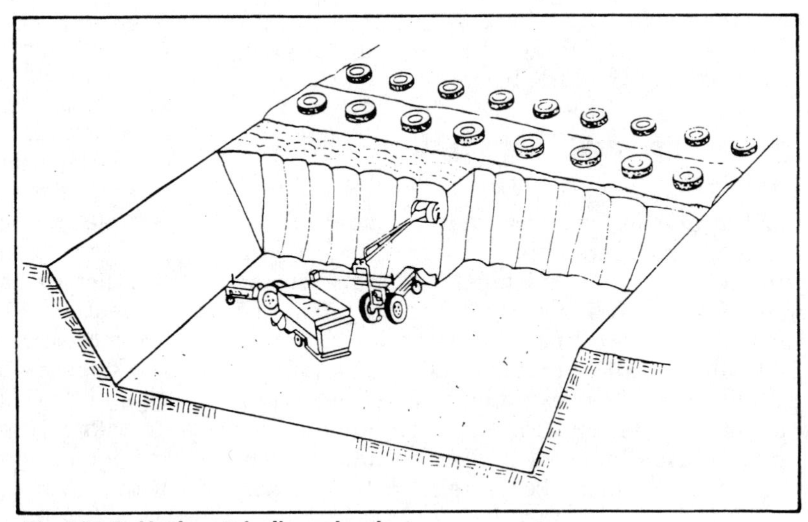

Fig 535-2. Horizontal silo unloader.

material may not compact adequately and may not "pickle" and tends to scorch. To reduce the risk of spontaneous fire, store haylage in upright silos at 40% or more moisture.

Safety with Silos and Silage

If you are going to operate silos, take time to learn about silo fires and gasses—if trouble comes, it may be too late to start learning. Contact the safety specialist at your College of Agriculture's Agricultural Engineering Department.

Fire hazard: To avoid spontaneous heating, do not ensile roughages below 40% moisture content. Once fire starts it is almost impossible to put out. It may not be possible for unloaders to empty the storage. There is danger of explosive gasses especially when water is added. Fires can burn for 6 months. Get expert help for fire control.

Silo gas is always a potential danger with roughages in conventional upright silos. Be especially cautious during filling and for the next few weeks—run the blower for some time before entering. Beware of suspicious smells or sight of gas in the silo, chute, or feedroom. Get a third person to help before you enter a silo to help a victim in trouble—you also may be overcome!

Don't enter a gas-tight silo without ventilating the open storage space thoroughly unless you wear an air pack. Do wear a safety harness attached to a line, with **two** people outside that can lift you out if necessary—one person cannot lift another in such circumstances, and one can go for help while the other gives first aid.

Equip a **silo unloader** with a safety switch to prevent its being started while repairs are made in the silo. Lock or flag the main control switch.

Keep **silo steps and chute** as clean as possible. Clean the steps as you enter the silo—you're more apt to slip coming down because you can't see the condition of the steps below you.

Tractor tipping risk has been mentioned. Set tractor wheels as wide as possible on the packing tractor. Pack especially well close to the walls as you fill—a firm pack reduces wall spoilage and prevents one wheel from sinking. As the silo fills, be aware of the added danger of a side slip or pitch, and the extra pressure the tractor puts on the wall.

Silage Storage and Distribution Systems

Horizontal silos may be most economical for very small producers, upright silos for the intermediate operation, and big horizontals for the large operator. Operator preference and ration and feeding circumstance affect selection.

The very small operator (under 200 tons or about 50 feeders) has few options. With an upright silo, a silo unloader and a mechanical bunk or feed wagon are usually needed. Mechanical conveyor feeding, if used, must be simple. With a small horizontal silo, hand or self-feeding is possible, but filling fenceline bunks with a tractor scoop is preferred. Storage losses may be high, but mechanization investment is small.

Upright gas-tight or conventional silos fit best with mechanical bunk feeding systems. They are typical in lots of 200-700 head near the feed center, and mostly family labor. Equipment investment can be efficient, and pushbutton feeding eliminates vehicle use in bad weather, on weekends, etc. Mechanical top or bottom unloaders match bunk conveyor capacities well. Conveyor lengths are limited to about 100' for augers and 200' for belts. Bunks are usually fed from both sides. With increasing size and number of lots, conveyors to the bunks may be longer than the bunks themselves. Accurately measuring daily feed use is difficult with continuous conveyor feeding unless weighers are added to the system.

Conventional upright silos can be combined with one or more gas-tight units. Use the unsealed silos for bulk storage, such as corn silage or corn and cob meal silage. Use a sealed unit to ensile roughage and grains added during the summer and early fall, and then fill the silo with wet grain or silage for winter storage. It can also be refilled from other silage storages to use its mechanical capacity. Filling the silo twice a year halves the use cost per ton.

Silo unloader delivery rates are usually too slow for well designed wagon feeding systems. The 10-20 minute wait to fill a load may be acceptable for only one load per feeding, but for 2 or more loads/feeding, it can be a serious time loss.

A surge bin or accumulator box can eliminate the delay. The box, mounted overhead on a scale or with a high capacity transfer conveyor, Fig 535-3, accumulates a load while the first load is being delivered. Size the accumulator to the feed wagon box; select one with fast delivery of non-free flowing materials. A surge bin installation is expensive and usually limited to larger feeders.

Feed wagons or trucks are more flexible than mechanical bunks. They can service a combination of feeding locations, different enterprises (cow-calf on pasture, feed lots), get feed from off- and on-farm sources, and permit feedlot locations to fit topography, use existing facilities, etc. Vehicle feed distribution also adapts to changes in the number of cattle and permits knowing how much feed is going to each lot, either by weighing or estimating the volume of each load.

Feed wagons or trucks require maintaining all-weather roads, drivers, and operation in all kinds of weather. Fenceline bunks fed from only one side are more expensive than comparable mechanical bunks ready to receive the conveyor. Bunk length along lot perimeter affects lot layout and drainage.

Many farms combine wagon and conveyor feeding for some of the convenience and flexibility of both systems.

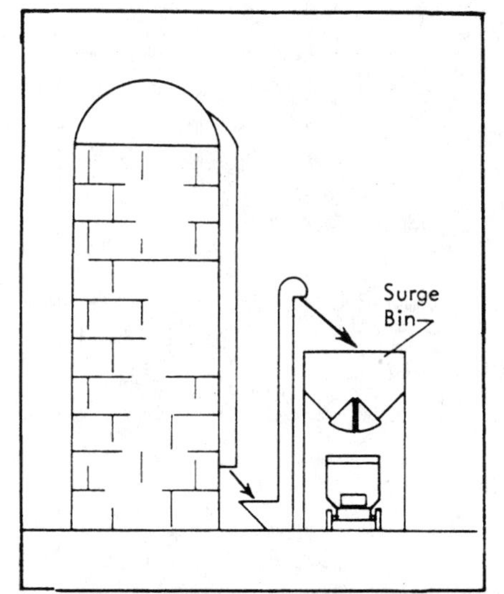

Fig 535-3. Overhead silage surge bin.

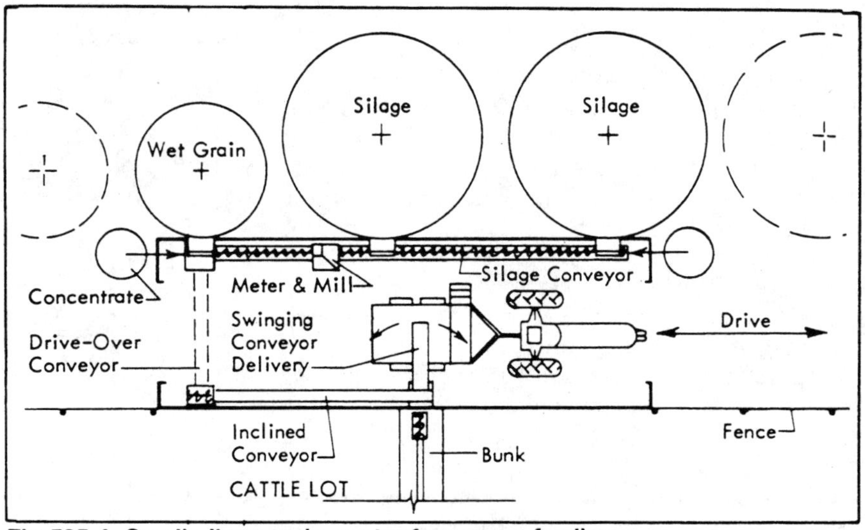

Fig 535-4. Small silage grain center for wagon feeding.

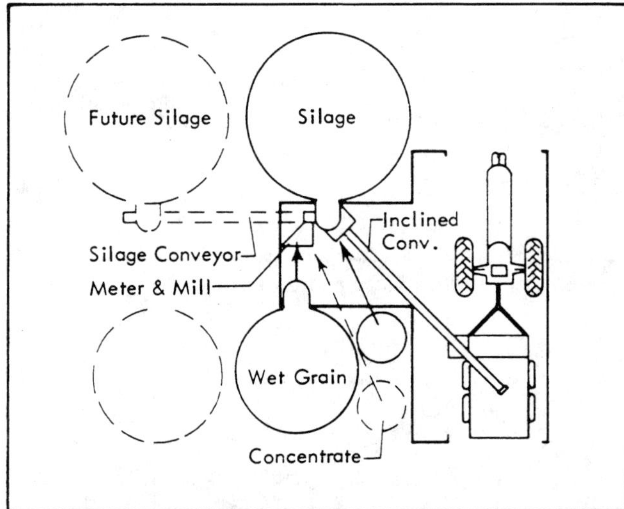

Fig 535-5. Silage grain center, bunk and wagon feeding.
Center building is about 20′x34′ plus 12′x16′ offset.

Consider a scale-mixer trailer or wagon for 300-1500 cattle and scale-mixer trucks to feed 1500-5000 head. One truck is needed for each 5000 head on feed.

Drive-through feed center buildings protect loading from wind, storms, and cold. Because a feed vehicle (Figs 535-4 and 535-5) is usually sheltered somewhere, the cost of shelter in the feed center alley may be little more than equivalent space elsewhere.

Conveying either wet or dry grain is usually easier and with cheaper, smaller equipment than with silage. First, locate silos for efficient silage handling. Locate grain storage to allow post-storage processing and for handling with silage.

Combination dry grain-silage feed centers may be needed on farms that sell grain and feed cattle. Grain and silage can be stored separately, especially with wagon feeding—route the wagon past the silo and then to the dry feed center.

Processed feed can be conveyed from a dry grain center to a bulk tank at the silage feed center. Or, with large dry grain volumes, put one grain storage at the silage-feed center and move dry grain for processing from the grain center. The dry grain center on many farms also processes feed for other livestock, such as hogs. Putting all dry grain at the silage center may limit growth and flexibility. Farmers with small volumes can combine silage and dry storage in one facility to reduce the number of building sites and permit some dual use of equipment.

Purchased feed reduces storage needs unless bought and stored at harvest time.

Hay Storage and Distribution Systems

Hay is handled as conventional square or round bales (60-80 lb), big round bales (1000 or 2500 lb), or as stacks (1-8 tons). Half-ton and larger bales and stacks are handled with special equipment. With small square bales, handling, storage, and feeding can be mechanized with a bale accumulator or automatic bale wagon system with ground level storage. Cornstalks and other crop residues can be handled as hay.

Small round bales are stored under roof or left in the field for winter grazing by the cow herd. Nutrient losses due to weather may be 15-20%, but handling labor is eliminated. Feeding losses from unrestricted pasture grazing may be as high as 50-60%; reduce losses to 20-30% with strip grazing. Small square bales are stored under roof or field stacked and covered with plastic or other material.

Feed baled hay in racks at or near the storage, Fig 535-6, in permanent or portable feeders filled with hay moved from storage, or just spread on the ground, Fig 535-7. Feeding losses are usually no more than 5%.

Big round bales and large hay stacks are designed to be stored outside. Nutrient losses due to weather are usually 10% or less with careful storage.

Select a well-drained area: divert all surface water away. Provide access that will be dry and hard during the feeding period.

Select a location with some wind protection for stacks, and unload so the fronts of the stacks face the prevailing wind. If possible, select a level storage area. If the area is sloped, arrange the stacks or bales so that the tractor and mover are backing downhill when loading for feeding.

Store large packages so they don't touch, so water runs off to the ground before soaking the hay. Allow for settling. Usually, 12"-18" between large bales and 24" between stacks is enough.

Store large bales on a slope, with their round sides uphill; flat ends uphill soak up water. Storing stacks in rows end-to-end without touching makes it easier to load them on a mover. Each stack prevents snow from drifting in front of the one behind it, and each stack pushes against the next one as the mover is backed under it.

Big bales and stacks can be stored in an open structure. Half-ton bales can be stacked 2 or 3 high under a roof with a front-end loader, Fig 535-7.

Handling big bales on a front-end loader is hazardous. The heavy bale makes the tractor more apt to tip if a rear wheel hits an object or a hole. When the loader is raised, the bale can roll down the loader arms onto the driver. Rear-mounted or trailing movers are safer.

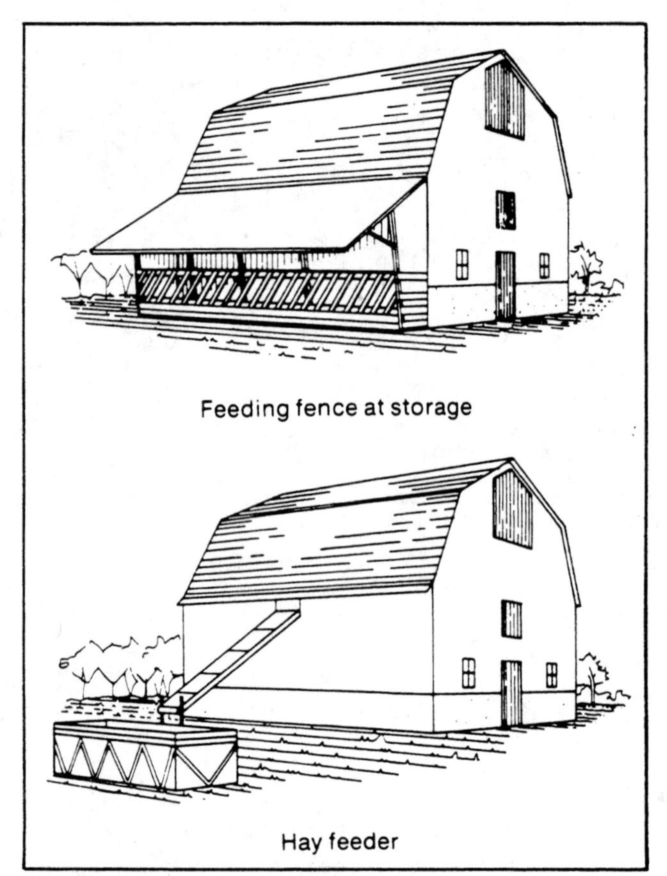

Fig 535-6. Hay feeding from roofed storage.

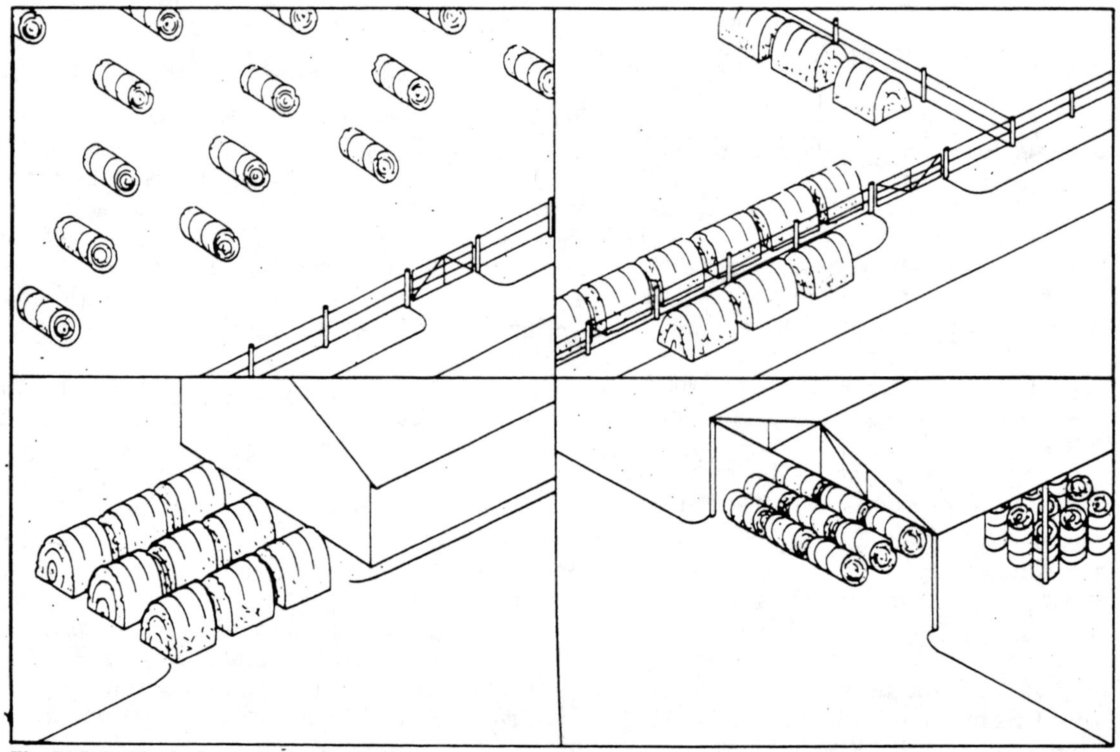

Fig 535-7. Big bale and stack storage.
From upper left: random storage in the hay field; along the edge of the field or a lane; in a concentrated stack yard; under-roof storage.

Bales and stacks are sometimes left scattered in the field for self-feeding, or are moved to the feeding area. Feeding waste can be 50%. Trampling also creates mud and usually kills vegetation in the area.

Reduce waste to 5%-10% by limiting animal access to the hay or by restricting the amount of hay exposed to the animals.

- The packages can be fenced on 3 sides; use a hot wire along the fourth side to control access and hay consumption, Fig 535-8. The hot wire system is difficult to manage and mud is usually a problem.
- Line large packages up 15' apart along a fence. Move temporary fencing to expose one package at a time. This technique is not recommended for large stacks.

Better ways to feed big hay packages are:

- Self-feed whole bales or stacks in circular feeders, feeder panels, or portable feeders, Fig 535-9.
- Three-sided fenceline feeders or a push-up feeding fence, Fig 535-10. Limit-feed with a bale unroller or a feeder attachment on a trailed stack mover. Drop the hay on the ground or into bunks, Fig 535-11.

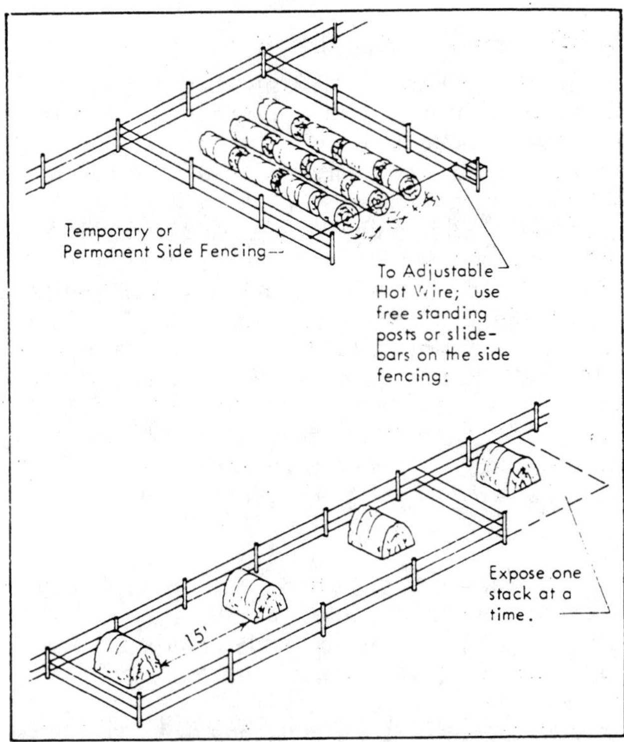

Fig 535-8. Rationing hay with an electric fence.
Big bales are stored close together in rows and columns. Animals feed under the hot wire, which is moved as required to control consumption. Or, expose 1 stack at a time.

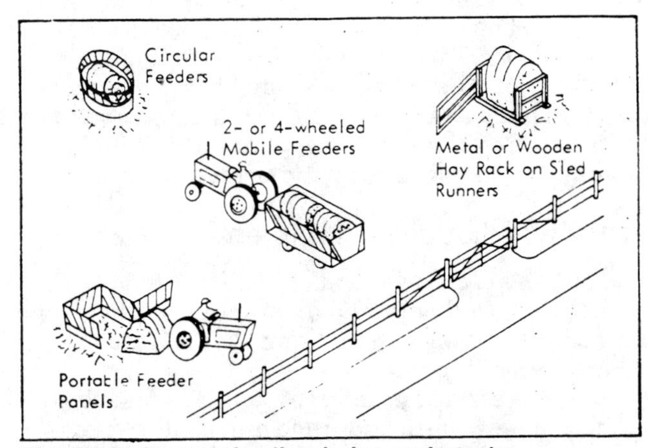

Fig 535-9. Pasture feeding bales and stacks.
Fill mobile feeders with a front-end loader with grapple fork. Move feeding site as desired.

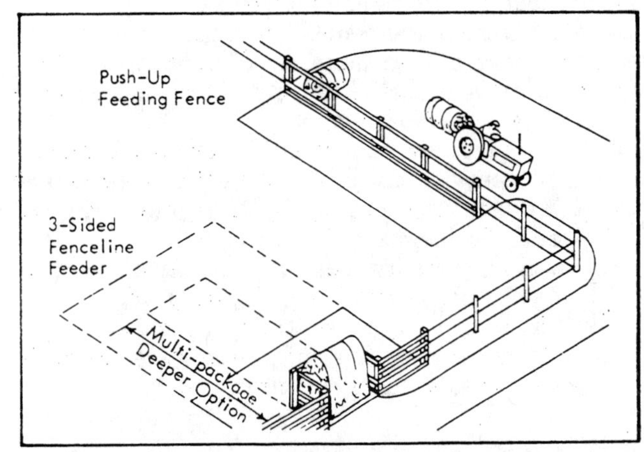

Fig 535-10. Fenceline feeding of big bales and stacks.
Push the hay closer, as necessary. A strip of hay in the center that cannot be reached is pushed back when new packages are moved in.

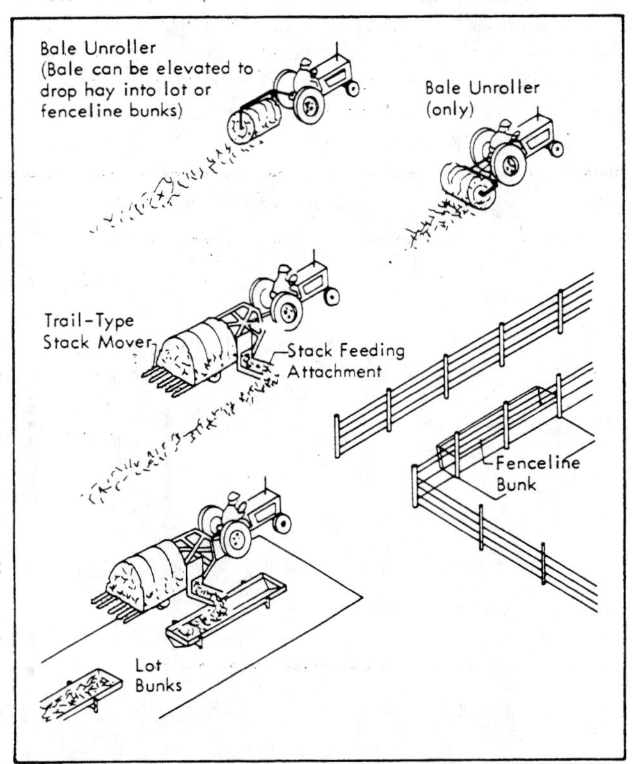

Fig 535-11. Limit-feeding big bales and stacks.
Feed on the ground or in lot or fenceline bunks.

540 FARM SHOPS

Most of this chapter is by W.H. Friday, D.D. Jones, S.D. Parsons, and R.M. Strickland. It was first published in *Planning Farm Shops for Work and Energy Efficiency,* Cooperative Extension Service Bulletin AE-104, Purdue University.

An energy efficient maintenance shop is needed on a modern farm. It provides a place to assemble, service, repair, adjust, and modify machinery and equipment for field and farmstead operations. It is also a place to work on farm, family, and even recreation and hobby vehicles and projects.

A well planned and well equipped shop encourages preventive maintenance of equipment, which not only extends its useful life, but also reduces the chances of costly failure.

The shop can be headquarters for farm management, employees and daily callers. Provide a heated office with washroom and space for farm records, tool catalogs and service manuals.

How well shop facilities are utilized depends on:

- Shop size and how well it is equipped.
- Skill and ability of those who use the shop.
- Extent that service and maintenance are coordinated with total farm operations.

Pre-Planning

Warm Weather Vs. Year-Round Shops

A shop in the corner of a machinery storage building and not partitioned from the storage area, Fig 540-1,is generally not heated, so it is used only during favorable weather. This type of warm weather shop, rather common on smaller farms, has minimal insulation, ventilation, or heat.

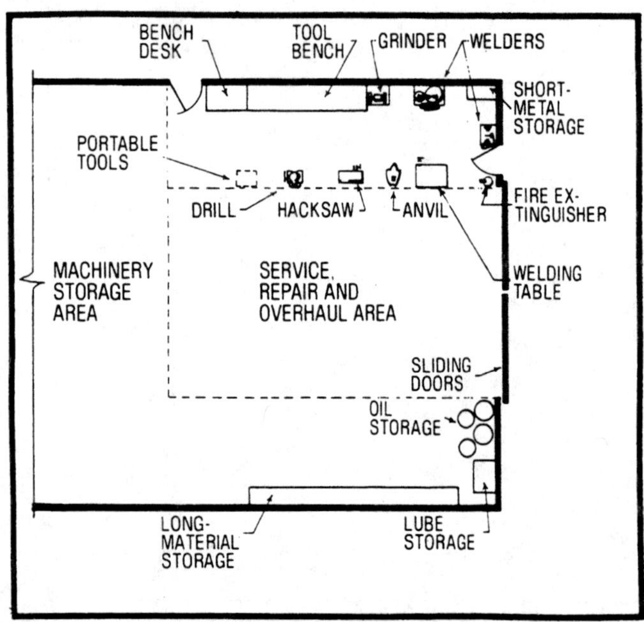

Fig 540-1. Warm weather shop layout.

Year-round shops have environmental control features, especially heating for maintenance and repair during winter. They can be a separate building, or they can be in the machinery storage structure but partitioned from the storage area. They are self-contained units, designed exclusively for use as shops.

This chapter emphasizes year-round shops, but much of the information also applies to warm weather ones.

Building New Vs. Remodeling

Many older shops are remodeled garages, corn cribs or other farm buildings not well suited to today's large farm equipment. In fact, any shop more than 10 years old is probably too small, unless it was built with unusual foresight.

Because of limited door size, most existing farm structures cannot economically be converted to modern farm shops. Therefore, consider building new shop facilities instead of remodeling.

However, a conventional two story barn, if the roof and foundation are in good condition, can sometimes be remodeled. Remove the mow floor and raise the ceiling height to about 14′ in the shop service area. Install a large service door in the end of the building or in the sidewall if the plate is high enough.

Site

A farmer is apt to go in and out of his shop more times than any other building on the farmstead except the house. Therefore, locate the shop convenient to the house, the machinery storage, and main farm traffic routes.

Because of the high value of shop tools, consider security in site selection. Yard lights help, but the best solution is for the shop to be accessible only from a lane which passes directly by the farm home. Locate large access doors so that valuable tools and equipment are not visible from the main road.

Put the shop on high ground with good drainage and with floors at least 12″ above the existing grade. Slope the ground surface 5% away from the shop. If possible, face doors away from prevailing winds to minimize wind damage. See Chapter 501 for site planning recommendations and details.

Farm Shop Layout Principles

Major shop activities are:

- Repairs, overhauls, and other major mechanical work, including annual preventive maintenance and reconditioning.
- Routine servicing and maintenance, including oil changes, lubrication, and minor adjustments.

For larger shops, make repairs in the larger area and do routine servicing in the smaller area. On smaller farms, repair and service equipment in the same area.

Consider the layout in Fig 540-2; modify it for your particular needs. Arrange the workbench and tool areas around the repair/overhaul area. The tool areas should be flexible; allow for large power tools and for implements which "fold out." Provide storage for bolts and parts, and office space for equipment manuals and plans, as well as service, repair, and inventory records.

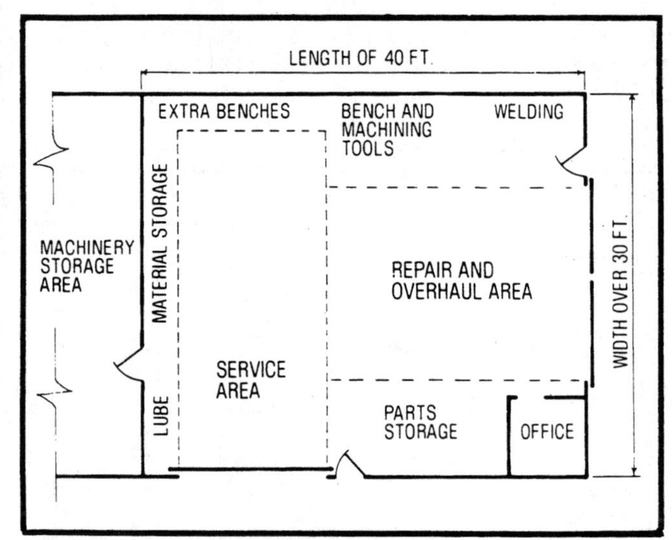

Fig 540-2. Basic farm shop floor plan.
Layout shows relative locations of repair and service areas and other work areas.

If additional service areas are needed, insert them at the end of the repair area, Fig 540-3.

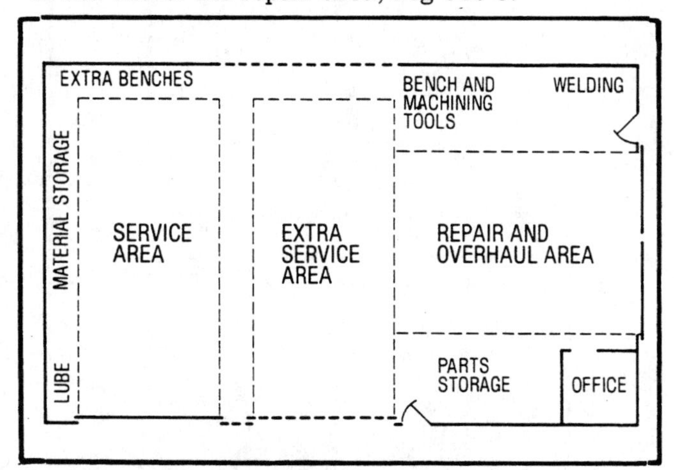

Fig 540-3. Basic plan with added service areas.

Repair/Overhaul Area

Make the repair/overhaul area large and well lighted. Size it for the length and width of the largest equipment to be repaired, Table 540-1. Accommodate implements with a tractor attached during repair by extending the work area into the service area, Fig 540-2. Extra-wide fold-out machinery can overlap the tool areas along the sides of the area.

Table 540-1. Shop repair/overhaul area dimensions.

Machinery matched to planter size	Repair area— Width	Length
	ft.	ft.
Less than 4-row	12-14	24
4-row, 38''	16	24*
6-row, 30''	20	28*
8-row, 30''	24	32
12-row, 30''	24**	32

* Self-propelled combines may exceed this length.
** Additional width required for fold-out equipment.

Service Area

The service area is for most day-to-day shop activity during busy field work periods. Put oil, grease, air and water near the door for servicing inside or on the outside apron.

The service area is a convenient garage for easily moved and often used equipment such as the farm pickup, a tractor for snow removal or a feed truck. You can also do emergency work there when the repair area is full.

Tool and Bench Areas and Parts Storage

Organize tools and supplies so repairs can be done efficiently and safely. Designate specific areas for welding, machining, and lubrication, and for carpentry, plumbing, and electrical work. Fig 540-4 shows efficient work area arrangements for four farm shop sizes; Table 540-2 recommends dimensions for these work areas.

Put 3′ high (depending on owner's height) work benches along the wall, with the large, free-standing power tools around the perimeter of the repair area. For shops wider than 32′, plan for 8′-12′ wide work areas on both sides of the repair area and for a 4′ wide area along the service area wall. For smaller shops that use somewhat smaller power tools, 6′ perimeter work areas may be enough.

Table 540-2. Welding, bench, and office sizes.

Shop size	Welding area	Bench length	Office space
	ft.	ft.	ft.
24 × 32 (Fig. 4a)	6 × 8*	8	Bench**
32 × 40 (Fig. 4d)	8 × 10	12	Bench**
40 × 48 (Fig. 4c)	10 × 10	16	8 × 8 (encl.)
48 × 56 (Fig. 4b)	12 × 12	24	10 × 12 (encl.)

* Portable welding table stored against the wall and moved into open area when in use.
** This is a small bench or desk located near the walk door entry to the shop.

Regardless of shop size, anchor benches and tools securely to the wall. Provide stable bases, but do not fasten tools to the floor around work areas; allow flexibility for large equipment.

The welding area is near an outside door and is at least 10′ long and 8′-12′ wide, depending on the equipment size. Provide for welders, grinder, anvil, and storage for short metal pieces and welding rods. The welding table should be at least 6" out from the wall and as far as possible from the lube area to reduce fire hazard.

Line the wall in the welding area with ½" fire retardant treated plywood or ⅝" exterior plywood. Or cover flammable wall lines with: 0.014" galvanized steel, 0.032" aluminum, ½" cement plaster, ½" fire-rated gypsum board, or ¼" asbestos board.

The machining area has cutting and drilling tools and bench space. It should be 8'-24' long and 8'-12' wide, depending on tool sizes. Put machining next to welding so that the grinder, anvil and welding table are available to both.

The machining area can include roll-about tool boxes, hoists, and movable tools like bearing presses when they are not in use. They can be moved when the area is needed for machining or to widen the service area temporarily.

For carpentry, plumbing and electrical work, have a wall-mounted wood bench at least 8' long. Line the walls with ⅝" exterior plywood or ¾" particle board at least 8' high to hang tools. Separate carpentry from electrical and welding areas to reduce equipment deterioration and fire hazards.

Put lubrication next to the service area entry. Locate oil, lube equipment, and supply storage along the service area wall. Space for three oil barrels permits one for truck and car oil, one for tractor oil, and one for waste. Oil storage can be wall-mounted for gravity flow discharge and pump filling.

The lubrication area is often about 8' long, includ-

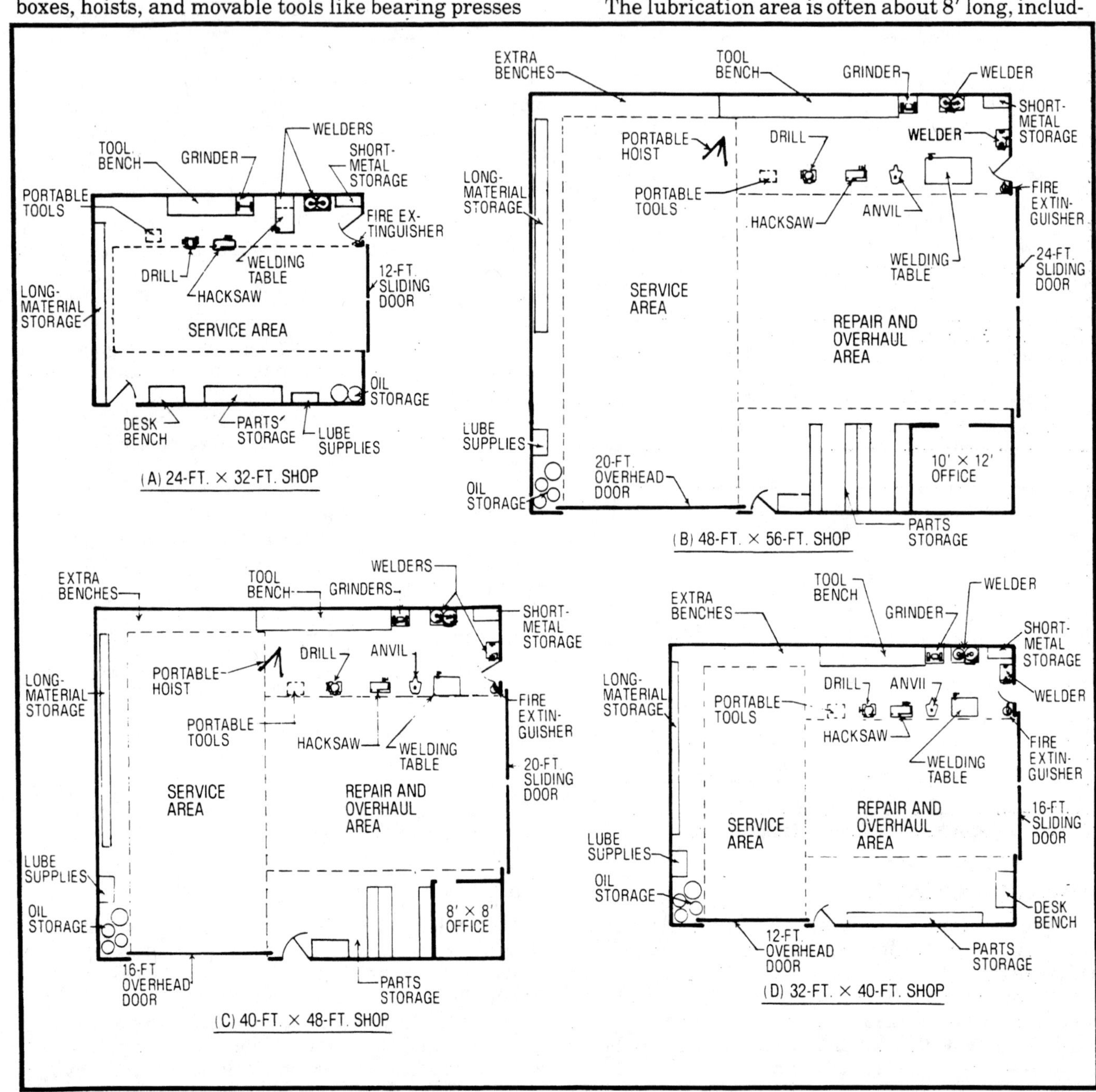

Fig 540-4. Four shop layouts from the basic plan.

ing tire service equipment and supplies.

Wall-mount the air compressor near the service area entry to save floor space and to service equipment on the apron. If noise is a concern, the air compressor can sometimes be in an attached machinery storage or in the shop attic.

Storage for parts, bolts, and screws occupy the remaining shop space. Use free-standing units for flexibility. Overflow supplies of lubricants and parts can be stored in the machinery storage building.

Wall-mounted racks along the service area wall, 12"-18" wide and at least 20' long, can store metal, pipe, and lumber.

Office

Provide convenient storage for repair and service records and machinery manuals—a small room in one corner of the shop or, in a small shop, merely a section of the work bench. Larger offices are needed on some farms for employee conferences and a communication headquarters.

Put offices larger than about 10'x12' in attached, single-story additions rather than within the shop. Storage over the office requires expensive, heavy-duty ceiling framing that cannot be justified in most cases.

Fig 540-5 illustrates some farm shop offices.

Restroom

Many farms, especially those with fulltime or seasonal labor, provide a small restroom in the shop, usually near the office. Although the water heater and septic system for a lavatory and toilet are expensive, a restroom improves working conditions and is a

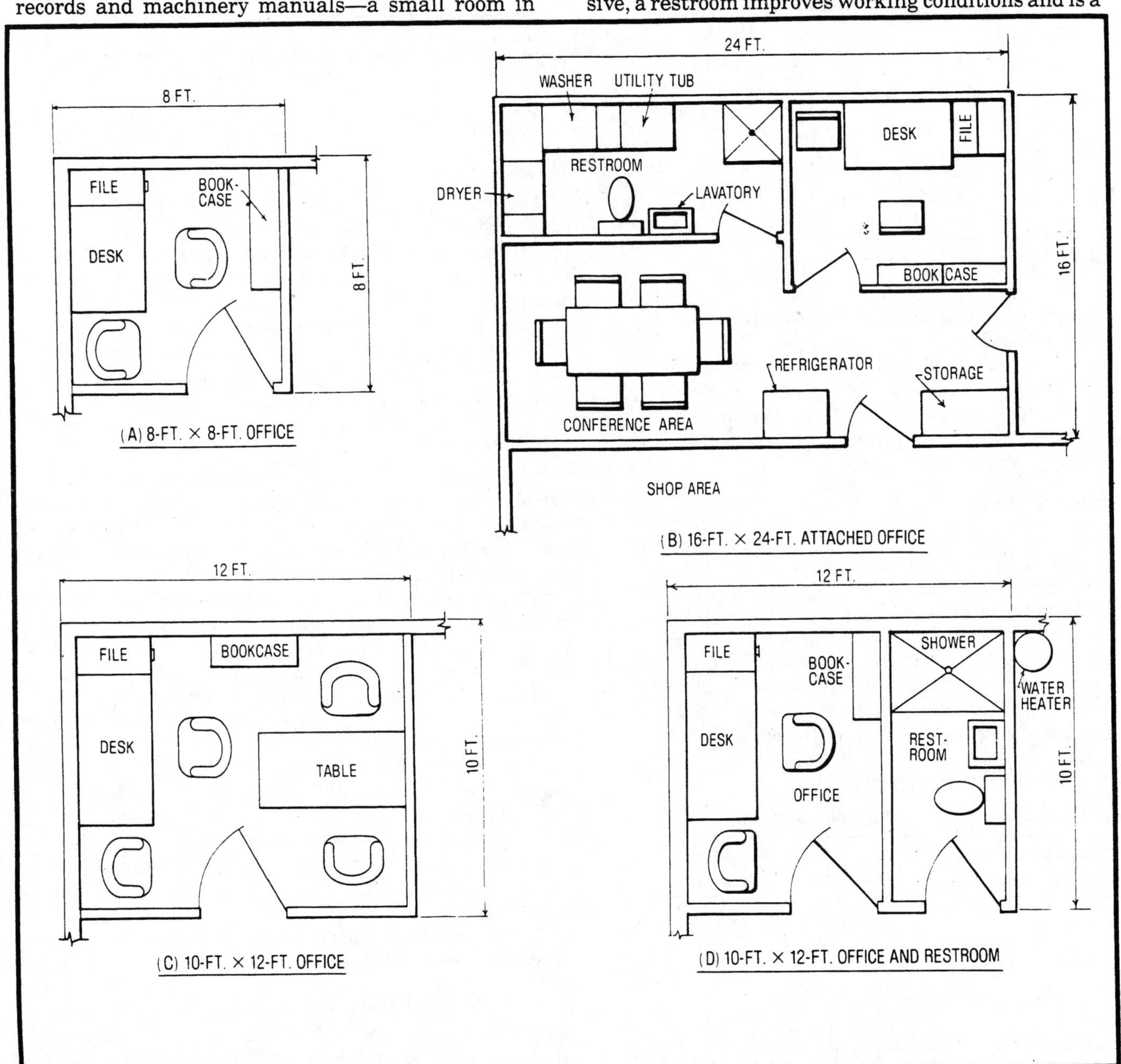

Fig 540-5. Farm shop offices.

convenient cleanup area. A shower, laundry facilities and clothes storage are also desirable for farms with off-farm labor.

At least a lavatory should be available in the shop for cleanup and safety. A frost-free hydrant and drain may be sufficient in smaller shops.

Farm Shop Size

Size your shop for the equipment on hand today but allow for tomorrow. Consider purchases over the next several years, such as a larger tractor or a combine with a bigger head. Enlarging a farm shop may not be economical. If your needs outgrow your shop, consider a new one.

Table 540-3. Shop length vs. equipment size.

Machinery matched to planter size	Length
	ft.
Less than 4-row	32
4-row, 38"	40
6-row, 30"	48
8-row, 30"	56
12-row, 30"	56
Other equipment	
4-WD tractor with plow	56
Semi-trailer with tractor	64

Building Length

Shop length depends on equipment length and how much space you want for convenience and work at the front and rear. See Table 540-3.

Because extra-long equipment can extend into the service area during repair, a separate service area need not add to shop length. Table 540-3 refers to overall shop length: repair area length plus service area width plus 4' storage space.

Building Width

Minimum building width (with repair area entry in the endwall, Fig 540-2) is twice the equipment-entry door width. This width provides support for sliding doors as well as work space on each side of the repair area.

If building width exceeds 50', enter the repair bay from the sidewall instead of the endwall, Fig 540-6. Fig 540-6 is a 'flip-flop' of Fig 540-2; the shop layout principles discussed earlier still apply. The rearrangement keeps the work area from getting too large and retains efficiency.

Table 540-4. Repair area door dimensions.

Machinery matched to planter size	Repair bay door— Width	Height
	ft.	ft.
Less than 4-row	12-14	12
4-row, 38"	16	13
6-row, 30"	20	14
8-row, 30"	24	15
12-row, 30"	32*	16

* Only 24 feet required for fold-up or end-trucked equipment.

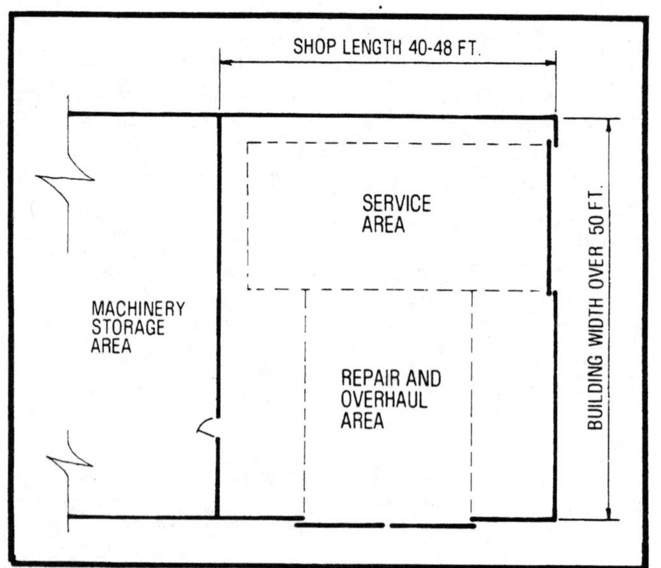

Fig 540-6. Layout for buildings wider than 50'.

Construction

Walk Doors

Locate 3' wide doors next to each machinery door for convenience and to prevent heat loss from opening large doors in cold weather. In combination shop-storage buildings, put a 3' door between the two sections.

In small shops with only one machinery door, a 3' door at the opposite end is a fire exit and improves summer ventilation, Fig 540-4a. Attached offices need an outside door for convenience, Fig 540-5.

Machinery Doors

Size doors for your machinery. For equipment on a modern grain farm, 13'-15' height is usually needed. Cabs, exhaust stacks, and bin extensions may require more clearance, Table 540-4. A 24' wide by 14' high door meets most needs.

The service area entry door is often 4' narrower than the repair area door, and about 1' lower, due to sidewall framing. Therefore, it is easier to open and close. Because this door is used often, consider a mechanically operated overhead door. Overhead doors can open into a recess in the ceiling so all the ceiling need not be raised 2'.

The repair area is entered less often than the service area; a less costly sliding door will do. Endwall sliding doors are up to half the shop width to leave space on each side of the repair area for benches and tools. For example, a 20' sliding door for a building 40' wide permits a 10' wide work area on each side of the repair area.

When possible, locate machinery doors away from prevailing winds to minimize wind damage and heat loss.

Provide machinery maneuvering space outside shop doors: at least 40' in front of entry doors; and at least 60' for semi-trailers and long, tractor-drawn machinery.

Protect all machinery-access doors both inside and outside with concrete-filled metal standpipes (at least 3″ diameter and 4′ high), embedded in concrete 12″ out from and in line with the door frame.

Floor and Drainage

Use 4″-6″ of compacted sand and pea gravel under the shop floor. Compact low- and moderate-clay content soils used as fill or allow them to settle several months. Avoid high-clay soils. Poor a 5″ thick concrete floor for all but exceptionally heavy machinery, which need 6″ floors.

Provide a central drain in each work area or slope floor 0.1″/ft toward the entry door.

A long, narrow, central slotted floor drain reduces total variation in floor elevation and eliminates the door freezing that can occur with one-way slopes. A floor drain with an oil and sludge trap performs well, if the trap is properly constructed and cleaned periodically, Fig 540-7.

Grates must not plug, yet they must be strong enough to support heavy equipment. See Fig 540-8.

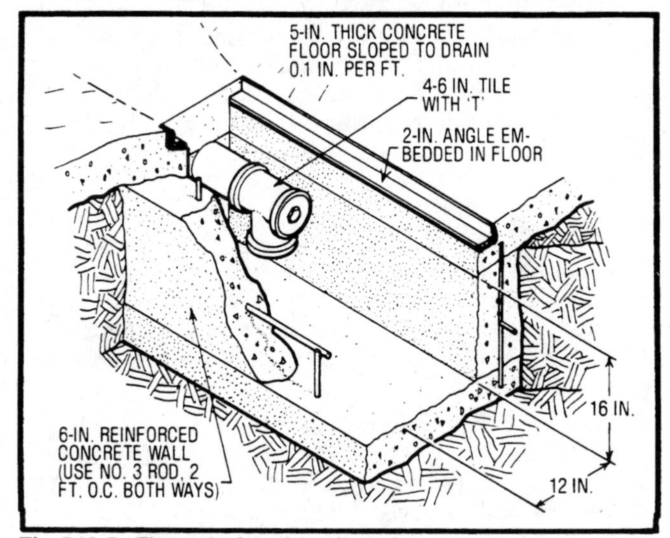

Fig 540-7. Floor drain with oil and grease trap.

Aprons

Provide an apron with a 1½″ raised shoulder at each machinery door. Make it as wide as the door and extend it 2′ inside and 3′ outside the door opening, Fig 540-9. Make the apron 5″-6″ thick, and reinforce it with No. 3 bars. With sliding doors, place a metal wedge in the center of the apron to force the doors tight against the shoulder when closing.

An apron 4′ wider than the door and extending 16′ out from it is for parking, servicing, emergency repairs, and clean up, Fig 540-9b. Concrete (4″ thick) is preferred, but well compacted gravel 4″-6″ thick also works. Slope the apron away from the building at least ⅛″/ft (¼″/ft for gravel) for drainage.

Hoists

A portable floor hoist avoids the extra framing required by ceiling hoists, is more maneuverable, and allows more space flexibility.

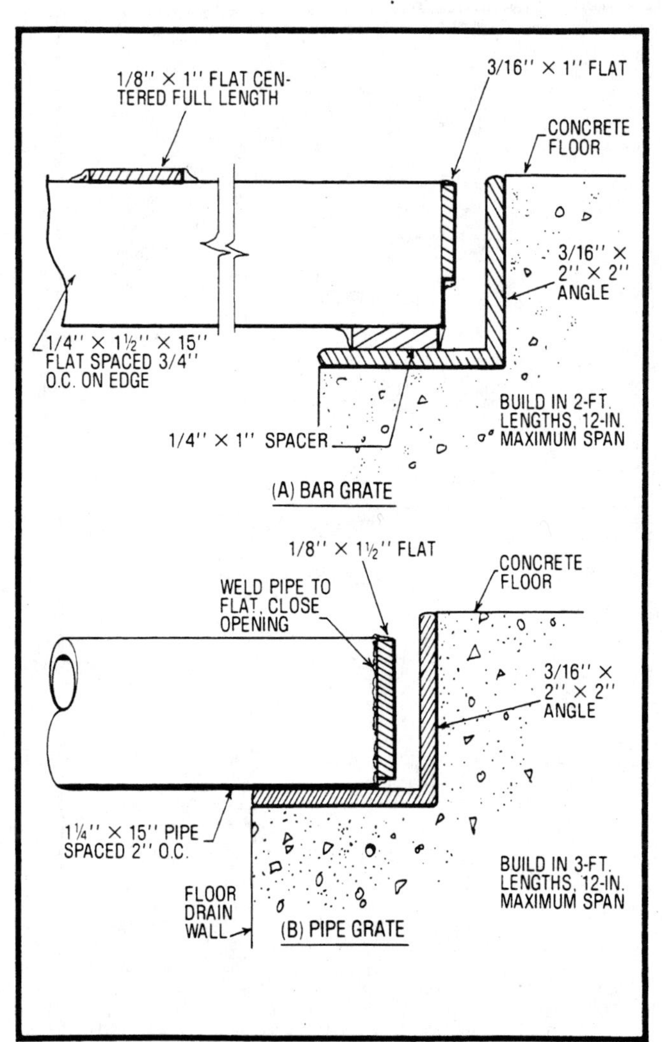

Fig 540-8. Grates for 12″ wide floor drains.

The framing for single-point and trolley-mounted hoists depends on truss span and the weight to be lifted. Single-point ceiling hoists are not very versatile. Trolly-hoist units, while more versatile, are much more expensive than portable floor units.

Environment and Utilities

Insulation

A warm-weather shop needs about R = 5 insulation in the ceiling or roof to relieve summer heat and minimize winter condensation. But a heated shop should be well insulated, including foundation, sidewalls and ceiling. (Insulating the ceiling takes 15% less insulation and gives 15% better heat loss protection than insulating the roof area.)

Insulate sidewalls to R = 13 and the ceiling to R = 20. See Chapter 631 for perimeter insulation and how to protect insulation from fire.

Ventilation

Ventilation removes smoke and fumes. Most important is a wall-mounted 1,000 cfm exhaust fan in the welding area. Although exhaust hoods are more

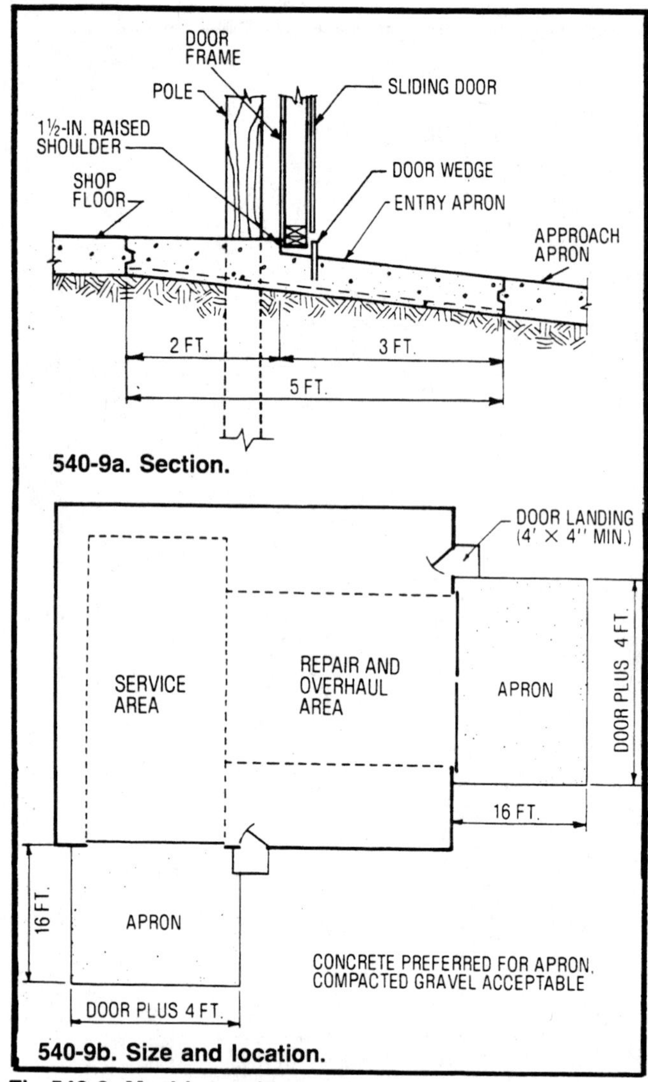

540-9a. Section.

540-9b. Size and location.

Fig 540-9. Machinery door aprons.

effective, they are also more expensive and difficult to work around. Open a door or window slightly (preferably one across the room from the fan) for fresh air to increase the fan's effectiveness.

If you do a lot of engine work, consider a permanent wall or ceiling-mounted flexible duct, with its own exhaust fan to remove fumes directly from an engine's exhaust pipe. Size the fan and ducts for:

CFM = CID x RPM ÷ 3230 = ft^3 air/min
CID = in^3 displacement of engines running
RPM = engine speed.

In summer, open endwall and sidewall doors.

Heating

Consider heating your shop to 45-60 F (40-45 F when not in use). Especially for electric heaters, consider several smaller units instead of one large furnace to heat work centers.

Ceiling-mounted, forced air space heaters work well because the furnace blast helps keep hot air from stagnating near the ceiling. Size the furnace at about 50 Btu/ft^2 of shop floor area. Use a minimum of 70,000 Btu/hr if the shop is fan-vented at 1,000 cfm.

Unvented fuel-fired heaters are for temporary use only. A uniformly-distributed minimum ventilation rate of 4 cfm per 1,000 Btu/hr furnace capacity is essential to prevent buildup of poisonous and odorless carbon monoxide. Operate a 1,000 cfm fan with a 10 minute timer set at 2½ minutes to provide ventilation for the unvented heater; leave a door open slightly

Radiant heaters are fast, can be localized, and are easily changed and relatively safe. They are efficient at low shop temperatures, because radiation heats the surfaces it strikes, providing comfortable equipment and surface temperatures. Overhead radiant heat reduces heat loss from doors. But radiant heaters may be more expensive than space heaters, and floors under vehicles are cold without floor heat.

Radiant heat makes a room at 60 F almost as comfortable as one at 70 F with forced air heat. Ceiling-mounted radiant heaters are less expensive to install and more dependable than floor heat systems. Size radiant heaters at 40 Btu/hr-ft^2 of shop area.

For floor heat, place heat cables (15 watt/ft^2) or water pipes (50 Btu/ft^2) near the bench area and extend the heating unit into the repair or service area about 4'. Maximum floor surface temperature is about 80 F. Reinforce heated concrete slabs with 6x6-10/ 10 wire mesh to prevent damage from heavy vehicle traffic. Solar heat can be stored under a shop floor.

Electricity and Lighting

Size the electrical service entrance to the size of shop. Minimum size is 200 amps and 240 volts. Locate the service entrance near a walk door for emergency shut-off. Fig 540-10 is a typical plan of convenience outlets and lighting.

Locate convenience outlets:

- uniformly along each work bench under the front edge to keep cords off the bench.
- at every motor-driven tool.
- at 10' intervals around the rest of the shop perimeter, 4' above the floor.

Use plug mold with outlets 1' apart or duplex outlets 4' apart along the workbench. Use overhead retractable drop feeds for large, free standing power tools and as convenience outlets around the perimeter of the repair and service work areas. Locate weatherproof duplex outlets on the outside wall of the shop near each machinery door.

Ground fault interrupters (GFIs) are recommended on all 20-amp convenience outlet circuits and are required for outside receptacles. The GFI trips the circuit if a 'leak' is detected, before a short can harm the operator.

Provide 20 foot-candles of lighting at floor level. That is a minimum of ½ watt fluorescent or 2 watts incandescent per ft^2 of typical shop floor area. Light-colored ceilings and upper walls help. A single incandescent light inside walk doors allows access until cold fluorescent or other lights start.

Provide 48" double-tube fluorescent fixtures

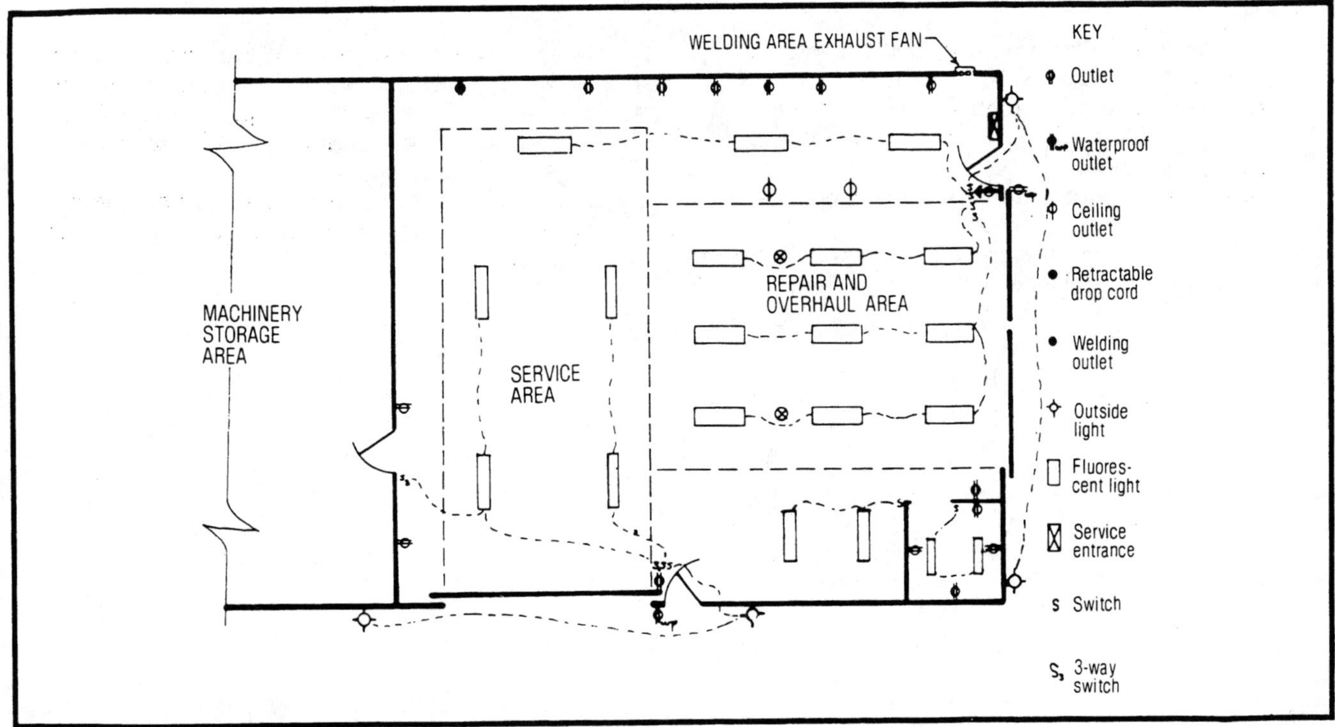

Fig 540-10. Typical shop electrical plan.

mounted 50" above benches. Mount them away from the wall over the front ⅓ of the bench.

For outdoor work, use a 400 watt mercury refractor, 25′ high for every 2,000 ft^2 of work area. Light doorways with 150 watt weatherproof flood lamps 8′ away from the entry doors. Place lights next to, rather than directly over, doors to attract flying insects away.

Water

For fire protection, equipment clean up, and general convenience, place a frost-free hydrant on each approach apron. An indoor hydrant is needed for radiator servicing and for floor washing. If there is no restroom, put a wash basin near the office for clean up.

Fuel Storage

Locate fuel near both the main farm traffic flow and the shop. Keep above-ground fuel tanks at least 40′ from any building, while below-ground tanks (with at least 2′ of soil cover) can be within 1′ of a building. If the tank will be driven over, it must have 3′ of soil cover.

Fuel storage next to the shop is very convenient but contributes to traffic congestion and can be a fire hazard.

Accident Prevention

Shop accidents do not just happen—they are caused. About 90% of shop accidents are caused by "human error," and the remaining 10% are caused by unsafe working conditions. Develop safe working habits, be alert to unsafe working conditions, and take the necessary precautions listed here to avoid accidents.

Minimize Errors

- Know how to use shop tools properly.
- Instruct and supervise new workers.
- Take frequent breaks, especially from tedious tasks.
- Use power tools only when you are reasonably calm and rested.

Protect Yourself

- Remove jewelry and loose clothing before starting.
- Wear eye protection when operating equipment, handling caustics or acids, and using a welder.
- Protect hearing from loud blasts and continuous noise.
- Protect your lungs from exhaust, paint, glue, chemical, and welding fumes.

Maintain Safe Conditions

- Clean up oil spills.
- Keep traffic areas clear.
- Block movable objects. Make sure equipment elevated with jack stands or hoists is secure.
- Use the right tool and the right-sized tool.
- Provide adequate lighting.
- Keep a clean, orderly work space: return tools to storage and dispose of greasy rags and other waste materials promptly.
- Keep foam insulation covered to reduce fire hazard.
- If chemicals must be kept in the shop, provide a locked cabinet or storeroom.

First Aid and Fire Safety

- Provide a portable first aid kit near the lavatory for minor cuts, abrasions, and burns.
- Keep an eye wash bottle (clean, fresh water) near the building entry and/or near where caustic chemicals are used.
- Place emergency phone numbers near shop and home phones.
- Place multi-purpose, dry chemical, class ABC fire extinguishers in conspicuous and accessible places around the shop, such as near doors, welding areas, etc. This is the only satisfactory extinguisher for all common types of shop fires.
- Never use water on grease or oil fires; water causes such fires to spread.
- In case of fire, first warn and evacuate people, then call the fire department, and only then try to put it out.

600 ENVIRONMENTAL FUNDAMENTALS

Environment means surroundings—all those materials and conditions that surround and affect us. In livestock housing, environmental control includes more than ventilation or temperature moderation. It includes feed and water and their quality, quantity, and availability. It includes building surfaces and their temperature, roughness, wetness, and cleanliness. It can include noise and other stress-producing influences. Waste management, health care, and accident-free facilities and management are also part of an animal's environment. In crop handling and storage, environment includes handling methods; building type and construction; drying, curing, or processing systems; and the effects of these factors on quality.

Ventilation is the method of environmental modification most often considered for agricultural structures. It is more than just pushing or pulling air through a building space or a pile of grain. Climate (long-term, average weather conditions) and weather (today's temperature, humidity, and rainfall) determine the properties of outdoor air used in ventilation. Indoor conditions are determined by the housed animals or stored crops; weather; building type and construction; equipment for ventilating, heating, or cooling, and their management; and systems for handling feed, water, wastes, and produced products (eggs, milk), and their management.

Escalating **energy** costs have increased our awareness of techniques to reduce our energy use for food and fiber production. Understanding the basic principles of environmental modification, ventilation, and grain drying is needed in order to evaluate new techniques and operating strategies.

The ASHRAE tables in this section were printed with permission of the American Society of Heating, Refrigerating and Air-Conditioning Engineers, Atlanta, Georgia.

Definitions

Heat is a form of energy transferred from a warmer to a colder body because of their temperature difference.

Temperature is a measure of a body's ability to transfer or receive heat from matter in contact with it or from surrounding surfaces. It is an indication of molecular activity.

The Fahrenheit scale was developed with 0 F as the minimum temperature of an ice and salt mixture, 32 F as the freezing point of water, and 212 F as the boiling point of water. On the Celsius scale, 0 C is the freezing point and 100 C is the boiling point of water. The relationship between the two temperature scales is:

Eq 600-1.

$$t_F = 1.8 \times t_C + 32$$

or

$$t_C = (t_F - 32) \div 1.8$$

The Rankine (absolute Fahrenheit) and Kelvin (absolute Celsius) scales are both based on zero equal to the lowest attainable temperature.

Eq 600-2.

$$T_R = t_F + 460$$

$$T_K = t_C + 273$$

Ambient temperature is the temperature of the medium surrounding a body or system, e.g. room air temperature.

Quantity of heat is measured in British thermal units (Btu); 1 Btu raises the temperature of 1 lb of water at atmospheric pressure 1 F at a temperature of 59 F. For the energy value of feed, one kilocalorie (kcal) raises 1 kg of water 1 C. One kcal = 3.968 Btu.

Specific heat capacity of a material is the quantity of heat needed per unit weight of the material to increase its temperature 1 degree; units are Btu/lb-F or kcal/kg-C.

Sensible heat is a measure of the energy that accompanies a temperature change:

Eq 600-3.

$$q_s = Mc(t_1 - t_2)$$

q_s = total sensible heat, Btu
M = weight of material, lbs
c = specific heat of the material, Btu/lb-F
t_1 = temperature of material at condition 1, F
t_2 = temperature of material at condition 2, F

Latent heat is absorbed or released when a material changes phase without a change in temperature. The three phases are solid, liquid, and gas.

Latent heat of fusion is required for a solid-liquid phase change. For example when ice melts or water freezes, 144 Btu/lb is absorbed or released without a temperature change.

Latent heat of vaporization is the heat required to change a liquid to vapor. The latent heat of vaporization of water varies with temperature as shown in Table 600-1. The common value for designing animal ventilating systems is 1044 Btu/lb of water. As a gas cools it returns to the liquid phase, or **condenses,** and gives up the same amount of heat required to vaporize it. The **heat of condensation** lost per unit weight equals the heat of vaporization. The sensible and latent heat for water to change phase and temperature are shown in Fig 600-1.

Mechanical equivalent of heat. From experiments, 778 ft-lb of mechanical energy, when converted to heat, raises the temperature of 1 lb water 1 F, or equals 1 Btu of heat. Other useful conversions from power to heat are:

1 hp = 2545 Btu/hr
1 kW = 3413 Btu/hr

Quantity of refrigeration. Refrigeration systems are rated in Btu/hr, or tons; one ton of refrigeration is the latent heat of fusion of 1 ton of ice melted in 24 hr (288,000 Btu/24 hr or 12,000 Btu/hr).

Table 600-1. Latent heat of water vaporization.
At standard atmospheric pressure.

Temperature, F	Latent heat, Btu/lb
32	1075
40	1071
60	1059
80	1048
87	1044
100	1037
120	1025
140	1014
160	1002
180	990
200	978
212	970

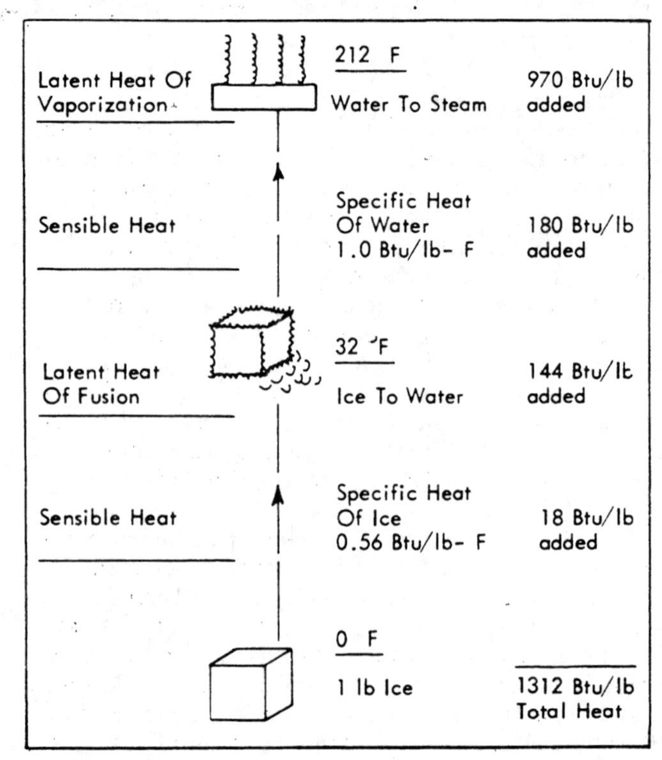

Fig 600-1. Sensible and latent heat required to vaporize ice.
Heat required to change 1 lb of ice at 0 F to vapor at 212 F.

601 HEAT TRANSFER THEORY

Every body contains some internal energy. One component of internal energy is thermal (heat). Heat will transfer from one body to another if there is a temperature difference between the bodies. Heat flows from the warmer to cooler body. When designing an agricultural building, an engineer is often required to accelerate or slow this heat transfer.

Steady state (constant with time) heat transfer problems can usually be solved with relatively simple equations. Non-steady state (varies with time) heat transfer problems require more complicated tools and are not discussed here.

There are three modes of sensible heat transfer: conduction, radiation, and convection.

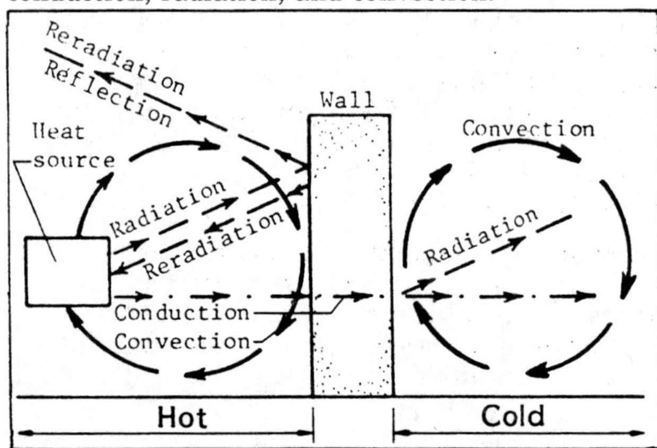

Fig 601-1. Modes of sensible heat transfer.

Conduction

Heat conduction is the exchange of heat between contacting bodies that are at different temperatures. The flow is from higher to lower temperatures.

Rate of Conduction

The rate of heat conduction through a substance is given by Fouriers Equation:

Eq 601-1.

$$q = -Ak(dt/dx)$$

q = heat conduction rate, Btu/hr
A = cross-sectional area normal to the direction of heat flow, ft^2
k = thermal conductivity of material, Btu/(hr-ft^2-F/in.)
dt/dx = temperature gradient, F/in.

Eq 601-1 shows that the rate of heat conduction is directly proportional to the cross-sectional area, A; the temperature gradient, dt/dx; and the thermal conductivity of the substance, k. For positive heat flow, the temperature gradient is negative. In practical applications, thermal resistivity, R, is often used instead of thermal conductivity. Thermal resistivity equals 1/k and has the dimensions hr-ft^2-F/Btu-in.

The thermal conductivity of a material is affected by density, size of fibers or particles, the degree of bond between particles and their arrangement, moisture content, and temperature. The ability of many materials to conduct heat has been determined experimentally. Experimental values for k vary slightly with temperature, but that difference can be ignored in agricultural buildings. Table 601-1 illustrates the wide range in k values of common building materials.

Table 601-1. Thermal conductivity varies widely with material.

Material	Thermal conductivity Btu/(hr-ft²-F/in.)
Glass wool	0.29
Wood (Douglas fir)	0.80
Concrete, stone	12.5
Steel	314
Aluminum	1536

Steady-State Conduction through Flat Bodies

We are usually concerned with steady state heat flow through flat bodies in which the direction of heat flow can be assumed normal to the surface. The steady-state heat flow rate through a homogeneous plane wall is determined from the solution of the differential equation, Eq 601-1. Separating variables:

Eq 601-2.

$$q\int_0^L dx = -Ak\int_{t_1}^{t_2} dt$$

Which upon integration and substitution yields:

Eq 601-3.

$$q = Ak(t_1 - t_2)/L$$

Note that t_1 is the higher temperature at $x = 0$ and that t_2 is the lower temperature at $x = L$.

Eq 601-3 applies only to a wall of homogeneous material with surface temperatures of t_1 and t_2. Fig 601-2 illustrates that with these conditions, the temperature of the wall material decreases uniformly from the warm side to the cold side. Computing heat flow through walls of more than one material is covered in Chapters 631 and 632.

Steady-State Conduction through Cylindrical Bodies

To calculate heat flow through the walls of a cylindrical structure, such as a circular tank, water pipe, silo, or storage bin, Eq 601-1 can be written in cylindrical coordinates:

Eq 601-4.

$$q = -Ak(dt/d\gamma)$$

Where γ is the radius of the cylinder wall.

Assuming steady-state conditions, negligible heat loss from the ends of the cylinder, and heat flow outward, the equation becomes:

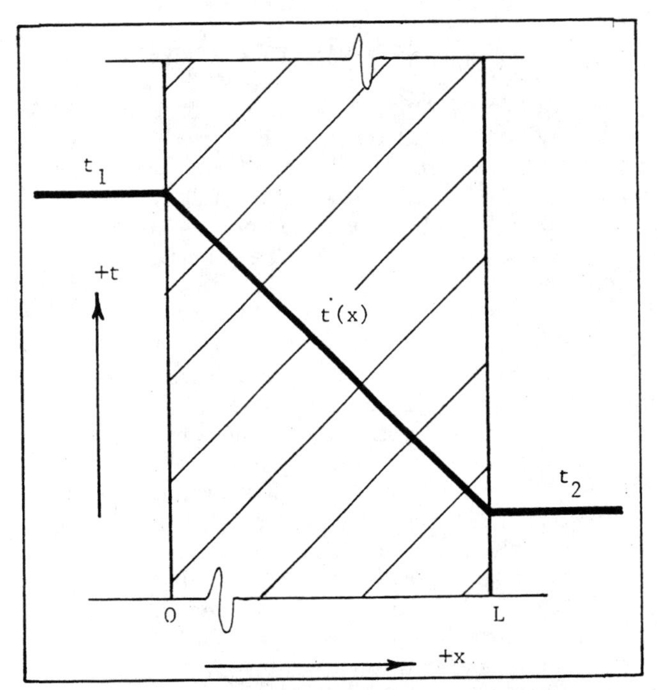

Fig 601-2. Temperature gradient for steady-state heat flow.
Through a plane wall.

Eq 601-5.

$$q = -2\pi\gamma hk(dt/d\gamma)$$

h = cylinder height
γ = radial distance from the cylinder axis

Separating variables and integrating between the temperature t_1 at γ_1 inside the wall and the lower t_2 at γ_2 outside the wall:

Eq 601-6.

$$q\int_{\gamma_1}^{\gamma_2} d\gamma/\gamma = -2\pi hk\int_{t_1}^{t_2} dt$$

$$q = 2\pi hk(t_1 - t_2)/\ln(\gamma_2/\gamma_1)$$

If heat is flowing inward, and using the same variables (t_1 = the lower temperature inside the cylinder), then $(t_1 - t_2)$ and q are negative.

Radiation

The Nature of Radiation

Thermal radiation is the exchange of thermal energy between objects by electromagnetic waves. The sun's energy radiates to earth; a flame radiates warmth to one side of your hand. Radiant energy transfer follows the same laws as light: it is transmitted through space, it follows a straight-line path, and it can be reflected, absorbed, or transmitted. It must be absorbed to be converted to heat.

All bodies emit heat in the form of radiant energy in all directions. The rate of emission varies with the temperature of the object (not just temperature difference) and the character of the body. Radiant energy is transferred between two bodies in both directions and not just from a warmer to a cooler body.

The net transfer, however, is from warm to cool bodies.

Bodies are transparent, translucent, or opaque to radiation, depending on the nature of the body and the wave length of the radiant energy. A glass windowpane is practically transparent to radiant energy from the sun, but is almost entirely opaque to radiation emitted at a longer wavelength by objects within an enclosure.

Black Body

Objects vary in their capacity to absorb, reflect, and emit radiation. A "black" body absorbs all radiation it receives and reflects none. It also emits radiation at the maximum rate for a given temperature. There are no true black bodies; real (gray) bodies reflect some of the incoming radiation and emit heat at a rate less than that of a black body.

Highly absorptive bodies generally radiate readily, and highly reflective ones radiate poorly. A ball covered with lamp black is nearly a black body; it receives and emits radiation at a high rate. Brightly polished aluminum is a good reflector, a poor absorber, and a poor radiator.

Emissivity-Absorptivity

Emissivity, ϵ, is the ratio of the emissive power of a real body to that of a black body at the same temperature. Absorptivity, α, is the fraction of incoming radiation that is absorbed.

The emissivity and absorptivity of all real bodies are functions of absolute temperature and of the radiation wavelengths being absorbed or emitted. For real bodies, ϵ and α can be average values over a range of wavelengths or for a specific wavelength. The monochromatic emissive power of a black body is determined from Planck's law:

Eq 601-7.

$$E_{b\lambda} = \frac{1.187 \times 10^{8} \lambda^{-5}}{e^{25896/\lambda T} - 1}$$

E = emissive power, Btu/hr-ft²-micron
b = black body
λ = energy wavelength, micron
T = absolute temperature of radiating body, R

The Stefan-Boltzmann equation is obtained directly from Planck's law by integrating over all wavelengths of radiation. The total emissive power of a black body, E_b, is:

Eq 601-8.

$$E_b = \int_{\lambda=0}^{\lambda=\infty} E_{b\lambda} d\lambda$$

$$E_b = 0.1714 \times 10^{-8}(T^4)$$

$$E_b = \sigma(T^4)$$

E_b = total emissive power of black body, Btu/ft²-hr
σ = 0.1714×10^{-8} Btu/ft²-hr-R⁴
T = absolute temperature, R

Calculating Radiation Heat Transfer

Radiant heat transfer is proportional to the fourth power of the absolute temperature difference. There are no true black bodies; the emissive power of a "gray" body is:

Eq 601-9.

$$E = \epsilon E_b$$

ϵ = emissivity
E_b = emissive power of a black body

The net heat q_r transferred by radiation between two bodies at temperatures T_1 and T_2 is:

Eq 601-10.

$$q_r = F_E F_A A \sigma(T_1^4 - T_2^4)$$

The values for A, F_E, and F_A are given in Table 601-2 for some common situations. F_A is a geometric factor allowing for size, slope, and orientation of the two bodies. F_E is a radiation factor allowing for part of the radiation being reradiated to the body it came from.

Convection

At ordinary temperatures, most heat transfer in liquids and gases is by convection. In convective heat transfer, heat is transferred to or from an object by the mass movement of either a liquid or gas.

The rate of heat transfer by convection is:

Table 601-2. Radiation between solids.
Factors A, F_A, and F_E for Eq 601-10.

Radiating surfaces	Area, A	F_A	F_E
Infinite parallel planes	A_1 or A_2	1	$1/(1/\epsilon_1 + 1/\epsilon_2 - 1)$
Completely enclosed body, 1, small compared with enclosing body.	A_1	1	ϵ_1
Completely enclosed body, 1, large compared with enclosing body.	A_1	1	$1/(1/\epsilon_1 + 1/\epsilon_2 - 1)$
Flat surface, 1, and sun	A_1	*$(r/R)^2\cos\phi$	ϵ_1

*r = radius of sun, 432,050 miles; R = distance from sun to surface, 92,900,000 miles; and ϕ = angle between the sun's rays and the normal to the surface; $(r/R)^2 = 2.16 \times 10^{-5}$.

Eq 601-11.

$$q_c = hA(t_s - t_\infty)$$

q_c = convective heat transfer rate, Btu/hr
h = heat transfer coefficient, Btu/hr-ft²-F
A = surface area, ft²
t_s = surface temperature, F
t_∞ = free stream fluid temperature, F

The laws governing the rate of heat transfer by convection are very complex, because complex fluid-dynamic phenomena are involved. The value of the heat transfer coefficient, h, is determined from relationships that have been found experimentally. Normally, experimental data provide the Nusselt number:

Eq 601-12.

$$h = N_{Nu}k/d$$

h = heat transfer coefficient, Btu/hr-ft²-F
N_{Nu} = Nusselt number
k = the conductivity of the fluid, Btu/(hr-ft²-F/in.)
d = the dimension of interest (e.g. tube diameter), in.

Heat transfer by convection is usually classified as either natural convection or forced convection. Natural convection involves fluid motion resulting from differences in fluid density and the action of gravity. Forced convection involves fluid motion created by a fan or pump. Heat transfer coefficients for natural convection are generally much lower than for forced convection.

Natural Convection

As a typical example, consider heat transfer by natural convection between a cold fluid and a hot surface. The fluid in immediate contact with the surface is heated by conduction, expands and becomes lighter, and rises due to the density differences. The motion is resisted by the viscosity of the fluid. Section I of Table 601-3 shows that the Nusselt number, N_{Nu}, is a function of the product of the Prandtl number, N_{Pr}, and Grashof number, N_{Gr}. These numbers depend upon the fluid properties, the temperature difference between the surface and the fluid, Δt, and the characteristic length of the surface, L. The constant, c, and the exponent, n, depend upon the physical configuration and the nature of the flow.

The entire process of natural convection cannot be represented by a single value of the exponent n. The process may be divided into three regions: turbulent natural convection for which n equals 0.33; laminar natural convection for which n equals 0.25; and a region having $(N_{Gr}N_{Pr})$ less than for laminar natural convection, for which the exponent, n, gradually diminishes from 0.25 to lower values.

Use the correct characteristic length as indicated in the table. Because the exponent, n, is 0.33 for a turbulent boundary layer, the characteristic length cancels out of the turbulent range equations in Section III of Table 601-3. Turbulence occurs when the length or the temperature difference is large.

Convection from horizontal plates facing downward when heated or upwards when cooled are special cases. Because the hot air is above the colder air, theoretically there is no reason for convection. However, some convection is caused by secondary influences such as temperature differences on the edges of the plate. As a rough approximation, a coefficient of somewhat less than half of the coefficient for a horizontal plate facing upwards is recommended.

Because air is often the heat transport fluid, simplified equations for air are in Section III of Table 601-3.

Use the heat transfer coefficients, h, from Table 601-3 in Eq 601-11 to predict heat transfer.

Forced Convection

When a fluid flows over a flat plate, a boundary layer forms adjacent to the plate. As shown schematically in Fig 601-3, the velocity of the fluid at the surface of the plate is zero and increases to its maximum free-stream value just past the edge of the boundary layer. The mechanism of boundary layer formation is important, because the temperature change from the plate to the fluid, and therefore thermal resistance, are concentrated here. Where the boundary layer is thick, thermal resistance is large and the heat transfer coefficient small. At the leading edge of the plate, the boundary layer thickness is theoretically zero and the heat transfer coefficient is infinite. Flow within the boundary layer immediately downstream from the leading edge is laminar. Farther along the plate, the laminar boundary layer gets thicker until a critical value is reached. Then, turbulent eddies develop within the boundary layer, except for a thin laminar sublayer adjacent to the plate. Beyond this point is turbulent forced convection. Between the end of the laminar boundary layer and the turbulent boundary layer is a transition region. Because turbulent eddies greatly increase heat transport into the main stream, the heat transfer coefficient increases rapidly through the transition region.

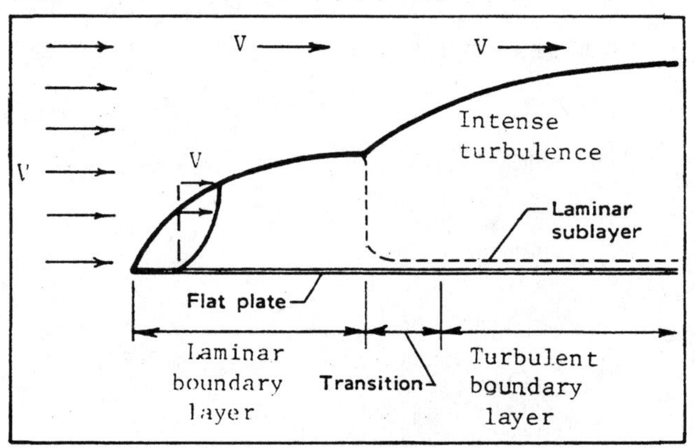

Fig 601-3. Boundary layer buildup on a flat plate.
Vertical scale is magnified for clarity.

For flow in long tubes or channels of small hydraulic diameter, the laminar boundary layers on each wall grow until they meet, if the velocity is low

Table 601-3. Natural convection heat transfer coefficients.

Description	Equation
I. General	$N_{Nu} = c(N_{Gr}N_{Pr})^n$
Characteristic length, L, for:	
Vertical plates or pipes	L = height
Horizontal plates	L = length
Horizontal pipes	L = diameter
II. Plates and pipes	
Horizontal or vertical plates, pipes, rectangular blocks, and spheres (excluding horizontal plates facing downward for heating and facing upward for cooling).	
(a) Laminar range: $(N_{Gr}N_{Pr})$ between 10^4 and 10^8	$N_{Nu} = 0.56(N_{Gr}N_{Pr})^{0.25}$
(b) Turbulent range: $(N_{Gr}N_{Pr})$ between 10^8 and 10^{12}	$N_{Nu} = 0.13\ (N_{Gr}N_{Pr})^{0.33}$
III. With air	
$(N_{Gr}N_{Pr}) = 1.6\text{x}10^6 L^3(\Delta t)$, (70 F, L in ft, Δt in F)	
(a) Cylinders:	
Small cylinder, laminar range	$h = 0.27(\Delta t/L)^{0.25}$
Large cylinder, turbulent range	$h = 0.18(\Delta t)^{0.33}$
(b) Vertical plates:	
Small plates, laminar range	$h = 0.29(\Delta t/L)^{0.25}$
Large plates, turbulent range	$h = 0.19(\Delta t)^{0.33}$
(c) Horizontal plates, facing upward when heated or downward when cooled:	
Small plates, laminar range	$h = 0.27(\Delta t/L)^{0.25}$
Large plates, turbulent range	$h = 0.22(\Delta t)^{0.33}$
(d) Horizontal plates, facing downward when heated, or upward when cooled:	
Small plates	$h = 0.12(\Delta t/L)^{0.25}$

Table 601-4. Equations for forced convection.
Unless otherwise indicated, all units are lb, hr, ft, F, Btu.

Description	Equation
I. General	
(a) Turbulent flow inside tubes	$\frac{hd}{k} = c(\frac{Gd}{\mu})^m(\frac{\mu c_p}{k})^n$
(1) Using fluid properties based on bulk temperature t	$\frac{hd}{k} = 0.023(\frac{Gd}{\mu})^{0.8}(\frac{\mu c_p}{k})^{0.4}$
(b) Laminar flow inside tubes	
(1) For large d or high Δt, the effect of natural convection should be included.	$\frac{hd}{k} = 1.86[(\frac{Gd}{\mu})(\frac{\mu c_p}{k})(\frac{d}{L})]^{0.33}(\frac{\mu}{\mu_s})^{0.14}$
II. Simplified equation for water, turbulent flow inside tubes	
(a) At ordinary temperatures, 40 to 220 F.	$h = \frac{150(1 + 0.011t)(V)^{0.8}}{(d')^{0.2}}$
III. Simplified equations for air	
(a) Vertical plane surfaces, V of 16 to 100 ft/sec (room temp)	$h' = 0.5(V)^{0.8}$
(b) Vertical plane surfaces, V < 16 ft/sec (room temp)	$h' = 0.99 + 0.21V$
(c) Single cylinder cross flow (film temp = 200 F) $1000 < Gd/\mu_f < 50{,}000$	$h = 0.026(\frac{G^{0.6}}{d^{0.4}})$

c = constant
c_p = specific heat at constant pressure, Btu/lb-F
d = tube diameter, ft
d' = tube diameter, in.
G = mass velocity, lb/hr-ft²
h = heat transfer coefficient, Btu/hr-ft²-F
h' = heat transfer coefficient, Btu/(hr) (ft²) (F initial temperature difference)
k = thermal conductivity, Btu/(hr-ft²-F/ft)
L = length, ft
t = temperature, F
V = linear velocity, ft/sec
μ = absolute viscosity, lb/ft-hr
μ_s = absolute viscosity at surface temperature, lb/ft-hr

enough. Beyond this point, the velocity distribution does not change and no transition to turbulent flow takes place. This condition is referred to as fully developed laminar flow.

At higher velocities or for tubes of large diameter, transition to turbulence takes place and fully developed turbulent flow is established as shown in Fig 601-4. The length dimension which determines the critical Reynolds number is the hydraulic diameter of the channel. For smooth circular tubes, flow is laminar for Reynolds numbers below 2100 and turbulent above 3000.

Forced convection equations are in Table 601-4. As indicated by the generalized, dimensionless formula in Section I, heat transfer is determined by the flow conditions (Reynolds number, $\mu c_p/k$) and by the fluid properties (Prandtl number, Gd/μ).

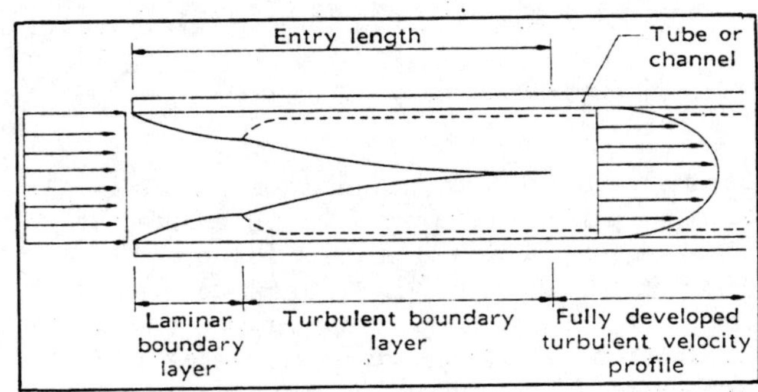

Fig 601-4. Boundary layer buildup.
At the entry of a tube or channel.

The characteristic length, d, is the outside or inside tube diameter or the plate length.

For convenience, simplified equations for water and air under normal operating conditions are in Sections II and III of Table 601-4.

602 AIR-WATER VAPOR MIXTURES—PSYCHROMETRICS

This chapter is from material prepared by Dr. Duane Mangold, Iowa State University

The earth's atmosphere is a mixture of gases and water vapor. The composition of dry air adopted by the International Joint Committee on Psychrometric Data is given in Table 602-1. An understanding of physical and thermodynamic properties of air-water vapor mixtures—psychrometrics—is fundamental to the design of environmental control systems for plants, crops, animals, or humans.

Table 602-1. Composition of dry air.

Gas	Molecular weight		Mole-fraction composition		Partial molecular weight
Oxygen (O_2)	32.000	x	0.2095	=	6.704
Nitrogen (N_2)	28.016	x	0.7809	=	21.878
Argon (A)	39.944	x	0.0093	=	0.371
Carbon dioxide (CO_2)	44.01	x	0.0003	=	0.013
					28.966

Properties of Moist Air

Pressure, volume, weight, and thermal properties are related by laws developed for a "perfect" gas, but are accurate enough for the air-water vapor mixtures in environmental control problems.

Perfect gas law. The thermodynamic state of moist air is established by pressure and two other independent properties. But most psychrometric charts are for a constant pressure, and only the two other independent properties are used. The relationship between the properties of a perfect gas is:

Eq 602-1.

$$P \times V = M \times R \times T$$

P = absolute pressure, lb/ft^2
V = volume, ft^3
M = mass, lb
R = gas constant, ft-lb/lb-R
T = absolute temperature, R (F + 460)

The gas constant for air, R_a = 53.35 ft-lb/lb-R, and the gas constant of water vapor, R_w = 85.78 ft-lb/lb-R, are functionally accurate for ventilation temperatures and pressures.

Dalton's Law states that each component in a mixture of gases exerts its own partial pressure. For a mixture of air, a, and water vapor, w:

Eq 602-2.

$$P = P_a + P_w$$
$$P = (M_a \times R_a \times T_a \div V_a) + (M_w \times R_w \times T_w \div V_w)$$

Assuming each component gas is uniformly diffused throughout the mixture, both components have the same volume and temperature, and Eq 602-2 becomes:

Eq 602-3.

$$P = (T \div V) \times (M_a \times R_a + M_w \times R_w)$$

Because the volume and temperature of the mixture are equal, from Eq 602-1:

Eq 602-4.

$$(P_w \times V) \div (P_a \times V) = (M_w \times R_w \times T) \div (M_a \times R_a \times T)$$
$$(P_w \div P_a) = (M_w \times R_w) \div (M_a \times R_a)$$

Partial pressures may be calculated from the total pressure and the weight of water vapor.

Humidity ratio, W, is the weight of water vapor in lb/lb of dry air (da) or gr/lb da (7000 grains = 1 lb). The base of one pound of dry air is constant for any change of condition, making calculations easier.

From Eq 602-1, M = (P x V) ÷ (R x T), and Dalton's Law, the humidity ratio for an air-water vapor mixture can be written:

Eq 602-5.

$$W = M_w \div M_a$$
$$W = [P_w \times V \div (R_w \times T)] \div [P_a \times V \div (R_a \times T)]$$
$$W = (P_w \times R_a) \div (P_a \times R_w)$$
$$W = (P_w \times R_a) \div [(P - P_w) \times R_w]$$

Substituting numerical values for the gas constants (R_a/R_w = 53.35/85.78 = 0.622), the humidity ratio becomes:

$$W = M_w \div M_a = (0.622 \times P_w) \div (P - P_w)$$

M_w = mass of water vapor in the mixture, lb
M_a = mass of air, lb

Relative humidity, RH, expressed as a percent, is the ratio of the actual water vapor pressure (P_w) to the vapor pressure of saturated air at the same temperature (P_{ws}):

Eq 602-6.

$$RH\% = 100 \times (P_w \div P_{ws})$$

P_w = actual water vapor pressure, lb/ft^2
P_{ws} = vapor pressure of saturated air at the same temperature, lb/ft^2

Vapor pressures at saturation are shown in Table 602-2.

Degree of saturation, μ, expressed as a percent, is the ratio of the actual weight of water vapor per pound of dry air to the saturated weight of water vapor per pound of dry air at the same dry bulb temperature:

Eq 602-7.

$$\mu = (W \div W_s) \times 100$$

μ = degree of saturation, %
W = humidity ratio of air, lb/lb da
W_s = humidity ratio of air at saturation, lb/lb da

Differences in the numerical values of the degree of saturation and relative humidity are small for ordinary atmospheric conditions and frequently are neglected.

Specific volume, V, of a gas or mixture is the space occupied by a given mass. In air conditioning work, the specific volume is the ft^3 of mixture/lb da. The base of one pound of dry air is again used because the pounds of dry air entering and leaving an air conditioning unit in a given time are constant after steady-state flow is established. Specific volume is related to absolute temperature and pressure in the equation:

Eq 602-8.

$$V_a = R_a \times T \div P_a$$

V_a = specific volume of dry air, ft^3/lb da
T = absolute temperature, R
P_a = pressure of dry air, lb/ft^2 absolute
R_a = gas constant, dry air, ft-lb/lb-R

Determine the specific volume of water vapor by substituting values for the gas constant of water vapor and water vapor pressure in Eq 602-8.

Standard air at atmospheric pressure of 14.7 psi, 70 F, and zero water vapor content has a specific volume of 13.34 ft^3/lb da. Specific volume increases as temperature increases.

The volume, V, of moist air per pound of dry air is calculated from:

Eq 602-9.

$$V = V_a + \mu \times V_{as}$$

V_a = specific volume of dry air, ft^3/lb da
V_{as} = $V_s - V_a$, ft^3/lb da
V_s = specific volume of moist air at saturation, ft^3/lb da
μ = degree of saturation

Calculate volume with the thermodynamic properties of moist air, Table 602-2, and Eq 602-9.

Temperature. Three different temperatures describe an air-water vapor mixture: dry-bulb (db), wet bulb (wb), and dewpoint (dp) temperatures.

Dry-bulb temperature is usually measured with a thermometer, thermocouple, or thermistor.

Table 602-2. Thermodynamic properties of air-water vapor mixtures.
From the *1981 ASHRAE Handbook of Fundamentals.*

Temperature F	Saturation vapor pressure lb/in²	Vapor weight lb/lb dry air	Volume ft³/lb da			Enthalpy Btu/lb da		
			V_a	V_s-V_a	V_s	h_a	h_w	h_a+h_w
-20	0.006	0.00026	11.073	0.005	11.078	-4.804	0.277	-4.527
-15	0.008	0.00035	11.200	0.006	11.206	-3.603	0.368	-3.235
-10	0.011	0.00046	11.326	0.008	11.335	-2.402	0.487	-1.915
-5	0.014	0.00060	11.453	0.011	11.464	-1.201	0.640	-0.561
0	0.019	0.00079	11.579	0.015	11.594	0.0	0.835	0.835
5	0.024	0.00102	11.706	0.019	11.725	1.201	1.085	2.286
10	0.031	0.00132	11.832	0.025	11.857	2.402	1.402	3.804
15	0.040	0.00169	11.959	0.032	11.991	3.603	1.801	5.404
20	0.051	0.00215	12.085	0.042	12.127	4.804	2.303	7.107
25	0.064	0.00273	12.212	0.054	12.265	6.005	2.930	8.935
30	0.081	0.00346	12.338	0.068	12.406	7.206	3.711	10.917
35	0.100	0.00428	12.464	0.085	12.550	8.408	4.603	13.010
40	0.122	0.00522	12.591	0.105	12.696	9.609	5.624	15.233
45	0.148	0.00633	12.717	0.129	12.846	10.810	6.843	17.653
50	0.178	0.00766	12.844	0.158	13.001	12.012	8.295	20.306
55	0.214	0.00923	12.970	0.192	13.162	13.213	10.016	23.229
60	0.256	0.01109	13.096	0.233	13.329	14.415	12.052	26.467
65	0.306	0.01327	13.223	0.281	13.504	15.616	14.454	30.071
70	0.363	0.01583	13.349	0.339	13.688	16.818	17.279	34.097
75	0.430	0.01883	13.476	0.407	13.882	18.020	20.595	38.615
80	0.507	0.02234	13.602	0.487	14.089	19.222	24.479	43.701
85	0.597	0.02643	13.728	0.581	14.310	20.424	29.021	49.445
90	0.699	0.03120	13.855	0.692	14.547	21.626	34.325	55.951
95	0.816	0.03676	13.981	0.823	14.804	22.828	40.515	63.343
100	0.950	0.04322	14.107	0.976	15.084	24.031	47.730	71.761
105	1.103	0.05074	14.234	1.156	15.390	25.233	56.142	81.375
110	1.277	0.05949	14.360	1.367	15.727	26.436	65.950	92.386
115	1.473	0.06968	14.486	1.615	16.101	27.639	77.396	105.035
120	1.693	0.08156	14.613	1.906	16.519	28.842	90.770	119.612

V_a = specific volume of dry air (da), ft^3/lb da
V_s = specific volume of saturated air-vapor mixture, ft^3/lb da
h_a = specific enthalpy of dry air, Btu/lb da
h_w = enthalpy of water vapor in a saturated mixture, Btu/lb da

Wet-bulb temperature is measured with a common mercury thermometer or thermocouple, with the bulb or junction covered with a water-moistened wick and in a moving stream of ambient air. Evaporation from the wick attains a steady state in which sensible heat from the surroundings provides heat of vaporization. Air flow past the bulb is high enough to prevent significant change in the ambient air temperature. Evaporation of water cools the bulb; the drier the surrounding air, the greater the rate of evaporation, and the lower the wet-bulb temperature.

Wet-bulb depression is the difference between dry-bulb and wet-bulb temperatures.

Dewpoint temperature is the temperature at which moisture starts to condense from air cooled at constant pressure and humidity ratio. Dewpoint temperature is determined by bringing moist air into contact with a polished metal surface whose temperature can be both controlled and measured. As the metal surface is slowly cooled, the mixture in contact with it is also cooled. When the mixture reaches its dewpoint temperature, condensation fogs the metal surface.

Enthalpy is the heat energy content of an air-water vapor mixture. The energy is both sensible heat (indicated by dry-bulb temperature) and latent heat of vaporization (energy content of the water vapor). For convenience in air conditioning problems, enthalpy is expressed as Btu/lb da. Measures of heat content are not necessarily absolute. It is convenient to express heat content as a difference in energy between a fixed condition or datum and the condition being considered. The usual datum for dry air is 0 F and for water is 32 F.

The enthalpy, h Btu/lb da, of an air-vapor mixture is calculated from:

Eq 602-10.

$$h = c_p \times t_{db} + W \times h_w$$

c_p = specific heat of dry air, 0.24 Btu/lb-F
t_{db} = dry bulb air temperature, F
W = humidity ratio of air, lb/lb da
h_w = enthalpy of water, Btu/lb water

or:

Eq 602-11.

$$h = 0.24 \times t_{db} + W \times (1061 + 0.444 \times t_{db})$$

1061 = latent heat of vaporization, Btu/lb
0.444 = specific heat of water vapor, Btu/lb-F

Adiabatic saturation. An adiabatic system is perfectly insulated; no heat is added or removed externally. Unsaturated air placed in an adiabatic container and exposed to a water surface becomes saturated by evaporation. If the initial and final water temperatures and the final air temperature are all the same, then all the heat of vaporization comes from the sensible heat of the air. The final temperature of the saturated air is the thermodynamic wet-bulb temperature.

Fig 602-1 illustrates an adiabatic process between states 1 and 2.

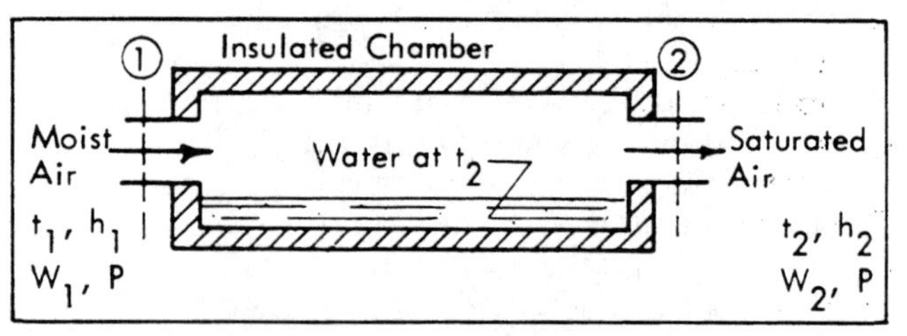

Fig 602-1. Adiabatic saturation of moist air.

The heat balance for one pound of dry air is:

Eq 602-12.

$$h_2 = h_1 + (W_2 - W_1) \times h_{w2}$$

h_1 = enthalpy of air, state 1, Btu/lb da
h_2 = enthalpy of saturated air, state 2, Btu/lb da
W_1 = humidity ratio of air, state 1, lb/lb da
W_2 = humidity ratio of saturated air, state 2, lb/lb da
h_{w2} = enthalpy of liquid water at t_2, Btu/lb

Enthalpy at saturation, h_2, is greater than the enthalpy of the entering air, h_1, by the small heat content of the water added, $(W_2 - W_1) \times h_{w2}$. Below 110 F, insignificant error results from assuming adiabatic saturation processes are at constant wet-bulb temperature and enthalpy.

Psychrometric Chart

A psychrometric chart, Figs 602-2 to 602-4, presents physical and thermal properties of moist air. It enables you to quickly trace changes in air properties through various air conditioning processes. Charts for air conditioning are at standard atmospheric pressure, 29.92 in. Hg = 14.7 psi.

Coordinates of the charts are dry-bulb temperature along the x-axis and moisture content or humidity ratio on the y-axis. Other properties of moist air are, Fig 602-5, wet-bulb temperature, enthalpy, dewpoint temperature, relative humidity, and specific volume. The intersection of any two property lines establishes a given state, and all other properties can be read.

Example 1

Fig 602-6 illustrates the properties of air at 73 F db and 20% RH: wet-bulb temperature, 52 F; humidity ratio, 0.0035 lb water vapor/lb da; enthalpy, 21.3 Btu/lb da; volume, 13.5 ft^3/lb da; and dewpoint temperature, 30 F. Verify these values in Fig 602-3.

Air-Water Vapor Mixture Processes

Air conditioning is more than just summer cooling. Air conditioning processes include heating, cooling, humidifying, or dehumidifying air-water vapor mixtures and mixing of various air volumes. A number of processes in agricultural engineering involve conditioning air-water vapor mixtures. Livestock housing usually includes some degree of environmental modification.

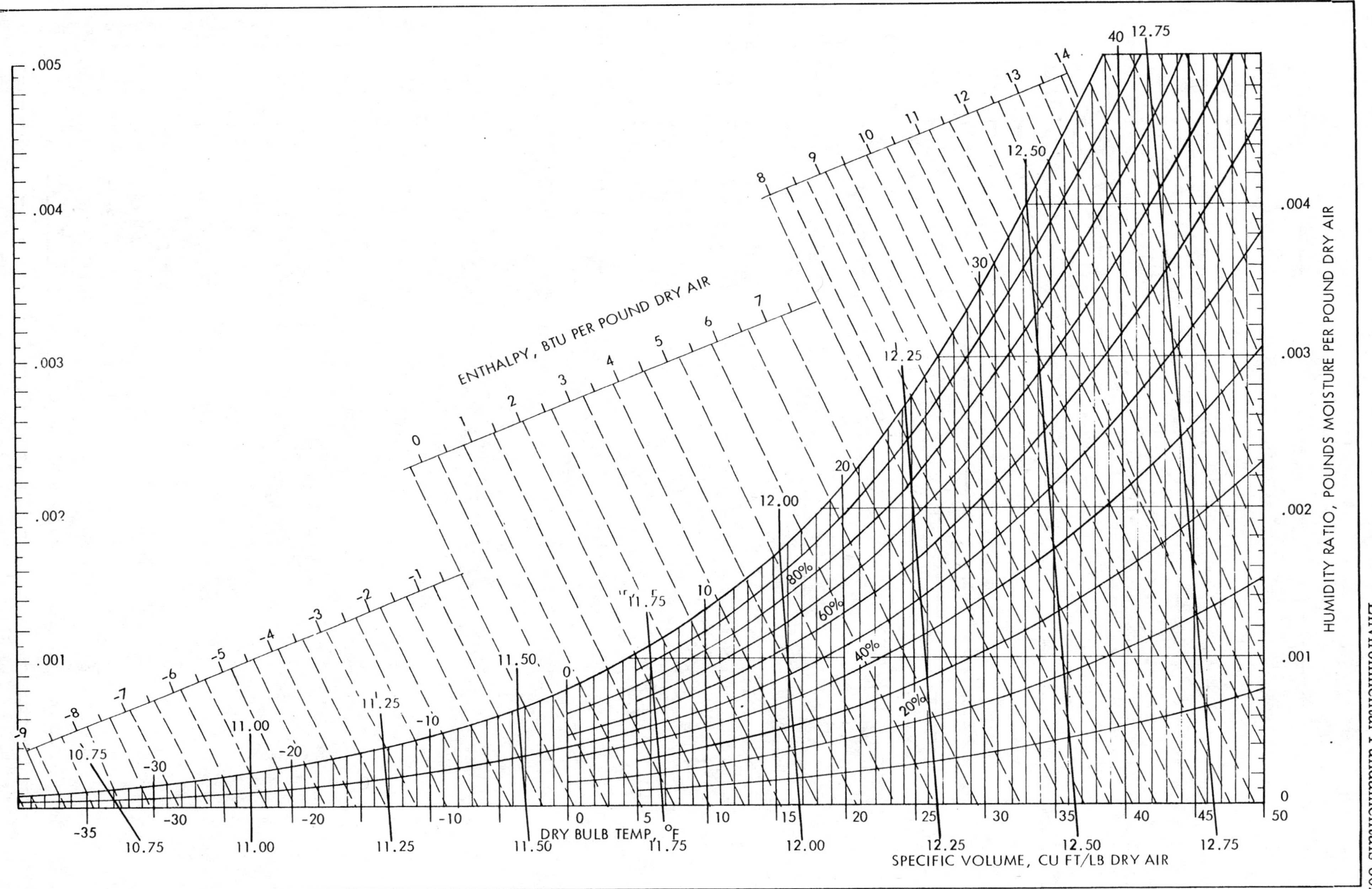

Fig 602-2. Low temperature psychrometric chart.

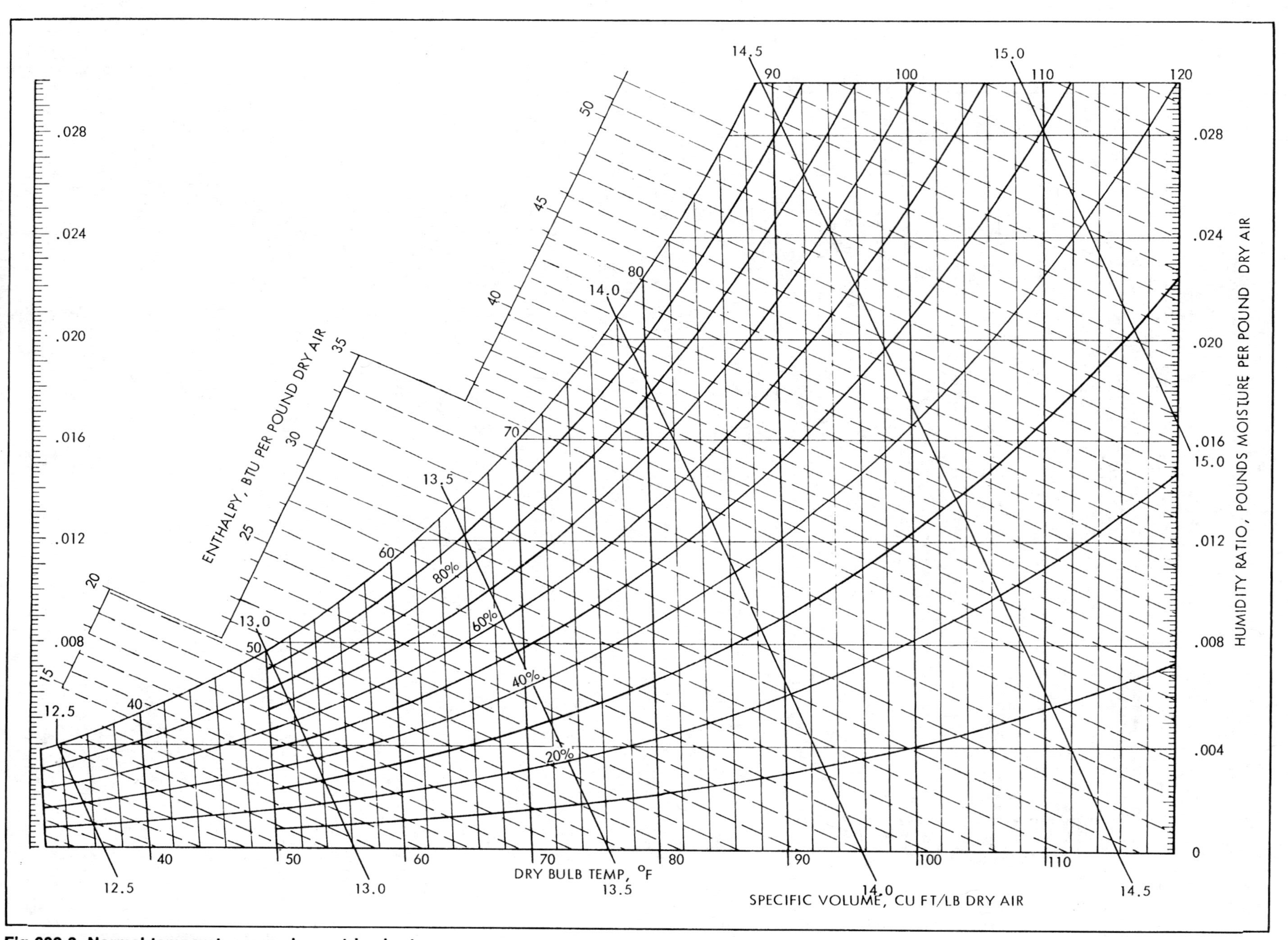

Fig 602-3. Normal temperature psychrometric chart.

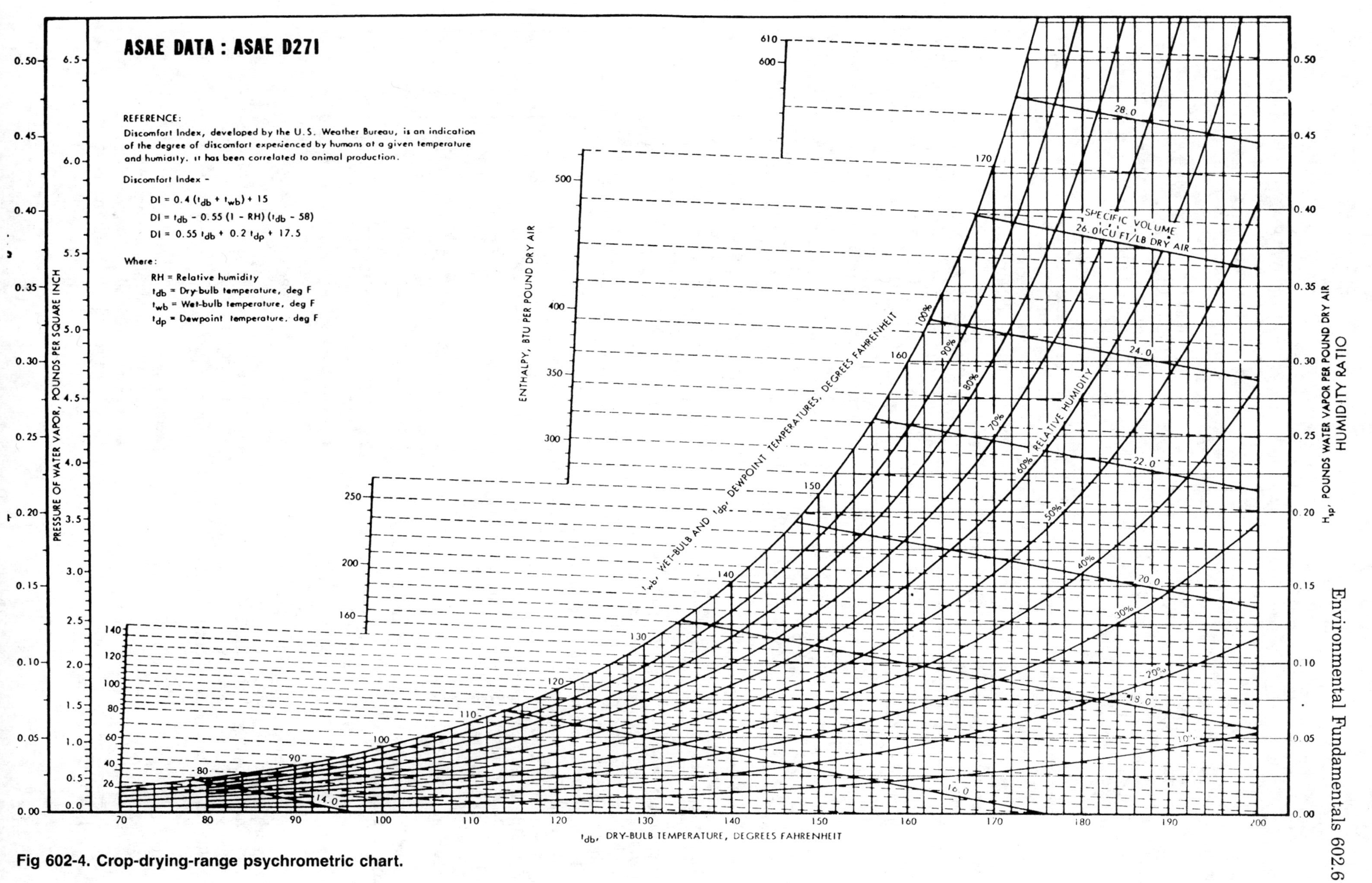

Fig 602-4. Crop-drying-range psychrometric chart.

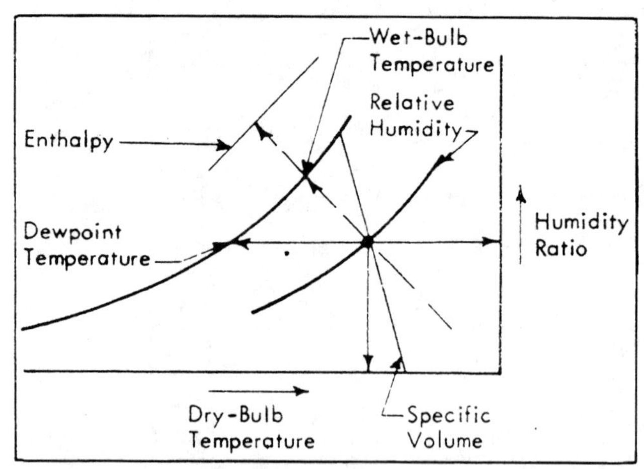

Fig 602-5. Properties of moist air on a psychrometric chart.

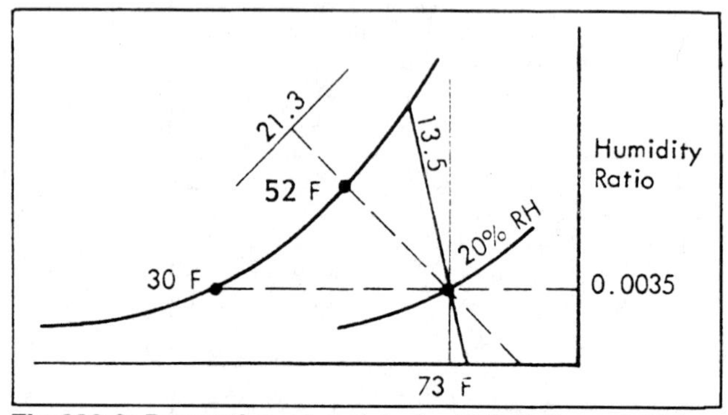

Fig 602-6. Properties of moist air at 73 F db and 20% RH.

Crop drying is adiabatic wetting of the drying air. Design of any fogging process, whether of water, insecticide, herbicide, or others, requires knowledge of air conditioning processes.

Also of interest in agricultural engineering are the plant, soil, and air dynamics relationships involving evapotranspiration.

Sensible heating is adding heat to air without changing its humidity ratio. Applications of sensible heating include heated-air grain drying and winter heating of room air in animal buildings. This process follows along a horizontal line and moves from left to right.

Calculate sensible heating from the mass and enthalpy change of the air:

Eq 602-13.

$$q = M_a \times (h_2 - h_1)$$

q = heat added, Btu
M_a = mass of dry air, lb da
h = enthalpy of air, Btu/lb da

For a continuous flow process, such as ventilation, sensible heat loss can be expressed per unit of time:

Eq 602-14.

$$q_t = (M_t \div V) \times (h_2 - h_1)$$

q_t = heat added, Btu/min
M_t = airflow volume, ft³/min
V = specific volume of air, ft³/lb da
h = enthalpy of air, Btu/lb da

The sensible heat raises the temperature of the air ($c_p \times t_{db}$) and of the moisture in the air ($W \times h_w$), as in Eq 602-11.

$$h_1 = 0.24 \times t_1 + W_1 \times 1061 + W_1 \times 0.444 \times t_1$$
$$h_2 = 0.24 \times t_2 + W_1 \times 1061 + W_1 \times 0.444 \times t_2$$
$$(h_2 - h_1) = 0.24 \times (t_2 - t_1) + 0.444 \times W_1 \times (t_2 - t_1)$$
$$= (t_2 - t_1) \times (0.24 + 0.444 \times W_1)$$
$$q = M_a \times (h_2 - h_1)$$
$$q = M_a \times (t_2 - t_1) \times (0.24 + 0.444 \times W_1)$$

Note that because the process is sensible heating (no moisture added), $W_1 = W_2$; there is no new water evaporated and therefore no increase in the heat of vaporization in the air-water mixture.

Example 2

Determine the heat required to raise 1250 ft³ of air at condition (1), 34 F db and 80% RH to (2), 80 F db.

Sensible heating in Fig 602-7 follows a horizontal line from 34 F db and 80% RH, to 80 F db. The relative humidity at state 2 is about 15%. At condition (1) the volume is about 12.5 ft³/lb da; enthalpy is about 11.7 Btu/lb da; the dewpoint temperature is 28 F; and humidity ratio is 0.0033 lb water/lb-da. The enthalpy at (2) is about 22.8 Btu/lb. To determine the heat added:

$$M_a = 1250 \text{ ft}^3 \div V$$
$$q = (1250 \div 12.5) \times (80 - 34) \times (0.24 + 0.444 \times 0.0033)$$
$$= 1{,}110$$

Or from a psychrometric chart:

$$q = M_a \times (h_2 - h_1)$$
$$q = (1250 \div 12.5) \times (22.8 - 11.7)$$
$$= 1{,}110 \text{ Btu}$$

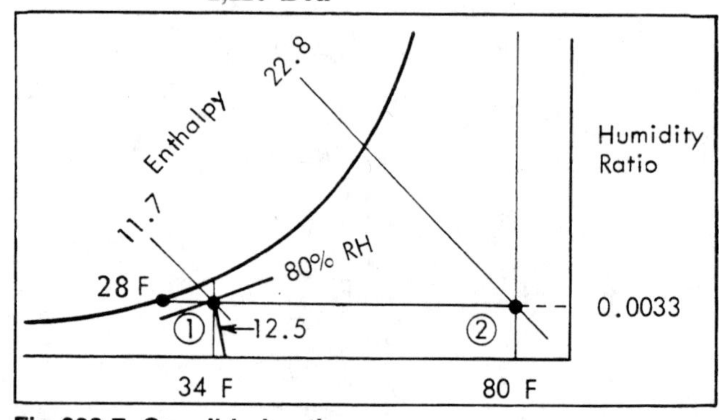

Fig 602-7. Sensible heating.

Sensible cooling is cooling at constant humidity ratio. An example is air passing over a cooling coil having a surface temperature above the dewpoint temperature of the air. Sensible cooling follows a horizontal line from right to left, Fig 602-8. The final temperature cannot be below the initial dewpoint temperature, or water vapor condenses and the process removes latent heat.

Example 3

Determine the amount of heat removed from 1400 ft³ of air at 90 F db and 64 F wb, which is cooled to 60 F without changing the moisture content (sensible cooling).

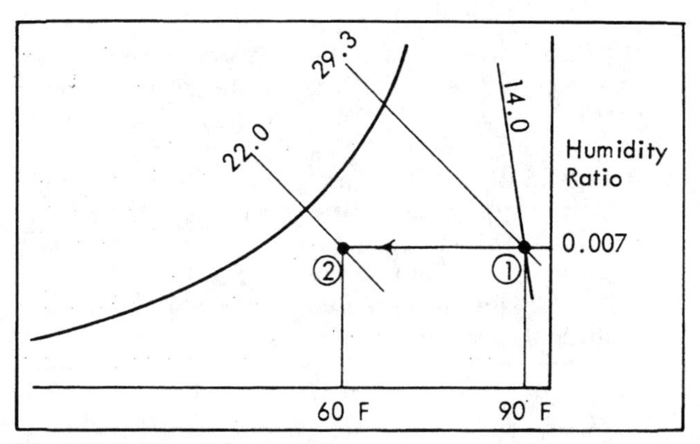

Fig 602-8. Sensible cooling.

The properties of the mixture from the psychrometric chart are in Fig 602-8. The specific volume (1) is 14.0 ft^3/lb da, and the enthalpy is 29.3 Btu/lb da. The enthalpy at (2) is 22.0 Btu/lb da. The humidity ratio remains constant at 0.007 lb water vapor/lb da. The sensible heat, q, removed is:

$$_1q_2 = M_a \times (h_2 - h_1)$$
$$= (1400 \div 14) \times (22.0 - 29.3)$$
$$= -730 \text{ Btu}$$

Note: A negative q means cooling.

The heat removed can also be calculated by:

$$_1q_2 = M_a \times [c_p \times (t_2 - t_1) + 0.444 \times W \times (t_2 - t_1)]$$
$$= (1400 \div 14) \times [0.24 \times (60 - 90) + 0.444 \times 0.007 \times (60 - 90)]$$
$$= -729.3 \text{ Btu}$$

Evaporative cooling is an adiabatic saturation process (no heat gained or lost) and follows upward along a constant wet-bulb temperature line. Air to be cooled is brought in contact with water at a temperature equal to the wet-bulb temperature of the air. The sensible heat of the initial air evaporates the water, lowering the air's dry-bulb temperature. Sensible heat is converted to latent heat in the added vapor, so the process is adiabatic. Evaporative cooling is most effective in hot dry climates where wet-bulb depression is large, and where the disadvantage of increased humidity is more than offset by a relatively large temperature drop.

Example 4

Air at 86 F db and 20% RH is saturated by spraying recirculated water into it. What is the dry-bulb temperature of the saturated air, and how much moisture is added per pound of dry air in the evaporative cooling process?

In Fig 602-9, the wet-bulb temperature at (1) is 60 F. Because the process is adiabatic and nearly follows a line of constant wet-bulb, the dry-bulb temperature at saturation is also 60 F. The difference in humidity ratio between the two points is: $W_2 - W_1 = 0.01100 - 0.00525 = 0.00575$ lb/lb da, the quantity of water vapor added to the air.

Heating and humidifying of ventilating air occur as the air moves through livestock buildings. Animals produce heat, vapor, and water; both sensible heat and water vapor are added to ventilating air,

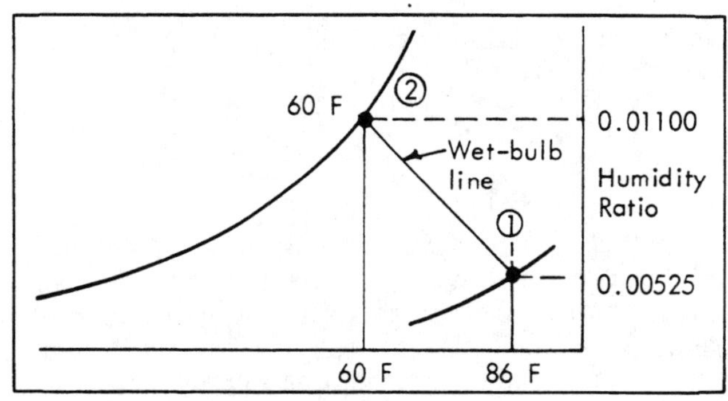

Fig 602-9. Evaporative cooling.

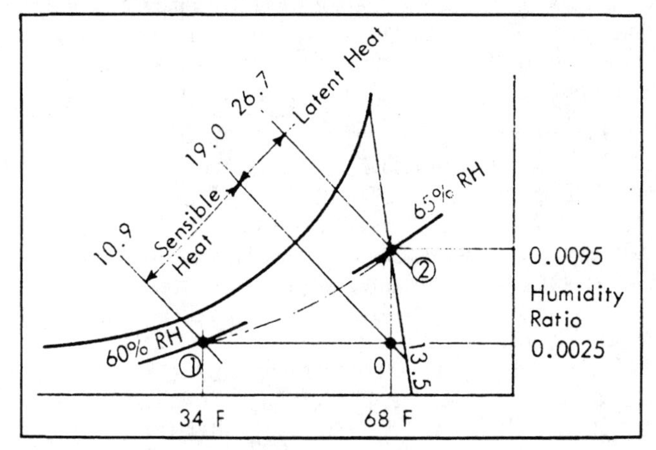

Fig 602-10. Heating and humidifying.

as in Fig 602-10. The process is along the dashed curve between the initial (1) and final (2) conditions.

Example 5

Moist air enters a farrowing building at 34 F db and 60% RH, replacing air removed from the space by an exhaust fan. If air leaves at 2700 cfm, 68 F db, and 65% RH, how much latent heat and sensible heat are added per hour to the air inside the building?

From Fig 602-10, the specific volume removed by the exhaust fan at state (2) is 13.5 ft^3/lb da. To separate the total enthalpy (energy) change into sensible and latent portions, locate state (0), which divides the process into a sensible heating process from state (1) to (0) and a humidifying process from state (0) to (2). State (0) has the same humidity ratio as state (1) and the same dry-bulb temperature as state (2). Thus h_1 = 10.9 Btu/lb da, W_1 = 0.0025 lb water/lb da, h_0 = 19.0 Btu/lb da, and h_2 = 26.7 Btu/lb da. The rate at which air flows through the building is:

Eq 602-15.

$$M_a = \text{cfm} \times 60 \text{ min/hr} \div V \text{ ft}^3\text{/lb da}$$

$$M_a = 2700 \times 60 \div 13.5$$
$$= 12{,}000 \text{ lb da/hr}$$

Sensible heat added to the air is:

$$_1q_0 = M_a \times (h_0 - h_1)$$
$$= 12{,}000 \times (19.0 - 10.9)$$
$$= 97{,}200 \text{ Btu/hr}$$

Latent heat added to the air is:

$_0q_2 = M_a \times (h_2 - h_0)$
$= 12{,}000 \times (26.7 - 19.0)$
$= 92{,}400$ Btu/hr

Total heat added is:

$_1q_2 = 189{,}600$ Btu/hr

Cooling and dehumidifying lowers both dry-bulb temperature and humidity ratio. The process path depends on the type of equipment used. In summer air conditioning, air passes over a cold finned-type evaporator coil of a refrigeration unit. The air is cooled below the dewpoint temperature and moisture condenses. Unless reheated or initially saturated, the final relative humidity of the moist air is always higher than at the start. Both sensible heat and latent heat are removed from the air in this process.

Example 6

How many tons of refrigeration are required to cool 700 cfm of air at (1) 86 F dry-bulb and 61 F dewpoint to (2) 53 F dry-bulb and 100% RH? How much moisture is removed? What is the ratio of latent heat to sensible heat removed in this process? (A ton of refrigeration is a cooling capacity of 200 Btu/min or 12,000 Btu/hr.)

Required properties from the psychrometric chart for the problem are in Fig 602-11. At state (1) enthalpy is 33.2 Btu/lb da, volume is 14.0 ft³/lb da, and humidity ratio is 0.0115 lb water vapor/lb da.

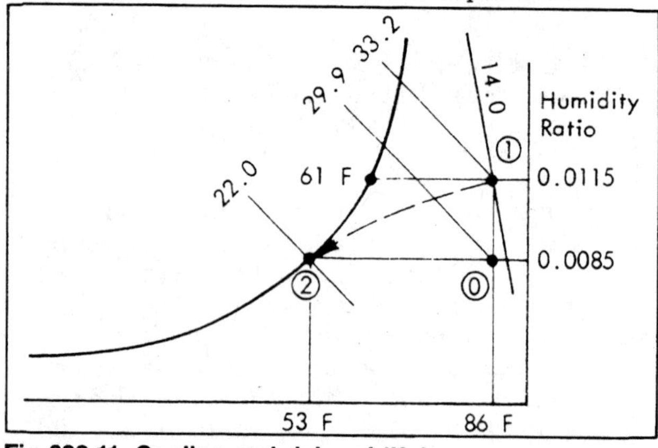

Fig 602-11. Cooling and dehumidifying.

Enthalpy at (2) is 22.0 Btu/lb da and humidity ratio is 0.0085 lb water vapor/lb da. State (0) divides the process into dehumidifying and sensible cooling segments; enthalpy is 29.9 Btu/lb da. The refrigeration required is:

$q_T = M_a \times (h_2 - h_1)$
$= (700 \div 14.0) \times (22.0 - 33.2)$
$= -560$ Btu/min or 2.8 tons of refrigeration

The water removed is:

$M_w = (700 \div 14.0) \times (0.0115 - 0.0085)$
$= 0.15$ lb/min

The ratio of latent heat to sensible heat removed is:

Eq 602-16.

$$q_L/q_s = (h_1 - h_0) \div (h_0 - h_2)$$
$$q_L/q_s = (33.2 - 29.9) \div (29.9 - 22.0) = 0.42$$

Adiabatic mixing of moist air. A common process in air conditioning is mixing moist air at two conditions to obtain air at a third. If the mixing is adiabatic as in Fig 602-12, it is governed by the following balances, and the state point of the air-vapor mixture after adiabatic mixing of air at two different states falls on the straight line connecting the two initial states. From Eq 602-20, the final state is the point dividing the line into two parts proportional to the weights of dry air being mixed.

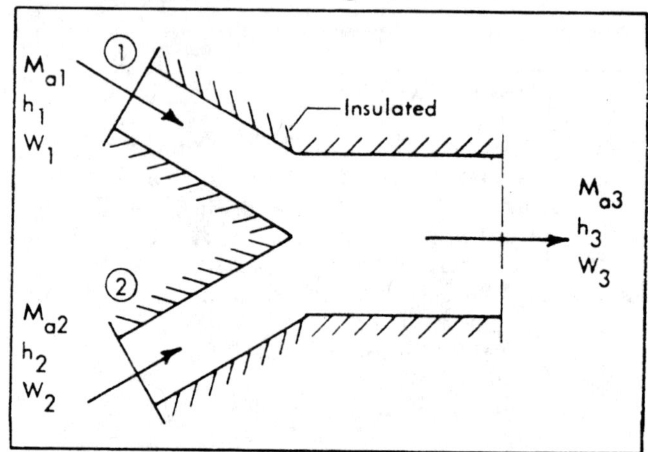

Fig 602-12. Adiabatic mixing of moist air.

Air mass balance:

Eq 602-17.

$$M_{a1} + M_{a2} = M_{a3}$$

Heat balance:

Eq 602-18.

$$M_{a1} \times h_1 + M_{a2} \times h_2 = M_{a3} \times h_3$$

Water mass balance:

Eq 602-19.

$$M_{a1} \times W_1 + M_{a2} \times W_2 = M_{a3} \times W_3$$

Substitute for M_{a3} from Eq 602-17:

$M_{a1} \times W_1 + M_{a2} \times W_2 = (M_{a1} + M_{a2}) \times W_3$

Transpose:

Eq 602-20.

$$M_{a2} \times (W_2 - W_3) = M_{a1} \times (W_3 - W_1)$$

Similarly:

Eq 602-21.

$$M_{a2} \times (h_2 - h_3) = M_{a1} \times (h_3 - h_1)$$

and, by division:

$(h_2 - h_3) \div (h_3 - h_1) = (W_2 - W_3) \div (W_3 - W_1)$
$(h_2 - h_3) \div (h_3 - h_1) = M_{a1} \div M_{a2}$

Also, it can be shown that:

Eq 602-22.

$$t_3 = [t_1 \times h_2 - t_2 \times h_1 + h_3 \times (t_2 - t_1)] \div (h_2 - h_1)$$

603 ANIMAL THERMAL ENVIRONMENT

This chapter is from material prepared by Dr. Duane W. Mangold, Iowa State University.

An animal's environment is the total of all external conditions that affect it. This chapter considers only the **thermal** environment. It describes the effects of air temperature, moisture, air velocity, and solar radiation on the regulation and balance of animal heat, and their influence on production, growth, feed conversion, and health.

Homeothermic Systems

Livestock and poultry are **homeothermic,** which means they maintain a relatively constant body temperature during environmental temperature changes. Body temperatures only a few degrees from those in Table 603-1 can be fatal.

Table 603-1. Normal rectal temperatures.

Animal	Temperatures Average	Temperatures Range
Dairy cow	101.5 F	100.4-102.8 F
Beef cow	101.0	98.0-102.4
Pig	102.5	101.6-103.6
Sheep	102.3	100.9-103.8
Chicken	107.1	105.0-109.4

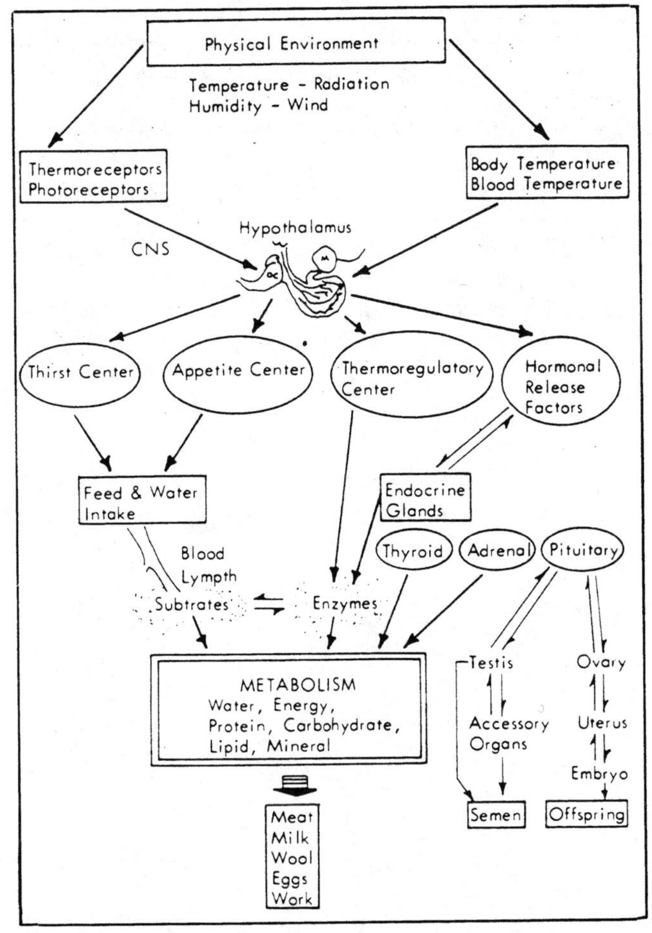

Fig 603-1. Interactions—environment and animal productivity.

Temperature regulation is an example of **homeostasis,** the maintenance of a bodily state within narrow limits. Small corrections in body temperature are controlled by the **hypothalamus gland** (Fig 603-1), which is a very sophisticated thermostat. The hypothalamus regulates body temperature by detecting internal body temperature, integrating it with neural signals from thermoreceptors at the skin surface, and then controlling neural and hormonal responses that alter heat production or heat loss.

Internal body temperature is controlled by a dynamic equilibrium between heat produced internally and heat gained from or lost to the environment. The animal produces heat when transforming the chemical energy of feed into work or body tissue. In its simplest form the thermal balance is:

Heat production = ± heat loss ± heat storage

The rate of internal heat production varies with size, body weight, breed, health, growth stage, feed type and intake rate, production level, gestation, age, degree of activity, and environmental conditions. Of the gross feed energy consumed by an animal in production, 25%-40% is converted into heat and lost to the environment. Animal performance is greatly affected by heat exchanged between it and the environment. In a cold environment where heat loss is greater than normal heat production, net energy normally used for growth and production must maintain body temperature, resulting in increased feed requirements per unit of production.

Fig 603-2 illustrates the uses of feed energy.

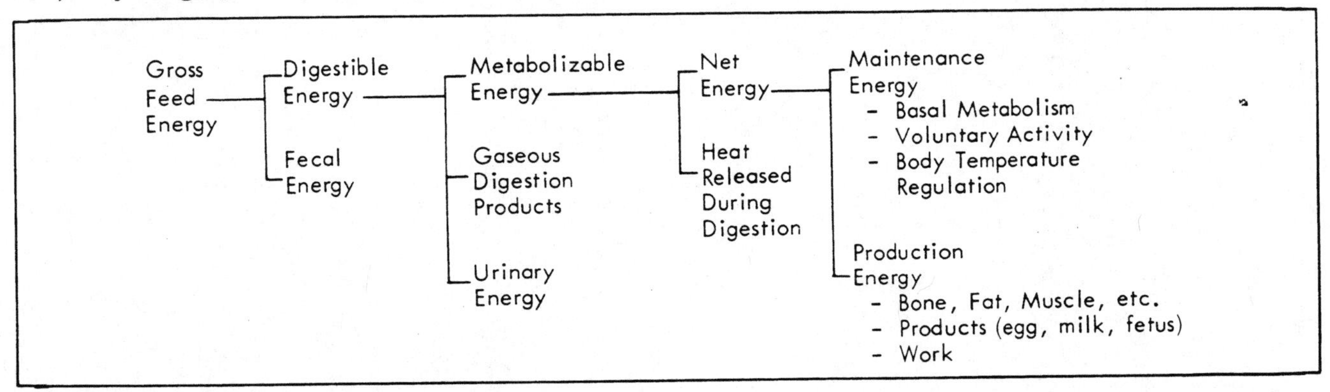

Fig 603-2. Feed energy utilization in animals.
Feed intake (gross feed energy) minus fecal energy is Digestible Energy. Deduct the energy in urine and digestive gases to determine Metabolizable Energy. Deduct the heat released during digestion to get Net Energy, which is the energy used by the animal.

Example 7

2540 cfm of outside air at 42 F db and 70% RH is adiabatically mixed with 1360 cfm of air at 70 F db and 70% RH. Find the dry-bulb and wet-bulb temperatures of the resulting mixtures. See Fig 602-13. Locate states (1) and (2) on the psychrometric chart: V_1 = 12.7 ft^3/lb da:

$$M_{a1} = cfm_1 \div V = 2540 \div 12.7 = 200 \text{ lb da/min}$$

$$M_{a2} = cfm_2 \div V = 1360 \div 13.6 = 100 \text{ lb da/min}$$

Since M_{a1} is twice M_{a2}, the line segment 1-3 is ½ line 3-2, or ⅓ the entire line 1-2. Locate state (3) with a ruler, and read the temperatures of 51.5 F db and 48 F wb.

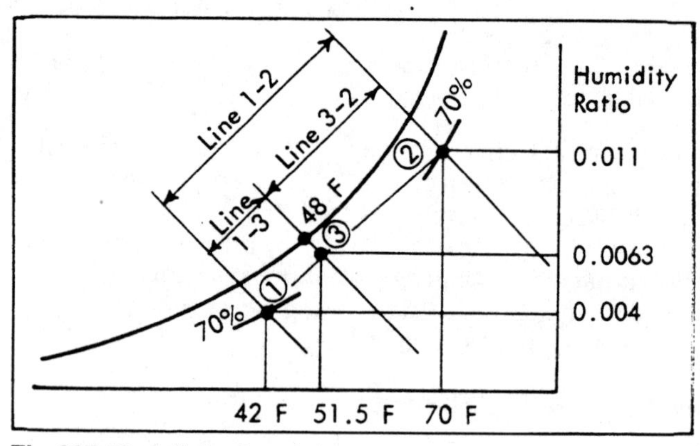

Fig 602-13. Adiabatic mixing.

Methods of Heat Transfer

The total heat exchanged per hour between an animal and its surroundings, q_T, consists of both sensible heat, q_s, and latent heat, q_L. Sensible heat transfer is by conduction, radiation, and convection. All animals dissipate significant amounts of latent heat by vaporization (evaporation) from the respiratory tract and body surface. The total heat exchange between animal and environment is:

Eq 603-1.

$$q_T = q_L \pm q_s = q_e \pm q_d \pm q_r \pm q_c \text{ Btu/hr}$$

q_T = total heat
q_L = latent heat
q_s = sensible heat
q_e = evaporation
q_d = conduction
q_r = radiation
q_c = convection

In Eq 603-1 sensible heat is positive if heat is transferred from the animal. If ambient temperature is greater than the animal's surface temperature, the animal gains sensible heat and the sensible terms are negative.

Conduction is heat transfer between contacting bodies at different temperatures without gross movement of the material. Heat transfer from the body core to the skin surface occurs by conduction through body tissue and also by convection associated with blood flow.

Eq 603-2.

$$q_d = U \times A \times (t_{sa} - t_{se})$$

q_d = heat transfer by conduction, Btu/hr
U = overall conductive heat transfer coefficient between the animal and the environment, Btu/hr-ft^2-F
A = contact area for conductive heat flow, ft^2
t_{sa} = surface temperature of the animal, F
t_{se} = surface temperature of the material in contact with the animal, F

Many animals adjust conductive heat loss simply by changing contact area—huddling with another animal or standing up from a cold floor.

Thermal radiation is the exchange of thermal energy between objects by electromagnetic waves. The rate depends on their temperatures and the nature of their surfaces. Radiation can pass through a vacuum. It warms the receiving body.

The radiant heat transfer equation is:

Eq 603-3.

$$q_r = A \times F_A \times F_E \times \sigma \times (T_{sa}^4 - T_s^4)$$

q_r = rate of radiant heat transfer, Btu/hr
A = area of radiating object, ft^2
F_A = shape factor
F_E = emissivity factor
σ = Stefan-Boltzmann constant, 0.1714×10^{-8}, Btu/ft^2-hr-R^4
T_s = absolute temperature (t_F + 460) of the surroundings, R
T_{sa} = absolute temperature (t_F + 460) of the animal surface, R

Evaluating F_A and F_E is complicated and detailed; values have been developed for some livestock shelter configurations.

The surroundings include the sun, sky, shade, fences, buildings, or other animals—any object or surface that an animal can "see." Sun radiant heat load can be reduced 30%-50% with shade.

Convective heat is transferred to or from an object by the mass movement of a fluid. Natural or free convection results from differences of density caused by temperature differences. Fans or pumps (including the heart) produce fluid motion and heat transfer known as forced convection.

The rate of heat exchange by convection is:

Eq 603-4.

$$q_c = h \times A_c \times (t_{sa} - t_a)$$

q_c = heat transferred by convection, Btu/hr
A_c = effective convective surface, ft^2
t_{sa} = surface temperature of the animal, F
t_a = ambient air temperature, F
h = convection coefficient, Btu/hr-ft^2-F = $C \times V^n$
V = air velocity, ft/min
C, n = constants

Values for h, C, and n have not been widely determined for livestock, but the form of the equation shows the important variables. The combined effect of high air velocity and cold ambient temperature results in high heat loss from livestock. When ambient air temperature equals the bodys surface temperature, convective heat transfer from the skin is zero.

Evaporation. Little moisture is lost through the skin surface of most domestic animals (swine, sheep, poultry); evaporative loss is largely from their upper respiratory tracts. Expired air has been heated nearly to body temperature and saturated with vaporized water in the upper respiratory tract. Little water evaporation or air warming takes place in the lungs.

Nonsweating species adjust to high temperature stress by greatly increased respiration—panting. Wetting an animal's skin maximizes cooling by external evaporation.

The evaporation heat transfer equation is:

Eq 603-5.

$$q_e = K \times A_e \times V^n \times (P_s - P_a)$$

q_e = evaporation heat loss, Btu/hr
K = evaporation constant
A_e = wet surface area of the animal, ft^2
V = air velocity, ft/min
n = velocity exponent
P_s = vapor pressure of water at the animal's internal or external surface, lb/ft^2
P_a = vapor pressure of water in the air, lb/ft^2

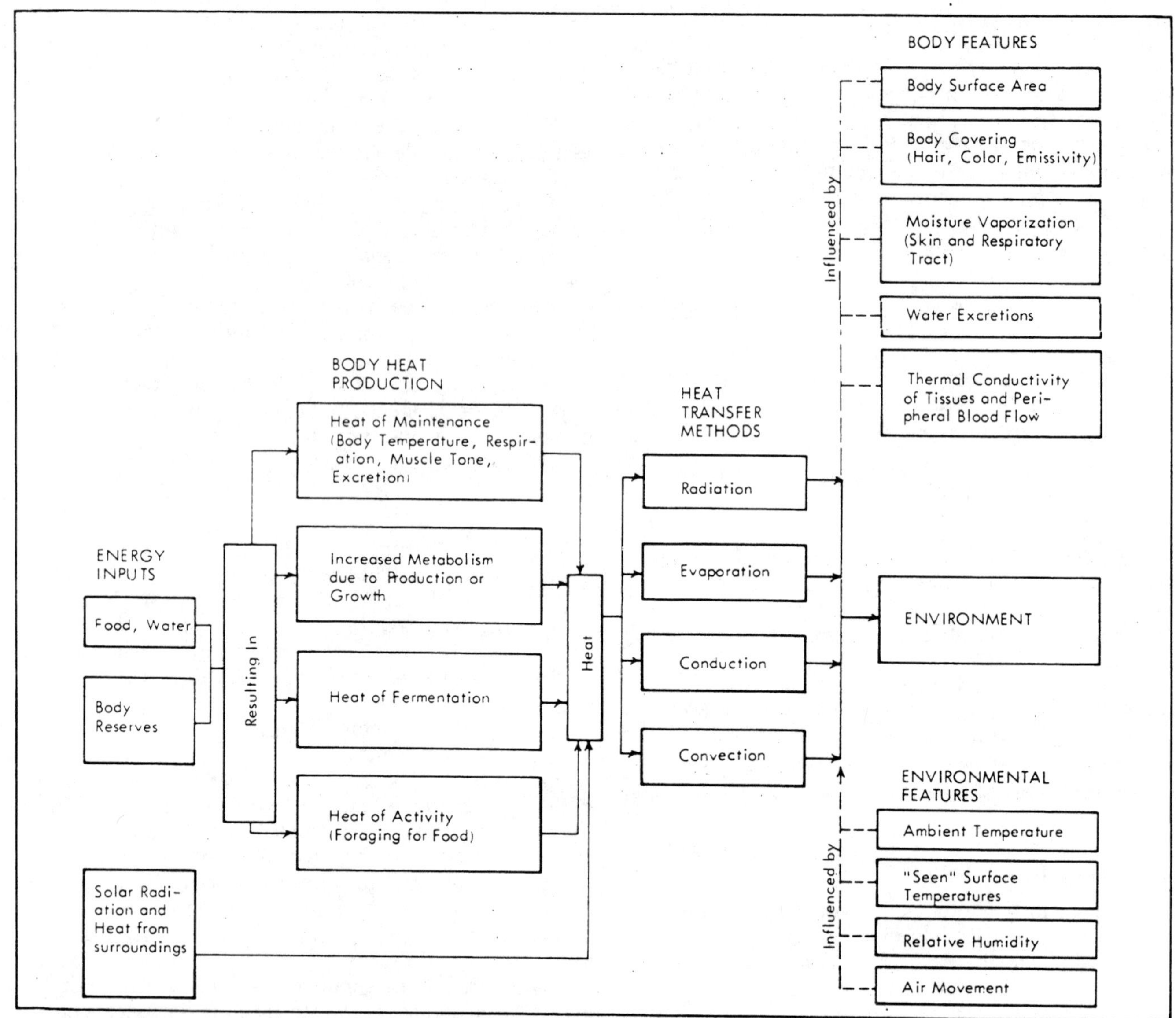

Fig 603-3. Factors influencing heat balance for homeothermic animals.
Animals usually have heat to dissipate to the environment. The rate and relative amount of heat transfer by each of the four methods depend on the amount of heat to be dissipated and on the animal and environmental factors listed.

Heat Production

Homeothermic animals maintain nearly constant body temperature by balancing internal heat production and heat loss to the surroundings. The factors involved are illustrated in Fig 603-3.

Animal Response to Temperature

The comfort zone (A-A′ in Fig 603-4) is not easily defined for animals, but it can be approximated:

- Blood vessels in the skin are neither all dilated nor all constricted (vasodilation, vasoconstriction).
- Moisture evaporation from the skin and respiratory tract is minimal.
- Hair and feathers are not erected (minimum piloerection).
- Behavioral response to heat or cold is not observed.

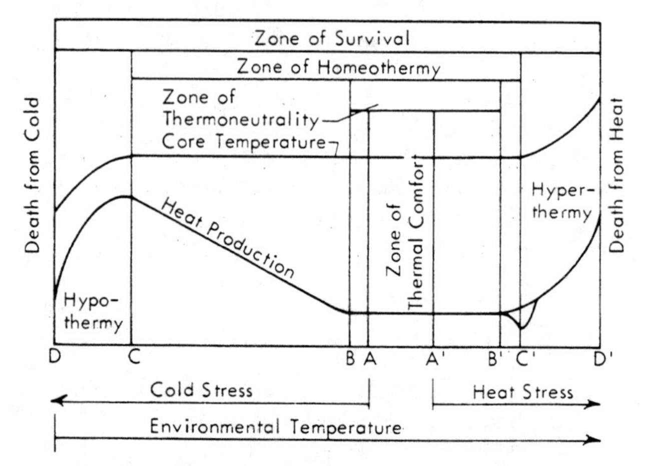

Fig 603-4. Homeothermic heat production and body temperature.
As affected by environmental temperature.

When environmental temperature drops below point A, vasoconstriction and erection of hair or feathers start to conserve metabolic heat. Vasoconstriction reduces blood flow to the skin reducing heat transfer by blood from the body core to the surface and the coefficient of thermal conductivity in that area. Piloerection increases the insulation value by holding more air in the coat cover.

If the environment cools below point B, the lower critical temperature, heat production rises. The added heat protects the core temperature from falling, and homeothermy is maintained. Increased feed intake sustains the increased metabolic rate.

At point C the extra heat produced cannot balance the heat lost, homeothermy fails, and the core temperature and heat production decline. This pattern can lead to death from cold.

Fig 603-4 lacks a time scale. Time is very important—a body can stand short periods of stress severe enough to kill it if extended.

Heat stress begins when a rising environmental temperature reaches point A′. The regulation above this point is physical, in contrast to the chemical regulation that increases metabolism to combat cold stress. The surface vascular vessels enlarge (vasodilation), increasing blood flow to the surface, increasing heat transfer, and raising the skin surface temperature, which increases the temperature difference between the skin surface and the environment. Sweating, increased respiration, or both begin in some species. In other species, water vapor diffusion through the skin increases. Appetite is depressed to reduce body heat production.

The upper critical temperature, point B′, is the threshold to radical change in heat production. At point B′, the evaporative mechanisms increase in intensity, and heat production may decrease, partly due to decreased feed intake.

The organism loses homeothermy at C′, because vapor loss from the skin can no longer increase. The respiration rate may still increase, but its cooling potential is also limited. Evaporation is the most powerful mode of heat loss as environmental temperature approaches skin surface temperature. When evaporation reaches its maximum rate, the core temperature begins to rise. A rise in body temperature increases biochemical reactions which further increase heat production (van't Hoff effect). If unrelieved, the cycle leads to death, although heat stress for a few hours probably causes no lasting harm in most animals.

Fig 603-4 illustrates the critical zones of temperature and stress. The importance of appetite to livestock production is clear. Below thermoneutrality, increased feed intake helps maintain core temperature without an increase in productivity. Above thermoneutrality, appetite depression reduces feed intake needed for maximum production of milk, meat, eggs, and fiber.

Livestock heat production

A general equation developed by Brody estimates basal heat production of most homeotherms:

Eq 603-6.

$$q_b = K \times W^{0.734}$$

q_b = basal heat production, Btu/hr
K = coefficient
W = animal weight, lb

The basal heat production of an animal is a standardized measurement of metabolism for comparisons among species, age, sex, size, etc. The basal metabolic rate is measured at thermoneutrality on a resting animal in a post-absorptive state.

Brody found K to be about 6.5 for the basal condition. For normal activity, K is about 2½ times basal, 16.25. The equation does not relate heat production to ambient temperature. The empirical exponent, 0.734, is partly explained by the fact that surface area (L^2) is dimensionally the ⅔ power of volume (L^3). In a uniformly dense body, weight is directly proportional to volume.

The ratio of surface area to animal body weight varies considerably, decreasing as weight increases. For example, the surface area of a 4-lb chicken is about 1.9 ft^2, a surface/weight ratio of 0.48; for a 1000-lb cow with a 49.5 ft^2 surface, the ratio is 0.05.

Because heat production is a function of body weight, and heat transfer from an animal depends largely on surface area, heat transfer per unit weight is higher for smaller animals, Fig 603-5.

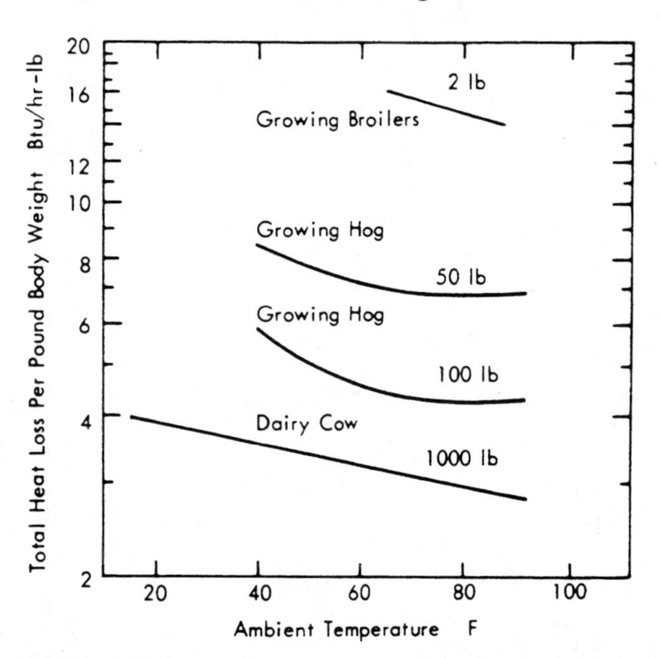

Fig 603-5. Effect of ambient temperature on total heat loss.

Air temperature and livestock production efficiency

Air temperature affects all four methods of heat transfer. Fig 603-6 relates the proportions of latent and sensible heat production to air temperature for several farm animals. As ambient temperature increases the sensible heat loss by conduction, radiation, and convection decreases and the latent heat loss by evaporation must increase if the animal is to maintain homeothermy.

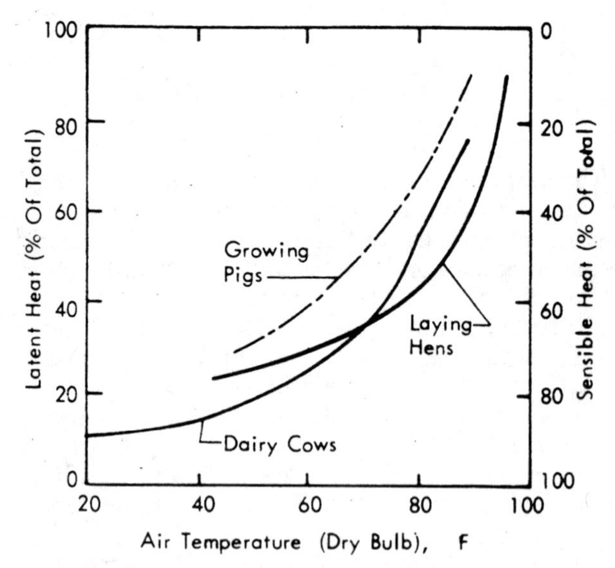

Fig 603-6. Percent of latent and sensible heat production.
As affected by air temperature.

The temperature range for maximum animal production efficiency is shown on the lower part of Fig 603-7. Between t_1 and t_2 is the thermal comfort zone, where production efficiency changes only slightly.

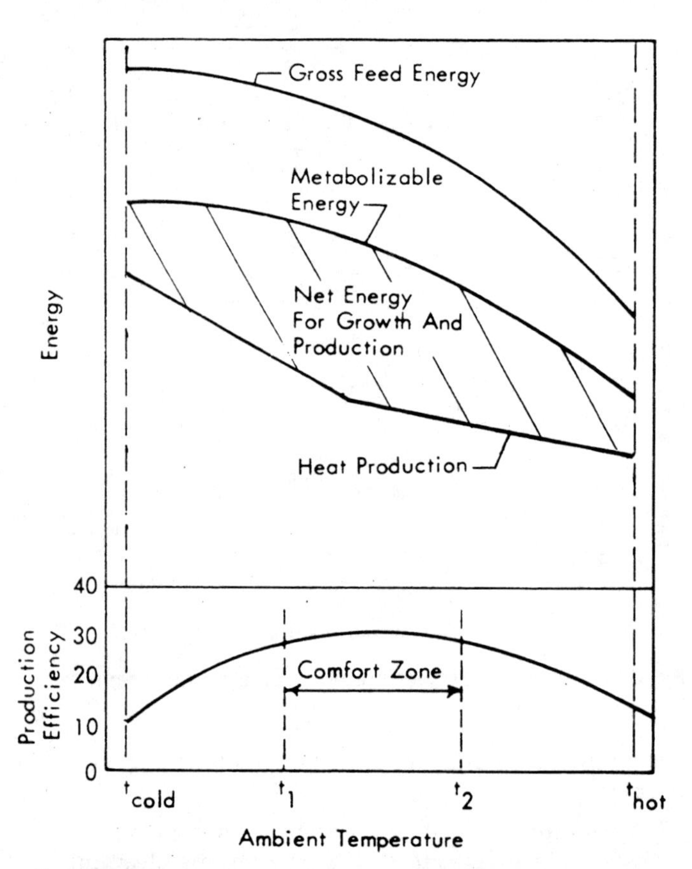

Fig 603-7. Proportion of food energy available for production.
As influenced by air temperature.

The effect of air temperature on the productivity of some classes of livestock and poultry is shown in Fig 603-8. The optimum temperature zone varies with type and age of animal, changes in air velocity, and to a lesser extent, relative humidity. Most temperature response data are observed at moderate humidity and at air velocities of 50 fpm or less ("still" air).

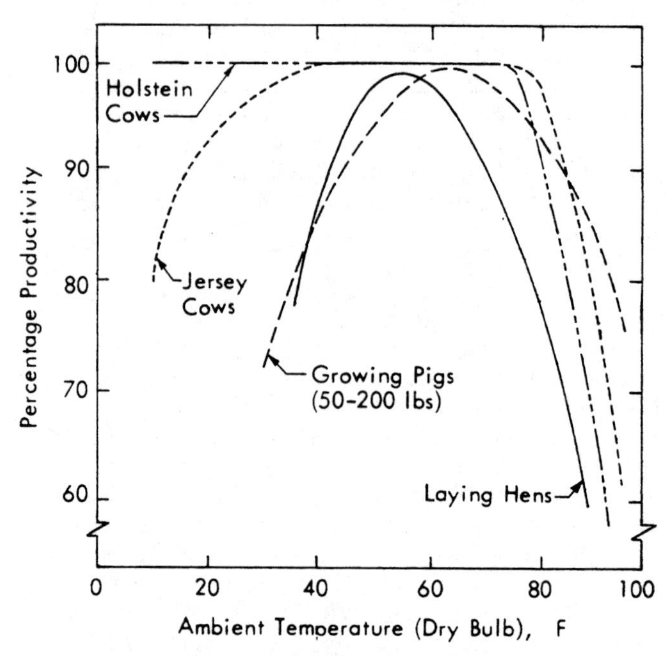

Fig 603-8. Percentage productivity of livestock and poultry.
As affected by air temperature.

604 ANIMAL HEAT AND MOISTURE PRODUCTION

Swine

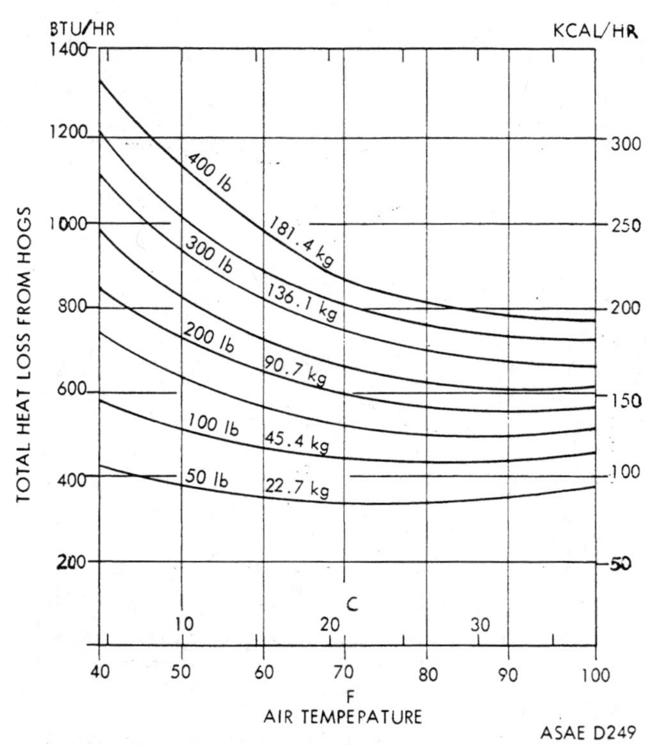

Fig 604-1. Total heat loss of swine.
Air velocity 20 to 30 ft/min; 50% RH; room, wall, and air temperatures the same. (Bond, Kelly, and Heitman, 1959).

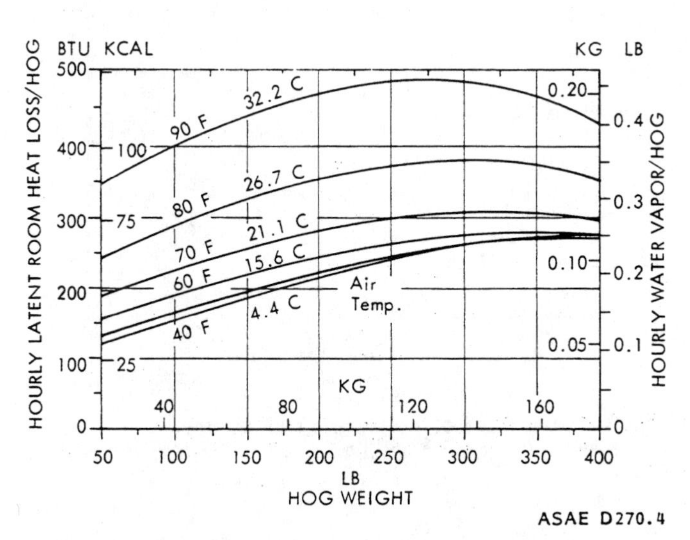

Fig 604-2. Room latent heat in a swine building.
With solid concrete floor scraped daily.
No bedding, air velocity 20 to 30 ft/min, and RH about 50%. (Bond, Kelly, and Heitman, 1959).
Water vapor from a totally slotted floor building is ½ as much as from a solid floor building. For a partly slotted building, moisture removed is in proportion to the percent of floor that is slotted. If bedding is used on solid floor, increase chart values by up to ⅓. The total heat production remains unchanged. (Harmon, Dale, and Jones, 1966).

Table 604-1. Partitioned swine heat loss.
Air velocity is 35 ft/min. (Bond, Kelly, and Heitman, 1959).

Air temperature F	C	Percent heat loss by: Radiation %	Convection %	Conduction %	Evaporation %
40	4.4	34.9	37.8	12.8	14.5
50	10.0	33.0	38.7	12.8	15.5
60	15.6	32.9	38.7	11.8	16.6
70	21.1	27.0	34.3	10.7	28.4
80	26.7	23.0	32.0	7.7	37.3
90	32.2	17.2	20.7	7.4	54.7
100	37.8	2.6	5.0	2.8	89.6

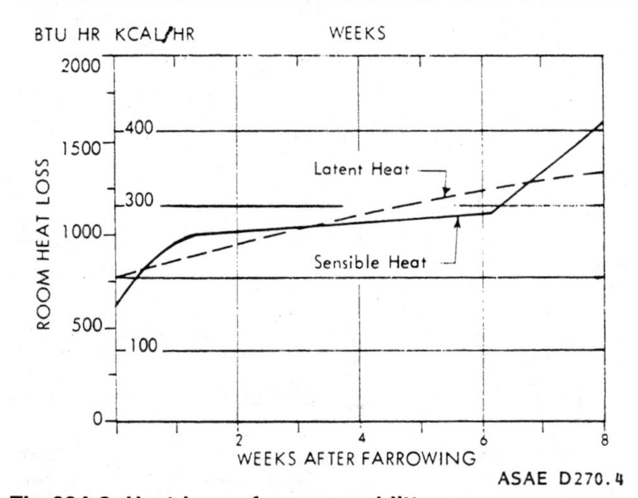

Fig 604-3. Heat loss of a sow and litter.
RH about 50%. Solid concrete floor, scraped daily, no bedding. (Bond, Kelly, and Heitman, 1959).

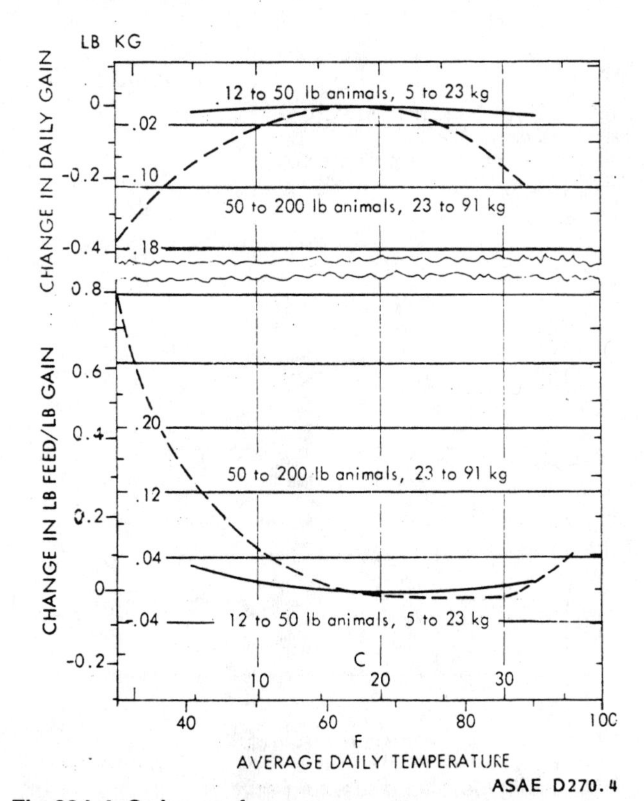

Fig 604-4. Swine performance.
Deviation from performance at 60 F in daily gain and feed efficiency of swine exposed to various average daily temperatures. (Hazen and Mangold, 1960).

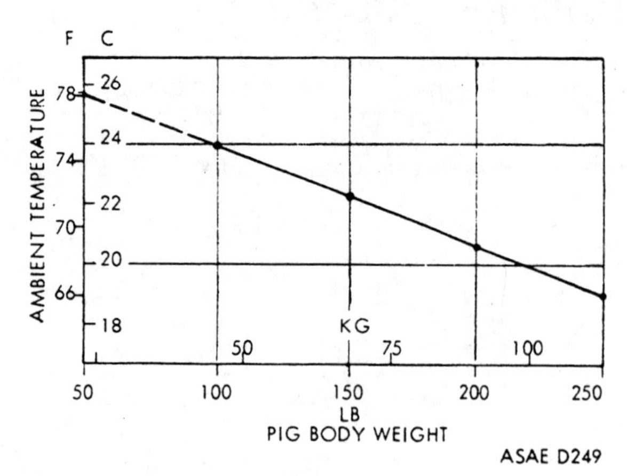

Fig 604-5. Temperatures for maximum swine growth.
At 50% RH.
(Heitman, Kelly, and Bond, 1958).

Dairy

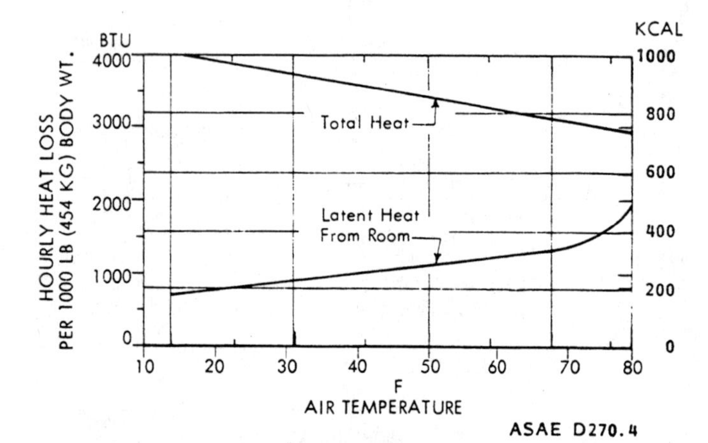

Fig 604-6. Barn heat and moisture loss.
Of stanchioned dairy cattle. RH = 55%-70%.
Total barn heat loss declined rapidly above 80 F.
(Yeck and Stewart, 1959).

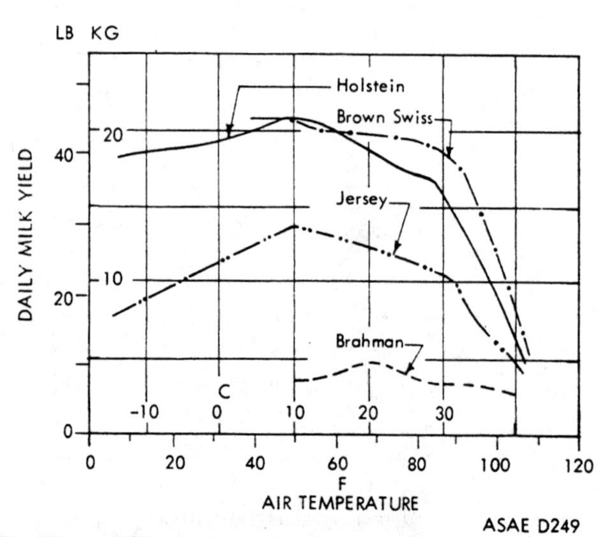

Fig 604-7. Milk yield vs. air temperature.
Cows at constant temperature, RH about 50%.
(Johnson, 1965).

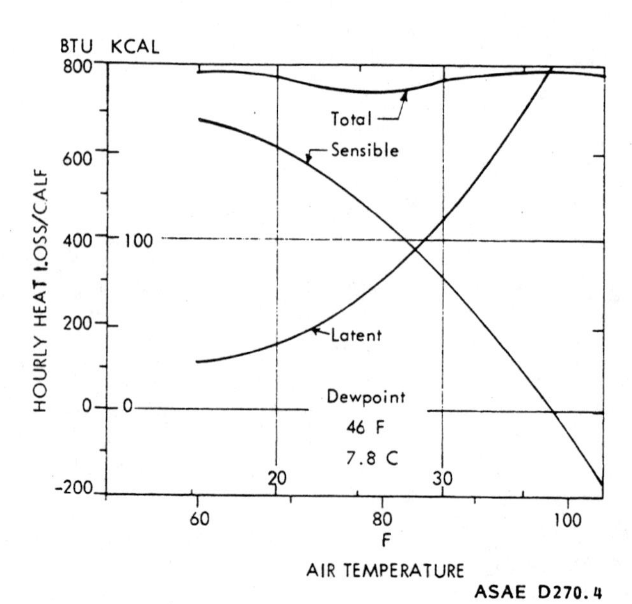

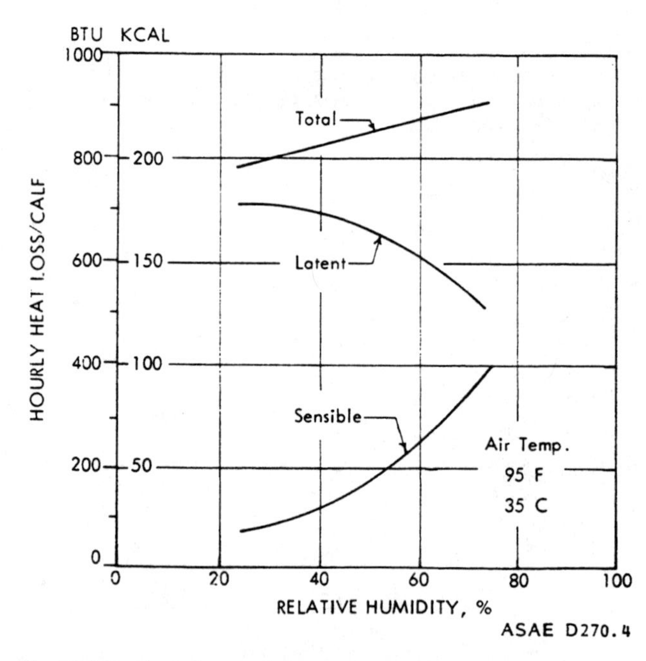

Fig 604-8. Heat loss of 3 Ayrshire bull calves.
Calves were 6 to 10 months old. Vapor pressure = 0.155 lb/in². (McLean, 1963).

Beef

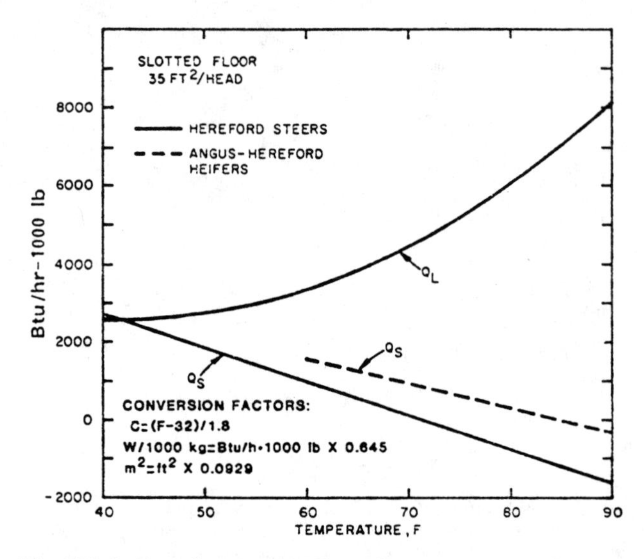

Fig 604-9. Heat loss of beef cattle.
Hereford and Angus-Hereford steers.
Q_s = sensible heat loss.
Q_L = latent heat loss.
From the *1981 ASHRAE Handbook of Fundamentals.*

Sheep

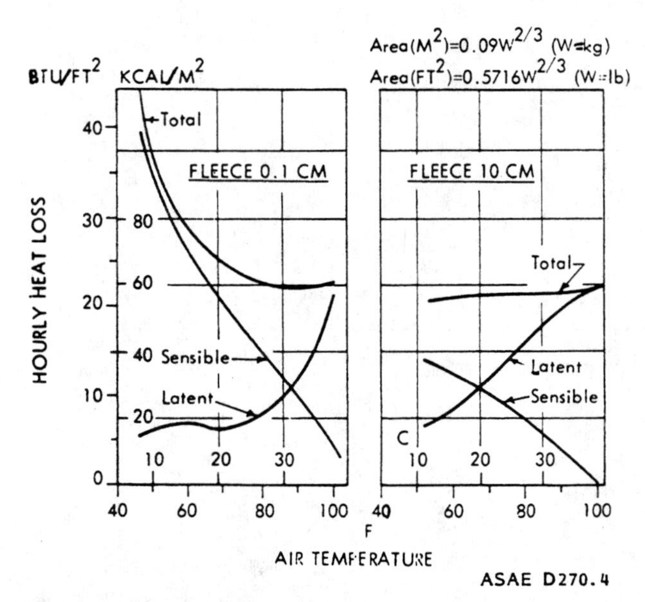

Fig 604-10. Heat loss of sheep.
Half bred x Dover-Cross Wethers.
RH: 45%-54%.
Air flow: 4 ft³/min. (Blaxter, McC. Graham, and Wainman, 1959).

Poultry

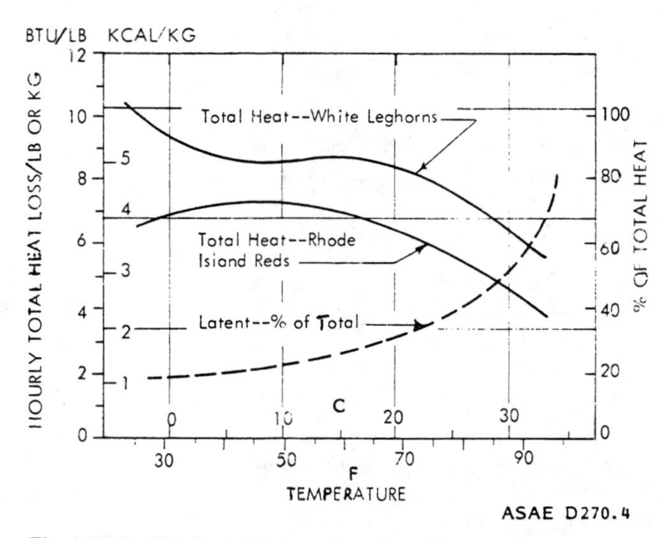

Fig 604-11. Total and latent hen heat losses.
Of Rhode Island Red and White Leghorn laying hens.
(Ota and McNally, 1961).

Table 604-2. Hourly hen moisture production.
Of 1000-4 lb White Leghorn laying hens. (Ota, 1966).

Temperature		Respired		Defecated		Total*	
F	C	lb	kg	lb	kg	lb	kg
25	- 3.9	6.3	2.9	14.5	6.6	22.8	10.4
35	1.7	8.3	3.8	14.5	6.6	24.8	11.3
45	7.2	8.4	3.8	12.9	5.9	23.7	10.8
60	15.6	11.4	5.2	12.7	5.8	26.4	12.0
80	26.7	14.3	6.5	14.4	6.4	31.6	14.3
95	35.0	20.0	9.1	10.3	4.7	33.7	15.3

*Includes wasted drinking water estimated at 10% of water consumed at 25-80 F and 15% at 95 F.

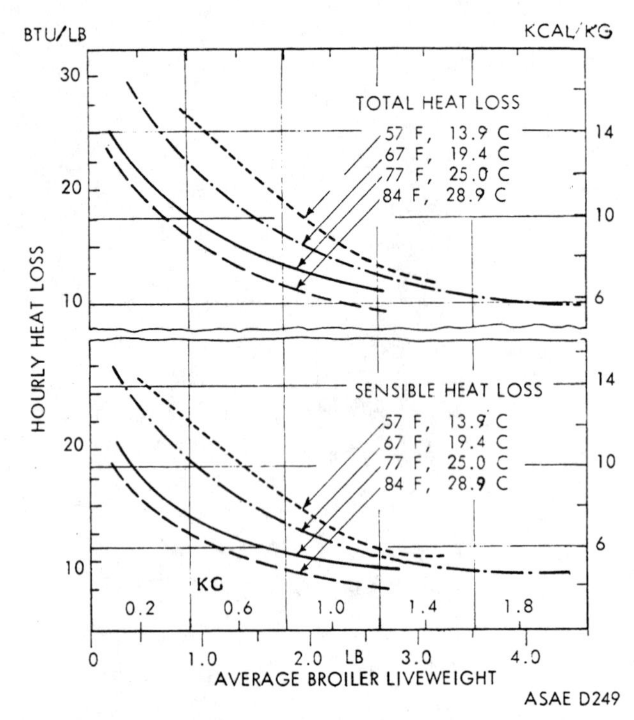

Fig 604-12. Heat loss of growing broilers.
At 75% relative humidity.
(Ota and McNally, 1966).

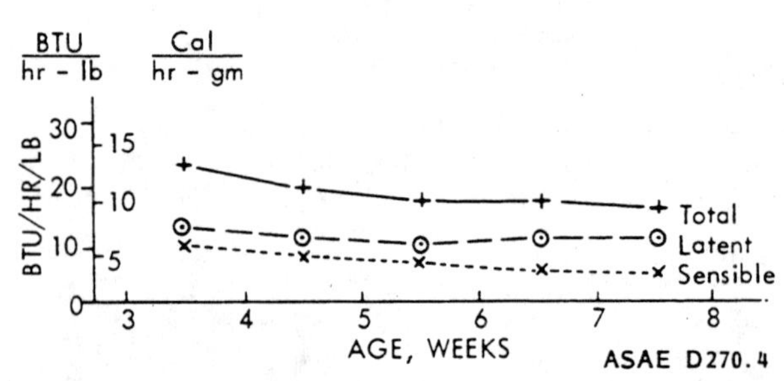

Fig 604-13. Specific total, sensible, and latent heat for broilers.
Data from a litter floor, insulated, windowless broiler building operated at an average temperature of 82 F and 87 F at the beginning and end of the test respectively. Heat loss through ceiling, floor, and walls was negligible. (Reece and Deaton, 1970).

Table 604-3. Total heat and moisture loss of turkeys.

				Heat loss			
				Light		Dark	
Sex	Temp F	R.H. %	Wt. lb	Total heat Btu/hr-lb	Latent Btu/hr-lb	Total heat Btu/hr-lb	Latent Btu/hr-lb
M	95	27	0.2	25.7	16.3		
M	90	31	0.5	20.0	10.2		
M	85	37	0.9	15.8	7.5		
M	80	42	1.3	13.4	4.3		
M	75	51	2.1	12.2	2.5		
Large White turkey (DeShazer, Olson, and Mather, 1974)*							
M	70	42	5.8	12.3		8.6	
F	70	42	6.8	12.0		9.1	
M	70	42	8.3	12.1		8.7	
Wrolstad White turkey (Buffigton, Jordan, Junnila, and Boyd, 1974)							
M	65	76	19.7	5.9	1.9	4.9	1.5
F	64	80-90	9.6	6.1	1.8	4.0	1.0
M	77	52	19.7	5.1	2.3	4.3	1.8
F	79	63	9.6	6.5	4.6	4.3	3.2
Beltsville White turkey (Ota and McNally, 1961)							

*Average heat losses of 4 turkeys under continuous light.

630 ENVIRONMENTAL CONTROL SYSTEMS

631 INSULATION AND VAPOR BARRIERS

Insulation is any material that reduces heat transfer from one area to another. Some building materials, such as wood, are good insulators, while others, such as concrete and metal, are poor insulators.

Although all building materials have some insulation value, the term "insulation" usually refers to materials with a relatively high resistance to heat flow. The resistance of a material to heat flow is indicated by its **R-value,** with good insulators having high R-values. See Table 631-1.

Heat is transferred from one place to another by conduction, radiation, and convection. Conducted heat moves through the material that separates a warm area from a cold area. Radiated heat passes through a space without warming the space, such as the sun heating the earth. Convected heat is transmitted by a moving fluid, such as circulating air or water.

In homes and farm buildings heat is lost or gained by conduction through construction materials, by radiation from warm surfaces, by air currents moving over the inside or outside surfaces, and by exhausting warm air which is replaced by cold outdoor air. In walls, the transfer of heat is complex, but simple measures can economically and effectively insulate the wall to resist heat flow.

Why Insulate

There are several reasons for insulating. First, insulation conserves heat in an enclosure (Fig 631-1), which reduces the amount of supplemental heat required. In non-heated buildings, the heat saved can evaporate water from the floor or warm additional ventilating air.

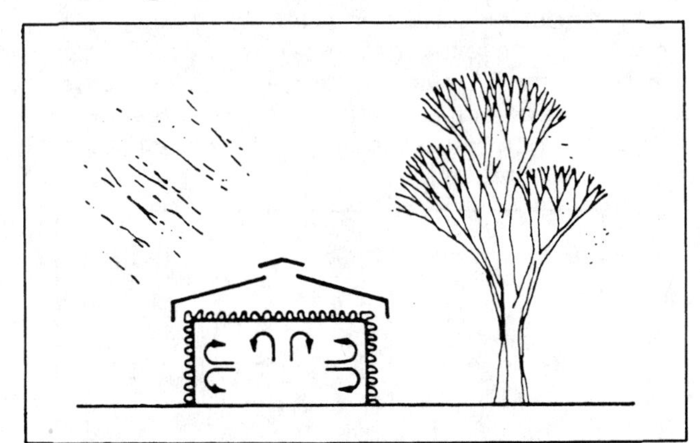

Fig 631-1. Insulation reduces heat loss in cold weather.

Second, insulation reduces heat gain during warm summer months, which improves comfort and reduces cooling costs, Fig 631-2. The temperatures of the walls and roofs of buildings exposed to direct sunlight can be as much as 50 F above air temperature.

Fig 631-2. Insulation reduces heat gain in hot weather.

Third, insulation provides warmer surface temperatures on the inside of exterior walls during cold weather. Warmer surface temperatures reduce the amount of condensation or "sweating" and lower the amount of radiant heat loss from objects in the building to cold walls. In a poorly insulated building, the inside ceiling and wall surfaces become cold in the winter. When the surface temperature is cold enough, the air next to the surface becomes saturated and moisture condenses, Fig 631-3. If the surface temperature is below freezing, frost occurs.

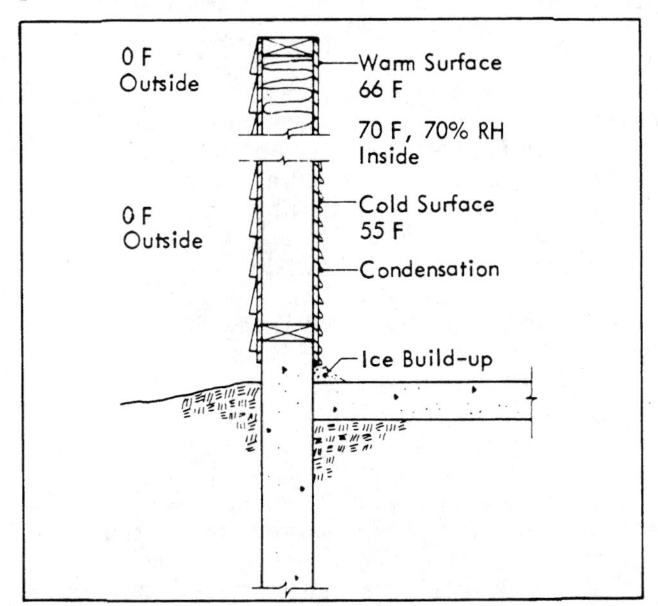

Fig 631-3. Warm, moist air condenses on a cold surface. Insulation helps to control sweating by making the wall and ceiling surfaces warmer.

Radiant heat loss affects the comfort level even though temperatures within the building are correct. Animals and people lose heat by radiant heat transfer when they are surrounded by a surface that is lower in temperature than their body temperature. Radiant heat loss causes you to feel cool when sitting in front of a window on a cold winter day, even though the room temperature may be 75 F.

Insulating Materials

Common insulations are bulky, porous, lightweight materials with countless tiny air spaces. Generally, the more air pockets in a material, the better it insulates.

Six basic types of insulating materials are available for use on the farm or home. They are available in more than one form: loose fill, batt, rigid board, etc.

Cellulose or wood fiber (R = 3.5/in. approximately). This is a fill-type insulation made from paper or pulp products chemically treated for fire and insect resistance. It is commonly blown into walls of existing buildings and is also widely used for ceiling insulation. The process of blowing the material fluffs the insulation and directs it through a hose to place it in wall cavities or over ceilings. This material can be partly fluffed with a garden rake or pitch fork before or after placing, but blowing is usually best.

Fiberglass or glass wool (R = 3.0-3.8/in.). This material is long, thin glass fibers or filaments intertwined and fluffy in appearance. It is available in batts or blankets and in loose bags for blown-in or pour installations.

Polystyrene (R = 4-5/in.). Polystyrene is usually in rigid boards, but it is also available as a granular or pour-type insulation. Extruded polystyrenes (styrofoam) have the higher insulating value and are more resistant to moisture absorption. Molded polystyrenes (bead board) have lower R-values and are not as moisture resistant. To tell the types apart, look at the particles making up the insulation. Bead board breaks up into many small ball bearing-type beads or pellets; extruded polystyrene is finer grained. Extruded polystyrenes often have a higher R-value when new because of the gases trapped during forming. The R-value decreases in time as the gases escape. Polystyrenes are flammable and should be covered with a fire-resistant material or used in locations where flammability is of little concern, such as perimeter foundation insulation. Check with your fire insurance company before using these insulations.

Polyurethane (R = 6.0/in. aged). Urethane is available in rigid sheets or can be sprayed onto walls, ceilings, or other structural components. Urethane has the highest R-value of common building insulations. New urethane has a higher insulating value because of the gas used for foaming; but as the gas escapes during aging, the R-value decreases. Cover urethane with a fire-resistant material, such as plywood or dry wall, to prevent rapid flame spread and to reduce the release of fumes in case of fire. Before using, check with your fire insurance company for special requirements.

Polyisocyanurate (R = 7.2-8.0/in.). Isocyanurate is readily available in rigid sheets and is used as foam cores of special prefabricated wall sections. It is more resistant to flame spread than many other rigid board insulations, but it must be used in strict accordance with manufacturer's directions. Aluminum foil faced sheets, meeting specification HHI 1972-1 Class 2 with a minimum of 1 mil thickness, reduce aging. Check with your insurance company for special requirements.

Urea formaldehyde. This foamed-in-place insulation is sold under a variety of trade names. It is banned by the Consumer Product Safety Commission for use in new residential installations due to formaldehyde emissions.

Vermiculite (R = 2.2/in.). This is a granular, free flowing, and extremely fire- resistant material. Use it to fill the cores of concrete blocks and odd holes or spaces.

Forms of Insulation

Manufactured insulation can be purchased in several common forms. There are four common types of insulation for the farm and home.

Batt and blanket insulation (Fig 631-4) is available in 1"-8" thicknesses and in widths to fit 16", 24", and 48" stud spaces. Batts are 4' or 8' long, and blankets are up to 100' long. Materials are fiberglass, mineral wool, or cellulose fibers. The batt or blanket may have a paper or aluminum face to serve as a partial vapor barrier. An additional vapor barrier is required.

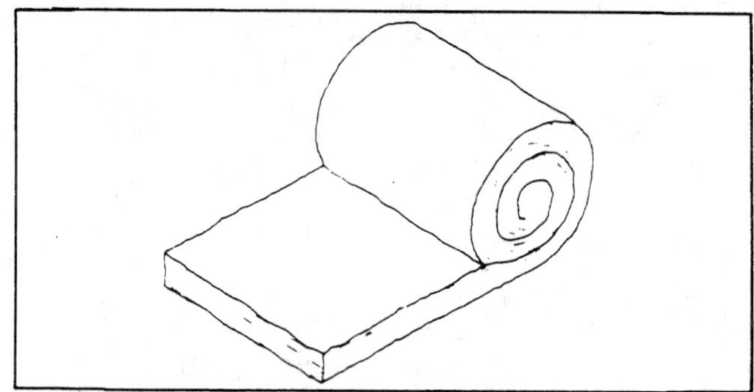

Fig 631-4. Batt or blanket insulation.

Loose-fill insulation (Fig 631-5) is packaged in bags and can be mineral wool, cellulose fiber (wood fiber), vermiculite (expanded mica), granulated cork, and/or polystyrene. It is easy to pour or blow above ceilings, in walls, and in concrete block cores. Poor quality insulation can settle in walls, leaving the top inadequately insulated.

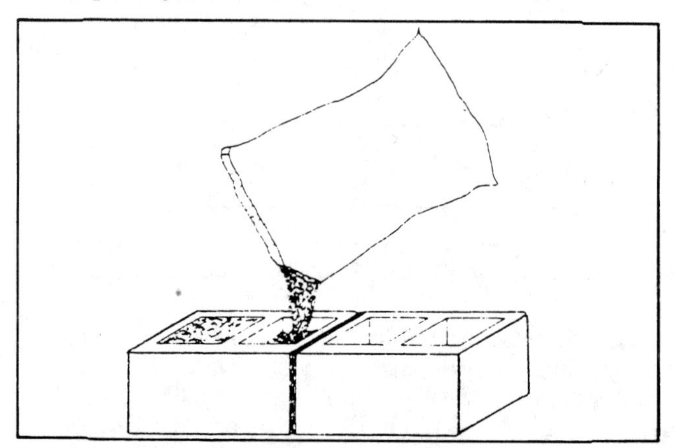

Fig 631-5. Loose fill insulation.

Rigid insulation (Fig 631-6) is made from cellulose fiber, fiberglass, polystyrene, polyurethane, polyisocyanurate, or foam glass and is available in ½"-2" thick by 4' wide panels. Some types have aluminum foil or other vapor barriers attached to one or both faces. Rigid insulation can be used for roofs and walls or as a ceiling liner. It can also be used along foundations (perimeter insulation) or buried under concrete floors (if it is waterproof and protected from physical and rodent damage).

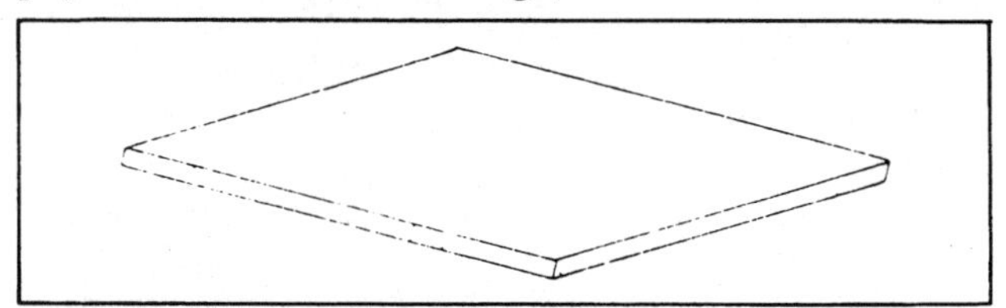

Fig 631-6. Rigid insulation.

Support rigid insulation at least 2' o.c. Seal the joints with caulk or tape to prevent moisture from passing through the joints, even if the panels are tongue-and-grooved. Check for flammability and toxic gas production if burned. Check if your insurance company requires rigid insulation to be protected with fire-resistant materials.

Foam or foamed-in-place insulation is made by foaming organic materials with air or inert gases. The most common materials are polystyrene and polyurethane. Some foams are available in either pre-cut board stock or foam in place while others are available as foam in place only. Foams may or may not have good moisture resistance, depending on their cellular structure. Use "open celled" foams only with a separate vapor barrier. Pre-cut foam boards are common directly under metal roofing or as interior liners for farm buildings. Protect them from mechanical damage by animals or birds. Some plastic foam insulations are flammable and release toxic gases when burned: check the building's insurability against fire. Many foams must be covered with fire-resistant materials, such as plywood or fire-rated gypsum board.

Sprayed-on insulation, applied to inside or outside surfaces, is difficult to protect with an adequate vapor barrier. Exterior application must also be protected from sunlight. As a result, improperly installed insulation may peel off.

Other. Concrete additives, reflective metallic foil, and air spaces can also help insulate.

Some materials (for example, expanded shale, ground corncobs, or wood shavings) when substituted for some of the concrete aggregates increase the insulation value **but** at a considerable reduction in the strength of the concrete. No additive has yet been proven to insulate concrete without reducing strength.

A reflective material like aluminum foil, placed so there are air spaces around it, reflects almost all of the radiant heat that strikes it. Because only part of the heat to be retarded is radiant heat, reflective insulations need several air spaces to resist heat flow by conduction and convection. Reflective insulation loses its value rapidly when covered with dust and corrosion and is not recommended in livestock facilities.

The effectiveness of an air space as insulation depends on its position and thickness. A dead air space at least ¾" but no more than 4" thick has a small insulating value. Where possible, fill air spaces with insulation.

Selecting Insulation

Consider the following factors:

- **R-Value.** The higher the R-value, the better the insulation.
- **Ease of installation.** Are you planning to do it yourself? Can it be done without tearing off the siding or inner wall surface? Some materials are harder to handle or take more time to install, which increases labor. Others are irritating to eyes and skin, requiring protective clothing and masks.
- **What is to be insulated?** Will you be insulating a ceiling where many inches of material may be needed, or a wall or roof with thickness limitations?
- **Fire resistance.** Will the material require a fire-resistant liner to prevent rapid flame spread? Check with your insurance company before choosing insulation.
- **Animal contact.** Will the insulation be exposed to livestock, rodent, or bird damage, requiring a protective covering which increases the cost?
- **Cost.** What will the different types of insulation cost considering preparation, installation, protection, etc., as well as purchase price? Variation in material and installation costs can be significant.

Recommended Insulation Levels

All building materials have some insulation value, but the amount varies considerably among different materials. For example, it takes 15.6" of concrete (R = 0.08/in.) to equal the insulation value of 1" of plywood.

Today there are many different insulations on the market, each with its own characteristics. Table 631-1 lists common insulating and building materials with their R values, either per inch of material thickness, or for the total thickness. Some manufacturers quote as-installed R-values, which allows them to include assumed values for other components of the ceiling, wall, or floor section. When comparing costs based on insulating value, be sure you are comparing R-values per inch of thickness or total R-value installed.

Table 631-2 shows some common wall sections in farm buildings and the relative insulation value of each. These show that a small amount of insulation can make a big difference in heat flow resistance. Table 631-3 shows insulation values for other types of construction, including roofs and ceilings.

Table 631-1. Insulation values.
From *1981 ASHRAE Handbook of Fundamentals*. Values do not include surface conditions unless noted otherwise. All values are approximate.

Material	R-value Per inch (approximate) 1/k	R-value For thickness listed 1/C
Batt and blanket insulation		
Glass or mineral wool, fiberglass	3.00-3.80*	
Fill-type insulation		
Cellulose	3.13-3.70	
Glass or mineral wool	2.50-3.00	
Vermiculite	2.20	
Shavings or sawdust	2.22	
Hay or straw, 20″		30+
Rigid insulation		
Exp. polystyrene,		
extruded, plain	5.00	
molded beads, 1 pcf	5.00	
molded beads, over 1 pcf	4.20	
Expanded rubber	4.55	
Expanded polyurethane, aged	6.25	
Glass fiber	4.00	
Wood or cane fiberboard	2.50	
Polyisocyanurate	7.04	
Foamed-in-place insulation		
Polyurethane	6.00	
Building materials		
Concrete, solid	0.08	
Concrete block, 3 hole, 8″		1.11
lightweight aggregate, 8″		2.00
lightweight, cores insulated		5.03
Brick, common	0.20	
Metal siding	0.00	
hollow-backed		0.61
insulated-backed, 3/8″		1.82
Softwoods, fir and pine	1.25	
Hardwoods, maple and oak	0.91	
Plywood, 3/8″	1.25	0.47
Plywood, 1/2″	1.25	0.62
Particleboard, medium density	1.06	
Hardboard, tempered, 1/4″	1.00	0.25
Insulating sheathing, 25/32″		2.06
Gypsum or plasterboard, 1/2″		0.45
Wood siding, lapped, 1/2″x8″		0.81
Asphalt shingles		0.44
Wood shingles		0.94
Windows (includes surface conditions)		
Single glazed		0.91
with storm windows		2.00
Insulating glass, 1/4″ air space		
double pane		1.69
triple pane		2.56
Doors (exterior, includes surface conditions)		
Wood, solid core, 1¾″		3.03
Metal, urethane core, 1¾″		2.50
Metal, polystyrene core, 1¾″		2.13
Air space (¾″ to 4″)		0.90
Surface conditions		
Inside surface		0.68
Outside surface		0.17

*The insulation value of fiberglass varies with batt thickness. Check package label.

Ways of expressing the value of insulation are:

R = thermal resistance, hr-ft^2-F/Btu.
It is the resistance to heat flow of 1 ft^2 of material when the temperature difference between the two sides is 1 F.
R is an additive quantity; 2″ of a material has twice the R-value of 1″. Also the individual R-values for all materials in a given section of a structure can be added together to obtain a total R-value.

R_T = total thermal resistance.
It is the total resistance of an entire wall, ceiling, etc. section, including the air film coefficients.

U = overall coefficient of heat transmission, Btu/hr-ft^2-F = $1/R_T$.
It is the heat in Btu/hr that passes through an entire wall, ceiling, etc. section of 1 ft^2, in one hour per 1 F temperature difference between the air on the warm side and the air on the cold side.

k = thermal conductivity, Btu-in./ft^2-F-hr.
It is the heat in Btu/hr that passes through a piece of material 1″ thick and 1 ft^2, when the temperature difference between the two sides is 1 F.

C = thermal conductance, Btu/ft^2-F-hr.
C is like k, except it is given for the total thickness: k for glass wool = 0.29; C for 3″ glass wool = 0.10. By convention, C does not usually include the effects of boundary layer resistances.

In the following discussion, R is used because the insulation value of a wall is easier to calculate, and many insulations are marked with their R-value.

Example 1:

Given the wall in Fig 631-7, find the total R-value. From Table 631-1 find the R-values for each material.

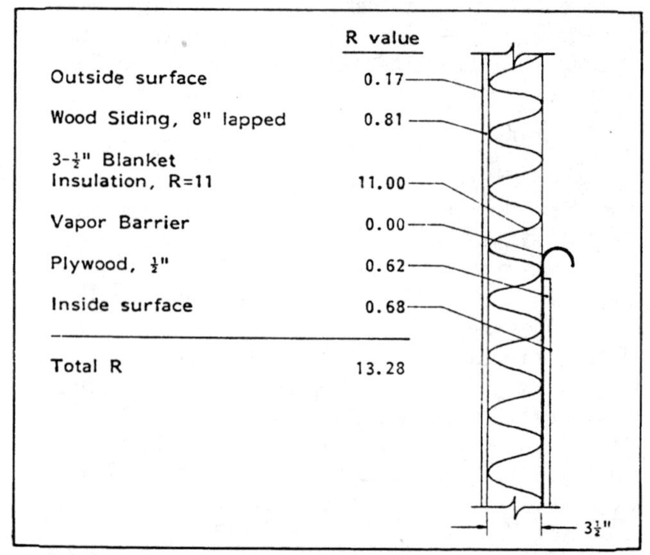

Fig 631-7. R-value of a wall section.

By adding the individual R-values, we find the wall has a total R-value of 13.28. Note that the blanket insulation provides more than 80% of the total R. Example calculations for other wall constructions are shown in Table 631-2.

Table 631-2. Insulation values for wall construction.

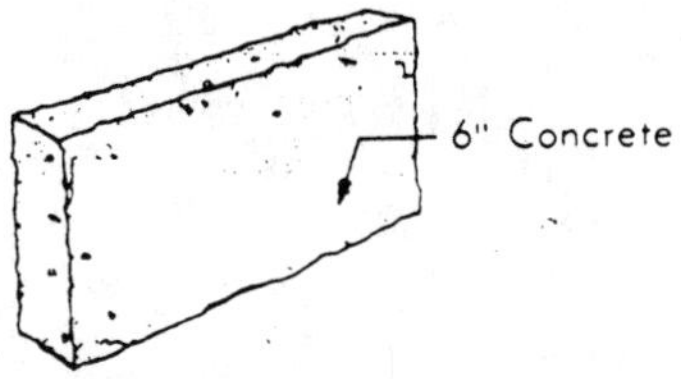

Outside surface (15 mph wind)	0.17
Concrete (6")	0.48
Inside surface (still air)	0.68
Total resistance, R_T	1.33

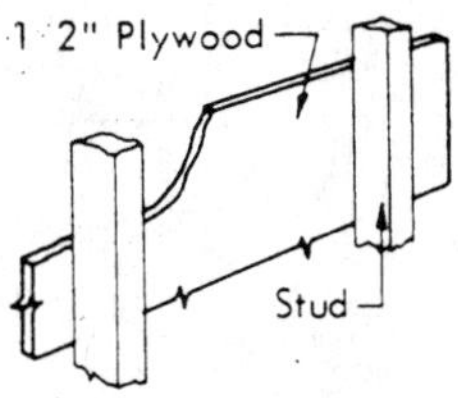

Outside surface (15 mph wind)	0.17
Plywood (½")	0.62
Inside surface (still air)	0.68
Total resistance, R_T	1.47

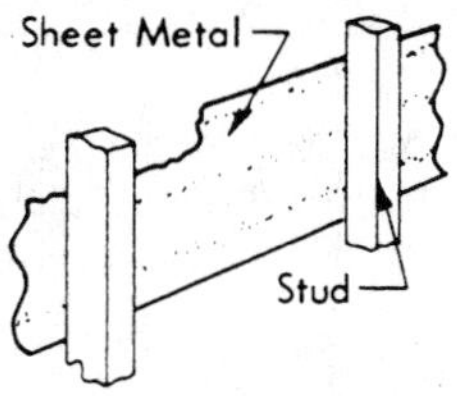

Outside surface (15 mph wind)	0.17
Sheet metal	0.00
Inside surface (still air)	0.68
Total resistance, R_T	0.85

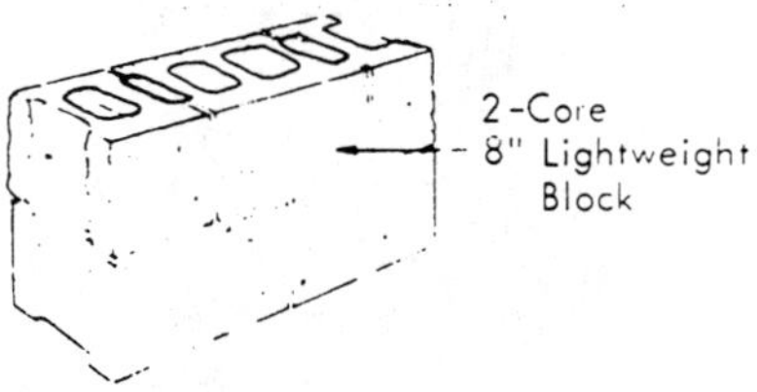

Outside surface (15 mph wind)	0.17
8" lightweight concrete block	2.00
Inside surface (still air)	0.68
Total resistance, R_T	2.85

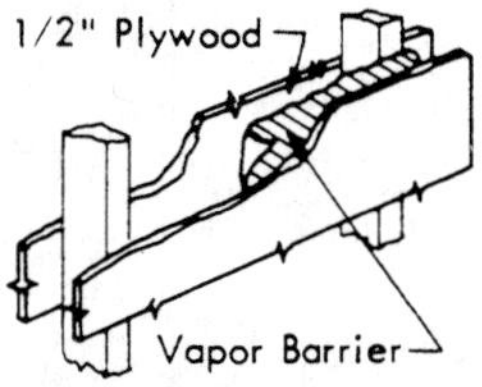

Outside surface (15 mph wind)	0.17
Plywood (½")	0.62
Air space	0.90
Supplemental vapor barrier	0.00
Plywood (½")	0.62
Inside surface (still air)	0.68
Total resistance, R_T	2.99

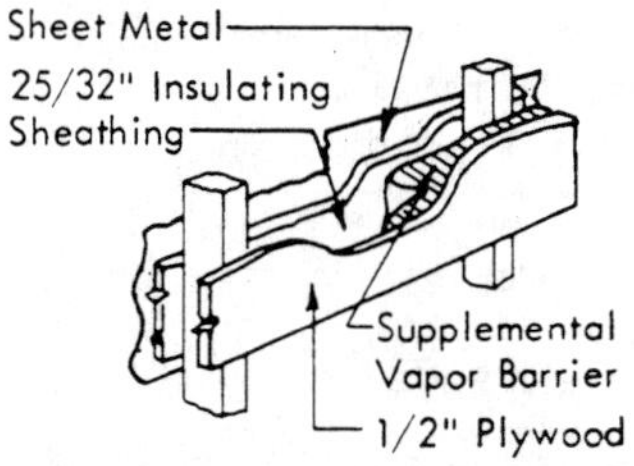

Outside surface (15 mph wind)	0.17
Sheet metal	0.00
Fiber board insulating sheathing, (25⁄32")	2.06
Air space	0.90
Supplemental vapor barrier	0.00
Plywood (½")	0.62
Inside surface (still air)	0.68
Total resistance, R_T	4.43

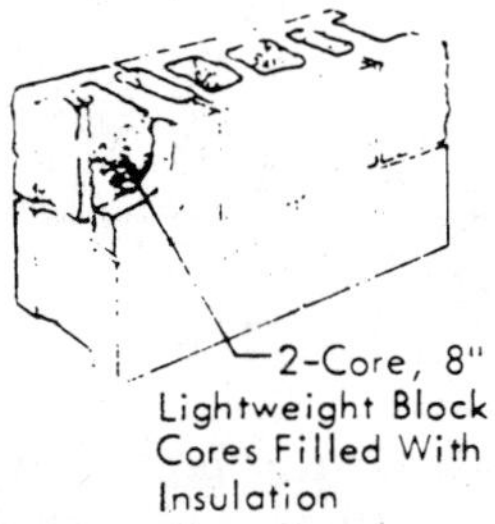

Outside surface (15 mph wind)	0.17
8" lightweight concrete block, cores filled with vermiculite	5.03
Inside surface (still air)	0.68
Total resistance, R_T	5.88

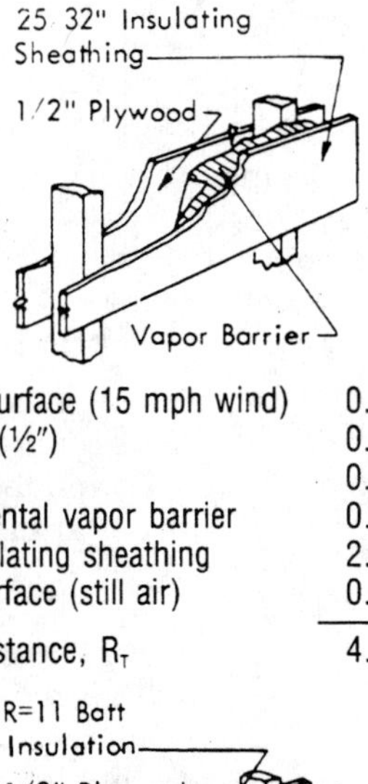

Outside surface (15 mph wind)	0.17
Plywood (½")	0.62
Air space	0.90
Supplemental vapor barrier	0.00
25⁄32" insulating sheathing	2.06
Inside surface (still air)	0.68
Total resistance, R_T	4.43

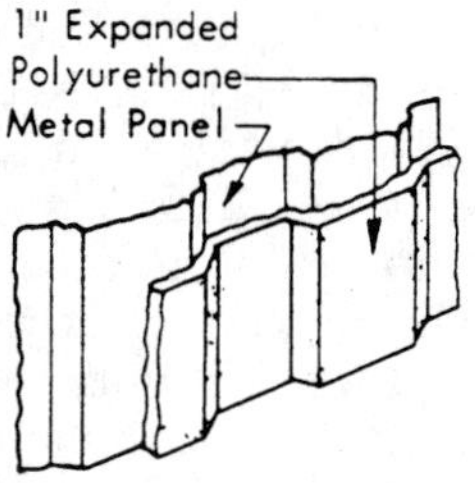

Outside surface (15 mph wind)	0.17
Sheet metal	0.00
Expanded polyurethane (aged)	6.25
Inside surface (still air)	0.68
Total resistance, R_T	7.10

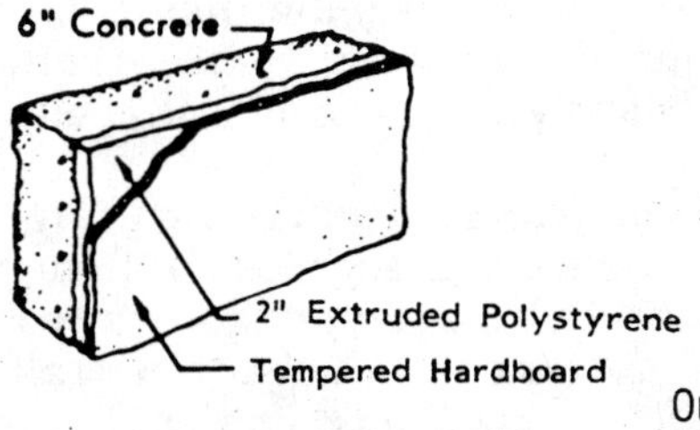

Outside surface (15 mph wind)	0.17
Concrete (6")	0.48
Extruded polystyrene (2")	10.00
Hardboard, tempered	0.25
Inside surface (still air)	0.68
Total resistance, R_T	11.58

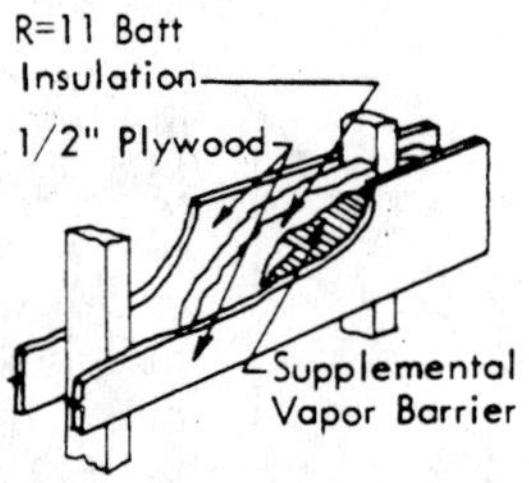

Outside surface (15 mph wind)	0.17
Plywood (½")	0.62
Batt insulation, R=11	11.00
Supplemental vapor barrier	0.00
Plywood (½")	0.62
Inside surface (still air)	0.68
Total resistance, R_T	13.09

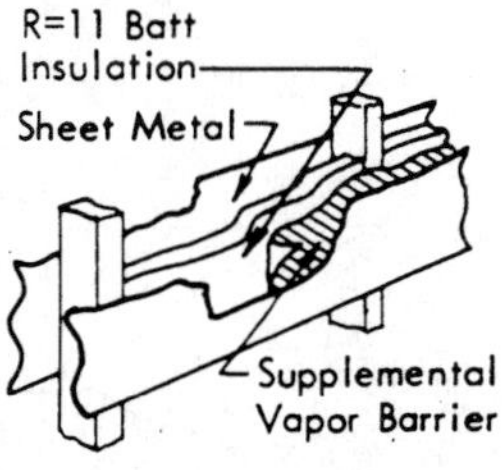

Outside surface (15 mph wind)	0.17
Sheet metal	0.00
Batt insulation	11.00
Supplemental vapor barrier	0.00
Plywood (½")	0.62
Inside surface (still air)	0.68
Total resistance, R_T	12.47

Table 631-3. Insulation R-values for other construction.

Roofs	R_T
Metal roofing, 25⁄32" insulating sheathing	2.91
Metal roofing, 0.4" expanded polyurethane	3.35
Metal roofing, 1" molded polystyrene, 1 pcf	5.85
Ceilings	
2" expanded polyurethane	13.35
½" plywood, 4" glass or mineral wool fill insulation	13.47
Metal roofing, R-19 blanket insulation	19.85
½" plywood, 8" glass or mineral wool fill insulation	25.47
Doors	
½" plywood, R-2 blanket insulation, ¾" air space, ½" plywood	4.99
½" plywood, 1" polystyrene molded, 1 pcf ¾" air space, ½" plywood	7.99
Floor perimeter (per foot of exterior wall)	
Concrete	1.23
Concrete, with 2"x24" of rigid insulation around perimeter	2.22

Residences

Unlike some farm buildings, homes **must** be insulated. The amount of insulation for a home depends primarily on economics and climate. Consider insulation as an investment against reduced heating and cooling bills. Often in new construction, installing the proper amount of insulation also reduces the size of the heating and cooling system.

Recommended insulation levels are often based on the heating season's degree day total. "Winter degree days" estimates winter season severity by comparing weather data with 65 F. A large number indicates average temperatures farther below 65 F and/or a longer season. The number of degree days for each day is calculated as the difference between 65 F and the day's average temperature. The number of degree days during the heating season is the total of the daily differences for the season. See Fig 631-8 for the number of degree days for your area. To qualify for financing by the Farmers Home Administration (FmHA), new residences must meet the minimum requirements in Table 631-4. Note that these R-values are minimums.

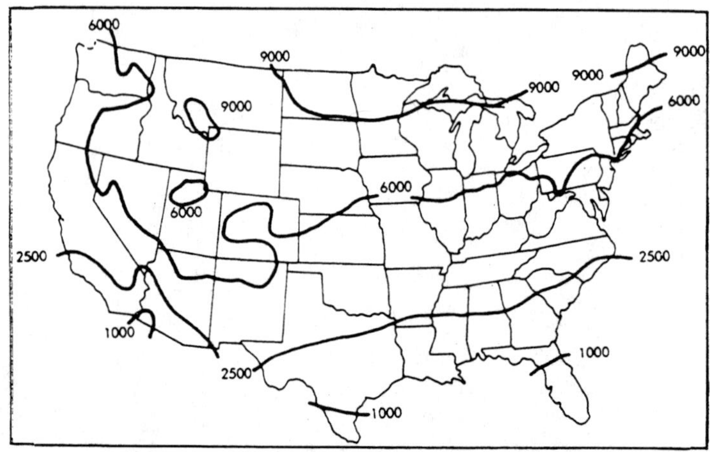

Fig 631-8. Winter degree days.
Accumulated difference between 65 F and average daily temperature for all days in the heating season.

Table 631-4. Home insulation levels.
Minimum R-values for ceiling, wall, and floor sections. R-values are not adjusted for framing.

	Minimum R-values		
Winter degree days	**Ceiling**	**Walls**	**Floors***
1000 or less	20	12	12
1001 to 2500	25	14	14
2501 to 6000	33	20	20
6001 or more	38	20	20

*For floors of heated spaces over unheated basements, unheated garages or unheated crawl spaces.

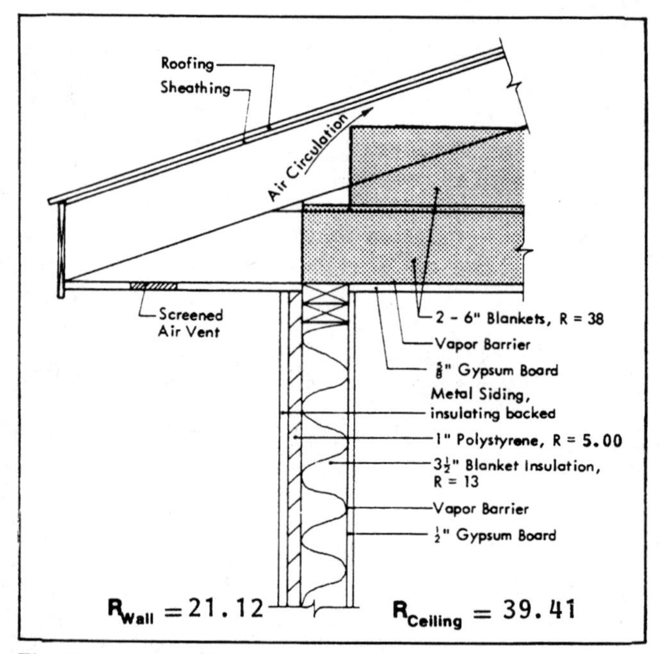

Fig 631-9. Example of residence insulation.

Farm buildings

The amount of insulation needed in farm buildings depends on many factors, such as the expected outside temperature (degree days), number and size of animals housed, desired inside temperature, and economics.

"**Cold**" **buildings** have indoor conditions about the same as outside conditions. Examples are machinery storages, cold free stall barns, and open-front livestock buildings. Minimum insulation is frequently recommended in the roof of these buildings to reduce solar heat gain in summer and condensation in winter.

Modified environment buildings rely on animal heat and controlled, natural ventilation to remove moisture and maintain the desired inside temperature. Examples are warm free stall barns, poultry production buildings, and swine finishing units.

Supplementally heated buildings require extra heat to maintain the desired inside temperature. Examples include farrowing buildings, some calf buildings, farm shops, and offices. Cold and modified buildings requiring supplemental heat in a small area, such as brooders in an open-front building, are **not** classified as supplementally heated.

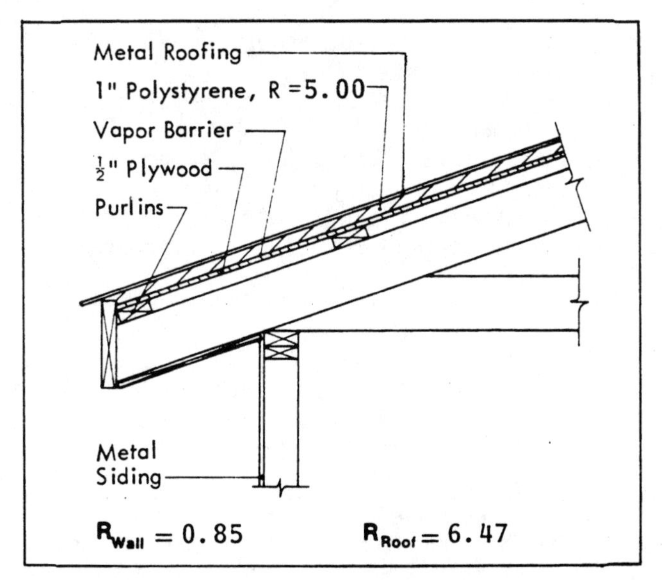

Fig 631-10. Example of "cold" building insulation.

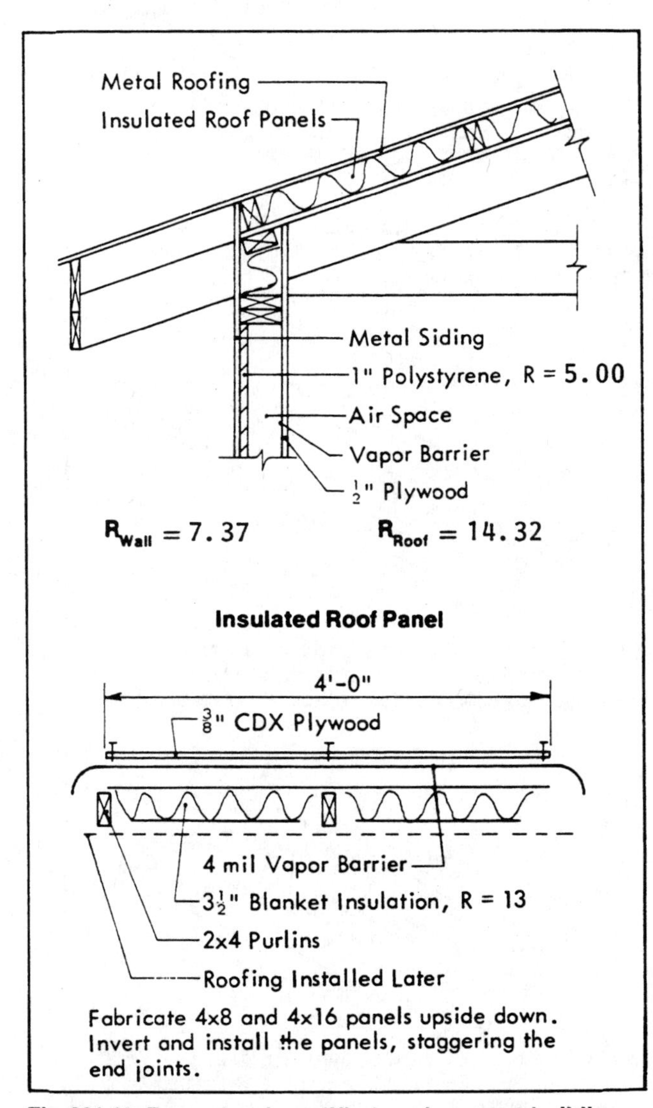

Fig 631-11. Example of modified environment building insulation.

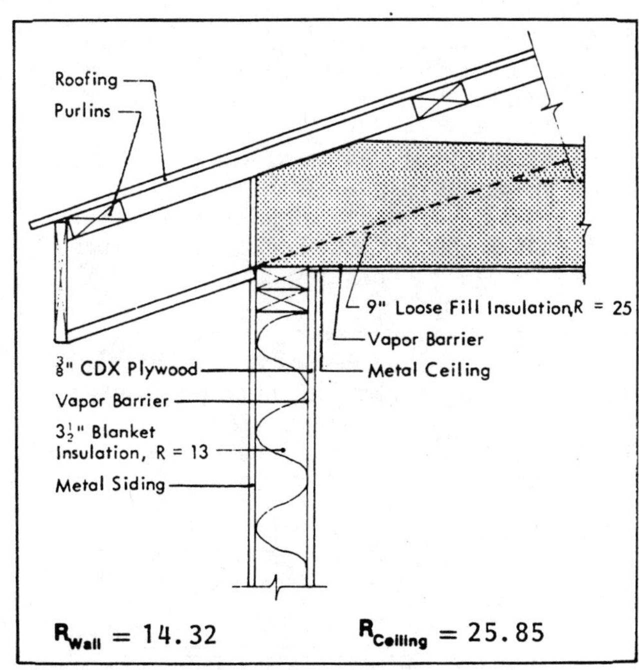

Fig 631-12. Example of supplementally heated building insulation.

Table 631-5 lists recommended minimum insulation levels for farm buildings. More insulation may be justified with increasing energy costs in supplementally heated buildings.

Table 631-5. Minimum insulation levels for animal buildings.
R-values are for building sections.

	Recommended minimum R-values					
	Open front		Modified open front		Environment controlled	
Winter degree days	Walls	Roof	Walls	Roof	Walls	Ceiling
2500 or less	—	6	6	14	14	22
2501-6000	—	6	6	14	14	25
6001 or more	—	6	12	25	20	33

Where to Use Insulation

Think of insulation as a blanket completely surrounding the living or housing area—all spaces that are heated in winter or cooled in summer. Fig 631-13 illustrates areas of a house that require insulation. Refer to the numbers in the house cross sections.

1. **Exterior walls.** Sections sometimes overlooked are the wall between living space and an unheated garage or storage room, dormer walls, and the portion of wall above the ceiling of an adjacent section of a split-level home. Pack insulation in narrow spaces between jambs and framing.
2. **Ceilings** with cold spaces above and dormer ceilings. An attic access panel can be insulated by stapling a piece of mineral wool blanket to its top.
3. **Knee walls,** when attic space is used as living quarters.

4. **Ceilings** above a finished attic, leaving open space above for ventilation.
5. Around the **perimeter** of a concrete slab on grade.
6. **Floors** above vented crawl spaces. When a crawl space is used as a plenum, insulate the crawl space walls instead of the floor above.
7. **Floors** over an unheated or open space, such as a garage or a porch. The cantilevered portion of a floor.
8. **Basement walls,** when below grade space is finished for living purposes. Mineral fiber sill sealer on top of the foundation provides an effective wind infiltration barrier.
9. Behind **band** or **header joists.**

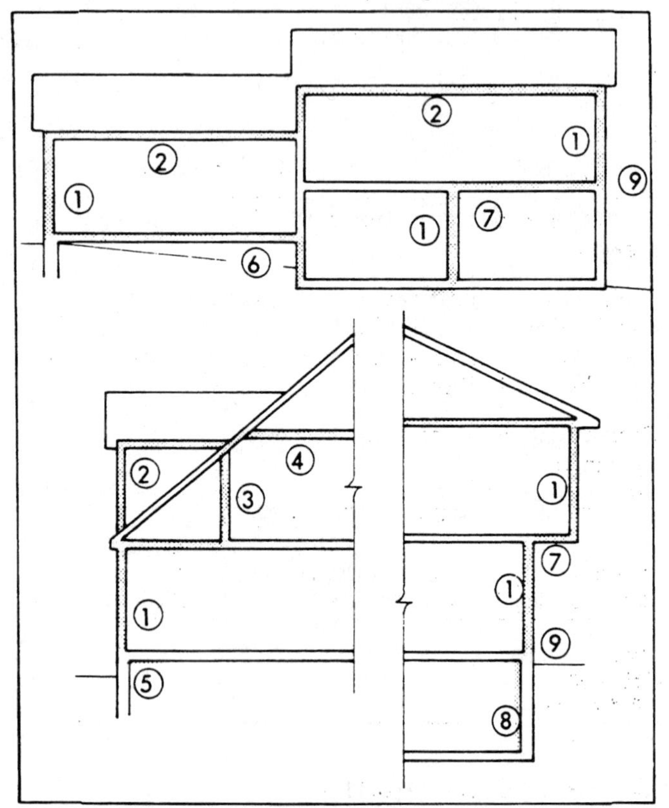

Fig 631-13. Residence areas requiring insulation.

In livestock housing the above areas also need to be insulated. In addition, be sure to insulate:

- Under metal roof surfaces in cold housing units to reduce radiation to animals in warm weather and to reduce moisture condensation.
- Under heated brooder areas where pigs or chickens are brooded.
- Heating ducts and heat pipes going through unheated areas.
- Floors of raised buildings with unheated space below them.

Installing Insulation

Fig 631-14 to 631-17 show common construction methods for insulated roofs, ceilings, walls, and foundations. Cracks around window and door frames, pipes, and wires result in cold spots, drafty areas, or condensation, and reduce ventilation inlet effectiveness. Caulk all cracks and joints.

Perimeter insulation reduces heat loss through the foundation and eliminates cold, wet floors. One way to insulate concrete foundations is by covering the foundation exterior below the siding to a minimum of 16″ below ground line. Use 2″ rigid insulation and protect it with an impact-, moisture-, and rodent-resistant covering.

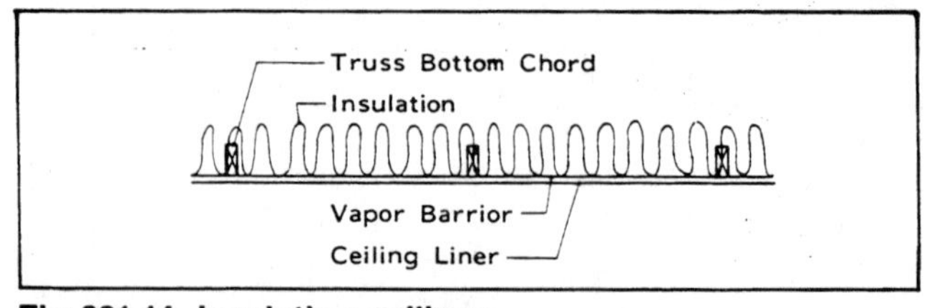

Fig 631-14. Insulating ceilings.
Loose-fill, batt, and blanket insulation. Recommended for insulating ceilings in environmentally controlled buildings.

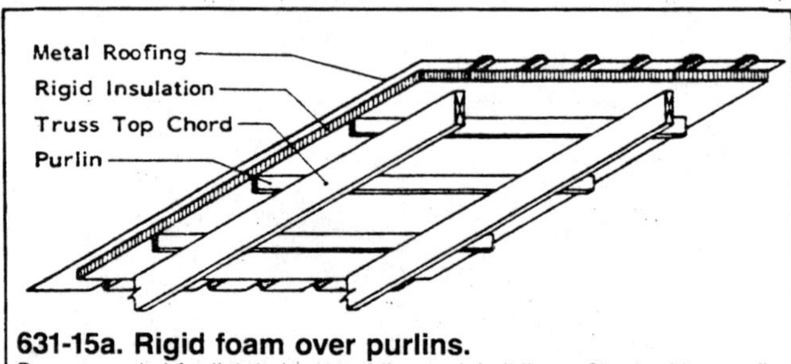

631-15a. Rigid foam over purlins.
Recommended for lightly-insulated livestock buildings. Check with your fire insurer before installing exposed insulation.

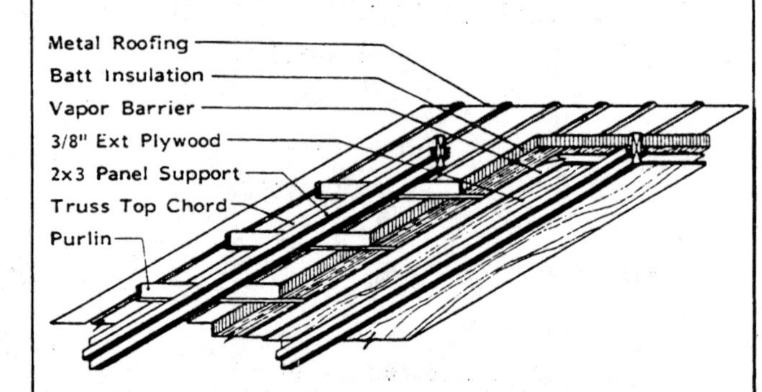

631-15b. Insulated roof panels between trusses.
Works well in modified open-front buildings. Fabricate on the ground and place between trusses. Apply roofing only after panels are nailed in place.

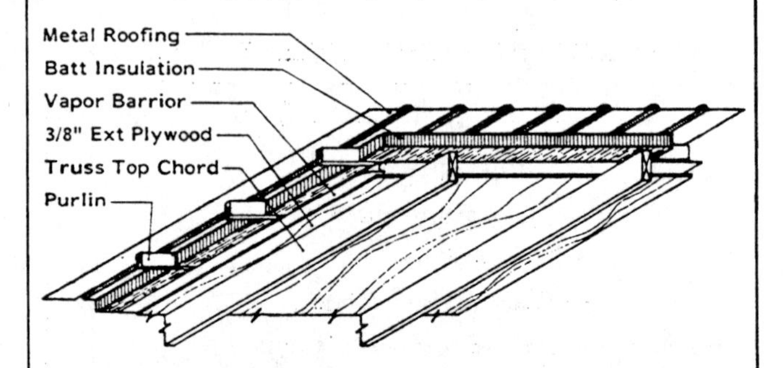

631-15c. Insulated roof panels over trusses.
Common in modified open-front buildings. Fabricate on the ground and lift into place—then apply roofing.

Fig 631-15. Insulating roofs.

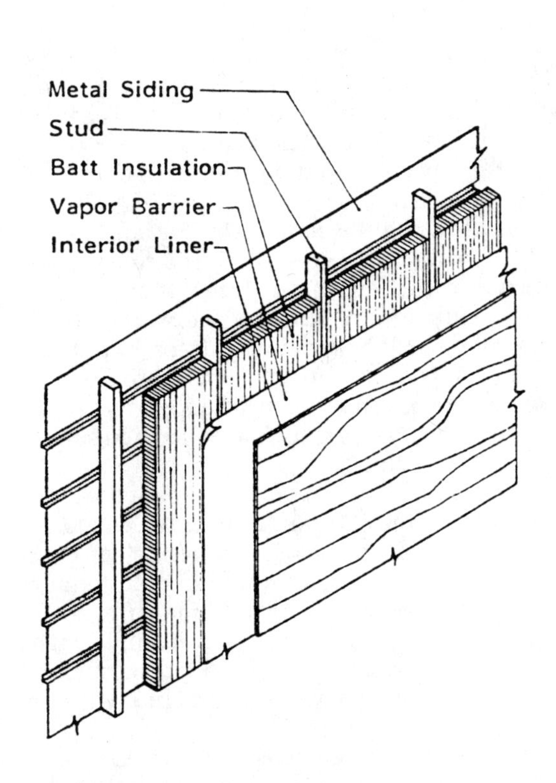

631-16a. Stud wall insulation.
(R = 12 if 2x4 studs, 20 if 2x6 studs). Common in supplementally heated livestock buildings.

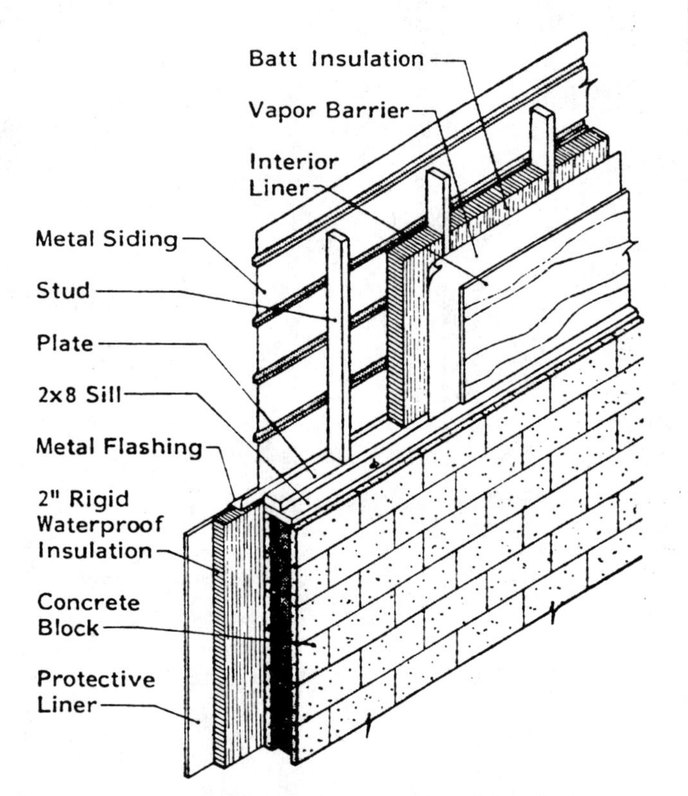

631-16b. Concrete block and stud wall insulation.
(Upper wall R = 12 if 2x4 studs, 20 if 2x6 studs. Lower wall R = 12 if insulation and standard blocks, 16 if insulation and lightweight blocks with cores filled). Concrete block wall (32" high) with stud wall above is popular for large swine. Although blocks provide a "pig-proof" wall, they lose much heat and "sweat" unless insulated. Providing adequate insulation levels is difficult with this wall.

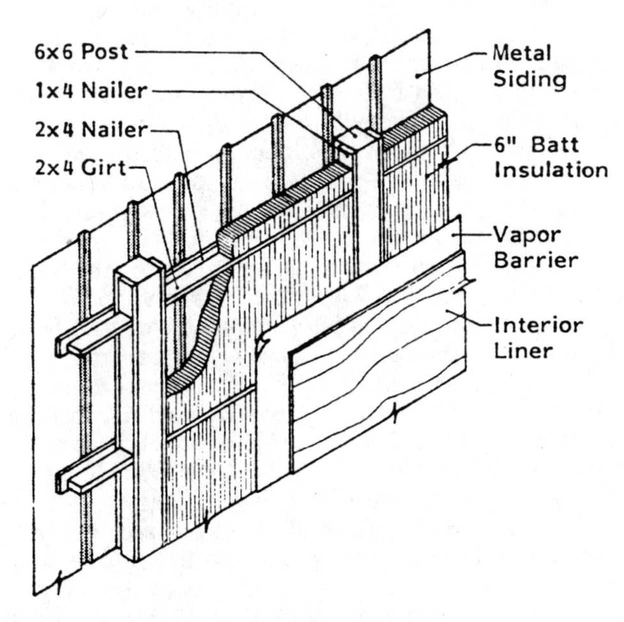

631-16c. Post wall with 6" batt insulation.
(R = 21)

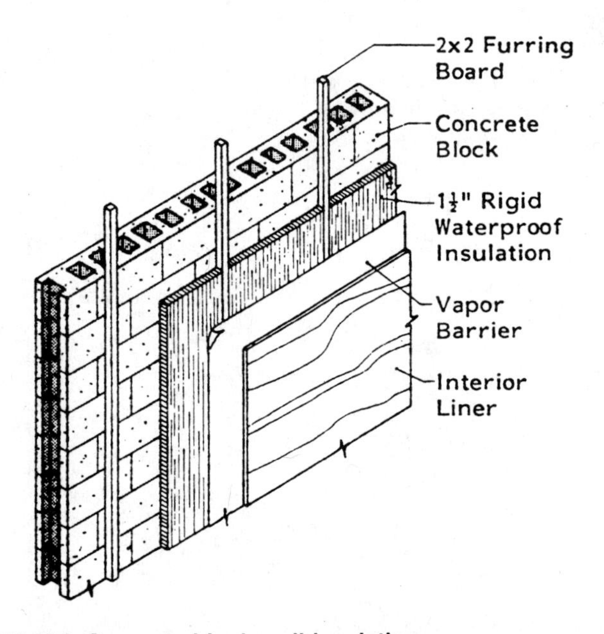

631-16d. Concrete block wall insulation.
(R = 10 if standard blocks, 14 if lightweight blocks with cores filled).

Fig 631-16. Insulating walls.

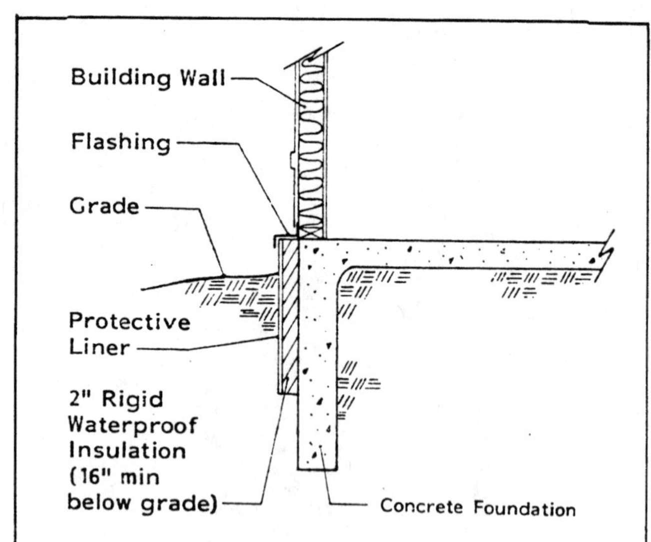

631-17a. Foundation perimeter insulation—outside.
Protect insulation from damage with a rigid, waterproof covering. High density fiberglass reinforced plastic or ¼" cement asbestos board are preferred. Tempered hardboard (¼") or ⅜" foundation grade plywood resist physical and moisture damage but are not rodentproof.

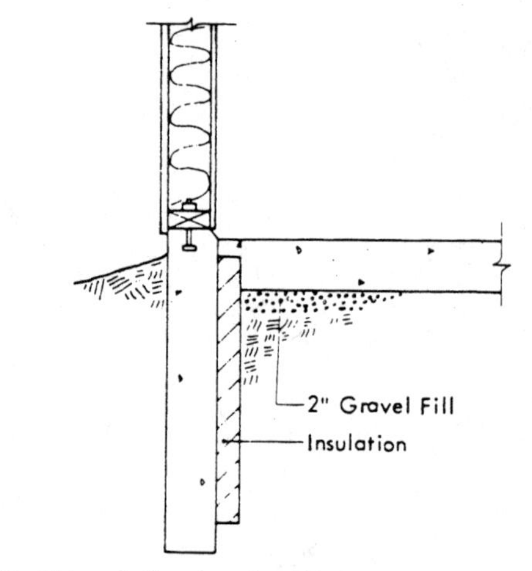

631-17b. Foundation perimeter insulation—inside.

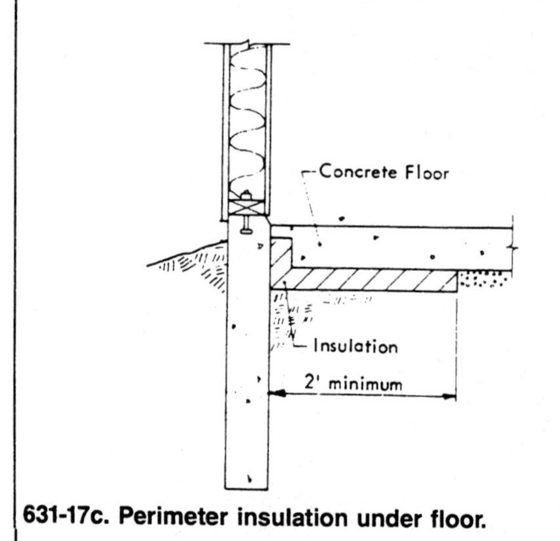

631-17c. Perimeter insulation under floor.

Fig 631-17. Insulating foundations.
Use waterproof insulation.

Maximizing Insulation Effectiveness

Moisture Protection

To prevent moisture problems, it helps to understand what causes them. All of the air in farm buildings contains some moisture in vapor form. Warm air can hold more water vapor than cold air. In fact, the vapor holding capacity about doubles with every 20 F increase in temperature.

Condensation on indoor building surfaces results from warm, moist inside air contacting a cold surface. The water vapor condenses into free water (just as it does on the outside of a glass of ice water). If the surface is cold enough, the water freezes, and the condensation appears as frost. As cold weather arrives, condensation appears first on surfaces that have low insulation values, such as the inside surface of a single-glazed window.

The amount of condensation depends on the amount of moisture in the air. Adequate ventilation to exhaust stale moist air, more frequent cleaning of livestock buildings, and good waterer maintenance help reduce the amount of vapor in indoor air.

Adequate insulation, including multiple-glazed windows and insulated doors, keeps indoor surfaces warm enough that indoor vapor is less likely to condense.

A more difficult and serious problem can result from water vapor getting into a building material or section. Water vapor on the warmer, more humid side of a building section tries to get to the other side. It exerts a force on the building section called vapor pressure (expressed in inches of mercury, in. Hg). If there is a difference in vapor pressure between two areas, water vapor moves from the high pressure area to the low pressure area until the pressures equalize.

If moisture movement through a wall section is not restricted, it can condense within the building section, if exposed to its dewpoint temperature. Condensation can saturate insulation, lowering its insulating ability, promote decay in structural members, and cause paint problems on exterior walls.

Prevent moisture problems in building sections by installing a vapor barrier which keeps water vapor from entering areas where it is likely to condense into free moisture. Install vapor barriers on the warm side of all insulated walls, ceilings, and roofs. Use vapor barriers underneath concrete floors and foundations to control soil moisture. Use waterproof rigid insulation if in contact with soil. Polystyrene and polyurethane boards have good vapor barrier ratings, but you need to seal the board joints to prevent moisture transfer through the joints. Refer to Fig 631-14 to 631-17 for proper vapor barrier locations.

The ability of a material to allow water vapor to pass through it is called **permeability** and is measured in units called **perms.** One perm equals one grain of water per hour per square foot per inch of mercury pressure difference.

Most building materials are highly permeable and **are not** good vapor barriers. Table 631-6 lists the perm ratings for several common construction materials and vapor barriers.

Table 631-6. Permeability of materials.
Based primarily on the *1981 ASHRAE Handbook of Fundamentals*. A vapor barrier should have a perm rating less than 1.0, and preferably less than 0.5.

Material	Perms*
Vapor barriers	
Aluminum foil, 1-mil	0.0
Polyethylene film, 6-mil	0.06
Kraft and asphalt laminated building paper	0.3
Two coats of aluminum paint (in varnish) on wood	0.3-0.5
Three coats exterior lead-oil base on wood	0.3-1.0
Three coats latex	5.5-11.0
Common building materials	
Expanded polyurethane, 1"	0.4-1.6
Extruded expanded polystyrene, 1"	1.2
Molded beads polystyrene, 1"	2.0-5.8
Tar felt building paper, 15-lb.	18.2
Structural insulating board, uncoated, ½"	50.0-90.0
Exterior plywood, ¼"	0.7
Interior plywood, ¼"	1.9
Tempered hardboard, ⅛"	5.0
Gypsum, ⅜"	50.0
Brick masonry, 4"	0.8
Poured concrete wall, 4"	0.8
Glazed tile masonry, 4"	0.12
Concrete block, 8"	2.4

*1 perm = 1 grain of water/hr/ft²/in. of mercury pressure difference.

One of the best vapor barrier materials for farm structures is polyethylene film. It is low cost, easily installed, and not affected by corrosive agents normally found in the farm building environment.

Doors and Windows

Use insulated and weatherstripped entry doors and locate them on the downwind side if possible. For upwind entries, a 4' air lock between outer and inner doors prevents cold wind from blowing directly into the building when you enter or leave. Be careful to not impede or restrict the flow of materials, feed, and animals, if an air lock is provided. To conserve floor area, consider making the air lock entrance part of an office, storage area, or wash room.

Windows or skylights provide little or no benefit in heated livestock buildings. The insulation values of windows (R=1-2.5) are well below an insulated wall (R=13-15) and increase heating requirements substantially.

During remodeling, consider replacing all windows with insulated removable panels or permanent wall sections. Hinged or removable panels can be opened for summer ventilation.

Birds and Rodents

To prevent rodent damage, cover exposed perimeter insulation with a protective liner and maintain a rodent bait program.

To prevent bird damage, construct buildings so birds cannot roost near the insulation and consider screening all vent openings. Use ½" hardware cloth for air intake vent openings and ¾" hardware cloth for air outlet vent openings. Screened vent **outlets** are very susceptible to freezing shut in cold climates, and you may need to knock ice off the screen regularly during cold spells.

An aluminum foil covering is not sufficient protection but may make the insulation less attractive to birds.

Fire Protection

The rate at which fire moves through a room depends on the interior lining material. Many plastic foam insulations have high flame spread rates. Urethane and styrene foam plastic insulations commonly used in farm buildings have extremely high flame spread rates. If foam plastic insulations are not protected suitably from potential fire, your insurance company may refuse to provide coverage on the structure. To reduce risk with these materials, protect them with fire-resistant coatings. Materials that provide satisfactory protection include:

- ½" thick cement plaster.
- Fire rated gypsum board (sheet rock). Do not use in high moisture environments such as animal housing.
- ¼" thick sprayed-on magnesium oxychloride (60 lb/ft³) or ½" of the lighter, foam material.
- Mineral asbestos board ⅛"-⅜" thick.
- Fire rated ½" thick exterior plywood.

Insulation Do's and Don'ts

1. **Do not underinsulate.** Insulation is an economical part of the construction in a new building.
2. **Do compare R-values of insulation alone when you shop.** Do not let values inflated by salesmen's claims or quotes on an installed basis affect your comparisons.
3. **Do compare installed costs per R-value.** One insulation may cost less than another for material, but it may cost much more to install than the other.
4. **Do a good job of installation.** For best results, follow manufacturer's directions to the letter. He knows his product and how to make it perform best. If it is done wrong, it adds cost every day you operate the facility.
5. **Do install a good vapor barrier.** Moisture is public enemy number one to insulating materials. It can be controlled with a properly installed vapor barrier.
6. **Do make sure insulation fits snugly.** Fully insulated walls and ceilings have the lowest heat loss. Cracks and voids reduce the insulating value and add substantially to air leakage into the structure.
7. **Do check with your insurance company before installing a foam plastic insulation.** They may refuse to insure it unless fire retardant treatment is incorporated.

Calculating Heat Loss

The rate of heat loss through each building component is proportional to its area and the difference between the inside and outside temperatures. The rate of heat flow is also determined by the R_T value of the building component; the higher the R_T value, the lower the rate of heat flow. The rate of heat loss from each building component, q, is given by:

Eq 631-1.

$$q = (A/R_T) \times (t_i - t_o)$$

q = rate of heat loss from the building component, Btu/hr
A = area of the building component, ft^2
R_T = total resistance to heat flow of the component, F-ft^2-hr/Btu
t_i = inside temperature, F
t_o = outside temperature, F

The floor perimeter is a special case and the R_T value in Table 631-3 is given per foot of length. In this case the area, A, is replaced by the length of the exterior wall.

To obtain the total heat loss from a building, the losses through each building component are simply added together. The following sample problem illustrates the procedure.

Sample Problem

Find the amount of heat loss that will occur in a 24'x36' building constructed as illustrated in Fig 631-18. The inside design temperature is 60 F and the outside design temperature is -10 F. The building has two 3'x7' doors, insulated with 1", 1 pcf of molded polystyrene.

Step I

List the length, width, wall height, and foundation height. Calculate the perimeter, frame wall area (excluding windows and doors), concrete wall area, ceiling area, window area, and door area. The resistance of the frame wall is 12.47 from Table 631-2. The R_T for a 6" concrete wall with 2" polystyrene insulation is 11.58 from Table 631-2. The R_T for the ceiling is 13.47 and for the doors is 7.99 from Table 631-3.

Step II

The heat losses, q, for the ceiling, walls, and perimeter are found by placing the appropriate values from Step I into the heat loss equation.

The R_T value of 2.22 in the perimeter equation assumes 2"x24" polystyrene perimeter insulation, Table 631-3.

The total heat loss, q_b, from the building is the sum of the ceiling, wall, perimeter, window, and door losses.

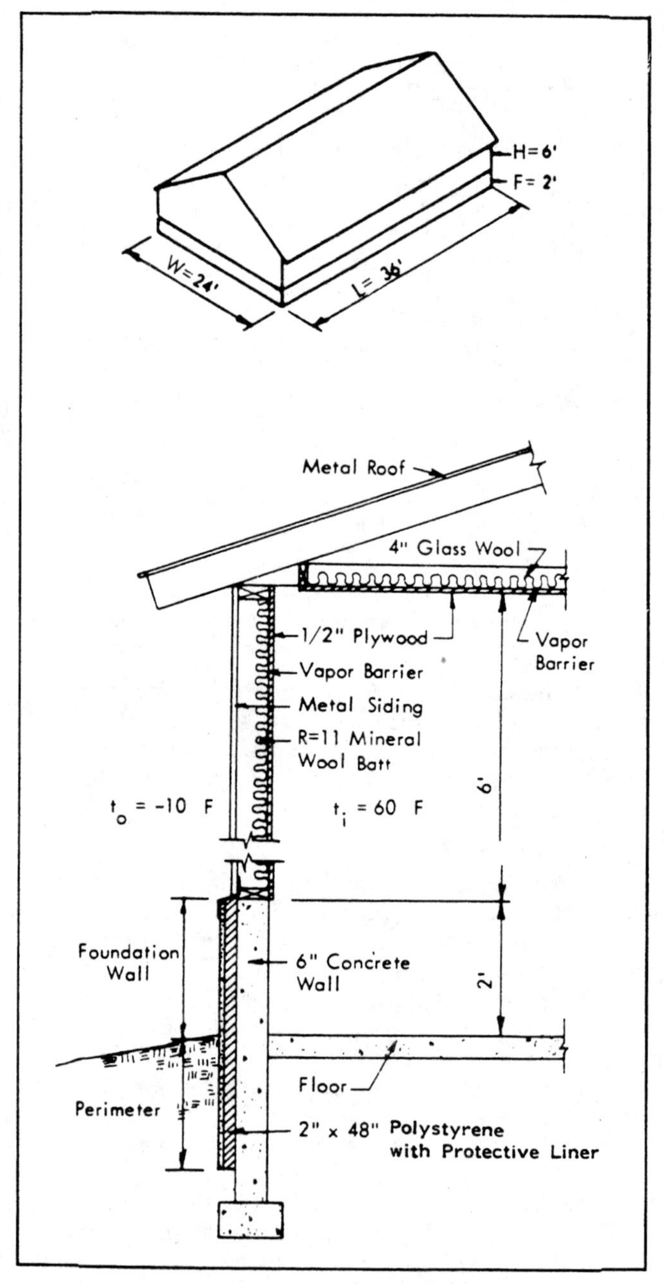

Fig 631-18. Sample problem building wall.

WORKSHEET—HEAT LOSS

Step I

Building dimensions	(ft)	Surface area	(ft^2)	R_T Values	
Length (L)	36	Ceiling area	864	Ceiling	13.47
Width (W)	24	Window area	0	Window	—
Frame wall height (H)	6	Door area	42	Door	7.99
Concrete wall height (F)	2	Frame wall area less		Frame wall	12.47
Perimeter	120	window & door area	678	Concrete wall	11.58
		Concrete wall area	240	Perimeter	2.22

Design temperatures (F)

t_o (outside temp) = -10 t_i (inside temp) = 60

Δt = 70

Step II

Heat loss from building, q_b

Ceiling $q_c = \dfrac{\Delta t \times \text{ceiling area}}{\text{ceiling } R_T}$ $q_c = \dfrac{70 \times 864}{13.47} = 4490$ Btu/hr

Windows $q_{wi} = \dfrac{\Delta t \times \text{window area}}{\text{window } R_T}$ $q_{wi} = 0$

Doors $q_d = \dfrac{\Delta t \times \text{door area}}{\text{door } R_T}$ $q_d = \dfrac{70 \times 42}{7.99} = 368$ Btu/hr

Frame walls $q_w = \dfrac{\Delta t \times \text{frame wall area}}{\text{frame wall } R_T}$ $q_w = \dfrac{70 \times 678}{12.47} = 3806$ Btu/hr

Concrete walls $q_f = \dfrac{\Delta t \times \text{concrete wall area}}{\text{concrete wall } R_T}$ $q_f = \dfrac{70 \times 240}{11.58} = 1451$ Btu/hr

Perimeter $q_p = \dfrac{\Delta t \times \text{perimeter}}{\text{perimeter } R_T}$ $q_p = \dfrac{70 \times 120}{2.22} = 3784$ Btu/hr

$q_b = q_c + q_{wi} + q_d + q_w + q_f + q_p = 13{,}899$ Btu/hr

---

Or:

Building heat loss, q_b can be expressed in terms of the inside-outside temperature difference, Δt:

Eq 631-2.

$q_b = A/R \times \Delta t$

A/R = sum of all (area/resistance) ratios of the building

Using the above sample problem:

Building (A/R) = ceiling (A/R) + frame wall (A/R) + concrete (A/R) + perimeter (A/R) + window (A/R) + door (A/R)
= 864/13.47 + 678/12.47 + 240/11.58 + 120/2.22 + 0 + 42/7.99
= 64.14 + 54.37 + 20.73 + 54.05 + 5.26
= 198.55 Btu/hr-F

Therefore, the heat loss from the building is:

q_b = 198.55 x Δt
= 198.55 x 70
= 13,899 Btu/hr

Effects of Changing Insulation Values

Note three important items in the example that can affect the quantity of heat loss:

1. Use enough insulation.

Suppose the insulation in the frame walls is decreased from R = 11 batt insulation to 25/32" insulating sheathing (see Table 631-2). This is a decrease in R_T value from 12.47 to 4.43. Recalculating the heat loss from the frame walls and building yields:

Frame wall:

q_w = Δt x frame wall area ÷ frame wall R_T
= 70 x 678 ÷ 4.43
= 10,713 Btu/hr

Therefore, the new building heat loss is:

q_b = 20,806 Btu/hr

This represents an increase in heat loss through the frame wall of 180%. The total building heat loss increases 50%.

2. Insulate all areas.

The use of insulation on concrete walls and around the building perimeter often went unnoticed until recent years. To illustrate the effectiveness of insulating the concrete wall and perimeter, assume the 2"x48" insulation covering the concrete wall and perimeter in Fig 631-18 is neglected. Recalculation of the heat loss from the concrete wall, perimeter, and building yields:

Concrete wall:

$R_T = 1.33$ (Table 631-1)

Perimeter:

$R_T = 1.23$ (Table 631-3)

Concrete wall:

$q_f = \Delta t \text{ x concrete wall area} \div \text{concrete wall } R_T$
$= 70 \text{ x } 240 \div 1.33$
$= 12{,}632 \text{ Btu/hr}$

Perimeter:

$q_p = \Delta t \text{ x perimeter} \div \text{perimeter } R_T$
$= 70 \text{ x } 120 \div 1.23$
$= 6{,}829 \text{ Btu/hr}$

Therefore, the new building heat loss is:

$q_b = 28{,}125 \text{ Btu/hr}$

The heat loss through the concrete wall increases 771%; and the perimeter, 80%. Not insulating the concrete wall and perimeter causes the total heat loss of the building to almost double.

3. Install windows only when necessary.

Single pane glass is poor insulation. If windows are required, use thermopane or windows with storms to cut heat loss. To illustrate heat loss through windows, assume there are eight in this building. The windows are single thickness having an area of 8.75 ft^2 (2′-6"x3′-6"). Recalculating the building heat loss yields:

Windows:

$q_{wi} = \Delta t \text{ x window area} \div \text{window } R_T$
$= 70 \text{ x } (8.75 \text{ x } 8) \div 0.91$
$= 5385 \text{ Btu/hr}$

Frame walls:

$q_w = \Delta t \text{ x (frame wall area} - \text{window area)} \div \text{frame wall } R_T$
$= 70 \text{ x } (678 - 70) \div 12.47$
$= 3413 \text{ Btu/hr}$

Therefore, the new building heat loss is:

$q_b = 18{,}891 \text{ Btu/hr}$

The heat loss through the eight windows is over 1½ times the heat loss of the four frame walls. The total building heat loss increases 36%.

WORKSHEET—HEAT LOSS

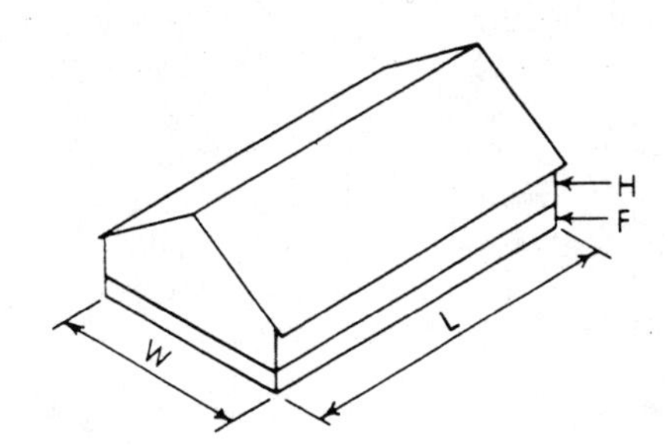

Step I

Building dimensions	(ft)
Length (L)	______
Width (W)	______
Frame wall height (H)	______
Concrete wall height (F)	______
Perimeter	______

R_T values	
Ceiling	______
Window	______
Door	______
Frame wall	______
Concrete wall	______
Perimeter	______

Surface area	(ft^2)
Ceiling area	______
Window area	______
Door area	______
Frame wall area less window & door area	______
Concrete wall area	______

Design temperatures, F

t_o (outside temp) = ______ , t_i (inside temp) = ______ ,
Δt = ______ .

Step II

Building heat loss, q_b

Ceiling $q_c = \dfrac{\Delta t \times \text{ceiling area}}{\text{ceiling } R_T}$ $\qquad q_c = \dfrac{(\quad) \times (\quad)}{(\quad)}$ = ______ Btu/hr

Windows $q_{wi} = \dfrac{\Delta t \times \text{window area}}{\text{window } R_T}$ $\qquad q_{wi} = \dfrac{(\quad) \times (\quad)}{(\quad)}$ = ______ Btu/hr

Doors $q_d = \dfrac{\Delta t \times \text{door area}}{\text{door } R_T}$ $\qquad q_d = \dfrac{(\quad) \times (\quad)}{(\quad)}$ = ______ Btu/hr

Frame walls $q_w = \dfrac{\Delta t \times \text{frame wall area}}{\text{frame wall } R_T}$ $\qquad q_w = \dfrac{(\quad) \times (\quad)}{(\quad)}$ = ______ Btu/hr

Concrete walls $q_f = \dfrac{\Delta t \times \text{concrete wall area}}{\text{concrete wall } R_T}$ $\qquad q_f = \dfrac{(\quad) \times (\quad)}{(\quad)}$ = ______ Btu/hr

Perimeter $q_p = \dfrac{\Delta t \times \text{perimeter}}{\text{perimeter } R_T}$ $\qquad q_p = \dfrac{(\quad) \times (\quad)}{(\quad)}$ = ______ Btu/hr

$q_b = q_c + q_w + q_d + q_w + q_f + q_p$ = ______ Btu/hr

632 ADVANCED HEAT AND VAPOR FLOW CALCULATIONS

Chapter 631 described methods to estimate heat loss from farm structures. This chapter presents refinements in these methods to account for variations in surface and air space heat transfer resistances. We also examine the effects of framing members and of ceiling, gable, and roof combinations on heat loss from buildings.

Condensation problems in farm structures are also discussed in more detail.

Surface Resistance

Heat transfer from a surface is affected by the surface temperature and emissivity, air velocity, and temperature difference between the surface and air. The surface resistances combine the effects of conduction, radiation, and convection. Values for surface resistances are in Table 632-1. Note the increase in surface resistance with a decrease in emissivity. The resistance for inside surfaces is usually for still air. For moving air, the R-value is independent of surface slope and direction of heat flow, because heat transfer by forced convection becomes predominant. For design, base outside surface resistances on a 15 mph wind for winter and a 7½ mph wind for summer.

Table 632-1. Surface heat transfer resistances.
R_1 hr-ft^2-F/Btu. Adapted from the *1981 ASHRAE Handbook of Fundamentals*.

Description of Surface	Direction of heat flow	Surface emissivity, ε 0.90	0.20	0.05
		- - - R - - -		
Horizontal (still air)	up	0.61	1.10	1.32
Horizontal (still air)	down	0.92	2.70	4.55
Sloping, 45° (still air)	up	0.62	1.14	1.37
Sloping, 45° (still air)	down	0.76	1.67	2.22
Vertical (still air)	horizontal	0.68	1.35	1.70
Any position (15 mph wind)	any	0.17		
Any position (7.5 mph wind)	any	0.25		

Air space resistance. In an air space, heat flows by convection and radiation and depends on:

- Mean surface temperature.
- Temperature difference of the two surfaces bounding the air space.
- Surface emissivities.
- Width, surface roughness, and orientation of the air space.

Thermal resistances, can be selected from Table 632-2 if you know the effective emissivity, E, of the two surfaces forming the air space. The effective emissivity is:

Eq 632-1.

$$E = 1/(1/\epsilon_1 + 1/\epsilon_2 - 1)$$

where ϵ_1 and ϵ_2 are the emissivities of the two surfaces forming the air space. Emissivities for typical building surfaces are given in Table 632-3.

Parallel Heat Flow Paths

In many installations, dissimilar components are arranged side-by-side which results in parallel heat flow paths with different thermal conductances. If there is no lateral heat flow between the paths, each path is considered to extend from inside to outside, and the transmittance is calculated for each path. The average transmittance is:

Eq 632-2.

$$U_{av} = a(U_a) + b(U_b) + \ldots + n(U_n)$$

where a, b, . . . , and n are the respective areas and U_a, U_b, . . . , and U_n are the respective coefficients of transmittance of the different components.

Adjustment for Framing Members

To adjust for parallel heat flow through wall sections with framing and insulated areas:

Eq 632-3.

$$U_{av} = U_s \times S/100 + U_i \times (1 - S/100)$$

U_{av} = average U-value for building section.
U_s = U-value for area backed by framing members.
U_i = U-value for area between framing members.
S = percentage of area backed by framing members.

The percentage of wall areas backed by framing members (studs, joists, plates, furring, headers, sills, etc.) varies widely with construction type. For buildings having 16" or 24" o.c. framing, typical values for S are 20% and 15%, respectively.

Combined Ceiling, Gable, and Roof Resistances

When estimating heat loss through the ceiling of structures having unheated, closed attic spaces, include the heat transfer characteristics of the roof and gable construction. Disregard the effect of the attic as an air space because convection currents make its thermal resistance very low; the situation is analogous to the simple electrical network shown in Fig 632-1.

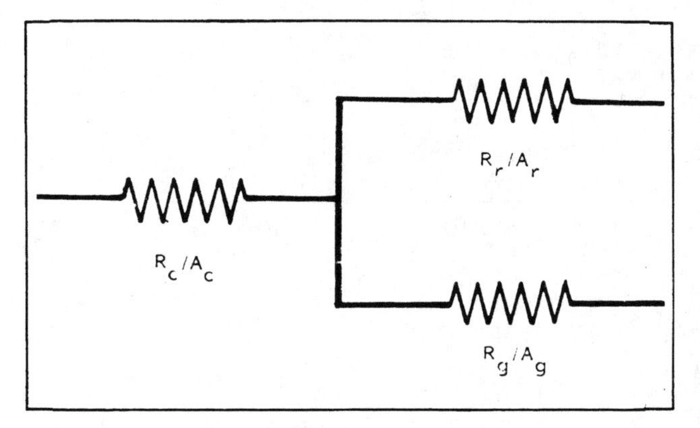

Fig 632-1. Electric analog.

The subscripts c, r, and g refer to the ceiling, roof, and gable, respectively. The total resistance of the network is:

Eq 632-4.

$$R_T/A = R_c/A_c + (R_r \times R_g) \div (R_r \times A_g + R_g \times A_r)$$

where A_c, A_r, and A_g are the areas of the ceiling, roof, and gable, respectively. The area, A, can be set equal to any of the three areas. Letting $A = A_c$ and multiplying:

Eq 632-5.

$$R_T = R_c + (R_r \times R_g) \div (m \times R_r + n \times R_g)$$

where $m = A_g/A_c$ and $n = A_r/A_c$ and R_T is the combined overall heat transmission coefficient for the ceiling, roof, and gable per unit ceiling area.

Condensation

Condensation on windows is a nuisance, but on wall or ceiling surfaces or in attic spaces it can be damaging to the structure or its contents. Condensation within walls can cause paint blistering, severe structural damage, and can greatly reduce the value of some insulation materials.

In farm buildings, condensation problems can be separated into two categories—condensation on the wall surface and condensation within a wall. To examine these problems we will look at determining the temperature gradient through a wall, the problem of surface condensation, determining water vapor flow through a wall, and the problem of condensation within a wall.

Estimating Temperature Within Walls

The flow of heat through a nonhomogeneous wall of unit area is:

Eq 632-6.

$$q = (t_i - t_o) \div R_T$$

Fig 632-2 illustrates a wall of two dissimilar solids with total resistance = R_T. The overall resistance from the indoor air to a point X in the wall is R_x; the temperature at that point is t_x.

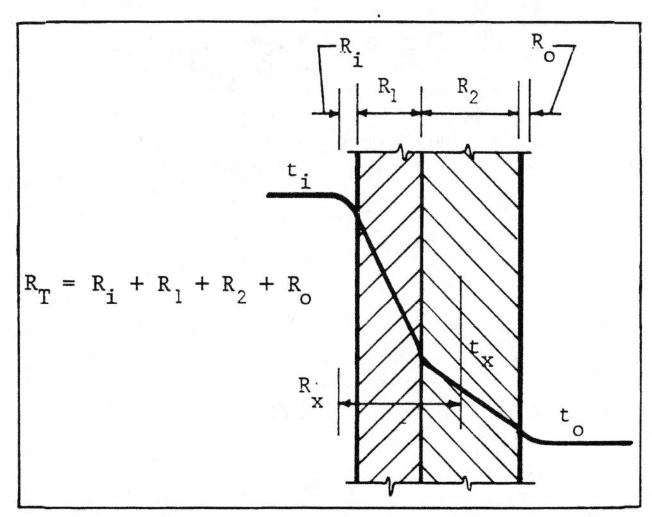

Fig 632-2. Temperature gradient in a nonhomogeneous wall.

The **rate** of heat flow through the entire thickness of the wall is the same as that through the wall to point X. The equations for heat flow through a unit wall area can be set equal to each other:

Eq 632-7.

$$q = (t_i - t_o) \div R_T = q_x = (t_i - t_x) \div R_x$$

Solving for t_x, the temperature at any point X is:

Eq 632-8.

$$t_x = t_i - (R_x \div R_T) \times (t_i - t_o)$$

Surface condensation. Condensation frequently occurs on inside wall surfaces. Estimate the surface temperature with Eq 632-8 by substituting R_i (indoor surface resistance) for R_x.

The indoor wall surface of a heated room is at a lower temperature than the room air. Condensation occurs whenever the wall surface temperature is less than the dewpoint temperature of the room air.

Fig 632-3 illustrates the temperature gradient in an uninsulated wall. Eq 632-8 can be used to find the indoor surface temperature, t_{is}:

$$t_{is} = 60 - (0.68 \div 2.67) \times (60 - 12) = 48\text{ F}$$

The ventilation-range psychrometric chart in Chapter 602 shows that condensation occurs if the indoor relative humidity is greater than 64%, Fig 632-4.

To prevent surface condensation or "sweating" on an interior wall or surface, the indoor relative humidity can be lowered or the surface temperature can be increased. To decrease the indoor relative humidity, reduce the moisture production or increase the ventilating rate. The indoor surface temperature can be increased by increasing R_T or decreasing R_i in Eq 632-8.

If 3.5" of insulation (R=11) is added to the wall section in Fig 632-3, the total R_T increases to 12.77 hr-ft^2-F/Btu and the wall surface temperature is:

$$t_{is} = 60 - (0.68 \div 12.77) \times (60 - 12) = 57.4\text{ F}$$

Table 632-2. Thermal resistance, R, for plane air spaces.
Adapted from the *1981 ASHRAE Handbook of Fundamentals.*

Air space orientation	Heat flow direction	Width in.	Mean temp F	Temp diff. F	Effective emmissivity, E 0.82	0.50	0.20	0.05
					- - - - - - - - -	- R -	- - - - - - - -	
Horiz.	Up	0.75	50	10	0.87	1.16	1.70	2.21
			0	20	0.93	1.16	1.52	1.79
			0	10	1.02	1.31	1.78	2.16
		3.5	50	10	0.93	1.28	1.95	2.66
			0	20	1.03	1.32	1.79	2.18
			0	10	1.12	1.47	2.07	2.62
	Down	0.75	50	10	1.02	1.45	2.41	3.59
			0	10	1.31	1.82	2.87	4.02
		1.50	50	10	1.15	1.73	3.27	5.90
			0	10	1.51	2.22	4.00	6.66
		3.5	50	10	1.24	1.93	4.09	9.27
			0	10	1.64	2.52	5.08	10.32
Sloping 45°	Up	0.75	50	10	0.94	1.29	2.00	2.75
			0	20	1.00	1.28	1.72	2.07
			0	10	1.12	1.47	2.08	2.62
		3.5	50	10	0.96	1.34	2.10	2.95
			0	20	1.06	1.38	1.90	2.35
			0	10	1.16	1.54	2.23	2.87
	Down	0.75	50	10	1.02	1.45	2.40	3.57
			0	20	1.26	1.72	2.63	3.57
			0	10	1.30	1.80	2.81	3.91
		3.5	50	10	1.08	1.57	2.73	4.36
			0	20	1.27	1.74	2.66	3.63
			0	10	1.34	1.88	3.02	4.32
Vert.	Horiz.	0.75	50	10	1.01	1.43	2.35	3.46
			0	20	1.18	1.58	2.32	3.02
			0	10	1.26	1.73	2.64	3.59
		3.5	50	10	1.01	1.42	2.32	3.40
			0	20	1.14	1.51	2.17	2.78
			0	10	1.23	1.67	2.50	3.33

Table 632-3. Surface and air space emissivities.
For various surfaces and effective emissivities of air spaces.
Adapted from the *1981 ASHRAE Handbook of Fundamentals.*

			Effective emissivity E, of air space	
Surface	Reflec-tivity %	Emis-sivity ε	1 surface with ε and other 0.90	2 surfaces with ε
Al foil, bright	92 to 97	0.05	0.05	0.03
Al sheet	80 to 95	0.12	0.12	0.06
Al coated	75 to 84	0.20	0.20	0.11
Steel, galv	70 to 80	0.25	0.24	0.15
Al paint	30 to 70	0.50	0.47	0.35
Building materials*	5 to 15	0.90	0.82	0.82

*Wood, paper, glass masonry, nonmetallic paints

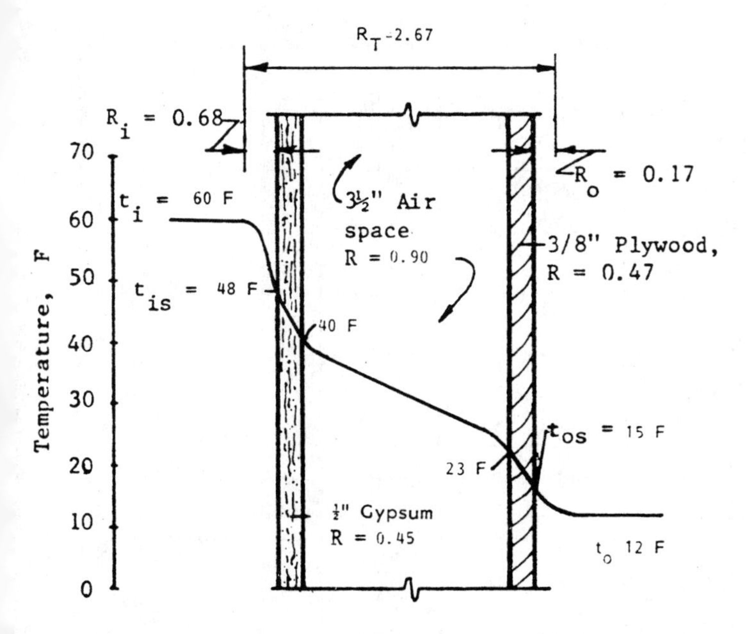

Fig 632-3. Temperature gradient in an uninsulated wall.

The added insulation increases surface temperature from 48 F to 57.4 F. This prevents surface condensation as long as the indoor relative humidity is below 91%.

Another method to increase wall surface temperature is to decrease the boundary layer resistance. By increasing air velocity over the surface to about 15 mph with a fan, R_i can be reduced from 0.68 to 0.17. If this is done on the wall section shown in Fig 632-3, R_i equals 0.17, but R_T is reduced to 2.16 hr-ft²-F/Btu. The resulting surface temperature, t_{is}, is:

$$t_{is} = 60 - (0.17 \div 2.16) \times (60 - 12) = 56.4\,F$$

The surface temperature is increased from 48 F to 56.4 F and condensation does not occur as long as the indoor relative humidity stays below 88%.

Water vapor flow through walls. Water vapor flow is analagous to heat flow. The rate of flow is a function of the vapor pressure difference and the vapor transmission characteristics of the wall.

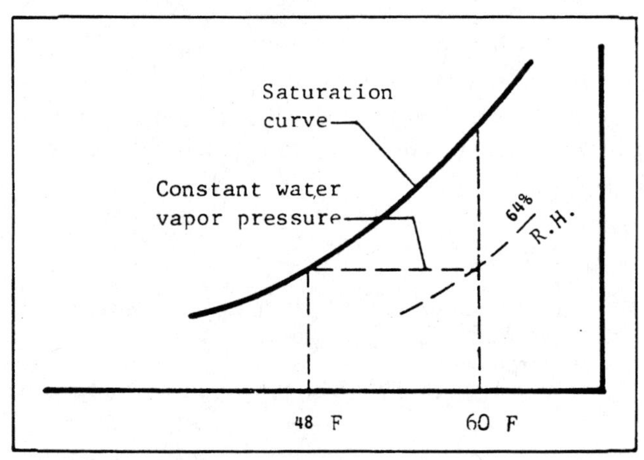

Fig 632-4. Condensation conditions on a psychrometric chart.

When temperature differs on opposite sides of a wall, the warm side usually has the higher vapor pressure. If the wall is vapor-permeable, vapor diffuses in the direction of the lower pressure, and the lower temperature may cause it to condense.

Water vapor transfer through materials. The transfer rate of water vapor through materials is based on Fick's law:

Eq 632-9.

$$m = -\mu \times A \times (dp/dx)$$

m = vapor transfer rate, grains/hr (7000 grains = 1 lb)
μ = water vapor permeability, grains-in./hr-ft-in. Hg
A = cross-sectional area normal to the direction of vapor flow, ft²
dp = vapor pressure gradient in the direction of flow, in. Hg
dx = distance of flow, in.

The transfer of water vapor through materials is complex and not well understood. The permeability, μ, can vary. It is dependent primarily on relative humidity (RH), and moderately on temperature (often neglected).

In the simple case of steady-state vapor transfer through a plane wall of homogeneous material subjected to constant vapor pressure, the vapor transfer rate can be obtained by separating variables and integrating Eq 632-9.

Eq 632-10.

$$m = \mu \times A \times (p_1 - p_2) \div L$$

Where p_1 and p_2 are the vapor pressures at each end of the flow path, L is the length of the flow path (material thickness), and the mean value of the permeability, μ', is:

$$\mu' = 1/(p_2 - p_1)\int_{p_1}^{p_2} dp$$ **Eq 632-11.**

Table 632-4 lists average water vapor permeabilities of some common building materials tested at the RH differences shown in the last column. Note that vapor permeability increases with relative humidity, as in Fig 632-5. It increases only moderately at low RH, but at an increasing rate at higher levels.

Water vapor permeabilities are determined with the wet-cup or dry-cup methods. In these tests the specimen is sealed over the top of a cup containing desiccant or water, placed in a controlled atmosphere, and weighed periodically. The steady rate of weight gain or loss is the water vapor transfer. When the cup contains a desiccant, the procedure is called the **dry-cup method.** When it contains water, it is called the **wet-cup method.** Usually, the surrounding atmosphere is held at 50% RH, thus providing about the same vapor pressure difference for either method. The results obtained by the two methods for the same specimen are likely to be very different, with the wet-cup method producing the higher values.

Table 632-4. Water vapor permeability.
For some common building materials.
k = grains/hr-ft²-in. Hg
D = hr-ft²-in. Hg/grain

	Thick. in.	Perm. K	Resistance D=1/K	Relative humidity diff. %
Construction materials				
Concrete	1	3.20	0.31	100-50
Brick masonry	4	0.80	1.25	
Concrete block	8	2.40	0.42	
Asbestos-cement board	0.12	8	0.13	
Gypsum wallboard	3/8	50	0.02	
Structural insulation board	1	50	0.02	
Hardboard	1/8	5	0.20	
Wood, sugar pine	1	5.4	1.85	
Ext plywood	1/4	0.7	1.43	
Int plywood	1/4	1.9	0.53	
Thermal insulation				
Still air	1	120	0.01	
Mineral wool	1	116	0.01	100-0
Polyurethane, blown	1	1.6	0.63	50-0
Polystyrene, extruded	1	1.2	0.83	50-0
Polystyrene, bead	1	5.8	0.17	50-0
Plastic and metal foils & films				
Aluminum foil, 1 mil		0.0		50-0
Polyethylene, 2 mil		0.16	6.25	50-0
PVC, plasticized, 4 mil		1.4	0.71	50-0
Building papers, felts, roofing papers				
Duplex sheet, asphalt laminated, aluminum foil, 1 side, 43 lb		0.18	5.56	100-0
Saturated and coated rolled roofing, 326 lb		0.24	4.17	100-0
15 lb asphalt felt		5.6	0.18	100-0
Liquid-applied coating materials				
Paint--2 coats				
Enamel on smooth plaster		1.5	0.67	
Primer & sealer on interior insulation board		2.1	0.48	
Primer & flat oil paint on plaster		3.0	0.33	
Paint--3 coats				
Exterior oil base paint on wood siding		1.0	1.0	50-0
Latex paint, 2 oz/sq ft		11	0.09	50-0
Latex paint, 4 oz/sq ft		5.5	0.18	50-0

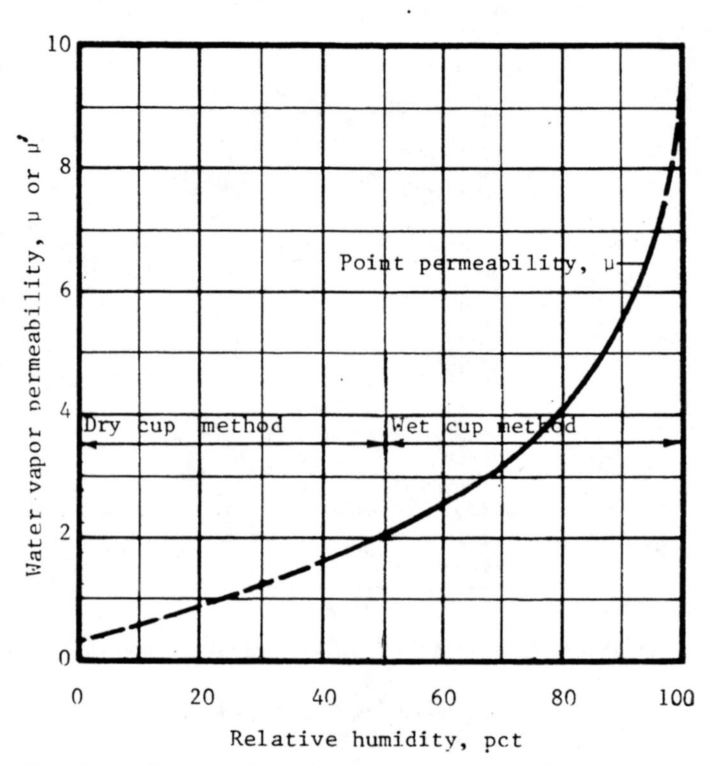

Fig 632-5. Water vapor permeability for plywood.

Water vapor transfer through walls. The transfer rate of water vapor, m, through a material is analogous to the transfer of heat and can be obtained by rewriting Eq 632-10:

Eq 632-12.

$$m = K \times A \times (p_1 - p_2)$$

or

$$m = A \times (p_1 - p_2) \div D$$

- m = rate of water vapor transmitted, grains/hr
- A = area normal to flow path, ft²
- K = $\mu' \div L$, permeance, grains/hr-ft-in. Hg
- p_1 = vapor pressure at beginning of flow path, in. Hg
- p_2 = vapor pressure at end of flow path, in. Hg
- D = 1/K, water vapor resistance, hr-ft²-(in. Hg)/grains

The overall vapor resistance (which is similar to the overall heat transfer resistance) in a nonhomogeneous wall is:

Eq 632-13.

$$D_T = D_1 + D_2 + \ldots + D_n$$

where, D_T = the overall wall vapor transmission resistance, and $D_1, D_2, \ldots D_n$ = the vapor resistance for each wall component.

The vapor transfer rate, m, through the whole wall is:

Eq 632-14.

$$m = A \times (p_i - p_o) \div D_T$$

where p_i and p_o are the vapor pressures on the inside and outside of the wall respectively.

Vapor pressure gradient. Consider two cases when estimating the vapor pressure gradient in a wall:

- Vapor flow through the wall without condensation.
- Vapor flow with condensation at a point within the wall.

Vapor flow without condensation. Vapor pressure at any point in a wall can be estimated by an equation similar to that for wall temperatures. It assumes an equal flow rate at all points in the wall. Vapor flow through the full thickness of the wall equals that through any point, X, within the wall. By analogy to Eq 632-8:

Eq 632-15.

$$p_x = p_i - (D_x \div D_T) \times (p_i - p_o)$$

p_x = vapor pressure at the point X
p_i = inside vapor pressure
p_o = outside vapor pressure
D_T = overall wall vapor resistance, $D_T = 1/K_T$
D_x = the wall resistance from p_i to point X, $D_x = 1/K_x$.

Example 1:

A residence wall cross section is illustrated in Fig 632-6. Assume the indoor air temperature is 70 F at 30% RH. Outdoor air is at 10 F and at 80% RH. What is the vapor pressure and temperature at the wall side of the outer sheathing, plane X, in Fig 632-6?

From Chapter 631 and Table 632-4, the thermal and vapor resistances for each component of the wall are in Table 632-5.

Table 632-5. Thermal and vapor resistances for example wall.

Component	Thermal resistance (R)	Vapor resistance (D)
Indoor surface	0.68	—
Gypsum sheathing, ½", with enamel paint	0.45	0.67
Air space, 3.5"	0.90	—
Structural insulating sheathing, 25⁄32":	2.06	0.02
Wood lapped siding, ½"x8" with 3 coats exterior oil base paint	0.81	1.00
Outdoor surface	0.17	—
TOTALS	5.07	1.69

In Table 632-6 the saturation vapor pressure at 70 F is 0.363 lb/in², and at 10 F it is 0.031. From the definition of relative humidity the vapor pressures p_i and p_o are:

Indoor vapor pressure = p_i = 0.30 x 0.363 = 0.109 lb/in²

Outdoor vapor pressure = p_o = 0.80 x 0.031 = 0.025 lb/in²

Solving for p_x by substituting in Eq 632-15:

$$p_x = 0.109 - (0.67 \div 1.69) \times (0.109 - 0.025) = 0.076 \text{ lb/in}^2$$

The vapor pressure gradient can be obtained by calculating p_x at each interface in the wall and then plotted as the line $p_i \ldots p_o$ in Fig 632-6.

Table 632-6 shows 0.076 lb/in² is the saturation vapor pressure for 28 F. Moisture will condense at plane X if the temperature is 28 F or less.

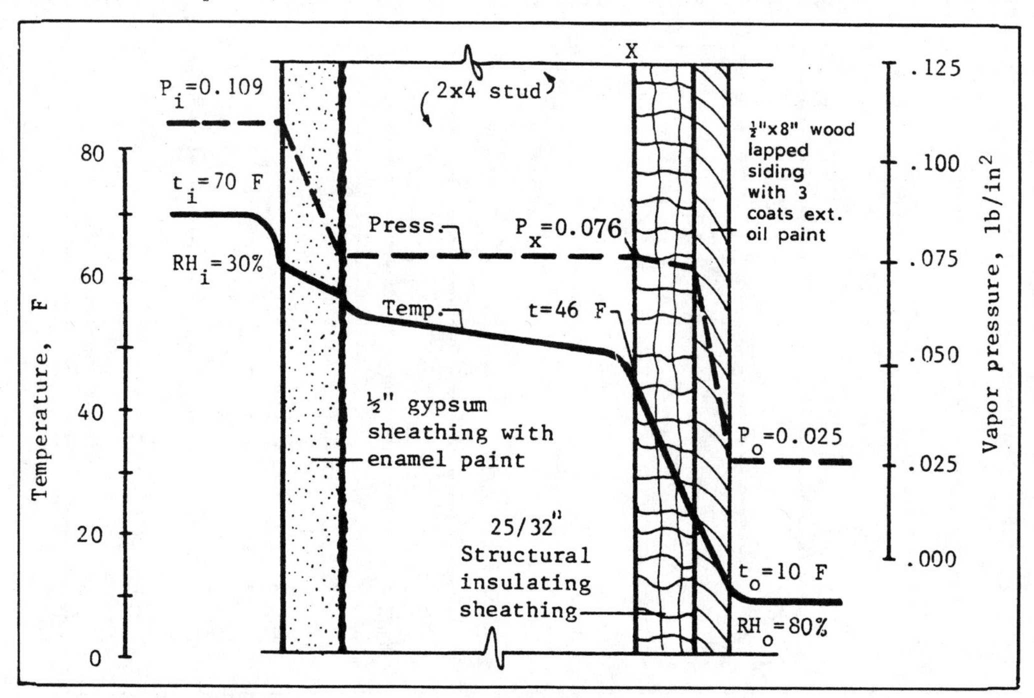

Fig 632-6. Temperature and vapor pressure gradients.
In a wall without fill insulation.

Table 632-6. Saturation vapor pressures.

Temp. F	Saturation vapor pressure lb/in^2	Temp. F	Saturation vapor pressure lb/in^2
-20	0.006	60	0.256
-15	0.008	65	0.306
-10	0.011	70	0.363
- 5	0.014	75	0.430
0	0.018	80	0.507
5	0.024	85	0.596
10	0.031	90	0.698
15	0.040	95	0.816
20	0.050	100	0.950
25	0.064	105	1.102
30	0.081	110	1.275
35	0.100	115	1.472
40	0.122	120	1.693
45	0.148		
50	0.178		
55	0.214		

The temperature at plane X can be calculated using Eq 632-8. The total thermal resistance of the wall in Fig 632-6 is about 5.07, and the resistance from the inside to plane X is 2.03. Because the temperature of the sheathing at plane X is calculated as 46 F, condensation will not occur.

If the temperature at plane X in Fig 632-6 had been less than 28 F, the vapor pressure would not be 0.076 lb/in^2. It would be the saturation vapor pressure for the wall temperature. If condensation occurred at X, Eq 632-15 would be valid only for the portion of the wall from p_i to p_x, and only if the saturation pressure for the sheathing surface temperature were substituted for p_x.

Vapor flow with condensation within the wall. The vapor pressure in the wall above was calculated assuming no condensation. Temperature and vapor pressure cannot always be treated independently.

Filling the air space in the wall in Fig 632-6 with insulation as shown in Fig 632-7 alters the vapor pressure-temperature relationship at the sheathing surface.

Example 2:

Calculate the vapor pressure and temperature at plane X for the wall in Fig 632-7.

The 3½" of fiberglass insulation has a thermal resistance of R = 11, and from Table 632-4 a vapor resistance of 3.5 x 0.01 = 0.04. The values for heat and vapor resistance for the wall are now:

$$R_T = 15.17: R_x = 12.13\text{—heat resistance}$$
$$D_T = 1.73: D_x = 0.71\text{—moisture resistance}$$

Substituting in Eq 632-8, the temperature at plane X is:

$$t_x = 70 - (12.13 \div 15.17) \times (70 - 10) = 22\text{ F.}$$

Table 632-6 gives the saturation vapor pressure for 22 F as 0.056 lb/in^2. A vapor pressure at plane X higher than 0.056 lb/in^2 causes condensation.

Using Eq 632-15, the vapor pressure at the sheathing surface, if estimated independently of the sheathing temperature, is:

$$p_x = 0.109 - (0.71 \div 1.73) \times (0.109 - 0.025)$$
$$p_x = 0.075\text{ lb/in}^2$$

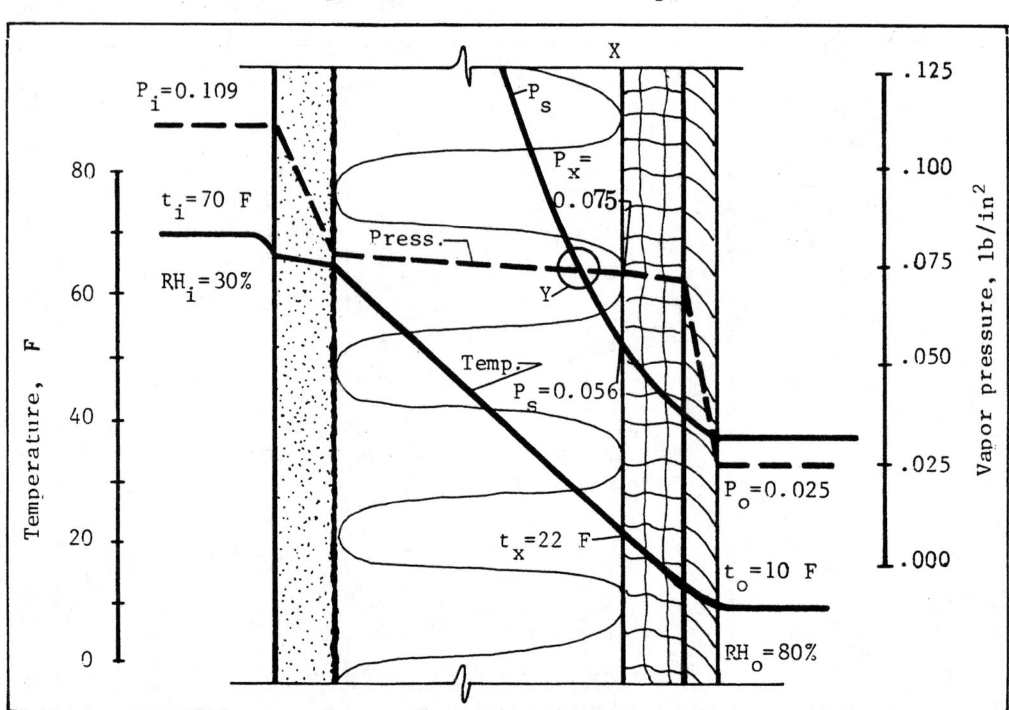

Fig 632-7. Temperature and vapor pressure gradients.
In a wall with fiberglass insulation.

The computed vapor pressure, p_x = 0.075 lb/in^2, cannot exist because it is greater than the saturation vapor pressure for 22 F, 0.056 lb/in^2. Condensation will occur within the wall.

The relationships between temperature and vapor pressure for this example are illustrated in Fig 632-7. The temperature gradient is the line t_i . . t_o; the line p_s represents saturation vapor pressures. The line p_i. . p_o shows vapor pressures calculated without regard to temperature gradient, but including vapor resistance of each component of the wall.

These examples show why insulating an older home with no vapor barrier on the inside wall surface can introduce moisture problems.

Based on this analysis, several recommendations can be made to prevent moisture problems within walls:

1. Maintain a low indoor relative humidity by reducing moisture production or by increasing ventilation. This will lower the vapor pressure at all points within the wall.
2. Install or apply an impermeable membrane (vapor barrier) to the interior wall surface. This will lower the vapor pressure behind the barrier where wall temperatures are coldest.
3. The resistance to vapor flow at the exterior wall surface should be as low as possible. This will permit any vapor within the wall to escape readily and maintain lower vapor pressures in the cold portion of the wall. Note that an impermeable exterior paint film increases possible moisture problems.
4. Do not install insulation, which decreases exterior wall temperatures, without installing an adequate vapor barrier on the inside surface of the wall.

633 HEAT AND MOISTURE BALANCES

Ventilation is a process to control temperature, humidity and odors. As air moves through a building its temperature may rise or fall, and the amount of water it carries (humidity ratio, W) may increase or decrease.

During the winter, a livestock shelter tends to be cold and damp; during the summer it tends to be hot and humid. With controlled ventilation designed to maintain desirable heat and moisture balances, a near-optimum environment can be maintained.

An energy balance relates sensible heat gains and heat losses to ventilating rate. A moisture balance relates latent heat sources and losses to ventilating rate. Fig 633-1 illustrates an overall energy and moisture balance between outside air, conditions in the building, and exhausted ventilating air.

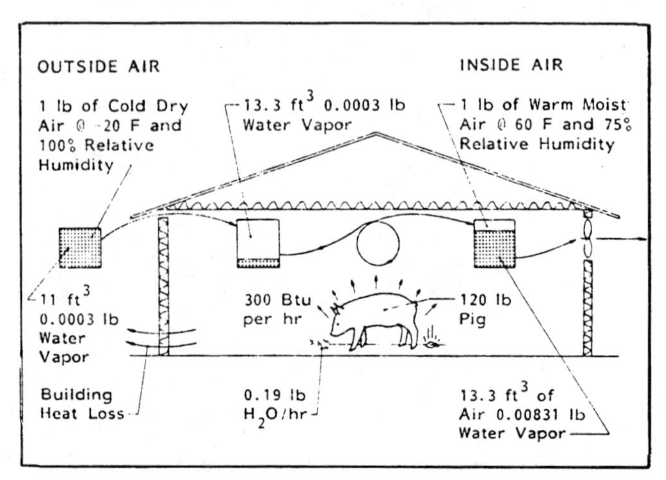

Fig 633-1. Winter ventilation in a controlled environment.

In the following discussion, most comments refer to winter conditions in a livestock building.

Sensible Heat Balance

The ventilation design procedure assumes an inside condition within the comfort zone of the animals, an outdoor condition within the anticipated weather range, and heat and moisture production and losses within the ventilated space. Heat sources include, Btu/hr:

q_s = sensible heat gain from animals
q_m = mechanical heat from lights, motors, etc.
q_{sr} = supplemental heat from furnaces or lamps

Heat losses include, Btu/hr:

q_b = building heat loss through doors, walls, etc.
q_v = sensible heat lost in ventilating air
q_e = sensible heat used to evaporate water

Sensible heat raises temperatures. If the building temperature is to remain constant, sensible heat produced in the building must be removed at the rate it is produced. If too much is removed, temperatures drop; if not enough is removed, temperatures rise.

A sensible heat balance in Btu/hr is:

Eq 633-1.

$$q_s + q_m + q_{sr} = q_b + q_v + q_e$$

Often q_m is neglected because it is relatively small. Most of the values for sensible and latent heat production, q_s and q_L, given in this book account for the fact that some of the sensible heat produced by the animals is used to evaporate water within the structure. Consequently, the term q_e is usually neglected.

This leaves either the supplemental heat, q_{sr}, or the amount of heat that must be removed in the ventilating air, q_v, as the unknown.

For a given building and animal population, the ventilating rate is assumed and the needed furnace size is calculated. Or, the ventilating rate is calculated if either the supplemental heat (such as brooder lamps) is known or no supplemental heat is assumed. Note that only the sensible heat portion of total animal heat loss is used.

Calculating sensible heat gains

Animal sensible heat gain is the sensible heat production of each animal at the assumed indoor temperature, times the number of animals.

Mechanical heat gain can be estimated from:

3.4 Btu/hr-watt incandescent lighting.
4.1 Btu/hr-watt fluorescent lighting.
4000 Btu/hr-horsepower, fractional hp motors.
(Motor heat gain does not apply to exhaust fans, because the heat is immediately exhausted.)

Supplemental heat needed is usually computed from the sensible heat balance. Enough heat is supplied with a furnace, floor heating, or brooders to make up for heat losses.

Calculating sensible heat losses

Building heat loss is the sensible heat lost through walls, ceilings, floor, windows, and doors. It is calculated as discussed in Chapter 631 on insulation, and is dependent on the construction type, insulation type and amount, the surface areas, and the indoor-outdoor temperature difference.

Sensible heat lost in the ventilating air, q_v, is the amount of heat needed to raise the temperature of incoming air to the exhaust air temperature:

Eq 633-2.

$$q_v = M_a \times c_p \times \Delta t$$

or

$$q_v = M_a \times (h_2 - h_1) \text{ Btu/hr}$$

M_a = amount of ventilating air, lb/hr
c_p = specific heat of air, 0.24 Btu/lb-F
Δt = difference between outside air temperature and exhaust air temperature, F
h_1 = enthalpy of incoming air, Btu/lb da
h_2 = enthalpy of exhaust air at the exhaust temperature, but at the entering humidity ratio, Btu/lb da

Moisture Balance

As ventilating air moves through a building it tends to evaporate moisture from floors, pits, and other wet surfaces. As animals breathe, moisture is evaporated from their upper respiratory tracts. To maintain desirable indoor conditions, enough moisture is removed from the building to keep the relative humidity below 70%-80%.

Moisture sources

Animal water vapor production is calculated from latent heat production. Moisture evaporation from wet building surfaces is usually included in the animal latent heat production data. Floor type and bedding can greatly affect water evaporation.

Moisture losses

Water vapor losses occur almost exclusively in the ventilating air as cool, dry air replaces warm, moist air. Moisture loss from the air by condensation on cold surfaces is small and is usually neglected. Surface condensation is usually not desirable in animal housing.

Calculating Moisture Balance

To determine the ventilating rate required to maintain a desired indoor relative humidity, a moisture balance calculation is performed. If the indoor humidity ratio is to remain constant the rate of moisture loss must equal the rate of moisture production.

The overall moisture balance is:

Rate of moisture loss = rate of moisture production

Eq 633-3.

$$M_a \times (W_2 - W_1) = W_a + W_e$$

M_a = ventilating rate, lb da/hr
W_1 = water vapor in incoming air, lb water/lb da
W_2 = water vapor in exhaust air, lb water/lb da
W_a = water vapor from animals, lb water/hr
W_e = water vapor from surface water evaporation, lb water/hr
da = dry air

Latent heat, q_L, from the animals is tabulated in Table 633-1, and the hourly rate of moisture production, W_a, can be calculated:

$$W_a = q_L \div 1044 = \text{lb water/hr}$$

Moisture produced by evaporation, W_e lb/hr, can be estimated, but it is usually included in W_a.

Moisture removed by the ventilating air is the gain in humidity ratio of the air as it moves through the building, $M_a \times (W_2 - W_1)$.

Ventilating Rates

Using energy and moisture balance concepts, two ventilating rates can be calculated that will maintain desirable indoor conditions. One of these calculated ventilating rates is the rate necessary to maintain the indoor temperature at the desired level (energy balance). The other ventilating rate is the rate necessary to maintain the desired indoor relative humidity (moisture balance).

Calculating Ventilating Rates

The ventilating rate, Q, is in units of ft^3 air/minute, cfm:

Eq 633-4.

$$Q = M_a \times V \div 60$$

Q = ventilating rate, cfm
M_a = amount of ventilating air, lb da/hr
V = specific volume of air, ft^3/lb da

Use V at inside conditions for exhuast systems, and V at outside conditions for pressure systems.

The ventilating rate required to maintain a desired indoor temperature can be calculated from the sensible heat balance equation. If q_m, q_{sr}, and q_e in Eq 633-1 are all assumed to be zero or negligible, then the animal heat production rate must equal the rate of heat loss through building surfaces and ventilating air.

Eq 633-5.

$$q_s = q_b + q_v$$

Substituting Eq 633-2 for q_v, q_b = A/R x Δt, and Eq 633-4 for M_a, the equation for the ventilating rate required to maintain the indoor temperature is:

Eq 633-6.

$$Q = [V \div (60 \times c_p \times \Delta t)] \times (q_s - A/R \times \Delta t)$$

Q = ventilating rate, cfm
V = specific volume of air, ft^3/lb da
c_p = specific heat of dry air, 0.24 Btu/lb-F
Δt = $t_2 - t_1$ = (indoor-outdoor) temperature difference, F
q_s = sensible heat production of animals, Btu/hr
A/R = building heat loss factor, Btu/hr-F

The units of Eq 633-6 are:

ft^3 = [(ft^3/lb da) ÷ (min/hr x Btu/lb da-F x F)] x Btu/hr

or

= ft^3/lb da x hr/min x lb da-F/Btu x 1/F x Btu/hr = cfm

The ventilating rate to maintain a moisture balance is computed using Eq 633-3 and 633-4. Solve Eq 633-3 for M_a. Substitute M_a into Eq 633-4 and calculate the ventilating rate:

Eq 633-7.

$$Q = (V \div 60) \times [W \div (W_2 - W_1)]$$

Q = ventilating rate, cfm
V = specific volume of air, ft^3/lb da
W = $W_a + W_e$ = moisture to be removed, lb water/hr

Table 633-1. Sensible and latent animal heat production.
ASHRAE values. The latent heat Btu/pig-hr column is based on 1044 Btu/lb water evaporated.

Livestock	Temp. F	Latent heat, q_L		Sensible heat, q_s
		lb H_2O/ pig-hr	- - - Btu/pig-hr - - -	
10-lb pig	85	0.03	18	130
20-lb pig	75	0.04	45	95
30-lb pig	65	0.07	70	165
50-lb pig (solid floors)	40	0.12	126	304
	50	0.13	137	243
	60	0.145	152	208
	70	0.18	189	151
	80	0.235	247	83
100-lb pig (solid floors)	40	0.14	147	443
	50	0.15	158	352
	60	0.18	189	281
	70	0.22	231	199
	80	0.27	284	136
200-lb pig (solid floors)	40	0.20	210	650
	50	0.21	220	520
	60	0.225	236	414
	70	0.265	278	322
	80	0.33	346	224
		lb H_2O/ sow-hr	- - - Btu/sow-hr - - -	
Sow and litter				
390-lb (0 wk)	60-80	0.703	733	785
400-lb (2 wk)	60-80	0.958	1000	1050
410-lb (4 wk)	60-80	1.066	1113	1079
440-lb (6 wk)	60-80	1.191	1243	1161
500-lb (8 wk)	60-80	1.301	1359	1627
		lb H_2O/ hd-hr	- - - Btu/hd-hr - - -	
Broiler				
0.22-lb	84.2	0.0009	0.9208	4.1
1.5-lb	60.8	0.0108	11.3	14.3
	77.0	0.0046	4.8	16.7
	86.0	0.0154	16.1	14.3

Livestock	Temp. F	Latent heat, q_L		Sensible heat, q_s
		lb H_2O/ hd-hr	- - - Btu/hd-hr - - -	
Broiler				
2.4-lb	60.8	0.0109	11.4	13.5
	66.2	0.0044	4.6	26.3
	86.0	0.0243	25.3	15.0
3.5-lb	66.2	0.0071	7.4	31.1
4.4-lb	66.2	0.0062	6.4	32.8
		lb H_2O/ lb-hr	- - - Btu/lb-hr - - -	
Leghorn (night)	33	0.0020	2.1	6.9
	54	0.0024	2.5	5.5
	64	0.0022	2.3	5.3
	82	0.0034	3.5	3.8
	94	0.0045	4.7	1.7
(day)	35	0.0022	2.3	8.4
	54	0.0032	3.3	6.6
	63	0.0034	3.5	6.6
	72	0.0034	3.6	6.5
	82	0.0041	4.3	5.9
	92	0.0051	5.3	0.0
Turkeys 1 lb	75	0.0059	6.2	10.8
2 lb	75	0.0025	2.6	9.5
		lb H_2O/ cow-hr	- - - Btu/cow-hr - - -	
1000-lb dairy cow	30	0.77	800	2950
	40	0.91	950	2650
	50	1.05	1100	2300
	60	1.28	1340	1900
	70	1.34	1400	1700
	80	1.82	1900	1000

= (animal latent heat production + water evaporated) ÷ (1044 heat of vaporization)
W_2 = indoor humidity ratio, lb water/lb da
W_1 = outdoor humidity ratio, lb water/lb da
60 = minutes/hr

The units of Eq 633-7 are:

$$\begin{aligned} ft^3/min &= (ft^3/lb\ da \div min/hr) \times (lb/hr \div lb/lb\ da) \\ &= ft^3/lb\ da \times hr/min \times lb/hr \times lb\ da/lb \\ &= cfm \end{aligned}$$

To determine the minimum winter ventilating rate at t_o = 97½% winter design temperature, compare the ventilating rates required for moisture and temperature control. Provide the larger ventilating rate. If more air is needed for moisture than for temperature control, the increased airflow increases sensible heat losses and supplemental heat is needed to prevent indoor temperatures from dropping below the desired level. If the ventilating rate for temperature control is higher, room moisture levels will be less than assumed, which is usually acceptable.

It is usually assumed that the minimum ventilating rate for moisture removal will also control odors within warm confinement buildings. However, if the calculated rate is lower than the recommended value in Table 633-2, removal of stale, odorous air may be the primary concern. Increasing the minimum winter ventilating rate above that required for moisture removal ensures that stale air is exhausted. It also increases operating costs because more supplemental heat is required.

Ventilation Curves

Use the two ventilating rate equations, Eq 633-6 for heat and Eq 633-7 for moisture, to plot required ventilating rate vs. outside temperature.

Once the ventilation curves are plotted, the ventilating system and the necessary supplemental heat

Table 633-2. Recommended ventilating rates.
The rate for each season is the total capacity needed.

Animal type		Cold weather rate	Mild weather rate	Hot weather rate
Swine				
For swine, circulation fans can replace exhaust fans for one-half of the hot weather ventilating rates. Use 300 cfm as the hot weather rate for sows in a breeding facility.				
		- - -	cfm/hd	- - -
Sow and litter	400 lb	20	80	500
Prenursery pig	12-30 lb	2	10	25
Nursery pig	30-75 lb	3	15	35
Growing pig	75-150 lb	7	24	75
Finishing pig	150-220 lb	10	35	120
Gestating sow	325 lb	12	40	150
Boar	400 lb	14	50	300
Dairy	**Unit**			
An alternative cold weather rate = Room volume, ft³/15.				
An alternative hot weather rate = Room volume, ft³/2.				
0-2 mo	cfm/hd	15	50	100
2-12 mo	cfm/hd	20	60	130
12-24 mo	cfm/hd	30	80	180
Mature cow	cfm/1000 lb	35	120	335
Milkroom	cfm			600
Milking parlor	cfm/stall		100	400
Beef				
Cattle in warm building	cfm/1000 lb	15	100	200 (min.)
Sheep				
Warm barn	cfm/1000 lb	25	100	200
Horse				
Warm barn	cfm/1000 lb	25	100	—
Poultry				
Chick	cfm	0.1/bird	0.5/bird	1/lb
Layer, pullet breeder, broiler	cfm/bird	0.50	2	4 (min.)
Turkey poults	cfm/lb	1/6	1/4-1/3	

can be determined. For a warm confinement building, the following information is needed to select ventilating equipment:

- Minimum winter ventilating rate for moisture or odor control.
- Maximum ventilating rate for temperature control as the outside temperature increases to within 5-15 F of the inside temperature. As outside temperatures increase above this point, ventilation requirements increase dramatically and natural ventilation is usually provided.
- Maximum supplemental heating rate needed to maintain the desired indoor temperature when ventilating at the minimum winter outside design conditions.

Before selecting the ventilating equipment and components, consider the adverse conditions that can develop during operation and that pose the strictest requirements for each piece of equipment. Both the minimum winter ventilating rate and the amount of supplemental heat required depend on animal size and stocking density, which vary throughout the lifetime of the building. It is often economical to allow the indoor design temperature to fluctuate from the optimum production level by ±10 F. Consequently, it is often desirable to prepare ventilation curves for three inside design temperatures and for both low and high stocking densities.

Figs 633-2 and 633-3 illustrate typical ventilation curves.

Example 1:

Plot the ventilating rates needed in a swine finishing building with a solid floor and 140 head of 150-lb pigs for outside temperatures from -10 F to 50 F. Assume inside conditions of 60 F and 80% RH and outside RH of 80%. Use 212 Btu/hr latent and 348 Btu/hr sensible heat production. These values were obtained by interpolation between the values for 100 and 200 lb pigs at 60 F in Table 633-1. The same values can be obtained from Chapter 604.

Moisture balance

W_a = (212 Btu/hr-pig, latent heat) ÷ (1044 Btu/lb water)
= 0.203 lb water/hr-pig
W = 0.203 lb water/hr-pig x 140 pigs
= 28.4 lb water/hr

The latent heat production, 212 Btu/hr, was obtained from pig production facilities so no additional allowance is needed for water being evaporated.

At 50 F, 80% RH outside, W_1 = 0.0061 lb water/lb air.

At 60 F, 80% RH inside, W_2 = 0.0089 lb water/lb air.

From the psychrometric chart at 60 F, 80% RH inside, V = 13.3 ft³/lb da, (exhaust system). Using these values, one point on the ventilation curve for moisture control can be calculated from Eq 633-7.

$$Q = (13.3 \div 60) \times [28.4 \div (0.0089 - 0.0061)]$$
$$= 2250 \text{ cfm at 50 F outside}$$

Heat balance

Using Eq 633-6, the ventilating rate required to maintain the indoor temperature can also be calculated. Sensible heat production, q_s, is:

q_s = (348 Btu/hr-pig, sensible heat) x (140 pigs)
= 48,720 Btu/hr

Assume the pigs are in the building described in the sample heat loss problem in Chapter 631. The A/R value was found to be 198.55 Btu/hr-F.

The building A/R value, the sensible heat production, q_s, and the outdoor temperature, 50 F, are substituted into Eq. 633-6:

$$Q = [13.3 \div (60 \times 0.24 \times (60 - 50))] \times [48{,}720 - 198.55 \times (60 - 50)]$$
$$= 4316 \text{ cfm at 50 F outside}$$

Similar procedures for other outside temperatures yield points on the curve as follows:

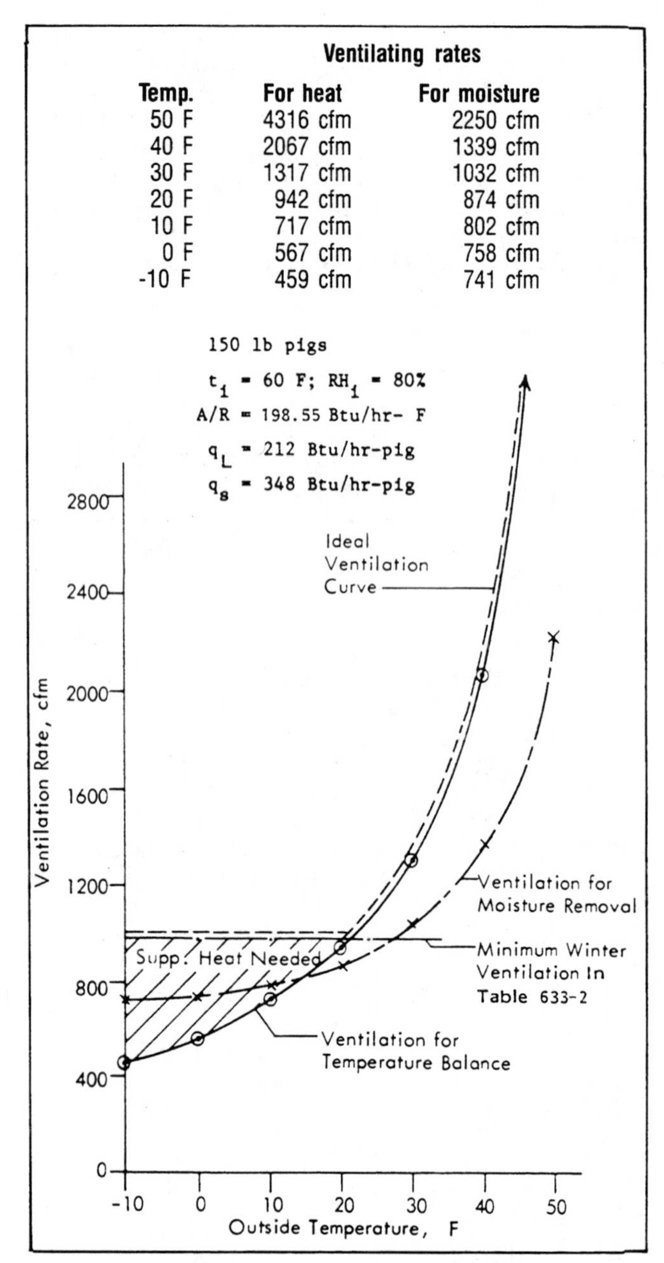

Ventilating rates

Temp.	For heat	For moisture
50 F	4316 cfm	2250 cfm
40 F	2067 cfm	1339 cfm
30 F	1317 cfm	1032 cfm
20 F	942 cfm	874 cfm
10 F	717 cfm	802 cfm
0 F	567 cfm	758 cfm
-10 F	459 cfm	741 cfm

Fig 633-2. Ventilation curves for growing swine.

Theoretically, the minimum ventilating rate is the larger of the two rates, given an outdoor temperature. For example, at -10 F you could ventilate at 741 cfm. However, most ventilating systems cannot vary the minimum ventilating rate with changes in outside temperature. Therefore, the minimum ventilating rate is usually taken at the intersection of the moisture and temperature ventilation curves, in this case about 830 cfm. Compare the calculated minimum winter ventilating rate with the recommendations in Table 633-2. For 150-lb pigs, the recommendation is 7 cfm/150 lb pig. For 140 head, the total minimum ventilating rate is 980 cfm.

The actual ventilating rate follows the minimum winter ventilating line to about 20 F, then it follows the heat removal curve for higher temperatures. At outside temperatures below the intersection of the minimum winter ventilating rate and the temperature balance lines, supplemental heat is required to maintain 60 F inside the building. At outside temperatures above the intersection, the inside relative humidity will be below the design RH of 80%.

Supplemental heat requirements

If it is -10 F outside and you are ventilating at 980 cfm, supplemental heat must be provided to maintain an indoor temperature of 60 F.

The amount of supplemental heat needed to maintain the inside temperature is determined from the basic heat balance equation.

Eq 633-8.

$$q_{sr} = q_v + q_b + q_e - q_s - q_m$$

Neglecting q_e and q_m:

Eq 633-9.

$$q_{sr} = q_v + q_b - q_s$$

q_{sr} = supplemental heat required, Btu/hr
q_v = ventilating air heat loss, Btu/hr
q_b = building heat loss, Btu/hr
q_s = sensible heat produced by animals, Btu/hr

$q_v = M_a \times c_p \times \Delta t$
$= Q \div V \times c_p \times \Delta t$
$= 980 \times 0.24 \times (60 - (-10)) \div 13.3$
$= 1237.9$ Btu/min = 74,274 Btu/hr

$q_b = A/R \times \Delta t$
$= 198.55 \times (60 - (-10))$
$= 13{,}899$ Btu/hr

$q_s = 48{,}720$ Btu/hr

$q_{sr} = 74{,}274 + 13{,}899 - 48{,}720$
$= 39{,}453$ Btu/hr

If only 70 pigs are in the building instead of 140, q_s is 24,360 Btu/hr and the amount of supplemental heat required to maintain t_i = 60 F is:

$q_{sr} = 74{,}274 + 13{,}899 - 24{,}360$
$= 63{,}813$ Btu/hr

If the minimum winter ventilation is selected to provide moisture removal at -10 F (741 cfm), the supplemental heat requirement for 140 pigs is:

$q_v = 741 \times 0.24 \times (60 - (-10)) \div 13.3$
$= 936$ Btu/min = 56,160 Btu/hr

$q_{sr} = 56{,}160 + 13{,}899 - 48{,}720$
$= 21{,}339$ Btu/hr

Note that reducing the minimum winter ventilating rate from 980 to 741 cfm significantly reduces energy consumption on cold days; however, the final system selected must have the capacity to prevent the buildup of obnoxious odors. Base the final decision on the equipment available, the amount of flexibility needed or desired, and a careful examination of the alternatives.

Calculations for equipment selection

To specify the ventilating equipment for this proposed building, it is advantageous to answer the following questions:

1. If the inside temperature is allowed to drop to 50 F with 80% relative humidity during extremely cold conditions (t_o = -10 F; RH_o = 80%), what ventilating rate is necessary for moisture removal?

   W = (189 Btu/hr-pig) x (140 pigs) ÷ (1044 Btu/lb water)
   = 25.34 lb water/hr

   Using Eq 633-7:

   Q = (12.95 ÷ 60) x [25.34 ÷ (0.0061 − 0.00037)]
   = 955 cfm

2. What maximum winter ventilating rate is needed to maintain the inside temperature at or below 70 F as the outside temperature increases to 60 F?

   Δt = 70 − 60 = 10 F

   q_s = (260.5 Btu/hr-pig) x (140 pigs) = 36,470 Btu/hr

   A/R = 198.55 Btu/hr-F

   V = 13.6 ft³/lb da

   Using Eq 633-6:

   Q = [(13.6 ÷ 60) x 0.24 x (70 − 60)] x (36,470 − 1,985.5)
   Q = 3257 cfm

   (The maximum winter ventilating rate can also be estimated using Table 633-2.)

3. If a minimum winter ventilating rate of 1000 cfm is selected, and t_i is allowed to decrease to 50 F, how much supplemental heat is needed?

   To anticipate the maximum needs, you may wish to assume that the building is only half stocked with 70 pigs weighing 100 lb. Note that this increases the heater size needed; the assumptions actually made should be realistic for each building analyzed. For 100 lb pigs at 50 F, q_s is 352 Btu/hr-pig (Table 633-1).

Eq 633-10.

$$q_{sr} = q_v + q_b - q_s$$
$$q_v = M_a \times c_p \times \Delta t = Q \div V \times c_p \times \Delta t$$

q_v = (1000 ft³/min x 60 min/hr ÷ 12.95 ft³/lb da) x (0.24 Btu/lb da-F x 60 F)
= 66,718 Btu/hr

q_b = A/R x Δt
q_b = 198.55 Btu/hr-F x (50 − (−10) F)
= 11,913 Btu/hr

q_s = 352 Btu/hr-pig x 70 pigs
= 24,640 Btu/hr

q_{sr} = 66,718 + 11,913 − 24,640
= 53,991 Btu/hr

Based on these calculations the following equipment specifications are realistic, although other equipment choices may be adequate and appropriate.

One 1000 cfm fan operating continuously.
One 1200 cfm fan to turn on at about 64 F.
One 1200 cfm fan to turn on at about 68 F.
One 60,000 Btu/hr heater to turn on at about 50 F.

Hot Weather Ventilating Rate

The number of fans indicated above would be adequate only for mild weather. Additional ventilation is required for hot weather ventilation. In many cases this additional capacity is most economically obtained by manually opening large ventilating panels for natural ventilation. If mechanical ventilation is desired, provide additional fans. Additional fan capacity is calculated using the sensible heat balance equation. The summer design temperature is used for the outside temperature and the desired inside temperature is assumed to be 3-5 F higher. Install thermostat controlled fans to provide this additional capacity.

For example, if the summer outside design temperature was 90 F for this swine building, the summer ventilating rate can be calculated using Eq 633-6.

t_o = 90 F, 97½% summer design temperature
t_i = 93 F, 3 F temperature difference to be maintained
q_s = 15,000 Btu/hr, estimated sensible heat production for 140 pigs weighing 150 lb at 93 F
V = 14.5 ft³/lb da, for T_{db} = 93 F and RH = 80%

Consequently the required summer ventilating rate is:

Q = [14.5 ÷ (60 x 0.24 x (93 − 90))]
x [15,000 − 198.55 x (93 − 90)]
Q = 4835 cfm

Compare this value to the standard recommendation of 75 cfm/pig, in Table 633-2. The table value is higher than needed for sensible heat removal, in order to provide high air velocities around the pig.

Example 2:

What ventilating rate is needed in a poultry confinement building with hanging cages over pits and 690, 4-lb laying hens, for outside temperatures from -10 to 50 F? Assume inside conditions of 60 F and 70% RH and an outside RH of 80%. Use 12.7 lb/hr moisture produced per 1000, 4-lb birds, and 6.6 Btu/hr sensible heat per lb of bird, Table 633-1.

Moisture balance

W = (12.7 lb/hr ÷ 1000 birds) x 690 birds
= 8.8 lb water/hr

At 50 F, 80% RH outside:
W_1 = 0.0061 lb water/lb da.

At 60 F, 70% RH inside:
W_2 = 0.0077 lb water/lb da.

At 60 F, 70% RH inside:
V = 13.3 ft³/lb da (exhaust system).

Using Eq 633-7:
Q = (13.3 ÷ 60) x [8.8 ÷ (0.0077 − 0.0061)]
= 1219 cfm at 50 F outside

Sensible heat balance

$q_s = 6.6$ Btu/hr-lb x 4 lb/bird x 690 birds

$q_s = 18{,}216$ Btu/hr

Because the number of layers in this example is about the capacity for the building in Example 1, the same A/R-value will be used to obtain the ventilating rate to maintain a sensible heat balance.

Using Eq 633-6:

$$Q = [13.3 \div (60 \times 0.24 \times (60 - 50))] \times [18{,}216 - 198.55 \times (60 - 50)]$$
$$= 1499 \text{ cfm at 50 F outside}$$

Similar procedures for other outside temperatures yield points on the curve as follows:

Temperature	Ventilating rates For heat	For moisture
50 F	1499 cfm	1219 cfm
40 F	658 cfm	557 cfm
30 F	377 cfm	398 cfm
20 F	237 cfm	325 cfm
10 F	153 cfm	293 cfm
0 F	97 cfm	275 cfm
-10 F	57 cfm	267 cfm

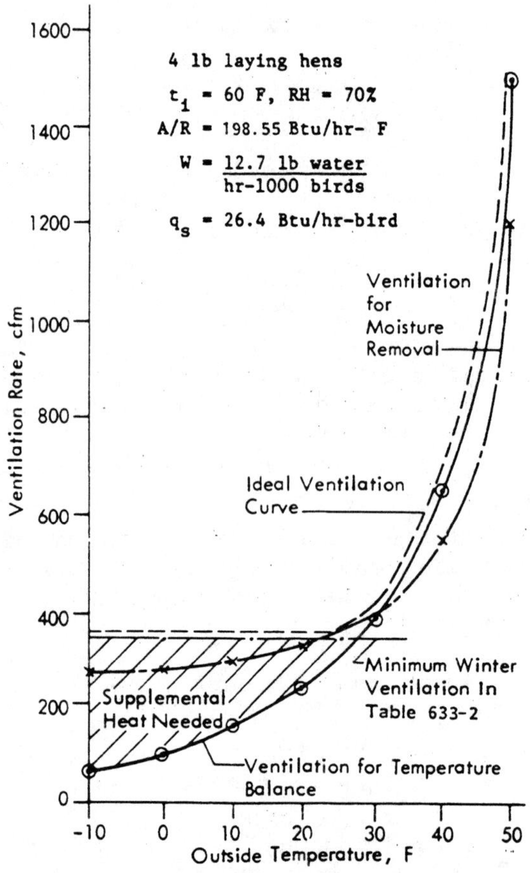

Fig 633-3. Ventilation curves for laying hens.

Compare the calculated minimum winter ventilating rates with the recommendations in Table 633-2. For poultry, the recommendation is ½ cfm/bird. For 690 layers, the total minimum ventilating rate is 345 cfm.

The actual ventilating rate must meet the maximum demand—it should follow the minimum winter ventilating rate line for low temperatures to about 25 F, the moisture removal curve from 25 F to about 30 F, and the heat removal curve for higher temperatures. At outside temperatures below the point of intersection of the heat and moisture removal curves, supplemental heat is required to maintain 60 F inside the building. At outside temperatures above the intersection, the inside relative humidity is below the design RH = 70%.

Determine the maximum amount of supplemental heat required if the minimum ventilating rate is 345 cfm and t_o = -10 F. From the sensible heat balance equation:

Eq 633-11.

$$q_{sr} = q_v + q_b - q_s$$

$q_v = M_a \times c_p \times \Delta t$

$q_v = [345 \text{ cfm} \times 0.24 \text{ Btu/lb-F} \times (60 - (-10) \text{ F})] \div 13.3 \text{ ft}^3\text{/lb da}$

$= 435.8$ Btu/min or about 26,150 Btu/hr

$q_b = A/R \times \Delta t$

$q_b = 198.55 \text{ Btu/hr-F} \times (60 - (-10) \text{ F})$

$= 13{,}899$ Btu/hr

$q_s = 18{,}216$ Btu/hr

Thus, the supplemental heat required is:

$q_{sr} = 26{,}150 + 13{,}899 - 18{,}216$

$= 21{,}833$ Btu/hr

Note that this rate changes if the minimum ventilating rate, the A/R value, or the stocking density changes.

634 VENTILATING SYSTEMS

Ventilation is a continuous process to:
- Remove moisture from inside the building.
- Provide fresh air for animals.
- Remove excess heat in hot weather.
- Remove odors and gases from animal manure.
- Remove air-borne dust and disease organisms.
- Remove combustion products of unvented heaters.

Good air movement reduces moisture condensation on walls and roofs in cold weather and helps keep inside temperatures acceptable in hot weather.

The Ventilating Process

Fresh air is made up of different gases as shown in in Table 634-1.

Table 634-1. Composition of fresh air.

Nitrogen (N_2	78%
Oxygen (O_2)	21%
Carbon dioxide (CO_2) and rare gases	Less than 1%

If air is not replaced in livestock housing, the air composition changes, and the concentration of carbon dioxide and other harmful gases increases to the danger level. Carbon dioxide, while essential to trigger breathing processes, becomes noxious at high concentrations. The critical level of some gases in air for sustaining the life of chickens is listed in Table 634-2.

Table 634-2. Lethal gas levels for chickens.

Gas	Symbol		Lethal		Tolerable
Carbon dioxide	CO_2	Above	30%	Below	2%
Methane	CH_4	Above	5%	Below	5%
Ammonia	NH_3	Above	0.05%	Below	0.004%
Hydrogen Sulfide	H_2S	Above	0.10%	Below	0.004%
Oxygen	O_2	Below	6%	Above	16%

Source: University of Kentucky

According to these data, the minimum safe biological rate of ventilation for layers is 1 cfm/50 birds. By comparison, 4 cfm/person is required in a crowded room. Greater rates are desirable for optimum performance.

Air exchange

As the ventilating system exchanges air it brings in oxygen needed to sustain life and carries out harmful gases and undesirable odors. It also dilutes air-borne disease organisms and odors.

Odor levels depend on type of animal, the manure handling system, air temperature, and the manager's attitude. Except during the coldest part of the year, ventilating enough to remove excess animal heat and moisture also controls odor. A cold weather ventilating rate that just prevents moisture buildup may not give good odor control.

It is very important that you vent gas, liquid, or solid fuel heaters to the outside with a fire-safe chimney or flue, to remove combustion byproducts. Or, ventilate at 4 cfm/1000 Btu heater capacity in addition to recommendations in Table 633-2. With exhaust ventilation, install a fan-powered flue instead of a gravity-powered flue so the exhaust fans do not draw the combustion products down the flue.

It is desirable to ventilate indoor manure storage pits to reduce gases at the animal level. Consider exhausting the cold weather ventilating air from the pit. Use high volume ventilation while agitating or emptying indoor storage pits, or remove the animals from the building and stay outside yourself.

Air also absorbs and removes sensible heat and moisture.

Moisture exchange

A primary purpose of winter ventilation is to remove excess water vapor from the building.

During cold weather, ventilation brings cold, relatively dry air into the building. The air is warmed by energy from animals, electrical equipment, and supplemental heat. As the air temperature rises, it can hold more moisture and its relative humidity decreases. The moisture holding capacity of air nearly doubles for every 20 F rise in temperature. This ventilating air picks up moisture (which increases its humidity ratio) and exhausts it from the building. These properties are shown schematically in Figs 634-1 and 634-2.

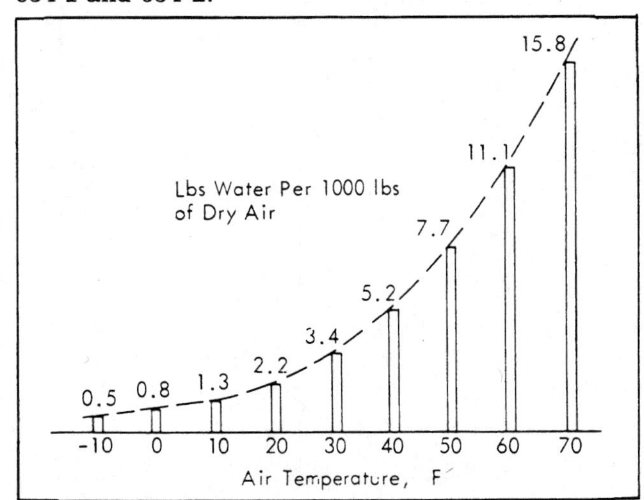

Fig 634-1. Moisture holding capacity of air.

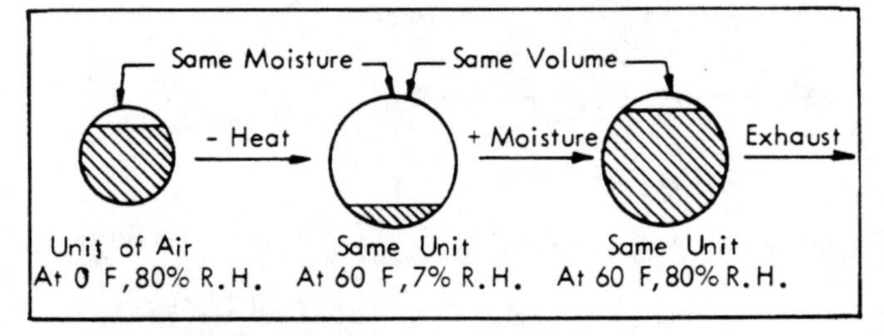

Fig 634-2. Heated air holds more moisture.
Cold air expands as it is heated and can absorb more moisture.

The amount of moisture to be removed depends on animal type, animal size, and manure handling system. Generally, larger animals give off more moisture and require more moisture control ventilation. With totally slotted floors, most liquids drain through instead of evaporating, reducing the moisture load on the ventilating system by about 50%, compared with solid floors. Partly slotted floors have moisture evaporation rates between solid and totally slotted floors.

Design ventilating rates to maintain room air at 50%-80% relative humidity. Higher humidities increase condensation; lower humidities increase dust levels. Also, a 50%-80% relative humidity is detrimental to the airborne bacteria found in livestock buildings.

Heat exchange

To maintain a constant room temperature, the heat **produced** by animals and heaters has to equal heat **lost** through building surfaces and ventilation. If heat loss exceeds animal heat production, provide supplemental heat. If heat production exceeds heat loss, increase ventilation.

When ventilating air entering a building is cooler than inside air, it removes sensible heat from the building. Each 12 to 14 ft^3 of ventilating air absorbs ¼ Btu of heat for each degree it is warmed. In warm weather, move more air through the building to remove excess heat, because each ft^3 of the warmer air absorbs less heat than cooler air. Conversely, in cold weather, move less air to prevent over-cooling.

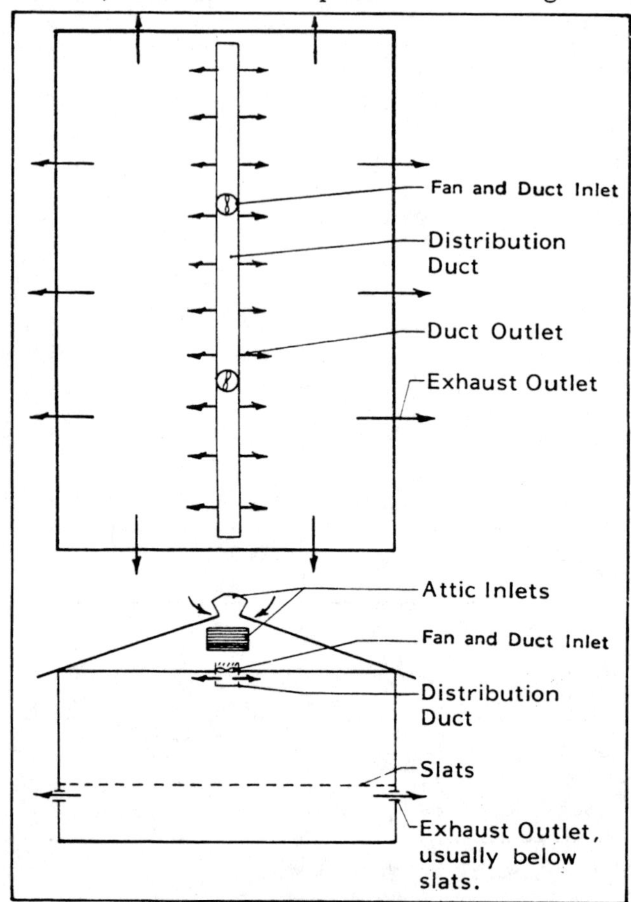

Fig 634-3. Positive pressure ventilating system.

Ventilating System Types

A mechanical ventilating system has fans, controls, and air inlets or outlets. It provides more control over room temperature and air movement than natural ventilation. It works especially well with young animals because they are susceptible to low temperature, sudden temperature changes, and drafts.

Positive pressure systems

Fans force fresh air into the building, creating a positive indoor pressure that pushes air out of the building through the outlet openings, Fig 634-3. The air is distributed with a baffle in front of the fan or a duct having carefully placed outlets, Fig 634-4. Use Table 634-6 to size the duct. Insulate the duct to prevent condensation.

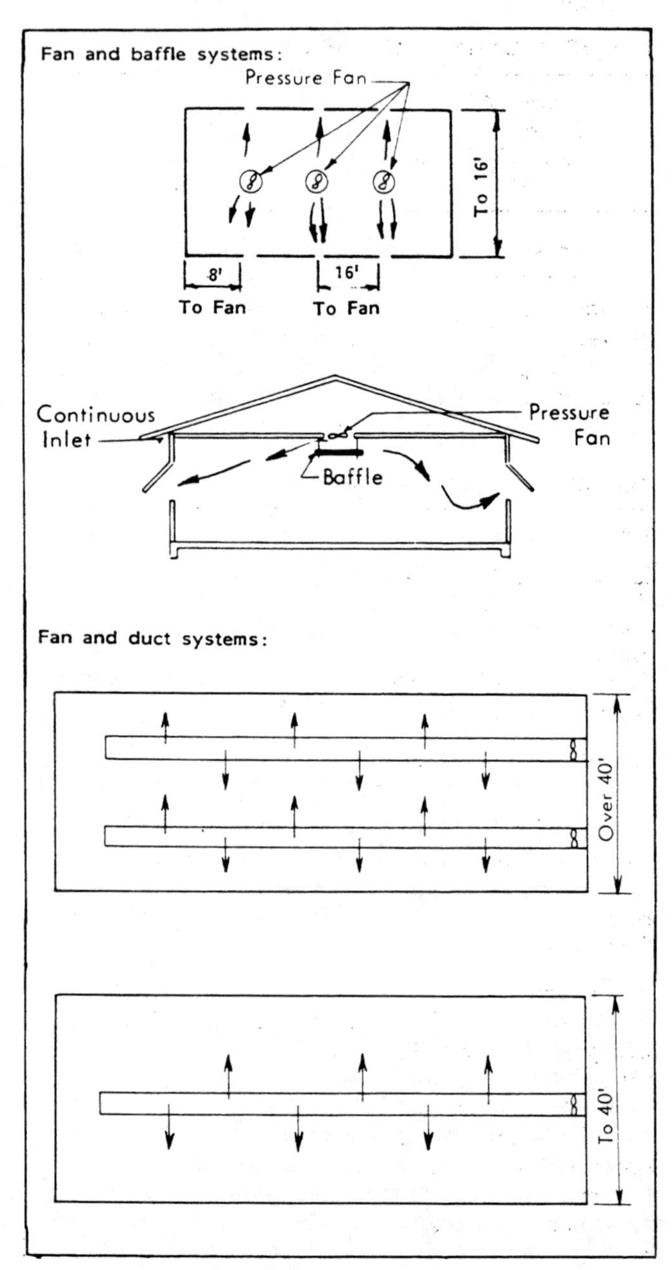

Fig 634-4. Positive pressure fan and duct locations.
For small buildings, positive pressure fans with baffles work. Use ducts spaced about 20′ apart for larger buildings.

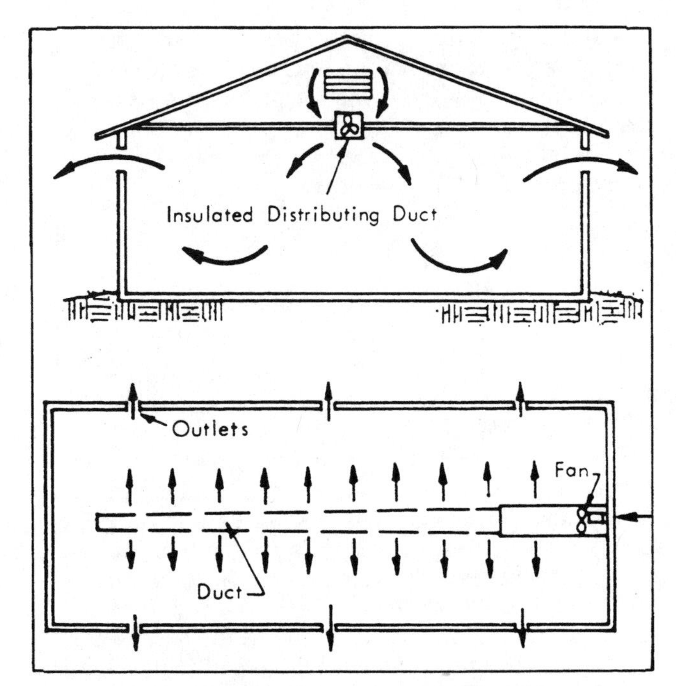

Fig 634-5. Positive pressure tapered insulated duct.
To provide for variable airflow, a multi-speed fan is required.

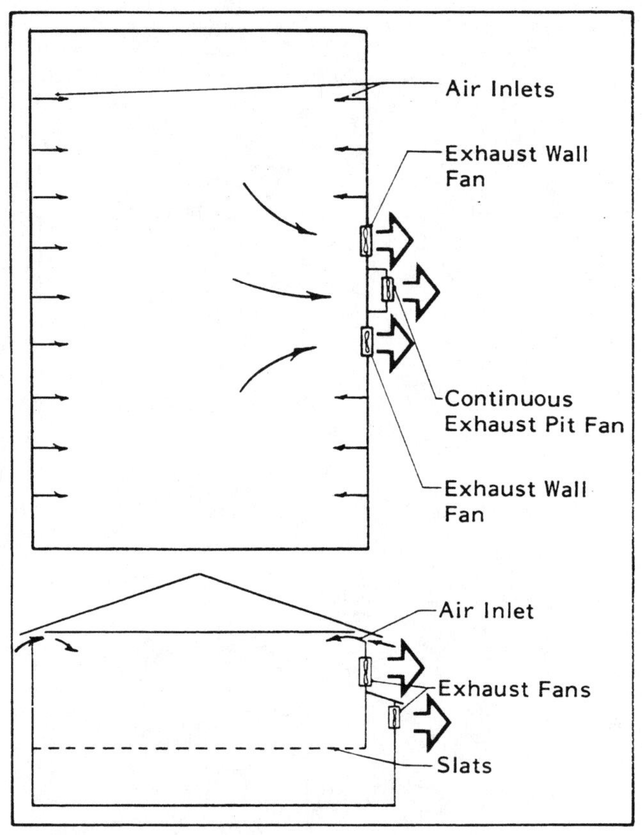

Fig 634-6. Exhaust ventilating system.
Negative pressure system.

Size positive pressure exhaust outlets at 300 in²/1000 cfm of fan capacity. For example, provide one pair of 2" diameter holes for each 20 cfm or one pair of 3" diameter holes for each 45 cfm of airflow through the duct. Base the number of holes on the highest airflow through the duct, and block off part of each hole when operating at lower airflow rates. Space holes uniformly along the duct.

Positive pressure systems tend to force moist air into the walls and attic, so a good vapor barrier is extremely important. Also, frost forms at air leaks around doors, freezing them shut. Positive pressure systems are often installed in remodeled buildings, because pressure outlets are easier to construct than inlets for exhaust ventilation.

Exhaust systems

Exhaust fans pull air out of the building, creating a negative pressure. The pressure difference between outside and inside forces fresh air into the building through baffled inlets, Fig 634-6. If the inlets are sized and located properly, fresh air is distributed uniformly throughout the room. Exhaust is by far the most common ventilating system in animal housing and is discussed in detail later.

Neutral pressure systems

Neutral pressure systems include inlet shutters, an air distribution duct (or tube), a duct fan, and exhaust fans, Fig 634-7. In cold and mild weather, ventilating air is distributed through the duct and exhausted through wall fans. A motorized shutter, located in the wall near the duct fan, allows fresh air to enter the duct from the outside or from the attic, Fig 634-8. The duct fan usually moves more air than is required in cold weather, so the shutter opens intermittently to let in the required amount of fresh air. An exhaust fan operates when the shutter opens to draw fresh air in through the shutter. When the intake shutter closes, the duct fan circulates only room air through the distribution duct. If unit space heaters are required, locate them at the duct intake so the duct will distribute the heat.

Design the duct and duct fan to handle at most the mild weather ventilation. Additional inlets and exhaust fans are required to supply the hot weather ventilation. Locate the hot weather inlets (usually motorized shutters) at the end of the building opposite the exhaust fans. Locate young animals so incoming air will not chill them.

While this system generally provides good air distribution at low ventilating rates, there are problems that can result from recirculating air:

- Dust collects in the duct.
- The intake shutters can freeze.
- Increased air movement can create a draft.
- Use of the same air duct from one animal group to another can transfer dust and bacteria.

The only solution to dust buildup is periodic cleaning of the duct. This may mean partial disassembly, so consider cleaning ease when constructing homemade ducts. Shutters can freeze if no supplemental heat is provided or if the duct fan is too far from the shutters to move warm air past them. Shutter freeze-up is a serious problem because it restricts ventilation. If you have this problem, keep the shutters warm with a heat lamp. Correct draft problems by installing a lower capacity, multiple-speed, or variable-speed duct fan; by raising room temperature; or by installing hovers.

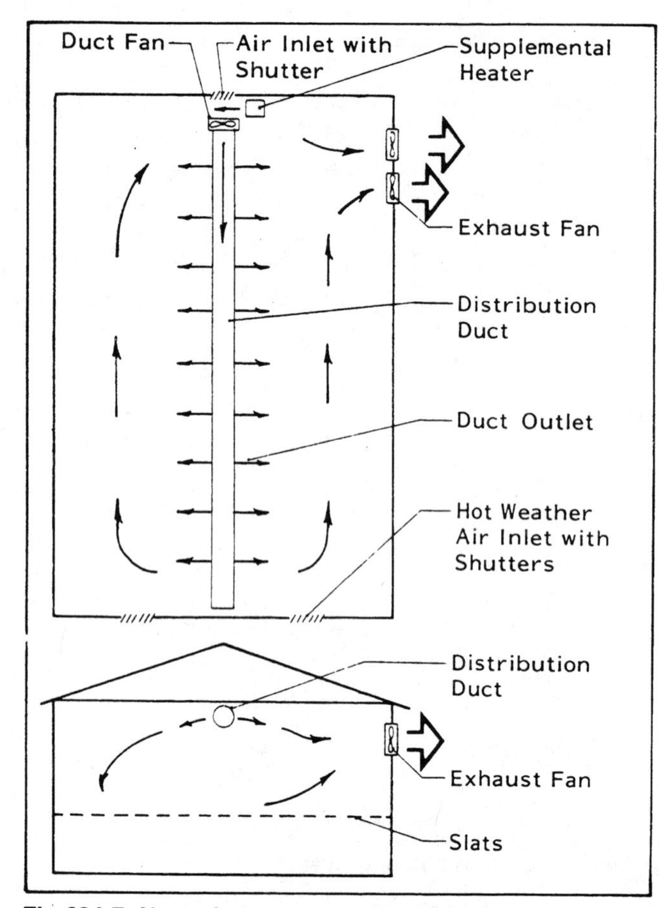

Fig 634-7. Neutral pressure ventilating system.

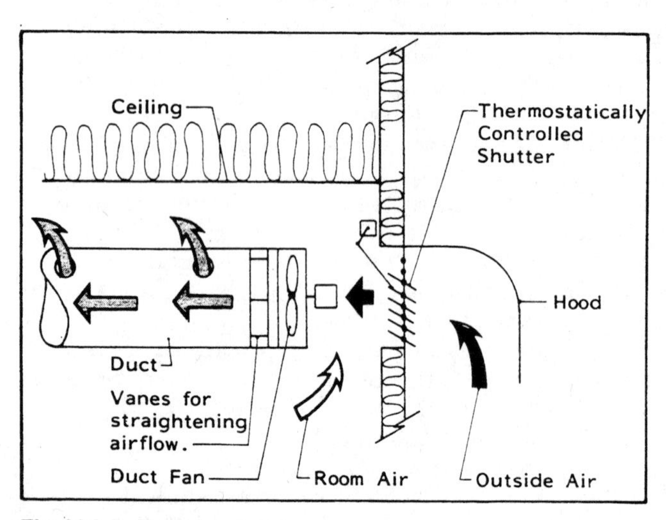

Fig 634-8. Duct fan inlet.
Remove the top one or two leaves from the shutter to still provide some air during a power or freezing failure.

Ventilation Design

Airflow Through Inlets and Outlets

The rate of air change is determined by fan delivery, but the uniformity of air distribution depends on the location, design, and adjustment of the air inlets. This is true of both exhaust and pressure systems, assuming that the longest path of air from inlet to outlet does not exceed about 75′.

When air flows through any opening of any shape, the cross-sectional area of the issuing jet is reduced to 60%-80% of the total free area of the opening—60% is a reasonable design value. This phenomenon, the Vena-Contracta effect, increases the velocity of the air emerging from the opening.

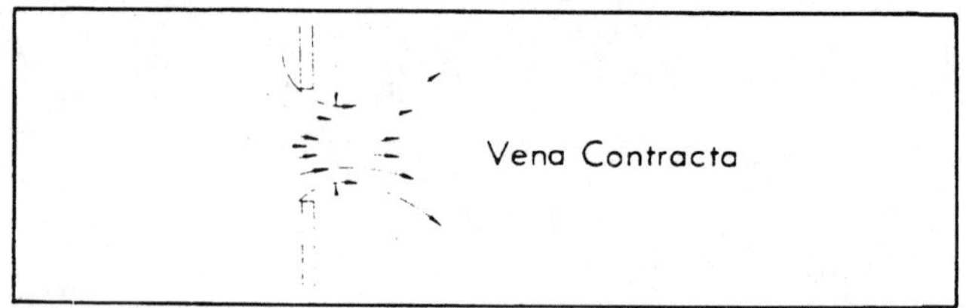

Fig 634-9. Vena contracta.

Tests at Penn State University related the dimensions of slot inlets and volume of air admitted at various static pressures. The airflow rate through a slot at a constant static pressure is proportional to the width of the slot. The average rate through an entire 4″ wide slot at 0.04″ water gauge (W.G.) is 200 cfm—four times the flow through a 1″ slot at the same W.G. This relationship holds at the higher static pressure of 0.125″ W.G. However, increasing the static pressure across the slot about three times (0.04″ to 0.125″) results in only doubling the airflow rate through the slot.

Table 634-3. Airflow through slots 1′ long.

Inches Slot width	Cfm Static pressure 0.04″	0.125″
1	50	100
2	100	200
3	150	300
4	200	400

Source: Pennsylvania State University

Make the total area of the air inlet proportional to the total fan capacity. A common rule of thumb sizes air inlets at 1 ft^2 (144 in^2) of area for each 600 to 1200 cfm of fan capacity. In the following discussion, air velocity through an inlet is considered to be the **average** velocity through the full slot width, i.e. Vena Contracta is not considered. For the above rule of thumb, the average inlet velocity through the full slot width is $V = Q/A = 600$ fpm at 0.04″ and 1200 fpm at 0.125″, Table 634-3.

Do not design an air inlet for the full static pressure rating of the fan. With use, inlets may fill with dirt and debris which constrict the opening. This may overload the fans if they are not initially slightly oversized.

Example 1:

Size a slot air inlet for a maximum fan delivery of 50,000 cfm at 0.04″ W.G.

Assume a maximum slot width of 3″ from Table 634-3. Each foot of length of a 3″ wide slot admits 150 cfm at 0.04″ W.G. The needed slot length is 50,000 ÷ 150 = 334′. If the static pressure is increased to 0.125″ then only 167′ of 3″ slot is needed (50,000 ÷ 300 = 167), or 334′ of 1½″ slot.

Usually the slot is made a little wider than calculated, and the opening width is adjusted for desired pressure differential and distribution.

Fans are most efficient if fan housing resistance is low and inlet area (or outlet area for positive pressure) is large. The maximum resistance of housings and inlets in livestock buildings should not exceed 0.2" W.G. and is often less than 0.1". Air heaters and filters in the airflow path further increase the resistance.

Circulation fans improve air distribution in winter to help dry or heat corners. In summer, they improve animal comfort by increasing air velocity over the skin. Circulation fans can be 1750 rpm unrated units, but should have sealed motors.

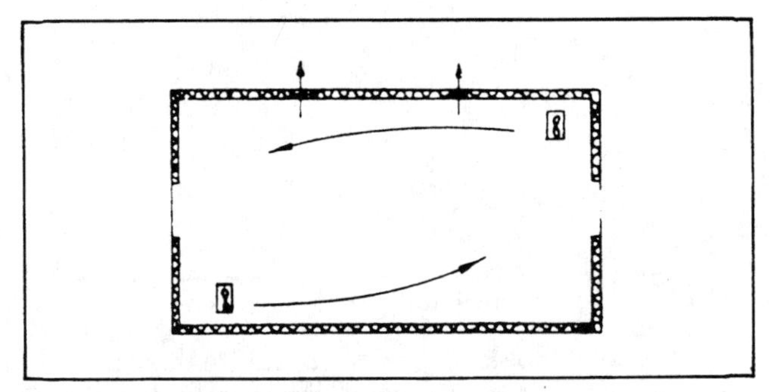

Fig 634-10. Circulating fans.

Air Inlets for Exhaust Ventilation

With proper inlet design and location, there are no damp corners, drafts, or dead air spots. One of the best air inlets is a continuous slot with an adjustable baffle, Fig 634-11a. The baffle restricts the inlet opening to increase air velocity and improve air mixing. Use rigid baffles to avoid warping and resulting uneven air distribution. Tightly seal all doors, windows, and unplanned openings, especially when operating at low ventilating rates.

Intermittent air inlets are common in remodeled buildings because they are easier to build than continuous slots, Fig 634-11b. Space intermittent openings uniformly in the room, but place them at least 8′ from building walls and at least 16′ apart. Intermittent inlet size depends on required airflow. Construct the baffle to extend 4" beyond all sides of the inlet opening.

While most exhaust ventilating systems with continuous slot inlets force air across the ceiling, Fig 634-11, a "down-the-wall" pattern also works, Fig 634-13. Air flows directly into the animal zone—control drafts with hovers over sleeping areas.

Inlet location

For buildings up to 35′ wide, place continuous slot inlets at the ceiling along both sidewalls; for wider buildings, add one or more interior ceiling inlets, Fig 634-14.

Air should travel less than 75′ from the inlet to the fan. During cold weather, close and seal air inlets within 8′ of the winter exhaust fans to prevent short circuiting.

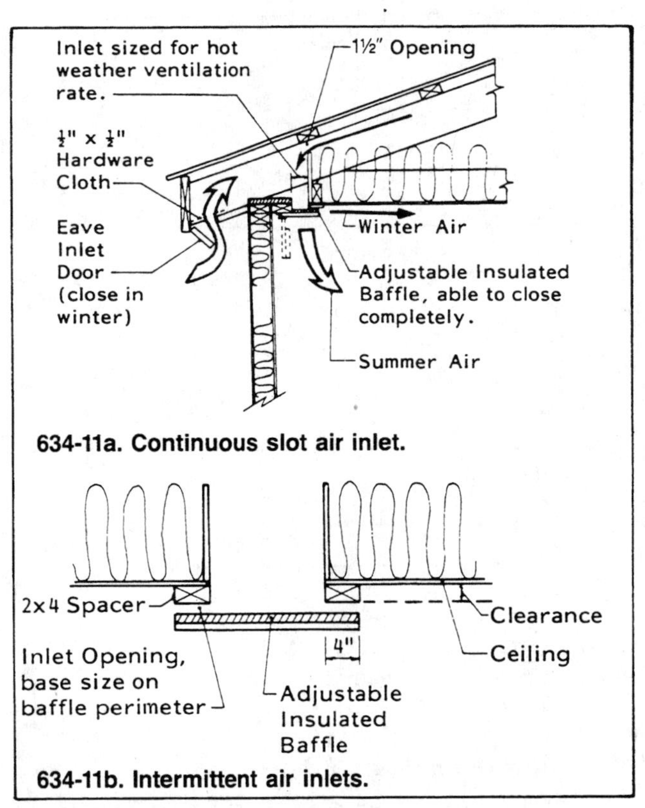

Fig 634-11. Baffled air inlets for "across-the-ceiling" airflow.
Keep augers, fluorescent lights, etc. at least 4′ away from inlets. Avoid even pipes, conduits, etc. within 4′ of the inlet.

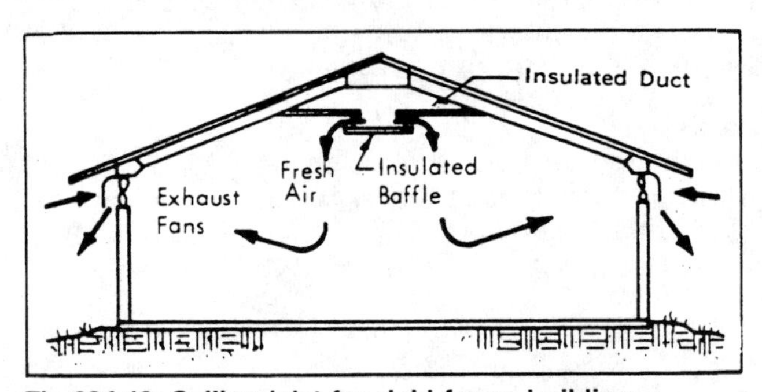

Fig 634-12. Ceiling inlet for rigid frame building.
An insulated duct can be built under the ridge. Air inlets to the duct are endwall louvers. Do not pull summer air from the ridge because of high heat gain from the roof.

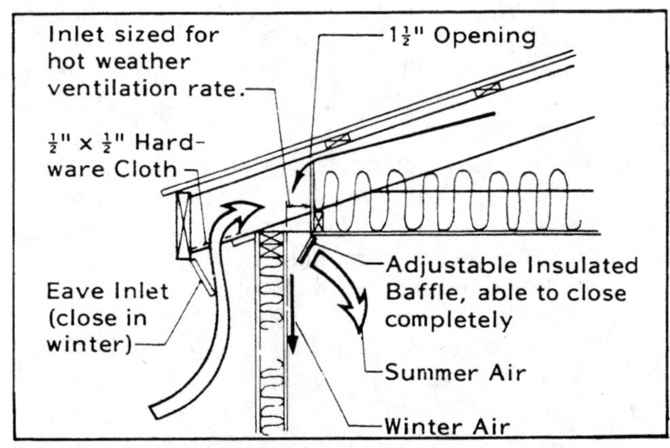

Fig 634-13. Baffled air inlet for "down-the-wall" airflow.

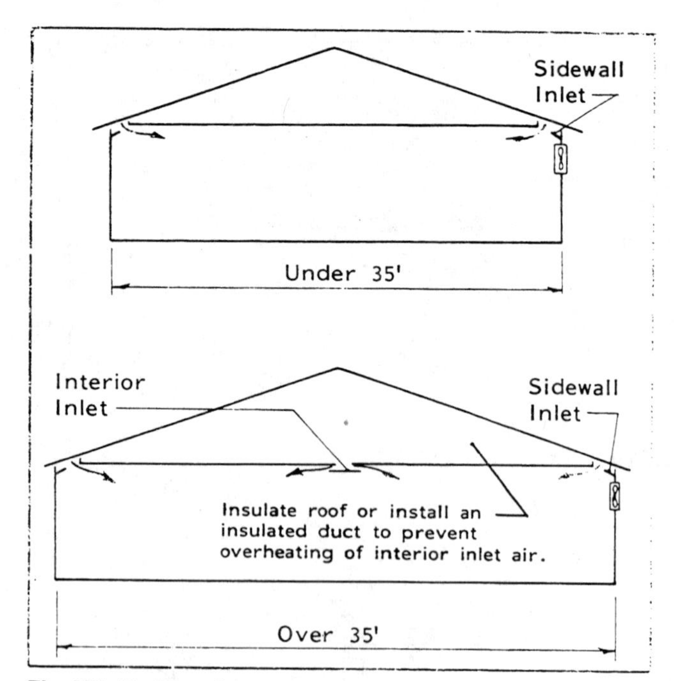

Fig 634-14. Locating slot inlets.

Inlet size

Size inlets to maintain a high average velocity (600 to 1200 fpm) where air leaves the baffle to get good mixing of cold and warm air. If the velocity is less, cold air settles too rapidly and may chill the animals; if greater, high static pressure decreases fan capacity and efficiency, Fig 634-15. Usually, inlets are sized for the hot weather rate, then reduced with a baffle for lower ventilating rates.

Example 2:

Calculate slot openings for a 24'x84', 30-sow farrowing building. Group exhaust fans in the center of the south wall and locate inlets at the ceiling on the long sides of the room. Design for 0.04" W.G. static pressure.

1. Total slot inlet length is: (building length x 2):
   (84' x 2) = 168'.
   In cold weather, close the inlets over the continuous exhaust fans to prevent short circuiting of air. **Winter** slot opening is about 150'.
2. Maximum ventilating rate per foot of inlet is: (hot weather rate, from Table 633-2 x number of sows) ÷ total slot inlet length:
   (500 cfm x 30 sows) ÷ 168' = 89.3 cfm/ft
3. A 1" wide slot provides 50 cfm per foot of length with a 0.04" static pressue drop (Table 634-3), so the hot weather slot width is:
   (89.3) ÷ (50 cfm/in.) = 1.8" or about 2" wide
4. In cold weather, the ventilating rate is much less: (cold weather rate, from Table 633-2 x number of sows) ÷ total slot inlet length:
   (20 cfm x 30 sows) ÷ 150' = 4.0 cfm/ft
5. To provide this cold weather rate, partly close the slot with a baffle, to:
   (4.0) ÷ (50 cfm/in.) = 0.08".
   In practice, a 0.08" opening is too small to control accurately. Close every other 4' section of inlet and double the slot opening in the alternate 4' sections.

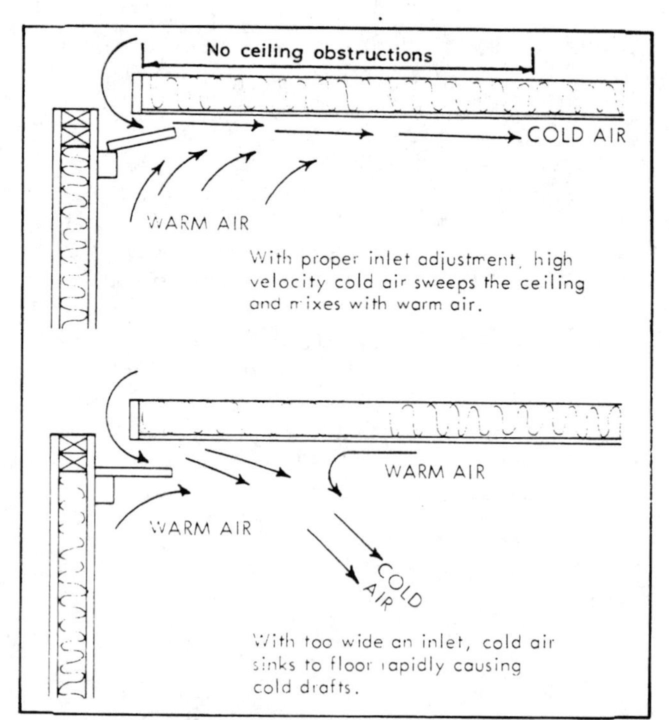

Fig 634-15. Proper winter inlet width.

Unplanned air leaks make it difficult to maintain adequate air velocity at low ventilating rates. Unless your building has very tight construction, decrease the inlet opening by 30%-50% to account for air leaks.

Cold weather inlets

In winter, bring fresh air from the attic. Run moisture control fans **continuously,** because when fans are off, warm room air rises through the slot, condenses on the underside of the cold roof, and drips on attic insulation. Avoid timers or install backdraft curtains on the inlets. Air enters the attic through downwind soffit openings, gable louvers, or ridge ventilators, Fig 634-16.

Hot weather inlets

In hot weather, pull fresh air directly from the outside instead of the attic. To do this, open the doors in both eaves, Fig 634-11a. Some air entering the eaves ventilates the attic to lower the ceiling temperature. Screen attic openings with ½" hardware cloth to keep birds out. Smaller mesh screen may plug with dust and restrict airflow.

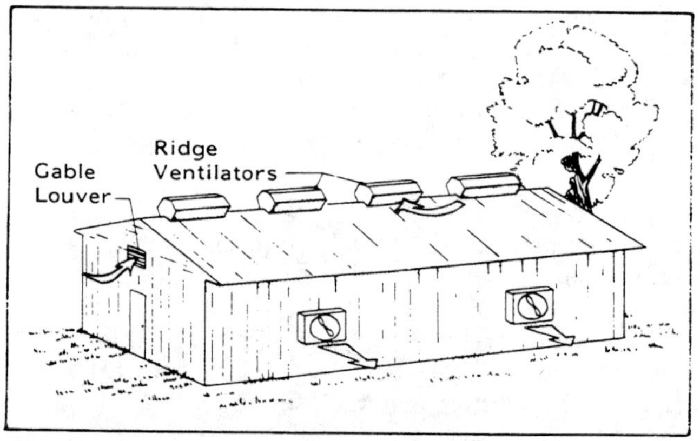

Fig 634-16. Gable louvers and ridge ventilators.
Screen attic openings with ½" hardware cloth. Total attic opening, ft² = (mild weather ventilation, cfm) ÷ 200.

Inlet control

Air inlet control is critical to good ventilation. The size of the slot inlet should change each time the ventilating rate changes, preferably automatically. Operate manually-adjusted baffles from one location with a winch and cable system. Install a manometer next to the winch for more accurate baffle adjustments.

A manometer measures static pressure, which is the difference in atmospheric pressure between the inside and the outside of the building measured in inches of water on a column or water gauge (W.G.). It indicates the resistance that fans must overcome to move air through the building, Fig 634-17. Ventilating systems for farm buildings are usually designed for 0.04" to 0.125" W.G. static pressures.

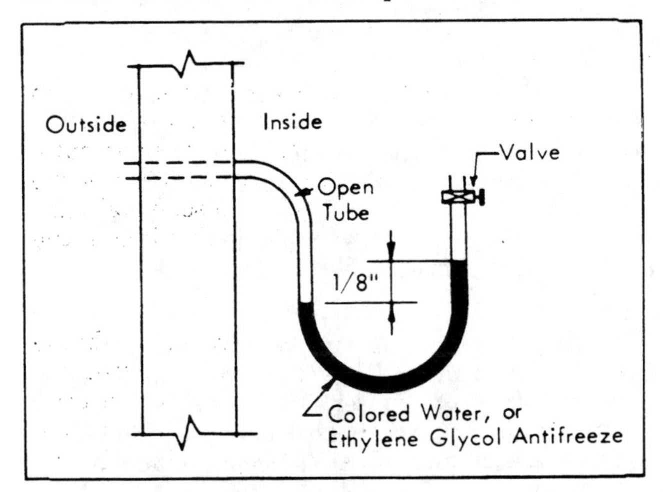

Fig 634-17. Manometer.
This water column illustrates a ⅛" pressure difference between inside and outside when exhaust fans are operating.

For continuous slot inlets in exhaust systems, a slot width sized to deliver air at 600 to 1200 fpm creates a static pressure of 0.04"-0.125" W.G. Keep doors and windows closed when the ventilating system is running, or little air is drawn in through the slot inlets.

Curtain-controlled slot inlets are self-adjusting but are not as precise as rigid baffles, Fig 634-18. They tend to have low inlet velocity and to deflect the cold air downward. Condensation and frost on the curtain and ceiling are often a problem. Plastic curtains tend to become less flexible with age, so check them at least once a year and replace as needed.

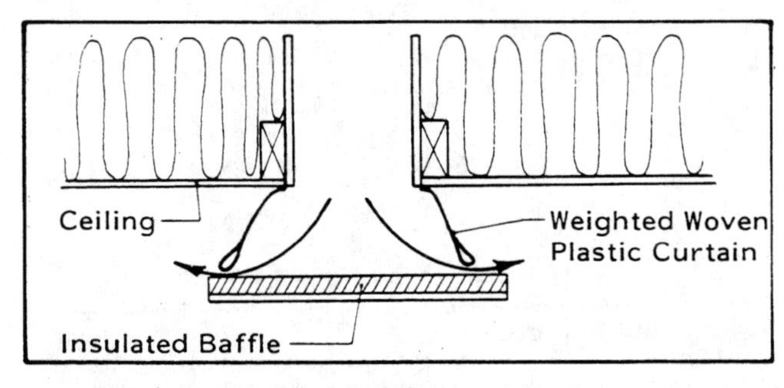

Fig 634-18. Gravity curtain inlet.

Fans

Fan types

Fans move the required ventilating air through the building. There are two main types—axial and centrifugal, Fig 634-19.

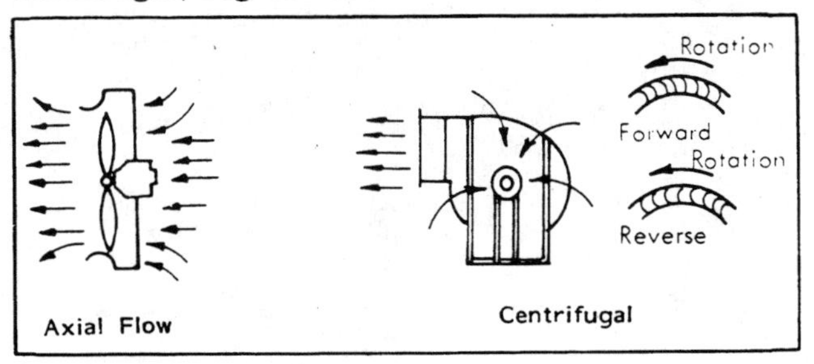

Fig 634-19. Types of fans.

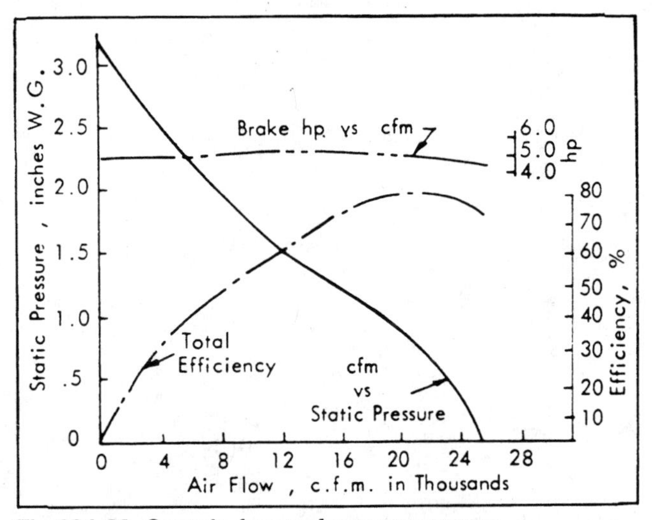

Fig 634-20. Sample fan performance curves.
For an axial flow fan.

Motors and drives

Fan motors should be:

- Totally enclosed for dust protection.
- Able to operate continuously.
- Equipped with overload protection.

There are several types of fan motors, including capacitor-start, two-speed, and variable-speed.

Fans can be direct- or belt-driven. Speeds of belt driven fans may be varied with a step pulley or a variable sheave pulley. Direct-driven fans generally require less maintenance and have less power transmission loss than belt driven fans.

Selection

The most important feature for ventilating fans is **airflow capacity** at given **static pressures,** usually specified in cfm vs. inches water. Cfm/watt is a measure of fan efficiency; consider it after airflow and static pressure needs are met. Diameter, speed, horsepower, shape of blade, and other features are less important to you. Some typical fan performance data is shown in Table 634-4.

Select a fan capable of moving the required amount of air against at least ⅛" static pressure. Look

Table 634-4. Typical fan performance data.

Fan Speed rpm	Fan Size in.	Fan capacity, cfm Free air	Fan capacity, cfm ⅛" W.G.	Motor h.p.	Type drive
Single-speed					
1100	10	560	380	1/20	Direct
1725	10	910	840	1/6	"
1150	12	1240	780	1/8	"
1725	12	1640	1520	1/6	"
1750	12	1760	1575	1/8	"
1050	14	1250	1020	1/20	"
1725	14	2100	2005	1/6	"
1140	16	1950	1780	1/8	"
1150	16	2630	2190	1/6	"
1725	16	2800	2680	1/4	"
1750	16	3000	2770	1/4	"
1140	18	2700	2380	1/8	"
1150	18	3110	2620	1/6	"
1725	18	2845	2680	1/4	"
1725	18	4000	3860	1/3	"
1750	18	4130	3720	1/2	"
1150	20	3620	3240	1/4	"
1725	20	2920	2485	1/4	"
1150	20	4430	3850	1/3	"
1725	20	4200	4120	1/3	"
860	24	4800	3770	1/4	"
860	24	5630	4950	1/3	"
1140	24	4400	4100	1/4	"
1140	24	6500	6200	1/2	"
1725	24	3975	3725	1/3	"
690	30	8800	7780	1/2	"
690	30	9250	8280	3/4	"
860	30	11000	10150	1	"
650	30	8570	7520	1/8	Belt
545	30	9950	8200	1/2	"
615	30	11200	9700	3/4	"
650	30	8570	7520	1/2	"
670	30	12100	10700	1	"
816	30	10850	10000	1	"
476	36	10900	9550	1/2	"
545	36	12450	11400	3/4	"
505	36	14200	11900	3/4	"
367	42	13400	10830	1/2	"
420	42	15300	13330	3/4	"
462	42	16800	15200	1	"
Two-speed					
1725	16	2534	2353	1/3	Direct
1140		1675	1374		
1140	18	2686	2395	1/6	"
855		2015	1573		
1725	18	4065	3880	5/8	"
1140		2686	2395		
1100	20	2920	2760	1/4	"
800		1930	1630		
855	24	4691	4180	1/3	"
570		3127	2105		
1100	24	5520	4800	1/3	"
800		4010	2800		

at the low-flow characteristics of your minimum-rate fan. It should deliver some air against about ¼" static pressure to avoid running backwards in high wind. Variable-speed fans have poor pressure ratings at low speeds and may not deliver enough air against wind.

Use fans designed specifically for animal housing and that meet the test standards of the Air Movement and Control Association (AMCA). AMCA fan ratings and performance charts certify a fan will deliver the volume of air specified for the static pressure and with the accessories in place as listed or illustrated in the certification. Also consider:

- Corrosion resistant blades and housing.
- A smooth streamlined ring around the blade tips.
- Size and type of shutters.
- Type of motor.

Installation

Protect each fan with a fused switch at the fan to protect against burned-out motors and electrical overloads, especially when the equipment is automatically controlled. Size the switch at 25% over the fan amperage and use a time-delay fuse. Install fans on separate branch circuits, so if one circuit fails, the other fans will still operate.

Fan location

With a properly designed and operated rigid baffle slot inlet, fan location has little effect on air distribution. If possible, place winter fans downwind from prevailing winds. To prevent air from becoming too stale, place fans no more than 75′ from the farthest part of the inlet.

If fans must exhaust against prevailing winter winds, protect them with a windbreak 5′-10′ from the building, Fig 634-21. Fan hoods can reduce wind effects. Put anti-backdraft shutters on all non-continuous fans. Place shutters on the inside (animal side) of fans to decrease freezing problems. Space mechanically ventilated buildings at least 35′ apart, so fans do not blow foul air into the intakes of adjacent buildings.

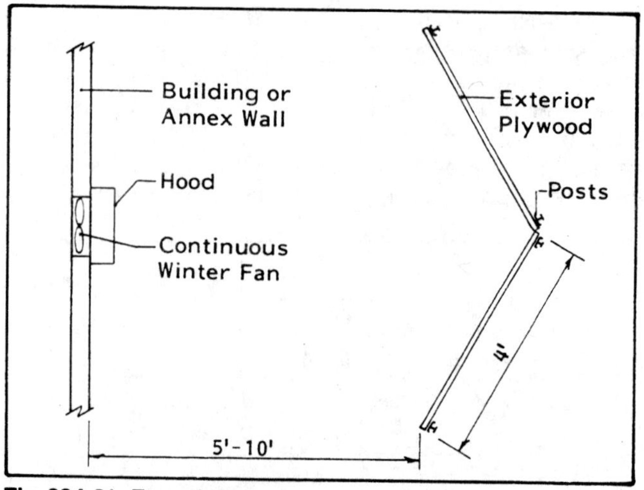

Fig 634-21. Fan windbreak.
Windbreaks for continuous fans help maintain more uniform ventilation. They are especially useful for variable-speed fans operated at the lower half of their capacity range.

Fan controls

Choose thermostats designed especially for livestock housing with an on-off range of 5 F or less. Locate the thermostat:

- At or near center of building width.
- Out of animal reach.
- Away from cold walls and ceilings.
- About 4'-5' off the floor.
- Out of the path of furnace exhausts, inlet air, and direct sunlight.

Several control circuits are illustrated in Fig 634-22 to 634-26.

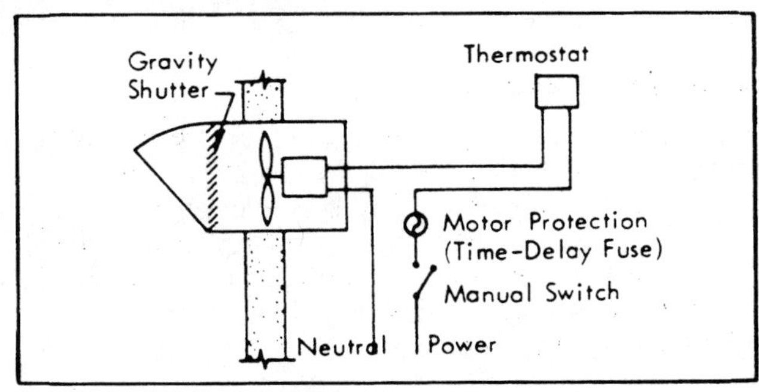

Fig 634-22. Single-speed fan and thermostat.

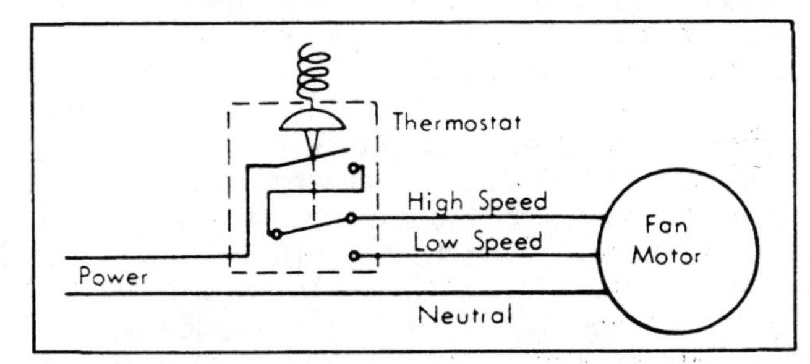

Fig 634-23. Two-speed fan and thermostat.

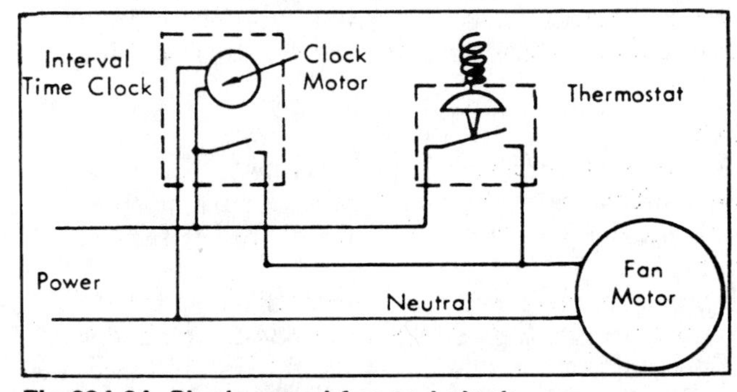

Fig 634-24. Single-speed fan and clock.
An interval timer is set so the fan delivers the minimum volume needed. A thermostat overrides the timer so the fan is on continuously at higher indoor temperatures.

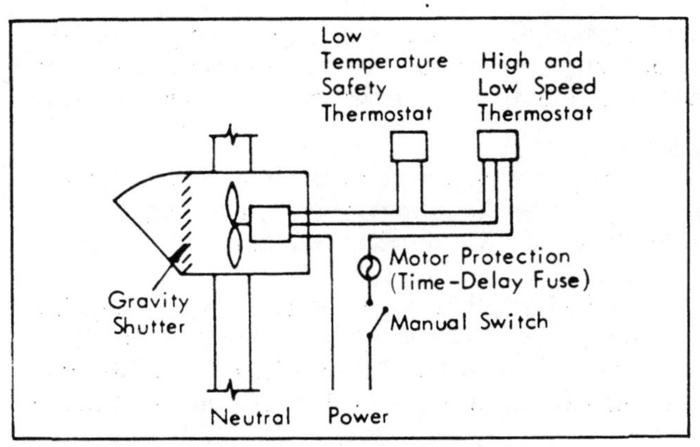

Fig 634-25. Two-speed fan thermostats.
Showing high-low and low temperature on-off thermostats.

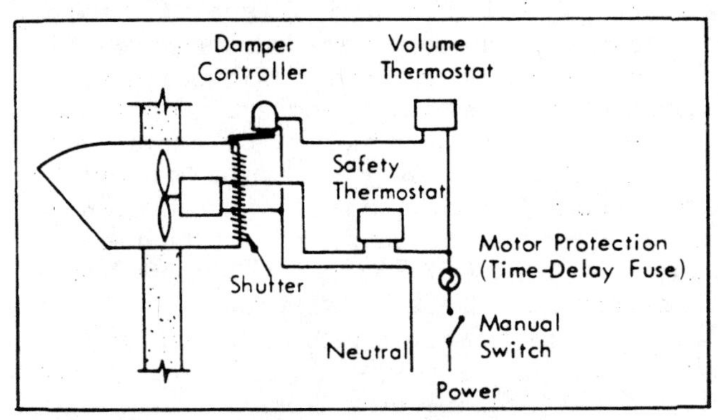

Fig 634-26. Variable air intake system.
A single-speed 2-volume fan is controlled with a thermostat on the shutter-damper and with a low temperature on-off safety thermostat.

Fan sizing and thermostat settings

Required ventilation varies with outside temperature. Ventilating air requirements vary from the minimum winter capacity to many times this value for summer. Fan capacity must change as outside temperature changes to meet changing ventilation needs.

- **Cold weather.** This lowest ventilating rate is continuous. It provides minimum air quality.
- **Mild weather.** Additional capacity controls indoor air temperature.
- **Hot weather.** Much higher rates in hot weather limit indoor air temperature rise and increase air velocity to reduce animal stress.

Select fan capacities and thermostat settings for a **gradual** airflow increase with an outside temperature increase. Use several fans, multi-speed fans, or variable-speed fans. Occasionally a multi- or variable-speed fan can provide both the mild and hot weather ventilating rates.

A single-speed fan with an interval timer allows a wide range of capacities, but requires manual setting of the timer for each adjustment. A 1,000 cfm fan **on 3** minutes and **off 7** minutes provides an average airflow of 300 cfm, but may cause relatively rapid temperature changes. Provide anti-backdraft curtains on inlets from an attic if fans are on timers.

Two or more single-speed fans or multi-speed fans provide a range of capacities.

Variable-speed fans modulate airflow from a minimum of about 20% capacity. A solid-state control regulates the voltage going to the capacitor-start, capacitor-run motor.

Find the recommended per head ventilating rates in Table 633-2, and multiply times the number of animals in the building or room to determine the ventilating rates. Select fan sizes from product literature.

Most thermostats have a 3-5 F range between when they turn on and when they shut off, so fans are normally staged to come on in 2-4 F increments.

Example 3:

For the 30-sow farrowing building of Example 2, Table 633-2 gives the following ventilating rates:

- Cold weather: 30 sows x 20 cfm/sow = 600 cfm total
- Mild weather: 30 sows x 80 cfm/sow = 2400 cfm (Additional capacity = 2400 − 600 = 1800 cfm)
- Hot weather: 30 sows x 500 cfm/sow = 15000 cfm (Additional capacity = 15000 − 2400 = 12600 cfm)

Choose the desired room temperature (70 F). Set heater thermostats about 2 F below the desired room temperature (68 F). With a 5 F range on the furnace thermostat, the furnace turns on at 68 F and shuts off at 73 F. The cold weather fan runs continuously so it does not have a thermostat. Set the thermostat of the smallest mild weather fan at about 3 F above the furnace-off temperature (76 F). Make sure the heater does not operate after the mild weather fans turn on. Set the other warm weather fans to turn on at increasing increments of 2-4 F (78 F, 80 F, etc.). The more fans in the room, the smaller the increments.

If variable-speed fans are used for cold weather ventilation, adjust the minimum speed to the cold weather rate (600 cfm). Set the fan controller to begin increasing speed at about 3 F above the furnace-off temperature (76 F). Most variable-speed fans increase speed over a range of about 5 F. The cold weather variable speed fan runs continuously so it does not have an on-off thermostat (although it has a variable-speed controller which controls the fan speed but never stops the fan). Except for totally slotted floors over storage pits, an emergency thermostat can be installed to shut the fan off at 35 F. Set the other fans to come on in 2-4 F increments, starting at about 5 F higher than the variable-speed fan setting (81 F, 83 F, etc.).

Table 634-5. Supplemental heat per animal.
Sized for twice the cold weather ventilating rate in a moderately well insulated building in the North Central U.S. Additional zone heat may be needed for young animals.
These heat requirements are peak loads for sizing heaters.

Animal type	Inside temp, F	Supplemental heat, Btu/hr-hd Slotted floor	Bedded/ scraped floor
Swine			
Sow and litter	80	4000	—
	70	3000	—
	60	—	3500
Prenursery pig (12-30 lb)	85	350	—
Nursery pig (30-75 lb)	75	350	—
	65	—	450
Growing-finishing pig (75-220 lb)	60	600	—
Gestating sow/boar	60	1000	—
Dairy			
Calf housing		1,000 Btu/hr	
Double 4-HB parlor		50,000	
-6 or -8 HB parlor		70,000	
Milkhouse		10,000	
Sheep			
Lambing		400 Btu/hr-ewe Plus heat lamps	
Horse			
Warm barn		5,000 Btu/hr per stall	

Manure Pit Ventilation

Properly designed pit ventilation reduces manure gas in the animal area, reduces odor levels, and helps warm and dry the floor. Totally and partly slotted floor buildings may benefit from properly designed pit ventilation, even though manure may be removed frequently by flushing or scraping.

There are two types of systems:

- **A duct** the length of the pit with small openings spaced to draw air uniformly from the pit.
- **An annex** outside the pit wall with a fan to draw air directly out of the pit.

Regardless of the system, allow at least 12" clearance between the bottom of slat support beams and the manure surface. Variable-speed fans are not recommended unless the lowest rate needed is at least 40% of the fan's maximum rated capacity or unless the fan is protected with a windbreak, Fig 634-21.

Pit fans encounter very corrosive conditions—obtain fans that are constructed from corrosion resistant materials like stainless steel or plastic. Obtain high static pressure fans (0.2"-0.6") to draw air through pit ducts, especially rigid plastic pipe ducts.

Ducts

Ducts are expensive but perform better than annexes. They work especially well with wire mesh or other slotted floors that have a high percentage of open area.

Ducts can be under the center alley of totally slotted buildings, Figs 634-27 and 634-29, or under the floor adjacent to the pit of partly slotted buildings, Fig 634-28. Larger ducts are usually constructed from concrete or plywood, while smaller ducts can be rigid plastic pipe. Select ducts and fans to ventilate at the cold weather rate. Try to limit distance between fan and the farthest point in the duct to 60′.

Large ducts

Size ducts and inlet openings with Tables 634-6 and 634-7. Place a duct inlet at each pen or stall. In long buildings, place a fan at each end and/or at the center of the duct to reduce duct airflow, and permit a smaller duct cross-sectional area.

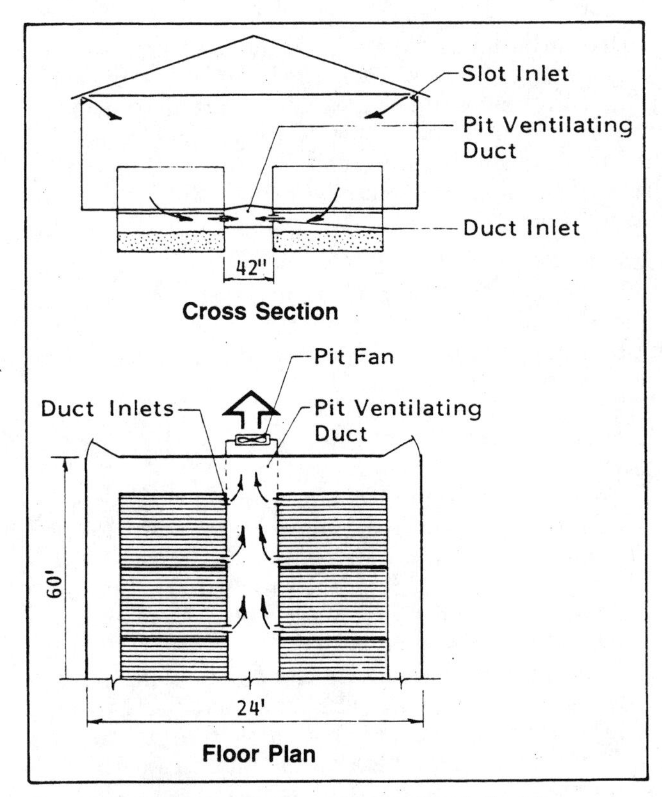

Fig 634-27. Pit ventilation with totally slotted floor.

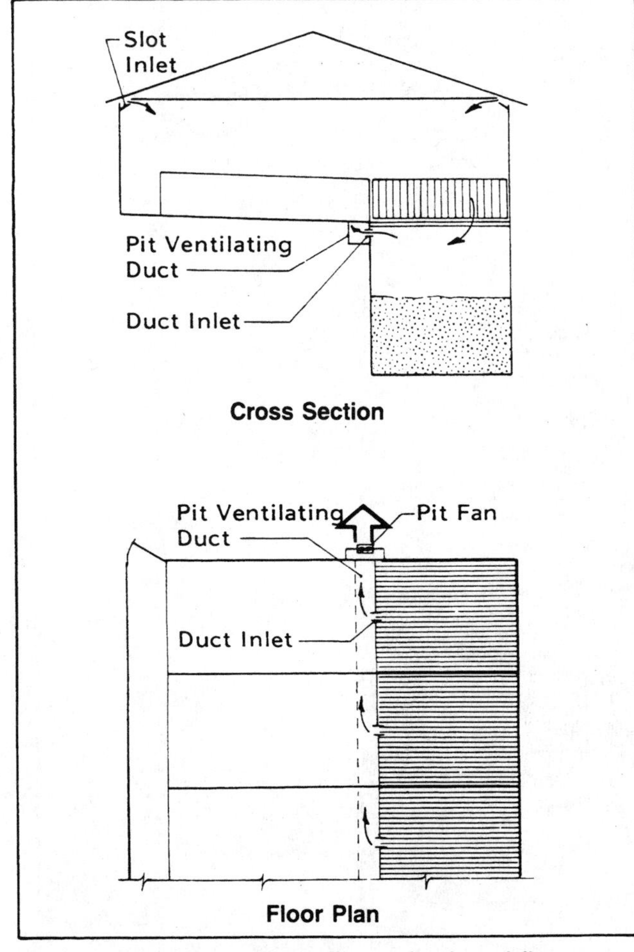

Fig 634-28. Pit ventilation with partly slotted floor.

Table 634-6. Ventilating duct sizes.
Based on a duct air velocity of 600 fpm.

Airflow rate cfm	Minimum duct size Area in²	Inside dimensions WxH, in.	WxH, in.	diam, in.
200	48	6x8	4x12	8
250	60	8x8	6x10	9
300	72	8x9	6x12	10
350	84	9x10	6x14	10
400	96	10x10	6x16	12
500	120	10x12	6x20	12
600	144	12x12	6x24	14
700	168	12x14	8x22	15
800	192	14x14	8x24	16
900	216	14x16	8x28	18
1000	240	16x16	8x30	18
1250	300	18x18	10x30	
1500	360	18x20	12x30	
2000	480	20x24	14x36	
2500	600	24x26	16x38	
3000	720	24x30	18x40	
3500	840	28x32	18x48	
4000	960	30x32	20x48	
5000	1200	34x36	24x50	
6000	1440	36x40	24x60	
7000	1680	40x42	28x60	
8000	1920	44x44	32x60	
9000	2160	46x48	36x60	
10000	2400	48x50	40x60	
12000	2880	54x54	48x60	
15000	3600	60x60	50x72	

Table 634-7. Inlets to pit ventilating ducts.
Put inlets at each pen or stall to improve room air distribution. These sizes are critical—do not vary. Based on an air velocity of 800 fpm through the opening.

Airflow cfm/pen	Opening area per pen in²	Net inlet sizes per pen HxW, in.	diam, in.
20	3.75	2x2	2¼
25	4.5	2x2¼	2½
30	5.5	2x2¾	2¾
35	6.25	2x3¼	3
40	7.25	2x3¾	3½
60	10.75	3x3¾	3¾
80	14.5	3x5	
100	18	4x4½	
125	22.5	4x5¾	
150	27	4x6¾	
200	36	4x9	
250	45	4x11¼	
300	54	2-4x6¾	
350	63	2-4x8	
400	72	2-4x9	
450	81	2-4x10¼	
500	90	2-4x11¼	
550	99	2-4x12½	
600	108	2-4x13½	
700	126	2-4x15¾	

Example 4:

Design duct pit ventilation for a 24'x60' slotted floor, 20-stall farrowing building that has 42" between pits, Fig 634-27.

1. From Table 633-2, pit ventilation = (20 cfm/sow x 20 sows) = 400 cfm.
2. Select a fan for 400 cfm at ⅛" static pressure.
3. From Table 634-6, 400 cfm requires a 96 in^2 duct. If the duct width is 42", the depth has to be at least (96 in^2 ÷ 42") = 2¼"; but an 8" depth could be used to simplify construction.
4. From Table 634-7, a per-stall ventilating rate of 20 cfm requires a duct opening of 3¾ in^2 or about a 2¼" diameter hole.

Rigid pipe ducts

(This section prepared by P.D. Bloome and R.L. Huhnke, Oklahoma State University.)

Rigid plastic pipe makes excellent ducts for small buildings. Space ducts across the building width so no portion of the slotted floor is more than 12' from a duct. Use Figs 634-30 and 634-31 to determine the required pipe size. Use Table 634-8 to determine the required hole size and spacing. We have assigned a duct identification number to each combination of pipe diameter, inlet hole diameter, and inlet spacings—see Table 634-8.

Drill inlet holes with a hole saw to ensure a uniform opening with **smooth edges.** Drill a ½" diameter hole in the bottom of the duct at several locations to drain any moisture. Dust tends to accumulate in the duct, especially at an elbow or annex. Construct the duct so a cleaner can be drawn through it to remove dust.

Example 5:

Design a rigid plastic ventilating duct for a 24'x66' totally slotted wire mesh floor, 24 sow farrowing building, Fig 634-29. Maximum distance of any slotted floor from a center duct is:

(24' ÷ 2) = 12'

which is equal to the 12' limit so one center duct is okay.

1. Select fan. The pit fan must supply the desired airflow at a static pressure of 0.2"-0.6". From Table 633-2, 24 sows require (20 cfm/sow x 24 sows) = 480 cfm for cold weather ventilation. Assume the fan selected can deliver 480 cfm at 0.21" static pressure.
2. Determine static pressure loss due to background pressure. Background pressure is created by airflow resistance at the building ventilating inlets. The pit fan pulls air into the building as well as through the pit duct. This pressure loss usually

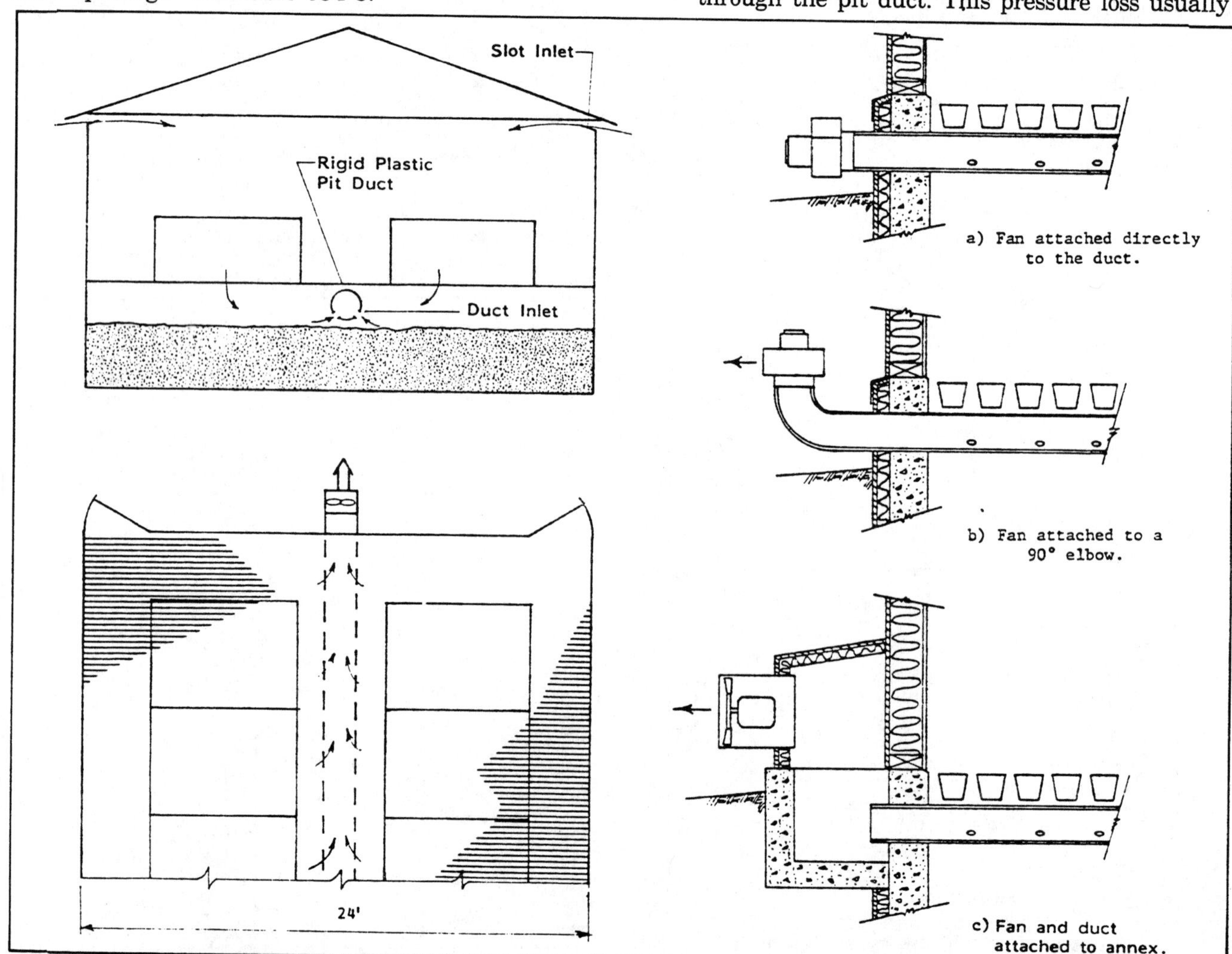

Fig 634-29. Rigid plastic tube pit duct.

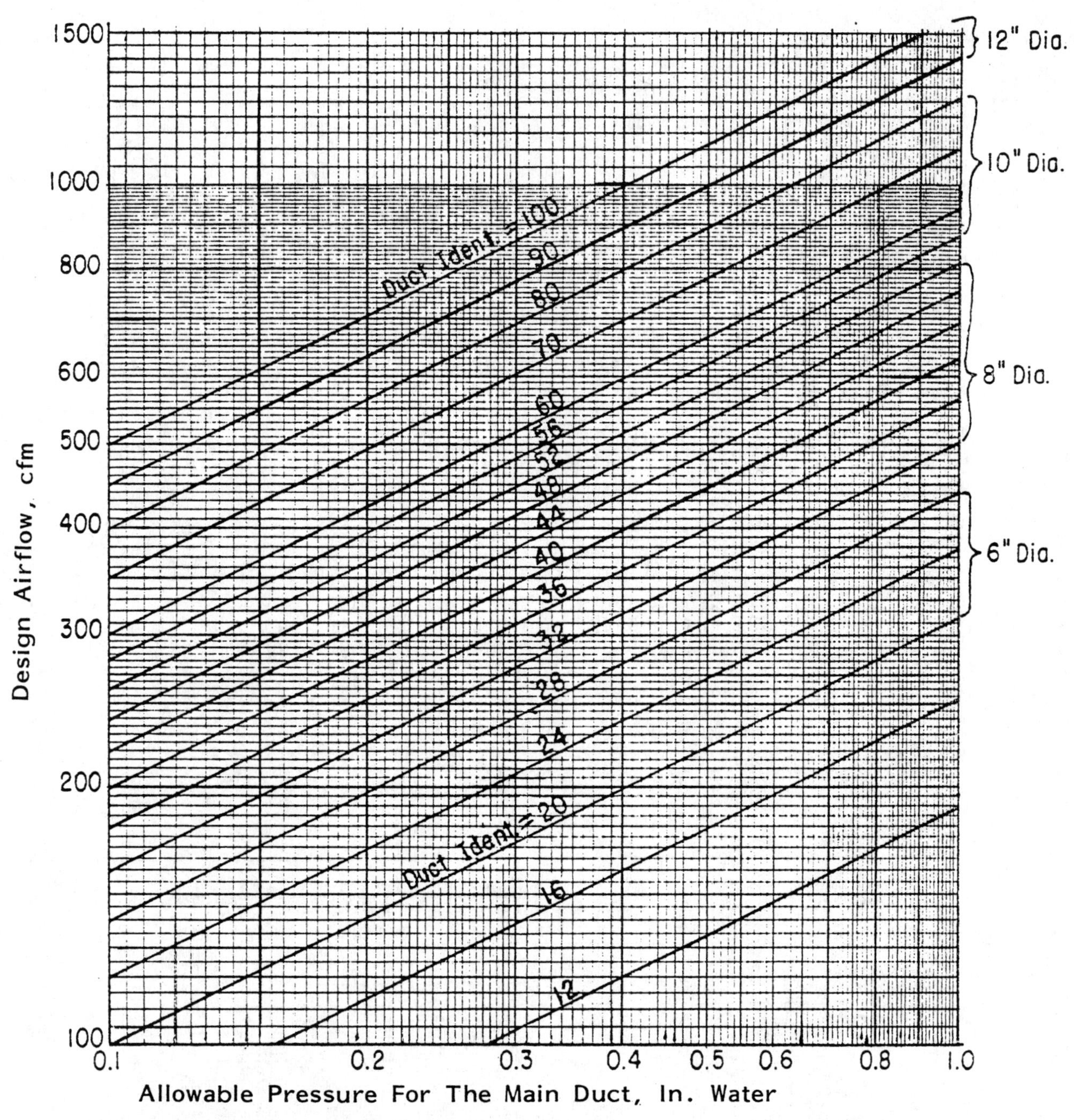

Fig 634-30. Rigid plastic pipe duct selection.

ranges from 0.04″-0.06″. Assume 0.05″, so total pressure available for the duct equals:

Fan pressure (in.) − Background pressure (in.) = (0.21″ − 0.05″) = 0.16″

3. First, base duct diameter on the **main** duct pressure loss and airflow. The **main** duct is the section between the first and last inlet holes. From Fig 634-30, a pressure loss of 0.16″ and an airflow of 480 cfm yield a #80 duct. The duct size may increase after step 4.
4. Determine pressure loss due to friction in the **attachment** duct. The **attachment** duct is the connecting duct between the main duct and the fan, Fig 634-32.

a. Calculate the **effective** attachment duct length (Eff ADL):

Eff ADL, ft = (number of 90° elbows x duct diam, in.) + (actual ADL, ft)

In the example:

Eff ADL = (1 elbow x 10″ diam.) + 5′ length = 15′.

b. From Fig 634-31, determine the pressure loss per foot of Eff ADL. For 480 cfm airflow and a 10″ diameter duct, pressure loss is 0.11″/100′ of Eff ADL.

c. Calculate total pressure loss across the attachment duct:

Table 634-8. Inlet hole size and spacing.

For rigid plastic pipe ducts. We have assigned a duct identification number to each combination of pipe diameter, inlet hole diameter, and inlet spacings. Multiply the table value (% of main duct length to inlet from first inlet) by the main duct length to locate each inlet from the first inlet.

Duct ident.	20	24	28	32	36	40	44	48	52	56	60	70	80	90	100
Duct dia. (in.)	6	6	6	8	8	8	8	8	8	10	10	10	10	12	12
Dia. of inlets (in.)	1	1	1	1	1	1¼	1¼	1¼	1½	1½	1½	1½	1¾	1¾	1¾
	% of main duct length to inlet from first inlet														
Inlet															
1	0.0	0.0	0.0	0.0	0.0	0.0	0.0	0.0	0.0	0.0	0.0	0.0	0.0	0.0	0.0
2	10.4	8.3	7.3	6.6	5.8	7.9	7.4	6.7	8.5	8.1	7.8	6.4	7.8	7.1	6.1
3	20.3	16.1	14.1	13.0	11.4	15.5	14.5	13.2	16.4	15.9	15.4	12.6	15.2	14.0	11.9
4	30.0	23.7	20.7	19.3	16.9	22.8	21.4	19.4	23.8	23.6	22.8	18.6	22.2	20.6	17.6
5	39.4	31.0	26.9	25.6	22.4	30.0	28.0	25.4	30.7	31.0	29.9	24.4	28.8	27.1	23.1
6	48.5	38.0	32.8	31.7	27.7	36.9	34.4	31.2	37.2	38.3	36.9	30.0	35.1	33.4	28.4
7	57.4	44.7	38.5	37.8	33.0	43.7	40.7	36.7	43.3	45.4	43.8	35.4	41.0	39.6	33.5
8	66.1	51.3	43.9	43.8	38.1	50.3	46.7	42.1	49.1	52.4	50.4	40.6	46.7	45.6	38.5
9	74.6	57.6	49.0	49.7	43.2	56.7	52.6	47.3	54.6	59.3	57.0	45.7	52.1	51.5	43.4
10	83.0	63.8	54.0	55.5	48.3	63.0	58.3	52.3	59.7	66.1	63.4	50.7	57.2	57.3	48.2
11	91.3	69.8	58.8	61.3	53.2	69.2	63.9	57.1	64.7	72.8	69.8	55.5	62.1	62.9	52.8
12	100.0	75.8	63.4	67.0	58.1	75.4	69.3	61.9	69.4	79.4	76.0	60.2	66.9	68.5	57.4
13		81.6	67.8	72.7	63.0	81.4	74.7	66.5	73.9	86.0	82.2	64.8	71.4	74.0	61.8
14		87.3	72.2	78.4	67.8	87.4	80.0	71.0	78.3	92.5	88.3	69.3	75.8	79.4	66.2
15		93.0	76.8	84.0	72.5	98.3	85.2	75.4	82.5	100.0	94.4	73.8	80.1	84.7	70.4
16		100.0	80.4	89.6	77.3	100.0	90.4	79.8	86.6		100.0	78.2	84.2	90.0	74.7
17			84.4	95.2	82.0		95.2	84.0	90.6			82.5	88.3	95.3	78.8
18			88.4	100.0	86.6		100.0	88.2	94.6			86.8	92.3	100.0	82.9
19			92.2		91.3			92.4	100.0			91.0	96.2		87.0
20			96.1		95.9			96.5				95.2	100.0		91.0
21			100.0		100.0			100.0				100.0			95.1
22															100.0

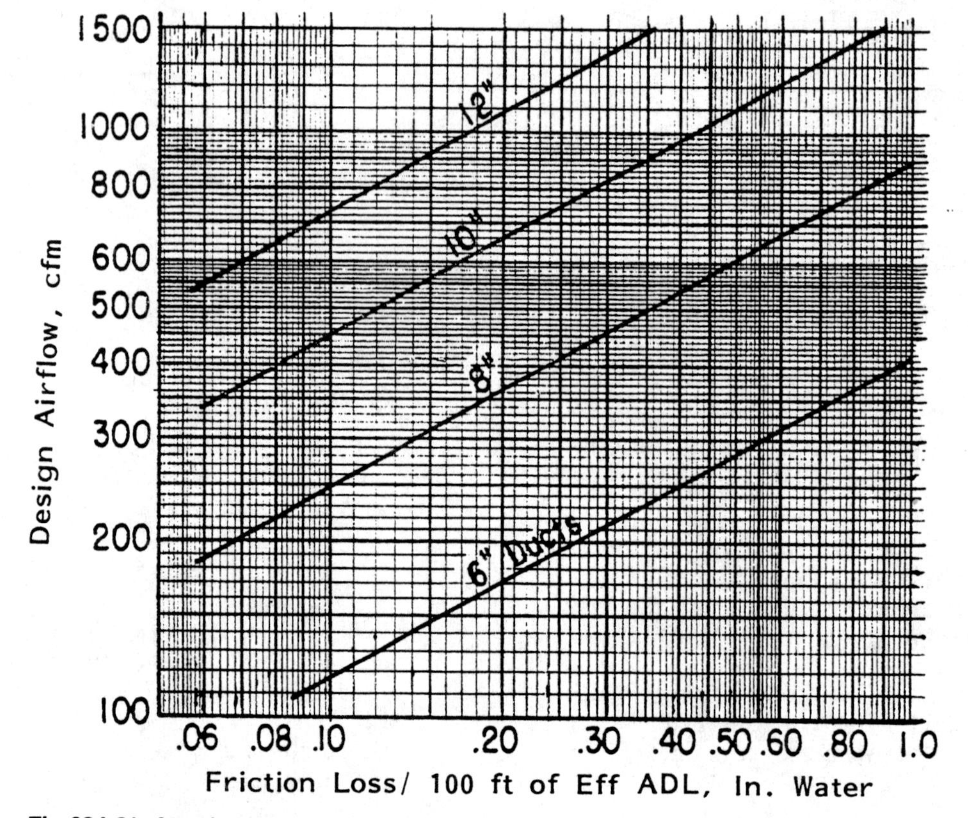

Fig 634-31. Attachment pressure loss.

Attachment pressure loss, in. = Eff ADL, ft x pressure loss/100′ duct, in/100′.

Attachment pressure loss = 15′ x 0.11″ ÷ 100′ = 0.0165″

5. Calculate the pressure available for the main duct.

   Available pressure for main duct, in. = total pressure across duct, in. − pressure loss from attachment duct, in.

   Available pressure for the main duct = (0.16″ − 0.0165″) = 0.1435″

Background pressure	= 0.05″
Attachment duct pressure	= 0.0165″
Main duct pressure	= 0.1435″
Fan pressure capacity	= 0.21″

6. From Fig 634-30, determine if the available pressure for the main duct is sufficient for the duct determined in step 3. A pressure drop of 0.1435″ and an airflow of 480 cfm requires a #80 duct. If this check indicates #80 is too small, repeat steps 3, 4, 5, and 6 with the next larger duct or a higher static pressure fan.

7. Determine inlet hole size and spacings from Table 634-8. Locate the first and last holes about 2′ from the pit endwalls. Duct length between the first and last holes of the example buildings is 62′. Each inlet hole location has two holes—drill as shown in Fig 634-32.
   The table indicates 1¾″ diameter inlet holes for a #80 duct. Multiply the duct length between the first and last holes, Fig 634-32, by the values in Table 634-8 to determine the space of each set of inlet holes from the first holes, Table 634-9.

Table 634-9. Locations of inlet holes for example building.

Inlet	% duct length	Inlet location
1	0 x 62′	0′
2	0.078 x 62′	4.8′
3	0.152 x 62′	9.4′
4	0.222 x 62′	13.8′
5	0.288 x 62′	17.9′
6	0.351 x 62′	21.8′
7	0.410 x 62′	25.4′
8	0.467 x 62′	29.0′
9	0.521 x 62′	32.3′
10	0.572 x 62′	35.5′
11	0.621 x 62′	38.5′
12	0.669 x 62′	41.5′
13	0.714 x 62′	44.3′
14	0.758 x 62′	47.0′
15	0.801 x 62′	49.7′
16	0.842 x 62′	52.2′
17	0.883 x 62′	54.7′
18	0.923 x 62′	57.2′
19	0.962 x 62′	59.6′
20	1.000 x 62′	62.0′

Annexes

An annex is common in totally slotted buildings, Fig 634-33. It does not provide as uniform air distribution through the floor as a duct, but it is cheaper.

Annexes work best for concrete slats with only 15%-20% open area because of the duct effect created by the slats. Air distribution under wire or metal flooring with 50%-60% open area is poor.

Design annex pit ventilation to supply at least the cold weather rate but no more than the mild weather rate. Table 634-10 shows the minimum pit wall opening for various fan capacities. Larger openings will not significantly affect performance. Locate annexes so no point in the pit is farther than 50′ from an annex.

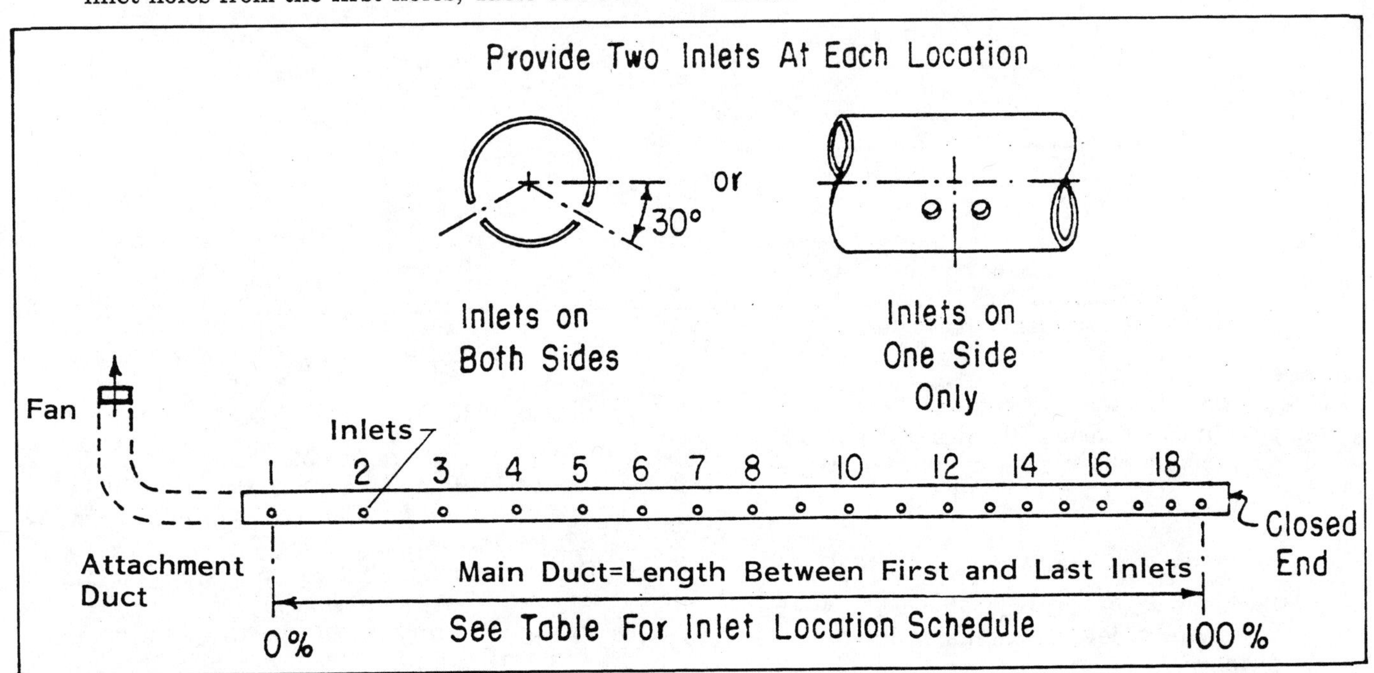

Fig 634-32. Inlet hole locations.
For rigid plastic pipe ducts.

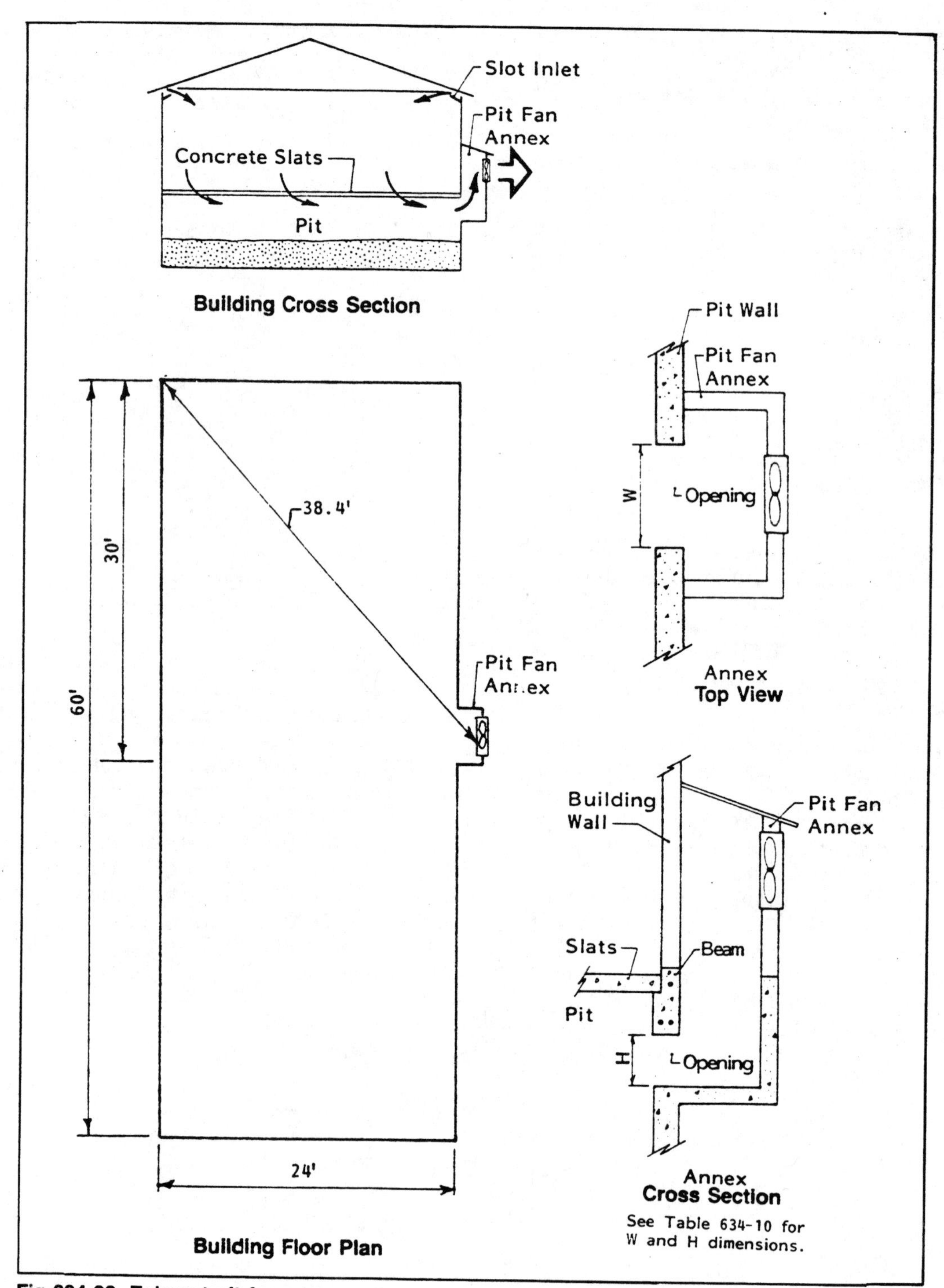

Fig 634-33. Exhaust pit fan annex.

Example 6:

Design annex pit ventilation for a 24′x60′ totally slotted pig-nursery housing 400 pigs, Fig 634-33.

1. From Table 633-2, the cold weather rate for nursery pigs is 3 cfm/pig. The pit ventilation is:

   (3 cfm/pig x 400 nursery pigs) = 1200 cfm.

2. With one annex in the center of the winter downwind sidewall, the maximum air-pull distance is:

   Square root[(60′ ÷ 2)² + (24′)²] = 38.4′,

   which is less than the 50′ limit.

3. Table 634-10 indicates a pit opening of 216 in² into the annex for 1200 cfm. Minimum opening for this rate is 6″x36″.
4. Select a pit fan that can provide 1200 cfm against ⅛″ static pressure.
5. Provide the mild and hot weather ventilation with wall fans.

Table 634-10. Minimum opening through pit fan annex.
Based on an air velocity of 800 fpm through the opening.

Pit fan capacity cfm	Opening area in²	Inside dimensions WxH, in.	diam, in.
400	72	4x18	10
500	90	4x23	12
600	108	4x27	12
700	126	4x32	13
800	144	4x36	14
900	162	6x27	15
1000	180	6x30	16
1100	198	6x33	16
1200	216	6x36	
1300	234	6x39	
1400	252	6x42	
1500	270	8x34	
1600	288	8x36	
1800	324	8x42	
2000	360	12x30	
2500	450	12x38	
3000	540	12x45	
3500	630	18x36	
4000	720	18x40	
5000	900	24x38	

Pit Fan Installation

Insulate and use corrosion resistant construction for pit fan housings. To prevent "short-circuiting" between fans, put each pit fan in its own annex. Shutters are not necessary on continuous fans.

If odors around the building are of concern, elevate the discharge to disperse them faster. Either vent the air high on the sidewall or add a chimney to exhaust even higher, Fig 634-34. Construct the chimney about 50% larger than the fan diameter to limit chimney air velocity to 1000 fpm or less. Put a drain hole at the base of the chimney for condensation or rainwater. Protect metal siding from corrosion with exterior plywood next to the building. Insulate the chimney liner to reduce condensation and frost.

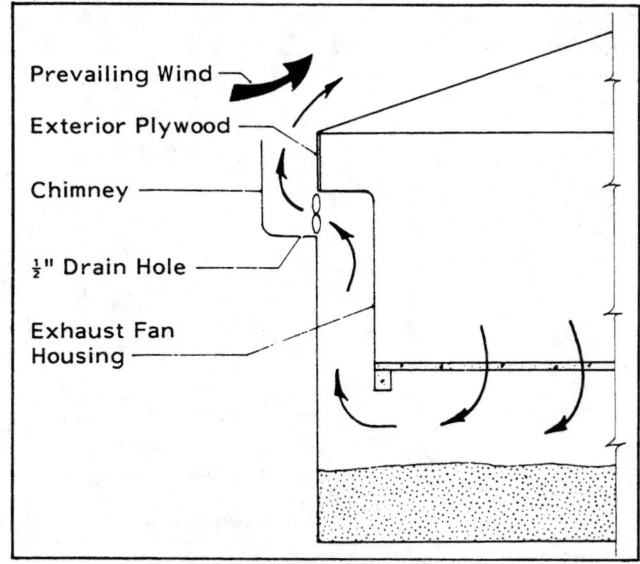

Fig 634-34. Chimney exhaust vents.
Locate vent on upwind side of building.

Chimneys also protect variable-speed fans from wind, allowing a more uniform ventilating rate. Exhausted materials may collect on metal roofs, causing corrosion. Check roofing around the chimney each year and retouch with a zinc base paint. Other solutions include extending the chimney 5′-6′ above the roofing for better dispersion or using asphalt base roofing.

System Maintenance

Environmental control systems require periodic maintenance that is conscientious and thorough. Develop a maintenance schedule from the following:

Every month:

- Clean fan blades and shutters. Dirty fan shutters can decrease fan efficiency by 25%. Shut off power to thermostatically controlled fans before servicing them.
- Check fans with belt drives for proper tension and correct alignment. If too tight, belts may cause excessive bearing wear; if too loose, excessive slippage and belt wear may result.
- During the heating season, remove dust from heater cooling fins and filters, and check gas jets and safety shut-off valves for proper operation.
- Test emergency ventilation and alarm systems.

Every 3 months:

- Make certain that shutters open and close freely. Apply graphite (not oil or grease) to fan shutter hinges.

Every 4 months:

- Clean motors and controls. Dirty thermostats do not sense temperature changes accurately or rapidly. Dust insulates fan motors and prevents proper cooling. If dust is allowed to build up, the motor can overheat.

Every 6 months:

- Add a few drops of SAE No. 10 non-detergent oil to oilable fan bearings. Never overlubricate. (Note: Most ventilating fans have sealed bearings and do not require lubrication.)

Every year:

- Clean and repaint chipped spots on fan housings and shutters to prevent further corrosion.
- During winter, cover hot weather fans (not cold or mild weather fans) with plastic or an insulated panel on the warm (animal) side of the fan and disconnect the power supply, or operate briefly every few weeks.
- Check slot air inlets for debris.
- Check gable and soffit air inlets for blockages.
- Check plastic baffle curtains. They can become brittle with age and require replacement.
- Check recirculation air ducts for dust accumulation.

Emergency Ventilation

Provide for emergency ventilation in all environmentally controlled buildings because of the danger of animal suffocation. The system can be as simple as several manually-opened sidewall doors or as sophisticated as an electric generator that starts automatically to power fans in case of electrical failure. Magnet-locked ventilating doors, which drop open when electrical power is cut off or room temperature rises sharply, are available commercially. Consider installing an alarm system to let you know when electrical power is off. Test your emergency ventilation and alarm systems monthly or according to manufacturers' instructions.

If fan failure is known, steps can usually be taken to prevent livestock losses. Fan failures may be mechanical and/or electrical. Electrical failure may be caused by such things as blown fuses, loose connections, broken wires, or power outage.

Warning systems cannot protect against all possible causes of fan failure, but the more common causes have been considered. Alarm systems are available commercially, Figs 634-35 to 634-38.

Battery maintenance. Some systems require a battery. Replace dry-cell batteries periodically and charge wet-cell batteries with a trickle-charger. Check water level in wet-cell batteries frequently.

Response time. The alarm shown in Fig 634-37 does not sound until conditions within the confined area are critical. The others respond to power failure.

Battery operated relay-controlled alarm

In Fig 634-35, a magnetic relay is connected to the fan circuit, so the relay is energized when there is power to the fan. The contacts of this normally closed relay are a switch for a battery operated alarm. Should the power fail, the contacts under spring tension close to sound the alarm. A type of magnetic relay rated at fan voltage for continuous operation with a single-pole, single-throw normally closed contacts may be used. A test switch (normally closed) should be included unless the fan is connected to a disconnect switch or circuit breaker.

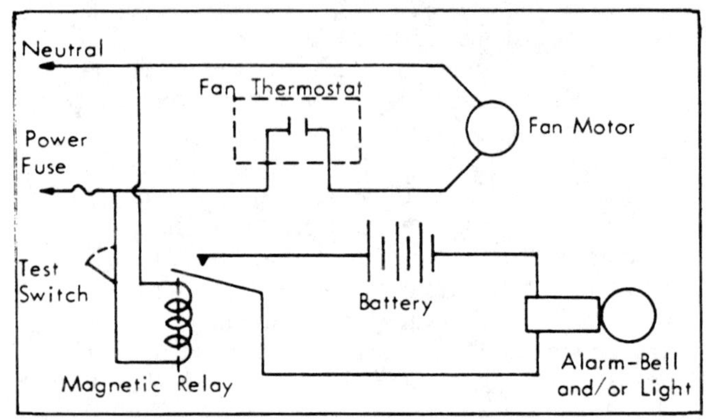

Fig 634-35. Battery operated relay-controlled alarm.

Solenoid valve-controlled compressed gas horn

In Fig 634-36, an air horn powered by an aerosol can (compressed gas), commonly used as fog horns or distress signals for small boats, is used with a normally open solenoid valve. The solenoid valve will be closed as long as there is power to the fan. With an interruption of power, the spring within the valve causes it to open, allowing the compressed gas to operate the horn. Again, include a test switch.

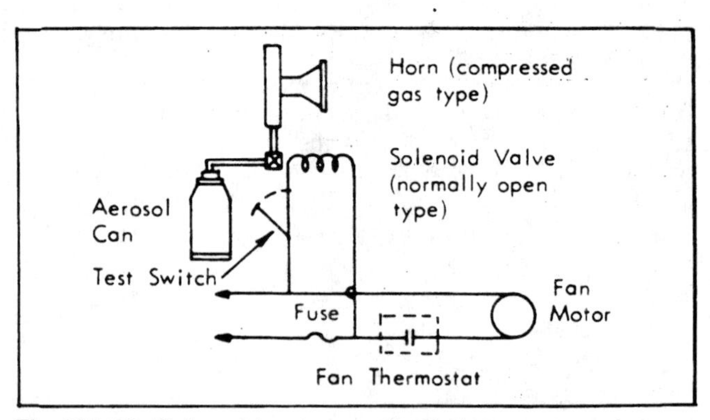

Fig 634-36. Solenoid valve-controlled compressed gas horn.

Battery operated alarm with thermostat

In Fig 634-37, a cooling thermostat (contacts close on temperature rise) is the sensing element, giving the alarm on a preset temperature, indicating the air temperature is above normal. In some cases the thermostat should be reset to adjust for seasonal changes—between 80-90 F under summer conditions, and lowered to 70-75 F in winter. This system may also indicate when a water spray or sprinkler should be turned on. Test the alarm circuit by changing the thermostat setting.

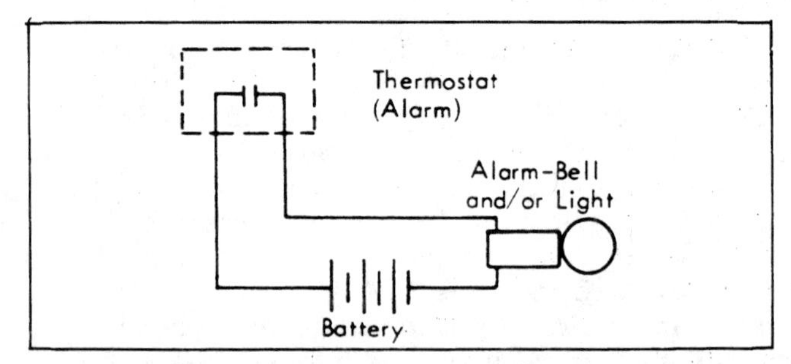

Fig 634-37. Battery operated alarm with thermostat.

Alarm for multiple fans

Wiring an alarm system for several fans requires a relay for each fan, but only one alarm, Fig 634-38. If the combination system is used, a relay is required for each fan and a thermostat for each pen.

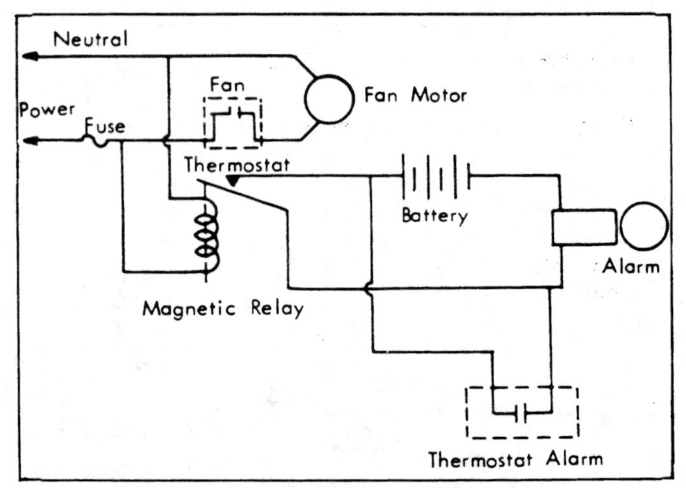

Fig 634-38. Temperature- and power-sensitive alarm.

Combination alarm system

An alarm combining circuits shown in Figs 634-35 and 634-37 gives added protection. Failure of ventilation due to broken or slipping belts, plugged air ducts, etc., is detected through increased air temperature, while a power failure to the fans is detected immediately, Fig 634-39.

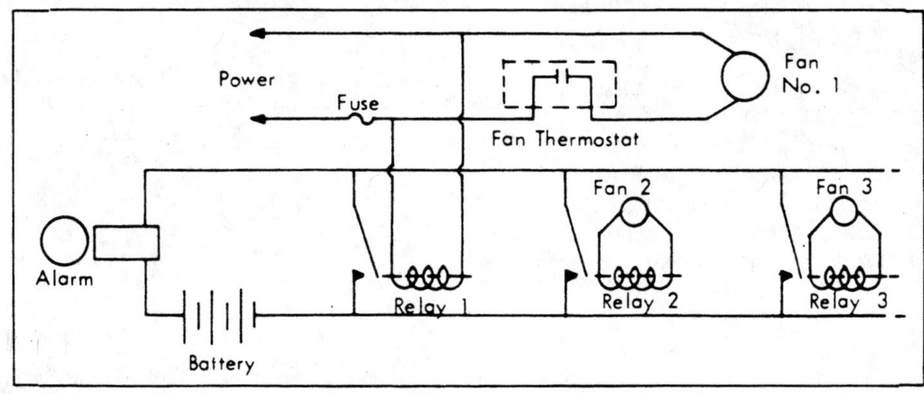

Fig 634-39. Alarm on multiple-fan system.

635 EVAPORATIVE COOLING

Principles

Evaporative coolers use heat from air to vaporize water. This increases the relative humidity but lowers the air temperature.

The lower the relative humidity of incoming air, the more effective is evaporative cooling. Therefore, evaporative coolers are more useful in dry western states, but they are also effective in the Midwest. Humidity drops as air temperature rises and is usually lowest during the hottest part of the day. Expected temperature drop during midsummer is shown in Fig 635-1.

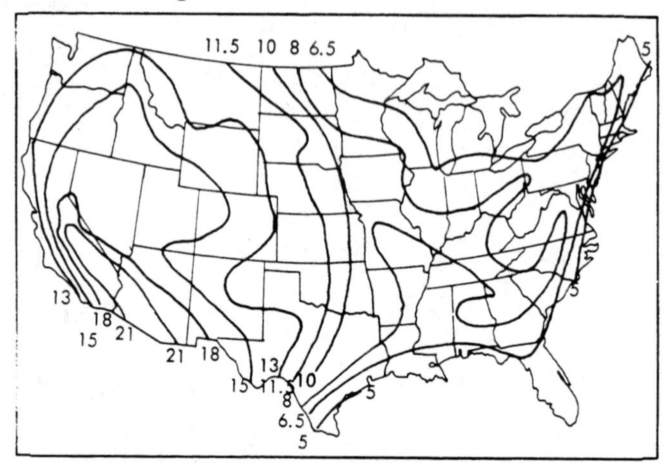

Fig 635-1. Effectiveness of evaporative coolers.
Average temperature drop (F) from the average daily outdoor temperature during July.

Commercial evaporative coolers are available for livestock buildings. Air is drawn through wet pads or foggers into the animal area. Most units pump water from a sump over a fibrous pad. One commercial unit rotates a drum-mounted pad through a pan of water.

Provide airflow through the evaporative unit of at least one half the hot weather rate, Table 633-2. On hot, humid days evaporative cooling is not very effective, so install circulation fans sized at one-half the hot weather rate, or supply the full hot weather rate on those days.

Design

Pad area (ft^2) for a wall evaporative cooler, Fig 635-2, is the airflow through the pad (cfm) divided by 150 for aspen pads or by 250 for cellulose pads. Provide insulated doors or covers to close off the pads during winter.

Supply water to the pads with an overhead system. Rigid plastic pipes or open rain gutters with spaced holes allow water to drip uniformly over the pads. Pipe size, hole size, and spacing depend on water flowrate and are sized for each system. For best evaporative efficiency, much more water is supplied than is used. To conserve unevaporated water, reuse it.

Install a water sump of at least ½ gal/ft^2 of aspen pad or ¾ gal/ft^2 of cellulose pad. Provide at least ⅓ gpm/linear ft of aspen pad or ½ gpm/linear ft of cellulose pad to the distribution pipe over the pads. A gutter beneath the pads, sloped at 1″/10′, collects and conveys unevaporated water back to the sump. Control make-up water to the sump with a shut-off float valve. Up to 1 gpm/100 ft^2 of pad evaporates on hot, dry days.

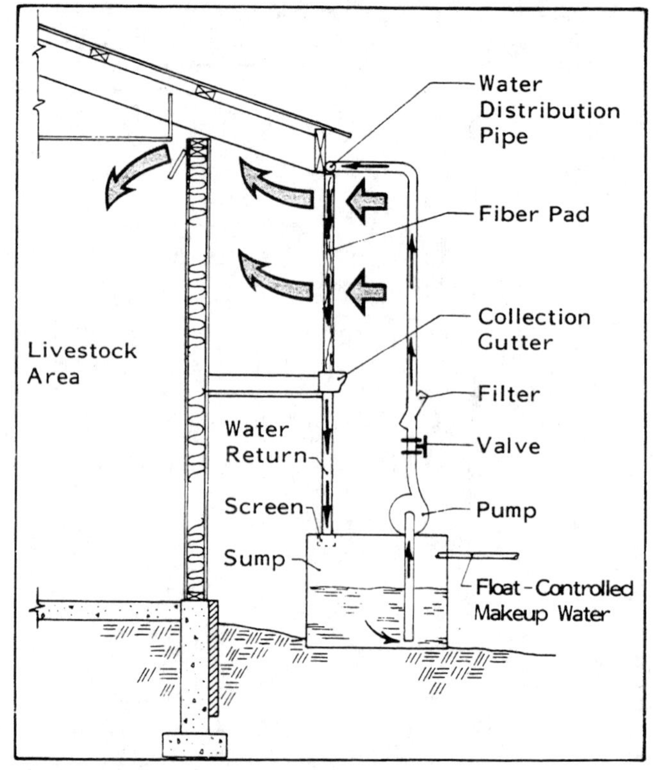

Fig 635-2. Evaporative cooling system.
Ventilating exhaust fans pull hot air through wet fiber pads. Air heat evaporates the water and lowers air temperatures.

Protect the water distribution system from insects and debris. Screen recirculated water before it returns to the sump. Also, install a filter between the pump and the distribution pipe or gutter.

Set the thermostat to begin wetting pads at about 85 F. To reduce algae growth, wire the pump to stop several minutes before the fans so the pads dry out after each use.

Maintenance

Most pads are woven aspen fibers and must be replaced annually. Cellulose pads with a life of about five years are available with some units. Pads are mounted on the sidewalls, endwalls, or roof. Wall-mounted units are easier to maintain than roof units. If the pad settles, fill in the opening so air does not short-circuit.

Hose pads off at least every other month to wash away dust and sediment. Control algae buildup in the water with a copper sulfate solution. Light-tight enclosures around pads and water sump help control algae. Water is constantly evaporated, so salts and other impurities build up. Bleed off 5%-10% of the water continuously to flush salts from the system as they are formed or flush out the entire system every month.

636 NATURAL VENTILATION

Natural ventilation is usually more economical for mature animals than mechanical ventilation. Natural ventilation can be used for young animals on bedded floors or in well-insulated modified environment buildings with zone heat.

Natural ventilation is accomplished by natural air movement through adjustable and fixed openings. Adjustable openings include windows, wall panels, vents, and ridge ventilators. Fixed openings include open fronts of buildings and continuous eave and ridge slots.

Building Types

There are basically two types of naturally ventilated buildings:

- **"Cold" buildings.** These are open-front or lightly insulated, enclosed buildings which are designed to maintain winter indoor tempertures within a few degrees of the outdoor temperature. They remain cold in winter so they are normally used only for large animals, such as cattle or finishing and gestating swine. Control of the airflow is very limited in these buildings.
- **"Modified-environment" buildings.** Modified open-front (MOF) buildings are well insulated and are designed to maintain higher winter indoor temperatures than cold buildings. They provide better winter performance for finishing swine and gestating sows and have fewer problems with manure freezing. Ventilation control is critical to prevent large temperature fluctuations and drafts.

Principles

The difference between inside and outside wind pressure and temperature moves air through the building.

Winter ventilation. Wind across the open ridge creates suction which draws warm, moist air out through the ridge and fresh air in through the eave openings. The downwind eave openings may occasionally act as air outlets. Some ventilation occurs even on calm days because warm air rises, but this chimney effect is only about 10% of the winter ventilation.

Summer ventilation. Large openings (typically one-third to one-half of the sidewalls) in each sidewall allow a crossflow of air. The ridge opening has little effect in summer.

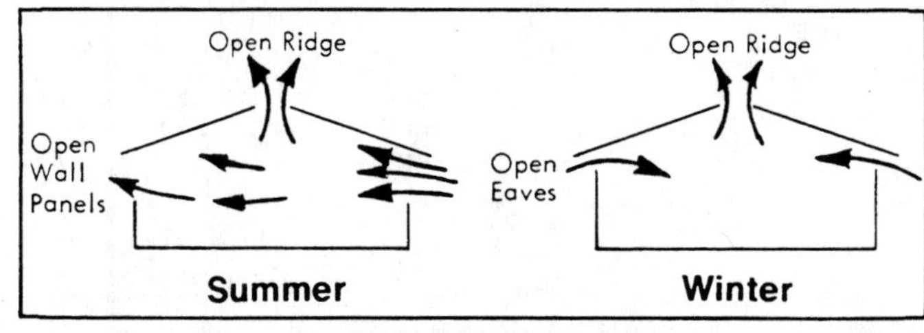

Fig 636-1. Airflow with natural ventilation.

Site selection

Trees, tall silos, grain bins, and other structures disturb airflow around adjacent buildings for 5 to 10 times their height downwind. Locate naturally ventilated buildings on high ground to expose them to wind and at least 50′ (in any direction) from other structures and trees. If naturally ventilated buildings are in a building complex, place them on the west or south side of mechanically ventilated buildings, Fig 636-2.

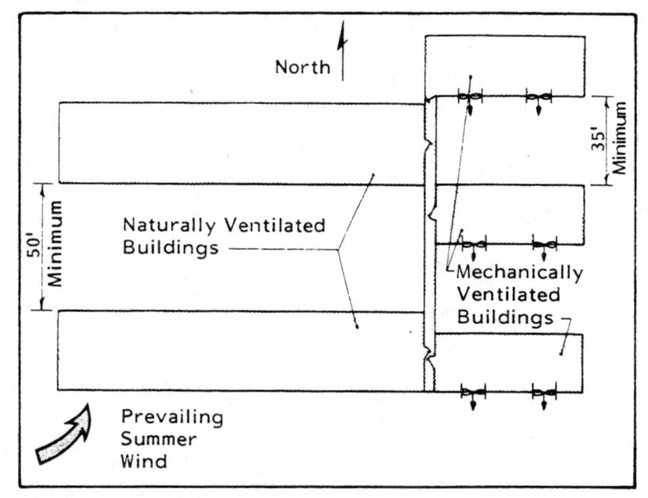

Fig 636-2. Layout of naturally and mechanically ventilated buildings.
Locate naturally ventilated structures on the west or south side so mechanically ventilated buildings do not interfere with prevailing wind.

Building orientation

Place open-front buildings with the open side away from prevailing winter winds. If possible, construct open-front buildings with an east-west long axis for best summer shade and winter sun penetration.

The greater solar heat load on north-south oriented roofs is not a major concern for well-insulated MOF buildings—try to orient these buildings perpendicular to winter and summer prevailing winds for better ventilation.

Insulation

In cold buildings, insulation is added primarily to control condensation and frost formation on roof surfaces and to resist high summer temperatures. An inch of asphalt-impregnated fiber board is sufficient. Unprotected styrene or urethane may not be insurable for fire and is susceptable to bird damage. Protect with a fire resistant liner on the inside.

MOF buildings require high levels of insulation to help maintain higher indoor temperatures. See Table 631-5.

Design

Provide two sizes of adjustable inlets: large openings for summer ventilation and smaller slot-type openings for winter ventilation. Summer air inlets are often 3′x6′ or 4′x8′ sliding, pivoting, or hinged doors in the sidewalls for good cross ventilation.

Winter openings generally consist of air **inlets** at the eaves of the building and air **outlets** at the ridge or high point of the roof. Winter air inlets are commonly between rafters or under overhangs to prevent drafts at animal level. In severe climates, make these openings with a baffle to be closed during storms to prevent snow from sifting in.

Winter air outlets are usually slots which can be baffled to control airflow. A variety of roof ventilators are also used. Some rotate from wind forces, and others have adjustable openings to suit weather conditions. Consider initial cost and number needed. Roof ventilators on livestock buildings usually fill with dust, birds' nests, or frost, which reduces air movement. They are not as satisfactory as an open ridge, but they do keep snow and rain out.

Ridge Openings

Make the ridge opening at least 3″ wide to prevent frost buildup. For cattle, provide a continuous slot sized at about 2″ of width for each 10′ of building width, Table 636-1.

Table 636-1. Natural ventilation openings for cattle.
Ridge openings are sized for a 10 mph wind.
Sidewall openings are sized for a 2 mph wind and are a minimum. Opening more of the back sidewall increases summer cooling.

Building width (ft)	30	40	50	60	70	80
Eave & ridge opening (in.)	6	8	10	12	14	16
Sidewall opening (in.)	28	36	44	52	60	68

For swine in the Midwest, size ridge openings to provide 5 to 6 in²/finishing pig and 7 in²/gestating sow. If swine are uniformly distributed in the building at recommended space allowances, use Table 636-2 to determine opening sizes.

For cold buildings, construct the ridge opening by removing the ridge cap and covering trusses with flashing, Fig 636-3a. Provide closures on MOF ridge openings or use commercial ridge ventilators with adjustable openings, so airflow can be reduced during cold periods, Fig 636-3b.

Air exiting through a properly sized ridge opening prevents most precipitation from entering. Raised ridge caps are not recommended—they disturb airflow and may even trap snow. If birds are a problem, consider installing ¾″ hardware cloth over the ridge openings. Screened ridge openings are very susceptible to freezing shut in cold climates. You may need to knock ice from the screen regularly during cold spells.

Eave Openings

Construct continuous eave openings along both sides of the building. Locate them high on the sidewall so incoming air is warmed before reaching the animals. Screen them with ½″ hardware cloth to keep birds out. Size **each** eave inlet to have at least as much open area as the ridge opening. Eave openings can be provided at the open spaces between trusses or with slightly opened sidewall doors, Fig 636-4.

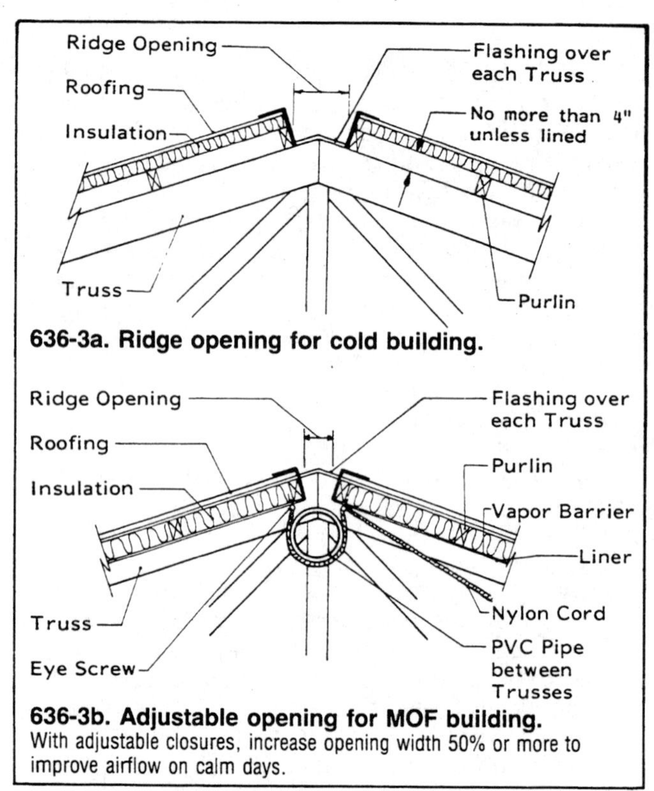

636-3a. Ridge opening for cold building.

636-3b. Adjustable opening for MOF building.
With adjustable closures, increase opening width 50% or more to improve airflow on calm days.

Fig 636-3. Ridge openings.

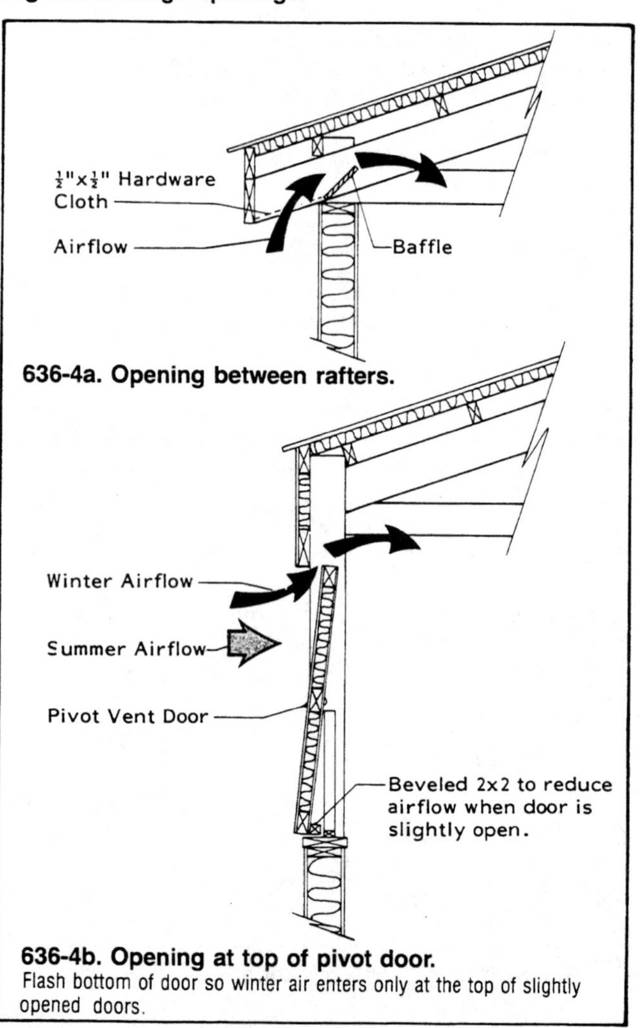

636-4a. Opening between rafters.

636-4b. Opening at top of pivot door.
Flash bottom of door so winter air enters only at the top of slightly opened doors.

Fig 636-4. Typical eave openings.

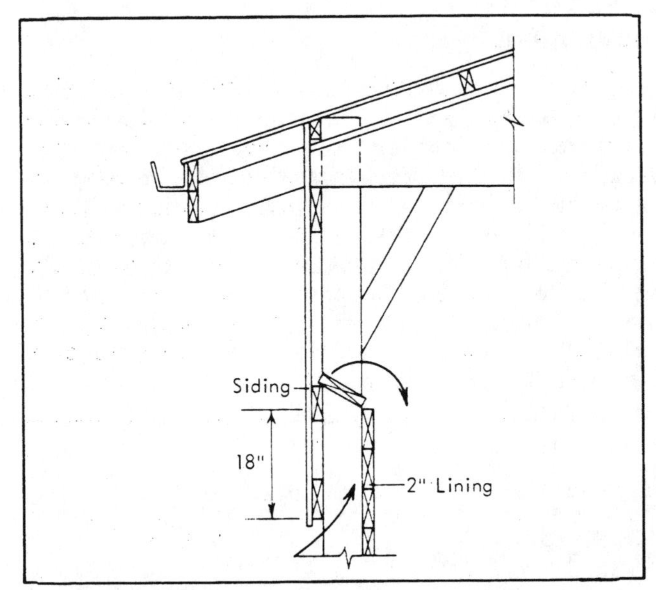

Fig 636-5. Alternate back wall winter vent.

Another winter air inlet is a hinged plank at the top of a splashboard wall lining in pole buildings. It is easily adjusted and prevents drafts at the roof. It will not provide enough air movement for totally confined units during hot weather, Fig 636-5.

Provide eave baffles so airflow can be reduced, but never completely close them. Protect eave openings from direct wind gusts with facia boards to reduce drafts on the animals.

Sidewall Openings

Several types of summer vent doors work well, including pivot doors, top- or bottom-opening doors, and plastic curtains, Figs 636-4b and 636-6. Generally, make ⅓ to ½ of the sidewall open for good airflow at animal level. For swine buildings, size per Table 636-2. Construct vent doors to open to a full horizontal position so none of the opening is constricted, or increase door size so the recommended open area is still provided.

Example 1:

Design the natural ventilation openings for a 30′ wide gable roof swine finishing building.

1. From Table 636-2a, **each** continuous eave inlet is 1.5″ high (HI).
2. Assume a 3″ wide ridge opening (WO). Divide the table constant by the ridge opening width to get the length of 3″ opening required for each 16′ of building length (LO). From Table 636-2a, the table constant (TC) = 24.
   LO = 24 ÷ 3″ = 8′ (i.e. 8′ of opening for each 16′ of building length)
3. From Table 636-2b, install a continuous 42″ high opening along **each** sidewall.

Roof

The steeper and smoother the roof underside, the better the airflow. Avoid roof slopes less than 3/12 for gable roofs and 2/12 for monoslope roofs. Avoid exposed purlins deeper than 4″. See section on insulation for recommended insulation values.

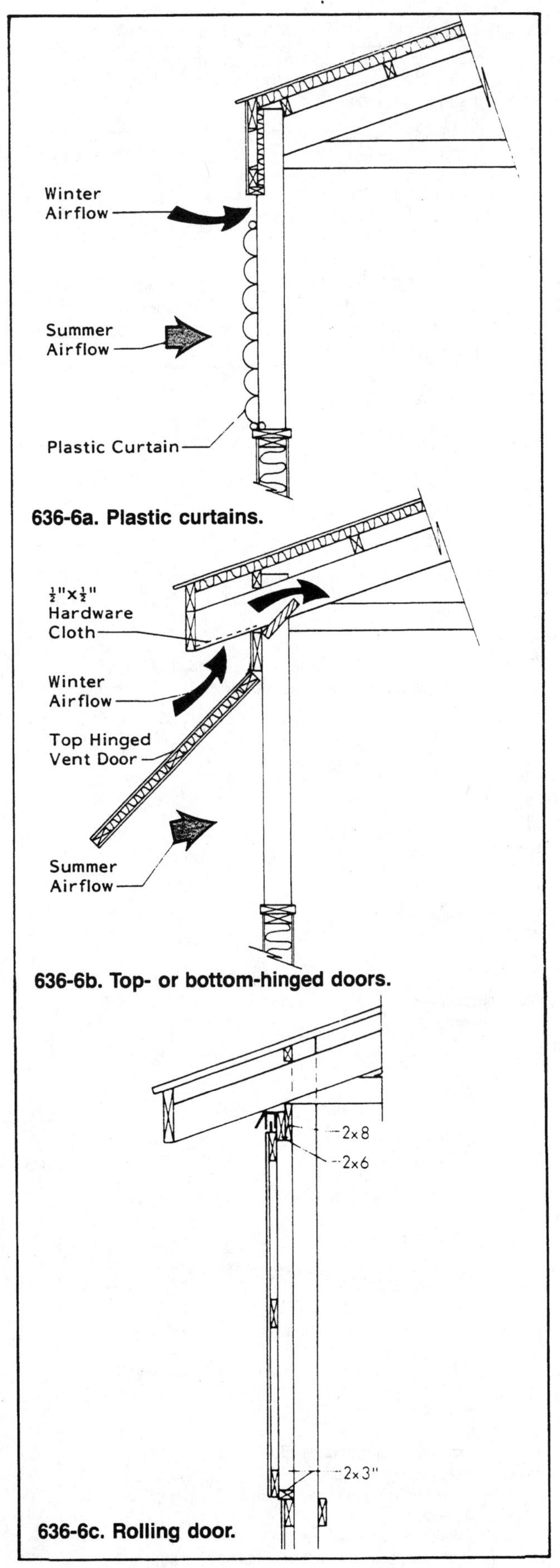

Fig 636-6. Sidewall openings.
See also Fig 636-4b.

Table 636-2. Natural ventilation openings for swine.

636-2a. Slot widths for cold and mild weather.
Install a continuous **winter air inlet** along each side of the building. Make **each** inlet the height (HI) in this table.

Make **winter air outlets** at least 3" wide to prevent them from freezing shut. It is usually necessary to install several shorter vents that are 3" or wider, instead of a continuous narrow opening. Provide an outlet at least 16' o.c. To determine the length of the outlet opening required per 16' of building length (LO), divide the table constant (TC) by the outlet opening width (WO = 3" or wider).

Building width	Winter air inlet height (HI)		Winter air outlet constant (TC)	
	Finishing	Gestation	Finishing	Gestation
ft	in.	in.	–	–
10-15	0.75	0.50	12	8
16-20	1.00	0.75	16	12
21-25	1.25	1.00	20	16
26-30	1.50	1.25	24	20
31-35	1.75	1.50	28	24
36-40	2.00	1.50	32	24

636-2b. Sidewall openings for warm weather.
Heights given are for one opening the entire building length. Provide an opening in **each** sidewall.

Building width	Gable roof	Monoslope roof	
	Each sidewall	Backwall	Frontwall
ft	in.	in.	in.
10-15	24	12	36
16-20	30	16	48
21-25	36	18	60
26-30	42	24	66
31-35	48	—	—
36-40	60	—	—

Winter Vent Openings:

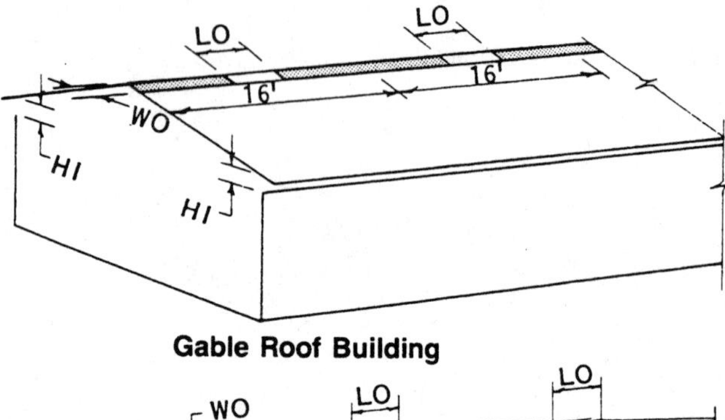

Gable Roof Building

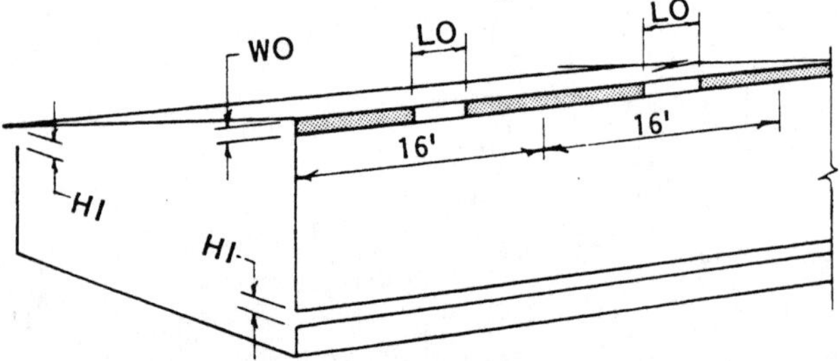

Monoslope Roof Building

HI = Height of winter air inlet.
Air inlets are continuous along both sides of the building
WO = Width of winter air outlets.
LO = Length of air outlet per 16' of building length.
LO = TC/WO

Management

Adjust vent openings several times a day during changing weather to avoid wide temperature fluctuations. Install a winch and cable system to adjust several doors from one convenient location. Consider winches on a thermostat for automatic door adjustment. Locate the controller at the center of building. Connect main cables from both directions to reduce stress on the controller supports. Use prestretched cable to minimize future adjustments. During cool weather, disconnect unwanted doors and control only the doors needed at that time.

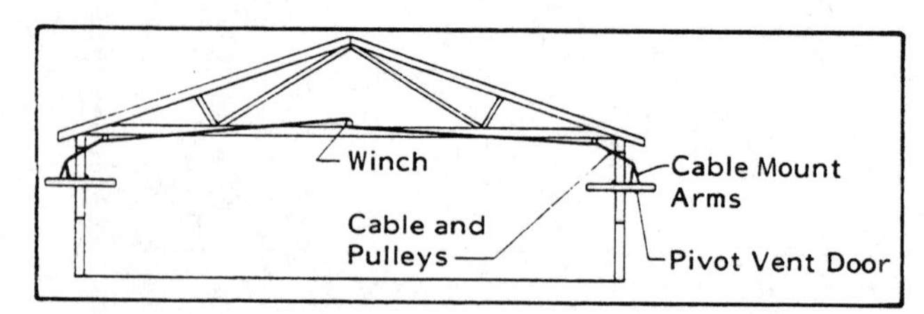

Fig 636-7. Cable-controlled pivot doors
Mount cables on outside arms so doors open to a full horizontal position. Hinge pivot doors just above center so they close when cable tension is released.

Gable Roof Buildings

MOF Buildings

In cold weather, adjust only the eave doors to maintain desired room air conditions. Keep the ridge fully opened except during very cold periods or blowing snow. Keep the large, sidewall vent doors closed unless they are used for the eave openings.

As the weather warms, gradually open the downwind vent doors first. When downwind doors are about half open, start opening the upwind doors. Completely open all the vent doors in hot weather. Reverse this sequence as temperature decreases.

Open-Front Buildings

Open fronts have limited ventilation control, and animals must adjust to rapidly changing temperatures. Open the back sidewall doors in spring and close them in fall; close eave vents when temperatures drop below freezing. Ridge vents are seldom closed, regardless of the weather.

Monoslope Buildings

Monoslopes have no ridge so both winter and summer ventilation are accomplished by a crossflow of air.

Open-front monoslopes are operated like open-front gable buildings. Close the back wall openings in winter and open them in summer.

MOF monoslopes have large, adjustable openings in the front (high) and the back sidewalls. The back sidewall opening is usually an insulated tilt door and the front sidewall is a plastic fabric curtain, a fiberglass panel, or an insulated tilt door in cold climates. These buildings also have small, adjustable openings along the eaves of both sidewalls and along

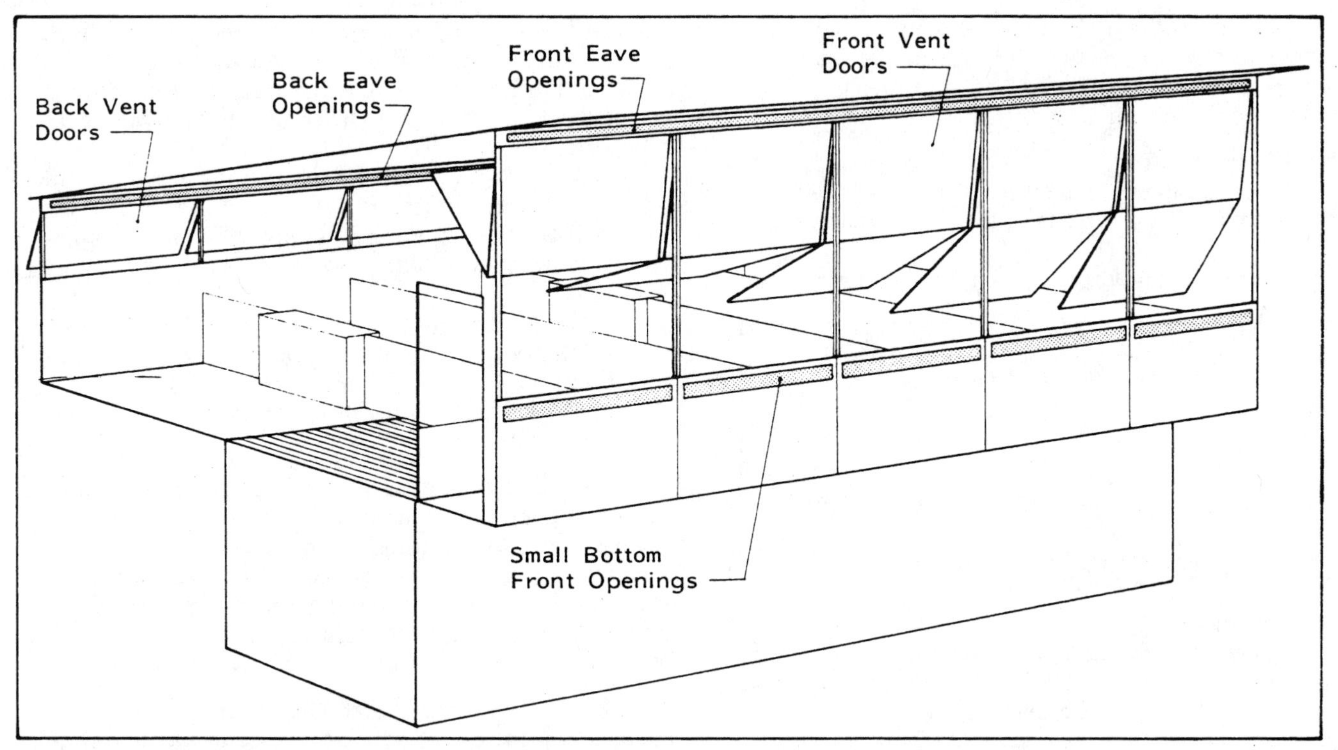

Fig 636-8. Modified open-front (MOF) monoslope building.

the bottom of the front sidewall, Fig 636-8. Size the small openings the same as the eave openings of gable roof buildings, Table 636-2.

When outside temperature gets below 20 F, close the back eave openings. Ventilating air enters through the small bottom front opening and moves 10′-15′ back into the animal area. As the air is warmed by the animals, it rises and travels up the sloped roof to exit through the front eave opening. The back sleeping area is not ventilated directly during extremely cold periods.

As outside temperature rises above 20 F, start opening the back eave openings and eventually the large front doors. In summer, open all the vent doors.

Snow and Wind Control

Wind against a building endwall tends to force air in the upwind end and out the downwind end, creating drafts along the building length, Fig 636-9. In MOF buildings, make every fifth pen cross partition solid (about every 50′). In open-front buildings, make every third cross partition solid (about every 25′). Extend the solid partitions up to the bottom of the sidewall vent doors and make them completely solid, even over the slats. If drafts are caused by wind swirling around the ends or over the top of an open-front building, close up 2′-3′ the top and 16′ near each end of each open wall.

Do not block summer cross ventilation—make alley partitions open or "porous." Open mesh partitions below the front vent doors of a MOF building are ineffective—if used, cover them in winter.

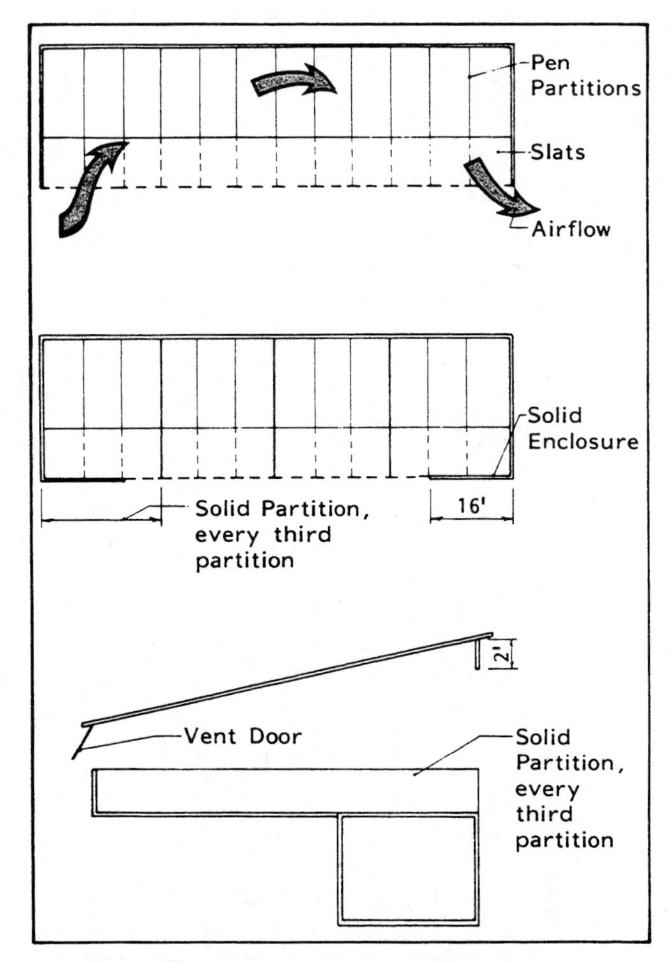

Fig 636-9. Draft control in open-front buildings.
Solid partitions and corners, and top extensions, reduce drafts.

Snow drifting can be reduced inside open-front shelters. Use local practice and experience to face the open front away from major winter storms.

See section on farmstead planning.

Cold, windy region

Reduce front wall opening:

- Side part way down the wall to reduce snow blow-in.
- Provide 16′ square snow swirl chambers at each end.

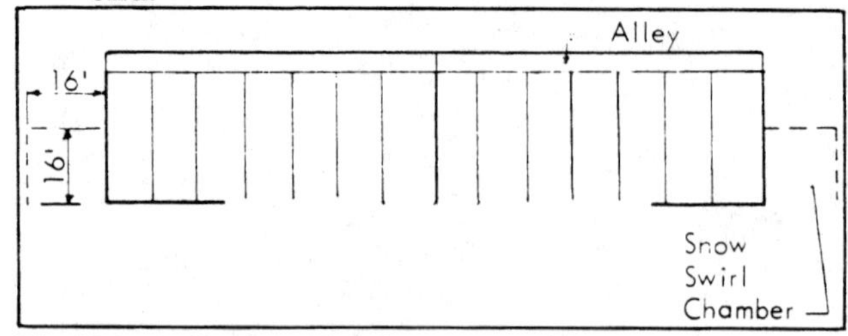

Fig 636-10. Snow swirl chamber.

Moderate region

Limit roof overhang, leaving front open as high as possible, because sunshine along the open front is of more value than reduced snow blow-in.

Problems

Ventilation and/or moisture problems occur in cold housing when:

- Buildings do not provide natural ventilation, usually because of inadequate openings.
- Ventilation needs are ignored. Buildings are closed in attempt to increase inside temperatures.
- Attached sheds and roof extensions prevent good natural ventilation.
- Sanitation and housekeeping leave extra moisture sources (manure, wet bedding, leaky waterers).

Factors that reduce open-ridge performance are:

- High silos, trees, or other buildings close to the barn.
- Too few animals in the building.
- Low roof slope.

Do not totally enclose a cold building in an effort to raise inside temperatures. The result is usually a severe moisture buildup, condensation on the roof and walls, animal respiration illness, and only a slight temperature rise. If winter moisture buildup is a problem in an existing building, open the ridge and provide an inlet under the eave as discussed earlier.

If heat buildup during the summer is a problem, open ⅓-½ of the closed wall area to allow air to circulate at animal level. Cable, wire, or pipe fences permit better air circulation than plank partitions. Steep roof slopes tend to increase air movement.

650 VENTILATION APPLICATIONS

651 VENTILATION OF SWINE BUILDINGS

Fans and Inlets

See Chapter 634 for inlet design and fan selection.

Heaters

Supplemental heat is added to swine buildings to maintain inside temperature within the range of animal comfort and to help control humidity.

Table 634-5 gives supplemental heat requirements for typical livestock housing in the North Central U.S. Swine heating systems include radiant, floor, unit space, and air make-up heaters.

Radiant Heaters

Radiant heaters work well in creep and sleeping areas. They allow the pig to find comfort by moving closer to or farther from the heater. Energy from a radiant heater passes through air without warming it—upon striking the pig or other surface, the energy is absorbed and the pig or surface is warmed.

Infrared heaters produce radiant heat from gas or electricity. Many types are available, from heat lamps to sophisticated fan-driven pipe units which can heat an entire building.

With floor heat in the farrowing creep area, provide an additional 250 watts (852 Btu/hr) with overhead heat lamps for the first few days after farrowing. If no floor heat is used, provide 2200 Btu/hr of overhead radiant heat per litter.

Heat lamps are a potential fire hazard. Suspend them on chains and make the lamp cord 1′ shorter than the floor-to-ceiling height. Mount lamps at least 30″ above pen floors and 18″ above creep floors. Place no more than seven 250 watt heat lamps on one 20 amp circuit.

Catalytic radiant heaters are flameless and have relatively low surface temperatures. No electricity is required and they are relatively unaffected by dust and moisture. Catalytic heaters are usually not thermostatically controlled, which makes them less efficient.

They consume LP gas but require no special venting in a properly ventilated building. With some ventilating systems, you may have to increase the ventilating rate 4 cfm/1000 Btu/hr of unvented heater capacity to prevent buildup of poisonous combustion gases.

Floor Heat

Floor heaters can be electric resistance cables or hot water pipes buried in concrete, Figs 651-1 and 651-2, or electric heating coils in fiberglass pads placed over an existing floor. Floor heat evaporates liquids from the floor surface which increases the relative humidity of the building. Arrange resting areas, waterers, and dunging areas so the heated area is seldom wet. Do not heat the floor under the sow in farrowing buildings.

Heat the correct amount of floor area, Table 651-1. Too much heated area wastes heat; too little encourages pigs to pile. Use waterproof insulation under a heated floor but do not use insulation or excessive bedding on heated sleeping areas. Do not lay the heating elements directly on plastic insulation. Install perimeter insulation if the heated slab is near an outside wall. Uniform concrete thickness above hot water pipes or electric heat cables prevents hot spots.

Table 651-1. Heated floor design criteria.

Pig weight lb	Heated floor space ft²	Floor surface temperature F	Hot water pipe spacing in.	Electric heat W/ft²
Birth-30	6-15/litter	85-95	See Fig 651-1	30-40
30-75	1-2/pig	70-85	5	25-30
75-150	2-3/pig	60-70	15	25-30
150-220	3-3½/pig	50-60	18	20-25

Locate the temperature sensing bulb about 1″ below the heated floor surface and about 4″-6″ from a heat pipe or 2″ from an electric heat cable. Install the temperature sensing bulb and connecting tube in a 1″ conduit or pipe so they can be removed for maintenance. Use a large radius bend in the conduit or pipe so the bulb can be inserted easily.

For farrowing buildings, provide about 95 F slab surface temperatures at farrowing. Gradually lower to 85 F when pigs are 3 weeks old. Heat lamps or hovers are common in addition to floor heat for 3 to 5 days after farrowing. Floor heat is occasionally installed in nursery buildings and naturally ventilated, growing-finishing buildings. See Table 651-1 for desired floor temperatures for larger pigs.

To calibrate the thermostat (which indicates internal instead of surface slab temperatures), adjust it to obtain desired surface temperatures, using a thermometer laid on the floor.

Start heating the floor at least 2 days before it is needed, to allow time for the concrete to warm up. Floor heat alone seldom provides the total heat requirements of a building.

Hot water floor heat uses wrought iron, black iron, or high temperature (160 F) plastic pipe. Do not use cold water plastic or galvanized pipe. Fig 651-1 shows the necessary equipment for a hot water floor heating system.

An input of 50 Btu/hour for each foot of pipe keeps the floor temperature at about 95 F. Multiply the linear feet of pipe times 50 Btu/hr to get the required output of a water heater or boiler. Multiply the rated capacity of gas heaters by 0.75 to approximate actual heat output.

Space pipes per Table 651-1. Place pipes within 6″ of the concrete slab edge and cover with 2½″-3″ of concrete. If a heated pipe passes under a farrowing

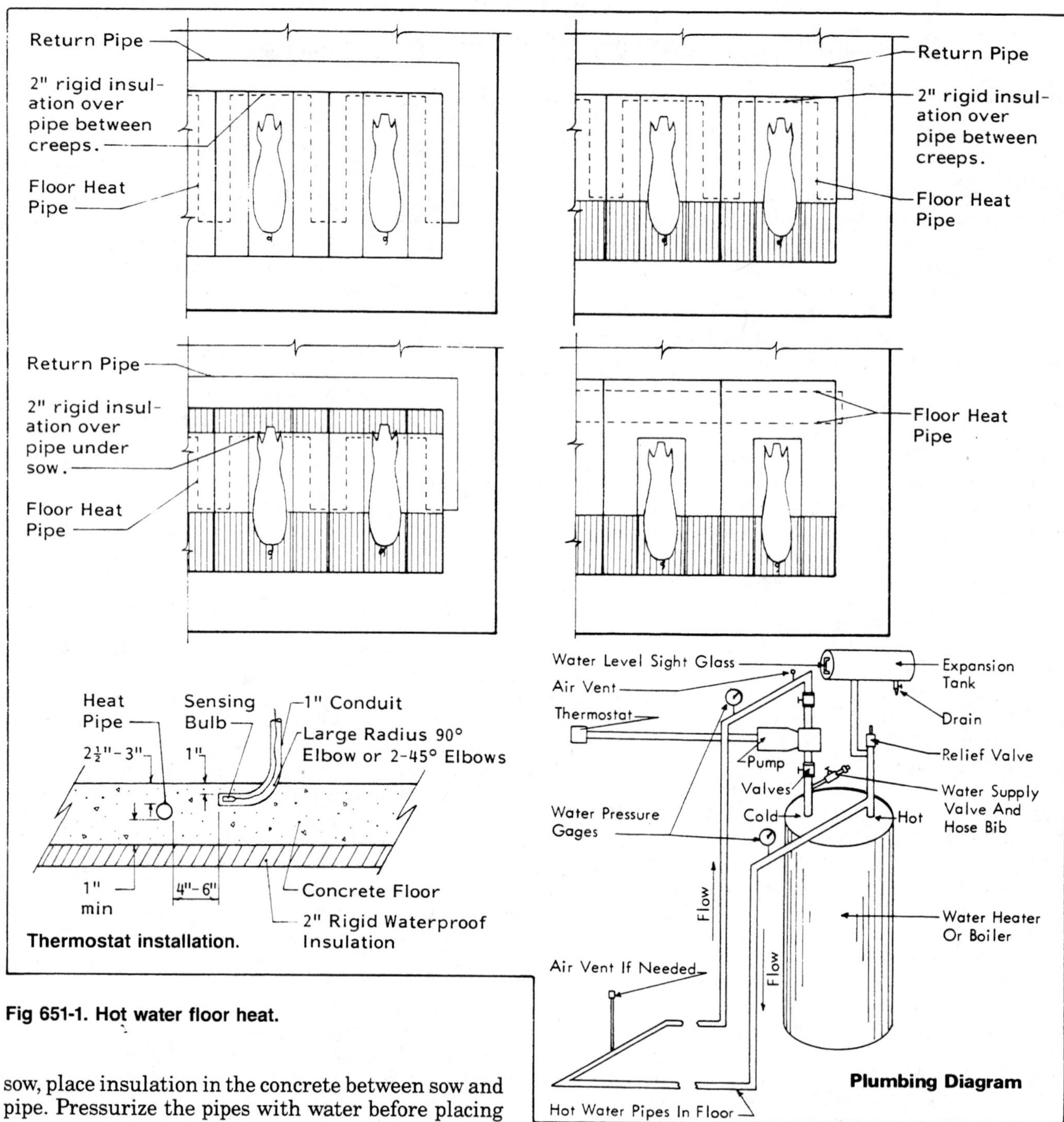

Fig 651-1. Hot water floor heat.

sow, place insulation in the concrete between sow and pipe. Pressurize the pipes with water before placing concrete, to check for leaks. Contact an experienced plumbing contractor to design and install hot water systems.

Electric heat cable usually has a lower installation cost than hot water pipes, avoids freezing problems, and permits individual control for each pen or stall. Electric heat has a higher operating cost than hot water heat.

Use cable approved for use in concrete. The most common type of heating cable is covered with polyvinylchloride (PVC) and is rated from 2 to 7 watts per linear foot. Prefabricated pads are available that have the cable already spaced in a plastic mesh to simplify installation.

Before grouting the heating cable in place, perform a continuity test to ensure the cable will operate. Embed the cable 1½"-2" into the concrete and space uniformly. Make sure it is completely surrounded by concrete, to prevent burnout. Use a thermostat for every one to five pens or stalls. A fused switch on each pen or stall permits disconnecting them when they are empty. Electrically ground all steel stalls and waterers.

Commercial plastic heat pads are set on the floor and plugged into electric outlets. Protect electric cable from animal abuse. Thermostatic control is recommended.

Unit space heaters

Unit space heaters are located within the room and heat room air directly. Unfortunately, they recirculate dusty, wet, corrosive air through the furnace, often resulting in high maintenance. Clean and lubricate unit heaters once a month during the heating season and more often in a very dusty environment such as a growing-finishing building. If possible, place heaters to blow along the coldest wall to reduce cold drafts and radiant heat losses to the wall. With more than one heater, arrange them to create a circular air pattern within the room.

Air make-up heaters

Air make-up heaters are unvented units mounted outside the building. They heat only the incoming ventilating air, so they require less service than units that recirculate room air. Fuel flow to the burner is modulated to maintain a constant exit air temperature.

Air make-up heaters supply ventilating air, so close down the air inlet accordingly. Size air make-up heaters so they provide no more air than the winter exhaust fan.

Air make-up heaters exhaust the products of combustion into the building which initially makes them more efficient than vented heaters. However, water vapor is one of the gases produced (about one lb water/lb propane burned), so the ventilating rate must be increased to remove this extra moisture, which tends to offset the initial efficiency gain. Allow 4 cfm extra ventilation for each 1000 Btu/hr of heater capacity.

Other heating systems

Solar heat can provide some of the energy needed. Management with solar preheated ventilating air is the same as with other systems. With space and floor heating, solar energy supplements heat from other sources. With underfloor warm air heating, you must prevent circulating cold air which could cool the floor instead of warming it. See Chapters 671 and 672 for more information.

Reusing exhaust air from one livestock building to ventilate an adjacent one is not recommended. Unless the first building is overventilated, its exhaust air is nearly saturated, so it has little capacity to remove moisture from the second building. If you overventilate the first building, drafts are created

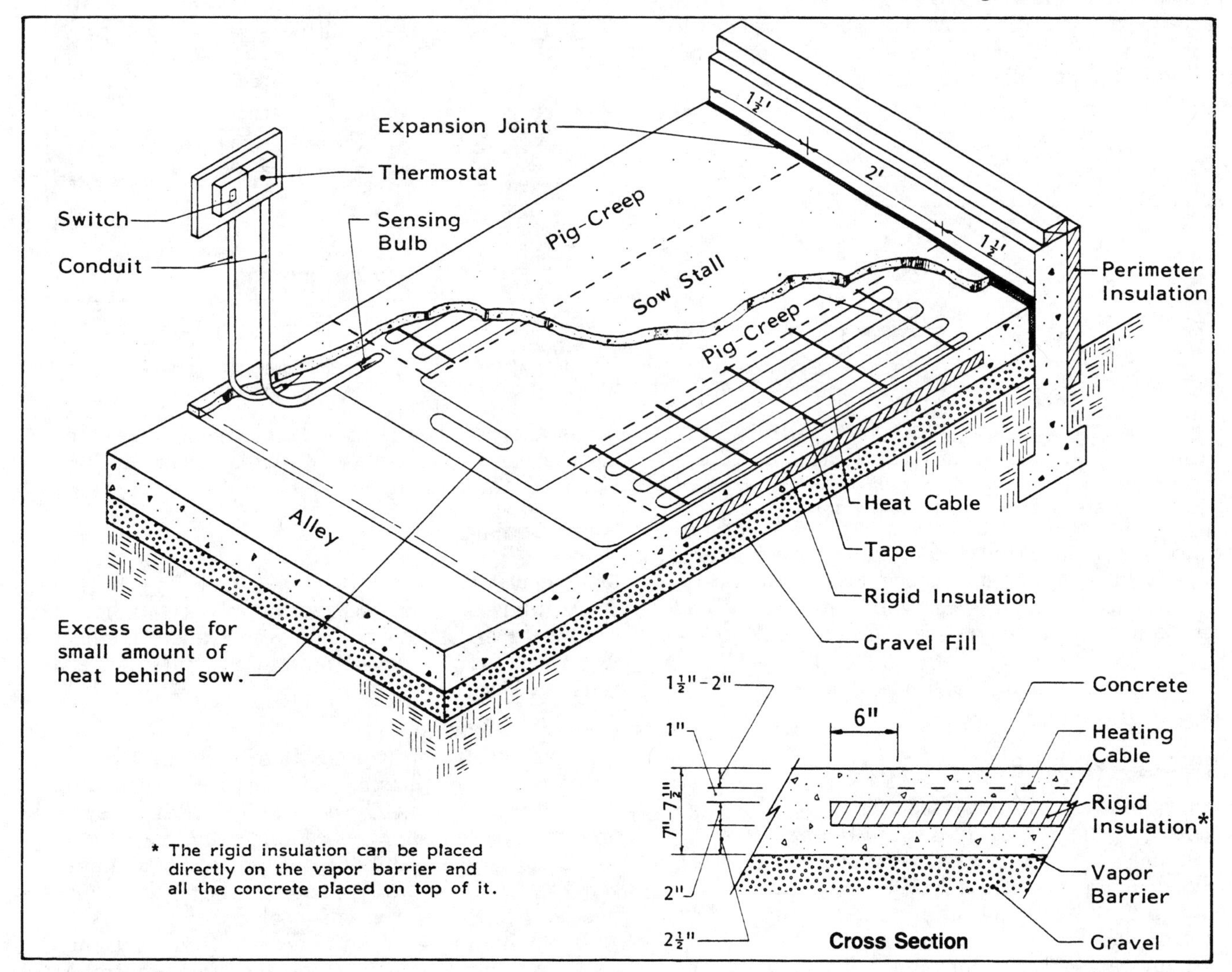

Fig 651-2. Electric floor heat.

and fuel savings are almost eliminated. Veterinarians and animal scientists discourage reusing ventilating air because of increased possibility of disease spread, particularly for farrowing and nursery pigs.

Heat exchangers extract heat from exhaust air which otherwise would be lost. This concept has a great deal of merit, but dust and condensation may cause equipment problems and decrease efficiency. See Chapter 673 for more information.

Earth tempering systems extract heat from the soil through pipes buried in the ground. The pipes handle cold and mild weather ventilating rates. Additional hot weather ventilation must be provided by a conventional ventilating system.

Draft Control

Hovers reduce vertical drafts better than solid covers placed over slotted floors. Hovers need not be insulated—try tempered hardboard, sheet metal, or exterior plywood. Heavy clear plastic on a frame is excellent and allows you to observe the animals.

Provide enough hover space for all animals in a pen to be under it at one time—about half of a farrowing creep area or a third of a nursery pen area. Locate hovers at no more than twice the animal height and preferably next to solid pen partitions.

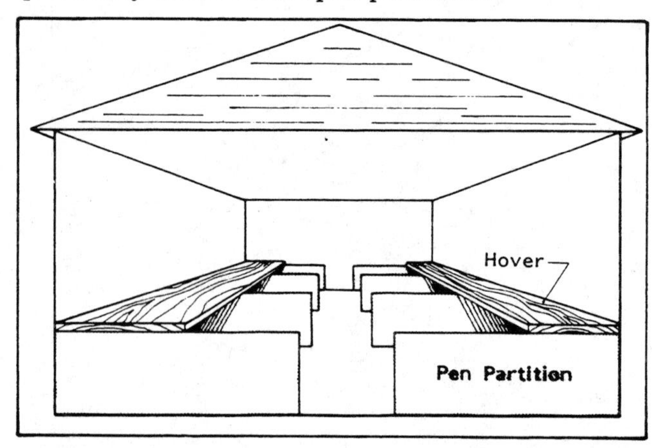

Fig 651-3. Hovers over animal resting areas.

Fire safety is essential for hovers with heaters. Put heaters beyond animal reach. Protect combustible materials with fire-resistant materials. Floor heat works well under hovers and reduces fire hazard.

Solid pen partitions also break up drafts, Fig 651-4.

Cooling Systems

Hot weather reduces feed intake of growing-finishing swine and decreases the reproductive capability of breeding stock. Swine over 50 lb start feeling heat stress at about 70-80 F, and temperatures above 85 F may cause substantial losses unless they are kept cool.

Shades are effective for cooling livestock in pastures and outside lots. Build shades at least 8′ high

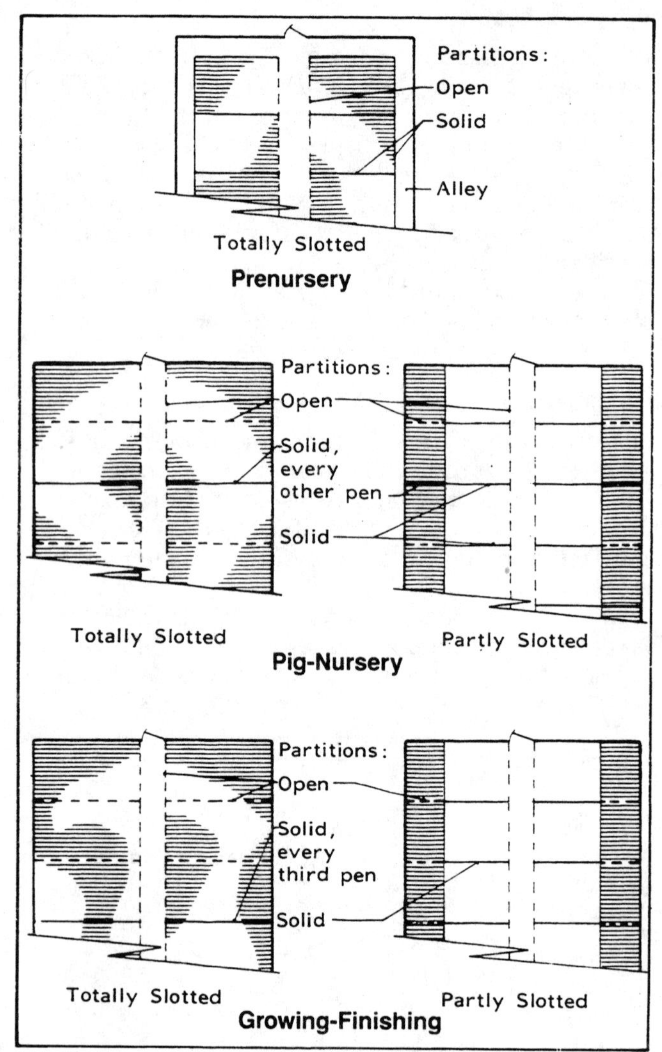

Fig 651-4. Draft control in environmentally controlled buildings.
For partly slotted buildings, open partitions in alternate pens are required to establish good dunging habits.

(preferably 10′). High shades increase exposure to the "cool" areas of the sky, which increases radiant heat loss from the animal.

Water Cooling

Drinking water

Animals drink a lot of water in hot weather to replace water lost due to increased evaporative heat loss. Cool (not cold) drinking water provides the most relief.

Wet-skin cooling

Substantial cooling is possible by wetting the animal's skin, then allowing the moisture to evaporate. A slight breeze across the animal increases the evaporation rate and improves cooling even more.

Pasture wallows provide wet-skin cooling. Shaded wallows are more effective because they reduce solar heat and the water remains cooler.

The best sprinkler systems wet the animal, then allow it to dry. A thermostat- and timer-controlled sprinkler wets the pigs for 2 to 3 minutes out of each hour that temperatures are above 85 F. Tables 651-2 and 651-3 suggest water nozzle and line sizes. Select non-corrosive nozzles designed to furnish a solid cone of water droplets, **not** a fog. A fogger cools the air; a sprinkler cools the animal directly.

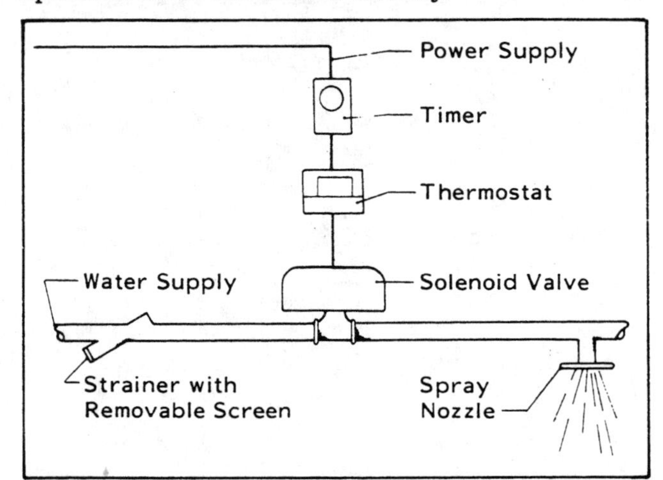

Fig 651-5. Sprinkler with solenoid valve control.

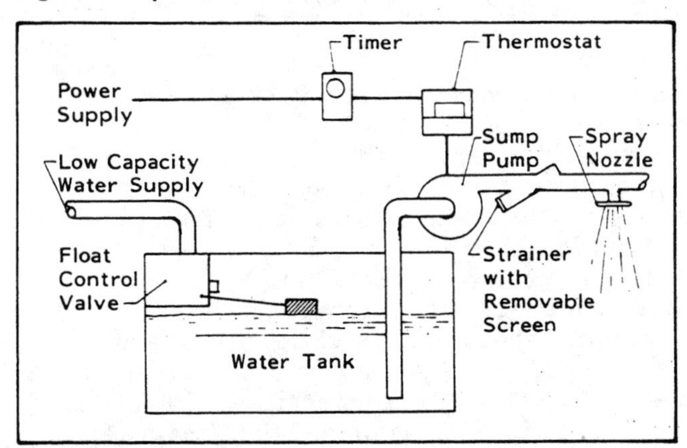

Fig 651-6. Sprinkler with float valve control.
A float valve and tank are often less costly and less complicated than a solenoid valve and manifold piping. Select a sump pump for at least 30 psi pressure.

Ventilation Cooling

Hot weather ventilation cools swine by:

- Replacing hot air with cooler air.
- Taking high-humidity respirated air away from the animals.
- Increasing air velocity around the animals.
- Using heat from the air to evaporate moisture from the building surfaces, creating an evaporative cooling effect.

Hot weather ventilation maintains inside temperatures no more than 3 F higher than outside conditions. Hot weather rates in Table 633-2 are about twice as high as needed to remove animal heat from a well-insulated building. They are designed for high air velocities around the animals. At least 150 fpm is necessary for appreciable hot weather cooling of larger animals. With some systems, you can direct fresh air onto the animals. One half of the total hot weather rate can be obtained with circulation fans to increase air velocity, while the other half must be exhausted from the building to control temperature.

Table 651-2. Nozzles for swine sprinklers.
One nozzle per pen about 6′ above floor.

Pigs per pen	Nozzle size
10	0.45 gpm
20	0.90 gpm
30	1.35 gpm

Table 651-3. Water lines for swine sprinklers.
Plastic pipe: Based on 5 psi/100′ maximum pressure drop and 4 feet/second maximum velocity.

Pipe size in. o.d.	Maximum flow, gpm
½	3
¾	6
1	10
1¼	18
1½	25
2	40
2½	50
3	90
4	140

Air Conditioning Systems

Livestock rooms are seldom air conditioned because it is expensive. Air conditioners cannot reuse cooled air from livestock buildings because of corrosive gases and dust, so they continually cool hot air in a one-pass process.

Zone Cooling

In hot weather, a hog loses 60%-70% of its heat through evaporation from the respiratory tract, so it helps to zone cool the area around its head. Zone cooling is generally used for stall- or tether-restrained animals and occasionally for animals in individual pens. In farrowing buildings, zone cooling cools the sow while allowing higher temperatures in the pig creep.

Zone cooling does not satisfy all of the hot weather ventilation needs. You still need to provide one-half the hot weather ventilating rate with conventional ventilation.

A zone cooling system has a main air duct and downspouts (or drop ducts) located as needed for the animals, Fig 651-7. Locate downspouts close to the animals' head; if spouts are too far away, the cooled air mixes with room air, decreasing its effectiveness. If the outlet is within the animals' reach, make it pig-proof. Provide dampers to close downspouts when stalls are empty.

Table 651-4 recommends ventilating rates for zone cooling systems. Tables 634-6 and 651-5 recommend main duct and downspout sizes. The main duct may be slightly larger, but not smaller. Downspout sizes are critical, but downspouts with dampers can be slightly larger. Insulate ducts to prevent excessive heat gain and condensation.

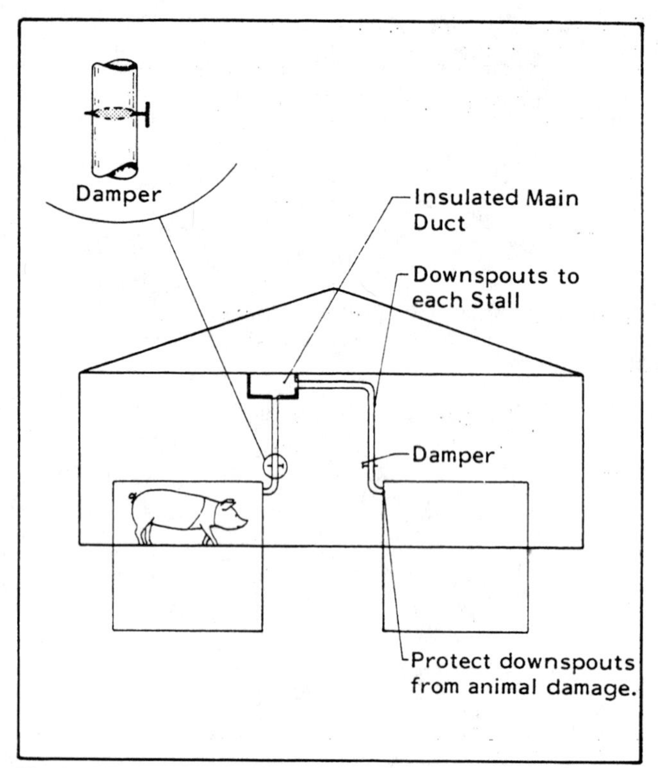

Fig 651-7. Zone cooling system.
Zone cooling provides only part of the total ventilation but is very effective, because high velocity air is ducted directly to the animal.

Table 651-4. Airflow for zone cooling swine.
In addition to zone cooling provide at least one-half of the hot weather rate in Table 633-2 with conventional ventilation.

	Airflow for:		
Type of animal	Uncooled air	Evaporative cooled air	Conditioned air
	- - - - - -cfm/animal- - - - - -		
Farrowing sow	70	40	35
Gestating sow	35	20	15
Boar	55	30	20

Example 1:

Determine the airflow and duct size requirements of the three zone cooling systems for a 20-sow farrowing building.

In addition to the zone cooling, ventilate the room with a conventional ventilating system. Size at one-half the hot weather rate in Table 633-2: (250 cfm x 20 sows) = 5000 cfm.

Uncooled air: From Table 651-4, recommended airflow is 70 cfm in each downspout and (70 cfm x 20 sows) = 1400 cfm in the main duct. From Tables 651-5 and 634-6, 70 cfm requires a 4" diameter downspout and 1400 cfm requires a 18"x20" main duct.

Evaporative cooling: From Table 651-4, recommended airflow is 40 cfm in each downspout and (40 cfm x 20 sows) = 800 cfm in the main duct. From Tables 651-5 and 634-6, 40 cfm requires a 3" diameter downspout and 800 cfm requires a 14"x14" main duct.

Air conditioning: From Table 651-4, recommended airflow is 35 cfm in each downspout and (35 cfm x 20 sows) = 700 cfm in the main duct. From Tables 651-5 and 634-6, a 3" diameter downspout and a 12"x14" main duct is required.

Size a zone air conditioning system at about one ton (12,000 Btu/hr) of cooling capacity per each 100 cfm (e.g. about one-third ton per sow and litter). Some of the cooling effect is lost because the cooled air gains heat as it passes through the duct to the animal. Insulate the duct to at least R=6.

Table 651-5. Downspout sizes for zone cooling.
Sized for 800 fpm air velocity.

Airflow cfm	Area in²	Inside dimensions WxH, in.	diam, in.
10	1.8	1½x1½	1½
15	2.7	1¾x1¾	2
20	3.6	2x2	2½
25	4.5	2x2¼	2½
30	5.4	2x2¾	3
35	6.3	2x3¼	3
40	7.2	2x3¾	3
45	8.1	3x2¾	4
50	9.0	3x3	4
55	9.9	3x3½	4
60	10.8	3x3¾	4
65	11.7	3x4	4
70	12.6	3x4¼	4
75	13.5	3x4½	4
100	18	4x4½	6
125	22.5	4x5¾	6
150	27	4x6¾	6

652 VENTILATION OF BEEF BUILDINGS

"Cold" housing (indoor temperatures about the same as outdoors) is adequate for beef cattle. Ventilation is by natural air movement and supplemental heating is limited to occasional use of heaters for newborn calves in cold weather.

Insulation

Add insulation to cold cattle barns to control condensation and frost formation on inside wall and roof surfaces and to resist high summer temperatures. See Chapter 631.

Ventilation Openings

Leave the front wall open and provide a 4"-6" continuous slot, or equivalent, under the eave or at the top of the back wall for cold weather air inlets. Add baffles on the back wall inlets to reduce snow blow-in, but do not close them completely as snow may be drawn in at the front of the building. The top of the open-front sidewall can be closed down to about 10′ in winter to reduce snow blow-in.

Install large adjustable openings, doors, or large panels in the back wall for warm weather air movement. Open the back wall vent doors in summer so air circulates through at animal level. Size summer back wall openings per Table 652-1.

Provide a continuous open slot in the roof at the ridge, Table 652-1. Protect exposed trusses at the slot from moisture with metal flashing.

Table 652-1. Natural ventilation openings for cattle.
Ridge openings are sized for a 10 mph wind.
Sidewall openings are sized for a 2 mph wind and are a minimum. Opening more of the back sidewall increases summer cooling.

Building width (ft)	30	40	50	60	70	80
Eave & ridge opening (in.)	6	8	10	12	14	16
Sidewall opening (in.)	28	36	44	52	60	68

Do not totally enclose a cold building in an effort to raise inside temperatures. The result is usually a severe moisture buildup, condensation on the roof, and only a slight temperature rise. If moisture buildup is a problem in an existing building, provide larger openings under the eave and at the ridge.

Drafts

Place a beef building with the open side away from prevailing winter winds. If drafts and snow blow-in are problems, close part of the front wall (up to ⅙ of the building's length) at each end and adjacent to solid partitions. Use spaced-board windbreaks at the ends of barns. Install solid cross partitions from floor to ceiling (or as high as the open-front wall) in long open sheds if indoor drafts continue to be a problem.

See Chapter 636 for more information on natural ventilation.

653 VENTILATION OF DAIRY BUILDINGS

Research at the University of Missouri Psychroenergetic Laboratory has indicated no change in production for Holsteins housed at temperatures between 10 F and 70 F. A decline in production occurred above 75 F. Jersey cows showed no change in production between 30 F and 80 F with declines above and below these temperatures.

Most controlled environment dairy buildings are kept at 40-60 F from November through April. Relative humidities of 60%-80% are common in ventilated dairy buildings.

Recommended Ventilating and Heating Rates

Cows

Provide about 35 cfm/1000 lb of body weight continuously, even during extremely cold weather. As temperatures increase, increasing amounts of air are needed. To provide this range, select one fan to run continuously and control the other fans with thermostats.

Mild weather fan capacity is about 120 cfm/1000 lb of body weight. For summer ventilation of buildings where cows are inside during the hottest part of the day, you need a fan capacity of about 335 cfm/1000 lb of body weight, Table 633-2.

Example 1:

Determine the fan capacities for 60-1400 lb cows.

Continuous cold weather fan:

60 cows x 1400 lb x (35 cfm ÷ 1000 lb) = 2,940 cfm

Mild weather fans:

60 cows x 1400 lb x (120 cfm ÷ 1000 lb) = 10,080 cfm

Warm weather fans:

60 cows x 1400 lb x (335 cfm ÷ 1000 lb) = 28,140 cfm

Select:

One continuous 2,940 cfm fan.

2-3,570 cfm fans (Total = 2 x 3,570 + 2,940 = 10,080 cfm).

3-6,020 cfm fans (Total = 3 x 6,020 + 10,080 = 28,140 cfm).

Calves

Warm housing is often selected for 15 calves or more. Maintain 40-50 F temperatures. Partition the calf area from the milking herd. Separate calf facilities are recommended for large herds.

Provide minimum continuous winter ventilation of 10 cfm/100 lb of calf or 4 air changes/hour, whichever is greater. Provide 25 cfm/100 lb of calf for mild weather and 50 cfm/100 lb of calf for warm weather ventilation.

To maintain a desirable inside temperature of 50 F, supplemental heat is needed. Do not simply shut down ventilating fans, because high humidities and condensation result. See Chapter 631 to estimate the amount of supplemental heat in Btu/hr needed in a well insulated building.

Install the heater and exhaust fans on opposite sides of the barn. Set the heater thermostat to turn the heat on at 45 F. The smallest fan should run continuously; set the thermostat for the next fan to come on at 55 F.

Calf pen areas in stall barns are difficult to ventilate and keep dry. In cold climates, locate the pen area to the east or south end of the barn.

Milking Center

Heating and ventilation

Heat the milking parlor, milk room, office, toilet, and utility room to prevent freezing and to dry the floors. Maintain 50 F during milking. Use heat from the compressor, water heater, and furnace. Exhaust equipment heat outside in warm weather. Insulate the milking center well to aid heating and to control condensation.

Use heat supplied by cooling milk. Cooling 100 lb of milk from 98 F to 40 F supplies about 4800 Btu of heat. This added to about 2500 Btu from the compressor motor is equal to a 2 kw electric heater running for 1 hr. A system to use this heat can be built at little extra cost. Locate the compressor, vacuum pump, water heater, and furnace together in a utility room. Connect the furnace fan to come on when the compressor starts. Connect the furnace to a thermostat in the milking parlor so when milk heat is insufficient the furnace comes on. Open an insulated door to the outside in summer to exhaust compressor heat outside or install a 2000 to 3000 cfm exhaust fan. See Fig 653-1.

Milking parlor

Use an exhaust ventilating system in the milking parlor—100 cfm/stall is suggested during cold weather. Control this fan with a time clock to provide some ventilation between milkings and a manual switch for continuous operation during milking and cleanup. For summer operation, 400 cfm/stall is suggested with the fan controlled by a manual switch.

Additional summer operator comfort is possible by providing increased air movement in the pit. Ceiling fans mounted horizontally to move air down across the operator is one option.

Provide screened air inlets with a total area of 1 ft^2/600 cfm of fan capacity. Vent attics or closed spaces above the milk room and milking parlor with screened louvers and/or eave vents.

Heat in the milking parlor during the winter dries the floor and provides operator comfort. Install at least 50,000 Btu/hr (15 kw) for a double 4 herringbone and 70,000 Btu/hr (20 kw) for a double 6 or 8 herringbone milking parlor. In northern states these heat capacities may need to be doubled for existing lightly insulated milking parlors.

Put the furnace of a central heating system in a utility room. A down draft furnace with a filter located for easy changing is recommended. Duct hot air

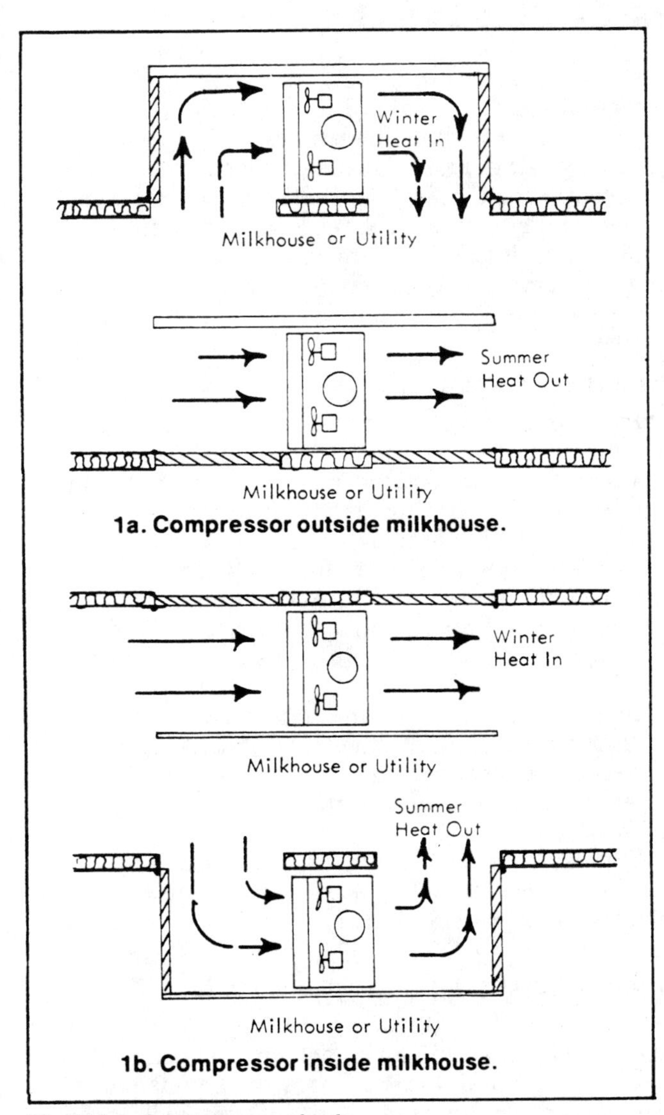

Fig 653-1. Compressor sheds.
The doors permit you to retain compressor heat in cold weather and remove it in warm weather. Make the cross-sectional area of all air passages 1½ times the condenser area.

to the milker's pit, hospital, milk room, and office. Underfloor ducts to the milker's pit enter the pit at least 8" above the pit floor, with the hot air directed toward the floor. The ducts should be vitrified bell tile or cement asbestos pipe 8"-12" in diameter and insulated to prevent heat loss. Locate cold air returns high in the wall. Run them from the milking parlor, milk room, office, and hospital to the utility room. The cold air return duct area should be equal to or larger than the warm air discharge area. Fit them with furnace filters. See Fig 653-2.

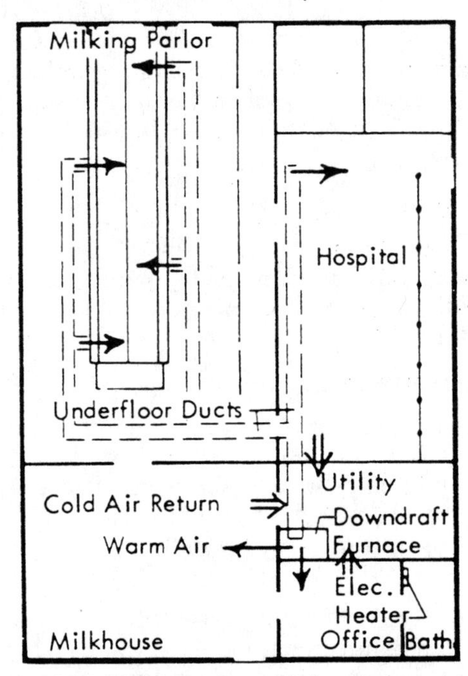

Fig 653-2. Milking center warm air heating system.

Electric heating cable or hot water pipe is sometimes used in the milker's pit floor and on cow ramps where ice and snow are a problem. Heating cable on a manual switch and providing 20 to 30 watts/ft^2 is recommended.

Overhead radiant heaters can heat the milking parlor. Hang them over the front of the stalls directed toward the cow's udder and the milker's hands. Control radiant heaters with a manual switch and turn on only during milking.

Milk room

Heat the milk room with at least a 10,000 Btu/hr thermostatically controlled unit space heater or a central furnace. Set the thermostat to keep temperatures above freezing except when higher temperatures are needed for chores.

Ventilate the milk room with a 600 to 800 cfm pressurizing fan to ensure that air is not drawn from the milking parlor. A larger fan is needed if compressors are in the room. Locate the fan to draw air from the cleanest and most dust-free side of the building or attic. Provide a filter to remove dust.

Office and toilet

If not heated with a central furnace, heat the office and toilet room with a small space heater. A small exhaust fan removes excess moisture created by a shower.

654 VENTILATION OF SHEEP BUILDINGS

Generally, ewes and feeder lambs are in cold housing (inside temperatures about the same as outside). Warm housing is common in cold climates for lambing.

Insulation

Insulating cold sheep barns helps control condensation, frost, and high summer temperatures. Insulate warm buildings well to minimize heating costs. See Chapter 631.

Cold Housing Ventilation

Open-front cold buildings are common in moderate climates, but ventilating doors in **both** long walls reduce drafts and snow blow-in in more severe climates. Place open-front barns with the open side away from prevailing winter winds.

Openings

Provide a continuous ridge opening on gable and shed roof barns as an air outlet. Make the slot about 1″ wide for each 10′ of barn width (4″ minimum in a cold climate to prevent freeze-up).

Provide continuous 4″-6″ openings along the eave of both side walls for cold weather air inlets. Adjustable baffles on the openings reduce snow blow-in. Protect lambing area with temporary walls or drop curtains during cold weather lambing.

Provide large adjustable openings—doors or large panels—in sided walls for warm weather ventilation. Open about half (4′ minimum) of the walls in summer; circulate air at animal level.

Warming a cold barn by tightly closing it during very cold weather results in severe moisture buildup, condensation on the roof and walls, and only a small temperature rise. It is much more important in sheep housing to maintain a dry and draft-free barn than to raise the temperature a few degrees.

See Chapter 636 for more information on natural ventilation.

Warm Housing Ventilation

In warm housing, inside temperatures are maintained above freezing and often at 45-55 F.

Warm barns are designed for fan winter ventilation. Natural ventilation is often used in summer through adjustable windows and wall panels.

Provide some air movement at all times. Select fans to provide 25 cfm/1000 lb animal weight continuously in winter. Provide an additional 75 cfm/1000 lb animal weight to turn on in steps as temperature rises.

In hot weather, increase fan capacity to 200 cfm/1000 lb, or open summer ventilating doors for natural air movement. Use large circulation fans to increase air speed over the animals. Consider evaporative cooling.

Supplemental heat (in addition to heat lamps for mothering pens) may be needed for adequate inside temperatures in warm barns.

655 VENTILATION OF HORSE BUILDINGS

As a rule, horses in mild or moderate climates can be housed in uninsulated buildings. However, wood or masonry sidewalls and tight sheathing under the roof are desirable.

In northern regions, insulated buildings and supplemental heat are more common to protect the animals and attendants from severe winter weather. Heated barns also help show horses stay in show condition throughout the year.

Enclosed buildings are usually tightly constructed with snug-fitting doors and windows. They may be uninsulated or insulated. Uninsulated and some insulated barns are used for cold housing. Use only well-insulated buildings for warm housing. As the move is made from cold to warm housing, the ventilating system requires more refinements.

Provide 25 cfm/1000 lb animal in very cold weather, and 100 cfm/1000 lb animal in milder winter temperatures.

Computing Supplemental Heat

Heat helps control moisture and maintain indoor temperatures above freezing. Because the density in a horse barn is low, more heat is needed than the horses can provide. Consequently, the insulation requirement for a warm horse barn is greater than for other mature animals, and supplemental heat has to be supplied. For an all-stall barn, insulate as recommended in Table 631-5. Estimate the supplemental heat needed for different conditions by adding the "building heat loss" and the "ventilation heat loss" for the selected minimum inside winter temperature. This simplified procedure assumes that animal heat is sufficient to control moisture. Additional heat is needed with higher ventilating rates.

Building heat loss (Btu/hr)

Select the ceiling and wall multipliers from Table 655-1 for your desired inside conditions and your climate.

Building heat loss = ceiling area (ft^2) x ceiling multiplier + wall area (ft^2) x wall multiplier

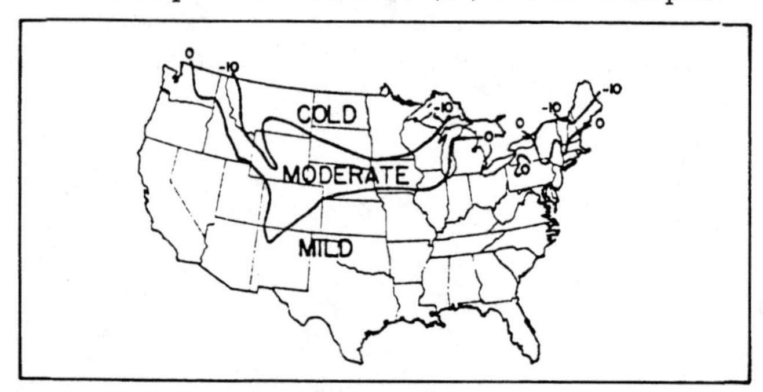

Fig 655-1. Temperature zone map.

Table 655-1. Heat loss multipliers.

	35 F inside			45-50 F inside		
	cold	moderate	mild	cold	moderate	mild
Ceiling	2.4	2.8	2.9	3.0	3.7	4.2
Wall	3.9	3.8	3.9	5.0	5.0	5.5
Ventilation	3.0	2.4	1.9	3.9	3.2	2.7

Example 1:

A 36′ x 36′ barn with 5 box stalls is to be kept at 45-50 F in southern Indiana.

Ceiling area = 36′x36′ = 1296 ft^2
Ceiling multiplier (mild zone) = 4.2
Wall area = 4 walls x 8′ high x 36′ long = 1152 ft^2
Wall multiplier (mild zone) = 5.5

Building heat loss = 4.2 x 1296 + 5.5 x 1152 = 5443 + 6336 = 11,779 Btu/hr.

Ventilation heat loss

Select the ventilation multiplier for your inside temperature and climate.

Ventilation heat loss = weight of horses (lb) x multiplier

Example 2:

Assume 5-1000 lb horses in the barn of Example 1.

Ventilation heat loss = 5 x 1000 lb x 2.7 = 13,500 Btu/hr

Total heat loss = building loss + ventilation loss

Supplemental heat needed = total heat loss = 11,779 + 13,500 = 25,279 Btu/hr.

NOTE: If there is no ceiling, treat the roof area as the ceiling area and the ends of the building as part of the total wall area.

Set the thermostat to shut the heater off below the temperature at which the main winter fans start to operate (usually 50-55 F).

656 VENTILATION OF POULTRY BUILDINGS

Broiler Chickens and Young Turkeys

Air temperature: 85-95 F for the first 3 to 5 days after hatching. Then decrease 1-2 F per day, with a 65-75 F minimum after 4 weeks of age.

Humidity: 60% minimum during first 2 to 3 weeks, then 30%-80%.

Ventilation. The ventilation requirement for broiler chicks and young turkeys depends on age and size, waste management system, building construction, and outside moisture content and temperature.

When the outside temperature does not exceed 40 F, 0.1 cfm/chick from 0 to 4 weeks of age is required. A ventilating rate of 5 cfm/bird is needed for summer.

Turkeys may need an increase in minimum ventilating rate to ⅙ cfm/chick. Because of limited data, test your turkey building ventilating system to determine the proper minimum ventilating rate needed for moisture removal and heat removal.

Laying Hens

Air temperature: 45 F minimum, 85 F maximum.

Maximum egg production usually occurs at or near 55 F.

Relative humidity: 50%-80%.

Heat and moisture production

The heat loss and moisture production of poultry is given in Chapter 604.

The ventilation requirements for layers depends on layer weight, waste management system, building construction, and outside moisture content and temperature. A minimum of ½ cfm/4 lb layer is recommended during the summer.

660 FRUIT AND VEGETABLE STORAGE

661 CONTROLLED ATMOSPHERE FRUIT STORAGE

From Selders and Ingle, West Virginia, 1967.

The atmosphere normally contains 20%-21% oxygen, 0.03% carbon dioxide, traces of inert gases, and the balance nitrogen. Apples can be stored for prolonged periods in atmospheres of 3%-5% oxygen and 0%-5% carbon dioxide, with the balance nitrogen. The term "modified atmosphere" is sometimes used rather than "controlled atmosphere," or CA, the term used here. Regular storage is conventional refrigerated storage.

With CA, chemical reaction rates that usually occur in fruits stored in normal atmospheres at comparable temperatures are reduced, so desirable quality is maintained over longer storage periods. You can see the effects only after about 90 days; therefore, place part of the crop in regular storage for use during the first 90 to 125 days after harvest, and the remainder in CA for later use.

Storage Scald Control

Apples in CA storage are very susceptible to storage or superficial scald and **must** be treated with scald inhibitors. Scald control materials and methods are described in West Virginia Agricultural Experiment Station Current Report 47. Treatment of regular stored fruit is recommended.

Atmosphere Modification and Control

Product generated atmosphere (Conventional CA)

As apples respire, they absorb oxygen from the surrounding atmosphere and produce carbon dioxide. In a gastight room, the oxygen level lowers and carbon dioxide accumulates. Fruit respiration modifies the atmosphere in about two weeks, the period called "pull-down." Some oxygen is added to the room periodically, and excess carbon dioxide is removed.

Determine the oxygen and carbon dioxide concentrations, usually with an Orsat gas analyzer, at least once a day. The volume of a gas sample is measured, and the oxygen and carbon dioxide are removed by chemical absorbers. The decreased sample volume measures the oxygen and carbon dioxide originally present.

Carbon Dioxide Scrubbers

Excess carbon dioxide can be removed from the storage in several ways.

Caustic soda scrubber. Although the first CA rooms used sodium hydroxide, or caustic soda, it is not used extensively now because it is very corrosive and difficult to handle. Pumps and plumbing must be replaced often. Spray mist injures fruit.

Water scrubbers. Water (or brine) can also remove enough carbon dioxide. Water or brine to defrost evaporator coils is circulated through an aeration system to remove the absorbed carbon dioxide.

One of the most common aerators is the so-called "stovepipe." Water is pumped from the collecting pan underneath the evaporator coils, taken outside the storage room, and sprayed into a length of 6"-12" stovepipe. As water falls through the pipe, carbon dioxide is released. The water is collected in a sump and pumped back over the evaporator coils. For a 12,000 bu room, about 70 gpm of water are circulated.

For rooms with dry evaporator coils, blow air through a closed chamber into which the water or brine is sprayed. Again, the water is aerated to remove the absorbed carbon dioxide.

Generally, water or brine alone cannot remove all excess carbon dioxide during the pull-down. Caustic soda can be used for 2 to 4 weeks after the room is sealed and then be replaced with water or brine.

An alternate to caustic soda is fresh hydrated, high-calcium spray lime. It is placed in the room where air is circulated over it or in a small "lime room." Use 0.10 to 0.15 lb lime/bu of fruit. If the lime is packaged in paper bags without plastic film liners, it is not necessary to open the bags.

Externally generated atmospheres

Generators outside the storage rooms produce a gas mixture with a low oxygen, and, if desired, a relatively high carbon dioxide concentration. This modification burns air and propane or natural gas with a catalyst. Undesirable impurities are removed with scrubbers. A water-cooled exchanger removes the heat of combustion. The gas is flushed through the room continuously or only during pull-down.

After pull-down, circulation may be continuous or intermittent. Use a carbon dioxide absorber and a fresh air intake for oxygen only when needed. Systems that recirculate the atmosphere allow the fruit to do much of the work of maintaining the modified atmosphere. Continuous flushing guarantees constant atmosphere but requires a large amount of fuel and electricity.

For externally generated atmospheres, the pull-down is less than five days as compared with 7 to 14 days for product-generated CA, and the rooms may not have to be as gastight. "Leaky" rooms are more expensive to operate than the cost of properly sealing a room. With externally generated atmospheres, the room can be opened, partially emptied, and resealed. Conventional CA rooms cannot be resealed; operating them partly full is more expensive. Consider smaller rooms to be emptied in a month or so.

Liquid nitrogen can flush conventional CA rooms and hasten pull-down. It also has a refrigerating effect to accelerate cooling. No major equipment investments are needed. Nitrogen generators are relatively small units on pallets or skids that are moved as successive rooms are filled.

Precautions

CA storages do not have enough oxygen to support human breathing. To enter a room for repairs during storage, wear breathing equipment. Have at least one person outside the room for each one in the room.

When the room is opened, remove fruit when gas analyses show 18%-20% oxygen concentration or enter for a very short time until there is enough oxygen.

Construction of CA Storages

Structural requirements are the same as for regular refrigerated storage rooms. There is one additional requirement—a good gas seal completely around the room.

Room Size

There is probably no "best" size for CA rooms; they range from 1000 to 50,000 boxes. The size of a room depends partly on the rate at which the room can be filled and the rate the fruit can be marketed once the room is opened. Some state laws specify time limits for filling the room, the atmosphere established, and that the apples must be in CA for a certain period (usually 90 days) before they can be marketed as CA apples.

Because of varietal responses to different atmospheres and temperatures, certain varieties are not stored in the same room. If five or six varieties are to be in CA, provide at least two rooms.

Insulation

In general, insulation is the same for CA as for regular storages. Provide an adequate gas seal in CA rooms. Some foam-type insulations applied with proper mastics and coatings provide both a vapor barrier and gas seal.

Wood Frame Structures

Wood frame structures for CA rooms are usually stud construction with the vapor barrier and siding on the outside, and a 28 ga sheet metal gas seal on the inside. Fill the wall space with insulation. Nail metal lining to either the lumber wall used to confine the insulation, or to the studs or other nailing strips to serve as the confining wall. Wall framing is rigid—any movement in the wall can break the gas seal. In gable roof construction, the sheet metal is under the ceiling and insulation is above it.

Aluminum corrodes from the salt brine spray from refrigeration units. Apply protective coatings to the aluminum, particularly near the cooling unit.

Seal nail holes and metal joints with a high-grade non-drying caulk. Lap metal sheets about 2"; lay caulking under the sheet before nailing. Then lay another bead on top of the first sheet and fasten the second sheet with large-headed galvanized nails spaced 2"-3". Caulking that oozes out is smoothed over the overlap and all nail heads. Leave ⅛"-¼" between butted sheets. Caulk under the sheets before they are fastened. Slightly overfill a metal moulding with caulk and fasten with screws 6"-8" o.c.

Floors and Doors

Pour concrete subfloor on sand or gravel fill. Mop two layers of 55 lb asphalt-impregnated roofing paper in hot asphalt for a gas seal. Stagger joints. Or, two or three layers of roofer's felt are made into a typical built-up roof over the subfloor. Tie the wall seal to the floor seal. Flare wall metal 12"-18" out on the floor and hot-mop into the felt or roofing paper. Coat any metal in contact with concrete with hot asphalt to check corrosion. Most engineers recommend at least 2" of floor insulation for CA rooms. Apply rigid insulation strong enough to support floor loads over the gas seal. If the floor is not insulated, extend 2" or more of perimeter insulation 1'-2' below the subfloor.

Pour a reinforced concrete wearing slab, at least 5" thick, over the insulation to carry the loads imposed on the floor. Fill a ¾" space between the floor and the wall with hot asphalt to make an expansion joint and to ensure a good seal.

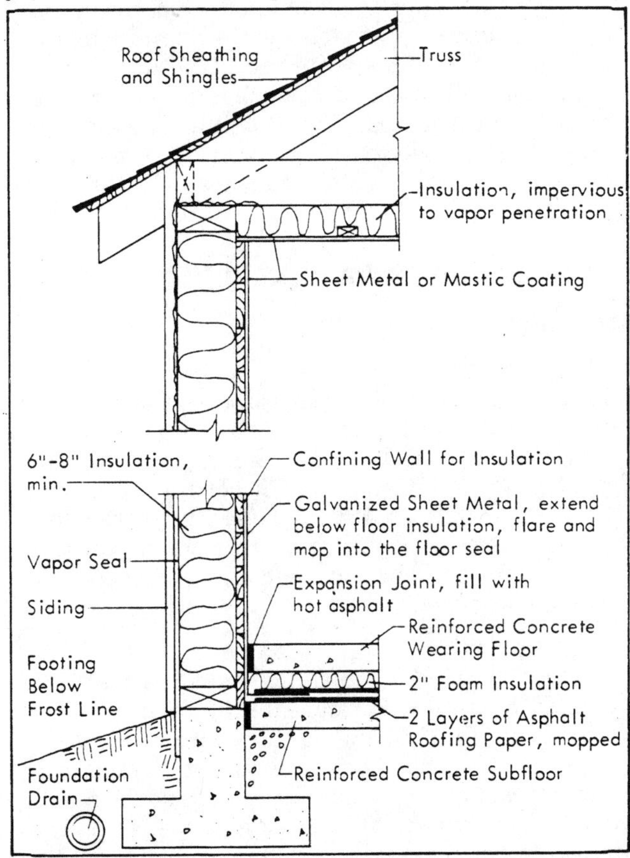

Fig 661-1. Construction of CA storage.

Make doors for CA rooms large enough for a fork lift truck—6'-8' wide by 8'-10' high. Have only one door per room if possible. Doors for CA rooms must be gastight.

Steel-clad insulated doors or closure panels bolted to the walls from the inside are used in CA rooms. Seal doors with heavy galvanized sheet metal bolted to the inside of the door opening. Use a standard hinged cooler door on the outside. Provide a second opening for a man to crawl through with a 6"-8" porthole for removing fruit samples and for observation.

Testing for Gastightness

Test CA rooms for gastightness before storing fruit, preferably in the morning and with about equal indoor and outdoor temperatures. Leave one man in the storage with a ladder or scaffolding and caulking; he will hear most leaks. Blow air into the storage to a pressure of at least 1½" water. Read air pressure loss on a manometer sensitive to 0.1" pressure. Small storages must be tighter than large ones; pressure should not drop to zero in less than an hour for a 2,000 bu room, or less than ½ hr in a 10,000 bu room. A gastight room continues to lose pressure over several hours.

662 BULK POTATO STORAGE

The main purpose of the storage is to provide conditions that will maintain high tuber quality, control diseases and provide a marketable raw product with minimal weight loss. The layout of the storage should be amenable to mechanized, high volume loading and unloading systems while minimizing the magnitude of handling impacts to the potato. Potatoes for different end uses should be stored at different temperatures. Five possible phases of a storage season in sequential order are: 1) removal of field heat, 2) suberization and wound healing, 3) cool-down, 4) holding, and 5) reconditioning.

Desired Environmental Conditions

Temperature

Suberization period

Immediately after harvest, suberization is most commonly accomplished in 10 to 14 days using temperatures of 50 F-60 F. This period may be lengthened if chipping potatoes do not appear to have the desired low internal sugar levels. In years when certain diseases or field frost are severe, potatoes should be brought down to the holding temperature as soon as possible.

Cool-down period

Potatoes for chip processing should be cooled at a rate of 0.5 F per day. Potatoes for all other uses can be cooled at a rate of 1.0 F per day if weather conditions permit. If sufficient cooling capacity is not available and final holding temperatures can not be maintained, cool-down should be done in stages so the potatoes do not reheat during warm spells.

Holding period

When cooling has been completed, the holding temperature should be:

Seed use: 38 F-40 F
Table use: 40 F-45 F
French fry use: 45 F-50 F
Chip storage (over 4 months): 50 F-55 F
Short-term chip storage: 55 F

Reconditioning period

Usually temperatures are increased above the holding temperature for at least a week before unloading a bin to reduce incidence of handling damage, or possibly several weeks if sugar accumulation has occurred and there is a dark color of the subsequent processed product. These temperatures should be:

45 F-50 F: to reduce handling injury
60 F: to stimulate rapid seed sprouting
55 F-60 F: to reduce internal sugar balance

Once the reconditioning temperature has been reached and product quality is satisfactory, it is recommended that a bin be emptied within one week to limit quality degradation during the bin unloading.

Relative Humidity

Relative humidities of at least 90-95% are recommended for all phases of storage. High relative humidities reduce weight loss and reduce the incidence of pressure flattening in deep, bulk piles. Maintaining high relative humidity during the suberization period is most important for reducing weight loss. Where exterior temperatures may be 0 F to -20 F for several weeks, it may not be economically feasible to insulate storages to maintain relative humidities over 90%.

If late blight, leak or frozen potatoes are present, the relative humidity should be reduced to approximately 85% to remove moisture that is freed with cell breakdown. Once the problem is controlled, the relative humidity level may be maintained at 90% or greater.

Controlled Atmosphere

Long term storage atmospheres with reduced oxygen content are detrimental to potatoes. A level of 1.0% carbon dioxide is normally considered the upper threshold. A common level of carbon dioxide in ventilated bins is 0.2%-0.3%

Cross Contamination/Sensitivities

Potatoes should always be stored in the dark. Light causes greening of the potato surface due to chlorophyll development and also initiates glycoalkaloid synthesis that gives a bitter off- flavor. Potatoes should be stored separately from other fruits and vegetables to avoid transfer of odor or off-flavors.

Chemical Treatments

The only current chemical treatment that is routinely applied in the storage is the sprout inhibitor isopropyl-N-(3-chlorophenyl) carbonate (CIPC). This controls the sprouting of table and processing potatoes which normally occurs two to three months after harvest. It is volatilized and distributed through the ventilation system after suberization and cool-down have been completed. All exposed structural surfaces and air contact surfaces must be thoroughly steam cleaned if seed potatoes are subsequently stored in a bin that has been treated with CIPC.

Facilities and Operation

Structural

Wind and snow loads

Design loads should be those specified in the American Society of Agricultural Engineers Standard EP288.4.

Product loads

Design loads for potatoes should be based on information in Chapter 104 and ASAE D446.

Floor loads

The maximum floor load is due to loaded field trucks. If specific information on size of trucks is not available, use ASAE EP378.3 for design loads.

Insulation

Thermal requirement

The amount of insulation that is specified for ceilings and exterior walls is based on the need to avoid condensation at high relative humidities in the storage. Based on interior storage temperature and relative humidity and exterior design temperatures, common ranges of thermal resistance (R) factors are:

Walls: (R) 20-40 hft^{2}F/Btu
Ceilings: (R) 30-40 hft^{2}F/Btu

Fire retardant

Exposed plastic foam insulations require a protective fire coating unless exempted from fire code.

Vapor retarder

A vapor retarder shall be installed to form a continuous seal on the warm side of the insulation. Six-mil, (0.006 in.) polyethylene plastic sheets are commonly used. All seams and edges need to be securely sealed. Normal procedure is to roll and tape all seams and seal the connection between the vapor retarder and structural members with a sealant. Other vapor retarders are also used such as foil or sprayed coatings. The installed permeance should not exceed 0.1 perm (0.1 gr/hft^2 in. Hg) wet-cup test. Installed performance is extremely important to preclude wood decay, steel corrosion and metal deterioration.

Moisture removal

A vented air space at least 1" wide is required between the outside face of the insulation and the exterior sheathing. This provides exterior air movement to remove moisture that penetrates the vapor retarder. Current practice is to provide 1/150 of building ceiling area as attic vent area. One quarter of this area is at each eave and one half of area is at the roof ridge. The storage should be ventilated after the potatoes are removed to dry the structural materials.

Mechanical

Ventilation system

Recommended capacities vary based on hours of available cooling temperatures, product respiration rate, and potential for losses due to disease or frozen potatoes. Maximum rates are used during the cool-down period and for the control of disease outbreaks or cell break down of field frozen product. Ranges in air flow rates are:

Seed or table use: 0.5-1.0 cfm/cwt
Process storage: 1.0-1.5 cfm/cwt

If thin wall, relatively rough ducts (corrugated metal) with hole outlets are used, an operating static pressure of 1.0" water column is common. With thick walled smooth ducts (wood, concrete) and slotted outlets, an operating static pressure of 0.75" water column is common for the ventilation system. Once holding temperatures are reached, sufficient ventilation capacity to maintain a 2 F difference between the bottom and top of the pile is required. Operating capacity is often reduced by one-half by utilizing intermittent operation, two speed fans or reducing the number of operating fans.

Based on duct roughness and physical dimensions, the maximum velocity should not exceed 1,000-1,500 ft/min. Maximum duct spacing should not exceed 80% of the pile depth. Spacings of 8'-10' on centers are common.

For slotted ducts, the effective discharge area when covered with potatoes should be 75%-100% of the duct inlet area. Potatoes will block approximately 65%-75% of the constructed slot area. For ducts with holes or slots that are protected, approximately 10% of the discharge area may be blocked by potatoes.

The plenum cross-sectional area is sized so velocities do not exceed 800 ft/min.

Intakes and exhausts are sized so that the velocity does not exceed 800-1,000 ft/min for the gross area. In colder areas, these are often sized to handle 80% of the total fan capacity. Successful operation in extremely cold outside temperatures requires that the moist exhaust air stream never comes in contact with the intake port. An air-lock type chamber or insulated covers have been successfully used.

Heating system

Supplemental heat may be required to warm the storage structure, warm incoming ventilation air or raise potato temperatures during reconditioning. Typical rated output may range from 1.5-5.0 Btu/cwt depending on storage fill and operating conditions. In a well constructed, completely filled storage, there is sufficient potato respiration heat so that heaters will not operate. All direct combustion heaters must be vented and have ignition systems that will not flameout in turbulent air conditions. All heaters shall be installed in accordance with prevailing safety and fire codes. Heaters are installed in the fan plenum or in the bin in the overhead space above the potatoes.

Humidification system

Humidification is required during the storage period. Centrifugal humidifiers, pneumatic nozzles or high pressure water spray nozzles are used. These are usually installed between the heaters and the duct inlets. Provision should be made for containing discharge fallout and channeling it to a floor drain. In certain situations, humidification systems are used for simultaneous cooling and humidification. System capacity varies with local climatic conditions and the amount of outdoor air being introduced. Common capacities may be as great as 1.0-3.0 gal/hr/1,000 ft^3/min ventilation capacity.

Refrigeration system

Generally, refrigeration systems are sized based on spring and summer holding conditions. Direct expansion freon systems, with evaporation coils that are mounted on individual bin ceilings, are used in multiple bin storages. In large single bin storages a central compressor and evaporator are used. All units should operate at a 5 F incoming air to incoming refrigerant temperature difference. Often this is integrated in a humidification system. Size varies according to local operating conditions. Sizes frequently range from 6-8 Btu/hr cwt.

Electric, hot gas or room air defrost are common in systems that require defrosting.

Electrical Service

The electrical service shall conform to the National Electrical Code and local codes where applicable. Service size should be sufficient for lighting, ventilation system, heating/refrigeration system and potato handling systems. Where available, three phase power should be used. All wiring should be surface mounted (exposed) to protect the integrity of the structural vapor retarder. All boxes should have a drip hole to prevent water accumulation. Plastic conduits should be used.

Control systems

The intake and exhaust dampers/louver are often controlled with an automatic system that contains a differential thermostat to sense bin-outdoor temperature differences and a proportional thermostat that controls the damper/louver drive motor. In very cold locations, manual damper/louver control is common. If the intake control motor is exposed to exterior air, it should have an internal crankcase heater. All systems should contain a low limit or safety thermostat that shuts the fan off if there is a damper/louver or heater failure and the duct air temperature becomes too cold. Often a 24-hr interval timer is incorporated for intermittent fan operation. Motor controllers that vary frequency and voltage are occasionally used to reduce operating fan capacities. All control systems should have a manual override provision with a mechanical lockout for use during servicing. Locate control panels in the driest possible location.

Heaters are thermostatically controlled. If more than one heater is installed in a bin a 2 F difference is maintained between their thermostat settings so heater operation is staged.

Some automatic control systems contain humidistats to control bin relative humidity. Care is required in selecting humidistats because of the high desired relative humidity operating levels and moisture and dirt particle contamination of the sensors. Long-term drift should not exceed 2% RH. All humidistats should be checked periodically with a reliable psychrometer. Many systems are manually operated until water droplets start forming on the ceiling of the bin. Then the humidifiers are temporarily turned off.

In automatic control systems for refrigeration systems, the operator has two selections: 1) the refrigeration system will come on only when outdoor air is too warm for cooling, otherwise outdoor air is used; 2) refrigeration is under continuous time clock control because there are no opportunities during the day for outdoor cooling. In manual control systems, a manual on-off switch is used.

All sensors located in air ducts should be mounted at least 3' beyond the duct entrance to insure adequate air mixing. This distance also reduces the risk of sensors being coated with water from the humidifier discharge in the air plenum. All sensors need to be protected from radiant energy and should be in well protected areas that receive a uniformly mixed representative air stream. In a well ventilated storage, the highest potato pile temperature is usually located 12"-16" below the potato pile surface.

Often automatic control systems are switched to manual control in winter to preclude damaging the damper/louver motors in case the damper/louver becomes frozen shut. All control systems and sensors should be tightly covered with plastic bags during the application of sprout inhibitor to the bin.

Operation/Management

Ventilation air distribution systems are normally operated as pressurized (divided) flow systems rather than suction (combining) flow systems and the ducts are designed accordingly. Unless cooling or humidification is required, fan operating time should be minimal during the suberization period. Unless very tight dampers or louver are used, the wind action on a storage will often keep CO_2 levels from becoming excessive. If CO_2 monitoring equipment is not available, operating the fans for several hours per day with the dampers or louver closed should preclude any excessive CO_2 accumulation. If the potatoes are wet, have suffered field frost, or there is a high risk of disease breakdown, fans should be run continuously with no humidification until suitable holding conditions are reached. When extremely cold weather conditions and subsequent ceiling condensation occur, several options may be used individually or in combination. Ventilation fans are often run continuously in storages that have a single exterior structural wall with a vertical air wall which contains discharge slots along the ceiling. Also fans in the overhead unit evaporators may be run or the heaters in the overhead bin space may be switched on. The subsequent air movement across the ceiling reduces condensation problems.

Storage Inspection

All bins should be inspected at least once a day to check the bin environment, appearance of the potatoes, operation of the mechanical system and sensor outputs.

670 ENERGY

671 SOLAR ENERGY FUNDAMENTALS

Solar energy comes from thermonuclear reactions in the core of the sun. The energy is mostly shortwave radiation emitted into space in all directions. When it strikes a material, the radiation can be reflected, transmitted, or absorbed. The absorbed fraction causes the material to heat.

Solar energy is nonpolluting—it has no undesirable byproducts. The energy itself is free, but the equipment required to collect and use it is not. The relative costs of solar collection equipment compared with fossil fuel (coal, gas, and oil) costs have slowed solar development. But solar energy collection is already economically feasible for a few applications, and if fossil fuel costs continue to increase, more solar applications may become feasible. Solar energy can reduce our dependence on fossil fuels, although its availability is too variable and too limited to completely replace fossil fuels.

This discussion of solar energy fundamentals gives basic information on solar energy and on collecting and using it as heat energy. Agricultural applications are emphasized. A study of solar energy fundamentals will help you:

- Determine how much solar energy is available in your area.
- Determine if solar energy is realistic for your operation.
- Select the best type of commercial or home-built collector for your situation.
- Calculate the size of solar collector you need.
- Determine the size and type of solar energy storage system best suited to your needs.
- Plan or design a collector to meet your needs.

Available Solar Energy

The quantity of solar energy, or solar radiation, reaching a surface decreases with greater distance from the sun. The sun is not at the center of the earth's orbit, so the earth's distance from the sun varies during the year, Fig 1. The average value for the solar radiation reaching the outside of the earth's atmosphere—called the solar constant—is about 428 Btu/hr-ft^2. Total average daily radiation received outside the earth's atmosphere is 24 hr x 428 Btu/hr-ft^2 = 10,272 Btu/day-ft^2. Only a fraction of this energy is available to solar collectors on the earth's surface. The actual amount of solar energy available to collectors on the earth depends on time of day, time of year, latitude, collector tilt angle, and weather.

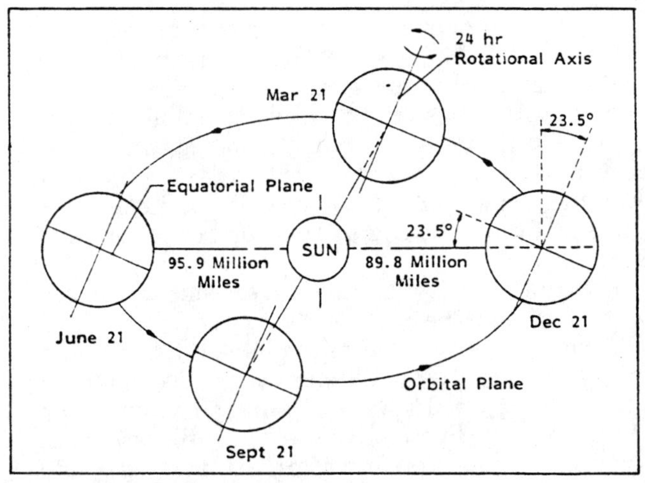

Fig 1. Earth's motion around the sun.
Solar year is one revolution of 365¼ days.

Day-Night Cycle and Season

Because of the earth's rotation, solar collectors on the earth's surface do not receive energy 24 hr/day. On sunny days, solar radiation increases from zero just before dawn to a maximum at solar noon and decreases to zero again at dusk, Fig 2.

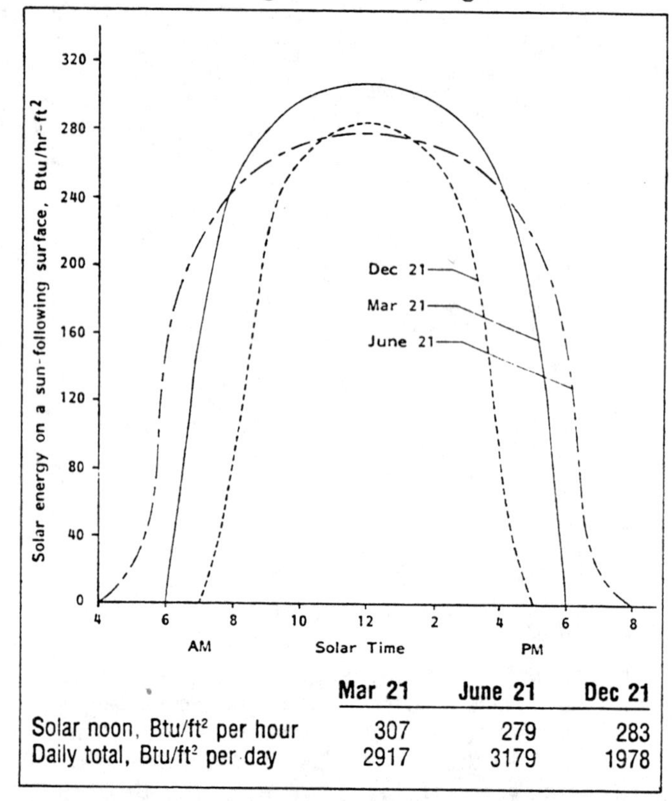

	Mar 21	June 21	Dec 21
Solar noon, Btu/ft² per hour	307	279	283
Daily total, Btu/ft² per day	2917	3179	1978

Fig 2. Hourly clear day radiation.
At 40° north latitude on a surface that follows the sun across the sky.

Note that solar time (used in Fig 2 and Table 672-4) is not the same as local clock time. At solar noon, the sun reaches its highest point in the sky for the day. Solar noon varies from local clock noon depending on time of year, where you live in the time zone (longitude), and whether Daylight Savings Time is in effect.

The number of hours of daylight changes with the season, Fig 2. Dec. 21 (winter solstice) is the first day of winter and the shortest day of the year. June 21 (summer solstice) is the first day of summer and the longest day of the year. March 21 (spring equinox) is the first day of spring and one of two days in the year (the other is Sept. 21, fall equinox) when the number of hours of daylight equals the number of hours of darkness.

The changing day lengths and seasons are caused by the fact that the earth's rotational axis is tilted 23.5° with respect to its orbital plane, Fig 1. We have summer in the United States when the northern hemisphere is tilted toward the sun and winter when it is tilted away. In summer, the sun is higher in the sky than in winter (greater solar altitude, Fig 3) and it rises and sets further to the north (greater solar azimuth, Fig 3). Because the sun is higher in the sky

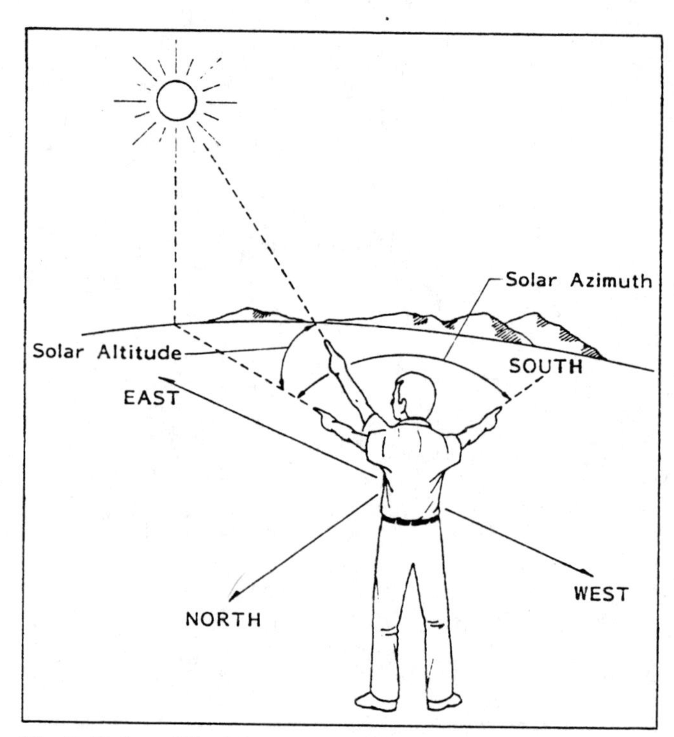

Fig 3. Solar altitude and solar azimuth angles.

(see Atmospheric Effects) and there are more hours of daylight, more daily total solar energy is received on surfaces exposed to the sun in summer. See the sun-following surface curve in Fig 4.

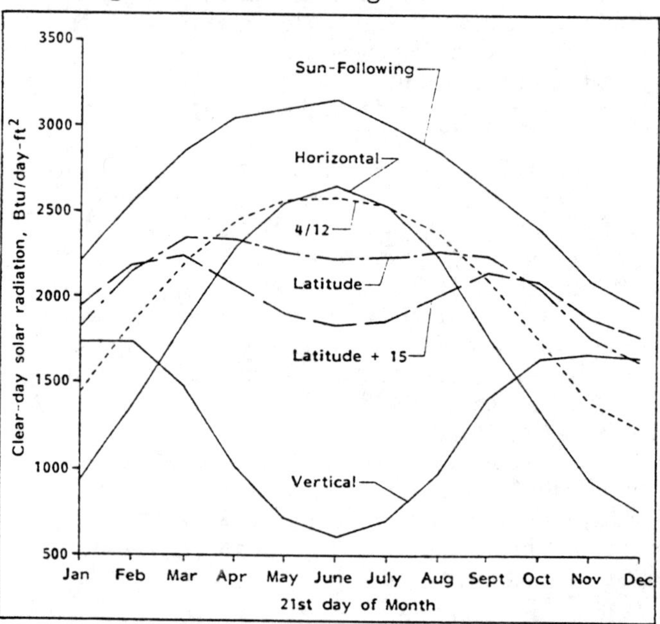

Fig 4. Clear day solar radiation.
South-facing surface. 40° north latitude. Curve labels are collector tilt angles measured from horizontal.

Surface Orientation

A sun-following surface—one that tracks or follows the sun's movement across the sky—receives more solar energy than surfaces with any other orientation, Fig 4. The sun-following surface is always perpendicular to the sun's rays, so the angle of inci-

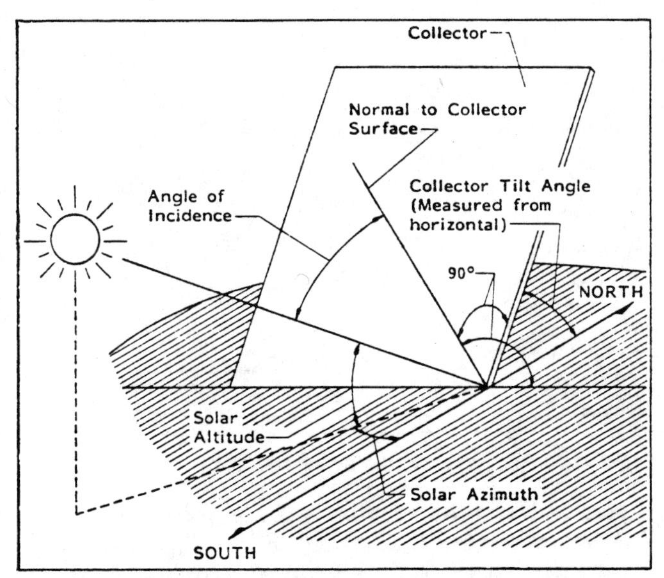

Fig 5. Solar angle of incidence on south-facing surface.
The angle of incidence is measured from the sun to normal to surface.

dence, or angle between the sun's rays and a line drawn normal to the surface, Fig 5, is 0°. ("Normal to" means perpendicular to or at right angles—90°— to a surface.) Utilization of collector surface area is maximized and reflection off the collector is minimized, Fig 6.

Sun-following surfaces must pivot both horizontally and vertically to track the sun's changes in altitude and azimuth. Tracking requires a device to sense the sun's position, a mechanism to move the collector, and a fairly complex support structure. Tracking increases collector costs and maintenance requirements and is seldom cost-effective for agricultural solar collectors.

A support and control system to track the sun in only one direction—either altitude or azimuth—is much simpler. Fixed collectors that do not track the sun are even simpler and less expensive. Less solar energy is available to fixed collectors than tracking ones, but their cost is usually low enough that you can afford extra collector area (perhaps as much as 50%) to make up the difference.

In the northern hemisphere, fixed solar collectors generally receive the most solar energy per day if they face due south. But deviations up to 20° east or west of due south have little effect on daily solar energy collection. Deviations greater than 30° from south greatly reduce the amount of solar radiation received. Sometimes, southeast or southwest collector positioning may be desirable, such as when an obstruction blocks the morning or afternoon sun or when more heat is needed in the morning or afternoon or in areas with predominantly cloudy mornings or afternoons.

For maximum solar energy collection with a fixed collector, set the tilt angle so the angle of incidence is near 0° at solar noon at the time of year when you want to collect the most heat. The collector will perform almost as well as a sun-following collector in the

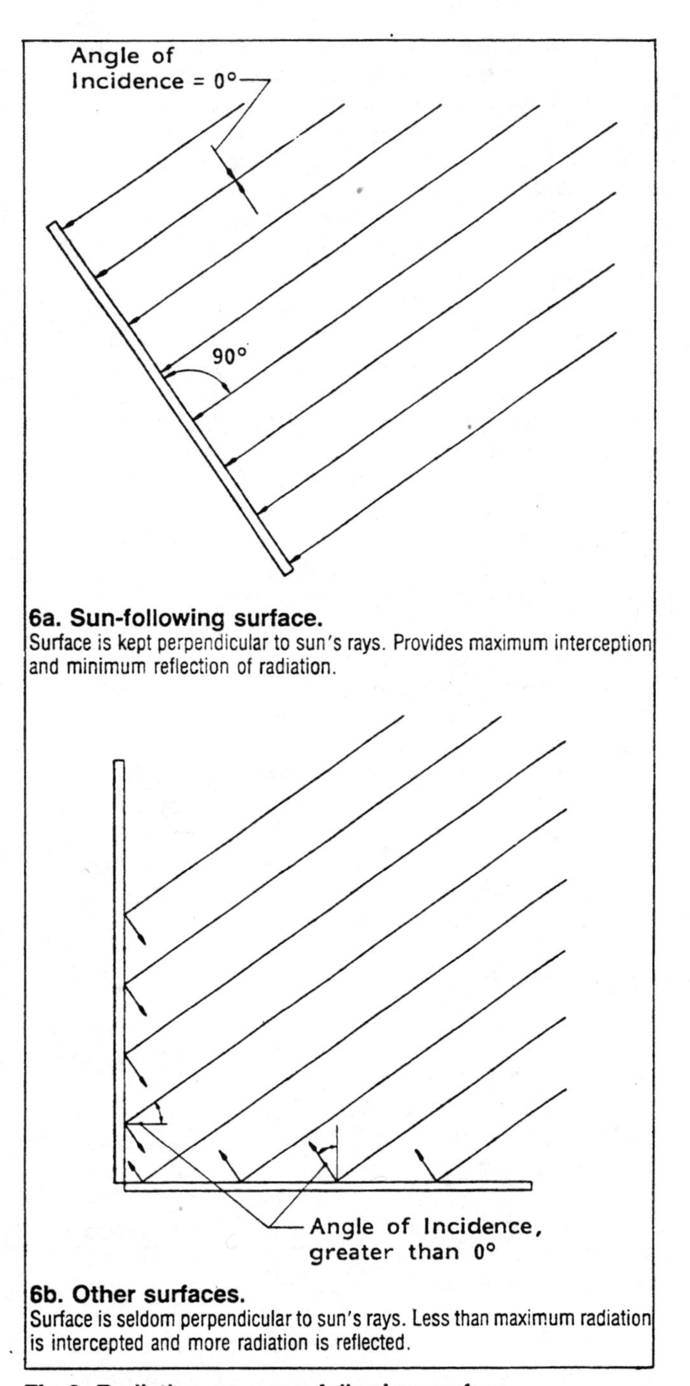

6a. Sun-following surface.
Surface is kept perpendicular to sun's rays. Provides maximum interception and minimum reflection of radiation.

6b. Other surfaces.
Surface is seldom perpendicular to sun's rays. Less than maximum radiation is intercepted and more radiation is reflected.

Fig 6. Radiation on a sun-following surface.

middle of the day, but in the early morning and late afternoon when the angle of incidence is large, much of the solar radiation will reflect off the fixed collector surface.

Because the sun's altitude at solar noon varies with latitude and time of year, the appropriate tilt angle for a fixed collector depends on when and where it will be used. Fig 4 shows the solar radiation received by a sun-following surface and several fixed, south-facing surfaces at 40° north latitude: horizontal (0° tilt angle); 4/12—such as the south slope of a building roof (18.4°); tilt angle equal to the latitude (40° in this case); latitude plus 15° (55°); and vertical—such as a south building wall (90°).

Of the fixed surfaces, the one with a tilt angle equal to the latitude has the most consistent energy reception throughout the year. For applications requiring heat year-round, choose a collector with the tilt angle equal to latitude. The surface with a latitude plus 15° tilt angle receives the most energy during the coldest months of the year and is the best surface for collectors needed most during the winter heating season.

Horizontal surfaces receive the most energy in summer and the least in winter. The 4/12 slope performs much like a horizontal surface. In the northern U.S., the 4/12 slope is usually not a very good surface for meeting winter heating needs—it collects too much energy in summer and too little in winter. Also, dust, frost, and snow accumulations on 4/12 slopes can greatly reduce energy collection.

Vertical surfaces receive about 10% less solar energy than latitude plus 15° surfaces in the coldest months, receive very little solar energy in summer months, and have no problem with snow accumulation. Vertical surfaces can be supported on an existing wall. These factors make vertical surfaces a frequent choice for solar collectors used for winter heating in the northern U.S.

The final choice of fixed collector tilt angle depends on when you need the most solar energy, where you have space available for the collector (on the south wall, on the roof, or along the south side of the building), and which collector surface is least costly to build per unit of energy returned.

Atmospheric Effects

Even at noon on sunny days, not all the solar energy reaching the outside of the earth's atmosphere is available on the earth's surface. At 40° north latitude on a sunny March 21, only about 307 Btu/hr-ft^2 is available to a sun-following surface on earth, Fig 2, compared with about 428 Btu/hr-ft^2 available outside the atmosphere. Part of the energy arriving from the sun is reflected back into space at the top of the atmosphere, Fig 7. Some is absorbed by ozone, water vapor, carbon dioxide, and other compounds in the atmosphere. Another portion of the solar radiation is scattered by dust particles and water vapor. The amount of solar radiation scattered and absorbed depends on the length of the travel path through the atmosphere and the concentration of water vapor, carbon dioxide, dust, smog, etc., in the atmosphere.

When the sun is high in the sky, the length of the solar radiation's travel path through the atmosphere is fairly short. When the sun is low in the sky in early morning, late afternoon, and in the winter, the travel path is longer and more of the solar radiation is scattered and absorbed with less reaching solar collectors. Because of this, you might expect the noontime solar radiation to be greatest on June 21 when the sun is highest in the sky. Fig 2 shows that the solar radiation is actually greater at noon on Dec. 21.

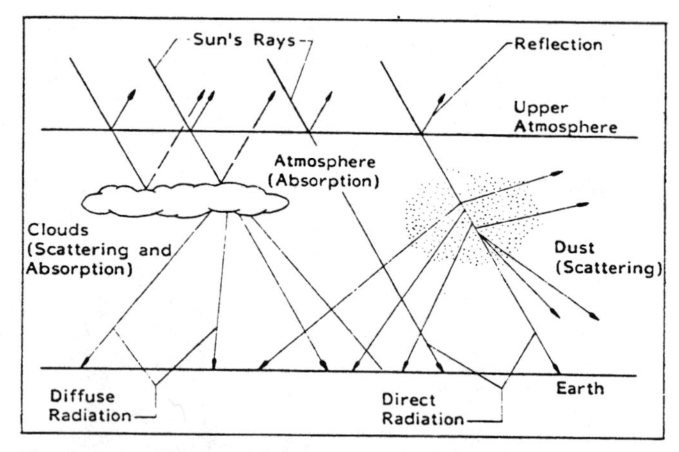

Fig 7. Atmospheric effects on solar radiation.

The solar radiation is greater in December because the earth is closer to the sun then, Fig 1. The earth is a little farther away from the sun in March than in December, but the shorter travel path for solar radiation through the atmosphere gives rise to a greater noontime solar intensity in March.

Solar radiation that passes directly from the sun to a collector surface without deflection is called direct or beam radiation. The scattered fraction is called diffuse radiation. The amount of diffuse radiation increases with the concentration of scattering materials in the atmosphere. On clear days, solar radiation is often about 85% direct and 15% diffuse. On overcast days, only diffuse solar radiation reaches the collector.

The quantity of solar radiation available on clear days is mainly a function of latitude (distance from the equator) and time of year, and is fairly predictable. Table 672-4 lists the clear day solar energy available at each hour of the day on the 21st of each month for one latitude. Locations at the same latitude receive about the same amount of solar energy on clear days, except when there are large differences in elevation above sea level and/or average atmospheric conditions.

Clear day insolation values in Table 672-4 include both direct and diffuse radiation, but not radiation reflected off the ground or surrounding surfaces. The values are for average clear days. Actual insolation taken on any given clear 21st of the month could differ from the table value by 15%.

On the average, the amount of solar energy available to a solar collector is not as great as that shown in Table 672-4 because not every day is clear. Clouds scatter and absorb solar radiation and greatly reduce the amount available beneath them. Cloud cover varies from one location to another and varies greatly from one year to the next at a given location. Because cloud cover varies so much and is so hard to predict, the best we can do is average the solar radiation received at a location over a number of years and assume that, on the average, the same amount will be received in the future. Table 672-5 lists the average solar radiation received each month at Lemont, Illinois. The insolation values include direct and diffuse radiation and radiation reflected off the ground, with average snow cover for each month taken into account. (Snow reflects more solar radiation than grass or soil.)

Data for other latitudes and cities are in MWPS-23, *Solar Livestock Housing Handbook*. Data for Columbus, OH (40° north latitude), illustrate the effect of cloud cover on solar energy availability. On a clear December day, the expected insolation on a vertical collector is 1644 Btu/day-ft^2. But on the average, only 686 Btu/day-ft^2 is available on a vertical collector in December. However, Lincoln, NE, which is also located at about 40° north latitude but has less cloudy weather, receives an average of 1346 Btu/day-ft^2 on a vertical collector in December.

Tables 672-4 and 672-5 list the solar energy available **to** a solar collector and do not indicate the amount of heat energy actually available **from** a solar collector. To estimate heat output, you need to multiply the solar energy available at the collector surface by the collector efficiency. Values from Table 672-4 allow calculation of the maximum collector heat output. The peak heat output and temperature rise from a solar collector will occur shortly after solar noon on sunny days unless concrete or other massive materials are used in the collector. Then, peak temperature rise will occur later in the day due to storage effects. Values from Table 672-5 allow calculation of average collector heat output over cloudy and sunny days or the total heat available over a selected season.

Collector Shading

The heat output from a solar collector is reduced by shading. The extent of the reduction depends on when and how long the shading occurs, how much of the collector surface is shaded, and the nature of the obstruction causing the shading.

Although total day length is usually greater than six hours, most of the usable solar energy is received in about a 6-hr period from 9 a.m. to 3 p.m. (solar time). Collector shading that occurs outside this time interval will have little effect on overall collector performance. Any shading between 9 a.m. and 3 p.m. has a much greater effect. The longer the shading period is and the closer to noon it occurs, the greater the reduction in performance.

Shading by surrounding objects is most likely to occur in December and January when the sun is lowest in the sky (the lowest point is reached on Dec. 21). This is also the coldest time of the year, when maximum solar collector area is required to meet heating needs and shading is least tolerable. The higher a solar collector is above ground, the less likely it will be shaded.

The worst shading problems are caused by long, east-west buildings located south of the collector. Short buildings, single trees, silos, and obstructions southeast or southwest of a collector can also cause shading, but usually for shorter time periods. Winter shade from deciduous trees (trees that shed their leaves in fall) reduces collector performance 10%-40%, but reduced performance might be more acceptable than removing the trees.

Separation distance required to prevent solar collector shading depends on month, time of day, and height of the obstruction. Calculate separation distance by multiplying the solar angle factor from Table 1 by the obstruction height. Measure the calculated distance straight north from an imaginary east-west line drawn under the obstruction's highest point. The solar angle factors in Table 1 were chosen to prevent shading between 9 a.m. and 3 p.m. solar time.

Table 1. Solar angle factors (SAF).

Date	Latitude, degrees north 24	32	40	48
1a. For winter, SAF = cos(9 a.m. azimuth)/tan (9 a.m. altitude). See Fig 8.				
To prevent winter shading:				
	------SAF------			
Dec 21	1.5	2.0	3.0	5.4
Jan 21 and Nov 21	1.2	1.7	2.4	3.8
Feb 21 and Oct 21	0.8	1.0	1.4	1.9
Mar 21 and Sept 21	0.4	0.6	0.8	1.1
1b. For summer, SAF = 1/tan (noon altitude). See Fig 9.				
To produce summer shading:				
Apr 21 and Aug 21	0.2	0.4	0.5	0.7
May 21 and July 21	0.1	0.2	0.4	0.5
June 21	0.0	0.1	0.3	0.5

Example 1:

Find the separation required to prevent shading of a collector at ground level by a long, 20′ high, east-west building between 9 a.m. and 3 p.m. during the winter heating season at 40° north latitude, Fig 8.

Answer:

Maximum separation is required on Dec. 21 when the sun is lowest in the sky. From Table 1a, the solar angle factor, SAF, for Dec. 21, 40° north latitude, equals 3.0.

d = SAF x h
= 3.0 x 20′ = **60′**
d = separation distance, ft
SAF = solar angle factor
h = obstruction height, ft

Example 2:

Find the length of overhang needed to completely shade a vertical collector at noon, May through July, at 40° north latitude. Collector bottom is 8′ below the eave, Fig 9.

Answer:

Use the solar angle factors in Table 1b to calculate required overhang lengths. The shortest overhangs are required in June when the sun is highest in the sky. To shade a collector during May, June, and July, you need a longer over-

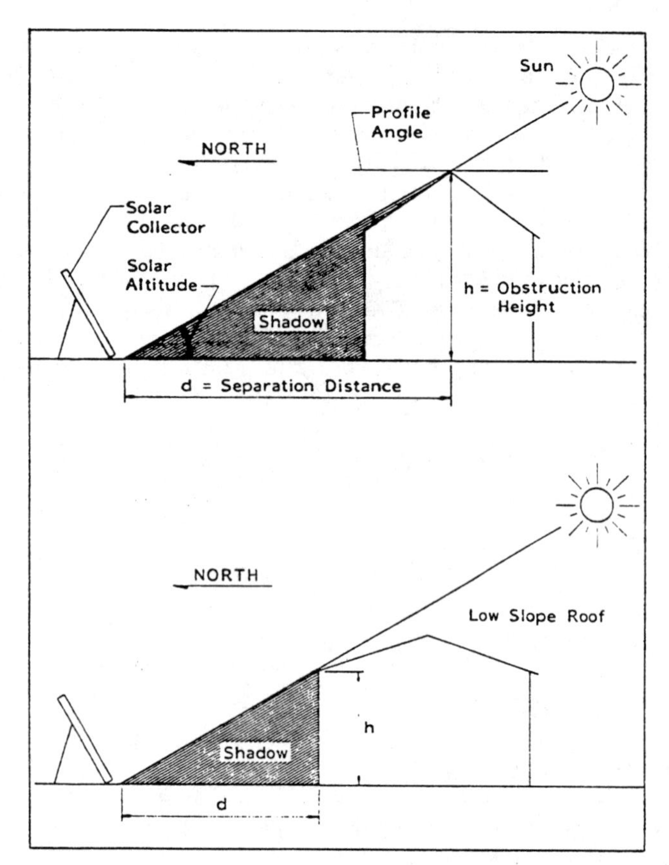

Fig 8. Separation distance to prevent shading.

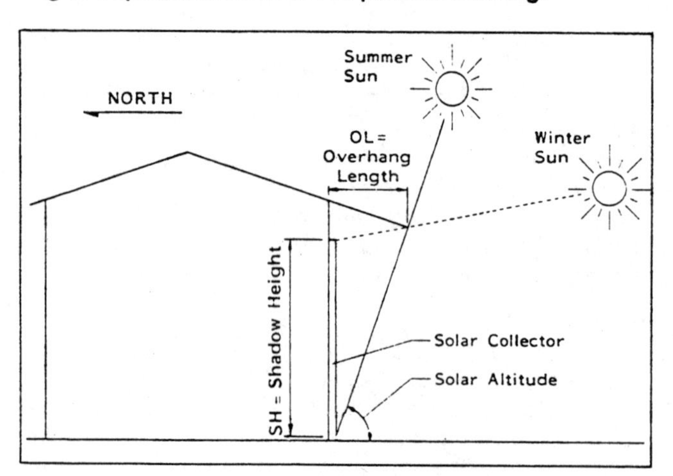

Fig 9. Overhang length to shade collector.

hang. Find the solar angle factor for May and July, 40° north latitude, from Table 1.

SAF = 0.4
OL = SAF x SH
= 0.4 x 8′ = **3.2′**
OL = overhang length, ft
SAF = solar angle factor
SH = shadow height, ft

A 3.2′ overhang will completely shade the collector at noon May 21 through July 21, but will also cause some shading at noon during the heating season. It is probably better to use a shorter overhang (e.g., 2′) to reduce winter shading.

Solar Collectors

Basics

The purposes of the solar collectors discussed in this handbook are to:

- Intercept radiation from the sun.
- Convert this solar energy into thermal (heat) energy.
- Transfer the heat energy to a fluid (air or liquid) that carries it to the point of use or heat storage.

Most solar collectors include the following components: an absorber; a glazing or cover; a support structure; and a heat transfer fluid. The absorber is heated by the sun and transfers the heat to the fluid moving over or through it. The glazing or cover plate transmits the solar radiation to the absorber, and reduces heat losses. The support structure is the mount for the collector components. The heat transfer fluid moves the heat from the collector to storage or use.

Basic types of solar collectors are flat plate and concentrating. A flat plate collector with reflectors has some characteristics of both, Fig 10.

Flat Plate Collectors

An essential part of a flat plate collector is the absorber. It absorbs solar energy, heats up, and then transfers the heat energy to a fluid moving over or through it. It is really a device that converts solar radiation into heat energy.

A number of area terms are used to describe flat plate collectors. **Gross collector area** is the surface area calculated from outside dimensions (e.g. a 4′x8′ collector has a gross area of 32 ft^2). In this book, gross collector area is considered to be the **energy intercepting area** and is the area used in efficiency and heat collection equations. In flat plate collectors, energy intercepting area equals absorbing area, while in concentrating collectors, intercepting area exceeds absorbing area.

Net collector area is the gross area minus the area of cover supports and framing members that shade the absorber. Many air-type flat plate collectors use corrugated, crimped, or finned absorbers to improve heat transfer from the absorber to the airstream. Such absorbers may have a material surface area greater than the gross collector area and may improve collector efficiency, but they do not increase the energy intercepting area used in calculations.

Most flat plate collectors are fixed (do not track the sun). They collect both direct and diffuse radiation, so they may produce small amounts of heat even on overcast days when all solar radiation is diffuse.

In air-type flat plate collectors, a moving air stream picks up heat from one or both sides of the absorber. In the simplest type, bare plate, an air channel is formed behind the absorber with back and side plates. The air moves through the channel and picks up heat conducted through the absorber, Fig 11a. A covered plate collector, Fig 11b, is usually more efficient. A transparent cover over the absorber forms the air channel. Air passes between the absorber and the cover to pick up heat. The transparent cover, or glazing, reduces heat losses from the absorber. A suspended plate version is even more efficient. The absorber is between a transparent cover and a back-plate, and air is directed either behind or on both sides of it, Figs 11c and 11d.

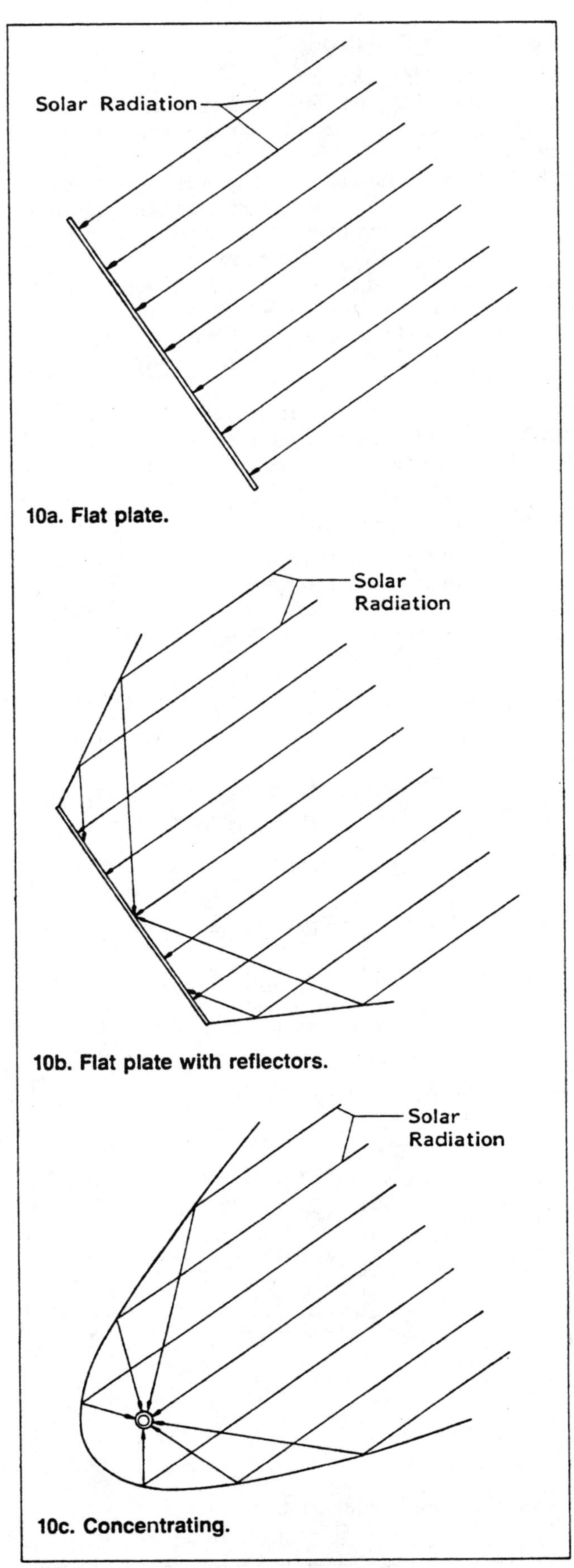

Fig 10. Basic types of solar collectors.

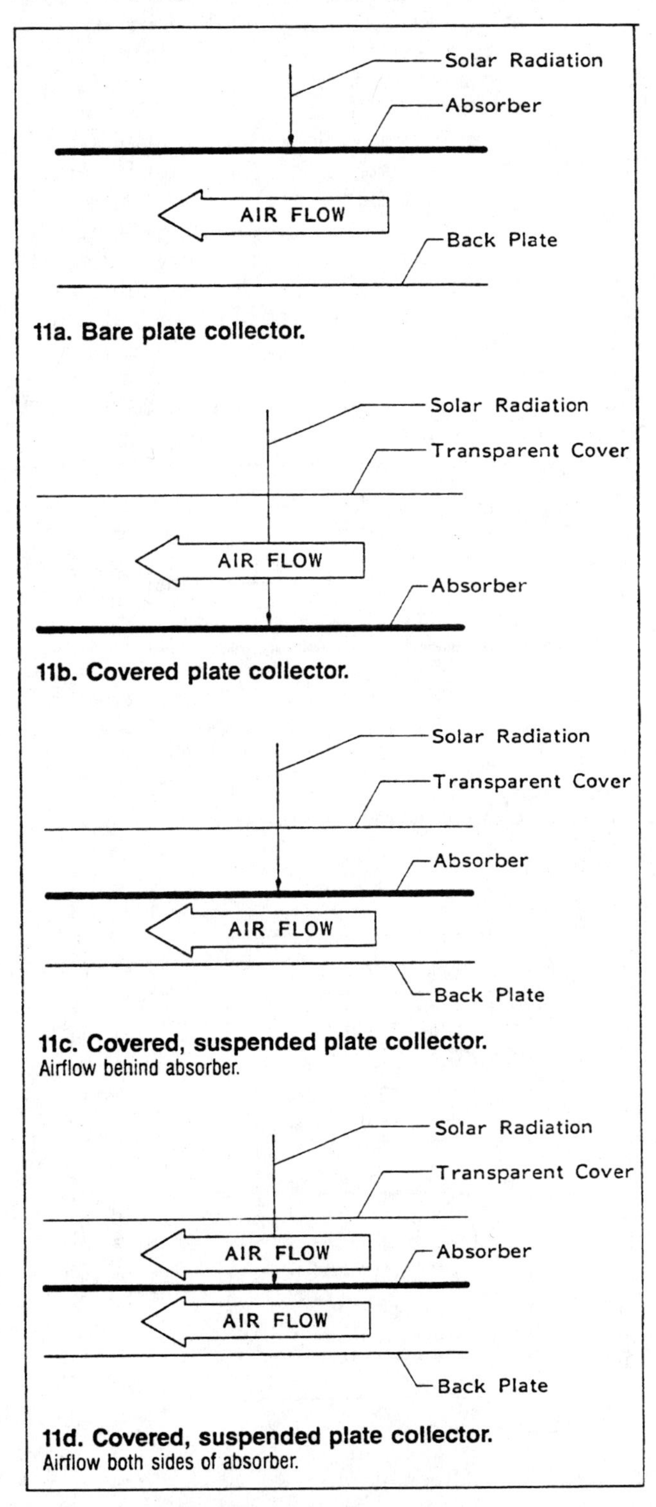

Fig 11. Air-type flat plate solar collectors.
Cross sections of collectors.

In liquid-type collectors, a liquid (usually water or water plus antifreeze) removes heat from the absorber. In many liquid-type flat plate collectors, the liquid flows in tubes bonded to or part of the absorber, Fig 12. The tubes are in a serpentine pattern, Fig 12c, or a parallel pattern with headers or manifolds on each end, Fig 12d. Covered plate liquid-type collectors are much more common than bare plate ones. In some liquid-type collectors, cool liquid enters at the top of a sloped absorber and trickles down open channels on the absorber surface. The heated liquid is collected at the base of the absorber.

On farms, liquid-type collectors are used mainly to provide service hot water or heating water for farm buildings or the farm home. (Service hot water is potable, or drinkable, water for bathing, cleaning, and cooking. Heating water is circulated through floor pipes or radiators in livestock housing or homes with hot water heating systems.) Most agricultural operations, such as building ventilation and grain drying, use air-type collectors. A heat exchanger is required to heat air with a liquid-type collector, which increases costs and reduces system efficiency.

Concentrating Collectors

Concentrating collectors have large reflecting surfaces (usually parabolic) or lenses that concentrate solar energy from a large area onto a relatively small absorbing area, Fig 13. Some concentrating collectors heat the fluid in the absorber to over 2000 F, usually to generate steam for industrial processes, electric power production, or to run engines.

Concentrating collectors use only direct radiation and must be pointed toward the sun to keep the rays focused on the absorber. At least a partial tracking system or frequent focus adjustment is required. Because concentrating collectors do not use diffuse radiation, no heat is produced on cloudy days. Because most concentrating collectors are more expensive and harder to build than flat plate collectors, few concentrating collectors have been used on farms.

Flat Plate Collectors with Reflectors

These solar collectors have some features of both flat plate and concentrating collectors. Reflectors increase the energy intercepting area and concentrate,

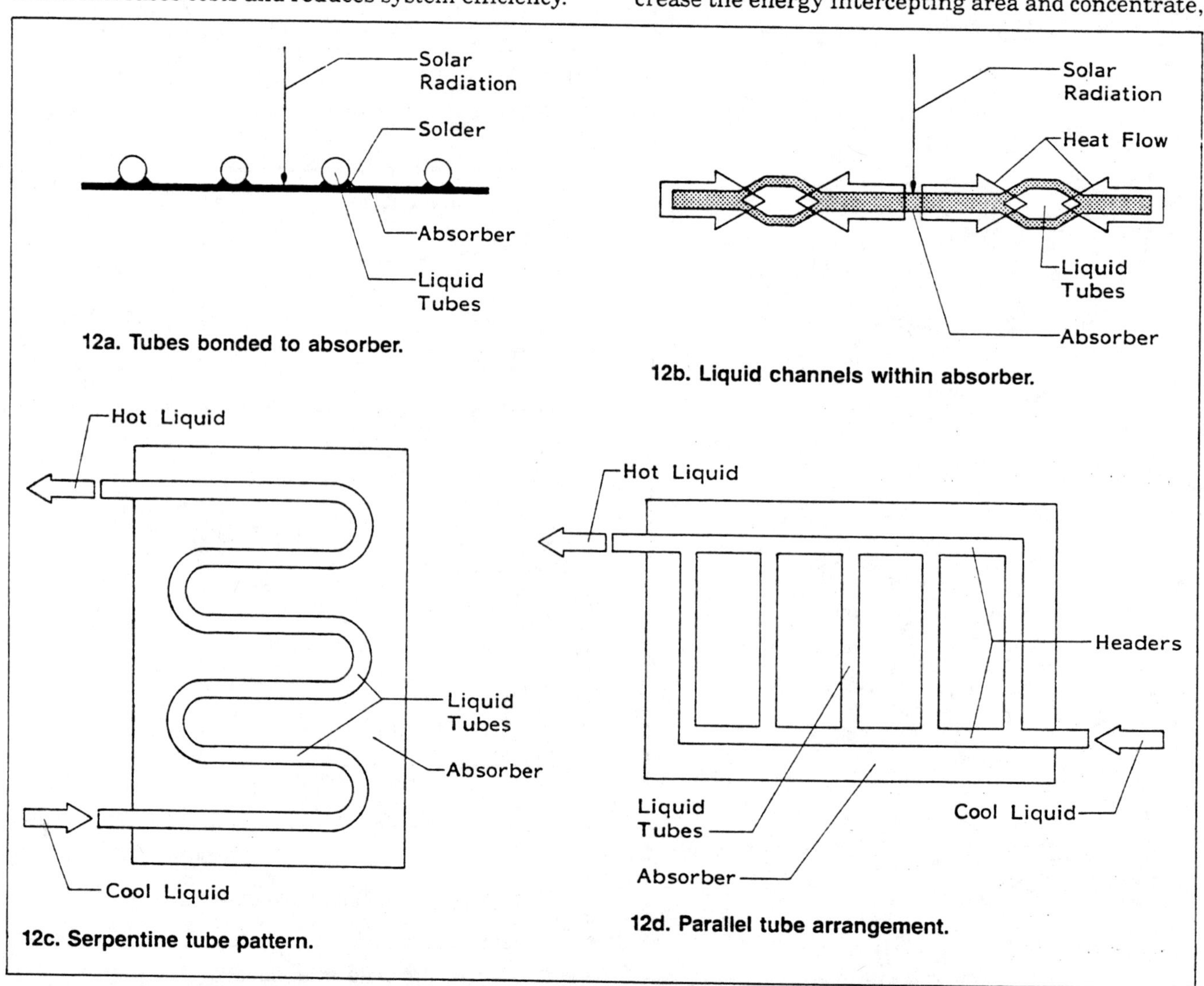

Fig 12. Liquid-type flat plate collectors.

to an extent, the energy available to a flat plate collector. With reflectors, temperatures in a flat plate collector can be increased. However, reflectors can use only direct radiation and work best in sunny regions.

Reflectors include plywood wings painted white or lined with aluminum foil, concrete slabs painted white, and parabolic sheets lined with aluminized film, Fig 14. Consider the value of the extra energy collected over the life of the reflectors against the cost of the materials and labor required to build and maintain them. Fresh snow reflects solar radiation fairly well and slightly increases the energy available to flat plate collectors at no cost to you.

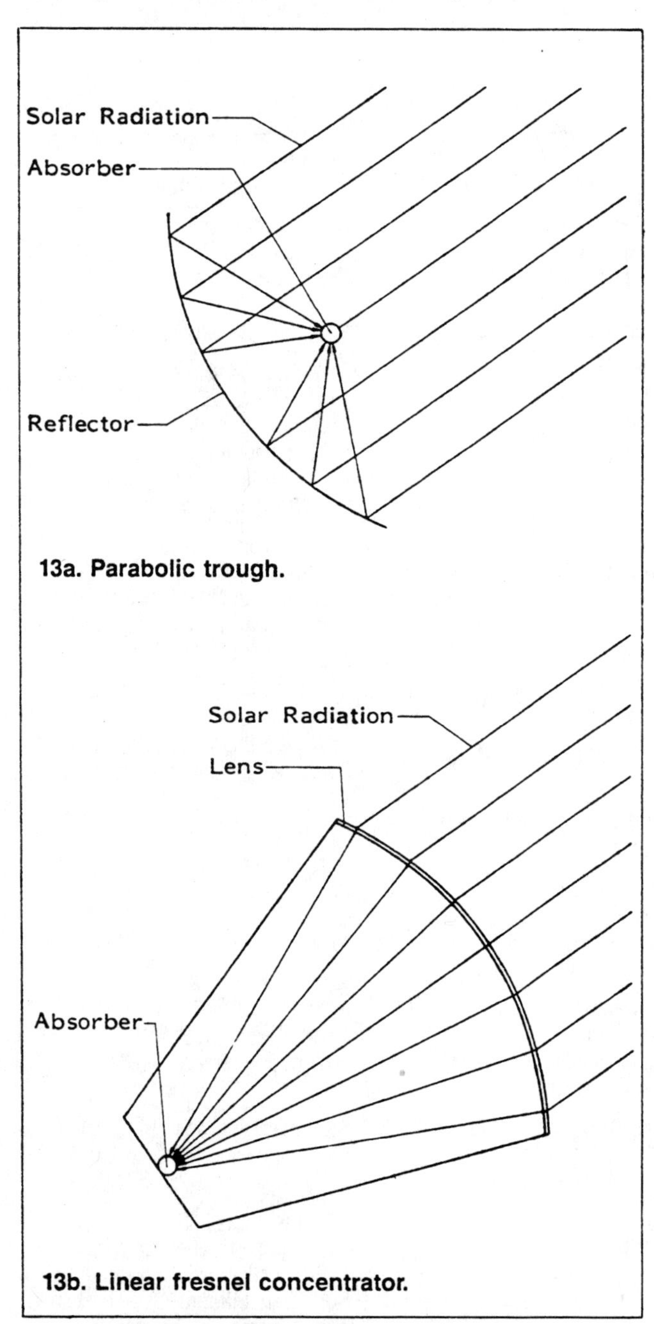

13a. Parabolic trough.

13b. Linear fresnel concentrator.

Fig 13. Types of concentrating solar collectors.

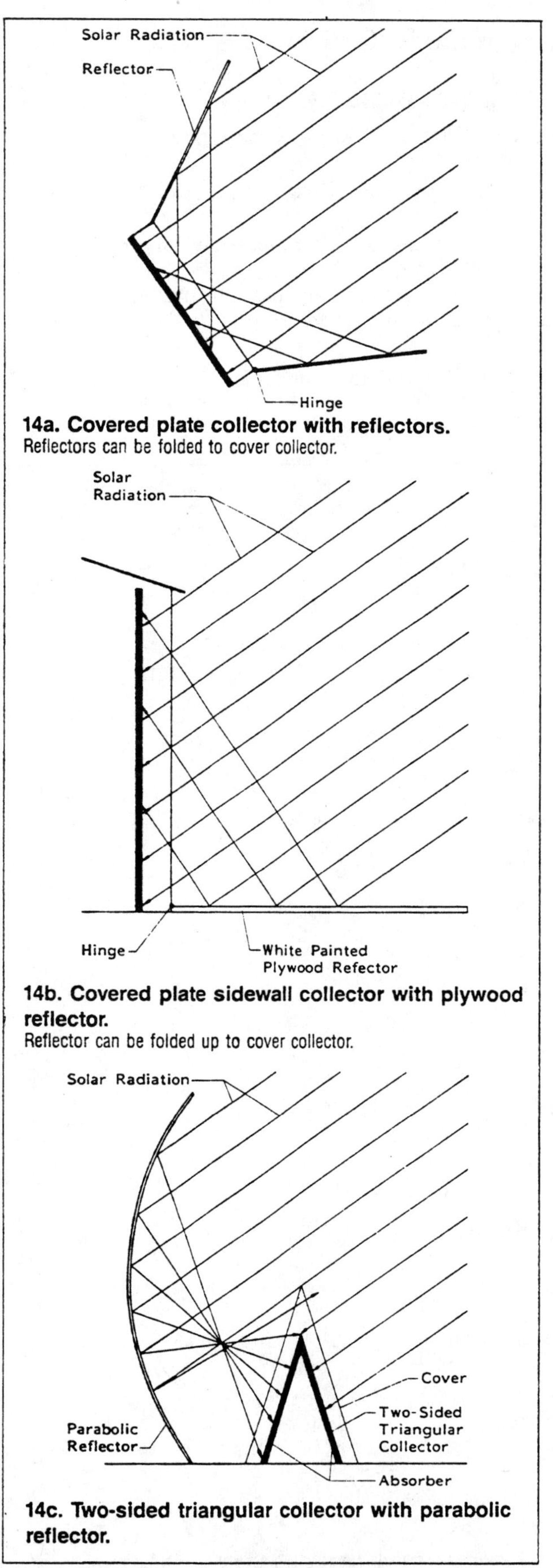

14a. Covered plate collector with reflectors.
Reflectors can be folded to cover collector.

14b. Covered plate sidewall collector with plywood reflector.
Reflector can be folded up to cover collector.

14c. Two-sided triangular collector with parabolic reflector.

Fig 14. Flat plate collectors with reflectors.

Flat Plate Collector Materials

Cover Materials (Table 2)

Solar collector covers or glazings have several basic purposes:

- To admit solar or shortwave radiation to the absorber.
- To reduce convection heat losses by shielding the absorber from the wind.
- To reduce radiation heat losses by preventing the escape of longwave radiation from the collector.

All objects that are warmer than their surroundings lose energy by radiation. Very hot objects, like the sun, radiate mostly high energy waves with short wavelengths (shortwave radiation). Objects at expected temperatures and only slightly warmer than surrounding temperatures, such as absorber plates in solar collectors, radiate or emit lower energy waves with long wavelengths.

When an absorber is not covered (as in bare plate collectors), much of the absorbed energy is lost to the wind. When an absorber is covered (as in covered plate collectors), the losses are reduced. A good cover

Table 2. Properties of cover materials.
Solar transmittances are daily averages for incident angles from 0° to 67°. Costs are relative to polyethylene. Example: acrylics cost 43 times as much as polyethylene, per ft².

Cover material type*	Solar transmittance	Longwave transmittance	Relative cost	Advantages	Disadvantages
Glass ⅛" double strength	0.84 to 0.91	0.03	21	Neat appearance Cleans easily Abrasion resistance High heat tolerance (400 F) Excellent weathering resistance	Heavy Breaks easily Hard to install Sharp edges Poor thermal stress resistance
Fiberglass reinforced plastic (FRP) 40-mil Greenhouse grade Flat or corrugated	0.77 to 0.90	0.01 to 0.10	12	Easy to install Can be fastened with screws Lightweight Good availability Tough, durable	Solar transmittance deteriorates with age (5-15 yr life) Protective coatings may come off or degrade over time Cannot tolerate long exposure to temps over 200 F
Polycarbonate ⅛", single cover	0.88	0.03	21	Easy to install Lightweight, very tough	Combustible Scratches easily High expansion & contraction Becomes brittle with age Solar transmittance deteriorates Low heat tolerance (170 F)
Acrylics	0.80 to 0.92	0.01	43	Easy to install Lightweight Nice appearance Long life (25 yr) Good weathering resistance	Expensive Low heat tolerance (170 F) Bends when one side heated
Polyester films	0.80 to 0.89	0.10 to 0.20	7	Low cost Easy to install Strong Fairly long life (10-15 yr) Fair weathering resistance	Scratches easily
Polyvinyl fluoride film	0.90 to 0.94	0.45	9	Light weight Very strong Excellent weathering resistance Scratch resistant	Becomes brittle with age Shrinks upon heating Low heat tolerance (225 F) Hard to install
Fluorocarbon film	0.96	0.58	12	Resistant to deterioration Good inner glazing High heat tolerance (400 F)	Stretches with age (5-15 yr) High longwave transmittance
Polyethylene 4-mil	0.89	0.80	1	Tough Lightweight	Punctures easily Susceptible to wind damage when collector is deflated Solar transmittance deteriorates Annual replacement required

*Formulations used by manufacturers vary greatly and consequently, so do material properties. Use this table only as a general guide. See manufacturer's literature for more accurate numbers.

material has a high transmittance to shortwave radiation and a low transmittance to longwave radiation.

Cover materials that allow shortwaves from the sun to pass but block longwaves from the absorber, cause the "greenhouse effect." The absorber is warmed by solar energy and the heat is trapped within the collector. An ideal cover material has a shortwave or solar transmittance of 1.0 (100% of the solar radiation passes through the coverplate) and a longwave transmittance of 0 (none of the longwave radiation escapes). But, with real materials, part of the incident solar radiation is reflected and part is absorbed by the cover, and part of the longwave radiation escapes. Short- and longwave transmittances for a number of materials are listed in Table 2.

The transmittance of a material is not constant. It varies from a maximum at a 0° incident angle (radiation normal to surface) to 0 at a 90° incident angle (radiation parallel to surface), Table 3. Less radiation is transmitted through a material at large incident angles because more is reflected off the surface—like a rock skips off a water surface. The transmittance decreases rapidly at incident angles larger than 60°. Because the solar angle of incidence on fixed, flat plate collectors usually exceeds 60° before midmorning and after midafternoon, most of the energy is collected by fixed collectors during the middle part of the day.

Adding cover layers reduces energy losses from a collector, but also reduces the radiation transmitted to the absorber. Each additional layer absorbs and reflects another fraction of the solar radiation.

In addition to good transmittance, other desirable properties of collector cover materials are:

- Resistance to deterioration by ultraviolet radiation, moisture, atmospheric pollutants (weathering resistance), and wide temperature ranges.
- Little expansion and contraction with temperature changes, or a rate of expansion that matches that of the framework holding the cover.
- High impact strength against hailstones, rocks, birds, and livestock.
- Low flammability.
- Economy. An expensive but durable material may be more economical than a cheaper one requiring annual replacement.
- Ease of installation and maintenance.
- Resistance to wind damage and abrasion at supports.
- Resistance to deterioration at high temperatures. When the fluid is stagnant and the sun is shining, temperatures can exceed 300 F.
- Lightweight.
- Resistance to static electricity charges which attract dust (reducing solar transmittance) and make the material difficult to handle.
- Ease of sealing against leakage.

Glass is most common on collectors with recirculated heat transfer fluids. (See Solar Systems.) It has excellent transmittance that does not change over time. It is rigid and has good abrasion resistance, so it does not sag or wear through at supports. Single-strength (0.085" to 0.100" thick), double-strength (0.115" to 0.133" thick), tempered (high-strength), and low-iron glass are available. Low-iron glass has a higher solar transmittance than other glass. The edge of low-iron glass appears water-white rather than the usual greenish-blue.

Table 3. Solar transmittance through double-strength window glass.

Incident angle degrees	Solar transmittance One layer	Solar transmittance Two layers
0-20	0.87	0.77
30	0.87	0.76
40	0.86	0.75
50	0.84	0.73
60	0.79	0.67
70	0.68	0.53
80	0.42	0.25
90	0.00	0.00

The biggest disadvantage of glass is brittleness. Rocks, hail, and even stresses caused by thermal expansion and contraction may break it. Vertical mounting or a wire-screen cover reduce hail damage but also reduce the amount of solar energy reaching the collector. The thermal expansion of glass is less than that of wood or steel. Glass attached directly to a wood or steel frame may break when subjected to temperature extremes. Use expansion gaskets between the glass and frame.

A number of transparent plastics can be used as collector covers. One of these, polyethylene, has a high initial solar transmittance and very low first cost, and was used on many early collectors. But it traps little longwave radiation and usually requires annual replacement due to ultraviolet degradation. More durable materials are probably more cost-effective in the long run.

Fiberglass reinforced plastics (FRP) are popular cover materials for low temperature agricultural collectors. The type used for collectors is sometimes called "greenhouse grade fiberglass." Its solar transmittance is less than glass but much better than skylight material. It is clear and almost see-through.

Not all materials sold as "greenhouse grade fiberglass" possess all the characteristics desirable in a cover material. The FRP suitable for solar collectors is **not** skylight and patio cover material.

Because FRP is tougher than glass and expands at nearly the same rate as wood and steel, it is easier to install and can be connected directly to a collector frame. However, even with FRP, keeping joints sealed is difficult. FRP is more flexible than glass and can be fastened to curved surfaces, but it requires more support. Corrugated FRP is more rigid than flat FRP and is harder to seal against air leaks. Use caulked, corrugated wood strips or closed cell foam strips for sealing.

Heat, ultraviolet radiation, and moisture cause FRP to gradually deteriorate. Some FRP is guaranteed for 15 to 20 years on greenhouses, but not yet on solar collectors where temperatures can be much higher. FRP develops fiber bloom as the binding plastic resin breaks down, exposing the glass fibers. Dirt and fungi trapped among these fibers reduce solar transmittance. FRP can be treated or coated to prevent fiber bloom, but the coating may peel off or degrade over time and require reapplication. Some plastics are ultraviolet resistant.

Other cover materials and their advantages and disadvantages are listed in Table 2.

Absorber Coating Materials

A good absorber in a solar collector:

- Absorbs a high percentage of incoming solar radiation.
- Loses minimum energy to the collector's surroundings.
- Efficiently transfers absorbed energy to the collector fluid.

All solar radiation striking an opaque (nontransparent) surface is either reflected or absorbed. The fraction that is absorbed (always a decimal between 0 and 1) is called the absorptance. The reflected fraction, the reflectance, is 1 minus the absorptance. A good absorber has an absorptance near 1.

Dark surfaces have high absorptances—so paint collector absorbers black if they are not naturally dark. Flat black paints reflect less energy than glossy ones. Any flat black paint that can withstand temperatures up to 300 F without cracking, peeling, or breaking down can be used. Use an appropriate primer or surface treatment before painting.

Let the paint dry completely before installing the cover. Some paints give off vapors (called outgassing) as they dry, which can condense on the collector cover and reduce transmittance. Also, the vapors from some paints (particularly latex) have unpleasant odors—an important consideration for collectors on shops or homes.

To avoid painting and paint adherence problems, you can buy metal that already has a baked enamel, flat, very dark surface.

Absorptances for a number of common surfaces are shown in Table 4. The solar absorptance of a surface varies with angle of incidence, just as transmittance does. At high angles of incidence a greater percentage of the insolation is reflected, Table 5.

Most surfaces with high solar absorptance also have a high longwave emittance. The longwave emittance of a surface (always a decimal between 0 and 1) is its tendency to radiate longwave energy. A solar absorber heats to a temperature above its surroundings and then radiates or emits energy as longwave radiation. An ideal absorber has a longwave emittance near 0.

"Selective surfaces" have both high solar absorptance and low longwave emittance. Many are special factory-applied coatings and are not suitable for field applications. Selective paints are available, but the cost per gallon is high. Selective surfaces can operate at higher temperatures because they lose less energy by radiation. Properties of several commercial, selective coatings are given in Table 6. Select one guaranteed to withstand high temperature exposure.

Well-weathered galvanized metal with a rough, dull gray surface is a natural mildly selective surface, Table 4. Painting the metal flat black increases the solar absorptance, but destroys the selective surface. In some cases, the unpainted surface is probably more cost-effective.

Table 4. Absorptance and emittance of common surfaces.
Normal solar incidence (0° incident angle).

Material	Solar absorptance	Longwave emittance
Flat very dark paint	0.95-0.99	0.95-0.99
Dark concrete and stone	0.65-0.80	0.85-0.95
Colored paints; brick	0.50-0.70	0.85-0.95
Bright aluminum paint	0.30-0.50	0.40-0.60
Dull metals: copper, brass, aluminum	0.40-0.65	0.20-0.30
Weathered galvanized steel	0.80	0.28
White paint	0.23-0.49	0.92

When preparing a **new** galvanized metal surface for use as an absorber plate, wear rubber gloves to protect yourself from the chemicals and the metal from skin oils. Clean the metal with a solvent, such as acetone, to remove manufacturing oils, and then wash the surface with detergent. Etch the metal with vinegar or a weak acid (e.g. 6:1 water/muriatic acid mixture) and rinse. Apply a galvanized metal primer and two coats of flat black paint.

Table 5. Solar absorptance of flat black paint.

Incident angle degrees	Solar absorptance
0-20	0.96
30	0.95
40	0.94
50	0.92
60	0.88
70	0.82
80	0.67
90	0.00

The solar collector absorber must transfer its absorbed energy to the heat transfer fluid. In liquid-type collectors, heat moves by conduction through the flat absorber plate to the tubes containing the liquid. Both the absorber and tube materials need high thermal conductivities, and the bond between the tubes and plate must provide a good thermal path. Metal absorber plates are best, and the most common in order of increasing conductivity and cost are steel, aluminum, and copper. The metals are painted black or coated with a selective surface material. Higher absorber conductivity permits greater tube spacing. Plastic and rubber absorbers are also available. Because of the low conductivity of plastic and rubber, the

Table 6. Absorptance and emittance of selective surfaces.
From *Solar Heating Materials Handbook* by Homann, Hilleary, and Darnall.

Absorber coating	Solar absorptance	Longwave emittance
Aluminum oxide and molybdenum dioxide	0.9	0.1-0.4
Black chrome (Chromium metal and chromium oxide on nickel-plated metals, copper, or steel)	0.91-0.96	0.07-0.16
Black copper (Cupric oxide and cuprous oxide on copper, nickel, or aluminum)	0.81-0.93	0.11-0.17
Black nickel (Nickel oxide on nickel, iron, or steel)	0.89-0.96	0.07-0.17
Iron oxide on iron or steel	0.85	0.08
Lead oxide	0.98-0.99	0.22-0.40
Stainless steel oxide on stainless steel	0.89	0.07
Selective paint	0.90	0.30

tubes must be very close together to obtain good efficiency.

The best bonds between metal tubes and absorber plates are solid beads of weld or solder or a thermal bonding mastic. Use only silver solder inside the collector itself. A 50:50 tin/lead solder is acceptable for the rest of the plumbing. Clamps, wires, press fits, and spot welds do not provide as good a thermal conducting joint. Bonding is avoided in tube-in-sheet absorbers—the liquid tubes are part of the absorber plate, Fig 12.

Chemical reactions between dissimilar metals can be a serious problem, especially with copper and aluminum or galvanized steel. Consider corrosion problems before installing a liquid solar system.

In air-type solar collectors, the heat transfer fluid passes over the entire absorber and picks up heat directly from the surface. In bare and suspended plate collectors, heat must be conducted to the back side of the absorber plate, so it should be made of metal. Absorber plates for other collector types can be of any convenient material: glass, plastic, concrete, or high temperature insulation board. The material must withstand sunlight and collector stagnation temperatures without sagging, deteriorating, melting, or burning. Paint the surface black if it isn't naturally dark.

Cautions about using wood and/or fiberglass in solar collectors:

- Extended exposure of wood to temperatures above 200 F can reduce its strength and ignition temperature.
- Wood exposed to temperatures above about 400 F may spontaneously ignite.
- A proposed UL standard suggests 194 F as the maximum temperature for wood in collectors.
- Exposing fiberglass cover sheets to temperatures above 200 F accelerates degradation.
- Cover, ventilate, or shade solar collectors that are not used during summer months.

Absorber-to-air heat transfer depends on flowrate per unit area, turbulence, and absorber material surface area. Up to a point, increasing air velocity improves heat transfer. Increasing the absorber surface area also improves heat transfer. Increasing the absorber surface area also improves heat transfer and often improves collector efficiency by decreasing reflective losses. Corrugated, crimped, and finned absorbers, which provide greater surface area than simple flat surfaces, have all been tested, Fig 15. With typical metal roofing and siding sheets, airflow direction over corrugated surfaces (parallel vs. perpendicular) makes little difference in collector performance. Make airflow parallel to the deeper corrugations of crimped and finned absorbers.

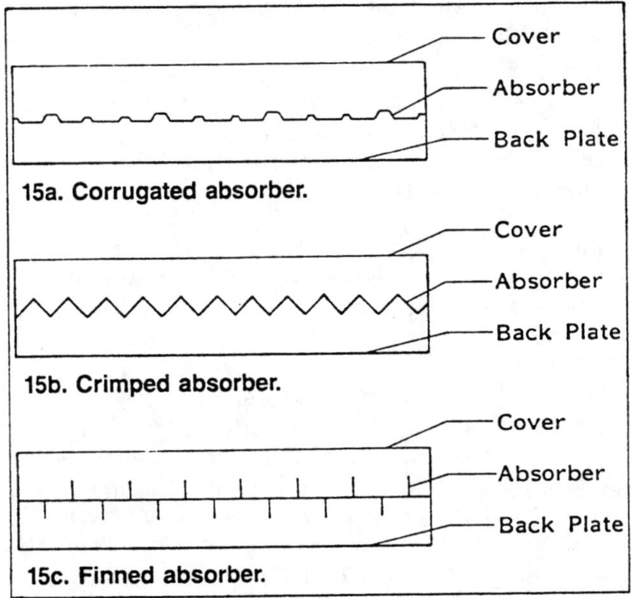

Fig 15. Air-type suspended plate solar collectors.
Airflow toward the viewer.

Insulation

(See Table 631-1 and Table 7)

In addition to losing heat by radiation from the absorber, solar collectors lose heat by conduction through the side and back plates. Insulate collectors to reduce conduction heat losses. Collectors that preheat ventilation air for livestock buildings operate at temperatures only slightly higher than outdoor temperatures and need to be insulated to about R=6. Solar collectors that heat recirculated air or water usually operate at temperatures much higher than outdoor temperatures and need to be insulated to about R=11. Insulate ducts and pipes passing through unheated spaces to the same R-value used in the collector.

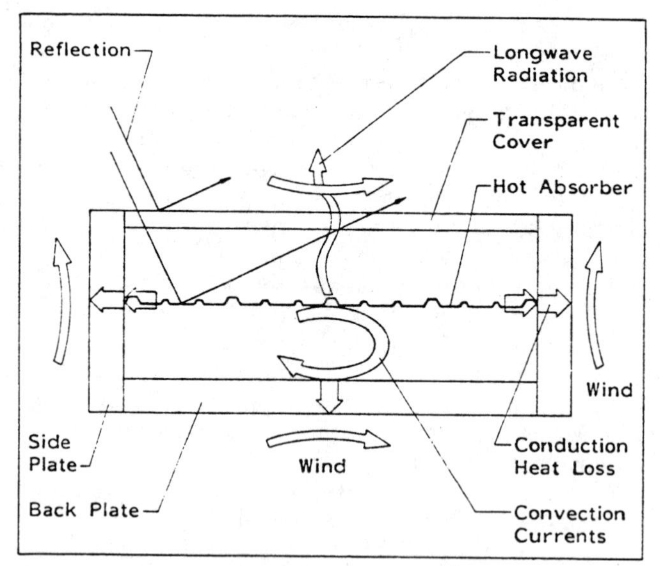

Fig 16. Heat losses from a solar collector.

Collectors mounted directly on insulated building walls may not need additional insulation.

In addition to R-value, consider these insulation properties:

- Flammability. Some insulating materials (especially plastics) are highly flammable and produce toxic gases when burning. If used, they must be covered with a fire-resistant material. Consult your insurance company and local codes.
- Upper temperature limit. Heat tolerance of insulating materials varies, but all of them have some upper temperature limit beyond which they melt, burn, or chemically deteriorate. Make sure insulation used in a solar collector has an upper temperature limit greater than collector stagnation temperatures (about 300 F). Polystyrene melts at about 165 F. Fiberglass insulation that is held together with an organic binder and some other types of insulation give off gases (outgas) that can condense on and cloud the collector cover at temperatures below the upper limit. Use Table 7 as a guide, but consult the manufacturer for information on a specific insulation brand.
- Durability. If insulation will be exposed to harsh conditions—sunlight, moisture, rubbing or chewing by livestock, birds, and rodents—either protect the insulation or select a type that can withstand the conditions. Polystyrenes and some other plastic insulations decay in sunlight. All materials lose their insulating value when wet—closed cell plastic foams may be a good choice for wet environments.
- Cost. When comparing insulation costs, include first cost of the insulation and any protective coverings, installation labor, and life for installations with equal R-values.

Foil-faced glass fiber board is usually a good insulation inside or outside solar collectors.

Table 7. Insulation upper temperature limits.
Consult the manufacturer for information on specific products—some materials can emit poisonous gases when overheated.
From *Solar Heating Materials Handbook* and UL Standard. Most conservative values were used.

Insulation type	Outgassing begins (F)	Approximate upper temperature limits (F)
Glass fiber building insulation	302	350
Glass fiber board	212	450
Glass fiber		
with low binder content	302	850
with no binder	482	850
Polystyrene		165 (melts)
Polyurethane		200
Isocyanurate	122	250

Efficiency

Converting solar energy to heat energy is not 100% efficient. Tables 672-4 and 672-5 show the amount of solar energy available at a collector surface—not the amount of heat available from the collector. Efficiency quantifies how well a collector converts available solar energy into heat energy. Collector efficiency helps you choose among collector types or predict the amount of energy or the temperature rise from a collector.

In this handbook, collector efficiency is the heat energy output over a time period divided by the solar energy available on the gross collector surface over the same time period. Because of energy losses from the collector, efficiency is always less than 100%.

Eq 1.

$$EFF = 100 \times CE \div AE$$

EFF = collector efficiency, %
100 = gives efficiency in % (For efficiency as a decimal, do not multiply by 100).
CE = collected energy, Btu per unit of time
AE = available energy, Btu per unit of time

Available solar energy is at the collector's tilt angle. Collector efficiency is for the time period over which the energy was collected. It is possible to calculate instantaneous, all-day, and average efficiency over a season for a collector. Because collector efficiency varies with time of day and weather conditions, these three efficiency values usually differ. Know which type of efficiency is being considered when a value is given. See Collector Sizing for examples using efficiency values.

Factors that affect collector efficiency include:

- Difference between the average collector fluid temperature and the outdoor temperature. The greater this difference, the greater the heat loss from the collector. Average fluid temperature approximately equals the average of inlet and outlet temperatures, i.e. (inlet temperature + outlet temperature) ÷ 2. Temperature rise is affected by the available solar energy and the fluid flowrate per unit of collector area. Collectors that heat outdoor air generally operate with relatively low temperature differences and high efficiencies.

- Fluid flowrate/unit area, which affects exposure of the fluid to the hot absorber and the cold glazing. High flowrates reduce exposure time, fluid temperature rise, and heat losses from the collector. High flowrates also increase fluid velocity and improve absorber-to-fluid heat transfer. So, up to a point, higher flowrates improve collector efficiency. Maximum collector efficiency is limited by the cover and absorber properties; once this efficiency is reached, higher flowrates cannot improve it. Also, pressure drop in the collector increases as flowrate increases. Pressure drop measures fluid friction or resistance to flow and requires extra fan or pump power. The selected flowrate must balance these factors. Air-type collectors heating livestock ventilating air are usually operated at 2 to 3 cfm/ft² of collector surface area and 500 to 1000 fpm air velocity past the absorber.
- Wind speed. Wind increases heat loss and reduces collector efficiency.
- Insulation reduces heat loss and increases efficiency.
- Length of fluid flow path through the collector. Up to a point, the longer the fluid is in the collector, the hotter it gets. As length is added to a collector, heat losses increase and in the extreme can equal the solar energy added by the extra area. If a collector is too long, the fluid temperature reaches its maximum part way through the collector and the remaining area is useless.
- Collector type and tilt angle. With the same insulation and fluid flowrates, air-type collectors (Fig 11) in order of increasing efficiency are:
  1) Bare plate.
  2) Covered plate, single channel.
  3) Covered, suspended plate—airflow above the absorber.
  4) Covered, suspended plate—airflow under the absorber.
  5) Covered, suspended plate—airflow both sides of the absorber.

  Collectors operating at a poor tilt angle (large angle of incidence) have poor efficiency, because much of the solar radiation reflects off the cover.
- Cover transmittance. High shortwave radiation transmittance and low longwave transmittance increase efficiency. See Cover Materials.
- Number of covers. Multiple covers reduce heat loss, but add cost and reduce solar transmittance. Most collectors in recirculating systems have two covers; most in single pass systems have one.
- Absorber properties. Increase efficiency with high solar absorptance, low longwave emittance (selective surfaces), and absorbers with corrugated, finned, or crimped surfaces. See Absorber Materials.
- Care in construction. Poorly-built collectors with many air leaks have very low efficiency. Collector components must fit together and be well sealed.

Maximum efficiency can be achieved only at high cost; multiple low-iron glass covers, selective absorber coatings, high insulation levels, etc. Select only those features needed to maintain reasonable efficiency in your application. Solar collectors that provide service hot water or space heat for homes must operate at low fluid flowrates and high temperature differences. Such collectors are relatively costly. Most agricultural collectors have higher fluid flowrates and lower temperature differences, so they can be simple and fairly inexpensive and still operate with good efficiency.

Solar **collector** efficiencies should not be confused with solar **system** efficiencies. Solar system efficiencies are based on energy supplied to the point of use and are always less than collector efficiencies because of losses in ducts, pipes, fans, pumps, and storage.

Solar Energy Storage

Install heat storage to save, and in most agricultural installations, to effectively utilize solar energy for night or cloudy days. Energy is stored by passing solar-heated air or liquid through a heat storage material (active systems) or by allowing direct sunlight to fall on the material (passive systems). Energy is recovered by passing cool air or liquid through the warm storage material or by direct radiation from the material. Choose storage materials that can store a lot of energy without taking up too much space (high volumetric specific heat). See Table 8.

Table 8. Heat storage materials.
Specific heat is for sensible heat storage.
Glauber's salt melts at 91 F and has a latent heat of fusion of 108 Btu/lb.

Material	Density lb/ft³	Specific heat Btu/lb-F	Volumetric specific heat Btu/ft³-F
Water (8.33 lb/gal)	62.4	1.0	62.4
50/50 water/glycol (8.8 lb/gal)	65.8	0.8	52.6
Clay bricks	135.0	0.2	27.0
Sand	95.0	0.2	18.0
Rock (3/4"-3" diameter)	100.0	0.2	20.0
Concrete (solid)	150.0	0.2	30.0
Glauber's salt			
Solid	100.0	0.5	50.0
Liquid	70.0	0.8	56.0

Specific heat is the number of Btus required to raise the temperature of 1 lb of material 1 degree F. Volumetric specific heat is the number of Btus required to raise the temperature of 1 ft³ of material 1 degree F. Storage material weight is an important consideration, because extra floor support—even under concrete slabs—may be required. Besides weight and volume, consider cost, availability, life of the material, and compatibility with your solar system.

Sensible Heat Storage

You can store solar energy as sensible heat in a large mass of water, rock, concrete, etc. (Sensible heat raises the temperature of a material.) The energy is regained as the material cools. To calculate the amount of energy stored:

Eq 2.

$$Qs = WT \times SH \times TD$$

or

Eq 3.

$$Qs = VOL \times SHv \times TD$$

Qs = quantity of sensible heat, Btu
WT = weight of storage material, lb
SH = specific heat of storage material, Btu/lb-F
SHv = volumetric specific heat, Btu/ft^3-F
VOL = volume of material, ft^3
TD = temperature difference, F

Water is the usual storage material in liquid-type solar systems. Use only softened or low mineral content water to prevent mineral deposits. Storage tanks must be leak free, corrosion resistant, and able to stand the heat and weight of the hot water. Steel, fiberglass reinforced plastic, and waterproofed concrete have been used. Glass or ceramic-lined tanks are common in service hot water systems. Steel tanks must be lined inside to prevent corrosion and must also be lined outside if they will be buried. Insulate tanks on all sides to at least R = 11 to reduce heat loss to the air or the soil.

Example 3:

How much energy is stored when a 5'x5'x4' rectangular water storage tank is heated from 50 F to 120 F?

Answer:

From Table 8, specific heat of water is 1 Btu/lb-F.

WT = VOL x DEN
= 5x5x4 x 62.4
= 100 x 62.4
= 6240 lb
Qs = WT x SH x TD (Eq 2)
= 6240 x 1 x 70
= **436,800 Btu**
WT = weight of storage material, lb
VOL = volume of storage material, ft^3
DEN = density of storage material, lb/ft^3

With an air-to-water heat exchanger, heat from air-type collectors can be stored in water. Rock or concrete is generally the heat storage material in air systems. Concrete floors and walls can store heat in passively heated homes or livestock buildings. For storing heat in rocks, select a rock size between ¾" and 2". Rocks smaller than ¾" cause too much resistance to airflow, and rocks larger than 2" have a poor surface area/mass ratio—heat does not transfer into the rock centers very well. To minimize airflow resistance, use smooth rock of uniform size and construct the bed with a large face area and relatively short length in the airflow direction. Use Table 9 to estimate pressure drops through rock beds. Remember that most agricultural ventilation fans are designed to operate at a static pressure of ⅛" (0.125") water, or less. Solar systems usually require fans that operate at higher pressures.

Table 9. Pressure drop in rock beds.
Face velocity is airflow divided by face area.
Underlined numbers refer to Example 5.

Face velocity fpm	Rock diameter, in. ¾	1	1½	2
	Pressure drop -in. of water/ft of bed length-			
10	0.01	0.0065	0.004	0.0025
15	0.02	0.012	0.007	0.005
20	0.03	0.02	0.01	0.0075
30	0.055	0.04	0.02	0.015
40	0.09	0.06	0.035	0.025
50	0.13	0.09	0.05	0.04
60	0.2	0.13	0.08	0.06
70	0.25	0.17	0.1	0.08
80	0.3	0.22	0.15	0.1
90	0.4	0.28	0.17	0.13
100	0.45	0.33	0.2	0.15

Values were interpolated from graph by Blaine Parker, University of Kentucky agricultural engineer.

Example 4:

How much energy is regained from a 100 ft^3 bed of 1½" rock cooled from 120 F to 50 F?

Answer:

From Table 8, rock has a volumetric specific heat, SHv, of 20 Btu/ft^3-F.

Qs = VOL x SHv x TD (Eq 3)
= 100 x 20 x (120 - 50)
= **140,000 Btu**
WT = 100 ft^3 x 100 lb/ft^3 = 10,000 lb

Note that rock stores less heat per cubic ft and per lb than water. Potential problems with rock beds include dust accumulations, water condensation in the bed, mold, odors, and algal growth on the rocks.

The storage container is usually made of lumber or concrete. It must be airtight and able to support the weight and sidewall pressure of the rock bed. Insulate rock beds to at least R = 11 on all sides.

Example 5:

Estimate the pressure drop caused by 1000 cfm of air moving through a 5'x5'x10' bed of 1½" rock.

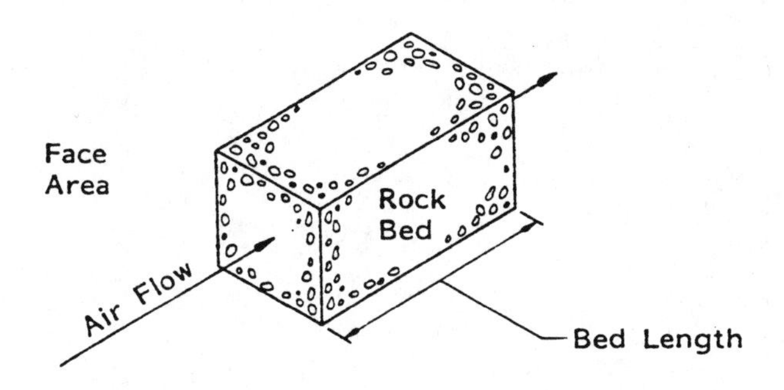

$$\text{Face Velocity} = \frac{\text{Air Flow}}{\text{Face Area}}$$

Answer:

Assume air enters the 5'x5' end.

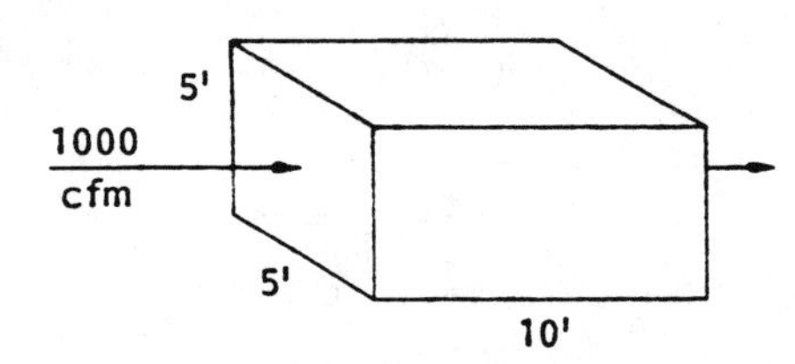

Volume = 5'x5'x10' = 250 ft^3
Face velocity = 1000 cfm ÷ 25 ft^2 = 40 fpm
From Table 9, the pressure drop is 0.035" of water per ft of bed length.
Pressure drop = 10' x 0.035"/ft = **0.35"** of water.

Assume air enters the 5'x10' top.

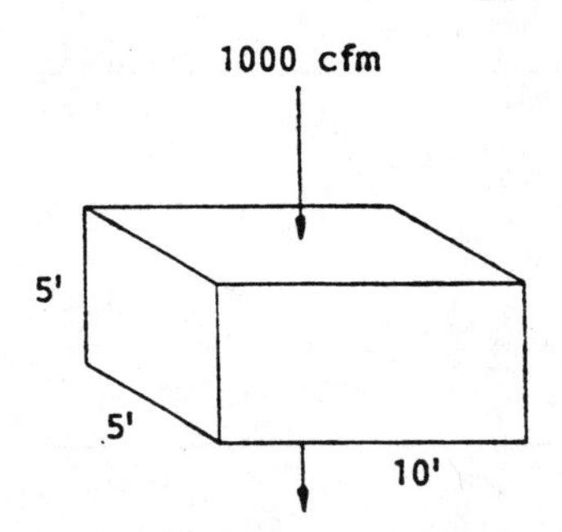

Volume = 5'x5'x10' = 250 ft^3
Face velocity = 1000 cfm ÷ 50 ft^2 = 20 fpm
From Table 9, the pressure drop is 0.01" of water per ft of bed length.
Pressure drop = 5' x 0.01"/ft = **0.05"** of water.
Note that the pressure drop is 7 times as high with 10' instead of the 5' air path.

As you can see, rock bed shape and airflow direction have a large effect on pressure drop. Design your rock storage bed with reasonable air velocity and air travel distance through the bed.

Latent Heat Storage

When a solid is heated (sensible heat), its temperature rises until it reaches the melting point. At the melting point, any added energy (latent heat of fusion) changes the material from the solid phase to the liquid phase without raising its temperature. The same quantity of heat is released when the material freezes from a liquid back into a solid. Phase change salts have melting points and heats of fusion that make them useful for storing energy in solar heating systems. One of these, Glauber's salt ($Na_2SO_4 \cdot 10H_2O$, sodium sulfate decahydrate), melts at 91 F and has a heat of fusion of 108 Btu/lb. Its other properties are listed in Table 8.

The amount of heat that can be stored in these salts can be calculated from the following equations:

Eq 4.

To warm a solid:
Qs = WT x SH x TD

Eq 5.

To melt a solid:
QL = WT x LH

Eq 6.

To warm a liquid:
Qs = WT x SHL x TD

Qs = quantity of sensible heat, Btu
QL = quantity of latent heat, Btu/lb
WT = weight of storage material, lb
SH = specific heat of storage material, Btu/lb-F
SHL = liquid specific heat, Btu/lb-F
LH = latent heat of melting, Btu/lb
TD = temperature difference, F

Stacks of small, plastic containers or rows of tubes of phase change salt make up a latent heat storage bed. The heat transfer fluid passes through spaces between the containers.

Example 6:

How much energy is stored when 3500 lb of Glauber's salt is heated from 50 F to 120 F? How much space does the salt bed occupy?

Answer:

Properties of Glauber's salt are listed in Table 8. At 50 F, Glauber's salt is a solid. It stores sensible heat as it is warmed to 91 F.

Qs = WT x SH x TD (Eq 4)
= 3500 x 0.5 x (91 - 50)
= 71,750 Btu

At the melting point, the salt absorbs 108 Btu/lb without an increase in temperature until all the solid is melted.

QL = WT x LH
= 3500 x 108
= 378,000 Btu

Once all the solid is melted, the liquid stores additional sensible heat as an increase in temperature from 91 to 120 F.

Qs = WT x SHL x TD
= 3,500 x 0.8 x (120 - 91)
= 81,200 Btu

Total heat energy = 71,750 + 378,000 + 81,200 = **530,950 Btu.**

Note that most of the heat stored is latent heat.

A salt bed occupies about twice the volume of the liquid salt, allowing for space between the salt containers.

$$VOL = 2 \times WT \div DEN$$
$$= 2 \times 3500 \div 70$$
$$= 100 \text{ ft}^3$$

VOL = volume of storage material, ft^3
WT = weight of storage material, lb
DEN = density of storage material, lb/ft^3

More energy can be stored in a cubic ft of phase change salts than in rock or water. See examples and Table 8. Much heat energy can be removed from or added to storage without changing the storage temperature. With sensible heat storage, the storage temperature and heating system performance drop as heat is removed. The advantages of phase change salts are offset by their high cost and uncertain life. **Some salts tend to break down and do not completely freeze after a number of freeze-thaw cycles.**

Utilizing Heat Storage

Many well-insulated, fully occupied livestock buildings need little supplemental heat during daytime hours when the sun is shining and temperatures are at their maximum. They frequently need heat at night when outside temperatures reach their daily minimum. Heat storage allows solar energy use when it is needed most and increases the collector system's efficiency.

A solar system designed to meet 100% of a building's heating load in all weather conditions (including extreme cold at the end of an extended cloudy period), would be very large and expensive and have excess solar capacity most of the time. It is generally more cost-effective to design a solar system with storage which provides enough heat for about 36 hr. For heating needs beyond that, provide a backup heater. Although the heater runs less often when storage is included in the solar heating system, the heater must be sized for the maximum calculated heat loss. On cold, cloudy days after stored heat is depleted, the backup heater must handle the full heating load.

In solar energy systems that preheat ventilating air for a livestock building, one of the important functions of the heat storage is to produce a time lag in the delivery of solar heat. During the day, outdoor air plus solar energy warm the storage material. Later, the cool night air removes heat from storage to help maintain the building temperature. The heat storage unit has two effects:

1. It smooths out the wide fluctuations in solar-heated air temperatures, Fig 17. A solar collector without storage increases natural day-night temperature fluctuations.
2. It increases the number of days that solar energy can heat the building. In winter, air leaving a solar collector may be below the desired building temperature, even on sunny days. But on sunny spring and fall days, a solar collector can overheat the building, so

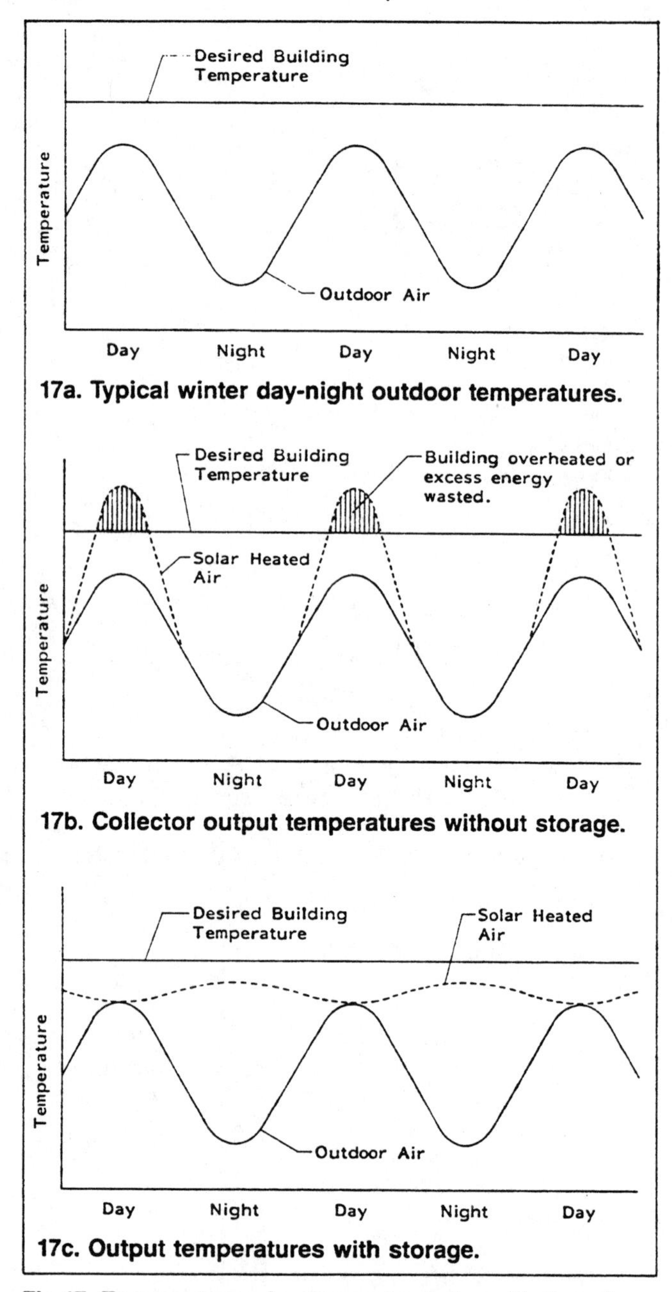

Fig 17. Temperature of solar preheated ventilating air.

the collector must be bypassed. With properly sized heat storage, however, the excess daytime heat is stored until evening, and the solar system can operate all day without overheating the building. A greater total amount of energy is utilized, more fuel is saved, and solar system efficiency is increased with storage.

Solar Systems

There are two basic types of solar systems—passive and active.

Passive Systems

In passive solar systems, heat is generally transferred without pump or fan power. Natural convection, conduction, and radiation distribute the collected solar energy.

The Trombe wall is one example of a passive solar heating system, Fig 18. The heart of the system is a massive, energy-storing concrete or masonry wall on the south side of the building. The outside of the wall is painted black and has a transparent cover. Solar radiation warms the wall during the day. Natural convection moves heated air past the front of the wall and into the building. Heat is also radiated into the building from the inside wall surface. Variations are possible—columns of water can replace the solid wall, or in combination passive and active systems, fans rather than natural convection can move the air past the wall.

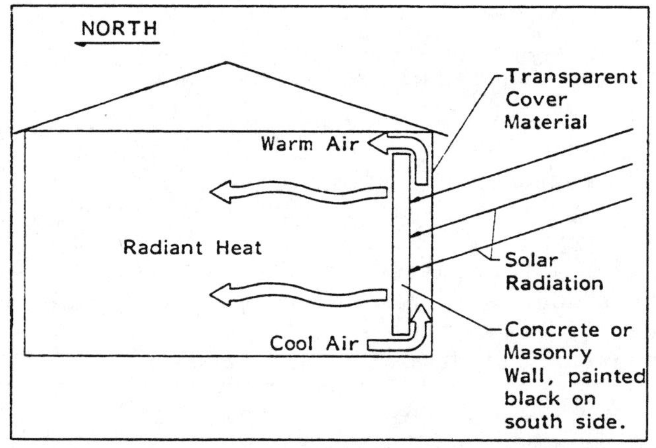

Fig 18. Passive solar heating with a Trombe wall.

Direct gain solar heating is also passive, Fig 19. Solar energy entering the building through south windows is absorbed by the materials inside. At night, these materials help keep the building warm. Heating is more effective if the back wall and floor of the structure are concrete (or other good heat storage materials). Window area is based on the heating needs for an average winter day. With direct gain heating, insulate the south windows at night or you can lose at least as much heat as you gained during the day. South facing, open-front livestock buildings utilize simple direct gain solar heating with a roof overhang that allows the winter sun to penetrate and keeps the summer sun out, Fig 20.

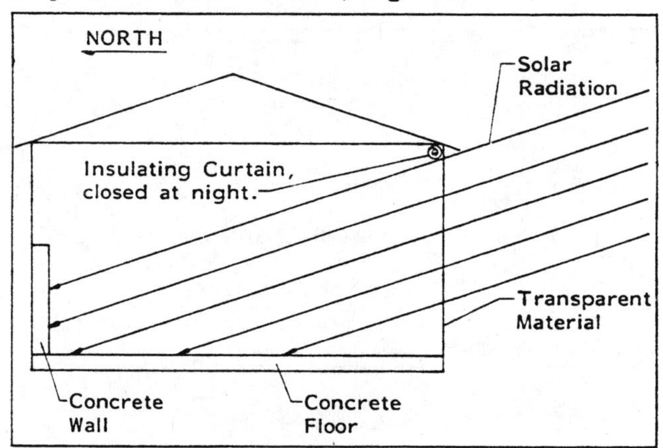

Fig 19. Passive solar heating by direct gain.

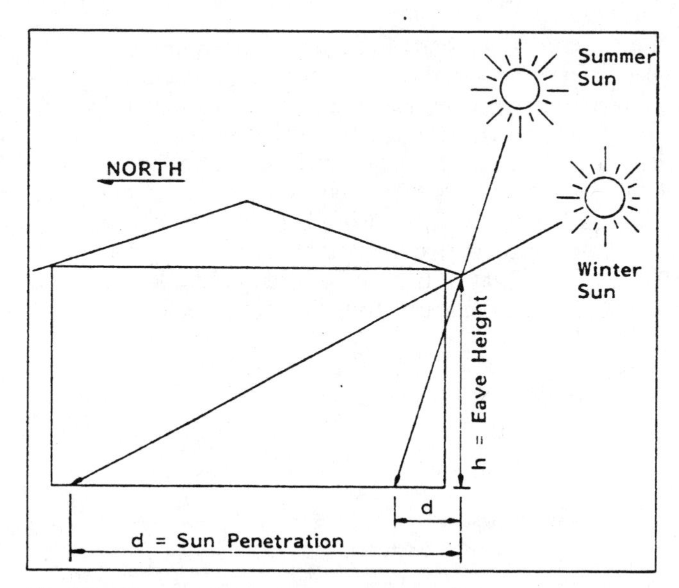

Fig 20. Direct solar gain in an open-front building.

You can estimate sun penetration with the solar angle factors in Table 1. Although sunlight can go deep into a building in early morning and late afternoon, little significant heating occurs.

Example 7:

Find the maximum sun penetration into a south-facing, open-front building with a 10′ eave height between 9 a.m. and 3 p.m. on both June 21 and Dec. 21 at 40° north latitude.

Answer:

Maximum 9 a.m. to 3 p.m. penetration occurs at 9 a.m. and 3 p.m. on Dec. 21 and at solar noon on June 21. Use the solar angle factors, SAF, from Table 1. SAF for June 21 is 0.3 and SAF for Dec. 21 is 3.0.

Pen = SAF x h

Pen = sun penetration, ft
SAF = solar angle factor
h = eave height, ft

Summer penetration
= 0.3 x 10′
= **3.0′** (at solar noon)

Winter penetration
= 3.0 x 10′
= **30.0′** (at 9 a.m. and 3 p.m.)

Passive solar systems can be simple and can provide heat at low operating and maintenance costs. Disadvantages include poor temperature control with wide temperature fluctuations in the heated space.

Active Systems

In active solar systems, pumps or fans move the heat transfer fluid. Some applications require extra pumps or fans, but others such as preheating livestock building ventilating air, use existing fans. The airflow resistance in the solar collector slightly reduces system airflow.

The heat transfer fluid in solar systems on livestock buildings is usually fresh, or non-recirculated, air. Ventilating air removes dust, gases, odors, disease organisms, and moisture. Although heat energy is lost, ventilating air is not recirculated through the collector to avoid bringing undesirable elements back into the building and to avoid fouling the collector.

Solar collectors in nonrecirculating systems preheat the heat transfer fluid, Fig 21. Heat storage can smooth out temperature fluctuations in the solar-heated air, Fig 17, and, if necessary, a backup heater can boost the temperature to a desired level.

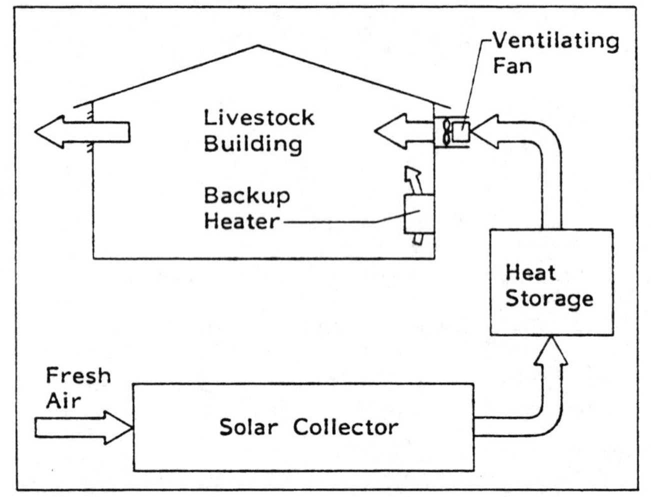

Fig 21. Solar preheated nonrecirculating system.

Solar air preheaters can be quite simple and still operate at high efficiency, because the difference between the heat transfer fluid temperature and the outdoor temperature is usually small. Collector efficiency is highest if all the ventilating air passes through the collector. If only part of the air is solar-heated:

$$TD = Scfm \times STD \div Tcfm$$

TD = temperature rise of total system airflow, F
Scfm = solar heated airflow (cfm)
STD = solar temperature rise, F
Tcfm = total system airflow (cfm)

Some active systems recirculate the same air or liquid through the solar collector. Recirculating systems, Fig 22, are used mainly to heat shops and homes and can be used for either service water or heating water. Fluid entering the collector is usually above room air temperature. Output temperatures can be high, but only high quality collectors are efficient. Seal air recirculating systems tightly to prevent cold air infiltration and efficiency loss. Also, avoid running the collector if the inlet fluid's temperature is higher than the absorber's, or heat will be lost from the system. Install a differential thermostat to sense fluid and absorber temperatures and switch on the pump or fan in the collector loop only when the absorber temperature is high enough.

Water is the usual storage material for liquid-type solar systems. In drain-down systems, water from storage is circulated through the collector, Fig 23a. To keep the water from freezing and rupturing the collector tubes, the collector is automatically drained

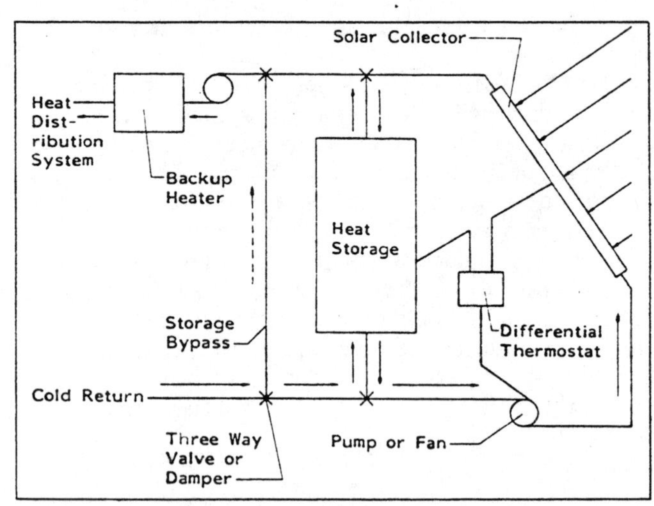

Fig 22. Active recirculating solar heating system.

every time the circulating pump stops. A more reliable but less efficient system uses a closed collector/storage loop that contains antifreeze, Fig 23b. Heat is transferred from the solar collector to the storage water through a heat exchanger in the storage tank. In solar service hot water systems that use antifreeze, install a double-wall heat exchanger to reduce chances of a leak contaminating the service water or use a non-toxic antifreeze, e.g. propylene glycol.

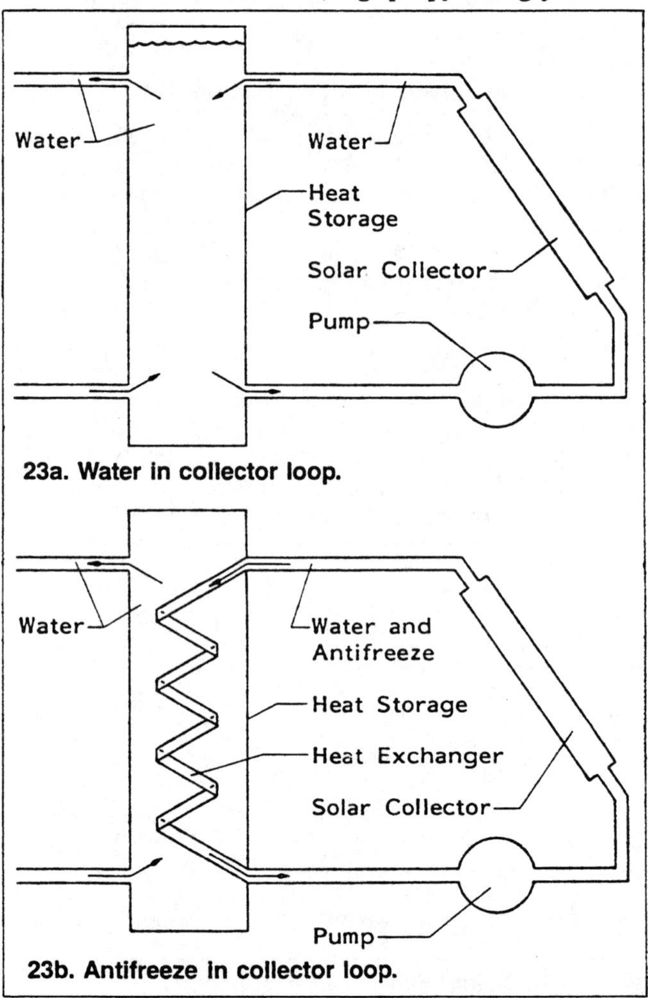

Fig 23. Liquid solar systems.

On sunny days, when the collector is not in use and the liquid is stagnant (not moving), temperatures inside the collector can exceed 300 F. Protect against damage from boiling liquid by draining the collector or providing a pressure relief valve. In addition to boiling and freezing problems, liquid-type systems are prone to corrosion, scale buildup, fluid leakage, and algal growth.

Flat Plate, Air-Type Collector Construction

There are many variations of air-type solar collectors for new or retrofit building construction and for freestanding and portable units. These guidelines will help you build efficient, long-lasting collectors:

- Select materials that withstand ultraviolet radiation, heat, wind, snow, rain, and hail; are easy to install; and have long life and low cost. Many common building materials (sheet metal, chipboard, plywood, dimension lumber) are adequate.
- In collectors with distinct air channels, keep collector air velocity between 500 fpm and 1000 fpm. Air velocity (fpm) equals airflow through the collector (cfm) divided by the cross sectional area of the collector (ft^2). At air velocities under 500 fpm, simple collectors may be inefficient; above 1000 fpm, pressure drop is usually too high, and fans require too much power.

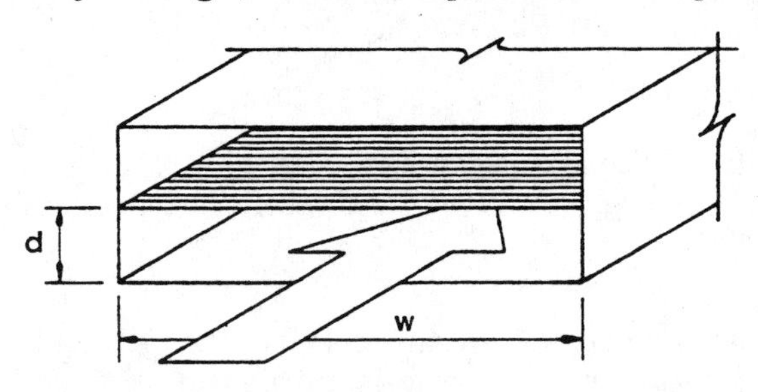

Cross-sectional Area = d x w

$$\text{Air Velocity (fpm)} = \frac{\text{Air Flow (cfm)}}{\text{Cross-sectional Area (sq ft)}}$$

- To prevent excessive pressure drop in ducts, keep ducts short and straight and keep air velocity below 800 fpm. Size ducts for about 800 cfm/ft^2 of duct area.
- Keep static pressure drop in collector and ducts less than ⅛". Make sure that your fan will deliver adequate air in solar systems with higher air resistance, such as with heat storage in rock. Greater pressure drops reduce fan performance too much. If pressure drop is excessive, bypass the solar system with part of the air, rebuild the collector and/or ducts, or install centrifugal fans designed for higher pressure. (Pressure drop is the difference between the inlet and outlet air pressure. In nonrecirculating systems where the fan draws outdoor air through the collector, the inlet pressure is 0—just measure the negative pressure at the entrance to the fan.) **Keep total solar system (collector and ducts) pressure drop less than ½" of water for grain drying air.**
- Cover air inlets with ½" hardware cloth to keep out birds, rodents, and trash.
- Use pressure preservative treated lumber for components in contact with the soil. Use CCA treatments to avoid the volatiles in oil borne preservatives.
- Seal the collector to prevent air leaks. Caulk joints and cracks with silicone or butyl rubber and seal cover-sheet laps with clear silicone. Other types of caulking are cheaper but do not last as long or work as well on solar collectors. Seal the edges of corrugated materials with corrugated wood or closed-cell foam strips.
- Fasten FRP with screws rather than nails.
- Position rows of portable or freestanding collectors so they do not shade one another. Maintain adequate distance between buildings to prevent shading.
- If collectors are not used during the summer, shade them from the sun and/or open vents (1 ft^2 of vent opening per 100 ft^2 of collector surface area) to keep them cool. Locate vent openings to cool the collector with prevailing winds or natural convection.

Basic Equations for Collectors

Use the following equations to size solar collectors or predict collector performance. (Note: collector area equals gross collector area based on outside dimensions.)

Definitions follow for Eqs 7 through 11:

1.1 = a constant to allow for air density and specific heat of air, Btu/F-hr-cfm
A = collector area, ft^2
Ccfm = collector airflow, cfm
CE = collected energy, Btu/hr
EFF = collector efficiency, % (For efficiency as a decimal, do not multiply by 100.)
RAD = solar radiation, Btu/hr-ft^2
TD = temperature rise, F

Eq 7.

Temperature rise through a collector:
TD = EFF x RAD x A ÷ (1.1 x Ccfm)

Eq 8.

Collector airflow to get temperature rise:
Ccfm = EFF x RAD x A ÷ (1.1 x TD)

Eq 9.

Collector efficiency:
EFF = 1.1 x Ccfm x TD ÷ (A x RAD)

Eq 10.

Collector energy output, by airflow:
CE = 1.1 x Ccfm x TD

Eq 11.

Collector energy output, by efficiency:
CE = EFF x A x RAD

The temperature rise used in the equations equals outlet temperature minus inlet temperature and can be either maximum or average. Maximum or peak temperature rise, for collectors without integral storage (no rock or concrete in the collector), occurs near solar noon on sunny days. Use noon hour, clear day solar insolation from Table 672-4 and noon hour efficiency when working with peak temperature rise. Measure peak temperature rise with the fan or pump running and thermometers shaded.

Twenty-four hour average temperature rise cannot be measured directly. It is the temperature rise that would occur if daily solar radiation arrived at a constant rate over 24 hours. When working with average temperature rises, use average daily insolation from Table 672-5 and average collector efficiencies.

Example 8:
Find the collector airflow required to produce 110 F output temperatures from a 96 ft^2 solar collector at noon on a sunny Dec. 21 at Lemont, IL. Collector inlet temperature is 65 F; the tilt angle is 60°; and noon hour efficiency is 45%.

Answer:
Lemont is at 41.4° north latitude, so use data for 42°. A 60° tilt angle is at about latitude plus 15° at 42° north latitude. From Table 672-4, solar insolation equals 287 Btu/hr-ft^2 at noon.

TD = 110 F - 65 F = 45 F

Using Eq 8,
Ccfm = 0.45 x 287 x 96 ÷ (1.1 x 45)
= **250.5 cfm**

Select a fan to deliver 250 cfm at about ¼" pressure.

Example 9:
Find the noon hour rate of energy output for the collector in Example 8.

Answer:
Solution 1: Calculate energy output from Eq 10:

CE = 1.1 x 250.5 x 45 F
= **12,400 Btu/hr**

Solution 2: Knowing solar input and collector efficiency, calculate collector output from Eq 11:
CE = 0.45 x 96 x 287
= **12,398 Btu/hr**

Example 10:
Find the expected 24-hr average December temperature rise for the collector in Example 9. Average collector efficiency is 40%; collector airflow is 250 cfm.

Answer:
From Table 672-5, the average December insolation on a 60° surface (about latitude + 15°) in Lemont is 964 Btu/day-ft^2.
RAD = 964 ÷ 24 hr/day
= 40.2 Btu/hr-ft^2 (average)

Using Eq 7,
TD = 0.4 x 40.2 x 96 ÷ (1.1 x 250)
= **5.61 F**

Example 11:
Find the expected average December daily energy output for the collector in Example 10.

Answer:
Solution 1 using Eq 10:
CE = 1.1 x 250 x 5.61 x 24 hr/day
= **37,000 Btu/day**

Solution 2 using Eq 11:
CE = 0.4 x 96 x 964
= **37,000 Btu/day**

Example 12:
What is the December daily fuel oil equivalent of the energy produced by the collector in Example 11?

Answer:
From Table 672-1, we see fuel oil contains 138,000 Btu/gal and is typically burned at 65% efficiency.
Oil = CE ÷ (Qoil x Beff)
= 37,000 ÷ (138,000 x 0.65)
= **0.41 gal/day** in December
Oil = fuel oil equivalent, gal/day
Qoil = energy in fuel oil, Btu/gal
Beff = burner efficiency, decimal

SYSTEM FLOW DIAGRAMS

This section shows some of the options for collecting, storing, and distributing solar energy for farm buildings. The schematic diagrams show the major components and the movement of air or liquid to transport the heat. The examples illustrate some of the ways the components are put into buildings.

Preheating Ventilating Air

Relatively simple and inexpensive solar systems can preheat ventilating air. For livestock buildings, do not recirculate room air. Liquid-type collectors are avoided because they require heat exchangers and are relatively expensive.

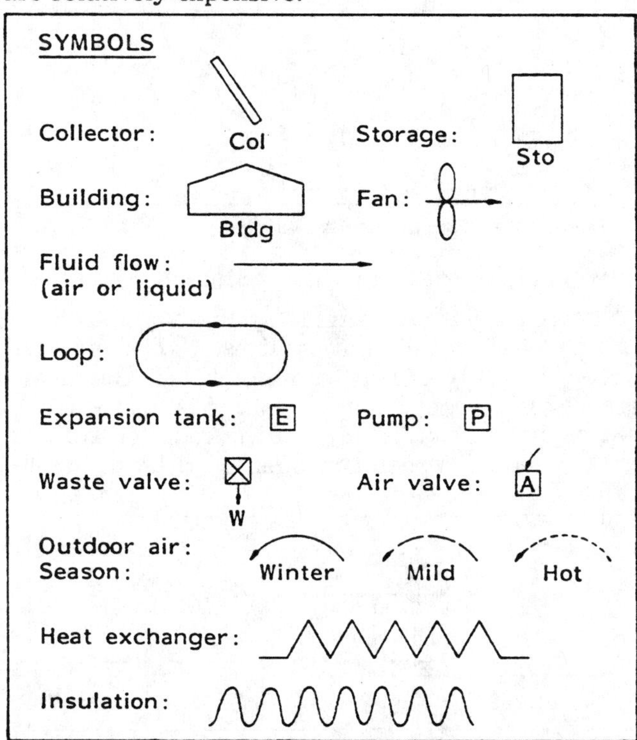

One Pass System without Storage (Fig 24)

In winter, air passes through the collector, where it is warmed, and goes directly to the heated space. In mild weather, some fresh air is blended into the warmed air to prevent overheating of the building. In summer, bypass the collector, but protect it from overheating by covering or venting it.

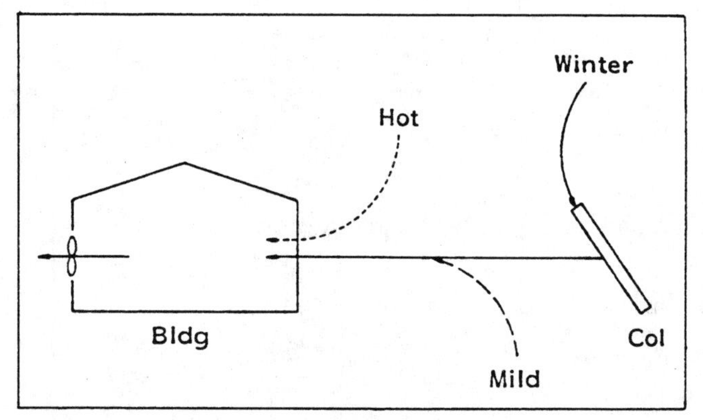

Fig 24. One pass system without storage.

This simple system has limited value, because it can overheat the building in warm weather unless the collector is undersized. See Fig 17. It is useful for retrofitting buildings with little insulation and in other cases where daytime heat is needed. (Adding insulation can be more cost-effective.) Applications include some swine growing buildings, colder climates, and retrofitting poorly insulated buildings for smaller animals. The system is also applicable to nursery and brooder buildings requiring summer heat.

If solar heating is added to a building that could be naturally ventilated, the fan energy needed to move the solar heat can exceed the useful energy gained.

Examples of one pass systems without storage include a solar attic without storage, a covered plate collector on the south wall, and some installations with portable or commercial collectors.

One Pass System with Storage (Fig 25)

Storage increases use of solar heat, levels out temperature cycles in the warmed space, and permits some summer cooling. These systems permit using less solar heat in the daytime and saving that heat for nighttime, when it is more typically needed. Well insulated buildings are desirable.

For summer cooling, pass night air through the collector and into the storage to cool it and exhaust it outdoors, Fig 25c. During the day, and until the storage warms up, bring ventilation air through the storage to the building space, Fig 25d.

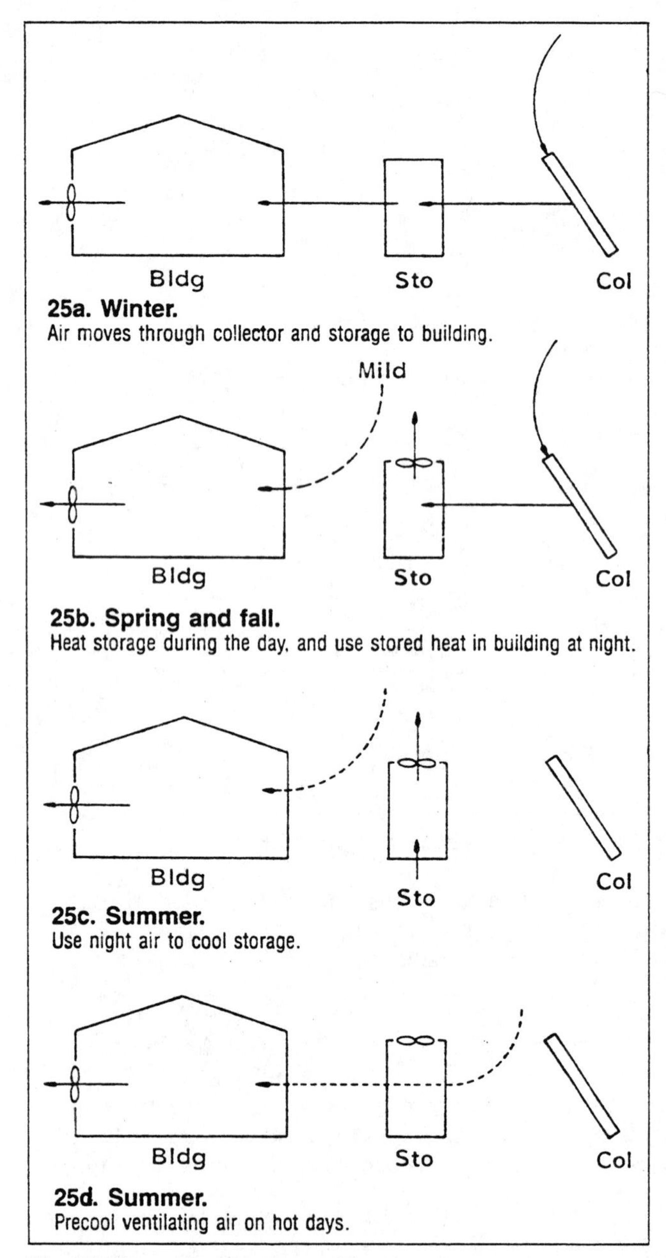

Fig 25. One pass system with storage.

Collect/Store Loop, One Storage/Bldg Pass (Fig 26)

In this system, the temperature of the storage is raised by recirculating air between the collector and the storage during daylight hours. In mild weather, heat can be stored while the ventilating air is bypassing the storage, permitting more heat collection during mild weather in anticipation of cold nights.

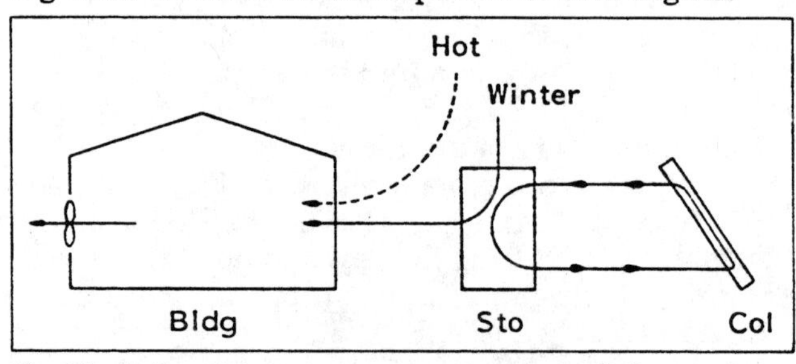

Fig 26. Collect/store loop, one storage/building pass.

Space Heating

Space heating is supplied by solar energy that is not added to the ventilating air.

Higher temperature collection and storage permits space heating of homes, shops, offices, milking centers, etc. These systems are suitable for buildings that do not require the high ventilating rates of typical livestock housing and that have relatively clean air that can be recirculated through storage. High temperature recirculating collectors are also useful for zone heating portions of livestock buildings.

Passive solar heating of livestock housing is another example of solar space heating.

Passive Solar Livestock Buildings (Fig 27)

Part of the south wall and/or roof allows sunlight to warm and dry the building directly. This is the simplest solar system, because no separate equipment or controls are needed. Some heat is stored in the building materials, especially in concrete walls and floors.

Some management is needed to conserve captured energy and to prevent seasonal overheating. Cover the solar glazing at night to reduce heat loss through it. Night losses can exceed the daytime gains unless the glazing is insulated at night. Shade the glazing in summer to prevent overheating.

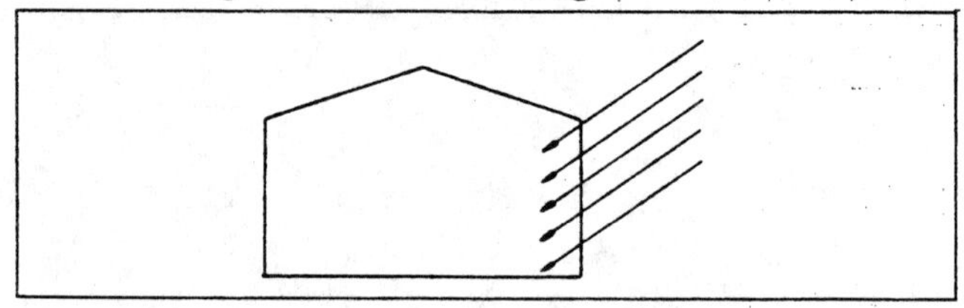

Fig 27. Passive solar livestock buildings.

Recirculating Air Heating without Storage (Fig 28)

This is a relatively simple system suitable for heating homes or shops but not livestock housing. It collects heat only during sunlight hours. Because there is no storage, it does not contribute heat at night or during cloudy days. If the collector is too large for the space being heated, overheating will result.

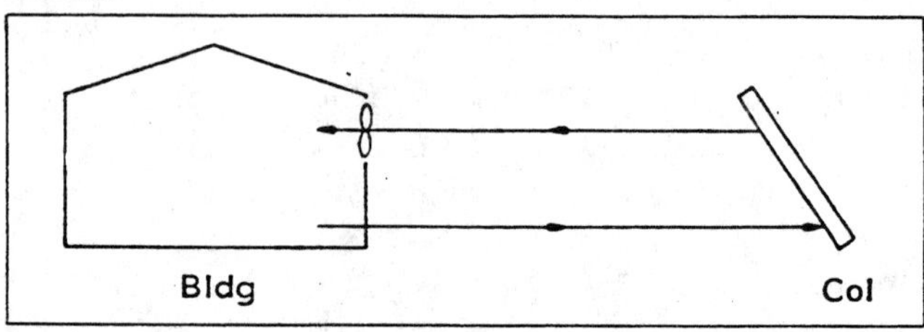

Fig 28. Recirculating air heating without storage.

Indirect Space Heating from Storage (Fig 29)

Heat from the collector can be stored in a rock bed or other storage beneath a floor. The room is heated directly by radiation, or resting animals are heated by conduction from the warm floor surface.

Circulate air through the collector and storage only when the temperature of the air coming out of the collector is higher than the temperature in the storage. Turn off the collector when storage reaches maximum temperature, or when the room is warm enough.

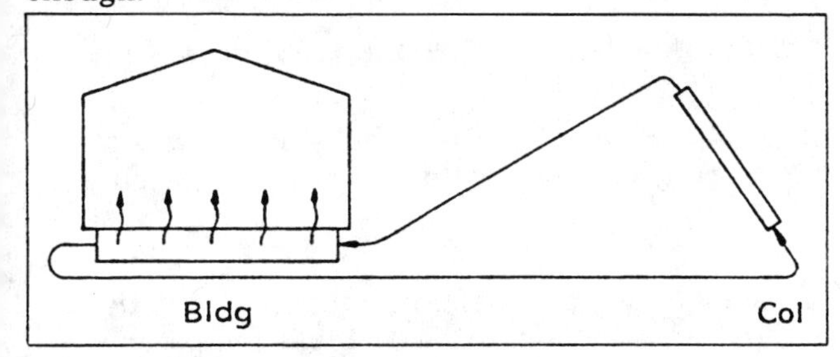

Fig 29. Indirect heating from storage.

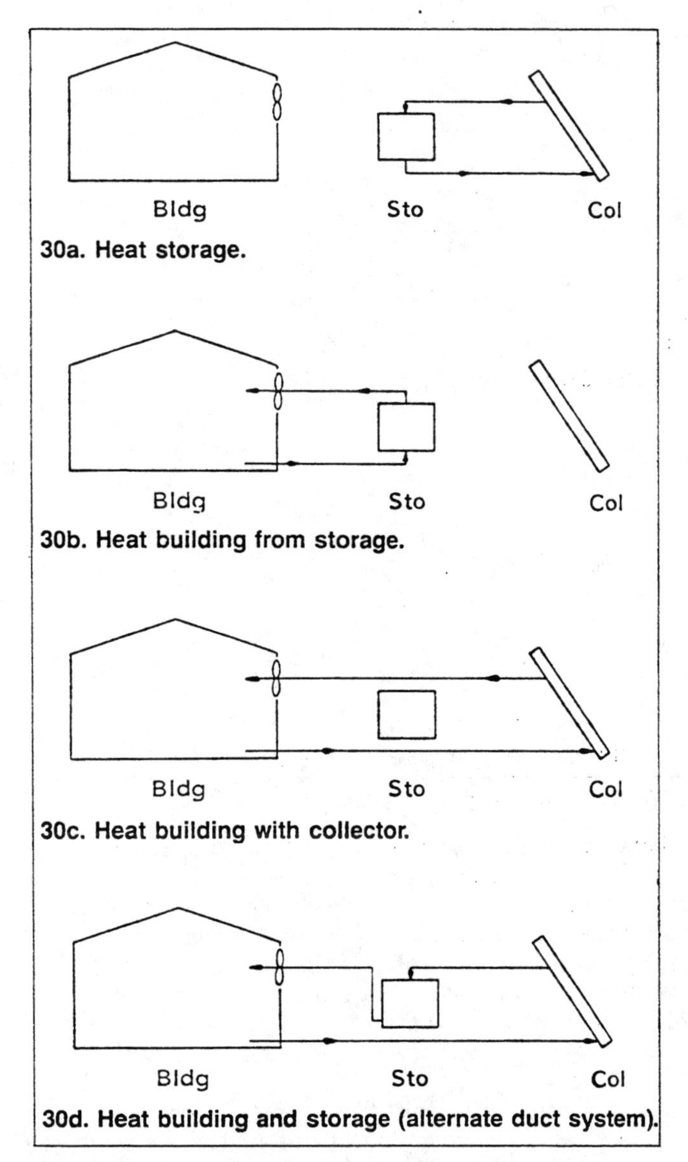

Fig 30. Recirculating air system with storage and bypass.
Not suitable for livestock housing.

Recirculating air system with storage & bypass (Fig 30)

In this system, the building is heated by circulating room air through the storage or the collector.

Circulate air through (see Fig 30):

a. Collector and storage when the sun is shining and the building needs no heat.
b. Building and storage when the building needs heat but the sun is not shining.
c. Collector and building when the building needs heat and the sun is shining.
Or,
d. Fig 30d shows a preferred system that circulates air through both the collector and storage when the sun is shining and the building needs heat. This mode requires more complicated ducts and dampers. Drawing heat from storage provides more uniform ventilating air temperature.

Usually air from the collector goes through the storage in one direction, while air for the building goes through in the other direction. More uniform temperatures in the ventilating air result.

Liquid-Type Collectors for Space Heating

One loop carries heat in the pumped fluid from the collector to the storage. Another loop carries heat from the storage to the building.

Heat from storage can be distributed through the building space with fin tubes, pipes in floor, fan and radiator, heat exchanger in the ventilating duct, or other liquid/air heat exchangers. This system is useful for residential, greenhouse, or milkhouse air heating and for some swine buildings.

Insulate the top, bottom, and all sides of the storage tank to at least R=11. Provide good drainage around buried storage tanks to prevent flotation and to reduce heat loss. Also provide an expansion tank in any closed system to allow for volume changes in the liquid caused by temperature changes.

Liquid system with drain-down (Fig 31)

A collector may freeze when there is no solar energy. One solution to the problem of freezing is to drain the collector when it starts to cool down.

In an open-loop system (Fig 31a), one pump delivers hot water to the building and a separate pump circulates fluid through the collector loop. There is an air gap between the surface of the liquid in the heat storage tank and the end of the collector loop return line. When the collector pump turns off, the air inlet valve opens, and the collector loop drains into the heat storage tank.

Install an air inlet valve at the high point of the collector loop so the whole loop drains. Also, locate the valve where it will not freeze shut—for example, inside the collector case. Because air enters the system, collector tube corrosion could be a problem.

In a closed-loop system (Fig 31b), one pump distributes hot water from heat storage and another circulates water through the collector loop. Install a drain-down valve in the lines to and from the collector. When the pump shuts off, the drain-down valve isolates the collector loop from the rest of the system. The air inlet valve opens, and the collector loop drains to waste.

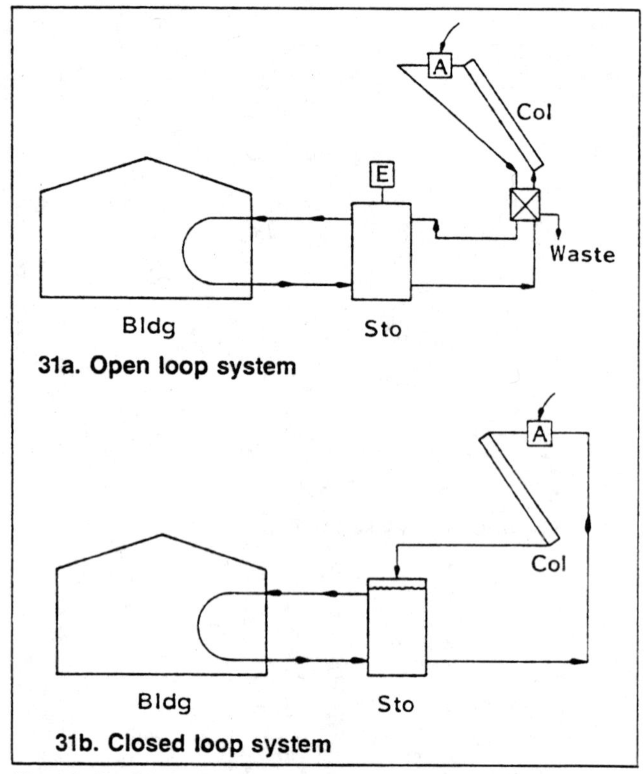

Fig 31. Drain-down liquid system.

Liquid system, antifreeze in collector loop (Fig 32)

By adding a heat exchanger in the collector loop, it can have antifreeze while the rest of the system uses plain water. The heat exchanger must be guaranteed to protect against antifreeze leakage, the antifreeze must be nontoxic, or the liquid in the rest of the system cannot be used for drinking or other pure-water purposes. Expect lower efficiencies than for the system in Fig 31.

Some systems with antifreeze mix the collector-loop fluid and the room-loop fluid in the storage. Mixing avoids heat exchangers but requires that all the fluid be protected with antifreeze.

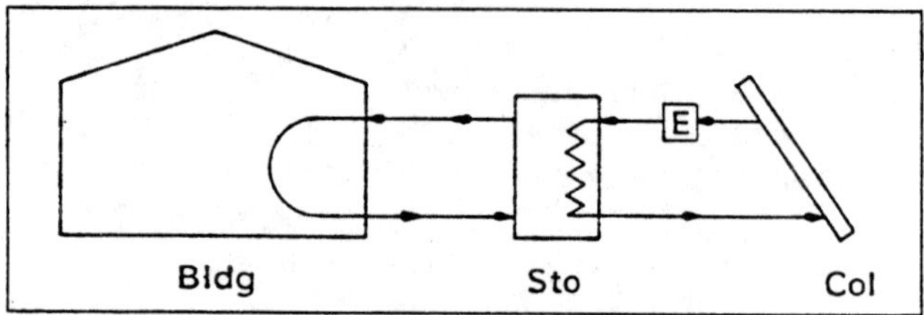

Fig 32. Liquid system with antifreeze in collector loop.

Service Water Heating

Solar energy can heat water for the home, milk-house, swimming pool, etc. Packaged commercial systems are available.

Because solar water heating is not the most cost-effective way to use solar energy, consider other heat sources first, such as compressor heat from a milk cooler. Also consider other uses for the solar energy.

Water Heating with Drain-Down (Fig 33)

Drain-down water heaters prevent freezing by draining the collector. When the storage is hot enough, during cloudy weather, and after the sun sets, the collector pump turns off, and the drain-down valve opens and the water drains out.

This system avoids antifreeze problems, but the whole system may have to be suitable for potable water.

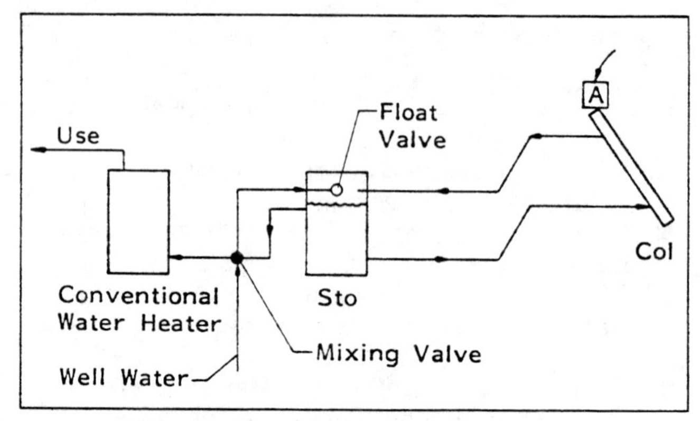

Fig 33. Service water heating with drain-down protection.
Hot water should be taken from top of tank, cold into bottom.

Water Heating with Antifreeze (Fig 34)

Heat from the collector warms water in the insulated preheater through a heat exchanger. The pre-warmed water supplies a conventional water heater.

Commercial units are available which tend to be about 20% efficient. Use non-toxic antifreeze or a double-wall heat exchanger to assure no contamination of potable water if storage fluid is delivered directly to the water heater.

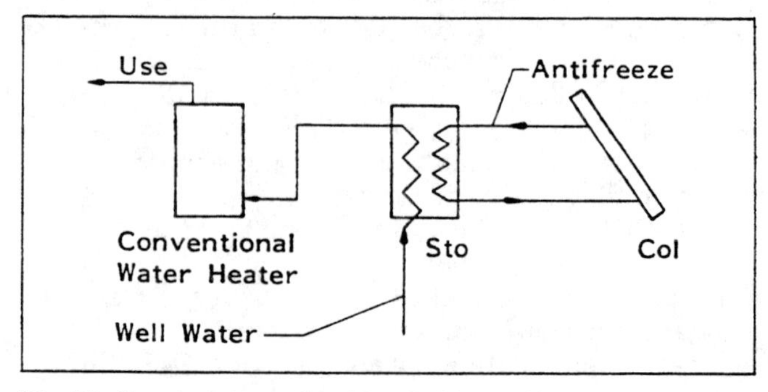

Fig 34. Service water heating, with antifreeze protection.

CONTROLS

The following is a summary of the types of controls that are useful in solar systems on farm buildings. Most of the components are commercially available. Look for quality, economy, cleanability, accuracy, and availability of adequate service.

Thermostats

Thermostats sense temperature and then open or close a switch to make something happen or stop happening. A **heating thermostat** closes a switch as the temperature falls, as for starting a furnace. A **cooling thermostat** closes a switch as the temperature rises, as for starting an air conditioner or ventilating fan.

Select a thermostat that matches the voltage of the system—12, 24, 120, or 240—and the amperage. Select one with a sensitivity range that matches needed accuracy—one or two degrees of error usually won't matter, but you will want to sense differences as little as 5 F.

Differential thermostats sense the difference between two temperatures. They are available with both fixed and adjustable differentials, but some are not very reliable. One application is to compare absorber plate temperature with storage temperature. If the absorber plate is at least 10 F above the actual storage temperature, the circulating pump turns on to move the working fluid through the collector which moves heat into storage. The thermostat turns off the pump when the temperature difference is down to about 6 F.

Thermometers

Thermometers in water and air lines and in storages help you learn how your system works, if it is working, and how it responds to various weather and building conditions.

Fittings are available for inserting a thermometer into a water pipeline. An indoor/outdoor thermometer can show the temperature within a storage. Remote sensing thermometers are available at electrical supply stores. Recorders are available that make a paper tape of temperature, or other readings, at several locations. They are expensive but can be helpful in a large installation.

Valves (Fig 35)

- **Flow control** valves balance the various parts of the system. They are manually set to limit flow below some maximum. Gate and ball valves contribute less to pressure loss in the system than do globe valves.
- **Solenoid** valves respond to a voltage. An electromagnet opens or closes a valve when energized by a circuit from a thermostat (or other control switch).

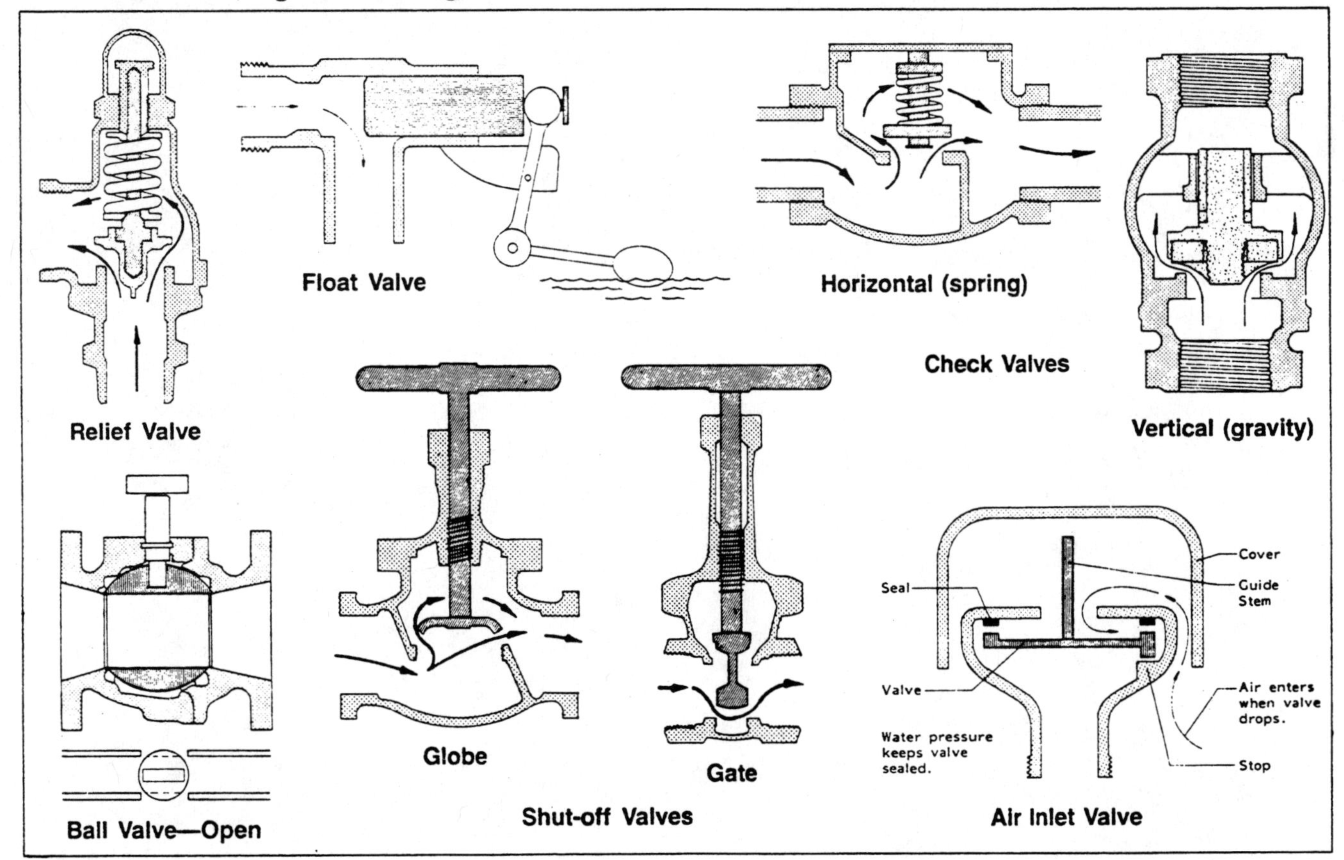

Fig 35. Typical valves.

- **Mixing valves** blend hot and cold liquids, like a typical domestic faucet. A type useful in solar installations responds to a thermostat to automatically blend cold water if a storage gets too hot (about 140 F).
- **Drain-down system valves.** Air inlet valves open when there is not water pressure against them. They let air into the high point of the collector loop so the lines can drain. In an open-loop system, the air inlet valve opens when the pump turns off. In a closed-loop system, a mechanically operated drain-down valve opens when the circulation pump turns off. The air inlet valve also opens, and the collector loop drains to waste. Solenoid valves are recommended because they are more likely to work in bad weather.
- **Check valves** prevent liquid from flowing backwards in a line.
- **Pressure relief valves** permit fluid to escape from the system to relieve excess pressure. They are spring controlled and usually are adjustable. Install a relief valve or an expansion chamber in any closed hot water system.

Expansion Chamber

Install an expansion chamber in any closed-loop liquid system. It allows for the change in volume of the liquid as temperatures change.

Air Controls

- **Dampers** open or close a duct and can also control flow in partly-open positions. They can be motor driven in response to a thermostat or manually set.
- **Diverters** at a "Y" in a duct cause flow down one leg or the other of the Y.
- **Shutters** are louvered dampers. They can be power driven in response to a thermostat or other signal, or closed by gravity and opened by a moving airstream. Gravity shutters can serve as a check valve by reducing backward flow.

Alarms

Consider a high temperature alarm in a water storage, especially with a recirculating collector loop which can create high temperatures. A high temperature alarm in a collector warns you of needed summer ventilation to prevent overheating. A low temperature alarm warns you of heating system failures in housing for young animals or that the building or solar system may freeze.

A general power failure alarm is important for livestock installations because othe damage that can result from the loss of fans, heaters, pumps, etc.

672 SOLAR LIVESTOCK HOUSING

Design

Solar collectors can provide part of the supplemental heat needed in livestock buildings. This section tells how to determine heat needed and how to size a solar collector. Detailed procedures and an example are given in Chapter 671. A detailed example is in MWPS-23, *Solar Livestock Housing Handbook*.

Calculate maximum heat loss at a low winter temperature: -10 F is often assumed in much of the Midwest. Maximum heat loss helps in sizing heaters (Btu/hr capacity).

Calculate average heat loss for each month using the average monthly temperature. Size solar collectors and estimate supplemental fuel needs with average heat loss for January.

Supplemental heating capacity is installed furnace size. It allows for unusually cold weather and the building being empty or only partly full of animals. Calculating maximum heat loss ensures adequate capacity for your climate and building. Buy a furnace for its rated output, not the input rating quoted in typical sales literature.

Calculate supplemental heat needed to maintain building temperature by subtracting heat produced by the animals from the building and ventilating heat loss. Some animal heat is wasted, because it is produced during the warm part of the day when building and ventilating heat losses are low. More animal heat is utilized if you provide solar heat storage. To calculate heat needed for solar heated livestock housing, use 50% of the animal heat production values from Chapter 633.

Collector Size

It is seldom practical or economical to provide all supplemental heat with solar energy, especially for collectors without storage, which provide heat only during the day when heating needs are typically lowest. Fig 672-1 illustrates the effect of collector size on utilization of solar energy. Much of the heat from a large solar collector is wasted in spring and fall, and collector pay-back is slow.

A number of factors and compromises affect final collector size. We suggest you estimate collector size by each of the appropriate methods below, and then compare the results. Seriously consider engaging an engineer experienced in designing solar systems and livestock ventilating systems to help you size your collector.

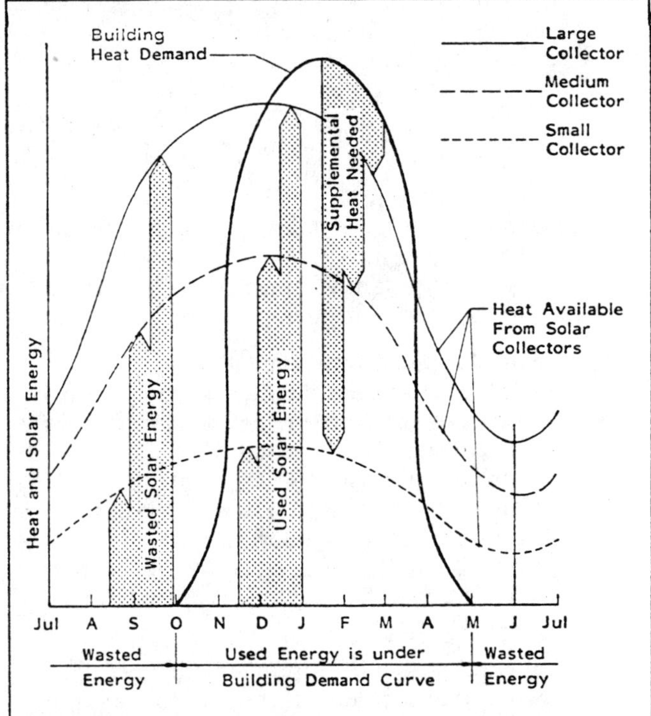

Large Collector
Small proportion of total energy collected is utilized. Collector cost pay-back very slow. Even with large heat storage, furnace is needed for midwinter and for cloudy fall and spring periods.
Medium Collector
Compromise. Much, but not all, furnace operation is replaced with solar energy over several months.
Small Collector
High proportion of available solar energy used. Pay-back rapid. Little effect on furnace operation.

Fig 672-1. Collector size and solar energy used.
Vertical surface, 40° N latitude.

1. Collector size by physical limitations.
   - Collector size is often limited by the size of the south wall, or south roof slope, of a building.
   - Collector size can be limited by the amount of unshaded area on a building. The collector must be unshaded during the season you need the heat. Avoid trees, silos, and other buildings.
   - Portable or other remote collectors need space away from shading, traffic, and animals. The amount of space available can limit collector size. Ducts get in the way, tend to have high heat loss, leak air, and increase cost, so minimize their length.
2. Collectors sized for minimum winter ventilation. For preheating ventilating air, the maximum collector size is about 1 ft^2 of collector area for each cfm minimum winter ventilating rate; 0.3 to 1.0 ft^2 collector area per cfm is a common design range. Collectors are more efficient at higher airflows, so smaller collectors produce more energy per ft^2.
3. Collector size by heat needs.
   - For heating water, provide 1 ft^2 collector area per gal of hot water needed per day.
   - For heating a floor in winter (e.g. swine nursery), size the collector for 50% (or 25% without heat storage) of the January floor heat needed. Allow 100 to 130 Btu/hr per ft^2 of floor area for farrowing and 85 to 100 Btu/hr-ft^2 for nursery pigs. See Chapter 651.

- For preheating winter ventilating air, select collector size to provide:
  Up to 50% of the January supplemental heat needed, with heat storage.
  Up to 25% of the January heat needed, without heat storage.

Eq 672-1.

$$A = 50\% \text{ (or } 25\%) \times Qsup \times 24 \text{ hr/day} \div (RAD \times EFF)$$

A = collector area
Qsup = supplemental heat demand for January, Btu/hr
RAD = average January insolation, Btu/ft^2-day
EFF = collector efficiency

Sizing collector for preheating ventilating air

Assuming a south sidewall collector with storage and an efficiency of about 30% near Lemont, Illinois, calculate the collector area needed.

Assume January supplemental heat needed is 25,000 Btu/hr. January insolation from Table 672-5 is 1159 Btu/ft^2-day on the average for the whole month.

Using Eq 672-1, collector area is:

$$A = 0.50 \times 25{,}000 \times 24 \div (1159 \times 0.30) = 862.81 \text{ ft}^2$$

If the collector is about 8′ high, it would need to be 862.81 ÷ 8, or about 108′ long. If the collector were at an angle of about 60° instead of vertical, the daily energy would be 1251, instead of 1159, Btu/ft^2-day, which only decreases the needed length to about 100′.

Estimating solar energy use

Estimate the total solar energy used from Eq 671-11. Assume adequate solar energy storage.

$$CE = EFF \times A \times RAD = 0.30 \times 108' \times 8' \times RAD = 259.2 \times RAD$$

Energy used equals energy collected, except that the maximum energy used is the demand for supplemental heat. Only 50%-100% of the solar heat collected will reduce costs.

Average daily solar energy use is 50% x 259.2 x average daily insolation.

For seasonal use of solar energy, multiply the average daily use for each month by the number of days in that month. Fossil fuel use is supplemental heat required minus solar energy use.

Miscellaneous Design Data

- Storages:
  -For liquid systems, allow about 25 lb water (about 3 gal) per ft^2 of collector.
  -For air systems, allow about 1 ft^3 or 100 lb rock per ft^2 of collector. At 1 to 2 cfm/ft^2 of collector, airflow through storage will be about 1 cfm/ft^3 of rock, which gives a 12-hr lag between time of maximum energy at the collector and when that energy is apt to be released to the ventilating air. One ft^3 rock provides about 20 to 30 Btu storage per degree F temperature difference in storage. Expect about a 30 F temperature rise in storage temperature during a sunny day; 30 F rise is 600 to 900 Btu stored per ft^2 of collector.
- Design airflow through the collector for the minimum ventilation needs of the building. Recommended airflow is 1 to 3 cfm/ft^2 of collector.

When Not Enough Solar Energy

Solar energy probably will not supply all the supplemental heat needed. Reasons include not enough collector and/or storage, extended sunless periods, or extended or extreme cold periods. It is not economic to totally solar heat most applications.

Solutions include:

- Provide supplemental heat, which is the only acceptable solution.
- Allow room temperatures to fluctuate with outside temperatures. Beware of freezing and causing stress on livestock.

Beware of reducing ventilation rates to save heat. Recommended rates are based on years of experience and research and are intended to maintain production and animal health. Problems resulting from underventilating include:

- Condensation, which corrodes equipment and electrical and electronic components. It can increase respiratory stress and bacterial growth.
- Increased concentration of airborne diseases.
- Too little oxygen (asphyxiation) and/or too much harmful gas (ammonia, hydrogen sulfide, CO, CO_2, etc.).

Economic Justification

Many factors influence how economical your solar energy system will be. Some of these factors are:

- **Collected solar energy** is energy obtained at the collector.
- **Solar energy storage** saves some collected energy for later use. Without storage, energy can be wasted during periods of peak solar collection and no solar energy is available for nights and cloudy days.
- **Utilized collected energy** is the proportion of collected energy actually used. Higher room temperatures and higher ventilation rates increase utilization of solar energy. An unused building has zero utilization. The more solar energy collected, relative to heating needs, the lower will be the portion of that energy that is utilized, Fig 672-1.

Utilized solar energy substitutes for fossil fuel normally used to heat the livestock building. The money saved on conserved fuel pays for the solar energy collection system. An efficient, well-sized, and well-maintained, low-cost solar system can be paid for in a reasonable time—perhaps 5 to 7 years.

Income from a Solar Collection System

The total energy saved per season is the solar energy used. To find the value of that energy, convert your cost for conventional fuel replaced by solar energy to cost per British thermal unit (Btu).

Eq 672-2.

$$EV = FC \div UE$$

EV = energy value, $/Btu
FC = your fuel cost, $/unit
UE = usable energy, Table 672-1

If your system is multiple use (e.g. grain drying or summer water heating), include savings from that application.

Table 672-1. Usable energy for four fuels.
Values for natural and LP gas and fuel oil are for vented heaters. Although usable energy for unvented heaters is relatively high, the moisture they release into the building must be removed by ventilation.

Fuel type	Total energy	Combustion efficiency	Usable energy UE
LP gas	93,000 Btu/gal	0.80	74,400 Btu/gal
Natural gas	100,000 Btu/ccf	0.75	75,000 Btu/ccf
Fuel oil	138,000 Btu/gal	0.65	89,700 Btu/gal
Electricity	3,413 Btu/kwh	1.00	3,413 Btu/kwh

If the money saved is less than the annual cost of the solar collection system, solar collection will not pay for itself. Estimate the annual cost of your collection system as outlined below.

Increase return on solar investment by increasing your use of the solar equipment. Consider other applications for the heat from your solar system. See the *Low Temperature & Solar Grain Drying Handbook*, MWPS-22, for a discussion of solar grain drying. Note that it is difficult to design a solar energy collection system that works well for both high airflow/low temperature (e.g. grain drying) and low airflow/high temperature (e.g. home heating) applications.

Estimating Solar System Costs

Table 672-2 is a convenient form for summarizing your cost data.

First cost

First cost is the extra investment needed to collect and use solar energy. Include everything extra needed for solar: collector, vents, ducts, storage, fans, controls, shutters, wiring, caulking, fasteners, labor, etc. Deduct any applicable investment and energy tax credits—check with a tax specialist.

For estimating costs before construction, use plans and specifications and talk to those who have built solar systems. Note that first cost is the cost of adding a solar collector to a conventional building. Solar glazing, for example, may also serve as siding; consider only the cost difference over normal siding.

Yearly fixed cost

Annual capital recovery allows for the principal and interest payments on money used to pay for the system.

Eq 672-3.

$$CR = [(Vf - Vs) \div SPWF] + (Vs \times i)$$

CR = capital recovery, $/yr
Vf = first cost of the solar system, $
Vs = salvage value of the system after "N" years, $. Can be zero.
SPWF = series present worth factor, Table 672-3
i = interest rate paid on borrowed, or earned on saved, money

Most of the components of a solar collection system should last 10 years, which is a reasonable value for N. Because polyethylene covers last only a year, add their replacement cost to the annual repair cost.

Example:
A $10,000 solar system will be worth about $2,000 after 10 years. At 10% interest, the annual capital recovery cost is (Eq 672-3):

CR = [($10,000 - $2,000) ÷ 6.145] + ($2,000 x 0.10)
CR = [1,302] + (200) = $1,502/yr

Table 672-2. Cost summary.

First cost		
1. Materials	______	
2. Additional equipment (Ducts, controls, etc.)	+ ______	
3. Labor	+ ______	
4. Total construction cost	=	______
5. Less tax credits	- ______	
6. First cost:	=	______
Yearly fixed costs		
Expenses		
7. Capital recovery	______	
8. Property taxes, if any	+ ______	
9. Insurance	+ ______	
10. Total expenses	=	______
Tax savings		
11. Line 10 x tax bracket	______	
12. Yearly fixed cost (10 - 11)	=	______
Yearly variable costs		
Expenses		
13. Repairs, maintenance	______	
14. Energy to operate, if any	+ ______	
15. Total expenses	=	______
Tax savings		
16. Line 15 x tax bracket	______	
17. Yearly variable cost (15 - 16)	=	______
Total Net Yearly Cost (12 + 17)	=	______

Real estate and property tax information is available locally. Some states exempt solar installations from property taxes. You can roughly estimate these taxes at 1% of first cost.

Insurance information is available from your agent. Some companies do not insure roof-mounted liquid collectors. You can roughly estimate insurance costs at 1% of first cost.

Tax savings can be estimated by multiplying your annual expenses, which are usually deductible, by your income tax bracket. Tax savings apply to use in agricultural buildings but not in a home.

Yearly variable costs

Repairs and maintenance can be estimated at about 2% of first cost. Add the replacement cost for polyethylene and other short-lived materials.

Extra energy is required if extra fans are needed to run the solar system. Estimate the cost of the extra energy.

Eq 672-4.

FEC = W x HS x FC x 0.024

FEC = fan energy cost, $/season
W = watts used by fan
HS = heating season, days/year
FC = cost of electrical energy, $/kwh
0.024 = 24 hr/day ÷ 1000 watts/kw

Example:

If a fan draws 165 watts, runs 150 days/yr, and electricity costs $0.06:

FEC = 165 x 150 x 0.06 x 0.024
= $35.64 fan operating cost per heating season

Net yearly cost

Net yearly cost is the total annual cost of owning the solar system and will be as accurate as you were when estimating costs and expenses. If this cost is less than fuel savings, solar collection will pay for itself within the assumed 10-yr life of the system.

Annual fuel cost

You can estimate fossil fuel cost. First calculate energy value, EV, with Eq 672-2.

Eq 672-5.

FF$ = FFU x EV

FF$ = fossil fuel cost
FFU = fossil fuel used, Btu/season
EV = energy value Btu/unit, Eq 672-2

Table 672-3. Series present worth factors (SPWF).
Factors for computing annual cost of investment over "N" years of life at the interest rates shown.

N	6%	8%	10%	12%	14%	16%	18%	20%
1	0.943	0.926	0.909	0.893	0.877	0.862	0.847	0.833
2	1.833	1.783	1.736	1.690	1.647	1.605	1.566	1.528
3	2.673	2.577	2.487	2.402	2.322	2.246	2.174	2.106
4	3.465	3.312	3.170	3.037	2.914	2.798	2.690	2.589
5	4.212	3.993	3.791	3.605	3.433	3.274	3.127	2.991
6	4.917	4.623	4.355	4.111	3.889	3.685	3.498	3.326
7	5.582	5.206	4.868	4.564	4.288	4.039	3.812	3.605
8	6.210	5.747	5.335	4.968	4.639	4.344	4.078	3.837
9	6.802	6.247	5.759	5.328	4.946	4.607	4.303	4.031
10	7.360	6.710	6.145	5.650	5.216	4.833	4.494	4.192
11	7.887	7.139	6.495	5.938	5.453	5.029	4.656	4.327
12	8.384	7.536	6.814	6.194	5.660	5.197	4.793	4.439
13	8.853	7.904	7.103	6.424	5.842	5.342	4.910	4.533
14	9.295	8.244	7.367	6.628	6.002	5.468	5.008	4.611
15	9.712	8.559	7.606	6.811	6.142	5.575	5.092	4.675
16	10.106	8.851	7.824	6.974	6.265	5.668	5.162	4.730
17	10.477	9.122	8.022	7.120	6.373	5.749	5.222	4.775
18	10.828	9.372	8.201	7.250	6.467	5.818	5.273	4.812
19	11.158	9.604	8.365	7.366	6.550	5.877	5.316	4.843
20	11.470	9.818	8.514	7.469	6.623	5.929	5.353	4.870

Table 672-5. Average solar energy at Lemont, Illinois.
South-facing collectors.
Data are from Table 2-1, *Introduction to Solar Heating and Cooling Design and Sizing*, DOE/CS-0011, 1978.
Table columns are:
0° Horizontal.
20° Approximates 4/12 roof slope.
_° Approximate latitude—usually best for year-round heat collection.
_° Approximate latitude plus 15°—usually best for winter heating.
90° Vertical.

Lemont, IL (41.40° North Latitude)

Month	Avg. Temp. °F	Collector tilt angle 0°	20°	40°	60°	90°
		- -Radiation, Btu/day-ft²- -				
Jan	24.8	629	937	1153	1251	1159
Feb	28.4	854	1124	1288	1326	1144
Mar	35.6	1200	1412	1494	1436	1110
Apr	50.0	1436	1520	1476	1308	876
May	59.0	1830	1818	1667	1390	830
Jun	69.8	2036	1964	1752	1418	800
Jul	73.4	1940	1897	1713	1405	812
Aug	71.6	1789	1851	1756	1515	953
Sep	64.4	1414	1602	1644	1532	1116
Oct	53.6	975	1239	1386	1399	1169
Nov	39.2	578	801	950	1007	908
Dec	28.4	482	718	885	964	901

Table 672-4. Clear day solar energy at Lemont, Illinois.
Data are for average clear days—on exceptionally clear days, values 15% greater are possible. Values include direct plus diffuse radiation, Btu/hr-ft², on south-facing surfaces. This table was developed by MWPS with technique from *1978 ASHRAE Applications Handbook.* Measure collector tilt angles from horizontal. Surfaces with a latitude + 15° tilt angle receive the most energy during the space heating season. Use this table to calculate sunny day collector output at your latitude. Table times are solar times.

42° North Latitude

Date	AM	PM	Collector tilt angle 0°	4/12	Lat.	Lat. + 15°	90°
			Hourly clear day radiation, Btu/hr-ft²				
Jan 21	8	4	21	42	62	70	71
	9	3	74	121	163	177	166
	10	2	116	178	231	246	221
	11	1	143	213	272	286	252
	Noon		152	225	285	300	262
	Daily total Btu/day-ft²		863	1336	1744	1861	1685
Feb 21	7	5	5	9	12	14	14
	8	4	63	92	115	120	105
	9	3	120	165	199	204	168
	10	2	165	221	262	265	213
	11	1	193	257	301	303	241
	Noon		202	269	314	316	250
	Daily total Btu/day-ft²		1299	1761	2095	2133	1734
Mar 21	7	5	43	50	53	50	34
	8	4	109	130	138	132	91
	9	3	166	200	214	206	141
	10	2	210	254	273	262	181
	11	1	238	288	310	298	206
	Noon		248	300	323	311	214
	Daily total Btu/day-ft²		1785	2149	2305	2213	1524
Apr 21	6	6	22	17	8	7	4
	7	5	87	84	69	54	13
	8	4	150	157	144	124	55
	9	3	204	221	211	188	97
	10	2	246	271	265	239	131
	11	1	272	303	299	271	153
	Noon		281	313	310	282	160
	Daily total Btu/day-ft²		2248	2422	2306	2052	1072
May 21	5	7	1	0	0	0	0
	6	6	52	38	15	13	9
	7	5	115	104	75	51	13
	8	4	175	171	143	113	28
	9	3	225	230	205	171	65
	10	2	264	277	253	218	95
	11	1	289	306	285	247	115
	Noon		297	316	295	257	121
	Daily total Btu/day-ft²		2546	2573	2254	1892	777
Jun 21	5	7	7	4	4	3	2
	6	6	63	46	18	16	10
	7	5	124	109	76	49	14
	8	4	182	174	141	109	19
	9	3	231	231	200	164	53
	10	2	269	276	247	208	82
	11	1	292	304	277	236	100
	Noon		300	314	287	246	106
	Daily total Btu/day-ft²		2643	2609	2219	1822	672
Jul 21	5	7	1	0	0	0	0
	6	6	52	38	16	14	9
	7	5	114	103	75	51	14
	8	4	173	169	141	112	29
	9	3	223	227	201	169	64
	10	2	261	272	249	214	94
	11	1	285	301	280	243	113
	Noon		293	311	290	253	119
	Daily total Btu/day-ft²		2515	2537	2221	1865	769
Aug 21	6	6	21	16	8	7	5
	7	5	85	82	68	53	14
	8	4	147	153	139	120	54
	9	3	200	215	205	182	94
	10	2	240	264	257	231	127
	11	1	266	295	290	263	148
	Noon		275	305	301	274	155
	Daily total Btu/day-ft²		2198	2360	2242	1993	1043
Sep 21	7	5	41	47	48	46	31
	8	4	104	123	131	125	86
	9	3	160	192	205	196	135
	10	2	203	244	262	251	173
	11	1	230	278	298	286	197
	Noon		240	289	310	298	205
	Daily total Btu/day-ft²		1723	2061	2201	2110	1452
Oct 21	7	5	4	7	9	10	10
	8	4	60	86	107	112	97
	9	3	116	159	190	194	160
	10	2	160	214	252	255	205
	11	1	187	248	290	293	232
	Noon		197	260	303	305	241
	Daily total Btu/day-ft²		1256	1692	2004	2038	1653
Nov 21	8	4	20	40	59	66	67
	9	3	72	118	159	172	161
	10	2	114	175	226	240	216
	11	1	141	210	267	281	247
	Noon		150	222	280	294	257
	Daily total Btu/day-ft²		849	1309	1704	1815	1641
Dec 21	8	4	7	18	29	33	36
	9	3	55	99	141	156	153
	10	2	96	157	212	230	215
	11	1	122	192	254	273	249
	Noon		131	204	268	287	260
	Daily total Btu/day-ft²		693	1139	1544	1675	1570

GLOSSARY OF SOLAR TERMS

ABSORBER PLATE—Collector part that absorbs incident solar radiation, converts it to heat energy, and transfers the heat to the working fluid or radiates it to the heated space.

ABSORPTANCE—Measure of incident radiation absorbed by a material; fraction between 0 and 1.

ACTIVE SYSTEM—Solar system requiring pumps or fans to move the heat transfer fluid.

ALTITUDE, SOLAR—Angle between the sun's rays and a horizontal surface.

ANGLE OF INCIDENCE—Angle between the sun's rays and a line perpendicular to the surface on which sunlight is falling.

AZIMUTH, SOLAR—Angle between a north-south line and the horizontal projection of the sun's rays.

BACKPLATE—Back of a solar collector; part farthest from sun.

BTU—British thermal unit; unit of energy; amount of energy required to warm one lb of water one degree Fahrenheit.

CFM—Cubic feet per minute; unit of airflow.

COEFFICIENT OF THERMAL EXPANSION—Change in length of a material for each degree change in temperature.

COLLECTOR—Device to receive and absorb solar energy and convert it to heat.

COLLECTOR EFFICIENCY—Ratio of the useful energy gain for a given time period to the solar energy incident on the surface during the same time period.

CONCENTRATING COLLECTOR—Device which focuses direct radiation; obtains energy at higher temperatures than attainable with a flat plate collector.

CONDENSATE—Liquid formed when a vapor condenses.

CONDUCTION—Heat transfer through or between bodies in physical contact—involves no fluid motion.

CONDUCTIVITY—Measure of how readily heat moves through a material.

CONVECTION—Heat transfer by fluid motion.

COVER—Collector part that admits solar radiation to the absorber, shields the absorber from heat losses to the wind, and reduces longwave radiation losses.

DENSITY—Weight of a unit volume of a substance.

DIFFERENTIAL THERMOSTAT—Switch that makes or breaks contact when the temperature difference between two points exceeds or falls below the setpoint.

DIFFUSE RADIATION—Solar energy scattered by particles in the atmosphere; solar energy available on a cloudy day.

DIRECT GAIN—Solar heating by direct exposure to sunlight.

DIRECT RADIATION—Solar energy arriving without diffusion or scattering; also called direct beam radiation.

EMITTANCE—Measure of the tendency of a material to radiate or emit energy of a specified wavelength; fraction between 0 and 1.

ENERGY INTERCEPT AREA—For flat plate collectors with reflectors, the area of an imaginary rectangle perpendicular to the sun's rays, which contains all rays striking the collector and useful reflector area.

EQUINOX—Date when the earth's axis of rotation has a 0# tilt angle toward the sun; day and night are equal length all over the earth; about March 21 and September 21.

EUTECTIC SALTS—See phase change salts.

FIBER BLOOM—Exposure of glass fibers at the surface of FRP—caused by deterioration of the binding resin.

FIXED COLLECTOR—One that does not follow the sun.

FLAT PLATE COLLECTOR—Basic collector in solar heating systems; consists of a dark-colored plate, usually insulated on the bottom and edges, and often covered by one or more transparent covers.

FLUID—Any liquid or gas.

FOSSIL FUELS—Natural fuels formed from prehistoric plants and animals; e.g. coal, petroleum, natural gas.

FPM—Feet per minute; unit for air velocity.

FREESTANDING—Self-supporting; not mounted on or part of another structure.

FRP—Fiberglass reinforced plastic; materials with glass fibers imbedded in a polyester resin.

GLAUBERS SALT—Sodium sulfate decahydrate; latent heat storage material; melting point = 91 F and heat of fusion = 108 Btu/lb.

GLAZING—Transparent solar collector cover; transmits solar radiation and blocks longwave radiation.

GREENHOUSE EFFECT—Trapping heat inside a glass or plastic enclosure, or trapping heat by the earth's atmosphere, by reducing convective and radiative heat loss.

HEADER—Manifold; a larger diameter pipe connecting smaller tubes through a solar collector absorber or other piping system.

HEAT EXCHANGER—Device that transfers heat between fluids without direct fluid contact; usually metal tubes with one fluid inside and the other outside them.

HEAT OF FUSION—Latent heat required to melt a material or the heat released when it freezes; Btu/lb.

HEAT TRANSFER FLUID—Gas or liquid used to move heat within a solar system.

HEATING WATER—Water (or water plus antifreeze) that transfers heat in liquid heating systems.

INCIDENT ANGLE—Angle of incidence.

INFILTRATION—Inward air leakage through cracks and joints, e.g. at windows and doors, caused by wind pressure and differences in indoor and outdoor temperatures.

INSOLATION—Shortwave or solar energy; includes ultraviolet, visible, and infrared radiation; Btu/hr-ft^2.

LATENT HEAT—Energy absorbed or released by a material when it changes phases (e.g. from solid to liquid); no temperature change is involved.

LATITUDE—Distance north or south of the earth's equator; degrees.

LONGITUDE—Distance east or west of the prime meridian (Greenwich, England); degrees.

LONGWAVE RADIATION—Low energy radiation with wavelengths longer than 3 micrometers; the type of radiation emitted by solar collector absorber plates; thermal radiation.

LOW GRADE HEAT—Energy in materials at low temperatures or only slightly warmer than their surroundings.

NATURAL CONVECTION—Heat transfer caused by the density difference between hot and cold fluids.

NORMAL—Perpendicular; a line at right angles (90°) to a surface.

OPAQUE—Not transparent; does not let light through.

OUTGASSING—Release of gas or vapor from organic materials as they deteriorate.

PASSIVE SYSTEM—Solar system that relies on natural convection of the heat transfer fluid or direct exposure to sunlight.

PHASE CHANGE MATERIAL—Material that stores energy as latent heat.

PHASE CHANGE SALTS—Phase change materials that melt and freeze at temperatures that make them useful for storing heat in solar heating systems.

POTABLE—Fit for drinking.

PRESSURE DROP—Measures fluid friction or resistance to flow and affects required fan or pumping power.

REFLECTANCE—Measure of incident radiation that reflects off the surface of a material; fraction between 0 and 1.

RETROFIT—Adaptation of a building for a technical innovation, such as solar heating.

R-VALUE—Resistance of a material to heat flow; good insulators have high R-values; hr-ft^2-F/Btu.

SELECTIVE SURFACE—One for which longwave emittance is much less than shortwave absorptance; a high fraction of incoming solar energy is absorbed, but little heat energy is lost by longwave radiation.

SENSIBLE HEAT—Energy applied to raise the temperature of a material or the energy removed to cool it.

SERPENTINE—Snake-like; having a curving, winding shape.

SERVICE HOT WATER—Potable hot water for washing, cleaning, or cooking.

SHORTWAVE RADIATION—High energy radiation with wavelengths shorter than 3 micrometers; radiation emitted by very hot objects.

SIDEPLATE—The side of a solar collector; supports absorber and cover.

SOLAR CONSTANT—Average amount of solar energy available on a sun-following surface just outside the earth's atmosphere; about 428 Btu/hr-ft^2.

SOLAR NOON—Time at which the sun reaches its highest point in the sky (zenith) for the day. Solar noon varies from local clock noon depending on season, longitude, and whether Daylight Savings Time is in effect.

SOLAR TIME—Time based on the position of the sun.

SPECIFIC HEAT—Amount of heat required to raise the temperature of a unit mass of material one degree.

STAGNANT—Not running or flowing.

STAGNATION—No fluid movement through a solar collector; high temperatures can develop when the sun is shining.

SUMMER SOLSTICE—Longest day of the year in the northern hemisphere; the first day of summer; about June 21; the northern hemisphere has a maximum tilt toward the sun.

SUN-FOLLOWING—Tracks or follows the sun so the solar angle of incidence is always 0°.

THERMAL BREAK—Insulating material between two heat conductors which reduces conduction heat transfer.

THERMAL ENERGY—Heat energy.

TILT ANGLE—Angle between a horizontal surface and a collector surface.

TRACKING—Able to follow the movement of the sun.

TRANSMITTANCE—Measure of incident radiation that can pass through a material; fraction between 0 and 1.

TROMBE WALL—Concrete, brick, or adobe wall on the south side of passively heated solar structures; solar energy heats the wall during the day and heat is released to the structure at night.

UL—Underwriters' Laboratories.

ULTRAVIOLET—Radiation band with wavelengths from 0.001 to 0.4 micrometers; a high energy component of solar energy; breaks down some rubber and plastic materials.

WINTER SOLSTICE—Shortest day of the year in the northern hemisphere; first day of winter; about December 21; the northern hemisphere has a maximum tilt away from the sun.

673 SINGLE-PASS HEAT EXCHANGER DESIGN

Chapter prepared by R.L. Fehr, University of Kentucky, and D.P. Stombaugh, The Ohio State University.

Heat exchangers for recovering energy from barn exhaust ventilating air can effectively reduce heating costs. Several companies are now marketing heat exchangers for livestock buildings. Industrial units have been available for years.

The information needed to design a heat exchanger has been condensed into a series of simplified equations in this chapter. These equations can be solved in various sequences to determine any one of the critical design parameters. This chapter concentrates on determining the **plate area** required to obtain the desired heat exchanger performance.

Heat Exchanger Design

The equation for heat transfer through a heat exchanger plate is:

Eq 673-1.

$$Q = U \times A \times \Delta T_{lm}$$

Q = heat flow across the exchanger plate, Btu/hr
U = overall heat transfer coefficient, Btu/hr-ft²-F
A = exchanger plate area, ft²
ΔT_{lm} = log-mean temperature difference, F

To design or evaluate a heat exchanger, determine the design variables U, A, and ΔT_{lm}.

Overall Heat Transfer Coefficient

The overall heat transfer coefficient for heat transfer through a heat exchanger plate is:

Eq 673-2.

$$U = 1 \div (1/hce + \Delta x/Kw + 1/hci)$$

hce = convective heat transfer coefficient on the exhaust side, Btu/hr-ft²-F
Δx = exchanger plate thickness, ft
Kw = thermal conductivity of exchanger plate, Btu/hr-ft-F
hci = convective heat transfer coefficient on the intake side, Btu/hr-ft²-F

In most heat exchangers, plate conductance is high and does not significantly alter the overall heat transfer coefficient (U), so it is usually ignored. The problem is reduced to determining the convective heat transfer coefficient on each side of the plate (hce and hci). Then use hce and hci in Eq 673-2 to determine U. The hce and hci coefficients are usually determined with Nusselt's and Reynold's numbers.

For a given heat exchanger, calculate U as follows:

1. Calculate Reynold's number (Eq 673-3).

Eq 673-3.

$$Re = V \times Hd \times \rho \div \mu = V \times Hd \div \upsilon$$

Re = Reynold's number
V = air velocity, ft/sec
Hd = hydraulic diameter = 4 x area ÷ wetted perimeter, ft
ρ = air density, lb/ft³
= 0.086 − 0.00016 x T (for 0 F<T<100 F)
μ = dynamic air viscosity, lb/ft-sec
= (1.1012 + 0.0019 x T) x 10^{-5}
(for 0 F<T<100 F)
υ = kinematic viscosity, ft²/sec
= (12.75 + 0.0519 x T) x 10^{-5}
(for 0 F<T<100 F)

2. Calculate Nusselt's number from the Reynold's numbers (Eqs 673-4 to 673-7):

Eq 673-4.

$$N_{Nu} = 0.344 \times Re^{0.35} \text{ (for } 100<Re<2100)$$

Eq 673-5.

$$N_{Nu} = 0.000000168 \times Re^{2.25} \text{ (for } 2100<Re<2850)$$

Eq 673-6.

$$N_{Nu} = 0.00255 \times Re^{1.04} \text{ (for } 2850<Re<5650)$$

Eq 673-7.

$$N_{Nu} = 0.0198 \times Re^{0.8} \text{ (for } 5650<Re<100{,}000)$$

The relationship between Nusselt and Reynold's numbers is also shown in Fig 673-1.

3. Calculate the convective heat transfer coefficients (Eq 673-8):

Eq 673-8.

$$hc = N_{Nu} \times k \div Hd$$

hc = convective heat transfer coefficient, Btu/hr-ft²-F
N_{Nu} = Nusselt's number
k = air thermal conductivity, Btu/hr-ft-F
= 0.0138 + 0.00002 x T
(for 0 F<T<100 F)
Hd = hydraulic diameter = 4 x area ÷ wetted perimeter, ft

4. Calculate the overall heat transfer coefficient, U (Eq 673-2).

Log-Mean Temperature Difference

Given a heat exchanger, the log-mean temperature difference can be used to calculate the rate of heat transfer through the plate with Eq 673-1. It is a function of the air temperature differences between chambers at each end of the heat exchanger:

Eq 673-9.

$$\Delta T_{lm} = (\Delta T_2 - \Delta T_1) \div \ln(\Delta T_2/\Delta T_1)$$

ΔT_{lm} = log-mean temperature difference
ΔT_1 = temperature difference at end 1 of the heat exchanger, e.g. $t_i - t_E$ of Fig 673-2
ΔT_2 = temperature difference at end 2 of the heat exchanger, e.g. $t_x - t_o$ of Fig 673-2

When **designing** a heat exchanger, the temperatures at both ends are usually not known, so base the design on the heat exchanger's effectiveness, E. Effectiveness indicates actual heat transfer relative to the maximum possible heat transfer:

E = Qactual ÷ Qmaximum

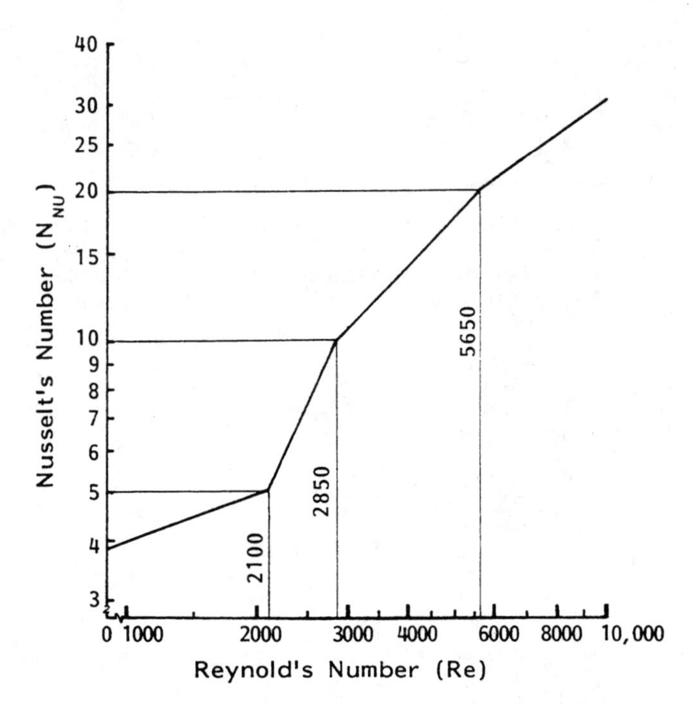

Fig 673-1. Relationship between Reynold's and Nusselt's numbers.

Effectiveness is usually calculated as the actual temperature change divided by the maximum possible temperature change. If intake < exhaust mass flowrate, effectiveness is:

Eq 673-10.

$$E = (t_E - t_o) \div (t_i - t_o)$$

- E = effectiveness, decimal
- t_E = air temperature to building from exchanger, F
- t_o = outside air temperature, F
- t_i = inside air temperature, F

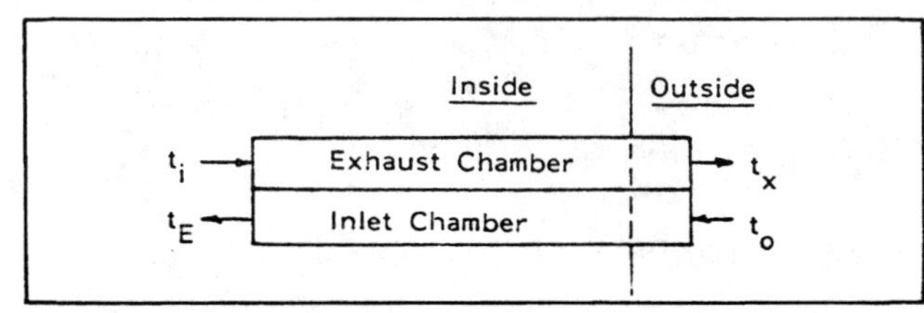

Fig 673-2. Terms for Eqs 673-10 and 673-11.

If exhaust < intake mass flowrate, effectiveness is:

Eq 673-11.

$$E = (t_i - t_x) \div (t_i - t_o)$$

- t_x = air temperature to outside from exchanger, F

If the mass flowrates are equal, Eqs 673-10 and 673-11 are equal. Effectiveness is generally considered equal to heat exchanger efficiency. However, effectiveness calculated from temperatures does not account for condensation. With condensation, the ratio of enthalpy changes in the exhaust and intake air streams does not equal the ratio of air temperature changes.

When designing a heat exchanger, the temperature differences at each end of the heat exchanger are usually not known, so E is determined by a parameter called the **number of heat transfer units (NTU).**

Eq 673-12.

$$NTU = A \times U \div Cmin$$

- NTU = number of heat transfer units, dimensionless
- Cmin = minimum mass flowrate times specific heat, Btu/hr-F

If NTU is known, calculate effectiveness from the equations in Table 673-1. If the desired effectiveness is known and plate area is unknown, then determine NTU from the other equations in Table 673-1.

Table 673-1. Formulas for E and NTU.

For parallel, counter, and crossflow heat exchangers. The crossflow equation assumes that each air stream is divided into a large number of separate channels passing through the heat exchanger with no cross mixing.
E = effectiveness
NTU = number of heat transfer units
Cmin = minimum mass flowrate x specific heat, Btu/hr-F
Cmax = maximum mass flowrate x specific heat, Btu/hr-F
CC = Cmin ÷ Cmax
ln = log base e

Parallel flow:

$$E = [1 - e(-NTU \times (1 + CC)] \div (1 + CC)$$
$$NTU = -\ln[1 - E \times (1 + CC)] \div (1 + CC)$$

Counterflow:

$$E = \{1 - e[-NTU \times (1 - CC)]\} \div \{1 - CC \times e[-NTU \times (1 - CC)]\}$$
$$NTU = -\ln[(1 - E) \div (1 - E \times CC)] \div (1 - CC)$$
$$NTU = E \div (1 - E), \text{ if } CC = 1$$

Crossflow:

$$E = 1 - e\{(CC \times NTU^{0.22}) \times [e(-NTU^{0.78} \times CC) - 1]\}$$

NTU = Solve by iteration of the equation for E.

The equations in Table 673-1 relate effectiveness and heat transfer units for any ratio of flow on either side of the plate for counter, parallel, and crossflow exchangers. In livestock buildings, fans are usually needed on both the exhaust and intake air chambers due to building air leakage.

For equal flow on both sides of an exchanger plate, Fig 673-3 relates desired effectiveness and the required NTUs. For units with low effectiveness, flow direction has little effect on the required number of NTUs. Crossflow and counterflow heat exchangers have similar effectiveness when E < 0.50. To achieve a 10% effectiveness increase, the number of heat transfer units, and consequently the plate area, must be increased dramatically when E > 0.5 is desired. Effectiveness of a parallel flow unit cannot be greater than 0.5.

Plate Area

When determining required plate area, NTU cannot be calculated from Eq 673-12. Estimate a value for E and calculate NTU from the equations in Table 673-1. Invert Eq 673-12 to calculate plate area:

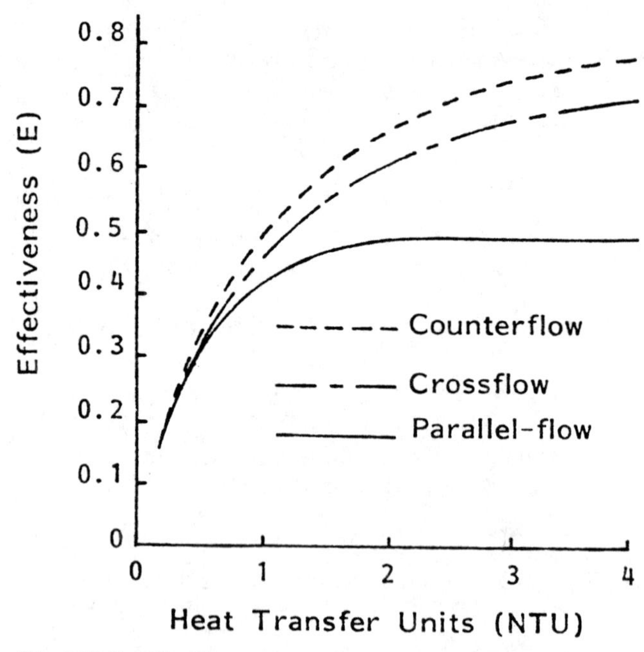

Fig 673-3. Effectiveness vs. heat transfer units.
For parallel, counterflow, and crossflow heat exchangers having equal flow on both sides of the exchanger plate.

Eq 673-13.

$$A = NTU \times Cmin \div U$$

To calculate required plate area for a heat exchanger:

1. Select the desired effectiveness, flow directions, exhaust flow rate, intake flowrate, width and height of flow channels, and approximate average air temperature on each side of the plate.
2. Calculate the required NTU (Table 673-1).
3. Calculate Re (Eq 673-3).
4. Calculate N_{Nu} (Eqs 673-4 to 673-7).
5. Calculate hc for each side of the plate (Eq 673-8).
6. Calculate U (Eq 673-2).
7. Calculate the required plate area, A (Eq 673-13).
8. Iterate; i.e. if the area required is unrealistically high, change the configuration to increase the heat transfer coefficients or reduce the desired effectiveness.

This procedure can also determine factors other than plate area, such as the effectiveness of a given heat exchanger.

Example 1:

Determine the required length (plate area) of a flat plate, counterflow, heat exchanger to provide an effectiveness of 0.7. The collector is 4′ wide, 7″ high (3.5″ each for the inlet and exhaust channels) and has inlet and exhaust flowrates of 400 cfm, Fig 673-4. Based on the desired effectiveness, the average temperature of the inlet and exhaust airflows is assumed to be 25 F and 45 F, respectively.

First calculate the required NTU from Table 673-1 for a counterflow unit and Cmin = Cmax:

$$NTU = E \div (1 - E) = 0.7 \div (1 - 0.7) = 2.33$$

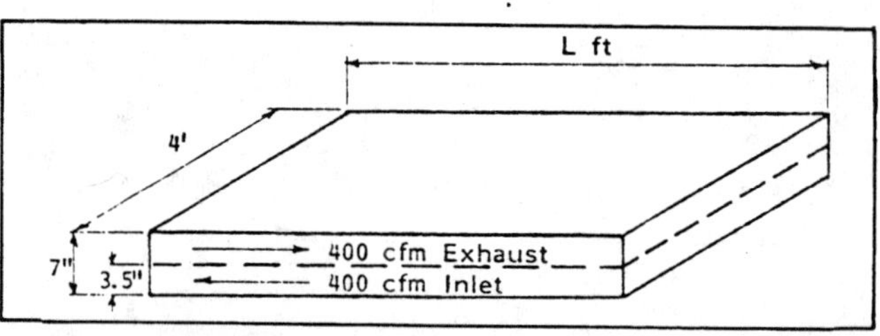

Fig 673-4. Heat exchanger in Example 1.
E = 0.7. Average temperature of inlet chamber = 25 F. Average temperature of exhaust chamber = 45 F.

Or read NTU directly from Fig 673-3.

Next calculate Reynold's number for both the inlet, Re_I, and exhaust, Re_E, flows:

$$Re = V \times Hd \div \upsilon$$

$$V = Q \div A = 400 \text{ cfm} \div (4' \times 0.2917')\text{ ft}^2 = 343 \text{ ft/min} = 5.714 \text{ ft/sec}$$

$$Hd = 4 \times (0.2917' \times 4') \div (0.2917' + 4' + 4' + 0.2917') = 0.544'$$

$$\upsilon_I = (12.75 + 0.0519 \times T) \times 10^{-5} = (12.75 + 0.0519 \times 25 \text{ F}) \times 10^{-5} = 14.05 \times 10^{-5} \text{ ft}^2\text{/sec}$$

$$\upsilon_E = (12.75 + 0.0519 \times 45 \text{ F}) \times 10^{-5} = 15.09 \times 10^{-5} \text{ ft}^2\text{/sec}$$

Thus:

$$Re_I = (5.714 \text{ ft/sec} \times 0.544') \div (14.05 \times 10^{-5} \text{ ft}^2\text{/sec}) = 22{,}124$$

$$Re_E = (5.714 \text{ ft/sec} \times 0.544') \div (15.09 \times 10^{-5} \text{ ft}^2\text{/sec}) = 20{,}600$$

From Re, determine the inlet and exhaust N_{Nu} numbers using Eq 673-7.

$$N_{Nu} = 0.0198 \times Re^{0.8}$$

$$N_{NuI} = 0.0198 \times (22{,}124)^{0.8} = 59.23$$

$$N_{NuE} = 0.0198 \times (20{,}600)^{0.8} = 55.94$$

Next, calculate the heat transfer coefficients, hci and hce, from Eq 673-8:

$$hc = N_{Nu} \times k \div Hd$$

$$hci = 59.23 \times (0.0138 + 0.00002 \times 25 \text{ F}) \div 0.544' = 1.56 \text{ Btu/hr-ft}^2\text{-F}$$

$$hce = 55.94 \times (0.0138 + 0.00002 \times 45 \text{ F}) \div 0.544' = 1.51 \text{ Btu/hr-ft}^2\text{-F}$$

Calculate the overall heat transfer coefficient, U, with Eq 673-2. Assume the plate resistance is negligible (i.e. $\Delta x/Kw = 0$).

$$U = 1 \div (1 \div hce + 1 \div hci) = 1 \div (1 \div 1.51 + 1 \div 1.56) = 0.767 \text{ Btu/hr-ft}^2\text{-F}$$

Finally, calculate the required plate area with Eq 673-13.

$$A = NTU \times Cmin \div V$$

$$Cmin = 400\ ft^3/min \times 60\ min/hr \times (1\ lb/13.3\ ft^3) \times 0.24\ Btu/lb\text{-}F = 433\ Btu/hr\text{-}F$$

Thus: $$A = 2.33 \times 433\ Btu/hr\text{-}F \div 0.767\ Btu/hr\text{-}ft^2\text{-}F = 1315\ ft^2$$

Because the plate is 4′ wide, the required length of the heat exchanger is 1315 ft² ÷ 4′ = 329′, which is unrealistic.

To reduce required plate area, increase flow velocity (by reducing flow channel height e.g. from 3.5″ to 1.5″) or reduce the desired effectiveness.

If the desired effectiveness remains at 0.7, but the flow area is reduced to 1.5″ x 4′:

$$Hd = 4 \times (0.125' \times 4') \div (0.125' + 4' + 0.125' + 4') = 0.242'$$

$$V = 400\ ft^3/min \div (0.125' \times 4')\ ft^2 = 800\ ft/min = 13.33\ ft/sec$$

$$Re_I = (13.33\ ft/sec \times 0.242') \div (14.05 \times 10^{-5}\ ft^2/sec) = 22{,}960$$

$$Re_E = (13.33\ ft/sec \times 0.242') \div (15.09 \times 10^{-5}\ ft^2/sec) = 21{,}377$$

$$N_{NuI} = 0.0198 \times (22{,}960)^{0.8} = 61.02$$

$$N_{NuE} = 0.0198 \times (21{,}377)^{0.8} = 57.63$$

$$hci = 61.02 \times (0.0138 + 0.00002 \times 25\ F) \div 0.242' = 3.61\ Btu/hr\text{-}ft^2\text{-}F$$

$$hce = 57.63 \times (0.0138 + 0.00002 \times 45\ F) \div 0.242' = 3.50\ Btu/hr\text{-}ft^2\text{-}F$$

$$U = 1 \div (1 \div 3.61 + 1 \div 3.50) = 1.78\ Btu/hr\text{-}ft^2\text{-}F$$

$$A = 2.33 \times 433\ Btu/hr\text{-}F \div 1.78\ Btu/hr\text{-}ft^2\text{-}F = 567\ ft^2$$

Required length = 567 ft² ÷ 4′ = 142′, a significant reduction. Reducing the flow area also increases power requirements and initial fan investment.

Reducing effectiveness from 0.7 to 0.4 can further reduce required plate area. With the same average temperatures as before and a flow area of 1.5″ x 4′:

$$NTU = E \div (1 - E) = 0.4 \div (1 - 0.4) = 0.67$$

$$A = NTU \times CMIN \div U = 0.67 \times 433\ Btu/hr\text{-}F \div 1.78\ Btu/hr\text{-}ft^2\text{-}F = 163\ ft^2$$

Required length = 41′, which reduces construction costs substantially. Further iterations to optimize design are desirable.

Power Requirements

To compare heat exchanger designs, also consider power requirements for their operation. The friction loss due to airflow between parallel plates is a function of length, velocity, and plate spacing.

Also consider pressure losses at the exchanger entrance and exit. These losses depend on the duct entrance and exit configurations and are often small compared with the pressure losses between the plates. Once the flowrates and pressure losses are estimated, select two fans (one for inlet air and one for exhaust air) and estimate their power requirements.

Eqs 673-3 through 673-8, show that you can change the convective heat transfer coefficient (hc) primarily by changing the velocity and the hydraulic diameter (Hd). For a given velocity, increasing Hd decreases hc. One method to increase Hd is to keep the same cross-sectional area but subdivide it with more plates. Increasing velocity decreases the thickness of the air surface film on the exchanger plate, which increases hc.

Condensation and Dust

Condensation in a heat exchanger slightly increases efficiency. As condensation occurs, one of three conditions exists:

- All the heat released from condensation is transferred across the plate.
- All the heat released from condensation remains on the exhaust side of the plate and helps maintain the temperature of the air stream.
- Some combination of the above two conditions.

Condensation may alter the exchanger's overall heat transfer coefficient, but it is more likely that increased efficiency results from the higher temperature being maintained in the exhaust air stream. The higher the temperature, the larger the temperature difference, and the higher the heat flow across the exchanger plate.

Evaluating the effect of condensation on heat exchanger efficiency requires fixed inside and outside air conditions. Inside air moisture content affects when condensation starts. Lower outside air temperatures cause the exhaust air to cool more rapidly to its dew point temperature. The greater the exchanger plate area where condensation occurs, the larger the increase in efficiency.

The convective heat transfer equations do not allow for dust accumulation or obstructions placed in the air stream to increase turbulance. A dust layer adds another thermal resistance term to Eq 673-2, which decreases heat transfer.

The effects of condensation and dust accumulation tend to offset each other. Until research yields better analysis techniques, they are often ignored in the theoretical equations.

Once an exchanger is designed, the expected seasonal energy savings for a given effectiveness can be obtained using the bin method. An example calculation for a heat exchanger with an effectiveness of 0.4 is given in Chapter 674, Example 4.

674 ESTIMATING SEASONAL ENERGY CONSUMPTION

By D.P. Stombaugh, The Ohio State University.

Reliable methods to estimate a building's seasonal energy use are needed to evaluate expected operating costs and possible savings from energy conservation methods. For example, the value of additional insulation or preheated ventilating air depends on the expected reduction in energy use during the heating season.

Methods to predict energy use during the heating season include the degree-day procedure, the bin or temperature frequency method, and more complex simulation methods. This chapter summarizes the degree-day and bin methods, but not simulation. Before describing the degree-day and bin methods, the concepts of infiltration and balance point temperatures are discussed.

Infiltration

Chapter 633 described energy balances to calculate supplemental heat for an animal building. The total instantaneous heating load is:

Eq 674-1.

$$q_{sr} = q_v + q_b - q_s$$

q_{sr} = supplemental heat required, Btu/hr
q_v = heat lost in ventilating air, Btu/hr
q_b = heat lost through the building envelope, Btu/hr
q_s = internal sensible heat production, Btu/hr

q_v = CFM x (60 ÷ 13.33) x 0.24 x (t_i − t_o)
= CFM x 1.1 x (t_i − t_o)
CFM = ventilating rate, ft³/min
0.24 = specific heat of dry air, Btu/lb da-F
13.33 = assumed specific volume, ft³/lb da
60 = conversion factor, 60 min/hr
t_i = inside design temperature, F
t_o = outside design temperature, F

q_b = (A/R) x (t_i − t_o)
A/R = building heat loss factor, Btu/hr-F

You can use Eq 674-1 to calculate residence heat load, except q_s is usually ignored and ventilation heat loss is replaced by air infiltration. Residences are not ventilated, but cold air infiltrates through doors, windows, and cracks in the house.

A rough, quick method to estimate air infiltration is to assume an air change rate. For most room types, an air volume/hr equal to ½ to 3 times the room volume enters by infiltration, while an equivalent volume of air leaves. Allow at least ½ air change/hr for most well constructed residences. A few new residences constructed with special techniques may have only ¼ air change/hr.

A more precise method is to determine the individual room air infiltrations from Table 674-1. The residence infiltration rate is the sum of the individual room air change rates. Construction quality, weather conditions, room use, and the number of people entering and leaving all affect the air change rate. Adjust the values in Table 674-1 to account for these conditions.

Convert the air change/hr value to cfm. For example, for a residence volume = 20,000 ft³ and one air change/hr the infiltration rate = 20,000 ft³/hr. This is equivalent to a ventilating rate of 20,000 ft³/hr ÷ 60 min/hr = 333 cfm.

Table 674-1. Residential infiltration rates.
For rooms with weatherstripped windows and doors, or storm sashes and storm doors, multiply these values by ⅔.

Room type	Air changes/hr
No windows or exterior doors	0.5
Windows or exterior doors in one wall	1.0
Windows or exterior doors in two walls	1.5
Windows or exterior doors in three walls	2.0
Entrance halls	2.0

Once the infiltration rate is determined, the house design heating load can be calculated by:

Eq 674-2.

$$H_L = A/R \times (t_i - t_o) + CFM \times 1.1 \times (t_i - t_o)$$

H_L = design heating load, Btu/hr
CFM = infiltration rate, ft³/min

Balance Point Temperature

The balance point temperature (t_b) is the outdoor temperature at which energy losses from a building equal the energy gains in the building. When it is colder than the balance point temperature, supplemental heat is required to maintain the indoor design temperature. Above the balance point temperature, internal heat production (lights, equipment, occupants, animals, solar heat gain) maintains the indoor temperature at or above the indoor design temperature.

At t_b, heat loss by conduction and ventilation (or infiltration) equal internal heat input:

Eq 674-3.

$$q_s = A/R \times (t_i - t_b) + CFM \times 1.1 \times (t_i - t_b)$$

or

$$q_s = [A/R + (1.1 \times CFM)] \times (t_i - t_b)$$

Solve for t_b, from Eq 674-3:

Eq 674-4.

$$t_b = t_i - [q_s \div (A/R + 1.1 \times CFM)]$$

t_b = balance point temperature, F
t_i = indoor design temperature, F
q_s = internal sensible heat from lights, equipment, animals, etc., Btu/hr
A/R = conduction heat loss term, Btu/hr-F
CFM = ventilation or infiltration rate, ft³/min

Example 1:

Calculate t_b for a farrowing building with 400 cfm ventilation and 22,000 Btu/hr sensible heat from animals and creep heaters. The A/R value for the building is 190 Btu/hr-F and the indoor temperature is 70 F. From Eq 674-4, t_b is:

$$t_b = 70 - [22{,}000 \div (190 + 1.1 \times 400)]$$
$$= 70 - 34.9$$
$$= 35.1 \text{ F}$$

Supplemental heat is not required until the outdoor temperature falls below 35.1 F. Below 35.1 F, the supplemental heat needed increases linearly with the difference between indoor and outdoor temperatures.

Residences usually have much less internal heat production than animal buildings, so t_b is typically much higher (50-65 F).

Degree-Day Procedure

Tests conducted in the 1930s and 1940s showed that fuel consumption in residences and public buildings varied directly with the difference between average daily outside temperature and 65 F. When the outside temperature was 65 F or above, practically no heat was required and fuel consumption approached zero. It was also found that fuel consumption doubled if an average daily outside temperature of 55 F (10 F difference) changed to 45 F (20 F difference). The average daily outdoor temperature was calculated as the average of the maximum and minimum daily temperatures. These averages are available from weather stations throughout the country.

The difference between 65 F and the average daily outside temperature is an index of heating requirements and is the basis for the degree-day method. One **degree-day** accumulates for every degree the average daily (24 hr) outside temperature is below 65 F. If the average outside temperatures for 5 days are 35, 40, 50, 68, and 55 F, then 80 degree-days (30 + 25 + 15 + 0 + 10) accumulates for the 5 day period.

Degree-days vary greatly from place to place and slightly from year to year. Average heating season degree-days for various locations are shown in Fig 674-1. A normal heating season in the United States is considered to be from October 1 to May 1 (212 days or 5,088 hr), although it varies throughout the country.

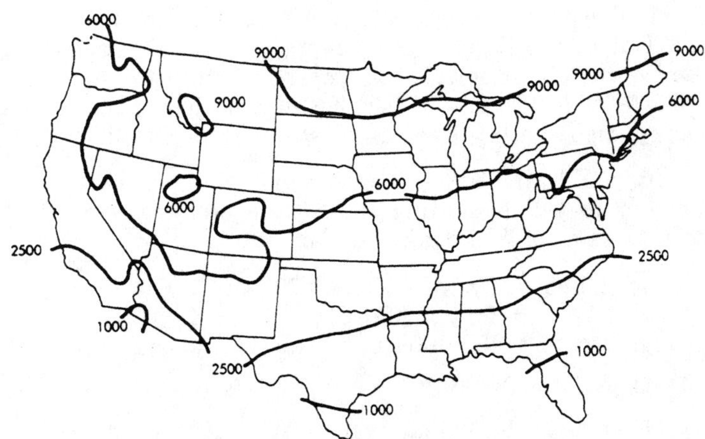

Fig 674-1. Winter degree days.

The degree-day procedure is simple, but gives only a rough estimate. Inaccuracies result from minor variations in degree-days from year to year, local wind conditions, unusual exposures, and wide variations in internal heat sources. Do not use this method for animal buildings, which have large internal heat production and low balance point temperatures.

Estimate the energy required during the heating season or a shorter period (one month, etc.) from:

Eq 674-5.

$$E = C_D \times [(H_L \times D \times 24) \div (\Delta t \times k \times V)]$$

- E = energy consumption for the season, Btu/season
- H_L = design heating load (Eq 674-2), Btu/hr
- D = number of degree-days for the season or period, F-day
- 24 = hr/day
- $\Delta t = t_i - t_o$ = design temperature difference, F
- k = correction factor to account for heating unit inefficiency
- V = conversion factor to obtain H_L in other units. See Table 674-2.
- C_D = correction factor to account for differences between the actual balance point temperature and the degree-day base temperature of 65 F.

The correction factor, k, varies dramatically with the equipment type, the control system, and whether the heating system is operating at full or partial load. For electric resistance heating systems, use k = 1.0. With natural gas and oil-fired heaters, k = 0.55 is typical.

Table 674-2. Unit conversion factors, V.
To determine energy use in the desired units. For example, V = 1000 Btu/ft³ of natural gas gives H_L in ft³ of natural gas.

Desired units	V
ft³ of natural gas	1,000 Btu/ft³
gal of LP gas	93,000 Btu/gal
gal of fuel oil	138,000 Btu/gal
kw-hr of electricity	3,413 Btu/kw-hr
$	No. of Btu/$

The degree-day base for most residences was set at 65 F in the 1930s and 1940s. Today, residences have much higher insulation levels and sensible heat production has increased dramatically. Balance point temperatures are now substantially below the original degree-day baseline temperature of 65 F.

There are two ways to adjust for this difference. The first is to use Eq 674-5 with C_D = 1.0, but determine the number of degree-days (D) with a base temperature equal to the balance point temperature. However, it is difficult to accurately calculate the balance point temperature for a residence.

The second is to use the empirical correction factor, C_D. The correction factor roughly accounts for internal and solar heat gains and varies with the severity of the climate. Typical values for C_D are in Fig 674-2.

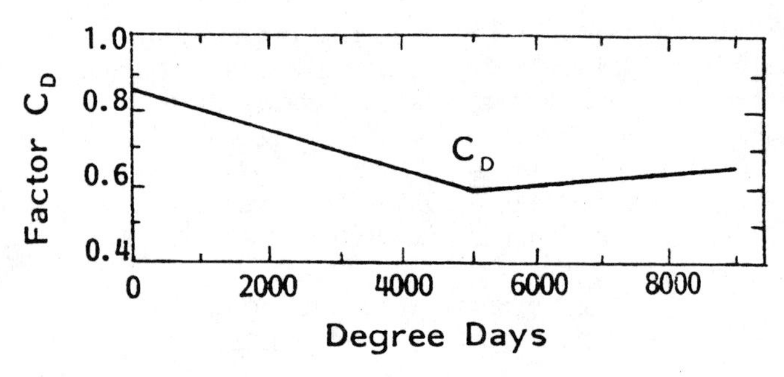

Fig 674-2. Correction factor, C_D, vs. degree-days.

Example 2:

Find the energy consumption of a residence in Columbus, Ohio both before and after installing storm windows and doors. The house has a conventional, natural gas, forced air furnace; initial A/R = 450 Btu/hr-F; initial air infiltration rate = 400 cfm; and interior volume = 16,000 ft^3. After installing storm doors and windows, A/R = 405 Btu/hr-F and infiltration rate = 267 cfm.

The initial energy consumption is:

$$H_L = A/R \times (t_i - t_o) + CFM \times 1.1 \times (t_i - t_o)$$
$$= 450 \times (70 - 5) + 400 \times 1.1 \times (70 - 5)$$
$$= 29{,}250 + 28{,}600$$
$$= 57{,}850 \text{ Btu/hr}$$

D = 5,660 degree-days, for Columbus, Ohio
$\Delta t = t_i - t_o$
= (70 − 5) F, 97½% design temperature for Columbus, Ohio
k = 0.55, for a standard, natural gas furnace
V = 1.0, for answer in Btu
C_D = 0.6, from Fig 674-2

$$E = 0.6 \times [(57{,}850 \times 5{,}660 \times 24) \div ((70 - 5) \times 0.55 \times 1.0)]$$
$$= 131{,}888{,}291 \text{ Btu/season}$$

If the cost of natural gas is 150,000 Btu/$, the cost to heat this residence throughout the heating season is:

Cost = 131,888,291 Btu ÷ 150,000 Btu/$ = $879

By installing storm doors and windows, the only term that changes is H_L.

$$H_L = A/R \times (t_i - t_o) + CFM \times 1.1 \times (t_i - t_o)$$
$$= 405 \times (70 - 5) + 267 \times 1.1 \times (70 - 5)$$
$$= 26{,}325 + 19{,}091$$
$$= 45{,}416 \text{ Btu/hr}$$

Seasonal energy use after installing storm doors and windows is:

$$E = 0.6 \times \{(45{,}416 \times 5{,}660 \times 24) \div [(70 - 5) \times 0.55 \times 1.0]\}$$
$$= 103{,}540{,}858 \text{ Btu}$$

With natural gas cost = 150,000 Btu/$, the revised winter heating cost is:

Cost = 103,540,858 Btu ÷ 150,000 Btu/$ = $690

Degree-day energy estimates are much more accurate if past operating expenses are known. For example, if the home energy costs in Example 2 are known for the preceding winter, use Eq 674-5 to get a better estimate of C_D. Exclude any natural gas for clothes drying, cooking, and hot water heating from these calculations.

Bin Method

In the bin method, steady-state energy balance calculations yield supplemental heat needs at several outdoor temperature increments (bins). Seasonal energy use is the sum of energy use times hours of occurrence for each bin temperature. This method is reliable for animal buildings and requires only a few calculations. This procedure can also be modified to analyze the performance of partly loaded heating equipment, weather related variations in capacity and performance of heat pumps, and other energy saving techniques.

The bins usually have a 5 F range. For example, calculate supplemental heat for t_o = 32 F. Assume the same heating load whenever t_o is between 29.5 F and 34.5 F. Multiply this supplemental heat value (Btu/hr) by the number of hours the temperature is between 29.5 and 34.5 F, to get the energy required during these hours. The number of hours in each bin are shown in Table 674-5.

Example 3:

Calculate the seasonal heating requirements of a farrowing building near Columbus, Ohio. Indoor temperature is 70 F, A/R = 240 Btu/hr-F, and minimum winter ventilating rate is 400 cfm. Average sensible heat from pigs, creep heaters, and lights is 15,000 Btu/hr.

First calculate the balance point temperature, Eq 674-4:

$$t_b = 70 - [15{,}000 \div (240 + 1.1 \times 400)]$$
$$= 47.9 \text{ F}$$

When t_o is above 47.9 F, no supplemental heat is required. Therefore, consider only the 47 F bin (44.5 to 49.5 F) and below, Table 674-5. The 47 F bin occurs for 603 hr, but heat is needed only when t_o is less than 47.9 F. The outside temperature is between 44.5 and 47.9 F about [603 x (47.9 − 44.5) ÷ 5] = 410 hr. During this 410 hr period, the average temperature is [(47.9 + 44.5) ÷ 2] = 46.2 F. Estimate total winter energy use from Eq 674-1 as shown in Table 674-3.

Divide E by heating equipment inefficiency to get fuel purchased. Assuming a 70% efficient heating unit, the Btu of fuel consumed is:

Btu consumed = 37,162,640 ÷ 0.70
= 53,089,486 Btu

If energy costs 150,000 Btu/$, the seasonal cost is:

Cost = 53,089,486 Btu ÷ 150,000 Btu/$ = $354

If heating unit efficiency varies with outdoor temperature, divide E for each bin by the efficiency,

Table 674-3. Seasonal energy use, E, for Example 3.
For a farrowing house using the bin method.
Example calculations:
$t_i - t_o = 70 - 46.2 = 23.8$ F
$q_v = 1.1 \times CFM \times (t_i - t_o) = 1.1 \times 400 \times 23.8 = 10{,}472$ Btu/hr.
$q_b = A/R \times (t_i - t_o) = 240 \times 23.8 = 5{,}712$ Btu/hr.
$q_s = 15{,}000$ Btu/hr (given).
$q_{sr} = q_v + q_b - q_s = 10{,}472 + 5{,}712 - 15{,}000 = 1184$ Btu/hr.
$E = q_{sr} \times$ Hours (from Table 674-5) $= 1184 \times 410 = 485{,}440$ Btu/season

Bin F	$t_i - t_o$ F	q_v Btu/hr	q_b Btu/hr	q_{sr} Btu/hr	Occurrence hrs	E Btu/season
46.2	23.8	10,472	5,712	1,184	410	485,440
42	28	12,320	6,720	4,040	658	2,658,320
37	33	14,520	7,920	7,440	730	5,431,200
32	38	16,720	9,120	10,840	772	8,368,480
27	43	18,920	10,320	14,240	502	7,148,480
22	48	21,120	11,520	17,640	280	4,939,200
17	53	23,320	12,720	21,040	169	3,555,760
12	58	25,520	13,920	24,440	94	2,297,360
7	63	27,720	15,120	27,840	40	1,113,600
2	68	29,920	16,320	31,240	20	624,800
-3	73	32,120	17,520	34,640	10	346,400
-8	78	34,320	18,720	38,040	4	152,160
-13	83	36,520	19,920	41,440	1	41,440
					Total =	37,162,640

rather than the total E. Heat pumps are evaluated in this way.

The effects of changing number of animals in the building, A/R value, or minimum winter ventilating rate can be examined by changing the appropriate variable. Complex energy conservation techniques can also be evaluated, such as preheating ventilating air in earth tubes.

Example 4:

Evaluate the building of Example 3 with a homemade heat exchanger. The flowrates of both inlet and exhaust air through the exchanger are 400 cfm. The building is near Columbus, Ohio; A/R = 240 Btu/hr-F; q_s = 15,000 Btu/hr; and t_i = 70 F.

The heat exchanger is designed for an effectiveness of 0.40. From the known effectiveness, you can estimate the temperature of the warmed, incoming air for any outdoor temperature. With equal mass flowrates entering and leaving through the heat exchanger, the effectiveness, E, is:

Eq 674-6.

$$E = (t_E - t_o) \div (t_i - t_o)$$

E = heat exchanger effectiveness
t_E = temperature of air entering the building after passing through the exchanger, F
t_i = inside design temperature, F
t_o = outdoor temperature, F

With E = 0.4, the entering temperature is:

Eq 674-7.

$$t_E = 0.4 \times (t_i - t_o) + t_o$$
$$t_E = (0.4 \times t_i) + (0.6 \times t_o)$$

At t_b, internal heat input equals heat loss by conduction and ventilation.

Eq 674-8.

$$q_s = A/R \times (t_i - t_o) + (1.1 \times CFM) \times (t_i - t_E)$$

or

$$q_s = A/R \times (t_i - t_o) + (1.1 \times CFM) \times [t_i - (0.4 \times t_i + 0.6 \times t_o)]$$

Substitute t_b for t_o and solve for t_b (the t_o at which balance occurs).

Eq 674-9.

$$q_s = A/R \times (t_i - t_b) + (1.1 \times CFM) \times [t_i - (0.4 \times t_i + 0.6 \times t_b)]$$

or

$$q_s = A/R \times (t_i - t_b) + (1.1 \times CFM) \times 0.6 \times (t_i - t_b)$$

$$t_b = t_i - q_s \div [A/R + 0.6 \times (1.1 \times CFM)]$$
$$= 70 - 15{,}000 \div [240 + 0.6 \times (1.1 \times 400)]$$
$$= 40.2 \text{ F}$$

The heat exchanger drops t_b from 47.9 F (Example 3) to 40.2 F, because the temperature of the ventilating air entering the building is higher than the outdoor temperature.

Using Eq 674-7 to obtain t_E, calculate supplemental heat, q_{sr}, as shown in Table 674-4. Note that t_E replaces t_o in the term for ventilating air heat loss.

Eq 674-10.

$$q_{sr} = (1.1 \times CFM) \times (t_i - t_E) + A/R \times (t_i - t_o) - q_s$$

If heat exchanger effectiveness varies with outdoor temperature, calculate t_E for each bin using the appropriate effectiveness.

For the 42 F bin (39.5 to 44.5 F), which occurs for 658 hr/year, heat is required only when $t_o < 40.2$ F. The outside temperature, t_o, is below 40.2 F for about [658 x (40.2 − 39.5) ÷ 5] = 92 hr. During this 92 hr period, the temperature averages [(40.2 + 39.5) ÷ 2] = 39.8 F.

If the heating unit efficiency is 70%, the fuel energy required is:

Btu consumed = 15,053,968 ÷ 0.70
= 21,505,669 Btu

If the cost of this energy is 150,000 Btu/$, the seasonal energy cost with the exchanger is:

Cost = 21,505,669 Btu ÷ 150,000 Btu/$ = $143

Comparing Examples 3 and 4, the exchanger saves $354 − $143 = $211/yr.

You can estimate the heating reduction at any outdoor temperature. For example, if t_{bin} = 7 F, the heat exchanger reduces the heating load [100 x (27,840 − 16,752) ÷ 27,840] = 40%. (These values are from Tables 674-3 and 674-4 at bin = 7). Combine this with the effect of changing t_b from 47.9 to 40.2 F, and the total reduction in energy use is [100 x (53.1 − 21.5) ÷ 53.1] = 60%.

Temperature data for the bin method are also available separated into three daily 8 hr periods with mean coincident wet-bulb temperature data. With this data, additional refinements in the method are possible, such as estimating energy savings with solar systems. Refer to the *1981 ASHRAE Handbook of Fundamentals* for further details.

Table 674-4. Seasonal energy use, E, for Example 4.

For a farrowing building with a heat exchanger (E = 0.4).
Example calculations:
$t_i - t_o = 70 - 39.8 = 30.2$ F
$t_i - t_E = 70 - 51.9 = 18.1$ F
$q_v = 1.1 \times CFM \times (t_i - t_E) = 1.1 \times 400 \times 18.1 = 7{,}964$ Btu/hr.
$q_b = A/R \times (t_i - t_o) = 240 \times 30.2 = 7{,}248$ Btu/hr.
$q_s = 15{,}000$ Btu/hr (given).
$q_{sr} = q_v + q_b - q_s = 7{,}964 + 7{,}248 - 15{,}000 = 212$ Btu/hr.
$E = q_{sr} \times$ Hours (from Table 674-5) $= 212 \times 92 = 19{,}504$ Btu/season

Bin F	t_E F	$t_i - t_o$ F	$t_i - t_E$ F	q_v Btu/hr	q_b Btu/hr	q_{sr} Btu/hr	Occurrence hrs	E Btu/season
39.8	51.9	30.2	18.1	7,964	7,248	212	92	19,504
37	50.2	33	19.8	8,712	7,920	1,632	730	1,191,360
32	47.2	38	22.8	10,032	9,120	4,152	772	3,205,344
27	44.2	43	25.8	11,352	10,320	6,672	502	3,349,344
22	41.2	48	28.8	12,672	11,520	9,192	280	2,573,760
17	38.2	53	31.8	13,992	12,720	11,712	169	1,979,328
12	35.2	58	34.8	15,312	13,920	14,232	94	1,337,808
7	32.2	63	37.8	16,632	15,120	16,752	40	670,080
2	29.2	68	40.8	17,952	16,320	19,272	20	385,440
-3	26.2	73	43.8	19,272	17,520	21,792	10	217,920
-8	23.2	78	46.8	20,592	18,720	24,312	4	97,248
-13	20.2	83	49.8	21,912	19,920	26,832	1	26,832
							Total =	15,053,968

Table 674-5. Hourly weather occurrences.

Adapted from *1980 ASHRAE Handbook of Systems.* The numbers in the table body are hr/yr, within ± 2.5 F of the temperatures given for each city. Example: for Albany and 72 F, the Albany temperature is at 69.5 F to 74.5 F about 588 hr/yr.

LOCATION	OUTDOOR TEMPERATURE, F																		
	72	67	62	57	52	47	42	37	32	27	22	17	12	7	2	–3	–8	–13	–18
Albany, NY	588	733	740	708	652	625	647	769	793	574	404	278	184	110	63	32	10	5	4
Albuquerque, NM	767	831	719	651	687	734	741	689	552	346	154	66	21	4	1	1			
Atlanta, GA	1185	926	823	784	735	676	598	468	271	112	44	19	8	2					
Bakersfield, CA	831	898	966	977	908	746	541	247	77	7									
Birmingham, AL	1138	908	805	742	668	614	528	433	292	143	69	17	6	3					
Bismark, ND	454	566	614	606	563	520	518	604	653	550	474	371	338	292	278	208	131	77	80
Boise, ID	492	575	643	702	786	798	878	829	522	307	148	53	26	14	6	2			
Boston, MA	676	819	804	781	766	757	828	848	674	429	256	151	74	35	4	9	1		
Buffalo, NY	646	772	760	700	666	624	647	756	849	602	426	267	170	81	5	24	2		
Burlington, VT	573	670	703	694	655	603	637	716	752	561	491	336	272	216	135	81	39	17	8
Casper, WY	423	532	592	642	606	670	782	831	806	683	495	324	200	116	73	45	30	15	5
Charleston, SC	1267	1090	889	787	651	576	434	321	192	79	27	5							
Charleston, WV	912	949	767	689	661	667	607	633	630	356	252	135	73	22	7	1			
Charlotte, NC	1115	908	839	752	730	684	634	515	360	166	64	23	5	2					
Chattanooga, TN	1021	895	775	722	713	679	642	553	414	228	113	45	4	4	2				
Chicago, IL	762	769	653	592	569	543	591	800	822	551	335	196	117	85	59	25	12	3	
Cincinnati, OH	879	843	726	639	611	599	627	698	711	460	249	131	68	44	18	8	2		
Cleveland, OH	763	831	732	641	638	607	620	754	806	578	355	201	111	47	22	11	2		
Columbus, OH	774	820	720	648	622	603	658	730	772	502	280	169	94	40	20	10	4	1	
Corpus Christi, TX	1175	1041	748	551	444	302	180	83	27	9	3								
Dallas, TX	831	795	693	656	629	576	504	371	231	91	34	17	4	1					
Denver, CO	549	684	783	731	678	704	692	717	721	553	359	216	119	78	36	22	6	1	1
Des Moines, IA	707	751	681	600	585	512	510	627	747	557	405	281	211	152	104	59	23	8	1
Detroit, MI	721	783	695	633	592	566	595	808	884	618	377	248	131	61	17	4	1		
El Paso, TX	933	839	749	760	687	611	494	369	233	104	34	10	2						
Ft. Wayne, IN	728	777	699	608	569	552	601	725	905	596	381	205	124	69	40	19	6	1	
Fresno, CA	709	803	921	1006	1036	952	673	426	168	34									
Grand Rapids, MI	634	739	712	647	571	565	554	742	938	690	469	293	172	78	31	10	1	1	
Great Falls, MT	407	520	636	754	822	830	832	813	698	533	355	218	167	136	118	101	68	51	62
Harrisburg, PA	807	824	737	692	635	659	722	888	749	427	222	125	52	18	4	1			
Hartford, CT	617	755	751	752	649	575	683	807	825	552	370	233	153	77	33	11	3	2	
Houston, TX	1172	980	772	681	570	452	291	141	64	18	4	2							
Indianapolis, IN	821	815	722	585	586	579	605	712	791	551	293	152	97	60	35	13	3	2	
Jackson, MS	1169	922	790	677	618	605	484	367	224	103	41	6	2	2	1				
Jacksonville, FL	1334	975	879	692	530	355	288	154	83	24	2								

Continued on next page.

Table 674-5. Hourly weather occurrences, continued.

LOCATION	*OUTDOOR TEMPERATURE, F*																		
	72	67	62	57	52	47	42	37	32	27	22	17	12	7	2	−3	−8	−13	−18
Kansas City, MO	761	723	601	572	553	562	628	625	591	407	265	175	99	51	21	4			
Knoxville, TN	1056	889	746	675	672	689	648	590	456	217	101	41	21	7	2				
Las Vegas, NV	651	644	699	786	769	716	591	396	194	44	7	1							
Little Rock, AR	940	803	725	672	638	669	605	509	363	172	50	25	5	1					
Los Angeles, CA	881	1654	2193	1904	1054	428	107	10											
Louisville, KY	869	758	693	654	619	634	649	703	631	332	169	97	45	25	8	3	1		
Lubbock, TX	833	829	688	700	642	618	620	546	490	346	180	86	33	7	5	1			
Memphis, TN	977	798	715	690	618	633	614	532	374	196	74	25	10	4					
Miami, FL	1705	810	452	277	147	71	26	4											
Milwaukee, WI	597	753	749	634	585	591	611	774	913	659	421	285	176	116	83	47	18	4	3
Minneapolis, MN	621	690	695	602	588	482	500	560	632	609	514	383	311	246	186	119	62	31	16
Mobile, AL	1411	1038	882	698	609	506	377	214	109	49	7	3							
Nashville, TN	933	838	738	697	637	619	627	565	463	263	132	67	28	9	3	1	1		
New Orleans, LA	1189	987	850	692	621	449	282	128	47	9	2								
New York, NY	926	877	754	745	722	796	838	858	603	330	188	2	26	10	1				
Oklahoma City, OK	881	769	717	643	645	611	641	570	468	287	173	77	36	12	3	1			
Omaha, NB	726	721	606	558	539	543	543	655	663	511	390	287	189	135	93	40	15	1	
Philadelphia, PA	863	809	735	710	663	701	758	818	654	335	189	100	32	9					
Phoenix, AZ	762	776	767	769	659	540	391	182	57	8									
Pittsburgh, PA	722	910	799	678	637	587	631	688	774	569	360	233	159	60	30	7	1		
Portland, ME	407	627	780	808	760	748	772	839	820	599	408	293	190	109	60	29	15	5	1
Portland, OR	373	581	1001	1316	1274	1271	1238	772	343	123	40	10	4	1					
Raleigh, NC	1087	937	848	762	707	672	638	527	410	236	103	38	11	1					
Reno, NV	418	477	572	690	845	909	890	829	733	530	387	227	101	37	15	4	1		
Richmond, VA	953	850	784	745	690	673	699	632	478	285	138	67	19	2	1				
Sacramento, CA	630	773	1071	1329	1298	1049	701	355	93	8									
Salt Lake City, UT	569	615	614	635	682	685	755	831	798	564	328	158	80	41	16	2			
San Antonio, TX	1086	943	789	669	569	445	387	190	94	31	11	4	1	1					
San Francisco, CA	285	665	1264	2341	2341	1153	449	99	10										
Seattle, WA	258	448	750	1272	1462	1445	1408	914	427	104	39	20	3						
Shreveport, LA	1063	886	772	679	619	609	516	361	200	72	23	6	2						
Sioux Falls, SD	566	684	669	605	522	498	501	625	712	585	520	448	293	208	152	102	59	43	18
St. Louis, MO	823	728	646	575	585	578	620	671	650	411	219	134	77	40	15	7	1		
Syracuse, NY	627	735	723	717	656	641	651	720	830	547	392	282	190	102	55	23	5	2	2
Tampa, FL	1387	1187	877	570	345	216	137	48	10	1									
Waco, TX	909	830	701	622	651	558	501	354	216	84	24	3	1						
Washington, DC	960	766	740	673	690	684	790	744	542	254	138	54	17	2					
Wichita, KS	758	709	641	603	589	592	611	584	607	426	273	161	85	45	14	3	1		

700 WASTE MANAGEMENT

A complete waste disposal system is no longer a luxury in a livestock business—it is a necessity. Careful waste management is needed to:

- Maintain good animal health through sanitary facilities.
- Avoid pollution of air and water.
- Comply with local, state, and federal regulations.
- Balance capital investment, labor, and nutrient use.

Animal producers generally have a choice of waste handling methods. Wastes are handled as either solids or liquids. Liquids can be handled with slotted floors, flushing systems, gravity flow gutters, or scraper systems. Solids can be handled with manual or mechanical scrapers or front-end loaders.

The disposal method chosen depends primarily on whether the waste is liquid or solid. Solids are usually spread on fields with conventional spreaders. Liquids are usually spread on fields with tank wagons, applied with irrigating equipment, or digested in a lagoon before field spreading.

701 ANIMAL WASTE CHARACTERISTICS

Manure refers to feces and urine. Waste refers to manure with added bedding, rain, soil, etc. It also refers to milkhouse or washing waters not particularly associated with manure. Livestock wastes also typically include hair, feathers, and other debris.

The quantity and composition of wastes produced influence livestock waste facility design. The properties of livestock waste depend upon several factors: animal species; ration digestibility, protein, and fiber content; animal age, environment, and productivity; and quantity of added bedding, soil, water, etc.

Manure

Manure with up to 15% solids has liquid handling characteristics. Extra water must often be added to dilute the manure for handling as a liquid. Wastes with up to 4% solids can be handled with irrigation or flushing equipment. Manure with 4%-15% solids is a semi-liquid slurry that can be handled as a liquid but may need special equipment for pumping.

A general range of solids content for manure without bedding is:

20-25% solids: stiff; some drying or liquid separation has taken place for most manures.
15-20% solids: semi-liquid, quite thick slurry.
5-15% solids: liquid slurry.
0-5% solids: irrigation or flushing consistency.

Solid Manure Handling

Wastes with 20% or more solids can usually be handled as a solid. Solid manure results from catching and holding manure in bedding, or from allowing the liquids to run off, leaving the solids to be handled separately.

Handling solid manure requires:

- solid floors that can be bedded or drained.
- a minimum of equipment.
- an area on which to spread the solids.

Suggestions:

- Install sloping floors; locate waterers where manure accumulation is desired; and keep pens full.
- Haul manure directly to fields whenever possible, but avoid spreading on frozen fields.
- When a stockpile is necessary, locate it for convenient access to a spreader, out of natural drainageways, and away from any water source. Divert surface water away from the storage area.
- Control runoff from stockpiles or lots.

Liquid Manure Handling

Livestock producers handle manure as a liquid for one or more of the following reasons:

- Liquid manure usually requires minimal time and labor, as it can be treated in lagoons or stored in pits, tanks, or earth storages until spread. Pits are concrete under-building storage and tanks are outside and usually separate from the building.
- Disposal of liquid manure can be postponed to fit field schedules, soil conditions, and expected rainfall, if the storage unit is properly sized.
- Objectionable odors, unsightliness, and fly problems can be controlled when wastes are stored in a covered storage. (However, odors that occur from spreading can be more objectionable than those from solid manure.)

Handling liquid manure requires the following facilities and equipment:

- scrapers, gutters, slotted floors, or drains to move the wastes into storage or treatment.
- a storage or treatment unit to which needed water can be added.
- pumps, agitators, or augers to stir and remove the liquid manure.
- tanker or irrigation equipment and land to dispose of the manure.

Table 701-1. Manure production and characteristics.
Table values are approximate because of differences in ration, animal age, and management.

Animal	Size pounds	Total manure production lb/day	cu ft/day	gal/day	Water %	Density lb/cu ft	TS lb/day	VS lb/day	BOD_5 lb/day	Nutrient content N lb/day	P lb/day	K lb/day
		(1)	(2)	(3)	(4)	(5)	(6)	(7)	(8)	(9)	(10)	(11)
Dairy cattle	150	12	0.19	1.5	87.3	62	1.6	1.3	0.26	0.06	0.010	0.04
	250	20	0.32	2.4	"	"	2.6	2.1	0.43	0.10	0.020	0.07
	500	41	0.66	5.0	"	"	5.2	4.3	0.86	0.20	0.036	0.14
	1000	82	1.32	9.9	"	"	10.4	8.6	1.70	0.41	0.073	0.27
	1400	115	1.85	13.9	"	"	14.6	12.0	2.38	0.57	0.102	0.38
Beef cattle	500	30	0.50	3.8	88.4	60	3.5	3.0	0.80	0.17	0.056	0.12
	750	45	0.75	5.6	"	"	5.2	4.4	1.2	0.26	0.084	0.19
	1000	60	1.0	7.5	"	"	6.9	6.0	1.6	0.34	0.11	0.24
	1250	75	1.2	9.4	"	"	8.7	7.4	2.0	0.43	0.14	0.31
cow[b]		63	1.05	7.9	"	"	7.3	6.2	1.7	0.36	0.12	0.26
Swine												
Nursery pig	35	2.3	0.038	0.27	90.8	60	0.20	0.17	0.07	0.016	0.0052	0.010
Growing pig	65	4.2	0.070	0.48	"	"	0.39	0.31	0.13	0.029	0.0098	0.020
Finishing pig	150	9.8	0.16	1.13	"	"	0.90	0.72	0.30	0.068	0.022	0.045
	200	13	0.22	1.5	"	"	1.2	0.96	0.39	0.090	0.030	0.059
Gestate sow	275[b]	8.9	0.15	1.1	"	"	0.82	0.66	0.27	0.062	0.021	0.040
Sow & litter	375[b]	33	0.54	4.0	"	"	3.0	2.4	1.0	0.23	0.076	0.15
Boar	350[b]	11	0.19	1.4	"	"	1.0	0.84	0.35	0.078	0.026	0.051
Sheep	100	4.0	0.062	0.46	75	65	1.0	0.85	0.09	0.045	0.0066	0.032
Poultry												
Layers	4	0.21	0.0035	0.027	74.8	60	0.053	0.037	0.014	0.0029	0.0011	0.0012
Broilers	2	0.14	0.0024	0.018	"	"	0.036	0.025	0.0023	0.0024	0.00054	0.00075
Horse	1000	45	0.75	5.63	79.5	60	9.4	7.5	-	0.27	0.046	0.17

(1) lb/day = animal wt x RM in lb/day per 100 lb / 1000
(2) cu ft/day = lb/day ÷ density
(3) gal/day = 7.5 x cu ft/day
(4) Water % = 100 - % RM of TS from Table 3.
(5) density = best estimate, not ASAE data.
(6) TS = lb/day x TS ans % RM from Table 3.
(7)-(11) lb/day of element = TS x % TS of element.

[b]Not ASAE data: assumptions:
cow = 1.05 cu ft/day
gestating sow = ½ of ASAE data for her weight because she's limit fed.
sow & litter = ASAE data for her weight + 8 pigs @ 1.0 lb/day.
boar = ½ of ASAE data for his weight because he's limit fed.

Table 701-2. Percent cattle manure on yard.
If animals have free access to yard, manure in yard = proportion of manure on yard in the table x total manure/day. If animals are released to yard for controlled periods, manure on yard = hr/day on yard ÷ 24 x total manure/day.

Type of housing	Proportion of manure on yard percent
Free stall dairy	
Inside feeding	50
Yard feeding	60
Manure pack bedded area and yard feeding	
Dairy cattle and heifers	50
Beef cows	70
Beef feedlot cattle	70

Bedding

Some livestock wastes include bedding. Table 701-3 lists common types of bedding by density and water absorbing capacity. To estimate bedding use, weigh the amount of bedding added to each pen per week and multiply by the number of pens and weeks between cleanout.

Estimate the total weight and volume of bedding and manure. Drained liquids can usually be neglected if the floor is well bedded. Bedding volume reduces during use.

Eq 701-1.

$$WWt = MWt + BWt$$

WWt = total weight of waste, lb or ton
MWt = manure weight, lb or ton
BWt = bedding weight, lb or ton

Eq 701-2.

$$WV = MV + \frac{1}{2} \times BV$$

WV = total volume of waste, ft^3
MV = manure volume, ft^3
BV = bedding volume, ft^3

Some livestock waste management systems require that the manure be solid—a consistency which does not flow but stays in a pile. Add bedding material such as straw or woodchips to manure to produce a solid waste. Use Fig 701-1 to estimate the amount of bedding needed to raise the solid or lower the moisture content. Use Tables 701-3a and 701-3b to estimate the absorption capabilities of your bedding.

Table 701-3. Bedding materials.
Approximate water absorption and density of dry bedding (typically 10% moisture).

701-3a. Water absorption of bedding.

Material	lb water absorbed per lb bedding
WOOD	
Tanning bark	4.0
Dry fine bark	2.5
Pine chips	3.0
sawdust	2.5
shavings	2.0
needles	1.0
Hardwood chips, shavings or sawdust	1.5
CORN	
Shredded stover	2.5
Ground cobs	2.1
STRAW	
Flax	2.6
Oats, threshed	2.8
combined	2.5
chopped	2.4
Wheat, combined	2.2
chopped	2.1
HAY, chopped mature	3.0
SHELLS, HULLS	
Cocoa	2.7
Peanut, cottonseed	2.5
Oats	2.0

701-3b. Bedding material density.

Form	Material	Density lb/cu ft
Loose	Alfalfa	4-4.4
	Non legume hay	3.3-4.4
	Straw	2-3
	Shavings	9
	Sawdust	12
Baled	Alfalfa	6-10
	Non legume hay	6-8
	Straw	4-5
	Shavings	20
Chopped	Alfalfa	5.5-7
	Non legume hay	5-6.7
	Straw	5.7-8

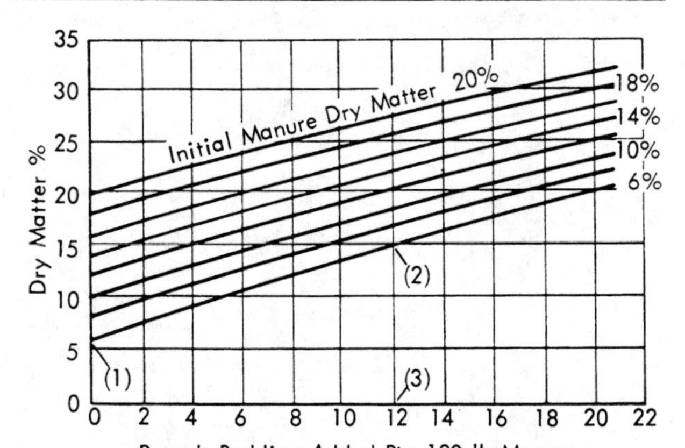

Fig 701-1. Changing manure dry matter by adding bedding.
Bedding is assumed 10% moisture.

Example 1:

(1) refers to example point on chart:

(1) to increase solids content from 6% to
(2) 15%, add
(3) 12 lb of bedding per 100 lb manure.

Or:

(3) Adding 12 lb bedding to 100 lb manure raises solids content from
(1) 6% to
(2) 15%.

Diluting Manure

It takes a lot of water to significantly dilute manure.

Eq 701-3.

$$WA = [(1 - SM) \div (1 - FM)] - 1$$

WA = amount of water added
SM = starting moisture content, decimal (e.g. 80% = 0.8)
FM = final moisture content, decimal

Example 2:

Dilute wastes from 80% (0.8) to 90% (0.9) water content:

Water added $= [(1 - 0.8) \div (1 - 0.9)] - 1$
$= (0.2 \div 0.1) - 1 = 1$

Or, add 1 unit (gal, ft^3, tank) of water to 1 unit of manure to dilute waste from 80% water content to 90% water content.

Stored Liquid Manure Gases and Odors

Gases from liquid wastes stored inside a building can be hazardous and create undesirable odors. Accumulating odors and gases are detrimental to animals and operators. The most serious gas problems occur when manure is agitated or when ventilation fans fail. Most (99% or more) of the gas produced is methane, ammonia, hydrogen sulfide, and carbon dioxide.

Animals asphyxiate because methane and carbon dioxide are heavier than air and displace oxygen. Oxygen content in the air below 10% is critical. Ammonia can irritate respiratory tracts and make animals more susceptible to disease.

Ventilating Manure Storages

The primary hazard from manure gases occurs with inadequate ventilation.

Choose a mild or warm day and provide maximum building ventilation when agitating or pumping wastes from a pit. Manure gases can be deadly so provide plenty of fresh air for workers and animals.

Exhaust some ventilation air from directly above stored manure. Even a low-volume continuous fan pulling air from above one corner of a pit helps reduce gas concentrations.

Vent engines and heaters to the outside to prevent toxic carbon monoxide (CO) accumulations. CO is a colorless and odorless gas about the same weight as air. It is exhausted from gas engines and from gas, oil, and coal heaters.

Provide an alarm system (loud bell or readily noticed light) to warn of power failures in totally enclosed buildings. Tightly closed buildings can have a very rapid gas buildup at animal level if ventilation is stopped.

Manure Gases Can Kill

Hydrogen sulfide poisoning or lack of oxygen kills. Enter a manure storage only after it has been well ventilated; wear self-contained breathing tanks; and have a safety rope attached and at least two people standing by to pull you out at the first sign of dizziness.

Methane gas can accumulate in covered pits, creating explosive conditions. Do not introduce a spark or flame unless the pit has been well ventilated.

Controlling Odors

To control odors, locate buildings and manure storage where prevailing summer winds and air drainage downhill on calm nights will carry odors away from you or your neighbor. In most of the Midwest, locate manure storages to the north or east of residences. There is no "safe" separation distance, but try to locate your facilities at least ½ mile from neighbors or farther if you have a large operation, open lot, or lagoon.

Open lots have more odor than environmentally controlled buildings. Frequent scraping reduces odor production.

Locate ventilation fans high on walls or place buildings on high ground to disperse ventilation exhaust odors. Frequent removal of manure and proper pit ventilation reduce odor within the building. Filtering dust particles, which carry odors, also reduces some noxious odors.

Lagoon odors can be reduced with proper dilution and loading but can still be severe during spring and fall when lagoon contents change temperatures and turn over.

Cover wastes to minimize odor release. Covered storage may be a floating crust or a concrete lid. Immediate plow down or chisel injection of wastes decreases odors from land applications.

Chemical treatment can reduce odors. However, many treatments require relatively large amounts of chemicals, have some disadvantageous results, and are at best temporary cures to odor problems.

702 COLLECTION

Several collection methods are possible. Some systems combine collection and storage, such as a built-up manure pack or slotted floors over a liquid pit. In selecting a collection system, consider:

- Facility type.
- Labor requirements.
- Investment.
- Total waste handling system.

Slotted Floors

Slotted floors rapidly separate an animal from its manure. Opening size and slat width depend on manure properties and experience with slipping, foot injury, and other animal responses. See Table 702-1.

Tapered slats (greater top than bottom width) tend to pass wastes better than uniform-width slats. For easy commercial slat installation, build the opening for the slat ½"-1" wider than the length of the slats to allow for variation in construction work. Where slats are parallel to a wall or partition, leave a 2" (3" for cattle) space to avoid solids buildup.

In partly slatted buildings, place slats parallel to the long dimension of rectangular pens. Try to place slats parallel to natural traffic flow patterns, so animals walk along rather than across slats to minimize foot injury.

Concrete slats are the most durable and work well for large animals. They can be purchased precast, precast at the building site, or cast in place. Wood slats have the lowest initial cost and shortest life. Wood wears, warps, and is chewed by hogs, leaving irregular spacings, and can become slick with use. Manufactured slotted floor systems of steel, aluminum, and plastic are more uniform for more accurate spacing and smoother for easier cleaning. They may be easier to handle, install, and replace than concrete, but can be more expensive. When selecting slats, consider initial cost, predicted life, intensity of use (wear rate), strength, corrosion, fire, noise, replacement cost, and most importantly, the type of animal housed. Younger animals require cleaner floors. Make slats stable; twisting, flexing, or shifting may cause cracks in the slat material or slat coating, or may catch or pinch animals.

Table 702-1. Slat size and spacing.

Animal type	Slot width in.	Concrete slat width, in.
Beef		
Cows	1½-1¾	4-8
Feeder cattle	1½-1¾	4-8
Calves	1¼	4-8
Dairy		
Cows	1½-1¾	4-8
Calves	1¼	4-8
Swine		
Sow and litter	⅜	4
Prenursery pig	⅜	Not recommended
Nursery pig	1	4
Growing-finishing pig	1	6-8
Gestating sow or boar		
pens	1	6-8
stalls	1	4
Sheep	½-¾	3-4

Other Floors

Woven Wire and Expanded Metal

Woven wire and steel rod mesh work well for farrowing and nursery facilities. Use an overlay in the pig creep the first few days after farrowing if foot injuries are a problem.

Flattened expanded metal (¾", 9 to 11 gauge) works for pigs under 30 lb, but does not last as long as concrete when subjected to concentrated traffic by heavier pigs. Avoid expanded metal with sharp edges. Flattened expanded metal is not recommended for sows with litters less than 14 days old because of foot, leg, and teat injuries.

Rough, unflattened, ¾" No. 9 expanded metal is recommended for sheep. Sheep are more sure-footed and hooves wear more evenly on expanded metal than on other materials. Put older lambs or dry ewes on the floor first to wear off small burrs before putting ewes with full udders on new expanded metal. Expanded metal sheets with 4-½" long openings can damage lambs' feet, so use it only for larger animals.

Removable steel grates of smooth round rods are best for covering the gutter behind dairy cows in stalls.

Sloped Solid Floors

Sloped floors help move manure toward gutters or scraped or slotted areas. Animal traffic tends to work manure down slopes of about ½"/ft (4%) or more. Slopes over about 1"/ft (8½%) may cause footing problems. Limit slopes of paved storages, lots, and drives to below 1'/10' (10%); 1'/20' is recommended. Steeper slopes limit maneuvering and traction.

Scrapers and Blades

The main advantage of mechanical manure scrapers is labor reduction. Use them on solid floors or under slotted floors in gutters, channels, or troughs to move wastes to storage, treatment, or spreading equipment.

Automatic scrapers are available commercially in a variety of weights and designs. A scraper system consists of one or more scraping blades, a cable or

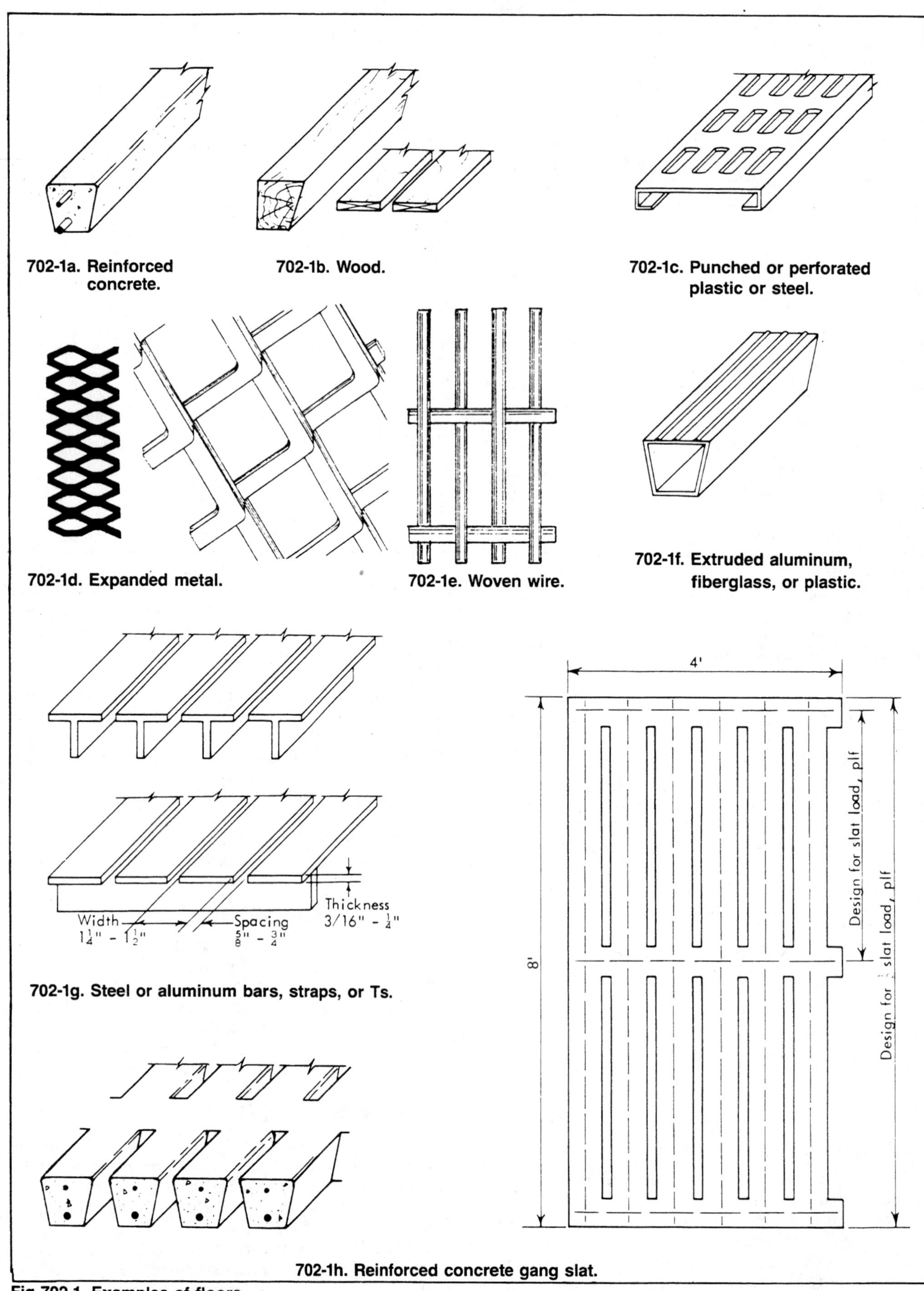

702-1a. Reinforced concrete.

702-1b. Wood.

702-1c. Punched or perforated plastic or steel.

702-1d. Expanded metal.

702-1e. Woven wire.

702-1f. Extruded aluminum, fiberglass, or plastic.

702-1g. Steel or aluminum bars, straps, or Ts.

702-1h. Reinforced concrete gang slat.

Fig 702-1. Examples of floors.

chain to pull the scraper, and a power unit with controls.

Control mechanical scrapers with a time clock. Two to four passes per day are usually sufficient to keep manure from drying and adhering to the floor. More frequent or continual operation may be necessary if the temperature is below freezing. In cold barns, use floor heat under the scraper rest position. Less frequent cleaning is needed under poultry cages because the moisture content of poultry manure is less than that of livestock waste.

Use a shallow manual gutter behind short rows of stalls or farrowing pens. Because a manual gutter usually takes space from an alley, make it no more than 10″ wide. This gutter is cleaned by pushing the manure to a sump or deeper gutter at the end of the alley.

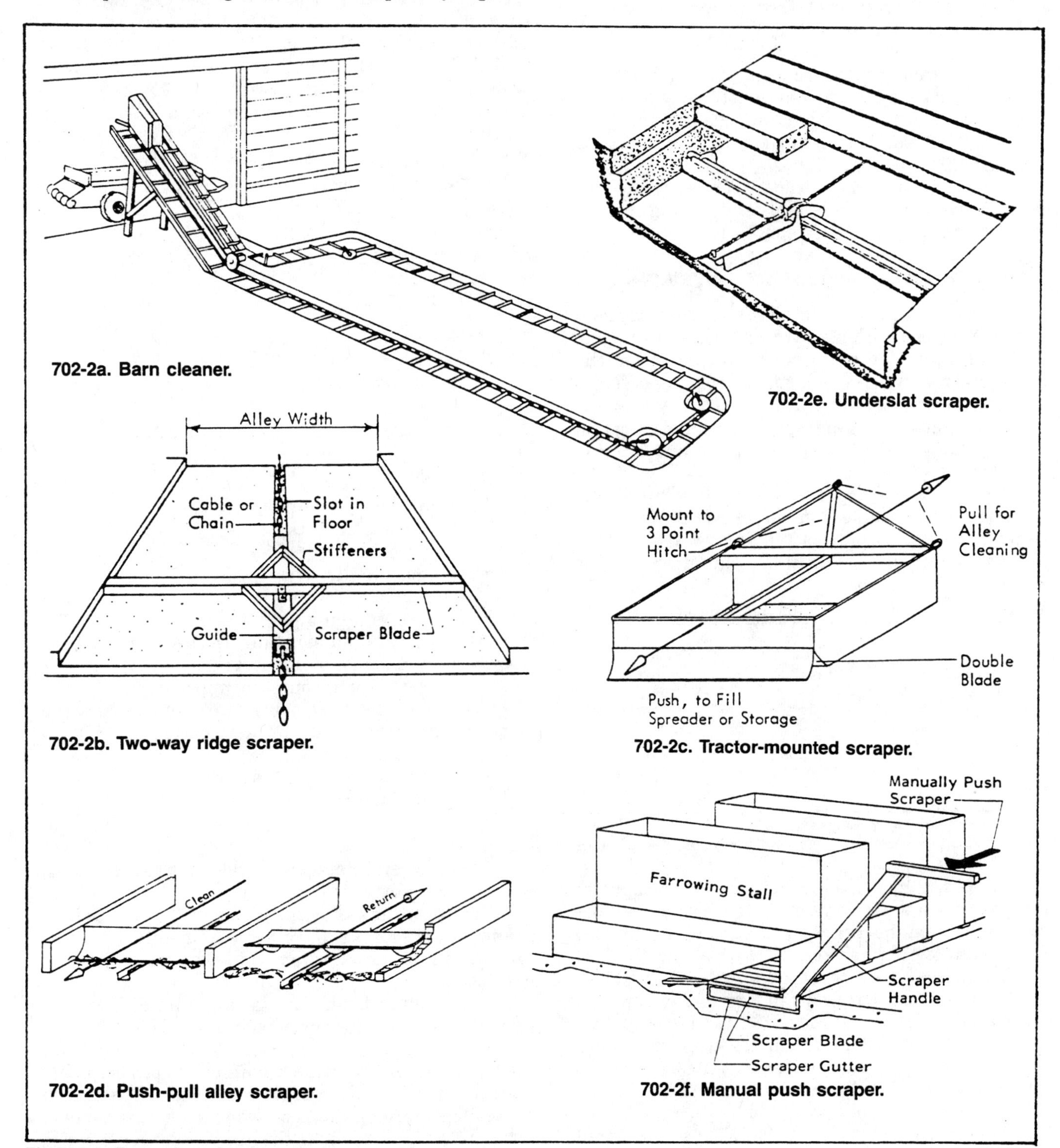

702-2a. Barn cleaner.

702-2e. Underslat scraper.

702-2b. Two-way ridge scraper.

702-2c. Tractor-mounted scraper.

702-2d. Push-pull alley scraper.

702-2f. Manual push scraper.

Fig 702-2. Examples of manure scrapers.

Front-end Loaders

Front-end loaders remove solid manure from open lot surfaces, storages, or building floors.

Conventional agricultural tractor-mounted loaders are typically available in capacities of 1,000 to 4,000 lb. Wide front tractor wheel spacing is desirable for stability. A loaded bucket reduces rear-wheel traction, limiting use to areas with slopes less than 1′/10′. Avoid building layouts requiring backing down long alleys or restricted turning areas.

Small self-propelled loaders (skid-steer) can clean in cramped areas, greatly reducing hand labor. Most are low and have a turning radius of their own length, so they work easily in tight quarters. Some have relatively low load lifting capacity and are designed primarily for pushing.

Larger industrial-type loaders can be justified for cleaning extensive open feedlots and moving stacked manure where traction is poor.

Livestock Building Flushing Systems

In flushing systems, a large volume of water flows from one end of a building to the other, down a sloped, shallow gutter. The water scours manure from the gutter and removes it to a lagoon. Two types of flushing systems are used in livestock facilities:

- Open-gutter flushing for finishing and gestation buildings, dairy free stall alleys, and holding areas.
- Underslat flushing for swine and beef buildings.

Gutter slopes range from flat to over 2% and have either no cross slope or a very slight crown. Use Table 702-2 to select gutter slopes for initial water flow depths.

Table 702-2. Minimum slope for flush gutters.

Initial depth of flow in.	Open gutter %	Underslat gutter %
1.5*	2.0	—
2	1.5	—
2.5	1.25	—
3**	1.0	1.25

*Recommended depth for open-gutters.
**Recommended minimum depth for underslat gutters and dairy alleyways and holding areas.

For hogs, a good recommendation for open gutter design is to allow a minimum of 1 ft² of gutter area per 150 lb of liveweight. For dairy, the entire 10′-12′ wide free stall alley is flushed.

For underslat systems, at least ⅓ of the floor should be slatted. For underslat gutters wider than 4′, divide all but the first 20′ of their length into gutter widths of 2′-2½′. This helps keep flush water from channeling around waste deposits as it moves down the gutter. Use inverted concrete slats as channel dividers.

For open gutters shorter than 125′, use a constant slope gutter. For gutters 125′-250′, slope both ends of the gutter so they flush toward the middle of the building length. For gutters longer than 250′, consult a knowledgeable building engineer.

Flush Volume

Determine two flush volumes: total volume required per day and volume required per flush. Select total daily flushing volume from Table 702-3. Divide the total daily volume by the volume per flush to determine the number of flushes per day. Total volume of flush water required per day for adequate cleaning is about 100 gal/1,000 lb liveweight, Table 702-3. Flush tanks should release the entire water volume in 10 to 20 seconds. Size the tank to deliver 30 gal/ft of gutter width for open gutters and 50 gal/ft of gutter width for underslat gutters per flush. Size the recycle pump for the total flush volume.

Table 702-3. Minimum volume of flush water.

Animal type	Flush volume gal/hd/day
Sow and litter	35
Prenursery pig	2
Nursery pig	4
Growing pig	10
Finishing pig	15
Gestating sow	25
Dairy cow	130
Beef feeder	100

Flush Tanks

Five types of flushing devices are:

- Automatic siphon tank.
- Tipping bucket.
- Trap door tank.
- Manual flush tank.
- Large volume pump.

An automatic siphon tank is the only flushing device that has no moving parts. As the tank slowly fills with water, an air bubble trapped under the bell is forced out the siphon pipe until it triggers the siphoning action. Place an automatic siphon tank above the pens to save floor space. See AED-17, *Siphon Flush Tank,* for construction details.

A tipping bucket tank dumps as it fills when the center of gravity of the water overbalances the pivot point. Because the water discharges all at once, flushing velocity and depth are high and can cover an 8′-10′ wide gutter if necessary.

Trap door tanks have more moving parts, but both tank volume and trap door can be modified to meet individual needs. A watertight seal around the door is needed.

The manual flush tank is best suited to twice-a-day flushing in underslat systems. Fig 702-3 shows a small but effective manual flush tank suitable for farrowing/nursery swine buildings. Another system uses a dividing wall across an underslat pit to dam up several hundred gallons of water in the high end of the gutter. The gutter is flushed by opening a sluice gate or by pulling one or more large standpipe overflows located in the water storage tank.

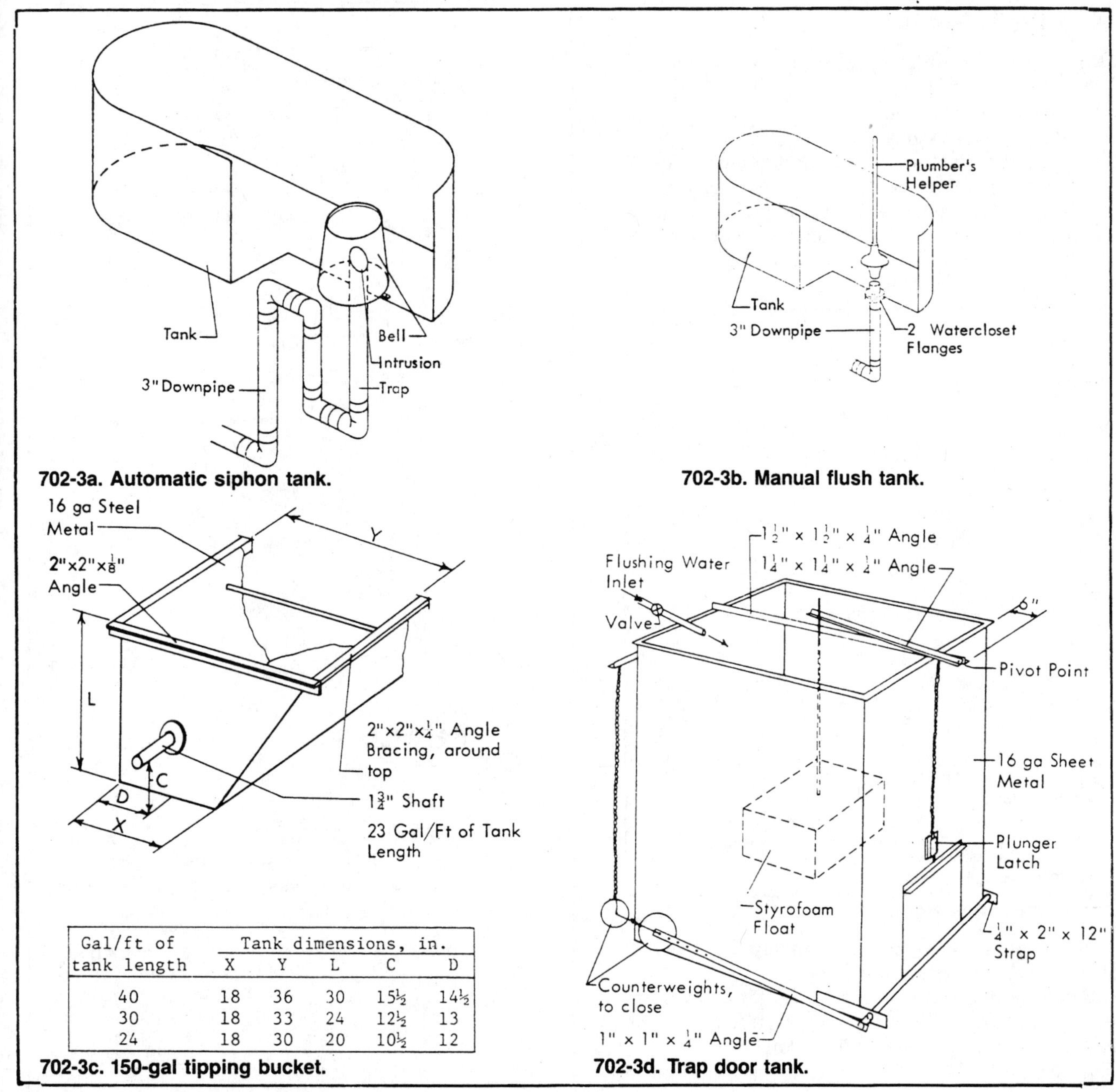

702-3a. Automatic siphon tank.

702-3b. Manual flush tank.

Gal/ft of tank length	Tank dimensions, in. X	Y	L	C	D
40	18	36	30	15½	14½
30	18	33	24	12½	13
24	18	30	20	10½	12

702-3c. 150-gal tipping bucket.

702-3d. Trap door tank.

Fig 702-3. Types of flush tanks.

Large volume pumps are becoming more common in large operations to flush underslat gutters. Size these pumps for 80 gpm/ft of gutter width. Pumps are typically operated once or twice a day. Total water use with pumps is often greater than with flush tanks.

Flushing Flumes

Flumes are underfloor channels that carry flushed manure out of a building. They are used in confined beef operations.

Typical flume channels are 12" wide under a 2" slot for manure passage and 8'-14' apart. Flumes are PVC pipe embedded in the floor or U-shaped precast concrete. Concrete is more popular and easier to construct than the plastic pipe flume. Supply flush water to the flume with flush tanks or timer-controlled pumps.

Precise design methods which account for various flow depths and flume roughness have not been determined. Guidelines based on successful systems are in Table 702-4.

Table 702-4. Design guidelines for flushing flumes.

Daily flush volume	15 gal/1000 lb liveweight/day (minimum)
Flushing frequency	1 to 6 hr
Flushing duration	5 min to 1 hr (continuous in freezing temperatures)
Gutter slope	0.5%-1%
Number gutters/bldg	3 to 5

Gravity Drain Gutters

Gutters for handling liquid manure include:
- deep narrow gutter.
- gravity drain system.
- gravity flow trenches.

Deep narrow gutters collect and remove liquid manure from solid floor pens. They are self-draining, allowing little time for heavy feed particles to settle out. Self-draining reduces odorous gas production. A 6″ width holds two days manure accumulation. Restrict length to 90′ maximum.

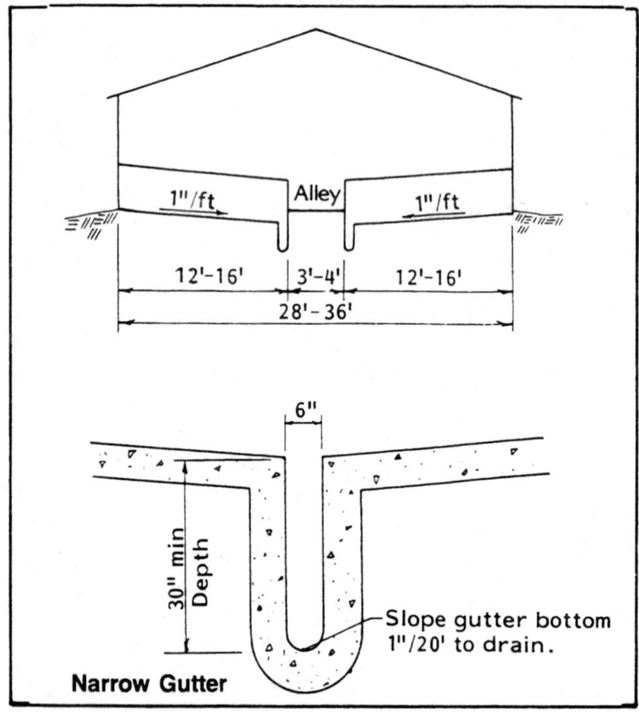

Fig 702-4. Deep narrow gutters.

The gravity drain system is a shallow, U-, V-, or Y-shaped trench under slotted floors.

Manure accumulates in the trench for 4 to 7 days. Then a drain plug, Fig 702-5, is released and the manure drains to storage or a lagoon.

The bottom of the gravity drain must be above the maximum liquid level of storage. Raise the building on fill or build the storage at a lower elevation. Or, pump the manure from a sump to the storage.

Gravity flow trenches are used in comfort stall dairy barns and under slats in swine and beef buildings, usually with an outdoor concrete storage, Fig 702-7.

Trench width is not critical. But, these trenches depend on biological activity, so do not add milkhouse wastewater containing disinfectants.

Collecting Runoff

Feedlot runoff, whether from rainfall or snowmelt, contains manure and is therefore handled in the livestock waste system.

Runoff from roofs, drives (but not animal alleys), and grassed or cropped areas without livestock ma-

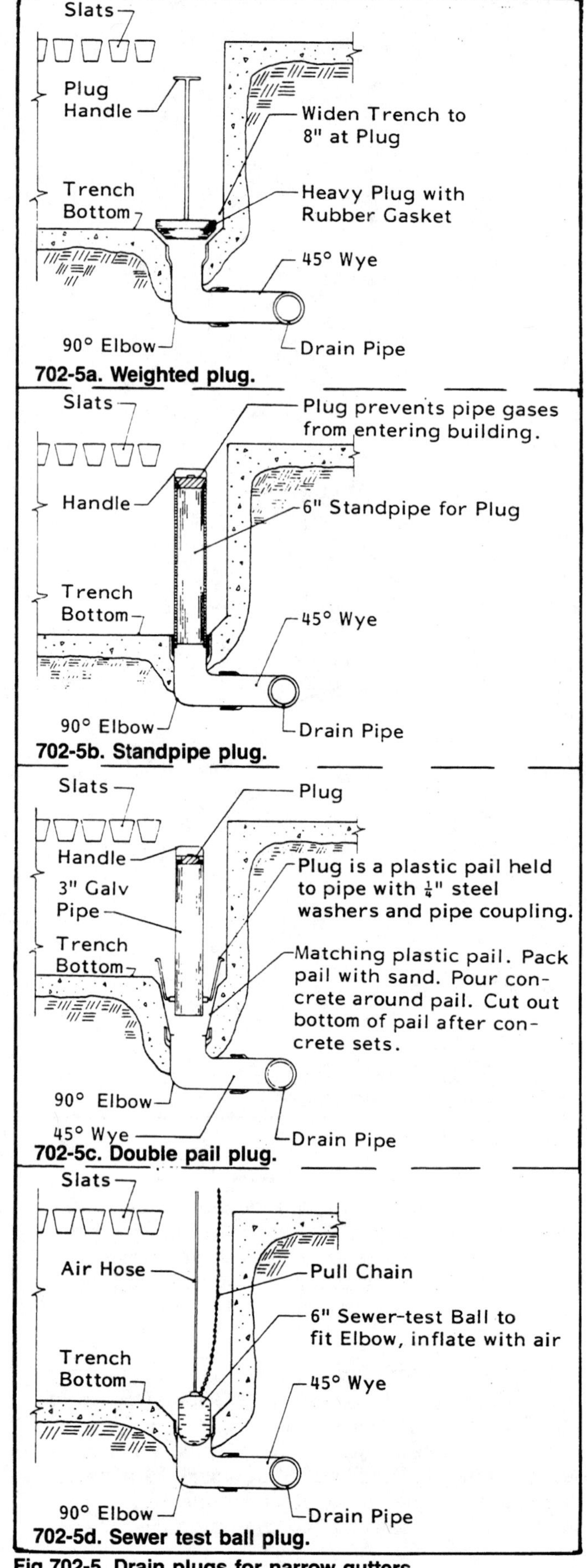

Fig 702-5. Drain plugs for narrow gutters.

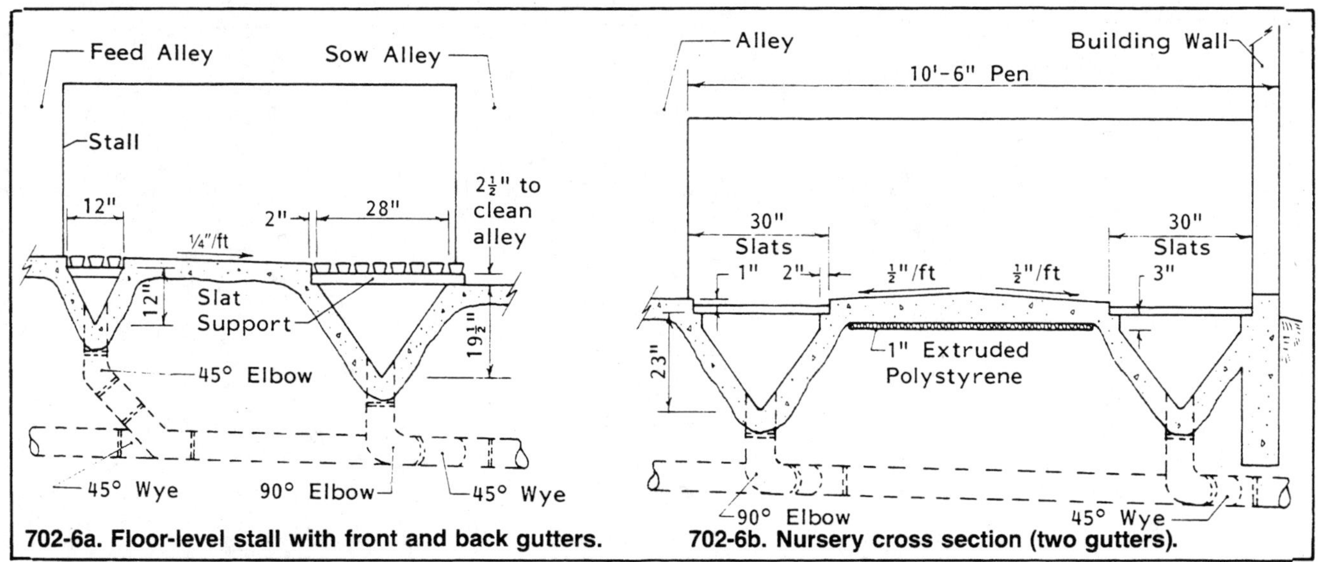

702-6a. Floor-level stall with front and back gutters.

702-6b. Nursery cross section (two gutters).

Fig 702-6. Gravity drain gutters

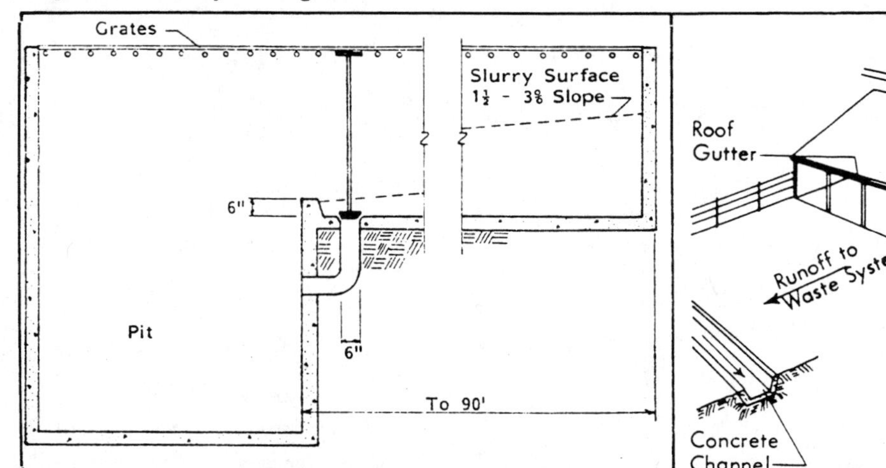

Fig 702-7. Gravity flow channel.
Biological activity helps liquefy the manure so it passes over the top of the bulkhead.

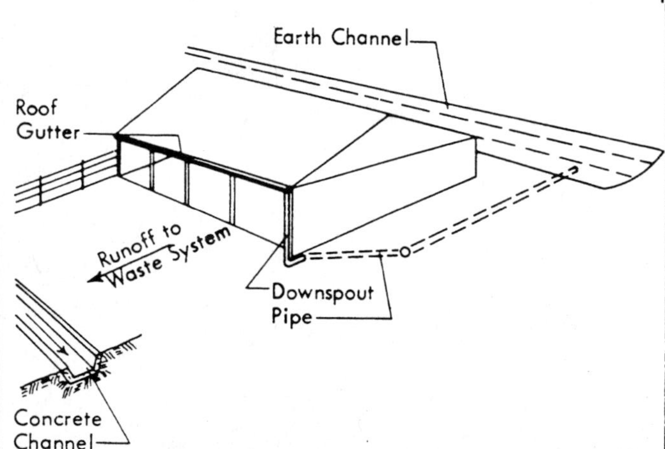

Fig 702-8. Gutters and open channels.

nure is relatively clean and need not be handled as a livestock waste. Divert clean runoff either away from the waste system to reduce the waste volume or into the waste system to dilute a lagoon. Use curbs, dikes, culvert pipes, and terraces to divert clean water away from a manure area.

The following design sections help you select appropriate channels, gutters, pipes, or culverts for liquids when solids content is less than about 5%. Thicker liquids flow slower and require larger conduits at steeper slopes.

Channel design is based upon the rainfall intensity of a 10-yr, 5-min storm. See Climate section. The capacity of a channel, ft^3/s, is determined from the rainfall intensity, in/hr, and the ground or roof area, ft^2.

$$Q(cfs) = [\text{area } (ft^2) \times \text{intensity (in/hr)}] \div [12 \text{ in/ft} \times 3600 \text{ s/hr}]$$
$$= ft^2 \text{ (in/hr)} \div 43{,}200$$

Note that because $43{,}560 \ ft^2 = 1$ acre, the following equation gives quantities within 1%.

$$Q \text{ (cfs)} = \text{area (acre)} \times \text{intensity (in/hr)}$$

Open Channels

The common design equation for fluid flow in open channels (of any configuration) is Mannings Formula. To find channel capacity in ft^3/sec (cfs):

$$Q = 1.486 \times S^{0.5} \times A^{1.667} \div (n \times P^{0.667})$$

- Q = channel capacity, ft^3/sec
- S = slope, ft/ft, in/in, etc.
- A = water cross sectional area, ft^2
- n = coefficient for surface roughness, Table 702-5
- P = wetted perimeter, ft

To find liquid velocity in the channel in ft/sec (fps):

$$V = Q \div A = 1.486 \times R^{0.667} \times S^{0.5} \div n$$

- V = liquid velocity, ft/sec
- R = hydraulic radius, ft
  = A ÷ P = area ÷ wetted perimeter
  = DW ÷ (2D + W)
- D = liquid depth, ft
- W = liquid width, ft

See Table 702-5b for the wetted perimeter and hydraulic radius of common configurations.

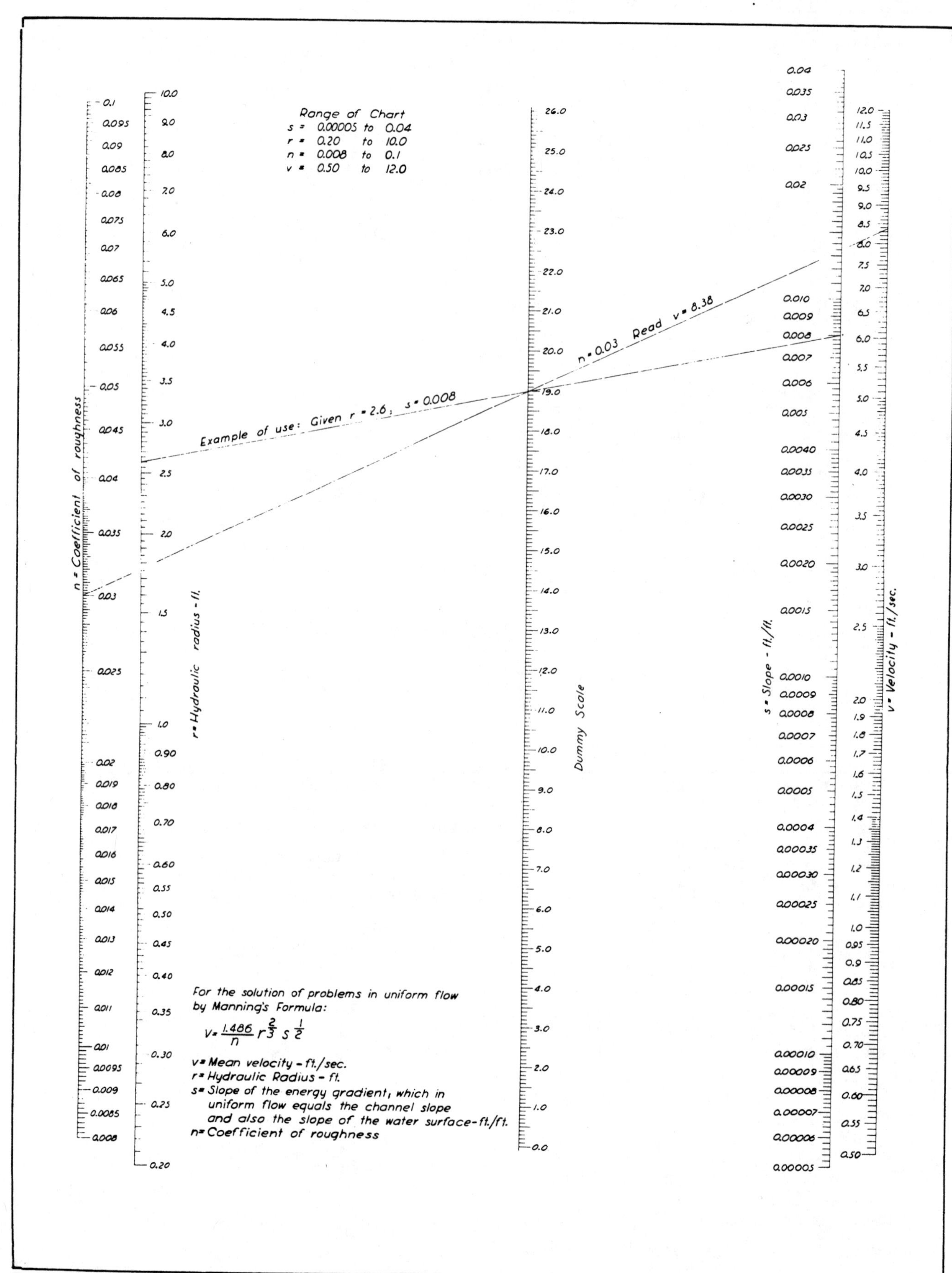

Fig 702-9. Mannings formula chart.

Fig 702-9 is used for solving Mannings equation. Table 702-5 lists coefficient n for various channel linings.

Example 1:

To solve Mannings equation with Fig 702-9:

1. Determine or select the hydraulic radius (R), slope (S), and coefficient of roughness (n).
2. For this example, select R = 2.65, S = 0.008, and n = 0.03.
3. Connect R = 2.65 and S = 0.008 with a straight line that establishes the pivot point of 19.0 on dummy scale.
4. Pass a straight line from n = 0.03 through pivot point (19.0) to the velocity scale.
5. Read the velocity, V = 8.38 fps. If too great or too small, select another R and solve for V. Continue until desired V is achieved. S and n could also be changed.

Channel velocity affects settling and erosion. Keep velocities above about 1.5 fps to prevent settling. In settling channels, design for velocities below about 1.5 fps. Maximum velocity depends on channel surface conditions. Velocities above 4 fps may erode unlined channels.

Open concrete channels

Where a paved drive or similar channel carries runoff, select width and depth required from Table 702-6. Although developed for vertical sides, the channels carry at least the listed capacities if the sides are sloped out.

Example 2:

Two 94′x800′ gable roof buildings are separated by a 12′ wide paved alley. The alley drains roof water and about 2 acres above the buildings. How deep must the alley be?

Compute: 2 acres = 2 x 43,560 = 87.120 ft^2
One half of each roof + the alley = 47 + 12 + 47 = 106′ wide; 106 x 800 = 84,800 ft^2
Total area drained = 87,120 + 84,800 = 171,920 ft^2
If intensity = 7 in/hr, Q = 171,920 x 7 ÷ 43,200 = 27.9 cfs

Select: A 6″ deep, 12′ wide channel carries 25.6 cfs, which is not quite enough; use 6½″ depth.
Check velocity: V = cfs ÷ area
Area = 12′ x 6.5″ ÷ 12 = 6.5 ft^2
V = 27.9 ÷ 6.5 = 4.3 fps; little settling occurs.

Open earth channels

Diversion ditches to carry clean runoff around a building, farmstead, or lot are usually grassed waterways with a roughly trapezoidal cross section. Base capacity on a 10-yr, 5-min storm.

Channel maintenance includes mowing, solids removal, occasional dike rebuilding, and culvert care.

Contact your district SCS engineer for help designing waterways draining large areas of cropland, pastures, or woodlots. The capacity of those channels is not included here.

Table 702-5. Channel properties.

702-5a. Mannings formula coefficient, n.

Type and Description of Channels	Design n values
Channels, lined	
Asphalt	0.014
Concrete	0.015
Shotcrete	0.017
Smooth metal, lumber, concrete	0.012
Channels, earth	0.040
Channels, dense grass	
Grass more than 10" tall	0.04 -0.23
Grass less than 2.5" tall	0.034-0.16

702-5b. Elements of channel sections.

Section	Area A	Wetted perimeter P	Hydraulic radius R	Top width T
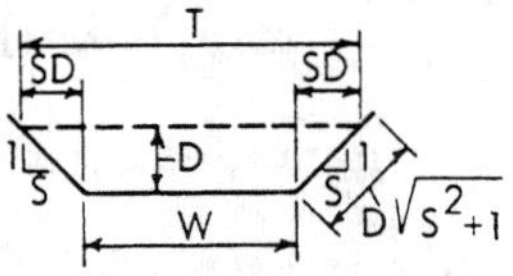 Trapezoid	$WD+SD^2$	$W+2D\sqrt{S^2+1}$	$\frac{WD+SD^2}{W+2D\sqrt{S^2+1}}$	W+2SD
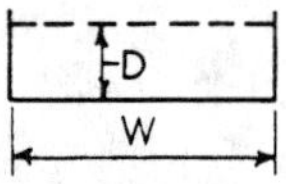 Rectangle	WD	W+2D	$\frac{WD}{W+2D}$	W
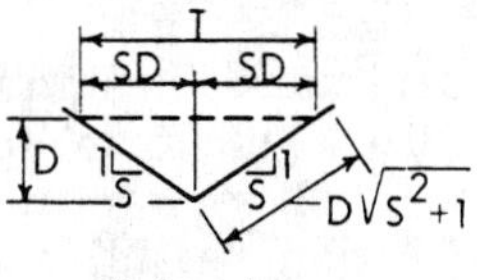 Triangle	SD^2	$2D\sqrt{S^2+1}$	$\frac{SD}{2\sqrt{S^2+1}}$	2SD

Table 702-6. Capacity of open rectangular concrete channels.
n = 0.015; S = 1/16 in/ft = 0.5%
Dashed lines refer to example.

Depth inches	Width, feet 2'	4'	6'	8'	10'	12'	14'	16'
	Capacity in cfs							
2"	0.6	1.4	2.1	2.8	3.5	4.2	5.0	5.7
4"	1.9	4.1	6.4	8.7	11.0	13.2	15.5	17.8
6"	3.4	7.8	12.2	16.6	21.1	25.6	30.1	34.6
8"	5.2	12.0	19.1	26.2	33.4	40.6	47.9	55.1
10"	7	17	27	37	48	58	68	79
12"	9	22	35	49	63	77	91	106
14"	11	27	45	62	80	98	117	135
16"	13	33	54	76	99	121	144	167
18"	15	39	64	91	118	145	173	200
20"	17	45	75	106	138	171	200	240
22"	20	51	86	122	159	197	230	270
24"	22	57	97	138	181	220	270	310

Low slope channels can settle solids from runoff. See Liquid-Solid Separation.

Example 3:

Assume the same runoff as in Example 2, Q = 27.9 cfs.

Select: A 1′ deep, 6′ wide channel with a 1% slope.
Velocity = 2.9 fps.
Velocity is greater than 1.5 fps.
Channel selected is OK.

Alternate solution: For a 3% slope, select a 1′x2′ channel and increase the depth slightly. To reduce velocity below 4.5 fps, make the channel wider so flow will be less than 1′ deep.

Roof gutters and small concrete channels

Where water is collected from relatively small areas, there is little time delay between peak rainfall and peak channel flow. Base design on 10-yr, 5-min peak rainfall, in/hr, and surface area drained, ft^2.

To divert roof runoff away from the waste control system, design roof gutters to carry all water to one or both ends of the building. Select adequate downspouts and a pipe or gutter to carry the water away from the site. Where an open channel will not become polluted, eave gutters can be omitted and roof water permitted to fall to a surface channel.

Table 702-8 gives maximum capacities of small open channels of rectangular cross section, such as roof gutters and small concrete gutters.

Example 4:

What size gutter is needed where each downspout drains a section of roof 30′x300′? Rainfall intensity for a 10-yr, 5-min storm = 7 in/hr.

Compute: Q = 30′x300′ x 7 in/hr ÷ 43,200 in-sec/ft-hr = 1.46 cfs

Select: 8″ wide x 8″ deep. Additional depth may be desired for silting, and if overflow is particularly damaging, a safety factor.

Example 5:

What size channel is needed to divert runoff from a one acre farm courtyard? 1 acre = 43,560 ft^2.

Table 702-7. Capacity of earth channels.
Select a velocity below 5 fps unless liner material is added to reduce erosion.
n = 0.4 bottom slope as shown; side slope is 3:1.
Dashed lines refer to example.

Longitudinal slope = 1%

Depth feet		Width, feet 1'	2'	4'	6'	8'	10'	12'
0.5'	fps	1.4	1.8	1.9	2.0	2.1	2.1	2.1
	cfs	1.4	3.5	5.7	8.0	10.3	12.6	14.9
1.0'	fps	2.3	2.6	2.8	2.9	3.0	3.1	3.2
	cfs	9.2	15.6	22.4	29.4	36.5	43.7	50.9
1.5'	fps	3.0	3.3	3.5	3.7	3.8	3.9	4.0
	cfs	27.1	39.6	52.7	66.1	79.8	93.7	107.7
2.0'	fps	3.7	3.9	4.1	4.3	4.5	4.6	4.7
	cfs	58.4	78.5	99.4	120.8	142.6	164.7	187.0
2.5'	fps	4.2	4.5	4.7	4.9	5.0	5.2	5.3
	cfs	106.0	135.0	165.0	195.8	227.0	258.7	290.8
3.0'	fps	4.8	5.0	5.3	5.4	5.6	5.7	5.8
	cfs	172.4	211.7	252.1	293.4	335.4	377.9	421.0

Longitudinal slope = 2%

Depth feet		Width, feet 1'	2'	4'	6'	8'	10'	12'
0.5'	fps	2.0	2.5	2.7	2.8	2.9	3.0	3.0
	cfs	2.0	5.0	8.1	11.3	14.6	17.8	21.1
1.0'	fps	3.2	3.7	4.0	4.2	4.3	4.4	4.5
	cfs	13.0	22.1	31.7	41.6	51.6	61.8	72.0
1.5'	fps	4.3	4.7	5.0	5.2	5.4	5.5	5.6
	cfs	38.3	56.0	74.5	93.5	112.9	132.5	152.3
2.0'	fps	5.2	5.6	5.9	6.1	6.3	6.5	6.6
	cfs	82.6	111.0	140.5	170.8	201.6	232.9	264.5
2.5'	fps	6.0	6.4	6.7	6.9	7.1	7.3	7.5
	cfs	149.9	191.0	233.4	276.8	321.1	365.9	411.2
3.0'	fps	6.8	7.1	7.4	7.7	7.9	8.1	8.3
	cfs	243.8	299.4	356.6	414.9	474.3	534.5	595.3

Longitudinal slope = 3%

Depth feet		Width, feet 1'	2'	4'	6'	8'	10'	12'
0.5'	fps	2.5	3.0	3.3	3.5	3.6	3.6	3.7
	cfs	2.5	6.1	9.9	13.9	17.8	21.8	25.8
1.0'	fps	4.0	4.5	4.9	5.1	5.3	5.4	5.5
	cfs	15.9	27.0	38.8	50.9	63.2	75.6	88.2
1.5'	fps	5.2	5.7	6.1	6.4	6.6	6.8	6.9
	cfs	46.9	68.6	91.3	114.6	138.3	162.3	186.6
2.0'	fps	6.3	6.8	7.2	7.5	7.7	7.9	8.1
	cfs	101.2	136.0	172.1	209.2	247.0	285.2	323.9
2.5'	fps	7.3	7.8	8.2	8.5	8.7	9.0	9.2
	cfs	183.5	233.9	285.9	339.1	393.2	448.1	503.6
3.0'	fps	8.3	8.7	9.1	9.4	9.7	9.9	10.1
	cfs	298.6	366.7	436.7	508.2	580.9	654.[illegible]	729.[illegible]

Rainfall intensity for a 10-yr, 5-min storm = 7 in/hr.

Compute: Q = 43,560 x 7 ÷ 43,200 = 7.06 cfs

Select: 12″ deep x 18″ wide; cfs = 7.6

Check for velocity (may get settling below about 1.5 fps).
V = cfs ÷ channel area
Area = 12″ x 18″ = 1′ x 1½′ = 1½ ft^2
V = 7.06 cfs ÷ 1.5 ft^2 = 4.7 fps
Velocity greater than 1.5 fps and less than 5 fps.
Channel selected is OK.

Liquid-solid Separation

Runoff from lots carry organic matter and other solids. Settling facilities intercept all lot runoff, settle most solids, and release liquids to holding ponds, lagoons, or soil infiltration areas.

The most critical situation is high intensity, short duration rain. The 10-yr, 1-hr storm is commonly used for settling design. When a more intense storm occurs, the percent of manure solids removed is reduced. A system can receive short overloads with little loss in treatment efficiency.

Table 702-8. Capacity of roof and concrete gutters.
Standard roof gutters are too small for large roof areas. Capacities given are for smooth metal, smooth concrete, or planed lumber gutters or channels. Dashed lines refer to examples.
n = 0.012 S = 1/16 in/ft = 0.5%

Gutter depth inches	Gutter width, inches						
	6"	8"	10"	12"	14"	16"	18"
	Capacity in cfs						
2"	0.16	0.23	0.30	0.37	0.44	0.52	0.59
4"	0.41	0.60	0.80	1.02	1.24	1.46	1.68
6"	0.68	1.02	1.38	1.77	2.17	2.58	3.00
8"	0.96	1.46	2.00	2.58	3.19	3.82	4.46
10"	1.24	1.90	2.64	3.42	4.25	5.12	6.00
12"	1.5	2.4	3.3	4.3	5.4	6.5	7.6
14"	1.8	2.8	4.0	5.2	6.5	7.8	9.3
16"	2.1	3.3	4.6	6.1	7.6	9.2	10.9
18"	2.4	3.8	5.3	7.0	8.8	10.7	12.7
20"	2.7	4.2	6.0	7.9	9.9	12.1	14.4

Settling Basins

Settling basins remove 50%-85% of the solids from lot runoff, so fewer solids settle in the holding pond or soil infiltration area. Solids removal reduces odors from holding ponds and makes liquids easier to pump with smaller irrigation equipment.

A good settling basin is large and shallow. Make the basin at least 2′ deep to avoid disturbing previously settled solids. A concrete bottom permits solids cleanout with a front-end loader. Slope basin access ramps no more than 1″ fall/1′ run. Slope basin bottoms 1″-5″/100′ with a uniform grade to prevent pooling. Earthen basins should have a 3:1 (horizontal:vertical) side slope; if erosion is a problem, use a 4:1 slope on the inlet side. Recommended settling basin capacity in most of the Midwest is 10 ft^3/100 ft^2 of feedlot area. Design for at least 30 min of detention to settle most runoff solids.

If liquid manure equipment is available, consider handling the semi-solids in the settling basin as liquid manure. Periodically, agitate and pump out the basin. Cleaning frequency depends on basin size and type, lot slope, amount of lot manure, and storm runoff characteristics. In most of the Midwest, 1′ of basin depth/month can be used to determine cleaning frequency.

In dewatering basins, perforated or slotted pipe outlets with expanded metal screens drain liquids to allow solids to dry, Fig 702-10b. Remove solids after each major storm. Dewatering-type concrete basins should have at least one vertical wall for cleaning with front-end loaders. Another alternative is a weir notch in the basin endwall, Fig 702-11. With the weir notch, liquid remains in the basin after a storm. Clean with liquid manure equipment 3 or 4 times a year.

Settling Basin Design Considerations

In general, determine:

- Peak flow rate (Q) using 10-yr, 1-hr rainfall intensity.
  Q, ft^3/hr = rainfall intensity, in/hr x runoff area, ft^2 ÷ 12 in/ft
- Basin surface area (SA). Provide 1 ft^2 of settling area for each 4 ft^3/hr loading rate.
  SA, ft^2 = Q, ft^3/hr ÷ 4 ft^3/hr-ft^2
- Basin depth (BD).
  BD, ft = 4 ft^3/hr-ft^2 x detention time, hr + desired solids storage.
  Minimum depth is 2′ and maximum depth is 4′.
- Depth of solids (DS).
  DS, ft = lot area, acre x solids volume, acre-ft/acre x 43,560 ft^2/acre ÷ SA, ft^2
  For solids volume, use ½ acre-ft/acre per year off of unpaved lots and 1/12 acre-ft/acre per year off of paved lots.
- Settling basin volume (V).
  V, ft^3 = BD, ft x SA, ft^2

For weir notch settling basins

- Length of outlet weir (WL). Provide 1′ of weir length for each 110 ft^3/hr loading rate.
  WL, ft = Q, ft^3/hr ÷ 110 ft^3/hr-ft

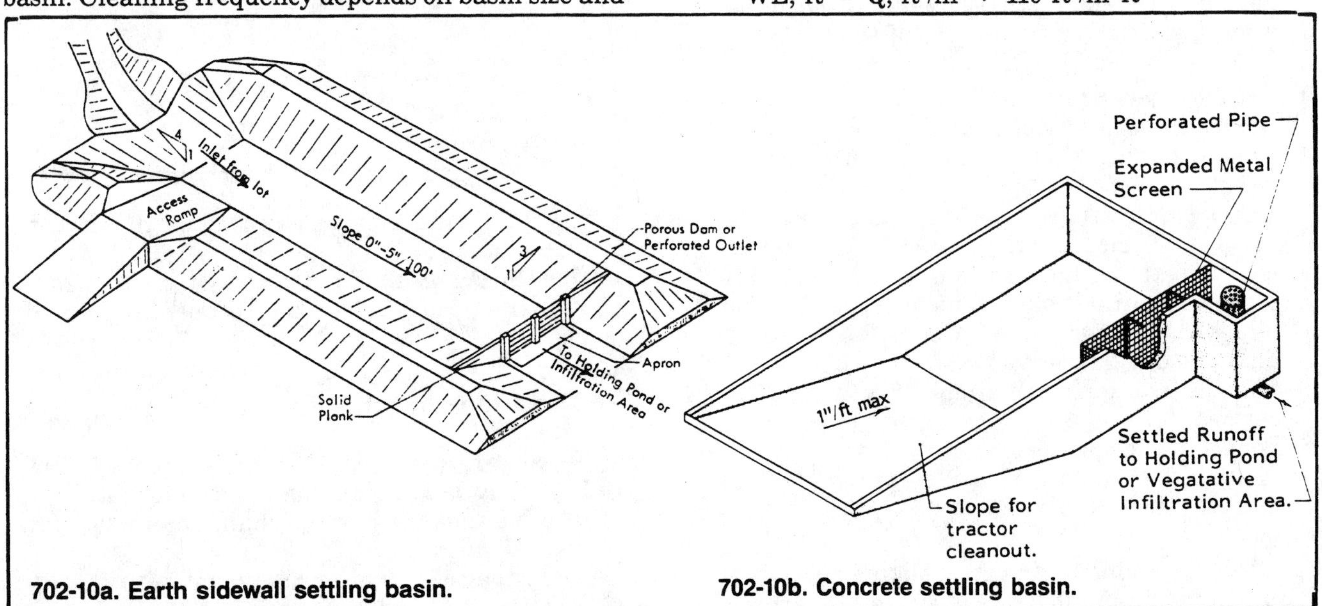

702-10a. Earth sidewall settling basin.

702-10b. Concrete settling basin.

Fig 702-10. Settling basins.
Expanded metal screens in 4′ lengths for removal during basin and screen cleaning.

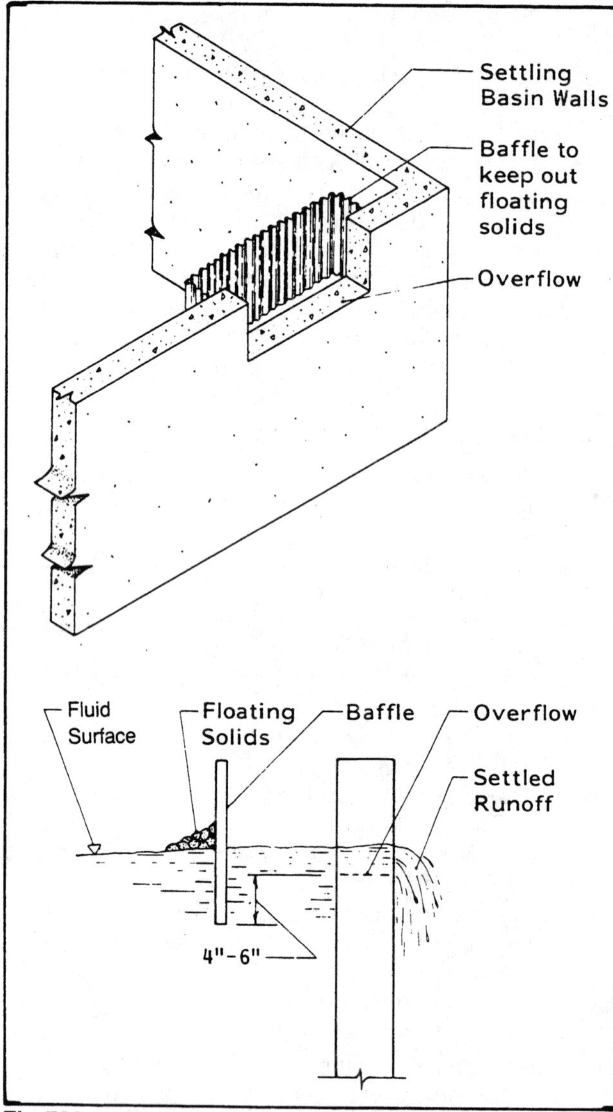

Fig 702-11. Weir notch in settling basin.

- Settling tank length (TL). Assume the outlet weir extends nearly the width of the tank.
  TL, ft = SA, ft^2 ÷ WL, ft

For dewatering settling basins

- Determine outlet opening area requirement using maximum flowrate, basin depth, and Fig 702-12.
- Select earth settling basin dimensions from Table 702-9. Because the rules of thumb for selecting settling basins assume vertical sides, compute required volume and select a somewhat larger basin. Assume square ends, 3:1 side slope, and a nearly flat bottom.
  Volume, ft^3 = length, ft x depth, ft x (width, ft + 3 x depth, ft)
  For 4:1 side slope, use:
  Volume, ft^3 = length, ft x depth, ft x (width, ft x 4 x depth, ft)

Example 6:

A 2′ deep settling basin with a slotted pipe outlet has a maximum flowrate of 40 cfm. Determine the outlet opening area required for the slotted pipe outlet.

From the curves in Fig 702-12, a 40 cfm flowrate with a 2′ head requires 10 in^2 of opening per ft of riser.

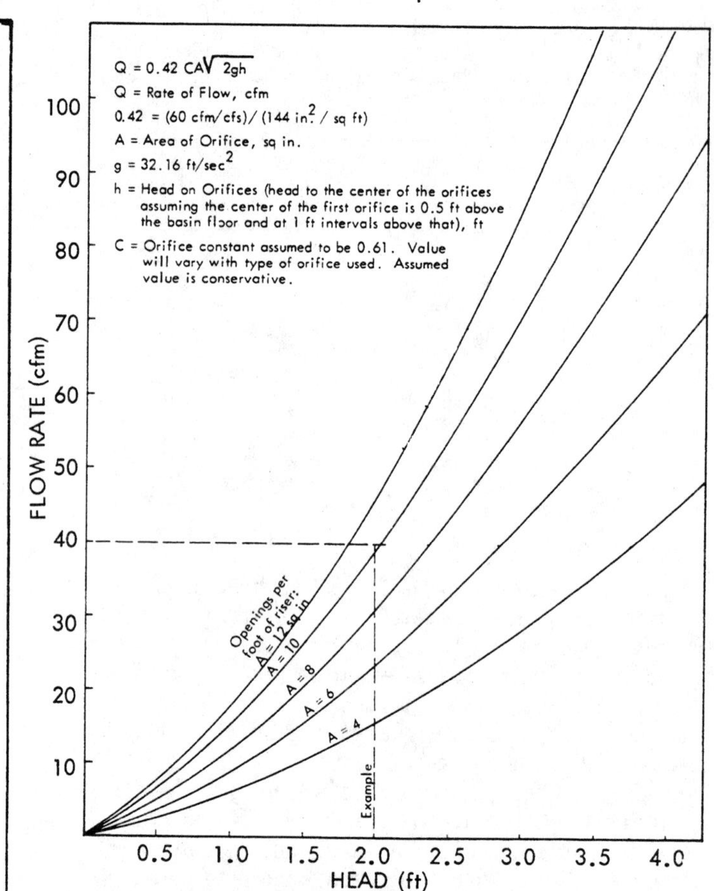

Fig 702-12. Settling basin outlet design.
Dashed lines refer to example.

Table 702-9. Volumes of earth settling basins.
Tabulated volumes are per 10′ length of basin and 1/100 of actual volume.

Width feet	Depth, feet 2'	4'	8'	12'	16'	20'
	100's of cu ft/10' basin length					
10'	3.2	8.8	27	55	93	140
15'	4.2	10.8	31	61	101	150
20'	5.2	12.8	35	67	109	160
25'	6.2	14.8	39	73	117	170
30'	7.2	16.8	43	79	125	180
40'	9.2	20.8	51	91	141	200
50'	11.2	24.8	59	103	157	220
60'	13.2	28.8	67	115	173	240

Example 7:

Assume the same runoff as in Example 2, Q = 27.9 cfs.

Select an appropriate channel from Table 702-10. A 0.1%, 1′x25′ channel is adequate. Determine the length of the channel for a 30 min detention time.

At 1 fps velocity, the channel must be:
30 min x 1 fps x 60 sec/min = 1800′

Table 702-10. Capacity of trapezoidal settling channels.
Construct channel at least ½′ deeper for solid waste storage.
n = 0.04; bottom slope as shown; side slope 3:1.
Dashed lines refer to example.

Bottom slope	Depth feet		10'	15'	20'	25'	30'	35'
Earth channels								
0.1%	0.5'	fps	0.6	0.7	0.7	0.7	0.7	0.7
		cfs	3.7	5.5	7.4	9.2	11.1	12.9
	1.0'	fps	0.9	1.0	1.0	1.0	1.1	1.1
		cfs	12.2	17.9	23.6	29.4	35.2	41.0
	1.5'	fps	1.2	1.2	1.3	1.3	1.3	1.4
		cfs	25.1	36.0	47.1	58.4	69.7	81.1
	2.0'	fps	1.3	1.4	1.5	1.5	1.6	1.6
		cfs	42.8	60.0	77.7	95.6	113.7	131.9
0.2%	0.5'	fps	0.9	0.9	1.0	1.0	1.0	1.0
		cfs	5.2	7.8	10.4	13.0	15.6	18.2
	1.0'	fps	1.3	1.4	1.5	1.5	1.5	1.5
		cfs	17.3	25.3	33.4	41.6	49.8	58.0
0.3%	0.5'	fps	1.1	1.2	1.2	1.2	1.2	1.2
		cfs	6.4	9.6	12.8	15.9	19.1	22.3
Concrete channels								
0.1%	0.5'	fps	1.6	1.9	1.9	1.9	1.9	1.9
		cfs	9.9	14.7	19.7	24.5	29.6	34.4

Bottom width, feet: 10' to 35'.

703 STORAGE

Storage design varies by state because of climate and pollution control regulations. Use recommendations on size, location, construction, and maintenance with care.

Avoid locating unlined storages over shallow creviced bedrock or below the water table. If shallow bedrock is present, consult a qualified engineer or contact your board of health. Also avoid storage in gravel beds or other areas where serious leakage can cause ground water pollution.

Allow at least 100′ between a water supply and the nearest part of a storage. Locate manure storages at least 50′ from a milkhouse or milking parlor. Check with milk and health authorities for minimum spacing requirements. Consider all farmstead operations, building locations, and prevailing winds when planning storages.

Locate, size, and construct storages for convenient filling and emptying and to keep out surface runoff. Provide all-weather access.

Storage Capacity

Storage capacity depends on regulations, number and size of animals, amount of dilution by spilled and cleaning water, amount of stored runoff, and desired length of time between emptying. Provide enough storage to spread manure only when field, weather, and local regulations permit. Post-harvest spreading saves time during busy spring planting activities. It also provides a chance for freezing and thawing during winter to lessen the effects of soil compaction from the spreading operation.

Plan for 10 to 12 mo storage capacity. Provide extra capacity for dilution water, rain, snow, and milking center waste. If the storage receives only animal manure (see Table 701-1), add dilution volumes of 10% for cattle and 20% for swine to account for waterer wastage, rain, and snow. Add up to 60% dilution volume to the animal manure production if milking center waste, sow wash, or farrowing stall wash water enter the storage. If the amount of additional water is unknown, plan for a total volume (manure plus dilution) equal to twice the animal manure production. Provide at least 1′ of freeboard.

From 20%-60% of the storage volume may be needed for dilution water if the manure is for irrigation.

Storage capacity =
(number of animals x daily manure production x desired storage time in days) + cleaning or leakage water + freeboard volume above lowest inlet opening

Filling

Top loading is suitable for a storage that will not crust over (e.g. one for swine wastes). It is also suitable or necessary for some of the methods listed below.

Use **bottom loading** for pumped wastes, such as cattle wastes. It reduces fly and odor problems and conserves nitrogen because the top surface crust is not disturbed. Keep the inlet pipe about 1′ above the bottom to prevent blockage.

Top-loaded solids pile up around the loading point, because the manure does not flow away, particularly in cold weather. Bottom loading pushes solids away from the inlet and distributes them more evenly.

Loading methods:

- **Barn cleaner extension.** Locate the storage next to the barn and top load. Freezing of machinery and manure can be a problem.
- **Tractor scraper or loader.** The storage is adjacent to the barn and top-loaded. Manure pileup at the push-off point can be a problem in freezing weather.
- **Conventional chopper-type liquid manure pumps.** Chopper pumps with a small sump or tank collect and pump the manure to the storage basin. The pump chops long bedding material for easier pump-out later. If the sump is outdoors, pump freezing can be a problem.
- **Piston manure pumps.** Piston pumps move manure through 6″-15″ diameter PVC or steel pipe. The pump is usually in the barn and the storage is up to about 300′ from the barn.
- **Pneumatic systems.** Compressed air forces manure through 12″ diameter PVC pipe to the bottom or side of the storage.
- **Centrifugal and other liquid manure pumps.** Liquids from swine barns, milkhouses, etc., can be pumped for top or bottom loading.
- **Gravity loading.** Storages below the level of the barn can sometimes be gravity loaded. For manure with solids less than about 5% (e.g. swine), a 6″-8″ pipe at 1% slope can fill or empty storage. For manure with higher solids content (e.g. cattle), a 24″-36″ pipe at 4% slope can fill or empty the storage. Avoid bottom loading by gravity. The inlet pipe liquid level is at the manure surface elevation in the storage. When manure flowing from the barn reaches the liquid level, its velocity drops, and solids settle out and plug the filling pipe.

Emptying

Use the following guidelines for emptying storage basins.

- If there is a crust, break it up before pumping any liquid from the basin. A high-volume, chopper-type, liquid manure pump is best for agitation. For best results, direct the agitator to swirl the crust. Slurry pumped on top of the crust will help to wet and sink floating dry material.
- If the pump is backed down a ramp, consider using a long hose connected to a stationary

standpipe outside the basin. Fill tank wagons without backing them down the ramp.
- Provide a paved equipment apron next to concrete below-ground storages.

Below-Ground Storage

Tanks and pits fall into three general groups:
- Shallow ones requiring relatively simple design.
- Large ones requiring complete structural design.
- Covered ones under slats or point-loaded by scrapers or liquid flushing. Tank lids, whether solid or slotted, need complete structural design.

In this chapter, pits are under buildings, and tanks are outside and usually separated from the barn. Use below-ground storages for slurry and liquid manure. Manure with up to about 15% solids can be agitated and pumped with conventional equipment.

Storage depth may be limited by soil mantle depth over bedrock, water table elevation, and possibly, effective lift of a pump. Columns and beams to support a floor or roof are usually spaced 8′-12′ apart. Provide agitation access ports no more than 40′ o.c.

Design and Construction

Tank design criteria are in the Concrete Pavements and Manure Tanks chapter. Tanks must be designed to withstand all anticipated earth, hydrostatic, and live loads, plus uplift if a high water table exists.

In cold climates, insulate the upper 2′-3′ of exterior tank walls with waterproof insulation. When construction is completed, clean out nails, lumber and other foreign material that could damage pumps. Before filling with manure, add water to keep solids submerged and counteract the uplifting forces caused by external pressures.

Evaluate site and soil conditions carefully to avoid contaminating ground and surface waters. Consider the soil characteristics to a depth of at least 5′ below the proposed storage basin. A soil survey is helpful in locating a site. Contact your county extension office or SCS for help in site evaluation.

Provide a firm base for the tanker loading area. Design for drive-through loading to avoid backing or maneuvering the tank wagon. When using a standpipe discharge, locate the outlet at the highest point on the line so the line drains after the pump is shut off.

Consider agitation requirements when selecting storage dimensions. Check manure pump manufacturer specifications. Locate agitation sites no more than 40′ o.c. for chopper-type manure pumps and 20′ o.c. for vacuum pumps. **Keep surface runoff out of the storage.** For dilution water, add roof runoff.

Protect openings with grills and/or covers and enclose open-top tanks with a fence at least 5′ high to prevent humans, livestock, or equipment from accidentally entering the tank. Provide removable grills over openings used for agitation and pumping. Install railings around all pump docks and access points for protection during agitation and cleanout. Provide wheel chocks and tie downs for pumps and tractors.

Storage Gutters

Narrow gutters are used primarily in swine facilities. Wastes from solid floors are washed or scraped to a gutter for short holding periods. The outlet is usually a 6″ bell tile with a plug. When the gutter is full (usually 3 days to 1 week), the plug is pulled and the manure flows by gravity to an outside storage or lagoon. See gravity gutters in the Collection chapter.

Wide gutters are used with swine or dairy. Most of the feces and urine are deposited in the gutter. Wastes are frequently removed from the gutter by scraping or flushing. Large gutters have greater capacity and are emptied periodically by gravity or pumping.

Earth Storage Basins

Outside storages can extend the capacity of indoor pits or provide total storage needs. Earth basins provide long-term storage at low to moderate investment. They are designed and constructed to prevent ground and surface water contamination. Check with your SCS office for help in evaluating site suitability, dike construction, and bottom sealing.

Locate storages to minimize odor nuisances and avoid eyesores. During construction, use the excavated material to build a dike around the basin. Dikes help screen the contents from view. If a basin is properly located, fenced, and its banks are well cared for, it should not be a public nuisance. Locate storages downwind and as far as possible from all homes.

Loading and unloading methods affect basin construction:
- Basin with one side vertical: rectangular basin with a vertical wall along one longer side, usually of concrete, Fig 703-1a. An 8′-10′ wide strip along the wall is paved to prevent the bottom from scouring during pumping.
- Basin with pumping platform: circular or rectangular basins with pumping platform(s) or dock(s), Fig 703-1b. A 10′-20′ square concrete pad under the pumping site prevents the bottom from scouring and keeps rocks and debris out of the pump. If a pad is not used, dig the pump-out area about 2′ deeper than the bottom of the pump.
- Basin with ramp: circular or rectangular basin with ramp(s) or driveway(s), Fig 703-1c. Pave the ramp with rough-surfaced concrete sloped no steeper than 10:1. If a self-loading wagon is used, pave the basin bottom.

Designing Outside Storage

- Decide how the storage will be emptied. Include items such as docks, ramps, pump access, and safety railings in your design.
- Keep the bottom of the storage at least 5′ above bedrock and 3′ above the water table or higher, depending on local pollution control regulations.

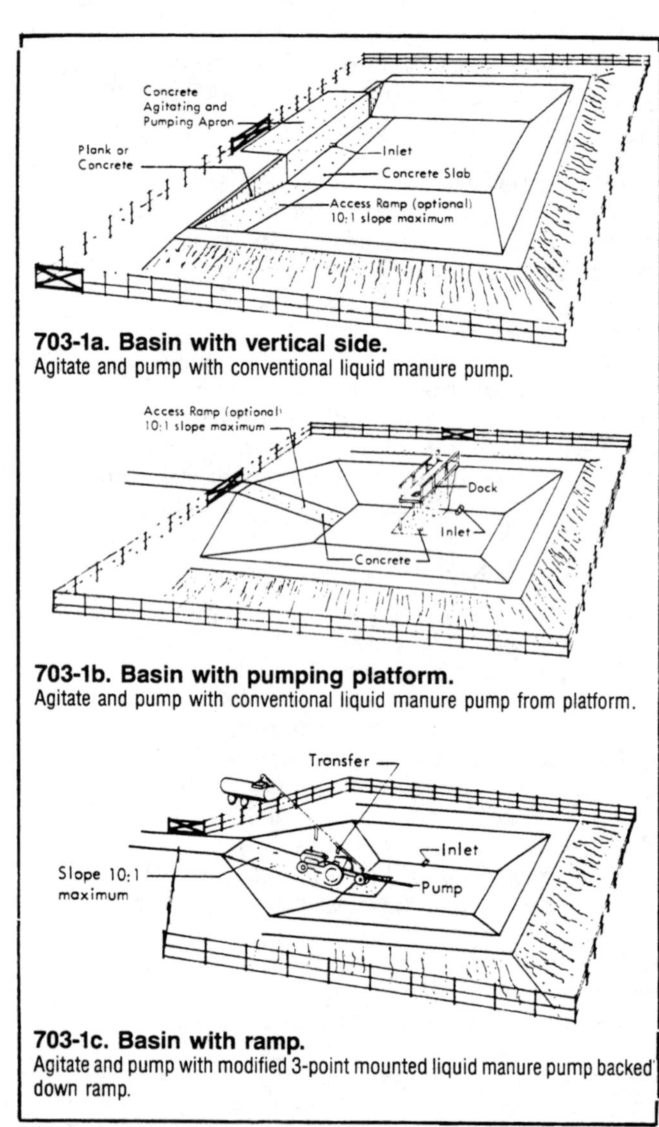

703-1a. Basin with vertical side.
Agitate and pump with conventional liquid manure pump.

703-1b. Basin with pumping platform.
Agitate and pump with conventional liquid manure pump from platform.

703-1c. Basin with ramp.
Agitate and pump with modified 3-point mounted liquid manure pump backed down ramp.

Fig 703-1. Earth storage basins.

- Study soil characteristics and check with your SCS office to determine basin wall side slopes. Follow accepted construction methods for sealing, building dikes, and bank seeding. In general, steeper banks conserve space, reduce rainfall runoff entering the basin, and leave less manure on banks as the basin is emptied. Inside bank slopes of 2:1 to 3:1 (run:rise) are common for most soils. Outside side slopes should not be steeper than 3:1 for maintenance.
- Determine the basin depth after considering the site, storage volume needed, groundwater conditions, and emptying equipment. Basins like Figs 703-1a and 703-1b usually have a maximum depth of 12′. Basins like Fig 703-1c can be deeper if site conditions permit, because the pump is backed down into the basin as cleanout progresses. A deeper basin requires less area for the same capacity. See Table 703-1.
- Use a slope no steeper than 10:1 for access ramps; 20:1 is preferred for tank wagons. Provide a rough surface for improved traction when wet. Form deep groves or ridges across the ramp with a garden rake or silage fork before the concrete sets.
- Maintain sod and mow vegetation.

Earth storage basin example

Size an earth basin to store manure for one year from 1,000 beef cattle averaging 1,000 lb each.

From Table 701-1, manure production per feeder is 1.0 ft^3/day.

1,000 feeders x 1.0 ft^3/feeder-day x 365 days = 365,000 ft^3

Adding 10% for dilution water = 36,500 ft^3

Total storage volume = 401,500 ft^3

From Table 703-1, determine the dimensions of the earth basin. Several designs are available. If 15′ deep, a 150′x324′ storage holds 425,000 ft^3.

For earth storages not in Table 703-1, determine the volume from:

Eq 703-1.

$$V = (W \times L \times D) - (S \times D^2) \times (L + W) + (4 \times S^2 \times D^3 \div 3)$$

W = width, ft
L = length, ft
D = depth, ft
S = side slope, ft
= amount of run for 1′ fall

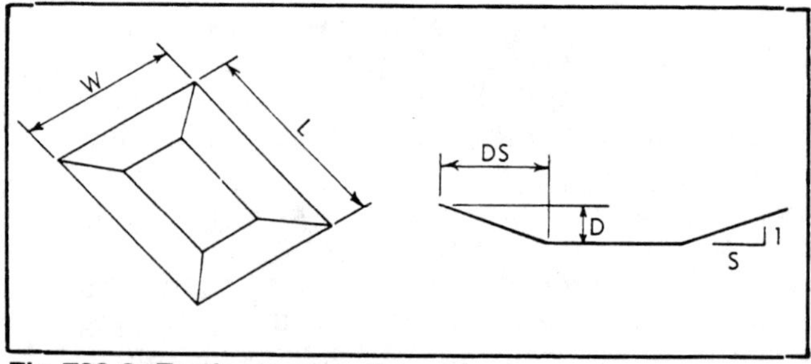

Fig 703-2. Earth storage dimensions.

Above-Ground Storage

Above-ground circular manure tanks are short, large-diameter tanks resembling silos. They are expensive compared to earth basins and are usually not used to store runoff or dilute wastes. But they are a good alternative where basins cannot be used because of space constraints, high groundwater, shallow creviced bedrock, or where earth basins are not aesthetically acceptable.

Above-ground liquid storages are from 10′-25′ high (15′ is most common) and 40′-100′ in diameter. Determine the required storage capacity and use Table 703-2 to size the storage tank. Estimate generously and allow for future expansion.

Construction

Construction options include:
- concrete stave.
- reinforced monolithic concrete.
- lap or butt joint coated steel.
- spiral-wound coated steel.

Leaks from joints, seams, or bolt holes can be unsightly, but most small leaks seal quickly with manure. The joint between the foundation and the

Table 703-1. Earth basin, holding pond, and lagoon capacities.

Bank slope = 2.5:1. Volume allows for 1′ of freeboard. All dimensions are in feet.

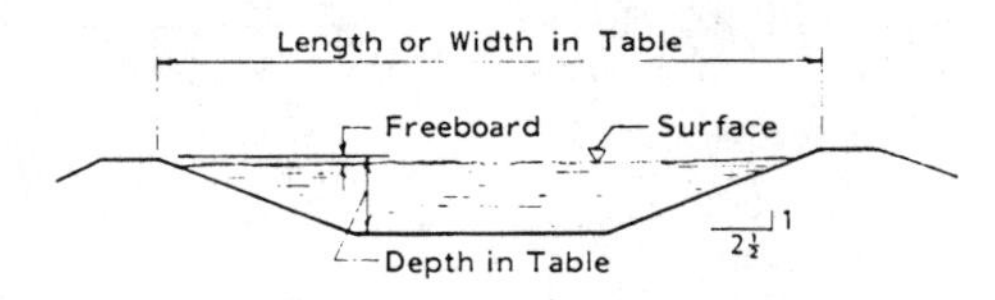

Volume, thousands of ft³	Depth 8, Interior width 50	75	100	150	200	Depth 10, Interior width 50	75	100	150	200	Depth 12, Interior width 50	75	100	150	200	Depth 15, Interior width 50	75	100	150	200
	Interior length, ft																			
4	49	37	-	-	-	48	38	-	-	-	48	-	-	-	-	37	-	-	-	-
6	62	44	-	-	-	61	44	-	-	-	61	45	-	-	-	57	46	-	-	-
8	75	50	42	-	-	73	49	42	-	-	74	50	-	-	-	78	51	-	-	-
10	89	57	46	-	-	86	55	46	-	-	88	55	-	-	-	98	55	-	-	-
15	122	74	57	-	-	117	69	55	-	-	121	67	-	-	-	149	67	56	-	-
20	155	90	68	51	-	148	83	64	50	-	154	80	63	-	-	201	79	63	-	-
30	222	124	91	64	-	211	110	82	60	-	221	105	78	-	-	303	103	77	-	-
40	289	157	113	78	63	273	138	100	71	-	288	130	93	68	-	406	126	90	68	-
50	355	190	135	91	72	336	166	117	81	66	354	155	109	77	-	508	150	103	75	-
60	422	224	157	104	82	398	194	135	91	73	421	180	124	85	70	611	174	117	82	70
70	489	257	180	118	91	461	221	153	102	81	488	205	139	94	76	714	197	130	89	74
80	555	290	202	131	101	523	249	171	112	88	554	230	155	103	82	816	221	143	96	79
90	622	324	224	144	110	586	277	189	123	95	621	255	170	111	88	919	245	157	104	84
100	689	357	246	158	120	648	305	207	133	103	688	280	186	120	94	-	268	170	111	89
110	755	390	268	171	129	711	333	225	143	110	754	305	201	129	100	-	292	184	118	94
120	822	424	291	184	139	773	360	242	154	117	821	330	216	138	106	-	316	197	125	99
130	889	457	313	198	148	836	388	260	164	125	888	355	232	146	113	-	339	210	132	104
140	955	490	335	211	158	898	416	278	175	132	954	380	247	155	119	-	363	224	139	109
150	-	524	357	224	167	961	444	296	185	140	-	405	263	164	125	-	387	237	147	114
160	-	557	380	238	177	-	471	314	196	147	-	430	278	172	131	-	410	250	154	118
170	-	590	402	251	186	-	499	332	206	154	-	455	293	181	137	-	434	264	161	123
180	-	624	424	264	196	-	527	350	216	162	-	480	309	190	143	-	458	277	168	128
190	-	657	446	278	206	-	555	367	227	169	-	505	324	198	149	-	481	291	175	133
200	-	690	468	291	215	-	583	385	237	176	-	530	339	207	155	-	505	304	182	138
225	-	774	524	324	239	-	652	430	263	195	-	592	378	229	170	-	564	337	200	150
250	-	857	580	358	263	-	721	475	289	213	-	655	416	251	185	-	623	371	218	162
275	-	940	635	391	286	-	791	519	315	231	-	717	455	272	200	-	683	404	236	175

Volume, thousands of ft³	Depth 10, Interior width 100	150	200	300	400	Depth 15, Interior width 100	150	200	300	400	Depth 20, Interior width 100	150	200	300	400	Depth 25, Interior width 100	150	200	300	400
	Interior length, ft																			
300	564	341	250	168	131	438	254	187	131	106	410	223	165	120	101	435	212	158	119	103
325	608	367	268	180	139	471	272	199	138	111	441	238	175	126	105	468	225	166	124	107
350	653	393	287	192	148	505	290	211	146	117	472	253	184	132	109	502	239	174	128	110
375	698	420	305	203	156	538	308	223	153	122	503	267	194	137	113	535	252	182	133	113
400	742	446	323	215	165	571	325	236	161	128	534	282	204	143	117	569	265	190	138	116
425	787	472	342	226	173	605	343	248	168	133	565	296	213	149	121	602	278	199	142	120
450	832	498	360	238	182	638	361	260	176	138	596	311	223	154	126	636	291	207	147	123
475	876	524	378	249	190	672	379	272	183	144	626	326	232	160	130	669	304	215	152	126
500	921	550	397	261	199	705	397	284	190	149	657	340	242	166	134	702	318	223	156	130
600	1100	654	470	307	232	839	469	333	220	171	781	399	280	188	150	836	370	256	175	143
700	-	758	544	354	266	973	540	382	250	192	904	457	319	211	166	970	423	289	194	156
800	-	862	617	400	300	1107	612	431	280	214	1028	516	357	234	182	1104	476	322	212	169
900	-	966	691	446	334	1240	683	480	310	235	-	574	395	256	198	-	528	355	231	182
1000	-	1071	765	492	367	1374	755	529	340	257	-	633	433	279	214	-	581	387	250	195

sidewall can be a problem with improper construction. The reliability of the dealer and construction crew are as important as the tank material in ensuring satisfaction.

Filling and Emptying

Above-ground storages are loaded with the same equipment as below-ground earth storages.

Commonly, agitation is with a centrifugal pump mounted on the storage foundation, Fig 703-3. The pump is connected with a large (12″x12″) slide valve to an inlet at the base of the silo storage. A tractor PTO operates the pump. For agitation, the pump discharges over the top into the storage. With large diameter tanks, a center agitation nozzle can be used.

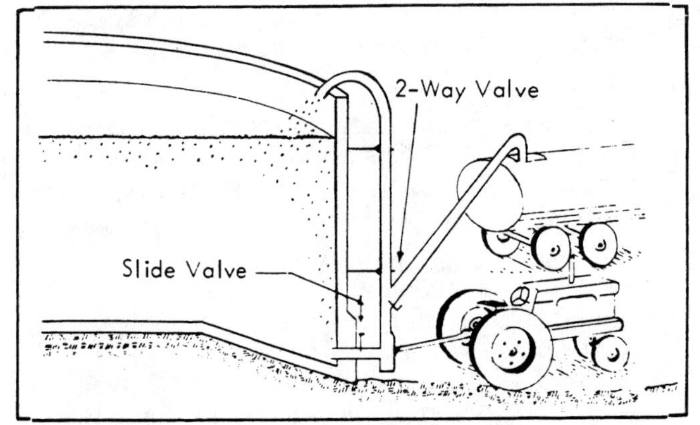

Fig 703-3. Side-mounted liquid manure pump.

After agitation, a valve on the discharge pipe is changed to pump into tank wagons. This recirculation system seems to be the simplest and most trouble-free method of agitating and unloading above-ground storages, although slide valves sometimes freeze.

An external sump and conventional chopper-type liquid manure pump can also agitate and unload, Fig 703-4. A large pipe or duct (about 12") with a valve drains the storage to the sump. The sump, underdrain piping, and valving add to the cost, but only one pump is needed for filling and emptying. Take care during pumping and agitating so the gravity drain fills the sump at the same rate as the pump empties it. The manure at the bottom of the tank must be liquid enough to flow to the sump for the recirculation process to start.

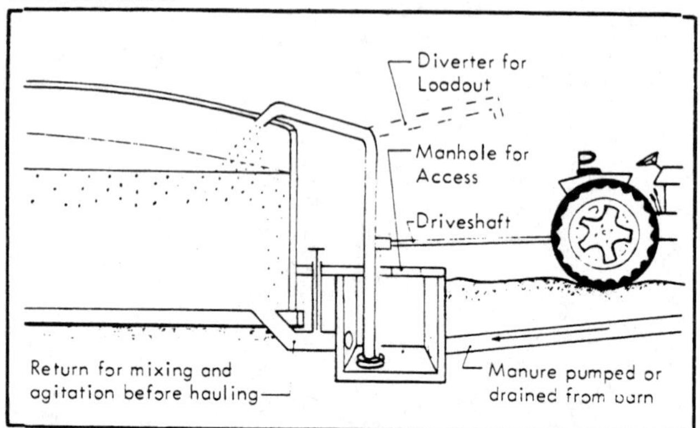

Fig 703-4. Chopper liquid manure pump.

Agitation can be a problem if the storage is too full, has too much bedding, or does not have enough dilution water. If sand or ground limestone get into the storage, hard sludge forms that is difficult or impossible to pump out.

Management

Put into the storage only material that can be agitated and pumped out. Keep out excess water unless dilution water is needed. Be sure adequate storage space is available for long periods, such as winter or cropping.

Remove ladders or other climbing devices to protect uninformed people. Maintain access to the storage for unloading. Follow manufacturers' instructions for draining pumps, pipes, and valves during cold weather. Do not allow runoff from the loading area or storage into streams or road ditches.

Above-ground storage example

Size an above-ground storage to provide 180 days of storage for 350 pigs averaging 150 lb/pig.

From Table 701-1, daily manure production per 150 lb pig is 1.13 gal/day.

350 pigs x 1.13 gal/pig-day x
180 days = 71,190 gal
Adding 20% for dilution water = 14,238 gal
Total storage volume = 85,428 gal

Select tank dimensions from Table 703-2. Several alternatives are available, such as 40′ diameter x 10′ high (94,200 gal).

Table 703-2. Round above-ground storage capacity.

Storage dimensions	Gal	ft³	Storage dimensions	Gal	ft³
40′D x 10′H	94,200	12,600	60D x 10′H	212,000	28,300
x 15′H	141,400	18,800	x 15′H	318,100	42,400
x 20′H	188,500	25,100	x 20′H	414,100	56,500
x 25′H	235,600	31,400	x 25′H	530,100	70,700
45′D x 10′H	119,200	15,900	72′D x 10′H	305,300	40,700
x 15′H	178,900	23,800	x 15′H	458,000	61,100
x 20′H	238,500	31,800	x 20′H	610,700	81,400
x 25′H	298,200	39,800	x 25′H	763,400	101,800
50′D x 10′H	147,200	19,500	80′D x 10′H	377,000	50,200
x 15′H	220,900	29,400	x 15′H	565,500	75,400
x 20′H	294,500	39,300	x 20′H	754,000	100,500
x 25′H	368,100	49,100	x 25′H	942,500	125,600
55′D x 10′H	178,100	23,700	100′D x 10′H	589,000	78,500
x 15′H	267,300	35,600	x 15′H	883,600	117,800
x 20′H	356,300	47,500	x 20′H	1,178,100	157,100
x 25′H	445,400	59,400	x 25′H	1,472,600	196,300

Above-Ground Roofed Storage

This system was developed for dairy comfort stall barns where large amounts of bedding are mixed with manure. Typically, piston pumps transfer manure to storage. Manure in storage crusts over, reducing odor and fly breeding problems.

Above-ground roofed storages have a concrete floor up to 3′ below existing grade. These storages usually have 12′ post and plank or concrete walls with earth backfill to withstand the internal fluid pressure. The roof protects collected manure from rainwater.

Unload by lifting out reinforced horizontal planks from a door in one end of the storage. Manure then flows over the wall onto a concrete slab outside the storage where a front-end loader fills a conventional spreader.

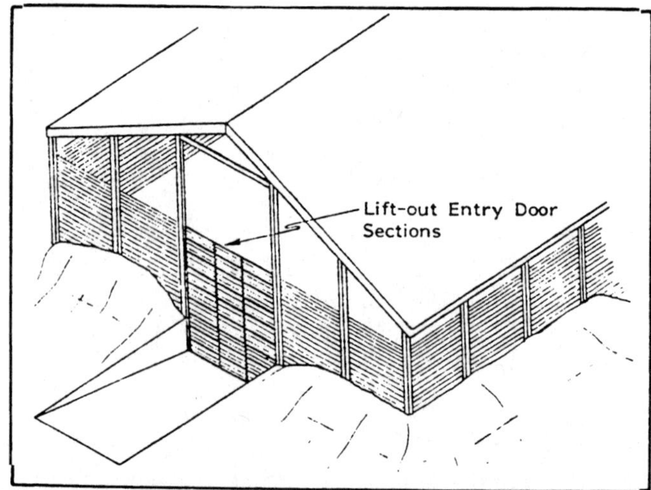

Fig 703-5. Roofed above-ground storage.

Holding Pond

A holding pond temporarily stores runoff water from a settling basin, Fig 703-6. Design for at least the storage time required in your state (usually 90 to 180 days). Base capacity on inches of rainfall on a drained area, such as a 25-yr, 24-hr rainfall event.

With additional capacity, emptying the pond can be delayed to fit labor and cropping schedules without fear that another runoff event will cause overflow.

Features of a holding pond:

- The bottom and sides are essentially watertight to avoid possible groundwater contamination.
- It receives runoff from a lot, usually after it has been through a settling unit, or from a lagoon's overflow.
- The interior banks of earth dams are usually no steeper than 2.5:1, depending on soil type. To maintain sod and mow weeds, limit exterior slopes to no steeper than 3:1.
- Empty the pond by pumping, usually through irrigation equipment.
- Pump before the storage is full and when wastes will infiltrate into the soil. Avoid spreading on frozen or wet ground.

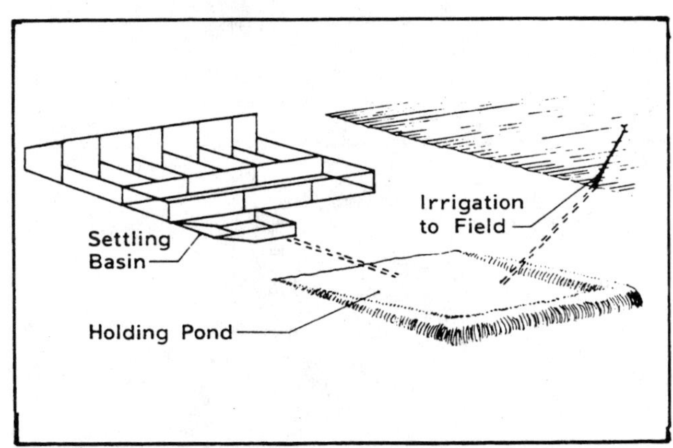

Fig 703-6. Holding pond.

Inlets

Inlets are discussed in the section on anaerobic lagoons. Because the liquid velocity in an inlet slows at the water line, solids can settle and plug inlets. An inlet pipe above the maximum water surface helps prevent plugging. A combination inlet helps prevent plugging and freezing problems. Pipes supported to discharge over the flat bottom of the pond are preferred, but freezing and thawing disturbs most pipe supports.

Storage capacity

Although part of needed capacity is difficult to estimate, e.g. sludge buildup, most of the major elements can be predicted with adequate accuracy. Fig 703-7 illustrates the volumes to be stored.

To determine needed storage capacity, consider:

- **Lot runoff** estimated from waste and climate data, and anticipated management practices.
- **Runoff solids** which can be a ½" layer of solids over the area drained. Settle out solids to reduce odors, required storage volume, and handling problems. Unless these are settled out, design to store and remove then.
- **Other liquid sources** and their required storage volume.
- **Sludge buildup** which depends on settling facilities and lot surface.
- **Net rainfall** estimated from weather data; use 25-yr, 24-hr rainfall. Estimate rainfall on storage surface and holding pond banks.

Add the total capacity required for the above five factors. Select storage size from Table 703-1. Use Eq 703-1 to determine the capacity of storages not given in Table 703-1.

Solid Manure

Cattle manure can be stored and handled as a solid or semi-solid if additional bedding is used and water is excluded. By handling all the manure as a solid, conventional equipment can be used.

Keep runoff and excess water out of the manure storage. Fill the storage with a tractor-mounted manure loader or scraper, elevator stacker, blower stacker, or piston pumping system. Unload with a tractor-mounted bucket or manure bucket.

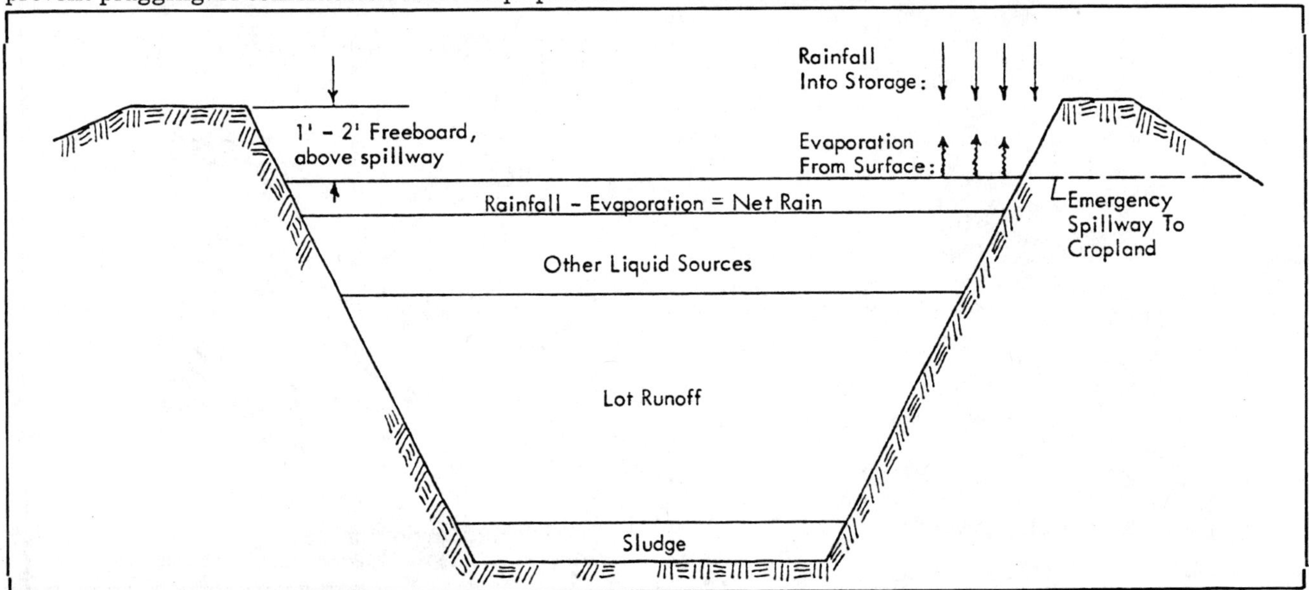

Fig 703-7. Liquids stored in holding ponds.
For capacity of earth storages, see Table 703-1.

Construction

Provide convenient access for unloading and hauling equipment. Slope entrance ramps upward to keep out surface water. Provide a load-out ramp at least 40′ wide with a 10:1 slope (20:1 preferred). A roughened ramp improves traction. Angle grooves across the ramp to drain rainwater. Concrete floors and ramps are recommended. Slope the floor ¼″/ft (2%). Use a 4″ concrete floor over 4 mil plastic over 6″ coarse gravel or crushed rock (up to 1½″ aggregate size). Two inches of sand can replace gravel over undisturbed or compacted soil.

Walls are usually concrete or post and plank. Provide one or two walls to buck against for unloading.

If liquids are to be retained in the storage, slope the floor toward a closed end.

To drain off liquids:

- Slope the floor toward a picket dam. A picket dam with continuous vertical slots about ¾″ wide between planks holds manure solids and allows liquids to drain. Vertical slots work much better than horizontal slots. See Fig 703-8.
- Provide floor drains with removable grills. Install underground noncorrosive 6″ pipe to carry away liquids.
- Provide a gutter to carry runoff away. Use removable covers of pressure-treated wood planks spaced about ¼″ apart.

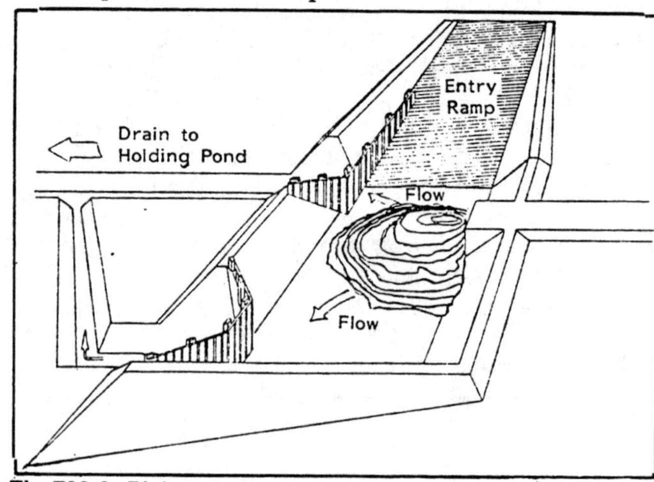

Fig 703-8. Picket-drained storage.

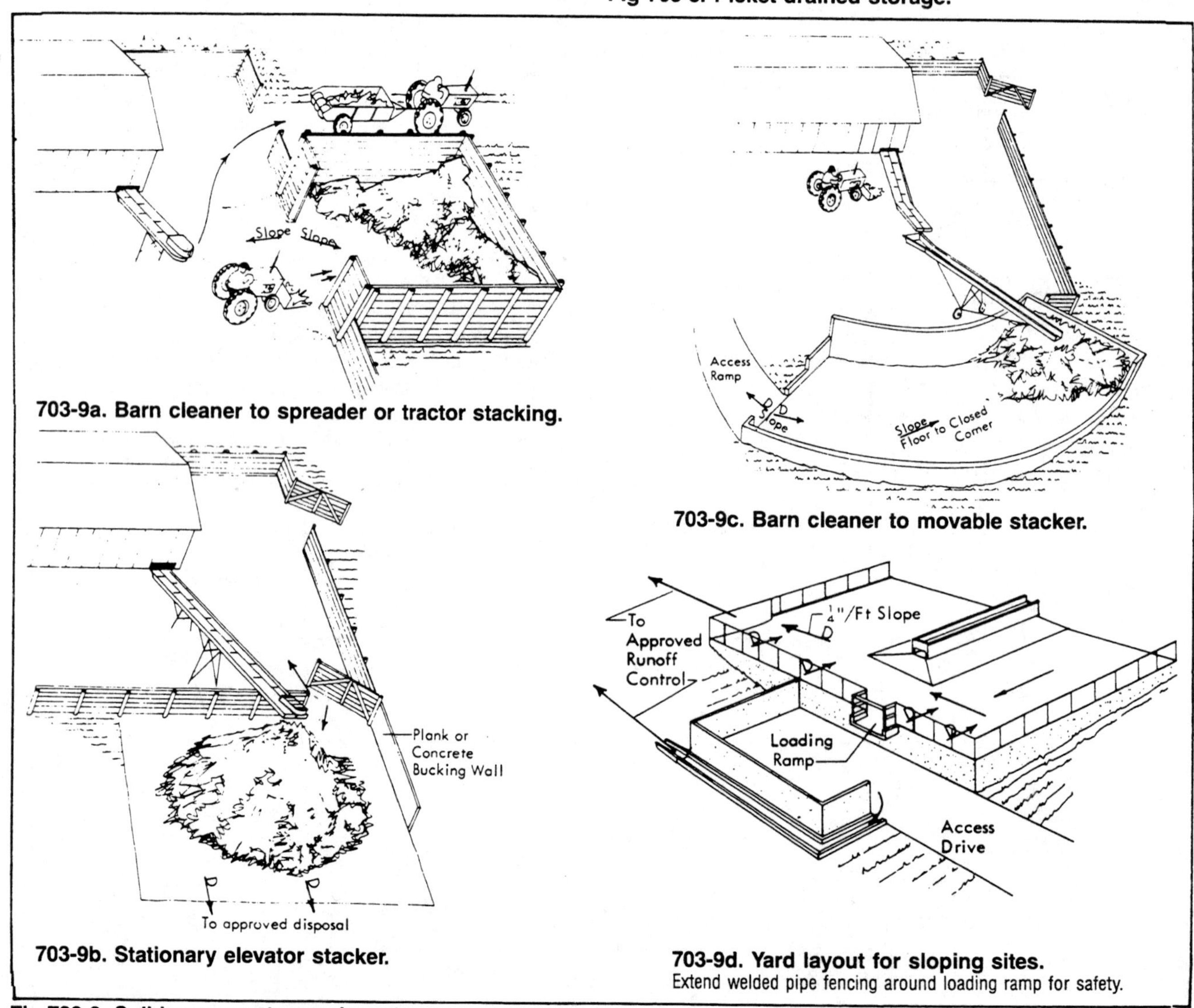

703-9a. Barn cleaner to spreader or tractor stacking.

703-9c. Barn cleaner to movable stacker.

703-9b. Stationary elevator stacker.

703-9d. Yard layout for sloping sites.
Extend welded pipe fencing around loading ramp for safety.

Fig 703-9. Solid manure storage layouts.
Divert surface drainage around lots and storage. Fence cattle away from storage. Divert lot drainage away from storage to approved disposal.

704 BIOLOGICAL TREATMENT

A biological treatment system is designed and operated for biodegradation—converting organic matter (feed, bedding, body byproducts) in animal manures to more stable end products.

Anaerobic processes occur without free oxygen and liquefy or degrade high BOD (biochemical oxygen demand) wastes.

Aerobic processes require free oxygen, and although aerobic lagoons require an oxygen supply, their odor-free operation can be a valuable advantage.

Facultative microorganisms can function either anaerobically or aerobically, depending on their environment.

Ideal conditions are not required because many varieties of microorganisms are adaptable. Anaerobic and facultative bacteria in animal excreta multiply rapidly. However, extreme environmental changes alter microbial activity. When microorganisms are stressed by the environment, waste treatment processes malfunction and odors can become a problem.

Anaerobic Lagoons

Anaerobic processes can decompose more organic matter per unit volume than aerobic ones. Therefore, use the anaerboic process for initial stabilization of strong (high BOD) organic wastes.

Anaerobic lagoons are more common than aerobic lagoons for treating livestock wastes. They can handle high loading rates, but give off some foul odors.

Advantages

- Long storage times permit labor flexibility while bacteria break down solids into liquid and gas.
- The high degree of stabilization can reduce odors during spreading.
- No oxygen is required, so no aeration system is needed.
- Nitrogen reduction is an advantage for disposal on small areas.

Disadvantages

- Odors are produced if environmental or management changes reduce biological activity—lagoons are sensitive to sudden changes in temperature or loading rate.
- Where winter water temperatures are near or below freezing, lagoons experience spring and fall inversion—that is, bottom water rises and top water falls. After a winter of little bacterial action because of low temperatures, odorous material from the bottom of the lagoon rises to the surface. Further, higher spring water temperatures increase microbial action, and foul odors are generated during bacterial buildup.
- Because higher temperatures improve manure decomposition, anaerobic lagoons work best above 70 F and in areas without cold winters.
- Fertilizer value reduction is a disadvantage if lagooon effluent is used to fertilize crops. Up to 80% of N is lost in a lagoon. Phosphorous may precipitate to the bottom but can be recovered when sludge is removed.

Function

Anaerobic lagoons liquefy and break down manure solids. Nondegradable wastes settle to the bottom as sludge. Sludge accumulation depends on management, environment, waste characteristics, and loading rate. Lagoons usually fill with sludge after about 10 to 15 years.

In areas with high evaporation rates, add extra water to maintain the lagoon's design volume. Mineral buildups, resulting from evaporation, require periodic diluting and pumping.

Because complete treatment is not practical, occasional pumping and draining are necessary management practices. Provide sufficient land area to properly dispose of lagoon effluent and sludge.

Well designed and managed lagoons have a musty odor. Continuous foul odors indicate malfunctioning. Do not use an anaerobic lagoon where odors may be a nuisance.

Design

Anaerobic lagoon design results from experience with successfully operating lagoons. Lagoons require more volume in cold than in warm climates because bacterial activity nearly stops at freezing.

Recommended design loading rates for the U.S. vary from 0.001 to 0.01 lb of volatile solids per ft^3 per day. Use the average daily volatile solids contribution, Table 701-1.

Table 704-1. Anaerobic lagoon design volume.
Cold = considerable ice formed each winter; mild = some ice; warm = no ice. The volumes listed include no allowance for sludge or dilution.

	Livestock			
Climate	Swine	Beef	Dairy	Poultry
	ft^3/lb liveweight			
Cold	2.7	2.4	3.4	3.6
Mild	1.8	1.2	1.7	1.8
Warm	1.4	0.6	0.9	0.9

Table 704-1 presents the minimum **design** volume per pound of livestock relative to ice buildup. To find the lagoon design volume, multiply the minimum volume by the number and average weight of animals. The larger required volumes frequently cited include allowances for spilled water, sludge accumulation, precipitation, etc.

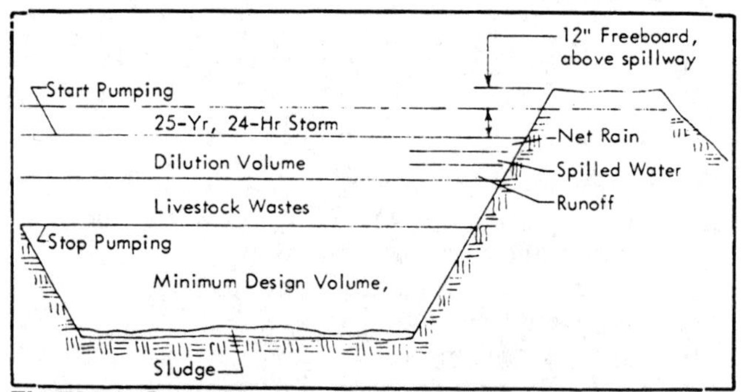

Fig 704-1. Capacity of one-stage anaerobic lagoon.
Wastes from inside a building.

Lagoon size

Minimum **design volume** provides enough space to maintain adequate bacterial populations. **Always** maintain this volume.

Livestock wastes enter the lagoon regularly—picture them occupying space above the minimum design volume.

Dilution volume is the water that must be added to the lagoon between each pumping period. Dilution water comes from wastewater, overflow from waterers, diverted rainfall from roofs and surface water, precipitation minus evaporation and, if necessary, pumped water from a well or water course. Livestock wastes are strong, and unless a lagoon is regularly diluted, it malfunctions because minerals and wastes become too concentrated.

Safety margin includes allowance for a 25-yr, 24-hr rain on the lagoon surface. Discharge from this or even a lesser storm is prohibited. It also includes freeboard above the emergency spillway so that if the lagoon should overflow, the dam will not be destroyed.

Procedure

Select the **minimum design volume** from Table 704-1 for your livestock and climate. Multiply it by the number of animals, and their average weight.

Decide on a **pumping cycle.** Usually, the lagoon is pumped in early spring before the water warms up and bacterial action starts again. Pump again in late fall while soil conditions and regulations still permit disposal.

- With twice-a-year pumping, select waste and dilution volumes for six months.
- With once-a-year pumping, select wastes and dilution volumes for one year. Greater storage volume is needed.

Compute the **livestock wastes** expected. Multiply the appropriate figure from Table 701-1 by the number of animals whose wastes will enter the lagoon and the number of days (usually 180) between planned pumpings. If the manure contains bedding, which is slow to decompose, allow for additional sludge accumulation.

The **dilution volume** is ½ the minimum design volume per year.

- Determine net rainfall—the difference between precipitation and evaporation—for your area. See the Climate section. Assume rain falls on the lagoon surface and on the banks that drain onto the surface. Assume evaporation only from the liquid surface.
- Prepare to divert roof water and surface drainage water into the lagoon if needed.
- After the lagoon is operational, keep the liquid level up to the dilution volume. The lagoon **cannot** operate properly without adequate and regular dilution.

Compute the **safety margin** for a 25-yr, 24-hr storm. See the Climate section. Apply this rainfall to all areas that drain into the lagoon, and compute the volume of water. Pump at least down to the **start pumping** level whenever the lagoon approaches spillway elevation, or another storm may cause overflow.

Total lagoon volume is the sum of minimum design volume, wastes volume, dilution volume (usually ½ design volume), and 25-yr, 24-hr storm volumes. Set the spillway height to provide this volume and build the dam 12″ higher for freeboard.

Size of two-stage lagoon

A two-stage lagoon needs more total volume than a one-stage lagoon. The effluent from the second stage is of higher quality. The first cell contains the minimum design volume and is kept at constant depth. Size the second cell for the volume of wastes expected, plus needed dilution volume, plus about ¼ the minimum design volume. Because only the first cell needs diluting, do not include net precipitation

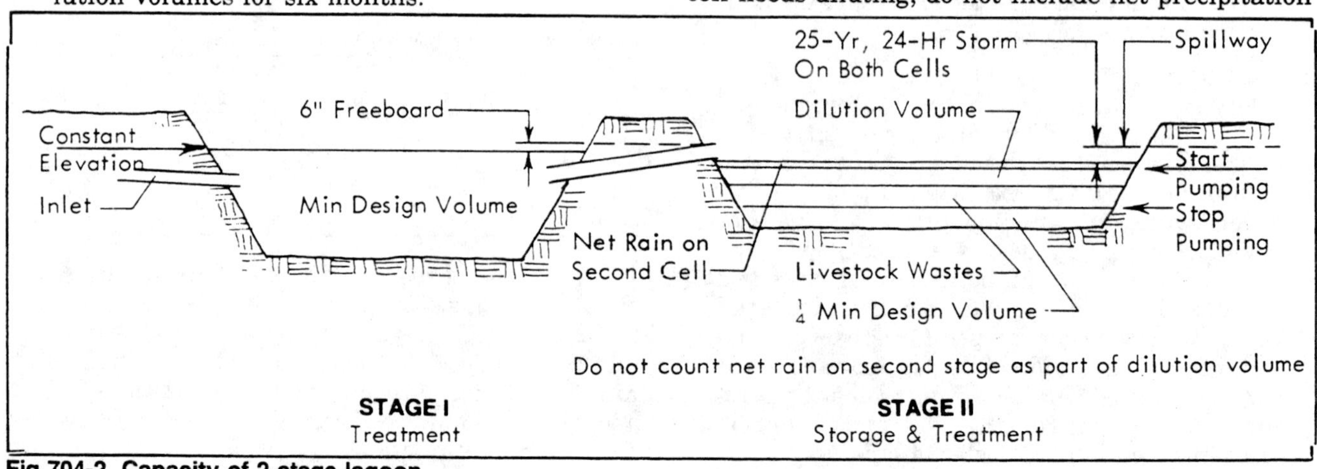

Fig 704-2. Capacity of 2-stage lagoon.

on the second cell as part of the required dilution volume. Do all pumping from the second stage.

Size of combined storage

The same earth basin can serve both as a lagoon to treat livestock wastes and as a holding pond to store liquid runoff for field spreading. Consider the runoff as stored on top of the lagoon.

Determine **minimum design volume,** Table 704-1, for all wastes from buildings and about half the wastes from lots (assuming other outdoor wastes are scraped and field-spread as solids). Determine the **waste volume** to be added between pumpings. Determine the **dilution volume** required; it is frequently less than the runoff to be stored. Determine the **runoff** and net precipitation to be stored.

Total lagoon volume is the sum of the minimum volume, the animal manure, and either dilution or runoff volume, whichever is greater. Provide at least 12″ of freeboard. If at any time the elevation approaches the spillway, pump within 20 days or as required by federal or state regulations.

Lagoon Construction

Location

Locate a lagoon as far from the farm home as practical, and where the prevailing winds will carry odors away from houses. Lagoon odors can be objectionable at distances of ½ mile and detectable at distances of a mile or more.

Locate the lagoon near the waste source. If the lagoon is downhill from the source, gravity can transport the waste. A sump and submersible sewage lift pump can elevate wastes into a lagoon if necessary.

Locate the lagoon over impervious soil, where bottom and sidewalls can be impervious. SCS and Agricultural Extension Service personnel can evaluate your soil. Avoid a site where the bottom of the lagoon would be less than about 20′ above limestone. Remove any field tile in the area.

Lagoons on some soils require sealing with liners, clay, or soil cement. Sealing is partially biological—animal waste solids are a good sealant in many soils. Use soil or soil cement to delay leaking while biological sealing can develop. Membrane sealing (plastic, vinyl, rubber, etc.) is positive and effective but expensive and difficult to install.

Consult your Agricultural Extension Service or state health or pollution control authorities for regulations governing lagoons near wells. If the lagoon must be built near a locally recharged shallow well, the bottom of the well should be higher than the top of the lagoon.

Inlets, trickle tubes, and outlets

When planning diversions for surface water, consider that some surface water may be needed for dilution. Locate pumping outlets for irrigation away from inlets to avoid unliquefied solids. Extend inlets to a lagoon beyond the cut slope of the embankment to reduce erosion and to uniformly distribute the waste load. See Fig 704-3.

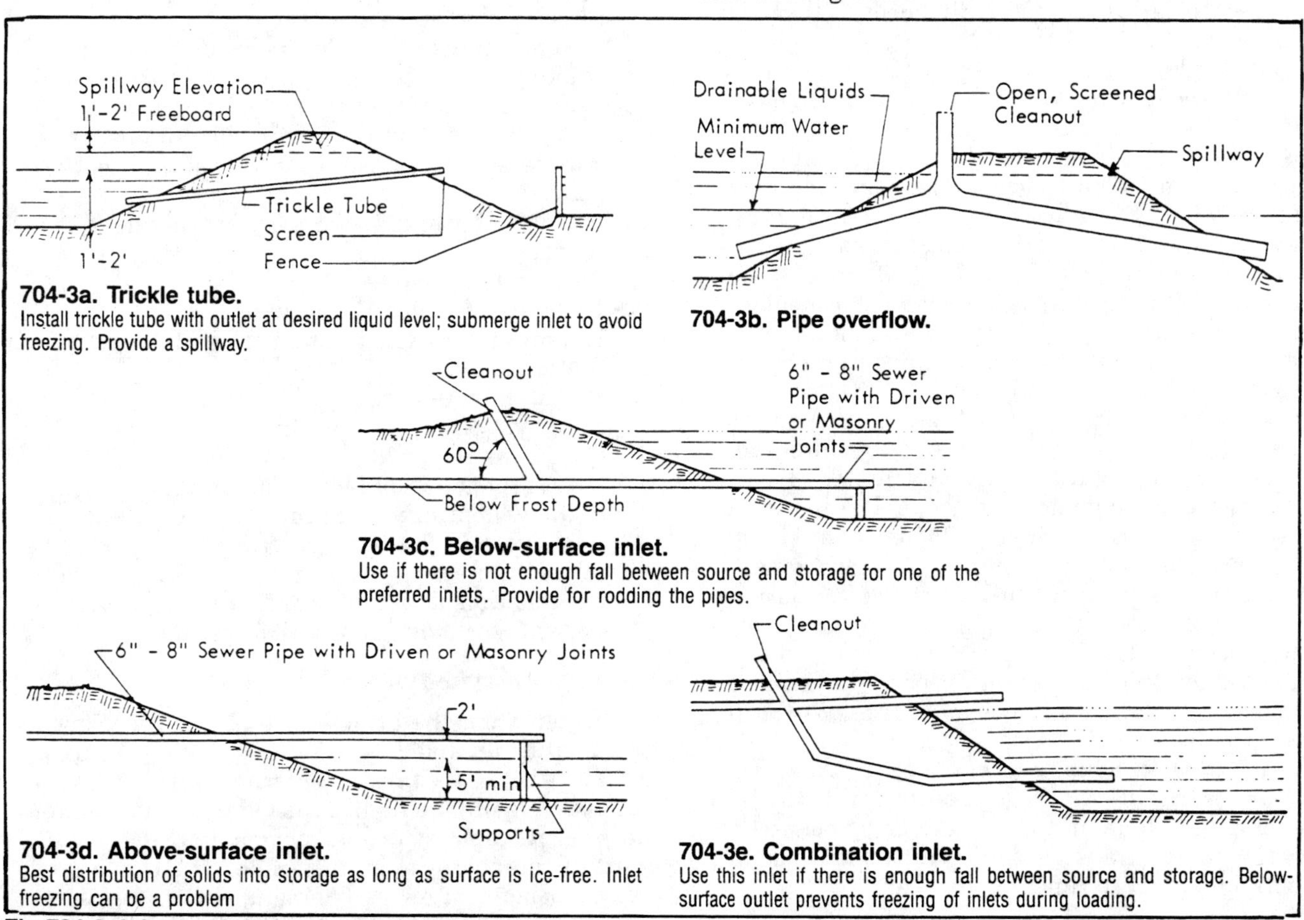

704-3a. Trickle tube.
Install trickle tube with outlet at desired liquid level; submerge inlet to avoid freezing. Provide a spillway.

704-3b. Pipe overflow.

704-3c. Below-surface inlet.
Use if there is not enough fall between source and storage for one of the preferred inlets. Provide for rodding the pipes.

704-3d. Above-surface inlet.
Best distribution of solids into storage as long as surface is ice-free. Inlet freezing can be a problem

704-3e. Combination inlet.
Use this inlet if there is enough fall between source and storage. Below-surface outlet prevents freezing of inlets during loading.

Fig 704-3. Inlets and drains for earth storages.
Slope inlet pipes so there is at least 2 fps velocity with dilute wastes (less than 5% solids).

A lagoon may freeze over in very cold weather. If this happens, either provide separate manure storage until the lagoon thaws or increase the inlet height to leave room above the ice for winter manure.

Shape

Build single-cell lagoons about square or round for uniform loading over the lagoon area and economical construction and maintenance. Build each cell of a multi-cell lagoon up to five times as long as wide. Install inlets and outlets at the ends to ensure transferring only wastes that have been in the cell for the design detention time. Consider limiting width to less than 75′ so a dragline can remove sludge if necessary. Make the lagoon as deep as practical to limit surface area, reduce odors, and increase vertical mixing of lagoon contents. A berm or dam width greater than 8′ is needed for machine operation, and 12′ is preferred.

Side slope is determined primarily by soil type, with more stable soils able to maintain steeper slopes.

Management

Fill a new or recently cleaned lagoon ⅓ to ½ full with water. Add manure frequently in increasing amounts. It is best to start loading in the spring, so several warm months of bacterial action can ensure good operation. If odors are a problem, take lagoon liquid samples and measure pH frequently. If pH is below 6.7, add 1 lb hydrated lime or caustic soda (lye) per 1000 ft^2 of lagoon surface each day until the pH is neutral (pH = 7.0).

After the lagoon is full, divert surface water away to prevent overflow. Maintain the lagoon depth at the designed water quantity.

It is very important to load lagoons continuously. Put manure in at least once a week; once a day is recommended. **Slug loading,** or loading large amounts of manure at extended intervals, causes rapid increases in volatile acids, lowered pH, and high odor. If slug loading is unavoidable, consider a holding tank that can be slug loaded and emptied gradually into the lagoon. Add hydrated lime or caustic soda to reduce pH changes.

Keep arsenic, copper, antibiotics, and other toxic feed additives out of the lagoon or at a minimum, because they can retard biological activity.

With good management and proper loading rates, lagoons usually establish equilibrium and operate consistently in about two years. Odors are usually stronger during spring warm-up. Even with a normal pH of 6.7-7.2, the lagoon can have excess odorous volatile acids.

During the winter, when digestion in the lagoon is very slow, store or haul as much manure as possible to reduce the load on the lagoon.

Irrigate when winds can carry odors away. The best time to irrigate with lagoon effluent is in the fall after crop removal and early in the spring before the lagoon temperature rises significantly. Odor should be minimal at these times of the year.

Use gravel rip-rap around the waterline on steep banks to prevent erosion. Control weeds around the water's edge to control mosquitos. Remove vegetation 3′ above and below the waterline. Use herbicides cautiously to destroy weeds. Insecticides applied weekly can control insect larvae.

Install a 5′ high fence to keep out animals, children, and trespassers. Post warning signs and keep the gate locked.

Keep the area neat and consider landscaping to provide for an attractive farmstead. Seed the bank with low-growing spreading grasses and keep it mowed. Sheep and goats can be grazed on slopes too steep to mow.

Remove sludge when it reaches a depth of more than 3′. Apply sludge to land. Plowing under or soil injection is recommended. Avoid spreading sludge on frozen ground where runoff from rain or spring thaw could pollute a waterway.

Aerobic Systems

In an aerobic treatment system, biological oxidation converts organic matter (mostly manure) to carbon dioxide, water, and microbe cells. Oxygen must be supplied either naturally or artificially to maintain an aerobic system. Microbiologists estimate that about 50% of the degraded organic matter becomes microbial cells. The remaining 50% becomes liquid or gas that escapes as ammonia or carbon dioxide. Cellulose, lignin, and inorganic solids are degraded very slowly in aerobic processes.

Aerobic organisms grow quite rapidly, permitting a treatment system to adjust to shock or sudden increases in loading. However, the additional cells present a different waste disposal problem. Cells accumulate as organic sludge and require eventual disposal.

The major advantages of aerobic treatment systems are:

- Relatively odor-free operation.
- Fast rate of biological growth.
- Rapid adjustment to changes in loading and temperature.
- Elevated temperatures not required.

The major disadvantages are:

- Oxygen is required.
- High production of biological sludge.
- Shallow depth requires large surface area.
- Relatively high space, maintenance, management, and energy requirements for artificial oxygenation. Cost makes this type of system unsuitable for most livestock operations.

Naturally Aerated Systems

Oxygen from the atmosphere enters the lagoon liquid naturally, aided by turbulent conditions. Algal growth utilizes waste nutrients and sun energy to produce oxygen. The higher the lagoon temperature, the lower its capacity to hold oxygen. In the Midwest, naturally aerobic systems are loaded at 20 to 45 lb of 5-day biochemical oxygen demand (BOD_5) per acre.

The loading rate decreases in northern latitudes because of colder weather conditions. Large volumes of high strength wastes from livestock operations require very large land areas and are not generally practical.

Table 704-2. Surface area for naturally aerobic lagoons.
Maximum depth 5′, with 3′ or 4′ preferred.

Livestock	Surface area/pound of animal sq ft/lb
Poultry	4.5
Swine	2.5
Dairy cattle	1.5
Beef cattle	1.5

Design depth for a naturally aerobic lagoon is 3′-5′. Bottom-growing aquatic weeds in lagoons shallower than 3′ limit oxygen transfer, reduce the treatment capacity of the system, and dead plants add to the lagoon loading. At about 5′ depth, the mixing of oxygen and water becomes limited and sunlight penetration for algal growth becomes negligible. At depths greater than 5′, natural thermal and wind currents do not mix dissolved oxygen from the surface down to the bottom of the pond. At least 1 to 2 milligrams per liter (mg/l) of dissolved oxygen is desirable throughout the entire system. To assume adequate oxygenation, aerobic design is based only on lagoon surface area, Table 704-2.

Mechanically Aerated Lagoons

Area requirements of naturally aerated lagoons can be reduced by mechanical aeration. Floating mechanical aerators pump or mix oxygen into the liquid.

Design criteria are:

- Surface area does not limit design, so depths up to 20′ are recommended. Recommended volumes are in Table 704-3.
- Oxygen requirements are about 2 lb oxygen/ lb 5-day BOD entering the lagoon.
- Mechanical aerators transfer about 2 lb oxygen/ hp-hr in animal waste lagoons.
- Lower temperatures reduce biological activity and winter freezing problems can be severe.

Table 704-3. Volume of mechanically aerated lagoons.

Livestock	Volume/pound of livestock cu ft/lb
Poultry	0.75
Swine	1.00
Dairy cattle	1.25
Beef cattle	0.75

Composting

Composting is a biological process to degrade organic matter (volatile solids) to a relatively stable humus-like material. Composting can be either anaerobic or aerobic, but is usually aerobic.

Objectives of composting are to:

- Stabilize organic matter.
- Kill pathogens and weed seeds.
- Conserve the nitrogen, phosphorus, potash, and resistant organic matter found in the raw material.
- Produce a uniform, sterile, odor-free, relatively dry end product that is a valuable and salable fertilizer and soil conditioner.

Particle size, moisture content, aeration, temperature, and initial carbon-nitrogen ratio determine the speed and extent of composting.

Smaller manure particles have a greater exposed surface available to organisms. Optimum moisture content for composting is 50%-60%, depending on particle size and aeration technique. Compost windrows are usually 6′-8′ high and naturally aerated. Turn windrows to reaerate the mixture at intervals of 1 to 60 days depending on the pile temperature.

High-rate composting is achieved in drums up to 12′ in diameter and 8′ long with forced aeration. The material rolls in these drums for 2 to 10 days.

The interior of a compost pile usually reaches elevated temperatures that destroy pathogens, weed and vegetable seeds, and insect eggs and larvae. The surface does not get hot enough, so frequent mixing is important.

Composting also depends on microbe food and nutrient supplies. The proportion of carbon to nitrogen is the C/N ratio. A C/N ratio of 30:1 is about optimum for rapid composting. If all other factors are near optimum, composting time varies with C/N ratio:

C/N ratio:	20:1	20-50:1	78:1
Days required:	12	14	21

With C/N ratios lower than 20:1, nitrogen escapes during composting as ammonia (animal manure has a C/N ratio of about 10-15:1). Composting animal manures with bedding, paper wastes, or other high carbon, low nitrogen wastes has been successful.

When a pile does not reheat after turning, the process is complete. Finished compost is dark brown to black in color, practically insoluble in water, has a slightly earthy to musty odor, and has a loose friable texture. Composted livestock manure contains about 0.5% nitrogen, 0.4% phosphorus, and 0.2% potassium. Adding compost improves soil moisture retention of light soils and pore volume of heavy soils. Compost-improved soils have a relatively stable structure and are resistant to erosion.

705 MILKING CENTER WASTES

Management greatly affects the concentration and volume of wastes from milking and cleaning operations.

Milking center wastewater includes:

- Dilute liquid manure with feed, bedding, and hoof dirt from cow and floor washing.
- Dilute milk and cleaners from equipment washing.
- Colostrum, medicated, or spilled milk.
- Clean water from the final pipeline rinse and from a water-cooled condensor.

Milk solids include fat, albumin, and lactose. The solids do not settle out and may plug soil absorption fields and dry wells. It is preferable to dispose of milk solids by field spreading or aerobic lagoon treatment. They can be diluted and treated in an anaerobic lagoon, but may produce strong odors.

Waste Volume

If cow udders are washed using paper towels and a bucket of disinfectant, little waste occurs. If cows are washed in an automatic prepstall, water use is about 9 gal/day for each cow milked. Floors can be washed with relatively little water and a stiff-bristle broom or with a high pressure hose using a large volume of water.

Table 705-1 includes data summarized from manufacturers' recommendations and field observations in several states. Overall, small operations use less total water but more per day per cow than larger ones. A 100-cow operation with automatic washing equipment can use over 800 gal/day or 3200 ft^3/month.

Table 705-1. Volume of milking center wastes.

Washing operation	Water volume
Bulk tank	
Automatic	50-60 gal/wash
Manual	30-40 gal/wash
Pipeline	
In parlor (Volume increases for long lines in a large stanchion barn.)	75-125 gal/wash
Pail milkers	30-40 gal/wash
Misc equipment	30 gal/day
Cow prep	
Automatic	1-4 1/2 gal/wash per cow
Estimated average	2 gal/wash per cow
Manual	1/4-1/2 gal/wash per cow
Parlor floor	40-75 gal/day
Milkhouse floor	10-20 gal/day
Toilet	5 gal/flush

Human Toilet Wastes

Handle human toilet wastes separately from milking center wastewater. Grade A Milk Rules require sanitary human toilet waste facilities. Toilet facility wastes must not be accessible to flies, pollute soil surface, or contaminate any water supply. In most states, human waste treatment system effluent should not be discharged to any milking center wastewater lagoon or animal manure storage facility. Mixing human wastes with livestock wastes and spreading it on the soil surfaces can cause a potential health hazard. Human toilet wastes are usually treated in a septic tank and soil absorption system. See section on Domestic Sewage Treatment for details.

Handling Methods

The amount and kind of suspended and settleable solids in a waste determines the wastewater treatment alternative. Milk solids and sanitizers are in solution or stay suspended. Wasted feed, manure solids, and foot dirt include fine particles in suspension and heavier particles that settle. Match milking center wastewater handling methods to site and management. Check with local health authorities for restrictions.

Septic Tank with Soil Absorption

This system is not a good solution. Use it only in well drained soils. The major problem is that the soil becomes plugged with manure and milk solids. However, many dairy farms use this method, so to minimize the amount of solids that reach the soil absorption bed:

- Scrape or sweep up wasted feed and manure.
- Avoid spilling or dumping milk into drains.
- Properly install inlet and outlet baffles in the septic tank.
- Pump out septic tank sludge every 3 to 6 months.

The life of the soil absorption system can often be extended:

- Alternate between two soil absorption systems, discharging for 2 to 5 months into one while resting the other.
- Feed unmarketable milk to animals or spread it on cropland.

Refer to the Domestic Sewage Treatment section for septic tank and absorption bed design.

Settling Tank with Surface Disposal

A settling tank removes solids that float or settle. The removal of solids reduces problems in pumping, distribution, and treatment. The settling tank must be compartmented, as shown in Fig 705-1, or have two (or more) tanks in series.

Settling tank capacity is at least five times the daily wastewater volume and no less than 2,000 gal. For example, a 50-cow herd using 8 gal/cow/day requires a 2,000 gal tank. Design the tank with a length to width ratio between 3:2 and 5:1 to aid settling. Pump out the tank when ¾ full of solids.

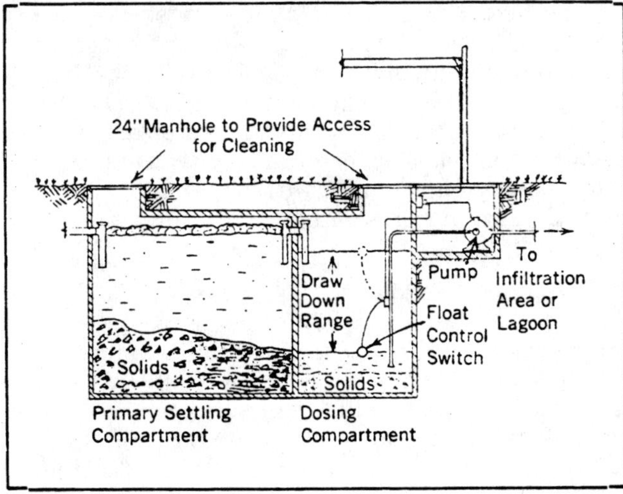

Fig 705-1. Settling tank.

Use a vegetative infiltration area or irrigation to dispose of settling tank effluent. Soils suitable for infiltration areas are silt loams to clay. Sandy soils are not suitable because infiltration is too rapid.

See the Transport and Disposal chapter for irrigation and vegetative infiltration area design procedures.

Handling With Liquid Manure

If you have a liquid manure system, the best option is to handle milking center wastewater with the manure. Adding water to dairy manure provides an effluent that can be easily agitated and pumped.

Other Methods

A holding tank provides short-term storage until wastes can be spread on nearby cropland, pastures, or woodlots. Use a storage volume of at least 15 days, preferably longer. See the Storage chapter for details.

A properly designed lagoon treats and stores wastes. Apply lagoon effluent to growing crops with irrigation equipment to add water and nutrients. See section on Biological Treatment for lagoon design procedures and requirements.

706 TRANSPORT AND DISPOSAL

Waste transport includes scraping, flushing, and runoff discussed in previous chapters, as well as hauling and pumping discussed in this chapter. Generally, solid manure is scraped or hauled to storage, or directly applied to fields. Liquid wastes are hauled or piped by gravity or pressure to storage, lagoons, or fields.

Handling Methods

Solid Manure

Wastes with 20% or more solids can usually be handled as a solid. Solid waste characteristics vary with the animal, ration, amount and type of bedding, time of year, and the amount of liquids that have been separated.

Most solid manure spreaders are box-type. Others include flail-type spreaders, dump trucks, earth movers, or wagons. A spreader should distribute wastes uniformly. Front-end loaders, scrapers and blades, and several mechanical systems transport solid wastes, but they are not usually used for spreading.

Flail-type spreaders usually have a shaft mounted near the open top and parallel to the main axis of the tank. Chain flails on this shaft throw the wastes out the side of the spreader as the shaft turns.

Box-type spreaders are pulled or mounted on trucks. Pull-types have 70 to 525 bu capacities. Flail-type spreader capacity is measured in tons—1 ton is about 27 bu. On spreaders under 110 bu capacity, the spreading mechanism is either ground or PTO driven. If ground driven, 2 or 4 drive wheels are used. PTO drives are used on larger spreaders. Truck-mounted box spreaders hold 175 to 525 bu. Spreader boxes are of steel or wood and should be watertight for road transport. Spreader mechanisms include paddles, flails, and augers. The feed apron, which moves the waste to the spreader, is often variable-speed. Some spreaders have moving front-end gates that push the wastes to the spreading mechanism.

Large spreader capacity reduces the number of trips to the field but may increase soil compaction. Calculate the number of trips from the amount of manure and spreader capacity.

$$NT = MS \div (SC \times 1.25) \text{ or}$$

$$NT = TS \div TL$$

MS = manure to be spread, ft^3
= AC x D
AC = area to be cleaned, ft^2
D = depth of manure, ft
NT = number of trips
SC = spreader capacity, bu
1.25 = ft^3/bu
TS = tons to be spread
TL = tons per load (1 ton = 27 bu)

Liquid Manure

Up to about 15% solids, manure has fluid handling characteristics. Up to about 4% solids, the waste can be handled as a liquid with irrigation or flushing equipment. From 4%-15% solids, the manure is semi-liquid and can be handled as a liquid, but equipment needs differ. Fibrous materials, such as bedding, hair, or feed, can hinder manure pumping. Chopper pumps can cut fibrous materials for improved pumping. Piston manure pumps handle manure with bedding.

Liquids—up to 4% solids

With proper management and screening, a liquid pump is satisfactory for handling 4% solids in liquids, but a slurry or trash pump is more trouble-free. If large quantities are handled, a pipeline may be preferred over tank wagons for transport. Prevent large solids such as ID tags, containers and lids, construction residue, rock, and animal teeth, tails, and hair from passing through the pump.

Settle out solids if possible. Provide an intake screen on the pump, with screen openings no larger than the smallest sprinkler nozzle, spile tube, or gate if irrigation is used. Use a large screen area to reduce velocities into the screen and to reduce plugging.

Required pump capacity is influenced by amount of wastes; time, labor, and power source available; where liquids are going; labor and equipment costs; and for cropland disposal, the rate at which soil and crops can receive water. A small capacity pump is less expensive but may require more labor and time than a larger pump.

Protect pumps and power units against malfunctions such as plugged lines or nozzles, loss of prime, overheating, and loss of lubricant. Match safety sensors and controls to pump, motor, and system types. Safety devices include temperature and pressure gauges, fuses, circuit breakers, and pressure switches.

Increase the size of power units rated for water capacity by at least 10% for increased friction losses and specific gravity of livestock wastes. Select pumps adequate for the flow and head of your system. Irrigation pumps can handle livestock wastes with a few solids. Use slurry pumps for liquids with low solids content.

Slurries—4% to 15% solids

Slurries with up to about 15% solids can be pumped. Agitate wastes before pumping. Select power source and capacity for pumping slurries using the same criteria as for pumping liquids, particularly if the slurries are pumped through pipelines. Open impeller chopper pumps are often used to agitate slurries in storage.

Pumps moving slurries through long pipelines and sprinklers operate against fairly high heads. Pumps for furrow irrigating with gated pipe or for filling tanks usually operate against much lower heads, which affects pumping needs.

Piston, helical rotor, submerged centrifugal, and positive displacement gear-type pumps can handle heavy slurries against high pressures. However, their performance is improved if solids are below about 10%. They do not require priming and therefore adapt to automation. Use submerged centrifugal, piston, auger, and diaphragm pumps to handle heavy slurries against low heads.

Pumping Livestock Waste

Pump Selection

Solids content is a major factor in selecting a waste handling pump. Settle solids or add dilution water to reduce pumping problems. Table 706-1 has estimates of solids content for various livestock waste systems.

Table 706-1. Solids content of liquid waste handling systems.

System type	Solids content (%)
Manure pit	
Swine	4-8
Cattle	10-15
Holding pond	
Pit overflow	1-3
Feedlot runoff	less than 1
Dairy barn washwater	less than 1
Lagoon	
Single or first stage	
swine	½-1
cattle	1-2
Second stage	less than ½
Flushing waste	½-2

Some installations, such as sprinkler irrigation systems, require high pressures and others only need to lift waste 10′-20′ to a manure tanker or earth storage. Table 706-2 and Fig 706-1 summarize the characteristics of waste handling pumps.

Centrifugal pumps

Centrifugal pumps are not positive displacement—the impeller can slip in the pumped liquid. This feature permits limited flowrate control by valving down the discharge side.

Performance of a centrifugal pump depends on impeller design, Fig 706-2. Closed impellers are efficient with water but cannot handle large solid particles. Closed impeller pumps develop high pressures and are useful in irrigating and recirculating water from a lagoon or pumping settled runoff.

Semi-open impellers have a plate or shroud on only one side of the impellers and handle water with some solids, such as pumping from a first stage cattle lagoon or unsettled lot runoff.

Open impeller pumps handle liquids with up to 15% solids. They are sometimes fitted with a sharp rotating blade at the pump inlet to chop material such as bedding, hay, or silage.

Pump size also affects the size of solids particles that can be pumped. Both open and semi-open impeller pumps are sometimes called trash pumps and may or may not develop enough pressure for sprinkler irrigation systems.

Most centrifugal pumps for livestock waste handling are self-priming. But with high solids liquids, foot valves can leak, requiring hand priming. Avoid priming by installing pumps below the level of the pumping reservoir.

Screw pumps

Positive displacement screw or rotary pumps handle fluids that are viscous but free from hard and abrasive solids such as nails, stones, and ID tags. The common rotary pump for livestock waste is the helical screw, Fig 706-3.

Because these pumps must not be operated dry, a small stream of fresh water is usually added directly into the pump casing during operation. Monitor line pressure when sprinkler irrigating directly out of liquid manure pits with positive displacement pumps—plugged irrigation nozzles cause burst pipes.

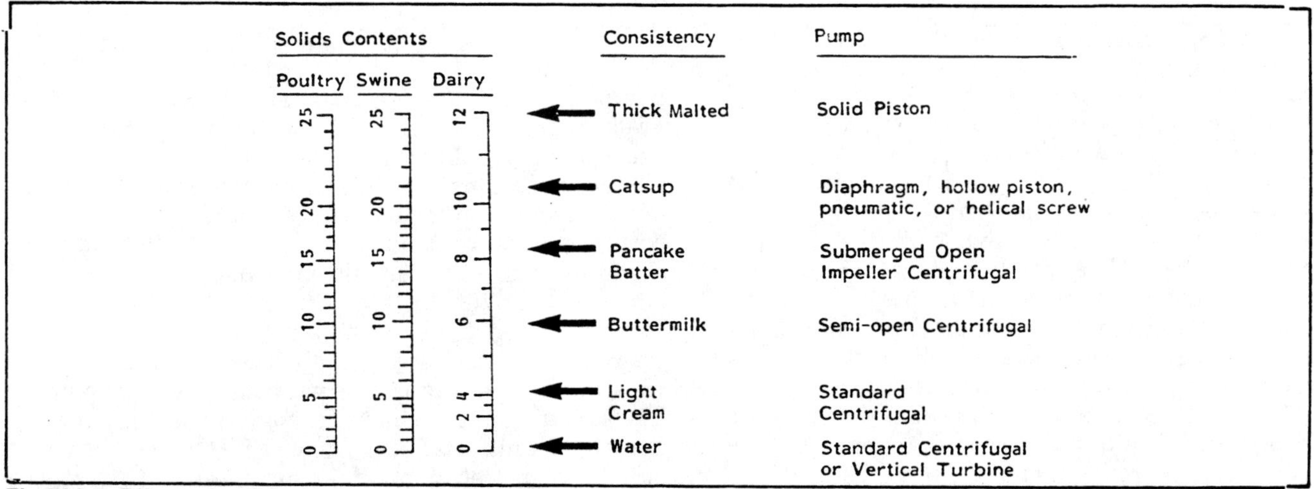

Fig 706-1. Effect of solids content on pump type.

Table 706-2. Waste handling pumps.

	Can be used for sprinkler irrigation	Solids content possible	Relative pump capacity	Applications
Centrifugal				
Open impeller	Usually not	High	Low	Moving waste from sump to lagoon Pit agitation Tanker filling
Semi-open	Sometimes	Moderate	Moderate	Lagoon recirculation Tanker filling
Closed impeller	Yes	Low	High	Recirculation Irrigation from lagoon
Rotary screw	Yes	High	High	Irrigation from lagoon Pit agitation Irrigation from pit Tanker filling
Reciprocating piston Diaphragm	No	High	Low	Tanker filling Transporting cattle waste from barn to outside storage
Air driven Vacuum tank Air piston	No	High	High	Tanker filling Transporting cattle waste from barn to outside storage

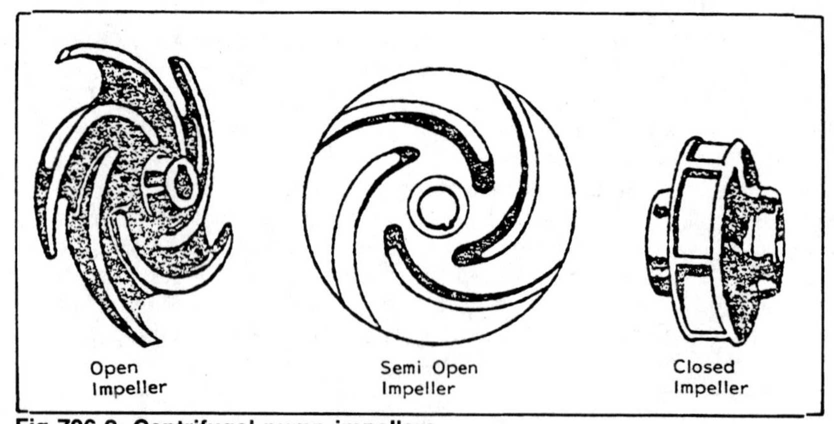

Fig 706-2. Centrifugal pump impellers.

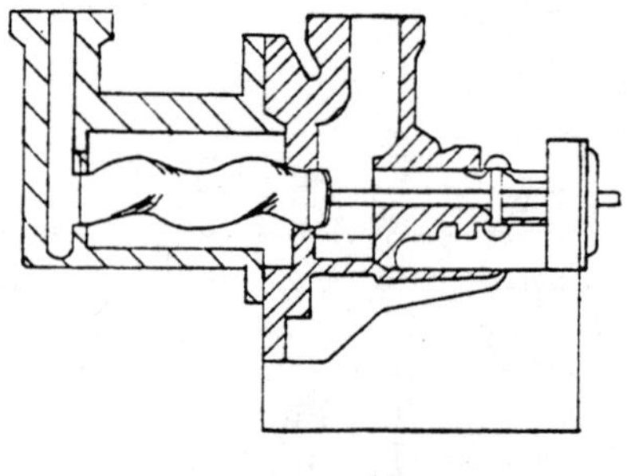

Fig 706-3. Screw-type rotary pump.

Piston pumps

Piston or reciprocating pumps have positive displacement and pulsating discharge. Except for leaks or bypasses, the piston or plunger capacity equals the piston area times the stroke length.

Piston pumps transport manure from cattle barns to outside storage. They can handle liquids with solids up to whatever can flow or be pushed into the pump inlet. Hollow pistons handle fairly dilute manure while the more expensive solid piston pumps handle manure with large amounts of bedding.

Piston pumps generate very high pressure if the discharge pipe is plugged. However, relief valves are seldom used to protect manure pumps, because discharge piping is large (9″-15″) and seldom plugs. The pumps are usually charged by gravity from a below grade hopper. One-way valves at the pump chamber exit control the direction of flow.

Hollow pistons are rectangular and from 8″x8″ to 8″x12″. The pump is at the base of a sloping hopper with the drive unit above floor level. A flap valve opens so manure passes through the piston during pumping, Fig 706-4a.

Solid piston manure pumps have round, square, or triangular pistons, Fig 706-4b. The piston is at the base of a hopper and in line with the transport pipe. The solid piston pulls back through the loading hopper and manure falls in front of it. On the return stroke, the piston pushes the manure into the cylinder and out through the underground pipe to storage.

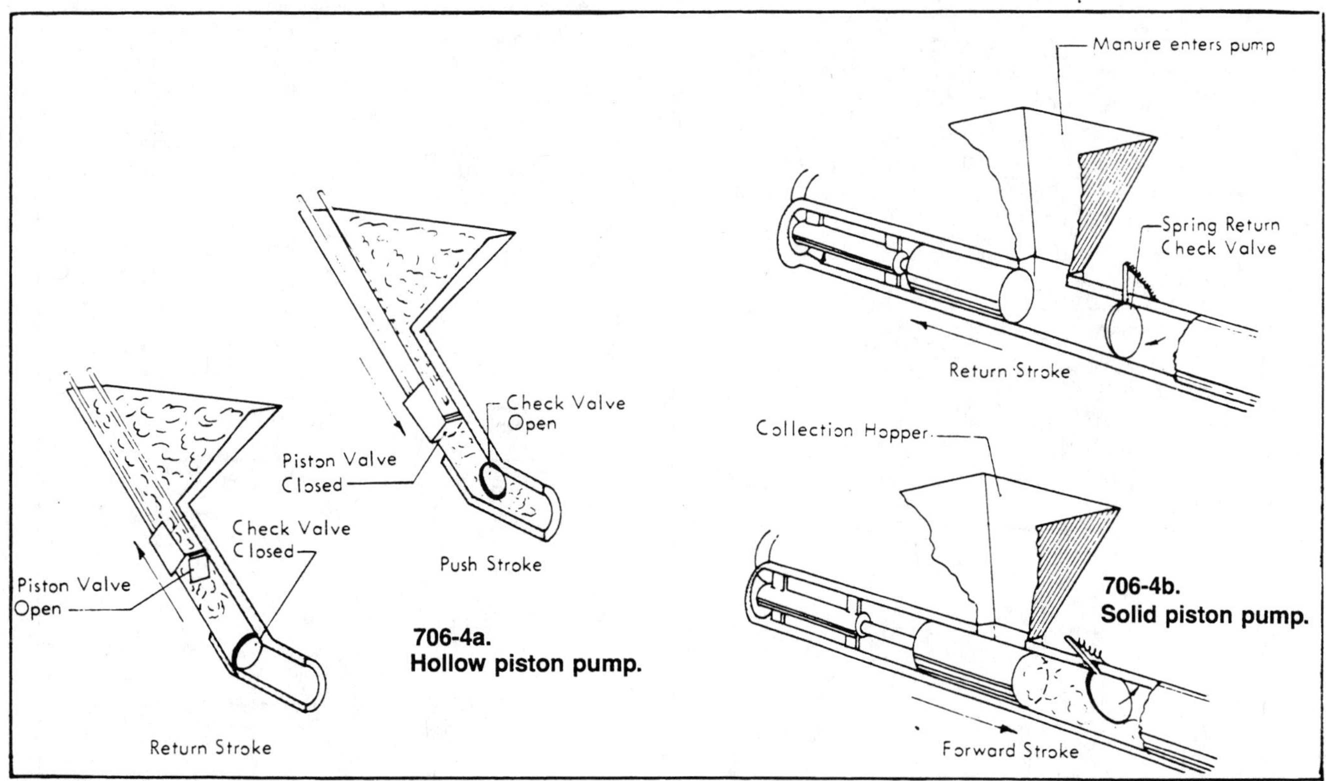

Fig 706-4. Pist‹ ›pumps.

A check valve prevents material from flowing back from the pipe. The close fitting piston creates a vacuum in the cylinder on the return stroke, helping to pull waste down into the cylinder.

Air driven pumps

Vacuum tanker wagons and air displacement transport pumps are two types of air driven pumps for livestock waste.

Tank wagons are common in 800 to 4500 gal capacities. Vacuum tanks pump liquid manure with up to about 10% solids. They are filled by suction developed by a PTO driven vacuum pump. Many vacuum tanks also develop pressure to improve unloading.

Tractor power for vacuum tanks depends on tank size and field spreading conditions. The vacuum pump needs only about 10 hp. Typically, the filling rate is about 500 to 700 gpm.

For air operated transport, enough compressed air is pumped into the top of a holding tank to force manure through a discharge pipe and to allow for leaks around the loading hatch. The tank is specially designed, usually underground, and typically about 1,500 gal. The discharge is at the bottom of the tank and fitted with a flap check valve. A large PVC pipe carries wastes to storage. See Fig 706-5.

Agitation pumps

Agitation pump types include open impeller centrifugal choppers, helical screws, inclined propellers, and vacuum tanker wagons. All but the vacuum tanker can pump liquid manure with fibrous solids up to about 15%.

How manure agitates varies with livestock ration, amount of drying before manure enters the tank, and

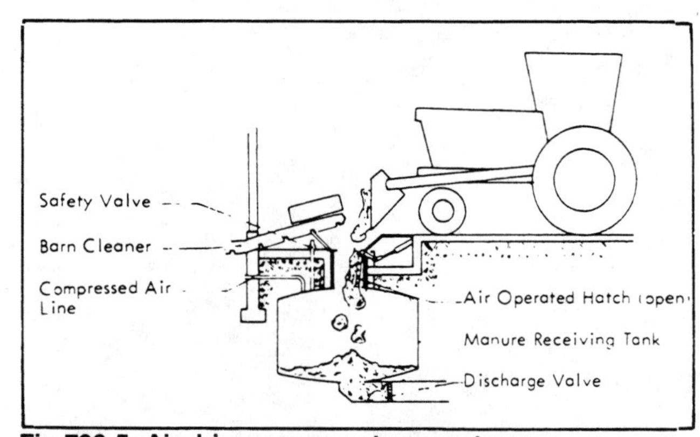

Fig 706-5. Air driven manure transport pump.

time of storage. Hog manure can be agitated easier than cattle manure. Solids tend to settle and pile up near where they are deposited, such as along feed bunks, or to float, dry out, and become caked.

Little agitation is needed for narrow gutters that are gravity drained about twice a week. Limit storages to be agitated to about 40′ across, Fig 706-7. Place agitation or access ports no more than 40′ apart in pits without partitions (20′ o.c. if agitated with vacuum wagon).

Agitate by cycling pumped liquid through an agitator nozzle. The stream from this nozzle breaks up the surface crust, stirs up settled solids, and makes the manure a more uniform slurry. Most chopper agitator pumps effectively agitate up to about 40′. Some commercial auger and propeller devices agitate but do not empty the storage.

Chopper agitators pump 100 to 3,000 gpm. Some of the smaller ones pump at high pressure for sprinkler irrigation. These dual purpose pumps are discussed further in the section on irrigation pumps.

Power requirements for chopper agitators depend on pumping pressures and capacities—20 hp for some to over 75 hp for larger capacity pumps.

Gases escaping from agitated manure are harmful to animals and humans. See section on gases and odors from stored liquid manure.

Irrigation pumps

Liquid manure slurries with 10% or less solids are satisfactorily pumped through long pipelines, sprinklers, and gated distribution pipe. Some agitation before or during pumping helps keep solids in suspension and break up larger solids that might clog nozzles, gates, or intakes. Irrigation pumps work against fairly high pressures, especially with sprinkler irrigation.

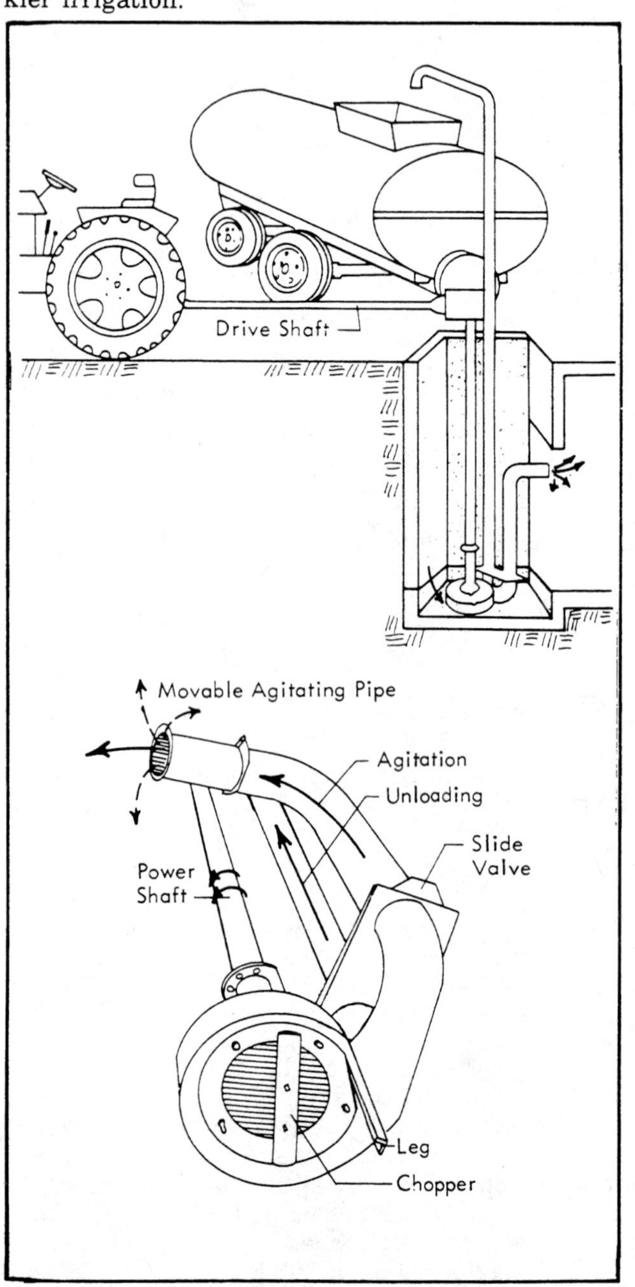

Fig 706-6. Agitating with submerged centrifugal pump.

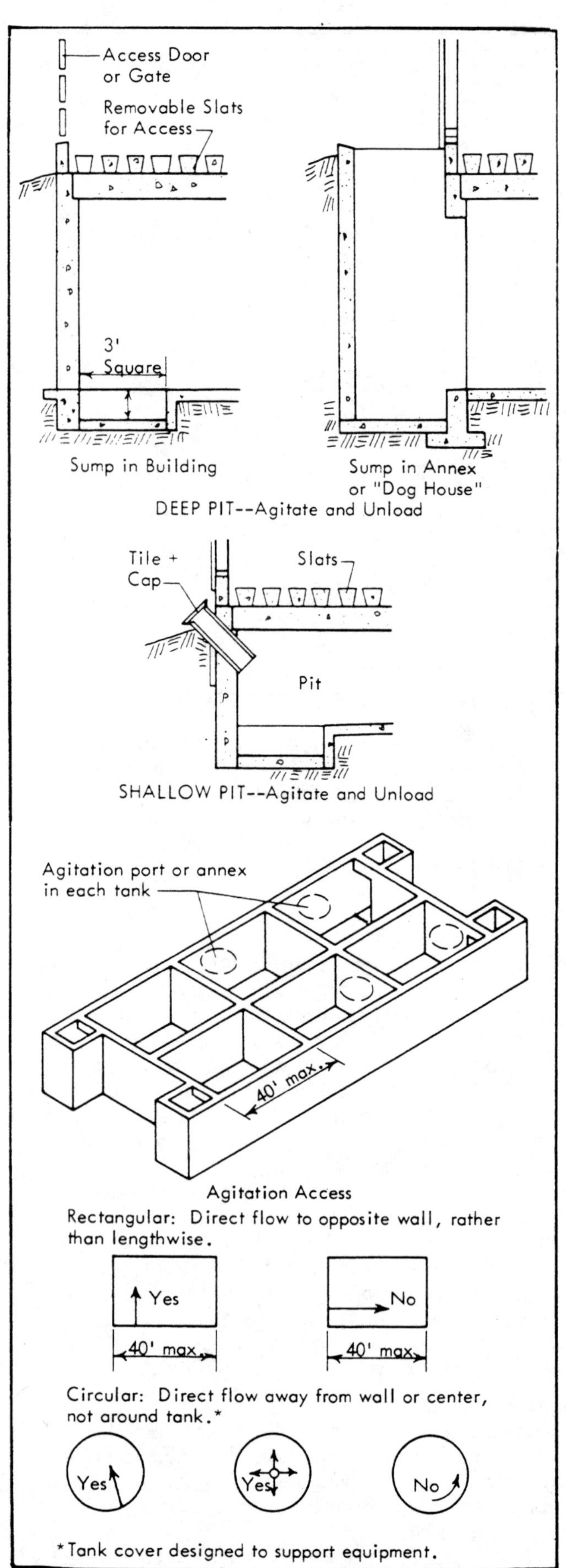

Fig 706-7. Agitation of manure storage pits.

For irrigating slurries, semi-open impeller centrifugal and screw-type pumps are usually used. Many of these pumps will satisfactorily handle slurries with more than 10% solids, but the high pipeline friction losses usually make pumping high solids slurries impractical. Sprinklers are usually the big gun type with ¾" and larger nozzles, requiring pressures of 80 psi or more and pumping rates of 100 to 800 gpm.

Effluents from waste lagoons, feedlot runoff holding ponds, and milk houses usually have 3% solids or less. These can often be handled by conventional irrigation pumps and equipment. Single stage, standard closed impeller centrifugal pumps work well. Semi-open impeller pumps are sometimes desirable to reduce clogging. Screen out large solids at the pump inlet.

Smaller irrigation systems can be used with effluents because of lower solids contents. Because there is less clogging, gated or perforated pipe distribution is also practical, allowing lower pumping pressures.

Flushing and recirculating pumps

Centrifugal and rubber impeller or roller-type pumps recirculate lagoon water for flushing and diluting pit storages. Centrifugal pumps with semi-enclosed or enclosed impellers work well for pumping from a swine waste lagoon.

A major disadvantage of recirculating pumps is the frequent service required when operated continuously. A plastic-fitted pump casing with high quality seals that can be easily serviced is recommended. Because of their short service life, stock several spare replacement seals, impellers, and a replacement motor.

Large capacity pumps to flush gutters are generally vertical turbine or closed impeller types. Design the pump and pipes to drain after flushing to minimize corrosion.

Sumps and sump pumps

Sumps are useful in many livestock waste handling applications where the fall is insufficient for gravity transport, Fig 706-8. Select the sump and pump size together—with the same inflow, a larger pump is needed with a smaller sump, Table 706-3. When handling livestock wastes, pumps are usually designed to empty sumps in less than 5 min to prevent solids from settling out. The type of pump depends on the amount of solids in the liquid. Sump pumps are usually low head pumps with a lift of 20'-30'. For high solids waste, make the discharge line the same diameter as the pump outlet to maintain high velocities and prevent settling.

Pipe

Pipe has many uses in a manure disposal system. Small pipes in flushing systems carry dilution water to a flush tank and other low-flow needs. Large pipe is for direct pump flushing of gutters, irrigation, roof and surface water control, conveying wastes to storage, and other high-flow needs.

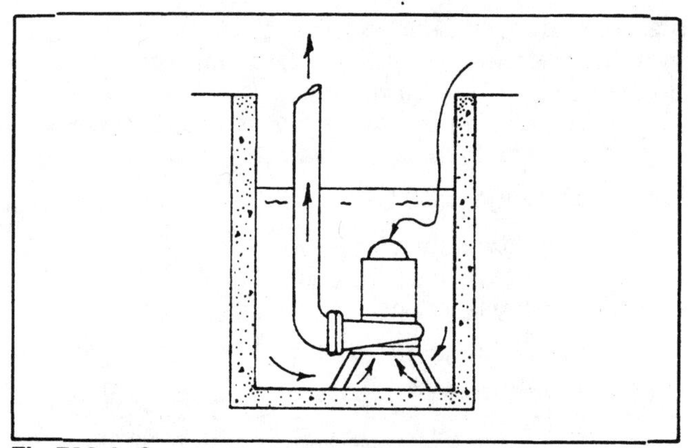

Fig 706-8. Sump with sewage lift pump.

Table 706-3. Volumes in round sump tanks.

Diameter of tanks (ft)	Volume per foot of depth (gallons)
1	6
1.5	13.25
2	23.5
2.5	36.75
3	53
3.5	72
4	94
4.6	119
5	147

Small pipe can be steel, copper, aluminum, or plastic. Large pipe is usually aluminum or plastic for irrigation and steel, cast iron, plastic, or asbestos-cement for culverts, drains, and other transport needs. Never bury aluminum pipe because of corrosion. Exposure to ultraviolet sunlight rays can make some plastic pipe brittle.

As water flows through a pipe, drag against the pipe wall reduces flow. The pressure needed to maintain flow is measured in feet—the pressure of a column of water of that height. Large pipe is more expensive, but has a lower pipe friction and therefore uses less pumping energy.

To determine the total head a pump must provide, consider friction losses due to pipe length, fittings, orifices, and elevation changes. Convert total head in feet to pressure, psi, by multiplying feet by 0.433.

Consider fluid velocity in pipelines. At a velocity below 2 fps, solids may settle in low spots. At high velocities, water hammer can be a problem. Design for velocities of 3 to 6 fps. Limit fluid velocity to 5 fps in pipe not buried or securely tied down. If possible, flush pipelines with clean water to remove any settled solids after pumping wastes.

Nozzles, Sprinklers

Sprinklers range in size up to large gun-type sprinklers with over 1" nozzles and over 1,000 gpm discharges. Discharge volume is controlled by nozzle size, number of nozzles, and operating pressure. Select a nozzle no smaller than the largest piece of waste to be discharged.

Sprinklers have 1 to 3 nozzles. Single nozzle sprinklers perform better where wind is a problem. One large nozzle is less likely to plug than two smaller ones with the same capacity. Minimum operating pressures range from 20 to 90 psi.

For approximate discharges from sprinklers with nozzle diameters from ⅛"-1¼", use the equation:

$$\text{Spray} = 29.84 \times d^2 \times P^{0.5}$$

Spray = nozzle discharge, gpm
d = nozzle diameter, in.
P = pressure, psi

See Table 706-5 for discharges from guns.

Table 706-4. Discharge of nozzles.

Pressure psi	Nozzle diameter, inches 3/16"	1/4"	5/16"	3/8"	1/2"	3/4"	1"	1¼"
	Discharge, gpm							
50	7.1	12.9	-	-	-	-	-	-
60	7.8	14.0	22.0	-	-	-	-	-
70	8.5	15.4	23.9	33.2	-	-	-	-
80	9.1	16.4	25.7	35.7	61.6	154	264	416
100	-	-	-	40.7	68.9	173	296	462
120	-	-	-	-	-	189	324	511

To convert nozzle discharge to in/hr of irrigation,

a. in/hr = gpm ÷ (acres x 453)
in/hr = application rate
gpm = total sprinkler discharge
acres = number of acres covered

b. in/hr = 96 x gpm ÷ ft^2
ft^2 = number of ft^2 sprinkled

Irrigation

Most irrigation systems can handle fluid wastes with up to 4% solids, which is typical of lot runoff and effluent from a lagoon, holding pond, or milk house. Crop flood irrigation and big guns can handle higher solids wastes. Irrigation equipment disposes of wastes and also adds water and fertilizer to crops.

Select a surface or sprinkler irrigation system adapted to your topography, soil, and crops. A well designed and managed system prevents runoff and erosion.

As with any waste management system, there are potential problems.

- Odor problems can be severe, depending on the waste and on management.
- Application of strong wastes to crops may adversely affect plant growth or utilization.
- Fine-textured and tight soils may not have enough permeability to receive liquids rapidly.

Surface Irrigation

Surface systems can be used on crops or forages, but require good management to avoid runoff and obtain uniform distribution. Do not use them on land with greater than 2% slope. Before attempting surface spreading, contact a soil and water engineer to inspect the site and help you determine the feasibility of such a system.

Advantages of surface irrigation are low cost, low power requirement, and few mechanical parts. Disadvantages are a high degree of management skill required, inflexibility with respect to land area, and moderate labor requirement.

Dissolved nutrients enter the soil with the liquids, but solids tend to settle out or be filtered out by grass near the field inlet. More nutrients and somewhat more liquid are absorbed at the high end of the slope.

Wastewater is piped to the field and spread on fields with gated irrigation pipe or through an open ditch with siphon tubes or turnout gates.

In **border irrigation,** an essentially level field is flooded with 3"-5" of water, which stands until it infiltrates. A small earth dike or berm boarders the field. In **furrow irrigation,** the water is spread in furrows between row crops—downhill with slopes up to 1% or along the contour for steeper slopes. In **corrugation irrigation,** shallow, closely spaced, V-shaped notches are tooled into a seedbed just before planting. Irregular and relatively steep land can be corrugation irrigated with grass or close-growing crops, especially if the corrugations are contoured.

Wild flooding is more or less random. It has low initial cost, adapts to a wide range of irrigation flows, and can be used on close-growing crops, rolling land, and shallow soils. However, wild flooding has a high labor requirement, uneven water and fertilizer distribution, and unless carefully managed, can cause water pollution.

Table 706-5. Discharge of irrigation guns

	Nozzle trajectory 24°								27°					
Taper bore, in:	.6		.7		.9		1.1		1.3		1.5		1.75	
Ring nozzle, in.			.86		1.08		1.26		1.41		1.74		1.93	
	GPM	Dia	GPM	Dia	GPM	Dia	GPM	Dia	GPM	Dia	GPM	Dia	GPM	Dia
50 psi	74	225'	100	250'	165	290'	255	330'						
60	81	240	110	265	182	305	275	345	385	390'	515	430'	695	470'
70	88	250	120	280	197	320	295	360	415	410	555	450	755	495
80	94	260	128	290	210	335	315	375	445	430	590	470	805	515
90	100	270	135	300	223	345	335	390	475	445	626	485	855	535
100	106	280	143	310	235	355	355	400	500	460	660	500	900	550
110	111	290	150	320	247	365	370	410	525	470	695	515	945	565
120			157	330	258	375	385	420	545	480	725	530	985	580
130									565	485	755	540	1025	590

Table 706-6. Flowrate for gated pipe.
Maximum recommended flowrates per gate with openings spaced at 30"-40".

Q, gpm	40	25	16	12	10	5
Slope, %	0.2	0.4	0.6	0.8	1	[illegible]

Sprinkler Systems

Sprinklers allow waste disposal on rolling and irregular land. Although initial and operating costs are generally higher for sprinklers than for surface systems, labor requirements are reduced, some systems can be automated, and application uniformity can be improved. Odors from sprinkled wastes can create nuisances. Avoid sprinkling on days with high humidity or when wind blows odors to areas of concern.

Select sprinklers and spacings to avoid runoff for the particular soil type, topography, crop, and time of application. Select equipment to handle anticipated waste particle sizes with minimum plugging and maintenance. Flushing the system with clean water after use prolongs equipment life by reducing corrosion.

Although there are other sprinkler systems, the following are those generally considered for waste disposal in the Midwest.

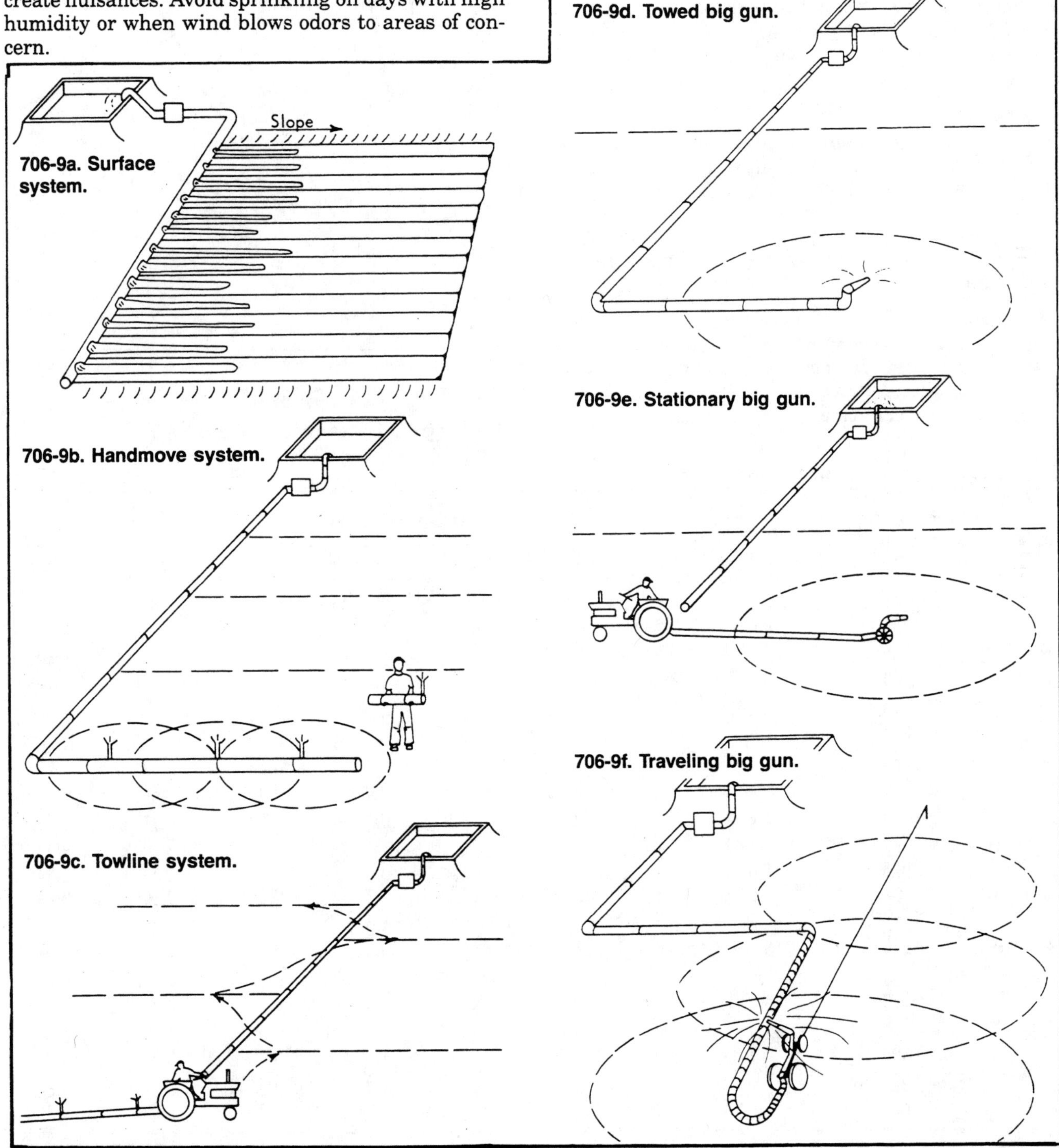

Fig 706-9. Irrigation systems.

Handmove system

Handmove sprinklers have a mainline and one or more laterals. The laterals are assembled from 20′, 30′, or 40′ sections of aluminum pipe; sprinklers are 30′ or 40′ apart and cover 60′-80′ circles.

As the pump runs, a strip is sprinkled as long as the lateral and as wide as the cover diameter of an individual sprinkler. That strip is one **set.** After that set is irrigated, the pump is shut off, the pipe is disassembled, moved to a new location, reassembled, and the new set is sprinkled.

With this system, a ¼ mile lateral covers 1.8 acres with each 60′ move; 32 sprinklers can discharge 10 gpm each for a total of 320 gpm pumped through 5″ pipe; the application rate = 0.4″/hr; and the power requirement is about 20 hp.

Advantages:

- Low initial investment; consider a used system.
- Few mechanical parts to malfunction.
- Low power requirements (50 psi at the sprinklers).
- Flexible with respect to land area. Different lengths can be set, and can be run almost any direction to get to isolated corners.

Disadvantages:

- High labor requirement; individual pipe sections are moved. With waste, this can be a very unpleasant task.
- Small sprinklers are prone to plugging.

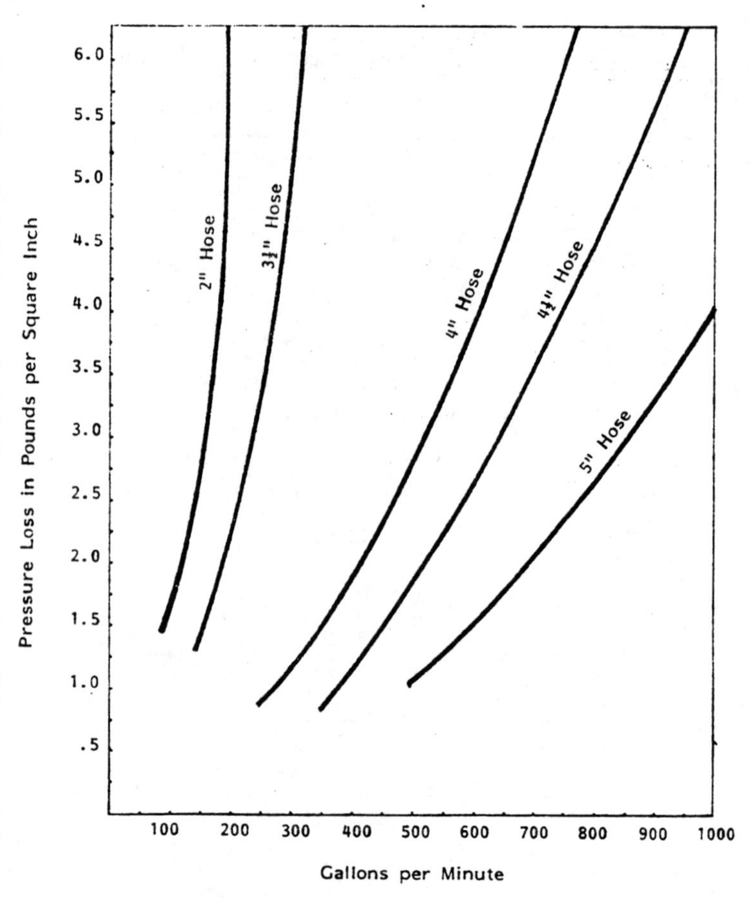

Fig 706-10. Irrigation hose pressure loss/100′.

Table 706-7. Water applied by traveling sprinkler.

Sprinkler G.P.M.	Travel lane spacing ft	Travel speed, ft/min 0.4	0.5	1	2	4	6	8	10
		Water applied, in.*							
100	165	2.4	1.9	1.0	0.5	0.24	0.16	0.12	0.09
200	165	4.9	3.9	2.0	1.0	0.5	0.32	0.24	0.19
	200	4.0	3.2	1.6	0.8	0.4	0.27	0.20	0.16
300	200	6.0	4.8	2.4	1.2	0.6	0.40	0.30	0.24
	270	4.4	3.6	1.8	0.9	0.4	0.30	0.22	0.18
400	240	6.7	5.3	2.7	1.3	0.7	0.44	0.33	0.27
	300	5.3	4.3	2.1	1.1	0.5	0.36	0.27	0.2
500	270	7.4	6.0	3.0	1.5	0.7	0.50	0.37	0.29
	330	6.1	4.9	2.4	1.2	0.6	0.4	0.3	0.24
600	270	8.9	7.1	3.6	1.8	0.9	0.6	0.45	0.34
	330	7.3	5.8	2.9	1.5	0.7	0.5	0.36	0.29
700	270	10.4	8.3	4.2	2.1	1.0	0.7	0.5	0.42
	330	8.5	6.8	3.4	1.7	0.8	0.6	0.4	0.34
800	300	10.7	8.5	4.3	2.1	1.1	0.7	0.5	0.43
	360	8.9	7.1	3.6	1.8	0.9	0.6	0.4	0.36
900	300	12.0	9.6	4.8	2.4	1.2	0.8	0.6	0.5
	360	10.0	8.0	4.0	2.0	1.0	0.7	0.5	0.4
1000	330	12.2	9.7	4.9	2.4	1.2	0.8	0.6	0.5
	400	10.0	8.0	4.0	2.0	1.0	0.7	0.5	0.4

*To convert table values to gal/acre, multiply by 27152.4.

Towline

Towlines are like handmove systems, except towlines have stronger, more permanent couplers between lateral pipe sections that are moved by tractor.

Advantages:
- Low initial investment.
- Lower labor requirement than handmove; do not have to work in mud to make moves.
- Few mechanical parts to malfunction.
- Low power requirement (50 psi at the sprinklers).

Disadvantages:
- Irregularly shaped fields can be a problem because of fixed lateral length.
- Small sprinklers can plug.
- Driving lanes for the tractor are required in tall row crops.

Stationary big gun

This system is applicable to many waste disposal systems. A single large sprinkler is moved as in handmove systems.

A big gun discharging 330 gpm sprinkles a 350′ diameter circle, an area of 2.2 acres, at an application rate of 0.33 in/hr at 90 psi instead of 50 psi. The power requirement is about 30 hp.

Advantages:
- Few mechanical parts to malfunction.
- Few plugging problems with large nozzles.
- Flexible with respect to land area.
- Pipe requirements are slightly less than with small sprinklers.
- Moderate labor requirements.

Disadvantages:
- Moderate to high initial investment.
- Higher power requirements (90 psi at the sprinkler).

Towed big gun

This system is essentially a towline with the small sprinklers replaced by a big gun. Or it can be a stationary big gun with the pipe couplers replaced by towline couplers. It has the nonplugging advantage of the big gun and the lower labor requirement of the towline.

Advantages:
- Low initial investment.
- Few mechanical parts to malfunction.
- Lower labor requirement than handmove or stationary gun systems.
- Few plugging problems with large nozzle.
- Pipe requirements are slightly less than with small sprinklers.

Disadvantages:
- High power requirement.
- Not as flexible with respect to land area.
- Driving lanes are required for tractor.

Traveling big gun

The traveling big gun is justifiable where it can be used several times each year on large acreages. A single large nozzle is mounted on running gear that is pulled across the field by cable.

Big guns travel either 660′ (⅛ mile) or 1,320′ (¼ mile) on each set and have variable speeds to control application rates. They are fed by dragging a flexible hose behind them. The hose is relatively expensive and constitutes a major part of the cost of a traveler.

Traveling guns are available in a wide range of sizes. A common size irrigates 10 acres per set, compared with 2 acres for a stationary gun.

Advantages:
- Lowest labor requirement of systems listed.
- No plugging problems with large nozzle.
- Flexible with respect to land area.

Disadvantages:
- Higher initial cost than the other systems.
- High power requirement.
- More mechanical parts than the other systems, especially with an auxiliary engine.

System Management

Water irrigation systems apply up to 20 in/yr in at least three separate, smaller applications. The equipment is large enough to cover the entire area one set at a time and be back to the starting point before the plants are out of water.

Waste irrigation systems may only apply 1″-2″ once a year, with crop water requirements not considered. Size is therefore much less critical, and is more affected by the time the producer wants to spend spreading wastes. To spend 3 days instead of 3 weeks, buy bigger equipment; costs are roughly proportional to size.

Table 706-8. Maximum water application rates.
Design the system for the rates suggested. Soils usually absorb water at a faster rate if applied in light applications (¾″-1½″) when soil is dry.

Soil characteristics	0-5% slope cover	0-5% slope bare
	----- in/hr -----	
Clay; very poorly drained	0.3	0.15
Silty surface; poorly drained clay and claypan subsoil	0.4	0.25
Medium textured surface soil; moderate to imperfectly drained profile	0.5	0.3
Silt loam, loam and very sandy loam; well to moderately well drained	0.6	0.4
Loamy sand, sandy loam, or peat; well drained	0.9	0.6

Reduce application rates on sloping ground:

Slope	Application Rate Reduction
0-5%	0%
6-8%	20%
9-12%	40%
13-20%	60%
over 20%	75%

Table 706-9. Guide for determining soil moisture.

Moisture Condition	Percent of available moisture remaining in soil	Soil Texture		
		Sands - sandy loams	Loams - silt loams	Clay loams - clay
Dry	0% Wilting point	Dry, loose, flows through fingers.	Powdery, sometimes slightly crusted but easily broken into powder.	Hard, baked, cracked; difficult to break into powder.
Low	50% or less	Loose, feels dry.	Forms a weak ball when squeezed but won't stick to tools.	Pliable, but not slick, balls under pressure, sticks to tools.
Fair	50 to 75%	Balls under pressure but seldom holds together when bounced in the hand.	Forms a ball somewhat plastic, sticks slightly with pressure. Doesn't stick to tools.	Forms a ball, ribbons out between thumb and forefinger, has a slick feeling.
Good	75 to 100%	Forms a weak ball, breaks easily when bounced in the hand; can feel moistness.	Forms a ball, very pliable, slicks readily, clings slightly to tools.	Easily ribbons out between thumb and forefinger, has a slick feeling, very sticky.
Ideal	Field capacity 100%	Soil mass clings together. Upon squeezing, outline of ball is left on hand.	Wet outline of ball is left on hand when soil is squeezed. Sticks to tools.	Wet outline of ball is left on hand when soil is squeezed. Sticky enough to cling to fingers.

The time required for a set is controlled by the application rate, soil intake rate, and the total amount of waste to be applied at that rate. For example, an application of 1.5" to a silt loam soil at 0.4 in/hr requires $1.5 \div 0.4 = 3.75$ hr. Approximate soil intake rates for water are in Table 706-8. These may have to be reduced for some liquid wastes.

To avoid saturating the soil, estimate the percent of soil moisture content from Table 706-9. Then, from Table 706-10, estimate the maximum field capacity. For example, if capacity is 1", and you estimate the soil at 50% of field capacity, you can add ½" to the top foot of soil. Usually, waste spreading proceeds until ponding and/or runoff starts. Then discontinue in that area.

Table 706-10. Soil moisture capacity.

Soil type	Moisture capacity Inches/foot
Light sandy	1.00
Medium	1.69
Heavy	2.39

Use Table 706-11 to estimate labor requirements for each system. The number of animals and the waste system determines the area to be sprinkled; the area determines the number of sets.

Once you have decided to irrigate for disposal and have an idea of the system you want, obtain additional help from the Extension Service, SCS, or consulting engineers. Irrigation equipment dealers and manufacturers usually provide free design service for their equipment.

Irrigating with Slurries

Big gun sprinklers can spread wastes with high solids content. Pipeline design is similar to that for dilute wastes, except that each component is selected to handle more solids. Big guns adapt to hand carry, permanent set, or traveling sprinkler systems. They handle thick milk or any thinner consistency. PVC plastic pipe with adequate pressure rating is suitable for pipelines. Flush with fresh water after irrigating to remove solids. Deposited solids can corrode aluminum pipe.

Manure solids can coat crop leaves, reducing photosynthesis and causing ammonia burning of leaves. Rinse-spray crops with clean water every hour or so if irrigating during the heat of the day. When the sprinkler is detached, liquid manure drains from the pipe and can kill the crop in that area. If possible, pump clean water for 10 to 15 min before moving the sprinkler to clean out the pipe, rinse the pipe near the sprinkler, wash solids off foliage, and dilute drained liquid manure.

Table 706-11. Labor and number of sets for 10 sprinkled acres.

System	No. of sets/ 10 acres	Labor/set minutes	Labor/10 acres hours
Hand move[a]	5.5	70	6.4
Towline[a]	5.5	23	2.1
Stationary big gun[b]	4.5	70	5.2
Towed big gun[b]	4.5	23	1.7
Traveling big gun[c]	1.0	60	1.0

[a]¼ mile lateral with 60 feet between sets

[b]350 feet wetted diameter

[c]350 feet wetted diameter and ¼ mile travel

Vegetative Infiltration Area

Feedlot runoff treatment by vegetative infiltration areas is practical and economical when grassland or other land that can be kept out of row crop production is available. A settling basin to remove solids is essential. The infiltration area can be a long, 10′-20′ wide channel or a broad, flat area. For uniform distribution, slope the first 50′ 1% to move the runoff rapidly away from the lot and settling unit before further settling occurs. The rest of the area should be nearly flat (less than 0.25%) with just enough drainage to prevent water from standing. The runoff must travel through the vegetated area at least 300′ before entering a stream or ditch.

The area needed depends on soil type. First, estimate the infiltration rate from Table 706-12. Then determine the design rainfall event—usually the 25-yr, 24-hr storm. Next, compute the runoff-receiving capacity as soil infiltration rate minus the rainfall rate, in in/hr. Then, size the infiltration area, ft^2:

Area = (Lot x Rain) ÷ (Cap x Time)

Area = infiltration area, ft^2
Lot = lot area, ft^2
Rain = design rainfall, in.
Cap = runoff receiving capacity, in/hr
Time = desired infiltration time, hr

The time to infiltrate all lot runoff is normally 24 hr, although other periods may be set by state regulations in some areas.

Table 706-12. Infiltration rates for soils.

Soil type	Infiltration rate, in/hr	
	Vegetated	No cover
Sand	1.0	1.0
Sandy loam	0.6	0.4
Silt loam	0.5	0.3
Silty clay loam	0.4	0.25
Clay	0.3	0.15

707 ANIMAL WASTE UTILIZATION

Regardless of the storage, treatment, and handling methods for animal wastes, some end products remain. They are valuable resources for benefit or return. Or, they are unwanted wastes to be disposed of economically and efficiently. The end use of wastes often dictates the waste disposal system. Consider the pollution and nuisance potential of the disposal method along with economy, fertilizer value, etc. The cheapest method may not meet regulations or be acceptable to your neighbors.

Most operators utilize wastes for an economic return, or at least to minimize costs. Animal manure is most commonly a fertilizer and soil conditioner.

Other established or potential practices include:

- Irrigation with waste effluent to supplement water needs for crop production.
- Reuse of liquids to flush and transport manure.
- Use of processed solids as bedding or litter.
- Use of processed solids for off-farm fertilizer, soil additive, or mulch.
- Salvage of energy from methane production.
- Reuse as feed ingredients for livestock, poultry, and aquatic life.

Land Application

Effects

Manure nutrients help build and maintain soil fertility. Manure also improves tilth, increases water-holding capacity, lessens wind and water erosion, improves aeration, and promotes beneficial organisms. When wastes include runoff or dilution water, they can supply water as well as nutrients to crop production.

The economic value of manure fertilizer is calculated from its available N, P, and K at commercial fertilizer prices. These values change with the costs of fertilizer and handling.

Excess wastes can harm crop growth, contaminate soil, cause surface and groundwater pollution, and waste nutrients.

While most soils have a tremendous capacity to absorb phosphorus, very high soil phosphorus levels can interfere with plant nutrition by inhibiting uptake of metallic trace elements such as iron, zinc, and copper. When plant residue or manure is added to soil, there is an immediate and marked drop in O_2 and an increase in CO_2 in the soil air which can inhibit plant growth.

The carbon-nitrogen ratio (C/N) of applied wastes affects both microbial and plant growth. If a waste having a high C/N ratio, such as manure with a lot of bedding, is added to a soil, organisms decomposing the organic matter grow until available mineral and nitrogen become limiting. All the immediately available nitrogen is bound by the microorganisms. In the short run, nitrogen is unavailable for plant use and more chemical fertilizer may have to be added than before the waste application.

Heavy manure applications can increase soil salinity (salts) especially in arid regions where little or no leaching occurs. Salts can inhibit plant growth and depress yields. If salinity becomes a problem, consult a crop specialist.

Sodium and potassium can alter soil structure and reduce water movement rates. Field equipment, such as heavy manure wagons, compacts wet soils, alters soil structure, and reduces water movement. Yield reduction can result.

Surface and Groundwater Quality

Several diseases that infect both animals and man can be transmitted in water-borne livestock wastes. Land application can successfully interrupt infection cycles if water pollution is prevented.

Surface runoff contains pollutants, including plant nutrients, oxygen-demanding material, and some infectious agents. Excessive nitrogen applications can cause nitrate pollution of water—the cause of infant cyanosis (blue babies) and perhaps chemical diarrhea.

Nitrogen in excess of crop requirements leaches through the soil once it is in the nitrate form. For local application rates to avoid groundwater pollution, consult your state extension crop and soil specialist or the SCS. Reducing excess nitrogen without pollution is difficult.

Excess nutrients in surface water can cause algae blooms, impaired fisheries, fish kills, odors, and increased turbidity. Nutrients in runoff from land where manure was applied and incorporated in the summer are less than in runoff where no manure was applied. But large nutrient losses can occur in spring runoff from land where manure was applied on frozen ground.

Nutrient Losses During Collection and Storage

Housing and waste handling systems affect the nutrient composition of wastes. Bedding and water dilute manure, resulting in less nutrient value per pound. Much nitrogen can be lost to the air as ammonia. Runoff and leaching in open lots can remove nutrients. But there is much less nitrogen loss from deep compost pits, liquid storage systems, or roofed feeding areas. See Table 707-1.

Phosphorus and potassium handling losses are negligible except for open lots or lagoons. About 20%-40% of the phosphorus and 30%-50% of the potassium can be lost by runoff and leaching in open lots. However, much of the P and K can be recovered by runoff control systems such as settling basins and holding ponds. Up to 80% of the phosphorus and nitrogen is lost in lagoons and becomes unavailable when sprayed on land.

Table 707-1. Nitrogen losses in handling and storage.
Average losses between excretion and land application adjusted for dilution in the various systems.

System	Nitrogen cost %
Solid	
Daily scrape and haul	15-35
Manure pack	20-40
Open lot	40-60
Deep pit (poultry)	15-35
Liquid	
Anaerobic pit	15-30
Above ground storage	10-30
Earthen storage	20-40
Lagoon	70-80

Application

Manure is usually:

- Broadcast (top dress) with plow down or disking.
- Broadcast without plow down or disking.
- Knifed (injection under the soil surface).
- Irrigated.

Table 707-2 shows average nitrogen losses by method of application.

The greatest nutrient response follows land application and immediate incorporation into the soil. Plow down solid manure as soon as possible to minimize nitrogen loss and to begin release of nutrients for plant use. Injecting, chiseling, or knifing liquids into the soil minimizes odors and nutrient losses to the air and/or to runoff.

Nitrogen loss as ammonia from land is greater during dry, warm, windy days than during humid or cold days. Ammonia loss is generally greater during the spring and summer months. Most losses occur in the first 24 hr after application, so incorporate manure into the soil as soon as possible.

Apply manure as near planting date as possible so more nutrients will be available to plants. While it may be more convenient to apply wastes in late fall or winter, 25%-50% of the total nitrogen can be lost from decomposition and leaching. With liquid manure, much nitrogen loss by leaching or denitrification can be stopped with a nitrification inhibitor. These products inhibit the action of certain soil bacteria that convert ammonium nitrogen to nitrate nitrogen. They are volatile compounds, so immediate incorporation in the soil is vital.

Table 707-2. Ammonium-nitrogen losses to the air.
Percent of the nitrogen applied that is lost within 4 days of application.

Method of application	Type of waste	Nitrogen %
Broadcast	Solid	15-30
	Liquid	10-25
Broadcast with immediate cultivation	Solid	1-5
	Liquid	1-5
Knifing	Liquid	0-2
Sprinkler irrigation	Liquid	30-40

Crop Nutrient Removal

Apply manure so nutrients added do not greatly exceed crop needs. See Table 707-3. Manure nutrients, especially nitrogen, are utilized more efficiently by grasses and cereals than by legumes. Legumes get most of their nitrogen from the air, so additional nitrogen is not usually needed.

For the greatest return, apply manure first to corn and small cereal grains, then to sorghum and forages, and finally to pasture.

With heavy manure applications, have your soil tested for fertilizer needs and nutrient imbalance. Adjust waste application rates for your soil conditions against soil tests for phosphorus and potassium to balance crop nutrient needs.

Example 1:

How much N. P_2O_5, and K_2O is utilized by a corn yield of 150 bu/acre?

From Table 707-3, this crop uses 185 lb/acre of N, 80 lb/acre of P_2O_5, and 215 lb/acre of K_2O.

Table 707-3. Nutrient utilization by crops.
Values for the total above-ground portion of the plants. When only grain is removed, much of the nutrients is left in the residues, but is temporarily tied up in them and not readily available. Estimate nutrient requirements for one crop year by assuming complete crop removal. Source: Potash Institute of America
Dashed lines refer to example.

Crop	Yield	N	P_2O_5	K_2O
		------lb/acre-------		
Corn	150 bu	185	80	215
	180 bu	240	100	240
Corn silage	32 tons	200	80	245
Soybeans	50 bu	257	48	120
	60 bu	336	65	145
Grain sorghum	8000 lbs	250	90	200
Wheat	60 bu	125	50	110
	80 bu	186	54	162
Oats, Barley	100 bu	150	55	150
Alfalfa	8 tons	450	80	480
Orchard grass	6 tons	300	100	375
Brome grass	5 tons	166	66	254
Tall fescue	3.5 tons	135	65	185
Bluegrass	3 tons	200	55	180

Nutrient Value per Animal Unit

The fertilizer value of manure depends upon livestock species and handling and disposal methods. Table 707-4 estimates the acres needed per 100 animal units for land application of manure. Table values consider nutrient losses from handling and disposal. Use Tables 707-3 and 707-4 to estimate the number of acres needed for a specific livestock and cropping operation.

Example 2:

A swine producer has a 50-sow farrow-to-finish operation. Liquid waste is stored in an anaerobic pit. The producer intends to broadcast and cultivate the liquid manure into the ground to produce 150 bu/acre corn. How many crop acres are needed?

Table 707-4. Land application of manure.
Values based on nitrogen application only.

Manure Handling method	Animal units: Sow + 8 pigs: Feeder pig production	Farrow to finish	Pigs fed 50 to 220 lb	Dairy 1 cow or 2 heifers	Beef feedlot	Sheep	Poultry: Broiler	Layer	Turkey
	(acres/100 animal units to yield 100 lb N/acre)								
Anaerobic pit									
Broadcast	31.2	100.0	10.6	84.9	43.4				
Broadcast/cultivate	40.0	125.0	14.2	115.7	57.2				
Injection	41.6	133.4	14.2	84.9	58.8				
Irrigation	22.0	71.4	7.6	63.6	31.2				
Deep pit (poultry)									
Broadcast								.40	
Broadcast/cultivate								.56	
Open lot									
Broadcast	15.0	48.8	5.4	45.4	21.2				
Broadcast/cultivate	22.8	74.0	8.0	67.0	32.2				
Manure pack									
Broadcast	15.0	46.6	4.0	43.9	31.5	3.8	.20	—	.68
Broadcast/cultivate	26.0	83.4	9.0	74.8	36.4	6.2	.32	—	1.16
Deep stack									
Broadcast						4.0			
Broadcast/cultivate						6.6			
Lagoon									
Irrigation	9.2	29.0	3.2	27.6	13.0				

1. From Table 707-4, 125 acres are needed to apply 100 lb N/acre from 100 animal units. For this example, 50 sows farrow-to-finish are 50 animal units.

   50 animal units x 125 acres ÷ 100 animal units = 62.5 acres fertilized to 100 lb N/acre.
2. From Table 707-3, 150 bu/acre corn requires 185 lb N/acre.
3. How many acres can be fertilized to 185 lb N/acre? Multiply the value from step 1 by the ratio of 100 lb N/acre to the required lb N/acre:

   62.5 acres x (100 ÷ 185 lb N/acre) = 33.8 acres fertilized to 185 lb N/acre.
4. Apply the liquid waste to 33.8 acres to fully utilize the N fertilizer value of the manure to grow 150 bu/acre corn. Applying manure to meet N requirements more than adequately meets crop P and K needs.

Nutrient Value for Crop Production

Tables 707-5 and 707-6 show estimated fertilizer compositions of waste applied to the land (wet basis). Nutrient contents vary widely, so these data are guidelines. For specific situations, have your manure analyzed. Conversion factors:

- Multiply P_2O_5 by 0.44 to convert to elemental P.
- Multiply K_2O by 0.83 to convert to elemental K.
- 1 bu manure spreader capacity = 35-40 lb
- 1 ton manure spreader capacity = 27 bu
- 1000 gal = 4 ton (approx)
- 27,152 gal = 1 acre-inch

Available N is nitrogen the plant can use. Total N is mostly organic and ammonium nitrogen. Organic N is slow releasing N. Ammonium N is equivalent to commercial fertilizer and except for that lost to the air, can be used by plants in the application year. (It is available nitrogen in Tables 707-5 and 707-6.) Organic nitrogen must be released for plant use. About 50% of the organic nitrogen is available the year of application, with the rest carried over to succeeding years. About 70%-80% of the phosphorus and 70%-90% of the potassium in animal wastes are available for plant use the year of application. After a few years of regular waste applications, the amounts available are about the same as one year's application.

Table 707-5. Nutrients in solid manure.
Approximate fertilizer value of manure—solid handling systems.

Specie	Bedding or litter	Dry Matter	Available N	Total N	P_2O_5	K_2O
		-- % --	-----lb/ton raw waste-----			
Swine	No	18	6	10	9	8
	Yes	18	5	8	7	7
Beef cattle	No	15[a]	4	11	7	10
	No	52[b]	7	21	14	23
	Yes	50	8	21	18	26
Dairy cattle	No	18	4	9	4	10
	Yes	21	5	9	4	10
Sheep	No	28	5	18	11	26
	Yes	28	5	14	9	25
Poultry	No	45	26	33	48	34
	Yes	75	36	56	45	34
	Deep pit	76	44	68	64	45
Turkey	No	22	17	27	20	17
	Yes	29	13	20	16	13
Horse	Yes	46	4	14	4	14

[a]Open concrete lot. [b]Open dirt lot.

How Much Animal Waste Can Be Applied?

After determining crop fertilizer needs and manure nutrient value, estimate how much manure to apply and determine if additional commercial fertil-

Table 707-6. Nutrients in liquid manure.
Approximate fertilizer value of manure—liquid handling systems.
Dashed lines refer to example.

Species	Waste handling	Dry matter	Available N	Total N	P_2O_5	K_2O
		%	---------- lbs/1000 gal raw waste ----------			
Swine	Liquid pit	4	[26]	36	[27]	[22]
	Lagoon*	1	3	4	2	4
Beef	Liquid pit	11	24	40	27	34
	Lagoon*	1	2	4	9	5
Dairy	Liquid pit	8	12	24	18	29
	Lagoon*	1	2.5	4	4	5
Poultry	Liquid pit	13	64	80	36	96

*Lagoon—including lot runoff water.

izer is needed for efficient crop production. Also see the section on irrigation for maximum liquid application.

Example 3:

A swine producer wants to apply liquid manure from a pit to his cropland to raise 150 bu/acre corn. How much manure should he apply? (See Table 707-7 for example values.)

1. What are the nutrient needs of the crop (Table 707-3)?
2. What is the fertilizer value of the manure? Use results of a recent laboratory analysis or values from Table 707-6 for liquid or Table 707-5 for solid. Divide table values by 1000 to get lb/gal.
3. How much waste is to be applied to meet crop nutrient needs? Divide the value from step 1 by the value in step 2.
4. Select the amount of manure to apply from step 3. Determine how much N, P_2O_5, and K_2O is supplied per acre. Is additional commercial fertilizer needed to meet crop nutrient needs? Use 9,770 gal/acre for this example.

To maximize the use of animal waste, select the lowest value from step 3 (2,960 gal) and supplement the manure with inorganic commercial fertilizer to meet crop nutrient needs. To use the animal waste as a complete fertilizer, select the highest value in step 3 (9,770) for the rate of application.

Other Management Considerations:

- Check state and local regulations concerning land application times and rates.
- Surface-apply animal wastes at least 100′ from streams, ponds, open ditches, and inlets to tile lines.
- While ground is frozen, apply animal wastes to relatively level land remote from surface water.
- Avoid spreading liquid manure on water saturated soils where runoff is likely.
- For odor control, spread raw animal wastes frequently, especially during the summer; spread early in the day as the air is warming up and rising rather than late when the air is cooling and settling; and avoid days when the wind is blowing toward populated areas or when the air is still and seems to hang.
- Agitate or mix liquid wastes for removal of settled solids and uniform waste nutrient application to the land.

Table 707-7. Values for Example 3.

Step 1:

Nutrient	lb/acre
N	185
P_2O_5	80
K_2O	215

Step 2:

Nutrient	Solid lb/ton	Liquid lb/gal
Available N	6	0.026
P_2O_5	9	0.027
K_2O	8	0.022

Step 3:

Nutrient	Step 1/Step 2	Solid ton/acre	Liquid gal/acre
N	185 / 6	31	
	185 / 0.026		7115
P_2O_5	80 / 9	9	
	80 / 0.027		2960
K_2O	215 / 8	27	
	215 / 0.022		9770

Step 4: a. Nutrients supplied

Nutrient	lb/gal	x	Application rate gal/acre	=	Nutrient applied lb/acre
N	0.026	x	9770	=	254
P_2O_5	0.027	x	9770	=	264
K_2O	0.022	x	9770	=	215

b. Is commercial fertilizer needed?

Nutrient Nutrient	Crop needs lb/ acre	-	Nutrients applied lb/acre	=	Fertilizer needs lb/acre
N	185	-	254	=	-69 (surplus)
P_2O_5	80	-	264	=	-184 (surplus)
K_2O	215	-	215	=	0

720 DOMESTIC SEWAGE TREATMENT

If access to a city or municipal sewer system is unavailable, a private sewage treatment and disposal system is needed. Select a system to properly handle all household wastes. Proper design, construction, operation, and maintenance of a treatment and disposal system will prevent any health hazard to the family or community.

In most areas, a permit is required for a proposed sewage disposal site. **Check your state, county, or city regulations and plumbing codes before starting construction.** Get information from your local health department, zoning administrator, or county extension office.

System Selection

There are several types of onsite sewage treatment systems to consider, Fig 720-1. To select the right system, carefully analyze the site. Table 720-1 lists available systems and the site criteria you must consider. An x shows that a system can be used. Select a system acceptable to local health authorities.

When a waste system needs maintenance or when ownership changes, an accurate sketch of the waste system is important. Make a plan showing:

- Septic tank size and location.
- Openings for pumping out sludge.
- Soil treatment area size and location.
- Direction and distance of the septic tank and treatment area from the house.
- Installer's name and address.

File the sketch in a place where it is readily available.

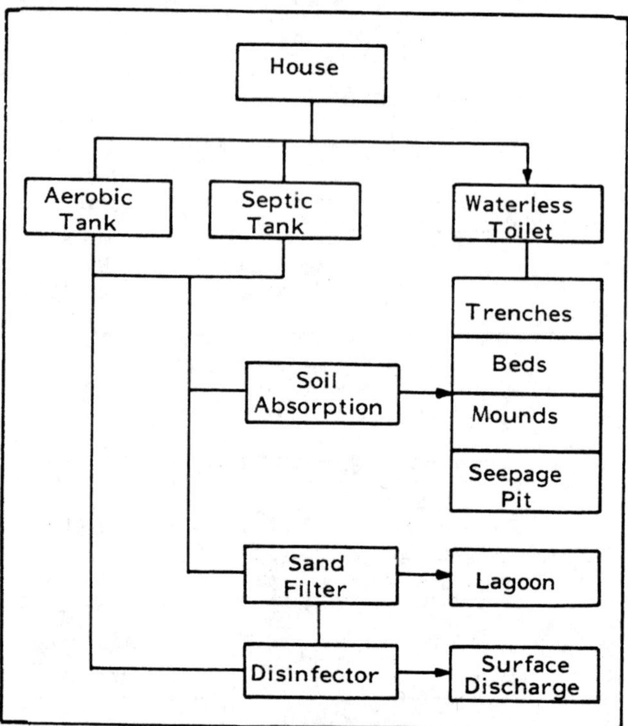

Fig 720-1. Alternative waste treatments.

Table 720-1. System selection chart.

Site criteria	System: Conventional Trench	Conventional Bed	Seepage pit	Mound	Treatment with surface discharge
Soil percolation rate in minutes per inch					
1- 30	x	x	x	x	
30- 60	x	x		x	
60-120				x	
More than 120				x	x
Depth to high water table or creviced bedrock					
More than 10′	x	x	x	x	
5′-10′	x	x		x	
2′- 5′	x	x		x	
Less than 2′					x
Depth to impermeable layer					
More than 10′	x	x	x	x	
5′-10′	x	x		x	
2′- 5′				x	
Less than 2′					x
Site slope					
More than 15%	x		x		x
5%-15%	x		x	x	x
Less than 5%	x	x	x	x	x

Septic Tanks

Location

Septic tank location with respect to the home is usually determined by the slope of the land and the major bathroom and kitchen plumbing. Locate septic tanks at least 10′ from foundation walls and straight out from the point where the sewer pipe goes through the wall. Check local codes for distance requirements between septic tanks, foundation walls, and property lines.

Locate the septic tank out of high vehicle traffic areas, because excessive loads can damage the tank. Avoid areas subject to flooding or discharge from downspouts. Locate the tank within 40′ of an access road so a tank truck can clean the tank and avoid driving over the disposal field. Consider future home improvements such as sidewalks, patios, garages, and storage buildings when selecting the tank site.

Installation

Septic tank depth depends on house plumbing and whether gravity flow from a basement sewer drain is provided. A sewage treatment system works better and is more easily maintained if the soil treatment area is near the ground surface. Gravity flow throughout the system is desirable.

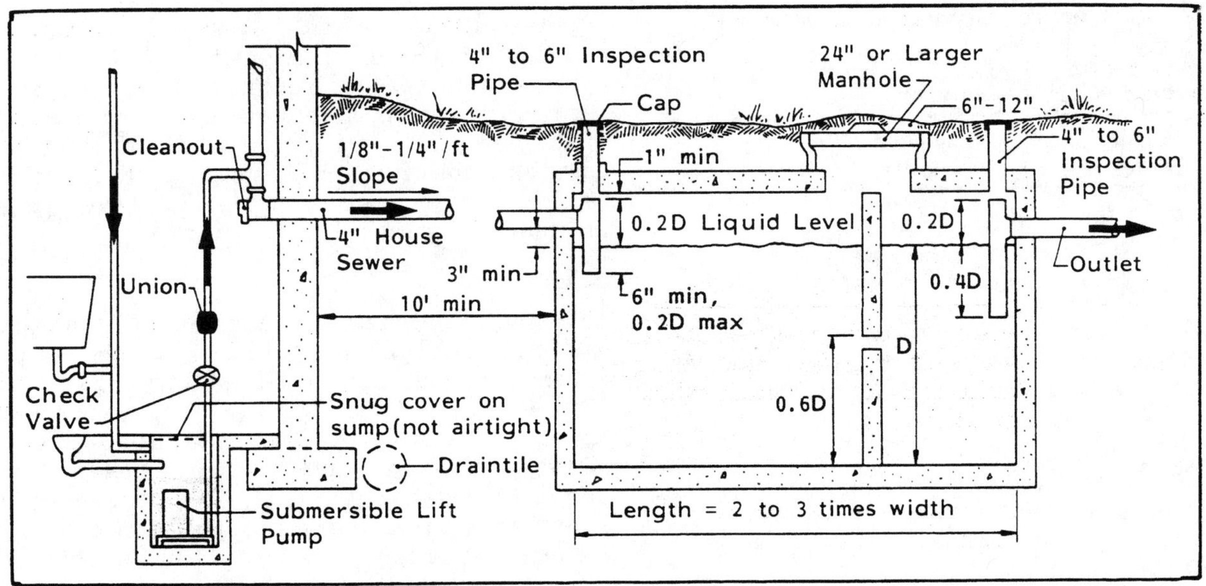

Fig 720-2. Shallow septic tank installation.

Provide a riser from the access port of the septic tank and pumping chamber. Grade the area away from the riser to divert surface water. Make the riser at least 24" wide and use a heavy cover to prevent children from removing it. In deep installations, check the strength of the tank with your supplier—it must support the extra weight of the riser.

Sizing

Check state and local codes for required septic tank size. If no requirements exist, use the values in Table 720-2. The minimum septic tank size for any installation is 750 gal. (See Graywater Systems for possible exceptions.) With garbage disposals, add at least 50% to the recommended size because of the extra load to the system. More frequent tank cleaning may be needed with a garbage disposal.

Septic tanks can be one- or two-compartment. A two-compartment septic tank, or two single-com-

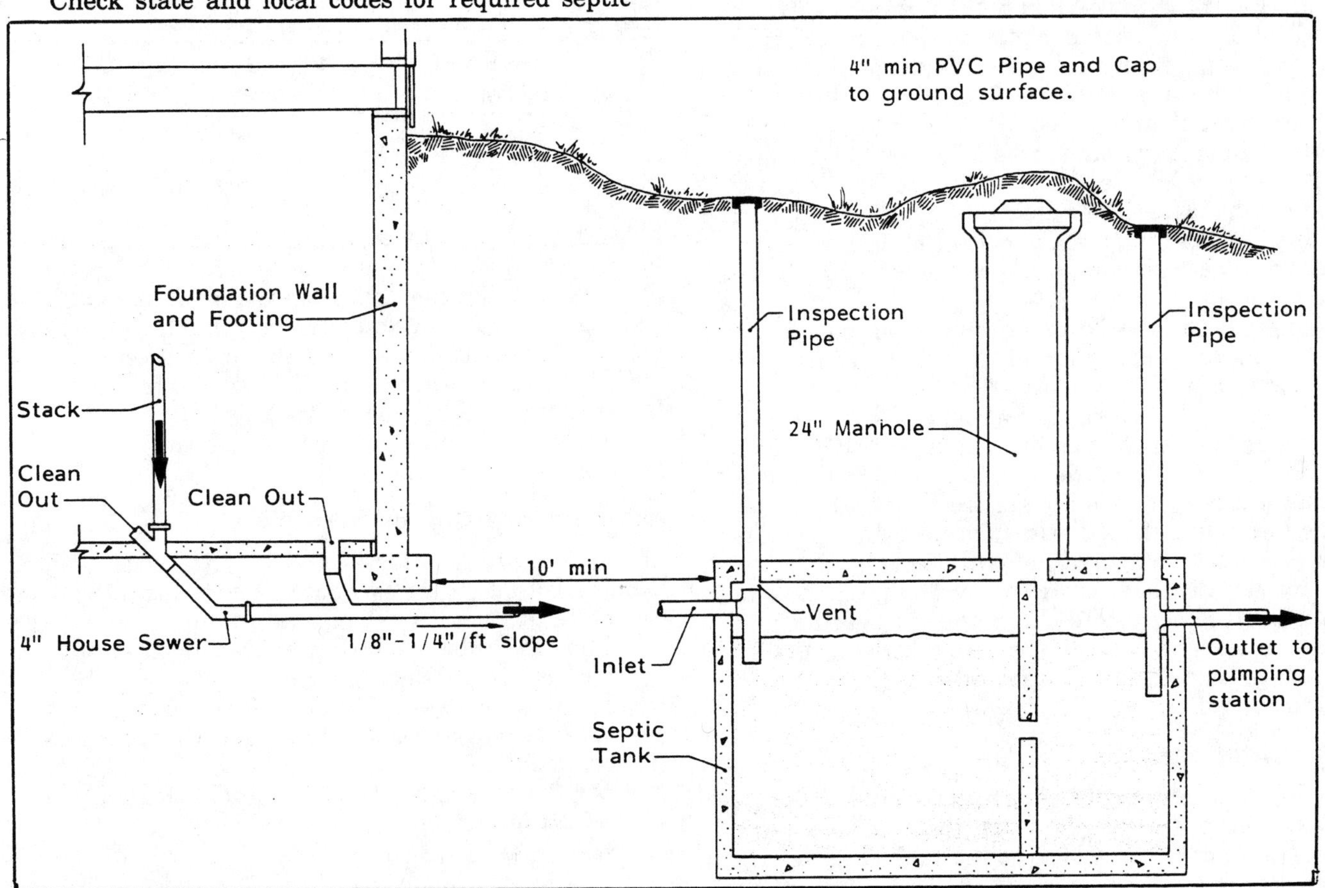

Fig 720-3. Deep septic tank installation.

Table 720-2. Minimum liquid capacities for residential septic tanks.
Increase capacity by 50% for kitchen sink garbage disposal. About 150 gal/day (gpd) of sewage/bedroom is assumed.

Number of bedrooms	Minimum liquid capacity, gal
2 or less	750
3	1000
4	1250
5	1500

partment tanks in series, are recommended. Make the first compartment at least 500 gal or ½ to ⅔ of the total capacity to minimize short circuiting and solids carry-over to the disposal field.

Construction

Septic tanks are constructed of the following materials unless dictated by state or local regulations:

- Precast reinforced concrete.
- Fiberglass.
- Cast-in-place reinforced concrete.
- Concrete blocks with concrete-filled cores. Insert reinforcing rods in the cores before filling them. Waterproof the tank inside and out.
- Asphalt-coated steel (useful life is 2 to 5 yr).
- Polyethylene.
- Other materials meeting state and local strength and durability criteria.

Place the tank on good quality fill or undisturbed earth. Level the tank, then backfill immediately or fill the tank with water to prevent flotation or shifting. For rectangular tanks, the length should be two to three times the width.

Arrange the inlet and outlet of each tank or compartment to retain scum and sludge, Fig 720-2. Inlet baffles can be sanitary tees at least 4" in diameter that extend at least 6" below the liquid surface and within 1" of the tank top. The bottom of the horizontal section of the inlet pipe should be at least 3" above the liquid level.

Put the outlet baffle, or sanitary tee, no more than 6" from the tank outlet. Extend it within 1" of the tank top and to a depth of at least 40% of the liquid depth.

Use a 4" plastic or rigid sewer pipe with watertight joints for the outlet pipe. Slope it at least 1% (⅛"/ft). Avoid running the outlet pipe under driveways to reduce the risk of crushing and freezing. If the outlet pipe does run under a driveway, protect the pipe from heavy loads by placing 6" of gravel around it.

To start septic action in a new tank, use it. The natural processes usually begin as the tank fills. It is not necessary to seed the system with commercial products. Late spring or summer is the best time to begin operation.

Tank Maintenance

The best designed and operated septic tank/disposal field eventually fails unless sludge is periodically removed from the septic tank. Inadequate maintenance can cause sewage to back up into the house and solids to overflow to the soil disposal area. When solids clog the soil, the disposal area must be abandoned and a new one constructed.

Most tanks need to be pumped every 3 to 5 years, depending on the septic tank size and the amount and quality of solids entering the tank. Have an experienced operator with appropriate equipment clean the septic tank and dispose of the pumpings (septage) in an approved location.

Safety

Septic tank gases are dangerous. Methane, an explosive, and hydrogen sulfide, a poison, are the major hazardous gases released during pumping. Torches or other flames near the septic tank opening can cause an explosion. **Never lean into or enter a septic tank, even to save someone else,** without proper breathing equipment. **You could be poisoned or asphyxiated.** Call a professional if tank maintenance is required.

Graywater Systems

A graywater sewer system does not contain any toilet or garbage waste, so the needed tank capacity is less. If possible, use a larger tank volume, Table 720-3, because the system may not always be a graywater system. If the system is upgraded to a sewage system, the absorption system must also be enlarged.

Table 720-3. Septic tank capacities for graywater systems.
Do not discharge toilet or garbage wastes into the septic tank.

Number of bedrooms	Minimum liquid capacity, gal
2 or less	450
3	600
4	750
5 or 6	900

For a graywater system, use at least 2" diameter lines sloped 1% to the tank from the home and treatment unit. Baffle submergence, length-to-width ratio, distance between inlet and outlet, access, and other septic tank features are the same as for a regular septic tank. Size the soil treatment unit as for a conventional system, but allow for the reduced sewage volume.

Non-residential Septic Tanks

Size septic tanks for establishments other than homes according to the estimated flow in Table 720-4. Calculate the septic tank size as follows:

- The volume below the liquid level for flows up to 500 gpd is **at least** 750 gal.
- For flows between 500 and 1500 gpd, the tank volume is **at least** 1½ times the estimated daily sewage flow.
- For flows greater than 1500 gpd, calculate the tank capacity below the waterline from:

$V = 1125 + 0.75\ Q$

V = tank volume below the waterline, gal

Q = daily waste flow, gpd.

- For restaurants, size the septic tank at least twice as large as the above values. Put tanks in series for cooling and coagulation of fats and greases.

Determine the soil absorption area requirements from Table 720-7.

Soil Disposal

Site Selection

Select the soil absorption field site before locating the house and any other facilities. Check local health or zoning regulations for minimum setback distances. Separate the absorption field at least 10′ from buildings, trees, and property lines, and 50′-150′ from lakes, streams, and wells, Fig 720-4.

Locate the absorption field where good grass cover is possible. Keep vehicles off the area before, during, and after construction. In the winter, keep all traffic off—including snowmobiles and foot traffic—to minimize frost penetration.

Divert all surface runoff water from roofs, patios, driveways, or other paved areas away from the absorption area. Grade the area so waterways, drainage ditches, etc., do not drain across the site. Never discharge water from basement footing drains or other sources onto the absorption system.

Evaluate prospective sites for percolation rate, depth to creviced bedrock, depth to high seasonal water table (saturated soil condition), depth to impermeable soil layer or rock, and slope. See Table 720-5.

Table 720-4. Quantities of sewage flow.

Commercial Sources	Unit	Wastewater flow Range	Typical
		---gal/unit/day---	
Airport	Passenger	2.1- 4.0	2.6
Auto service station	Vehicle served	7.9- 13.2	10.6
	Employee	9.2- 15.8	13.2
Bar	Customer	1.3- 5.3	2.1
	Employee	10.6- 15.8	13.2
Hotel•	Guest	39.6- 58.0	50.1
Industrial building (excluding industry and cafeteria)	Employee	7.9- 17.2	14.5
Laundry (self-service)	Machine	475. -686.	580.
	Wash	47.5- 52.8	50.1
Motel	Person	23.8- 39.6	31.7
Motel with kitchen	Person	50.2- 58.1	52.8
Office	Employee	7.9- 17.2	14.5
Restaurant	Meal	2.1- 4.0	2.6
Rooming house	Resident	23.8- 50.1	39.6
Store, department•	Toilet room	423. -634.	528.
Shopping center•	Parking space	0.5- 2.1	1.1
Institutional Sources			
Hospital, medical••	Bed	132. -251.	172.
Hospital, mental••	Bed	79.3-172.	106.
Prison••	Inmate	79.3-159.	119.
Rest home••	Resident	52.8-119.	92.5
School, day			
Cafeteria only	Student	10.6- 21.1	15.9
Cafeteria, gym, showers	Student	15.9- 30.4	21.1
No cafeteria, gym, showers	Student	5.3- 17.2	10.6
School, boarding	Student	52.8-106.	74.0
Recreational Sources			
Apartment, resort	Person	52.8- 74.0	58.1
Cabin, resort	Person	34.3- 50.2	42.3
Cafeteria•	Customer	1.1- 2.6	1.6
Campground (developed)	Person	21.1- 39.6	31.7
Cocktail lounge	Seat	13.2- 26.4	19.8
Coffee shop•	Customer	4.0- 7.9	5.3
Country club	Member present	66.0-132.	106.
	Employee	10.6- 15.9	13.2
Day camp, no meals	Person	10.6- 15.9	13.2
Dining hall	Meal served	4.0- 13.2	7.9
Dormitory, bunkhouse	Person	19.8- 46.2	39.6
Hotel, resort	Person	39.6- 63.4	52.8
Store, resort•	Customer	1.3- 5.3	2.6
Swimming pool•	Customer	5.3- 13.2	10.6
Theater	Seat	2.6- 4.0	2.6
Visitor center	Visitor	4.0- 7.9	5.3
•Add for employees	Employee	7.9- 13.2	10.6
••Add for employees	Employee	5.3- 15.9	10.6

Source: EPA Design Manual, Onsite Wastewater Treatment and Disposal Systems, October 1980.

Table 720-5. Soil evaluation for conventional and mound systems.
Depths are minimum and measured from the soil surface. "High water table" means high seasonal saturated soil level.

	Units	Conventional Trench	Bed	Mounds
Percolation rate	min/in	1-60	1-60	1-120
Depth to creviced bedrock	ft	3	3	2
Depth to high water table	ft	3	3	2
Depth to impermeable soil layer or rock	ft	3	3	2-5[a]
Maximum slope	%	24	6	12[b]

[a] Depends on soil permeability, frost depth, and system layout. Provide enough area to move effluent away. See section on mounds.
[b] For percolation rates of 1-29 min/in, maximum slope is 12%; for 30-120 min/in, 6%.

Absorption Trenches

Absorption trenches are the most common and effective soil absorption system. As the soil filters sewage effluent, a biological mat develops on the trench bottom and sidewalls that treats the sewage. Freezing rarely occurs in a properly constructed and continuously operated absorption trench.

Sizing

The trench bottom area required depends on (Table 720-6):

- The amount of sewage entering the system daily.
- Soil porosity as measured by the percolation test.
- The depth of rock below the distribution pipe.

Estimate 150 gpd/bedroom for sizing trenches. Base trench design on the number of bedrooms, because the number of people in a home changes.

Rock around and below the distribution pipe distributes the effluent over the trench bottom and provides a porous storage zone until the liquid can infiltrate into the soil. Greater rock depths require less trench bottom area because some water flows through the trench walls as well as the bottom.

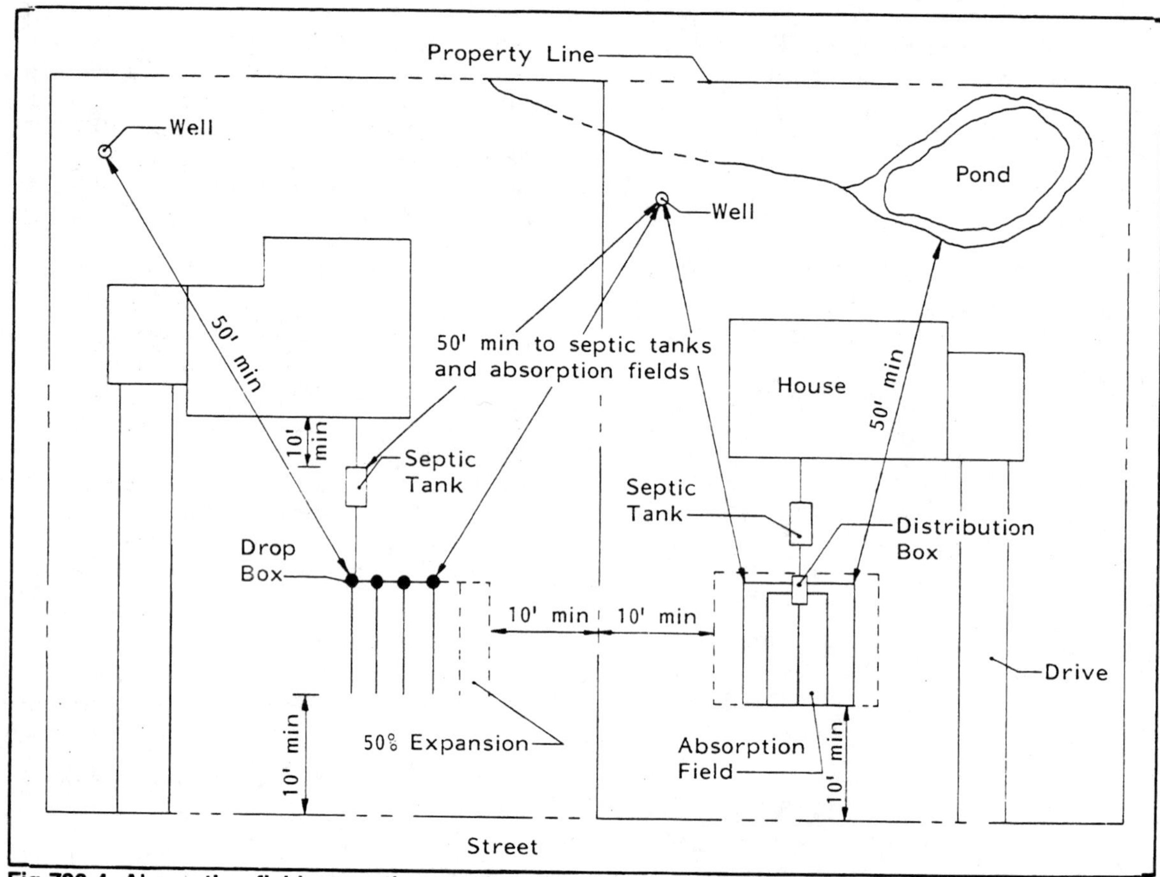

Fig 720-4. Absorption field separation requirements.
Consider neighbors' facilities, too. State, county, or local regulations may require greater separation distances.

Put the bottom of the trench 18"-36" below the soil surface. Minimum trench width is 18"; maximum width is 36". Limit single trenches to 100' long. Use at least two trenches, unless needed trench length is less than 50'. Leave 5' of undisturbed soil between trenches.

Table 720-6. Trench loading rates and bottom areas.
Provide enough area for at least two bedrooms in residences at 150 gpd/bedroom.
If the percolation rate is outside the range shown, the soil is unsuitable for standard soil treatment units. Refer to section on mounds and alternative systems.

Percolation rate MPI	Depth of rock below distribution pipe 6"	12"	18"	24"
	Trench loading rate, gal/ft²/day			
1 to 5	1.2	1.5	1.76	2.0
6 to 15	0.8	1.0	1.2	1.3
16 to 30	0.6	0.75	0.91	1.0
31 to 45	0.5	0.63	0.75	0.83
46 to 60	0.45	0.57	0.68	0.75
	ft² of trench bottom/bedroom			
1 to 5	125	100	85	75
6 to 15	190	150	125	115
16 to 30	250	200	165	150
31 to 45	300	240	200	180
46 to 60	330	265	220	200

Example 1:

Find the number of trenches required for a disposal field for a three-bedroom home.

1. Find the quantity requiring disposal.
   3 bedrooms x 150 gpd/bedroom = 450 gpd
2. Find the soil percolation rate. For this example, assume a percolation test showed 35 minutes per inch (MPI).
3. Find the loading rate for 35 MPI soil from Table 720-6. It is 0.75 gal/ft²-day for 18" of rock below the distribution pipe.
4. Divide the estimated flow rate by the loading rate to get the trench bottom area.
   450 gpd ÷ 0.75 gal/ft²-day = 600 ft²
5. Another way to find bottom area is to use the lower half of Table 720-6. For 35 MPI and 18" of rock, area is 200 ft²/bedroom. Multiply the area by the number of bedrooms.
   200 ft²/bedroom x 3 bedrooms = 600 ft²
6. The total area required is the trench bottom area from Step 4 or Step 5.
7. Find the trench length required for the desired width. Assume a 24" (2') width; the length is the required trench bottom area divided by the width in feet.
   600 ft² ÷ 2' = 300 lineal ft
8. Single trench length is limited to 100', so divide the 300' into three or more trenches.

Trench construction

Soil conditions and construction practices are important factors affecting how an absorption system will function. Construction when the soil is too wet can cause complete system failure in a very short time.

If the soil crumbles when you try to roll it into a ribbon, the soil is dry enough for construction, and little danger of compacting or smearing the soil exists.

Excavate absorption trenches with a backhoe. Depth can vary from 18"-48" but is typically 24"-30". A 30" depth allows for 12" of rock below the 4" distribution pipe and 2" of rock plus 12" of soil cover over the pipe.

Use rock in the trench that is clean and free of dust, dirt, and debris. It should be at least ¾", but no greater than 2½" in diameter. Do not use pea gravel or **pit run gravel,** which clogs in a short time. Use caution in selecting crushed limestone. Do not use calcite limestone—it is soft, deteriorates in contact with septic tank effluent, and clogs the soil.

Distribution pipes are commonly 4" rigid plastic, corrugated plastic, concrete, or clay tile. Plastic agricultural drain tubing is not suitable because the small holes clog easily. Corrugated plastic pipe is flexible, so tie it to a grade board. Install the distribution pipe on a uniform grade. Comply with local and state codes.

Cover the top of the distribution pipe with at least 2" of clean rock. Cover the top of the rock with untreated building paper (red rosin), 3"-4" of hay or straw, synthetic fabric, or other permeable material. This covering helps prevent the rock from becoming clogged by the soil backfill. With sandy loam or coarser backfill, use a 2"-4" layer of hay or straw over the rock and cover this with untreated building paper or a synthetic fabric.

Backfill over the rock with soil of the same texture as the original soil. Crown the backfill 4"-6" above finished grade to allow for settling and to shed rain and runoff. Establish a grass cover over the absorption area as soon as possible.

On sites with a high seasonal water table, use long trenches and space them as far apart as practical. Trenches can extend 200′ along a slope if a drop box supplies effluent at the center, so each trench is only 100′ long. Put a 4" observation pipe, extending to the trench bottom, at the end of each trench. If water is ponding at the end of a trench, take it out of service by plugging the distribution pipe at the box.

Absorption Beds

An absorption bed is a wide absorption system with more than one distribution pipe, Fig 720-6. Beds are usually rectangular, with the pipe parallel to the long side. They are not permitted in some states. If

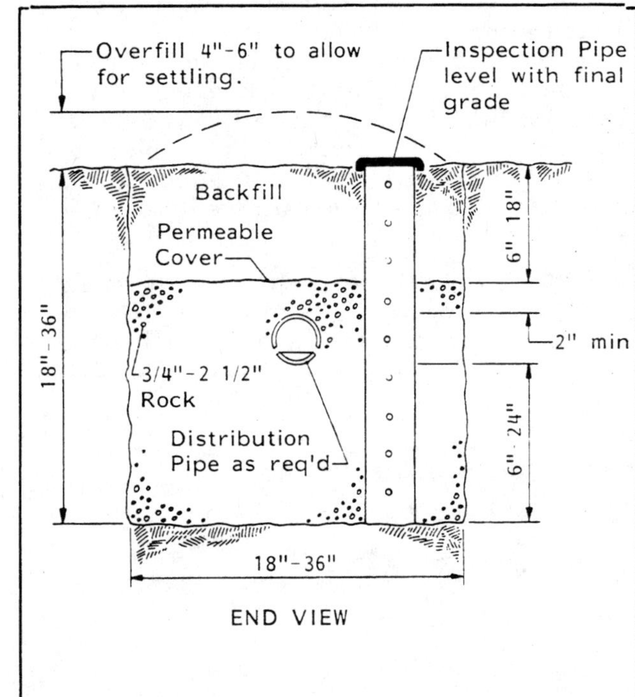

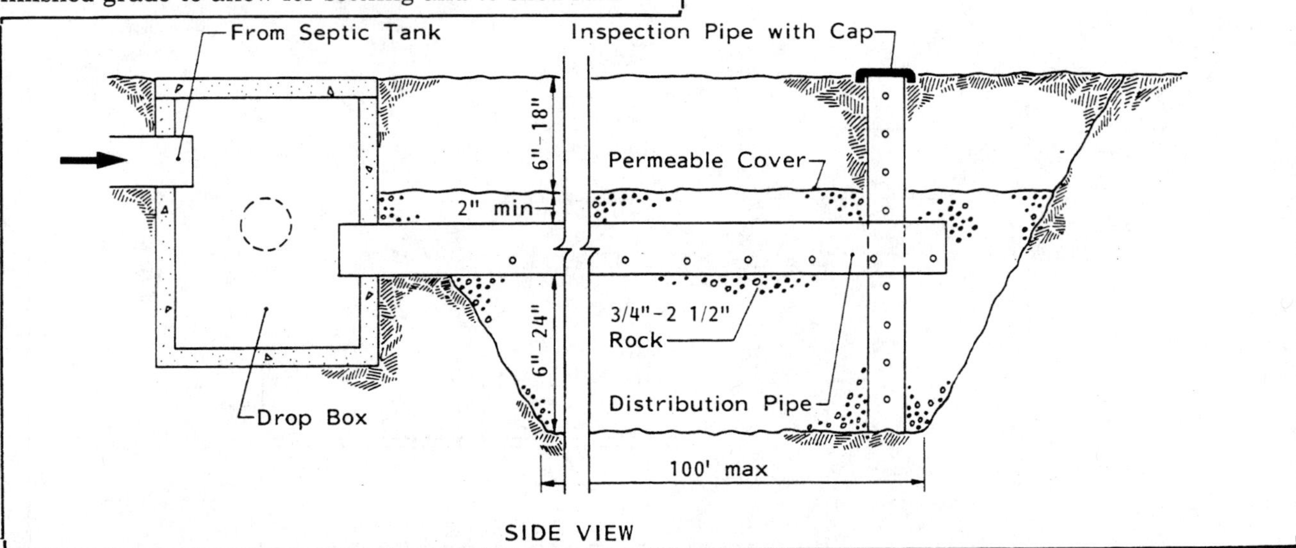

Fig 720-5. Trench construction details.
Do not slope bottom of trench. Slope pipe 2"-4"/100′. Permeable cover is untreated building paper or 3"-4" of loose hay or straw. Check local regulations for location and number of holes in the distribution pipe.

enough space is available, use absorption trenches because they have more sidewall area than a bed.

Locate absorption beds in areas with less than 6% slope. On steeper sites, excavation gets too deep on the upslope side. And on the downslope side, the bed may be too shallow, causing surface seepage as the effluent level rises in the bed. Put the long side perpendicular to the slope (along the contour) to minimize depth differences.

Construction

Absorption bed construction is similar to trench construction. Refer to the Trench Construction section for precautions, rock and material specifications, and construction procedures. The bed bottom must be level in all directions. Excavate with a backhoe or other equipment not requiring wheels or tracks on the bottom of the bed. Do not excavate with a front-end bucket loader or bulldozer.

Place distribution pipes on 6″ or more of rock; space them 4′-6′ apart and 18″-36″ from the sides of the bed. Backfill the absorption bed with 6″-18″ of topsoil. Crown the backfill about 12″ to allow for settling and to promote runoff away from the bed area. Do not mechanically compact the backfill.

Sizing

See Table 720-7 for the required bottom area of the bed. Note that more than 6″ of rock below the distribution pipe does not reduce bottom area, because the bed has relatively little sidewall area. But, in most cases, the total land area for effluent absorption is less for a bed than for a trench.

Table 720-7. Recommended absorption bed area.
Provide enough area for at least a two-bedroom residence (300 gpd). If the percolation rate is outside the range shown, the soil is unsuitable for standard soil treatment units. Refer to the sections on alternative systems. Underlined number refers to Example 2.

Percolation rate min/inch	Treatment bed area per bedroom ft^2	Loading rate $gal/ft^2/day$
1 to 5	125	1.2
6 to 15	190	0.8
16 to 30	<u>250</u>	0.6
31 to 45	300	0.5
46 to 60	330	0.45

Example 2:

Determine the bottom area for an absorption bed for a three-bedroom residence. In this example assume the percolation rate is 20 MPI.

Solution:

From Table 720-7, the required absorption bed bottom area for a percolation rate of 20 MPI is 250 ft^2/bedroom.

3 bedrooms x 250 ft^2/bedroom = **750 ft^2**

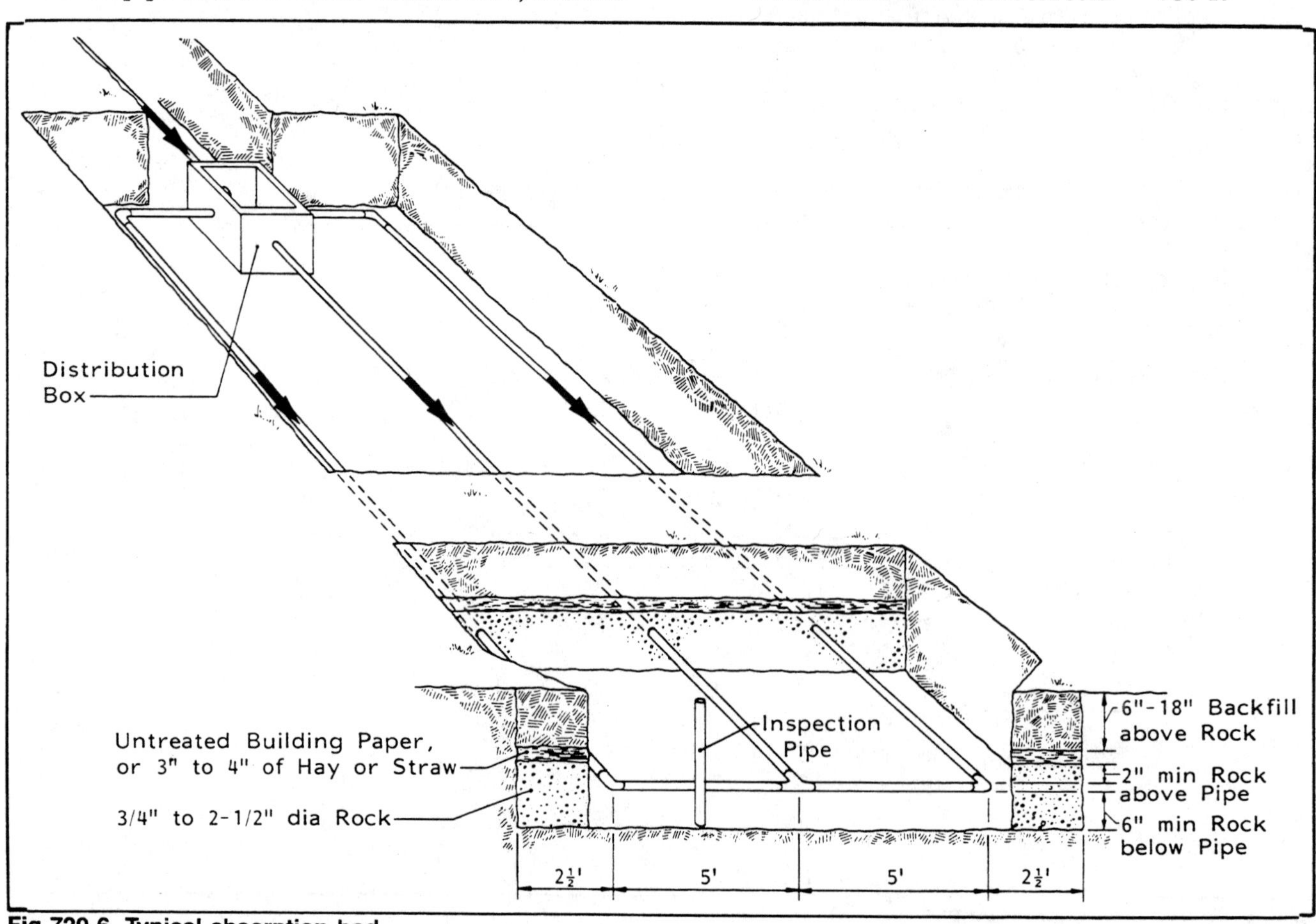

Fig 720-6. Typical absorption bed.
Crown the finished soil surface about 12″.

Seepage Pits

Seepage pits, or dry wells, are the least desirable soil treatment system. Do not use them:

- Where soil is suitable for absorption trenches, absorption beds, or mounds.
- Where the percolation rate is faster than 1 MPI or slower than 30 MPI.
- Where domestic water supply wells are shallower than 50′.
- Where creviced bedrock is closer than 50′ to the ground surface.
- Where the water table or an impervious layer is within 4′ of the bottom of the proposed pit.

Design seepage pits with a wall area equal to the bottom area of an absorption bed, Table 720-7.

Construct seepage pits with an inside wall diameter of at least 5′; use precast concrete or bricks, stones, or blocks at least 4″ thick. Make the wall watertight above the inlet and provide wall openings below the inlet for liquid passage.

If multiple pits are used, separate them by a distance of three times the diameter of the pit. All pits in a multiple pit system need not be the same size.

Place 6″-12″ of clean ¾″-2½″ rock between the outside of the pit wall and the excavated soil. Place 6″ of rock in the bottom of the pit to prevent removing the bottom soil if the pit is pumped.

The top of a pit is 6″-12″ below the ground surface and should be strong enough to support the earth cover and any other reasonable surface load. Provide an inspection pipe or a manhole for access. Beware—the pit may contain deadly gas. Do not extend the manhole to the surface, because children and others could fall into the pit.

Access for pumping can be through a 4″ inspection pipe. Provide a corrosion-proof and easily removable watertight cap.

Design and Construction of Mounds

Use mounds for onsite wastewater disposal on sites that have:

- Slowly permeable soils with or without high water tables.
- Shallow permeable soils with creviced or porous bedrock.
- Permeable soils with high water tables.

A septic tank and mound system includes a septic tank, pumping chamber, and mound. The pump elevates effluent to the mound and pressurizes the distribution system. The mound has fill material, an absorption area, a distribution system, a soil cap to shape the top surface, and topsoil. Effluent is pumped into the absorption area through the distribution system. Fill material treats the effluent before it passes into the natural soil. The topsoil provides frost protection, retains moisture for vegetation, and promotes runoff of precipitation.

Soil and site requirements

Table 720-8 summarizes soil and site limitations for mound construction. See Site Selection for other requirements.

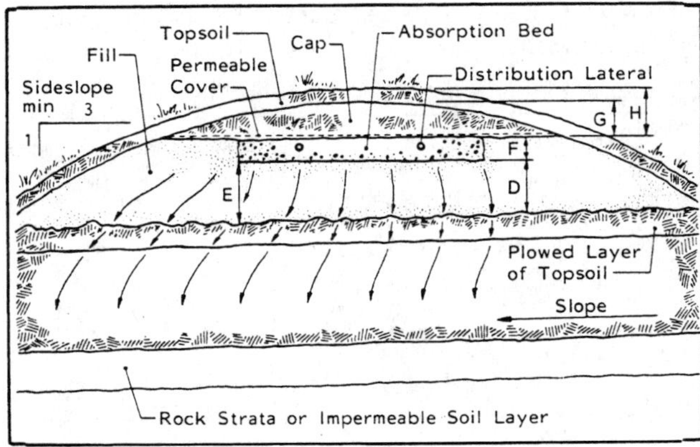

Fig 720-7. Mound system with absorption bed.
Arrows show effluent movement in permeable soil, with high water table and shallow permeable soil over creviced bedrock. D, F, and G are dimensions at the upslope side of the bed; H is at the crown.

Table 720-8. Soil and site restrictions on mound systems.

Restricting factor	Soil group: Slowly permeable soils	Permeable soils with pervious bedrock	Permeable with high water tables
Percolation rate, MPI	60-120	1-60	1-60
Test depth[a]	24″	16″	16″
Depth to pervious rock	24″	24″	24″
Depth to high water tables	24″	24″	24″
Depth to impermeable soil layer or rock strata	2′-5′[b]	2′-5′	2′-5′[b]
Depth to 50% by volume rock fragments:	24″	24″	24″
Slope	6%	12%[c]	12%[c]

[a] Percolation test depth, unless there is a more restrictive horizon above. If perched water is at 24″, hold test depth to 16″.
[b] See discussion in text. Depth depends on soil permeability, length, and climate.
[c] For percolation rates of 1-29 min/in, maximum slope is 12%; for 30-60 min/in, 6%.

Generally, sites with large trees, numerous smaller trees, or large boulders are unsuitable for a mound system because of difficulty in preparing the surface and the reduced infiltration area beneath the mound. Rock fragments, tree roots, stumps, and boulders reduce the amount of soil for proper purification.

Recommended separation distance between the toe of a mound (where the mound meets the original soil level) and buildings, property lines, and driveways is 10′ for 3:1 sideslope mounds. On slowly permeable soils, 30′ is recommended. More clearance permits mound repair if seepage occurs.

Fill material

The mound requires a **medium** sand texture fill beneath and beside the absorption area (trenches or bed). See Table 720-9. The sand need not be screened or washed, but select it with care to be sure that there are not more than 10% fines.

The topsoil and a medium or medium-fine textured soil cap provide higher water holding capacity for plant growth and minimize rainfall infiltration. Often, excavated soil from the site can be used. Place good topsoil 6″ deep over the entire mound to promote good vegetation.

Table 720-9. Medium texture for mound fill sand.

	More than 25%			Less than 50%		Less than 10%
Gravel	Very coarse sand	Coarse sand	Medium sand	Fine sand	Very fine sand	Silt and clay
2mm	1-2mm	.5-1mm	.25-.5mm	.1-.25mm	.05-.1mm	.05mm
Example of a good fill:						
5%	0%	30%	50%	13%	1%	1%

Table 720-10. Mound fill material characteristics.

Material	Characteristics % by weight	Design infiltration rate gpd/ft²
Medium sand	See Table 9.	1.2
Sandy loam	5-15% clay content	0.6
Sand/sandy loam mixture	88-93% sand 7-12% finer grained material	1.2
Dune sand		1.0

Absorption bed

Water table height and soil permeability determine absorption bed shape. Use a long absorption trench 4′-6′ wide on sites with a high water table and slowly permeable soils. For permeable soils, a rectangular bed up to 10′ wide can be used. Use either layout for shallow permeable soils over creviced or porous bedrock.

Orient the absorption bed along the contour so the effluent spreads out as it moves downslope.

The required absorption area is the daily sewage flowrate (gpd) divided by the absorption rate of the fill material (gpd/ft²), Table 720-10. The absorption rate for a medium sand fill is 1.2 gal/ft²-day. For a three-bedroom home and medium sand, the required absorption area is 3 bedrooms x 150 gpd/bedroom ÷ 1.2 gal/ft²-day = 375 ft².

Mound dimensions

Mound height (Fig 720-8b) is fill depth (D and E), plus bed depth (F), plus cap and topsoil depth (G and H). **Fill depth** depends on the treatment capabilities of the fill material. A minimum unsaturated flow depth of 3′, including natural soil fill, is needed for proper effluent treatment. When the water table is more than 2′ beneath the soil surface, use at least 1′ of fill (D). For shallow permeable soils over creviced bedrock, use at least 2′ of fill (D) when the natural soil depth is 2′.

Place at least 6″ of rock under, and 2″ above, the distribution pipe lateral. Use ¾″-2½″ washed rock or crushed stone. Do not use soft limestone.

Make the cap and topsoil at least 1.5′ deep at the center (H) and 1′ deep at the edges (G). For 3 or more parallel trenches, increase (H) to 2′ for enough slope

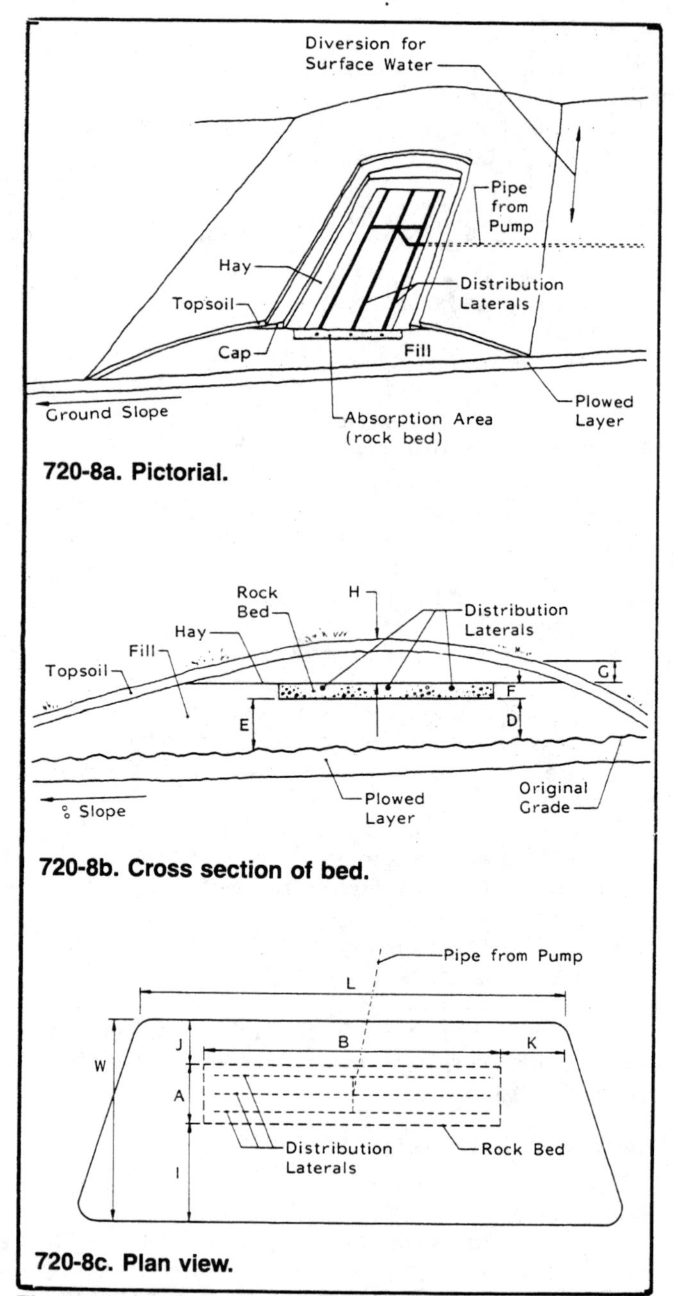

Fig 720-8. Mound with absorption bed.

for surface drainage. For sites with slopes greater than 6% or with mounds longer than about 60′ on a sloping site, consider sloping the cap and topsoil entirely to the downslope side to help shed water from the area.

Mound width and length depend on the absorption area needed, mound depth, and sideslopes. Limit sideslopes to 3:1 (4:1 often preferred).

Mound length (L) is the sum of the endslopes (K) plus the bed or trench length (B), plus 1′ of sand on each end of the bed to contain the rock for the bed. The mound width (W) is the sum of the upslope width (J), the bed or trench width (A), plus 1′ of sand on each side of the bed to contain the rock for the bed, the spacing between the trenches (C) if trenches are used, and the downslope width (I). As slope increases, downslope width (I) increases. Also, a slowly permeable soil usually requires more downslope width to get enough absorption area.

Basal area is the natural soil-fill interface. It accepts the effluent from the fill and transfers it to the subsoil.

The basal area required depends on the soil and slope conditions. For level sites, the total basal area is effective. For sloping sites, the only basal area considered for design is the area under and downslope of the bed. It includes the area enclosed by B x (A + I) for the bed system, Fig 720-8c. The upslope and endslopes transmit very little of the effluent. The percolation rate of the natural soil determines the area required. Find the required basal area by dividing the daily sewage flowrate (gpd) by the infiltration rate (gpd/ft²), Table 720-11.

Table 720-11. Mound subsoil infiltration rates.

Soil texture	Percolation rate MPI	Infiltration rate gpd/ft²
Sand/sandy loam	0- 30	1.2
Loams, silt loams	31- 45	0.75
Silt loams	46- 60	0.5
Clay loams, clay	61-120	0.25

Example 3:

Determine the required basal area for a three-bedroom house if the soil percolation rate is 50 MPI.

3 bedrooms x 150 gpd = 450 gpd = sewage flowrate.

From Table 720-11, infiltration rate = 0.50 gpd/ft².

450 gpd ÷ 0.50 gpd/ft² = **900 ft² basal area.**

After sizing the mound for absorption area and slope conditions on slowly permeable soils, add fill to make the mound wider if required for sufficient base soil area.

Distribution system

The effluent should spread uniformly over the entire absorption area. Gravity loading with 4″ perforated pipe is not adequate. Each trench has one lateral, while a rectangular bed up to 10′ wide has up to three laterals spaced 3′ apart. Pipe diameter depends on mound length. Table 720-12 gives allowable lateral lengths. For ¼″ holes 30″ apart in 1″ pipe, maximum length is 25′ from manifold to end of lateral. Tee-to-tee construction of manifold to laterals is preferred; cross-to-cross construction is satisfactory.

The laterals and manifolds must drain after every dosing. If the mound is downslope of the pumping chamber, put the manifold above the laterals so it drains or use cross-to-cross construction.

Site preparation

Lay out the trenches or beds perpendicular to the slope. Trench and lay the effluent pipe from the pumping chamber to the mound. Lay pipe sloping uniformly back to the pumping chamber so that it drains after dosing. Backfill and compact soil around the pipe to prevent back seepage of effluent and to reduce settling. Plow the site along the contour with a mold board or chisel plow 7″-8″ deep. Prepare the site when the soil is not too wet. Excess compaction and smearing reduce infiltration to the natural soil and can cause seepage.

Keep all traffic off the plowed site. Place fill immediately after site preparation with a backhoe or push it on from the upslope side with a lightweight, track-type tractor. Do not compact the soil below the mound or you may have seepage failure.

Table 720-12. Allowable lateral lengths.
For ¼″ perforation diameter. Underlined number refers to Example 4.

Perforation spacing	Pipe diameter 1 in	1¼ in	1½ in
in	Length, ft		
30	25	38	50
36	27	<u>42</u>	54
42	30	48	65
48	33	52	70

Mound preparation

Work from the ends and upslope side, especially on fine textured soils.

- Place the fill to the top of the bed. Shape sides to the desired slope.
- Form the bed with the tractor blade; hand level the bottom.
- Place 6″ of ¾″-2½″ rock in the bed.
- Place the distribution system on the rock. Connect the manifold to the pipe from the pumping chamber. Slope manifold slightly to the inlet pipe.
- Lay laterals.
- Place 2″ of rock over the distribution pipe.
- Place untreated building paper, permeable fabric, or 4″-5″ of material such as uncompacted straw or marsh hay over rock.
- Place cap soil over the bed—12″ deep in the center and 6″ at the edges of the bed.
- Place 6″ of good topsoil over the entire mound surface. The elevation of the mound is then at least 1.5′ (H) at the center and 1′ (G) at the outside edges of the bed.
- Plant grass or the best vegetation adaptable to the area.

Distribution Systems

Distribution systems convey septic tank effluent to the soil absorption system and distribute it. The successful operation of any soil absorption treatment system depends on proper distribution system choice and installation.

Gravity Distribution

When the total length of the absorption trench has been determined, trench layout and effluent distribution depend on topography. Gravity distribution methods are a distribution box or drop boxes.

Distribution boxes are primarily for level sites, but drop boxes can be used on level or sloping sites. Place distribution boxes and drop boxes on a firm base to prevent settling during or after backfilling.

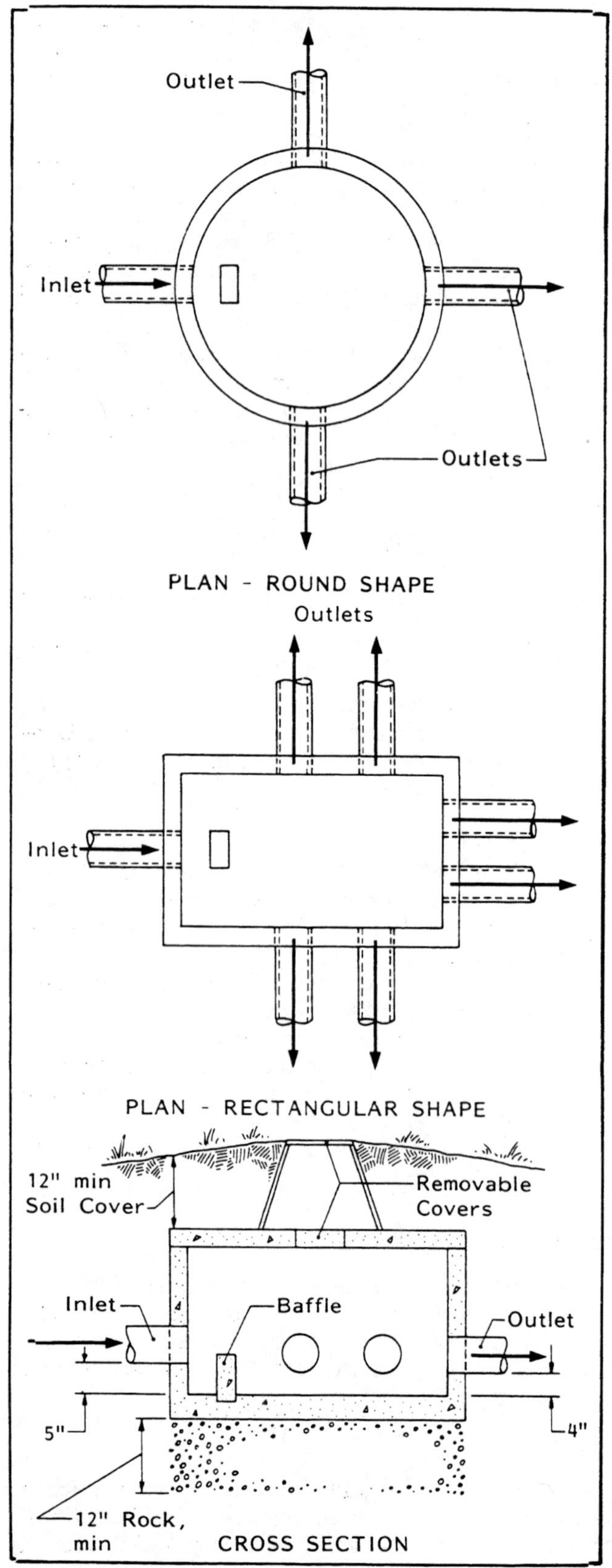

Fig 720-9. Distribution boxes.

Distribution box

A distribution box has an inlet from the septic tank and an outlet to each distribution pipe in the trench field or absorption bed, Fig 720-9. The boxes are watertight, have a removable cover, and are usually concrete or plastic. The inlet is 1″ higher than the outlets. The outlets to trenches are all at the same elevation and at least 4″ above the bottom of the box. Connect each distribution pipe separately and directly to the distribution box.

Drop boxes

Drop boxes are usually concrete, 12″-18″ in diameter or square, and about 18″ deep, Fig 720-10. Install a drop box at the inlet end of each trench.

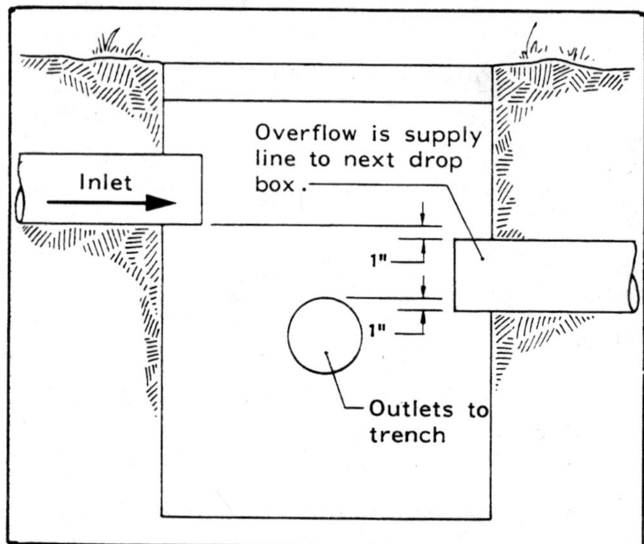

Fig 720-10. Drop box.

Effluent flows from the septic tank to the first drop box. An outlet near the bottom of the drop box connects to the distribution pipe of the trench. The outlets to two trenches supplied from one drop box are usually at the same elevation. A watertight overflow 1″ below the inlet leads to the next drop box. On level sites, place each row of trenches about 2″ deeper than the previous row.

The amount of effluent a distribution pipe receives depends on the elevation of the supply line leading to the next box. Place the bottom of the overflow pipe 1″ below the top of the trench pipe so most of the trench sidewall is used. Only that portion of the soil absorption area required to handle the effluent is used.

A drop box is a convenient inspection point. A check of the drop boxes shows how many of the trenches are being used. Every other year, rest the upper trench by plugging the distribution pipes at the drop box.

Pressure Distribution

Install a pressure distribution system where gravity cannot convey septic tank effluent to the disposal field, such as with mounds or a disposal field above the level of the septic tank discharge. Pressure distribution with a pump or siphon also provides more even distribution and therefore more uniform loading to the soil disposal area.

The Midwest Plan Service recommends dosing four times a day in sandy soils and mounds but less frequently in heavier soils. Manifold and distribution laterals should drain after each dose.

Table 720-12 gives the allowable lateral length by pipe size, hole size, and spacing. For larger systems, consult an engineer.

Example 4:

Determine the maximum allowable lateral length if you have a 1¼″ pipe with ¼″ diameter perforations spaced 36″ apart.

From Table 720-12, we find the allowable length to be 42′.

Dosing volume

Dosing volumes for homes are listed in Table 720-13. The quantity per dose is the design load divided by the desired dosing frequency. For a soil absorption area sized for three bedrooms, the design load is 450 gpd. Using a dosing frequency of four times/day, the dosing volume is 450 ÷ 4, or about 115 gal/dose.

Table 720-13. Recommended dosing volume for homes.
Assumes dosing 4 times/day. Check each system—make dosing quantity about 10 times the lateral volume, Table 720-14.

Home size no. bedrooms	Dosing volume gal/dose	Daily sewage flow, gpd
1	50	150
2	75	300
3	115	450
4	150	600
5	200	750

A dosing volume of about 10 times the lateral pipe volume minimizes differences in discharge volumes over the length of the discharge pipe. Estimate pipe volume from Table 720-14. Compare 10 times the pipe volume with the recommended dose from Table 720-13. Select the larger dose volume.

Table 720-14. Pipe volume for various pipe diameters.

Diameter in	Volume gal/ft of length
1	0.04
1¼	0.06
1½	0.09
2	0.16
3	0.37
4	0.66
6	1.47

Example 5:

Determine the dosing volume for a three-bedroom house using pressure distribution to a mound. The mound contains four 1¼″ lateral pipes, each 30′ long (120′ total length).

The design criteria require the dosing volume to be the larger of 10 times the lateral pipe volume or the value in Table 720-13.

1. Table 720-13 recommends a dosing volume of 115 gal/dose for a three-bedroom home.
2. Table 720-14 gives the volume of 1¼″ pipe as 0.06 gal/ft of length.
3. Dosing volume = 10 x (0.06 gal/ft x 120′) = 72 gal/dose.
4. 115 gal/dose is larger than 72 gal/dose, so use a dosing volume of 115 gal.

Pumping chamber

A pressure system includes a pumping chamber, pump, and alarm system. Size the pumping chamber to provide the desired dosing volume, space for the controls and pump, and extra volume for a malfunction and pipe draining after the pump shuts off. Table 720-15 gives pumping chamber sizes.

Table 720-15. Pumping chamber sizes for homes.

Home size no. bedrooms	Pumping chamber size, gal
1	250
2	250
3	500
4	500
5	750

The pumping chamber and septic tank must be watertight, so ground water is not pumped to the absorption system. Provide drain pipes around the chamber or a flotation collar to prevent the pumping chamber from floating out of position due to hydrostatic pressure. Vent the pumping chamber for proper pump performance. If possible, mount all electrical controls outside the tank. Provide an easy pump disconnect, such as a union, for pump removal in case of failure.

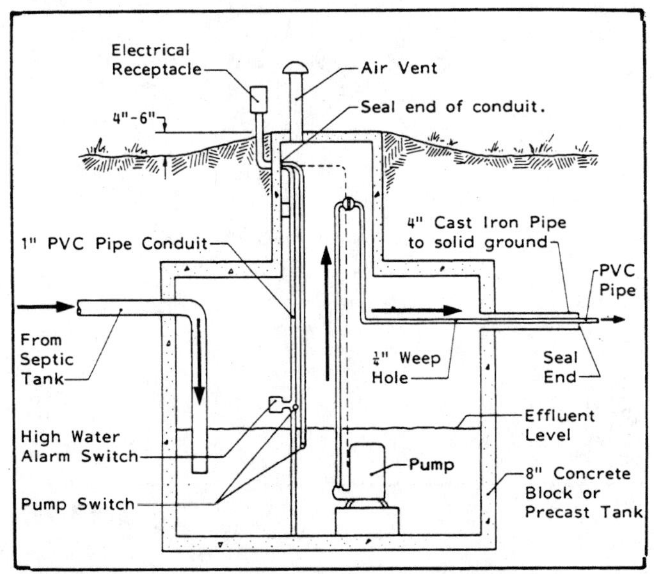

Fig 720-11. Cross section of a typical pumping chamber.
Keep pump submerged for cooling to prolong pump life.

Pump selection

Sufficient pump capacity and head are needed for good performance of a pressure distribution system.

Base pump selection on pump performance curves or tables. The total head is the sum of:

- The elevation difference between pump and lateral pipe.
- Friction loss in pipe between pump and manifold (Table 720-16).
- Recommended head pressure at the supply end of the perforated lateral (2.5′ or 1 psi).

Table 720-16. Friction loss, schedule 40 plastic pipe (C = 150).

Flow gpm	Pipe diameter (in) 1	1¼	1½	2	3	4
	ft/100 ft					
1	0.07					
2	0.28	0.07				
3	0.60	0.16	0.07			
4	1.01	0.25	0.12			
5	1.52	0.39	0.18			
6	2.14	0.55	0.25	0.07		
7	2.89	0.76	0.36	0.10		
8	3.63	0.97	0.46	0.14		
9	4.57	1.21	0.58	0.17		
10	5.50	1.46	0.70	0.21		
11		1.77	0.84	0.25		
12		2.09	1.01	0.30		
13		2.42	1.17	0.35		
14		2.74	1.33	0.39		
15		3.06	1.45	0.44	0.07	
16		3.49	1.65	0.50	0.08	
17		3.93	1.86	0.56	0.09	
18		4.37	2.07	0.62	0.10	
19		4.81	2.28	0.68	0.11	
20		5.23	2.46	0.74	0.12	
25			3.75	1.10	0.16	
30			5.22	1.54	0.23	
35				2.05	0.30	0.07
40				2.62	0.39	0.09
45				3.27	0.48	0.12
50				3.98	0.58	0.16
60					0.81	0.21
70					1.08	0.28
80					1.38	0.37
90					1.73	0.46
100					2.09	0.55
125						0.85
150						1.17
175						1.56

Other Systems

In recent years, several systems have been developed for handling home sewage that have had little evaluation by testing or actual use. They are considered experimental in many states and most require much more maintenance than septic tank systems.

Aerobic System

A septic tank treats wastes by bacteria which live in an anaerobic (without air or oxygen) environment, while an aerobic system treats waste with bacteria requiring air. An aerobic system usually treats waste more completely than a septic tank, but the effluent is still not of sufficient quality to be discharged directly on the soil surface or to a waterway. Depending on local or state regulations, discharge treated effluent from an aerobic system:

- To a conventional soil absorption field.
- To a sand filter.
- To the soil surface or a receiving stream after adequate disinfection (which is often difficult on a continuous basis).

An aerobic system has higher initial operating and maintenance costs. It is suggested that a service agreement be purchased with an aerobic unit. If effluent is discharged to a soil absorption area, design the absorption field the same size as for a septic tank.

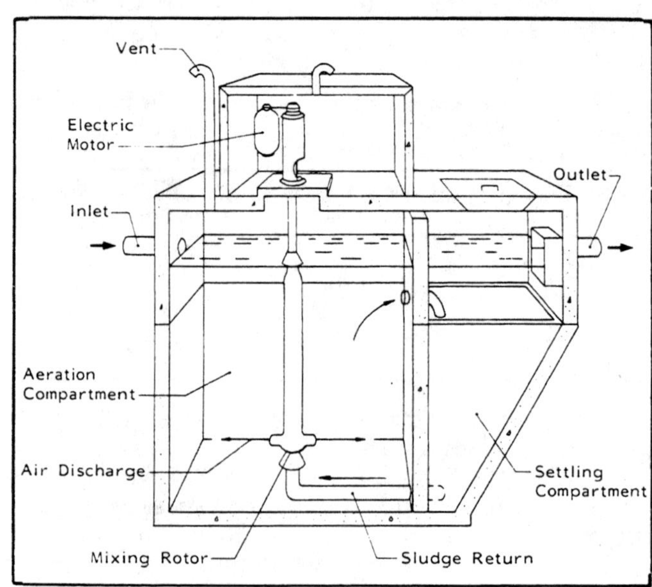

Fig 720-12. Aeration tank.

Evapo-transpiration System

Evapo-transpiration systems are advertised as systems which dispose of septic tank effluent by utilizing the normal function of vegetation. Plants require moisture to live and in doing so they release this moisture to the atmosphere through their leaves. These systems are still experimental and cannot be recommended at this time. They require water conservation and seem better suited to arid regions with 2″/month more evaporation than precipitation.

Lagoons

Some states require permits for and/or inspection of lagoon design and construction.

In remote locations with very tight soils, a lagoon can replace a soil absorption system. In many cases, the combination of evaporation and slow seepage is adequate to dispose of the wastewater from the septic tank. In other cases, periodic discharge or land application is needed. Septic tanks, although seldom used with lagoons, increase lagoon treatment and seepage rate.

When designing a lagoon:

- Seek professional help—size depends on local climate.
- Make the lagoon at least as large as a properly designed absorption field.
- Fence children and livestock away from the lagoon.
- Keep lagoons shallow for good seepage, evaporation, and oxygen transfer. Maximum depth is 4′ plus 2′ freeboard.
- Prevent pipes from freezing shut with a submerged inlet.

Experience suggests a lagoon surface of about 500 ft² per bedroom in most of the Midwest. Increase surface area to 650 to 750 ft²/bedroom in extreme northern climates.

Sand Filters

Where soils are too tight for an absorption field or mound, a sand filter can treat septic tank or aerobic system effluent before it is discharged into a stream, ditch, or ravine. Buried and recirculating sand filters are used in the Midwest. Design according to Table 720-17.

Table 720-17. Buried sand filter design.
Based on septic tank effluent and 150 gpd/bedroom. For a recirculation system, ½ the tabulated area is enough.

Number of bedrooms	Sewage flow gpd	Bed area ft²
1	150	200
2	300	400
3	450	500
4	600	600
5	700	700

Buried sand filters

Provide one collection line for each 10′ of filter width or fraction thereof, Fig 720-13. Slope the collection lines 6″/100′. Completely plug the upper end of the collection line. Lay the collection line on the bottom of the filter and place ¾″-2½″ diameter washed rock bedding around the collection line. Place 3″ of pea gravel over the rock.

The sand filter media should be 24″ deep; use 1.0 to 1.5 mm sand with a uniformity coefficient of less than 3.5. Use washed sand free of clay and silt. Lay the distribution pipes in 10″ of rock and then cover the rock with up to 12″ of porous soil.

Use 4″ clay tile, cement, or rigid or semi-rigid plastic pipe for distribution and collection lines. Space tile ¼″-½″ apart. Perforated tile or pipe needs two rows of ⅝″ holes 4″ apart. Lay the distribution lines level and space them 3′ apart.

Recirculating sand filter

This system has a recirculation tank and an open sand filter. It is similar to a buried sand filter but has many important differences. Size the recirculation

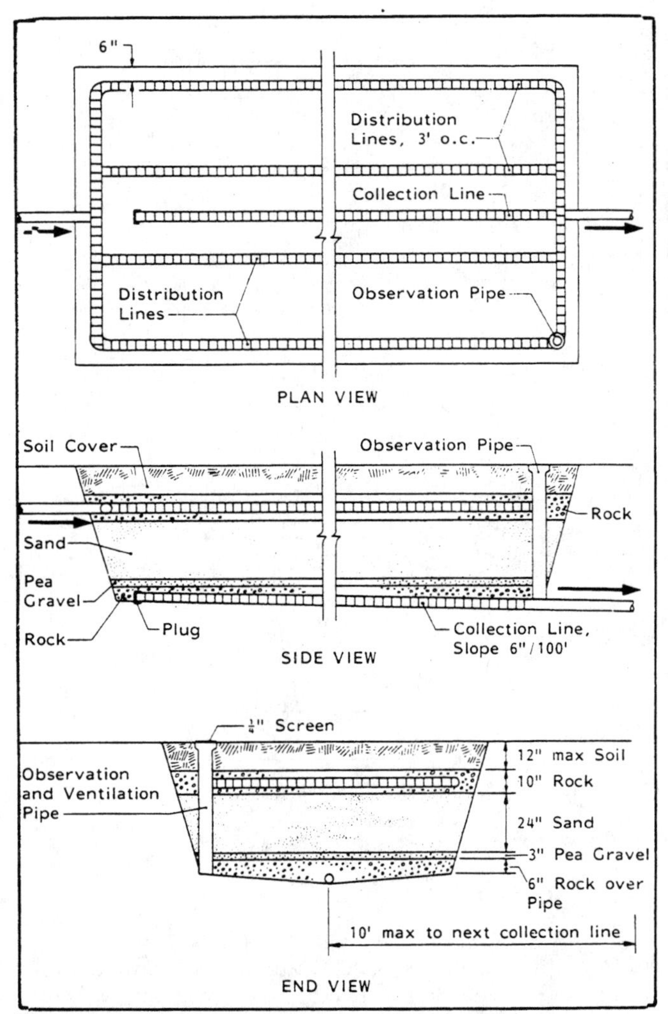

Fig 720-13. Buried sand filter.

tank for one day's volume of raw sewage (based on 150 gpd/bedroom) from the overflow of a septic tank.

A recirculation pump discharges from the recirculation tank to the sand filter. Some of the sand filter effluent returns to the recirculation tank, while the rest is discharged, usually after disinfection. Size the recirculation pump with a maximum pumping time of 10 min/half hour and a 4:1 recirculation rate.

For example, the tank size for a four-bedroom house is 4 x 150 = 600 gal or more. Average sewage flow is 600 ÷ 24 hr/day = 25 gal/hr. The pump must handle, at a recirculation rate of 4:1, 25 x 4 = 100 gal/hr. Because the pump operates 10 min/half hour, or ⅓ of the time, pump size is 3 x 100 = 300 gal/hr.

Construct the sand filter with a filter bed at least 30″ deep. Use 1.0 to 1.5 mm sand with a uniformity coefficient of less than 3.5. Put about 12″ of graded rock (¾″-2½″) under the sand for support and to surround the underdrain system. Space underdrains 10′ o.c. and level distribution pipes 3′-4′ apart. Size pressure distribution pipe as shown in previous sections.

Install the recirculation tank and sand filter so that effluent flows by gravity. Keep the filter surface free of debris and open for good air circulation. In areas subject to severe freezing, design the system with a bypass so filter effluent does not flow back to the recirculation tank during freezing periods.

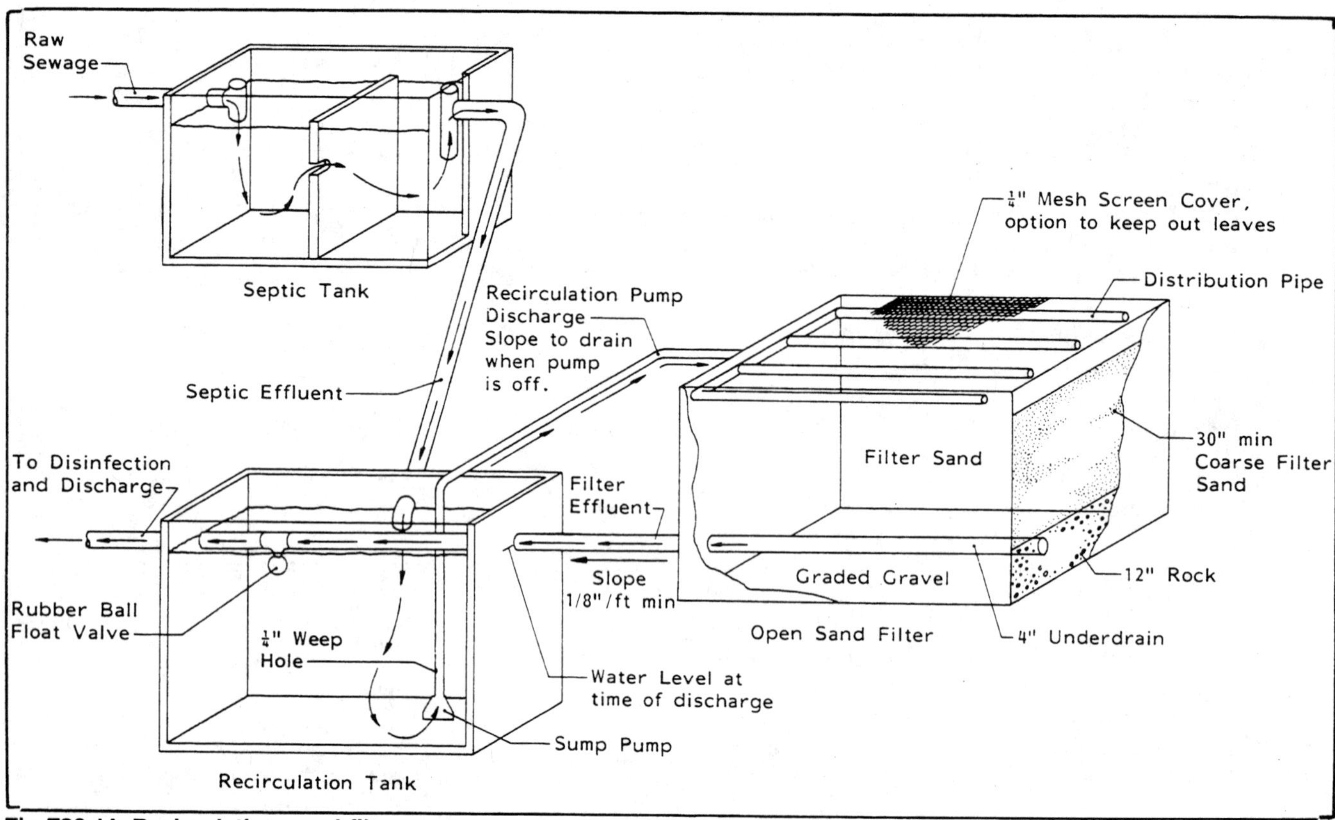

Fig 720-14. Recirculating sand filter system.

Waterless Toilets

Reduce family water use with a waterless toilet. Most will handle feces, urine, toilet tissue, and some other biodegradable materials. A separate system is needed to handle other sewage wastes from the home.

Consult local health officials when planning to install a waterless toilet to be sure that the type selected meets codes.

Composting toilets

Several companies manufacture composting toilets, which are large containers to hold toilet wastes and some biodegradable kitchen garbage. Some use no water or energy. Bacterial action produces heat to drive off excess moisture and reduces the waste to about 5%-10% of its original volume. A fan or roof vent carries away gases, odors, and moisture. Composting destroys harmful organisms and produces a residue that can be disposed of in a trash bin or garden if permitted by the health department. Some models have a heater to aid composting.

Incinerating toilets

Incinerators burn with oil, gas, or electricity until wastes are reduced to a sterile ash. Empty the ash box periodically. These units use a lot of energy and release some odors and gases into the air.

Biological toilets

Toilets are now available that decompose body wastes and toilet tissue with enzymes and aerobic and anaerobic bacteria. To maintain proper operation, regularly add enzymes and bacteria to the base of the toilet. Activated carbon filters remove odors and are replaced about every two years. These toilets operate on a principle similar to composting toilets. Germicidal cleansers inhibit the action of biological toilets.

Collector Systems

A collector system with a common soil treatment unit may be a satisfactory waste treatment system for two or more property owners, Fig 720-15. Each home has a septic tank. The effluent usually flows by gravity into a collector line to a pumping station. If adequate slope is not available, intermediate pump stations may be needed along the collector line. From the pump station the effluent is pumped to a suitable soil treatment unit.

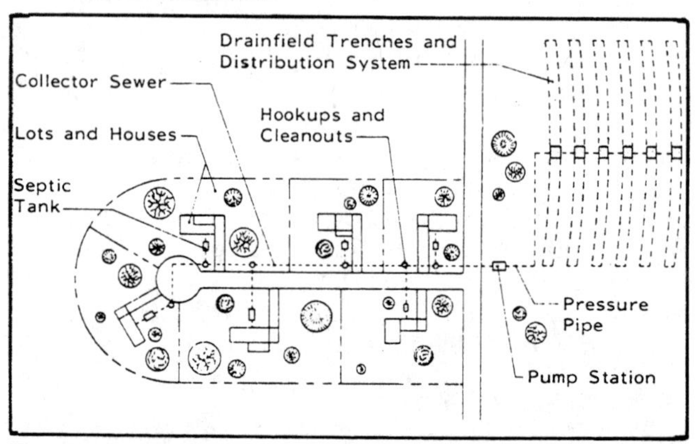

Fig 720-15. Collector sewer system with common soil treatment area.
From NCR Ext Pub 130.

Property owners **must agree** on the organizational and operational details before the treatment system can be designed. A detailed engineering design is necessary to ensure a satisfactory system. Cooperation, understanding, and proper installation and maintenance are necessary for a successful system.

Holding Tanks

For a residence on a small lot with no suitable soil treatment area and in an area too isolated for a collector system, a holding tank could be used. Sewage is discharged to a holding tank and later pumped and transported to a treatment and disposal system. Holding tanks are usually allowed only for an existing residence and not for a new home. Check state and local regulations to determine if they are allowed in your area.

The high cost of hauling sewage makes the holding tank undesirable. Minimize cost by conserving water.

Locate the holding tank so it is accessible to a pump truck under all weather conditions and where accidental spills during pumping will not create a nuisance. Check local regulations for required separation distances from wells, water pipes, buildings, and property lines.

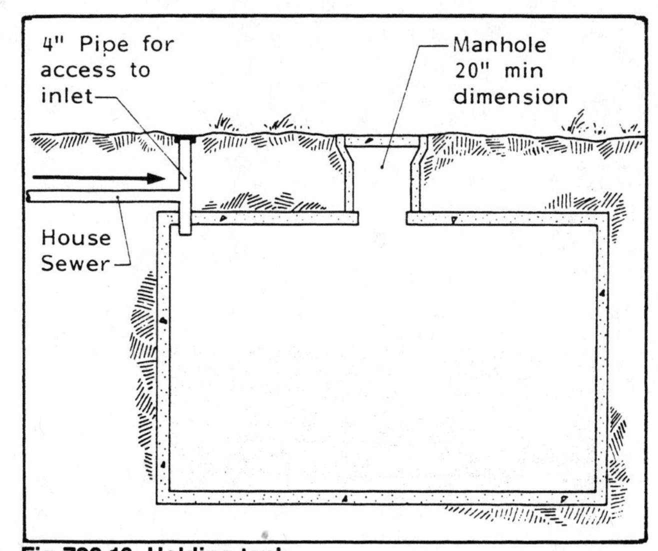

Fig 720-16. Holding tank.
Tank capacity = 1000 gal minimum or 400 gal/bedroom, whichever is greater. For commercial operations, use five times the daily flowrate. Meter water flowing into the system. Secure the manhole cover against unauthorized opening.

Install a water meter in your water system beyond the pressure tank and outside sillcocks to record all liquid wastes flowing into the holding tank. The charge for hauling the sewage should be based on the water meter readings.

800 UTILITIES
801 WATER SUPPLY

A correctly designed water system supplies the quantity of water needed every day all year. It delivers water at a rate that meets the temporary large demands that occur each day, and it provides enough water to fight small fires.

Typical water needs for many uses are listed in Tables 1 and 2.

Water systems must often meet the needs of many uses during short periods of time. These times, or **peak use periods,** usually last from 30 minutes to 2 hours. The flow rate during peak use periods is the **peak use rate.**

Peak use periods in the home are usually near mealtimes, during midmorning laundry periods, and shortly before bedtime.

A farm or farmstead water system must be able to supply the peak use rate continuously for 2 hours. For livestock operations, peak use periods usually occur during feeding and cleanup periods, when animals come in from day pastures and during the heat of the day.

If the peak use exceeds the maximum well yield, install intermediate storage to help supply water during peak use periods.

Use peak use rates to determine pump capacity, pipe size, temporary storage capacity and other water system features. Use average year-round water use to determine long-time storage capacity where the water source is a pond or cistern.

Reducing water consumption in the home and on the farm is important if: the capacity of your water system or water source is limited, it is difficult to dispose of waste water, expansion of a livestock enterprise is exceeding the water system capacity, or you want to conserve water and energy.

Water System Flow Rate

The water system flow rate is the quantity of water delivered in gallons per minute (gpm). It should at least equal the peak use rate, gpm. Estimate the peak use rate by determining the greatest water demand likely to occur at one time. The flow rate should be at least the peak use rate of the largest single fixture to prevent undersizing. Note that only approximate flow rates can be calculated.

Home water system

For family living, provide 50-60 gallons per day per person, of which about one-third is for flushing toilets. Peak daily use may reach 100 gallons per day per person in homes with several bathrooms and water-using appliances.

A home water system must be able to supply the peak use rate continuously for 1 hour.

The minimum flow rate for a home is about 6 gpm; 10 gpm is more desirable. Some water filters need 20 gpm for backwashing. Estimate future water requirements so an adequate flow rate will be available when needed. Use Table 3 as a guide for home water systems flow rates.

Farmstead water system

If water for livestock production is supplied through the home water system, increase the flow rate to meet those needs or provide intermediate storage. The minimum flow rate for a farmstead is 8 gpm, but 10 gpm is more desirable.

Flow rates for various types of water-using equipment are in Table 2c. The farmstead flow rate is the sum of those flow rates for all equipment to be in use at one time, added to the flow rate for the farm home.

Example

(For 600 market hogs, plus farm family)

10 automatic waterers, 2 watering spaces per waterer	@2 gpm	= 20 gpm
1 hose for hydraulic waste removal	@10 gpm	= 10 gpm
Flow rate for livestock production		= 30 gpm
Flow rate for farm home (4 br, 2 baths)		= 14 gpm
Flow rate for farmstead		44 gpm

	Gal/day
600 hogs @ 4 gal/day	2400
Hose for hydraulic waste removal (1 hr.)	600
Total	3000
From Fig 1, farmstead flow rate is	24 gpm

Water Storage

A water system's pressure tank also provides a small amount of storage. The storage is usually 10%-30% of the tank size, which provides small amounts of water without starting the pump. It also helps satisfy water demand during short peak use periods. (See Pressure Tanks.)

A well casing stores water at the source; see Table 4 for storage capacity. Water is also stored at the source in a pond, spring box, cistern or storage bag. See the sections on those facilities for their storage capacity.

When the water source and pressure tank cannot deliver the required flow rate, use an intermediate storage in a two-pump system.

The first pump, usually the well pump, has a low level cut-off and a capacity slightly less than the well yield so as not to pump the well dry. This pump fills an intermediate storage with water for peak use periods. A second pump draws the water from the intermediate storage and forces it into a pressure tank. Size this second pump to provide the peak use flow rate.

Intermediate storages are usually concrete or steel tanks. Protect the storage from contamination. If the intermediate storage is an elevated tank, the pressure tank and the second pump are not required. The water pressure for the system is gravity.

If your water source can deliver the required flow rate most of the time, size the intermediate storage for the 2 hours of peak flow. If your source tends to decrease or dry up for short periods, consider an intermediate storage with at least one day's total water needs. Additional storage capacity may be desirable for future expansion and emergencies, such as at least 1200 gal for fire protection.

Table 801-1. Home and outdoor living needs.

[a]Water restricting valves and shower heads reduce flow and water use by up to 50%.
[b]Ordinary toilet; low flow toilets will reduce water usage by 40%-90%.
[c]Water hardness, softener size, etc., affect water use.
[d]For limited fire fighting, at least 10 gpm with a ¼" nozzle at 30 psi for 2 hr/day. Preferred: 20 gpm at 60 psi—2400 gal.

Use	Flow rate gal/min	Total use gal
Adult or child		50 - 100 / day
Baby		100 / day
Automatic washer	5	30 - 50 / load
Non-automatic washer, hand tub	5	15 - 45 / load
Dishwasher	2	7 - 15 / load
Garbage disposer	3	4 - 6 / day
Kitchen sink [a]	3	2 - 4 / use
Shower or tub [a]	5	25 - 60 / use
Toilet flush [b]	3	4 - 7 / use
Bathroom lavatory	2	1 - 2 / use
Water softener regeneration [c]	5	50 - 100 / time
Backwash filters [c]	10	100 - 200 / backwashing
Outside hose faucet	5	—
Fire protection [d]	10	1200 / 2-hr period

Table 801-2. Farm water needs.

801-2a. Water use per animal.

	gal/day
Milking cow	35 - 45
Dry cow	20 - 30
Heifer	10 - 15
Calves (1-1½ gal/100 lb body weight)	6 - 10
Swine, finishing	3 - 5
Nursery	1
Sow & litter	8
Gestating sow	6
Beef animal	8 - 12
Sheep	2
Horse	12
100 chicken layers	9
100 turkeys	15

Table 801-2. Continued.

801-2b. Water use for milkhouses and parlors.

Washing operation	Water volume
Bulk tank	
Automatic	50 - 60 gal/wash
Manual	30 - 40 gal/wash
Pipeline	
In parlor (Volume increases for long lines in a large stanchion barn.)	75 - 125 gal/wash
Pail milkers	30 - 40 gal/wash
Miscellaneous equipment	30 gal/day
Cow preparation	gal/wash per cow
Automatic	1 - 4½
Estimated average	2
Manual	¼ - ½
Parlor floor	40 - 75 gal/day
Milkhouse floor	10 - 20 gal/day

801-2c. Water use flow rates.

Livestock water consumption is affected by air temperature, size of animal, species, age, milk or egg production, type of ration, dry matter consumed, and other variables.

Average summer values are listed—use 60% for cool weather. Also use 60% of the tabulated livestock consumption for pond storage if the average year-round temperature is about 50 F.

	Minimum	Preferred
Automatic waterers		
Cattle, hogs, or sheep (20-40 head per bowl)	½ gpm	2 gpm
Poultry (100-150 layers)	¼	1
Cleaning hose for milkhouse and dairy utensils	3	5
Cleaning and manure removal hose for milking barn or hog house	5	10
Outdoor hydrant for uses other than firefighting	3	5

Table 801-3. Flow rates for home water systems.

No. of bedrooms	Number of bathrooms in home: 1	1½	2	3
	Flow rate, gpm			
2	6	8	10	–
3	8	10	12	–
4	10	12	14	16
5	–	13	15	17
6	–	–	16	18

Table 801-4. Storage in well casing or pipe.
Casing storage capacity is usually negligible.

Well diameter In.	Storage per foot of depth Gal.	Well diameter Feet	Storage per foot of depth Gal.
2"	0.163 gal	1'	5.87 gal
3	0.367	2	23.50
4	0.653	3	52.87
5	1.02	4	94.00
6	1.47	5	146.87
8	2.61	7	287.86
10	4.08	9	475.86

Water Sources

A water supply may come from either surface or underground sources. The choice and use of these sources may be regulated by governmental agencies. Surface water is in lakes, streams, reservoirs, and surface depressions that capture and hold water. Ground water is in water-yielding geologic formations called aquifers, beneath the earth's surface.

It is sometimes impossible to get enough water from a single source. Consider water from two or more good water resources to solve problems such as low well yield, high pumping costs and poor water quality.

Caution. Never cross-connect an unsafe water source with a safe source.

Examples of water from several sources are:

- Low-yielding well for home and milkhouse. Pond, stream, or spring for livestock.
- Low-yielding well for cold water in home. Cistern for hot water in home and milkhouse. Pond, stream, or spring for livestock.

Ground Water

Ground water comes from wells or springs and is usually the safest and most reliable source of water. But, some shallow water-table wells and springs are easily contaminated and also may go dry in droughts.

Artesian wells are less subject to contamination than water-table wells, because the upper confining bed restricts the movement of contaminants into the artesian aquifer. An artesian well taps water under pressure from an aquifer lying below an impermeable layer; the pressure may drive the water to the surface in the well pipe.

State water agencies have information on the ground water available in your area and on codes that regulate their development.

Surface Water

Stream and lake water is usually contaminated. It is not recommended for domestic use in private water systems without extensive treatment.

Caution. Check state codes: some states prohibit the use of surface water for a domestic supply.

Surface water collected in farm ponds is usually suitable for livestock and for fighting fires, but consider it for home and milkhouse use **only** when ground water sources are inadequate or too costly. Filter and disinfect surface water for the home and milkhouse.

Surface water collected in cisterns, reservoirs and other man-made storages can supplement ground water sources and be for emergencies such as fires. Some water systems depend on these storages for all water needs. For home and milkhouse use, carefully construct the storages and protect them from additional contamination.

Well Location and Site Preparation

It is desirable to locate a well to avoid subsurface contamination to prevent, for example, any influence from a feedlot. Underground geologic formations, changes in water table, and other factors prevent general rules from being guarantees of safe water.

Contact county and state health departments for local regulations on well location and construction. Health department recommendations include

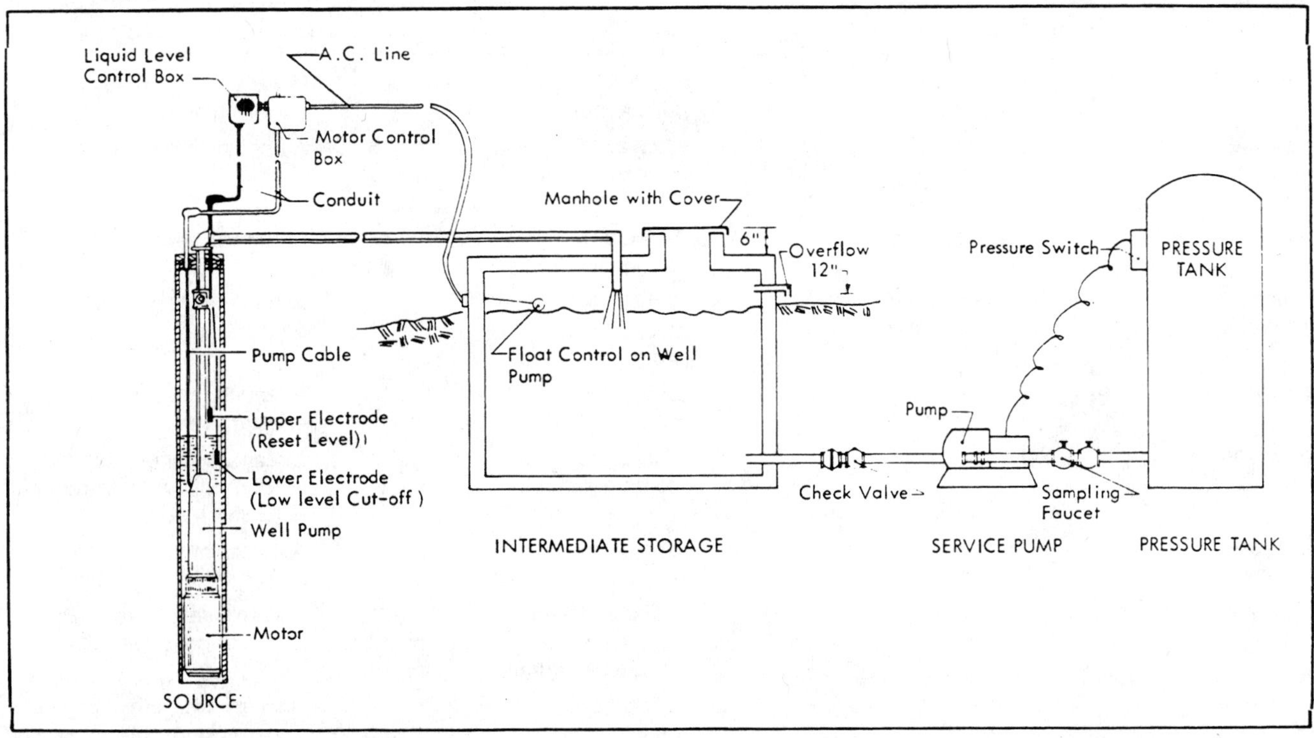

Fig 801-1. Intermediate storage with 2-pump system.

minimum distances between a new well and sources of contamination such as septic tanks, feed lots, fuel tanks and absorption fields.

Locate a water-table well on high ground with the water level in the well higher than any nearby source of contamination; or if the well must be located on low ground, put it at least 1000' from a source of aquifer contamination. Avoid flooding or surface water contamination.

Put a well where it will be accessible for maintenance, inspection, and pump or pipe replacement—at least 5' from a building. Locate a pump house at least 50' from any major building so a fire won't short out the water system. Wire the pump motor directly from the meter pole and protect its circuits from the effects of building fires.

Pumps

Total "Head"

To find the total head on a pump, find the "lift," "elevation," "pressure," and "friction loss" and then total them.

Total Head = Lift + Elevation + Friction Loss + Pressure at Outlets.

Pump manufacturers design and make their pumps to overcome specific heads at various flow rates and list these capabilities in tables and graphs. Typical catalog data are in Table 7 . Study data from several manufacturers before selecting your pump.

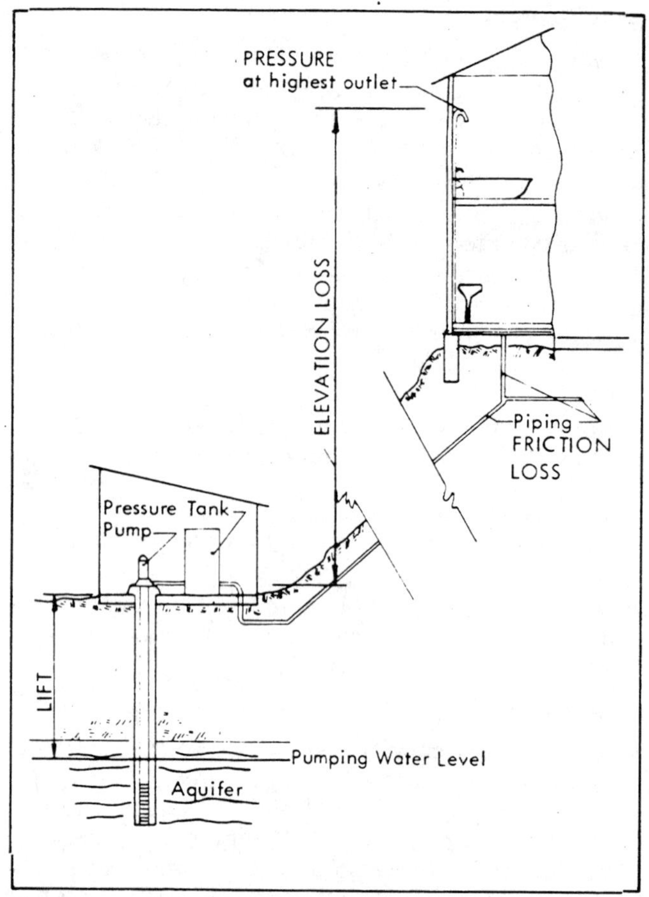

Fig 801-3. Elements of total head.

Fig 801-2 is on page 801.7.

Table 801-5. Safety characteristics of water sources.
Additional conditioning may be needed to remove undesirable minerals.

Source	Primary uses	Safety	Treatment usually recommended to make water safe for DOMESTIC use
Drilled well	Domestic and livestock	Usually best of all sources	None, unless subject to contamination
Dug, jetted, driven, or bored	Domestic and livestock	Subject to contamination	Automatic chlorination
Spring	Livestock and domestic	Subject to contamination	Automatic chlorination
Cistern (untreated water)	Domestic and fire protection	Subject to contamination from roof runoff	Automatic chlorination and filtration
Farm pond	Livestock and fire protection	Subject to contamination	Automatic chlorination and filtration
Stream or lake	Livestock	Subject to contamination	Stream not recommended for domestic use without extensive treatment
Controlled catchment	Livestock and domestic	Subject to contamination	Automatic chlorination and filtration
Hauled water	Domestic	Good if hauled from safe source in safe containers	None if safely stored, but chlorination is desirable
Community water system	Domestic and livestock	Good if properly maintained	None by user, but if stored underground, chlorinate

Table 801-6. Well characteristics.

	Drilled cable tool or percussion	Drilled, rotary hydraulic	Jetted
Depth, ft. **Diameter, in.**	To 1000' 4" - 18"	To 1000' 3" - 24"	To 50' 2" - 12"
Construction method	A heavy drill bit and stem is raised and dropped in the borehole. The bit breaks rock and loosens other material, which is mixed with water and removed by a bailer. In unconsolidated formations, the casing follows the bit closely to keep the borehole open.	A rotating bit breaks up material which is mixed with drilling fluid ("mud"), carries it out of the hole, supports the wall of the borehole to prevent caving, seals the wall of the borehole to reduce fluid loss, and cools and cleans the drill bit. The mud flows to a settling pit and is recirculated. After the hole is complete, a metal or plastic casing is inserted.	Water under pressure is forced down a riser pipe and loosens the material around the washing point. The loosened material is drawn up as the riser is lowered into the ground. The drilling water carries the cuttings in suspension upward in the space around the riser. The water and cuttings flow to a settling pit and are recirculated. Casing fitted with a driveshoe is usually sunk as drilling proceeds.
Precautions	Little trouble with surface contamination. Upper portion of drill hole made 4" larger than casing and grouted. Depth of grout and casing varies with geological formation.	Little trouble with surface contamination. Upper portion of drill hole made 2" larger than casing and grouted. Depth of grout and casing varies with geological formation.	Install a protective casing to a depth of at least 25' to protect against surface water contamination.
Limitations	Can be used in all type formations. Is usually slower than rotary drilling. Must case well while drilling in unconsolidated materials.	Sometimes difficult to recognize water-bearing formations because of the drilling fluid. Difficult to drill in rock formations.	Can be jetted in only fine unconsolidated formations. Best for small holes of about 4" diameter.

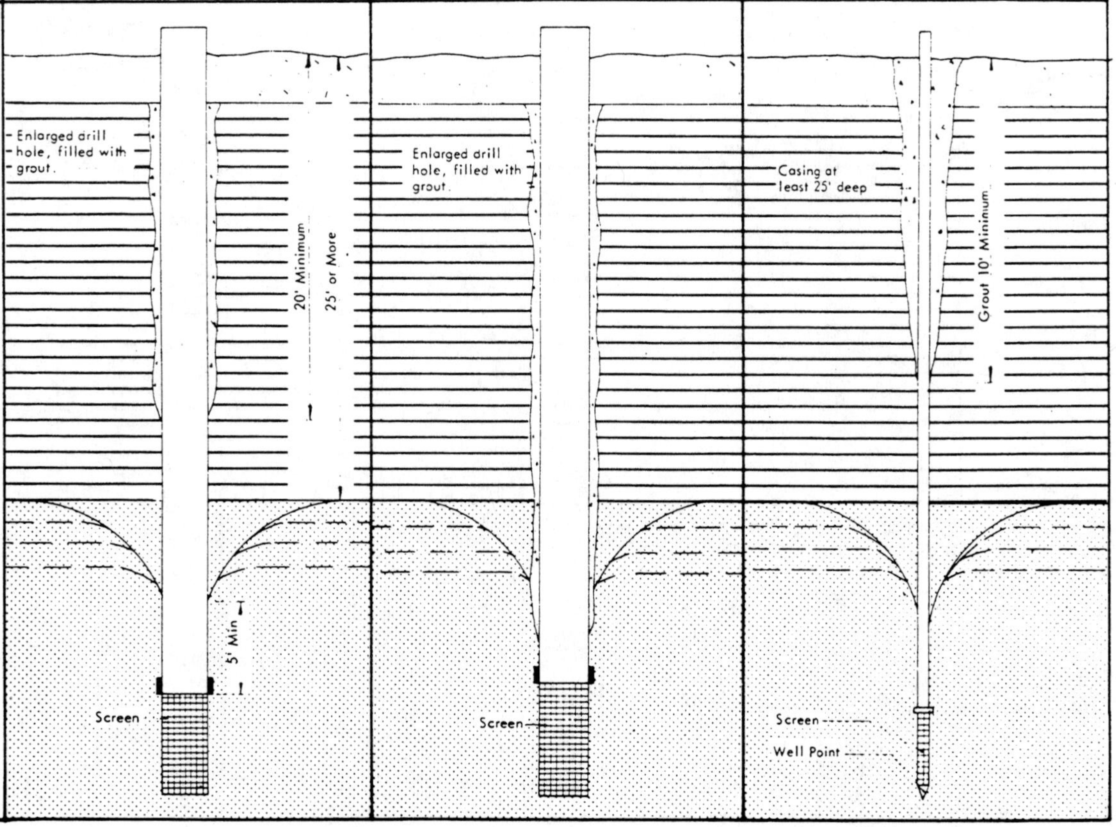

Table 801-6. Continued.

	Driven	**Bored**	**Dug**
Depth, ft. **Diameter, in.**	To 50' $1\frac{1}{4}$" - 12"	To 1000' 2" - 30"	To 50' 3 to 20 feet
Construction method	A special driving point with well screen on a series of short pipe sections is driven into the ground with a drive-block assembly or a post or pile driver. Pipe sections are added as the pipe is driven. A pilot hole, bored as deep as possible, reduces damage to the well screen during driving.	An earth auger, rotated by hand or power, bores the hole and carries the earth to the surface. Saturated sand must be jetted, not bored. The casing is usually steel, concrete, or plastic pipe.	The hole is dug the desired diameter and depth by hand or power. The hole may be shored; often casing is installed as digging progresses and is allowed to sink by its weight as the hole is excavated under the casing. The walls are often brick, stone, concrete or precast concrete pipe.
Precautions	Well screen can be damaged during driving. Maul not recommended for driving pipe—glancing blows may break or bend pipe. Turn riser pipe with wrench periodically to ensure that threaded fittings remain tight and that pipe is undamaged.	Seal casing against surface contamination by grouting the space around the casing. Somewhat easily contaminated.	Seal well lining tightly to prevent contamination. Case or shore the walls during construction. Needs careful backfilling. Ventilate deep holes while hand digging.
Limitations	Yields are small to moderate because of small pipe. Cannot be driven into coarse or cemented gravel, boulders, sandstone, limestone and dense rock. Cannot be grouted to prevent contamination from perched water tables.	Soils must have enough clay to keep the borehole open until casing is installed. Can be bored only a few feet into a watertable. Difficult to bore into boulders and dense rock. Cannot bore into cemented limestone or sandstone.	Many fail during droughts. Easily contaminated. Can be dug only a few feet below the water table. Cannot be dug in dense rock.
Top soil **Geological formations** **Water table** **Pumping level** **Water-bearing Formation**	20' - 25' 30' Minimum Screen Well Point	Clay Grout Casing 25' Minimum and Below Contamination Cast Concrete Concrete Pipe	Well Seal 8" Grout Seal Precast Concrete Pipe 6" Minimum Grout Seal, 25' Depth Minimum Perforated Pipe Screen

	Submersible Multistage	Jet (or Ejector) Shallow-Well or Deep Well	Centrifugal Shallow-Well
Practical Lift	To 1,000	To 22' for shallow-well jet and to 85' for deep-well jet. Deeper deep-well jets are possible but less efficient.	To 15'
How It Works	Operates like a shallow-well centrifugal pump except there are several impellers mounted close together on a single shaft. The impellers and motor are in a housing immersed in the water source. Each impeller and its diffuser (a guide to the next impeller) is called a stage. A 4" or larger casing is usually required.	Jet pumps consist of a pump (usually centrifugal) and a jet or ejector assembly. The assembly is in the pump for shallow-well units or in the well on deep-well units. The pump forces some water through the nozzle and venturi tube of the assembly and forces the rest of the water to the distribution system.	A rotating wheel or impeller develops a vacuum in the intake pipe. Water fills the vacuum; the impeller increases its velocity and forces it into a surrounding casing shaped to slow down the flow and convert the velocity to pressure.
Advantages	Produces a smooth, even flow. Easy to frost-proof. Short pump shaft to motor.	Few moving parts. Both shallow-well and deep-well jets can be offset from the well. High capacity at low heads.	Produces a smooth, even flow. The open-impeller, but not the closed-impeller, type pumps water with some sand. Usually reliable and with good service life.
Disadvantages	Repair to pump or motor requires pulling from well. Easily damaged by sandy water.	Easily damaged by sandy water. The amount of water returned to ejector increases with increased lift; 50% of the total water pumped at 50' lift and 76% at 100' lift.	Loses prime easily. Efficiency depends on operating under design heads and speed.
Remarks	These pumps usually operate at 3500 rpm, the fastest practical speed for a 60-cycle electric motor. Pump capacity depends on impeller design. Pressure depends on diameter, speed, and number of impellers.	Capacity depends on design and number of impellers in the jet. Pressure depends on diameter, speed, and number of impellers.	Very efficient for capacities over 50 gpm and pressures less than 65 psi. Ideal for a booster pump. Can be offset from the well.
	Discharge Nylon Rope Cable Diffuser Impeller Bowl Pump Water Inlet Screen Seals and Bearings Motor Bearing Oil Reservoir Well Casing Well Screen	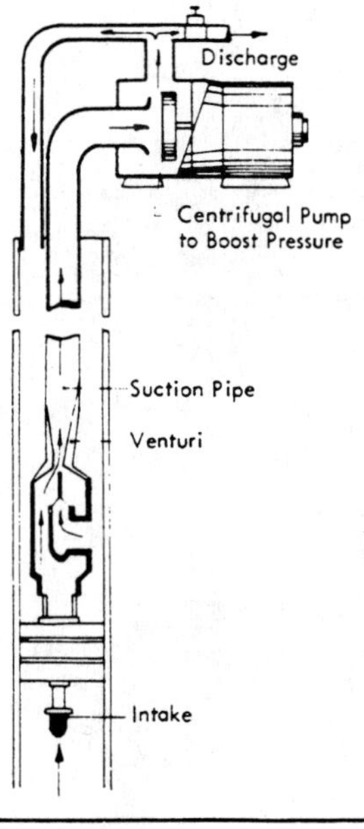 Discharge Centrifugal Pump to Boost Pressure Suction Pipe Venturi Intake	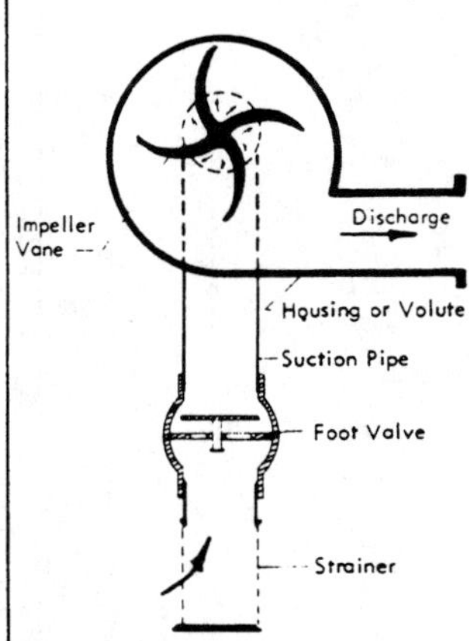 Impeller Vane Discharge Housing or Volute Suction Pipe Foot Valve Strainer

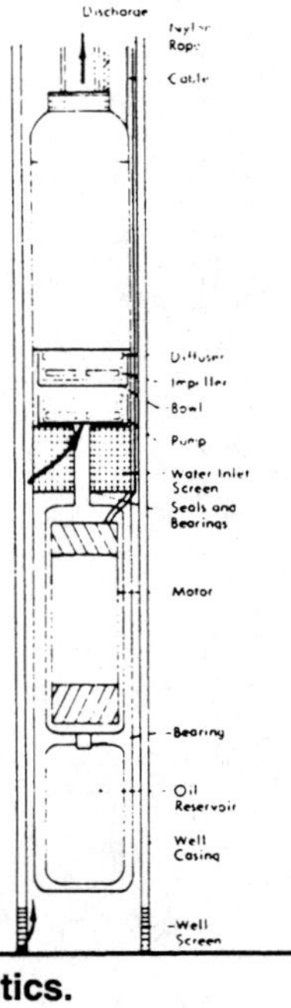

Fig 801-2. Well pump characteristics.

Reciprocating or Piston **Shallow Well**	**Deep Well**	**Deep Well Turbine** **Multistage**	**Submersible** **Helical Rotor**
To 22'	To 600'	To 1,500'	To 1,000'
A piston is driven from a chamber to develop a vacuum. Water fills the vacuum and is forced into the water system as the piston reverses direction.	A pump cylinder is attached to the bottom of the drop pipe. A piston is attached to a rod in the drop pipe. As the piston is forced up and down, it pumps water up through the drop pipe.	Operates like a centrifugal pump except there are one or more impellers mounted close together on a vertical shaft. The bowls (each bowl is one stage—an impeller and its diffuser) are below the pumping water level with the discharge pipe and shaft extending to the power unit at the surface.	A positive displacement pump mounted with a motor in a submersible housing.
Can pump small amounts of sand. Can be installed over small diameter wells. Positive displacement means a constant rate of yield. Adaptable to hand operation.	Same as for shallow-well. The open-type cylinder is easy to maintain.	Produces a smooth, even flow. Easy to frost-proof. Long drive shaft requires a straight and vertical well casing.	Produces a smooth, even flow. Easy to frost-proof. Short pump shaft to motor. Pumps sand with less pump damage than any other type.
Pulsating discharge. May cause vibration and noise.	Same as for shallow-well. The pump must be directly over the well.	Pump repair requires pulling from well.	Pump or motor repair requires pulling from well.
Pump capacity depends on cylinder size (displacement) and strokes per minute. Pressures are limited by the strength of the pumping equipment and the motor horsepower. Can be offset from the water source	Double-acting barrels pump 65% more water with 15% more horsepower.	Usually operates at 1,760 or 3,500 rpm, depending on kind of power used. Usually used for high capacity from deep wells. Capacity depends on design, diameter, and speed of the impellers. Pressure depends on diameter, speed, and number of impellers.	Capacity depends on rotor design. Can be used in 4" or larger wells.
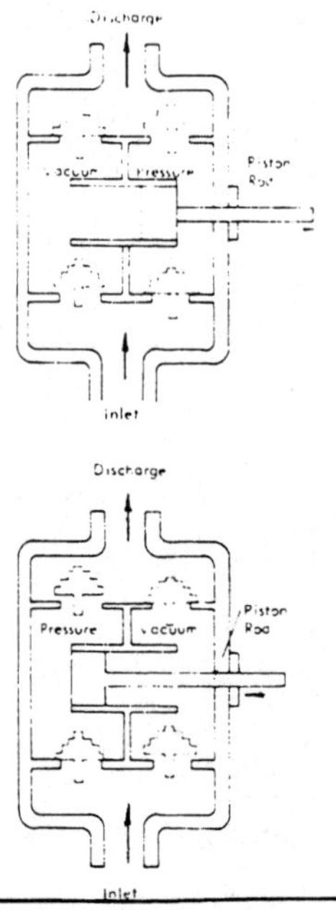 Discharge Vacuum Pressure Piston Rod Inlet Discharge Pressure Vacuum Piston Rod Inlet	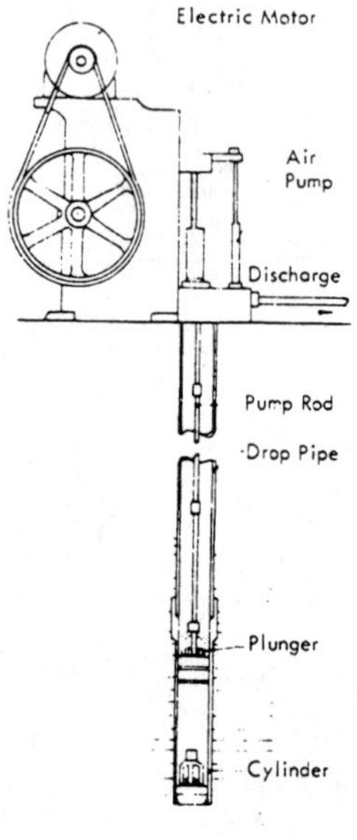 Electric Motor Air Pump Discharge Pump Rod Drop Pipe Plunger Cylinder	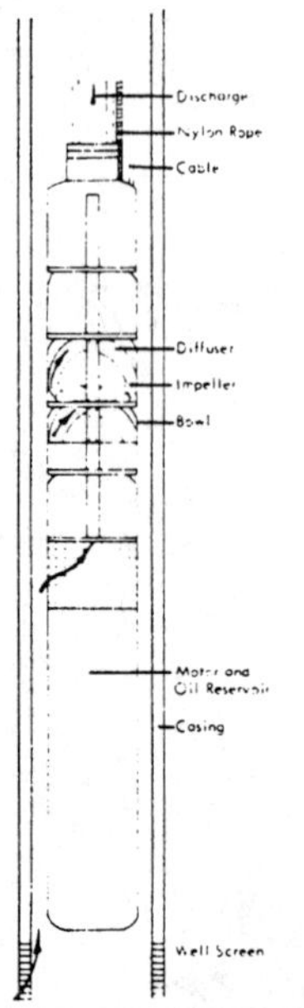 Discharge Nylon Rope Cable Diffuser Impeller Bowl Motor and Oil Reservoir Casing Well Screen	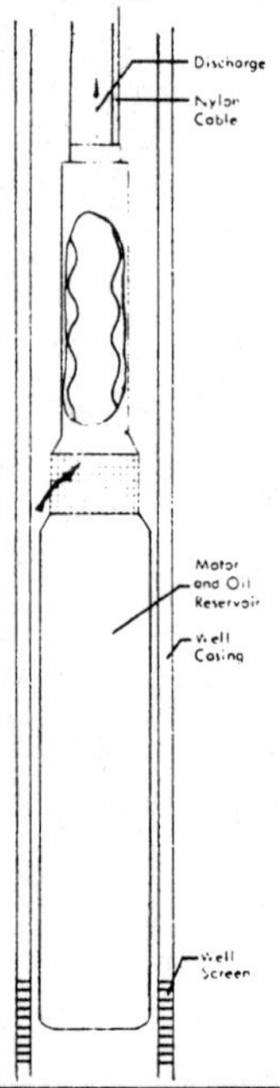 Discharge Nylon Cable Motor and Oil Reservoir Well Casing Well Screen

Fig 801-2. Continued.

Submersible multistage
Several impellers (small centrifugal pumps) act in series to force water up the drop pipe. A nylon rope permits pulling the pump if the drop pipe breaks. An electric cable provides electricity to the pump motor.

Jet
In the single-pipe jet, the well casing is the return, or pressure, line. The venturi is at the motor of a shallow-well jet, but is down in the well for a deep-well jet, requiring both a suction and a return, or pressure, line.

Centrifugal
The impeller is motor-driven to suck water into the inlet and force it out the high pressure outlet side.

Piston
The double-acting piston pump sucks water from the well during both strokes and forces the water out the pressure side.

The deep-well piston sucks water through the check valve on the upstroke and forces it past the piston on the downstroke.

Turbine
The deep well multi-stage turbine operates the same as a submersible centrifugal. Bowl, impeller, and diffuser design are modified for higher pressures.

Helical Rotor
The rotor operates like an auger to force water up through the pump.

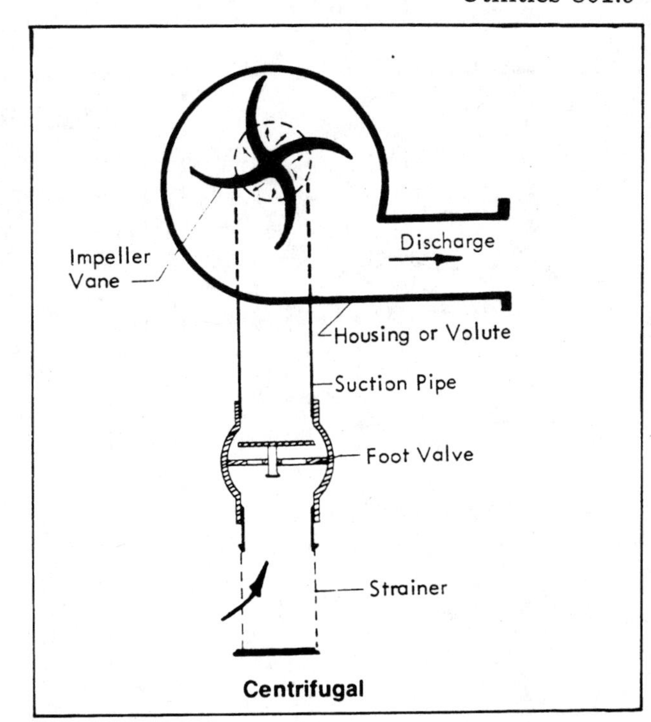

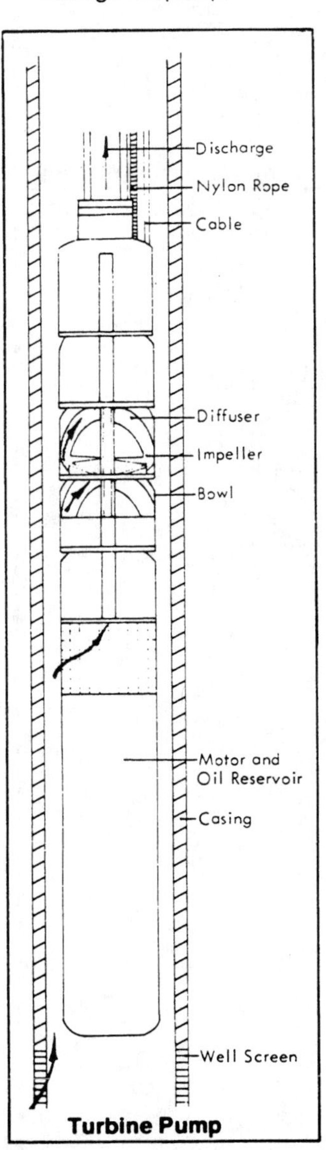

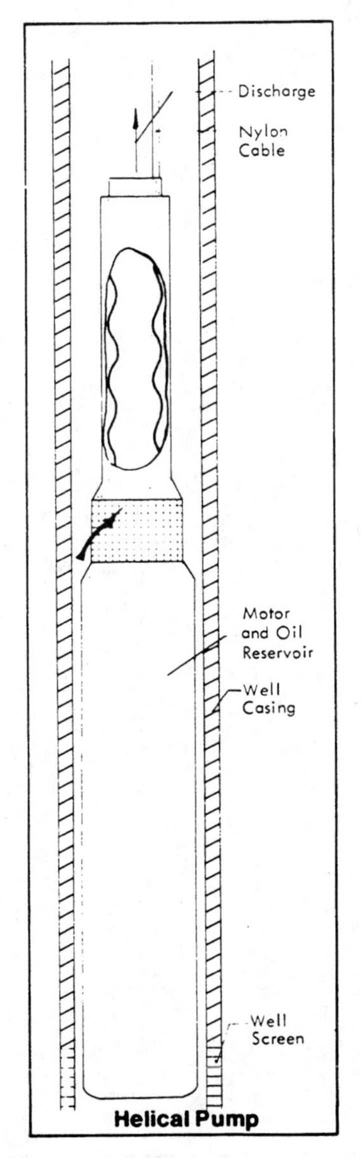

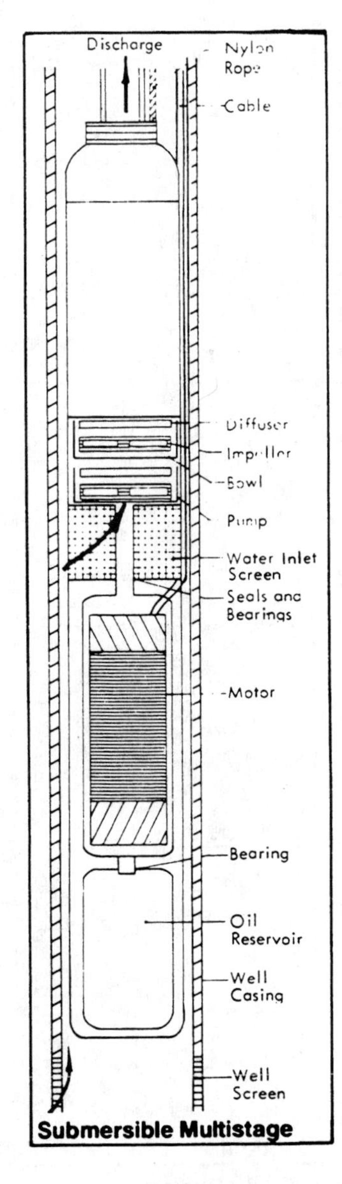

Fig 801-2a. Pump details.

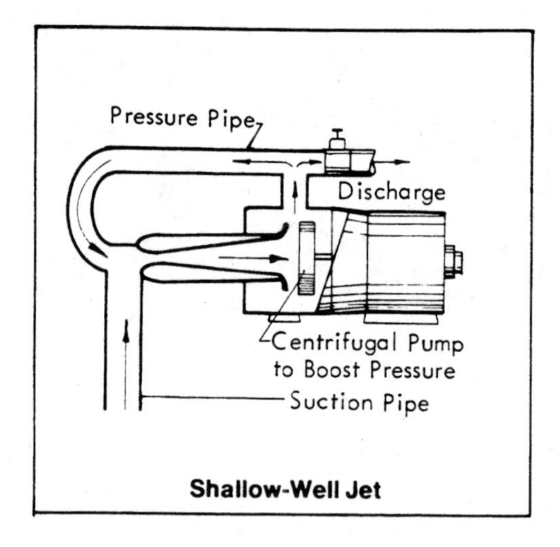

Shallow-Well Jet

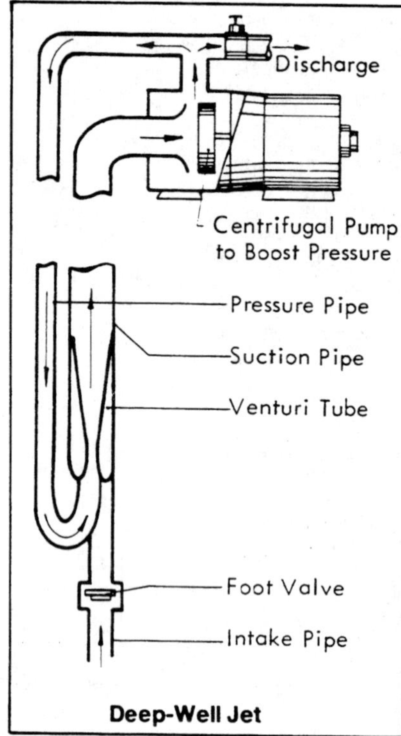

Deep-Well Jet

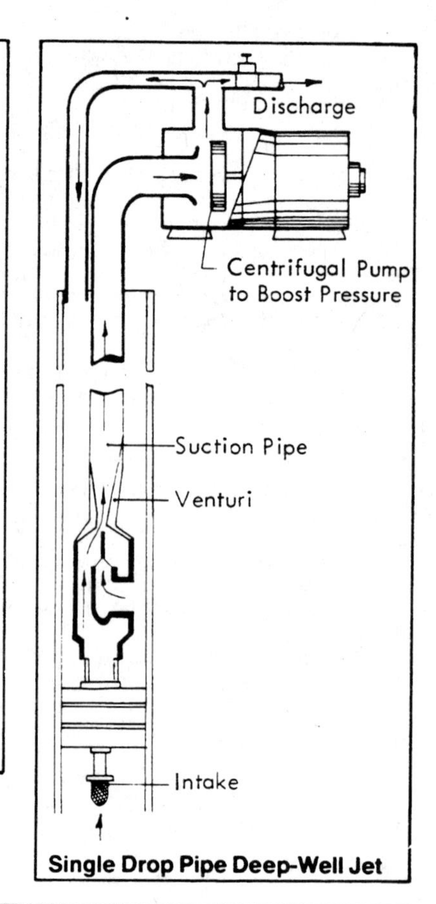

Single Drop Pipe Deep-Well Jet

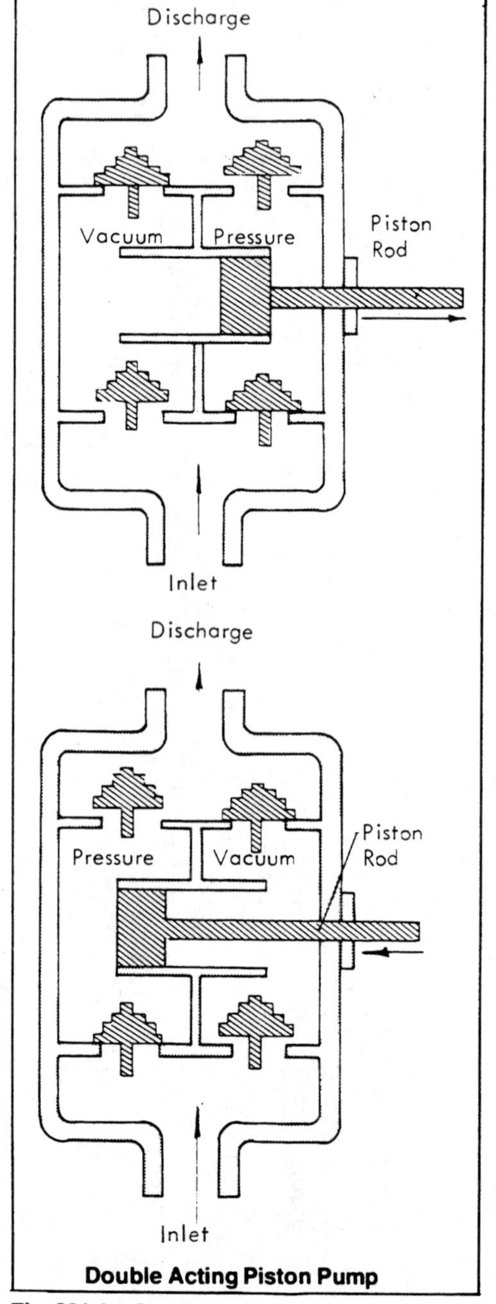

Double Acting Piston Pump

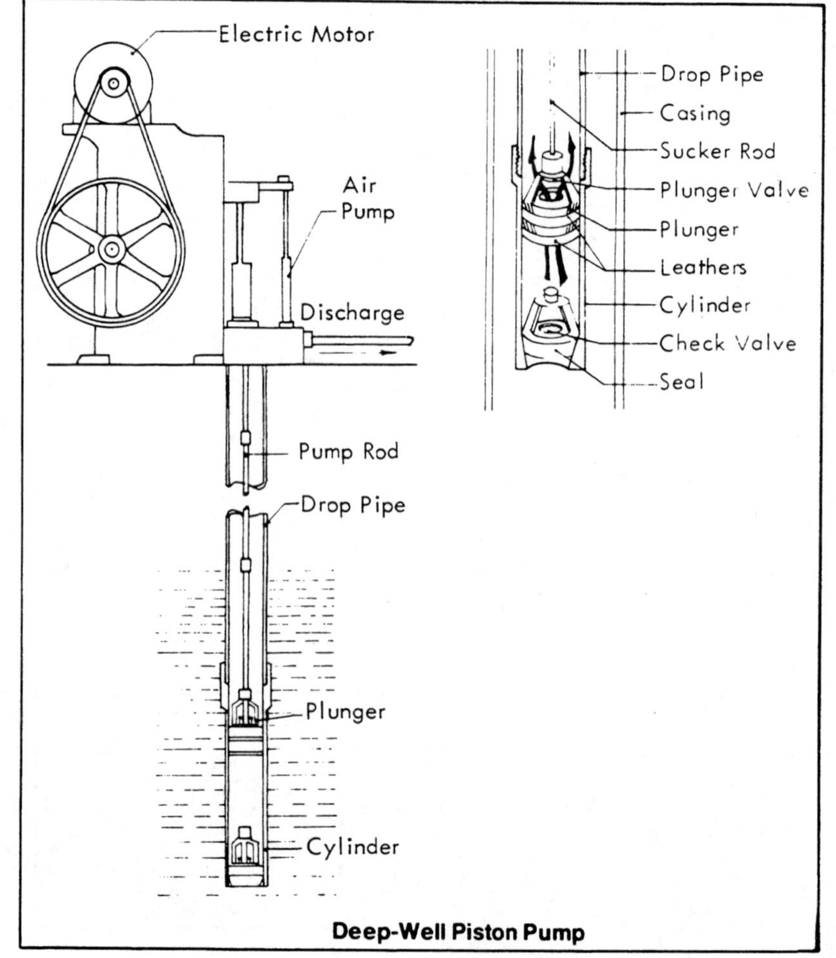

Deep-Well Piston Pump

Fig 801-2a. Continued.

Table 801-7. Typical shallow well jet pumps.
HP = horsepower; psi = pounds per square inch; gph = gallons per hour.

HP	Operating pressure psi	10'	15'	20'	25'
		At suction lift of:			
		Capacity in gph is:			
	20	525	420	320	250
1/3	30	515	420	320	250
	40	435	390	295	240
	20	900	825	650	490
1/2	30	750	660	600	490
	40	500	410	330	260

It is sometimes not possible to buy a pump with low enough capacity to match the well yield. Protect your pump to prevent loss of prime, even more reduced well capacity, and/or damage to the pump and motor.

For two-pipe jet pumps, install a 35' length of tailpipe between the ejector assembly and the foot valve so the water will never be lowered to the foot valve level. At some level below the ejector, the pump delivery will become equal to the rate at which water is flowing into the well as long as the pump is operating. For one-pipe jets and Eureka cylinders keep the jet 35' above the screen to prevent losing prime and damaging the well.

For submersible pumps, an adjustable gate or globe throttling valve between the check valve and the pressure tank can be adjusted to restrict the pumping rate to the well yield. Make this adjustment by trial and error unless the well yield is premeasured. Or, protect any type of pump with a liquid level control.

Installation

Install above-ground pumps in sheltered, dry, accessible locations. Protect against freezing.

Install shallow-well pumps as near the well or water source as possible to minimize "suction lift."

Deep-well reciprocating and turbine pumps are installed directly over the well. Submersible and helical rotor pumps are in the well. A deep-well jet may be offset from the well.

To protect a pump from freezing, install it below the frost line, in a heated above-ground pump house, or in some other heated and sanitary location.

Sanitary and Frost Protection

Provide both frost protection and the sanitary features required by well codes. Install either:

- Pitless well devices, or
- An insulated, heated well or pump house which covers the top of the well and allows the well casing to protrude above the ground level. The top of the well casing must be capped with a well seal.

Well pits and buried wells for frost protection are now prohibited by most well codes because they are easily contaminated.

Pitless Well Devices

Pitless devices provide frostproof, sanitary wells. Watertight connections to the well casing below ground and casing extensions above ground level prevent contamination near the ground surface.

Select a pitless device that is approved by your code.

The three basic types are:

- **Pitless adapters** connect to a well casing below the frost line. Usually a hole is cut in the casing at the desired level; the discharge part of the adapter is fastened over the hole by a clamp and gasket or by welding. The other part of the adapter is connected to the top of the drop pipe and seals into the first section.

Water from the well is diverted horizontally at the adapter while still below frost level.

Extend the casing above ground level, usually at least 8".

- **Pitless units** are similar to pitless adapters except they are manufactured assemblies which replace the upper portion of the well casing. The casing is cut off just below frost and the pitless unit is attached to it by threads, welding, or compression gaskets. The units are available in sizes to match well casing diameters and in lengths to extend below frost depths.

Internal parts are removable and attach to the drop pipe. When in the pitless unit, they form a watertight seal and divert the water horizontally below the frost line into underground water lines.

Pitless adapters and pitless units are for submersible, deep and shallow well jet, and reciprocating pumps. To prevent contaminated water from entering a leaking discharge pipe, it is always under pressure.

A combined pitless unit and pressure tank fits directly over the well. The pressure tank is outside the pitless unit so well parts can easily be removed for servicing.

A pitless buried tank is practical only with submersible and deep-well reciprocating pumps. It frees space in the house for other purposes, provides colder drinking water, eliminates the moisture condensation problem associated with tanks housed indoors and provides the pressure close to the well. Electrical supply and controls can be outside the building, permitting operation in case of fire. Disadvantages include increased corrosion and inaccessibility for maintenance, such as flushing out sediment.

- **Above-ground discharge adapters** cover the top of the well casing above ground level. For frost protection, a check valve in the discharge pipe permits it to drain by gravity, leaving only compressed air in the line. The well seal, check valve, and pipe connections are above ground for sanitary protection.

The adapter attaches to the top of the drop pipe and seats on the top of the well casing. The discharge pipe slopes for gravity flow to a pressure tank either buried next to the well or in a basement.

Well Pits—Not Recommended

Well pits (with a pump installed in a dug pit to prevent freezing) have been common but are unsanitary if located over the well, because of the danger of contamination entering the well from the pit.

Buried wells, with the well cap and suction line buried, are not recommended because of the danger of contamination.

Pressure Tanks and Controls

Most private water systems include a pressure tank to prevent the pump from running each time a small amount of water is used, and to smooth out water pressure in the system. The tank also stores a small quantity of water under pressure. Pressure is developed by forcing water into the tank until the air in the tank is compressed to a pre-set pressure.

When an outlet is opened, the pressure forces water out. As water leaves the tank, the pressure drops. When the pressure falls to another pre-set level, the pump starts, forces water into the tank, and recompresses the air.

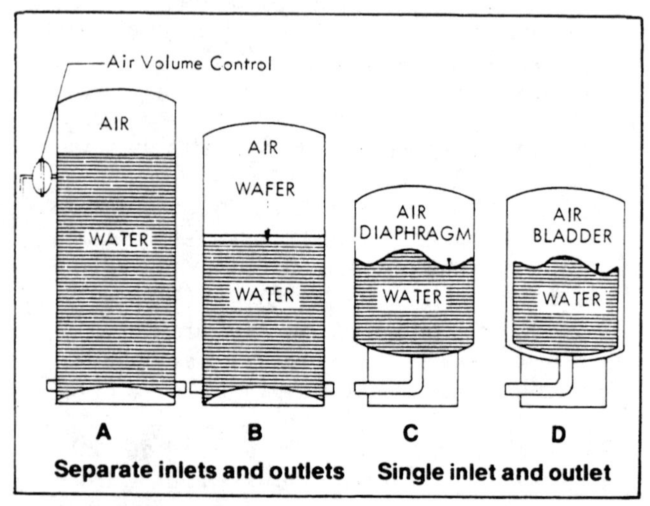

Fig 801-4. Pressure tanks.
A, plain steel tank; B, tank with floating wafer; C, diaphragm tank; D, bladder tank.

The pressure needed at a fixture depends on the type of fixture; provide at least 10 psi. See Table 8.

Table 801-8. Outlet water pressures.
Recommended minimum pressures.

Fixture	Pressure
Lavatory, sink or bathtub	10 psi
Shower	12
Tank toilet	15
Flush-valve toilet	20
Hose faucet	20
Hose faucet for fire control	30
Livestock waterer	15

Pressure Control Switch

A pressure switch automatically starts or stops a pump at set tank pressures. The switch has two adjustments: one to set the cut-in or "on" pressure, and the other to adjust the difference between the "on" and "off" pressure. Most switches are factory adjusted for specific pressures, but many can be readjusted within the limits of the switch rating.

Before the pressure switch can be properly adjusted, the outlet pressures and pressure losses must be determined.

Procedure:

1. Make a diagram of equipment and pipe locations. Determine and label differences in elevation between the pressure tank and outlets. Select needed flow rate: Tables 1, 2, or 3.
2. Select the largest minimum delivered pressure from Table 8 which the line is to supply.
3. Determine pressure loss due to pipe friction, Tables 13 to 22, in the Data Section.
5. To the needed delivered pressure add pressure losses due to elevation difference and pipe friction. (Total of steps 2, 3 and 4). Set the lower switch setting for at least this total.
6. Set the "pump-off" pressure switch setting about 20 psi higher than the "pump-on" setting.

Needed Tank Volume

Remember that the main reason for a pressure tank is to prevent the pump from cycling too frequently. A larger pressure tank permits longer pump cycles but costs a little more; it also has a little more water in storage.

For common water systems, a pressure tank about ten times the pump capacity is adequate (8 gpm pump, 80-gal tank). If the tank is a pre-charged type, six times pump capacity is adequate (8 gpm pump, 48-gal tank). For water systems requiring pumps larger than about 25 gpm, consult your pump supplier for recommended tank sizes.

The storage capacity of a pressure tank is usually small when compared to the daily water used.

Usable storage capacity is affected by air volume control location, operating pressures, and whether or not the tank is a precharged type.

Don't depend on the pressure tank to meet peak use rate. For a very few minutes, the pressure tank can supply water at any rate demanded, but as soon as the "pump on" capacity is reached, system flow rate is reduced to the pump's capacity.

Air Controls

When air and water in a pressure tank are in direct contact, water absorbs some of the air. If this absorbed air is not replaced, the tank becomes "waterlogged," causing the pump to start and stop frequently because there is no air to be compressed.

Install one of the following controls to maintain an adequate amount of air.

Constant-air devices are part of the pressure tank. One type uses a movable separator to prevent air absorption. (Type C, Fig 4) Others use an elastic air container to prevent air absorption. The constant-air devices provide more available water than add-air or release-air devices. They are precharged with air to the lower pressure setting through an air charging valve.

Add-air devices (Types A and B, Fig 4) are available in three types. The float-operated type is for systems with positive displacement shallow-well pumps. The diaphragm type is for jet pump systems. The displacement air control is for any system where the pump is above ground.

Air-release devices release excess air continuously added by certain types of pumps. Deep-well reciprocating and submersible pumps usually add air as they pump water. The extra air must be released or the tank becomes "air-bound" and holds little reserve water.

Air Chamber

Air chambers are often needed in water systems with reciprocating pumps. They are installed between the pump and pressure tank to lessen the pulsating delivery. Some pumps have this air chamber built in. They are also installed at the extreme end of each major supply line to counteract "water hammer" caused by rapidly closed faucets.

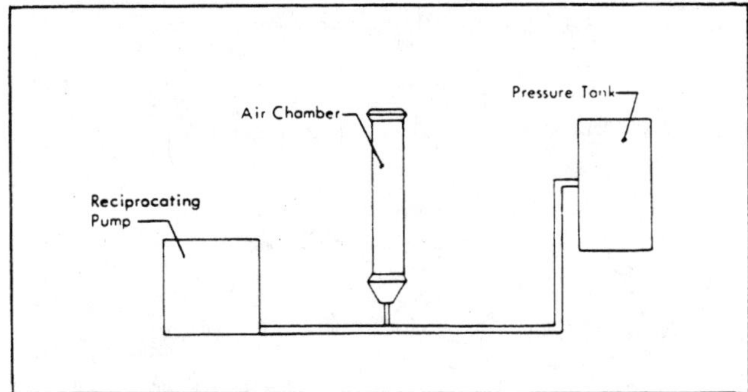

Fig 801-5. Air chamber.

Piping Types

The three types of pipe generally used in water systems are galvanized steel, copper, and plastic.

Galvanized Steel

Galvanized steel pipe is covered with a protective coating of zinc that greatly increases its life compared to black pipe.

It is made in standard 21' lengths with each end threaded. The pipe is cut and threaded to fit the job. Make joints by placing a small amount of pipe joint compound on the pipe threads and screwing on the fitting.

Galvanized steel pipe is suitable for all piping inside a building, but plastic and copper pipe are preferred for underground installation. Highly mineralized water greatly reduces the life of steel pipe.

Copper

Copper pipe is available in types "K," "L," and "M." Type K is heavy duty—pump suction lines and underground piping. Type L is standard weight—inside buildings. Type M is light-weight—use only behind walls inside buildings.

Types K and L are available in both "hard-tempered" or "soft-tempered" form. Hard-tempered pipe is rigid, comes in 10' and 20' lengths, and is usually used as exposed piping inside buildings. It needs little support compared to soft-tempered (flexible) tubing. Type M is available in hard-tempered form, but requires physical protection.

Soft-tempered tubing is excellent for underground use and for piping inside existing walls in old buildings. For underground piping, use the heavier Type K tubing.

Plastic

Plastic pipe is in flexible, semi-rigid, and rigid forms. Flexible pipe is common for underground water piping because of installation ease and economy. It is ½" or more in diameter and in coils of 100' or more.

Solvent welded, or glued, PVC semi-rigid pipe is now common because of better and more readily available quality, ditching equipment, and ease of joining. The joints of polyethylene pipe use nylon or brass fittings and stainless steel clamps and clamp screws.

Use only pipe and fittings that have the National Sanitation Foundation seal, NSF. It assures that non-toxic virgin materials were used in the manufacture and that the pipe is safe for drinking water.

There are five classes of plastic pipe and fittings that meet commercial standards established by the American Society of Testing Materials (ASTM) for potable household water:

Polyethylene (PE)

This flexible or semi-rigid pipe is for cold-water pipe, as its strength decreases as temperature rises. It has pressure ratings between 80 and 160 psi.

Polyethylene pipe is used for hot water heating in concrete floors; temperatures up to 100° F are common for foot comfort. Install the best brand of virgin plastic rated at 115°. Install mixing valves for water temperature adjustment and lay the pipe in straight lines to avoid strain.

Polyvinyl Chloride (PVC)

This rigid pipe is available with pressure ratings of 50 to 315 psi. PVC is for cold water only. It is used for some household cold water pipes and drains and in some permanent irrigation installations. Pipe for pressure water systems should be rated at least 80 psi.

PVC comes in 20' lengths and common diameters. It is joined with a coupling and solvent.

Acrylonitrile Butadene Styrene (ABS)

This semi-rigid pipe has pressure ratings between 80 and 160 psi. It is suitable for sewer pipe.

Chlorinated Polyvinyl Chloride (CPVC)

This material is similar to PVC but better for handling corrosive water at temperatures 40°-60° F above the limit of other vinyl plastics. It is suitable for hot or cold water lines.

Polybutylene (PB)

This pipe is for hot or cold potable water lines.

1. CPVC and PB were developed to handle hot water. Check the manufacturer's maximum stress and temperature limitations before installing. Also, check local plumbing codes to see if it is acceptable.
2. Within each of these classes, manufacturers sometimes make a number of pipes with different physical characteristics. Make sure that the one selected meets your needs.

Safety Measures

Install connections that will not allow backsiphoning of polluted water into the system. Disease-causing organisms often get into a system through backflow or backsiphoning from toilets, sinks, livestock waterers, flooded pump pits, flooded basements, and other places. Insecticides and herbicides can be backsiphoned into the system when the spray tank is filled with a hose that is allowed to become submerged. Follow the suggestions listed below.

- Eliminate all cross connections and interconnections between potable and non-potable water systems. Fig 7 . A source and all piping for non-potable water for such fixtures as stool and outside hose connections must not interconnect with a potable water pipe.
- Install only non-backsiphoning livestock waterers. Backsiphoning is caused by a loss of pressure in the supply causing the fixture to drain back into the piping. Standard household fixtures are non-backsiphoning. Be sure that the livestock waterer you install or build is non-backsiphoning. When the waterer is full there should be an air gap of twice the diameter of the outlet between the water surface and the outlet. Fig 8.
- Replace or reconstruct pump pits and pump installations to avoid flooding.
- Fill spray tanks so there is no possibility of backsiphoning.

There are anti-back siphoning connectors for hose bibs.

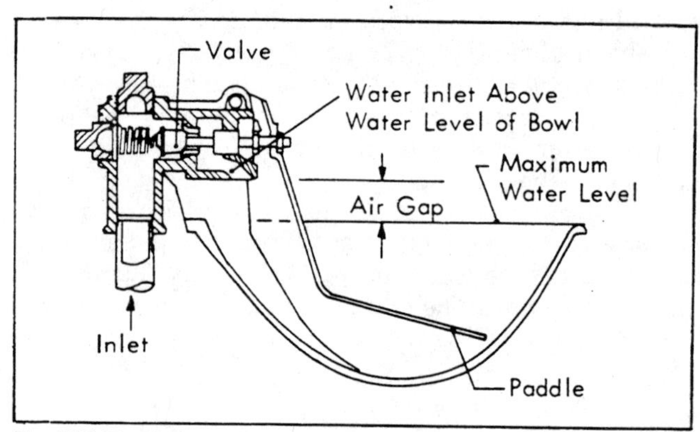

Fig 801-7. Air gap livestock waterer.

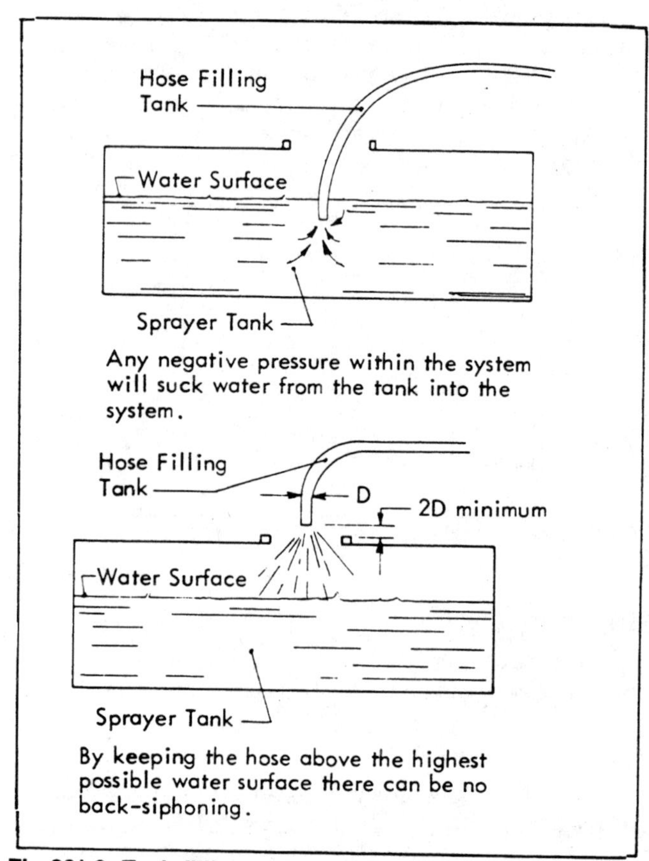

Fig 801-8. Tank filling without back-siphoning.

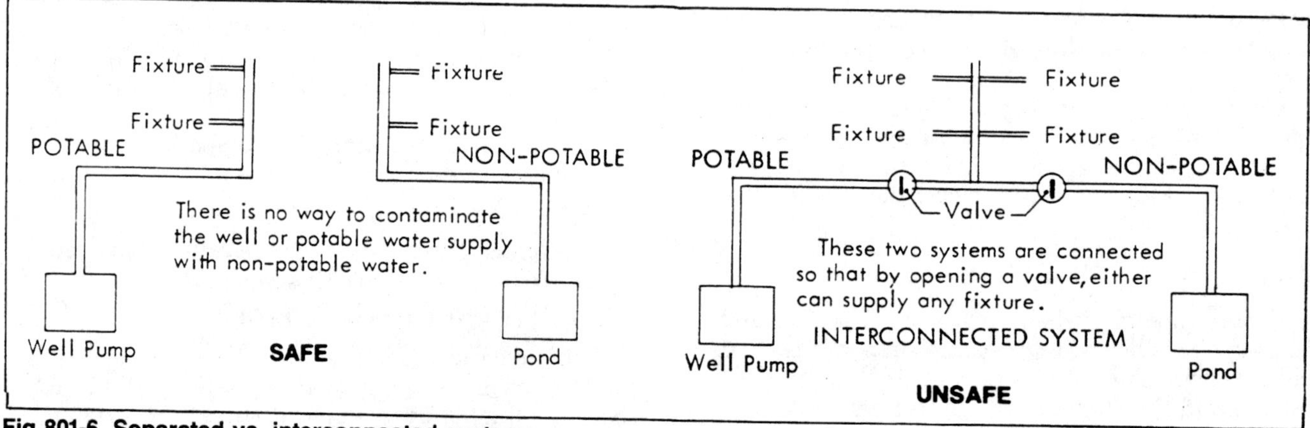

Fig 801-6. Separated vs. interconnected systems.

Water Treatment

All water from natural sources contains impurities. Some of these impurities adversely affect the usefulness and suitability of water, while others may improve its palatability.

Pure water is tasteless, colorless, and odorless. Because pure water is one of the best solvents available, it picks up impurities easily. Most impurities are picked up naturally, but some are added, accidentally or intentionally, by man.

Water may be cleansed or polluted as it flows over or filters through soil or other material; it may pick up or lose bacteria; it may dissolve or lose chemicals, minerals, and sediment.

The belief that flowing or soil-filtered water has purified itself is false and leads to an unjustified feeling about water safety. Clear water is not necessarily safe; colored water is not necessarily unsafe. To know whether your water is acceptable for its intended uses:

- Determine the quality needed for each use;
- Test the water for bacteriological quality;
- Analyze for chemical quality;
- Treat the water if needed.

Some of the procedures, processes, and equipment to do these jobs are described in this chapter.

Needed Quality

Acceptable quality depends on the ultimate use of the water. For example, drinking water must be of high quality—safe and palatable.

Potable = safe to drink

Palatable = nice to drink; acceptable flavor and, usually, appearance and odor.

Water for livestock does not usually need to be as free of impurities as water for humans. But water containing certain biological or chemical impurities can adversely affect the growth and health of livestock.

Safe water is free from disease-causing organisms and contains no chemical that exceeds the amounts listed in Table 10.

Table 801-9. Water qualities for various uses.
Softened water may not be recommended for a person on a low-salt diet. Consult your doctor.

Use	Quality			
	Safe	Palatable	Soft	Non-Staining
Drinking	X	X		
Food Preparation	X	X	X	X
Personal hygiene	X	X	X	X
Dish and dairy utensil washing	X		X	X
Clothes washing	X		X	X
Toilet flushing				X

Most disease-causing organisms—bacteria, protozoa, and viruses—are identified only by a complex laboratory examination.

Do not use water containing disease-causing organisms or excessive amounts of chemicals until corrective measures have been installed or until advice from a health authority is followed.

Water to be palatable must be non-offensive to the taste, smell, or sight, and contain no chemicals that exceed the amounts listed in Table 11.

Remove all visible and odorous materials, even though they may not cause disease, from water to be used in the home. These materials include silt, clay, precipitated minerals, colored matter, insects, worms and visible growths of algae.

Table 801-10. Chemicals in drinking water.
Consult local health authority for optimum and maximum amounts of fluoride for your area.

Chemical	Maximum Concentration in mg/1	Chemical	Maximum Concentration in mg/1
—Arsenic	0.05 mg/1	—Lead	0.05 mg/1
—Barium	1.00	—Mercury	0.002
—Cadmium	0.01	—Nitrate Nitrogen	10.0
—Chromium	0.05	—Selenium	0.01
—Fluoride	1.4 to 2.4	—Silver	0.05

801-10a. Secondary maximum levels.
Advisable maximums for water delivered to the user.

Contaminant	Level
Chloride	250 mg/1
Color	15 color units
Copper	1 mg/1
Corrosivity	Non-corrosive
Foaming agents	0.5 mg/1
Hydrogen sulfide	0.5 mg/1
Iron	0.3 mg/1
Manganese	0.05 mg/1
Odor	3 threshold odor number
pH	6.5-8.5
Sulfate	250 mg/1
TDS	500mg/1
Zinc	5 mg/1

Source: Federal Register

Table 801-11. Chemicals causing odor and/or taste.
Many individuals will accept greater concentrations.

Chemical	Concentration mg/l	Chemical	Concentration mg/l
Chlorides	100-250 mg/l	Copper	1 mg/l
Total dissolved solids	500-1000	Hydrogen Sulfide	0.1-0.2
		Iron	0.1-0.2
		Zinc	5
		A B S (detergent)	0.5
		Phenol	0.001

Determining Water Quality

Many tests and analyses can be made on water; select carefully those actually needed. Contact your county extension agent for help in selecting tests and locating testing laboratories. Also, water treatment equipment dealers can help determine the tests and analyses needed, and may even make some analyses. Some states have water testing services.

A sample of water should be checked:

- For bacteriological safety, following completion, reconstruction, or addition to a water supply and its distribution system.
- At least annually, as a check for continued bacteriological safety.
- For safety, whenever disease occurs that could be caused by unsafe water. (Diseases such as typhoid fever, gastroenteritis, dysentery, diarrhea, hepatitis, and leptospirosis).
- For chemicals, if a hazardous pollutant gets into the supply or distribution system.
- For safety, following unexplained changes in taste, odor, or appearance.
- Before installing treatment equipment designed to solve problems outlined in Table 12.

Table 801-12. Water quality problems and solutions.

Problem	Possible Cause	Test for	Typical Corrective Procedures (one or several procedures may be needed)
Diseases such as typhoid fever, dysentery, diarrhea, hepatitis, and others	Disease-producing bacteria, virus, protozoa, etc.	Coliform bacteria	Repair or reconstruct the complete water system to keep out pollutants; then disinfect, retest, and if the test results are satisfactory from 3 or more samples taken 2 weeks apart, use the water. If the results are unsatisfactory, either • Develop a new water source; disinfect; test, and if satisfactory, use; or • Install and properly operate continuous disinfection equipment.
Rust or black colored water. Stained clothes and fixtures. Reduced water flow. Discolored, unpalatable food and beverages.	Iron or Manganese	Iron and Manganese	• Aerate the water in a storage tank (pressurized or non-pressurized) and then run water through a sediment filter. • Continuously chlorinate the water and run it through a sediment filter and carbon filter. • Install a sediment filter and a softener. • Install a softener, but only if the water is clear. • Install a potassium permanganate regenerated oxidizing filter and a softener. If water pH is less than 7.0, the acid must be neutralized ahead of the filter.
Rust colored slime in toilet tank and pipes. Fuzzy particles in water. Staining. Reduced water flow.	Iron bacteria	Iron bacteria	• Shock chlorinate the well and system, blow out all pipes, and shock chlorinate again. Do this when needed. • Install a continuous chlorination system and carbon filter.
Clothes becoming gray even with good washing Scum on wash and bath water after using soap. Water heater takes long to heat water. Reduced water flow.	Hard water (dissolved minerals)	Calcium and Magnesium	• Correctly size, install, and operate a water softener. • Use small amount of packaged water softener in baths, washer, etc. • Clogged pipes, water heaters, and other equipment must be cleaned or replaced.
Smelly water—rotten egg odor. Rapid tarnishing of silverware. Staining, black color	Hydrogen sulfide	Hydrogen sulfide	• Install a continuous chlorination system and carbon filter. • Aerate the water. Install a sediment and/or carbon filter if needed. • Install a potassium permanganate regenerated oxidizing filter. • Clogged and corroded pipes and equipment must be cleaned or replaced.
Corroded metal parts of pump and water system. Staining, green color	Low pH (Acid water)	pH	• Install a continuous soda ash solution feeder. • Install a neutralizer tank containing limestone chips.
Cloudy water	Particles of suspended or colored matter	Turbidity and color color	• Install a sediment filter, and possibly a coagulant feeder ahead of the filter. • Use activated carbon to remove color due to organisms.
Gassy well and water	Decay of organic matter	Methane	• Install gas release valve on pressure tank and vent to outdoors. • Aerate water in a non-pressurized storage tank.

Friction Loss Tables

Recommendation:

"Friction loss" should not exceed 5 psi from the pressure tank to the service entry. It should not exceed 10 psi from the pressure tank to any isolated fixture. The velocity of the water within the pipe should not exceed 4 feet per second. Before using the tables:

1. Select kind of pipe
2. Estimate size of pipe
3. Determine length of pipe*—from diagram of complete water system.
   Friction loss for a pipe less than 100' long is Table value x pipe length / 100.
4. Estimate needed flow rate—from Figure 1.

Pipe Friction Loss Tables, Tables 13-21

The friction loss tables are based on the following equation:

$H = 4.67\,(cfs/C)^{1.85} L/D^{4.87}$

h = ft of head; L = ft of pipe length; D = ft of pipe diameter; C = coefficient for pipe material.

For 100 ft of pipe length, gpm, and diam in inches, the equation becomes:

$h = 1043.8\,(gpm/C)^{1.85}/d^{4.87}$

h, psi = 0.4332 h,ft

C = 140, plastic

130, copper

100, 17-yr old steel

(Multiply losses by 0.6 for new pipe, by 1.2 for 25-yr old pipe.)

Pipe diameters used are listed in the tables.

If possible, use friction loss data furnished by the manufacturer of the specific pipe you will use.

Table 801-13. Friction loss/100' of ½" pipe.

Gallons per Minute	Steel I.D. = 0.6222" Feet	psi	Copper I.D. = 0.545" Feet	psi	Plastic I.D. = 0.622" Feet	psi
1 gpm	2.1'	0.9 psi	2.5'	1.1 psi	1.1'	0.5 psi
2	7.6	3.3	8.9	3.8	4.1	1.8
3	16.1	7.0	18.8	8.1	8.6	3.7
4	27.3	11.8	32.0	13.9	14.7	6.4
5	41.3	17.9	48.4	21.0	22.2	9.6

- - - - V = 4 ft/sec - - - -

Table 801-14. Friction loss/100' of ¾" pipe.

Gallons per Minute	Steel I.D. - 0.824" Feet	psi	Copper I.D. = 0.785" Feet	psi	Plastic I.D. = 0.824" Feet	psi
1 gpm	0.5'	0.2 psi	0.4'	0.2 psi	0.3'	0.1 psi
2	1.9	0.8	1.5	0.7	1.0	0.4
3	4.1	1.8	3.2	1.4	2.2	0.9
4	6.9	3.0	5.4	2.3	3.7	1.6
5	10.5	4.5	8.2	3.5	5.6	2.4
6	14.7	6.4	11.5	5.0	7.9	3.4
7	19.6	8.5	15.2	6.6	10.5	4.5
8	25.0	10.9	19.5	8.5	13.4	5.8

Table 801-15. Friction loss/100' of 1" pipe.

Gallons per Minute	Steel I.D. - 1.049" Feet	psi	Copper I.D. = 1.025" Feet	psi	Plastic I.D. = 1.049" Feet	psi
2 gpm	0.6'	0.3 psi	0.4'	0.2 psi	0.3'	0.1 psi
3	1.3	0.5	0.9	0.4	0.7	0.3
4	2.1	0.9	1.5	0.6	1.2	0.5
5	3.2	1.4	2.2	1.0	1.7	0.8
6	4.5	2.0	3.1	1.4	2.4	1.1
8	7.7	3.3	5.3	2.3	4.1	1.8
10	11.7	5.1	8.0	3.5	6.3	2.7
12	16.4	7.1	11.3	4.9	8.8	3.8
14	21.8	9.4	15.0	6.5	11.7	5.1

Table 801-16. Friction loss/100' of 1¼" pipe.

Gallons per Minute	Steel I.D. - 1.380" Feet	psi	Copper I.D. = 1.265" Feet	psi	Plastic I.D. = 1.380" Feet	psi
4 gpm	0.6'	0.2 psi	0.5'	0.2 psi	0.3'	0.1 psi
5	0.9	0.4	0.8	0.3	0.5	0.2
6	1.2	0.5	1.1	0.5	0.6	0.3
8	2.0	0.9	1.9	0.8	1.1	0.5
10	3.1	1.3	2.9	1.3	1.6	0.7
12	4.3	1.9	4.0	1.8	2.3	1.0
14	5.7	2.5	5.4	2.3	3.1	1.3
16	7.3	3.2	6.9	3.0	3.9	1.7
18	9.1	3.9	8.6	3.7	4.9	2.1
20	11.1	4.8	10.4	4.5	5.9	2.6
22	13.2	5.7	12.4	5.4	7.1	3.1
24	15.5	6.7	14.6	6.3	8.3	3.6

Table 801-17. Friction loss/100' of 1½" pipe.

Gallons per Minute	Steel I.D. = 1.610" Feet	psi	Copper I.D. = 1.505" Feet	psi	Plastic I.D. = 1.610" Feet	psi
4	0.3	0.1	0.2	0.1	0.1	0.1
6	0.6	0.2	0.5	0.2	0.3	0.1
8	1.0	0.4	0.8	0.4	0.5	0.2
10	1.5	0.6	1.2	0.5	0.8	0.3
12	2.0	0.9	1.7	0.8	1.1	0.5
14	2.7	1.2	2.3	1.0	1.5	0.6
16	3.5	1.5	3.0	1.3	1.9	0.8
18	4.3	1.9	3.7	1.6	2.3	1.0
20	5.2	2.3	4.5	1.9	2.8	1.2
25	7.9	3.4	6.8	2.9	4.2	1.8
30	11.1	4.8	9.5	4.1	5.9	2.6
35	14.7	6.4	12.6	5.5	7.9	3.4

Table 801-18. Friction loss/100' of 2" pipe.

Gallons per Minute	Steel I.D. = 2.067" Feet	psi	Copper I.D. = 1.985" Feet	psi	Plastic I.D. = 2.067" Feet	psi
10	0.4	0.2	0.3	0.1	0.2	0.1
12	0.6	0.3	0.5	0.2	0.3	0.1
14	0.8	0.3	0.6	0.3	0.4	0.2
16	1.0	0.4	0.8	0.3	0.5	0.2
18	1.3	0.6	1.0	0.4	0.7	0.3
20	1.5	0.7	1.2	0.5	0.8	0.4
25	2.3	1.0	1.8	0.8	1.3	0.5
30	3.3	1.4	2.5	1.1	1.8	0.8
35	4.4	1.9	3.3	1.4	2.3	1.0
40	5.6	2.4	4.2	1.8	3.0	1.3
45	6.9	3.0	5.2	2.3	3.7	1.6
50	8.4	3.7	6.3	2.7	4.5	2.0

Table 801-19. Friction loss/100' of 2½" pipe.

Gallons per Minute	Steel I.D. = 2.469" Feet	psi	Plastic I.D. = 2.469" Feet	psi
20	0.7	0.3	0.3	0.2
30	1.4	0.6	0.7	0.3
40	2.3	1.0	1.3	0.5
50	3.5	1.5	1.9	0.8
60	5.0	2.2	2.7	1.2
70	6.6	2.9	3.5	1.5

Table 801-20. Friction loss/100′ of 3″ pipe.

Gallons per Minute	Steel I.D. = 3.068" Feet	Steel psi	Plastic I.D. = 3.216" Feet	Plastic psi
40	0.8	0.4	0.3	0.2
50	1.2	0.5	0.5	0.2
60	1.7	0.7	0.7	0.3
70	2.3	1.0	1.0	0.4
80	2.9	1.3	1.3	0.5
90	3.7	1.6	1.6	0.7
100	4.4	1.9	1.9	0.8
120	6.2	2.7	2.7	1.2

Table 801-21. Friction loss/100′ of 4″ pipe.

Gallons per Minute	Steel I.D. 4.026" Feet	Steel psi	Plastic I.D. = 4.134" Feet	Plastic psi
40	0.2	0.1	0.1	0.0
60	0.5	0.2	0.2	0.1
80	0.8	0.3	0.4	0.2
100	1.2	0.5	0.6	0.2
120	1.7	0.7	0.8	0.3
140	2.2	1.0	1.0	0.5
180	3.5	1.5	1.7	0.7
220	5.1	2.2	2.4	1.0

Table 801-22. Friction loss in fittings.

Equivalent lengths of straight pipe for fittings and equipment.

[1]Loss figures are based on equivalent lengths of indicated pipe material.

[2]Loss figures are for screwed valves and are based on equivalent lengths of steel pipe.

Type Fitting & Application	Pipe & Ftg[1] Material	Equivalent Length in feet Nominal Size of Fitting & Pipe 1/2"	3/4"	1"	1 1/4"	1 1/2"	2"	2 1/2"
Insert coupling	Plastic	3'	3'	3'	3'	3'	3'	3'
Threaded adapter Plastic or copper to thread	Copper	1'	1'	1'	1'	1'	1'	1'
	Plastic	3	3	3	3	3	3	3
90° standard elbow	Steel	2'	3'	3'	4'	4'	5'	6'
	Copper	2	3	3	4	4	5	6
	Plastic	4	5	6	7	8	9	10
Standard tee, straight flow thru run	Steel	1'	2'	2'	3'	3'	4'	5'
	Copper	1	2	2	3	3	4	5
	Plastic	4	4	4	5	6	7	8
Standard tee, turn flow thru	Steel	4'	5'	6'	8'	9'	11'	14'
	Copper	4	5	6	8	9	11	14
	Plastic	7	8	9	12	13	17	20
Gate or ball valve[2]	Steel	2'	3'	4'	5'	6'	7'	8'
Swing check valve[2]	Steel	4	5	7	9	11	13	16
Globe valve	Steel	15	20	25	35	45	55	65
30-gal vertical water heater	---	4'	17'	56'				
Water softener	---			Up to 10 psi loss				
Iron or sediment filter	---			Up to 20 psi loss				
Chlorinator	---							

802 ELECTRICAL WIRING OF FARM BUILDINGS

Glossary

A	ampere(s)
AC	alternating current
Ampacity	current carrying capacity of wire
AWG	American Wire Gauge
Cable	multi-strand conductor in same insulating sheath
CU	copper
DCO	duplex convenience outlet
DP	distribution panel
GFCI	ground fault circuit interrupter
hp	horsepower
NEC	*National Electrical Code®*
PVC	polyvinyl chloride
SPO	special purpose outlet
UL	Underwriters' Laboratories
V	volt(s)
Voltage drop	reduction in voltage from power supply to load due to conductor resistance
W	watt(s)
Wire	single strand conductor

This chapter outlines materials and methods for electrical equipment and wiring in agricultural buildings. It can also help you determine if existing wiring is adequate. It does **not** cover wiring from the power supplier to the building's service entrance. Have the farmstead electrical distribution and service entrance equipment installed by a qualified electrician cooperating with your power supplier.

You will also need to consult with the following persons or agencies to ensure proper wiring:

- Your local power supplier can help you plan and install the distribution system to your building.
- A licensed electrician can help you plan and install distribution panels and motor circuits, select conductors and fixtures, and verify compliance with state and local codes.
- Your electrical equipment supplier can supply the dust- and moisture-tight fixtures and wiring required for Damp buildings. This equipment is only available through electric wholesale supply stores. Plan ahead; you may need to order equipment.
- Your insurance company will help you meet insurance requirements. If the electrical system does not meet their standards, they may increase rates or refuse to insure the building.

Codes

The code referred to for electrical work in the United States is the NFPA 70-1981 *National Electrical Code®* (NEC) published by and registered trademarks of the National Fire Protection Association (NFPA), Quincy, MA 02269. Tables 802-3 and 802-11 are reprinted with permission from that code and are not the complete or official position of NFPA. The NEC is a guide to proper and safe materials and installation methods. Even though many farm buildings do not presently fall under code jurisdiction, it is a good idea to follow NEC. Also, your insurance company may require a "code" installation for approval. Before starting construction, check if a wiring permit is required.

This chapter occasionally makes a reference to the code (e.g. NEC 547 where 547 = article number).

Building Groups

This chapter applies **only** to agricultural buildings. An agricultural building is a structure for housing farm implements, hay, grain, poultry, livestock, or other agricultural produce. It is not for human habitation or a place of employment where agricultural products are processed, treated, or packaged. Also, it is not used by the public.

Agricultural buildings are divided into three groups:

- **Dry:** machine sheds, shops, and garages not attached to the residence. NEC 547-1 (c).
- **Damp:** animal housing (open or closed), milkhouses, ventilated manure pits, well pits, silos, silo rooms, and high humidity produce storages (e.g. potatoes and apples). It also includes buildings or areas that are washed periodically. NEC 547-1 (b).
- **Dusty:** fertilizer, dry grain, and dry hay storage buildings; and grain-feed processing centers.

Not included in this chapter are structures housing methane or alcohol production equipment or any other system that may produce explosive gas or dust. These buildings may require explosion-proof materials—refer to the NEC.

Different parts of the same building may fit different groups. For example, a wash down area in a machine shed fits the Damp group while the rest of the building fits the Dry group.

Materials

Use only equipment that is "UL Listed." Listed items include switches, plugs, connectors, and receptacles. The Underwriters' Laboratories (UL) is a nonprofit organization that tests and establishes standards for various products. The UL does not approve items but rather "lists" them as having met minimum safety standards. Use only 20 A rated materials, because they are safer and last longer than 15 A ones. Include a grounding wire in all circuits.

Dry Buildings

Dry buildings do not require special wiring materials, but use quality materials and practices as outlined in the NEC. Generally, they may be wired with the same type of materials used for residential wiring.

Surface wiring is common and saves materials and labor. Protect wiring and components from physical damage. Either type NM or NMC cable is acceptable, but use NM only indoors and in areas of permanent low humidity, Fig 802-1.

Most agricultural applications require dust- and moisture-resistant fixtures that resist physical damage. Metal or plastic boxes are acceptable. Avoid the less expensive type of surface mount fixtures shown in Fig 802-2.

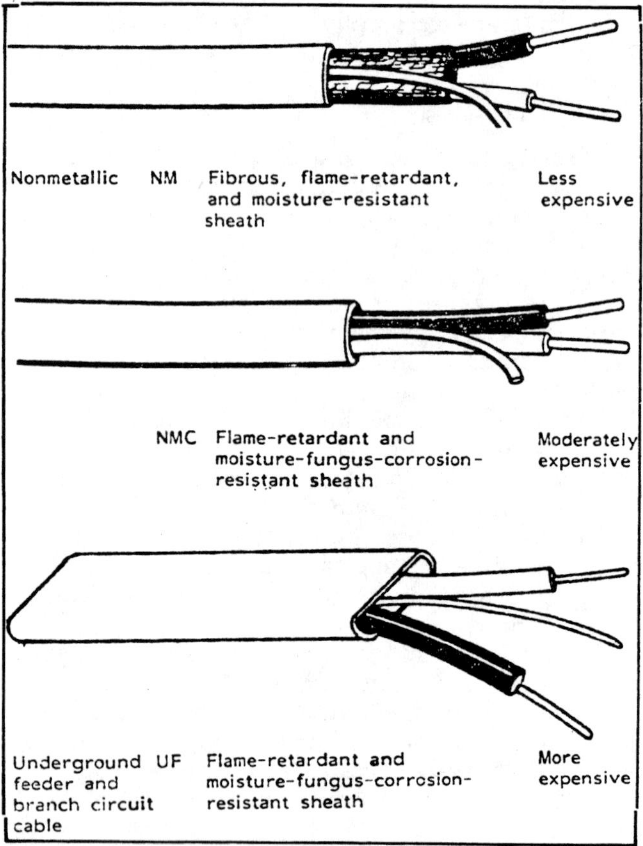

Fig 802-1. NM, NMC, and UF cable.
Protect cable from physical and rodent damage. Include a grounding wire in all circuits.

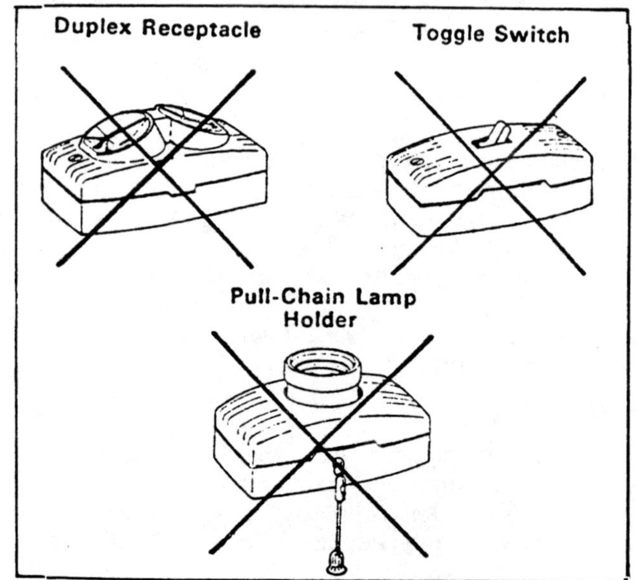

Fig 802-2. Do NOT use this type of fixture.
These fixtures generally do not endure agricultural applications.

Damp Buildings

Damp buildings require special materials and wiring methods, because high levels of moisture and corrosive dust and gas can quickly corrode standard electrical equipment. Dust and moisture can also lead to fire and safety hazards by creating short circuits or heat buildup in electrical components. All wiring hardware must be dust- and moisture-tight and made of corrosion-resistant materials.

Use dust- and moisture-tight incandescent light fixtures with a heat-resistant globe to cover the bulb. Use dust- and moisture-resistant fluorescent light fixtures with a gasketed cover. Figs 802-3 and 802-4.

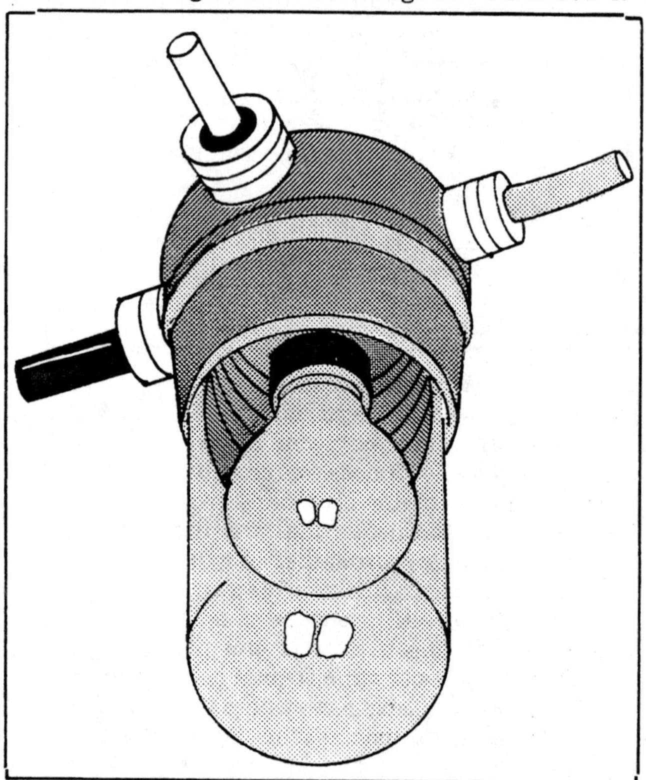

Fig 802-3. Incandescent fixture for Damp buildings.
Fixtures are nonmetallic, globed, and dust- and moisture-tight. Use fixtures with at least a 150 W rating.

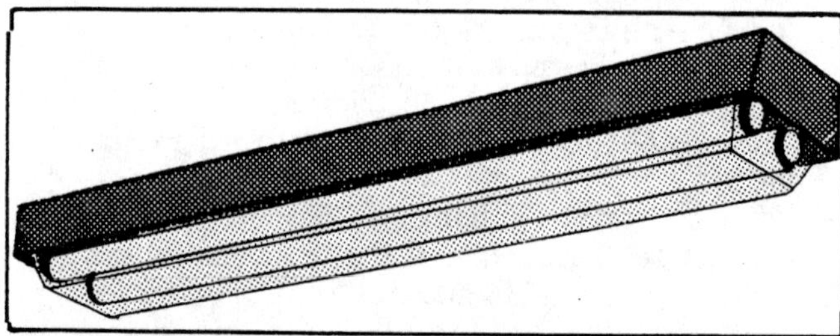

Fig 802-4. Fluorescent fixture for Damp buildings.
Fixtures are nonmetallic, enclosed, and gasketed.

Surface Mount Cable

Use type UF or NMC **(not NM)** cable **with a grounding wire,** Fig 802-1.

Mount cable on the inside surface of walls or ceiling with plastic or plastic coated staples or straps every 2′ and within 8″ of junction or fixture boxes. Install cable where it cannot be easily damaged. Avoid sharp bends in the cable—minimum radius is five times the cable diameter.

If the building has joists, try to run the conductor along a joist or a beam. If the conductor must run perpendicular to joists or ribs of metal liners, install it on a 1x2 running board.

Boxes

House every wire splice, switch, and receptacle in a box. Mount every fixture (lights, etc.) on a box. NEC 300-15. All boxes are to be noncorrosive and dust- and moisture-tight. Use molded plastic or cast aluminum boxes. Gasketed covers are required to seal all junction boxes, Fig 802-12. Use receptacle boxes equipped with gasketed, spring-loaded covers. Switch boxes can have spring-loaded covers, moisture-tight switch levers, or moisture-tight covers over the surface, Fig 802-5.

Do not crowd wires into boxes—crowding makes work difficult, increases work time, and makes shorts and grounds more likely. See Fig 802-6 and Table 802-1.

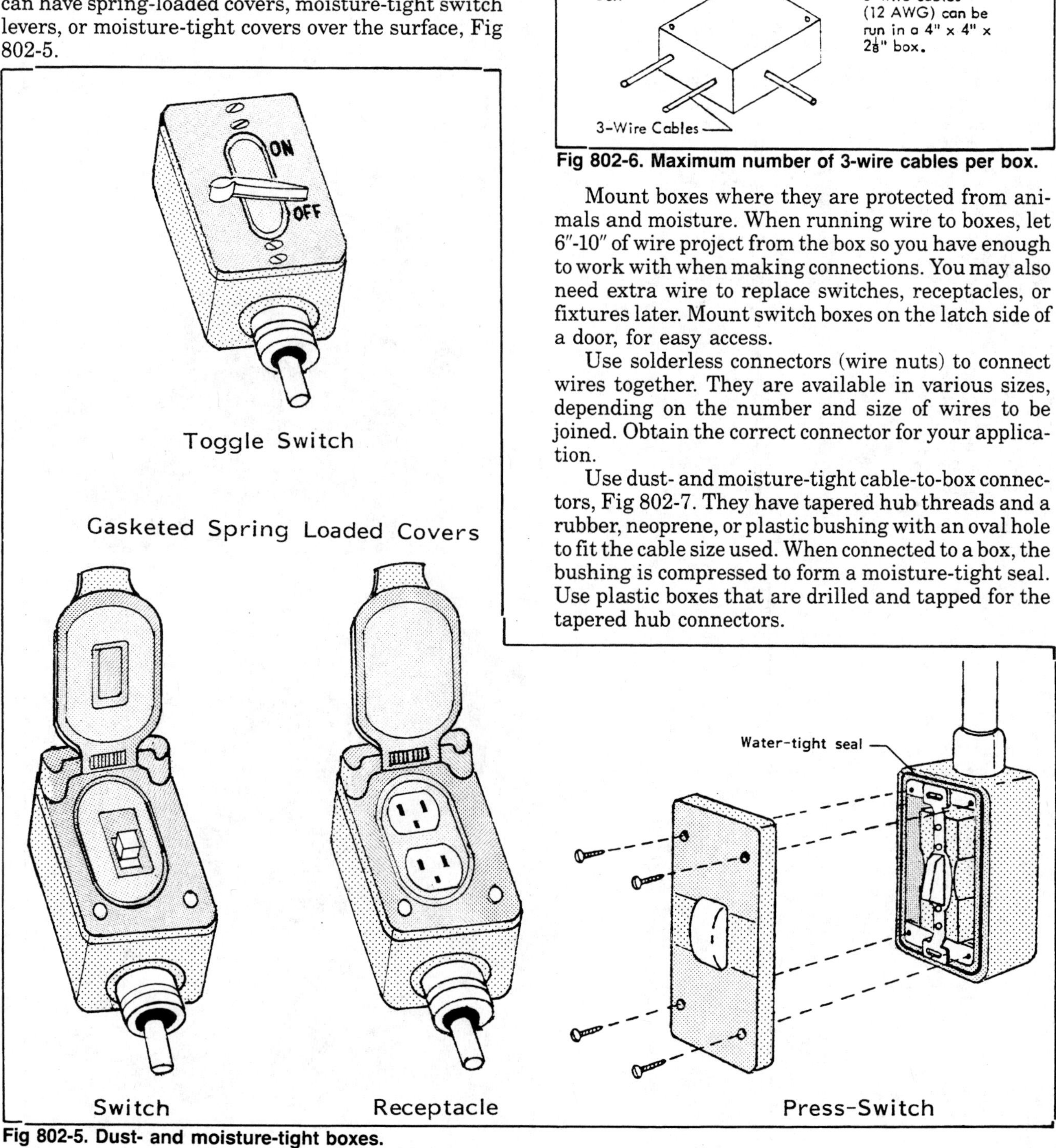

Fig 802-6. Maximum number of 3-wire cables per box.

Fig 802-5. Dust- and moisture-tight boxes.

Mount boxes where they are protected from animals and moisture. When running wire to boxes, let 6″-10″ of wire project from the box so you have enough to work with when making connections. You may also need extra wire to replace switches, receptacles, or fixtures later. Mount switch boxes on the latch side of a door, for easy access.

Use solderless connectors (wire nuts) to connect wires together. They are available in various sizes, depending on the number and size of wires to be joined. Obtain the correct connector for your application.

Use dust- and moisture-tight cable-to-box connectors, Fig 802-7. They have tapered hub threads and a rubber, neoprene, or plastic bushing with an oval hole to fit the cable size used. When connected to a box, the bushing is compressed to form a moisture-tight seal. Use plastic boxes that are drilled and tapped for the tapered hub connectors.

Table 802-1. Required box size.

Volume per conductor

Conductor size AWG	Box volume per conductor in³
# 14	2
# 12	2.25
# 10	2.5
# 8	3
# 6	5

Conductors per box

Box dimension, in. Trade size or type	Box volume in³	Conductor size #14	#12	#10	#8
		- - max./box - -			
4 x 1¼ round or octagonal	12.5	6	5	5	4
4 x 1½ round or octagonal	15.5	7	6	6	5
4 x 2⅛ round or octagonal	21.5	10	9	8	7
4 x 1¼ square	18.0	9	8	7	6
4 x 1½ square	21.0	10	9	8	7
4 x 2⅛ square	30.3	15	13	12	10
4¹¹⁄₁₆ x 1¼ square	25.5	12	11	10	8
4¹¹⁄₁₆ x 1½ square	29.5	14	13	11	9
4¹¹⁄₁₆ x 2⅛ square	42.0	21	18	16	14
3 x 2 x 1½ device	7.5	3	3	3	2
3 x 2 x 2 device	10.0	5	4	4	3
3 x 2 x 2¼ device	10.5	5	4	4	3
3 x 2 x 2½ device	12.5	6	5	5	4
3 x 2 x 2¾ device	14.0	7	6	5	4
3 x 2 x 3½ device	18.0	9	8	7	6
4 x 2⅛ x 1½ device	10.3	5	4	4	3
4 x 2⅛ x 1⅞ device	13.0	6	5	5	4
4 x 2⅛ x 2⅛ device	14.5	7	6	5	4

Plastic conduit

The most common conduit for Damp buildings is Schedule 40 PVC (polyvinyl chloride). Do not use metal conduit because it corrodes in these buildings. All fittings and boxes for PVC conduit must be non-corrosive. PVC conduit is available in 10′ and 20′ lengths and the most common diameters for agricultural buildings are ½″ and ¾″.

Obtain wires with a type W designation—TW, THW, THWN, RHW, RUW, and XHHW. NEC 310-8(a). See Table 802-2. Use bare or green covered copper wire for the grounding wires in PVC conduit. See Table 802-3 for the maximum number of wires per conduit.

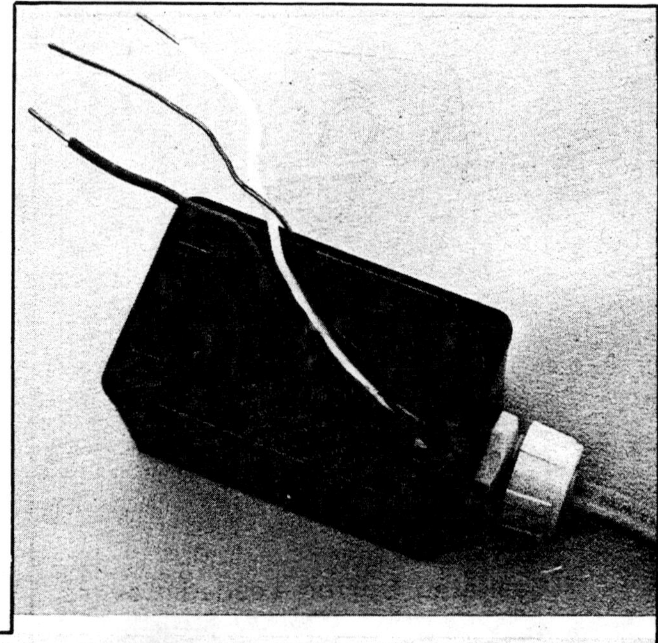

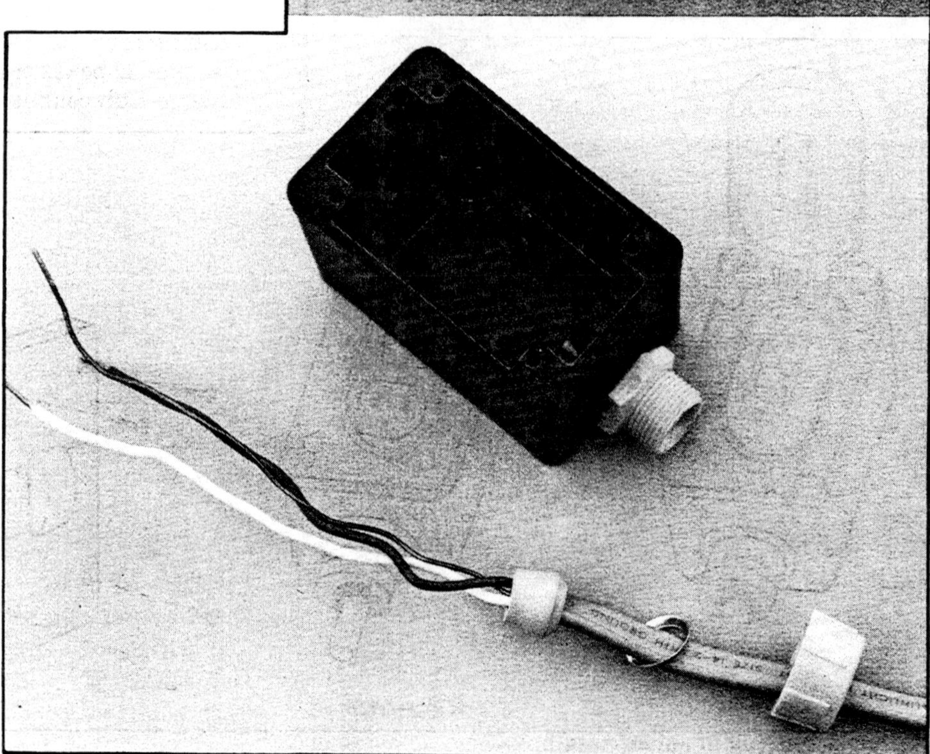

Fig 802-7. Moisture-tight cable-to-box connection.

Surface mount conduit on walls and ceiling. Avoid running the conduit through building surfaces—if conduit must pass through a wall or ceiling from a warm to cold area, seal the conduit with a sealing compound on the warm side of the wall or ceiling. This prevents moisture in the warm part of the conduit from moving into the cold part and condensing.

Support ½"-¾" diameter PVC conduit at least 3' o.c. with nonmetal fasteners, Fig 802-8. NEC 347-8. Where walls are washed frequently, space the conduit, boxes, and fittings ¼" out from walls. NEC 300-6(c). You can buy PVC conduit elbows and offsets, Fig 802-9, or you can heat the conduit and bend it by hand. Heat the conduit with "hot boxes" or a hot air blower—never use an open flame. Maintain the circular cross section of the conduit through the bend. Put no more than the equivalent of four 90° bends between junction boxes and/or access fittings. NEC 347-14.

Cut PVC conduit with a fine-toothed saw or a special plier-type cutter. Ream the ends smooth with a file after cutting. Permanent joints are made with PVC couplings and are glued togther with a solvent-cement. For nonpermanent joints, use threaded adapters sealed with rubber washers. See Fig 802-10.

Allow for thermal expansion in PVC conduit by leaving the bends unrestrained, Fig 802-11a. If the conduit has few corners or is exposed to a wide temperature range (difference of 120 F), install a 6" expansion joint every 200' or a 2" expansion joint every 65', Fig 802-11b.

Use PVC molded or cast aluminum junction and outlet boxes, Fig 802-12. Connect conduit to boxes as shown in Fig 802-13, or glue directly into boxes with socket fittings.

Flexible connections to motors and equipment must be moisture-tight, dust-tight, and corrosion-resistant, Fig 802-14. Use moisture-tight, plastic covered flexible metal conduit—ordinary flexible metal conduit is **not** suitable. Flexible connections to portable equipment can be hard service cord such as types SO and STO. Dust- and moisture-tight plugs and caps are required for locations such as feed rooms, grain dryers, and silos.

Table 802-2. Wire classifications.

Insulation material	Type letter	Description
Rubber	RHW	Moisture- and heat-resistant rubber
Latex rubber	RUW	Moisture-resistant latex rubber
Thermoplastic	TW	Moisture-resistant thermoplastic
	THW	Moisture- and heat-resistant thermoplastic
	THWN	Flame retardant, moisture- and heat-resistant, with nylon jacket outer covering
Cross-linked synthetic polymer	XHHW	Flame retardant
Hard service cord	SO	Thermoset insulated with oil-resistant thermoset cover, no fabric braid
	STO	Thermoplastic or thermoset insulated with oil-resistant thermoplastic cover, no fabric braid

Table 802-3. Maximum number of wires in PVC conduit.
From NEC Tables 3A, 3B, 3C.

Insulation type letter	Wire size AWG	Conduit size ½"	¾"	1"
TW	14	9	15	25
RUW	12	7	12	19
XHHW	10	5	9	15
	8	2	4	7
THW	14	6	10	16
RHW (without outer	12	4	8	13
covering)	10	4	6	11
	8	1	3	5
RHW (with outer	14	3	6	10
covering)	12	3	5	9
	10	2	4	7
	8	1	2	4
THWN	14	13	24	39
	12	10	18	29
	10	6	11	18
	8	3	5	9

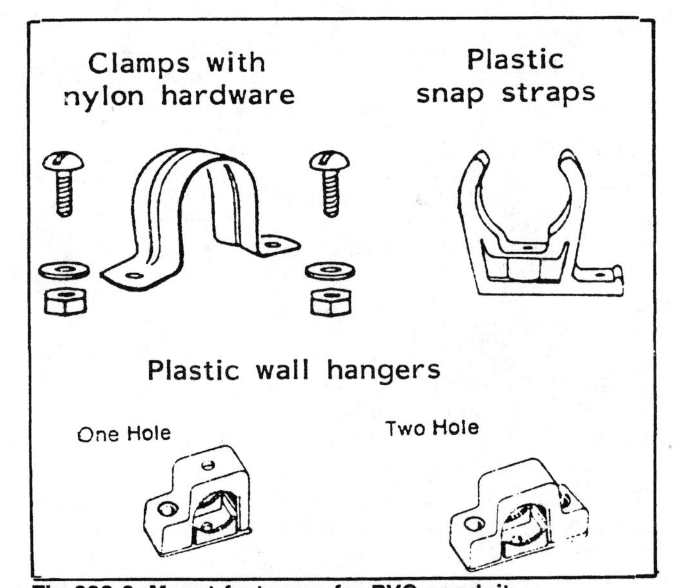

Fig 802-8. Mount fasteners for PVC conduit.

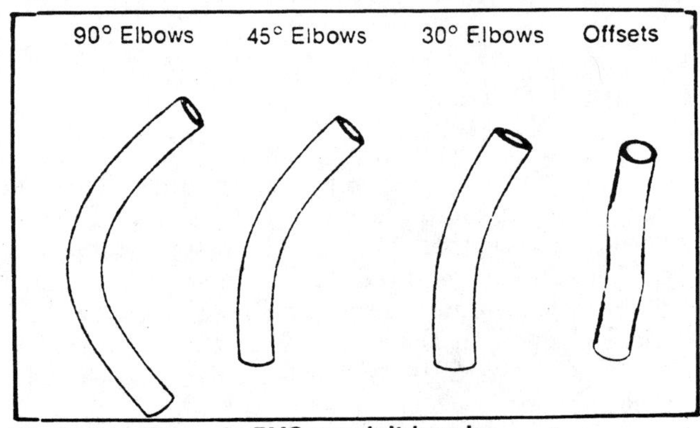

Fig 802-9. Premade PVC conduit bends.

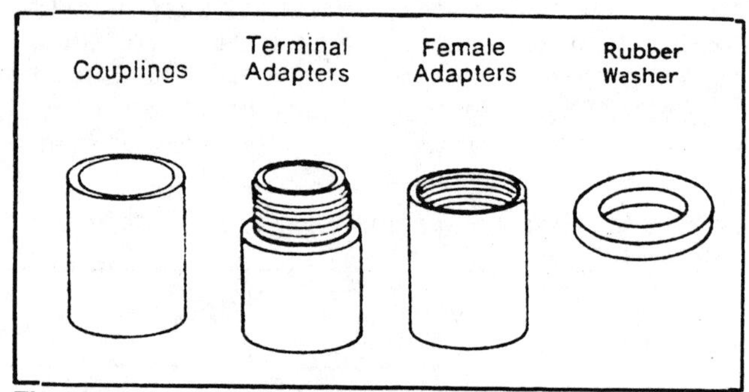

Fig 802-10. PVC couplings and threaded adaptors.

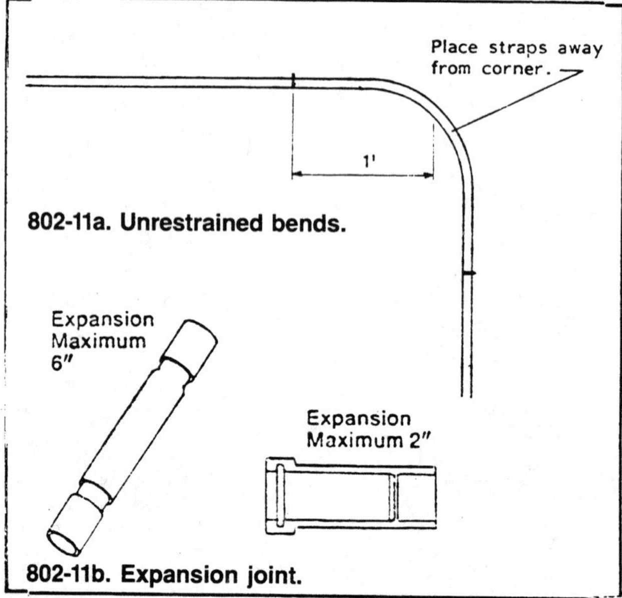

Fig 802-11. Allow for thermal expansion in PVC conduit.
Place straps about 1′ away from bends. If there are only a few bends, install 6″ expansion joints every 200′ or 2″ ones every 65′.

Dusty Buildings

Farm grain-feed centers are usually not considered to be **Dusty** buildings by local codes, so wire them like Damp buildings.

Commercial Dusty buildings that may have relatively high levels of explosive dust require "dust-ignition-proof" materials and wiring techniques as described in NEC 500-S (Hazardous Location, Class II, Division I). These structures are not covered in this book.

Planning

Before wiring a building, draw plans locating outlets, lights, boxes, switches, motors, and all other electrical equipment. Show the route, size, and junction of all conductors. This plan helps you prepare a list of materials required for the electrical system and helps you locate all the components properly. A typical plan is in the example at the end of this chapter. Fig 802-15 lists symbols for common electrical equipment.

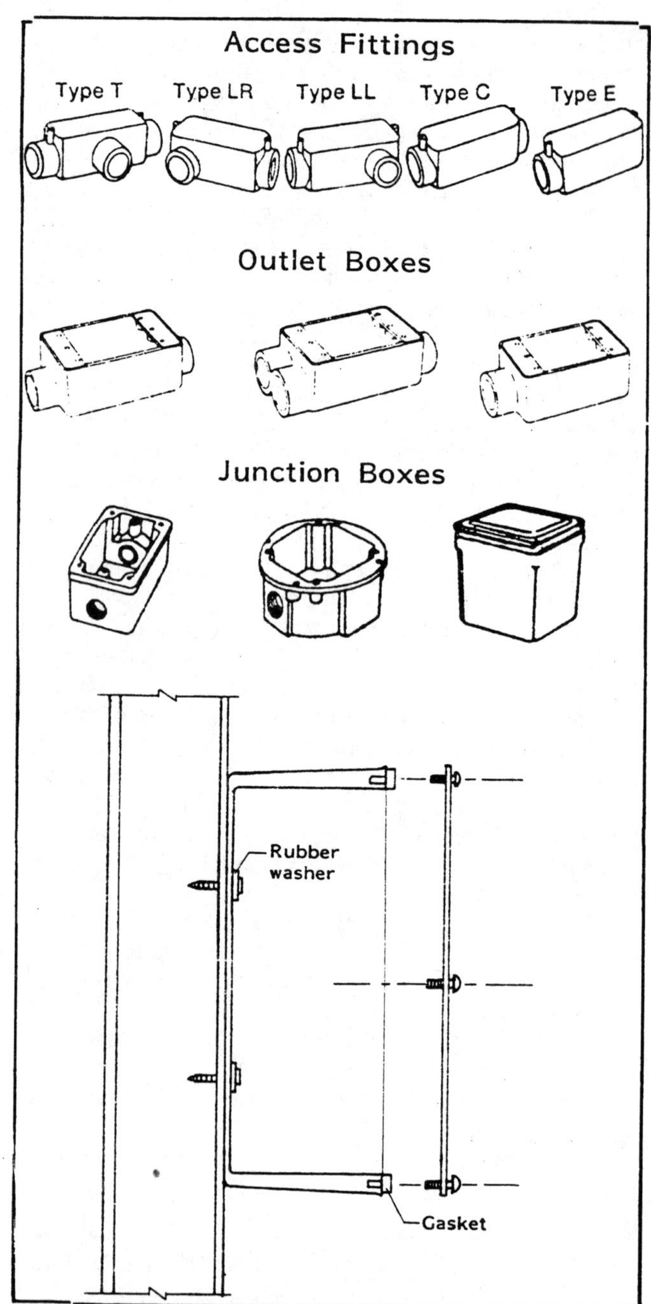

Fig 802-12. PVC boxes.

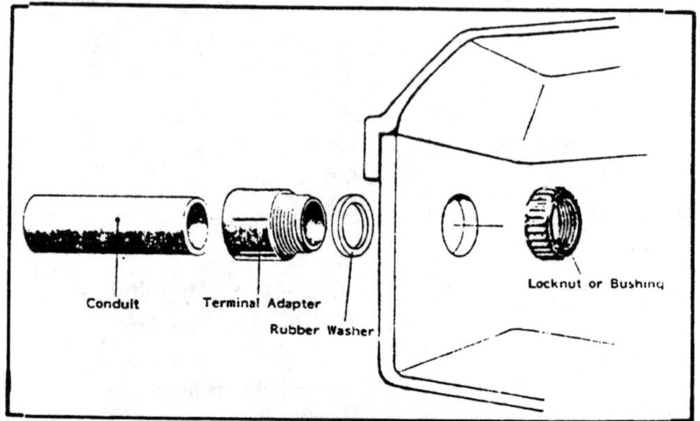

Fig 802-13. PVC conduit-to-box connection.
Place flat washer over the threads of the terminal adapter, securely against the shoulder. Insert the adapter threads through knock-out and fasten with a standard locknut or threaded bushing.

Fig 802-14. Moisture-tight flexible metal conduit.

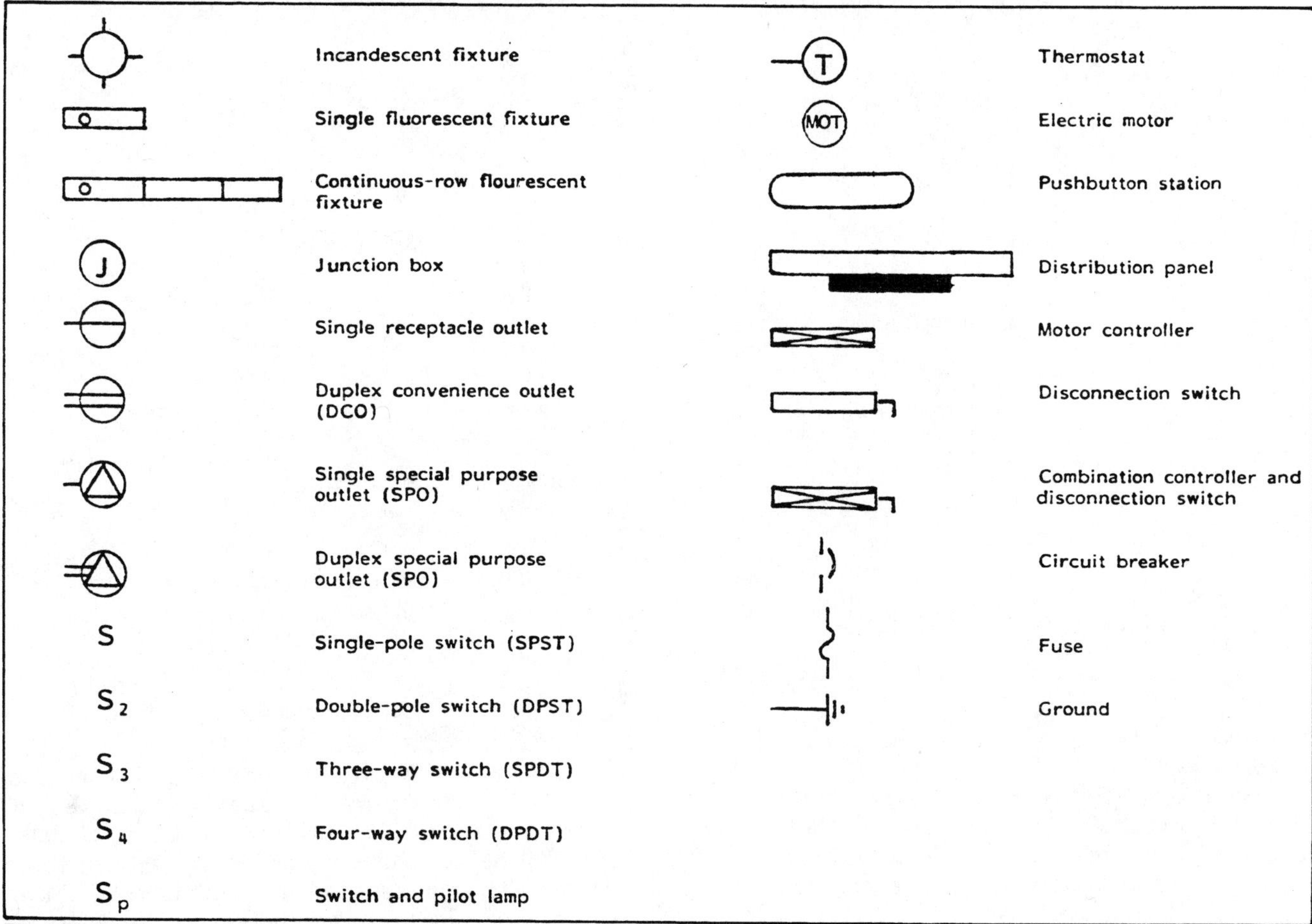

Fig 802-15. Electrical equipment symbols.

Lights

Light types

Three types of lights are common on farms:

- Incandescent
- Fluorescent
- High intensity discharge (HID)

Each type has individual properties of light output, maintenance, color, efficiency, and cost that affect selection for a particular task.

Incandescent

Consider incandescent fixtures when light is needed for short periods and when they are turned on and off frequently. Their initial cost is relatively low and they operate well in most conditions including low temperatures.

Bulbs up to 300 W have standard screw bases that fit ordinary medium base sockets. Porcelain sockets are recommended for incandescent lights over 100 W.

Incandescent light efficiency is relatively low, Table 802-4, so they are the most expensive type to operate. Note that their efficiency increases with wattage. Incandescent lights are short lived—only 750 hr for 100-150 W bulbs. Their light output decreases to 80%-90% of its initial value as the bulb approaches its rated life.

A 100 W bulb radiates 10% of the input energy as visible light and 72% in the infrared spectrum. This results in the low efficiency of incandescents. The high infrared output may also be detrimental in some applications.

Table 802-4. Characteristics of lights.
From *USDA Farmer's Bulletin Number 2243.* HG and HPS lamps operate longer than shown, but with greatly reduced output.

Lamp	Rated size	Size including ballast	Mean lumens	Lumens/W including ballast for fluorescents	Lifespan
Incandescent:	- - - - Watts - - - -		- - Number - -		Hours
Standard	15	15	110	7	2,500
	25	25	210	8	2,500
	40	40	410	10	1,500
	60	60	780	13	1,000
	100	100	1,580	16	750
	150	150	2,500	17	750
Fluorescent (cool or warm white):					
Tube length:					
18"	15	20	750	38	7,500
24"	20	25	1,100	44	9,000
48" (energy-saving)	34	39	2,700	69	20,000
48"	40	46	2,800	60	20,000
High intensity discharge (HID):					
Mercury (HG)	50	65	1,300	20	16,000
	100	120	3,600	30	16,000
	175	200	7,600	38	16,000
	250	280	11,000	39	16,000
	400	440	20,000	45	16,000
Metal halide (MH)	175	210	13,000	62	15,000
	250	290	18,500	64	15,000
	400	460	31,000	67	15,000
High-pressure sodium (HPS)	50	70	3,600	51	16,000
	70	90	5,220	58	16,000
	100	125	8,550	68	16,000
	150	180	14,400	80	16,000
	200	240	19,800	82	16,000
	250	295	27,000	91	16,000
	310	365	33,300	91	16,000
	400	470	45,000	96	16,000
Low-pressure sodium (LPS)	18	30	1,800	60	10,000
	35	55	4,800	87	18,000
	55	75	8,000	107	18,000
	90	120	13,500	112	18,000
	135	185	23,000	124	18,000
	180	230	33,000	143	18,000

Fluorescent

Fluorescent fixtures cost more than incandescents but produce 3 to 4 times more light/watt, Table 802-4. Turning these lights on and off frequently or leaving them on for only 2 to 3 hours at a time reduces lamp life. Such situations are best served by incandescents. For longer on periods, fluorescents are better because of their high efficiency.

Rated life ranges from 7,500 hr at 3 burning hr/start to 20,000 hr at 12 burning hr/start. Near the end of its life a typical fluorescent lamp emits only 60%-80% of its initial light output.

Fluorescent lights are used mainly indoors, because they are temperature sensitive. Standard indoor lights perform well down to 50 F; with special ballasts, down to -20 F. Humid air makes the fluorescent tubes difficult to start. Problems develop above 65% relative humidity and become severe at 100%.

There are four standard colors of fluorescent lamps:

- Standard cool white
- Deluxe cool white
- Standard warm white
- Deluxe warm white

Standard white gives the highest light output but is not desirable for color matching tasks. Deluxe whites are about 25% less efficient than the standard. The deluxe cool white produces light nearest to daylight. The deluxe warm white produces light that is closest in color to the incandescents. Fluorescents produce little infrared light.

High intensity discharge (HID)

HID lamps include mercury, metal halide, high pressure sodium, and low pressure sodium. They tend to have long lives and to be very energy efficient, Table 802-4. They operate well in cold temperatures. Light output is colored, i.e. mercury is greenish-blue and sodium is golden yellow.

HID lamps require 5 to 15 minutes to start and are not usable where lamps are turned on and off frequently. HID lamps are best where mounted at least 12' high and lamps are on for at least 3 hr. Common uses include feedlot and outdoor security lighting.

Light levels

Use enough lighting for inspection and work to be done efficiently. Provide at least two lighting circuits in each building. Consider two rows of lights on separate switches, so two light intensity levels are possible over pen areas.

Install enough switches so at least one light in each room is controlled from main access doors. Consider a separate switch for each row of lights and for lights over feeding stations for all-night feeding. Equip outdoor and haymow light switches with pilot lights to indicate when lights are on.

No light reflectors are necessary when ceiling and wall surfaces are white. Provide reflectors if surfaces are dark.

Table 802-5 shows lighting requirements for agricultural buildings. For example, a 24′ wide farrowing building could reach 15 foot-candles with (0.42 x 24) = 10.1 watts (W) of fluorescent light/ft of building length or (1.5 x 24) = 36 W of 150 W incandescent light/ft of building length. Table 802-6 gives light location guidelines.

Table 802-5. Light level for agricultural buildings.
Space lights uniformly throughout the work area. Some areas may need additional task lighting, e.g. the office.
These table values assume frequent cleaning of bulbs (once a month in animal buildings) and regular bulb replacement (at rated life). With infrequent cleaning and delayed replacement, increase table values 40%.
Dairy, poultry, and general illumination levels are from ASAE; others are comparable.

Design values based on ASAE *Farm Lighting Design Guide*, SP-0175. Assumed 8′ high ceiling and 70% ceiling and 50% wall reflectance. Incandescent: maintenance factor (MF) = 0.75 and coefficient of utilization (CU) = 0.69; fluorescent: MF = 0.70 and CU = 0.66.

Task	Level of illumination	Standard cool white Fluorescent 40 W	Standard Incandescent 100 W	Standard Incandescent 150 W
	foot-candles	- - - watt/ft² of building area - -		
Dairy				
Cow housing	7	0.20	0.80	0.70
Calf housing	10	0.28	1.15	1.00
General milking operation	20	0.55	2.29	2.00
Equipment washing area	100	2.75	11.44	10.00
Milk handling	20	0.55	2.29	2.00
Loading platform	15	0.42	1.72	1.50
Poultry				
Brooding and laying hens	20	0.55	2.29	2.00
Egg handling	50	1.38	5.72	5.00
Egg processing	70	1.93	8.00	7.00
Swine				
Farrowing	15	0.42	1.72	1.50
Nursery	10	0.28	1.15	1.00
Growing/finishing	5	0.14	0.58	0.50
Breeding/gestation	15	0.42	1.72	1.50
Animal inspection/handling	20	0.55	2.29	2.00
Sheep				
Lambing	15	0.42	1.72	1.50
Growing/finishing	5	0.14	0.58	0.50
Animal inspection/handling	20	0.55	2.29	2.00
Beef				
Housing	5	0.14	0.58	0.50
Along feed bunk	10	0.28	1.15	1.00
Animal inspection/handling	20	0.55	2.29	2.00
General				
Feed storage/processing	10	0.28	1.15	1.00
Office	50	1.38	5.72	5.00
Haymow	3	0.09	0.35	0.30
Machine storage	5	0.14	0.58	0.50
Machine repair area	30	0.83	3.43	3.00
Shop bench area	100	2.75	11.44	10.00
Rough bench work	50	1.38	5.72	5.00
Restrooms	30	0.83	3.43	3.00

Outlets

There are two types of receptacle outlets—duplex convenience outlets (DCO) and special purpose outlets (SPO). SPOs are for motors and equipment that are on individual branch circuits (see section on Branch Circuits). An SPO can have a receptacle for plug in equipment or the equipment can be hand-wired into the outlet box.

Install DCOs where portable equipment (heat lamps, power tools, etc.) is used often. Locate additional DCOs along alleys, near doors, etc. for convenient short-term use of portable equipment. Avoid long-term use of extension cords. Table 802-7 has guidelines for locating DCOs.

Install SPOs next to the fixed equipment that they serve. Fixed equipment stays in one place for long periods of time, e.g. feed handling equipment, ventilating fans, tank heaters, and gutter cleaners.

Refer to MWPS handbooks and plans, or the *Agricultural Wiring Handbook* (Food & Energy Council Inc., 909 University Ave., Columbia, MO 65201) for more guidelines on quantities and locations of lights and outlets.

Branch Circuits

Branch circuits carry current from the overcurrent protection device (fuse or circuit breaker) to the loads (lights, outlets, motors). They are either nominal 120 volt (V) or 240 V circuits. These circuits have wires for conducting current and for grounding. Grounding falls into two catagories:

- **System grounding** is the grounding of one of the current carrying conductors at the distribution panel. The grounded conductors for system grounding are called **circuit neutrals.**
- **Equipment grounding** is the grounding or bonding of noncurrent carrying equipment such as motor frames back to the distribution panel grounding bar. The equipment grounding wire serves only to ground the metal frames of electrical equipment. It does not carry current during normal operation, but will carry current in case of damage to or defect in electrical equipment or wiring. The conductors for equipment grounding are called **equipment grounds.**

The equipment ground must make a **continuous** connection from the ground busbar of the distribution panel to all receptacles, switches, metal fixtures, and motor frames. If the boxes are metal, they must also be grounded. Connect the equipment ground to the green hex head grounding screw of receptacles and switches. Use grounding type receptacles. Connect to the frame of metal fixtures or motors.

120 V circuits

Nominal 120 V circuits have one hot wire, one circuit neutral, and one equipment ground. The hot

Table 802-6. Guidelines for locating lights.

Dairy
(Contact your dairy inspector concerning regulations.)

Stanchion barn
Face out arrangements: Place incandescent lights along litter alley center line; one light directly behind every other stall divider. Or install a continous row of 2-lamp fluorescent lights along alley center line for even lighting of cow udders.
Face in arrangements: Place light fixtures about 1′ to rear of gutter and directly behind every other stall divider.
Install lights about 12′ o.c. along center line of feed alleys.

Free stall barn
Place a row of lights over center of each work/litter alley and over feed bunks. Add lights to aid specific tasks, brighten dark corners, and avoid long dark shadows.

Pens and stalls
Provide one light fixture per bull, maternity, or calf pen. Locate switch outside of pen. Provide one row of lights over each row of calf stalls.

Milking parlor
Provide one row of lights for the operator and one row for each row of stalls. Locate lights so operator does not work in any shadows. Install one light at each animal entrance.

Milk room
Provide one fluorescent light over the outer edge of each wash vat and an additional fluorescent light for each 100 ft^2 of floor area. Use fixtures with two 4′-40 W cool white fluorescent tubes. Locate fixtures away from bulk tank openings to keep broken glass out of the tank.

Poultry
In general, space lights 8′-12′ o.c. along building width and 10′-16′ o.c. along building length, with end fixtures 4′-5′ from walls. Refer to bulletins from your local agricultural college or power supplier for more specific information for your region and operation type.

Swine
Provide one row of lights over each row of pens or stalls and one row along feed alley center. Install at least one light over every other pen partition.

Sheep
Locate lights over pens and feeders to enhance inspection. Provide one light 12′-15′ o.c. down alley center line along lambing or feeding pens.

Beef
Provide a row of lights over the bunk. Install additional lights in pen or handling areas, especially for veterinary care. Place lights to aid chores, brighten dark corners, and eliminate long shadows.

Farm shop
Two 150 W reflector bulbs about 50″ above a bench provide excellent lighting. For continuous lighting, use fluorescents above benches. Position lamps above the front half of the bench. Provide lights over each concentrated work area, such as welding.

Table 802-7. Guidelines for locating DCOs.
DCO = duplex convenience outlet. Locate DCOs at least 4′ high—higher where they may be damaged by animals.

Dairy
(Contact your dairy inspector concerning regulations.)

Stanchion barn
Provide DCOs 20′ o.c. along litter alley. They can be on outside walls where cows face in. Install extra ones in feed alley for supplemental lighting.

Free stall barn
Provide DCOs wherever portable equipment (tools, circulation fans, spot heaters, etc.) may be used. Also provide one on inside wall near each major entrance.

Pens and stalls
Provide one DCO for each pen for equipment like heat lamps or groomers. In stall areas, install at least one on each wall and no more than 20′ apart.

Milking parlor
Provide one DCO at each end of operator's work area. Add SPOs for heating equipment.

Milk room
Provide a DCO for each work station. DCOs for radio, clock, and desk lamp may be desirable for recordkeeping room. A receptacle SPO on a manual switch may be required for electric pump on milk truck; check local regulations for location, type, and rating. Place outlets high to avoid splashed water.

Poultry
Provide DCOs for tools, water warming, ultraviolet lamps, etc. Install one per 400 ft^2 of floor area with at least one per pen.

Swine
Provide at least one DCO per two farrowing stalls or two pens. Locate over heat lamp locations. Provide one DCO 15′-20′ o.c. along alleys.

Sheep
Provide one DCO per two stalls or pens. Provide one DCO 15′-20′ o.c. along alleys. Provide extras near water troughs (for water warming), near shearing stations, and at each major building entrance.

Beef
Provide one DCO wherever portable equipment can be conveniently used. Also install one near each major entrance to the building.

Machine sheds
Provide one DCO per 40′ of wall perimeter.

Farm shops
Provide one DCO every 4′ along work bench. Provide at least one extra DCO on each shop wall. Locate an SPO at each permanent location of a motor-driven device.

wire is usually black or red, the circuit neutral is white, and the equipment ground is bare or green.

Do **NOT** connect the equipment ground to the circuit neutral. NEC 250-61. Even though these two wires are often connected to the same busbar in the distribution panel, a connection earlier in the branch circuit can lead to current in the equipment ground. See the section on Stray Voltage.

The circuit neutral and equipment ground must run without interruption to all 120 V equipment.

- Never protect circuit neutrals or equipment grounds with fuses or circuit breakers.
- Do not install switches on circuit neutrals or equipment grounds—switches can be installed on hot wires only, Fig 802-16.

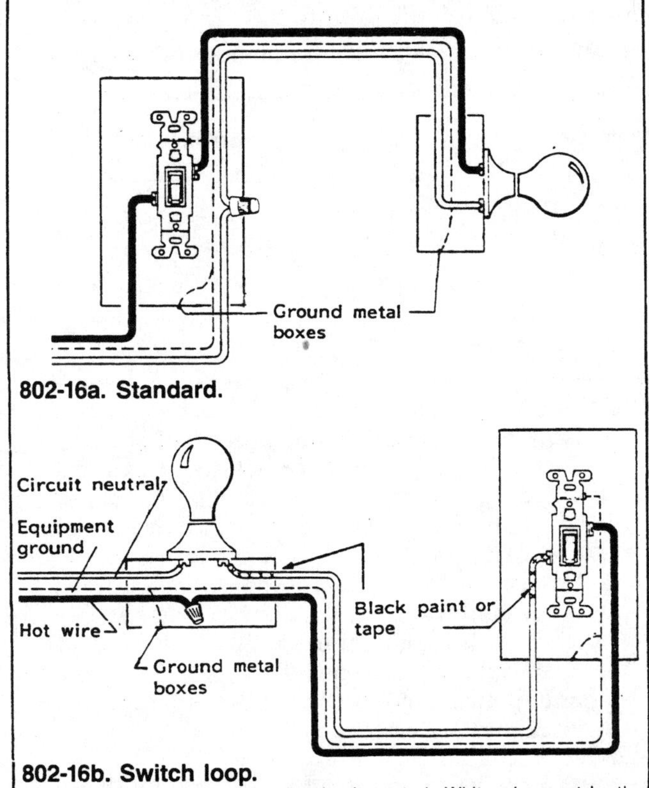

802-16a. Standard.

802-16b. Switch loop.
Note: The switch box contains no circuit neutral. White wire must be the return side. Paint or tape the ends of the white wire black to indicate that it is a hot wire.

Fig 802-16. Light with one switch.
In 120 V circuits, white wires may not be hot wires except on switch loops. Note that switch is installed in the hot wire—not the circuit neutral. Connect the equipment ground to all metal boxes, switches, and equipment.

240 V circuits

Nominal 240 V circuits have two hot wires and an equipment ground. Circuit neutrals are not required by most 240 V equipment. If an appliance has both 120 V and 240 V circuits (e.g. ranges, clothes dryers), both the circuit neutral and equipment ground are required.

Three wires run from the power supplier transformer to the building. One of the wires is grounded at the transformer and at the service entrance of the building. It is called the neutral (wire N of Figs 802-17a and 802-17b). The voltage between lines A and B of Figs 802-17a and 802-17b is 240 V, while the voltage between A (or B) and N is 120 V.

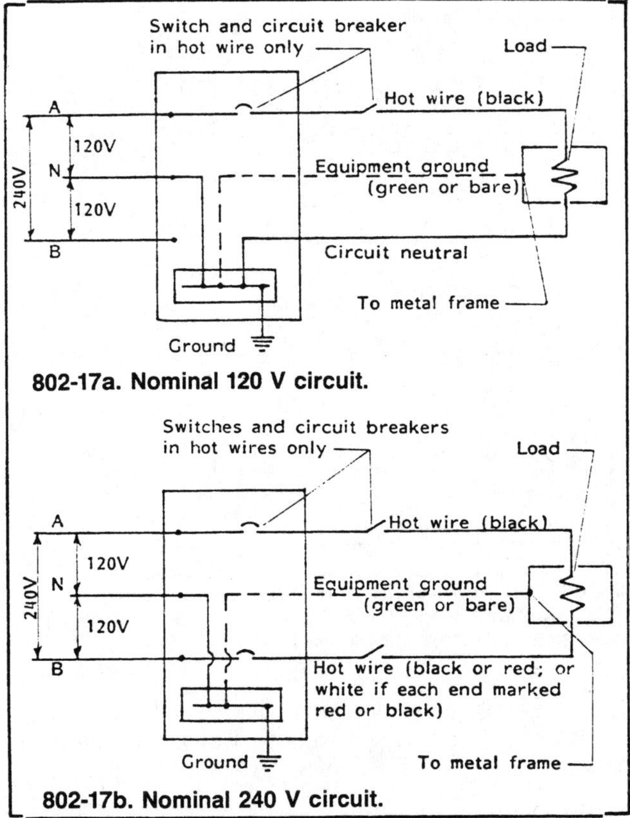

802-17a. Nominal 120 V circuit.

802-17b. Nominal 240 V circuit.

Fig 802-17. Circuit wiring schematics.
With a 240 V power supply, 120 V is also available.

Circuit types

Two types of branch circuits are discussed here:

- **General purpose,** which includes lights and duplex convenience outlets (DCO). The DCO can be used for portable appliances under 1500 W and portable motors ⅓ hp or less.
- **Equipment,** which includes fans, heaters, motors, and special purpose outlets (SPO). Use equipment branch circuits for fixed equipment and for appliances over 1500 W and motors over ⅓ hp.

A general purpose circuit has 120 V service with #12 AWG-CU (copper) wire. For farm buildings, it is recommended that no more than 10 DCOs and/or light fixtures be on one 20 A general purpose branch circuit, Fig 802-18.

Individual branch circuits are recommended for equipment such as fans, heaters, and augers. Use receptacles in the SPO for equipment with plugs—otherwise hard-wire the equipment into the SPO.

Wire only one or at most two fans per circuit. If possible, install at least two fan circuits in each room of environmentally-controlled animal buildings, so if one circuit fails a fan on the other circuit can ventilate the room.

Equipment branch circuits often require 240 V. They may also require higher capacity fuses or circuit breakers and larger wires.

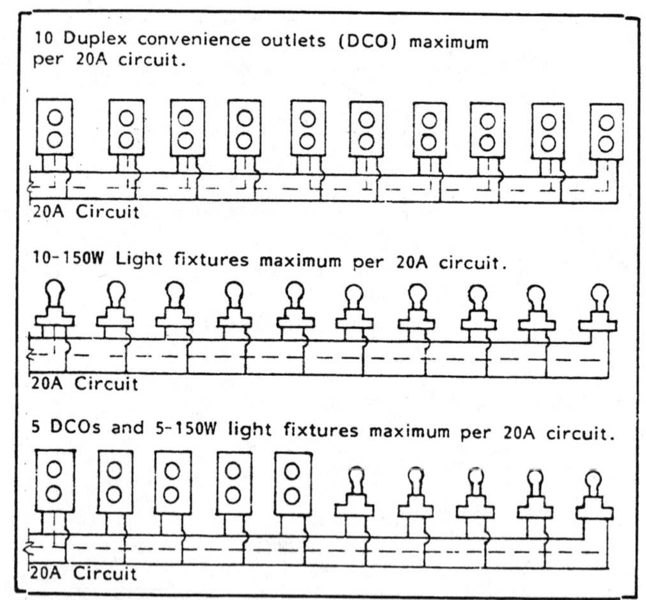

Fig 802-18. Maximum lights and DCOs per circuit.
DCO = duplex convenience outlets. For 20 A, 120 V circuits. Maximum amperage load per DCO and/or light fixture is 1.5 A (about 170 W on a 120 V circuit).

Motor circuits require special devices and design methods and are not discussed here. Contact an electrician to install motor circuits.

Branch Circuit Conductors

Conductors are wire or cable for conducting electricity. Wire usually refers to single strand conductors, whereas cable refers to multistrand conductors in the same insulating sheathing. Copper is the most common material, but aluminum is becoming increasingly popular.

Aluminum wire requires special wiring devices and techniques and is not discussed in this book.

Wires carrying electric current are like pipes carrying water—the larger the pipe (wire), the more water (current) it can carry. The more current required by the end use, the larger the wire must be. Wire is sized by the American Wire Gauge (AWG) system; the larger the wire the smaller the gauge number, Fig 802-19.

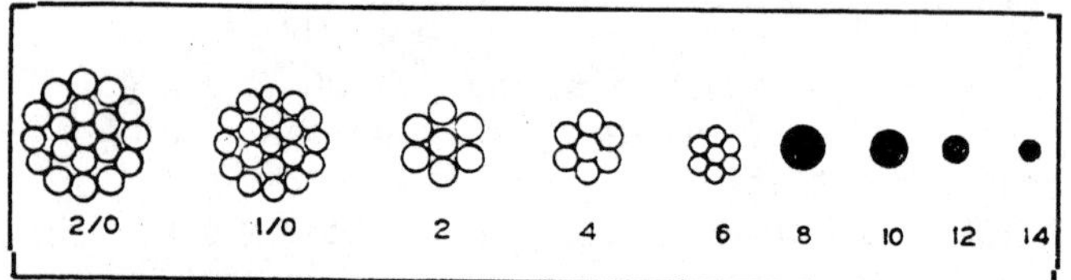

Fig 802-19. Wire sizes.
Diameters of typical wires without insulation. Note the larger the gauge number, the smaller the diameter.

Although the NEC allows #14 AWG-CU (copper) wire and 15 A circuits for branch circuits, #12 AWG-CU is highly recommended for farm buildings. Most **general purpose** branch circuits can be #12 AWG-CU, if the number of DCOs and/or lights does not exceed 10 and the maximum load of each DCO and/or light does not exceed 1.5 A.

However, check each **equipment** branch circuit conductor size for the following criteria. Wire size is based on two factors: ampacity and voltage drop. Ampacity is the safe current carrying capacity of a wire in amperes (A). Current flowing through a wire creates heat. If the amperage is too high, the wire may become hot enough to damage the insulation and start a fire.

Heat lost from the wire not only wastes energy but drops voltage. For example, with 120 V at the distribution panel, the voltage at a motor 100′ away drawing 14 A is about 115 V if wired with #12 AWG-CU wire. Voltage drop results in power or light loss and can result in inefficient appliance operation, Table 802-8. The larger the conductor, the less the voltage drop. Select branch circuit conductors large enough to limit voltage drop to 2%.

Table 802-8. Effect of voltage drop on power and light loss.

Voltage drop %	Light loss %	Power loss %
1	3	1.5
2	7	3.0
3	10	4.5
4	13	6.0
5	16	7.5
10	31	15.0
15	46	22.5

To size a conductor:

1. Determine the maximum amperage the conductor has to carry. Increase the maximum amperage by 25% if the branch circuit may carry sustained load for longer than 15 minutes. This includes all motor circuits.
2. Determine the conductor length (one way). It is the distance from the distribution panel to load via the conductor's route.
3. From Tables 802-9 and 802-10, select the minimum wire size to handle the amperage at a voltage drop less than 2%.

To determine amperage:

- **Voltage** is the pressure in the circuits (similar to water pressuse in your plumbing system) and is measured in volts (V). Common nominal voltages in buildings are 120 V or 240 V.
- **Amperage** is the current flowing through conductors and is measured in amps (A).
- **Wattage** is the power (work done over a period of time) and is measured in watts (W), or kilowatts (kW). 1 kW = 1,000 W.

For incandescent lights and electric resistance heaters, calculate amperage from:

Amps = Watts ÷ Volts

For example, a 600 W heater on a 120 V circuit uses (600 W ÷ 115 V) = 5.2 A of current. **To compute amperage, use the value of 115 V for nominal 110-120 V service, and 230 V for nominal 220-240 V service.**

Compute the amperage in each branch circuit as follows:

Table 802-9. Minimum copper conductor size for branch circuits.
For type TW and RUW wire or UF and NMC cable. Conductor length is one way. Use for equipment circuits and for general purpose circuits which do not meet the criteria in Fig 802-18. Based on conductor ampacity or 2% voltage drop, whichever is limiting.

Nominal 120 V service

Load A	30′	40′	50′	60′	75′	100′	125′	150′	175′	200′
	Conductor length									
	American Wire Gauge									
5	12	12	12	12	12	12	12	10	10	10
7	12	12	12	12	12	12	10	10	8	8
10	12	12	12	12	10	10	8	8	8	6
15	12	12	10	10	10	8	6	6	6	4
20	12	10	10	8	8	6	6	4	4	4
25	10	10	8	8	6	6	4	4	4	3
30	10	8	8	8	6	4	4	4	3	2
35	8	8	8	6	6	4	4	3	2	2
40	8	8	6	6	4	4	3	2	2	1
45	6	6	6	6	4	4	3	2	1	1
50	6	6	6	4	4	3	2	1	1	1/0
60	4	4	4	4	4	2	1	1	1/0	2/0
70	4	4	4	4	3	2	1	1/0	2/0	2/0
80	2	2	2	2	2	1	1/0	2/0	2/0	3/0
90	2	2	2	2	2	1	1/0	2/0	3/0	3/0
100	1	1	1	1	1	1/0	2/0	3/0	3/0	4/0

Nominal 240 V service

Load A	50′	60′	75′	100′	125′	150′	175′	200′	225′	250′
	Conductor length									
	American Wire Gauge									
5	12	12	12	12	12	12	12	12	12	12
7	12	12	12	12	12	12	12	12	10	10
10	12	12	12	12	12	10	10	10	10	8
15	12	12	12	10	10	10	8	8	8	6
20	12	12	10	10	8	8	8	6	6	6
25	10	10	10	8	8	6	6	6	6	4
30	10	10	10	8	6	6	6	4	4	4
35	8	8	8	8	6	6	4	4	4	4
40	8	8	8	6	6	4	4	4	4	3
45	6	6	6	6	6	4	4	4	3	3
50	6	6	6	6	4	4	4	3	3	2
60	4	4	4	4	4	4	3	2	2	1
70	4	4	4	4	4	3	2	2	1	1
80	2	2	2	2	2	2	2	1	1	1/0
90	2	2	2	2	2	2	1	1	1/0	1/0
100	1	1	1	1	1	1	1	1/0	1/0	2/0

- **Lights:** allow 1.5 A per light fixture for 120 V service. For heat or flood lamps larger than 170 W, use the actual figure. For example, a 250 W heat lamp on 120 V service, consumes: 250 W ÷ 115 V = 2.2 A.
- **Outlets:** allow 1.5 A/DCO on 120 V service, if not used for motors. Use the actual value of the load for SPOs that supply motors or other large power users.
- **Motors:** for a branch circuit with only one motor, size the wire for 125% of the motor full load current rating. For example, for one 6 A motor on a circuit, 6 x 1.25 = 7.5 A. For more than one motor on a circuit, rate the largest motor at 125% and the other motors at 100% of their full load current rating. If other loads, such as lights or heaters, are on this circuit, add 100% of their full load current to the above calculations. See Table 802-11 for full load motor amperages.

Circuit breakers

Most distribution panels (DP) contain one large 2-pole main breaker that protects the entire installation and disconnects it from the power source. However, there must also be breakers to protect the individual branch circuits—single pole breakers for 120 V and 2-pole for 240 V. The branch circuit breakers are purchased separately from the DP and plugged into the DP as needed. Only circuit breakers are discussed because of their many advantages over fuses.

Table 802-10. Minimum copper conductor size for branch circuits.
For type THW, THWN, RHW, and XHHW wire. Conductor length is one way. Use for equipment circuits and for general purpose circuits which do not meet the criteria in Fig 802-18. Based on conductor ampacity or 2% voltage drop, whichever is limiting.

Nominal 120 V service

Load A	30′	40′	50′	60′	75′	100′	125′	150′	175′	200′
	Conductor length									
	American Wire Gauge									
5	12	12	12	12	12	12	12	10	10	10
7	12	12	12	12	12	12	10	10	8	8
10	12	12	12	12	10	10	8	8	8	6
15	12	12	10	10	10	8	6	6	6	4
20	12	10	10	8	8	6	6	4	4	4
25	10	10	8	8	6	6	4	4	4	3
30	10	8	8	8	6	4	4	4	3	2
35	8	8	8	6	6	4	4	3	2	2
40	8	8	6	6	4	4	3	2	2	1
45	8	8	6	6	4	4	3	2	1	1
50	6	6	6	4	4	3	2	1	1	1/0
60	6	6	4	4	4	2	1	1	1/0	2/0
70	4	4	4	4	3	2	1	1/0	2/0	2/0
80	4	4	4	3	2	1	1/0	2/0	2/0	3/0
90	3	3	3	3	2	1	1/0	2/0	3/0	3/0
100	3	3	3	2	1	1/0	2/0	3/0	3/0	4/0

Nominal 240 V service

Load A	50′	60′	75′	100′	125′	150′	175′	200′	225′	250′
	Conductor length									
	American Wire Gauge									
5	12	12	12	12	12	12	12	12	12	12
7	12	12	12	12	12	12	12	12	10	10
10	12	12	12	12	12	10	10	10	10	8
15	12	12	12	10	10	10	8	8	8	6
20	12	12	10	10	8	8	8	6	6	6
25	10	10	10	8	8	6	6	6	6	4
30	10	10	10	8	6	6	6	4	4	4
35	8	8	8	8	6	6	4	4	4	4
40	8	8	8	6	6	4	4	4	4	3
45	8	8	8	6	6	4	4	4	3	3
50	6	6	6	6	4	4	4	3	3	2
60	6	6	6	4	4	4	3	2	2	1
70	4	4	4	4	4	3	2	2	1	1
80	4	4	4	4	3	2	2	1	1	1/0
90	3	3	3	3	3	2	1	1	1/0	1/0
100	3	3	3	3	2	1	1	1/0	1/0	2/0

Table 802-11. Full load motor currents.
From NEC Table 430-148 for single phase AC motors.
The following values are for motors running at usual speeds with normal torque characteristics. For low speed, high torque, and multi-speed motors, use the nameplate current ratings.

hp	115 V	230 V
	amperes	
1/6	4.4	2.2
1/4	5.8	2.9
1/3	7.2	3.6
1/2	9.8	4.9
3/4		6.9
1		8
1½		10
2		12
3		17
5		28
7½		40
10		50

Table 802-12. Ampacity of copper wire.
Column A = TW and RUW wire and UF, NM, and NMC cable.
Column B = THW, THWN, RHW, and XHHW wire.
#14 wire is only used for some equipment—not for branch circuits.

Wire size AWG-CU	Ampacity amperes A	Ampacity amperes B
14	15	15
12	20	20
10	30	30
8	40	45
6	55	65
4	70	85
3	80	100
2	95	115
1	110	130
1/0	125	150
2/0	145	175
3/0	165	200
4/0	195	230

Sizing circuit breakers

Circuit breakers are rated in amperes. Except for motor circuits, the circuit breaker must have a rating in amperes **not** greater than the allowable current carrying capacity (ampacity) of the wires it protects, Table 802-12. For example, #12 AWG-CU wire has an ampacity of 20 A, so use a circuit breaker with an amperage rating of 20 A or less.

Installing circuit breakers

For 120 V circuits, attach the black hot wire from the branch circuit cable to the terminal on the circuit breaker and attach the circuit neutral and the equipment ground to the grounding busbar. For 240 V circuits, attach one of the hot wires to one of the circuit breaker terminals and attach the other hot wire to the other terminal. Attach the equipment ground to the grounding busbar, Fig 802-20.

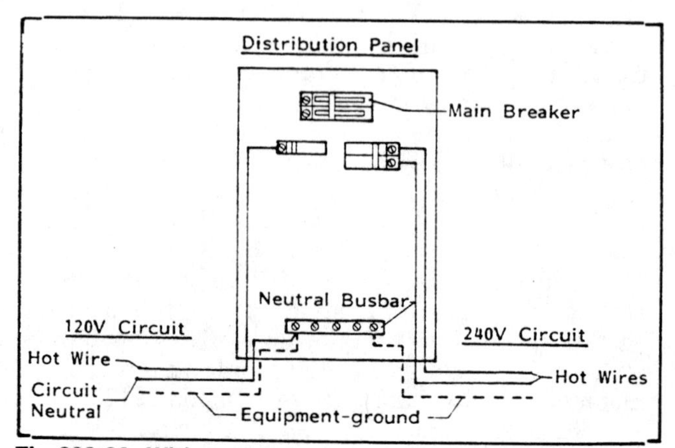

Fig 802-20. Wiring of branch circuit breakers in DP.

Ground fault circuit interruptors

There are special circuit breakers with ground fault interruption. These devices (referred to as GFCIs or GFIs) protect humans and animals from dangerous shock by cutting the circuit almost immediately upon detecting a ground fault (current is escaping somewhere in the circuit). A ground fault usually results from a breakdown in the insulation or wires of an electrical device. The GFCI monitors the current going out and the current returning. If these current levels are not equal, then a ground fault is detected. Fig 802-21.

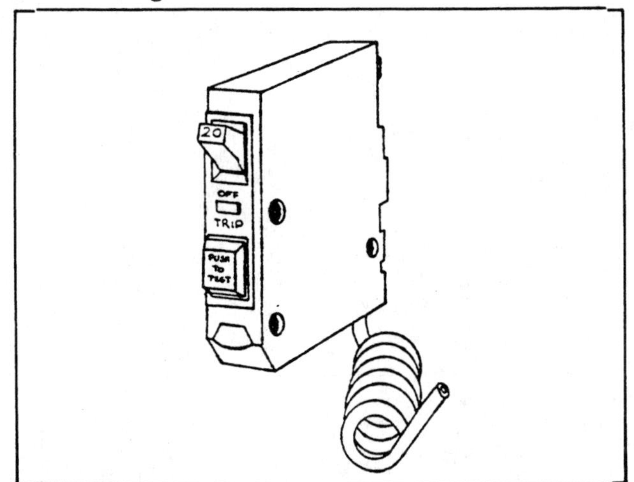

Fig 802-21. GFCI circuit breaker.
For 120 V circuits, both the circuit neutral and hot wire connect to the GFCI. The GFCI has a white lead to connect to the grounding busbar.

GFCIs break the circuit much quicker and at lower currents than do circuit breakers. In many cases, a dangerous or even fatal shock can occur before the circuit breaker opens. For farm buildings it is wise to use GFCIs for toilet facilities and in shop/office circuits, although they can be used anywhere.

Current "leaks" from all circuits and may be detected by the GFCI. Such "nuisance tripping" is most common when extension cords are used.

Distribution Panel (DP)

If possible, locate the DP outside of humid, dusty, and corrosive environments. If the building has an entry, office, or utility room, locate the DP there. If the DP must be placed in the same room as the animals, use a weathertight or raintight plastic cabinet.

Surface mount the DP with at least a 1" gap between the mounting panel and the wall.

Make all branch circuit cables enter and leave from the knockout holes at or near the bottom of the DP. This arrangement tends to keep water from running down the cable into the DP, Fig 802-22.

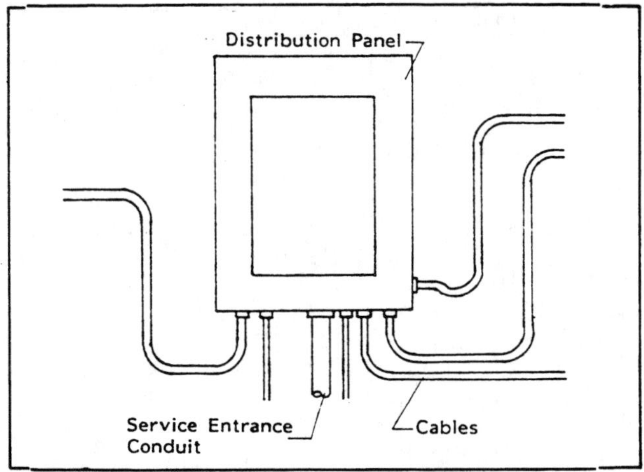

Fig 802-22. All cable and conduit enters DP at or near the bottom.

Stray Voltage

Stray voltage is a small voltage difference that exists between two surfaces (stanchion, waterer, floor, etc.) which an animal can contact. When an animal contacts both surfaces, current flows through its body, Fig 802-23. Voltage differences as low as 0.5 V can affect some animals. Stray voltage may stress the animal or cause it to avoid feeders and waterers, which reduces production.

Causes

Stray voltages usually are caused by problems in the grounded neutral network. As shown in Fig 802-24, the grounded neutral network consists of:

- The primary neutral from the power supplier.
- The secondary neutral to the building distribution panel.
- The branch circuit neutrals.
- The equipment ground.
- The ground wires and rods.
- The grounding of all metal equipment (stanchions, waterers, etc.) in the building.

Excessive stray voltages can be created by:

- Undersized conductors, poor or corroded connections, or improper wiring on the neutral and equipment grounding system.
- Unbalanced 120 V loads in the distribution panel.
- Voltage on a nonfaulty neutral grounding network due to the resistances and current flows inherent in the systems.
- Electrical conductors (e.g. cow trainers) which run near and parallel to an ungrounded metal pipe can induce a voltage in that pipe.

Solutions

The solution can be simple if the problem source or sources are clearly diagnosed and the alternatives evaluated. Good diagnosis requires a thorough understanding of electricity and farmstead wiring fundamentals. Often there is more than one cause, which further complicates the diagnosis. If you feel that you may have a problem, contact your electrician, power supplier, and state extension agricultural engineer to diagnose the problem. It is to everyone's benefit if the farmer, electrician, power supplier, equipment supplier, and others involved work together to attack this problem.

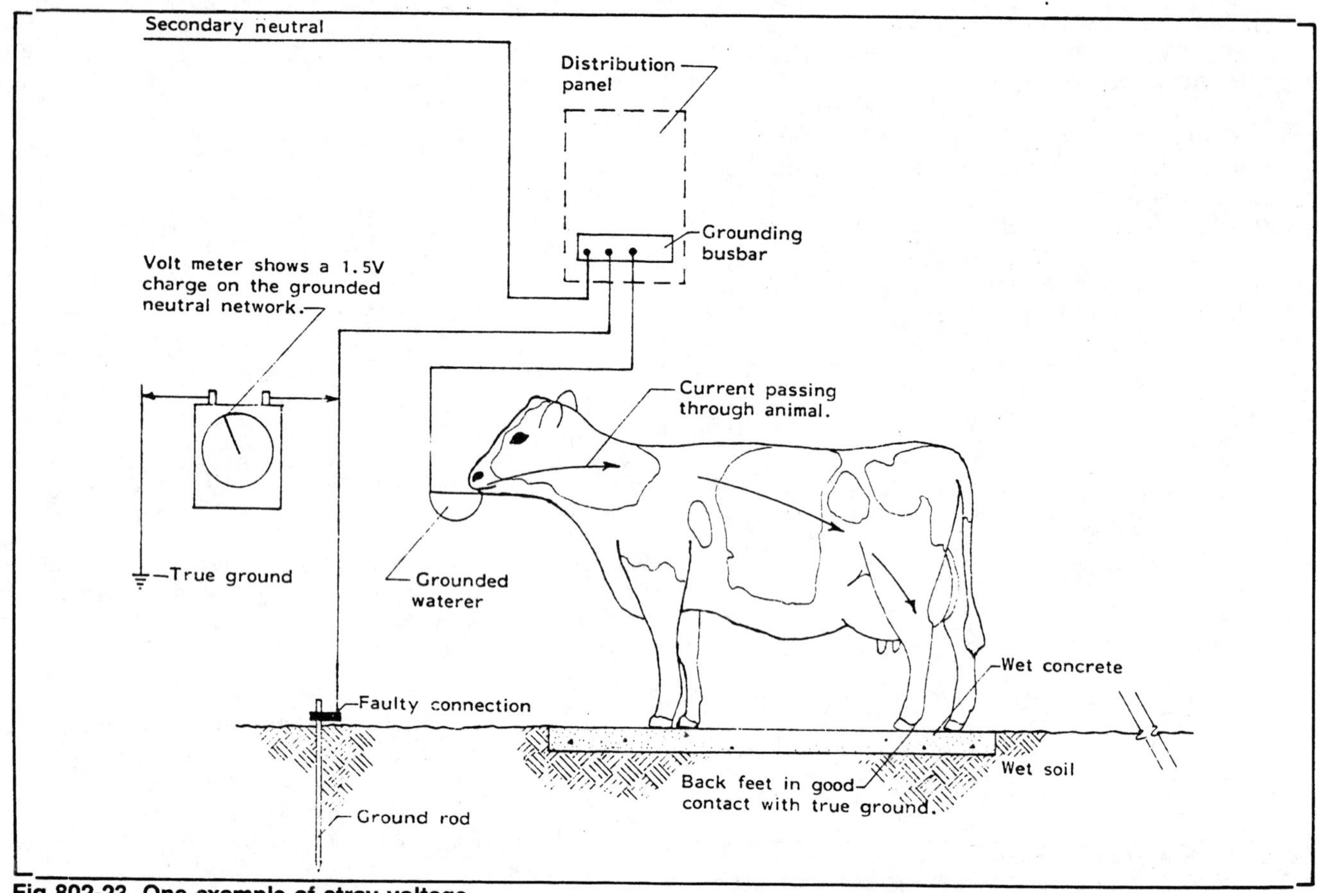

Fig 802-23. One example of stray voltage.
In this case, the grounded neutral network is at 1.5 V relative to true ground. This 1.5 V may be created by one or more sources such as the faulty ground connection shown here. The 1.5 V is accessible to the cow through the waterer since it is part of the grounded neutral network. When the animal touches the waterer and the wet concrete, she provides a path for current flow.

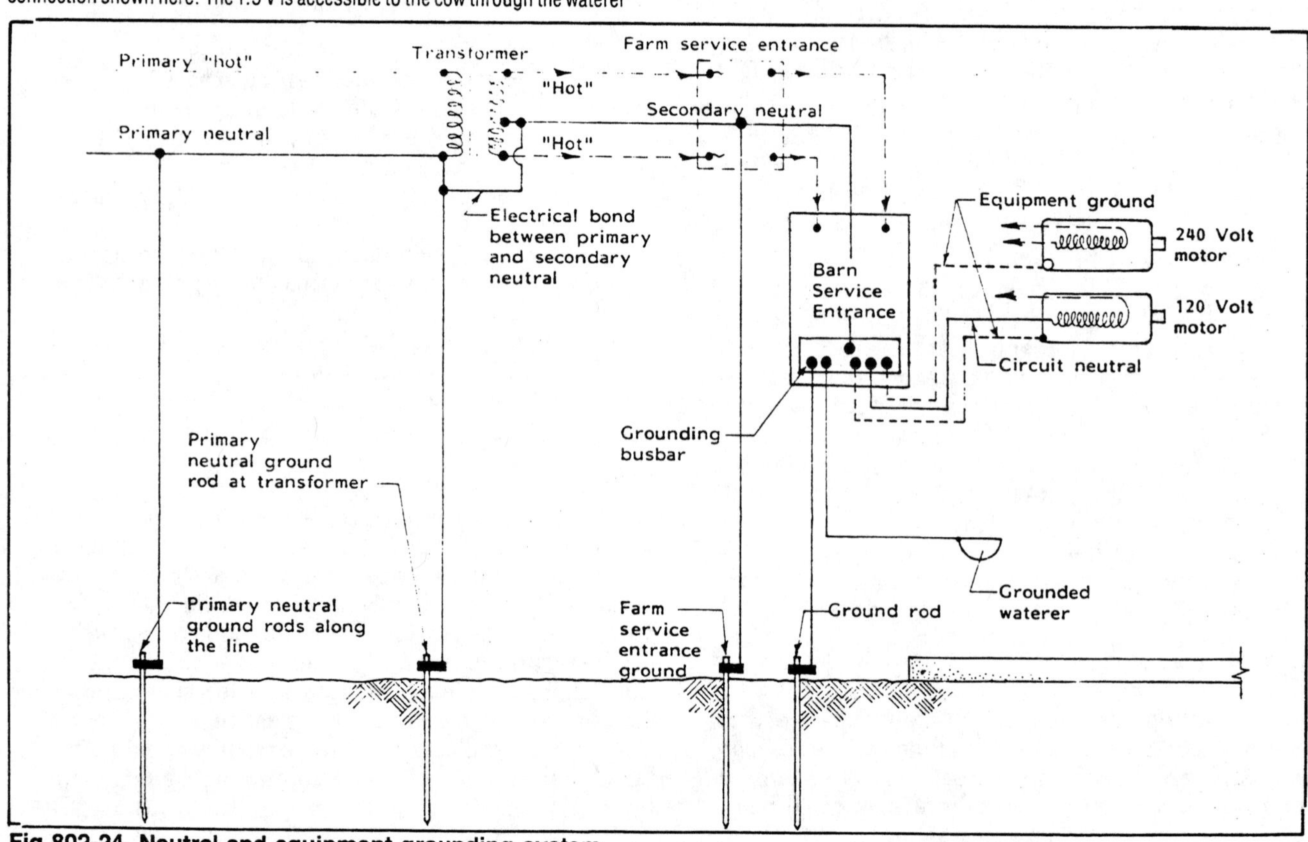

Fig 802-24. Neutral and equipment grounding system.
Single phase power.

There are three basic solutions to stray voltage problems:

1. Eliminate or reduce the voltage causing the problem.
2. Isolate the voltage from any equipment in the vicinity of the animals.
3. Install an equipotential plane to keep all possible animal contact points at the same voltage.

Eliminate or Reduce the Stray Voltage

If diagnosis indicates that current on the primary neutral system is a major contributor (due to either on-farm or off-farm loads), ask the power supplier to check the distribution system.

If the diagnosis indicates that a voltage drop on the secondary neutral system is a major contributor:

- Make sure all connections of the circuit neutral and equipment grounds are uncorroded and tight.
- Make sure all the circuit neutrals and equipment grounds are large enough in diameter and not excessively long.
- Make the 120 V loads from the distribution panel as balanced as possible.
- Convert 120 V motors to 240 V, especially those above ⅓ hp.

If the diagnosis indicates major contributions from the following, correct the wiring problem:

- Currents on the equipment ground.
- Improper use of the neutral as an equipment grounding conductor (the two wires should be separate in all branch circuits.)
- Improper interconnection of neutral and grounding conductors (the two wires should be separate in all branch circuits.)

To avoid these problems, install the secondary wiring system strictly by the requirements in the NEC, as outlined in this book.

Isolate Voltage

If the diagnosis shows a major contribution from the primary neutral, it is possible to isolate this voltage from electrically grounded equipment near livestock. One option is to disconnect the bond between the primary and secondary neutrals at the distribution transformer that is shown in Fig 802-24. Although the NEC may be interpreted as allowing this, many power suppliers prohibit it because of safety considerations.

Another option is an insulating transformer (240 V to 240/120 V) between the distribution transformer and the service entrance of the livestock facility. The "isolation" transformer can be at either the main farm or barn service entrance, Fig 802-25. If the isolation transformer is at the barn service entrance, it will also reduce secondary neutral contributions from imbalance currents.

To completely isolate the voltage, remove any above or below ground conductor that may bypass the isolation. Some common interconnections are telephone ground wires, metal water and propane lines,

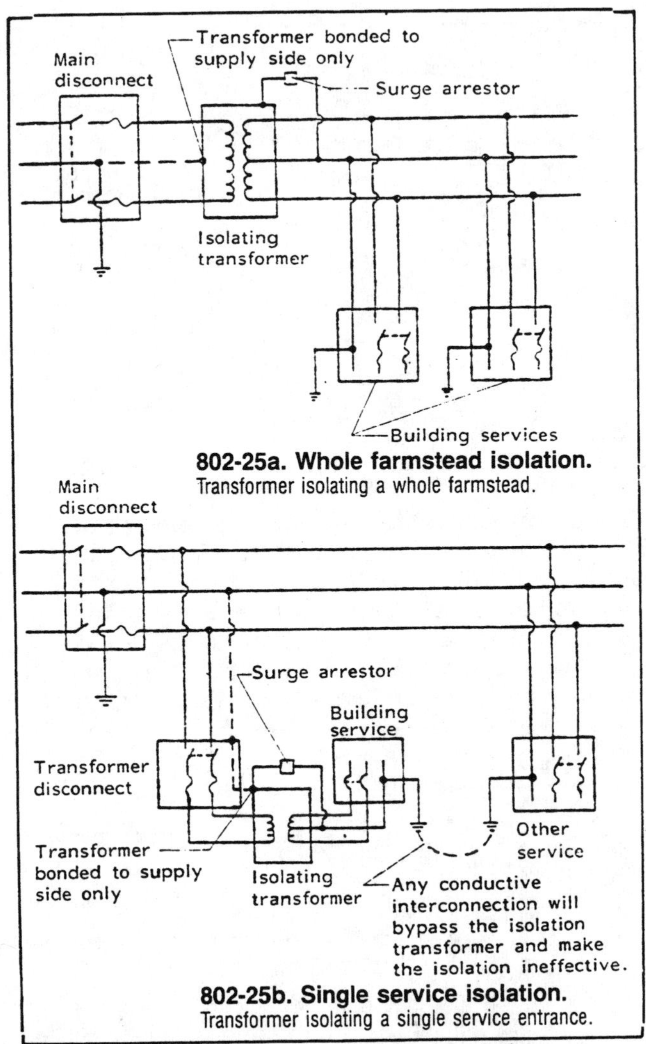

802-25a. Whole farmstead isolation.
Transformer isolating a whole farmstead.

802-25b. Single service isolation.
Transformer isolating a single service entrance.

Fig 802-25. Transformers for electrical isolation.

metal buildings, and feeding equipment between buildings. **Any** conductor reduces the effectiveness of isolation—test to verify the absence of metal interconnections.

Do not try to solve your stray voltage problem by isolating metal equipment (pipes, stanchions, etc.) from the electrical grounding system. **This creates a potential electrical hazard.** Any electrical fault to ungrounded equipment may create a lethal condition. Electrically ground all conductive equipment to the service entrance with an equipment ground, particularly if there is electrical equipment in the area.

Install Equipotential Planes

Equipotential planes eliminate stray voltage problems regardless of the source, if they are successful in maintaining the same voltage at all possible animal contact points. Consider them for all areas where electrically grounded equipment is near livestock.

Equipotential planes or grounding mats are an extension of good electrical wiring and grounding practice. The principal is that if all possible animal contact points are at the same voltage, there can be no current flow through an animal's body. Install a continuous metal grounding mat in the floor, bond it to

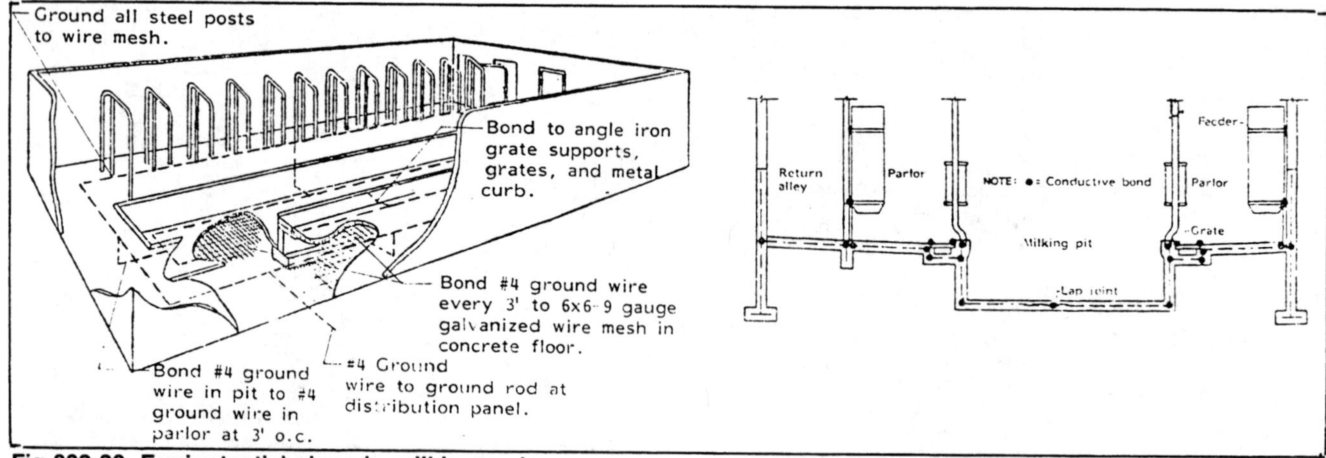

Fig 802-26. Equipotential plane in milking parlor.

all metal equipment in the area, and electrically ground the complete system at the distribution panel, Fig 802-26. Properly installed equipotential planes can be very effective in solving stray voltage problems in milking parlors. Animal access to equipotential planes should be through some type of voltage ramp installed in the access areas. Voltage ramps give a gradual voltage rise to the voltage of the equipotential plane.

Electrical Plan Example

This example shows an electrical system for a swine farrowing building (Damp building).

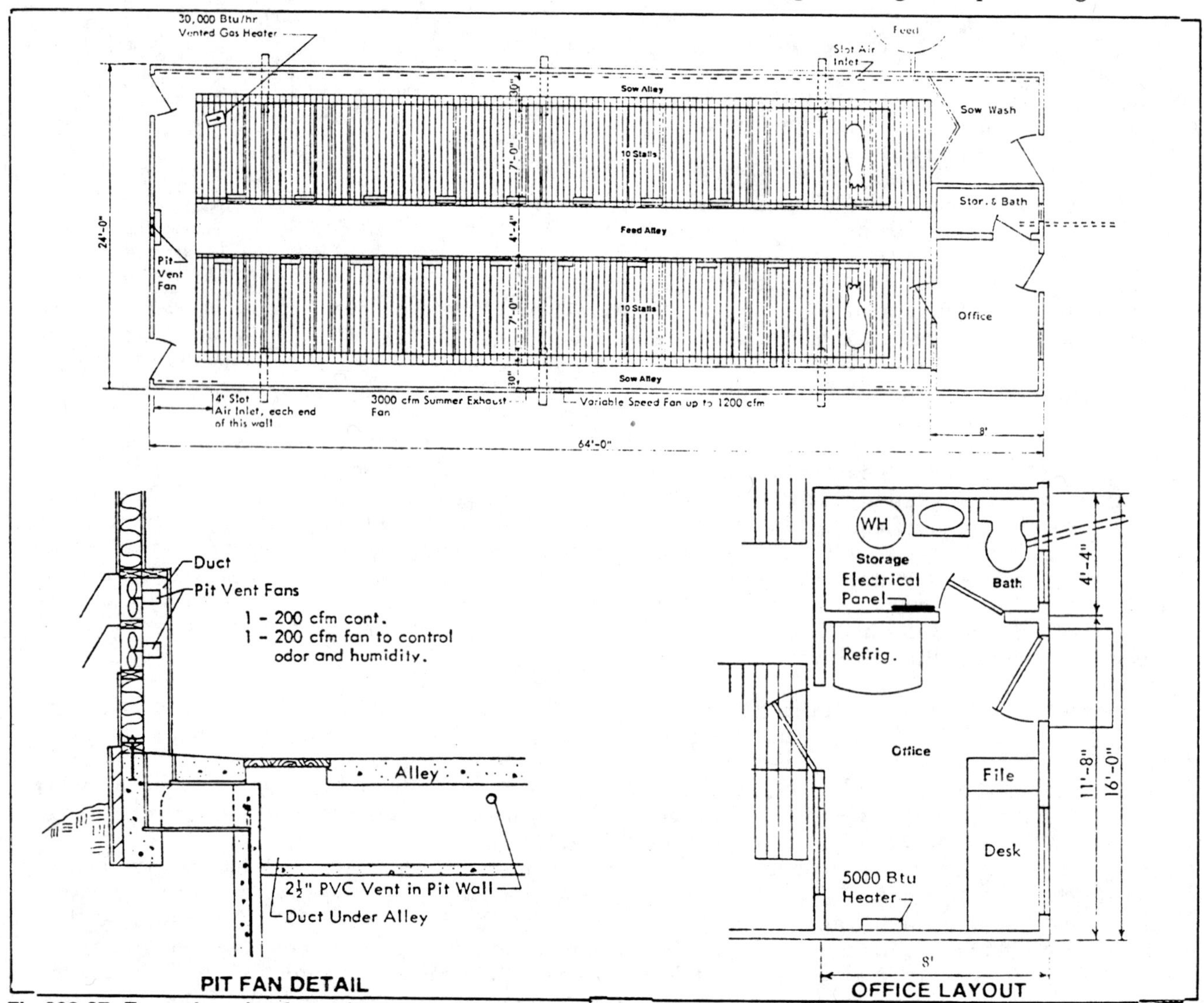

Fig 802-27. Example swine farrowing building layout.

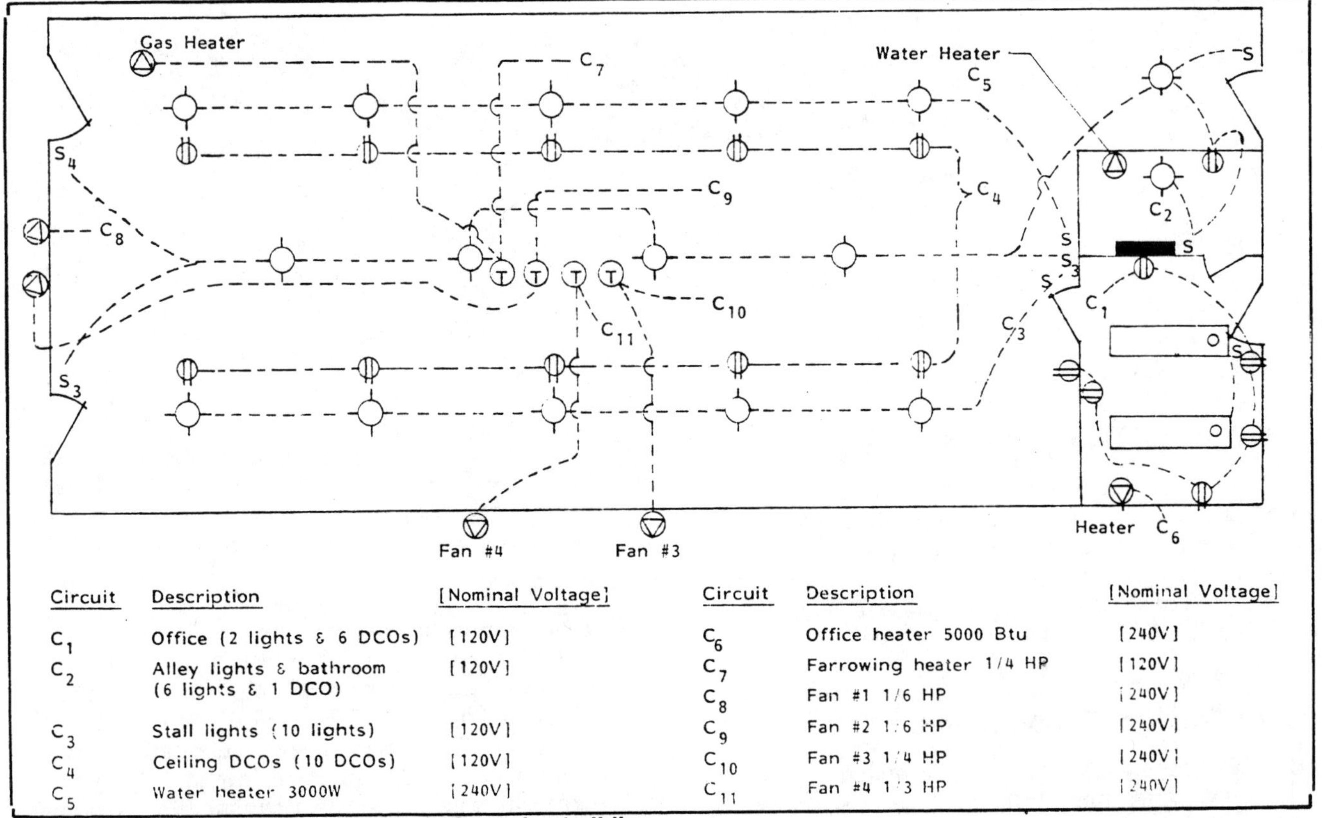

Circuit	Description	[Nominal Voltage]
C_1	Office (2 lights & 6 DCOs)	[120V]
C_2	Alley lights & bathroom (6 lights & 1 DCO)	[120V]
C_3	Stall lights (10 lights)	[120V]
C_4	Ceiling DCOs (10 DCOs)	[120V]
C_5	Water heater 3000W	[240V]
C_6	Office heater 5000 Btu	[240V]
C_7	Farrowing heater 1/4 HP	[120V]
C_8	Fan #1 1/6 HP	[240V]
C_9	Fan #2 1/6 HP	[240V]
C_{10}	Fan #3 1/4 HP	[240V]
C_{11}	Fan #4 1/3 HP	[240V]

Fig 802-28. Electrical plan for example farrowing building.

Farrowing area (Table 802-5 requires 15 foot-candles for lighting):

(24′ x 56′) x (1.5 W/ft²) ÷ (150 W/bulb) = 13.44 bulbs

Provide one 150 W incandescent light over each pair of farrowing stalls and four 150 W lights along the center alley. Wire each row of lights to a separate switch to give control over the area lighted.

Provide one overhead duplex convenience outlet (DCO)/2 stalls, one near the office, and one at each entry door. Because these DCOs are for 250 W heat lamps, check the wire and circuit breaker sizes.

(250 W/outlet x 10 outlets) ÷ 115 V = 21.74 A

Because these lamps will be on continuously for periods longer than 15 minutes, increase the amperage capacity by 25%:

21.74 A x 1.25 = 27.17 A

The first few DCOs are within 30′ of the distribution panel (DP), so #10 AWG-CU wire is required, Table 802-9. A 30 A circuit breaker is required to protect #10 AWG-CU wire in UF cable, Table 802-12.

Place each fan and the heater on a separate circuit. All the fan motors are 240 V and the largest is ⅓ hp with a full load amperage of 3.6 A, Table 802-11. Because motors are considered to be sustained loads, increase the full load current by 25% to determine wire size:

3.6 A x 1.25 = 4.5 A

None of the motors is over 75′ from the DP, so use #12 AWG-CU wire in UF cable, Table 802-9. Contact an electrician to size the circuit breaker and to install the other devices needed in motor circuits.

Sow wash area (20 foot-candles required):

(8′ x 8′) x (2.0 W/ft²) ÷ (150 W/bulb) = 0.85 bulbs

Use one 150 W incandescent light with a switch at the entry door. No DCOs are required.

Office (50 foot-candles required):

(7′ x 11′) x (1.38 W/ft²) ÷ (80 W/fixture) = 1.33 fixtures

Use two double 40 W fluorescent fixtures. Provide at least one DCO on each wall and two along the desk wall. A separate 240 V circuit is required for the electric heater. The heater draws 5000 Btu/hr ÷ 3.41 W/Btu/hr ÷ 230 V = 6.38 A. The heater is within 50′ of the DP, so use #12 AWG-CU wire, Table 802-9. A 20 A circuit breaker is required for a #12 AWG-CU wire in UF cable, Table 802-12.

Storage area (30 foot-candles required):

(4′ x 7′) x (3.43 W/ft²) ÷ (100 W/bulb) = 0.96

Provide one 100 W incandescent light and at least one DCO. Use an individual branch circuit for the 3,000 W, 240 V electric water heater, (3,000 W ÷ 230 V = 13.04 A). From Table 802-9, a #12 AWG-CU wire size is sufficient, and from Table 802-12, a 20 A circuit breaker will protect this size wire.

900 APPENDIX

901 CLIMATIC DATA

Indoor Design Temperatures

The correct inside temperature of a building depends on the product or animals housed, and varies widely. The inside design temperature for a residence may be 70 F; for an apple storage, 33 F; for a poultry house, 60 F. These are average room air temperatures. Table 634-5 has indoor design temperatures for swine.

Outdoor Design Temperatures

Local weather data provide recorded extremes, but these extremes are rarely used as design temperatures. It is customary to design structures for more moderate temperatures, because building heat storage reduces short-term extremes, equipment costs are reduced, and indoor temperatures slightly above or below their design range can usually be tolerated.

How far the outdoor design temperature is from the extreme recorded value is a matter of judgment. Consider the financial loss and physical discomfort of deviating from design temperatures, compared with the added construction and equipment costs incurred by selecting a more extreme design temperature.

Also consider the heat storage capacity of the building, possible extreme winds or other local weather disturbances, and the amount of insulation. These factors are all associated with thermal lag.

Winter

Outdoor temperatures for winter design are in Figs 901-1 and 901-2 and Table 901-2.

The winter temperatures in Table 901-2 have been equalled or exceeded by 99% or 97.5% of the winter hours. The 97½% temperature is adequate for designing most enclosed animal buildings, but design 35 F potato storages for at least the median extreme temperature.

Table 901-1. Hours of extreme temperatures.
In a normal winter, outdoor temperatures are below the 99% data for about 22 hr (54 hr at 97½%). In a normal summer, outdoor temperatures are above the 1% data for about 30 hr (75 hr at 2½%).

Season	Months	Total hours	Level	Extreme hours
Winter	Dec-Feb	2160	1%	22
			2½%	54
Summer	Jun-Sept	2928	1%	30
			2½%	75

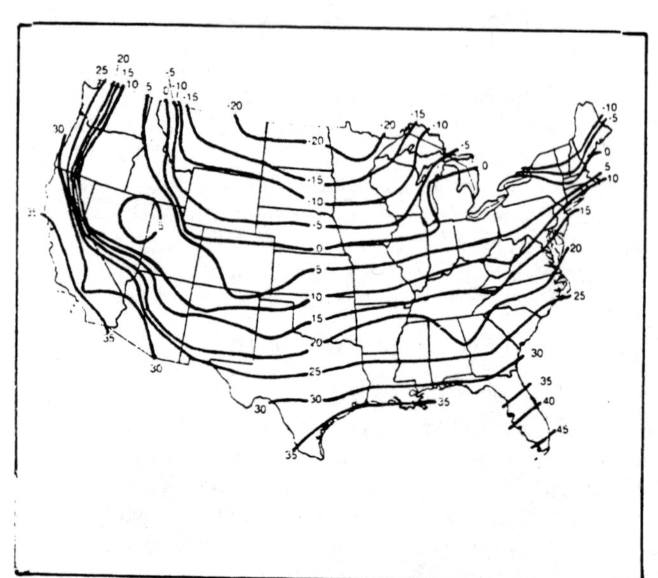

Fig 901-1. Winter design temperatures.
It is colder than this 2½% of the time (54 hrs) during December through February.

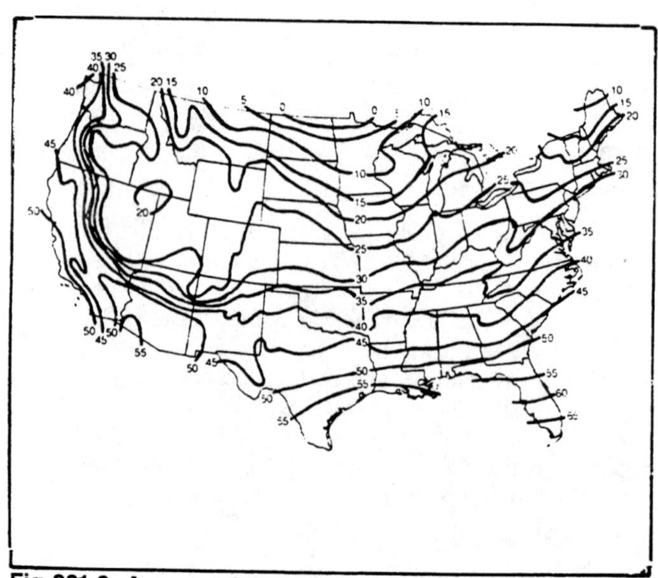

Fig 901-2. Average daily temperatures for January.

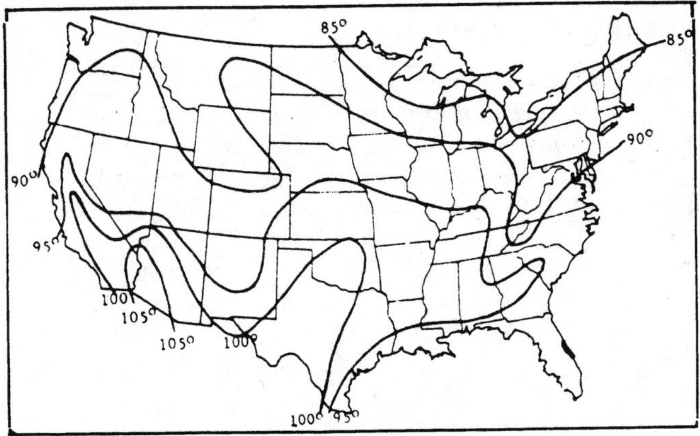

Fig 901-3. Summer design temperatures.
It is warmer than this 2½% of the time (75 hr) during June through September.

Summer

Fig 901-3 contains 2½% design temperatures for summer cooling.

The summer dry-bulb temperatures in Table 901-2 have been equalled or exceeded by 1% or 2.5% of the summer hours. A coincident wet-bulb temperature is listed with each dry-bulb temperature. The coincident wet-bulb temperature is the mean of all wet-bulb temperatures occurring at that dry-bulb temperature.

The mean daily range is the difference between the average daily maximum and average daily minimum temperatures in the warmest month. The design wet-bulb temperatures in Table 901-2 have been equalled or exceeded by 1% or 2.5% of the summer

Table 901-2. Weather design data.
From the *1981 ASHRAE Handbook of Fundamentals*. For agricultural structures, use the 97.5% (2.5%) values, unless the application is critical.

		Winter		Summer				
		Design dry-bulb		Design dry-bulb and mean coincident wet-bulb		Mean daily range	Design wet-bulb	
		99%	97.5%	1%	2.5%		1%	2.5%
AL	Birmingham	17	21	96/74	94/75	21	78	77
AZ	Phoenix	31	34	109/71	107/71	27	76	75
AR	Little Rock	15	20	99/76	96/77	22	80.	79
CA	Sacramento	30	32	101/70	98/70	36	72	71
CO	Boulder	-2	8	93/59	91/59	27	64	63
CT	Hartford	3	7	91/74	88/73	22	77	75
FL	Gainesville	28	31	95/77	93/77	18	80	79
GA	Atlanta	17	22	94/74	92/74	19	77	76
ID	Boise	3	10	96/65	94/64	31	68	66
IL	Freeport	-9	-4	91/74	89/73	24	77	76
IN	Indianapolis	-2	2	92/74	90/74	22	78	76
IA	Ames	-11	-6	93/75	90/74	23	78	76
KS	Dodge City	0	5	100/69	97/69	25	74	73
KY	Lexington	3	8	93/73	91/73	22	77	76
LA	New Orleans	29	33	93/78	92/78	16	81	80
ME	Caribou	-18	-13	84/69	81/67	21	71	69
MI	Lansing	-3	1	90/73	87/72	24	75	74
MN	St. Cloud	-15	-11	91/74	88/72	24	76	74
MS	Jackson	21	25	97/76	95/76	21	79	78
MO	Columbia	-1	4	97/74	94/74	22	78	77
MT	Great Falls	-21	-15	91/60	88/60	28	64	62
NE	Lincoln	-5	-2	99/75	95/74	24	78	77
NV	Reno	5	10	95/61	92/60	45	64	62
NJ	Trenton	11	14	91/75	88/74	19	78	76
NM	Albuquerque	12	16	96/61	94/61	27	66	65
NY	Albany	-6	-1	91/73	88/72	23	75	74
NC	Raleigh	16	20	94/75	92/75	20	78	77
ND	Bismark	-23	-19	95/68	91/68	27	73	71
OH	Columbus	0	5	92/73	90/73	24	77	75
OK	Oklahoma City	9	13	100/74	97/74	23	78	77
OR	Corvallis	18	22	92/67	89/66	31	69	67
PA	State College	3	7	90/72	87/71	23	74	73
SC	Charleston	24	27	93/78	91/78	18	81	80
SD	Rapid City	-11	-7	95/66	92/65	28	71	69
TN	Nashville	9	14	97/75	94/74	21	78	77
TX	Abilene	15	20	101/71	99/71	22	75	74
UT	Salt Lake City	3	8	97/62	95/62	32	66	65
VA	Norfolk	20	22	93/77	91/76	18	79	78
WA	Yakima	-2	5	96/65	93/65	36	68	66
WI	Madison	-11	-7	91/74	88/73	22	77	75
WY	Laramie	-14	-6	84/56	81/56	28	61	60

hours. These wet-bulb values are computed independently of the dry-bulb values; their coincident dry-bulb values are not given.

Wet-Bulb Temperatures

In drying high-moisture grain, the temperature of undried grain approaches the wet-bulb temperature. Higher wet-bulb temperatures reduce the allowable drying time. The mean wet-bulb temperature is lower than the values in Fig 901-4 for about 50% of the years. If one standard deviation is added to the values in Fig 901-4, the mean wet-bulb temperature is lower than the resulting sum about 84% of the years.

Wet-Bulb Depressions

In drying high-moisture grain, the rate at which drying can occur is determined in part by the wet-bulb depression. Greater wet-bulb depressions increase drying rates. The mean wet-bulb depression is greater than the values in Fig 901-5 for about 50% of the years. If one standard deviation is subtracted from the values in Fig 901-5, the mean wet-bulb depression is greater than the resulting difference in about 84% of the years.

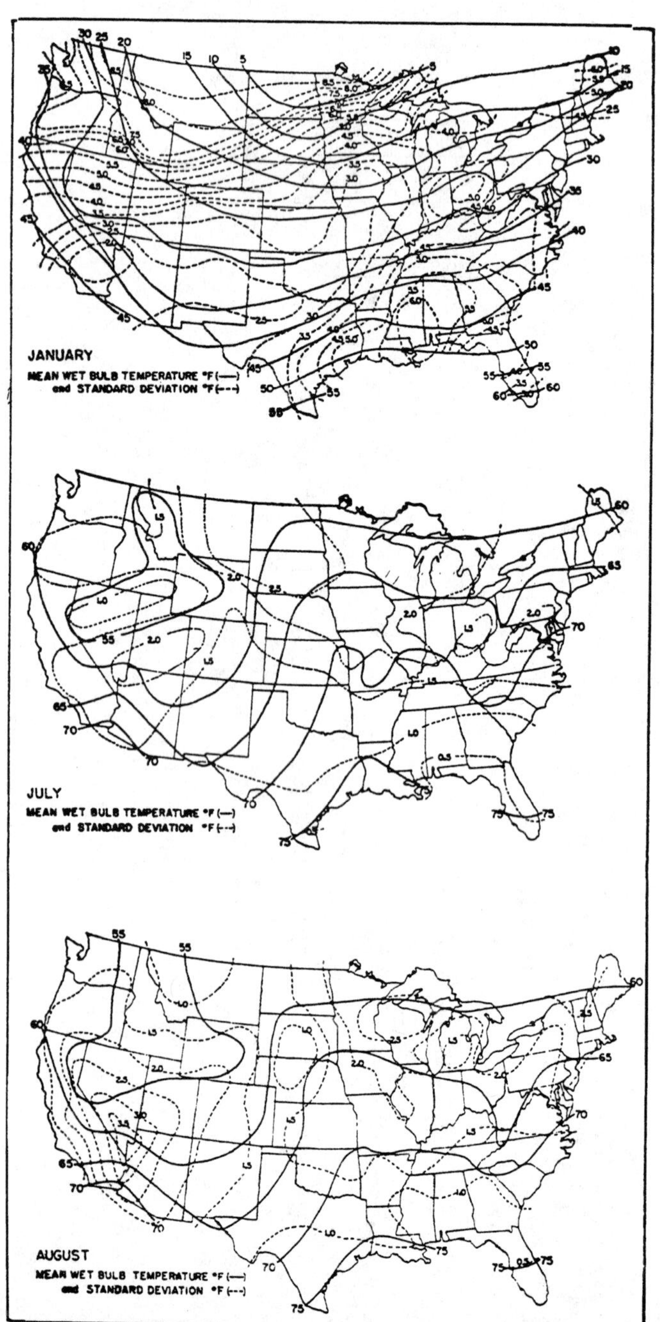

Fig 901-4. Wet-bulb temperatures.
Extracted from the *1982 Agricultural Engineers Yearbook.* ASAE Data: ASAE D309.

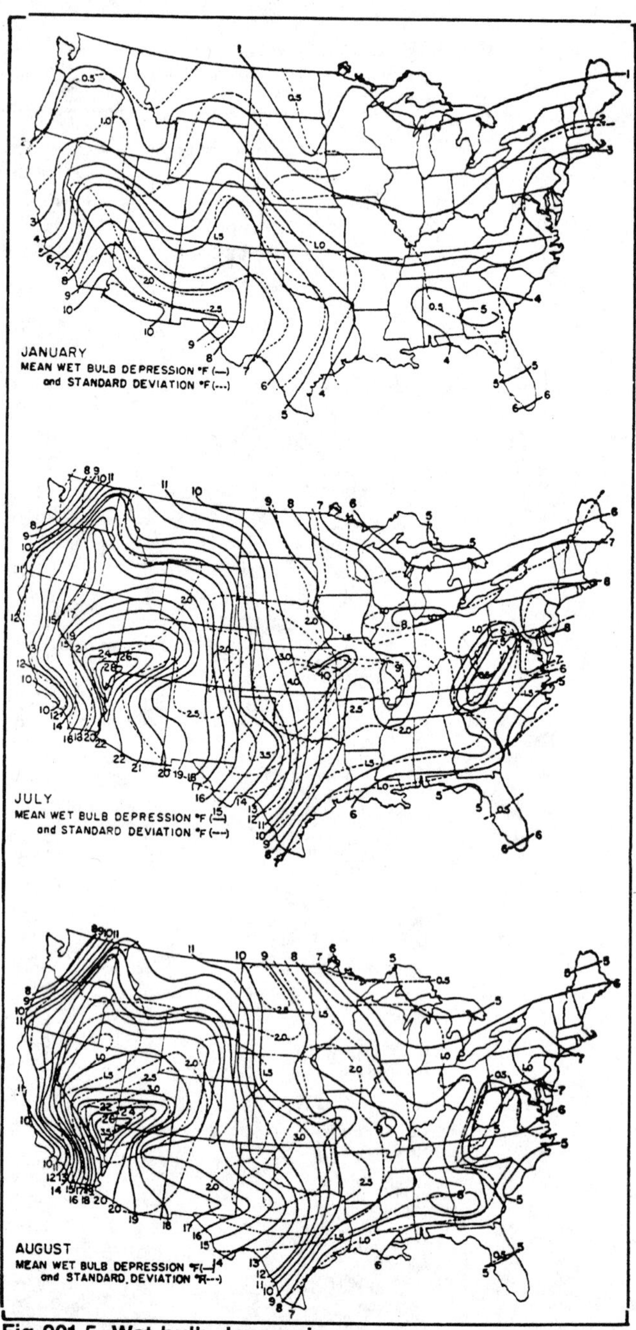

Fig 901-5. Wet-bulb depressions.
Extracted from the *1982 Agricultural Engineers Yearbook.* ASAE Data: ASAE D309.

Ground Temperatures

Design ground temperatures, for estimating heat flow through basement walls and floors laid on earth, depend on the locality, season, and depth below the ground surface. In northern climates, a winter temperature of 32 F is suitable for basement walls near the surface. At greater depths, use the local ground water temperature (about 50 to 55 F).

Soil Freezing

Figs 901-6 and 901-7 give average and extreme frost depths. An area free of snow freezes deeper than one protected with a snow blanket. Because even cold water provides some heat, a pipe with water always running is less apt to freeze than one with no water flow.

Frost heaving depends on a soil being saturated, subject to cyclical freezing-thawing, and underlain with a relatively impermeable soil so the upper soil stays saturated. Local experience with soil and drainage patterns is the best predictor for frost heaving. Drainage at the base of a foundation prevents saturation; insulation outside a foundation permits building heat to increase temperatures of the foundation and the soil just inside it. Banking a building

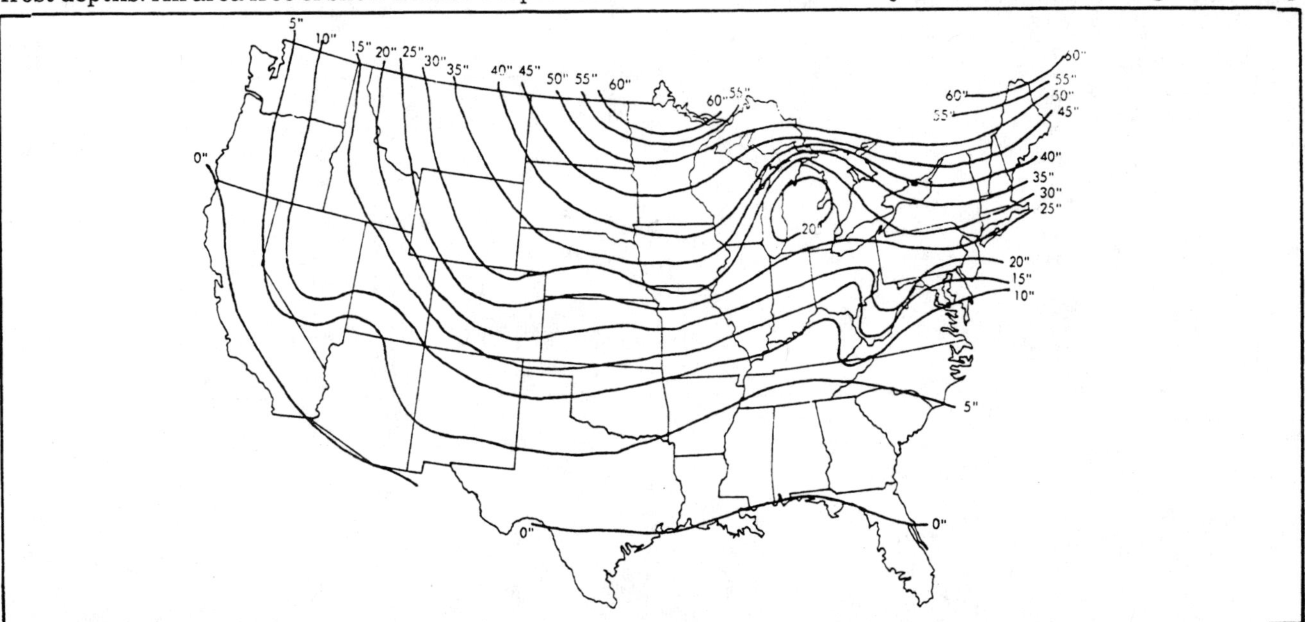

Fig 901-6. Average annual frost penetration.
Use average values in areas where frost heaving is uncommon, for foundations that will be drained and insulated against extreme frost, or where frost heaving or freezing presents no particular hazard.

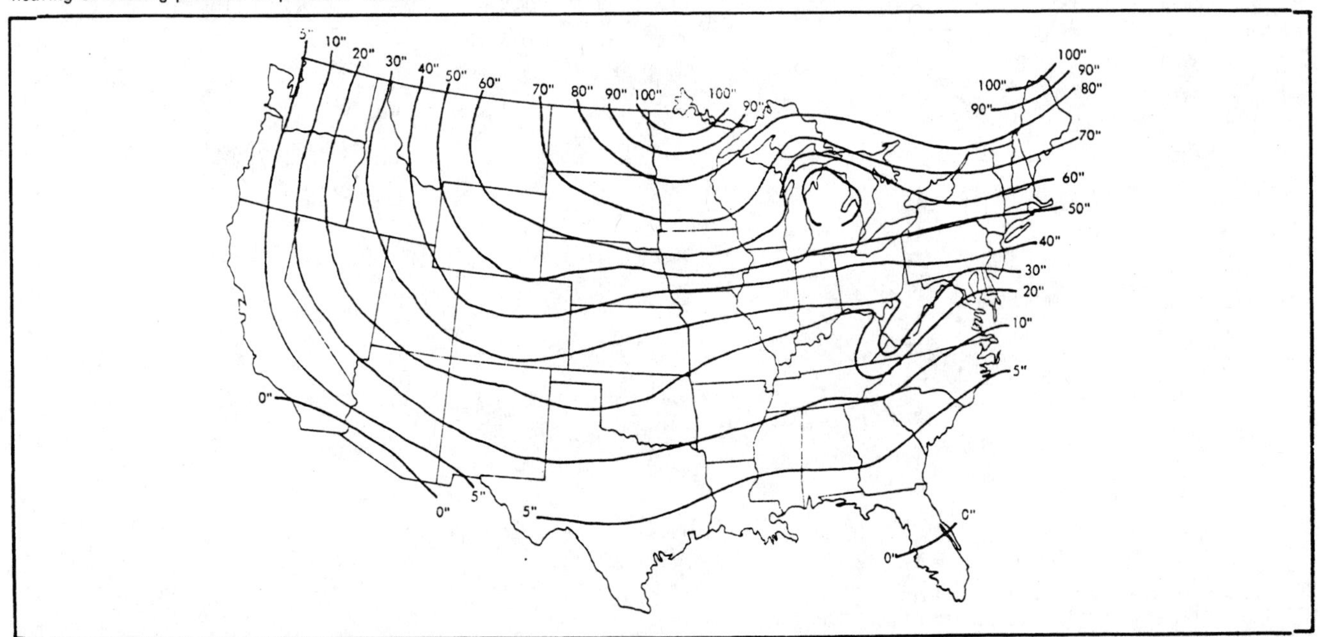

Fig 901-7. Extreme frost penetration.
Use extreme values where freezing is costly (as for water lines), in areas where soil and groundwater conditions cause frequent frost heaving problems, or for facilities that frost heaving would severely damage.

with straw, snow, or other insulation reduces freezing cycles and therefore frost heaving.

Wind Speed

A wind rose is a picture of prevailing wind directions, Fig 901-8.

Fig 901-9 has maps of wind experience; averages for 24 hourly readings per day for the given month for 10 years. At each location (e.g. Des Moines, Iowa; January), the inside of the circle is the percentage of hours (1%) of calm wind. Each line represents the direction and the percentage from that direction, each band being 5%. For Des Moines, about 7% of the time in January the wind blows from the south, about 8% from the north, and about 12% from the NW etc. For wind loads, see Chapter 101.

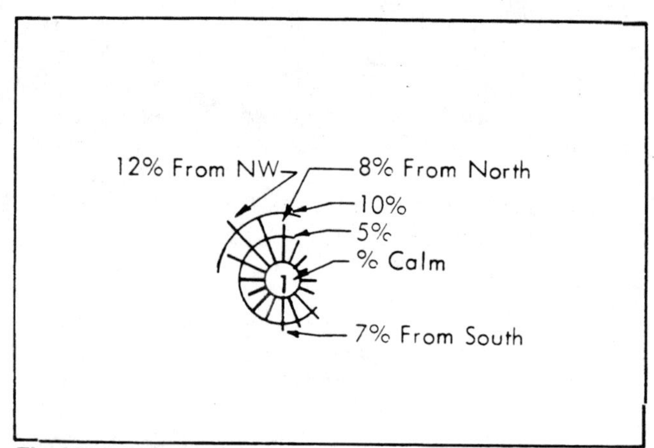

Fig 901-8. Wind rose.
Des Moines, Iowa; January.

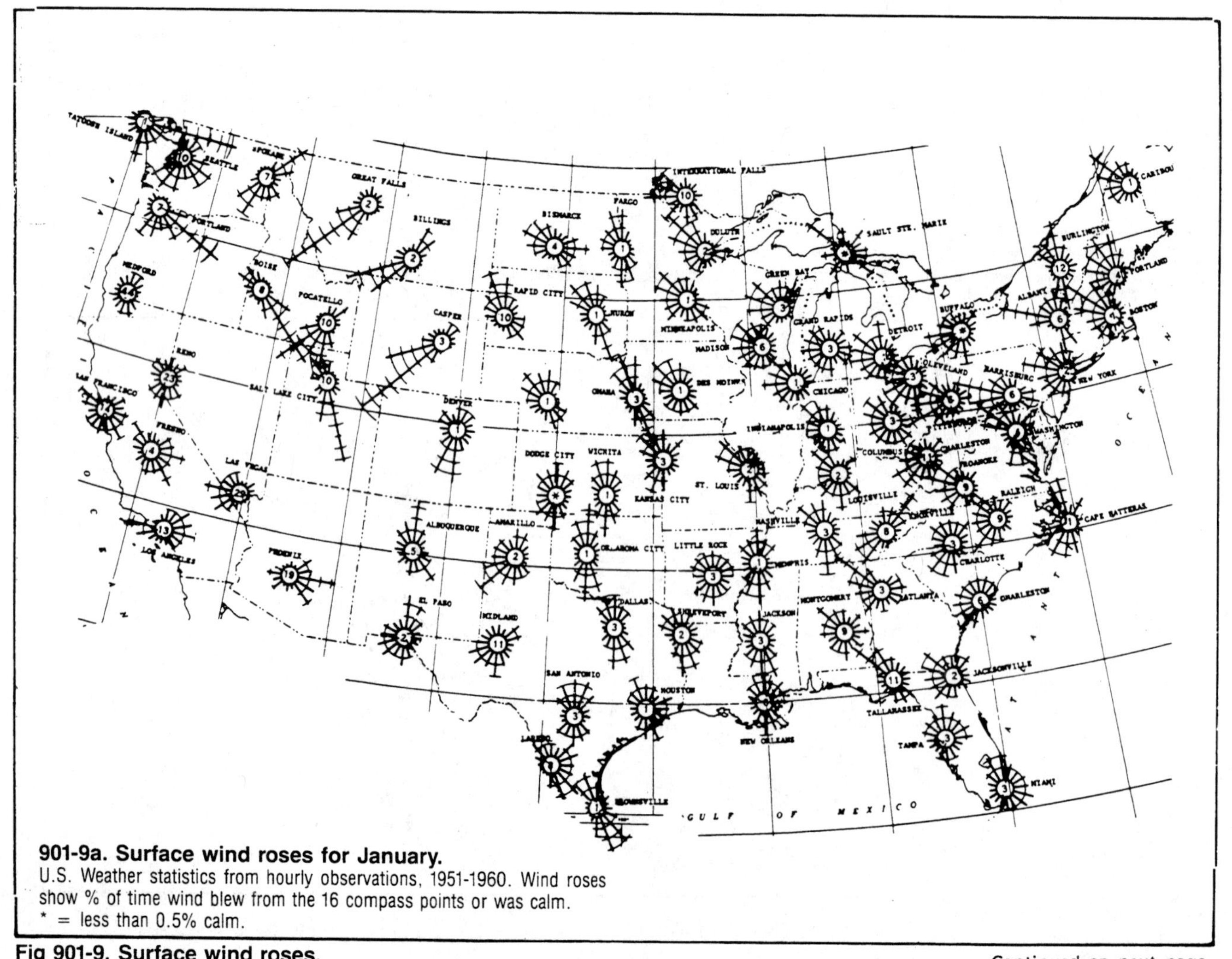

901-9a. Surface wind roses for January.
U.S. Weather statistics from hourly observations, 1951-1960. Wind roses show % of time wind blew from the 16 compass points or was calm.
* = less than 0.5% calm.

Fig 901-9. Surface wind roses.

Continued on next page.

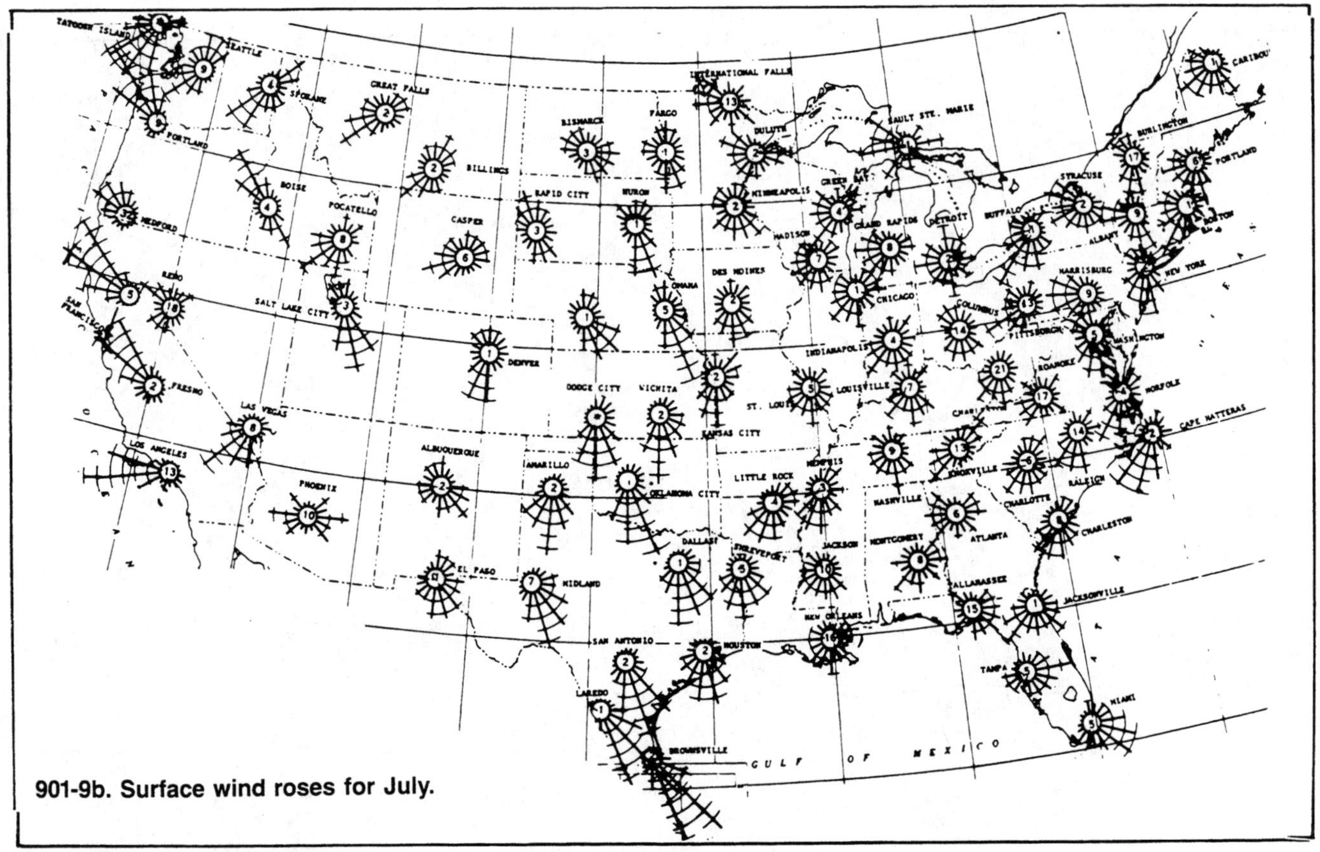

901-9b. Surface wind roses for July.

Fig 901-9. Surface wind roses, continued.

Rainfall

The climate data in Figs 901-10 through 901-14 are from or adapted from the following sources: *Climate Atlas of the U.S.*, 1968 Dept. Commerce—ESSA; and *Technical Papers 25, 37, 41, and 49,* U.S. Dept. Commerce—ESSA.

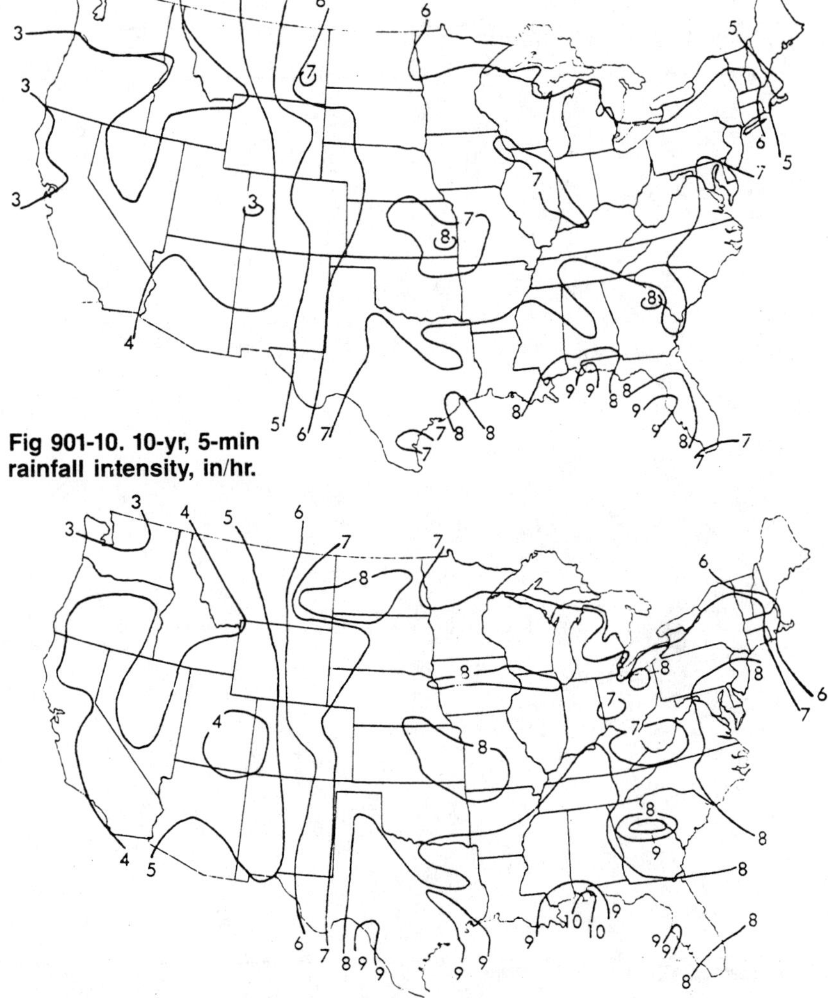

Fig 901-10. 10-yr, 5-min rainfall intensity, in/hr.

Fig 901-11. 25-yr, 5-min rainfall intensity, in/hr.

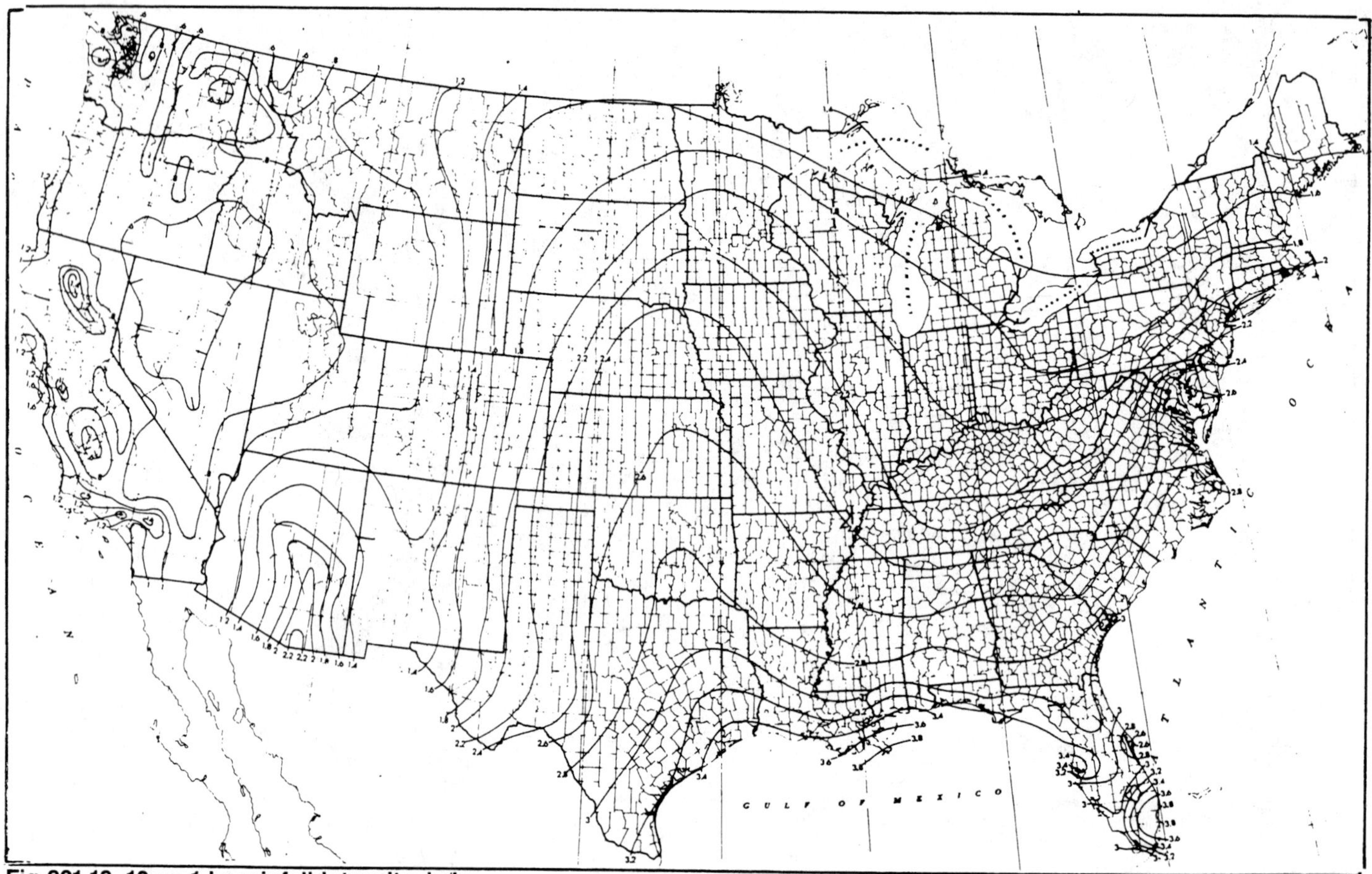

Fig 901-12. 10-yr, 1-hr rainfall intensity, in/hr.

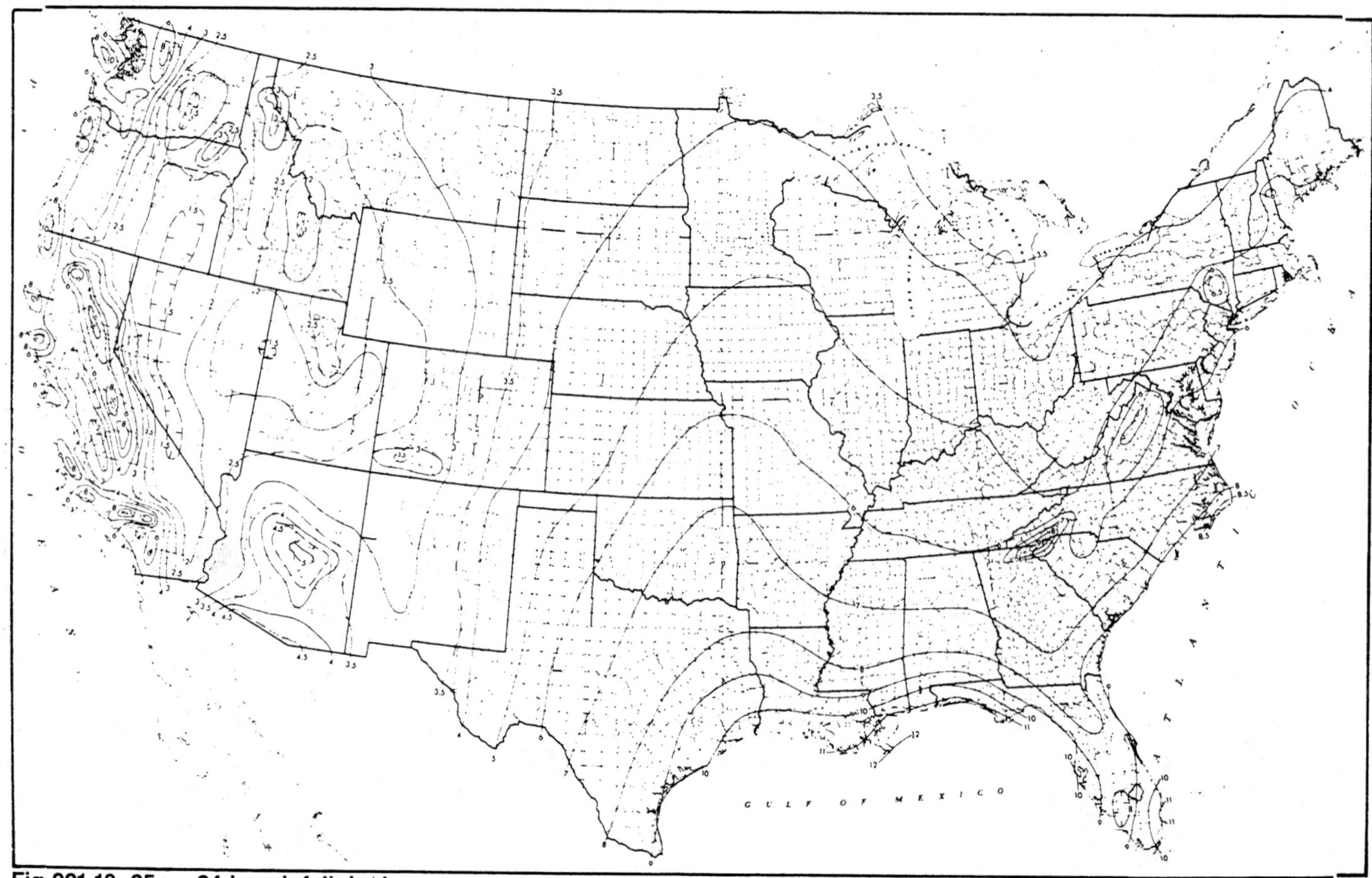

Fig 901-13. 25-yr, 24-hr rainfall, in/day.

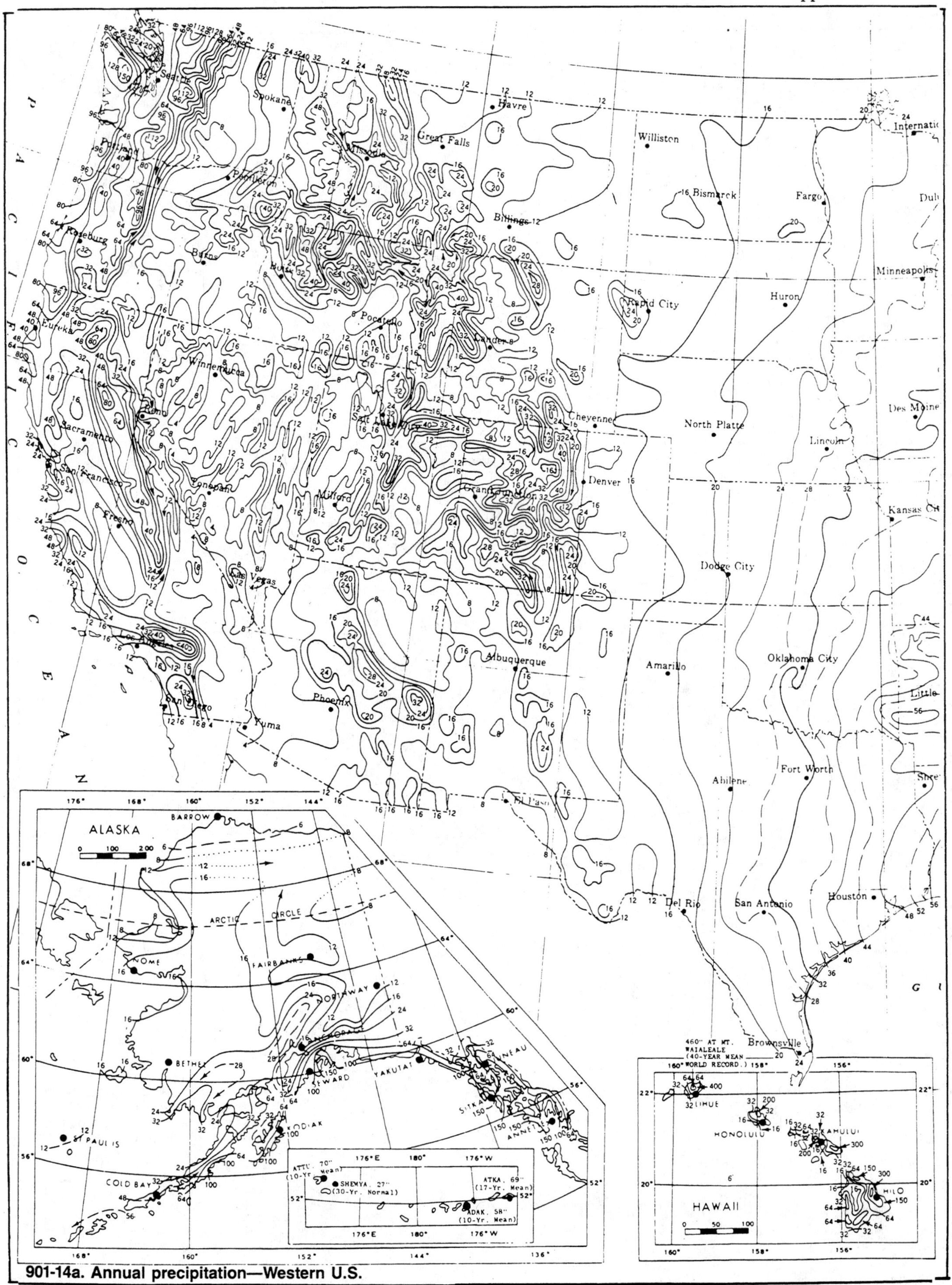

901-14a. Annual precipitation—Western U.S.

Fig 901-14. Normal annual precipitation, in.

Continued on next page.

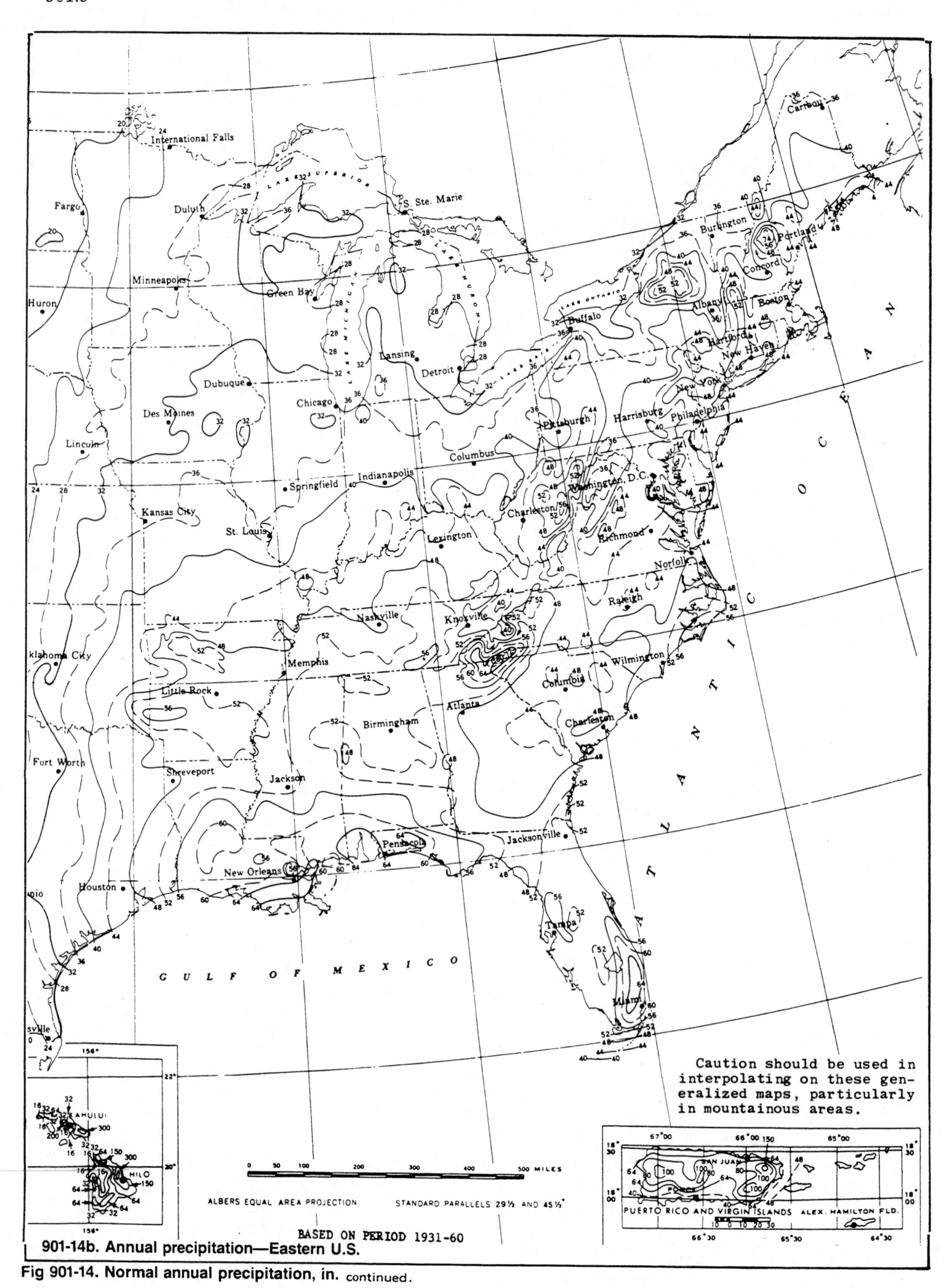

901-14b. Annual precipitation—Eastern U.S.

Fig 901-14. Normal annual precipitation, in. continued.

Evaporation

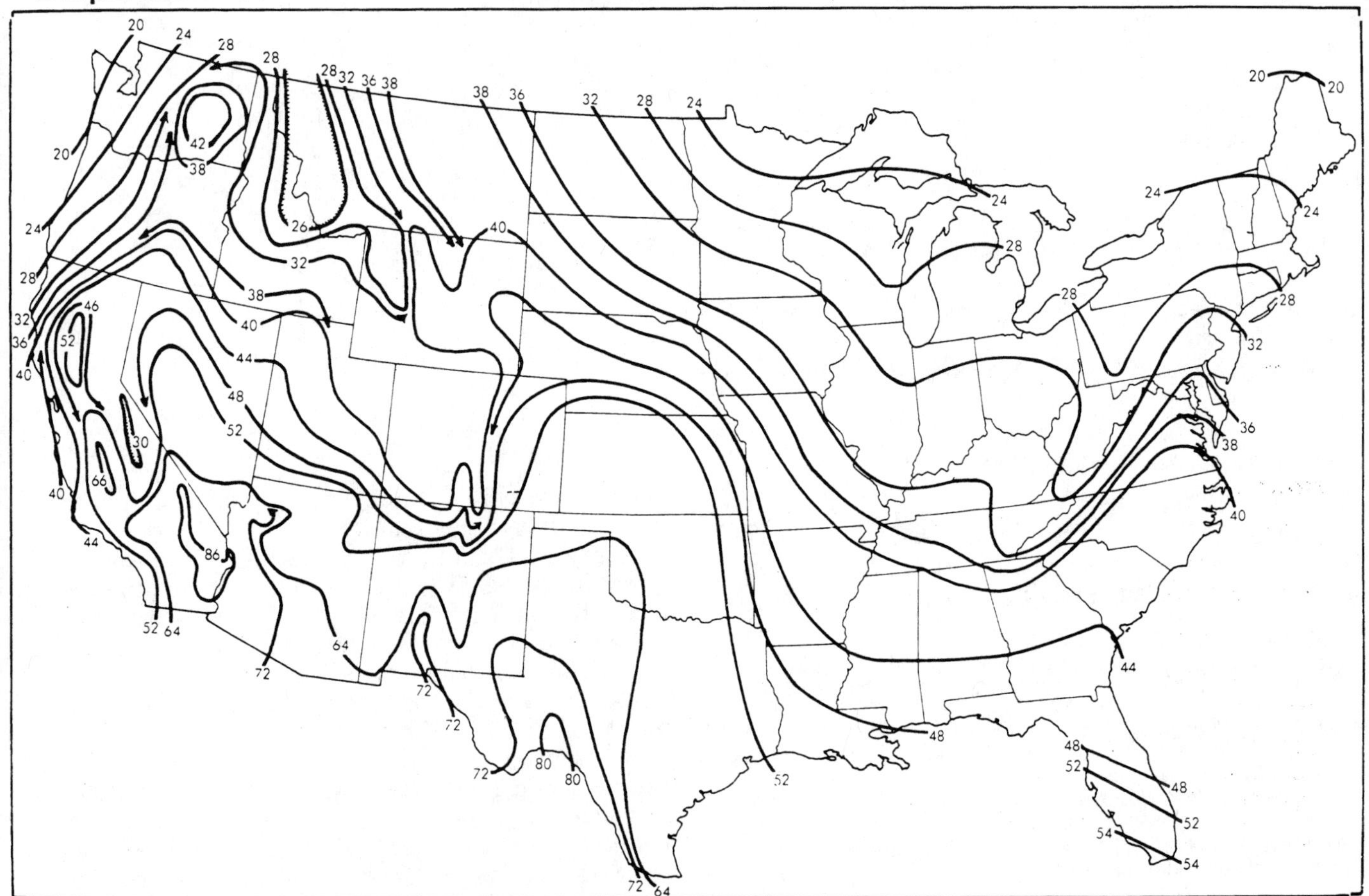

Fig 901-15. Mean annual lake evaporation, in.

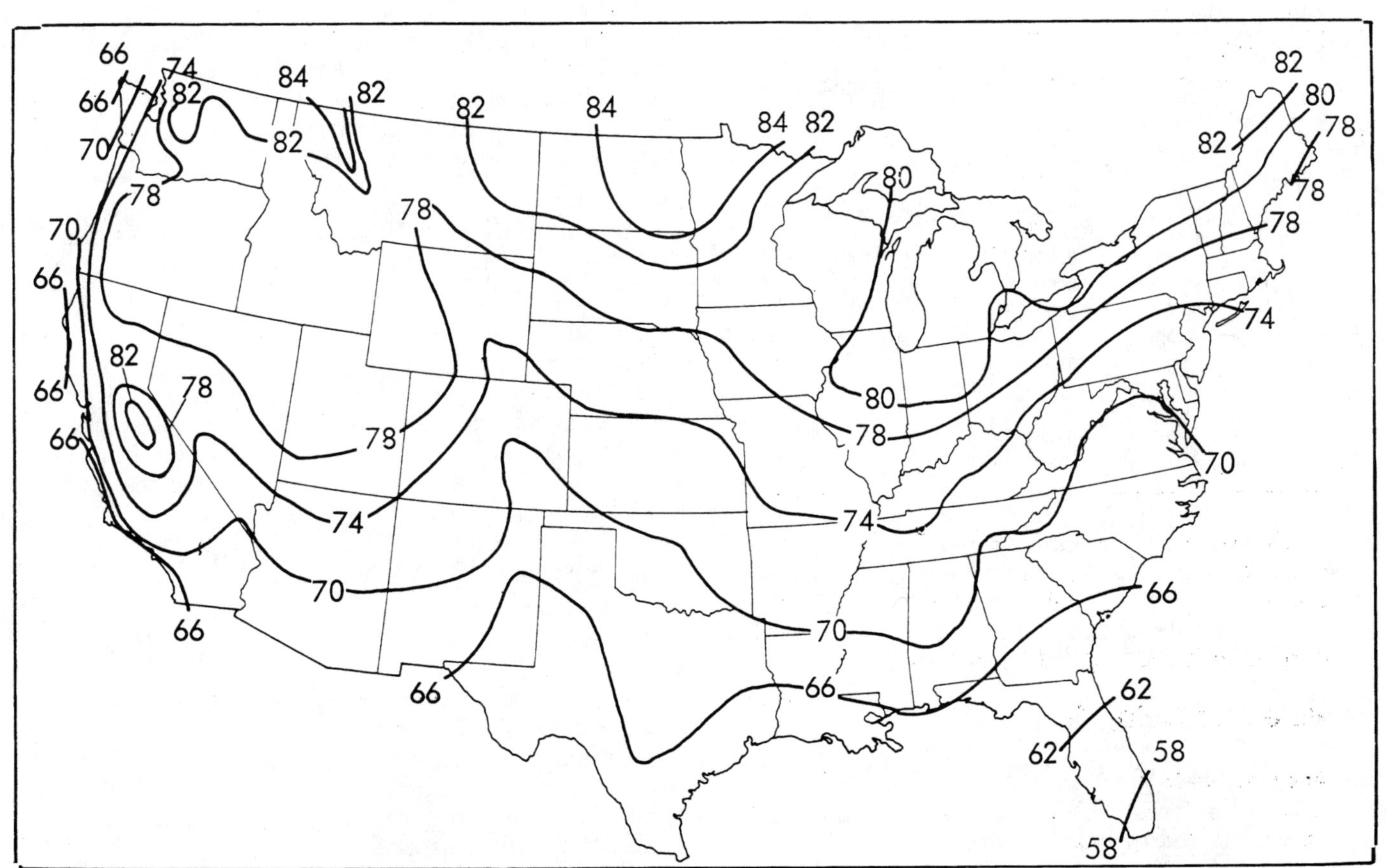

Fig 901-16. Mean May-October evaporation in % of total.

902 SI (METRIC) UNITS

This chapter is adapted from *ASAE Engineering Practice: EP285.6, 1982.*

Units

The International System of Units (SI) consists of seven base units, two supplementary units, derived units, and prefixes to form multiples of the units.

Table 902-1. SI units and prefixes.
Note especially which abbreviations are capitalized.

Base Units
meter (m) — unit of length
second (s) — unit of time
kilogram (kg) — unit of mass
kelvin (K) — unit for thermodynamic temperature
ampere (A) — unit of electric current
candela (cd) — luminous intensity
mole (mol) — amount of a substance

Supplementary Units
radian (rad) — plane angle
steradian (sr) — solid angle

SI unit prefixes

Multiple	Prefix	SI Symbol
10^{12}	tera	T
10^{9}	giga	G
10^{6}	mega	M
10^{3}	kilo	k
10^{2}	hecto	h
10^{1}	deka	da
10^{-1}	deci	d
10^{-2}	centi	c
10^{-3}	milli	m
10^{-6}	micro	u*
10^{-9}	nano	n
10^{-12}	pico	p
10^{-15}	femto	f
10^{-18}	atto	a

Derived units are combinations of basic units or other derived units as needed to describe physical properties, e.g., area. Some derived units are named; others are simply a combination of SI units. Some derived units are tabulated in Table 902-1.

Rules for Using SI

Applying prefixes

Use the prefixes to eliminate insignificant digits and decimals and to avoid powers of 10 as generally preferred in computation. For example: 12 300 m or 12.3 x 10^3m becomes 12.3 km.

Apply prefixes to the numerator of compound units, except when using kilogram (kg) in the denominator—kilogram is a base unit and is preferred to the gram. For example: 200 J/kg, not 2 dJ/g.

With higher order units such as m^2 or m^3, the prefix is also raised to the same order. For example: mm^2 is $(10^{-3}m)^2$ or 10^{-6} m^2.

Select prefixes so that numerical values lie between 0.1 and 1000 except where certain multiples are generally accepted. Use the same unit or multiple in tables even if the series exceeds the preferred range. Do not use double or hyphenated prefixes. For example: use GW (gigawatt), not kMW.

Capitalization

Symbols are capitalized when the SI unit is derived from a proper name: N for Isaac Newton. The units are not capitalized. Numerical prefixes and their symbols are not capitalized, except M (mega), G (giga), and T (tera).

Plurals

Form SI unit plurals normally. SI symbols are always written in singular form. For example: 50 newtons or 50 N.

Punctuation

For numerical values less than one, a zero precedes the decimal point. SI unit symbols do not have a period except at the end of a sentence. English speaking countries use a dot for the decimal point, others use a comma. Use spaces instead of commas for grouping numbers into threes (thousands).

For example: 6 357 831.376 88
not 6,357,831.376,88

Derived units

The product of two or more unit symbols is preferably indicated by a raised period or with a space if there is no risk of confusion with another unit symbol. For example: Use n·m or N m, but not mN. A solidus (oblique stroke, /), a horizontal line, or negative powers express a derived unit formed by division. For example: m/s, m ÷ s or $m \cdot s^{-1}$. Use parentheses to avoid ambiguity if needed.

Table 902-2. Derived units.

Quantity (unit)	SI Symbol	Formula
acceleration	—	m/s²
angular acceleration	—	rad/s²
angular velocity	—	rad/s
area	—	m²
density	—	kg/m³
electrical capacitance (farad)	F	A·s/V
electrical conductance (siemens)	S	A/V
electrical field strength	—	V/m
electrical inductance (henry)	H	V·s/A
electrical resistance (ohm)	Ω	V/A
energy (joule)	J	N·m
entropy	—	J/K
force (newton)	N	kg·m/s²
frequency (hertz)	Hz	(cycle)/s
illuminance (lux)	lx	lm/m²
luminance (candela/m²)	—	cd/m²
luminous flux (lumen)	lm	cd·sr
magnetic field strength	—	A/m
magnetic flux (weber)	Wb	v·s
magnetic flux density (tesla)	T	Wb/m²
magnetomotive force (ampere)	A	—
power (watt)	W	J/s
pressure (pascal)	Pa	N/m²
quantity of electricity (coulomb)	C	A·s
quantity of heat (joule)	J	N·m
radiant intensity	—	W/sr
specific heat	—	J/kg·K
stress (pascal)	Pa	N/m²
thermal conductivity	—	W/m·K
velocity	—	m/s
viscosity, dynamic	—	Pa·s
viscosity, kinematic	—	m²/s
voltage (volt)	V	W/A
work (joule)	J	N·m

Alternate alphabets

For correct usage in all capitals, etc., see International Organization for Standardization (ISO) 2955, *Information Processing—Representation of SI and Other Units for Use in Systems with Limited Character Sets.*

Non-SI Units

Certain units outside the SI are recognized by ISO because of their practical importance in specialized fields. These include units for temperature, time, and angle. Also included are some multiples of units such as liter (L, volume), hectare (ha, land measure), and metric ton (t, mass).

Temperature

The SI base unit for thermodynamic temperature is Kelvin (K). Celsius (formerly centigrade) may be used for temperature.

Eq 902-1.

$$1\ \mathrm{C} = 1\ \mathrm{K}$$
$$t = T - 273.15$$

t = Celsius temperature
T = Kelvin temperature

273.15 = absolute zero

Time

The SI unit, the second, is preferred, especially for technical calculations. Minute, hour, day, etc., are permissible.

Angles

The SI unit for plane angle is the radian. Arc degrees (°) and decimal, or minute (′) and second (″), submultiples are permissible. Express solid angles in steradians.

Logical conversions depend on the intended precision of the original quantity.

The number of significant digits shown implies precision. The implied precision is plus or minus one-half unit of the last significant digit, because the one-half unit is the limit of error from rounding. For example, 2.14 may have been any number from 2.135 to 2.145. Whether rounded or not, a quantity has this implication of precision: 2.14 m implies a percision of ± 0.005 m (the last significant digit is in units of 0.01 m).

Express quantities in digits that are significant. A common problem arises with fractions: in the dimension 1.1875 the fourth place may be significant, or it may be the decimal equivalent of 1³⁄₁₆, in which case too many decimal places were given.

Include significant zeros. The dimension 2 m means "about 2 m." Write 2.0000 m if that degree of precision is intended.

Rounding

Where feasible, round SI equivalents in reasonable, convenient, whole units. American National Standards Institute Z210.1-1973, *Metric Practice Guide,* outlines methods to ensure interchangeability of parts.

When a number is to be rounded to fewer decimal places proceed as follows:

Round to closest even number when first digit discarded is 5, followed only by zeros.

When the first digit discarded is less than 5, leave the last digit unchanged.

When the first digit discarded is greater than 5, or if it is a 5 followed by at least one digit other than 0, round up.

Round directly to the desired number of decimal places. Do not round by successive steps to fewer places.

For example:

3.4535 = 3.454 (5)
or = 3.45 (<5)
or = 3.5 (>5)
or = 3

Do not round 3.4535 to 3.5 and then round again, because it would round to 4 instead of 3.

Table 902-3. Metric conversion factors.

Unit	Times	Equals
Length		
foot (ft)	0.304 8	m
inch (in.)	0.025 4	m
in.	25.4	mm
in.	2.54	cm
mile	1 609.	m
rod	5.029	m
yard	0.914 4	m
Area		
acre	4.047	m^2
acre	0.404 7	hectare
ft^2	0.092 9	m^2
in^2	0.000 645	m^2
in^2	645.	mm^2
in^2	6.45	cm^2
$mile^2$ (U.S. statute)	2.59	km^2
yd^2	0.836 1	m^2
Volume (capacity)		
acre-inch	102.8	m^3
bushel	0.035 24	m^3
fluid ounce	0.000 029 57	m^3
fluid ounce	0.029 57	liter
ft^3	0.028 32	m^3
gallon	0.003 785	m^3
gallon	3.785	liter
in^3	0.000 016 39	m^3
in^3	0.016 39	liter
quart	0.946 4	liter
yd^3	0.764 6	m^3
Acceleration		
ft/sec^2	0.304 8	m/s^2
free fall, standard	9.807	m/s^2
Bending Moment, Torque		
in-lb	0.113	N·m
ft-lb	1.356	N·m
Energy, Heat, Power		
R (F-hr-ft^2/Btu)	0.176 1	k-m^2/W
U (Btu/F-hr-ft^2)	5.678	W/m^2-K
British thermal unit (Btu)	1 055.	J
Btu/lb-F	4 187.	J/kg-K
Btu/ft^2	11 360.	J/m^2
Btu/ft^2-hr	3.155	W/m^2
Btu/hr	0.293 1	W
Btu/lb	2 326.	J/kg
calorie (cal)	4.187	J
foot-pound (ft-lb)	1.356	J
ft-lb/sec	1.356	W
horsepower (550 ft-lb/s)	745.7	W
hp-hr	0.745 7	kWh
kilowatt-hour (kWh)	3 600 000.	J
watt-second (Ws)	1	J
Force		
ounce (oz)	0.278	N
pound (lb)	4.448	N
Light		
foot candle	10.76	lm/m^2
Mass		
ounce	0.028 35	kg
pound	0.453 6	kg
ton, 2000 lb	907.2	kg
ton	0.907 2	tonne
tonne, 15.5% grain	39.37	bu
Mass Per Area		
lb/acre	1.121	kg/ha
lb/ft^2	4.882	kg/m^2
lb/in^2	703.1	kg/m^2
Mass Per Length		
lb/ft	1.488	kg/m
Mass Per Time (flow)		
lb/sec	0.453 6	kg/s
ft^3/min	0.000 471 9	m^3/s
ft^3/min-bu	0.018 58	m^3/s-tonne
ft^3/sec	0.028 32	m^3/s
gal/min	0.000 063 09	m^3/s
Mass Per Volume (density)		
lb/ft^3	16.02	kg/m^3
lb/gal	119.8	kg/m^3
Moment of Area		
in^4	416 200.	mm^4
in^4	41.62	cm^4
Plane Angle		
degree	0.017 45	rad
Pressure, Stress (force/area)		
atmosphere (normal = 760 torr)	101 300.	Pa
bar	100 000.	Pa
in. of mercury (60 F)	3 377.	Pa
in. of water (60 F)	248.8	Pa
lb/ft^2	47.88	Pa
lb/in^2	6 895.	Pa
Velocity (speed)		
ft/sec	0.304 8	m/s
mile/hr	1.609	km/h
Volume Per Area		
bu/acre	0.087 08	m^3/ha

903 WEIGHTS AND MEASURES

Conversions

Multiply to the right: acres x 43,560 = ft^2
Divide to the left: ft^2 ÷ 43,560 = acres

Unit	Times	Equals
Acres	43,560	ft^2
	4,840	yd^2
	160	square rods
	1/640	square mile
Acre-ft	325,851	gallons
	43,560	ft^3
Acre-in	3,630	ft^3
Acre-in/hr	453	gpm
	1	cfs (approximate)
Bushels	1.25	ft^3
	2.5	ft^3 ear corn
ft^3	7.48	gallons
	1728	in^3
	62.4	lb water
	0.4	bu ear corn
	0.8	bu grain
cfs	448.8	gpm
	646,317	gal/day
Cubic yard	27	ft^3
concrete	81	ft^2 of 4″ floor
concrete	54	ft^2 of 6″ floor
Gallons	231	in^3
	0.134	ft^3
	8.35	lb water
Miles	5,280	ft
	1,760	yd
	320	rods
Pressure, psi	2.31	ft of water head
Rods	16.5	ft
	5.5	yd

Metric conversions are at the end of Chapter 902 SI (Metric) Units.

AREAS & VOLUMES

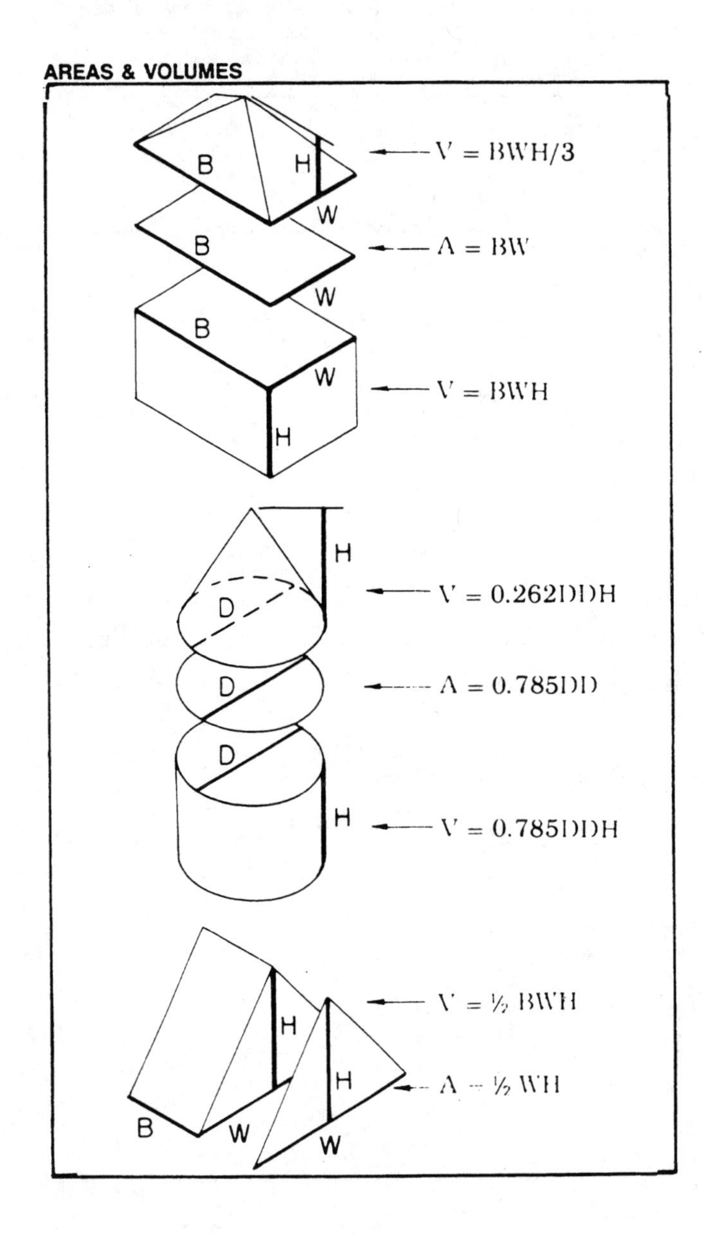

Symbols and Abbreviations

²	= square; e.g. ft^2 = sq ft (ft x ft)
³	= cubic; e.g. ft^3 = cu ft (ft x ft x ft)
/	= per; e.g. lb/ft = lb per ft
%	= percent
°	= degree of angle, e.g. 45°
′	= feet, e.g. 12′-3″
″	= inches
a-ft	= acre-ft
Btu	= British thermal unit
bu	= bushel
cfm	= cu ft per min
cfs	= cu ft per sec
cfh	= cu ft per hr
F	= degrees fahrenheit
fpm	= ft per min
fps	= ft per sec
ft	= foot
FRP	= fiberglass reinforced plastic
hr	= hour
ga	= gauge
gal	= gallon
gpd	= gal per day
gpg	= grains per gallon
gph	= gal per hour
gpm	= gal per minute
hp	= horsepower
hr	= hour
in.	= inch
lb	= pound
mg/l	= milligrams per liter
min	= minute
o.c.	= on center
ppm	= parts per million
psf	= lb/ft^2
psi	= lb/in^2
pt	= preservative treated
R	= insulation resistance
sec	= second
yd	= yard
yr	= year

Definitions

Hard Conversion: physical dimensions and mechanical properties have changed, and products are converted to modular metric increments.
Soft Conversion: physical dimensions remain unchanged, and products are converted to an appropriate degree of precision in metric designations.

Temperature

Fahrenheit (F); Celsius (C); Kelvin (K)
C = (F - 32) ÷ 1.8
F = (1.8 x C) + 32
K = C + 273.15

R - Values (insulation)

RSI = R x 0.176

Common	Metric
R8	RSI 1.4
R12	RSI 2.1
R14	RSI 2.5
R20	RSI 3.5

Vapor Barriers

Common	Metric
2 mil	50 um
4 mil	100 um
6 mil	150 um
10 mil	250 um

Slope (approximate)

	Common ratio	Metric ratio
Roof Slopes		
	1:12	1:12
	2:12	1:6
	3:12	1:4
	4:12	1:3
	6:12	1:2
	8:12	1:1.5
	12:12	1:1
Floor Slopes		
	1/16″/1′	1:200
	1/8″/1′	1:100
	1/4″/1′	1:50
	1/2″/1′	1:25
	5/8″/1′	1:20
	1″/1′	1:10

Roof Load (approximate)

psf	kN/m²
20	1
40	2
60	3
80	4
100	5
120	6

Length

ft-in.	mm
1/8″	3
1/4″	6
3/8″	10
1/2″	13
5/8″	16
3/4″	19
1″	25
1′-0″	300
4′-0″	1 200
6′-0″	1 800
8′-0″	2 400
10′-0″	3 000

Sawn Lumber and Poles

Nominal in.	Actual mm
1	19
2	38
3	64
4	89
5	114
6	140
7	165
8	184
9	210
10	235
11	260
12	286

Round Poles

in.	mm
4	100
5	125
6	150
7	175
8	200
9	225

Softwood Plywood

Modular sheet: 1 200 x 2 400 mm

in.	mm (actual)	
	sanded	sheathing
1/4	6	—
5/16	—	7.5
3/8	8	9.5
1/2	11	12.5
5/8	14	15.5
3/4	17	18.5

Modular Plywood Gussets

in.	mm
4	100
6	150
8	200
9 1/2	240
12	300
16	400
19	480
24	600
32	800
48	1 200

Doors

ft-in.	mm
1′-6″	455
2′-0″	610
2′-4″	710
2′-6″	760
2′-8″	810
2′-10″	860
3′-0″	910
6′-8″	2 030

Concrete Strength

(hard conversion)

psi	MPa
2 000	15
20	3 000
3 500	25
4 000	30
5 000	35
40	6 000
7 000	50